FUNCTIONAL METABOLISM

FUNCTIONAL METABOLISM: REGULATION AND ADAPTATION

EDITED BY

Kenneth B. Storey, Ph.D.
Institute of Biochemistry
Carleton University
Ottawa, Ontario
Canada

WILEY-LISS
A JOHN WILEY & SONS, INC., PUBLICATION

This book is printed on acid-free paper. ♾

Published by John Wiley & Sons, Inc., Hobocken, New Jersey.
Published simultaneously in Canada.

For general information on our other products and services please contact our Customer Care Department within the U.S. at 877-762-2974, outside the U.S. at 317-572-3993 or fax 317-572-4002.

Wiley also publishes its books in a variety of electronic formats. Some content that appears in print, however, may not be available in electronic format.

Library of Congress Cataloging-in-Publication Data:

Functional metabolism: regulation and adaption/edited by Kenneth B. Storey
p. ; cm.
Includes bibliographical references and index.
ISBN 0-471-41090-X (pbk. : alk. paper)
1. Metabolism–Regulation. 2. Homeostasis. 3. Adaption (Physiology) 4. Cellular signal transduction.
[DNLM: 1. Metabolism–physiology. 2. Cells–ultrastructure. 3. Metabolic Diseases–genetics. 4. Molecular Biology–methods. QU 120 F979 2004] I. Storey, K. B. (Kenneth B.)
QP171.F86 2004
572′.4–dc22

2003027397

Printed in the United States of America

10 9 8 7 6 5 4 3 2

CONTENTS

Preface

I thank the authors of this book who are, except for my wife, all former graduate students or postdoctoral fellows from my lab. We are a "serial scientific lab family" in that people come and then go, moving on to their new scientific futures over time. Each new arrival first learns from the lab, then enriches the lab with their ideas and hard work, and then moves on leaving a luminous beacon in the form of scientific achievements that allow the next students to find their way to the edge of knowledge and then plunge into the unknown. I have always operated as a "time vampire" — taking years from the young, asking long hours as well as physical and mental effort — and returning to them the immortality of adding to the totality of scientific knowledge. The authors of this book are only a few of the many students and fellows who have passed through my lab and who have succeeded at science and built productive careers. I am by far the big winner in this bargain of time for opportunity and I truly appreciate the chance the Cosmos has bequeathed me in allowing me to hold onto their coattails for a short time while they learned and applied the scientific method.

I must also thank the mentors who I have had throughout my career — there are several who have been key at different forks in the road but two who must be named. Shining above the rest are my father, Dr. Arthur G. Storey, who instilled in me a desperate work ethic and a need for accomplishment (regardless of the arena), and Dr. Peter W. Hochachka who gave direction and substance to my desire to succeed and introduced me to the elegance of minimalist experimentation and the power of "Synthetic Intuition." Two great fathers in one lifetime are more than I deserve.

In producing this book, I would also like to thank Luna Han, my editor at Wiley, for her support and encouragement, and Christine Punzo for keeping us on track during production and proofreading. Big thanks also to my wife, Jan, for her scintillating writing, and her "eagle eye" and attention to detail during the writing and editing process.

Ken Storey

Contributors

Tolga Bilgen, Department of Zoology, University of Washington, Seattle, Washington

Stephen P. J. Brooks, Nutrition Research Division, Health Canada, Ottawa, Canada

Thomas A. Churchill, Surgical Medical Research Institute, University of Alberta, Edmonton, Canada

Kyra J. Cowan, Epitomics, Inc., Burlingame, California

Steven C. Greenway, The Hospital for Sick Children, Toronto, Canada

Jean E. Grundy, Healthy Environments and Consumer Safety Branch, Health Canada, Ottawa, Canada

Marcelo Hermes-Lima, Departamento de Biologia Celular, Universidade de Brasilia, Brasilia, Brazil

Denis R. Joanisse, Department of Social and Preventive Medicine, Laval University, Sainte-Foy, Canada

Justin A. MacDonald, Department of Biochemistry and Molecular Biology, University of Calgary, Calgary, Canada

William C. Plaxton, Department of Biology, Queen's University, Kingston, Canada

Janet M. Storey, Institute of Biochemistry, Carleton University, Ottawa, Canada

Kenneth B. Storey, Institute of Biochemistry, Carleton University, Ottawa, Canada

William G. Willmore, Institute of Biochemistry, Carleton University, Ottawa, Canada

1

PRINCIPLES OF METABOLIC CONTROL

WILLIAM C. PLAXTON

KEY CONCEPTS

The ability to control the rates of metabolic processes in response to changes in the internal or external environment is an indispensable attribute of living cells that must have arisen with life's origin. This adaptability is necessary for conserving the stability of the intracellular environment (homeostasis), which is, in turn, essential for maintaining an efficient functional state. In the absence of such control, all metabolic processes would achieve a state of equilibrium with the external environment. For example, the intracellular storage of a fuel macromolecule such as glycogen would be impossible since there is an enzyme (glycogen phosphorylase) dedicated to catalyzing the breakdown of this storage polyglucan into its constituent glucosyl units. Obviously, the existence of glycogen as an important energy store in animal tissues implies that the activity of glycogen phosphorylase is carefully controlled so as to allow this fuel to be utilized as dictated by the needs of the organism. Indeed, elaborate regulatory mechanisms have been discovered that affect glycogen phosphorylase activity, thereby allowing wide variations in the rate of glycogen breakdown *in vivo*. The aim of this chapter is to outline the biochemical regulatory mechanisms that are believed to be the most important in metabolic control. Practical aspects for the study of metabolism and its control, as well as the advantages and disadvantages of qualitative versus quantitative approaches to metabolic control, will also be highlighted.

Metabolic Renaissance in Postgenome Era?

The remarkable advances in molecular genetics that have occurred over the past couple of decades have somewhat eclipsed areas of traditional biochemistry such as protein chemistry, enzymology, and metabolic control. With many genomes sequenced and others nearing completion, the next step is the less straightforward task of analyzing the expression and function of gene products (proteins), as well as more thoroughly elucidating metabolism and its control. The task of completing the picture of all cellular proteins, their actions and reactions, is one of the biggest challenges facing life science researchers today. Although molecular biology has generated many impressive techniques [e.g., protein overexpression, site-directed mutagenesis, metabolic engineering, complementary deoxyribonucleic acid (cDNA) microarrays, etc.] for assessing various aspects of protein/enzyme structure–function and regulatory control, one cannot deduce the properties of a functional protein or the kinetic and regulatory properties of an enzyme solely from genetic information. Moreover, recent animal, plant, and microbial genome sequencing projects have revealed a plethora of gene sequences that encode proteins having unknown functions. Furthermore, many organisms whose genomes have been sequenced have not had their metabolism extensively studied. Where feasible, their metabolic phenotype is determined using annotated genome sequence data. Thus, there appears to be a resurgence of interest in protein, enzymological, and metabolic research for understanding biological processes in the postgenome era. Efficient approaches are needed for determining: (a) the function of unknown gene products, (b) protein expression in different cells under various conditions, (c) covalent modifications of proteins in response to different stimuli, (d) protein–protein interactions, (d) the relationship between protein structure and protein function, and (e) the sophisticated mechanisms that serve to control the flux of metabolites through specific

Functional Metabolism: Regulation and Adaptation, edited by Kenneth B. Storey.
ISBN 0-471-41090-X

metabolic pathways *in vivo*. Novel methods are also being developed to map proteomes (i.e., the proteins encoded by the genome) and to discover new enzymes of interest.

Metabolic Engineering

Since it is now possible to manipulate nucleic acids and gene expression at will, an important goal of biotechnology is to modify (usually enhance) the output of specific biosynthetic pathways via the process of *metabolic engineering*. Contemporary genetic engineering techniques have created the potential to directly modify the metabolism of a target organism in a desired fashion. However, the ability to manipulate an organism's genetics has thus far transcended our ability to predict the effects of these manipulations on metabolism. Our lack of a thorough appreciation for, and understanding of, metabolic control is reflected by the meager results from most attempts to apply the powerful tools of molecular biology and genetic transformation to the practical goal of metabolic engineering. Many unsuccessful efforts at so-called shotgun metabolic engineering have been based upon the misguided assumption that if the expression of a gene encoding a particular enzyme is suitably manipulated, then there will be a corresponding change in both the *in vivo* activity of the encoded enzyme as well as the *flux* (or rate of movement) of metabolites through the pathway in which the enzyme functions. "Rational" metabolic engineering is a longer-term, but arguably more scientific and interesting process that involves the targeted and purposeful alteration of a specific metabolic pathway. It does not necessarily depend upon altering the concentration of an enzyme but could be carried out by introducing a mutant (or heterologous) enzyme with altered control properties. Apart from an ability to manipulate nucleic acids, rational metabolic engineering also requires a strong background in protein/enzyme and metabolic biochemistry.

Metabolic Regulation *versus* Metabolic Control

Although biochemists frequently employ the terms *regulation* and *control* interchangeably, the need to discriminate between these terms has been emphasized (Fell, 1997). *Metabolic control* refers to adjusting the output of a metabolic pathway in response to an external signal. By contrast, *metabolic regulation* occurs when an organism maintains some variable relatively constant over time, despite fluctuations in external conditions. Homeostasis is therefore a consequence of metabolic regulation, which itself may be a result of metabolic control. For example, the regulation of mammalian blood glucose is largely due to the secreted peptide hormones glucagon ("starved" signal) and insulin ("fed" signal) controlling intracellular metabolism within the liver. In this case, the concentration of blood glucose is regulated (kept constant) mainly by controlling (varying) fluxes of metabolic pathways (i.e., glycogen breakdown versus synthesis, glycolysis, gluconeogenesis) in hepatocytes. Regulation and control are properties of highly elaborate metabolic systems. An ongoing challenge is to link our knowledge of molecular, reductionist-based, enzyme control mechanisms to organism-level explan-ations of metabolic regulation.

Complexity of Metabolism and Concept of Biochemical Unity

As protein catalysts, enzymes vastly accelerate the rates of chemical reactions without themselves undergoing a permanent change. As each cellular reaction is catalyzed by its own enzyme, every cell contains a large number of different enzymes. Although a "simple" prokaryote, such as *Escherichia coli*, is only about 1/500th the size of a typical eukaryotic cell, each *E. coli* cell contains about 3000 different proteins, at least 90% of which are enzymes. The metabolic complexity of all cells is reflected by the many separate enzymatic reactions that make up the metabolic pathways that collectively constitute metabolism.

Despite its complexity, a general understanding of metabolism has been achieved because common solutions to the problem of biochemical design have been evolved. Thus, the types of substrates, cosubstrates, cofactors, fuels, and types of metabolic pathways used are common to most cells. This is the concept of *biochemical unity*. In general, biochemical unity also applies to metabolic regulation and metabolic control. Comparative biochemistry has revealed that the types of control mechanisms found in metabolic pathways are similar from one organism to the next. However, it is the implementation of these designs—the regulatory details—that cannot only differ widely from species to species but can differ widely for similar metabolic pathways in different cell types of a single organism, or even within different organelles of a single cell.

For example, citrate synthase, which catalyzes the reaction acetyl-CoA + oxaloacetate $\rightarrow$ citrate + CoA, is controlled in dissimilar manners in different cells. In respiring animal cells, a major function for this enzyme is in the citric acid cycle that operates in the mitochondria in conjunction with oxidative phosphorylation to produce adenosine 5′-triphosphate (ATP). Here, the overall end product, ATP, feeds back to inhibit citrate synthase. This regulatory mechanism is logical since at high ATP levels, the ATP-generating citric acid cycle will then be inhibited, but if ATP levels fall, substrate catabolism by the cycle will speed up. In *E. coli*, however, citrate synthase and the citric acid cycle have a different function. This bacterium lives a mainly anaerobic life, generating its ATP primarily

CHAPTER 228

Pelizaeus-Merzbacher Disease and the Allelic Disorder X-linked Spastic Paraplegia Type 2

Lynn D. Hudson

1. Pelizaeus-Merzbacher disease (PMD) (MIM 312080), together with the allelic disorder spastic paraplegia type 2 (SPG2) (MIM 312920), which also maps to Xq22, are leukodystrophies of widely varying severity. PMD is characterized chiefly by impaired motor development that presents within the first year and progresses throughout life: nystagmus, ataxia, spasticity, and mental retardation are usually encountered. SPG2 patients display spasticity of the lower limbs. In patients with the complicated form of SPG2, cerebellar ataxia, sensory loss, nystagmus, and optic atrophy may also be present.

2. A loss of myelin is evident by MRI or histopathology in both PMD and SPG2, and a loss of oligodendrocytes is obvious in the severe connatal form of PMD. Abnormal central nervous system conduction velocities, as analyzed by evoked potentials, reflect the myelin defect.

3. Mutations in the X-linked PLP gene (GenBank M27110), which encodes the major central nervous system myelin protein (PLP) and its alternatively spliced isoform DM20, are responsible for the pathogenesis of PMD and SPG2. Females are affected only infrequently in PMD, but in the relatively mild SPG2 the PLP mutation can act semidominantly.

4. Animal models exist that have the identical mutation found in the human form of the disease: the *jimpy^{msd}* mouse for the connatal form of PMD and the *jimpy^{rsh}* mouse for the complicated form of SPG2.

5. PLP is a gene subject to dosage control, as overexpression of PLP is the most common type of mutation in PMD. Transgenic mice carrying multiple copies of PLP mimic the phenotype of PLP duplications found in the classical form of PMD.

6. At a cellular level, the extent of pathophysiology in PMD/SPG2 can be correlated with the type of mutation in the PLP gene. In the most severe mutations (the connatal form of PMD), nascent PLP and DM20 accumulate in the rough endoplasmic reticulum of oligodendrocytes and trigger apoptosis. Less severe mutations spare DM20, which can traffic to the sites of myelin assembly and participate in myelin sheath formation. Abnormal PLP and DM20 proteins generate phenotypes that span the PMD/SPG2 spectrum, while either too much or too little of these lipoproteins leads to intermediate phenotypes, such as the classical form of PMD or the complicated form of SPG2.

A list of standard abbreviations is located immediately preceding the index in each volume. Additional abbreviations used in this chapter include: DM20 = an isoform of proteolipid protein with a MW of 20 kDa; PLP = proteolipid protein; PMD = Pelizaeus-Merzbacher disease; RER = rough endoplasmic reticulum; SPG2 = X-linked spastic paraplegia type 2.

HISTORY AND CLASSIFICATION

Friedrich Pelizaeus published the first clinical picture of Pelizaeus-Merzbacher disease in 1885, and managed to pinpoint the mode of inheritance by including a quote from the affected family "that the disease is passed on by the mother but does not hurt her."[1] Twenty-five years later, Ludwig Merzbacher documented a widespread loss of myelin in the cortical white matter of the original family examined by Pelizaeus.[2] A high degree of similarity of what is now known as the classical form of Pelizaeus-Merzbacher disease was found among the 12 male family members, with onset in the first months of life. Initially, the disease manifested itself by aimless, wandering eye movements, followed by nystagmus. Infants failed to develop normal head control, displayed tremors or shaking movements of the head, and may have had microcephaly. The disease was slowly progressive, with additional signs including bradylalia, scanning speech, ataxia and intention tremor of the upper limbs, spastic contractions of the lower limbs, athetotic movements, mild dementia, and death in the second decade.[1,2]

Franz Seitelberger reported a much more severe disorder, detectable at birth, in which the nearly complete absence of myelin sheaths was accompanied by a profound loss of myelin-forming cells (oligodendrocytes) in the central nervous system.[3] Seitelberger correctly identified both this connatal form (type II) and the classical form (type I) of PMD as leukodystrophies; that is, conditions in which the absence of myelin is primary. Although originally described as a "demyelinating" (destruction of myelin) disorder, the pathology of the connatal form of PMD is that of a "dysmyelinating" (inability to form normal myelin) disease. A transitional intermediate (type III) between the classical and connatal forms of PMD has also been described,[4,5] but the extensive overlap of this type with the classical and connatal forms, together with intrafamilial heterogeneity in the clinical findings of all types (particularly with respect to the age of death and the extent of cerebellar dysfunction), speaks for limiting the classification to the classical and connatal forms of PMD. Additional variants have been described,[5] but the absence of X-linkage in these disorders genotypically distinguishes them from PMD.

A distinct clinical entity, X-linked spastic paraplegia type 2 (SPG2), was recently discovered by Odile Boespflug-Tanguy and colleagues to be an allelic disorder with PMD.[6] A combination of studies led to this unexpected finding: (a) the mapping of SPG2 to the Xq21 region, where linkage studies had previously placed PMD; (b) commonalities in the clinical presentation of PMD and the complicated form of SPG2, namely spasticity, nystagmus, cerebellar ataxia, and a pyramidal syndrome; (c) the detection of hypomyelination in a complicated form of SPG2; and (d) the observation that several PMD families included a family member diagnosed as having the rare spastic paraplegia disorder.[7] The X-linked form of spastic paraplegia type 2 exists as either a complicated form or a pure one in which the clinical phenotype is confined to lower limb spasticity. Both the pure and complicated forms of SPG2 arise from mutations in the gene encoding the myelin proteolipid protein (PLP), the same gene affected in PMD.[6–12] The detection of families in which PMD and SPG2 coexist[7,8] emphasizes the clinical continuum of these disorders, all of which share a phenotype of spasticity and hypomyelination.

CLINICAL FEATURES

The common feature of lower limb spasticity unites the four clinically distinct categories of PMD/SPG2. Distinguishing features among the PMD/SPG2 subtypes include the presence or absence of nystagmus,cerebellar signs, optic atrophy, laryngeal stridor, dystonia, and the extent of developmental delays due to motor impairment.[7,14,15] The severity of the disease ranges from the most severely affected connatal form of PMD, the classical form of PMD, the complicated form of SPG2, and to the least affected pure form of SPG2.

Pelizaeus-Merzbacher: Connatal (Type II) Form

The connatal form of PMD is apparent within the first few months after birth, and affected males display a nearly complete lack of psychomotor development.[4,5,16–19] Progression of the disease is rapid, with death usually occurring in the first decade. Severe feeding problems and extrapyramidal dyskinesia are early indicators of the connatal form of PMD. Prominently featured are nystagmus (roving, horizontal, or rotatory), titubation, dysarthria, optic atrophy, seizures, ataxia, and spasticity that progresses to contractures in the extremities.[4,7,14–19] The movement disorder includes head-nodding, eye-rolling, involuntary jerking of the head, eyes, or limb, and continuous random uncoordinated movement. Laryngeal stridor is often seen in these patients.[17,18]

Pelizaeus-Merzbacher: Classical (Type I) Form

Onset of the classical form of PMD is usually noted within the first year of life, with motor handicaps as the outstanding feature.[1,2,7,14,15,17–19] Unlike the connatal form of PMD, classical types can improve in performance up to 10 to 12 years of age before a slow deterioration ensues with death occurring in mid adulthood.[7] Nystagmus is evident early, together with generalized hypotonia, slowness of motor development, head tremor, and trunk ataxia. Nystagmus may disappear on maturation, and bilateral optic atrophy may develop. Spasticity and pathologic reflexes of the lower limbs, ataxia, athetotic movements of the upper extremities, and dysarthria accompany the delayed motor development. Speech and motor deficits make it difficult to determine the extent of mental retardation.

In a mild variety of classical PMD associated with duplications of the PLP gene, the age of onset is unchanged, but the life expectancy is extended to 60 years, indicating that the progression of the disease is comparatively slow in this type of PLP mutation.[20] Nystagmus, pyramidal tract signs, and cerebellar involvement are all evident shortly after birth in these families, but features such as laryngeal stridor, optic atrophy, and dementia are not commonly encountered.[20]

SPG2: Complicated Form

All forms of the rare SPG2 display spasticity and weakness in the lower extremities, but the complicated form has additional signs in common with PMD, which may include nystagmus, optic atrophy, ataxia, and cerebellar syndrome of the upper limbs with dysarthria.[6,7,11,13,21,22] Unlike PMD, no hypotonia, choreoathetosis, or bobbing movements of the head or trunk are evident. Patients with the complicated form of SPG2 have slowly progressive weakness and spasticity of the lower extremities, and at the onset of the disease (2 to 5 years) may appear to have the pure form of SPG2. Nystagmus occurs at a later onset than PMD, and only half of the patients have optic atrophy. Patients may be delayed in the age of walking (15 to 36 months), but can walk unaided for a long period. Mental retardation, when present, is mild, and most patients attend school.[6]

SPG2: Pure Form

The pure form of SPG2 presents with spasticity of gait and increased deep tendon reflexes, all in the absence of other neurologic deficits.[9,11,12] The age of onset can range from toddler to teen years within a family. Like the complicated form of SPG2, the pure form is slowly progressive and the life span is nearly normal. Cognitive function is normal. More spasticity is seen in the legs than in the arms, and intention tremors may also be present.

NEUROIMAGING, NEUROPATHOLOGY AND NEUROPHYSIOLOGY

The myelin sheath in the central nervous system is a specialized extension of the oligodendrocyte plasma membrane that successively enwraps axons, thereby permitting fast saltatory conduction of nerve impulses. Myelin formation begins prenatally, but the most rapid changes in myelination occur within the first eight postnatal months.[23,24] In general, sensory pathways tend to myelinate prior to motor pathways (unlike the peripheral nervous system). While central nervous system myelination is largely complete at the age of 2, myelination in regions of the cerebral hemispheres involved in higher-level associative functions and sensory discrimination progresses over a decade.[24] Mutations in myelin proteins that compromise the synthesis of a normal myelin sheath are, therefore, expected to present at an early developmental stage as "dysmyelinating" (inability to form normal myelin) disorders. PMD patients, for the most part, exhibit the neuroimaging and neuropathologic features of dysmyelinating disorders, in which a reduction in the amount of myelin (hypomyelination) is apparent early and signs of myelin breakdown may be prominent subsequently. "Demyelinating" (destruction of myelin) disorders are ones in which problems arise not in the synthesis but in the maintenance of myelin sheaths. Mutations that have only subtle effects on myelin structure may permit an ostensibly "normal" myelin sheath to be synthesized, but this sheath may be structurally unsound and therefore more susceptible to turnover, which can present as a demyelinating disorder.

All forms of PMD/SPG2 have hypomyelination, a feature that can be detected by magnetic resonance imaging (MRI), pathologic exam of central nervous system tissue, and/or analysis of central evoked potentials.[4,7,14,15,17,18] MRI typically shows diffuse hyperintense white matter signal in T2-weighted images, as apparent in Fig. 228-1, PMD patients generally have a homogeneous pattern of myelin loss throughout the central nervous system, with the less-affected classical PMD patients displaying small scattered areas of normal signal resembling the "tigroid pattern" described in the histopathology of the disease.[25] The MRI abnormalities can be detected in PMD patients with the severe connatal form of the disease as early as 3 months of age.[26] MRI diffusion-weighted images can be used to distinguish between dysmyelinating and demyelinating lesions—dysmyelinating lesions maintain diffusional anisotropy, whereas demyelinating

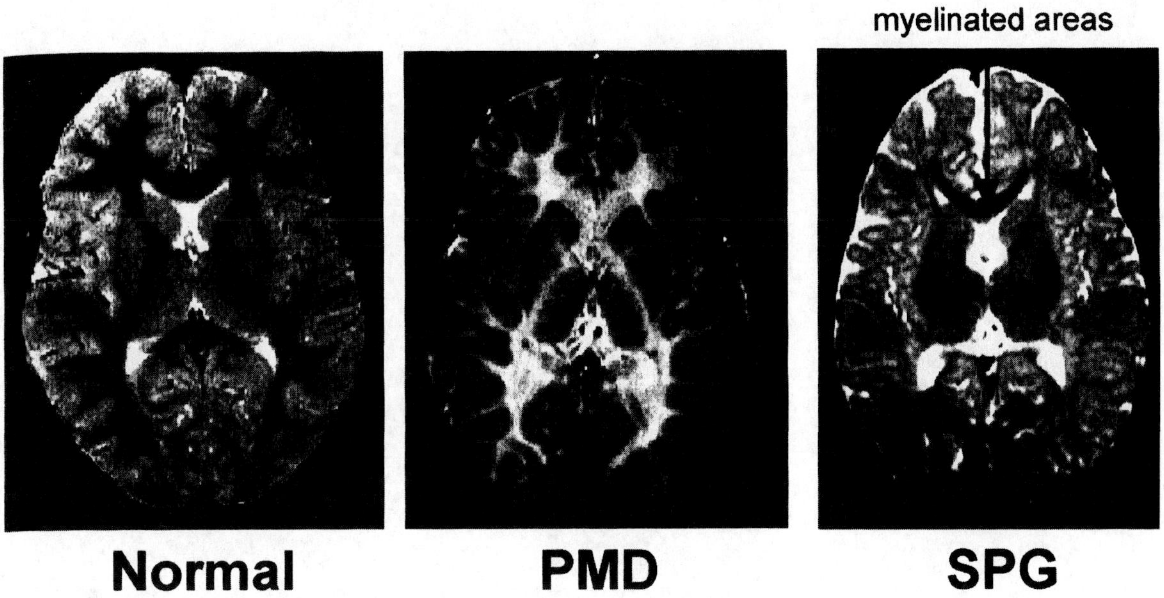

Fig. 228-1. Pathology of SPG2 and PMD detected by MRI. T_2-weighted cerebral MRI images of a normal (3 years of age), PMD-affected (3 years of age), and SPG-affected (10 years of age) boy. In the white matter of the PMD-affected boy, a diffuse high-intensity signal is present instead of a diffuse hyposignal (seen in the normal control) that occurs when myelination has been normally achieved. Some myelinated areas (arrow) are observed in the white matter of the SPG-affected patient. (*Provided by and used with permission of Dr. O. Boespflug-Tanguy, INSERM, Cedex, France*).

lesions lose it.[27] This technique has been applied to a PMD patient who displayed the diffusional anisotropy of a dysmyelinating lesion.[27] SPG2 patients often show a heterogeneous, patchy pattern consisting of occasional extensively myelinated regions (arrow in Fig. 228-1),[7] generally with a later onset and a more variable presentation than in PMD patients. The pure and complicated forms of SPG2 are not distinguishable by MRI, as both have a diffuse hyperintense signal in white matter, and the less clinically severe pure form of SPG2 can present with a global lack of myelination, sparing of only a few transverse fibers in the pons.[12]

On histopathologic exam, the striking loss of white matter in PMD can be seen even without the benefit of dyes that stain myelin. Connatal forms of PMD may have a nearly complete absence of myelin, with few degradation products apparent.[4] Islands of myelin are preserved in the classical form of PMD, giving rise to the characteristic "tigroid" appearance. The myelin sheaths in such islands appear abnormal and remnants of myelin are found in vacuoles of surrounding cells, including astrocytes. Fibrillary gliosis occurs in affected areas, and some hypertrophic astrocytes are evident. Axonal pathology and neuronal loss can occur in affected areas.[4,17] Oligodendrocytes can be reduced in number, and often show abnormal cytoplasmic inclusions and condensed chromatin granules in the nuclei characteristic of apoptotic cells.[28] The lack of myelin extends to the grey matter, where a small number of oligodendrocytes normally reside. The peripheral nervous system is normally myelinated, although a PMD family with a mild form of the disease was recently found to have a loss of peripheral myelinated fibers, detectable degradation of myelin sheaths, and a slowing of peripheral nerve conduction.[29]

As anticipated in disorders of defective central nervous system myelination, PMD patients have extensive conduction slowing involving the brain stem to the cerebrum.[7,14,15,17,18,30–33] Conduction blocks are prominent in motor rather than sensory systems, as seen by electrophysiologic findings with brain stem auditory, somatosensory, motor, and visual evoked potentials. The visual evoked potentials can be particularly informative in young infants.[30] SPG patients also have abnormal central nervous system conduction velocities.[6,7]

GENETICS AND MOLECULAR BIOLOGY OF PMD/SPG2

The Proteolipid Protein (PLP) Gene Encodes a Major Myelin Protein

The most abundant protein in central nervous system myelin is proteolipid protein (PLP). An integral membrane protein, PLP is one of nature's most hydrophobic proteins due to the high content of hydrophobic amino acids (50 percent) and an unusual degree of fatty acid acylation (Fig. 228-2).[34,35] Despite the difficulties of working with a protein that is soluble only in organic solvents, Stoffel and coworkers managed to pinpoint the location of the covalently linked long chain fatty acids and to elucidate the

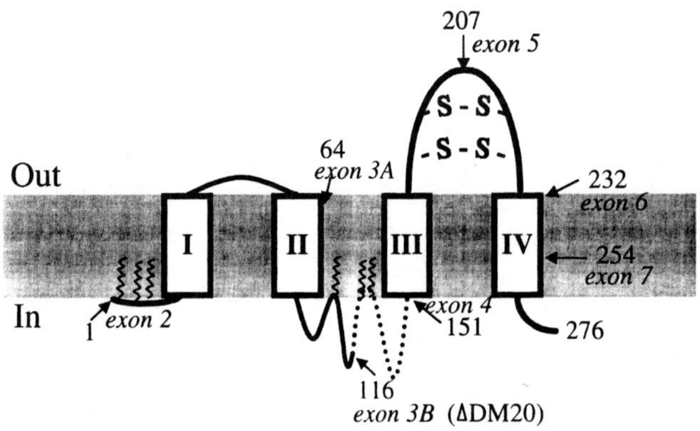

Fig. 228-2. Structure and topology of PLP and DM20. The four transmembrane domains are labeled with roman numerals. The first amino acid of each exon is numbered. Because exon 1 only encodes the initiator methionine, which is cleaved from the protein, the numbering begins with exon 2. Disulfide bridges are indicated in the large extracellular loop, and the fatty acids are denoted by squiggly lines. The exon 3B region that is deleted in DM20 is shown as a dotted line. During compaction of the layers of myelin membrane, the inside of the membrane becomes the major dense line and the outside becomes the intraperiod line of the myelin sheath. (*Adapted from Weimbs and Stoffel.*[36])

structure of PLP, which features four transmembrane domains with both the amino and C-termini located on the intracellular face of the membrane (Fig. 228-2).[34,36] Further structural studies have confirmed and extended the Stoffel model by demonstrating that the alternatively spliced isoform of PLP named DM20 shares the topology delineated for PLP.[37] DM20 is missing part of the intracellular loop that contains two of the six acylation sites (Fig. 228-2), which may account for the altered conformation and physical properties observed for DM20.[35,37,38] The overall structure and individual regions of PLP/DM20 resemble channel-forming regions of the nicotinic acetylcholine and glutamate receptor complexes,[39] and ionophoretic activity of PLP has been demonstrated in reconstituted systems.[35,38]

PLP is an exceptionally well-conserved gene: No amino acid differences exist between mouse, rat, and man, and no amino acid polymorphisms have been detected in any species. A single gene composed of 7 exons spread over 18 kb encodes PLP/DM20.[40–43] The first exon ends with the initiator methionine, which is cleaved off the nascent PLP/DM20 protein. The third exon contains an internal splice donor site, which is used to generate DM20 transcripts. The PLP locus is transcriptionally active in early prenatal development, at which time alternative splicing favors the DM20 transcript. Maturing oligodendrocytes orchestrate the expression of PLP/DM20 together with the other myelin proteins and lipids.[44] PLP/DM20 is synthesized in the rough endoplasmic reticulum and transported in vesicles to the sites of myelin assembly. Like PLP, DM20 is an abundantly produced myelin protein, with an estimated 1.6 ratio of PLP to DM20 protein.[45] DM20 has been proposed to facilitate, or chaperone, the movement of PLP to the cell surface.[46–49] In addition to being expressed in oligodendrocytes, the PLP gene is transcriptionally active in the myelin-forming cells of the peripheral nervous system, the Schwann cells,[50] where the predominant isoform expressed is DM20.[51] Schwann cell expression of PLP/DM20 is an order of magnitude lower than that observed in oligodendrocytes, and most of the proteins produced are not normally incorporated into the myelin sheath.[52] A low level of PLP expression also occurs outside of the nervous system, in the heart[53] and fetal thymus.[54]

Animal Models of Pelizaeus-Merzbacher Disease/SPG2

Several of the animal models of PMD/SPG2 have the same mutation as in the human disease: *jimpy*^{msd} for the connatal form of PMD[55,56] and *jimpy*^{rsh} for the complicated form of SPG2.[8,10,57] The *jimpy*^{msd} and *jimpy*^{rsh} are alleles of *jimpy*, a strain in which hemizygous males containing the jimpy mutation display ataxia, tremor, and ultimately seizures commencing with the general onset of central nervous system myelination.[58] All of the mutant PLP animals have varying degrees of hypomyelination.[59] One of the most severely affected mutants, the *jimpy* mouse, has only 1.6 percent of control myelin in the spinal cord.[60] The biochemical profiles reflect the lack of myelin, with dramatic reductions in all myelin proteins. The residual myelin is in the patchy pattern seen as the "tigroid" pathology of PMD patients. The occasional myelin sheath may appear well-compacted but the intraperiod line, which is formed when the extracellular faces of oligodendrocyte plasma membrane are fused, can be abnormally condensed.[59] Most of the point mutations in the PLP gene feature a reduced number of mature oligodendrocytes, accompanied by a proliferation of immature oligodendrocytes and noticeable oligodendrocyte cell death.[61] A unique oligodendrocyte abnormality, grossly distended rough endoplasmic reticulum (RER), is present in several of the animal mutants (the *myelin deficient* rat and *shaking* pup).[59] This swelling of the RER may represent a general block or retardation in protein transport from the RER due to the abnormal folding and accumulation of the abundantly synthesized mutant PLP/DM20 protein. Distended RER has also been detected in oligodendrocytes from a PMD patient.[62] The other cell types in the central nervous system are secondarily affected in animals with PLP point mutations. Astrocytic hypertrophy is predominant, with a peculiar invasion of astrocytic processes evident between the axon and the myelin sheath. Excessive numbers of microglia are found in the white matter during the period of the most cell death, a time consistent with a role for microglia in disposing of cell debris. Neuronal abnormalities, which appear subsequent to the myelin loss, include an accumulation of axonal organelles that are often segregated into giant spheroids.[59]

Moderate phenotypes predominate in the *rumpshaker* mouse (*jimpy*^{rsh}),[57] the animal model for SPG2,[7,8,10] and the *paralytic tremor* rabbit.[63] Animals are long-lived and reproduce normally. In *jimpy*^{rsh}, oligodendrocytes are not subject to premature death and do not display cytoplasmic abnormalities. The astrocyte and microglial populations are likewise normal. Much more myelin is present in *jimpy*^{rsh} mice,[57] but the myelin sheaths are structurally unstable.[64] The DM20 isoform is much less affected than PLP in *jimpy*^{rsh} mice, as the myelin sheaths contain mutant DM20 but no PLP.[57]

Transgenic mice that are engineered as null for PLP/DM20 expression are only mildly affected.[65–67] Oligodendrocytes do not undergo premature cell death as in *jimpy* and *jimpy*^{msd} mice. Nonetheless, the absence of PLP/DM20 noticeably alters compaction of the myelin sheath, an effect that is magnified with age.[65–67] The PLP-deficient mice suggest that neither PLP nor DM20 is essential for the normal development or differentiation of oligodendrocytes in rodents. A developmental role for PLP/DM20 had previously been indicated by the early expression of DM20 in wild-type mice and the paucity of mature oligodendrocytes in *jimpy* and *jimpy*^{msd} mice with PLP point mutations. It is likely that other lipoproteins can partially substitute for PLP/DM20 in the transgenic null mice. Indeed, a proteolipid protein gene family exists, as identified by the cloning of a homologous neuronal membrane glycoprotein named M6/EMA[68,69] and related proteins expressed in sharks and rays.[39] The deletion of the PLP gene in man appears to have more serious phenotypic consequences (i.e., the classical form of PMD)[70] than those observed in transgenic mouse models, perhaps because the compensatory mechanism of alternative lipoprotein expression is not as effective in man.

The overexpression of PLP/DM20 in transgenic mice, on the other hand, has pronounced effects on myelination.[71–73] Multiple copies of the PLP gene on an autosome create severe hypomyelination, astrocytosis, seizures, and premature death. While the impact on PLP/DM20 mRNA levels was minor, little PLP protein was incorporated into myelin. Thus, increased PLP dosage leads to a translational or posttranslational arrest of PLP expression. Although the animal model differs from the duplications of the PLP gene found in a majority of patients with the classical form of PMD[7,14,20,74–77] in that three to seven copies of the PLP gene were present in the mouse, a similar phenotype occurs in both cases of PLP overexpression.

X-linked Pattern of Inheritance

Two approaches led to the discovery of PLP mutations in PMD, and both were dependent on the assignment of PLP to the X chromosome in man[78,79] and mouse.[79,80] The first relied on the detection of a splicing defect in the PLP gene of *jimpy* mice[80–83] as the impetus for sequencing the PLP gene of PMD patients that had the same behavioral, biochemical, and morphologic profiles as *jimpy* mice.[43,84] A single base mutation substituting a charged amino acid for tryptophan in one of the transmembrane domains of PLP was found in this PMD family.[43] A large duplication of the Xq13-q22 region, which included PMD among the clinical phenotypes,[85] together with the mapping of PLP to the Xq21.2-q23 region[79] and linkage studies placing the PMD phenotype within that interval, enabled Hodes and coworkers to identify a point mutation in the PLP gene that segregated with the PMD phenotype.[86] Nearly 40 distinct point mutations in the coding regions or splice sites of the PLP gene have been subsequently identified in PMD or SPG families (Table 228-1), yet less than a quarter of the PMD/SPG patients have identifiable sequence

Table 228-1 Mutations in the Proteolipid Protein Gene

Mutation Type	Exon	DM20 Spared	Clinical Phenotype	Affected Females	Animal Disorder	References
Missense						
M0I	1		PMD: Classical	+		89
P14L	2		PMD: Connatal	+¶		86,91
F31V	2		PMD			103
H36P	2				*shaking* pup	106
H36Q	2				*paralytic tremor* rabbit	63
A38S	2				*jimpy*^{4J}	107
T42I	2		PMD			108
L45R	2		PMD			109
F50S	2		PMD			7
G73R	3A		PMD			110
T74P	3A				*myelin-deficient* rat	111, 112
T115K*	3A		PMD: Classical	+		90
H139Y	3B	+	SPG2: Complicated	+		6
K150N*	3B	+*	PMD			101, 102
T155I	4		PMD: Classical			113, 114
W162R	4		PMD: Connatal			43
V165E	4		PMD: Classical			92
S169F	4		SPG2: Pure			12
T181P	4		PMD: Connatal			115
I186T	4		SPG2: Complicated		*jimpy*^{rsh} mouse	10, 13, 57
D202N	4		PMD			7
D202H	4		PMD: Classical			110
Y206C	4		PMD			99
V208N	5		PMD			116
P210L	5		PMD			116
P215S	5		PMD: Classical			117
G216S	5		PMD: Connatal			118, 119
V218F	5		PMD: Classical			120
G220C	5		PMD: Connatal			100
L223P	5		PMD: Connatal			115
S225P	5		SPG2: Pure			9
F236S	6		SPG2: Complicated			121
A241P	6		PMD: Connatal			122
A242V	6		PMD: Connatal		*jimpy*^{msd} mouse	55, 56
A248P	6		PMD			123
Nonsense						
W144X	3B	+	SPG2/PMD	+		11, 93
E233X	6		SPG2/PMD	+		8
Frameshift						
del G3	1		PMD: Classical	+§		29
del/ins 583-9	4		PMD			124
insG772	7		PMD: Classical			125
Deletions						
T117-V165	3B, 4		PMD: Connatal			126
A170-P172	4		PMD: Classical			102
entire gene	1–7		PMD: Classical		transgenic knockouts	65–67, 70
Duplications						
entire gene	1–7		PMD: Classical		transgenic overexpressors	20, 71–77
Splice sites						
A438T, exon 3B	3	+?	PMD			127
G → T, intron 3	4		PMD			128
A → G, intron 4	5				*jimpy* mouse	81–83

The initiator methionine is numbered as amino acid zero, as the terminal methionine is removed from the protein following translation. The nucleotide sequencing numbering is according to Hodes et al.[15] Mutations that should spare DM20 are indicated by +. Deletions are abbreviated del and insertions, ins.

* Splicing is anticipated to be affected in these mutations, as the base change alters the donor splice site at the 3′ end of either exon 3A (T115K) or exon 3B (K150N).

? This mutation would create an ideal consensus splice site AG | GT in the region of exon 3 where intra-exonic splicing normally occurs to generate the DM20 spliced isoform.

¶ Only a single female was affected in this large pedigree.

§ With a peripheral nervous system neuropathy.

An updated database of mutations in the PLP gene is available at http://www.med.wayne.edu/Neurology/plp.html

alterations in these regions.[7,13,14] The discovery by Malcolm and coworkers that PLP is a gene subject to dosage control[75] led to the current estimate that over half of PMD patients have duplications of the PLP gene.[7,14,20,74–77] There is also a significant fraction of PMD patients with no detectable PLP mutations, although a tight linkage exists between the PMD phenotype and markers of the Xq22 region in these families.[7,13,14,87] Given that the PLP gene dosage is strictly controlled, the mutations in these families probably reside in the *cis* regulatory elements that influence the levels of PLP gene expression. Linkage studies also figured prominently in the discovery that SPG2 results from mutations in the PLP gene.[6,9] Two loci on the X chromosome are associated with spastic paraplegia, SPG-1 resulting from mutations in the L1 neuronal cell adhesion molecule at Xq28 and SPG-2, which mapped to the PLP gene.[6,7,9,88]

PMD is usually inherited as an X-linked recessive.[7,13,14] No clinical phenotype or abnormal neuroimaging has been found in any mother carrying a PLP duplication. However, an occasional family with point mutations in the PLP gene will include affected female relatives.[6–8,11,29,89–93] Females heterozygous for PLP mutations are mosaics: one population of oligodendrocytes expresses normal PLP and synthesizes normal myelin sheaths, while the other population expresses a mutant form of PLP which, if these abnormal oligodendrocytes survive, would generate a defective sheath. The lack of phenotype in the vast majority of PMD female carriers can be attributed to the same phenomenon observed in female heterozygous *jimpy* mice or *myelin-deficient* rats, in which the dying mutant oligodendrocytes are gradually replaced by normal oligodendrocytes.[59,94] Most of the remaining population consequently consists of genotypically normal oligodendrocytes by early adulthood. Adult *myelin-deficient/+* females are morphologically and phenotypically normal with the possible exception of regions such as the optic nerve, where the migration of oligodendrocyte progenitors into the nerve occurs during only a brief window of time.[94] If enough normal progenitors do not enter the optic nerve in early development, there may not be enough normal cells to compensate for the apoptotic *myelin-deficient* oligodendrocytes. Another scenario that accounts for a PMD phenotype in females is unfortunate lyonization of the affected X chromosome, yielding only a very small population of normal oligodendrocyte progenitors to seed the central nervous system. Nonrandom inactivation of the affected X chromosome (the Lyon effect) occurs infrequently, and therefore is not expected to account for families with a large number of affected females. A family where the Lyon effect is a plausible explanation for an affected female is the large, multigeneration Indianapolis kindred with the connatal form of PMD in which only a single female was affected.[86,91]

Families with mild cases of PMD/SPG2 may paradoxically have more females involved, as the mutant oligodendrocytes can survive in these patients. Because the mutant oligodendrocytes are not overrun by normal oligodendrocytes in early development, they compete with the normal oligodendrocytes for axons and proceed to elaborate myelin sheaths that are structurally flawed. Over time, these sheaths are susceptible to degradation, a situation yielding a progressive loss of myelin characteristic of a "demyelinating" disease. The resultant clinical phenotype is that of an adult-onset neurodegenerative disorder with variable symptoms of spastic paraplegia[7] and/or PMD.[6–8,11,90,92,93] Over half of the SPG2 families, as well as null mutations, which give rise to a mild classical form of PMD, have affected females,[8,11,13,14,89,92,93] indicating a semidominant pattern of X-linked inheritance in these families (Table 228-1).

Correlation of Phenotype and Genotype

Mutations at the PLP locus give rise to a clinical continuum ranging from a lack of motor development and early death in the severe connatal form of PMD to mild gait disturbances in the pure form of SPG2. While any sequence alteration in the PLP gene appears to yield a phenotype, the most dramatic disruption of myelination occurs with certain missense mutations (Table 228-1). A model accounting for the distress inflicted on oligodendrocytes by missense mutations in PLP/DM20 has been formulated by Gow and Lazzarini, based on their observations that the intracellular trafficking of mutant PLP proteins is disrupted in transfected cells or in affected oligodendrocytes.[47,95] Mutant PLP gene products accumulate at their site of synthesis in the RER,[47,95] previously detected as grossly distended RER in PMD and in the animal models.[59] Point mutations in PLP/DM20 that eliminate the export of both DM20 and PLP from the RER are ones that create the most severe, connatal form of PMD, whereas mutations that selectively arrest PLP transport while allowing mutant DM20 to traffic to the plasma membrane result in the less severe classical form of PMD or spastic paraplegia.[47] Excessive amounts of misfolded proteins in the RER can lead to apoptosis of differentiated oligodendrocytes following axonal contact and up-regulation of myelin protein synthesis.[95] The toxic effects of mutant PLP and DM20 proteins are further illustrated by the inability of normal PLP/DM20 to correct the dysmyelinating phenotype in transgenic models[96,97] and by the enhanced survival of cultured affected oligodendrocytes treated with PLP/DM20 antisense oligodeoxynucleotides.[98]

Missense mutations are distributed equally between the transmembrane sections of PLP and the remainder of the protein. Nonetheless, mutational hot spots are evident. For the regions of the protein not embedded in the membrane, 70 percent of the missense mutations occur in the large extracellular loop derived from exons 4 and 5 (schematized in Fig. 228-2 and listed as D202N through S225P in Table 228-1). This loop, which in compact myelin would be located within the intraperiod line, may participate in protein-protein interactions crucial for myelin maintenance. Only three of the missense mutations in the PLP gene represent conservative amino acid changes, one of which (A242V) creates the most devastating impact on myelin formation in mouse and man,[55,56] and another (H139Y) that presents at the other end of the clinical spectrum with spastic paraplegia.[6] The remainder of the missense mutations impart obvious changes to the PLP structure; examples include the Y206C[99] and G220C[100] mutations, which would introduce cysteine residues into the disulfide bridge-containing large extracellular loop (Fig. 228-2). The two missense mutations that least perturb the structure of PLP/DM20, as judged by the very mild phenotype of the pure form of SPG2 in these cases, are located in the third transmembrane domain (S169F)[12] and the large extracellular loop (S225P)[9] of PLP and DM20.

Missense and nonsense mutations that affect PLP transcripts but spare the alternatively spliced DM20 isoform of PLP (H139Y[6] and W144X[11], in Table 228-1) create the much less severe disorder of SPG2. The form of SPG2 in these patients is the complicated one, probably due to the presence of abnormal PLP proteins in oligodendrocytes. An additional missense mutation which is present in the exon exclusive to PLP transcripts, but at the very end of the exon 3B (K150N[101,102] of Table 228-1), also alters the donor splice site and is anticipated to affect the synthesis of both PLP and DM20. This mutation does, in fact, produce the more severe phenotype of PMD.

The overexpression (e.g., duplications) or underexpression (e.g., null mutations or deletions of the entire gene) of PLP/DM20 generally manifests with less extreme phenotypes than the expression of abnormal proteins (e.g., the missense, frameshift, and splice-site mutations). These intermediate phenotypes range from the classical form of PMD to the complicated form of SPG2. Both the mutation of the initiator methionine[89] and a complete deletion of the PMD gene[70] give rise to a long-lived classical form of PMD. These loss-of-function mutations reveal that PLP/DM20 performs an essential role in myelination that cannot be compensated by other proteolipid proteins in humans. The transgenic models of loss-of-function mutations with normal numbers of mature oligodendrocytes, on the other hand, show that PLP/DM20 is not required for oligodendrocyte development or

maturation, and reinforce the notion that many point mutations in PLP/DM20 poison oligodendrocytes through a general block in protein transport from the RER. Too much PLP can similarly undermine myelination, as seen by the duplication of the PLP gene in the majority of PMD patients.[7,14,20,74–77] While one extra copy of PLP yields the classical form of PMD, additional copies create the features of the connatal form of PMD, as characterized in a patient with a possible triplication of the PLP gene.[75–77] The correlation between copy number and severity of disease has also been documented in the transgenic mouse overexpressors.[71–73]

Myelination is a unique process that demands coordinate transcriptional and translational controls both to synthesize the vast quantities of myelin proteins and to assemble the appropriate proportions of each myelin protein into the myelin sheath.[44] It comes as no surprise that PLP, the most abundantly produced gene in central nervous system myelin, is a gene that is subject to dosage control. An analogous situation occurs in the peripheral nervous system, where duplication of one of the myelin genes expressed by Schwann cells, peripheral myelin protein 22 (PMP22), results in the myelin disorder of Charcot-Marie-Tooth disease type 1A.[105] Dosage control does not operate on all of the genes encoding myelin proteins, as no dysmyelination occurs when another major central nervous system protein (myelin basic protein) is overexpressed in transgenic mice.[104] Perhaps a single myelin gene, acting as a kind of pacemaker, is key to balancing the production levels of the other myelin constituents. Such a pacemaker role for myelination in the central nervous system would be carried out by the PLP gene, and for the PNS, PMP22 might be the regulator.

No change at the PLP locus, however subtle, goes unnoticed. The dramatic range of clinical phenotypes can be explained largely by the extent of apoptosis: mutations that highly induce the apoptotic pathway cause the most derangement. The loss of oligodendrocytes is much more debilitating than the loss of myelin sheaths or the manufacture of abnormal myelin, a phenomenon confirmed by other myelin gene mutants[104] that may point to nervous system functions of oligodendrocytes extending beyond myelin formation. Oligodendrocytes may supply the neurons that they ensheath with essential cytokines, and in the absence of these trophic molecules, neurons eventually degenerate. The cellular pathology of PMD/SPG2 illustrates how mutations that generate the mildest phenotypes in affected male patients are also the ones most likely to create problems for the "carrier" mothers. In these cases, the mutant oligodendrocytes survive in the female heterozygotes and contribute to myelin sheath formation. The abnormal myelin is unstable and breaks down over time, resulting in a late onset degenerative disorder with variable symptoms. The aversion of the PLP gene to change must be factored into therapeutic proposals for PMD/SPG2. With the exception of the loss of function mutations in the PLP gene, the PLP mutations behave dominantly within the oligodendrocyte population in which they are expressed, favoring strategies to disable the defective gene or eliminate expression of the extra copy of the PLP gene. Supplying a wild-type PLP gene can be especially problematic because of the strict dosage control displayed by PLP. The animal models of PMD/SPG2 are valuable resources for working through the special challenges posed by genes such as PLP whose expression is highly regulated. Indeed, the recent success by Duncan and coworkers in transplanting oligodendrocyte precursor cells in these animal models[129] highlights a promising approach for treating myelin deficiencies.

REFERENCES

1. Pelizaeus F: Über eine eigentumliche Form spastischer Lähmung mit Zerebralerscheinungen auf hereditärer Grundlage (multiple Sklerose). *Arch Psychiatr Nervenkr* **16**:698, 1885.
2. Merzbacher L: Eine eigenartige familiäre Erkrankungsform (Aplasia axialis extracorticalis congenita). *Z Gesamte Neurol Psychaitr* **3**:1, 1910.
3. Seitelberger F: Die Pelizaeus-Merzbacher Krankheit, Klinisch-anatomische Untersuchungen zum Prolbem ihrer Stellung unter den diffusen Sklerosen. *Wien Z Nervenheilkd* **9**:228, 1954.
4. Seitelberger F: Pelizaeus-Merzbacher disease, in Vinken PJ, Bruyn GW (eds): *Handbook of Clinical Neurology. Leucodystrophies and Poliodystrophies*, Vol 10. Amsterdam, North Holland, 1970, p 150.
5. Roizin L, Haymaker W, D'Amelio F, Adams RD, Willson N, Kaufman MA: Disease states involving the white matter of the central nervous system, in Haymaker W, Adams RD (eds): *Histology and Histopathology of the Nervous System*, Vol 1. Springfield, IL, C.C. Thomas, 1982, p 1375.
6. Saugier-Veber P, Munnich A, Bonneau D, Rozet JM, Le Merrer M, Gil R, Boespflug-Tanguy O: X-linked spastic paraplegia and Pelizaeus-Merzbacher disease are allelic disorders at the proteolipid protein locus. *Nat Genet* **6**:257, 1994.
7. Nave KA, Boespflug-Tanguy O: X-linked developmental defects of myelination: From mouse mutants to human genetic diseases. *Neuroscientist* **2**:33, 1996.
8. Bond C, Si X, Crisp M, Wong P, Paulson G, Boesel C, Dlouhy S, et al: Family with Pelizaeus-Merzbacher disease/X-linked spastic paraplegia and a nonsense mutation in exon 6 of the proteolipid protein gene. *Am J Med Genetics* **71**:357, 1997.
9. Cambi F, Tang XM, Cordray P, Fain PR, Keppen LD, Barker, DF: Refined genetic mapping and proteolipid protein mutation analysis in X-linked pure hereditary spastic paraplegia. *Neurology* **46**:1112, 1996.
10. Kobayashi H, Hoffman EP, Marks HG: The rumpshaker mutation in spastic paraplegia. *Nat Genet* 7:351, 1994.
11. Osaka H, Kawanishi C, Inoue K, Uesugi H, Hiroshi K, Nishiyama K, Yamada Y, et al: Novel nonsense proteolipid protein gene mutation as a cause of X-linked spastic paraplegia in twin males. *Biochem Biophys Res Commun* **215**:835, 1995.
12. Hodes ME, Hadjisavvas A, Butler IJ, Aydanian A, Dlouhy SR: X-linked spastic paraplegia due to a mutation (C506T; Ser169Phe) in exon 4 of the proteolipid protein gene (PLP). *Am J Med Genet* **75**:516, 1998.
13. Naidu S, Dlouhy SR, Geraghty M, Hodes ME: A male child with the *rumpshaker* mutation, X-linked spastic paraplegia/Pelizaeus-Merzbacher disease and lysinuria. *J Inherit Metab Dis* **20**:811, 1997.
14. Hodes ME: Pelizaeus-Merzbacher disease, in: Gilman S, Goldstein GW, Waxman SG (eds). *Neurobase*, 4th ed. San Diego, CA, Arbor Publishing, 1998.
15. Hodes, ME, Pratt VM, Dlouhy SR: The genetics of Pelizaeus-Merzbacher disease. *Dev Neurosci* **15**:383, 1993.
16. Cassidy S, Sheehan N, Farrell D, Grunnet M, Holmes G, Zimmerman A: Connatal Pelizaeus-Merzbacher disease: An autosomal recessive form. *Pediatr Neurol* **3**(5):300, 1987.
17. Zeman W, DeMyer W, Falls JS: Pelizaeus-Merzbacher disease: A study in nosology. *J Neuropathol Exp Neurol* **23**:334, 1964.
18. Boulloche J, Aicardi J: Pelizaeus-Merzbacher disease: Clinical and nosological study. *J Child Neurol* **1**:233, 1986.
19. Scheffer I E, Baraitser M, Wilson J, Harding B, Kendall B, Brett EM: Pelizaeus-Merzbacher disease: Classical or connatal? *Neuropediatrics* **22**:71, 1991.
20. Sistermans E, deCoo R, DeWijs I, Van Oost B: Duplication of the proteolipid protein gene is the major cause of Pelizaeus Merzbacher disease. *Neurology* **50**:1749, 1998.
21. Johnston AW, McKusick VA: A sex-linked recessive form of spastic paraplegia. *Am J Hum Genet* **14**:83, 1962.
22. Bonneau D, Rozet JM, Bulteau C, Berthier M, Mettey R, Gil R, Munnich A, et al: X-linked spastic paraplegia (SPG2): Clinical heterogeneity at a single gene locus. *J Med Genet* **30**:381, 1993.
23. Tanaka S, Mito T, Takashima S: Progress of myelination in the human fetal spinal nerve roots, spinal cord and brainstem with myelin basic protein immunohistochemistry. *Early Hum Dev* **41(1)**:49, 1995.
24. Volpe, J: Neuonal Proliferation, migration, organization and Myelination, in *Neurology of the Newborn*, 3d ed. Philadelphia, WB Saunders, 1995, p.78.
25. Caro PA, Marks HG: Magnetic resonance imaging and computed tomography in Pelizaeus-Merzbacher disease. *Magn Reson Imaging* **8**:8128, 1990.
26. Andre M, Monin P, Moret C, Brown M, Picard L: Pelizaeus-Merzbacher disease. Contribution of magnetic resonance imaging to an early diagnosis. *J Neuroradiol* **17**:216, 1990.
27. Ono J, Harada K, Mano T, Sakurai K, Okada S: Differentiation of dys- and demyelination using diffusional anisotropy. *Pediatr Neurol* **16(1)**:63, 1997.

via the fermentation of glucose by glycolysis. The main role of the citric acid cycle in *E. coli* is in the generation of biosynthetic precursors and reducing power [reduced nicotinamide adenine dinucleotide (NADH)]. *E. coli* citrate synthase is unaffected by ATP, but it is inhibited by one of the ultimate end products of the cycle, NADH. Germinating seeds contain a third type of citrate synthase, localized in the glyoxysome. This isozyme is insensitive to both ATP and NADH, and here the enzyme functions as part of the glyoxylate cycle, an indispensable component in the metabolic conversion of fatty acids derived from storage triacylglycerides into sucrose. Thus, the concept of biochemical unity tends to break down when individual metabolic controls are compared. Although the structure and products of a metabolic enzyme or a metabolic pathway may be identical in various organisms, the environment and function of that enzyme or pathway may not be the same. Nevertheless, all metabolic controls have a common basis, and certain regulatory strategies are ubiquitous.

BASIS OF METABOLIC CONTROL

Pacemaker Enzymes

It is self-evident that the flux of metabolites through any pathway must be closely coordinated with the needs of the cell, tissue, or organism for the final end product(s) of the pathway. The traditional view of metabolic control is that such regulation is accomplished by altering the activity of at least one pacemaker enzyme (or *rate-determining step*) of the pathway. Thus, a major focus of enzymology has been to characterize these key enzyme reactions—the pacemakers—that are believed to be most important in controlling pathway flux. Substantial efforts have been directed to identifying the pacemaker enzyme(s) of metabolic pathways, as well as the complex mechanisms that serve to modulate the activities of these key enzymes.

Identification of a Pacemaker Enzyme

Normally, the pacemaker enzyme(s) of a pathway has a low activity overall, is subject to control by metabolites other than its substrates, and is often positioned as the first committed step of a pathway, directly after major branch points, or at the last step of a "multi-input" pathway. However, such circumstantial evidence for a putative pacemaker enzyme-catalyzed reaction still requires confirmation by techniques that rely upon accurate quantification of the *in vivo* concentrations of the enzyme's substrate(s) and product(s).

The standard method for determining metabolite concentrations begins with ultrafast freezing of cells/tissues in liquid nitrogen at $-196°C$. This effectively quenches any reactions that could lead to artifactual alterations in metabolite levels. The frozen tissue is then extracted by homogenization at low pH in cold perchloric acid, which inactivates enzymes that would affect the metabolites of interest. Following centrifugation to remove cell debris and precipitated proteins, the supernatant is neutralized and analyzed for metabolites by appropriate enzymatic or chemical methods. More recently, ^{31}P nuclear magnetic resonance (^{31}P-NMR) has become widely used for determining the intracellular concentrations of phosphate-containing metabolites such as the hexose-phosphates, adenylates, phosphocreatine, and inorganic phosphate. It is now feasible to measure the concentrations of phosphate-containing metabolites in a tissue, perfused organ, or even an intact living organism inserted into the wide-bore magnet of a NMR spectrometer. Results obtained from ^{31}P-NMR are generally consistent with those obtained by acid extraction techniques. However, noninvasive NMR techniques continue to hold great promise for providing a detailed insight into metabolite levels of living cells, and how these levels may vary following perturbations such as oxygen deprivation, or the addition of metabolites or hormones. Whether obtained via classical or NMR methods, the amount of a metabolite (determined as micromoles per gram of tissue), can also be expressed as concentration [i.e., in millimolars (mM)] in the intracellular water if the water content of the tissue is known.

Accurate determinations of metabolite levels can be difficult, particularly if the same metabolite is distributed between several intracellular compartments. Nevertheless, the availability of metabolite data is quite relevant to determining how and where flux control of a specific metabolic pathway is exerted, that is, probable pacemaker enzyme(s). A major controlling enzyme should theoretically catalyze the slowest step in the pathway. Thus, a reaction that is far from equilibrium *in vivo* is likely to be catalyzed by a pacemaker enzyme (although under certain conditions, enzymes that catalyze reactions close to equilibrium may also be regulatory). The reason that equilibrium is not achieved is that owing to the action of various "fine" metabolic controls (such as feedback allosteric inhibition; see below) there is insufficient active enzyme present to bring the reaction to equilibrium *in vivo*. Hence, a metabolic biochemist interested in identifying important sites of pathway control initially searches for those enzymes that catalyze nonequilibrium reactions *in vivo* (i.e., catalyze irreversible reactions that have a highly negative overall free energy change). This requires measurement of the intracellular concentrations of the particular metabolites involved. Results obtained with a rat heart perfused with glucose are listed in Table 1.1.

For the glycolytic reaction catalyzed by 6-phosphofructokinase (PFK): $F6P + ATP \rightarrow FBP + ADP$, the ratio of its intracellular concentrations of products : substrates, known

TABLE 1.1 Intracellular Concentration of Several Metabolites Obtained with a Rat Heart Perfused with Glucose

Metabolite	Intracellular Concentration (mM)
D-Fructose-6-phosphate (F6P)	0.09
D-Fructose-1,6-bisphosphate (FBP)	0.02
ATP	11.5
ADP	1.3
AMP	0.17

Source: Price and Stevens (1989).

as the mass action ratio (Γ), is given by

$$\Gamma = \frac{[\text{FBP}][\text{ADP}]}{[\text{F6P}][\text{ATP}]} = 0.025$$

where F6P is fructose-6-phosphate, FBP is fructose-1,6-biphosphate and ADP is adenosine 5′-diphosphate. The equilibrium constant (K_{eq}) for a chemical reaction, a value that is independent of the presence or absence of a catalyst (enzyme), can be determined in separate experiments under physiological pH, temperature, and pressure. The K_{eq} for the aforementioned reaction is about 1200. Since this value is 48,000-fold greater than the reaction's corresponding Γ value in perfused rat heart, it is evident that the reaction *in vivo* is very far removed from equilibrium and, thus, is essentially irreversible. Therefore, PFK is regarded as a probable pacemaker enzyme of the glycolytic pathway. This is logical because PFK catalyzes the first committed step of glycolysis, that is, the first step that does not form part of other metabolic processes as well. By contrast, for the reaction catalyzed by adenylate kinase: ATP + AMP $\leftrightarrow$ 2ADP,

$$\Gamma = \frac{[\text{ADP}]^2}{[\text{ATP}][\text{AMP}]} = 0.85$$

where AMP is adenosine 5′-monophosphate. This is very close to the reaction's K_{eq} value of 0.44. This indicates that rat heart adenylate kinase is sufficiently active *in vivo* to allow the reaction to remain very close to equilibrium (or readily reversible).

Another method used to identify pacemaker enzyme(s) that has been employed is the "crossover theorem," which states that when pathway flux is enhanced, the substrate concentration of the pacemaker enzyme will decrease and its product concentration will increase (and vice versa for a flux decrease). A crossover in relative metabolite levels between two physiological states indicates where a regulatory signal has acted on a pacemaker enzyme to alter flux. For example, in yeast, as well as in many animal and plant tissues, oxygen deprivation greatly enhances glycolytic flux (owing to the need to generate ATP via anaerobic fermentation). This is associated with an immediate reduction in intracellular [F6P] and an increase in [FBP]. This results in a "positive crossover" between F6P and FBP and indicates that the perturbation (anoxia) has activated PFK *in vivo* so that the concentration of its substrate, F6P, is lowered whereas that of its product, FBP, is increased. An elevated [FBP] would directly cause an increased flux through the next enzyme in the glycolytic pathway, FBP aldolase (which catalyzes a reaction close to equilibrium and shows no crossover following the aerobic to anoxic transition). This flux increase would then be transmitted to the remainder of the glycolytic sequence. Crossover analysis clearly identifies PFK as a pacemaker enzyme whose activity can respond appropriately to facilitate the marked stimulation of glycolysis that accompanies the imposition of anoxia stress. This example also underscores an important principle in metabolic biochemistry research: namely, that the full elucidation of pathway control is often dependent on a thorough comparative analysis of a control versus perturbed (i.e., stressed) tissue. Correlation of alterations in metabolic fluxes and metabolite levels that ensue from a perturbation of cells/tissues generates indispensable information for the metabolic biochemist seeking to identify key sites of control in a particular pathway.

Enzyme Purification

From the various approaches described above it is possible to pinpoint the most important sites of control of a metabolic pathway. The next step is to examine the molecular and kinetic properties of the putative pacemaker enzyme(s), particularly those properties that might be involved in the control of pathway flux. Of particular interest are the following questions:

1. What is the enzyme's subunit structure? Pacemaker enzymes are invariably *multimeric*; that is, in their native state they consist of two or more subunits held together by noncovalent bonds. A more complex protein structure appears to be a prerequisite for a more complex protein function/regulation (i.e., allosterism, etc.).

2. How does the activity of the enzyme vary with alterations in assay pH and substrate(s) concentration(s)? Are hyperbolic (Michaelis–Menten) or sigmoidal (cooperative) substrate saturation kinetics observed? As discussed below, the sigmoidal substrate saturation plot observed for some pacemaker enzymes implies that over a certain range of substrate concentrations, the activity is more sensitive to

[S] than would be the case for enzymes that display hyperbolic kinetics.

3. Is the activity of the enzyme controlled by any metabolites (referred to as *effectors*) that are structurally distinct from its own substrate(s) or product(s)? If so, (a) what is the nature of the interaction (i.e., activation or inhibition) and (b) do these metabolite effectors significantly alter enzyme activity *in vitro* at concentrations that are known to exist *in vivo*?

4. Is the activity of the enzyme subject to control by covalent modification such as by reversible protein phosphorylation, and if so, what mechanisms serve to control the activities of the modifying enzymes that catalyze these changes?

It is clear that if we hope to gain a detailed understanding of the behavior of an enzyme in a complex biological system (such as an isolated organelle, intact cell, or entire organism) we must first attempt to understand its properties in as simple a system as possible. Thus, one cornerstone of metabolic biochemistry research has been the reductionist approach of enzyme purification and characterization. Enzyme purification eliminates metabolites and contaminating proteins that would otherwise confound kinetic and/or structural studies. By performing accurate and detailed analyses of the kinetic and regulatory properties of purified key enzymes *in vitro* and then combining these with knowledge of the enzyme's subcellular localization and of the *in vivo* concentrations of the enzyme's substrates, products, and metabolite effectors, one attempts to formulate a theory for the control of the enzyme *in vivo*. Such information can have broad practical applications and may be used to provide insights about various metabolic diseases, facilitate the identification of suitable targets for the disruption of a pathogen's metabolism with drugs, or generate key information for biotechnologists wishing to manipulate specific metabolic pathways via metabolic engineering. From *in vitro* studies of a purified enzyme we can also learn about its structure, specificity for substrates, and reaction mechanism. With sufficient quantities of purified enzyme, structural biochemists can employ powerful methods such as chemical modification and covalent labeling, peptide isolation and sequencing, X-ray crystallography, mass spectroscopy, NMR spectroscopy, and so on, to determine the relationship between an enzyme's structure and its function. This leads to the identification of key amino acid residues that are critical in substrate/product or allosteric effector binding, catalysis and covalent modification (i.e., phosphorylation) sites. Integration of biochemical (kinetic and structural data) and genetic information for a given enzyme ultimately provides a logical basis for site-directed mutagenesis by suggesting which amino acid(s) should be modified to produce a mutant enzyme with altered kinetic/regulatory properties.

Cautionary Note about Relating *In Vitro* Kinetic Studies of a Purified Enzyme to Its *In Vivo* Function and Control

The traditional approach to metabolic control outlined above is largely based upon the extrapolation of the *in vitro* kinetic/regulatory characteristics of purified enzymes to conditions within the intact cell. However, it is well-known that large multimeric regulatory enzymes such as PFK are susceptible to artefactual posttranslational modifications such as partial proteolysis by endogenous proteases (during purification), as well as dilution-dependent alterations in their oligomeric structure. Even very minor proteolytic clipping may obscure the allosteric properties of a purified regulatory enzyme, without markedly influencing its V_{max}. The diagnosis and prevention of unwanted proteolysis, through the addition of appropriate protease inhibitors to purification buffers, should be a major concern of all enzymologists.

The influence of protein concentration must also be considered because enzymes are present *in vivo* at far greater concentrations than they are during *in vitro* kinetic assays. Concentration dependence is believed to be particularly significant for enzymes important in metabolic control because their structure, and hence kinetic/regulatory properties, may be affected by protein–protein interactions. The interactions between enzyme subunits that normally exist at the high protein concentration prevailing *in vivo* can be specifically promoted *in vitro* by the addition of molecular crowding agents such as 10% (v/v) glycerol or 5% (w/v) polyethylene glycol to the reaction mixture. The mechanism involves exclusion of the protein from the aqueous solvent, thus increasing local enzyme concentration and favoring protein–protein interactions. The *in vitro* activity of many regulatory oligomers, including rat liver PFK, can be enhanced by the presence of such molecular crowding agents and can aid the examination of the catalytic and regulatory properties of enzymes in an *in vitro* environment that may be a closer approximation of the conditions prevailing *in vivo*.

Compartmentation of Metabolism

In the intact cell the individual enzymes of a metabolic pathway function to convert a starting material to end product(s) without necessitating the accumulation of elevated concentrations of the corresponding metabolic intermediates. In addition, there are many connections between the major metabolic pathways, with selected substrates, cofactors, regulatory molecules, and occasionally enzymes being common to more than one pathway. These complex interactions can only be fully understood when, in addition to studying the isolated enzymes, some knowledge has been acquired concerning the intracellular location and concentrations of the enzyme and metabolites involved, and of any permeability barriers that separate the individual components. Thus, the functions of many enzymes cannot be fully understood without knowledge of their subcellular location.

The appearance of eukaryotic cells during evolution was associated with the process of *compartmentation* of metabolism through the formation of specialized organelles such as lysosomes, mitochondria, and plastids, separated from the cytosol and from each other by selectively permeable membranes. Compartmentation concentrates enzymes of a metabolic pathway and their associated metabolites, and prevents the simultaneous occurrence of potentially incompatible metabolic processes. The integration of cellular metabolism necessitates controlled interactions between pathways sequestered in the various subcellular compartments. This is facilitated by the existence of numerous membrane transporter proteins that selectively translocate specific metabolites between subcellular compartments. Thus, a major advance in the biochemical study of eukaryotic cells was the development of methods for separating intact organelles from the cytosol and from each other. This has not only allowed analysis of the overall biological functions of isolated organelles (i.e., respiration in mitochondria, photosynthesis in chloroplasts, etc.) but has also facilitated the determination of the distribution of enzymes and metabolites within cells, as well as the various metabolite translocators of the organelle membranes. Understanding these transport processes is of great importance for the overall understanding of metabolism and its control.

Formulation of a Theory of Metabolic Control

Once the controlling enzyme(s) of a given pathway have been identified and their kinetic/regulatory properties and subcellular localization investigated, it should be feasible to postulate a theory for the control of pathway flux. The theory should give rise to predictions that can be tested experimentally. This often necessitates collecting data about the activities and state of covalent modification (i.e., phosphorylation status) of the key enzymes, and the concentrations of their substrates and effectors under a variety of physiological conditions. It is important to demonstrate that these parameters reflect the *in vivo* situation as closely as possible. For example, is the *in vivo* ratio of a pacemaker enzyme's [activators] : [inhibitors] proportional to pathway flux (i.e., following pathway stimulation, do the levels of the enzyme's activators and inhibitors, respectively, increase and decrease, and vice versa)? Similarly, are the variations in the *in vivo* concentrations of effectors sufficient to account for the observed changes in enzyme activity or pathway flux? If not, then the initial theory must be revised and retested. This may require a more detailed investigation of the properties of the isolated enzyme.

One problem with the traditional pacemaker approach to metabolic control is that the studies are for the most part *qualitative* rather than *quantitative*. For example, the demonstration that a particular enzyme catalyzes a nonequilibrium reaction *in vivo*, shows a positive crossover during pathway activation, and exhibits pronounced control properties *in vitro* provides a unequivocal indication that it is relatively important in controlling pathway flux *in vivo*. However, designation of an enzyme as a pacemaker is not based upon any direct measurement of the precise contribution of each enzyme in a pathway to the overall control of pathway flux. Where more than one enzyme appears to be "regulatory," there is no estimate as to their relative contributions, or how the degree of control exerted by each enzyme might vary under differing physiological circumstances. Furthermore, biological systems may display regulatory properties that are not possessed by their isolated components. In other words, the properties of biological systems tend to be greater than the sum of the properties of their isolated parts (i.e., the so-called Humpty-Dumpty principle). For instance, it would be impossible to understand how a mitochondrion functions in respiration by only studying purified mitochondrial enzymes and electron transport proteins in isolation from each other and from mitochondrial membranes. Thus, another important approach to the problem of metabolic control is to analyze the whole system.

METABOLIC CONTROL ANALYSIS

The metabolic control analysis (MCA) theory developed by Kacser in 1973 attempts to provide a quantifiable mechanism for probing intact biological systems and interprets the resulting data without preconceived notions as to which enzymes in a pathway are rate-determining steps or pacemakers (Fell, 1997). In fact, an important tenet of MCA theory is that metabolic control is shared among many, if not all, steps in a pathway. As discussed in more detail in a later chapter (and in references at the end of this chapter), Kacser has established the concept of the *flux control coefficient* (C_E^J) whose value specifies the change in metabolic flux (ΔJ) that results from a small change in the activity of any enzyme (ΔE) in the metabolic system as follows: $C_E^J = (\Delta J/\Delta E)$. For a linear pathway, the flux control coefficients of the component enzymes will lie between 0 and 1, with higher values indicating a greater contribution to metabolic control.

However, for complex pathways involving branches or substrate cycles, the flux control coefficients can hold any value, less than 0 or greater than 1. It should be emphasized that measurement of a flux control coefficient for a single step in a metabolic pathway may be difficult to interpret. The most satisfactory, yet highly challenging, way to apply MCA to a pathway would be to estimate the flux control coefficients for each component enzyme, as values for a single step may mislead.

Experimental determination of the magnitude of appropriate flux control coefficients apparently yields an unambiguous evaluation of the existing quantitative allotment of control among the various steps in a pathway, under specified conditions. Moreover, since the values of flux control coefficients can redistribute between enzymes according to physiological circumstances, any particular flux control coefficient applies only to the physiological state in which it was determined. Because metabolic control theory predicts that all enzymes in a pathway exert some control on pathway flux, all enzymes (in a linear pathway) should theoretically have flux control coefficients greater than 0. However, no single enzyme would be expected to have a flux control coefficient as high as 1.0 (which would be the case if pathway flux were entirely controlled by a single pacemaker enzyme). In fact, the *summation theorem* states that the sum of flux control coefficients for all component enzymes of a metabolic system should equate to 1.0.

The magnitude of any one flux control coefficient is not an intrinsic property of the enzyme per se but is a system property that depends upon the concurrent activities of all the other enzymes in the pathway. Thus, the value of a flux control coefficient cannot be determined by considering the properties of a purified enzyme since the characteristics and amounts of other enzymes in the system will influence the result. Individual flux control coefficients must therefore be determined experimentally from the intact system by measuring how pathway flux changes following alterations of the activity of a specific enzyme *in situ*. Advances in molecular biology now allow for direct manipulations of *in vivo* enzyme activities and continue to yield new information on the control of metabolism. There are a number of excellent publications and several Internet sites devoted to the quantitative MCA approach to metabolic control (see references). The various formulations and concepts of the mathematical models of MCA have given rise to considerable debate over the meaning and usefulness of flux control coefficients (see Text Box 1.1). However, most metabolic biochemists would likely agree that flux control of a metabolic pathway is generally dominated by a minority of its component enzymes (i.e., the pacemakers), although under different physiological conditions the degree of control exerted by the individual enzymes may vary.

TEXT BOX 1.1 THE PFK PARADOX

Phosphofructokinase (PFK) is generally considered to be an important pacemaker enzyme of the glycolytic pathway. It catalyzes the first unique step of glycolysis, a nonequilibrium reaction *in vivo* and shows a strong positive crossover concomitant with glycolytic stimulation. PFK is a multimeric enzyme that displays sigmoidal substrate (F6P) saturation kinetics as well as complex and potent allosteric regulation by numerous effectors, the levels which are controlled by the hormonal and/or nutritional status of the tissue. For example, the role of fructose-2,6-bisphosphate as a potent allosteric activator of animal and yeast PFK is well-established. However, the use of molecular genetic techniques for the selective overexpression of PFK in yeast, mammals, and plants has failed to yield significant increases in glycolytic flux or respiration that were expected. It appears that the elevated PFK concentration was somewhat compensated for *in vivo* by changes in the levels of PFK allosteric activators and inhibitors. PFK flux control coefficients were determined to be very small, leading to the surprising conclusion that PFK exerts very little or no control over glycolytic flux or respiration in vivo. Although proponents of MCA have challenged the traditional concept that PFK is a pacemaker enzyme of glycolysis, several MCA advocates have nevertheless agreed that there is little doubt that "control of PFK activity plays a part in glycolytic flux control" (Thomas and Fell, 1998) and "PFK makes an important contribution to the control of glycolysis in most cells" (Cornish-Bowden, 1999). A possible explanation for this "PFK paradox," is that MCA also indicated that significant flux control of glycolysis and respiration lies in the metabolism of key feedback inhibitors of PFK, namely ATP and citrate in yeast and mammalian cells and phospho*enol*pyruvate in plant cells. This would suggest the somewhat contradictory conclusion that, although the flux control coefficient for PFK may be low, it does indeed play an important role in the control of carbohydrate catabolism in most cells.

MECHANISMS OF METABOLIC CONTROL

The magnitude of metabolite flux through any metabolic pathway will depend upon the activities of the individual enzymes involved. It is possible to group mechanisms of metabolic control into two major classes on the basis of the relative lengths of time they take to bring about a change in the velocity of a particular enzyme. These are "coarse" and "fine" metabolic control.

Coarse Metabolic Control

Coarse metabolic control is a long-term (hours to days in eukaryotes; perhaps minutes to hours in rapidly growing prokaryotes), energetically expensive, response that is achieved through changes in the total cellular population of enzyme molecules. The total amount of a given enzyme is dependent upon the relative rates of its biosynthesis versus degradation. Thus, any alteration in the rates of gene expression [i.e., transcription, translation, messenger ribonucleic acid (mRNA) processing or degradation] or proteolysis can be considered as coarse metabolic control. Coarse control can be applied to one or all the enzymes in a particular pathway and most frequently comes into play in response to hormonal stimulation and/or during tissue differentiation or long-term environmental (adaptive) changes. The dynamic range of coarse metabolic control can be large, particularly when a previously absent enzyme is induced and rises to high levels in response to a stimulus. Coarse metabolic control might appear to be an inefficient and wasteful use of energy since each peptide bond formed requires the hydrolysis of several ATP equivalents, whereas protease activity is not coupled to the production of ATP (but as discussed below can also be ATP-dependent). However, coarse control may be particularly important when a cell must acclimate to changes in its environment, or it becomes necessary to remove abnormal enzyme molecules that have become damaged or arisen by errors in gene expression. In general, the longer the life of an individual cell the more important is the process of intracellular enzyme turnover. For example, in *E. coli* growing under optimal conditions, mitosis may occur every 20 min. Acclimation of *E. coli* to its environment occurs largely by the induction or repression of enzyme synthesis. For example, if lactose is added to the growth medium, rapid induction of β-galactosidase (needed to catabolize lactose) occurs. If lactose is then removed from the medium, the enzyme is not synthesized, and existing β-galactosidase molecules will be rapidly diluted out within the cytoplasm during the rapid division of cells. In contrast to rapidly dividing microbes, the average lifetime of a cell in a multicellular eukaryote may be several hundred days or more, but many enzymes need to be completely replaced every few days.

Gene Expression The regulation of transcription and translation is covered in Chapters 6 and 7 and will not be discussed in detail here. However, in the context of metabolic control it is important to note that an underlying assumption of many genomic studies is that the expression of a gene at the mRNA level is a quantitative indicator of function of the encoded enzyme. Thus, an n-fold increase in transcript levels (detected via Northern blots or gene chip screening) equates to n-fold more enzyme and hence n-fold more activity. However, it is becoming apparent that this assumption does not always reflect reality. For example, a study that determined fluxes through steps of a central metabolic pathway (glycolysis) in three parasitic protists found that these did not correlate proportionally with the concentration of the corresponding enzymes; that is, relative to various fine metabolic control mechanisms, gene expression alone exerts little control on glycolytic flux.

Protein Turnover Relative to gene expression, much less is known about the mechanisms governing protein degradation. Animal and plant enzymes that coexist in the same cellular compartment may exhibit vastly different turnover rates, ranging from several minutes to hundreds of hours. In general, larger, oligomeric proteins that display complex biological properties and significant hydrophobicity tend to show much shorter half-lives *in vivo* than do less complex monomeric (and/or less hydrophobic) proteins. It is clear that proteolysis of enzymes can be selectively targeted and may be initiated in response to specific stimuli.

How are Enzymes Selected for Intracellular Degradation? Many enzymes need to first become "tagged" before becoming susceptible to degradation by endogenous proteases. The types of covalent modification used for tagging enzymes for degradation include the formation of a peptide bond between the target enzyme and a protein called ubiquitin, or the modification of the protein by phosphorylation or by oxidation. Ubiquitin (M_r 9000) is so-called because of its widespread occurrence in eukaryotic cells. Its role in protein turnover has been well-established in animals and plants. Certain proteins destined for degradation are covalently bonded to ubiquitin via the NH_2 groups of lysine residues. A single protein may become tagged with many ubiquitin molecules. ATP is required for the ubiquitin conjugation, together with several enzymes. Another method of tagging an enzyme for protease degradation is by phosphorylation, which again is dependent upon the hydrolysis of ATP, in this case by the modifying protein kinase (see below). The ATP requirement of the ubiquitin and phosphorylation tagging systems reflects the bioenergetic cost for endowing the cell with proteolytic specificity. Although tagging methods appear to make many enzymes susceptible to proteolytic attack *in vivo*, it is not yet clear precisely which features of the target enzymes are recognized by the tagging machinery.

Fine Metabolic Control

Fine metabolic controls are generally fast (i.e., seconds to minutes), energetically inexpensive, regulatory devices that modulate the activity of the preexisting enzyme molecule. Operating primarily on the regulatory or pacemaker enzyme(s) of a pathway, fine controls allow the cell to prevent metabolic chaos. Fine controls can be thought of as *metabolic transducers* that "sense" the momentary metabolic needs of the cell and modulate flux through the various pathways accordingly. It is important to note that the fine metabolic controls discussed in detail below are not mutually exclusive but often interact with, or may actually be dependent upon, one another.

Fine Control 1. Alteration in Substrate Concentration The rate of an enzyme-catalyzed reaction is dependent upon [S] when [S] is subsaturating. Substrate concentrations for most enzymes are subsaturating *in vivo*. Often the *in vivo* [S] is less than or nearly equal to the K_m or $S_{0.5}$ value of the enzyme for that substrate. The main exception to this are enzymes such as nucleases, proteases, lipases, phosphorylases, or amylases that catalyze initial steps in macromolecule catabolism, cases where substrate reserves (i.e., glycogen, triglycerides deposits) are huge.

Can pathway flux be controlled by alterations in [S] for any of the enzymes that comprise the pathway? Following stimulation of a metabolic pathway, the concentration of its constituent metabolites may increase severalfold (frequently 2- to 5-fold). However, pathway flux may increase by as much as 100-fold under the same stimulation. For enzymes that obey hyperbolic substrate saturation kinetics (i.e., Michaelis–Menten kinetics where Hill coefficient or $n_H = 1$) (Fig. 1.1*a*), Table 1.2 shows that the increase in [S] must be about 80-fold in order to obtain a 9-fold change in the activity of the enzyme. Such an increase in [S] is rarely if ever seen and so variation in [S] cannot be the sole determinant of the *in vivo* activity of enzymes that follow hyperbolic saturation kinetics.

TABLE 1.2 Quantitative Influence of Hyperbolic versus Sigmoidal Substrate (*S*) Saturation Kinetics on an Enzyme's Response to Variations in Its Substrate Concentration[a]

Value of n_H	Required Change in [S] to Increase V_0 from 10 to 90% of V_{max}
0.5	6561-fold
1.0	81-fold
2.0	9-fold
3.0	4.33-fold
4.0	3-fold

[a]The parameter n_H represents the Hill coefficient, derived by fitting enzyme initial velocity (V_0) versus [S] data to the Hill equation.

Source: Adapted from Price and Stevens (1989).

Not all enzymes show simple Michaelis–Menten kinetics, however. Multimeric enzymes (i.e., consisting of two or more polypeptides in their native state) often contain more than one substrate binding site. Binding of substrate to one subunit can affect the conformation of other subunits and positively enhance the binding of substrate to them. The result of such cooperative binding of substrate is a sigmoidal relationship between enzyme activity and [S] (Fig. 1.1*b*). Enzymes of this nature have been termed *allosteric* because they can assume "other shapes" or conformations by the reversible, noncovalent binding of a specific metabolite. Sigmoidal substrate saturation kinetics has been referred to as *homotropic* allosterism since the allosteric modulator and the substrate are identical. Table 1.2 shows that the same 9-fold increase in activity that required an 80-fold increase in substrate concentration for a hyperbolic enzyme can be achieved with only a 3-to 4-fold increase in [S] for a sigmoidal enzyme (Table 1.2). The actual increase in [S] that is required is dependent upon the degree of cooperativity with which the enzyme binds its substrate. With increased cooperativity (i.e., higher values for n_H reflected by increasing sigmoidal *V* vs. [S] plots) smaller fold increases in [S] are required to give the same relative increase in enzyme activity.

In summary, changes in substrate concentrations that normally occur *in vivo* can alter the rate of pathway flux but do so most effectively for enzymes that show sigmoidal saturation kinetics (i.e., homotropic allosterism). Enzymes of this nature have been found in all phyla. Invariably such enzymes have been identified as pacemakers. Note, however, that not all multimeric or pacemaker enzymes necessarily display sigmoidal substrate saturation kinetics. Although sigmoidal kinetics allow a much more sensitive control of reaction rate by [S], it should be stressed that an alteration in [S] as a mechanism of fine control is often unimportant, relative to the other fine controls discussed below. The real metabolic "advantage" of sigmoidal substrate saturation kinetics is that this may allow metabolite activators (and/or covalent modification) to facilitate a marked enhancement in enzyme activity (at a relatively constant [S]) by promoting an allosteric transition that brings about a reversible shift between hyperbolic and sigmoidal saturation kinetics (see below).

Fine Control 2. Variation in pH Most enzymes have a characteristic pH at which their activity is maximal, that is, the pH optimum. Above or below this pH the activity normally declines, although to varying degrees depending upon the particular enzyme. Thus, enzymes can show pH versus activity profiles ranging in shape from very broad

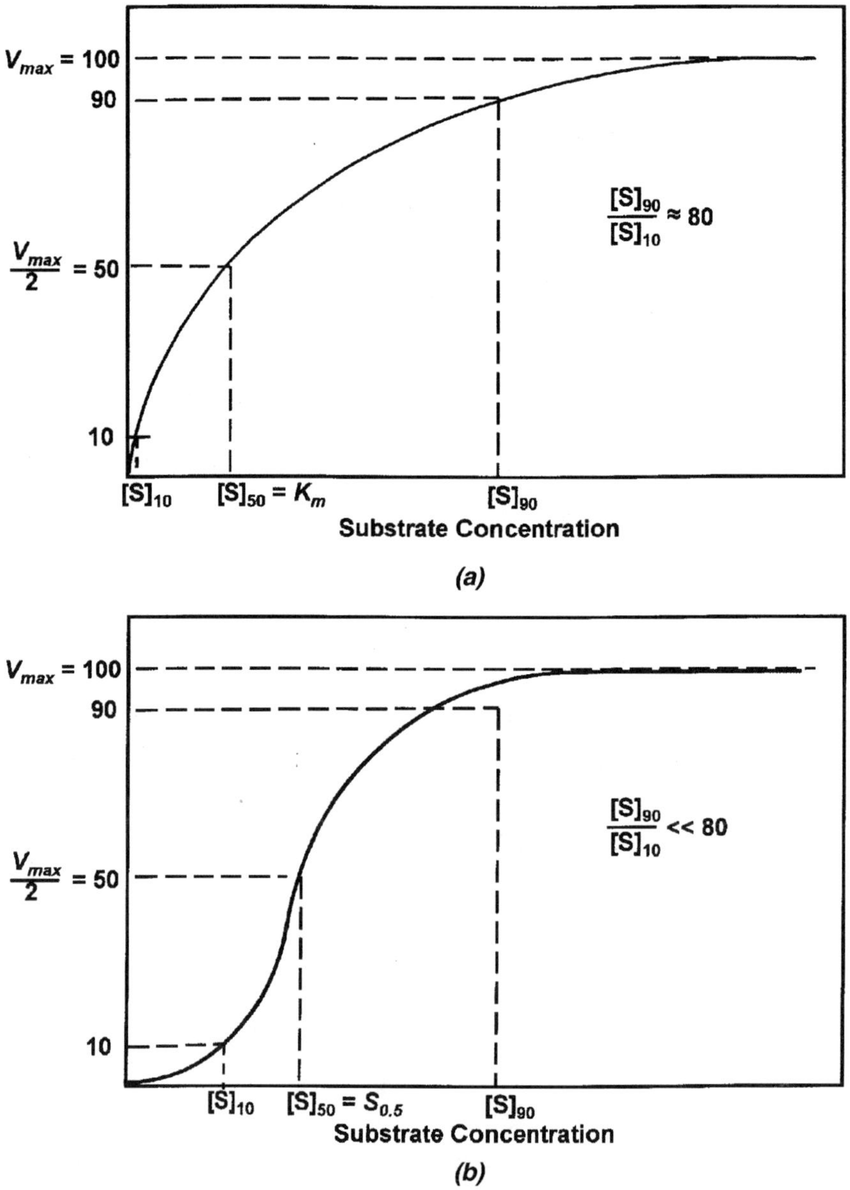

Figure 1.1 Relationship between substrate concentration and reaction rate for enzymes that follow (*a*) hyperbolic or (*b*) sigmoidal substrate saturation kinetics. $[S]_{90}$, $[S]_{50}$, and $[S]_{10}$ denote the respective substrate concentration that yields 90, 50, and 10% of V_{max}, respectively. To increase the activity of an enzyme that follows hyperbolic substrate saturation kinetics (*a*) from 10 to 90% of V_{max}, the increase in its [S] must be about 80-fold. A much lower increase in [S] is needed to achieve the same relative increase in the activity of an enzyme that follows sigmoidal substrate saturation kinetics (see text and Table 1.2 for details).

to very narrow. As the pH optimum of an enzyme is not always the same as the pH of its intracellular surroundings, this suggests that the pH dependence of enzyme activity may be one factor that determines its overall activity in the cell. As all cells contain thousands of enzymes, many of which show very different responses to pH, the intracellular pH (pH_i) may represent an important element of fine metabolic control.

The light-dependent activation of several of the enzymes of the reductive pentose–phosphate pathway (Calvin–Benson cycle) provides a well-documented example of how changes in pH_i can contribute to metabolic control

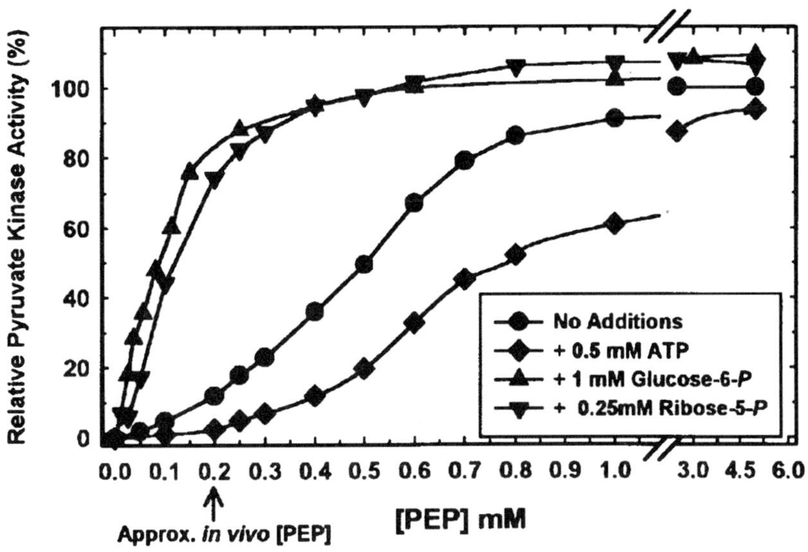

Figure 1.2 Influence of metabolite effectors on the substrate saturation kinetics of purified pyruvate kinase from the cyanobacterium *Synechoccous* PCC 6301 (adapted from Knowles et al., 2001). Note that at the approximate concentration of PEP (phosphoenolpyruvate) substrate that occurs *in vivo* (~0.2 mM), the addition of physiological concentrations of glucose-6-phosphate or ribose-5-phosphate can raise pyruvate kinase activity from ~10% of V_{max} to ~90% of V_{max}, whereas the inhibitor (ATP) can suppress activity to <2% of V_{max}.

within the chloroplast stroma of plants. Photosynthetic electron transport has been linked to H^+ uptake from the stroma into the thylakoid lumen. This establishes a proton gradient between the lumen and stroma that drives the photophosphorylation of ADP by P_i (inorganic phosphate) to form ATP. This ATP is used to help power the biosynthetic (CO_2 fixation) reactions in the stroma. The transport of H^+ ions into the lumen results in a light-dependent increase in stromal pH_i from about 7.0 to 8.0, and the high pH facilitates CO_2 fixation because several enzymes of the reductive pentose–P pathway have relatively sharp alkaline pH optima of between 7.8 and 8. In the dark, however, H^+ ions leak back into the stroma and stromal pH_i falls back to 7.0. Both the drop in stromal pH and the limitation of ATP supply in the dark combine to help suppress the reductive pentose–P pathway until the reintroduction of light makes the system fully operational once again.

Fine Control 3. Allosteric Effectors Multisubunit regulatory enzymes often contain allosteric sites that are separate from the active or catalytic site where specific inhibitor or activator metabolites can bind reversibly. By varying the concentration of these nonsubstrate effector molecules, it is possible to markedly alter the activity of an enzyme and thereby modify the flux of metabolites through an entire pathway. Allosteric effectors alter enzyme activity by binding to an allosteric site and eliciting a precise change in the enzyme's conformation. This conformational change has been termed *heterotropic* allosterism because, in contrast to homotropic allosterism, the allosteric modulator is a molecule other than the substrate. The amount of an allosteric modulator bound to the corresponding allosteric site of an enzyme is a reflection of the concentration of allosteric modulator and the enzyme's affinity for this particular ligand. Allosteric effectors themselves may show hyperbolic or sigmoidal (i.e., cooperative) saturation kinetics. As outlined above for sigmoidal substrate saturation kinetics, a sigmoidal relationship would allow a very sensitive response of enzyme conformation (and hence activity) to changes in effector concentration. The conformational change brought about by the binding of an allosteric effector thereby promotes (activators) or hinders (inhibitors) enzyme–substrate interactions such that some or all of the kinetic constants may be significantly altered including V_{max}, K_m or $S_{0.5}$, and n_H. This regulatory strategy allows metabolites that are remote from a specific reaction to function as feedforward (activators) or feedback (inhibitors) control signals on enzyme activity. The key significance of this interaction is that, at the low and relatively constant substrate concentrations that usually exist *in vivo*, the rate of a reaction can be dramatically increased or decreased by effector binding. This is illustrated graphically by the example illustrated in Figure 1.2 that shows that by addition of an activator or inhibitor it is possible to vary the rate of an allosteric enzyme reaction from less than 10% to nearly 100% of V_{max} with little or no change in [S].

Essentially all pacemaker enzymes are at least partially controlled by allosteric effectors. As indicated above, for an effector to have meaningful regulatory significance, the concentration at which it significantly activates or inhibits *in vitro* must be close to the effector's actual concentration range *in vivo*.

Enzyme Activation Activators are metabolites that reversibly interact at an allosteric activator site. When bound, they elicit an allosteric transition that typically elevates the reaction rate at subsaturating [S] by causing a reduction in the enzyme's K_m (or $S_{0.5}$) value. This may or may not be accompanied by a corresponding increase in V_{max}. This is illustrated by the example given in Figure 1.2, which concerns purified pyruvate kinase from the cyanobacterium, *Synechococcus*. Pyruvate kinase is an important control enzyme of glycolysis in all phyla and catalyzes the essentially irreversible reaction: phosphoenolpyruvate (PEP) + ADP → pyruvate + ATP. The cyanobacterial pyruvate kinase was fully purified and shown to be potently activated by glucose-6-phosphate and ribose-5-phosphate, the immediate products of glycogen breakdown. Low concentrations of either activator caused a slight increase in enzyme V_{max} but elicited a marked enhancement in PEP binding, which was accompanied by a shift from sigmoidal to hyperbolic PEP saturation kinetics (Fig. 1.2). The concentrations of glucose-6-P and ribose-5-P required for half-maximal activation (i.e., K_a values) were less than 25 μM, which suggests that the activating effects observed *in vitro* are relevant to pyruvate kinase (PK) and glycolytic flux control *in vivo*.

Enzyme Inhibition Any substance that reduces the rate of an enzyme-catalyzed reaction is called an *inhibitor*. Competitive (or specific) inhibition is the most common form of reversible enzyme inhibition and arises when the inhibitor and substrate compete for a common substrate binding site (i.e., the active site) so that when one binds the other cannot. The effect of a competitive inhibitor is to decrease the apparent affinity of the enzyme for its substrate without any effect on the reactivity of the enzyme–substrate complex once formed. Thus, competitive inhibition causes an increase in the K_m (or $S_{0.5}$) value, does not affect V_{max}, and is overridden by increasing the [S]. Although competitive inhibitors are not allosteric inhibitors (because they do not bind to an allosteric site, a site that is distinct from the active site), they nevertheless constitute an important aspect of enzyme control *in vivo*. For example, potent competitive inhibition of many microbial, yeast, and plant acid phosphatases by their product, inorganic phosphate (P_i), is in accord with the hypothesis that acid phosphatases are particularly active *in vivo* during nutritional P_i starvation when intracellular P_i levels become depleted. Conversely, any accumulation of cellular P_i due to P_i resupply would act as a tight regulatory control to prohibit further hydrolysis of certain phosphorylated compounds by the various intracellular acid phosphatases.

Mixed-type inhibition is a rarer, but important, form of metabolic control in which the inhibitor reversibly interacts with the enzyme, or enzyme–substrate complex, at a true allosteric site. In this instance V_{max} is reduced, whereas K_m is increased. Mixed inhibition is frequently misconstrued as noncompetitive inhibition, which theoretically reduces an enzyme's catalytic potential (i.e., V_{max}) without affecting substrate binding (i.e., K_m or $S_{0.5}$). However, it is unlikely for an inhibitor to interact with an enzyme and alter its V_{max} without having any effect on substrate binding. Moreover, most commonly quoted examples of noncompetitive inhibition prove, on reexamination of the original data, to be examples of mixed inhibition.

Interacting Effectors The activity of an allosteric enzyme *in vivo* is dependent upon the relative concentrations of its activators versus inhibitors. Often, the presence of an activator can override, or cancel, inhibitory signals. Thus, it is the ratio of [activator] : [inhibitor], rather than their absolute concentrations per se, that is believed to be a major factor in determining the overall activity of an allosteric enzyme *in vivo*.

Adenine Nucleotides as Metabolic Effectors Energy transduction and energy storage, involving the adenine nucleotides (ATP, ADP, and AMP), are fundamental features of metabolism in all organisms. Thus, it is not surprising that the adenine nucleotides (particularly ATP and AMP) are potent allosteric effectors of many enzymes, their effects having a major impact on the balance between ATP formation by catabolic pathways and ATP use by many other cell functions. In most tissues, the adenine nucleotides are maintained in equilibrium by the enzyme adenylate kinase (also known as myokinase in muscles), which catalyzes the interconversion of the three adenine nucleotides: ATP + AMP ↔ 2ADP. Since the concentration of adenine nucleotides in a healthy cell are [ATP] > [ADP] ≫ [AMP], and the reaction catalyzed by adenylate kinase is very close to equilibrium *in vivo*, any small change in the balance between ATP and ADP will be transmitted (by adenylate kinase) into a relatively large change in [AMP]. Thus, enzymes that are allosterically regulated by AMP can respond with high sensitivity to small changes in [ATP] and [ADP]. For example, in insect flight muscle, [ATP] falls by about 10% during the early seconds of flight, and this causes an immediate 250% increase in [AMP]. AMP is a potent allosteric activator of key control enzymes such as PFK, and via allosteric effects as well as the action of the AMP-activated protein kinase (see below), AMP functions as an "amplifier" of relatively small changes in ATP concentration. Although not sufficient to fully explain the approximately 100-fold increase in glycolytic flux during insect flight, it is believed that the increase in [AMP] makes a major contribution to the response.

The widespread importance of adenine nucleotides in metabolic regulation led Atkinson in 1977 to introduce the concept of the *adenylate energy charge* (AEC), which is defined as

$$\text{AEC} = \frac{[\text{ATP}] + \frac{1}{2}[\text{ADP}]}{[\text{ATP}] + [\text{ADP}] + [\text{AMP}]}$$

The AEC can vary theoretically between 0 (i.e., all adenylates present as AMP) and 1 (i.e., all adenylates present as ATP), but in healthy cells AEC is normally controlled at a value of between 0.7 and 0.9. Fluctuations in adenylate levels under different circumstances provide a basis for metabolic control by the AEC. Regulatory enzymes that occur in pathways in which ATP is consumed (i.e., anabolic pathways) respond to changes in AEC (via the action of ATP as an activator and/or AMP as an inhibitor) in the general way shown by curve 2 of Figure 1.3. Regulatory enzymes such as PFK that occur in pathways in which ATP is generated (i.e., catabolic pathways) respond to changes in AEC (via the action of ATP as an inhibitor and/or AMP as an activator) in the general way shown by curve 1 of Figure 1.3. It is clear from Figure 1.3 that any fall in AEC would be counteracted by a consequent increase in flux through ATP-generating pathways (curve 1) and by a decrease in flux through ATP-consuming pathways (curve 2).

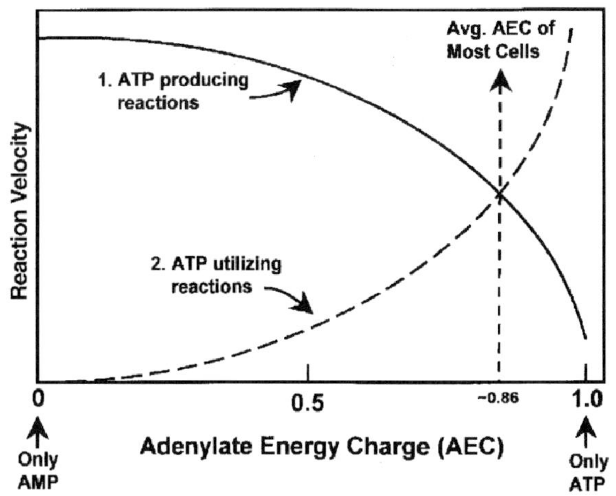

Figure 1.3 Responses to adenylate energy charge by regulatory enzymes in metabolic pathways that produce ATP (curve 1) or consume ATP (curve 2) (adapted from Atkinson, 1977).

Fine Control 4. Reversible Covalent Modification A wide variety of posttranslational modifications of proteins can have an influence on enzymatic activity. Some of these, such as the ubiquitin-tagging system mentioned above, are irreversible and are not mechanisms of fine metabolic control. Similarly, proteolytic cleavage of an inactive precursor protein (or *zymogen*) is not a mechanism of fine control, although it is an important mechanism for safely synthesizing and storing various enzymes (e.g., digestive enzymes) until they are released at their site of action (e.g., in the lysosome of animal cells).

Reversible covalent modification (also known as an *interconvertible enzyme cascade*) plays a dominant role in fine metabolic control. The general model is that an enzyme is interconverted between a less active (or inactive) and more active form because of the effects of the covalent modification on its substrate saturation kinetics and/or response to allosteric effectors (Fig. 1.4*a*). This interconversion is not the result of a reversible equilibrium, as is the case for regulation by allosteric effectors (Fig. 1.4*b*) but is usually governed by two thermodynamically favorable enzyme-catalyzed reactions that result in the formation of new stable covalent bonds on the surface of the target enzyme. Interconversion in either direction can be very fast (e.g., seconds to minutes) and very complete (up to 100% conversion). A change in enzyme conformation that is induced by covalent modification normally causes an alteration of enzyme–substrate interactions such that kinetic parameters such as $V_{\max}$, K_m (or $S_{0.5}$), and n_H are significantly elevated or reduced. The sensitivity of the target enzyme to allosteric activators or inhibitors can also be affected. Covalent modification can quickly provide the cell with an essentially "new" enzyme form whose kinetic properties are geared to the cell's momentary metabolic requirements.

Enzyme control by reversible covalent modification is the major mechanism whereby extracellular signals, such as hormones, nervous impulses, and numerous environmental stimuli, can coordinate the control of key enzymes of intracellular pathways. Reversible covalent modification of enzymes typically allows a very marked sensitivity (e.g., "amplification") to signals, much greater than is possible for an enzyme responding to allosteric effectors, and indeed, often regulates enzymes in a virtual on–off manner (Fig. 1.4).

Although at least 150 types of posttranslational modifications of proteins have been reported *in vivo*, very few appear to be important in the control of enzyme activity. The major kinds of reversible covalent modifications that reversibly control enzyme activity are summarized in Table 1.3. Protein phosphorylation was the first to be discovered and has emerged as the predominant method of reversible covalent modification used in eukaryotic metabolic control. Although reversible enzyme phosphorylation

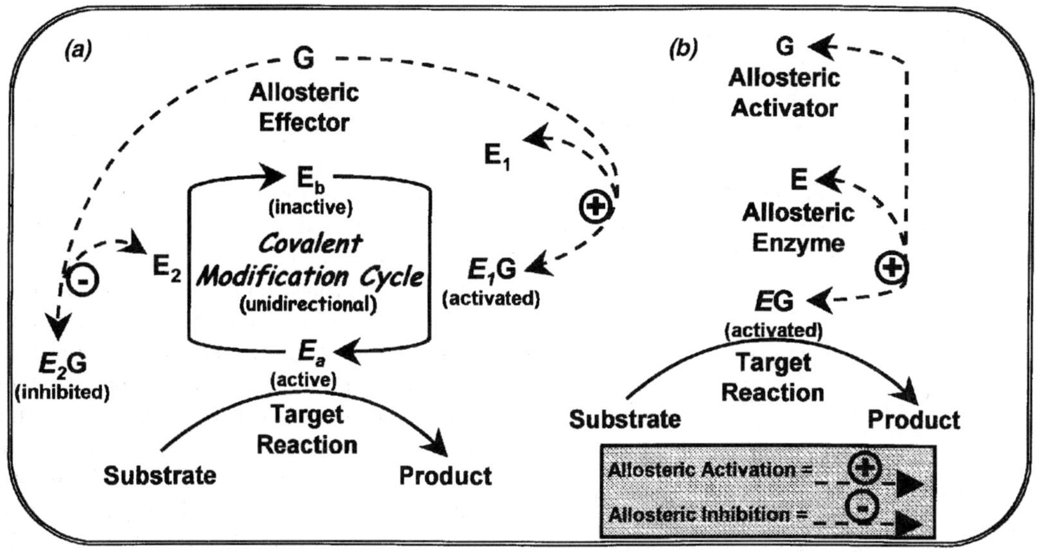

Figure 1.4 General model for enzyme control by reversible covalent modification and allosteric activation (*a*) In reversible covalent modification an effector G typically exerts reciprocal allosteric effects on the modifying enzymes (E_1 and E_2) that catalyze the covalent interconversion of a target enzyme between inactive (or less active) and active conformations. Additional substrates (such as ATP and water) are needed to make the respective interconversion reactions thermodynamically irreversible. (*b*) In conventional allosteric activation the activator G reversibly binds directly to the enzyme to enhance its activity (adapted from Fell, 1997).

does not appear to play as prominent a role in the control of prokaryotic metabolism, it does play a significant role in the control of some prokaryotic enzymes. Reversible ADP-ribosylation, methylation, and adenylation also contribute to the fine control of various enzymes in eukaryotes and/or microbes. In addition, enzyme dithiol–disulfide interconversion is important in the control of plant metabolism. To further illustrate the principles and critical importance of reversible covalent modification in fine metabolic control, phosphorylation–dephosphorylation, and to a lesser extent dithiol-disulfide interconversion will be discussed.

Phosphorylation–Dephosphorylation Enzyme modification by the reversible covalent incorporation of phosphate is a widespread phenomenon with considerable consequences for metabolic control *in vivo*. As outlined in Figure 1.5*a*, the protein phosphorylation reaction is catalyzed by a protein kinase that typically transfers the γ-phosphate of ATP to a hydroxyl group on a serine, threonine, or tyrosine residue of the target protein. The reverse reaction is catalyzed by a phosphoprotein phosphatase. Both protein kinases and phosphoprotein phosphatases may have wide or narrow substrate specificities and are themselves always subject to reciprocal fine metabolic

TABLE 1.3 Major Classes of Reversible Enzyme Covalent Modification Important in Metabolic Control

Type of Covalent Modification	Donor Molecule(s)	Amino Acids Modified	Distribution
Phosphorylation	Mainly ATP (also GTP)	Ser, Thr, Tyr, His	More widespread in eukaryotes than in prokaryotes
Disulfide–dithiol interconversion	Reduced thioredoxin	Cys	Vascular plants and green algae
Nucleotidylation	ATP, UTP	Tyr, Ser	Only in prokaryotes
ADP-ribosylation	NAD^+	Arg, Glu, Lys	Eukaryotes and prokaryotes
Methylation	*S*-adenosyl-methionine	Asp, Glu, Lys, His, Gln	Eukaryotes and prokaryotes

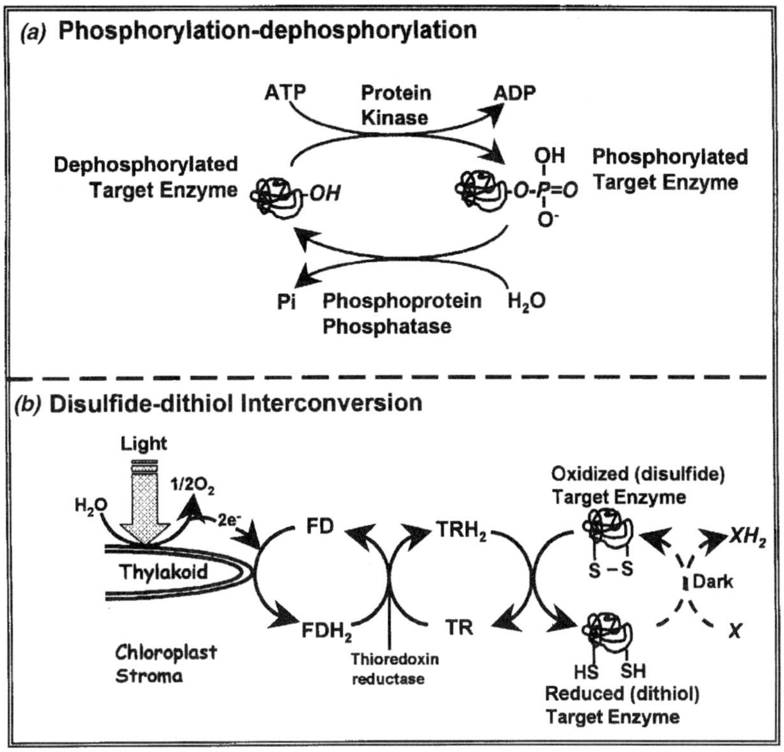

Figure 1.5 Examples of the control of enzyme activity by reversible covalent modification. For details refer to the text. (*a*) Phosphorylation–dephosphorylation and (*b*) disulfide–dithiol interconversion. Abbreviations: FD and FDH_2, oxidized and reduced ferredoxin, respectively; P_i, inorganic phosphate; TR and TRH_2, oxidized and reduced forms of thioredoxin, respectively; X, some oxidant normally kept reduced in the light.

controls. To prevent a futile cycle that wastes ATP, under conditions where a protein kinase is active, the corresponding phosphatase is usually inhibited and vice versa. In animals, plants, and yeast there are significant similarities in amino acid sequences among most of the known protein kinases and phosphoprotein phosphatases indicating that the interconverting enzymes belong to extended families of proteins that have probably evolved from common ancestors.

Since more than 10% of the 10,000 proteins in a normal mammalian cell are thought to be controlled by reversible phosphorylation, this aspect of metabolic control has received considerable attention. Abnormal protein phosphorylation is the basis for or the result of major human diseases including cancer, diabetes, and arthritis. Mutations of certain protein kinases and phosphatases underlie a number of hereditary diseases, including certain leukemias. Many drugs are specific inhibitors of protein kinases and phosphatases. For example, by potently inhibiting of a specific type of phosphoprotein phosphatase, the immunosuppressant drug cyclosporin plays a key role in the success of organ transplants. Likewise, it is probable that protein kinases and/or phosphatases will become critical targets for metabolic engineering.

The first example of enzyme control by reversible covalent modification was provided in the mid-1950s by Edwin Krebs and Edmond Fischer who were studying the control of muscle glycogen phosphorylase, the enzyme that cleaves glucose-1-phosphate units from the glycogen polymer. Phosphorylase *b*, the less active enzyme form, was found to be converted to the more active phosphorylase *a* by the covalent attachment of a phosphate group, derived from ATP, to a serine residue on the protein, catalyzed by a protein kinase that was named phosphorylase kinase. Conversely, a phosphatase was discovered that catalyzed the dephosphorylation and conversion of phosphorylase *a* back to the less active phosphorylase *b*. This seminal work led to the subsequent discovery by Earl Sutherland of the hormone-mediated production of cyclic 3′,5′-adenosine monophosphate (cAMP) by the enzyme adenylyl cyclase and, thus, the concept of the *second messenger* as an intracellular transducer of extracellular (i.e., hormonal) signals. It has since become clear that there are many other signals and second messengers apart from hormones and cAMP that influence protein phosphorylation status (Table 1.4). To complicate matters even further, although protein phosphorylation may lead to a change in enzyme

TABLE 1.4 Classification of Major Serine and Threonine Protein Kinases Important in Metabolic Control

Signal	Protein Kinase Class	Distribution
Cyclic nucleotides	cAMP-dependent protein kinase cGMP-dependent protein kinase	Animals and yeast, (plants?)
Ca^{2+}/calmodulin	Ca^{2+}/calmodulin protein kinases	Animals and yeast
Ca^{2+}	Ca^{2+}-dependent protein kinase	Plants
Diacylglycerol/Ca^{2+}	Protein kinase C	Animals and plants (yeast?)
AMP	AMP-dependent protein kinase	Animals and yeast (plants?)
Metabolic intermediates	Various target-specific protein kinases	Animals, plants, and yeast

activity, many enzymes are subject to multiple phosphorylations at different sites, and some of these phosphorylations appear to affect the ability of protein kinases and phosphatases to phosphorylate or dephosphorylate other sites on the same target enzyme.

Protein Kinases The central role of phosphorylation–dephosphorylation in cell biology and metabolic control is illustrated by the fact that the largest known protein family consists of the eukaryotic protein kinase *superfamily*. These enzymes are related by virtue of their similar *kinase domains*, which consist of about 250 to 300 amino acid residues. The rapid pace of discovery of members of the eukaryotic protein kinase superfamily can be attributed to the development of molecular cloning and genome sequencing technologies, as well as the advent of the polymerase chain reaction, which stimulated the use of similarity-based cloning strategies. It has been estimated that the mammalian and yeast genomes contain approximately 1000 and 120 different protein kinase genes, respectively, comprising 1 to 2% of all genes! Similarly, in the vascular plant *Arabidopsis thaliana* there are believed to be as many as 2500 different protein kinase genes (comprising more than 5% of its entire genome). Identifying the mechanisms that control the activity of each protein kinase, as well as the endogenous protein substrates, for the myriad of eukaryotic protein kinases and the influence of phosphorylation on their biological activities will remain a daunting task for many years to come. Apart from mammals and vascular plants, eukaryotic protein kinase superfamily members have been found in a wide range of other animal phyla, as well as fungi and protists. Tyrosine protein kinases appear to be associated with the receptors for certain hormones such as insulin, and although undoubtedly important, they do not appear to play as direct a role in the fine control of metabolic enzymes as do the serine/threonine protein kinases. As outlined in Table 1.4, each major class of protein kinase is allosterically stimulated by a specific *signal metabolite* that makes enzyme phosphorylation responsive to extracellular and/or intracellular signals.

Protein Kinases That Respond to Extracellular Signals Relative to direct allosteric control of a target enzyme, protein-kinase-mediated modulation of the activity of the same target enzyme is usually initiated by a much smaller change in allosteric effector concentration and can thus provide a much larger amplification of the initial signal. This is illustrated by the aforementioned regulation of glycogen phosphorylase, in which the conversion of the inactive *b* form to the active *a* form is catalyzed by phosphorylase kinase, which is, in turn, controlled by reversible phosphorylation. As outlined in Figure 1.6, a sequence of events can be traced back to the binding of a hormone (i.e., epinephrine in muscle or glucagon in liver) to a receptor on the plasma membrane. Hormone binding stimulates the synthesis of the second messenger, cAMP, and it has been demonstrated that there is only a very small increase (i.e., $<1\ \mu M$) in the concentration of cAMP upon treatment of muscle tissue with a hormone such as epinephrine. However, each molecule of cAMP-dependent protein kinase (also known as protein kinase A) that is activated can itself rapidly activate (phosphorylate) many molecules of phosphorylase kinase, which then activates (phosphorylates) an even larger number of glycogen phosphorylase molecules, and leads to a rapid increase in glycogen breakdown. It has been estimated that 50% of glycogen phosphorylase *b* could be converted to the active *a* form with only a 1% increase in the concentration of cAMP, the process occurring within about 2 s after hormone administration.

The same phosphorylation mechanism that activates glycogen phosphorylase also inactivates the key pacemaker enzyme of glycogen synthesis, glycogen synthetase, and vice versa (Fig. 1.7). This precludes futile cycling between glycogen and glucose-1-phosphate and illustrates the crucial point that when antagonistic (oppositely directed) enzymes in potential futile cycles are phosphorylated, it is invariable that one is activated whereas the other is inhibited. Although there are exceptions to the rule, in general it appears that the activities of catabolic and anabolic enzymes (such as glycogen phosphorylase and glycogen synthetase), which are subject to this mechanism of

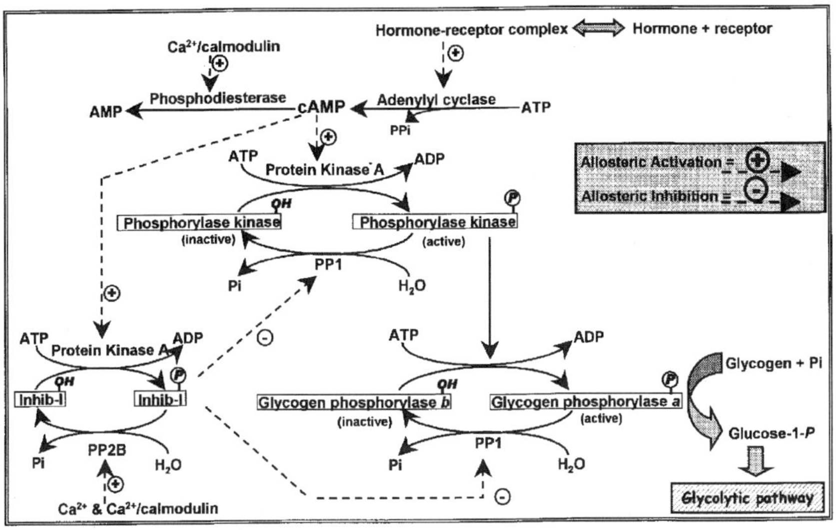

Figure 1.6 The glycogen phosphorylase activation cascade in mammalian muscle. Abbreviations: Inhib-1, Inhibitor 1; PP1 and PP2B, protein phosphatase 1 and 2B, respectively (adapted from Fell, 1997).

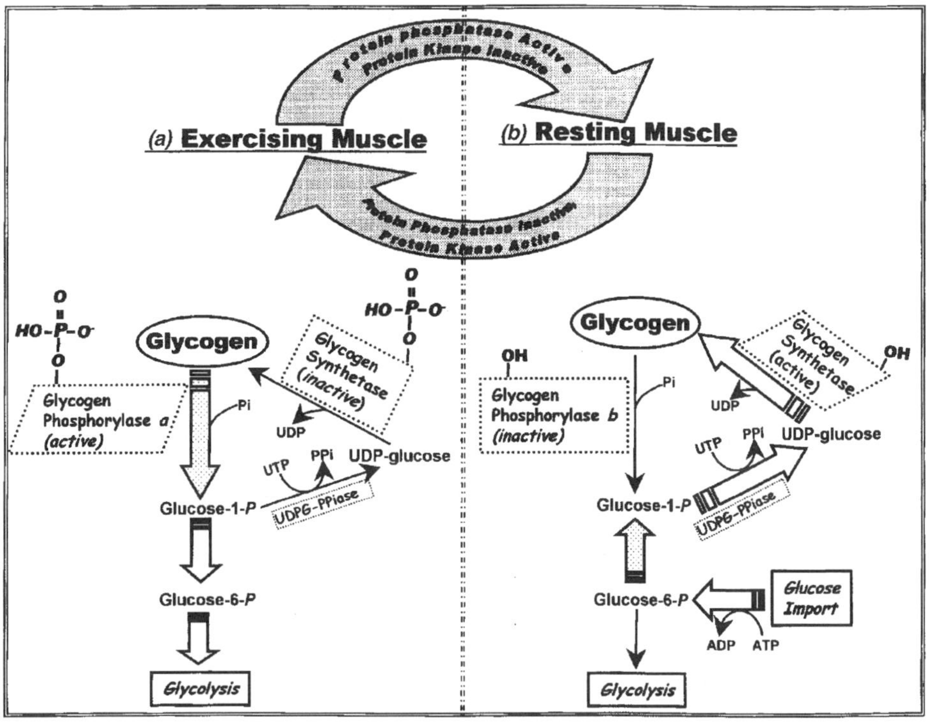

Figure 1.7 Reciprocal control of glycogen phosphorylase and glycogen synthetase by protein-kinase-mediated phosphorylation prevents futile cycling between glycogen and glucose-1-phosphate. (*a*) In exercising muscle hormonal or neuronal signals lead to second-messenger-mediated (i.e., cAMP or Ca^{2+}) protein kinase activation and the consequent phosphorylation/activation of glycogen phosphorylase and phosphorylation/deactivation of glycogen synthetase. (*b*) In resting muscle the corresponding protein phosphatase is activated to dephosphorylate both enzymes leading to inactivation and activation of glycogen phosphorylase and glycogen synthetase, respectively. Abbreviations: PPi, pyrophosphate; UDPG-PPase, UDP–glucose pyrophosphorylase.

fine control, are, respectively, activated and inhibited in their phosphorylated form.

Low cAMP concentrations are maintained via the action of a variety of phosphodiesterase isozymes that convert cAMP to AMP and are generally activated by Ca^{2+}/calmodulin (see Fig. 1.6 and below). cAMP is a true *signal metabolite* whose only role is to control enzyme activities; it is not a metabolic intermediate for the synthesis of any other compound. Various diseases and pathological states in humans arise through "incorrect" protein phosphorylations caused by either abnormally high or low intracellular levels of cyclic AMP (see Text Box 1.2). In contrast to animal systems, cAMP has no confirmed effect on the phosphorylation of plant proteins (although plant cells have been reported to contain cAMP).

TEXT BOX 1.2 CYCLIC NUCLEOTIDES—TOO MUCH OR TOO LITTLE IS UNHEALTHY

Many diseases and pathological states in humans arise from the incorrect phosphorylation of proteins brought about by abnormal levels of cellular second messengers. For example, the cholera toxin causes elevated cAMP levels in intestinal epithelial cells due to its continual stimulation of adenylyl cyclase in these cells (adenylyl cyclase converts ATP into cAMP). The cholera toxin-induced elevation in cAMP levels in intestinal epithelial cells leads to all of the symptoms associated with this debilitating disease. By contrast, impotence in human males is associated with abnormally low 3′,5′-cyclic guanosine monophosphate (cGMP) levels in erectile tissue. The drug Viagra is effective owing to its potent and specific inhibition of a cGMP-phosphodiesterase (enzyme responsible for converting cGMP to GMP) isozyme that is expressed in the male reproductive organ.

The Ca^{2+} ion is also widely recognized as having an important signal transduction role in all eukaryotic cells. Ca^{2+} functions as a second messenger that can transduce extracellular stimuli such as light, hormones, nervous impulses, or stress to produce a coordinated and appropriate metabolic response. In response to an external stimulus the rate at which Ca^{2+} ions are transported into or out of a specific subcellular compartment is modified. The resulting change in intracellular free Ca^{2+} concentration represents the "signal," which elicits appropriate metabolic responses, sometimes as a direct effect of Ca^{2+} on the target enzyme and sometimes via stimulation of a protein kinase that then phosphorylates the target enzyme. It is generally accepted that animal and plant cytosolic free Ca^{2+} concentrations are maintained in the range of 0.1 to 1 μM and that metabolic processes in the cytosol are activated by an increase in free Ca^{2+} concentration from 0.5 to 5 μM. The very low cytosolic Ca^{2+} concentration is maintained via the active transport of Ca^{2+} ions by various membrane-localized Ca^{2+} pumps that move the ion against its concentration gradient to return it to the extracellular space or to various subcellular compartments (i.e., the lumen of the endoplasmic reticulum). Several types of specific Ca^{2+} channels (that facilitate Ca^{2+} influx) and ATP-dependent pumps (for mediating Ca^{2+} efflux) are present in membranes of animal and plant cells and are tightly regulated by plasma-membrane-associated and intracellular control mechanisms. Thus, eukaryotic cells are endowed with a complex "machinery" with which to maintain or alter Ca^{2+} concentration gradients.

Many of the effects of Ca^{2+} on animal metabolism are mediated by Ca^{2+} acting through the highly conserved Ca^{2+}-binding protein, *calmodulin*, to activate Ca^{2+}/calmodulin-dependent protein kinases. There is ample evidence that plants also utilize Ca^{2+} and calmodulin to mediate protein phosphorylation, and that this is an important mechanism whereby various extracellular stimuli are coupled to intracellular metabolic events. Relative to animal systems, however, far less is known about the structure or control of plant protein kinases. A number of Ca^{2+}-dependent and calmodulin-independent protein kinases, which appear to be unique to the plant kingdom, have been purified and cloned from various plant sources. The enzymes contain both a protein kinase catalytic domain and a Ca^{2+}-binding domain similar to calmodulin. This explains the direct activation of the enzyme by Ca^{2+} and has established it as the prototype for a new class of protein kinase.

Protein Kinases That Respond to Intracellular Signals

Most kinases described to date respond to extracellular signals. However, there are at least two categories of protein kinases that respond to intracellular signals such as changes in AEC or concentrations of metabolic intermediates.

The mammalian AMP-dependent protein kinase (AMPK) was initially discovered as a regulator of various key pacemaker enzymes of anabolic and catabolic pathways. As previously discussed, due to the action of adenylate kinase, a sharp elevation in AMP concentration occurs whenever the cell is subjected to stresses that cause ATP depletion, for example, hypoxia, heat shock, glucose starvation, and (in muscle) exercise. The rapid rise in AMP activates AMPK and leads to the phosphorylation of target enzymes. Thus, the AMPK cascade acts as an intracellular "fuel gauge," monitoring the AEC of the cell. Once activated by a decrease in AEC, AMPK acts to conserve

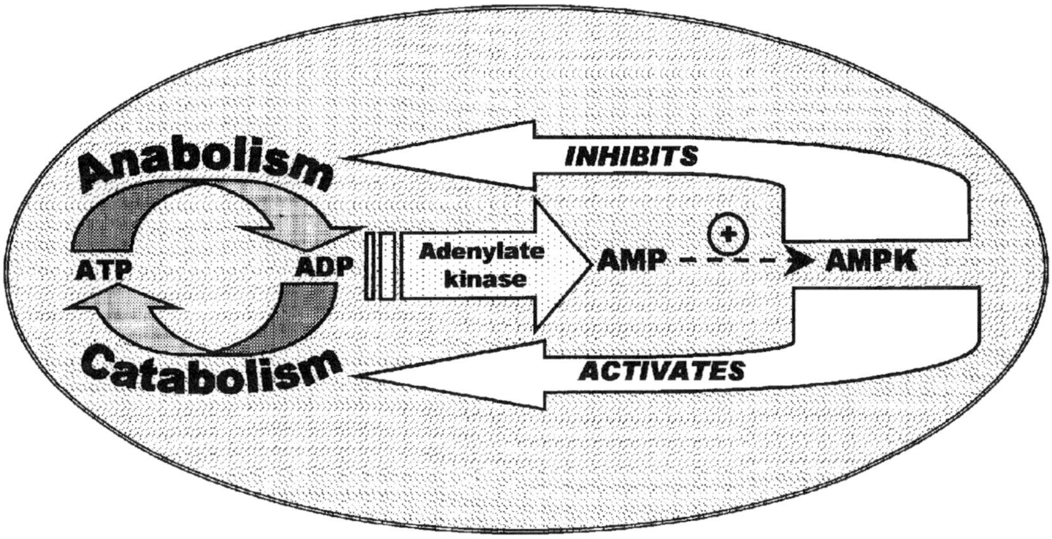

Figure 1.8 A general model for the role of the AMP-activated protein kinase (AMPK) as an "energy charge" monitor within the cell. If cellular stress causes the rate of an ATP-consuming pathway to exceed that of an ATP-producing pathway, adenylate kinase will catalyze the net conversion of ADP to AMP, which activates the AMPK. AMPK then catalyzes the phosphorylation of key enzymes of anabolism and catabolism.

ATP by inhibiting key enzymes of ATP-consuming anabolic pathways and promoting ATP production by stimulating key enzymes of ATP-producing catabolic pathways such as those involved in glucose uptake, glycolysis, and fatty acid oxidation (Fig. 1.8). AMPK is activated in two ways by AMP: first, AMP is an allosteric activator of the enzyme, and, second, AMP makes the enzyme more susceptible to activation via phosphorylation by another protein kinase (called a *kinase kinase*). The combined effects cause an up to 50-fold increase in the activity of AMPK in response to elevated AMP, providing further amplification of the effect of a reduction in ATP levels. AMPK now appears to be one of the primary intracellular sensors of cellular AEC, and most effects of ATP depletion may actually be mediated through the AMPK pathway, rather than via direct allosteric effects of the adenine nucleotides on metabolic enzymes. The AMPK cascade appears to be highly conserved between vertebrates, yeast, and higher plants and may have evolved from an ancient "stress response" system.

Protein kinases that are allosterically controlled by intracellular metabolites tend to specifically phosphorylate a single target enzyme to which they are usually tightly bound. The specific phosphatases that dephosphorylate the phosphoenzyme are also tightly bound to the target enzyme. A good example of this is the pyruvate dehydrogenase complex (PDC) of eukaryotic mitochondria that catalyzes the key reaction (pyruvate + CoA + NAD^+ $\rightarrow$ acetyl-CoA + NADH + CO_2) that links the glycolytic pathway with the Krebs cycle. PDC is subject to inactivation when phosphorylated by a specific PDC kinase. The PDC kinase is allosterically activated by metabolites such as acetyl-CoA and NADH that accumulate when mitochondrial respiration is suppressed (i.e., during hypoxia) or when mitochondrial oxidative capacity is at its limit. Conversely, PDC kinase is inhibited by metabolites such as ADP that signal a need for more respiratory metabolism.

Phosphoprotein Phosphatases The action of protein kinases is reversed by phosphoprotein phosphatases that tend to have very broad substrate specificities, making determination of their physiological functions difficult. Phosphoprotein phosphatases have been classified into four main groups (Table 1.5). These phosphatases are generally highly conserved proteins that appear to be ubiquitous to all eukaryotic cells.

The activity of phosphoprotein phosphatase 1 is specifically inhibited by a regulatory protein termed *inhibitor 1*, which complexes with the enzyme under conditions when the corresponding protein kinase is active. Inhibitor 1 only does so when it has been phosphorylated by protein kinase A. Thus, activation of protein kinase A leads to inhibitor-1 phosphorylation, which reduces the dephosphorylating activity of protein phosphatase 1, and thereby augments the increase in target enzyme phosphorylation due to protein-kinase-A activity (e.g., see Fig. 1.6). Phosphoprotein phosphatase 2B is activated by Ca^{2+} and

TABLE 1.5 Classification of Major Phosphoprotein Phosphatases Important in Metabolic Control

Type	Specific Properties	Typical Targets
1	a. Inhibited by phosphorylated inhibitor 1 b. Particulate (i.e., bound to glycogen)	Glycogen metabolism, muscle contraction
2A	Cytosolic	Glycolysis, gluconeogenesis, fatty acid synthesis
2B	a. Activity-dependent on Ca^{2+}/calmodulin b. Poor activity with metabolic enzymes c. May result in Ca^{2+} attenuation of the influences of cAMP	Protein kinases, protein phosphatases, and inhibitor 1 that have been phosphorylated by protein kinase A
2C	a. Cytosolic b. Mg^{2+} required as a cofactor	AMP-dependent protein kinase, and possibly the targets of AMP-dependent protein kinase

Ca^{2+}/calmodulin and specifically acts on the protein kinases, phosphatases, and inhibitor 1 that are phosphorylated by protein kinase A (and thus provides another link between the cAMP and Ca^{2+} signal transduction systems).

Disulfide–Dithiol Interconversion Reversible covalent modification by disulfide–dithiol exchange involves reactions that are chemically similar to those that occur in the formation of the disulfide bonds that stabilize the tertiary structure of certain proteins. However, unlike structural disulfide bonds, those that participate in the control of enzyme activity must be readily susceptible to reduction back to dithiol groups via the action of compounds such as reduced thioredoxin.

Thioredoxin is a 12-kDa heat-stable protein that in its reduced SH form can function as a protein–disulfide reductase. Thioredoxin has been found in all cell types and all phyla where it has been sought. The protein plays a variety of different roles. In bacterial and animal systems it may function as a general protein–disulfide reductant, or it may participate in the reduction of specific enzymes. In these organisms reduced thioredoxin is regenerated by thioredoxin reductase, at the expense of NADPH (reduced NAD phosphate).

Disulfide–dithiol enzyme interconversion has a much greater regulatory significance in photosynthetic organisms. This type of modification is extremely important in linking photosynthetic electron transport flow with light control of several key chloroplastic enzymes involved in photosynthetic CO_2 fixation and related processes. Reducing equivalents (electrons) are ultimately generated from water by the photosynthetic electron transport chain on the thylakoid membrane. The proteins ferredoxin, ferredoxin–thioredoxin reductase, and thioredoxin constitute the machinery that shuttles the reducing equivalents from the electron transport chain to selected target enzymes in the chloroplast stroma (Fig. 1.5*b*). Two different thioredoxins, designated *f* and *m* are part of the ferredoxin–thioredoxin system in chloroplasts. In the reduced state thioredoxin *f* selectively activates several Calvin cycle enzymes involved in photosynthetic CO_2 fixation, whereas thioredoxin *m* preferentially activates and inhibits NADP-malate dehydrogenase and glucose-6-phosphate dehydrogenase, respectively. The thioredoxin system thus plays a key role in coordinating the light-dependent photosynthetic electron transport chain with the control of several key enzymes within the chloroplast stroma. By mechanisms not yet fully understood, enzymes that are reduced by thioredoxin in the light are oxidized to their disulfide forms in the dark. This might be catalyzed by the action of either oxidized thioredoxin or low-molecular-weight oxidants such as oxidized glutathione, dehydroascorbate, or hydrogen peroxide.

Fine Control 5. Subunit Association–Disassociation Usually, a multimeric enzyme becomes less active or inactive when its constituent subunits are dissociated. This property has been exploited to facilitate the control of certain enzymes. Aggregation or dissociation of subunits is normally induced by the binding of a small molecular weight allosteric effector molecule but may also be instigated by reversible covalent modification. For example, acetyl-CoA carboxylase is a tightly regulated enzyme that catalyzes the irreversible first committed step of fatty acid biosynthesis as follows: acetyl-CoA + HCO_3^- + ATP $\rightarrow$ malonyl-CoA + ADP + P_i. In eukaryotic cells, the enzyme is subject to both allosteric regulation and reversible phosphorylation so as to control the flow of fatty acid precursors into malonyl-CoA. In its active, dephosphorylated form the enzyme's subunits polymerize into long filaments. As outlined in Figure 1.9, protein-kinase-mediated phosphorylation (triggered in vertebrates by the hormones glucagon or epinephrine) promotes dissociation of the acetyl-CoA carboxylase polymer into inactive monomeric subunits, thereby inhibiting fatty acid synthesis.

Another interesting example of enzyme control by subunit association–dissociation concerns the glycolytic enzyme, pyruvate kinase, from the mammalian pituitary gland. Researchers who isolated a clone for a thyroid

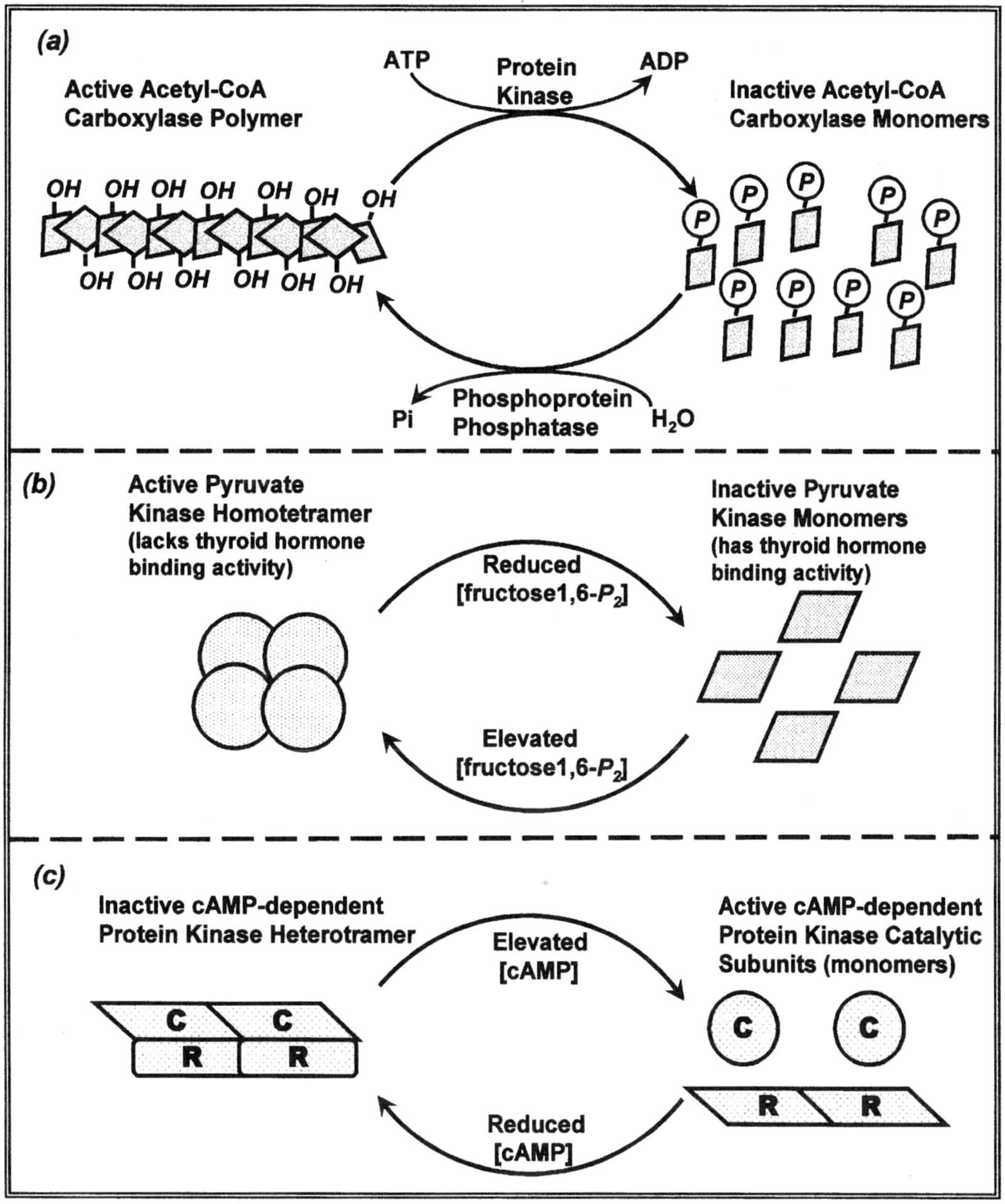

Figure 1.9 Examples of the fine control of enzyme activity by subunit association–dissociation. (*a*) Protein-kinase-mediated phosphorylation of mammalian acetyl-CoA carboxylase causes the enzyme to dissociate from its active polymeric form into inactive monomers. This process is reversed by the action of a phosphoprotein phosphatase. (*b*) Pyruvate kinase from pituitary glands dissociates from an active homotetramer into inactive monomers when levels of its allosteric activator (fructose-1,6-bisphosphate) are reduced, and vice versa. The inactive pyruvate kinase monomers acquire thyroid hormone binding activity. (*c*) In the inactive R2C2 heterotetramer of cAMP-dependent protein kinase the regulatory subunits (R) block the substrate binding sites of the catalytic subunits. (*c*) Cyclic AMP activates the enzyme by causing dissociation of the C subunits from the inhibitory R subunits. This allows the enzyme to phosphorylate various protein substrates that contain the cAMP-dependent protein kinase consensus sequence.

hormone binding protein from a pituitary cDNA library unexpectedly discovered that its sequence exactly matched that previously reported for pituitary pyruvate kinase. Further research revealed that pituitary pyruvate kinase is a bifunctional protein that interconverts between an active pyruvate kinase tetramer (lacking thyroid hormone binding activity) and inactive pyruvate kinase monomers (having thyroid hormone binding activity) (Fig. 1.9). This interconversion has been attributed to the potent allosteric activator of pyruvate kinase, FBP. Elevated [FBP] promotes association of the monomeric subunits into the active pyruvate kinase homotetramer and reduced [FBP] promotes dissociation.

There are exceptions to the general rule that multimeric enzymes become less active when dissociated. For example, in its inactive state the cAMP-dependent protein kinase exists as a heterotetramer formed of two regulatory subunits (R) and two catalytic subunits (C). Allosteric activation by cAMP involves binding of four cAMP molecules by the two regulatory subunits and release of the free catalytic subunits (Fig. 1.9). The two sites for cAMP per regulatory subunit interact cooperatively so that the Hill coefficient for kinase activation with respect to cAMP is about 1.6. As previously discussed, this allows a more sensitive response by the enzyme to small changes in cAMP concentration than would be possible otherwise.

Fine Control 6. Reversible Associations of Sequential Enzymes The existence of known stable multienzyme complexes provides favorable evidence for the interaction of soluble enzymes that are metabolically sequential (i.e., that catalyze the consecutive reactions in a metabolic pathway). Due to high intracellular protein concentrations, such interactions are far more likely to occur *in vivo* than in the dilute solutions that are used for *in vitro* enzymological studies. The limited solvent capacity of the cell also supports the idea of interactions between sequential enzymes of a metabolic pathway since the need for a large pool of free intermediates would then be eliminated. There is now convincing evidence for subcellular structuring of metabolic pathways that were once believed to be entirely soluble. For example, various workers have described associations between many of the so-called soluble enzymes of glycolysis and membrane fractions or structural proteins in animal and plant cells. Of regulatory significance is the fact that the extent of enzyme binding appears to be closely related to pathway flux. During anaerobic muscle contraction, for instance, a marked increase in the binding of key glycolytic enzymes to contractile proteins has been observed. Enzyme binding dissipated once contractile activity, and need for a high glycolytic flux, ceased. Of equal significance are the findings that sequential glycolytic enzymes can also show specific interactions with each other—interactions that have mutual kinetic effects on enzyme activities. Similar correlations between the degree of glycolytic enzyme binding and rate of glycolytic flux have been observed in other animal and plant systems. Enzymes that reversibly bind to intracellular structures such as membranes or the cytoskeleton have been termed *ambiquitous*.

The microcompartmentation of enzymes and metabolic pathways that can result from ambiquitous interactions could provide an effective means of metabolic control via: (i) direct transfer or channeling of intermediates between consecutive enzymes (Fig. 1.10), and (ii) altering enzyme kinetic properties due to allosteric transitions occurring during binding. It has been suggested, therefore, that metabolic control of pathways such as glycolysis can occur not only by allosteric regulation and covalent modification of key enzymes but also by altering the partitioning of enzymes between the soluble and bound phases. The molecular mechanisms that control the extent of enzyme binding *in vivo* are not yet fully understood. Binding of enzymes *in vitro* can be influenced by pH, concentrations of substrates, products and allosteric effectors, enzyme phosphorylation, and osmotic and/or ionic strength. For example, during muscle contraction PFK was found to be phosphorylated *in vivo*. However, there was no obvious influence of phosphorylation on the *in vitro* substrate saturation kinetics of PFK or its response to allosteric effectors. The physiological reason for covalent modification remained perplexing until it was discovered that PFK phosphorylation was a signal that promoted the association of the enzyme with contractile proteins (actin/myosin), and hence other metabolically sequential glycolytic enzymes.

Although there is compelling evidence for complex formation and/or binding of so-called soluble enzymes to structural elements in the cell and the direct transfer of substrates between sequential enzymes of animal cells, these possibilities have received relatively scant attention in plant systems. However, there is accumulating evidence that metabolically sequential enzymes in plant pathways may physically associate to form a multienzyme complex *in vivo*. For example, the chloroplastic ferredoxin, ferredoxin–thioredoxin reductase, and thioredoxin system used for disulfide–dithiol enzyme interconversion (Fig. 1.5b) may exist in the chloroplast stroma as an enzyme complex termed *protein modulase*. Protein–protein interactions may also play a significant role in addition to disulfide–dithiol exchange reactions in the light-dependent activation of several chloroplastic enzymes by thioredoxin. Various studies have also provided compelling evidence that there is direct channeling of substrates between Calvin cycle enzymes in illuminated chloroplasts and that the enzymes of this pathway exist as large supramolecular complexes that may reversibly associate with the thylakoid membrane in the light.

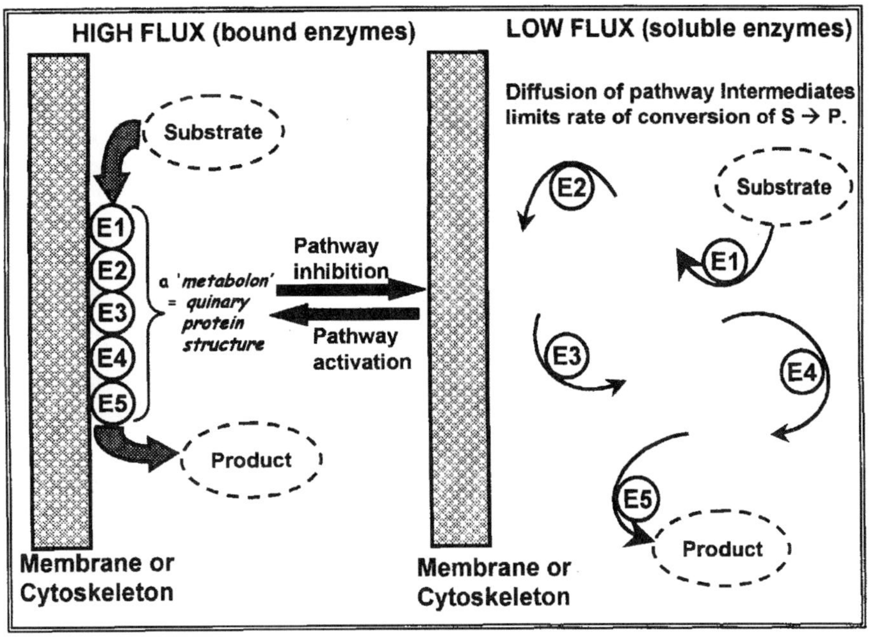

Figure 1.10 The control of enzyme activity by the reversible association of metabolically sequential enzymes into an organized multienzyme system, or metabolon (also termed quinary protein structure). E1, E2, E3, E4, and E5 represent the sequential enzymes that catalyze the consecutive reactions of a hypothetical metabolic pathway. Substrate (or S) and product (or P) depict the pathway's overall starting material(s) and end product(s), respectively.

The above studies indicate that before an overall understanding of metabolic control is achieved we must determine how, where, and when "soluble" metabolically sequential enzymes might be aggregated into microcompartments to form an organized multienzyme system or *metabolon*. It will also be of interest to determine how such associations may alter individual enzyme kinetic/regulatory properties. Equally significant will be an analysis of the extent to which the *reversible* formation and dissolution of metabolons contributes to the overall integration and control of metabolism.

CONCLUDING REMARKS

Enzyme activity is one of the major factors influencing the magnitude of metabolic fluxes in any cell. Metabolic control may occur at several levels, beginning with the gene and proceeding through various stages of protein synthesis and turnover. More rapid alterations in metabolic flux occur through activation and inhibition of key enzymes along the major metabolic pathways, particularly by mechanisms such as reversible covalent modification and by the actions of effector molecules that reflect the cell's AEC, oxidation/reduction potential, and/or the accumulation of metabolic end products. Discoveries concerning metabolic control continue to be made at a rapid rate, particularly in the field of signal transduction. Each discovery adds to the view that signal transduction and metabolic control networks have remarkable intricacy. Although great advances have been made in understanding the mechanisms that contribute to the control of enzyme activity, our comprehension of why metabolic systems behave as they do *in vivo* is incomplete. However, the tools to address these questions are rapidly evolving, and advances in the theory of metabolic control and in computing power to analyze metabolism have kept pace with experimental developments. This holds great promise for those molecular geneticists who wish to reap a bounty via the process of metabolic engineering.

LIST OF ABBREVIATIONS

AEC	adenylate energy charge
AMPK	AMP-activated protein kinase
cAMP	3′,5′-cyclic adenosine monophosphate
cGMP	3′,5′-cyclic guanosine monophosphate
CoA	coenzyme A
F6P	fructose-6-phosphate

FBP	fructose-1,6-bisphosphate
K_{eq}	equilibrium constant for a chemical reaction
K_a	activation constant = activator concentration yielding half-maximal activation
K_m or $S_{0.5}$	substrate concentration yielding $0.5V_{max}$ for enzymes that show hyperbolic or sigmoidal substrate saturation kinetics, respectively
MCA	metabolic control analysis
n_H	Hill coefficient
NMR	nuclear magnetic resonance
[S]	substrate concentration
PEP	phospho*enol*pyruvate
PDC	pyruvate dehydrogenase complex
pH_i	intracellular pH
P_i	inorganic phosphate
PFK	phosphofructokinase
V_{max}	maximal rate of an enzyme-catalyzed reaction
V_0	initial rate of an enzyme-catalyzed reaction

REFERENCES

ApRees, T., and Hill, S. A. (1994). Metabolic control analysis of plant metabolism. *Plant, Cell Environ* **17**:587–599. *A clearly written and concise introduction to the concepts underlying metabolic control analysis, focusing on its application to plant metabolism.*

Atkinson, D. E. (1977). *Cellular Energy Metabolism and Its Regulation.* Academic, New York. *A classic book introducing the concept of adenylate energy charge.*

Cornish-Bowden, A. (1999). *Fundamentals of Enzyme Kinetics.* Portland Press, London. *A discussion of the principles of enzyme kinetics, with an emphasis on principles rather than an accumulation of facts. More advanced.*

Fell, D. A. (1997). *Understanding the Control of Metabolism.* Portland Press, London. *An appealing introduction to the basic theories of metabolism and metabolic control analysis. More advanced.*

Knowles, V. L., Smith, C. S., Smith, C. R., and Plaxton, W. C. (2001). Structural and regulatory properties of pyruvate kinase from the cyanobacterium *Synechococcus* PCC 6301. *J Biol Chem* **276**:20966–20972.

Plaxton, W. C. (1996). The organization and regulation of plant glycolysis. *Annu Rev Plant Physiol Plant Mol Biol* **47**:185–214. *Helpful review on glycolytic compartmentation and control in plants. Compares and contrasts plant and animal glycolysis.*

Price, N. C., and Stevens, L. (1989). *Fundamentals of Enzymology*, 2nd ed. Oxford University Press, New York. *Excellent book that provides a comprehensive and readable account of enzyme properties in systems of increasing complexity (from purified enzymes to enzymes in the living cell).*

Thomas, S., and Fell, D. A. (1998). The role of multiple enzyme activation in metabolic flux control. *Advan. Enzyme Regul.* **38**:65–85.

2

ENZYMES: THE BASIS OF CATALYSIS

STEPHEN P. J. BROOKS

INTRODUCTION

Of the over 2000 reactions that can occur within a eukaryotic cell, virtually all are catalyzed by enzymes. This means that cellular metabolism is effectively directed by enzymes. In fact, nonenzymatic reactions are usually deleterious to cells, producing toxic products that must be dealt with effectively and quickly. Enzyme pathways are, therefore, analogous to the road system within a city, keeping traffic moving along defined paths. Like roadways, they must be regulated to ensure a proper flow that can respond to the needs of a cell.

ENZYME STRUCTURE

Enzymes are made from sequences of amino acids (protein catalysts) or nucleic acids (nucleic acid catalysts). Although this chapter will focus on protein-based catalysis, it is currently thought that during the early evolution of life on Earth, ribonucleic acid (RNA) molecules acted as both genetic template and RNA replicating enzyme, although only a limited number of RNA catalysts have been observed in the laboratory (Strobel, 2001). The importance of RNA catalysis to early life is suggested from the demonstration of self-splicing activity by a pre-rRNA intron from *Tetrahymena thermophila*. If RNA can act as a catalytic agent, then early forms of life may not have needed protein to catalyze the replication of genetic material. This solves one of the oldest dilemmas in evolution: Which came first, the protein or the genetic material?

Although the reader is referred to a more fundamental textbook for a thorough treatment of protein structure, an understanding of the basics of enzyme architecture is important for a fuller appreciation of enzyme function and regulation. This chapter will, therefore, start with a short review of this topic. The focus will be on protein enzymes as a model for all enzyme activity. This is because of the large amount of literature on protein enzyme-determined metabolic regulation and the importance of protein enzymes to cellular metabolism. Four levels of organization have been assigned to protein enzymes: primary, secondary, tertiary, and quaternary structures. The ordering of these four categories reflects the order of influence of one type of structure on the next. For example, the primary structure influences the degree of secondary structure present, but not the reverse.

The primary structure is defined as the sequence of amino acids that makes up a protein. This is determined by the sequence of nucleotide bases in the gene that codes for the protein. Translation of the mRNA (messenger RNA) transcript produces a linear chain of amino acids linked together by a peptide bond between the carboxyl carbon of the first amino acid and the free amino group of the second amino acid, as shown in Eq. (2.1). The first amino acid in any polypeptide sequence has a free amino group and the terminal amino acid has a free carboxyl group. This numbering convention is dictated by the direction of mRNA translation:

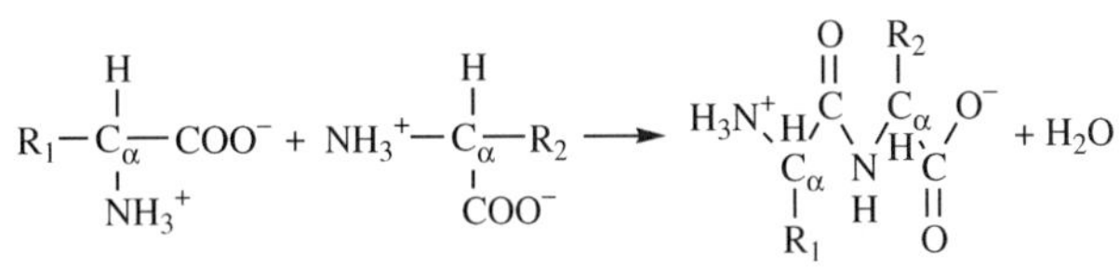

(2.1)

Free amino acids are usually zwitterionic at pH 7 with the carboxyl group deprotonated and the amino group pro-

Functional Metabolism: Regulation and Adaptation, edited by Kenneth B. Storey.
ISBN 0-471-41090-X

TABLE 2.1 Types of Amino Acids[a]

Class	Name	R group[b]
Nonpolar	Alanine	$-CH_3$
	Valine	$-CH(CH_3)_2$
	Leucine	$-CH_2-CH(CH_3)_2$
	Isoleucine	$-CH(CH_3)-CH_2-CH_3$
	Proline	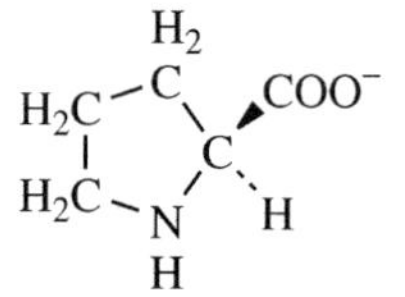
	Phenylalanine	$-CH_2-\phi$
	Tryptophan	$-CH_2-$indole
	Methionine	$-CH_2-CH_2-S-CH_3$
Polar, uncharged	Glycine	$-H$
	Serine	$-CH_2-OH$
	Threonine	$-CH(OH)-CH_3$
	Cysteine	$-CH_2-SH$
	Tyrosine	$-CH_2-\phi-OH$
	Asparagine	$-CH_2-CO-NH_2$
	Glutamine	$-CH_2-CH_2-CO-NH_2$
Polar, negative	Aspartic acid	$-CH_2-COO^-$
	Glutamic acid	$-CH_2-CH_2-COO^-$
Polar, positive	Lysine	$-CH_2-CH_2-CH_2-CH_2-NH_3^+$
	Arginine	$-CH_2-CH_2-CH_2-NH-C(NH_2^+)-NH_2$
	Histidine	$-CH_2$-imidazole$^+$

[a]The basic amino acid structure is $R-C_\alpha(H(NH_3^+)-COO^-$. Charges on amino acid R are measured at pH 7. The complete structure of proline is shown.
[b]The symbol ϕ represents benzene.

tonated. Combining two amino acids results in the elimination of water and produces a bond where the carbonyl carbon, amide nitrogen, and α carbon atoms all lie in the same plane. The chain can bend because rotation can occur around the α carbon atoms. The resulting chain of amino acids is called a polypeptide (many peptide bonds) and can be identified by its unique sequence of amino acids. The individual properties of the various amino acids and the relative amount of each amino acid in the protein play an important role in determining the secondary, tertiary, and quaternary structure of an enzyme.

Amino acids can be functionally divided into four different classes depending on the structure of their distinct side chains (called R groups): nonpolar, polar uncharged, negatively charged (at pH 7), and positively charged (at pH 7). This classification is based on R-group polarity and is the most useful for understanding the relationship between primary structure and higher order protein structure and functionality. Table 2.1 shows this classification for the 20 most common amino acids. The polar, negatively charged, and positively charged R groups tend to be on the outside of a protein, in direct contact with water. The nonpolar R groups are usually hidden in the interior of the protein away from water.

The secondary structure refers to highly ordered structures found within an enzyme. Two types are common: α-helices and β-pleated sheets. The α helix can be thought of as having a structure similar to a coil or spring. The β-pleated sheet can be visualized as a series of parallel strings laid on top of an accordion-folded piece of paper. The existence of these structures is a function of the protein primary structure. Relatively small, uncharged, polar amino acids readily organize themselves into α helices. Relatively small, nonpolar amino acids form β-pleated sheets. Both α helices and β-pleated sheets represent stable structures that minimize the free energy of a protein. In the case of the α helix, the structure is stabilized by hydrogen bonding between amino acids. These secondary structures may form a significant proportion of an enzyme's three-dimensional structure. For example, the muscle oxygen transport protein, myosin, contains 6 well-defined α-helical regions representing over

70% of the polypeptide chain. On the other hand, only about 25% of the lysozyme polypeptide chain is in the form of α-helical regions. Proline is a special amino acid because of its unique structure (Table 2.1). Introduction of proline into an amino acid sequence introduces a permanent bend at this position. For this reason, the presence of proline in an α helix or β sheet will disrupt the secondary structure at this point.

Tertiary enzyme structure refers to the overall configuration obtained when an amino acid chain is placed in water (or another solvating medium). Tertiary structure is determined largely by the interaction of R groups on the surface of the protein with the water and with other R groups on the protein surface. Polar R groups will tend to move to the outside of the protein, and nonpolar R groups will tend to move toward the inside of the protein (away from water). This movement gives a protein a characteristic three-dimensional shape. In addition, electrostatic interactions between the charged amino acid R groups and hydrogen bonding help form the tertiary structure. Posttranslational modifications, such as the formation of disulfide bridges between cysteine residues can also be present. The final structure represents a minimum energy point and is a compromise between all the forces acting on it. It is possible to alter the tertiary structure of an enzyme by changing the solvating medium to an organic solvent or by altering the pH, ionic strength, or temperature. These changes cause proteins to lose structure either by solvating nonpolar R groups (dissolving the protein in an organic solvent) or by causing R groups on the surface of the protein to become nonsoluble so that proteins stick together (pH changes). These changes can, if carried out to a significant extent, radically change protein structure, often to the point of irreversible denaturation. Despite the impression that the enzyme is locked into a single conformation by all these forces, one must not think of an enzyme's tertiary structure as static. The enzyme is continually "breathing" and subtle motion of the various regions always occurs. The extent of this motion depends on many factors including temperature, the properties of the solvating medium, the presence or absence of substrate and product, as well as the degree of quaternary structure (see next paragraph). As discussed below, theories of enzyme action rely on the fact that the enzyme is not a rigid ball but rather a pliable string of amino acids.

The fourth level of enzyme structural organization, quaternary structure, exists only in multisubunit enzymes. In fact, most enzymes are not single polypeptide chains but are dimers, tetramers, or polymers of several subunits (each polypeptide chain is termed a subunit in a multiple subunit enzyme). These subunits can be identical or can be different polypeptide chains. In the case of multiple subunit enzymes, the R groups on each polypeptide chain may interact with each other, as well as with the solvating medium to give a final enzyme structure that is different from that for the isolated subunit. These final structures can be quite varied. For example, collagen is a filamentous protein whereas most enzymes are globular.

THERMODYNAMICS AND KINETIC THEORY OF ENZYME FUNCTION

It is important to understand what governs chemical processes in order to understand how enzymes work since chemical processes and enzyme action are intimately linked. That is, enzymes possess no magical abilities and all reactions occur by well-known chemistry.

All biological processes, from the catabolism of glucose to produce cellular energy to the reactions driving muscle contraction, are chemical reactions and, as such, must obey the fundamental rules governing chemical processes. These rules determine (1) how fast a reaction happens, and (2) how far a reaction proceeds to completion. Enzymes can influence how fast a reaction occurs but not how far a reaction goes to completion.

Thermodynamic Aspects of Kinetic Function: Processes Not under Enzyme Control

In any chemical reaction, the degree to which a particular reaction proceeds (how much product is formed) is determined by the energy change of all the components of the reaction as well as those of the surrounding system. This is because the extent to which a reaction occurs is a function of the starting materials, the products, and the properties of the entire system. In 1878, Willard Gibbs derived a function relating the change in free energy of a system (ΔG) to the change in enthalpy (ΔH, related to the internal energy of the system) and entropy (ΔS, related to the degree of randomness of the system):

$$\Delta G = \Delta H - T\Delta S \tag{2.2}$$

The ΔG value is important because it defines how far a reaction will proceed. For example, if the ΔG for a reaction is negative, this means that energy is released during the conversion of starting material to product—that is, a negative ΔG means energy flows out of the system to the surroundings. This also means that the reaction will favor the accumulation of product rather than starting material. A positive ΔG value means that the reaction requires energy and that the reaction favors the accumulation of starting material over products. It is possible to drive a reaction toward product formation by adding extra energy, such as heat. This is predicted by Eq. (2.2) for reactions that involve an increase in entropy (as T increases ΔG becomes more negative).

Equation (2.2) also shows that a reaction can be driven by a decrease in the enthalpy of the system (a negative ΔH results when the products have a lower energy than the starting material) and/or by an increase in the randomness of the system (ΔS). An example of the first type of process is the burning of wood, which releases a large amount of energy because the heat of combustion of the products is much lower than that of the starting materials. An example of the second type of process is the dilution of salt in water because the Na^+ and Cl^- ions are now no longer in a highly ordered crystal.

It is common to think of a reaction, especially in metabolism, as a unidirectional process that converts starting material (S) completely into product (P) with a rate constant k:

$$S \xrightarrow{k} P \tag{2.3}$$

This is incorrect because all reactions are reversible, no matter how far the equilibrium lies to one direction. Equation (2.4) gives a correct illustration of a reaction:

$$S \underset{k_r}{\overset{k_f}{\rightleftharpoons}} P \tag{2.4}$$

Equation (2.4) demonstrates that there is a constant interconversion of starting material into product and product into starting material. These reactions are governed by their rate constants, k_f (forward direction) and k_r (reverse direction). Although there is a constant flow of material across the reaction, there will be a point at which the flux of starting material to product is exactly equal to the flux of product to starting material. At this point, the equilibrium condition has been achieved, and the ratio of product to starting material is defined as the equilibrium constant (K_{eq}) for the reaction. This value is a constant for any given set of conditions (molarity of reactants, temperature, solvating medium). It is independent of the path of molecular transformation as well as the time required to reach equilibrium and is defined as the ratio of the concentration of product to starting material at the equilibrium condition:

$$K_{eq} = \frac{[\text{Product}]}{[\text{Starting material}]} = \frac{k_f}{k_r} \tag{2.5}$$

Equation (2.5) is useful because it allows the calculation of the equilibrium concentration of product and starting materials when the K_{eq} and the product and starting material concentrations are known. Equation (2.5) can also be used to monitor how far a given reaction is from its equilibrium point by comparing the measured concentration of starting material and product with the values predicted by K_{eq}.

Because cellular metabolism is based on chemistry, Eq. (2.5) also applies to the chemical reactions that occur in the cell. Often in descriptions of these pathways the reactions are written as proceeding in one direction only with well-defined starting materials and products [see Eq. (2.3)]. It is important to note that the designation of a specific cellular metabolite as a product or as a starting material is somewhat arbitrary and depends entirely on how one views the reaction sequence. For example, the enzyme phosphoglyceromutase (PGlyM) is involved in the glycolytic pathway (responsible for glucose catabolism). It catalyses the transfer of a phosphate group from the 3 to the 2 position of glyceric acid:

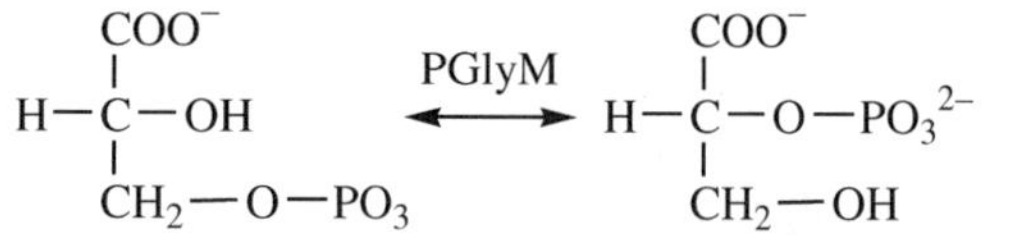

(2.6)

The equilibrium constant is usually written as

$$K_{eq} = \frac{[\text{2-phosphoglycerate}]}{[\text{3-phosphoglycerate}]} = 0.18 \tag{2.7}$$

with an equilibrium constant that favors starting material as shown by $K_{eq} < 1$. However, the reaction runs in the reverse direction during glucose synthesis in the liver so that one could also (correctly) write the inverse of Eq. (2.7) with a K_{eq} value of 5.6 favoring 3-phosphoglycerate formation. This shows the importance of convention in labeling reactions so that the proper values are correctly assigned. It also shows the importance of remembering that reactions are readily reversible and that conventions have no bearing on actual chemical processes.

The K_{eq} constant [Eq. (2.5)] can be related to the change in the total energy of the system (ΔG) and the energy change measured under standard conditions ($\Delta G^{\circ\prime}$; measured at pH 7, 25°C and 1 M aqueous solution concentration) by Eq. (2.8):

$$\Delta G = \Delta G^{\circ\prime} + RT \times \ln(K_{eq}) \tag{2.8}$$

where R is the gas constant and T is the temperature in degrees Kelvin. At equilibrium, $\Delta G = 0$ (by definition since the system is neither gaining nor losing energy), and Eq. (2.8) gives a relationship between the standard energy change of a system and the equilibrium constant (under standard conditions):

$$\Delta G^{\circ\prime} = -RT \times \ln(K_{eq}) \tag{2.9}$$

This equation expresses mathematically what we know intuitively—a reaction that releases energy goes much further toward completion than a reaction that requires energy. Equation (2.9) also restates mathematically what was stated at the start of this section: The equilibrium constant of a reaction is independent of any kinetic mechanism. Note that the K_{eq} value is defined under stated conditions. Its value will change if any of the reaction conditions change. This concept will be important when we consider how enzymes function later in this chapter.

It is possible to determine $\Delta G^{\circ\prime}$ values for any reaction simply by measuring the concentrations of the reactants at equilibrium. Tables of $\Delta G^{\circ\prime}$ values for several reactions have already been complied and can be found in many standard biochemistry and chemistry texts. These values are useful for calculating equilibrium concentrations of reactants.

Cellular Equilibria and Enzymes

As indicated above, it is possible to measure how far a cellular reaction is from its equilibrium point by comparing predicted and measured concentrations of starting materials and products. In the past, this type of analysis was used to obtain an idea of how enzyme activity is related to pathway flux and to identify potential metabolic "bottlenecks." Although there are many kinetic arguments against the validity of this type of analysis, it brings up an important point that one must consider when thinking about cellular metabolites and equilibria. It is common to think about cells as tiny bags containing mostly water with freely dissolved metabolites diffusing from one enzyme to another to react. However, total cellular protein concentration is high, on the order of 200 to 250 mg/mL and the concentrations of individual enzymes are also quite high (Table 2.2). As we shall see later, enzymes function by binding starting materials and products to their active sites. An active site is a specialized area or cleft on the protein surface where substrate binds and catalysis occurs. What is important for the present analysis is the fact that the high cellular enzyme concentrations (Table 2.2) mean that there are many proteins that can potentially bind metabolites and remove them from the "free" water of the cytosol. This has two effects on cellular reactions. First, as we saw above, the K_{eq} value for any reaction is dependent on the amount of free energy liberated (the $\Delta G^{\circ\prime}$). The change in $\Delta G^{\circ\prime}$ is a function of the energy of the reactants. The binding of the reactants to an enzyme will change their energies and, consequently, the K_{eq} for bound reactants will differ from that of free reactants. Second, in order to measure the proximity of a reaction to equilibrium, it is has been a relatively common practice to break open the cell and determine the concentration of metabolites. Once a cell has been broken (and diluted in an extraction medium), it is no longer possible to know the original (unbroken cell) distribution of metabolites; the enzyme–reactant binding distribution has been altered by the physical methods used to disrupt the cellular milieu. Since we can only know the K_{eq} value for the reaction under standard conditions measured *in vitro* (broken cells), it is virtually impossible to assess how close the reaction is to equilibrium *in vivo* (in the unbroken cell) unless we have some other knowledge that tells us the free concentrations of metabolites in the cytoplasm. An example of these difficulties is best illustrated by a consideration of the creatine kinase reaction. In a muscle cell, creatine kinase (CK) catalyses the transfer of a high-energy phosphate from adenosine triphosphate (ATP) to creatine to make phosphocreatine (PCr) and adenosine diphosphate (ADP). Theories of how CK interacts with ATP-producing reactions to capture the high-energy phosphate were developed because it was believed that the CK reaction was far from equilibrium. This conclusion was based on total ATP, ADP, creatine, and PCr concentrations measured in the cell. However, all this changed with the advent of nuclear magnetic resonance (NMR), which can be used to quantify phosphorylated compounds in living cells and was used to determine the amount of free versus bound ADP, ATP, and PCr. The free concentrations of these metabolites turned out to be very different from their total concentrations. New calculations, based on the free levels of these metabolites, revealed that the CK reaction was, in fact, close to equilibrium. This had a significant impact on how the working of muscle cells was viewed.

Reaction Rate Theories

Thermodynamic analysis shows that a reaction can go far to completion if ΔG is negative; however, it says nothing of how fast that reaction may occur. The factors determining the rate of the reaction are illustrated by considering the energy diagram for an idealized reaction scheme shown in Figure 2.1. For starting material (S) to react to form product (P), additional energy is required to break the chemical bonds of the starting material and/or produce the required reactant configurations. This energy is called the activation energy (E_a); and, once enough energy is supplied to overcome this barrier, the reaction can proceed to form products. The greater the height of the barrier, the slower the reaction since a smaller proportion of reactants will have enough energy to overcome the barrier. If enough energy is released by the formation of products, the reaction may even catalyze itself once started. For example, wood does not spontaneously combust without first lighting it, even though the burning of wood has a large negative ΔG. However, once lit, it burns to completion.

TABLE 2.2 Concentration of Specific Glycolytic Enzymes and Their Substrates in Mammalian Skeletal Muscle[a]

Enzyme	Muscle Tissue (nmol/g)	Active Site Concentration (μM)	Free Active Site (%)	Substrate[b]	Total [S] (μM)	Free [S] (μM)	Products	Total [P] (μM)	Free [P] (μM)	Active Site/Free S Ratio
Phosphofructokinase	1.2	6.4	25	F6P	1500	1495	N[b]	N	N	0.004
Fructose-1,6-bisphosphatase	0.4	2.1	14	$F16P_2$	80	8.9	N	N	N	0.23
Aldolase	40.6	216	50	$F16P_2$	80	8.9	DHAP	160	116	24
							Gald3P	80	17.8	
Triosephosphate isomerase	22.4	60	77	DHAP	160	116	Gald3P	80	17.8	0.5
Glyceraldehyde-3-phosphate dehydrogenase	70	373	73	Gald3P	80	17.8	1,3-diPGA	50	0.14	21
Phosphoglycero-kinase	26	35	82	1,3-diPGA	50	0.14	3-PGA	200	196	247
Pyruvate kinase	8.7	46	59	PEP	65	46.9	Pyruvate	380	360	1.0
Lactate dehydrogenase	14.8	79	52	Pyruvate	380	360	Lactate	3700	3655	0.22

[a]The active site concentration was calculated by multiplying the enzyme concentration in nanomoles per gram muscle tissue by the number of subunits and dividing by 0.75 mL water/gram muscle tissue. Concentrations of enzyme active sites and metabolites are adapted from Srivastava and Bernhard (1986). Free concentrations were determined by solving simultaneous equations relating the free and bound concentrations of all metabolites. The equilibrium constants for the binding reactions were assumed to be equal to the reciprocal of the Michaelis constants obtained from Barman (1969).

[b]Abbreviations: S, substrate; P, product; F6P, fructose-6-phosphate; $F16P_2$, fructose-1,6-bisphosphate; DHAP, dihydroxyacetone phosphate; Gald3P, glyceraldehyde-3-phosphate; 1,3-diPGA, 1,3-diphosphoglyceric acid; PEP, phosphoenolpyruvate; 3-PGA, 3-phosphoglyceric acid; N, negligible amounts.

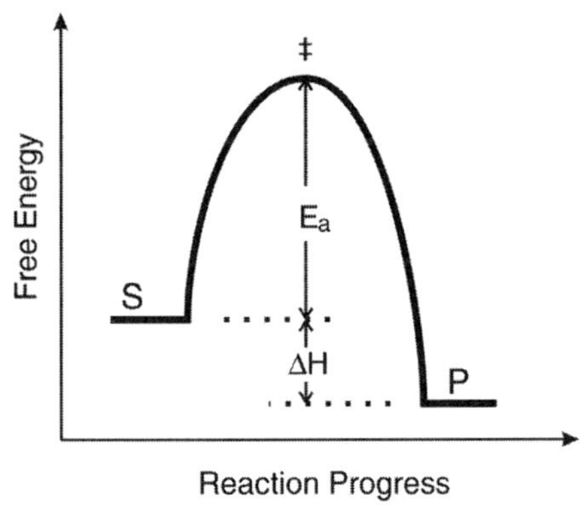

Figure 2.1 Idealized representation of a single-substrate reaction scheme. The free energy of the reactants is plotted versus hypothetical reaction coordinates (the reaction proceeds from left to right). S represents starting material and P products. The symbol ‡ represents the activated state, and E_a is the activation energy for the reaction.

The way in which the E_a influences the rate of a reaction is best described by transition state theory. The transition state is defined as a configuration of maximum potential energy through which the reactants must pass before continuing to form products. It is commonly denoted by a "‡" symbol (Fig. 2.1). The E_a barrier is thought of as a static barrier; it is fixed for a given set of starting parameters. The transition state is clearly distinguished from an intermediate that may form during a reaction sequence. An intermediate is a metastable minimum on the reaction profile, whereas the transition state is found at a point of maximum energy.

We can treat the formation of the transition state as an equilibrium problem. The equilibrium constant for the formation of the transition state can then be written as

$$K^{\ddagger}_{\text{eq}} = [\text{A}^{\ddagger}]/[\text{A}] \qquad (2.10)$$

where $[\text{A}^{\ddagger}]$ represents the concentration of the starting material that has formed a transition state. Using the thermodynamic principles developed above and quantum mechanics, it is possible to transform Eq. (2.10) into an equation relating the enthalpy ($\Delta H^{\ddagger}$) and entropy ($\Delta S^{\ddagger}$) associated with transition state formation to the rate of the reaction:

$$k = \frac{RT}{Nh}\exp\left(\frac{\Delta S^{\ddagger}}{R} - \frac{\Delta H^{\ddagger}}{RT}\right) \qquad (2.11)$$

where k is the reaction rate, R is the gas constant, N is Avogadro's number, and h is Plank's constant. Equation (2.11) explains much about reaction properties. First, it shows that a reaction is temperature-dependent, with rate increasing as temperature increases. Second, it shows that a reaction rate is directly influenced by the activation energy ($E_a = \Delta H^{\ddagger} + R\text{T}$), defined as the minimum energy that reactants must have in order to form products. A reaction with a larger E_a value has a smaller rate constant. Finally, it shows that the reaction rate is only dependent on the formation of the transition state (the transition state is regarded as the highest energy point on the reaction profile). Once the transition state is formed, the reaction quickly proceeds to form products.

The transition state theory is a powerful theory because it explains much of the behavior of chemical reactions as well as enzyme behavior. Other, simpler theories of catalysis have also been developed and are useful to know. One of the simplest is that described by the Arrhenius equation, which is based on collision theory. In this theory, only a fraction of the collisions between two molecules in solution produce a reaction:

$$k = Ae^{-E_a/RT} \qquad (2.12)$$

In Eq. (2.12), the factor A is called the preexponential factor and E_a is the activation energy (as defined above). Equation (2.12) is a useful relationship because it can be used to determine the E_a for a reaction and provides insight into the influence of temperature on the reaction rate. According to Arrhenius theory, increasing temperature increases potential collisions and, thus, increases the probability of a successful reaction. The use of Eq. (2.12) in determining the E_a for an enzyme reaction is described later in this chapter.

How Enzymes Influence Reaction Rates

The transition state theory outlined in the previous section has long been applied to the study of enzyme catalysis. Its power rests in the accuracy of its predictions of enzyme catalytic behavior and of enzyme structure–function relationships. Transition state theory shows that an enzyme can speed up catalysis by reducing the E_a in two different ways. Both of these ways have been experimentally confirmed. First, enzymes can act by partially stabilizing the transition state [lowering $\Delta H^{\ddagger}$ in Eq. (2.11)]. A more stable transition state means that less energy is required to produce the transition state, and this means that more starting material can react to form a transition state. Thus, more starting material will form product—in other words, the reaction proceeds at a faster rate. Enzymes that act in this manner either have specific amino acids within their active site (a specialized area or cleft on the protein surface where substrate binds and catalysis occurs) to stabilize the transition state or use other chemical species (called coenzymes) to aid in the catalytic process. Enzymes that act by stabilizing the transition state and that use specialized amino acids in their active site commonly catalyze general acid/base reactions. Enzymes that use specialized coenzymes as catalytic aids catalyze a variety of different reactions. For example, the enzyme–coenzyme complex can covalently bind to high-energy intermediates, stabilizing them by delocalizing electrons across extensive conjugated double-bond systems. These reactive species can then react with other substrates present in the active site. This is common for two substrate reactions. Coenzyme A, pyridoxyl phosphate, and thiamine pyrophosphate are good examples of coenzymes that, when bound to enzymes, stabilize high-energy intermediates.

The second way that enzymes can increase the reaction rate is by providing a favorable reaction entropy [decreased $\Delta S^{\ddagger}$, Eq. (2.11)]. This occurs when reactive groups on both substrates and enzymes are oriented to optimize catalysis. Enzymes that hydrolyse sugar polymers are good examples of this type of catalysis (see How Enzyme-Catalyzed Reactions Occur: Lysozyme as a Model).

Transition state theory explains many enzyme properties, but it is not the only theory of enzyme-catalyzed reactions. About the same time that Eyring popularized the transition state theory, Kramers developed a transient strain model. In this theory, the reaction proceeds by a process of diffusion over a potential energy barrier. The driving force is derived from random structural fluctuations in the enzyme protein molecule and the motion of the substrate (remember that the enzyme is not a fixed sphere but rather a flexible string of amino acids). At any given time, a certain fraction of the enzyme molecules produce the correct strain on the substrate and this allows catalysis to proceed. The relationship between the rate constant and the height of the potential energy barrier (ΔU) is given by

$$k = \left(\frac{1}{\tau}\right) \exp\left(\frac{-\Delta U}{RT}\right) \qquad (2.13)$$

where τ is the time constant of structural fluctuations and is proportional to the local viscosity of the enzyme. Changes in the viscosity of the local environment produce changes in enzyme fluctuations with a consequent change in the value of τ. Equation (2.13) is similar in form to the Arrhenius equation [Eq. (2.12)] with ΔU replacing E_a. Equation (2.13) is important because it is a framework for developing theories of enzyme-catalyzed hydrogen atom tunneling. Tunneling has been recently postulated because of discrepancies between predicted and observed kinetic isotope ratios and problems with the large energies needed to break stable C–H bonds. The kinetic isotope ratios are generated by measuring the rates of enzyme-catalyzed reactions with normal and deuterated or tritriated substituted substrates. Isotopes are useful because they can often provide mechanistic details of reactions and are useful for differentiating potential reaction mechanisms.

Tunneling theory (referred to as quantum tunneling) considers the wavelike properties of matter, rather than just focusing on the particlelike properties (as transition state theory does). One of the features of the wave-particle duality of matter is the ability of matter to pass through physical barriers than would normally be inaccessible. This can significantly lower the energy required to perform the reaction. This "tunneling" (so-called because it is like connecting two energy valleys by a tunnel rather than having to go over the intervening E_a mountain) arises because of the wavelike properties of matter. The de Broglie equation relates the wavelength (λ) of a particle to its kinetic energy (E) and mass (m):

$$\lambda = h/(2mE)^{1/2} \qquad (2.14)$$

Equation (2.14) shows that as m decreases, λ becomes longer and there is higher uncertainty in its physical location. Tunneling is a probabilistic event, and the probability of its occurrence increases with the uncertainty of the position of the particle. In order for an enzyme to catalyze a reaction via quantum tunneling, it must act to increase the probability of this event. The only theoretical treatment of H atom tunneling that incorporates the role of protein structural fluctuations is called vibrationally enhanced ground state tunneling. In this theory, the thermal fluctuations in protein structure shorten the distance that an H atom must tunnel. This, therefore, increases the probability of a tunneling event so that it can occur with a reasonable certainty. As the vibrational motion of a protein increases, the likelihood of a tunneling event increases. The energy barrier is, therefore, the energy required to jump to the next quantized vibrational state. The quantized vibrational states are comparable to the potential energy barrier in Eq. (2.13) (ΔU).

The transient strain theory is interesting in that it predicts different evolutionary stresses on enzyme development. Transition state theory predicts that enzyme architecture evolved to maximize the formation of the transition state. Thus, active site architecture would have evolved to stabilize the transition state and lower E_a values. Transient strain theory predicts that enzyme architecture would evolve to maximize "correct" thermal structural fluctuations to maximize tunneling probabilities. This would mean that the overall protein structure–active site structure architecture must vibrate correctly to allow the shortest tunneling distances to exist and evolutionary pressures would be on achieving productive vibrations. These pressures would apply only to enzyme reactions that transfer electrons or hydrogen atoms because only these entities are small enough to allow a tunneling event to occur.

How Enzyme–Catalyzed Reactions Occur: Lysozyme as a Model of Transition State Theory

As indicated above, enzymes have a specialized area or cleft on the protein surface called the active site. It is here that substrate binds, and it is here that the R groups responsible for binding substrate (in enzyme-catalyzed reactions starting material is usually referred to as substrate) and for stabilizing the transition state are located. The active site of an enzyme is a highly conserved region of the protein and is specifically configured to bind compounds with a particular geometry. Often, this means that the enzyme has a good affinity for a particular substrate but can also bind structurally similar compounds. This has implications for metabolic control and enzyme inhibition (see below). Obviously, how well a particular substrate binds the active site depends on (1) which amino acids are present, (2) the positioning of the amino acids within the active site, and (3) other environmental factors such as the ionic strength, pH, and ion composition of the solvating medium.

Originally, it was thought that the binding of substrate to active site could be described as a "lock-and-key" fit. We now know this to be incorrect. As noted in previous sections, enzymes are not rigid structures. Their final structure is always "breathing" and responds subtly to changes in the enzyme's environment. The plasticity of the enzyme structure also permits the more profound changes in conformation that accompany the binding of substrate. This dynamic binding mechanism is called *induced fit* and can be illustrated by measurements of enzyme structure in the presence and absence of substrate. X-ray crystallography shows that enzyme structure is different when substrate is bound to the active site. This is consistent with the idea that enzymes undergo changes in conformation when bound to substrate. The induced-fit model does not imply that the enzyme–substrate complex is a perfectly complementary structure. Enzymes could not operate efficiently if substrate binding was optimized since this would cause a minimum in the energy profile at this point. In other words, if enzymes were optimized to bind substrate, then the enzyme–substrate complex would be stable and would require extra energy to continue along the catalytic pathway. This would lead to an increase in the reaction E_a and slow catalysis—not a good thing for an enzyme.

The fact that enzymes, in general, are not optimized to bind substrates does not mean that enzymes are not selective in which substrates they bind. In fact, enzymes are highly selective, easily capable of distinguishing between stereoisomers that are difficult to distinguish chemically. However, it is the combination of active site geometry (orientation of reactive, charged, and nonpolar groups in the active site) and changes in enzyme conformation after substrate binding that gives the enzyme–substrate complex its specificity.

A good example of an enzyme that stabilizes a transition state is lysozyme, an enzyme found in mammalian tears and mucus and in egg white that hydrolyses sugar polymers found in bacterial cell walls (Fig. 2.2). These sugar polymers are composed of alternating residues of *N*-acetylmuramic acid (NAM) and *N*-acetylglycosamine (NAG). Lysozyme acts as an *N*-acetylmuramidase, breaking a NAM–NAG linkage within the sugar polymer while transferring the *N*-acetylmuramyl group to water at the sites indicated by the bold arrows in Figure 2.2. This acts to weaken the bacterial cell wall and leads to lysis of the cell from turgor pressure.

The enzyme has a long active site that binds six sugar residues at a time, three NAG and three NAM units. The active site is a well-defined cleft in the polypeptide, easily identified by X-ray crystallography. Upon substrate binding, a NAM residue occupies the fourth spot in the active site. After substrate binding, the enzyme undergoes a conformational change that puts a physical strain on the fourth NAM residue, forcing it to adopt an energetically unfavorable half-chair conformation. This destabilization means that the enzyme–substrate complex is at a higher energy than the free substrate, effectively lowering the activation energy (and increasing the rate of reaction).

Once substrate is bound to the active site, it is oriented so that amino acid R groups on the protein can directly participate in the catalytic mechanism. This is illustrated in Figure 2.2*c*, which shows how aspartate 52 and glutamic acid 35 of lysozyme act to catalyze the NAG–NAM cleavage. Both groups are important to catalysis; glutamic acid provides the proton for acid catalysis and aspartate stabilizes the resulting glycosyl-oxocarbonium ion (transition state). Lysozyme action demonstrates that enzymes catalyze reactions using simple chemical species (glutamic acid and aspartic acid in this case). This shows that the active site is not an isolated environment where specialized chemistry occurs. Nevertheless, enzymes are more efficient catalysts than simple chemical agents because: (1) they orient the substrate so that optimum efficiency between the catalyst (e.g., protons for lysozyme) and the susceptible bond is achieved, (2) they bind substrates with a high degree of specifically so that other molecules are not acted upon, and (3) the specific orientation of reactants in the active site means that only one stereo isomer is produced. This is in contrast to chemical catalysts that are frequently neither substrate-specific nor stereo-specific and, consequently, produce a significant proportion of undesirable products. Enzyme-catalyzed reactions are highly specific and so enzymes produce essentially 100% of a single product.

Lysozyme can also bind small oligosaccharides similar in structure to the NAG–NAM polymers. Because they

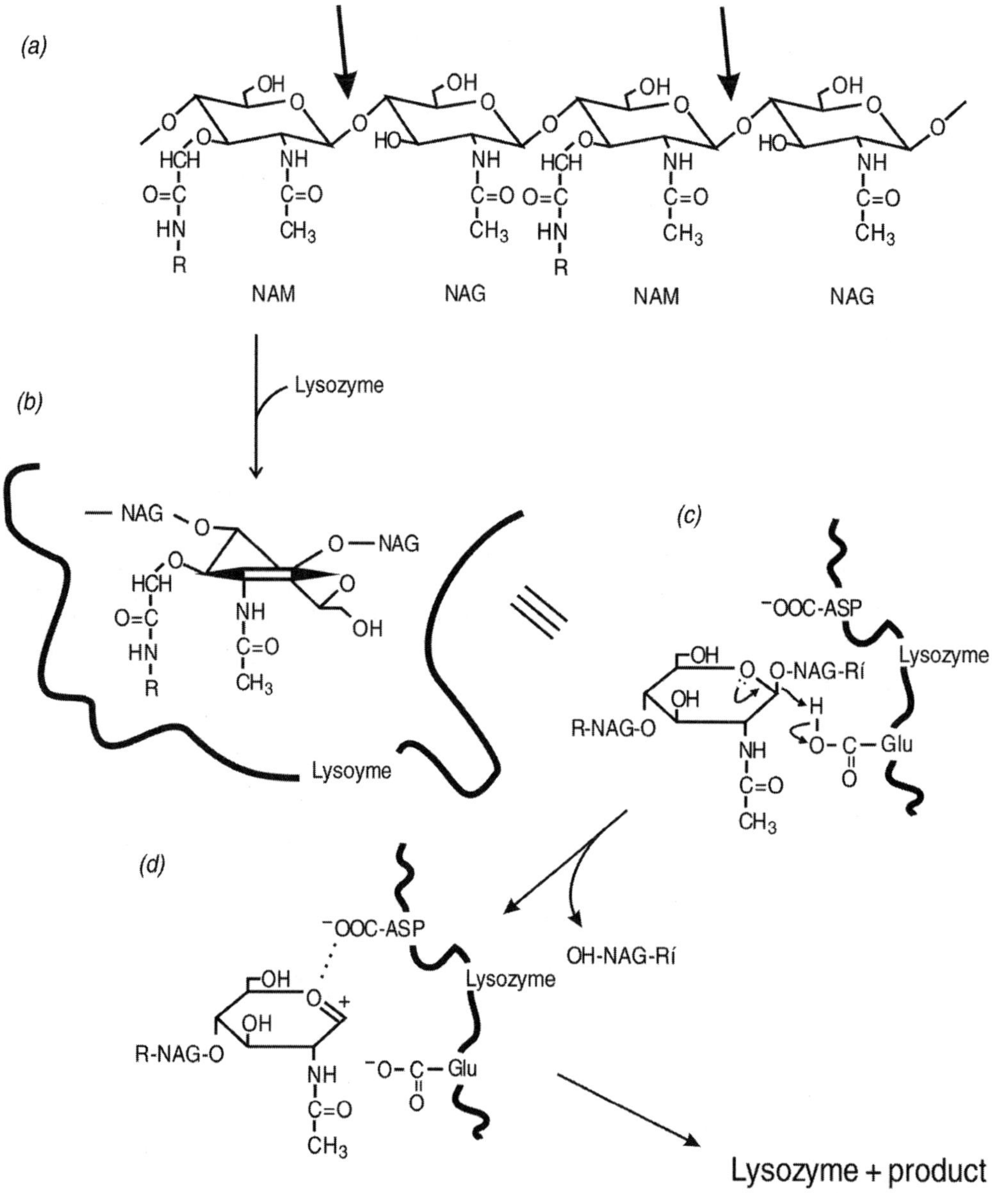

Figure 2.2 Diagrammatic representation of the lysozyme catalytic mechanism. (*a*) NAM–NAG polymer binds to lysozyme to force the fourth sugar residue to adopt (*b*) a half-chair conformation. Aspartate (Asp) and glutamate (Glu) participate directly in the reaction sequence. (*c*) Glutamate donates a proton, and (*d*) aspartate stabilizes the glycosyl-oxocarbonium ion (shown by dots). Arrows indicate bonds that can be lysed. Modified from Walsh (1979).

occupy the same physical location as the bacterial cell wall polymer on the lysozyme protein surface, they effectively compete with the bacterial cell wall polysaccharide for the enzyme active site. These oligosaccharides are not lysed by lysozyme and so do not undergo a normal catalytic cycle. This is a good example of how structurally similar chemicals can bind to an enzyme active site and influence activity. Inhibition by molecules that compete with the active site will be more fully developed in the Enzyme Inhibition section later in this chapter.

Transition state stabilization is a common catalytic strategy and many instances have been reported in the scientific literature. The enzyme CK provides a particularly striking example of this (Fig. 2.3). As previously discussed, CK functions by transferring the phosphoryl moiety of ATP to creatine. Both ATP and creatine substrates lie in the

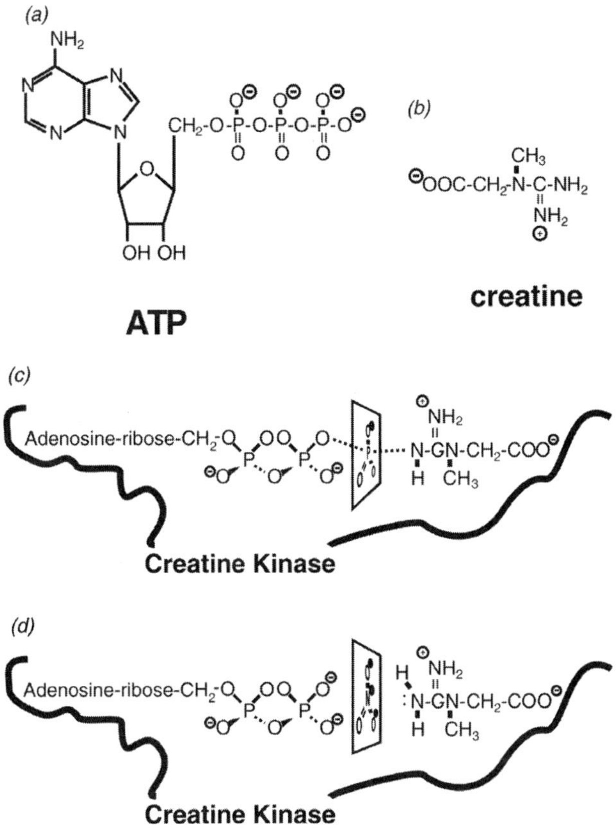

Figure 2.3 Transition state complex formation during the catalytic cycle of creatine kinase. Structures for the substrates (*a*) adenosine triphosphate (ATP) and (*b*) creatine are shown in (*c*) the enzyme active site. The transition state involves the formation of a planar phosphate (PO_3) intermediate that is shared by adenosine diphosphate (ADP) and creatine [square structure in (*c*)]. The transition state can be mimicked by adding nitrate [NO_3^-; in (*d*) square structure in active site], ADP, and creatine to a solution containing creatine kinase.

active site beside one another, and the transition state includes a flat phosphate intermediate (Fig. 2.3*c*) with the bond delocalized over the planar phosphate, ADP, and creatine. This active site geometry allows the creation of a tightly bound inhibitor when ADP, creatine, and nitrate ion (NO_3^-) are added because the flat nitrate ion mimics the transition state phosphate (Fig. 2.3*d*). This complex cannot react to give products. Knowledge of this transition state permitted a facilitated purification of CK that made use of ADP-derivatized columns. In the presence of creatine and nitrate, CK could be effectively immobilized to the ADP column because a tight transition state complex was formed. The specific nature of this binding meant that all other contaminating proteins could be separated from CK by washing the column matrix with a salt solution. The CK enzyme was then released from the column by adding free ADP to compete with the column-bound ADP. The enzyme was, therefore, obtained in a highly purified form because of prior knowledge of its transition state.

Transition state mimics are also commonly manufactured as enzyme inhibitors. This is a common strategy when specific enzymes in a mixture of enzymes must be inhibited. For example, many of the drugs that inhibit proteases are transition state analogs. All of these examples show that many enzyme structures are optimized to bind and stabilize transition states (but not substrates).

WHEN ENZYMES NEED EXTRA HELP: COENZYMES AND THEIR FUNCTIONS

Lysozyme, considered above, is an example of a group transfer catalyzed by simple acid–base hydrolysis. The enzyme itself contains the necessary reactive groups to catalyze these types of reactions so that no other reactive groups need be present. Although a great number of enzymes make use of simple acid–base catalysis, this represents only one of several different types of enzyme reactions. These can be loosely divided into four different classes: group transfer, oxidation/reduction, elimination, and carbon–carbon bond breaking.

When enzymes require specialized reactive groups not available on constituent amino acid R groups, they often make use of coenzymes as catalytic aids. Coenzymes are specialized compounds that can act to accept and donate chemical groups, hydrogen atoms, or electrons (see Table 2.3). Coenzymes can be broadly classified into three groups: compounds that transfer high-energy phosphate, compounds that accept and donate electrons, and compounds that activate substrates.

Coenzymes That Transfer High-Energy Phosphate

These compounds are highly phosphorylated derivatives of the nucleotide bases (e.g., adenosine and guanosine), which exist in mono-, di-, or triphosphate forms. Usually, the diphosphate acts as a phosphate acceptor and the triphosphate acts as a phosphate donor: ATP $\rightarrow$ ADP + P_i (inorganic phosphate). The monophosphate (AMP) form is produced during biosynthetic reactions requiring a pyrophosphate intermediate: ATP $\rightarrow$ AMP + PP_i. The adenylates are extremely important to metabolic and biosynthetic reactions since they act as the energy currency of the cell. ATP is by far the most common cellular high-energy phosphate coenzyme and occurs at concentrations approximately 100-fold higher than other high-energy phosphate coenzymes such as GTP, CTP, TTP, and UTP. Other high-energy phosphate compounds exist and are found at higher concentrations than ATP, such as creatine

TABLE 2.3 Common Coenzymes and Their Functions

Function	Types of Reactions	Coenzyme Common Name	Abbreviation	Vitamin Derived from
Transfer of "high"-energy phosphate	Activation of metabolic intermediates to push unfavorable equilibria	Adenosine triphosphate	ATP	
		Guanosine triphosphate	GTP	
		Uridine triphosphate	UTP	
Accept and donate electrons or H atoms	Two-electron acceptors and donors. With H^+ participation appear as 2 H atom reactions	Nicotinamide adenine dinucleotide	NAD	Niacin
		Nicotinaminde adenine dinucleotide phosphate	NADP	Niacin
		Flavin adenine dinucleotide	FAD	Riboflavin (B_2)
		Riboflavin 5′-phosphate	FMN	Riboflavin (B_2)
	Membrane-based electron acceptors and donors	Lipoamide		
		Coenzyme Q	CoQ	Ubiquinone
Activate acyl groups	Water-soluble reactions	Coenzyme A (pantothenic acid)	CoA	Pantothenic acid
Activate acyl groups on growing fatty acid chain	Associated with fatty acid synthase	Acyl carrier protein (phosphopantetheine)	ACP	
Carboxyl group carrier	β-carboxylation reactions	Biotin		Biotin
α-Condensation and α-cleavage	α-Keto acid decarboxylation and α-hydroxyketone reactions	Thiamine pyrophosphate	TPP	Thiamine (B_1)
α-Hydrogen loss, decarboxylation, side chain cleavage	Several different types of reactions—water soluble	Pyridoxyl phosphate	PLP	Pyridoxine and related compounds (B_6)
H-transfer reactions	Group migration via radical intermediate chemistry, methyl transfer	Coenzyme B_{12} (cyanocobalamin)	B_{12}	Cobalamin (B_{12})
One-carbon transfers	Methylation reactions	Tetrahydrofolate (folic acid)	THF	Folate
		S-Adenosyl methionine	SAM	

phosphate and arginine phosphate. Creatine and arginine phosphate do not, however, act as coenzymes in the cell but rather as repositories of high-energy phosphate. ATP, GTP, UTP, TTP, and CTP are designated as "high-energy phosphates" because hydrolysis of the γ-phosphorous yields a high amount of energy; for example, for ATP, the $\Delta G^{\circ\prime} = -30.6\,\mathrm{kJ/mol}$ corresponding to a K_{eq} value of approximately 10^8 in favor of the formation of ADP and inorganic phosphate. This means that many reactions can be forced to completion by coupling them to ATP hydrolysis. This is true because the overall $\Delta G^{\circ\prime}$ for a series of reactions is simply the sum of the individual $\Delta G^{\circ\prime}$ values for each reaction. Thus, the overall free energy can be made more negative [goes farther to completion, see Eq. (2.8)] if it is coupled to ATP hydrolysis.

Coenzymes such as ATP and ADP do not remain bound to an enzyme after participating in a reaction but dissociate and diffuse to other enzymes, participating in other enzyme reactions. This allows these coenzymes to shuttle between energy-generating and energy-utilizing reactions. It also allows them to be regenerated. For example, ATP may donate a phosphate to drive one reaction becoming ADP. The ADP can then dissociate and be rephosphorylated by another reaction to give ATP once again. In this way, the ATP generated by glycolysis, for example, can be used in any number of energy-requiring processes such as biosynthetic pathways, membrane ion pumps, and protein synthesis. Because they are freely soluble entities that reversibly bind to enzymes, these coenzymes are usually treated as enzyme substrates in kinetic equations. In general, enzymes that use ATP are highly regulated since they control the cellular concentration of ATP through their action. These enzymes are usually found at the start of metabolic pathways and are often inhibited or activated by ATP, ADP, and AMP, as well as by the end products of the metabolic pathway.

Coenzymes That Accept and Donate Electrons

Coenzymes that accept and donate electrons usually have delocalized π-ring systems that stabilize positive charges (oxidized coenzymes) and negative charges (reduced coenzymes). These coenzymes usually accept or donate electrons easily. Some reversibly bind enzyme active sites (similar to the way ATP acts), whereas others remain closely associated with one enzyme and do not dissociate. Nicotinamide adenine dinucleotide (NAD), nicotinamide adenine dinucleotide phosphate (NADP), and coenzyme Q (CoQ) are good examples of coenzymes that diffuse from enzyme to enzyme, whereas flavin adenine dinucleotide (FAD) and flavin mononucleotide (FMN) remain bound to one enzyme. Like the coenzymes that transfer high-energy phosphate, coenzymes such as NAD, NADP, and CoQ act as substrates in the reaction mechanism. NAD and NADP participate in water-based reactions and CoQ is lipid soluble. In the case of FMN and FAD coenzymes, the reduced riboflavin coenzyme usually interacts with a soluble coenzyme to complete its oxidation–reduction cycle. NAD and NADP can be considered to be high-energy electron carriers, much like ATP is a high-energy phosphate carrier. Nicotinamide coenzymes transfer electrons from substrates to either reduce and oxidize compounds during biosynthetic reactions or provide the electrochemical energy necessary for ATP synthesis in the mitochondrion. Similar to ATP-utilizing enzymes, nicotinamide-coenzyme-utilizing enzymes can be highly regulated. Good examples are glyceraldehyde 3-phosphate dehydrogenase and pyruvate dehdyrogenase of the glycolytic pathway and α-ketoglutarate dehydrogenase of the tricarboxylic acid cycle in the mitochondrion. However, unlike ATP-utilizing enzymes they are not usually regulated by coenzyme (NAD or NADP) concentrations but are sensitive to substrate and product concentrations.

Coenzymes That Activate Substrates

There are many different types of coenzymes that activate substrates. All act by reacting with substrate to produce new, more reactive products. For example, CoA and acyl carrier protein (ACP) covalently bind to acetyl residues (CH_3–CO–) to form acetyl-CoA and acetyl-ACP. Both CoA and ACP are good chemical leaving groups so that the reactivity of acetyl-CoA and acetyl-ACP is much higher than that of acetic acid alone. This means that the $\Delta G^{\circ\prime}$ for hydrolysis of these compounds is large and negative; thus, formation of these compounds effectively creates a high-energy acetate derivative.

Biotin is a highly reactive double-ring system that is covalently attached to an enzyme by a long, flexible arm. Biotin serves as a carboxyl group carrier, adding $-COO^-$ groups to many different substrates. The reaction is energetically unfavorable and is coupled to the hydrolysis of ATP to drive the reaction to completion.

Thiamine pyrophosphate and pyridoxal phosphate catalyze several different types of reactions including α-condensation and α-cleavage reactions (thiamine pyrophosphate), as well as transaminations, decarboxylations, eliminations, and aldol cleavages (pyridoxal phosphate). Both coenzymes sit in the enzyme active site, although neither is covalently attached. The thiamine pyrophosphate mechanism proceeds via a thiazolium dipolar ion that can attack ketone or imide compounds. The newly formed adduct reacts readily to give products (carbon dioxide is often a product). Pyridoxal phosphate also forms an adduct with substrate (called a Schiff's base), which decomposes to give product. These coenzymes increase the rate of catalysis by stabilizing the transition state to lower the overall activation energy. In both cases, the substrate–coenzyme adduct is stabilized by the conjugated π-ring systems of the coenzyme that act either as an electron sink or an electron source during the reaction.

Coenzyme B_{12} is another example of a coenzyme that catalyses a reaction by stabilizing the transition state. B_{12} is an unusual coenzyme because it contains a cobalt atom in the center of a corrin ring system coordinated to adenosine below and ribose above the plane of the corrin ring. The cobalt–carbon bond is very rarely found in nature. The cobalt is the reactive center of the coenzyme and can easily accept and donate a single electron; coenzyme B_{12} catalyses reactions using radical ion chemistry. It is this property that enables B_{12} to catalyze a wide range of reactions including group migrations, dehydrations, and oxidation–reduction reactions (often in conjunction with the oxidation–reduction coenzymes listed above).

S-adenosylmethionine (SAM) and tetrahydrofolate (THF) serve as methyl group carriers in much the same way that CoA and ACP serve as acetyl carriers. Since the chemistry of carbon–carbon bond breaking is complex, reactions involving THF and SAM often involve other coenzymes required for reduction or oxidation, as well as B_{12}. SAM and THF reactions are not common, but, because of their unique function, these coenzymes participate in some important reactions. For example, they are involved in the methylation of deoxyribonucleic acid (DNA) to regulate gene transcription, the methylation of homocysteine to form methionine, (see Table 2.1), and the production of creatine, phosphatidyl choline, and epinephrine.

KINETIC MECHANISMS OF ENZYME ACTION

All enzymes are catalysts that can increase reaction rates by one of three processes: stabilizing high-energy intermediates, orienting reactive groups so that they are in close proximity, or facilitating tunneling of electrons or hydrogen atoms. Although many different enzyme reactions exist, they can be effectively reduced to a handful of kinetic mechanisms. This is because, in general, the mathematics of kinetics deals with overall enzyme rates and does not consider the details of the individual catalytic mechanisms that operate at the active site. Having said this, of course, it is apparent that some consideration of detail is necessary to construct accurate representations of enzyme action. Up to now, we have considered the thermochemistry of reaction mechanisms. From this point on, we will consider the kinetic behavior of reaction mechanisms.

Consider the general reaction of Figure 2.1 where substrate S reacts smoothly to form product P. The introduction of an enzyme (E) catalyst to this process changes the details of the reaction mechanism because enzymes are physical entities that participate directly in the reaction mechanism. As discussed above, this means that enzymes bind substrates (and products) so that (at least) three equilibria are now required to depict a general one-substrate reaction:

$$S + E \underset{k_{-1}}{\overset{k_1}{\rightleftharpoons}} ES \underset{k_{-2}}{\overset{k_2}{\rightleftharpoons}} EP \underset{k_{-3}}{\overset{k_3}{\rightleftharpoons}} E + P \tag{2.15}$$

These three equilibria illustrate all the steps that are necessary to describe any one-substrate reaction, including the binding of substrate to enzyme, the conversion of enzyme-bound substrate to enzyme-bound product, and the release of product from the enzyme. Equation (2.15) does not formally include the formation of a transition state as a discrete step in the reaction mechanism. Its existence is, however, encompassed in the interconversion of enzyme-bound substrate into product (ES $\leftrightarrow$ EP; step 2). Equation (2.15) also demonstrates the temporal order for a simple enzyme-catalyzed reaction: (1) substrate binds to the enzyme to form an enzyme–substrate complex (ES), (2) ES reacts to give the enzyme–product complex (EP), and (3) EP dissociates to give free enzyme and product. This latter step regenerates the catalyst so that free enzyme is now capable of binding to new S to continue the reaction.

Each of the rate equations for the individual reaction steps of Eq. (2.15) could be written as a form of Eq. (2.11) with thermodynamic parameters that govern the formation of ES, the formation of EP, and the conversion of ES to EP. Thus, each step has its own associated rate constant defined by Eq. (2.11). Enzymes work by influencing one or more of these individual rate constants to lower the overall reaction rate. One must remember, however, that speeding up the forward process also means speeding up the reverse process so that the equilibrium constant for the reaction is unchanged.

The binding of reactant to an enzyme active site involves several different steps including the removal of water structured around both substrate and active site reactive groups, the correct orientation of all the reactive groups on both reactants and enzyme, and the breaking of electrostatic bonds that may exist between reactive groups and other free ion species present in the solvating medium (such as Na^+, K^+, and Cl^-). This means that there may be significant differences in the free energies of free and enzyme-bound substrates. This also means that all enzyme-catalyzed reactions have two distinct equilibria that describe the interconversion of substrate to product: the equilibrium between the free reactants and the equilibrium between the enzyme-bound reactants. The K_{eq} value commonly listed in published tables is always that of the free reactants—this value, therefore, represents the equilibrium constant describing the overall conversion of "free" substrate to "free" product. The "free" K_{eq} value may differ substantially from that of the enzyme-bound reactants because of the difference in the free energy of the free and bound reactants. The equations for the K_{eq} values [derived from Eq. (2.5)] are simply

$$K_{eq}(\text{free reactants}) = \frac{[\text{free product}]}{[\text{free substrate}]} \tag{2.16a}$$

$$K_{eq}(\text{bound reactants}) = \frac{[\text{enzyme bound product}]}{[\text{enzyme bound substrate}]} \tag{2.16b}$$

A simple relationship exists between the equilibria of the free and bound reactants. This is determined by the thermodynamics of the reaction and can be expressed as

$$\Delta G_{\text{free}} = \Sigma \Delta G_i \tag{2.17}$$

where the ΔG_i values represent the free energy losses or gains for each of the steps in a reaction. Thus, if an enzyme binds substrate with a higher affinity than product, the ΔG for the enzyme-bound reaction will be more positive than that of the free reactants since more substrate will be bound to the enzyme. Conversely, if the enzyme binds product with greater affinity, the enzyme-bound reaction will have a more negative ΔG (product formation being favored) than that for the free reactants. If the ΔG for the free reactants is negative but the enzyme binds substrate better than product, things still must balance out. Thus, substrate binding may proceed with a large negative ΔG and product binding with a positive ΔG (Fig. 2.4).

A consideration of the many equilibria present in an enzyme-catalyzed reaction illustrates the difference between transition state theory and enzyme kinetic modeling. Transition state theory was developed to describe solution kinetics of free reactants. In this theory, the reaction rate is only a function of the energies of the reacting species

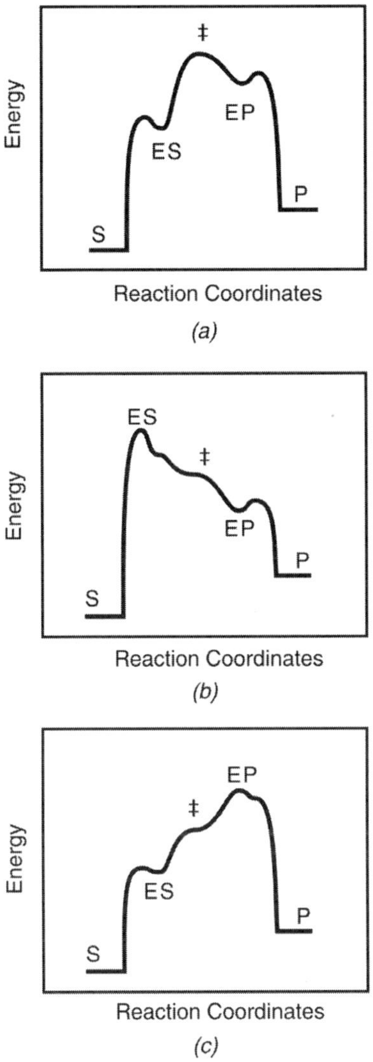

Figure 2.4 Energy diagram for idealized enzyme-catalyzed reactions. The energy level of the reactants and reacting species is plotted versus reaction coordinates. (*a*) Reaction sequence in which the transition state is the most energetically unfavorable species. (*b*) Formation of ES is the most energetically unfavorable step. S and ES represent the free and enzyme-bound substrate. P and EP represent free and enzyme-bound product, respectively. (*c*) Formation of EP is the most energetically unfavorable step.

and collision frequencies. Enzyme-catalyzed reactions have several steps, including substrate and product binding, each with their own rate constant. Many factors can influence the individual steps including the nature of the solvating medium, the ambient temperature (effect on enzyme conformation), the total enzyme concentration, and the presence of dissolved ions in the medium. It is important to consider all these phenomena when thinking about metabolic regulation under different physiological conditions. These considerations are especially important for substrate and product binding which are complex processes in and of themselves.

Development of Rate Equations: Simple Solution Kinetics

The reaction of Eq. (2.15) was drawn with arrows going in both directions to illustrate that all reactions are reversible; a finite flow of S to P and P to S always exists for any given reaction, even if the reaction is at equilibrium. It is possible to solve Eq. (2.15) to obtain an expression that relates the rate of the enzyme to the concentration of substrate, but some assumptions about the reaction help to simplify the analysis. A common assumption is that the overall reaction is at steady state. This means that the concentration of each of the enzyme species of Eq. (2.15) does not change over time:

$$\frac{d[E_{\text{free}}]}{dt} = \frac{d[\text{ES}]}{dt} = \frac{d[\text{EP}]}{[dt]} = 0 \qquad (2.18)$$

Another common assumption is that the concentrations of substrate and product are much greater than the concentration of enzyme. This means that

$$[\text{S}]_{\text{free}} \gg [\text{ES}] \quad \text{and} \quad [\text{P}]_{\text{free}} \gg [\text{EP}] \qquad (2.19)$$

so that

$$[\text{S}]_{\text{total}} \approx [\text{S}]_{\text{free}} \quad \text{and} \quad [\text{P}]_{\text{total}} \approx [\text{P}]_{\text{free}} \qquad (2.20)$$

Equations (2.19) and (2.20) imply that the enzyme concentration is much lower than the concentration of reactants. This is true under *in vitro* conditions where a reaction is monitored by adding a large excess of one reactant to a small catalytic amount of enzyme. It does not generally apply to *in vivo* conditions where the concentrations of reactants may be similar to or much greater than the concentration of enzyme (Table 2.2). The implications of this will be dealt with in greater detail in Chapter 3.

Finally, it is usually assumed that the changes in substrate and product concentrations over the time taken to measure the reaction are so small that they are essentially zero. This assumption is required to simplify the resulting differential equations. This condition is relatively easily engineered under *in vitro* conditions as both substrate and product concentrations are relatively large compared with the amount of product produced or substrate consumed over the relatively short time period that a reaction is monitored *in vitro*. With these assumptions, and the enzyme conservation equation

$$[\text{E}]_{\text{total}} = [\text{E}]_{\text{free}} + [\text{ES}] + [\text{EP}] \qquad (2.21)$$

it is possible to generate steady-state expressions for each of the enzyme species:

$$\frac{d[\mathrm{E}]_{\mathrm{free}}}{dt} = k_{-1}[\mathrm{ES}] - k_1[\mathrm{E}]_{\mathrm{free}}[\mathrm{S}] = k_3[\mathrm{EP}] - k_{-3}[\mathrm{E}]_{\mathrm{free}}[\mathrm{P}] = 0 \tag{2.22}$$

$$\frac{d[\mathrm{ES}]}{dt} = k_1[\mathrm{E}]_{\mathrm{free}}[\mathrm{S}] + k_{-2}[\mathrm{EP}] - [\mathrm{ES}] \times (k_{-1} + k_2) = 0 \tag{2.23}$$

$$\frac{d[\mathrm{EP}]}{dt} = k_2[\mathrm{ES}] + k_{-3}[\mathrm{E}]_{\mathrm{free}}[\mathrm{P}] - [\mathrm{EP}] \times (k_3 + k_{-2}) = 0 \tag{2.24}$$

Note that Eqs. (2.22) to (2.24) have been set equal to zero. This is the steady-state assumption, which states that, at steady state, the concentrations of the different enzyme species do not change. Combining and rearranging Eqs. (2.21) to (2.24) gives an equation of the form:

$$V = \frac{\left(\dfrac{V_f}{K_{m,\mathrm{S}}}[\mathrm{S}] - \dfrac{V_r}{K_{m,\mathrm{P}}}[\mathrm{P}]\right)}{\left(1 + \dfrac{[\mathrm{S}]}{K_{m,\mathrm{S}}} + \dfrac{[\mathrm{P}]}{K_{m,\mathrm{P}}}\right)} \tag{2.25}$$

In Eq. (2.25), V_f and V_r represent the maximal rates in the forward and reverse directions, respectively. The V_f and V_r values are a function of the relative magnitude of the rate constants of the individual steps and are intrinsic properties of individual enzymes. Because the maximal rates are intrinsic properties of enzymes, they are directly proportional to the enzyme concentration. This is demonstrated in Eq. (2.26), which shows the relationship between the maximal rates and the microscopic rate constants. Substituting Eq. (2.26) into Eq. (2.25) will show that the rate of the reaction is also a linear function of the amount of enzyme. This means that the rate of the reaction is doubled when the amount of enzyme in the reaction vessel is doubled. This is expected for a reaction where the catalyst is neither consumed nor created.

The $K_{m,\mathrm{S}}$ and $K_{m,\mathrm{P}}$ values are measures of the affinity of the enzyme for substrate and product. In a practical sense, they are the concentration of substrate or product that gives a rate equal to $V_f/2$ or $V_r/2$, respectively. The $K_{m,\mathrm{S}}$ and $K_{m,\mathrm{P}}$ values are the kinetic equivalent to enzyme–substrate and enzyme–product dissociation constants ($1/K_{\mathrm{eq}}$ values) but are usually made up of many individual rate constants and so are often not equal to the true K_{eq} value. In fact, all the parameters V_f, V_r, $K_{m,\mathrm{S}}$, and $K_{m,\mathrm{P}}$ are complex combinations of the individual kinetic constants of Eqs. (2.21) to (2.23):

$$V_f = [\mathrm{E}]_{\mathrm{total}} \times \frac{k_2 k_3}{k_{-2} + k_2 + k_3} \qquad V_r = [\mathrm{E}]_{\mathrm{total}} \times \frac{k_{-1} k_{-2}}{k_{-1} + k_{-2} + k_2} \tag{2.26}$$

$$K_{m,\mathrm{S}} = \frac{k_{-1}k_{-2} + k_{-1}k_3 + k_2k_3}{k_1(k_{-2} + k_2 + k_3)} \qquad K_{m,\mathrm{P}} = \frac{k_{-1}k_{-2} + k_{-1}k_3 + k_2k_3}{k_{-3}(k_{-1} + k_{-2} + k_2)} \tag{2.27}$$

Although Eq. (2.25) was developed from a steady-state treatment of detailed kinetic mechanism to give an equation with microscopic kinetic constants for each step, it should be regarded as the general equation for a reversible enzyme reaction as other deviations will also produce Eq. (2.25). For example, one can assume that k_2 and k_{-2} are rate limiting. This means that the free and enzyme-bound reactants are at equilibrium and the $K_{m,\mathrm{S}}$ and $K_{m,\mathrm{P}}$ values (generally called Michaelis constants after one of the researchers who studied enzyme kinetics in the early 1900s) represent true dissociation constants. In this case, the maximal velocities are a function only of the catalytic conversion of S into P:

$$K_{m,\mathrm{S}} = \frac{k_{-1}}{k_1} = \frac{1}{K_{\mathrm{eqS}}} \qquad K_{m,\mathrm{P}} = \frac{k_3}{k_{-3}} = \frac{1}{K_{\mathrm{eqP}}} \tag{2.28}$$

$$V_f = k_2 \times [\mathrm{E}]_{\mathrm{total}} \qquad V_r = k_{-2} \times [\mathrm{E}]_{\mathrm{total}} \tag{2.29}$$

Other limiting assumptions will also give Eq. (2.25) meaning that it is impossible to determine the kinetic meaning of the K_m and V values without a detailed examination of the microscopic rate constants. Equation (2.25), therefore, is written as an empirical equation that can and does serve as a description of many other enzyme mechanisms that are more complex than the simple reversible mechanism of Eq. (2.15) so that the overall form of Eq. (2.25) can be used to describe many different kinetic mechanisms. Note that the derivation of Eq. (2.25) assumes that the changes in substrate and product concentrations are small relative to the total product and substrate concentrations. This may occur *in vitro* but may not reflect the *in vivo* situation (the ramifications of this will be investigated later).

Experimental verification of Eq. (2.25) has been accomplished under *in vitro* conditions using relatively high substrate and low enzyme concentrations. This ensures that the total substrate concentration does not change during enzyme assay. When assayed *in vitro*, enzyme rates are measured as initial rates to ensure a negligible amount of product formation. The enzyme concentration is usually adjusted so that these rates are linear for a prolonged period of time to allow greater measurement precision. If the *in vitro* assay reaction vessel contains only substrate (the initial product concentration is zero and no product has been added to the reaction mixture), the reaction scheme of Eq. (2.15) can be simplified to

$$S + E \underset{k_{-1}}{\overset{k_1}{\rightleftharpoons}} ES \overset{k_2}{\longrightarrow} E + P \tag{2.30}$$

which can be described by an equation of the form

$$V = \frac{V_{max}[S]}{K_m + [S]} \tag{2.31}$$

Equation (2.31) is commonly referred to as the Michaelis–Menten equation, although it was derived by three different groups working on enzyme kinetics around the early 1900s. It contains the terms V_{max} and K_m instead of V_f and $K_{m,S}$ for comparison with other textbooks and treatises that commonly use these terms. It is common not to specify the substrate in the K_m term or the reaction direction in the V_{max} term when only one substrate is used because these designations are intuitive. The three groups differed in their assumptions about the relative magnitude of the rate constants. As previously discussed, Michaelis and Menten assumed that the binding was at equilibrium (i.e., k_2 was the slow step in the reaction sequence) and Van Slyke and Cullen assumed that substrate binding was irreversible (k_{-1} was much slower than k_2). Briggs and Haldane made no assumptions about the relative magnitude of the individual rates but assumed that the reaction was at steady state. Differences between the microscopic reaction rates ($k_1, k_{-1}, k_2, \ldots$) of a reaction can be graphically illustrated by energy versus reaction coordinate diagrams (Fig. 2.4); the slower steps require more energy. Regardless of a reaction's energy profile, its mechanism can still be adequately described by Eq. (2.31) [or Eq. (2.25) for reversible reaction kinetics]. It is the kinetic definition of the K_m and V_{max} values that differ. (See Text Box 2.1).

The Haldane Relationship

At the beginning of this chapter, it was mentioned that certain properties of a reaction are under enzymatic control and others are not. It was stated that the degree to which a reaction proceeds is not influenced by an enzyme, but the rate at which that reaction proceeds can be altered by the presence of a catalyst, such as an enzyme. While this is true, the equations developed in the preceding section show that a relationship between these two reaction properties exists. When a reaction is at equilibrium, the net velocity is zero and we can write [from Eq. (2.25)]

$$V = 0 \frac{\left(\frac{V_f}{K_{m,S}}[S]_{eq} - \frac{V_r}{K_{m,P}}[P]_{eq}\right)}{\left(1 + \frac{[S]_{eq}}{K_{m,S}} + \frac{[P]_{eq}}{K_{m,P}}\right)} \tag{2.32}$$

where $[S]_{eq}$ and $[P]_{eq}$ are the concentrations of substrate and product at equilibrium. Rearranging Eq. (2.32), one can show that

$$K_{eq} = \frac{[P]_{eq}}{[S]_{eq}} = \frac{V_f K_{m,P}}{V_r K_{m,S}} \tag{2.33}$$

Equation (2.33) relates the equilibrium constant of any reaction to the enzyme kinetic parameters. It applies irrespective of the details of the kinetic mechanism even though the V_{max} and K_m values will be made up of different combinations of the microscopic kinetic constants. More complex kinetic equations (e.g., those involving multiple substrates) will have more complex Haldane relationships but a relationship will still exist.

Because the Haldane equation relates the K_{eq} to the kinetic constants for a reaction, it places limits on the kinetic behavior of an enzyme. This is best demonstrated by examining enzymes that more effectively catalyze a reaction in one direction or another. A good example of this type of enzyme is methionine adenosyltransferase, an enzyme originally isolated from baker's yeast that catalyzes the transfer of adenosine from ATP to L-methionine:

$$\begin{aligned} &\text{ATP} + \text{L-methionine} + H_2O \leftrightarrow P_i + PP_i \\ &\quad + S\text{-adenosylmethionine} \end{aligned} \tag{2.34}$$

The reaction is thought to be essentially irreversible because of the enormous difference in kinetic activity between forward and backward rates: The forward rate is approximately 2×10^5 faster than the reverse rate. This is true, despite the fact that the equilibrium constant is near unity. Substituting these values into Eq. (2.33) shows that the K_m value for substrate ($K_{m,S}$) must be about 2×10^5 greater than the K_m value for product ($K_{m,P}$). This means that the enzyme has a 2×10^5-fold higher affinity for product than it does for substrate because the K_m values are inverse measures of the enzyme's affinity for reactant. Thus, under initial velocity conditions *in vitro*, when the concentration of product is zero, it is possible to measure a forward rate that is much greater than the reverse rate. This example brings to mind the discussion leading up to the development of Eq. (2.17) when it was noted that the overall free energy change for the reaction was proportional to the equilibrium free concentrations of reactants only and that the equilibrium condition is unaffected by the presence of a catalyst. It was also noted that two separate equilibria exist: an enzyme-bound equilibrium and a free solution equilibrium. The methionine adenosyltransferase example described here illustrates this principle. Whereas the equilibrium of the unbound reactants is approximately unity, the enzyme-bound equilibrium is approximately 2×10^5 since this many more enzyme–product complexes exist:

$$K_{eq}(\text{bound}) = [\text{EP}]/[\text{ES}] \approx 2 \times 10^5 \qquad (2.35)$$

Multisubstrate Reactions

The preceding analysis gave a simple equation relating substrate concentrations to reaction velocities. The result of this analysis, Eq. (2.31), applies directly to systems where one substrate reacts to give one product under initial velocity conditions. The analysis of one-substrate enzyme reactions is relatively simple, but, unfortunately, this describes only a very few enzyme mechanisms. It is much more common to find enzymes that catalyze reactions between two substrates. Examples of these reactions include group transfer reactions (e.g., all reactions involving ATP, transaminase reactions, acyl carrier protein reactions, reactions with SAM and THF; see Table 2.3), oxidation and reduction reactions (enzymes that use NAD, FAD, FMN, CoQ; see Table 2.3), many elimination reactions, condensation reactions, as well as alkylation reactions. Numerous equations of varying complexity exist to describe enzyme reactions involving two substrates. These equations are necessary because of the myriad catalytic mechanisms that exist. It is fortunate, however, that most two-substrate reactions can be described essentially by two equations: the ternary complex equation and the substituted enzyme equation. A ternary complex arises when two substrates bind to an enzyme simultaneously to form an enzyme–substrate$_1$–substrate$_2$ complex, which then goes on to react and form products. Two types of ternary complex mechanisms are common: the random-order and the compulsory-order mechanism (Fig. 2.5). In the first case, free enzyme can bind either substrate to produce two potential complexes. These enzyme–substrate complexes can bind the other substrate to give the reactive enzyme complex. In the second case, free enzyme can bind only one of the two substrates. The resulting enzyme–substrate complex can then bind the second substrate to give the reactive enzyme complex. Figure 2.5 details these two possibilities by illustrating the reaction sequence for a group transfer where the group X is transferred from A to B. The overall reaction is written as

$$\text{A-X} + \text{B} \rightleftharpoons \text{A} + \text{B-X} \qquad (2.36)$$

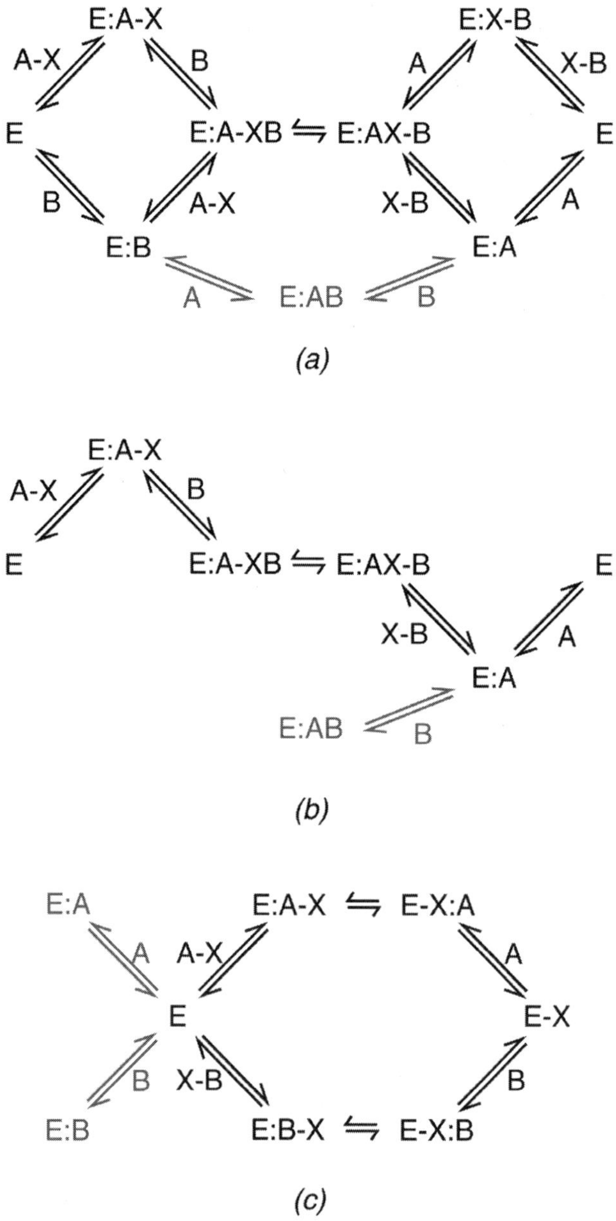

Figure 2.5 Common initial velocity two-substrate mechanisms. The reaction shows the mechanisms for random-order, compulsory-order, and substituted enzyme mechanisms that describe the catalysis of A-X + B ↔ A + B-X. E represents the free enzyme. Dead-end complexes (complexes that cannot react) are indicated in light gray. The mechanisms are adapted from Cornish-Bowden (Cornish-Bowden, 1995).

Both ternary complex mechanisms also incorporate "dead-end" complexes (shown in Fig. 2.5 in gray), which arise when one of the enzyme–substrate complexes binds one of the products. As shown in Figure 2.5, only specific substrate–product pairs can form dead-end complexes because the size and orientation of reactive groups in the enzyme active site can accommodate only certain pairs of reactants. In the random-order example of Figure 2.5, dead-end complexes can occur only when the product A and substrate B bind simultaneously to the enzyme. For the compulsory-order mechanism, the dead-end complex is shown on the product side of the reaction; however, this would be reversed if substrate B bound the free enzyme and substrate A-X bound second. Dead-end complexes can play a significant role in influencing a reaction rate if the product concentration is high because dead-end complexes remove enzyme from the reaction pathway. This is a type of enzyme inhibition known as product inhibition and will be discussed later.

The initial rate equation (at zero product concentration) for the random-order and compulsory-order ternary complex mechanisms is given by

$$v = \frac{V_f\,[\mathrm{AX}][\mathrm{B}]}{K_{i,\mathrm{AX}}K_{m,\mathrm{B}} + [\mathrm{AX}]K_{m,\mathrm{B}} + [\mathrm{B}]K_{m,\mathrm{AX}} + [\mathrm{AX}][\mathrm{B}]} \qquad (2.37)$$

where V_f is the maximal velocity of the forward direction, $K_{m,\mathrm{AX}}$ is a measure of the binding affinity of the first substrate (A-X) for the free enzyme and $K_{m,\mathrm{B}}$ is a measure of the binding affinity of the second substrate (B) for the free enzyme. Similar to the case for a one-substrate reaction, if the binding step is much faster than the rate of substrate–product conversion, the $K_{m,\mathrm{AX}}$ and $K_{m,\mathrm{B}}$ values are equal to the dissociation constants for these substrates. The $K_{i,\mathrm{AX}}$ value is a measure of the binding affinity of the first substrate for the "enzyme–second substrate" complex. For the random-order mechanism, the interpretation of the kinetic constants is straightforward, and the symmetrical binding of both substrates means that the term $K_{i,\mathrm{AX}}K_{m,\mathrm{B}}$ can be replaced by the term $K_{i,\mathrm{B}}K_{m,\mathrm{AX}}$. However, constants such as $K_{m,\mathrm{AX}}$ and $K_{i,\mathrm{B}}$ have no simple meaning for the compulsory-order mechanism.

The third general mechanism presented in Figure 2.5 is the substituted mechanism. This differs from the previous two mechanisms in that the enzyme cycles between two states: a native state and a covalently modified state. The first substrate binds and reacts with enzyme to form an intermediate (E-X, Fig. 2.3, bottom) that then reacts with the second substrate to regenerate the native enzyme. In essence, the mechanism is described by two single-substrate reactions, each dependent on the other to regenerate the enzyme form that can catalyze the other reaction. The equation for the substituted mechanism is very similar to Eq. (2.37). It simply lacks one term since no tertiary complex exists:

$$v = \frac{V_f\,[\mathrm{AX}][\mathrm{B}]}{[\mathrm{AX}]K_{m,\mathrm{B}} + [\mathrm{B}]K_{m,\mathrm{AX}} + [\mathrm{AX}][\mathrm{B}]} \qquad (2.38)$$

These equations show the complexity of normal cellular reactions but also illustrate how they may be simplified to Eq. (2.31) (the Michaelis–Menten equation) by an appropriate grouping of terms. For example, at a fixed concentration of the second substrate B, Eq. (2.37) becomes

$$v = \frac{\dfrac{V_f\,[\mathrm{B}]}{([\mathrm{B}] + K_{m,\mathrm{B}})}[\mathrm{AX}]}{\dfrac{K_{i,\mathrm{AX}}K_{m,\mathrm{B}} + [\mathrm{B}]K_{m,\mathrm{AX}}}{([\mathrm{B}] + K_{m,\mathrm{B}})} + [\mathrm{AX}]} = \frac{V_f^{\mathrm{app}}[\mathrm{AX}]}{K_{m,\mathrm{AX}}^{\mathrm{app}} + [\mathrm{AX}]} \qquad (2.39)$$

so that the kinetics resemble those of the simple Michaelis–Menten equation. Two important differences are apparent. First, the apparent $V_{\max}$ will change with changing concentrations of the second substrate, and this relationship will be a square hyperbola. Second, the enzyme will be subject to product inhibition because of the existence of dead-end complexes (although product inhibition is also a feature of any reversible enzyme mechanism as discussed below). Table 2.4 shows the equations that define the apparent kinetic constants for each mechanism.

For an enzymologist interested in defining an enzyme mechanism, the relationships defined in Table 2.4 are extremely useful because they illustrate how to plot apparent $V_{\max}$ and K_m values in order to determine the complete set of kinetic parameters. In addition, they predict specific relationships between the primary parameters ($V_{\max}^{\mathrm{app}}$ and K_m^{app}) and the fixed substrate. The inhibition patterns predicted by the products can also be extremely useful in elucidating potential mechanisms. For individuals wishing to grasp an understanding of how enzymes may work *in vivo*, these patterns also provide information on factors that can potentially influence enzyme activity in the cellular environment.

Cooperative Enzyme Mechanism

In the previous section, the compulsory-order mechanism illustrated a case in which binding of one substrate permitted the binding of a second, different substrate. This is one example of a situation where the binding of a substrate to an enzyme can alter the enzyme's conformation and affect its affinity for other substrates. Cooperative enzymes provide another example of this phenomenon. Cooperative enzymes are multisubunit enzymes where binding of one substrate molecule to one subunit causes conformational changes in the other subunits to alter the affinity of the other subunits for the same substrate. When the binding of one substrate increases the affinity of the enzyme for the second substrate, the enzyme is

TABLE 2.4 Kinetic Equations Describing Common Multisubstrate and Cooperative Enzyme Models[a]

Pattern	Variable [S]	V_{max}^{app}	K_m^{app}
Random-order and compulsory-order	[AX]	$\frac{V_f\,[B]}{([B]+K_{m,B})}$	$\frac{K_{i,AX}K_{m,B}+[B]K_{m,AX}}{([B]+K_{m,B})}$
	[B]	$\frac{V_f\,[AX]}{([AX]+K_{m,AX})}$	$\frac{K_{i,AX}K_{m,B}+[AX]K_{m,B}}{([AX]+K_{m,AX})}$
Substituted	[AX]	$\frac{V_f\,[B]}{([B]+K_{m,B})}$	$\frac{[B]K_{m,AX}}{([B]+K_{m,B})}$
	[B]	$\frac{V_f\,[AX]}{([AX]+K_{m,AX})}$	$\frac{[AX]K_{m,B}}{([AX]+K_{m,AX})}$
Cooperative	$[S^h]$	V_{max}	K_m^h

[a]The values for the random-order, compulsory-order, and substituted patterns were obtained by assuming that the binding reactions are at equilibrium (Michaelis–Menten assumption). The equation describing the cooperative pattern is an empirical relationship derived by Hill to describe oxygen binding to hemoglobin. In cooperative enzyme kinetics, h represents the Hill coefficient, an arbitrary measure of the cooperativity of the system that has no physical significance.

said to obey positive cooperativity. Cooperative enzymes do not show the "square hyperbolic" velocity versus substrate relationship typically shown by enzymes that obey Eq. (2.31) (see Fig. 2.6, top). Instead, they are sigmoid (or S-shaped) to reflect the increase in substrate affinity that occurs as a function of increasing substrate concentration (see Fig. 2.6, top).

Two specific models and one general model have been developed to describe cooperative enzyme kinetics. The diagram of Figure 2.7 schematically presents the general model for a tetrameric enzyme, although this scheme is incomplete as presented. The complete symmetrical general scheme contains 44 different states to satisfy all possible enzyme–substrate states. Equations can be developed to describe the complete model, but they are much too complex to be of any general use.

Simplifications of the general model have also been proposed. In fact, two of the most common simplified models were developed before the general model to explain the binding of oxygen to hemoglobin. These two simplified models are shown as gray-shaded areas in Figure 2.7. The Monod–Wyman–Changeux (MWC) concerted mechanism assumes that the enzyme exists in two separate states: a T form (tense form) and an R form (relaxed form). In the MWC model, the two forms are assumed to be in equilibrium, described by the relationship: $L = [T]/[R]$. The MWC model also assumes that all the subunits on a single enzyme tetramer are either in the T form or in the R form (no mixed forms of enzyme exist) and that all the binding sites in each state are equivalent and have identical binding constants for substrate. The MWC model is also known as the symmetry model to reflect the symmetry of the enzyme states and substrate binding. The T form has a lower affinity for substrate and the R form has a higher affinity for substrate. In the absence of substrate, the $T \leftrightarrow R$ equilibrium favors the T form (lower binding affinity). Addition of substrate drives the equilibrium toward the R form. This occurs simply by mass action principles—the substrate essentially pulls the equilibrium toward one side. This model is described by an equilibrium between the left and right columns of Figure 2.7 (light gray). The fractional saturation of the enzyme (y) can be described for an enzyme with n subunits by

$$y = \frac{L\dfrac{[S]}{K_T}\left(1+\dfrac{[S]}{K_T}\right)^{n-1} + \dfrac{[S]}{K_R}\left(1+\dfrac{[S]}{K_R}\right)^{n-1}}{L\left(1+\dfrac{[S]}{K_T}\right)^{n} + \left(1+\dfrac{[S]}{K_R}\right)^{n}} \tag{2.40}$$

This model easily explains the sigmoid form of positive cooperativity. It has several advantages. First, it describes the observed kinetic behavior using only four unknown parameters and the concentration of ligand. Second, the model is much simpler than the general scheme of Figure 2.7. Third, it is easily applicable to enzymes that are activated by substances that bind to sites different from the active site (enzyme activation will be discussed below). It is easier to understand Eq. (2.40) when considering some special cases that illustrate how the equation behaves. If the enzyme exists only in the R form, then $L = 0$ and Eq. (2.40) reduces to

$$y = \frac{[S]}{K_R + [S]} \tag{2.41}$$

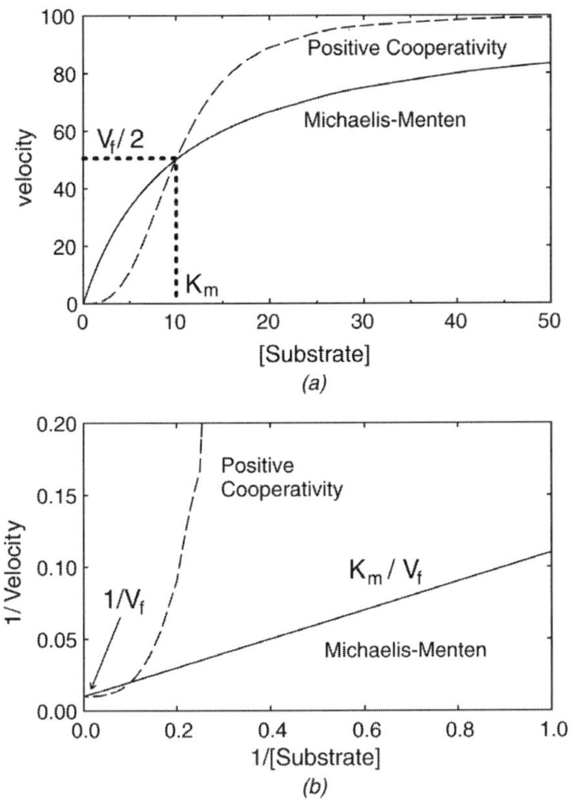

Figure 2.6 Reaction velocity of an enzyme-catalyzed reaction as a function of the substrate concentration: Michaelis–Menten and sigmoid mechanisms. (*a*) Normal plot of the velocity versus substrate. (*b*) Lineweaver–Burk plot of $1/v$ versus 1/[Substrate] traditionally used to graphically obtain estimates of the K_m and $V_{\max}$ values. The maximal velocity ($V_{\max}$) was set to 100 and the concentration of substrate that produces half-maximal saturation (K_m) is indicated in the upper graph. Enzyme reaction kinetics are most commonly graphed using the Lineweaver–Burk plot, even though some authors dislike this plot because it magnifies constant errors at low substrate concentrations (see Text Box 2.1). Sigmoid kinetic patterns are not always readily apparent from an inspection of velocity versus substrate plots (*a*) but are readily apparent when other plots, such as the Lineweaver–Burk plot, are employed (*b*). Other statistical and kinetic tests are required to confirm sigmoidal behavior.

which perfectly describes the fractional saturation of a single substrate enzyme. In this case, the K_m value would be low, reflecting the tighter binding of the R form. Multiplying Eq. (2.41) by the enzyme maximal velocity gives the Michaelis–Menten equation. A similar equation results if the enzyme exists only in the T state. In this case, $L = \infty$ and Eq. (2.40) simplifies to

$$y = \frac{[\mathrm{S}]}{K_{\mathrm{T}} + [\mathrm{S}]} \tag{2.42}$$

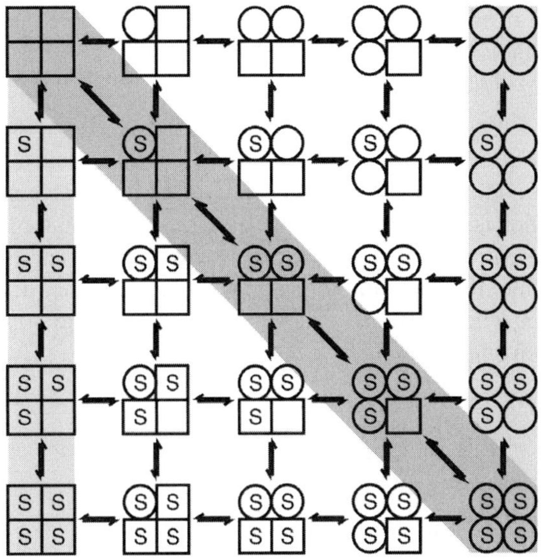

Figure 2.7 Model showing the various states possible for a symmetrical tetramer that obeys positive cooperatively. The MWC symmetry model is shown in light gray. In this model, the enzyme can exist in either two states: a T state (left column, boxes) or an R state (right column, circles). Binding of substrate to the R state is tighter so that increased substrate concentrations drive the equilibrium between T and R states to the R state. The sequential KNF model is illustrated by the dark gray diagonal. In this model, binding of substrate to any of the subunits induces a conformational change that makes the binding of the second substrate molecule more favorable.

If both forms of the enzyme bind substrate with equal affinity ($K_{\mathrm{T}} = K_{\mathrm{R}}$), then Eq. (2.40) gives Eq. (2.41) [or Eq. (2.42)] once again. This shows that cooperativity is only observed when two forms of the enzyme exist and when they bind with different affinities.

The second model is called the Koshland–Némethy–Filmer (KNF) sequential model. It is diagrammatically illustrated by the darker diagonal gray area of Figure 2.7. This model assumes a sequential progression of enzyme states from T to R as the enzyme binds substrate (hence the name sequential). In the absence of substrate, the enzyme exists only in one conformation (the T state) and no equilibrium exists between T and R forms. Binding of substrate to one subunit induces a conformational change in the subunit that is transmitted to the other subunits to promote substrate binding at their active sites. This model produces equations that are formally equivalent to the Adair equation, an equation developed to describe an enzyme with two active sites that bind substrate independently. The Adair model assumes that the substrate binding is at equilibrium and that the maximal velocity of the individual subunits is identical so that the maximal velocity of the enzyme can be described by (for a tetrameric enzyme)

$$V_{max} = V_{max1} + V_{max2} + V_{max3} + V_{max4} \tag{2.43}$$

In addition, the Adair model assumes different binding coefficients for each subunit. If we let K_1, K_2, K_3, and K_4 represent the dissociation constants for a tetrameric enzyme, it is possible to derive a relationship that defines the enzyme fractional saturation at any substrate concentration (y):

$$y = \frac{\dfrac{[S]}{K_1} + \dfrac{3[S]^2}{K_1K_2} + \dfrac{3[S]^3}{K_1K_2K_3} + \dfrac{[S]^4}{K_1K_2K_3K_4}}{1 + \dfrac{4[S]}{K_1} + \dfrac{6[S]^2}{K_1K_2} + \dfrac{4[S]^3}{K_1K_2K_3} + \dfrac{[S]^4}{K_1K_2K_3K_4}} \tag{2.44}$$

This model sacrifices some simplicity since the equation is more complex and it requires more constants (four in this case) to describe the binding. The model requires as many constants as there are subunits so that the equation for a dimer would be much simpler. In sacrificing simplicity, the Adair (or KNF) model is more general and probably is a better description of some enzyme mechanisms than the symmetry model.

General equations that describe cooperative kinetics have also been developed. The most popular of these is the Hill equation, an empirical description of the binding of oxygen to hemoglobin that has a form similar to the Michaelis–Menten equation [Eq. (2.31)]:

$$V = \frac{V_{max}[S]^h}{K_{0.5}^h + [S]^h} \tag{2.45}$$

The affinity of enzyme for substrate is measured by the term $K_{0.5}^h$, which represents the midpoint of the substrate versus velocity curve and cannot be absolutely related to the substrate dissociation constant. The exponent term, h, is the Hill coefficient, a purely empirical measure of the degree of cooperativity. It is widely believed that the Hill coefficient represents the upper limit for the number of active sites on the enzyme but this is not always true. For example, the Hill coefficient for hemoglobin is typically around 2.7, although the number of substrate binding sites on the hemoglobin molecule is 4. Equation (2.45) is useful for fitting sigmoid data, but it is best confined to velocity values that range between 10 and 90% of V_{max}. This is because the equation is, at best, an empirical approximation of a more complex process and tends to deviate from actual enzyme behavior at the ends of the curve.

The Hill equation has two advantages over the more complete MWC and KNF models. First, it is possible to use nonlinear regression to fit a kinetic curve having a reasonable number of individual velocity measurements. This is best illustrated by comparing the MWC symmetry model and the Hill equation in their ability to fit actual kinetic data (Table 2.5). The regression data show that it is difficult for the nonlinear algorithm to obtain reasonable estimates of the MWC model parameters (demonstrated by the large standard errors associated with the best-fit parameters) while the Hill equation parameters are easily obtained. Part of the problem occurs because the MWC model requires one additional parameter (V_{max}) in addition to the four contained in Eq. (2.40). Because of this, nonlinear routines have difficulty in minimizing the error structure of Eq. (2.40). This problem can be partly alleviated by constraining the estimates (e.g., all must be held >0), but this did not help with the regressions of Table 2.5. This means that other, nonkinetic data must be provided to lower the number of parameter estimates. However, even this does not always help as shown by fixing (holding constant) the V_{max} and n values in the first example of Table 2.5. This suggests that the parameters do not vary independently so that changing one forces another parameter to change as well. The Hill equation, on the other hand, provides reliable estimates of the Hill parameters and, as shown by the low standard deviation associated with the estimates, the fit is quite good.

Second, the Hill equation provides an estimate of the midpoint of the substrate–velocity curve. This is a measure of how much substrate is required to obtain half-maximal velocity. Changes in this parameter, which can occur, for example, as a result of enzyme modification (which often occur when animals are exposed to environmental stress as will be discussed in later chapters), allow some interpretation of the change in activity of an enzyme (although this may or may not be related to changes in pathway flux, as we shall see in the next chapter). Even though the $K_{0.5}$ is only a relative estimate of substrate affinity, the fact that these parameters may be obtained with reasonable accuracy allows relevant comparisons between enzymes isolated from cells or animals under different conditions. The reader is referred to later chapters for some concrete examples of this phenomenon.

Allosteric Enzyme Activation and Inhibition

The preceding discussion on positive cooperativity leads naturally into a discussion of enzyme activation because the models for the two processes are very similar. In the preceding section, the binding of one substrate molecule to a multisubunit enzyme influenced the binding of the next substrate molecule. In this section, we shall consider what happens when compounds other than substrates bind to enzymes and influence activity.

TABLE 2.5 Practical Applications of Kinetic Models to Kinetic Data. How Useful Is Symmetry Model in Defining a Kinetic Curve?[a]

Data Set	Hill Equation Parameter	Value	Symmetry Model Parameter	Value
Segal (1968)	V_{max}	250 (fixed)	V_{max}	250 (constant value)
	$K_{0.5}$	9.987 ± 0.006	K_R	0.766 ± 0.175
	h	2.991 ± 0.005	K_T	113.7 ± 22.6
			L	28165 ± 24324
			n	4 (constant value)
Weiker et al. (1970)	V_{max}	199.6 ± 1.7	V_{max}	207.2 ± 21.1
	$K_{0.5}$	1.902 ± 0.023	K_R	0.51 ± 1.14
	h	2.15 ± 0.05	K_T	6241213 ± 212191284
			L	60 ± 188
			n	2.89 ± 1.52

[a]Two different kinetic data sets (Segal, 1968; Weiker et al., 1970, were fit to the Hill equation [Eq. (2.45)] or the MWC model [Eq. (2.40)] by nonlinear regression. The table shows the best-fit value ± standard deviation from the nonlinear regression procedure (assuming constant error). The standard deviations of the individual parameters are good indicators of the goodness of fit. Note that the data easily fit the Hill equation but poorly fit the MWC model. This was not a function of having too few data points as the Segal data set contained 12 data points with 4 points $\geq 0.8V_{max}$ and the Weiker et al. (1970) data set contained 31 data points with 8 points $\geq 0.8V_{max}$. This shows that the MWC model requires some additional information provided by nonkinetic measurements to enable the complete determination of the parameters. The Hill equation, on the other hand, can be used to completely describe a kinetic pattern, although the constants have no exact physical meaning.

Binding of a molecule to a site other than the enzyme active site—with a subsequent influence on an enzyme's activity—is known as allosteric regulation. It is usually the result of a change in the tertiary structure of the enzyme to alter substrate binding or catalytic efficiency. If the enzyme is a multisubunit enzyme, binding to one subunit may also affect the activity of the other subunits. Allosteric inhibition is common in single and multisubunit enzymes and will be discussed in the following section (see noncompetitive and uncompetitive inhibition). Allosteric activation of enzymes is also a common phenomenon, but this is not adequately addressed by any common detailed mechanism other than models based on the MWC symmetry model presented in the preceding section. A closer look at this model shows why it lends itself easily to allosteric activation. It is relatively simple to envision a case where binding of a compound could influence the activity of a multisubunit enzyme that exists as a mixture of T (lower activity) and R (higher activity) forms (left and right columns of Fig. 2.7). Let us assume that the enzyme behaves exclusively as a Michaelis–Menten enzyme in the absence of activator and that the equilibrium favors the T form of the enzyme. One can envision a case where an activator could drive the enzyme into the R form by preferentially binding that form only. In this scenario, only the R form would exist in the presence of very high activator concentrations. The activated enzyme would also obey simple Michaelis–Menten kinetics, but the enzyme would exhibit only the properties of the R form. Intermediate states would show sigmoidal kinetics. The mathematics of this process has already been discussed [Eqs. (2.40) to (2.42)]. In practice, however, the enzymes do not exclusively exhibit Michaelis–Menten behavior in the absence of activator since the enzyme T and R forms are in equilibrium and respond to increasing substrate concentrations. Some examples of enzymes that are allosterically activated include pyruvate kinase (activated by fructose 1,6-bisphosphate), isocitrate dehydrogenase (activated by ADP), and phosphofructokinase (activated by AMP, fructose 2,6-bisphosphate, and phosphate when assayed at lower pH values). In all these cases, the velocity versus substrate concentration curve changes from sigmoid shape to square hyperbola in the presence of the activator (Fig. 2.8).

Enzyme Inhibition

Three general types of inhibitors exist; classified according to the form of the enzyme to which they bind [either E and/or ES in Eq. (2.30)]. This produces three distinct kinetic patterns as visualized by velocity versus substrate plots. Competitive inhibitors compete directly with substrate for free enzyme and so affect the apparent substrate affinity for enzyme (the K_m value increases with increasing inhibitor concentration). Uncompetitive inhibitors bind to the ES complex and stop it from reacting to form product. This type of inhibition, therefore, reduces the rate of product formation (the V_{max} value decreases) but does not affect the K_m value. Mixed inhibitors bind both free and substrate-bound enzyme to give increased K_m values and decreased

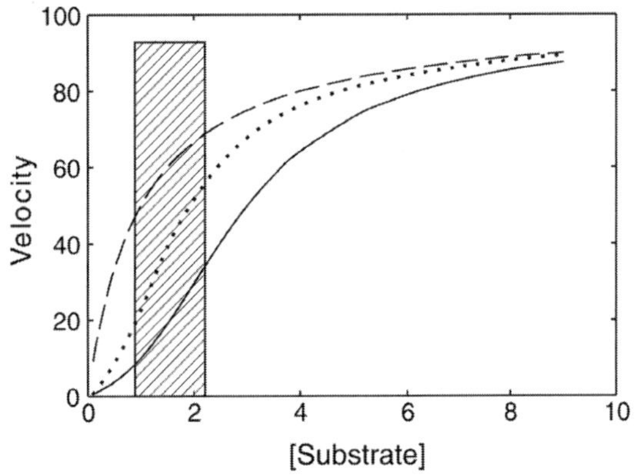

Figure 2.8 Graphical representation of allosteric activation of an enzyme as predicted by the symmetry model. The model assumed $K_R = 1\,\mu M$, $K_T = 20\,\mu M$, $V_{max} = 100\,\mu mol/min$, and $n = 4$. The three curves were constructed by assuming the addition of an activator that influences the equilibrium between T and R forms. The equilibrium constant ($L = [T]/[R]$) was set at $L = 100$ (solid line), $L = 20$ (dotted line), and $L = 0$ (dashed line). Thus, as L decreases, the enzyme becomes more active. The curves highlight the expected increase in activity for a system with substrate values around the midpoint of the velocity curve ($K_{0.5}$). For enzymes that obey positive cooperativity, it is meaningless to discuss K_m values since they have no absolute physical meaning. However, the concentration of substrate giving half maximal enzyme activity ($K_{0.5}$) is a practical measure of enzyme activity. In the present example, the curves show that, at a substrate value of $1\,\mu M$, the enzyme activity increases from 10.0 to 22.7 to $50\,\mu mol/min$ (solid to dotted to dashed lines) as the concentration of activator increases to drive the enzyme into the R form.

V_{max} values. The effects of inhibitors are best illustrated by considering a Botts–Moralis scheme [Eq. (2.46)]. The Botts–Moralis scheme is simply the Michaelis–Menten equation expanded to include the binding of an inhibitor (I):

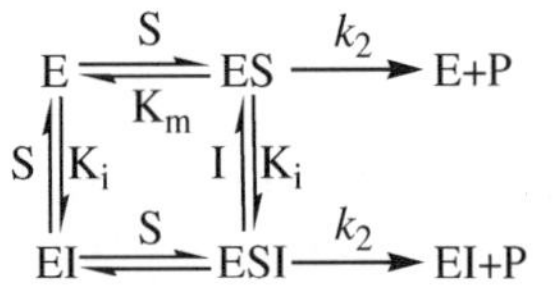

(2.46)

In Eq. (2.46), K_i corresponds to the inhibitor dissociation constant for competitive inhibitors, and K_i' to the inhibitor dissociation constant for uncompetitive inhibitors. The modified parameters K_m' and k_2' are the K_m value and turnover number of the enzyme when bound to inhibitor. Solving the equilibria of Eq. (2.46) gives a generalized equation that relates the observed rate to all the binding parameters. The scheme is easily solved by assuming that the binding reactions are in equilibrium. This is equivalent to the Michaelis and Menten assumption; that is, the enzyme turnover step (k_2 or k_2') is the slow step in the reaction mechanism. The resulting solution is an equation similar to the Michaelis–Menten equation with V_{max}^{app} replacing V_{max} and K_m^{app} replacing K_m. Equations (2.47) and (2.48) show how these new values relate to the individual binding constants of Eq. (2.46):

$$V_{max}^{app} = [\text{enzyme}] \times \frac{\left(k_2 + k_2' \frac{[I]}{K_i'}\right)}{\left(1 + \frac{[I]}{K_i'}\right)} \tag{2.47}$$

$$K_m^{app} = K_m \times \frac{\left(1 + \frac{[I]}{K_i}\right)}{\left(1 + \frac{[I]}{K_i'}\right)} \tag{2.48}$$

The V_{max}^{app} and K_m^{app} values are the apparent V_{max} and K_m values measured at finite concentrations of inhibitor. They are obtained graphically by extrapolating [S] to infinite concentration at fixed values of inhibitor. Table 2.6 lists some of the more commonly occurring inhibition patterns and illustrates how each inhibition pattern arises. For example, competitive inhibition occurs when an inhibitor competes with substrate for free E only (the inhibitor does not bind the ES complex). This is expressed mathematically as $K_i' = K_m' = \infty$. This means that the dissociation constants for the ES complex are infinitely large (the K_{eq} values are essentially 0) so that at equilibrium no complex forms. Similarly, uncompetitive inhibition occurs when inhibitor binds only to the ES complex and the ESI complex does not react ($K_m = K_m' = \infty$ and $k_2' = 0$).

Many cellular enzymes are subject to inhibition. These enzymes are normally found at the traditional control loci of pathways—the beginning and branch points of metabolic pathways. Traditionally, it was believed that this made these enzymes subject to metabolic regulation, providing the cell with an easy way to regulate flux through biochemical pathways. In accordance with this view, most inhibitors are cellular metabolites that serve as monitors of cellular energy status or pathway flux. Although it is compelling to think of inhibition as a mechanism that provides the cell with absolute control over activity (much like turning a tap off and on), in reality the influence of the inhibitor on pathway activity is a function of many factors and the final relationship between enzyme activity, pathway flux, and inhibitor concentration is quite complex. Thus, the current discussion permits an understanding of how inhibition may act on a single enzyme, but the next chapter will show how enzyme systems behave and give a much fuller picture of inhibition and activation.

TABLE 2.6 Common Inhibition Patterns for Single-Substrate Enzyme Reactions[a]

Pattern	Conditions	V_{max}^{app}	K_m^{app}
Competitive	$K_i' = K_m' = \infty$	V_{max}	$K_m \times \left(1 + \frac{[I]}{K_i}\right)$
Partially competitive	$K_m' > K_m$ $K_i' > K_i$ $k_2 = k_2'$	V_{max}	$K_m \times \frac{\left(1 + \frac{[I]}{K_i}\right)}{\left(1 + \frac{[I]}{K_i'}\right)}$
Mixed (noncompetitive)	$K_m = K_m'$ $K_i = K_i'$ $k_2' = 0$	$\frac{V_{max}}{\left(1 + \frac{[I]}{K_i'}\right)}$	K_m
Uncompetitive	$K_i = K_m' = \infty$ $k_2' = 0$	$\frac{V_{max}}{\left(1 + \frac{[I]}{K_i'}\right)}$	$\frac{K_m}{\left(1 + \frac{[I]}{K_i'}\right)}$
Partially noncompetitive	$K_i \neq K_i'$ $K_m \neq K_m'$ $k_2 = k_2'$	$[\text{enzyme}] \times \frac{\left(k_2 + k_2' \frac{[I]}{K_i'}\right)}{\left(1 + \frac{[I]}{K_i'}\right)}$	$K_m \times \frac{\left(1 + \frac{[I]}{K_i}\right)}{\left(1 + \frac{[I]}{K_i'}\right)}$

[a]The expressions for the apparent V_{max} and K_m values were derived from the Botts–Moralis scheme assuming the equilibrium condition. The three traditionally defined inhibition patterns (competitive, noncompetitive, and uncompetitive) are presented as well as other patterns that represent mixtures of these cases. In practice, pure uncompetitive inhibition is rarely observed; most enzymes obey competitive or mixed-type inhibition patterns.

TEMPERATURE EFFECTS ON ENZYME FUNCTION

The preceding sections dealt with enzyme activation and inhibition by cellular metabolites. Other potential controls on enzyme activity *in vivo* exist such as pH, temperature, and pressure (the latter of importance only for deep-sea organisms). As we shall note in later chapters, temperature is a major factor in the lives of all organisms, whether it be the impact of temperature change on the metabolism of poikilotherms (cold-blooded organisms) or the complex regulatory measures needed to sustain a constant body temperature by homeotherms (warm-blooded organisms). Hence, an understanding of the mathematics of temperature effects on enzyme activity is important. Temperature change can affect enzyme activity in two ways: (1) It can influence the flexibility of the protein molecule (change tertiary structure) and, (2) it has a direct effect on the thermodynamics of the enzyme reaction. Temperature-dependent changes in enzyme properties can be distinguished from the temperature-related thermodynamic effect by plotting a rearrangement of Eq. (2.9) as a function of temperature. Taking the natural log of each side of Eq. (2.9) gives an equation of the form $y = mx + b$ (straight line) with the slope of the line equal to $-E_a/R$.

$$\ln(k) = \left(\frac{-E_a}{R}\right) \times \left(\frac{1}{T}\right) + \ln(A) \qquad (2.49)$$

Equation (2.49) shows that a plot of the natural logarithm of the reaction velocity, ln(k), as a function of the reciprocal of the temperature in degrees kelvin (called an Arrhenius plot) gives a straight line with a slope equal to the activation energy divided by the gas constant, R. If the effect of temperature on the reaction is simply thermodynamic, one expects this plot to give a straight-line relationship. If temperature influences an enzyme's conformation or its ability to adapt to substrate/product binding, a break in the line appears so that two activation energies can be calculated for higher versus lower temperature ranges. The biphasic nature of the relationship results from the influence of temperature on the initial enzyme conformation and/or its ability to respond to substrate binding. Either situation changes the final enzyme–substrate active site configuration so that there is a change in the overall E_a (Fig. 2.9).

In older textbooks, one may find a reference to the Q_{10} value, which can be calculated from the E_a. The Q_{10} is a

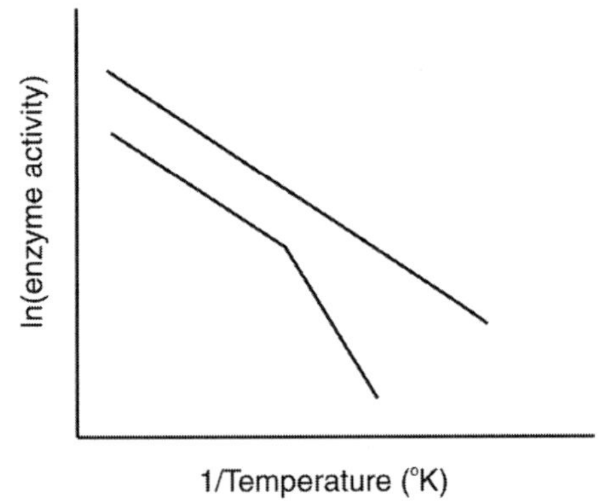

Figure 2.9 Theoretical Arrhenius plots showing the continuous straight-line relationship predicted by the thermodynamics of chemical reactions and a hypothetical discontinuous relationship observed when temperature influences an enzyme's conformation or its ability to adapt to substrate/product binding.

rule-of-thumb value defined as the relative change in reaction rate for every 10°C change in temperature. For most reactions, the Q_{10} value is approximately 2 as the reaction rate approximately doubles for every 10° rise in temperature. Values lower than 2 reflect a significant enzyme influence on the reaction rate, usually due to temperature influences on enzyme conformation. Values are not usually much higher than 2 since this is about what is expected from thermodynamic considerations. The Q_{10} value can be related to the E_a by the following equation:

$$Q_{10} = \exp\left(\frac{10 \times E_a}{T_1 \times T_2}\right) \tag{2.50}$$

ESTIMATING KINETIC PARAMETERS FROM ENZYME VELOCITY DATA (THE INFLUENCE OF ERROR STRUCTURE)

Several factors must be considered before one decides on the optimum method for determining kinetic constants from enzyme rate data. The most important of these factors is the structure of the error because error structure impacts on the choice of statistical model used for estimation as well as the design structure of the kinetic experiment. Two types of error can be envisioned. The first case is a constant error (ε) that may apply uniformly to all measurements of velocity:

$$V = \frac{V_{\max} \times [\mathrm{S}]}{K_m + [\mathrm{S}]} + \varepsilon \tag{2.51}$$

It is evident that, when the error is constant, rate measurements made at very low substrate concentrations will contain little information because the rate will be swamped by the error of the measurement. Many of the original computer programs designed to properly weight statistical estimates of $V_{\max}$ and K_m assume a constant error. However, it is difficult to envision a process that produces an error that is random in direction but equal in magnitude. This is required if the weighting schemes predicted by Eq. (2.51) are to be used. On the other hand, a constant, systematic error can arise at any stage of the analysis. For example, the pipette or instrument may not be properly calibrated, the original sample may contain an incorrect amount of substrate or enzyme, the temperature may not be well regulated, and so forth. These errors will give kinetic constants that have a constant bias but will not produce random error within a single kinetic run.

The second type of error is a relative error. This error is proportional to the measured parameter (enzyme velocity in the present case). The kinetic equation for this type of error can be written as

$$V = \frac{V_{\max} \times [\mathrm{S}]}{K_m + [\mathrm{S}]} + f(v) \tag{2.52}$$

The proportionality factor, $f(v)$ may be constant or a complex function. It is much easier to envision how relative errors can arise within a single kinetic experiment. For example, one may make a 10% error in measuring the slope of a line regardless of the rate of the reaction. A pipette used to add substrate may have an accuracy of ±5%. In both cases, the absolute error will increase with increasing velocity or substrate concentrations. Relative errors increase in absolute magnitude as the velocity increases even though they may be a constant percentage of the velocity.

Published analyses of the error structure of kinetic data have shown that it is not simple nor is it adequately described by Eqs. (2.51) or (2.52). It is apparent that, in general, the error increases with increasing velocity, but the exact form of the relationship is not clear. This is expected when one considers the many factors that may influence an experiment. For example, a fixed relative error of ±5% for a pipette has already been discussed in terms of relative error, but the actual error may increase or decrease as the volume is changed (the error may be 10% at 20 μL, 5% at 50 μL, 7% at 100 μL, and 10% at 150 μL). In order to take some of these considerations into account, the error has been described empirically as

$$\sigma^2(v) = Kv^{\alpha} \quad \text{or} \quad \sigma^2(v) = \sigma_0^2 + v^2\sigma_2^2 \tag{2.53}$$

where $\sigma^2(v)$ represents the variance of the velocity measurement, K and α are factors to be determined for each kinetic experiment, σ_0^2 is the constant variance term, and $v^2\sigma_2^2$ is the variance that is directly proportional to the velocity. Cornish-Bowden (1978), has argued that the equation on the left embodies the worst features of both simple and relative errors, whereas the equation on the right is a more sound mathematical treatment that encompasses both types of error. Whatever your preference, it is apparent that no single equation can adequately describe the error structure for all experimental setups. As already stated, it is easy to envision cases of relative error, but this does not mean that the variance can adequately be described by $\sigma^2(v) = v^2$. As any experimentalist knows, relationships that describe rapid changes in a measured parameter as a function of relatively small changes in experimental conditions are highly subject to errors. This situation occurs when velocity is changing rapidly with small changes in substrate concentration—a situation that occurs when the enzyme rate is between 0 and $0.6V_{max}$, and the enzyme mechanism obeys a Michaelis–Menten relationship. If this type of error is to be included in the analysis, the relative error structure may have an inverted "U" shape or may even decrease as enzyme rates increase, whereas the absolute error may increase as velocity increases.

It is, therefore, impossible to assign an error structure beforehand to a set of kinetic data. This is because no single error structure applies to all kinetic experiments and because no one has undertaken a complete analysis of the error structure of a kinetic experiment due to the large number of points required. A further problem arises when we consider the analyses that have been performed to date and try to apply their conclusions to modern experimental results. This is because modernization of laboratory equipment has changed the way in which kinetic experiments are performed and analyzed. For example, adjustable or automatic pipettes have replaced glass pipettes, spectrophotometers, and data recording devices are electronic, chart recorders no longer exist, and computer software performs much of the analysis. This means that the error structure is considerably different, although it may still depend on traditional sources of error (enzyme stability, pipetting reproducibility, adequate mixing, temperature regulation, etc.) in addition to operator error (incorrect use of computer programs, nonlinearity of assay progress curves, etc.).

An understanding of error structure is important because it impacts on both the optimal experimental design and on the method of analysis. Statistical modeling of Eq. (2.31) has shown that there are different optimal methods for measuring K_m and V_{max} values in a single experiment depending on the nature of the error. If one knows beforehand that the enzyme mechanism is described by Eq. (2.31) (is a Michaelis–Menten type of mechanism) and that the error is constant, the optimal design for estimating V_{max} and K_m is obtained by making multiple velocity measurements, half performed at the highest substrate concentration that can be reasonably added ($[S]_{max}$) and the other half at

$$[S] = \frac{[S]_{max} K_m}{[S]_{max} + 2K_m} \tag{2.54}$$

If the error is proportional to the velocity, the optimal design occurs when the substrate is kept as low as possible, consistent with the sensitivity of the analytical method. If the error is somewhere in between the constant and variable case, the substrate concentration is given by

$$[S] = \frac{[S]_{max} K_m}{2[S]_{max} + 3K_m} \tag{2.55}$$

These calculations are useful if one wishes to analyze kinetic parameters of traditional Michaelis–Menten enzymes. However, most metabolic research concerns itself with enzymes at the start of metabolic pathways and enzymes at metabolic branch points. Analyses usually proceed under various conditions to determine how enzyme activity may respond to changes in cellular conditions or how enzyme activity may be altered by hormonal or physiological changes in organism. These analyses require one to determine the complete shape of the substrate versus velocity curve. This is important since not all enzymes obey Michaelis–Menten kinetics under all conditions. In this case, it is important to have many different substrate concentrations, most of which bracket the apparent K_m value (or substrate concentration that gives a velocity half of the V_{max}) and some that are used to define the V_{max}.

Once an optimal experimental design has been selected and the rates recorded, one must estimate the kinetic constants from the data. The K_m and V_{max} values can be obtained easily from one of several graphical plots of velocity as a function of increasing substrate concentration, but one must be careful in selecting an appropriate plot. Figure 2.10 shows some of the more common plot types as well as the effect of proportional error on the ability of the graph to predict unbiased estimates of K_m and V_{max} values. The Lineweaver–Burk plot (Fig. 2.10, middle) is well-known to have severe problems when the error is constant because it gives too much weight to values at low substrate concentrations (high $1/[S]$ values). The effect is not so severe when the error is proportional but is still present. The Eadie–Hofstee plot (Fig. 2.10, bottom) distributes the error more evenly when the error is constant (data not shown) as well as when the error is relative. This means that linear estimation would provide a relatively unbiased estimate of V_{max} and K_m. Nonparametric plots, such as the direct linear plot (discussed below) are also available. However, the complexity of the plot limits its use for estimating K_m and V_{max} values in the absence of a computer algorithm.

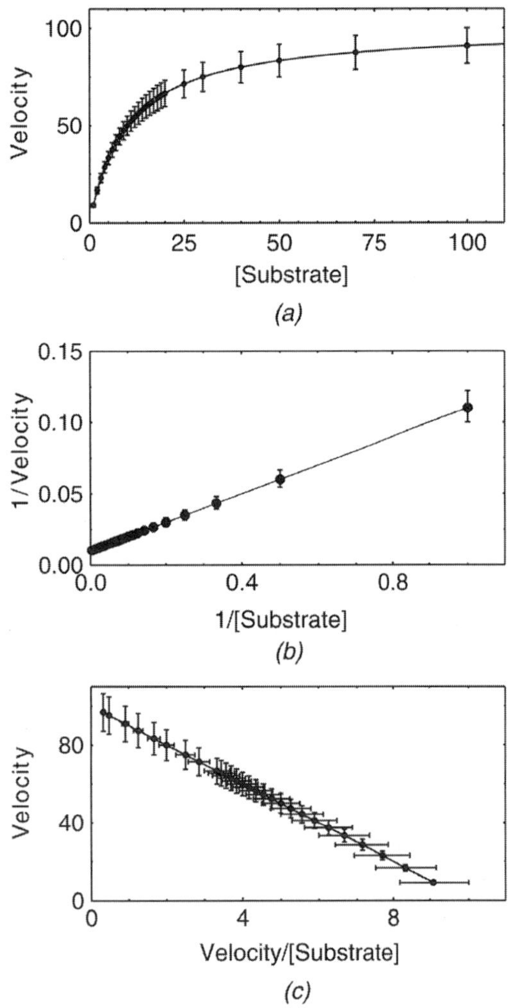

Figure 2.10 Error associated with graphical methods of analysis: effect of proportional error. Values plotted for a Michaelis–Menten enzyme mechanism [Eq. (2.31)] with $V_{\max} = 100\,\mu\text{mol/min}$ and $K_m = 10\,\mu\text{M}$. The error was assumed to be 7% of the calculated velocity value.

The practical question of how to estimate $V_{\max}$ and K_m values from kinetic data remains. Currently, most researchers obtain kinetic parameters through the use of computer programs. These programs either linearize the Michaelis–Menten equation prior to analysis or use nonlinear minimization algorithms to fit Eq. (2.31) directly. In either case, some sort of error structure must be assumed prior to parameter estimation. Assuming a constant, absolute error [$\alpha = 0$ in Eq. (2.53)] allows a direct fit of the data to Eq. (2.31) with the sum of squares given by Eq. (2.56):

$$\text{Sum squares} = \sum u\left(v - \frac{\hat{V}[\text{S}]}{\hat{K} + [\text{S}]}\right)^2 \tag{2.56}$$

where $\hat{V}$ and $\hat{K}$ are the best-fit estimates of $V_{\max}$ and K_m. When the error is constant, $u = 1$ in Eq. (2.56). If the error is exactly proportional to the velocity, $f(v) = \text{error} \times v$ in Eq. (2.52) [or $\alpha = 2$ or $\sigma_0 = 0$ in Eq. (2.53)] one can substitute $u = 1/v^2$ in Eq. (2.54). This has been shown to correspond better (but not perfectly) to the results of the few detailed analyses of error structure. Some authors prefer to use a weighting scheme that represents an intermediate between these two extreme cases since error analysis is not often performed, and this seems to be a reasonable compromise. In this case $u = 1/v$. A practical solution to the problem has also been suggested. Estimates of local variance should be derived from at least five observations from closely spaced substrate concentrations (not necessarily replicate observations at a single substrate concentration) to estimate a single value of α that can then be used to provide a reasonable summary of the total error behavior. Thus, when computing kinetic parameters from kinetic data using a nonlinear regression program, the equation to minimize should be defined by Eq. (2.56) with an appropriate weighting factor derived from estimates provided by the data itself.

In the absence of a complete statistical treatment for each kinetic run, it is possible to use another method to estimate K_m and $V_{\max}$ values. This is a computer application of the nonparametric direct linear plot (Fig. 2.11) that makes no assumptions about the error structure. The method uses nonparametric statistics to provide an estimate of K_m and $V_{\max}$. First, estimates of the K_m and $V_{\max}$ values are calculated from straight lines joining $(-[\text{S}], 0)$ and $(0, v)$ pairs of points that intersect in the first quadrant of a graph. Each intersection provides a single estimate of the K_m (x axis) and $V_{\max}$ (y axis) values (one must be sure not to count $K_{m(i,j)}$ and $K_{m(j,i)}$ as two estimates). Thus, for any two pairs of points, v_j, s_j and v_i, s_i, one can write

$$K_{m(i,j)} = \frac{v_j - v_i}{\left(\dfrac{v_i}{[\text{S}]_i}\right) - \left(\dfrac{v_j}{[\text{S}]_j}\right)} \tag{2.57}$$

$$V_{\max(i,j)} = \frac{[\text{S}]_j - [\text{S}]_i}{\left(\dfrac{[\text{S}]_j}{v_j}\right) - \left(\dfrac{[\text{S}]_i}{v_i}\right)} \tag{2.58}$$

It is difficult to resolve these estimates graphically but not computationally. To estimate the K_m and $V_{\max}$ values, individual $K_{m(i,j)}$ and $V_{\max(i,j)}$ estimates are ordered from lowest to highest. The middle value (median) provides an unbiased estimate of the true K_m and $V_{\max}$ values. This method appears to be quite robust to many types of error. The analysis does not provide a measure of how well the parameters fit the data (no estimates of parameter variance are possible from a nonparametric analysis). If desired, the parameter variances can be estimated by using the best fit K_m and

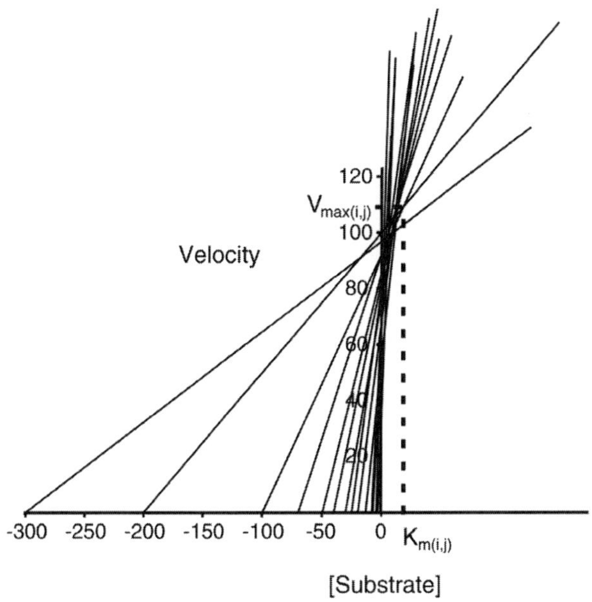

Figure 2.11 Example of direct linear plot analysis. The negative value of the substrate concentration is placed along the x axis and the associated velocity is placed along the y axis. A straight line is drawn through all substrate–velocity pairs. Intersections between lines give V_{max} (y axis) and K_m (x axis) estimates. An example is shown (dashed line). The V_{max} and K_m estimates are ordered from lowest to highest value and the median is used as the best estimate. The number of estimates is given by $\frac{1}{2}n(n-1)$ where n is the number of substrate–velocity pairs.

V_{max} values to calculate the experimental variance and then using this to calculate the K_m and V_{max} variance. Cornish-Bowden (1995) gives the equations for this procedure for those who are interested but, in general, these values are of little use. If one is interested in how well the parameters describe the data, visual inspection of a v vs. S plot should suffice. If one is following changes in enzyme parameters as a function of some environmental or physiological change affecting the whole animal, replicates of control and experimental animals will be needed to estimate the population variance (which is required for comparative purposes, not the variance of the individual measurements).

Although the direct linear plot represents the best method for obtaining unbiased estimates of the K_m and V_{max} values, it is computationally challenging for individuals who do not possess programming skills, and it requires development of different equations for each model. It is not possible to adapt this method to cooperative enzymes, and so one must rely on nonlinear fits of the data to a Hill equation as discussed in the text.

BIBLIOGRAPHY

Cornish-Bowden, A. (1995). *Fundamentals of Enzyme Kinetics*, rev. ed. Butterworths, Toronto.

Cornish-Bowden, A. (1978). In Endrenyi, L. (Ed.), *Kinetic Data Analysis. Design and Analysis of Enzyme and Pharmacokinetic Experiments*. Plenum, New York, pp. 105–119.

Fersht, A. (1977). *Enzyme Structure and Mechanism*. W.H. Freeman, New York.

Segal, (1975). *Enzyme Kinetics*. Wiley, New York.

Segal, I. H. (1968). *Biochemical Calculations. How to Solve Mathematical Problems in General Biochemistry*. Wiley, New York.

Suelter, C. H. (1985). *A Practical Guide to Enzymology, Biochemistry, A Series of Monographs*, Vol. 3, Wiley, New York.

Strobel, S. A. (2001). Biological catalysis: repopulating the RNA world. *Nature* **411**:1003–1006.

Walsh, C. (1979). *Enzymatic Reaction Mechanisms*. W.H. Freeman, San Francisco.

Weiker, H.-J., Johannes, K.-J., and Hess, B. (1970). A computer program for the determination of kinetic parameters from sigmoid steady-state kinetics. *FEBS Lett* **8**:178–185.

3

ENZYMES IN THE CELL: WHAT'S REALLY GOING ON?

STEPHEN P. J. BROOKS

INTRODUCTION

In Chapter 2, we reviewed how enzymes catalyze reactions and how they behave when studied *in vitro* under the artificial experimental conditions of very large dilutions and zero product concentration. This chapter will bring the reader from this basic understanding of enzyme behavior to an understanding of how enzymes may function in cells. It is important to stress the fact that it is very difficult to measure the activity of most enzymes *in situ* so that the ideas presented in this chapter are based on limited observational data. The difficulties arise from problems associated with measuring free substrate concentrations, product, inhibitor, or activator concentrations, manipulating substrate levels, and even measuring enzyme activity in an intact cell. Some of the following sections are, therefore, of a speculative nature. Nevertheless, they all provide insight into the workings of a cell and should stimulate much discussion.

BASIC ENZYME MECHANISM UNDER *IN VIVO* CONDITIONS

Two differences are immediately apparent when one compares the environment of an enzyme *in vitro* and *in vivo*. First, the concentration of product, negligible under the initial velocity conditions in the test tube, is rarely so low as to be considered negligible in the cell. Second, the enzyme/substrate ratio is rarely low enough to permit the assumption that the total substrate concentration is essentially equal to the free substrate concentration. Table 2.2 listed the actual *in vivo* measurements of enzyme (as active site) and metabolite (total and free substrate and product) concentrations from muscle cells. Examination of this data immediately shows the wide variability in the active site/free S ratio (right-hand column) that can exist within a single metabolic pathway (in this case, the glycolytic pathway). These differences limit the usefulness of the equations developed to describe enzyme kinetic data *in vitro* when describing enzyme activity under *in vivo* conditions. The present chapter will expand the equations used for *in vitro* work to develop a system of equations that more closely describe cellular events and give the reader a better perspective of the control of cellular metabolic activity. The reader may wish to return to the development of the general Michaelis–Menten equation contained in Chapter 2 in order to refresh his or her memory before reading this chapter.

Effect of Enzyme–Substrate Binding

The assumption that the free substrate concentration is approximated by the total substrate concentration was critical to the development of all enzyme mechanisms presented in Chapter 2 [see Eqs. (2.19) and (2.20)]. Eliminating this assumption means that one must take the ES (enzyme–substrate complex concentration into account when calculating the total substrate concentration. Thus, the total substrate concentration is the sum of free and bound substrate:

$$[S_T] = [S_{\text{free}}] + [ES] \tag{3.1}$$

Note that the relationship of Eq. (3.1) is exactly analogous to that relating free and bound enzyme to the total enzyme concentration that was developed in Chapter 2

Functional Metabolism: Regulation and Adaptation, edited by Kenneth B. Storey.
ISBN 0-471-41090-X

($[E_T] = [E_{free}] + [ES]$). Including Eq. (3.1) in the derivation of a Michaelis–Menten type of mechanism [Eq. (2.30)] gives a quadratic equation that does not yield a neat analytical solution. In order to calculate the enzyme rate, one must solve first for the concentration of enzyme-bound substrate [ES].

Before we proceed with the development of the equations, it should be noted that most of the equations in this chapter were derived using the dissociation constants of the substrate and product (P) from the ES or EP complexes, respectively. The dissociation constants are abbreviated as $\bar{K}$, to distinguish them from the Michaelis constants, K_m. It is possible to derive kinetic equations by assuming that the K_m values are equal to the $\bar{K}$ values. This is known as the equilibrium assumption and, in fact, it was the assumption used by Michaelis and Menten to derive their equation. However, the equilibrium assumption is not valid until tested by detailed mechanistic analysis. In order to emphasize this fact, the symbol $\bar{K}$ will be used throughout the text. The dissociation constants are used because they make the derivation simpler while giving correct kinetic expressions. In addition, they allow the calculation of free and enzyme-bound distributions of substrates. Using the dissociation constants, one can de-rive an equation that relates the $\bar{K}$ value and the total substrate and enzyme concentrations to the concentration of the enzyme–substrate complex (ES):

$$0 = [ES]^2 - [ES] \times ([S_T] + [E_T] + \bar{K}) + [S_T][E_T] \quad (3.2)$$

The velocity is simply the concentration of ES multiplied by the kinetic constant that defines how quickly it breaks down to form products (k_{cat}):

$$v = k_{cat} \times [ES] \quad (3.3)$$

One can show that if $[S_T] \gg [E_T]$ (and $[S_T][E_T] \gg [ES]^2$, by extension), Eq. (3.2) reduces to $[ES] = [S_T][E_T]/(K_m + [S_T])$ and Eq. (3.3) gives Eq. (2.31) once again. The $[E_T]/[S_T]$ ratio, therefore, becomes important in interpreting enzyme behavior under *in vivo* conditions. In Table 2.2, consideration of the $[E_T]/[S_T]$ ratio for some common enzymes of the glycolytic pathway showed that this can range from a value of 0.004 to 247. When the ratio is very low, as is the case for phosphofructokinase, the assumption that $[S_{free}] \approx [S_T]$ is valid. However, when the ratio approaches or exceeds a value of 0.4, serious deviations from the Michaelis–Menten equation can be observed. Figure 3.1 demonstrates the effect of including substrate binding in the derivation of the Michaelis–Menten equation. In Figure 3.1*a*, the predicted velocity is plotted as a function of increasing [S] for both the Michaelis–Menten equation [Eq. (2.31)] and

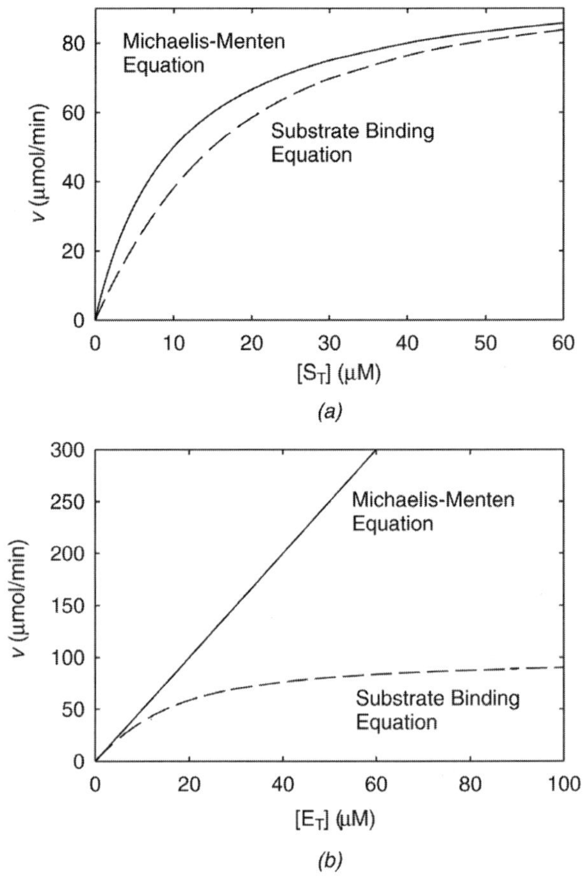

Figure 3.1 (*a*) Substrate concentration–enzyme velocity relationship and (*b*) enzyme concentration–velocity relationship predicted by the Michaelis–Menten equation [Eq. (2.31), solid lines] and by the substrate-binding equation [Eqs. (3.2) and (3.3), dashed lines). Arbitrary values were chosen for the kinetic constants in order to run the simulation. These values were: a K_m or $\bar{K}$ value of 10 μM, a total enzyme concentration of 10 μM [fixed for (*a*)], a k_{cat} value of 10 min^{-1}, and a total substrate concentration of 10 μM [fixed for (*b*)].

the substrate-binding equations [Eqs. (3.2) and (3.3)]. This graph shows that the effect of enzyme–substrate binding is to reduce the observed velocity at a given total substrate concentration. The effect is greatest around the K_m (or, more correctly, the $\bar{K}$ value), which is 10 μM in this simulation but disappears as the total substrate concentration increases to very high values. In addition to an effect on the velocity–substrate curve, the substrate binding equation also predicts a lower velocity when $[E_T]/[S_T]$ ratios are high. This is illustrated in Figure 3.1*b*, where the calculated velocity is plotted as a function of increasing total enzyme concentration at a fixed total substrate concentration. As the total enzyme concentration increases, the substrate-binding curve asymptotically approaches $[S_T] \times k_{cat}$. This is because, as defined in Eq. (3.3), the velocity is a function of k_{cat} and [ES]. With a fixed total [S], the maximal amount of [ES] that can be formed is equal to $[S_T]$. The difference between the two equations becomes apparent when $[E_T] = 4\,\mu M$. Here the difference between the predicted velocity is approximately 11% and increases with increasing $[E_T]$. For example, at an $[E_T]/[S_T]$ ratio of 1.0 ($E_T = 10\,\mu M$ in Fig. 3.1), the difference in the calculated velocity is 31%. Although Figure 3.1 gives an idea of the magnitude of the difference in velocity as a function of $[E_T]$, it does not give a complete picture because the difference between the Michaelis–Menten equation and the substrate-binding equation also depends on the $\bar{K}$ value as shown by

$$\text{Difference} = 0.5(\bar{K} + [S_T]) + \frac{[E_T](\bar{K} - [S_T])}{\bar{K} + [S_T]} - 0.5\sqrt{(\bar{K} + [S_T] + [E_T])^2 - 4[E_T][S_T]} \quad (3.4)$$

Modifications of the Michaelis–Menten equation have also been proposed that partially take into account the contribution of substrate binding to enzyme activity. One such equation assumes that the $[ES]^2$ term of Eq. (3.2) is much smaller than the other two terms. The resulting equation contains a term for total enzyme concentration in the denominator:

$$v = \frac{V_{max}[S_T]}{K_m + [S_T] + [E_T]} \quad (3.5)$$

Equation (3.5) gives much lower errors than the conventional Michaelis–Menten equation when the enzyme concentration is higher than the substrate concentration but still fails when both are approximately equal.

Effect of Reaction Reversibility

As already mentioned, the other important difference between *in vitro* and *in vivo* conditions is the presence of significant concentrations of enzyme product in the cell. Table 2.2 shows the *in vivo* total concentrations of substrate–product pairs for the enzymes of the glycolytic pathway. For some enzymes, such as pyruvate kinase, the concentration of product is approximately sixfold higher that the substrate concentration. An equation that takes product concentration into account has already been developed—it is the complete single-substrate reaction equation [Eq. (2.25)]. The presence of product has two effects. First, it competes with substrate for the enzyme active site so that, in a sense, it acts like a competitive inhibitor. This effectively lowers the rate of the forward direction by reducing the amount of ES complex that can form (Table 2.4). Second, because the reaction is reversible, product can

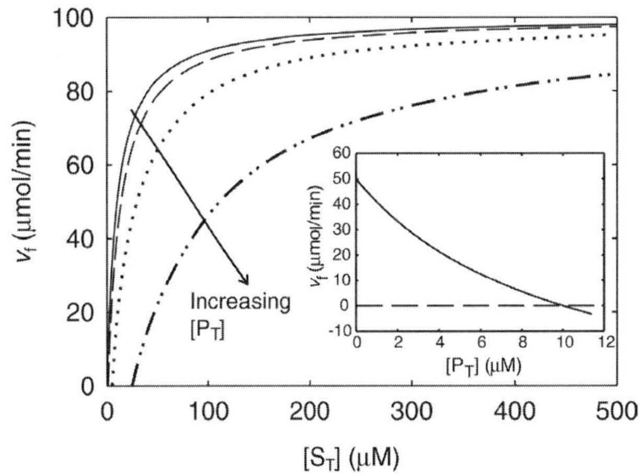

Figure 3.2 Effect of increasing product concentration on the rate of product formation using the reversible Michaelis–Menten equation [Eq. (2.25)] to calculate enzyme velocity. The rate of product formation (v_f) is plotted as a function of increasing, fixed $[S_T]$ for different fixed values of $[P_T]$. Inset: The rate of product formation is plotted as a function of increasing concentrations of product for a fixed concentration of substrate.

react to form substrate. This means that increasing product concentrations can actually generate negative velocity values when the substrate concentration is not high enough to force the reaction forward (Fig. 3.2), inset). This is a simple consequence of the reversibility of the enzyme reaction. Note that when the net enzyme rate is zero ($v = 0$ in Fig. 3.2, inset), the reaction is at equilibrium and the Haldane relationship [Eq. (2.33)] applies.

Effect of Substrate and Product Binding Combined with Reaction Reversibility

Figures 3.1 and 3.2 demonstrated the effect of substrate binding and reaction reversibility independently. For simple one substrate–one product reaction mechanisms [Eq. (2.15)], it is possible to derive an equation that includes both effects simultaneously. The resulting cubic equation is complex and must be solved algebraically to give the free enzyme concentration ($[E_{free}]$):

$$0 = [E_{free}]^3 + [E_{free}]^2 \times (\bar{K}_S + \bar{K}_P + [S_T] + [P_T] - [E_T]) + [E_{free}] \times (\bar{K}_S\bar{K}_P + [S_T]\bar{K}_P + [P_T]\bar{K}_S - [E_T](\bar{K}_S + \bar{K}_P)) - [E_T]\bar{K}_S\bar{K}_P \quad (3.6)$$

The value of $[E_{free}]$ is then used to calculate the concentrations of [ES] and [EP]. Remember that the K_m values are not used in Eq. (3.6) but are replaced by dissociation constants ($\bar{K}$ values). This is to emphasize the fact that the K_m values are not true dissociation constants but are "kinetic" dissociation constants—the $\bar{K}$ values represent actual binding constants, which are not always equal to K_m values. The two may be equal for some enzymes but this must be experimentally verified. Thus, using K_m values may give erroneous concentrations of ES and EP. Using the dissociation constants allows the concentrations of [ES] and [EP] to be determined by Eq. (3.7):

$$[\mathrm{ES}] = \frac{[\mathrm{S}_T][\mathrm{E}_{\mathrm{free}}]}{(\bar{K}_{\mathrm{S}} + [\mathrm{E}_{\mathrm{free}}])} \qquad [\mathrm{EP}] = \frac{[\mathrm{P}_T][\mathrm{E}_{\mathrm{free}}]}{(\bar{K}_{\mathrm{P}} + [\mathrm{E}_{\mathrm{free}}])} \tag{3.7}$$

Knowing [ES] and [EP] allows calculation of the reaction rate:

$$v = k_f \times [\mathrm{ES}] - k_r \times [\mathrm{EP}] \tag{3.8}$$

with k_f and k_r related to the maximal velocity of the reaction in the forward and reverse direction by

$$k_f = V_f/[\mathrm{E}_T] \qquad k_r = V_r/[\mathrm{E}_T] \tag{3.9}$$

Figure 3.3 gives an idea of how an isolated enzyme might behave when the effects of substrate and product binding are included. In this model, the enzyme reaction was set up so that the equilibrium constant is unity ($K_{\mathrm{eq}} = 1$). Thus, if the reaction were allowed to reach equilibrium, equal amounts of free product and free substrate would be present. The maximal rate in the forward direction (V_f) was set to 100 μmol/min and the maximal rate in the reverse direction was set to half this value ($V_r = 50$ μmol/min) so that one conceptualizes the reaction as going in the forward direction.

The top graph of Figure 3.3 shows the effect of increasing the total enzyme concentration on the forward rate of catalysis at constant substrate and product concentrations. Interestingly, there are concentrations of substrate and product where the reaction runs in the reverse direction at low enzyme concentrations and in the forward direction at higher enzyme concentrations. This is a function of the binding coefficients for substrate and product. Because product binds with higher affinity in the simulation, it effectively out-competes substrate for the enzyme active site at low enzyme concentrations (when $[\mathrm{S}_T] + [\mathrm{P}_T] > [\mathrm{E}_T]$). As enzyme concentration increases, there are more sites for substrate binding (i.e., there is more enzyme) and activity increases. The lines all tend to a final maximal velocity given by ($[\mathrm{S}_T]k_f - [\mathrm{P}_T]k_r$) because, in the case of increasing enzyme concentrations, the concentrations of substrate and product are limiting and not the concentration of enzyme. At infinitely high enzyme concentrations, all the reactants will be bound by enzyme ($[\mathrm{ES}] = [\mathrm{S}_T]$ and $[\mathrm{EP}] = [\mathrm{P}_T]$).

The effect of changing product concentrations (Fig. 3.3, bottom) is best illustrated by holding the total concentration of enzyme constant (think about drawing a vertical line through the graph of Fig. 3.3, top). When product concentration is low, substrate binding predominates and the reaction goes forward. However, as product concentration increases, the rate of the back reaction exceeds that of the forward reaction, and the rate of the forward direction becomes negative (reverse flux). This simple graph explains the classic metabolic interpretation of what happens in a pathway when an enzyme downstream is turned off by a regulatory event. As product backs up, activity in the forward direction is expected to decrease simply because of the presence of high amounts of product. While this is true for an enzyme in isolation

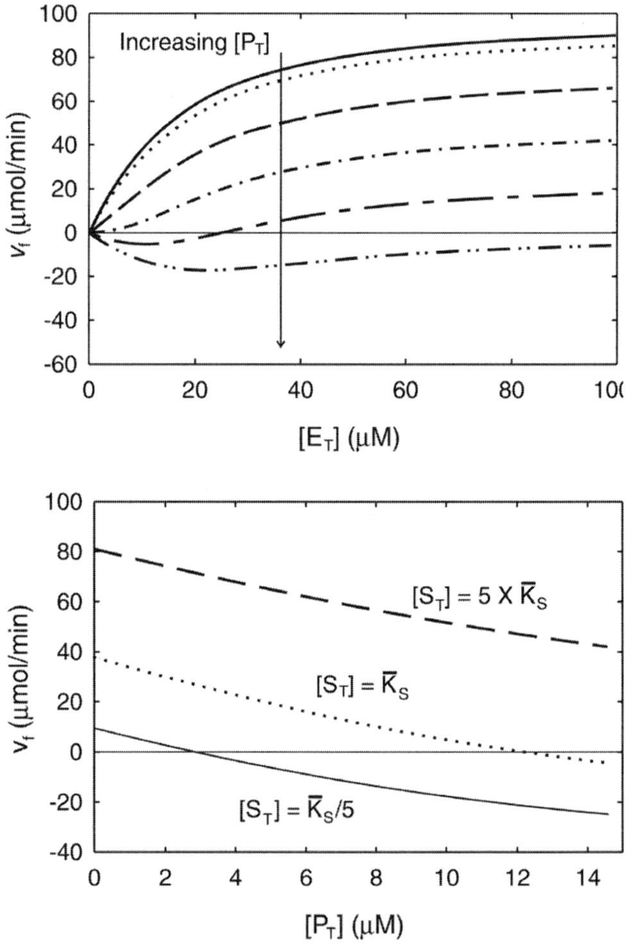

Figure 3.3 Effect of changing product and enzyme concentration on the rate of product formation according to the complete reversible single-substrate equation [Eq. (3.6)]. Top graph: The rate of product formation (v_f) is plotted as a function of increasing total enzyme concentration ($[\mathrm{E}_T]$) for a single fixed substrate concentration and different fixed concentrations of product ($[\mathrm{P}_T]$). Bottom graph: The forward velocity is plotted as a function of increasing total product concentrations for different fixed total substrate concentrations.

with high product concentrations, the situation is more complex for enzymes in a pathway. Equations showing how enzymes behave in a sequence will be presented later in this chapter.

Time Course of the Complete Enzyme Mechanism

The above simulations suffer from one major limitation—the total concentration of substrate and product were held at fixed concentrations so that they did not change as the reaction progressed. This means that these equations allow calculations of the enzyme rate at any given substrate/product concentration but do not describe the behavior of an enzyme over a complete reaction time course. It is possible to derive a complete equation for the time-dependent behavior of an enzyme, but the process is complex and the equation will apply only to single enzymes isolated in a reaction vessel with a fixed total amount of substrate + product. The first step is to define the total amount of substrate + product at any time:

$$A = [S_T] + [P_T] \qquad (3.10)$$

The value of A is then substituted into the algebraic solution to Eq. (3.6) (which is, itself, a complicated equation) to allow calculation of $[E_{free}]$. Equation (3.7) is then used to calculate values of [ES] and [EP], and these values are substituted into Eq. (3.8), rewritten in differential form:

$$d[P_T]/dt = k_f \times [ES] - k_r \times [EP] \qquad (3.11)$$

The new equation would then be integrated to obtain the change in enzyme rate as a function of time. Needless to say, the equation would be extremely complex and will not be presented here for one reason. The extra information provided by the integrated form of Eq. (3.8) can be easily obtained by substituting $[S_T] = A - [P_T]$ into Eq. (3.6) and choosing small enough time increments so that the changes in substrate and product are small (on the order of 1% of the total). This is equivalent to numerically integrating Eq. (3.6) and produces a workable result providing that the changes in $[S_T]$ and $[P_T]$ are small. The results of this process are shown in Figure 3.4, where three different reaction progress curves are presented: one based on Eq. (3.6) (solid lines), one based on the Michaelis–Menten equation [Eq. (2.25), dotted lines), and one based on an equation without a reverse reaction [Eqs. (3.2) and (3.3), dashed lines). When the reverse reaction is included, the reaction rates slow as the concentration of product accumulates (or the concentration of substrate decreases). This is due to the conversion of substrate into product, which serves to reduce the forward rate of reaction—the enzyme rate is defined as the amount of substrate converted into

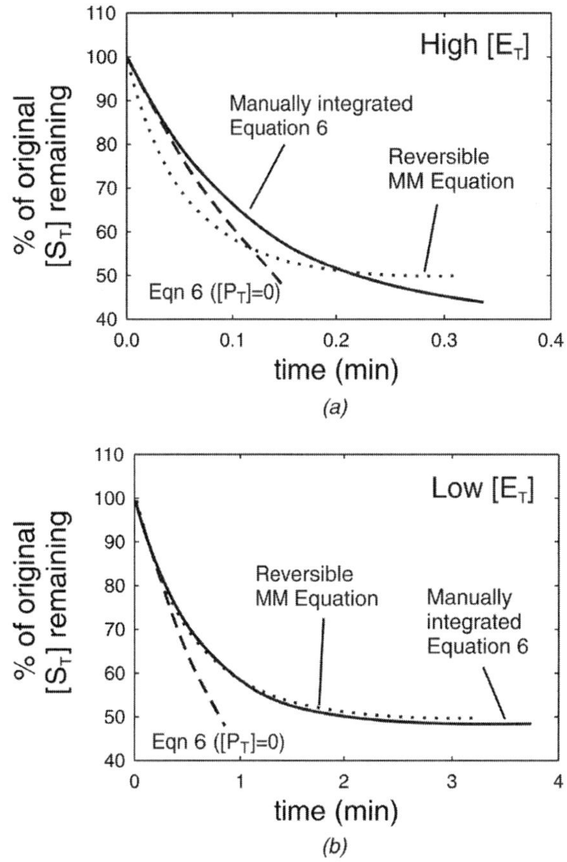

Figure 3.4 Complete enzyme time course for a reaction with a fixed total amount of $[S_T] + [P_T] = A$. The graphs show the time course based on Eq. (3.6) (complete equation with substrate binding; solid line), on the reversible Michaelis–Menten equation [Eq. (2.25); dotted line], and on Eq. (3.6) with no reverse reaction ($k_r = 0$). The calculations were performed with high (10 μM $[E_T]$; upper figure) or low (1 μM $[E_T]$; lower figure) amounts of total enzyme. When enzyme concentrations are high, there is significant deviation from Michaelis–Menten behavior.

product. Even though the rate in the forward direction decreases, it does not mean that the enzyme loses activity: There is a constant flow of [S] → [P] and [P] → [S] at any one time. It is only the *net* conversion of [S] into [P] that decreases over time.

When the enzyme concentration is high ($[E_T] = 10\,\mu M$, Fig. 3.4, top), the reaction rate is greatly influenced by the binding of substrate and product to enzyme. Under these conditions, the rate predicted by the manual integration of Eq. (3.6) is slower than that predicted by the reversible Michaelis–Menten equation, and the reaction apparently proceeds to a further extent (i.e., more substrate is used up). Both effects are due to the binding of the reactants to the enzyme. The reaction proceeds at an apparently slower rate because increasing enzyme concentrations apparently reduce the concentration of free substrate. In reality, this effect comes about because the reversible Michaelis–Menten equation does not take substrate binding into account.

In addition to an apparent slowing of the reaction rate, the extent of the reaction is also affected: The reaction proceeds further than would be predicted from the equilibrium constant. This is most apparent in the top graph of Figure 3.4 where the "percentage of the original [S_T] remaining" predicted by the manual integration of Eq. (3.6) dips below that predicted by the reversible Michaelis–Menten equation [the same equilibrium point as that predicted by the Haldane relationship, Eq. (2.33)]. This happens because the equilibrium constant, as developed in Chapter 2, applies only to the free reactants, whereas Figure 3.4 plots the change in total substrate over time. At high enzyme concentrations, a significant amount of substrate and product are associated with the enzyme. The difference between these two equations is due to the difference in enzyme-bound and free reactant equilibriums. As discussed in Chapter 2, a considerable difference between the enzyme-bound equilibrium and the free reactant equilibrium may exist depending on the differences in the enzyme binding constants for substrate and product. In the present case, this difference is a factor of twofold. Both these effects are much less prominent when the enzyme concentration is lowered to 1 μM (Fig. 3.4, bottom figure).

The effect of increasing enzyme concentration on the apparent equilibrium constant can be derived from the completely reversible substrate-binding equation. This is done by rearranging Eqs. (3.7) and (3.8) when the reaction is at equilibrium ($v = 0$):

$$K_{\mathrm{eq}}^{\mathrm{app}} = \frac{[\mathrm{P}_{T,\mathrm{eq}}]}{[\mathrm{S}_{T,\mathrm{eq}}]} = \frac{V_f(\bar{K}_\mathrm{P} + [\mathrm{E}_\mathrm{free}])}{V_r(\bar{K}_\mathrm{S} + [\mathrm{E}_\mathrm{free}])} \tag{3.12}$$

Equation (3.12) shows mathematically what has been described in the preceding paragraphs—that is, the apparent K_{eq} value is a function of the free enzyme concentration. Equation (3.12) can be reduced to the traditional Haldane relationship when $[\mathrm{E}_\mathrm{free}] \ll \bar{K}_\mathrm{P}$ and $[\mathrm{E}_\mathrm{free}] \ll \bar{K}_\mathrm{S}$ ($[\mathrm{E}_T]$ is small). The original Haldane equation does not take enzyme concentration into account because it is based on the reversible Michaelis–Menten equation, which does not include substrate and product binding. Thus, the original Haldane equation assumes that [E_T] is much lower than [S_T] and [P_T] so that [E_{free}] is small. Equation (3.12) can be expressed explicitly in terms of [S_T], [P_T], and the kinetic constants for the reaction by substituting the algebraic solution for Eq. (3.6). The resulting definition for $K_{\mathrm{eq}}^{\mathrm{app}}$ is, however, complex and does not give a solution containing only the $[\mathrm{P}_{T,\mathrm{eq}}]/[\mathrm{S}_{T,\mathrm{eq}}]$ ratio.

TEXT BOX 3.1 MEASURING EQUILIBRIA IN THE CELL AND ITS RELATIONSHIP TO FLUX

The classical method to identify potential control loci of enzyme systems involves examining a reaction's theoretical equilibrium position and comparing this to the measured reaction products in the cell. The reasoning behind this method was that enzyme reactions close to their theoretical equilibrium position could not be control points for a reaction because they have a large excess of enzyme capacity. The fact that there is a large catalytic discrepancy between different enzymes in a single pathway has intrigued biochemists for a long time.

Measuring a reaction's proximity to its theoretical equilibrium point requires measuring the cellular concentrations of substrates and products followed by a calculation of the mass action ratio for the reaction (designated by the capital Greek letter gamma, Γ):

$$\Gamma = [\mathrm{P}]/[\mathrm{S}] \tag{TB3.1}$$

To do this, cells or tissues are freeze-clamped to instantaneously stop metabolism. The metabolites are then extracted in a medium (usually acid) that kills enzyme activity, neutralized and quickly quantified. The mass action ratio for each enzyme is then compared to the theoretical equilibrium position for the reaction. The K_{eq} can be measured directly in an isolated reaction with known quantities of total substrate or can be calculated through the Haldane relationship [see Eq. (2.33)]. In the cell, a reaction can never be at equilibrium since this means that the whole pathway has no net flux passing through it. However, reactions can be close to equilibrium and this is measured by the ratio Γ/K_{eq}. Classically, potential control points were evaluated by the size of their Γ/K_{eq} ratios.

Two problems arise in calculating a reaction's proximity to its equilibrium point. First, contrary to what one would expect, the proximity of a reaction to equilibrium depends on many factors, including the kinetic parameters of the enzyme, the flux through the reaction step, and the degree of substrate saturation of the enzyme. This is different from the case of an isolated enzyme in a test tube where the substrate and product concentration are regulated only by the activity of an isolated enzyme that is usually present in very low concentrations. At steady state in the test tube, the reaction can be at equilibrium while the *in vivo* steady-state conditions may be far from equilibrium. As will be shown later in this chapter, a sequence of reversible enzymes obeying the Michaelis–Menten equation [Eq. (2.25)] can be resolved to give equations relating the free concentration of any intermediate in a pathway at steady state to the kinetic parameters of the enzymes [see Eqs. (3.24) and (3.25)]. Rearrangement of these equations gives the kinetic expression for the mass action ratio:

$$\Gamma = \frac{[\mathrm{S}_{\mathrm{ss,free}}]}{[\mathrm{P}_{\mathrm{ss,free}}]} = \frac{[\mathrm{I}_{(i+1),\mathrm{ss,free}}]}{[\mathrm{I}_{i,\mathrm{ss,free}}]}$$

$$= \frac{\bar{K}_i}{(V_i + J_{\mathrm{ss}})}\left(\frac{(V_i^+ - J_{\mathrm{ss}})}{\bar{K}_i^+} - \frac{J_{\mathrm{ss}}}{[\mathrm{I}_{i,\mathrm{ss,free}}]}\right) \quad \text{(TB3.2)}$$

The proximity to equilibrium is defined as the ratio Γ/K_{eq}. Using the Haldane relationship to calculate K_{eq} (for the free reactants) gives

$$\frac{\Gamma}{K_{\mathrm{eq}}} = \frac{V_i}{V_i^+(V_i + J_{\mathrm{ss}})}\left(V_i^+ - J_{\mathrm{ss}} - \frac{J_{\mathrm{ss}}\bar{K}_i^+}{[\mathrm{I}_{i,\mathrm{ss,free}}]}\right) \quad \text{(TB3.3)}$$

Equation (TB3.3) is interesting because it shows the conditions that must be fulfilled before an enzyme reaction approximates equilibrium conditions. First, it shows that the flux through the system must be much smaller than either the maximal forward or reverse rates ($V_i^- \gg J_{\mathrm{ss}}$ and $V_i^+ \gg J_{\mathrm{ss}}$). Under these conditions, Eq. (TB3.3) becomes

$$\frac{\Gamma}{K_{\mathrm{eq}}} \approx 1 - \frac{J_{\mathrm{ss}}}{V_i^+} \times \frac{\bar{K}_i^+}{[\mathrm{I}_{i,\mathrm{ss}}]} \quad \text{(TB3.4)}$$

Equation (TB3.4) illustrates the other requirement for a reaction to be close to equilibrium. This is that the fraction $\bar{K}_i^+/[\mathrm{I}_{i,\mathrm{ss}}]$ (i.e., the ratio of free to substrate-bound enzyme also given by the fraction $[\mathrm{E}_{i,\mathrm{free}}]/[\mathrm{E}_i\mathrm{I}_i]$) must be relatively small. This suggests that relatively large concentrations of $\mathrm{I}_{i,\mathrm{ss}}$ are required if the reaction is to be close to equilibrium.

The second problem with measuring the proximity of a reaction to its equilibrium is that the above equations apply only to free substrate and product. Thus, they do not apply to the case where a relatively large enzyme concentration can bind significant amounts of substrate and product. However, this is precisely the type of enzyme that is considered an equilibrium enzyme! Under conditions where enzyme binding is significant, two different equilibria exist. One is the equilibrium of the bound substrate/product pair. This is different from the solution equilibrium because of the effects of enzyme–substrate binding. The equilibrium position for free reactants is still given by the Haldane equation, but the bound equilibrium for a reaction catalyzed by a single, isolated reaction (in a cuvette) is given by

$$K_{\mathrm{eq}}^{\mathrm{bound}} = \frac{[\mathrm{EP}]}{[\mathrm{ES}]} = \frac{k_f}{k_r} = \frac{V_f}{V_r} \quad \text{(TB3.5)}$$

An equation that combines the limiting conditions of Eqs. (TB3.4) and (TB3.5) as well as describes the relationship between K_{eq} and the enzyme concentration for a single enzyme in a cuvette is given by combining Eqs. (3.7) and (3.8) and setting $v = 0$ (the equilibrium condition):

$$K_{\mathrm{eq}}^{\mathrm{app}} = \frac{[\mathrm{P}_{T,\mathrm{eq}}]}{[\mathrm{S}_{T,\mathrm{eq}}]} = \frac{V_f(\bar{K}_{\mathrm{P}} + [\mathrm{E}_{\mathrm{free}}])}{V_r(\bar{K}_{\mathrm{S}} + [\mathrm{E}_{\mathrm{free}}])} \quad \text{(TB3.6)}$$

When the concentration of free enzyme is much lower than the dissociation constants for substrate and product (free enzyme concentration is low), Eq. (TB3.6) gives the Haldane relationship again. When the concentration of free enzyme is high ($[\mathrm{E}_T] \gg$ dissociation constants), Eq. (TB3.6) gives (TB3.5). For a single enzyme, the concentration of free enzyme can be related to the kinetic parameters of the enzyme and the enzyme concentration by Eq. (3.6).

Equation (TB3.6) is useful for an isolated enzyme but not for an enzyme reaction in a metabolic pathway. This is because Eq. (TB3.6) takes into account the binding of substrate and product to only one enzyme. In a metabolic pathway, the product of one enzyme reaction is the substrate for the next reaction. Thus, binding of intermediate to the two surrounding enzymes must be taken into account. This is true for both the substrate and product of the enzyme of interest. It is possible to use Eq. (TB3.3) but only if we can either measure the free concentration of the intermediates or if we can calculate them. If the kinetic parameters of the enzymes around the reaction of interest are known and if the flux through the system is known, it is possible to use Eqs. (3.24) to (3.26) (see main text) to calculate the concentration of free intermediate. Because of the nonlinear nature of Eqs. (3.24) to (3.26), it is impossible to write a neat analytical solution for the equilibrium of an enzyme in a metabolic pathway. However, as one would expect from the above analysis, the proximity to equilibrium is dependent on many factors including the concentrations of the enzymes before and after the reaction being investigated and the kinetic parameters of the system.

Thus, as expected, the apparent K_{eq} value is a function of the concentrations of enzyme, total substrate and total product since changing these parameters changes $[E_{free}]$.

SIMPLE ENZYME SYSTEMS

The differences between Eq. (3.6) and the reversible Michaelis–Menten equation are important when following single enzymes in isolated reaction vessels, but they are not that important when studying systems of enzymes. This is because there is a constant flow of material into and out of a reaction sequence. Thus, the concentrations of substrate and product do not change as radically as they would if the enzyme were isolated in a test tube. In addition, while enzyme reactions tend toward equilibrium, and may in fact reach equilibrium in a test tube, reactions *in vivo* are usually not at equilibrium. Indeed, it is the desire of reactions to go toward equilibrium that drives metabolism.

All the enzymes of metabolism exist in pathways—enzymes do not function in isolation in the cell. This means that enzymes are always interacting by sharing products and substrates. When two enzymes act together, there is always some "adjusting" that occurs before the enzyme activities are coordinated. This adjusting occurs as intermediates (shared substrate/product pairs) take time to build up to their steady-state levels. Although enzymes are not at equilibrium in a cell, the metabolic pathways are believed to be at, or close to, steady state. This means that there is a constant flow of substrate into, and product out of, a metabolic pathway and that the concentrations of the intermediates of the pathway do not change appreciably.

While the concentrations of intermediates are coming to their steady-state levels, the activity of the enzymes in a pathway are not constant but increase or decrease in response to the changing concentrations of the intermediates. This adjusting can apply to two-enzyme systems or to multienzyme systems as we shall see later. This section will focus on very simple enzyme systems to see how they work. The understanding obtained from this section will help to build bigger models.

The simplest two-enzyme system is one where the first enzyme is irreversible and the second enzyme is present in excess. Indeed, enzymologists commonly use this system to measure the rate of an enzyme that does not produce an end product that is easily quantified. For example, pyruvate kinase (PK) catalyzes the reversible transfer of phosphate from phosphoenolpyruvate (PEP) to adenosine 5′-diphosphate (ADP). Its products, pyruvate and adenosine 5′-triphosphate (ATP), are not easily monitored so this reaction is normally coupled to the lactate dehydrogenase (LDH) reaction. Lactate dehydrogenase is another two-substrate reaction, catalyzing the conversion of pyruvate to lactate using reducing equivalents from its second substrate, reduced nicotinamide adenine dinucleotide (NADH; a compound that absorbs light at 340 nm). Combining both enzymes, therefore, allows one to monitor the progress of the first enzyme by following the disappearance of NADH from the second reaction. The reduction in [NADH] is measured as a decrease in absorbance at 340 nm in a spectrophotometer. The rate of the PK reaction can be determined because there is a 1 : 1 stoichiometry between the pyruvate that is produced and the NADH that is consumed. Enzymologists commonly use an activity of LDH that is very much higher than that of pyruvate kinase so that there is no lag time. The lag time is the time it takes for the second reaction to catch up to the first reaction (or the time before the system enters a steady state).

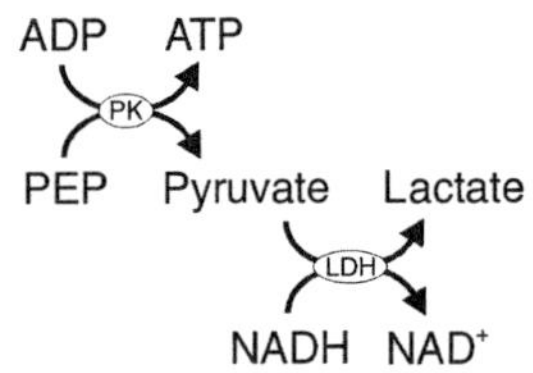

(3.13)

Under *in vitro* conditions there is little or no product so that the scheme of Eq. (3.13) can be simplified to an irreversible two-enzyme system with PK as the first enzyme, pyruvate as the intermediate (I), and LDH as the coupling enzyme:

$$\mathrm{S} \xrightarrow{v_1} \mathrm{I} \xrightarrow{\frac{V_{max}[\mathrm{I}]}{K_m + [\mathrm{I}]}} \mathrm{P} \tag{3.14}$$

The complete time-dependent equation for this system is complex:

$$\exp\left(\frac{-t\Phi^2}{K_m V_{max}}\right) = \frac{\Psi}{\Psi_0}\exp\left(\frac{\Phi[\mathrm{I}]}{K_m V_{max}}\right) \tag{3.15}$$

with $\Phi = (V_{max} - v_1)$, $\Psi = (v_1 K_m - [\mathrm{I}]\Phi)$, and $\Psi_0 = (v_1 K_m - [\mathrm{I}_0]\Phi)$. However, at steady state, it resolves to give a simple relationship between $[\mathrm{I}_{ss}]$ and the kinetic constants for the reaction:

$$[\mathrm{I}_{ss}] - [\mathrm{I}_0] = \frac{v_1 K_m}{V_{max} - v_1} \tag{3.16}$$

For coupling enzyme systems, $[\mathrm{I}_0]$ is normally zero and this value is dropped from the expression:

$$[\mathrm{I}_{ss}] = \frac{v_1 K_m}{V_{max} - v_1} \tag{3.17}$$

Equation (3.17) shows that a steady state is not attained

unless the maximal rate of the second enzyme (V_{max} for LDH) is greater than the rate of the first enzyme (the rate of PK, expressed as v_1). This is because the concentration of the intermediate $[I_{ss}]$ is negative when $V_{max} < v_1$, an impossible result. This confirms what is intuitively known: If the second enzyme cannot go fast enough to deal with the rate of formation of intermediate, the concentration of intermediate will always continue to increase so that a steady state will not be achieved. This result seems trivial but is important. It shows that when you add substrate (S) to an enzyme system such as that shown in Eq. (3.14), the system will adjust to this change by increasing the concentration of intermediate, which will speed up the second enzyme (according to the relationship predicted by the Michaelis–Menten equation). Once steady state is achieved, both enzyme rates are equal and the overall flux through the system is, therefore, equal to that of the first enzyme. This adjusting can occur as long as the rate of the first enzyme does not exceed the capacity of the second enzyme to catch up. Equation (3.17) applies, of course, only to a two-enzyme system without any back reaction. This specialized system exists in many initial velocity enzyme reactions assayed *in vitro* but does not exist in the cell.

It is possible to extend the system of Eq. (3.14) so that the second enzyme is reversible. In this case, the equation defining the second enzyme is the reversible Michaelis–Menten equation [Eq. (2.25)]. Under steady-state conditions (when [I] does not change) and at a fixed concentration of [P]:

$$[I_{ss}] = \frac{K_{m,S}}{K_{m,P}} \times \frac{[P_{fixed}](V_r + v_1) + v_1 K_{m,P}}{(V_f - v_1)} \tag{3.18}$$

Here V_f and V_r correspond to the forward and reverse maximal activities of the coupling enzyme. Once again, a steady state is only attained when the forward maximal velocity of the second enzyme is greater than the flux through the first enzyme. One may be tempted to think that the concentration of product will affect the system's ability to enter steady state. This may be expected because increasing $[P_{fixed}]$ will increase the rate of the back reaction, and the rate of a reversible reaction is the difference between the forward and reverse rates. However, a closer examination of Eq. (2.25) shows that both S and P compete for the enzyme and that high enough concentrations of either will effectively out-compete the other for the enzyme, dominating the final rate. What really happens is that increasing $[P_{fixed}]$ forces the system to adopt a higher $[I_{ss}]$ level. Equation (3.18) shows that $[I_{ss}]$ will increase until it is high enough to run the second enzyme at the steady-state flux. In these simple systems, the rate of the primary enzyme is constant (and irreversible) so that the steady-state flux through the system is always equal to v_1 regardless of the value of V_f (providing, of course, that $V_f > v_1$).

SIMPLE ENZYME PATHWAYS UNDER *IN VIVO* CONDITIONS

The analysis of the preceding section dealt with how isolated enzymes or simple two-enzyme systems may be expected to behave in the presence and absence of substrate and product. Here this concept will be expanded to include multiple enzymes in a pathway. It is easiest to understand the relationships using a linear sequence of enzymes so the model for this system is the linear set of glycolytic reactions from fructose-1,6-bisphosphate (Fru-1,6-P_2) to pyruvate (Fig. 3.5).

In this model, it is assumed that there are no branch points so that carbon flows uniformly from starting material (Fru-1,6-P_2) to product (pyruvate). This is not true of the glycolytic pathway as it exists *in vivo* because branch points exist at the glycerol-3-phosphate/dihydroxyacetone phosphate step. However, assuming a linear pathway will make the analysis simpler. The enzyme kinetic parameters

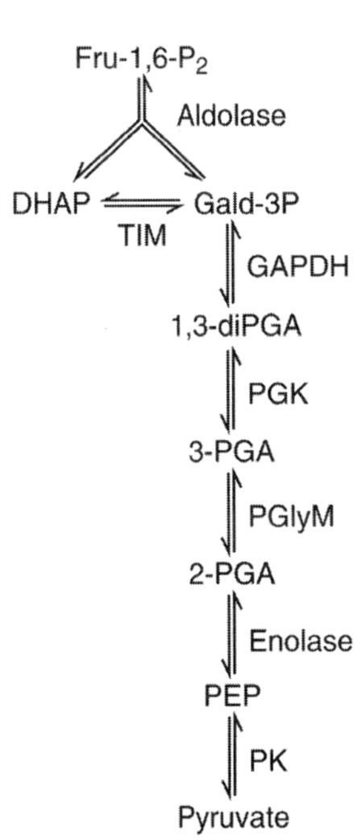

Figure 3.5 The linear sequence of glycolytic enzymes from aldolase to pyruvate kinase (PK). Enzyme abbreviations are triosephosphate isomerase (TIM), glyceraldehyde-3-phosphate dehydrogenase (GAPDH), phosphoglycerate kinase (PGK), phosphoglyceromutase (PGM). Metabolites are dihydroxyacetone phosphate (DHAP), glyceraldehyde-3-phosphate (Gald-3P), 1,3-diphosphoglycerate (1,3-diPGA), 3-phosphoglycerate (3-PGA), 2-phosphoglycerate (2-PGA), and phosphoenolpyruvate (PEP).

TABLE 3.1 Kinetic Parameters for Sequence of Enzymes Shown in Figure 3.5[a]

Species	Actual $[E_T]$ (μM)	Active Site $[E_T]$ (μM)	Substrate	K_m (mM)	$V_{\max}$ (mM/min)
Aldolase	54	217	FBP	0.014	0.17
			DHAP	0.45	0.49
			GAP	0.23	0.49
Triosephosphate isomerase	30	60	DHAP	0.39	14.9
			GAP	1.87	85
Glyceraldehyde-3-phosphate dehydrogenase	93	373	GAP	0.09	2.1
			1,3-diPGA	0.0008	0.011
Phosphoglycerokinase	36	36	1,3-diPGA	0.0022	4.3
			3PGA	1.2	0.46
Phosphoglyceromutase	16	63	3PGA	5	554
			2PGA	2	44
Enolase	72	144	2PGA	0.07	280
			PEP	0.1	63
Pyruvate kinase	12	46	PEP	0.07	1.2
			Pyruvate	10	0.007

[a]For most enzymes, the active site concentration was calculated by multiplying the enzyme concentration in nanomoles per gram muscle tissue by the number of subunits and dividing by 0.75 mL water/gram muscle tissue. Concentrations of enzyme active sites and metabolites are adapted from Srivastava and Bernhard (1986). For phosphoglyceromutase and enolase, the values were taken from Barman (1969).

and equilibrium constants used for the simulation are those measured *in vitro*, and the enzyme concentrations are those measured in skeletal muscle (Table 3.1). In addition, it is assumed that all enzymes are evenly distributed within the cell and the reactions are regulated only by diffusion and kinetic parameters. Thus, supramolecular enzyme organization has not been incorporated into these simulations.

When analyzing systems of enzymes, one usually assumes that the entire pathway is in steady state. This means that substrate and product concentrations do not change appreciably over time. This allows one to neglect the effect of substrate disappearance and product appearance on enzyme activity. The study of non-steady-state pathway kinetics is used to define the response of a system to a perturbation and its kinetic behavior as it progresses toward a new steady state. It is a subject unto itself and will not be considered here.

In order to resolve the system of Figure 3.5, it is treated as a sequential system of enzymes that obey the reversible single-substrate/single-product equation previously developed [Eq. (2.25)]. The effect of enzyme–substrate binding is easily incorporated into the system by setting $[S] = [S_{\text{free}}]$ and $[P] = [P_{\text{free}}]$ in Eq. (2.25). Thus, a system of reactions that obeys Eq. (3.6) will give exactly the same response as that predicted by Eq. (2.25). In addition to these changes, three slight modifications of Eq. (2.25) will be introduced: The forward direction will be denoted by a plus sign, the reverse direction by a minus sign, and substrate and product will be defined as intermediates (I_j). These changes make it easier to keep track of the enzyme reactions in a metabolic sequence such as

$$I_0 \overset{v_0}{\rightleftharpoons} I_1 \overset{v_1}{\rightleftharpoons} I_2 \overset{v_2}{\rightleftharpoons} \cdots \overset{v_{(j-1)}}{\rightleftharpoons} I_j \overset{v_j}{\rightleftharpoons} \cdots \overset{v_{(n-1)}}{\rightleftharpoons} I_n \tag{3.19}$$

With these modifications, the activity of each enzyme is given by

$$v_i = \frac{V_i^+ \bar{K}_i^- [I_{i,\text{free,ss}}] - V_i^- \bar{K}_i^+ [I_{(i+1),\text{free,ss}}]}{\bar{K}_i^+ \bar{K}_i^- + \bar{K}_i^- [I_{i,\text{free,ss}}] + \bar{K}_i^+ [I_{(i+1),\text{free,ss}}]} \tag{3.20}$$

Equation (3.20) is simply a rearrangement of the reversible Michaelis–Menten equation [Eq. (2.25)] with + and − signs replacing the P and S designations. Because it is based on the reversible Michaelis–Menten equation, Eq. (3.20) does not include the effect of enzyme–reactant binding, which can be significant *in vivo*. However, as mentioned above, it can be used to calculate free substrate and product concentrations. Enzyme–intermediate binding can be incorporated into the final analysis if desired, although the resulting equations are unwieldy. Before an examination of the effect of enzyme–intermediate binding on the kinetics of the system can be undertaken, equations using the free (rather than enzyme-bound) intermediate concentrations must be developed. If the system is in steady state, the rate of each enzyme (v_i) is equal to all the others, and this value can be represented by a single variable that represents the steady-state flux through the system, J_{ss}:

$$J_{ss} = v_1 = v_2 = v_3 = \cdots = v_n \tag{3.21}$$

Substituting the value for v_i from Eq. (3.20) into Eq. (3.21) gives a general equation that relates the steady-state concentration of the substrate and product pairs for any enzyme in a linear pathway:

$$[I_{i,\text{free}}] = \alpha_i[\text{I}_{(i+1),\text{free}}] + \beta_i \tag{3.22}$$

$$[\text{I}_{(i+1),\text{free}}] = ([\text{I}_{i,\text{free}}] - \beta_i)/\alpha_i \tag{3.23}$$

with α_i and β_i defined as

$$\alpha_i = \frac{\bar{K}_i^+(V_i^- + J_{ss})}{\bar{K}_i^-(V_i^+ - J_{ss})} \qquad \beta_i = \frac{J_{ss}\bar{K}_i^+}{(V_i^+ - J_{ss})}$$

Note the use of the "free" subscript to remind the reader that Eqs. (3.22) and (3.23) apply to free intermediates only. It is difficult to relate the results of Eqs. (3.22) and (3.23) back to the cell, where only total concentrations can usually be measured. It is possible to derive an equation that relates the total concentration of an intermediate to its free concentration but the equation is complicated. Rather than present this equation, it is more instructive to present a series of three equations that provide a link between the free and total intermediate concentrations. These three equations can be combined to give the "complicated" equation already mentioned or can be numerically solved using a computer algorithm designed for nonlinear matrices:

$$[\text{E}_{i,\text{Total}}] = [\text{E}_{i,\text{free}}] \times \left(1 + \frac{[\text{I}_{i,\text{free}}]}{\bar{K}_i^+} + \frac{[\text{I}_{i,\text{free}}] - \beta_i}{\bar{K}_i^- \alpha_i}\right) \tag{3.24}$$

$$[\text{I}_{i,\text{Total}}] = [\text{I}_{i,\text{free}}] \times \left(1 + \frac{[\text{E}_{i,\text{free}}]}{\bar{K}_i^+} + \frac{[\text{E}_{(i-1),\text{free}}]}{\bar{K}_{(i-1)}^-}\right) \tag{3.25}$$

$$[\text{E}_{(i-1),\text{Total}}] = [\text{E}_{(i-1),\text{free}}] \times \left(\frac{1 + [\text{I}_{i,\text{free}}]\alpha_i + \beta_i}{\bar{K}_{(i-1)}^+} + \frac{[\text{I}_{i,\text{free}}]}{\bar{K}_{(i-1)}^-}\right) \tag{3.26}$$

A close examination of these equations shows that the free intermediate concentration is affected by the concentrations of the two enzymes that it binds (E_i and $\text{E}_{(i-1)}$), by the kinetic properties of these enzymes, and by the flux through the system [expressed as α and β; see Eqs. (3.22) and (3.23)]. If the enzyme concentrations are small or the concentration of the intermediate is large, $[\text{I}_{i,\text{Total}}] \cong [\text{I}_{i,\text{free}}]$ and Eqs. (3.22) and (3.23) can be used to estimate flux and intermediate concentrations without worrying about the effect of enzyme–intermediate binding. When either of the enzyme concentrations becomes large enough to bind a significant amount of the intermediate, Eqs. (3.24) to (3.26) must be used to calculate what is happening in the cell when only the total concentrations can be measured. In all cases, Eqs. (3.22) and (3.23) apply since they deal with free concentrations only. It is only the relationship between these equations and the total cellular concentration that must be calculated with Eqs. (3.24) to (3.26).

Thought Experiment: Regulating Flux *In Vivo*

A close examination of Eqs. (3.22) and (3.23) illustrates some interesting points, but how you interpret these points depends on the system and the metabolic conditions at the time of analysis. It is apparent that two possibilities exist. In the first case, J_{ss} can be essentially fixed by an external reaction, and the pathway would then respond to this new flux by adjusting the intermediate concentrations to accommodate the increased flux. In the second case, the flux is determined by the concentrations of the pathway intermediates (i.e., the concentrations of the intermediates are fixed, not the flux) and no limiting condition exists.

An example of the first case would be a situation where a linear pathway responds to a strongly directional step upstream (there are no irreversible steps in metabolism, even if the reactions strongly favor product formation). The linear part of the glycolytic pathway, in an animal under postprandial conditions when insulin and carbohydrate levels are high, can be thought of in this way. The classical metabolic interpretation is that, under these conditions, glucose is being pumped into the cell and the glycolytic pathway must respond to the increased flux. Systems such as these are similar to the simple two-enzyme system with the second enzyme reversible, as was discussed in the previous section. When J_{ss} is essentially fixed externally, Eqs. (3.22) and (3.23) show that a steady state cannot be achieved if the V_{max} value for any enzyme in the sequence is less than the forward flux through the system (the quantity V_i^+ must be greater than J_{ss} for all enzymes). The same would also apply to this system when flux goes in the reverse direction (i.e., $J_{ss} < 0$). For reverse flux, the maximal velocity in the reverse direction (V_i^-) must be greater than the rate of the reverse flux. If these conditions are not met, the calculated intermediate concentrations become negative—a physical impossibility that means the system is no longer in steady state. The physical consequence of a negative $[\text{I}_{ss}]$ is the division of the metabolic pathway into two parts—the part before the point at which $J_{ss} > V_i^+$ and the part after this point. In the scheme of Figure 3.5, with an externally fixed J_{ss}, only the "after" part of the pathway would come to a new steady state dictated by the V_{max} of the limiting enzyme. The intermediates before the limiting enzyme would continue to rise.

A condition when $J_{ss} > V_i^+$ can be brought about through two different processes. Either the initial reaction can increase to very high values or the activity of an enzyme in the pathway can be reduced by one of many different factors. In either case, a relative reduction in the activity of an enzyme in the middle of a linear pathway would back up the substrate of the limiting enzyme, and this would constantly increase since there is nowhere else for the flux to go. The limiting enzyme would essentially become unidirectional because its substrate would keep increasing. This would fix the flux into the lower part of the pathway—it would essentially act as a unidirectional enzyme. This type of scenario does not occur *in vivo* because cellular pathways are usually highly branched, and their rates are influenced by several different cellular metabolites and processes. Decreasing the enzyme activity of one pathway would simply divert intermediate into other pathways and a new steady state will be achieved based on a function of pathway kinetic constants.

The above discussion is based on a condition that fixes the flux for a pathway (or part of a pathway). The other possibility is a system where the concentrations of the metabolites are influenced by many different reactions. Under these conditions, the flux is determined by the concentrations of intermediates in the pathway (i.e., intermediate concentrations are fixed, not the flux) and no limiting condition exists. In this situation, reducing the activity of an enzyme in the middle of the pathway will still allow the achievement of a steady state by increasing the concentrations of intermediates before the enzyme (to increase product concentrations and reduce net flux) and reducing the concentrations of intermediates after the limiting enzyme to reduce the flux through the after part of the pathway.

One may well wonder how often either of these situations is encountered in the cell. While it is possible to imagine that an irreversible step occurs at the start of a metabolic pathway, there are no essentially irreversible steps in metabolic pathways. Even when the equilibrium strongly favors formation of a product, there is always some back reaction. In addition, most strongly directional enzymes have a counterpart that catalyzes the reversible reaction, or are part of a cycle that can readily reproduce the starting material. For example, phosphofructo-1-kinase and fructose-1,6-bisphosphatase [Equation (3.27)], which will be discussed in later chapters of this book, coexist in the liver and catalyze two sides of the fructose-6-phosphate/fructose-1,6-bisphosphate locus in glycolysis.

In addition, there are enzyme reactions whose flux is fixed externally. Examples of these enzymes are most likely found, as pointed out by Cornish-Bowden (1995), in the middle part of pathways and only act to process intermediates as quickly or as slowly as they receive them. This type of situation may also occur in the glycolytic pathway of postprandial liver and skeletal muscle when these cells respond to increased plasma glucose and insulin concentrations although this situation is only temporary.

While a system that responds to an external (fixed) flux represents one extreme view of metabolism, the other extreme is the situation where cellular metabolite concentrations are fixed by virtue of the many different reactions in which they participate (e.g., ATP and NAD). The majority of reaction pathways lie somewhere between these two extremes so that transient cellular conditions promote strong directional flux to increase metabolite concentrations that, in turn, influence pathway flux. However, it is always important that some enzymes be regulated to reduce the extent of substrate cycling. For example, some regulation of phosphofructo-1-kinase (PFK) and fructose-1,6-bisphosphatase (FBPase) is required so that the fructose-6-phosphate (F6P)/fructose-1,6-bisphosphate ($F1,6P_2$) [Eq. (3.27)] does not continually turn over and exhaust all the cellular ATP (PFK is an ATP-dependent enzyme). This control may not lie at the PFK locus, since limiting substrate and product availability will also act to regulate flux through this step. It is interesting to contemplate whether regulation of the PFK/FBPase enzyme pair allows regulation of glycolytic/gluconeogenic activity or whether regulation of the glycolytic/gluconeogenic activity allows regulation of the PFK/FBPase enzyme pair. Metabolic control will be discussed in greater detail later.

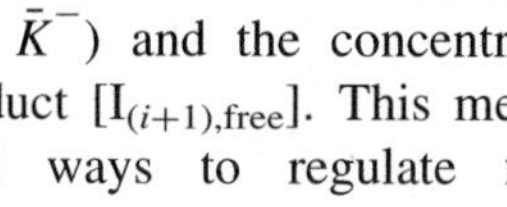

(3.27)

Equations (3.22) and (3.23) show that the flux through any enzyme along the sequence (which is, in fact, the flux through the whole pathway at steady state) is defined by many variables—the kinetic constants of the enzyme (V^+, V^-, $\bar{K}^+$, $\bar{K}^-$) and the concentration of substrate $[I_{i,free}]$ and product $[I_{(i+1),free}]$. This means that there are many potential ways to regulate flux through the pathway. For example, the kinetic binding constants can be altered by allosteric activators or by inhibitors, the total enzyme activity (V_{max} values) can be affected through changes in enzyme synthesis/destruction, covalent modification of the enzyme, or by allosteric effectors, the concentrations of product can be changed by changing the activity of other pathways that use an intermediate (although that does not apply to the linear pathway of

Fig. 3.5), or the flux itself can be affected by events outside the pathway such as allosteric inhibitors/activators from another metabolic pathway.

Mathematical Modeling of the Linear System

Equations (3.22) and (3.23) illustrate the interdependence of all the reactions in a sequence. Changing the kinetic constants of one reaction causes an adjustment in all the other reactions. This means that the system is flexible and can support many different rates of flux, concentrations of intermediates, and kinetic constants. These principles can be illustrated using the data of Table 3.1 to construct a mathematical model of the scheme of Figure 3.5. Figures 3.6 and 3.7 present the results of these simulations for different conditions. A cursory examination of Eqs. (3.22) and (3.23) shows that one can completely resolve the system if the kinetic parameters of the enzymes are constant (and are known) and one knows either (i) the concentration of any two intermediates or (ii) the flux through the pathway and the concentration of any one intermediate. Figures 3.6 and 3.7 demonstrate these relationships. In Figure 3.6, the steady-state flux through the system (J_{ss}) was held constant at 0.1 μmol $\cdot$ min^{-1} $\cdot$ mL^{-1} and the concentration of pyruvate (the terminal intermediate in the pathway) was varied. The relative concentrations of the intermediates was plotted as a function of varying free pyruvate concentrations to observe the effect of changing one intermediate on the concentrations of the other intermediates. Increasing the concentration of the reaction sequence product (pyruvate) acts to competitively inhibit the terminal reaction. However, as we saw with the two enzyme-coupled system, this can be overcome by increasing the substrate of the terminal reaction. This simulation thus shows that the concentration of pathway intermediates can change in response to an external stress even though the flux through the pathway does not change. In this particular example, the change is not large relative to the change in pyruvate concentration. A maximum relative increase of 1.2 times for intermediates closest to pyruvate was observed and the other intermediates were almost unaffected. This shows that some metabolic perturbations may have only local effects. The intermediate concentrations change only slightly because the flux does not change (in the absence of changes in V_{max}, enzyme activities are controlled by the concentration of their substrate/product pairs). Nevertheless, Figure 3.6 shows that there are many possible combinations of intermediate concentrations that can support the same pathway flux.

In Figure 3.7, the concentration of pyruvate was fixed, and J_{ss} was varied to observe the relationship between the concentrations of intermediates and the steady-state flux through the pathway. The graph shows increasing concentrations of intermediates as J_{ss} increases. Higher concentrations of intermediates are expected because higher enzyme substrate concentrations are needed to increase the enzyme rates [see Eq. (2.25)]. The lines of Figure 3.7 are parallel, showing that both substrate and product increase in parallel to maintain flux through the system. As the flux through the system of Figure 3.7 increases, deviations from the log-linear relationships appear because large increases in the concentrations of enzyme substrates are required to increase enzyme velocity by a

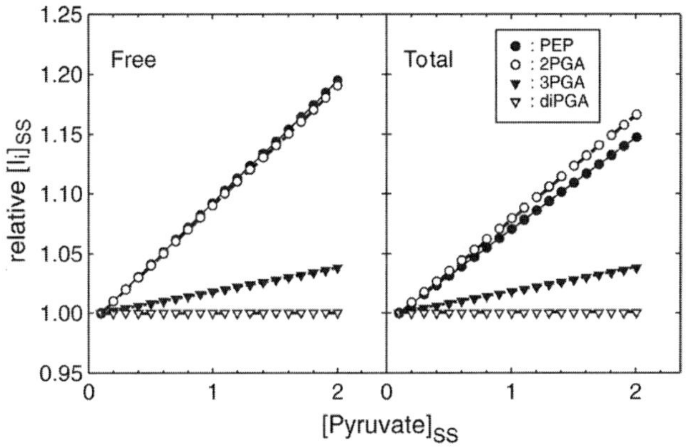

Figure 3.6 Mathematical modeling of the linear set of reactions depicted in Figure 3.5. The graph shows the relative changes in the concentrations of pathway intermediates relative to [pyruvate$_{free}$] = 0.1 mM for a fixed flux and variable free pyruvate concentrations (the terminal intermediate in the sequence). The simulations were run using *in vitro* measured kinetic constants and the total enzyme concentrations (see Table 2.2).

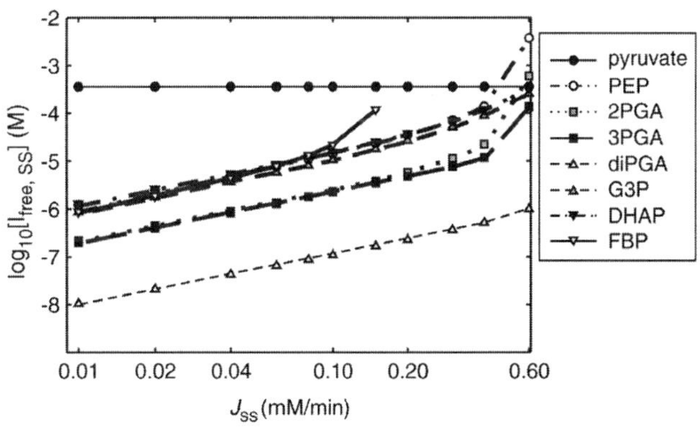

Figure 3.7 Mathematical modeling of the linear set of reactions depicted in Figure 3.5. The graph shows the free concentration of pathway intermediates (plotted on a log scale) as a function of changing J_{ss} values (plotted on a logarithmic scale). The curves were generated by fixing the total free concentration of pyruvate ([pyruvate$_{free}$] = fixed). The simulations were run using *in vitro* measured kinetic constants and the total enzyme concentrations (see Table 2.2).

small amount when enzyme–substrate binding is close to saturation. This means that the system is close to its maximal steady-state rate.

BRANCH POINTS AND IRREVERSIBLE ENZYMES IN PATHWAYS

The previous section dealt with linear pathways but not many metabolic pathways are strictly linear, even though sections of pathways may be linear themselves. Adding branch points to the pathways adds another level of complexity, but the equations can still be resolved algebraically. As before, any intermediate along a linear part of the pathway will obey Eqs. (3.22) and (3.23). The linear sequence of Eq. (3.19) is symmetrical, allowing both positive and negative flux so that the system can be resolved with either positive or negative flux. With a branched reaction sequence, such as that described by Eq. (3.28):

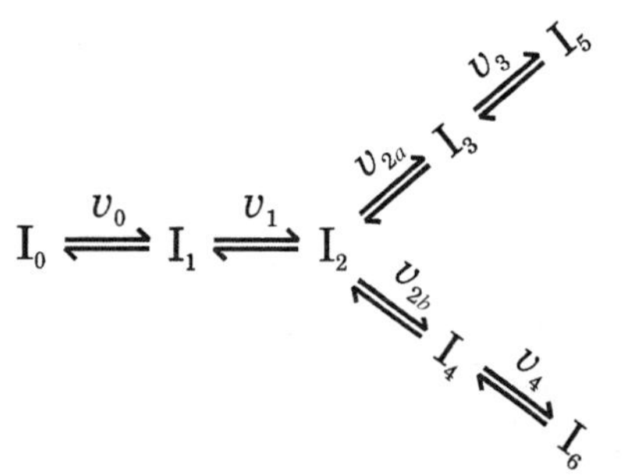

(3.28)

the total flux before and after the branch point must add up. Assuming that the reaction starts with the addition of I_0, one can write

$$J_{ss} = J_{ss,\text{before}} = J_{ss,a} + J_{ss,b} \tag{3.29}$$

with the subscripts a and b referring to different arms of the branch point. As with the linear system, flux can be positive or negative with the convention decided before the equations are written. In the final analysis, it makes no difference if the flux is positive or negative. The intermediate concentrations are calculated using Eqs. (3.22) and (3.23). In order to completely resolve the system, a combination of three fluxes or intermediate concentrations must be known. The equations are much easier to resolve when at least two fluxes and one intermediate concentration are known, but it is possible to use numerical methods to resolve the system of equations when other parameters are known.

The previous section examined the situation when one of the enzymes was limiting in a linear system. In a branched pathway, reducing the activity of an enzyme along one of the branches would simply increase the flux into the other branch of the pathway as long as it can be supported. This simple system illustrates the advantages of regulating enzymes at the termini of branched pathways to regulate the flow of carbon. This is a classical metabolic view of a pathway and of regulation of a pathway and has fallen out of favor with more recent theories of metabolic control. Nevertheless, it is a mathematically sound idea that may find application in cellular metabolism. Later in this book, the reader will be introduced to animals that can survive anoxia by successfully rearranging their metabolism to respond to the lack of oxygen. A perfect example of the control of branched metabolic pathways occurs when oxygen is limiting and the tricarboxylic acid cycle stops because of the lack of oxidizing equivalents. Under these conditions, the carbon that would normally flow into the cycle is diverted away from the pyruvate dehydrogenase reaction into formation of lactate or other anoxic end products. This shows what happens when one branch of a metabolic pathway is shut off—intermediate concentrations increase and flow down their energy gradient into another branch.

SIMPLE ENZYME PATHWAYS AND CROSSOVER PLOTS

Figures 3.6 and 3.7 show how an enzyme system may respond to changes in intermediate concentrations and to changes in J_{ss}. They also show how difficult it is to tell what is going on when only intermediate concentrations are monitored. This was highlighted by the data of Figure 3.6, which showed that flux was independent of the absolute value of the concentrations of the intermediates although the relative proportion of the intermediates was important.

In classical treatments of cellular data, the relative flux through various enzyme loci is monitored by plotting the ratio of intermediate concentrations under two different metabolic conditions. The resulting graph, termed a crossover plot, is used to identify potential control loci. The control points are identified by relative changes in the concentrations of enzyme substrates and products. Positive crossovers are observed when substrate decreases and product increases, negative crossovers are seen when substrate increases and product decreases. The crossover theorem was originally developed by Britton Chance and his colleagues who used it to identify control points in the mitochondrial electron transport chain. By following relative changes in the ratio of oxidized and reduced cytochromes and electron transport chain intermediates, they could tell what steps were affected by their experimental treatments. It has been proven that the theorem is valid when the total amounts of substrate and product are conserved at an enzyme locus. This is true of the cytochromes and intermediates of the electron transport system. However, when substrate flows into a system and product

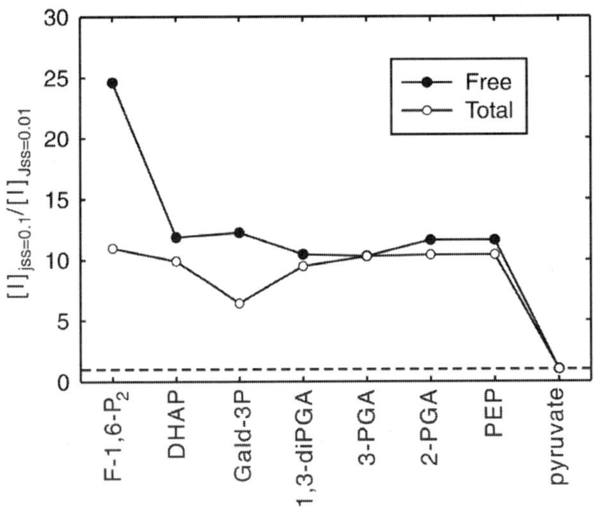

Figure 3.8 Crossover plot for identifying putative control points in a metabolic sequence. The ratio of the intermediates was plotted for two different fluxes (data from Fig. 3.7). The graph shows the relative concentration of each intermediate. Both free and total intermediate ratios are presented. The concentration of pyruvate was held constant in this simulation so that its ratio is 1. See Figure 3.5 for enzyme abbreviations and substrate/product pairs that correspond to each enzyme.

flows out of a system, as occurs for the glycolytic pathway, the total amounts of substrate and product are not conserved. In fact, as shown by Eqs. (3.22) and (3.23), the concentrations of all the intermediates along the pathway are functions of the other properties of the enzymes of the pathway.

A concrete example of the problems with crossover plots follows from an examination of the theoretically manipulated metabolic pathway of Figure 3.5 under the changing conditions of Figure 3.7. Plotting the results of this simulation under two conditions—$J_{ss} = 0.01\,\text{mM} \cdot \text{min}$ and $J_{ss} = 0.1\,\text{mM} \cdot \text{min}^{-1} \cdot /\text{mL}^{-1}$ (Fig. 3.8) showed some interesting patterns. First, the graph shows that the concentrations of the pathway intermediates all increase when flux through the system increases (except pyruvate, which was held constant). This was expected because increased substrate concentrations are required to drive more flux through the system. Figure 3.8 also shows that the increases are not uniform and that the free concentration of one intermediate (F-1,6-P_2) is 25-fold higher at high flux than at low, whereas most others are 10 to 12-fold higher at high flux versus low flux. This differential increase in intermediate concentrations is a function of the properties of the enzymes.

Looking at the relative change in the concentrations of the free intermediates between the low and high flux conditions, one would predict an enzyme activity inhibition at both the adolase and PK enzyme loci (although this is not, strictly, a negative crossover). This is because a large decrease occurs between the substrate ratio and the product at these loci; for example, the high/low flux ratio for F-1,6-P_2 (aldolase substrate) is ~25, whereas for DHAP and Gald-3P (aldolase products) it is ~12. This is a false conclusion because we know that the kinetic parameters of these enzymes were not changed during the simulation!

The corresponding crossover plot for total metabolites is also shown in Figure 3.8. This graph is included because it is often only possible to measure total metabolite concentrations even though free metabolite concentrations are the most important for determining pathway flux. There are significant differences between the free and total crossover plots. This occurs because some of the enzymes are present in high enough concentrations to bind a significant proportion of some of the intermediates. Figure 3.8 still suggests an inhibition at PK, but now there is no apparent inhibition of ALD. This shows another limitation of the crossover plot when applied with total metabolite concentrations. Analyses such as these show the importance of combining data from a number of sources before drawing conclusions about the relative activity at enzyme loci in a metabolic sequence and before making conclusions about an enzyme's importance in regulating a pathway. It also demonstrates the danger of using crossover plots on systems that do not have fixed total amounts of substrates.

BASICS OF METABOLIC CONTROL ANALYSIS

This section will deal with metabolic control analysis from a very simple perspective. Other, more comprehensive books should be consulted for a more thorough and mathematical rigorous treatment of the theory of metabolic control analysis. This section will focus on the fundamentals of the system and attempt to point out where it is at odds with more traditional views of metabolic regulation.

The regulation of flux through a metabolic pathway has been studied for many years. The traditional viewpoint seeks to identify rate-determining steps, which are believed to be responsible for controlling flux. Regulatory enzymes are those that can be inhibited or activated through many processes that are thought to link activity to the needs of the cell. It is relatively easy to imagine what happens to pathway flux when an enzyme is completely inhibited but the relationship between pathway flux and less dramatic changes in individual enzyme activity is not intuitive. We tend to think of an enzyme pathway as a series of containers with spigots arranged along a rising stairway. If we dump water into the top container, it will fill and drain at a rate dictated by its height and by the size of its spigot. The next container will fill and empty at a rate that depends on the rate of water flowing into it and the size of its spigot, and so forth. However, this model suffers from two critical problems. First, enzymes catalyze reactions in both directions so that there must be a way for water to move up the stairway in a manner that reflects the amount of water in the previous reaction. Second, cellular pathways are never isolated so that they are not properly represented by a single staircase. There are many compounds that are produced by one pathway that impact on the performance of other pathways.

Up to this point, we have dealt largely with simple linear systems of enzymes that are not regulated by any external or internal factors. The only things they have in common are substrates and products. However, enzymes *in vivo* can be regulated by many cellular compounds, including their own reactants or the reactant of another enzyme of the same pathway. It is the combination of reaction specificity and regulation of activity that helps to maintain cellular homeostasis under widely different metabolic conditions and permits cells to perform various tasks. In animal liver, for example, cellular and blood glucose concentrations are regulated by controlling the enzymes responsible for storing glucose (in the form of glycogen) or releasing glucose (breakdown of glycogen) depending on whether blood glucose levels are high (during eating) or low (in between meals). The liver and muscle tissue regulate these processes through control of the enzymes that act specifically on glycogen. How the changes in enzyme activity affect the flux through the various pathways is the subject of metabolic control analysis.

Classically, enzymologists assume that the important enzymes of the pathway are those regulating the entry of substrate into the pathway and those producing the final product. In order to determine how a pathway could be regulated, the first enzyme would be tested for sensitivity to inhibition by its own product, by any intermediate along the metabolic pathway and by the final product. The terminal enzyme in the pathway would be tested for activation or inhibition by any pathway intermediate as well as for inhibition by the final product. The enzymes would be investigated for possible covalent modification (such as protein phosphorylation) that can alter activity and make them sensitive to external signals (such as those derived from as hormone action). Finally, the substrates, intermediates, and products of competing metabolic pathways would also be tested for possible effects on the enzymes. As you can see, this requires the completion of an enormous number of experiments before the pathway can be adequately defined. However, once all this information had been gathered, the experimenter still had no precise method for determining how changes in enzyme rates observed *in vitro* would translate into control of cellular metabolic pathways *in vivo*.

Theoretical mathematical frameworks have been developed to try to answer, in a quantitative fashion, the relationship between pathway flux and enzyme activity. Three such frameworks have been developed, each with their own merits, but, recently, all three frameworks have been unified to present a single theory of metabolic regulation. The fundamental concept of metabolic control analysis (MCA) is that control of a metabolic pathway is shared among all the enzymes of the pathway. As David Fell (1997) pointed out, the theory essentially seeks to answer the question: How does metabolic flux vary as substrate or enzyme activity is changed? MCA was developed to be independent of the kinetic particulars of the enzymes in question. This allows almost universal application of the equations to any metabolic pathway. MCA theory defines three basic relationships. The two most important are the influence of enzyme concentration on pathway flux and the sensitivity of an enzyme's activity to changes in the concentrations of intermediates. Mathematically, these relationships are defined as the fractions $\delta J_{ss}/\delta[E_T]$ and $\delta v_i/\delta[I_i]$, respectively. These fractions represent the change in pathway flux as a function of changing enzyme concentration and the change in enzyme velocity as a function of changing intermediate concentrations, respectively. In metabolic control analysis, these fractions are normalized to represent the relative change in flux for a relative change in total enzyme concentration or the relative change in enzyme activity for a relative change in substrate or product concentration. The resulting variables are called the flux control coefficient (abbreviated as C) and the elasticity (abbreviated as ε). They are defined as

$$C_i^J = \frac{\delta J_{ss}}{\delta[E_{i,T}]} \times \frac{[E_{i,T}]}{J_{ss}} = \frac{\delta \ln J_{ss}}{\delta \ln[E_{i,T}]} \tag{3.30}$$

$$\varepsilon_S^v = \frac{\delta v_i}{\delta[I_i]} \times \frac{[I_i]}{v_i} = \frac{\delta \ln v_i}{\delta \ln[I_i]} \tag{3.31}$$

Flux control coefficients and elasticities are defined as normalized partial derivatives. Therefore, they represent the change in flux or enzyme activity while holding all other variables constant. The partial derivative of the flux with respect to total enzyme is the change in flux brought about by changing only the total concentration of one enzyme while keeping all the other parameters of the pathway constant. The flux control coefficient is, in reality, the slope of the flux versus enzyme concentration at the initial enzyme concentration. It should be remembered that these equations were developed to describe the effect of small changes in enzyme concentration on the flux through the system or changes in intermediate concentration on enzyme activity. As such, they do not generally apply to situations where large changes in these parameters occur. It should also be remembered that Eqs. (3.31) and (3.32) apply only to the original system under investigation. Changing any parameter in the system produces a new system. For example, one can determine the flux control coefficient for a particular enzyme in a pathway by genetically altering the expression of its gene to either increase or decrease the amount of enzyme protein in the cell. This will give the flux control coefficient for the enzyme under the original conditions only. If one wants to determine the flux control coefficient for the enzyme under the changed conditions, one must redesign the experiment (or reverse the flux control equations so that the changed condition represents the initial condition and vice versa). Thus, researchers usually express the activity of a pathway as a function of the change in the native amount of enzyme.

Application of MCA leads to the conclusion that no single enzyme controls the flux through a metabolic pathway. In fact, all enzymes of the pathway have the ability to control the pathway to some degree—flux control is shared. The flux control coefficients indicate the degree to which pathway flux may be regulated by each enzyme's concentration, and the elasticities indicate how an enzyme's activity may be affected by a change in the concentration of a particular intermediate.

Unless a reaction is at equilibrium, the flux control coefficients are never zero, although they may be small. Metabolic control analysis makes no assumptions about the relationship between enzyme control and pathway flux but does show that they are related in interesting ways. For example, reducing the rate of one enzyme will increase its flux control coefficient but will reduce the flux control coefficients of the other enzymes in the pathway since the sum of the flux control coefficients for the components of any single pathway is unity:

$$\sum_{i-1}^{n} C_i^J = 1.0 \tag{3.32}$$

Thus, the magnitude of the changes in the flux control of the other enzymes will depend on the flux through the system and on the activities of the enzymes of the system. The flux control coefficients and the elasticities are defined so that higher values indicate a higher degree of control. Consider a three-enzyme pathway that has flux control coefficients of $E_1 = 0.75$, $E_2 = 0.09$, $E_3 = 0.16$. In this system, small changes in the activity of enz_1 produce small relative changes in pathway flux that are $0.75 \times$ the change in E_1 activity (higher or lower, depending on whether enz_1 was increased or decreased). The relative changes in pathway flux that would result from changes in E_2 and E_3 activity are only 0.09 and 0.16 times the change in enzyme concentration, respectively. This shows that the majority of control of the pathway rests with enz_1, but that the control of flux is shared among all three enzymes.

Metabolic control analysis also provides a link between the flux control coefficients and the kinetic properties of the enzymes—the connectivity theorem. It is expressed mathematically as

$$\sum_{i=1}^{n} C_i^J \varepsilon_S^i = 0 \tag{3.33}$$

Equation (3.33) relates the flux control coefficients and elasticities of all the enzymes that share a common intermediate. For example, for the intermediate PEP of the pathway of Figure 3.5, one can write

$$0 = C_{\text{Enolase}}^J \times \varepsilon_{\text{PEP}}^{\text{Enolase}} + C_{\text{PK}}^J \times \varepsilon_{\text{PEP}}^{\text{PK}}$$

This theorem, therefore, provides the link between the enzyme's sensitivity to its substrate, which is a function of its kinetic mechanism, and the overall pathway flux. It is, perhaps, the most important relationship because it allows the calculation of flux control coefficients from elasticities (which can be calculated from rate equations or obtained through experimentation). It should be noted that Eq. (3.33) applies only to linear pathways. Special equations are needed for branched systems. Equation (3.33) and its corollaries are important because they allow the calculation of flux control coefficients from enzyme properties (remember to use free intermediate concentrations).

In addition to flux control coefficients, there are also concentration control coefficients. These are defined in analogous fashion to the flux control coefficients. They are the relative change in the concentration of a pathway intermediate as a function of the change in the concentration of an enzyme:

$$C_i^{\text{I}_i} = \frac{\delta \text{I}_i}{\delta[\text{E}_{i,T}]} \times \frac{[\text{E}_{i,T}]}{\text{I}_i} = \frac{\delta \ln \text{I}_i}{\delta \ln[\text{E}_{i,T}]} \tag{3.34}$$

There is a corresponding connectivity theorem for this parameter. However, the connectivity relationship for concentration control coefficients is more complex than the two already presented and has two different values, depending on what is being changed. When the concentration control coefficient and the elasticity refer to the same intermediate (say $i = 1$), then

$$\sum_{i=0}^{n} C_i^{\text{I}_1} \varepsilon_{\text{I}_1}^i = -1 \tag{3.35}$$

and when the intermediates are not the same, we have another relationship:

$$\sum_{i=0}^{n} C_i^{\text{I}_1} \varepsilon_{\text{I}_2}^i = 0 \tag{3.36}$$

Relationships between the flux control coefficients, the concentration control coefficients, and elasticities have also been derived. These relationships are useful in completely defining systems of equations:

$$C_i^{v_i} = 1 + \sum_{j=0}^{m} C_i^{I_j} \varepsilon_{I_j}^{i} \tag{3.37}$$

$$C_i^{v_k} = \sum_{j=0}^{m} C_i^{I_j} \varepsilon_{I_j}^{k} \tag{3.38}$$

Note that, when the pathway is at steady state, the flux though all enzymes is given by J_{ss} (all fluxes are equal).

The use of these equations to obtain flux control coefficients is best illustrated by following an example. The typical example is that of a linear three-enzyme system (Fell and Sauro, 1985):

$$I_0 \overset{v_0}{\rightleftharpoons} I_1 \overset{v_1}{\rightleftharpoons} I_2 \overset{v_2}{\rightleftharpoons} I_3 \tag{3.39}$$

The summation theorem [Eq. (3.32)] is written as

$$C_0^J + C_1^J + C_2^J = 1.0 \tag{3.40}$$

The connectivity theorem [Eq. (3.33)] provides the other two equations necessary to solve the system. One equation is written for each shared intermediate:

$$C_0^J \varepsilon_{I_1}^0 + C_1^J \varepsilon_{I_1}^1 = 0 \tag{3.41}$$

$$C_1^J \varepsilon_{I_2}^1 + C_2^J \varepsilon_{I_2}^2 = 0 \tag{3.42}$$

Equations (3.40) and (3.42) can now be solved for the unknown flux control coefficients. This can be done using algebra (three equations and three unknowns) or by writing the equations in matrix form and solving the system using matrix algebra or using numerical methods with a computer program. In matrix form, Eqs. (3.40) to (3.42) are written as

$$\begin{bmatrix} 1 & 1 & 1 \\ \varepsilon_{I_1}^0 & \varepsilon_{I_1}^1 & 0 \\ 0 & \varepsilon_{I_2}^1 & \varepsilon_{I_2}^2 \end{bmatrix} \begin{bmatrix} C_0^J \\ C_1^J \\ C_2^J \end{bmatrix} = \begin{bmatrix} 1 \\ 0 \\ 0 \end{bmatrix} \tag{3.43}$$

The matrix is built up from the three equations. There will always be as many rows and columns as enzymes in the left array. The first row will always be a series of 1s. The second, third, fourth (etc.) rows will be a series of elasticities corresponding to the first, second, third (etc.) intermediates. The second array (really a vector) is a column of the flux control coefficients. The right-hand side is a vector of the sums (the summation and connectivity theorems). Matrices are multiplied by vectors in a straightforward way. The top left element of the matrix is multiplied by the top element of the flux control vector, and this is added to the top middle element multiplied by the middle element of the flux control vector, and so on. In this way, Eqs. (3.40) and (3.42) can be reproduced to give the right-hand side vector (1 for the summation theorem and 0 for each connectivity theorem). Using matrix algebra, Eq. (3.43) can be written as

$$\mathbf{M} \times \mathbf{C} = \mathbf{A} \tag{3.44}$$

where **M** is the leftmost 3×3 matrix, **C** is the vector of control coefficients, and **A** is the right-side vector. This can be rearranged to give $\mathbf{C} = \mathbf{A}/\mathbf{M}$ and solved using either matrix algebra or a numerical algorithm to solve the division. The advantage of Eq. (3.43) is that flux control coefficients can be obtained from a knowledge of changes in flux for different metabolic conditions, providing the free concentrations of intermediates are known. Many experiments have been performed in this fashion because it is relatively easy to change overall pathway flux and then measure the concentration of metabolites. A plot of the log of the flux versus the log of the metabolite concentrations will give the elasticities for the individual enzymes. These are then used to populate the **M** matrix as detailed in Eq. (3.43). The values of **C** are then obtained by inverting Eq. (3.44).

Metabolic control analysis does not define the number of enzymes that constitute a pathway or their relationship. Thus, a pathway can be as few as two enzymes or as many as the whole cell. This is allowed because of the relative nature of the flux control coefficients and the elasticities. Flux control coefficients are relative terms and will vary depending on the length of the pathway and the properties of the enzymes within that pathway. It is up to the experimenter to define the metabolic pathway appropriately so that a meaningful analysis is obtained. Elasticities apply only to the enzymes themselves, that is, they are properties of the individual enzymes. Thus, their values are not dependent on the properties or number of enzymes in a pathway.

There is an interesting version of MCA that has been developed to analyze mitochondrial energetics called the "top-down" approach to MCA. This method looks at control coefficients and elasticities of segments of metabolism instead of isolated enzymes. In order to apply this version of MCA, the segments have to be isolated, and defining control parameters need to be identified for each segment. The control equations are then applied in the same fashion, but one obtains coefficients for linear blocks of enzymes (rather than for individual enzymes). This idea has found many applications because it allows one to group several enzymes into a block and analyze for the important metabolites that affect the activity of the block. Recently, this approach has been used to examine the contribution of ATP turnover, glycolysis, gluconeogenesis, oxidative phosphorylation, pyruvate oxidation, NADH oxidation, proton leak, and lactate production to hepatocyte energy balance and function (Ainscow and Brand, 1999a, 199b). The trick in the analysis is to identify intermediates that link the processes. In the analysis, these were identified as glucose-6-phosphate, the cytoplasmic $NADH/NAD^+$ ratio, ATP, the mitochondrial membrane potential, and pyruvate (Fig. 3.9). Interesting relationships came out of this analysis. In particular, the authors found that, in total, the ATP-consuming processes had only 35% control over the rate of ATP consumption. The processes that produce ATP, oxidative phosphorylation and glycolysis, had more control over ATP consumption: Glycolysis exerted 19% of the control and oxidative phosphorylation exerted 30% of the control. The analysis also allowed an estimation of the Pasteur and Crabtree effects. The Pasteur effect is the stimulation of glycolysis brought about by a large decrease in the rate of formation of ATP, classically brought about by making a system anoxic. The associated rapid increase in cellular concentrations of free P and adenosine 5′-monophosphate (AMP) is thought to stimulate glycolytic rates by turning on phosphofructokinase. The Crabtree effect is the slowing of glycolysis brought about through a direct competition of glyceraldehyde-3-phosphate dehydrogenase with mitochondrial oxidative phosphorylation for free P. Mitochondrial ATP production had a negative control over glycolysis (−26%; the Pasteur effect) and glycolysis had a small degree of control (−10%; the Crabtree effect) over mitochondrial ATP production.

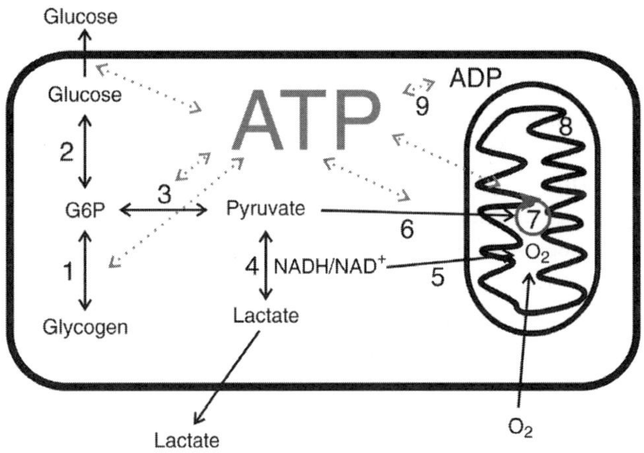

Figure 3.9 Top-down analysis: control of ATP concentration. The scheme represents the first stage of a top-down analysis where the flux control coefficients for processes that may affect cellular ATP concentration are determined. This diagram is a representation of a hepatocyte with blocks of metabolism represented by black arrows. The processes that can affect ATP concentration are defined as: (1) glycogen breakdown, (2) glucose excretion, (3) glycolysis, (4) pyruvate fermentation, (5) NADH oxidation, (6) pyruvate oxidation, (7) mitochondrial oxidative phosphorylation, (8) the mitochondrial proton leak, and (9) process that utilize ATP. The scheme is based on an actual analysis performed by Ainscow and Brand (1999a). The first step is to identify intermediates (between blocks) and end products that can be measured. These are used to measure flux and calculate elasticities, which are used to calculate flux control coefficients according to equations presented in the text.

MCA Analysis and Metabolic Pathways

As pointed out by Cornish-Bowden (1995), neither velocity nor intermediate concentrations can be manipulated independently of the other in the cell. This means that the two are intimately interconnected [as shown by Eqs. (3.22) and (3.23)], and it is impossible to change one without influencing the other. However, the degree of influence that each has over the other is a function of the metabolic pathway. For example, if an enzyme operates near maximal velocity, small increases in flux through the system may mean that large increases in the concentration of the enzyme's substrate are required in order to increase enzyme velocity enough to maintain steady state. This is seen in Figure 3.7: As flux increases, the concentrations of some metabolites begin to increase rapidly, signaling that the flux through the system is close to its maximum. On the other hand, Figure 3.6 demonstrates that changes may also be small and localized. In Figure 3.6, the concentration of pyruvate was manipulated while fixing the flux through the system. Small changes in PEP were observed but the concentration of fructose-1,6-bisphosphate was unaffected. This shows that parts of a metabolic pathway may change only very little even when relatively large changes occur in another part of a pathway if flux is relatively unaffected. These discussions are important to keep in mind while reading the next section, where control of cellular metabolism is discussed. They demonstrate that the control of a system will vary depending on the state of the system. While this conclusion should be intuitive, it does not easily fall out of MCA analysis because of the nature of the definition of flux control coefficients.

CONTROLLING PATHWAY FLUX UNDER *IN VIVO* CONDITIONS: DIRECT CALCULATIONS OF CONTROL COEFFICIENTS

The reaction pathway of Figure 3.5 will be used to examine the application of MCA to cellular metabolism but, for simplicity, only the pathway from GAPDH (Fig. 3.5) to PK will be considered. Using this part of the pathway ensures that there are no branched reactions at ALD to deal with.

In reality any linear system of enzymes will suffice, and we could make up kinetic constants for the enzyme parameters and observe the same relative trends. Figure 3.5 just happens to be a convenient system to illustrate the application of MCA. As shown earlier [Eqs. (3.22) and (3.23)], it is possible to calculate the flux through a system when the parameters of the individual enzymes as well as the concentrations of any two intermediates in the pathway are known. The truncated Figure 3.5 pathway, therefore, can be used to illustrate the effects of varying enzyme concentration (and, by extension, total enzyme activity) on the steady-state flux through the system. It also allows the luxury of fixing certain metabolites to determine their effect on pathway flux.

Metabolic control analysis is primarily concerned with determining the factors that have the greatest influence on a pathway's flux. For individual enzymes in a pathway, this amounts to determining the flux control coefficients and elasticities. Flux control coefficients show which enzymes regulate flux for a system under a given set of conditions. The higher the flux control coefficient, the more sensitive the pathway flux is to changes in this enzyme. The flux control coefficients can be determined in two different ways. The first involves changing the enzyme concentration and looking at the changes in pathway flux. The second involves changing the pathway flux and measuring changes in the concentrations of the intermediates (the elasticities). This section will examine the relationship between changing cellular enzyme concentrations and flux control coefficients.

Before proceeding with an examination of the experimental approach to accomplish this task, it is instructive to examine the relationship between flux control coefficients and the individual enzyme parameters. The flux control coefficients for each enzyme can be directly derived for the system of Figure 3.5 by using the definition of the flux control coefficient from Eq. (3.30) and the reversible Michaelis–Menten equation [Eq. (3.20)]. By definition, a partial derivative is obtained by treating all variables as constant, except the one in which we are interested. In this case, both [S] and [P] are treated as constant and only $[E_{i,T}]$ is treated as variable. Following this procedure gives the relationship: $C_i^{J_{ss}} = 1$! While this is true for an isolated enzyme, there is a problem with this relationship for an enzyme in a pathway. Equation (3.32) shows that the sum of all flux control coefficients for a pathway is 1, while the above relationship ($C_i^{J_{ss}} = 1$) shows that each enzyme's flux control coefficient is equal to 1. This difficulty arises because both [S] and [P] cannot be treated as constants in a metabolic pathway, even for the calculation of a flux control coefficient. Thus, in order to derive the flux control coefficient for an enzyme in a pathway, both [S] and [P] must also be treated as variables (i.e., as functions of $[E_{i,T}]$). The flux control coefficient for any enzyme is then defined as

$$\begin{aligned}\frac{E_i}{J_{ss}} \times \frac{d}{dE_i}(J_{ss}) &= C_i^J \\ &= 1 + \frac{\bar{K}_i^-[E_i]^2}{[I_i]}\left(\frac{k_i^+}{T} - \frac{1}{B}\right) \times C_i^{I_i} \\ &\quad - \frac{\bar{K}_i^+[E_i]^2}{[I_{i+1}]}\left(\frac{k_i^-}{T} + \frac{1}{B}\right) \times C_i^{I_{i+1}} \qquad (3.45)\end{aligned}$$

where

$$T = k_i^+\bar{K}_i^-[I_i] - k_i^-\bar{K}_i^+[I_{i+1}] \quad \text{and}$$

$$B = \bar{K}_i^+\bar{K}_i^- + \bar{K}_i^-[I_i] + \bar{K}_i^+[I_{i+1}]$$

Equation (3.45) makes sense in light of what is known about the system. It says that the flux control coefficient is increased by factors that increase the forward rate of the reaction multiplied by the concentration control coefficient for the forward reaction ($C_i^{I_i}$) and is decreased by factors that increase the reverse rate of the reaction multiplied by the concentration control coefficient for the reverse reaction ($C_i^{I_{i+1}}$). Equation (3.45) is not very useful for calculating the flux control coefficient for an enzyme since one must also calculate the concentration control coefficients (which are defined in terms of flux control coefficients), and so one ends up with quite a complicated calculation to perform. A much simpler equation has already been presented that relates the flux control coefficient for an enzyme to the properties of the system (the concentration control coefficients and the elasticities of the other enzymes): It is a version of Eq. (3.43) that includes the elasticities of all the intermediates of the system. This will be explored in the next section.

Although it is possible to use the above equations to calculate the flux control coefficients for each enzyme, the advantage of MCA is that one does not really need to know anything about the system to obtain flux control coefficients providing one can engineer an identical system with slightly lower enzyme activity (or change flux through the system and monitor the intermediate concentrations). Calculations using MCA require no prior knowledge of the kinetic mechanisms in the pathway. The flux control coefficients and the elasticities are simply relative measures of the degree of influence one parameter has on another.

The simulated pathway (Fig. 3.5) provides the ability to mathematically change the concentrations of the enzymes. It is possible, therefore, to conduct a computational experiment by varying each $[E_{i,T}]$ and observing what happens to the system. For this simulation, the concentrations of

Gald-3P and pyruvate were fixed and the total enzyme concentrations were varied. The flux was calculated by inverting Eq. (3.22) or (3.23). These mathematical manipulations are analogous to experimentally changing the total activity of an enzyme by a small amount either through direct genetic manipulation or by inducing an enzyme in a pathway by growing cells on appropriate media. The observed pathway flux is then plotted as a function of the total enzyme activity. Both can be measured experimentally by traditional enzymological methodologies.

Figure 3.10 shows the resulting graphs for each enzyme. By plotting the natural log of each parameter, the flux control coefficient can be obtained directly from the slope. Each graph was generated by changing the concentration of only one enzyme. The enzyme concentrations of the unperturbed system are listed in Table 3.1. By definition, the flux control coefficient for an individual enzyme is the tangent to the curve at the point where the enzyme concentration is at its initial value. This point has been marked in Figure 3.10 with an arrow for the reader's convenience. Thus, the experiment should be performed by: (i) measuring the flux through the system, (ii) changing $[E_{i,T}]$ by a small amount, and (iii) measuring the new flux. The slope of the graph is the flux control coefficient for this enzyme. The resulting flux control coefficients are listed in Table 3.2 (under the heading "Actual") and show that the majority of flux control rests potentially with GAPDH with a small amount of control resting with (PGK). This means that the flux through the system is most sensitive to changes in GAPDH concentrations and less so to changes in the concentrations of the other enzymes.

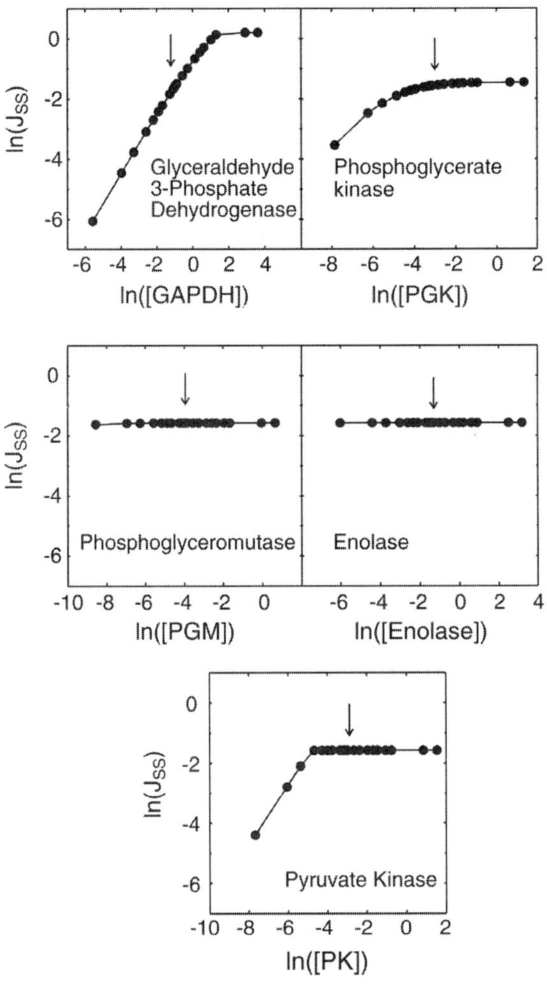

Figure 3.10 Plot for determining the flux control coefficients of the individual enzymes of a linear pathway. The data is for a linear sequence of enzymes from GAPDH to PK (see Fig. 3.5). The natural log of the pathway flux was plotted as a function of the natural log of the concentration of enzyme. The enzyme concentrations of the reference system were taken from Table 2.2 and are indicated by an arrow. The slope of the tangent to the line at the point indicated by the arrow is equal to the flux control coefficient for each enzyme. See Figure 3.5 for enzyme abbreviations.

The fact that control is held mostly by the first few enzymes of a linear pathway has been pointed out before by other authors analyzing linear systems of enzymes. In fact, one can show mathematically (using simpler enzyme mechanisms) that there is a tendency in a linear pathway of enzymes for the flux control coefficient to be largest near the start of the pathway. This will not always be true. For example, Figure 3.10 shows that there are conditions where the flux control coefficient for GAPDH (the first enzyme) is lower than that for other enzymes (the slope is smaller). It is interesting that there is a tendency in a linear pathway for flux to be controlled near the start of the pathway. As pointed out above, the classical view of cellular regulation identifies control enzyme as those at the beginning and the end of metabolic pathways. It is here, as well, where most enzymes that can be regulated are found. These are the allosteric enzymes that respond to changes in the concentrations of cellular metabolites. MCA, therefore, provides a mathematical framework to explain this phenomenon. However, it does not support the idea that all enzymes at the start of metabolic pathways exert a high degree of regulation.

The curves presented in Figure 3.10 are more complete than required to determine the flux control coefficients but are presented to show how flux control could vary depending on the enzyme concentration. For example, while the flux control coefficient for PK is virtually zero (meaning that the slope of the asymptote to the curve at the original PK concentration, $\ln[PK] = -3.0$, is very close to zero), its flux control coefficient would be much higher at a very low PK concentration, when $\ln[PK] = -8$. This is also true for PGK. In addition, while the flux control coefficient for GAPDH is high, it would be virtually zero if the concentration of GAPDH were higher, around $\ln[GAPDH] = 2$.

TABLE 3.2 Flux Control Coefficients for Mathematical Simulation of Metabolic Pathway from Glyceraldehyde-3-Phosphate Dehydrogenase to Pyruvate Kinase[a]

	$C_i^{J_{ss}}$ Value Determined after Multiplying Initial Enzyme Concentration by:				
Enzyme	Actual (1.001X)	0.1	0.5	2	10
GAPDH	0.9	0.95	0.92	0.86	0.74
PGK	0.1	0.25	0.14	0.08	0.05
PGM	≈0	0.002	0.001	0.0003	0.0002
Enolase	≈0	0.00001	0.000014	≈0	≈0
PK	≈0	0.23	0.0003	0.0001	0.00005

[a] The table shows the effect of relatively large changes in enzyme concentration on the measured $C_i^{J_{ss}}$ values. The $C_i^{J_{ss}}$ values were calculated by fixing all enzyme concentrations except for the indicated one. The initial enzyme concentration (see Table 3.1) was then multiplied by either 0.1, 0.5, 2, or 10, the flux calculated and the corresponding $C_i^{J_{ss}}$ value determined from the fraction: $\Delta \ln(J_{ss})/\Delta \ln(E_i)$. This corresponds to an experiment where cellular enzyme activity is changed to a value 10-fold lower (0.1), 2-fold lower (0.5), 2-fold higher (2) or 10-fold higher (10) by a genetic manipulation, through activation of a cellular pathway that can covalently modify an enzyme to alter its activity (such as a hormonal pathway) or through the addition of a chemical inhibitor (such as iodoacetate). The flux control coefficient would then be calculated by measuring the flux and taking the ratio: $\Delta \ln(J_{ss})/\Delta \ln(E_i)$. The table shows the $C_i^{J_{ss}}$ values measured for each change. In general, the error increases as the difference in enzyme concentration becomes large but the effect depends on the enzyme and on the direction of change (increase vs. decrease).

This provides some insight into the relationship between the enzyme activity (expressed as concentration) and its flux control coefficient.

By definition, the flux control coefficients should be determined for very small changes in enzyme concentration. Table 3.2 shows what happens when the changes are not small. Here, the $C_i^{J_{ss}}$ values were calculated for different relative changes in enzyme concentrations. Remember that flux control coefficients are measured by changing the concentration of one enzyme and measuring the new flux. The change in enzyme concentration and flux are used to calculate the flux control coefficient. When the changes in enzyme concentration are very small, flux control coefficients of 0.9 and 0.1 are obtained for GAPDH and PGK, respectively (see the second column in Table 3.2). As the magnitude of the difference between the initial condition and the modified system (with one of the enzyme concentrations altered) increases, the apparent $C_i^{J_{ss}}$ values either decrease (as more enzyme is added) or increase (as less enzyme is added). As an example, consider the values for GAPDH. When [GAPDH] is increased by a large amount (rather than a small amount, as required by the $C_i^{J_{ss}}$ definition), the measured flux control coefficient is lower than the true value. When [GAPH] is decreased by a large amount, the measured flux control coefficient is lower than the true value. Thus, designing an experiment to measure the $C_i^{J_{ss}}$ value of an enzyme by creating a situation where a cell has one-tenth the original activity of an enzyme may give an erroneously large or small flux control coefficient (depending on the system and how sensitive it is to changes in this enzyme concentration). The reason the apparent $C_i^{J_{ss}}$ values change as a function of the relative change in enzyme activity is that the actual flux control coefficients vary as the enzyme concentrations vary but this change in not constant. Thus, one obtains a flux control coefficient that partly reflects the new system, which is not the original intent. This emphasizes the importance of keeping the changes small when measuring flux control coefficients.

CONTROLLING PATHWAY FLUX UNDER *IN VIVO* CONDITIONS: INDIRECT CALCULATIONS OF CONTROL COEFFICIENTS

It is important to remember the definition of an elasticity before using elasticities to determine the flux control coefficients of a pathway. Cornish-Bowden (1995) provides a particularly good definition: An elasticity is a local property of an isolated enzyme that expresses how its rate varies with the concentration of any metabolite that affects it: This can be its substrate, product, or any other metabolite. It is relatively easy, therefore, to calculate the elasticity of an isolated enzyme that obeys the reversible Michaelis–Menten equation under *in vitro* conditions. When holding the concentration of product $[\mathrm{I}_{i+1}]$ constant, the elasticity for an enzyme with respect to its substrate $[\mathrm{I}_i]$ is

$$\varepsilon_{\mathrm{S}}^{v} = \frac{[\mathrm{I}_i] \times \delta v}{v \times \delta[\mathrm{I}_i]} = \frac{1}{1 - \Gamma/K_{\mathrm{eq}}} - \frac{\bar{K}_i^{-}[\mathrm{I}_i]}{\bar{K}_i^{+}\bar{K}_i^{-} + \bar{K}_i^{-}[\mathrm{I}_i] + \bar{K}_i^{+}[\mathrm{I}_{i+1}]} \tag{3.46}$$

and when holding the substrate constant, the elasticity for an enzyme with respect to its product is

$$\varepsilon_S^v = \frac{[I_{i+1}] \times \delta v}{v \times \delta[I_{i+1}]}$$

$$= \frac{-\Gamma/K_{eq}}{1-\Gamma/K_{eq}} - \frac{\bar{K}_i^+[I_{i+1}]}{\bar{K}_i^+\bar{K}_i^- + \bar{K}_i^-[I_i] + \bar{K}_i^+[I_{i+1}]} \qquad (3.47)$$

where Γ is the *mass action ratio*, defined as $\Gamma = \frac{[I_i]}{[I_{i+1}]}$. The K_{eq} value is that defined by the Haldane relationship [Eq. (2.33)]. Previously, it was pointed out that the explicit equation for the flux control coefficient of an enzyme in a pathway required that both substrate and product were variable. Because the elasticity is a property of an isolated enzyme, one can fix product or substrate to calculate the partial derivative. Equation (3.47) shows that the elasticity normally varies between 0 and 1, depending on the concentration of $[I_i]$ and $[I_{i+1}]$. This value has been defined using the free concentrations of substrate and product. Equations (3.46) and (3.47) are interesting because they show that the elasticity is a property of the system of enzymes and not just the enzyme itself, even though it is defined as an isolated enzyme property. This is because it is related to the concentrations of substrate and product, which Eqs. (3.22) and (3.23) show are a property of the system.

Theoretically, Eqs. (3.46) and (3.47) can be used to calculate elasticities providing all the parameters are known. However, one does not need to know all the enzyme parameters to obtain elasticities. Similar to the situation where the flux control coefficients were obtained by varying an enzyme's activity and measuring the change in flux (see preceding section), elasticities can be obtained from an experiment designed to slightly alter the flux through the system. Unlike the experiment for calculating flux control coefficients directly, a method called the *double modulation method* must be utilized. The method involves perturbing the flux through a system of enzymes, measuring the new flux and new intermediate concentrations and then repeating the process by perturbing the flux by a different method. The idea is based on Eq. (3.48), which says that a small change in flux can be approximated by the change in the concentration of an enzyme's substrate, multiplied by its sensitivity to that change added to the change in the concentration of an enzyme's product, multiplied by its sensitivity to that change:

$$\Delta J \approx \frac{\delta v_i}{\delta[I_i]}\Delta[I_i] + \frac{\delta v_i}{\delta[I_{i+1}]}\Delta[I_{i+1}] \qquad (3.48)$$

Normalizing with respect to flux and intermediate concentrations can be preformed, by dividing each side by J, which is equal to v_i, when the system is at steady state:

$$\frac{\Delta J}{J} \approx \varepsilon_i^i \frac{\Delta[I_i]}{[I_i]} + \varepsilon_{i+1}^i \frac{\Delta[I_{i+1}]}{[I_{i+1}]} \qquad (3.49)$$

where $\varepsilon_i^{E_{i+1}}$ will normally be negative. A total of three different measurements are required to perform the experiment: one under unperturbed conditions and two more under two different perturbations. This will give two different sets of ΔJ, $\Delta[I_i]$, and $\Delta[I_{i+1}]$ values (one for each perturbation compared to the control) and, therefore, two different versions of Eq. (3.49). These can then be solved to give the elasticities (two equations and two unknowns; see Text Box 3.2).

RELATING FLUX CONTROL COEFFICIENTS TO PATHWAY REGULATION

The $C_i^{J_{ss}}$ values obtained for very small changes in enzyme concentration showed that the majority of control in the linear sequence from GAPDH to PK is held by the first two enzymes, GAPDH and PGK (Fig. 3.10). This point is not trivial and represents the real advance in thinking brought about by the MCA revolution. The fact that flux control is shared by more than one enzyme means that there is no one rate-controlling enzyme in the pathway as it exists. The enzyme with the slowest rate does not necessarily regulate pathway flux. In fact, in the example of Figure 3.10, the activity of PK is lower than that of GAPDH. It is, rather, the interaction between the enzymes that regulates pathway flux. This result also shows that there is not a 1 : 1 correspondence between changes in a "regulatory" enzyme and activity. In the present example, a change in the activity of GAPDH would only change the activity of the pathway by 0.9 times this change. Thus, a $C_i^{J_{ss}}$ value of 0.9 means that if you change enzyme velocity by a very small amount, you will be able to change only 90% of the flux. Stated another way, this says that 90% of the flux is regulated by this enzyme under these conditions. In addition, in the present example, small changes in PGK activity could also regulate the pathway flux on a smaller scale than that of GAPDH.

Metabolic control analysis shows that control is a shared property of the system but says nothing about an enzyme's ability to control the flux through a pathway—it only deals with the potential for control and the magnitude of this potential. This means that we can talk about the relationship between GAPDH activity and pathway flux, but this says nothing of the cellular mechanisms that exist to allow the regulation of GAPDH activity. Although this point is not relevant to the mathematical arguments

TEXT BOX 3.2 CALCULATION OF ELASTICITIES USING THE DOUBLE MODULATION METHOD USING PGK AS AN EXAMPLE

It is possible to solve for flux control coefficients indirectly using *in vivo* measured elasticities. The method is a little involved but has the advantage that all the flux control coefficients in a single pathway can be obtained after a relatively minor number of experimental measurements. We say relatively because the experiments are still complicated but do not require the manipulation of each enzyme in the pathway, as would be needed if the flux control coefficients for all the enzymes in a pathway were to be measured directly. The first step is to perturb the system in some way and then measure the free concentrations of all the metabolites and the flux through the system. The system is then perturbed again and the measurements repeated. In the following example, the activity of GAPDH was doubled (Perturbation 1) and the PGK activity was halved (Perturbation 2) (Table TB3.1). The change in these values, relative to the unperturbed condition is then calculated.

TABLE TB3.1 Changes in Free Concentration of Pathway Intermediates after Two Perturbations

	Value		
Parameter	Original Condition	Perturbation 1: Double the GAPDH Activity	Perturbation 2: Halve the PGK Activity
J (steady-state flux)	0.2062	0.3732	0.1867
$[\text{Gald-3P}_{\text{free}}]$	0.0369	0.1135	0.0319
$[\text{diPGA}_{\text{free}}]$	0.0060	0.01846	0.00519
$[\text{3PGA}_{\text{free}}]$	0.0049	0.01051	0.0044
$[\text{2PGA}_{\text{free}}]$	0.00024	0.00047	0.00021
$[\text{PEP}_{\text{free}}]$	0.0280	0.0770	0.02420
$[\text{Pyr}_{\text{free}}]$	0.36	0.36	0.36

Application of Eq. (3.50) to these data gives a series of equations relating the relative change in flux to the relative change in intermediate concentration. These equations apply to each substrate/product pair for each enzyme. For example, for PGK, there would be two equations:

For perturbation 1: $0.810 = 0.992 \times \varepsilon^{\text{PGK}}_{\text{diPGA}} + 1.128 \times \varepsilon^{\text{PGK}}_{\text{3PGA}}$
For perturbation 2: $-0.094 = -0.104 \times \varepsilon^{\text{PGK}}_{\text{diPGA}} - 0.105 \times \varepsilon^{\text{PGK}}_{\text{3PGA}}$

which can be solved to give:

$$\varepsilon^{\text{PGK}}_{\text{diPGA}} = 1.64$$
$$\varepsilon^{\text{PGK}}_{\text{3PGA}} = -0.70$$

The pairs of elasticities would then be substituted into equations of the form given by Eqs. (3.43) and (3.44) for each of the intermediates. A matrix would then be built up, similar to that of Eq. (3.45), which would then be solved for the flux control coefficients. When the values from Table TB3.2 are used to construct the matrix, the flux control coefficients are not correct. For example, the flux control coefficient obtained for GAPDH is −9.9. This is clearly, an impossible result and reflects the sensitivity of this method to deviations from the rule that small changes must be used to measure flux control coefficients. As the change in flux becomes smaller, the flux control coefficients begin to approximate the correct values, but it is only when the relative changes in flux are on the order of 0.01% that accurate flux control coefficients are obtained. Part of the problem may be related to the system of Figure 3.5. Figure 3.5 was constructed so that it represented a linear system of enzymes with most of the flux determined by the first enzyme but with appreciable changes in the levels of intermediates when the concentration of the first or second enzyme is changed. This makes this system sensitive to small changes, a feature that may not always occur in normal metabolic pathways. Indeed, most of the work with cellular systems of enzymes suggests that the flux through a system is dependent on many factors—that is, it is not dominated by a single enzyme. Nevertheless, this analysis shows that caution is warranted when using the double perturbation method for measuring flux control coefficients.

TABLE TB3.2 Relative Changes in Flux and Free Intermediate Concentrations after Perturbation of the Theoretical System of Figure 3.5

Flux or Ratio of Change in Free Concentration	Perturbation 1: Double the GAPDH Activity	Perturbation 2: Halve the PGK Activity
$\Delta J/J$	0.810	−0.0943
Δ[Gald-3P]/[Gald-3P$_0$]	1.752	−0.1352
Δ[diPGA]/[diPGA$_0$]	0.992	−0.1038
Δ[3PGA]/[3PGA$_0$]	1.128	−0.1047
Δ[2PGA]/[2PGA$_0$]	2.075	−0.1356
Δ[PEP]/[PEP$_0$]	2.077	−0.1357

presented, it does have implications for cellular control. In particular, we must ask the question: What does MCA say about events in a cell where we might expect enzyme activity changes of large magnitude to occur as a result of a regulatory event? The realization that an enzyme has a $C_i^{J_{ss}} = 0.9$ says something about the existing interaction between the enzymes of the pathway but says nothing about what may happen after a regulatory event. This is because the flux control coefficients only measure the system, as it exists. In their strict definition, they say nothing about how flux control coefficients may change as actual flux changes.

As a general rule, one can state that the flux control coefficient for an enzyme increases as its activity decreases. Thus, an enzyme with a flux control coefficient of 0.9 will only become more regulatory as its activity decreases while the flux control coefficients of the other enzymes in the pathway will decrease. This is also true for an enzyme with a flux control coefficient of 0.01, but it may require quite a change in activity before the flux control coefficient becomes sufficiently large to have an impact on the flux. Once again, as the flux control coefficient of one enzyme increases, the flux control of all the other enzymes decreases [Eq. (3.32) will always hold true] so that an enzyme with a flux control coefficient of 0.9 under one condition may have a flux control coefficient of 0.3 under another condition, even though its activity was never altered.

While it is relatively easy to imagine what happens to the flux control coefficient of an enzyme as its activity decreases, it is virtually impossible to say what happens to flux control as its activity increases. Presumably, the flux control coefficients of all the other enzymes will increase, but it is difficult to predict the magnitude of this increase. In addition, it is difficult to predict the effect on flux. This is because there is no simple relationship between flux and enzyme activity contained within MCA. If the kinetic parameters of all the enzymes are known, a relationship can be calculated by Eqs. (3.22) and (3.23), but it is rare that kinetic constants determined *in vitro* are accurately reflected by the enzymes *in vivo*.

POTENTIAL OF AN ENZYME TO BE REGULATED

Although somewhat mathematically daunting, MCA provides a way to think about a metabolic system and provides a framework to model it. In essence, it provides a measure of the correlation between flux and changes in enzyme activity or the concentrations of intermediates. In a practical sense, it allows a test of the predictions of classical metabolic control, that is, that rate-controlling enzymes exist and that changes in the kinetic properties of controlling enzymes will regulate the flux through a pathway. Metabolic control analysis says that the classical view of metabolism is flawed and that control is shared.

Although MCA provides a convenient mathematical framework for examining metabolic pathways, there seems to be an important point left out by MCA theories: the potential for an enzyme to be regulated. It is, in fact, at this point that traditional biochemical approaches and MCA clash. It is certain that the measurement of flux control coefficients for individual enzymes along a pathway shows that control is shared by all members of a pathway. In fact, MCA makes the important point that the flux through a pathway is a function of all the enzymes in the pathway. This, however, says nothing about what enzymes in the pathway may actually be regulated. Enzymes are regulated through the actions of inhibitors and activators as well as by changes in activity. Changes in an enzyme's susceptibility to inhibitors and activators are often brought about by a posttranslational modification of the enzyme, such as by reversible phosphorylation carried out by protein kinases or protein phosphatases. Hence, total activity is regulated by the relative rate of enzyme synthesis and breakdown as well as by changes in the phosphorylation state of the enzyme. Only some enzymes are subject to inhibition or activation or to protein phosphorylation. As a consequence of this, there are comparatively few enzymes in the cell that have the potential to be regulated. Traditional approaches to the problem of cellular regulation have identified these enzymes because their total activity or their affinity for substrate can be altered by internal or even external (such as hormone) signals and because their activity is often the lowest in the pathway.

Traditional enzymology holds that, even if an enzyme's flux control coefficient is high for any given pathway flux, changes in its activity (such as increasing the affinity for substrate or reducing the V_{max} value) can effectively regulate pathway flux. Mammalian liver pyruvate kinase (PK) is a good example of an enzyme that is thought to be regulatory even though its flux control coefficient may not be large. PK activity can be altered by one of two mechanisms: (1) the enzyme's kinetic properties can be altered by reversible phosphorylation to change its kinetic pattern from square hyperbolic (described by the Michaelis–Menten equation) to sigmoid (cooperative or Hill equation). This leads to a decrease in the enzyme's affinity for PEP substrate, and (2) the concentration of PK can be reduced by changes in the ratio of enzyme synthesis to enzyme breakdown during prolonged food deprivation. The combination of these effects are thought to shut down the enzyme during times of starvation (when blood glucose concentrations are low and liver glycogen concentrations are depleted) so that the liver can make glucose from amino acids (via gluconeogenesis). However, MCA

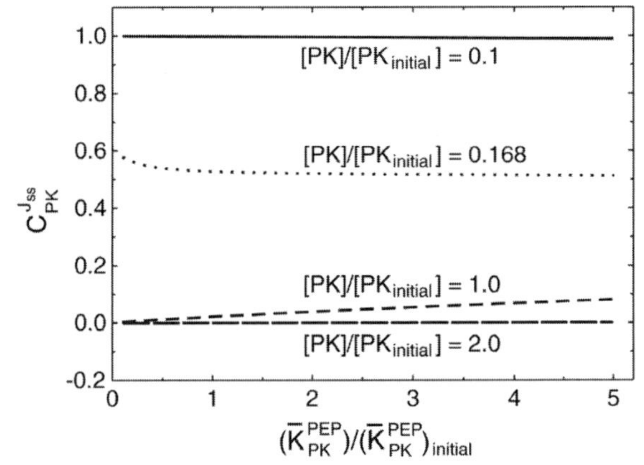

Figure 3.11 Flux control coefficient for pyruvate kinase (PK) in the linear pathway from GAPDH to PK: effect of a simulated phosphorylation event. The graph shows that only small changes in the flux control coefficient for PK can be obtained, even after a fivefold change in the binding affinity for substrate (the *X*-axis value, $\bar{K}^{PEP}_{PK}$, was varied from $0.2 \times \bar{K}^{PEP}_{PK}$ to $5 \times \bar{K}^{PEP}_{PK}$). Similar results were obtained when the total concentration of PK was decreased by 10-fold ($[PK]/[PK_{initial}] = 0.1$) or increased twofold ($[PK]/[PK_{initial}] = 2$), even though the flux control coefficient increased to approximately 1.0. The simulation was run by holding Gald-3P and pyruvate concentrations constant and calculating the flux using Eqs. (3.22) and (3.23).

predicts a different response, and this is borne out by direct mathematical simulation of the lower part of the pathway using Eqs. (3.22) and (3.23) as demonstrated in Figure 3.11. Under the reference conditions used to construct the simulated pathway, PK has very little control over the flux; its $C^{J_{ss}}_i$ value is essentially zero. What happens when a phosphorylation event is simulated by decreasing PK's affinity for its substrate, PEP, by as much as fivefold (phosphorylation *in vivo* significantly increases the K_m for PEP)? Figure 3.11 shows that there is no appreciable change in the PK flux control coefficient. This runs contrary to what is expected from a classical understanding of metabolism. It is tempting to think that the lack of response was due to the low initial $C^{J_{ss}}_{PK}$ value. However, changing substrate affinity had no effect when $C^{J_{ss}}_{PK} \approx 0.5$ or when $C^{J_{ss}}_{PK} \approx 1.0$; at these flux control coefficients, PK exerts a considerable control on pathway flux. Although this result is at odds with the classical interpretation, it should not be surprising in light of what was learned about linear systems and was discussed in previous sections of this chapter. When the kinetic parameters of the system are changed, the system can adapt by changing the concentrations of the intermediates. This point is illustrated by Figure 3.12, where the concentration of PEP is plotted for the conditions outlined in Figure 3.11. As the activity of PK decreases, the concentration of its substrate, PEP, increases dramatically. This will help to force flux

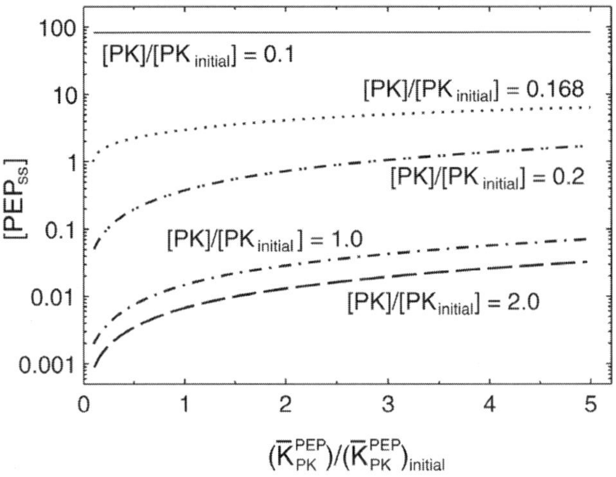

Figure 3.12 Changes in the concentration of phosphoenolpyruvate, the substrate of PK, as a function of a simulated PK phosphorylation event. The log of the concentration of PEP is plotted as a function of the relative PK binding affinity for PEP at various concentrations of total enzyme. The graph shows that the linear enzyme system adjusts to a reduction in PK activity and to an increase in the binding affinity for PEP by increasing the concentration of PEP.

through PK. This also occurs when the affinity of PK for its substrate decreases. Figure 3.12 shows that one can expect a 10- to 100-fold change in the concentration of an intermediate (like PEP) when the binding affinity is reduced by up to 5-fold and there is no change in flux.

Although such dramatic fold changes have been observed for some metabolites in the cell, this clearly never happens with intermediates such as PEP. In fact, cellular metabolite levels usually change only very little. For example, the glycolytic flux in working animal muscle can rise by 100- to 1000-fold compared with resting muscle with very little change in the concentration of the glycolytic intermediates. This means that other things must be going on to regulate the flux while maintaining the concentration of metabolites. Nevertheless, the fact that flux can remain constant while kinetic parameters of an enzyme are altered is a restatement of what MCA is telling us: A metabolic system can adapt to changes in flux (or intermediate concentrations) by altering its own properties. These changes are spread over the entire pathway so that control is not a property of any single enzyme; modeling shows that changes in one enzyme's parameters are not enough for the enzyme to take "control" of a pathway. This point has been made more clearly and completely by Thomas and Fell (1998) when they examined the role of multiple enzyme activation in metabolic flux control.

Some further discussion is warranted on the example of PK phosphorylation in the starving liver. It is known that limiting the glycolytic flux is not the only thing that happens when glucose availability is reduced. The starving liver is also actively involved in making glucose from amino acids. Reverse flux through the glycolytic pathway predominates when PK is "inhibited," including special pathways that circumvent selected enzymes such as PK and PFK. The mathematical simulations of Figures 3.11 and 3.12 show that inhibiting PK is not enough to force flux backward through the linear part of the glycolytic pathway. What MCA tells us, then, is that other things must be going on to promote reverse flux. One thing could be a change in the metabolite concentrations that were held fixed for the simulation (Gald-3P and pyruvate for this simulation). Other things are also happening. Carbon is being pushed into the PEP pool by amino acid catabolism through PEP carboxykinase, and glucose is being pulled backward through the glycolytic chain by the glucose transporter, which is pumping glucose into the blood. What role the phosphorylation of PK plays in this larger pathway cannot be determined by the mathematical simulation of Figures 3.11 and 3.12 but this determination has already been made for hepatocytes making glucose (Groen and Westerhoff, 1990). The flux control coefficient for PK, in the presence of glucagon—a gluconeogenic promoting hormone—was measured as 0.0, while PEP carboxykinase had a flux control coefficient of 0.83.

There is one other important point to consider when determining flux control coefficients: the cellular concentrations of inhibitors and activators that act on the enzyme under study. This consideration is important when studying the flux control of traditionally identified regulatory enzymes, such as 6-phosphofructo-1-kinase (PFK). PFK occupies an important place in the glycolytic pathway; it is the first true committed step of glycolysis because it catalyzes the phosphorylation of fructose 6-phosphate using ATP as a phosphate donor. The enzyme is inhibited by many different cellular metabolites at neutral pH, including citric acid and high concentrations of ATP. It is activated by many compounds, including adenosine monophosphate, fructose 2,6-bisphosphate, and phosphate and ammonium ions. In the classical view of metabolism, these inhibitors and activators are thought to link the activity of PFK to the energy state of the cell so that glycolytic flux increases during periods when cellular energy is being used and decreases when cellular energy supply is plentiful. However, the PFK flux control coefficient in yeast cells, obtained by genetic modification to decrease PFK expression (Heinisch, 1986), is approximately 0.3, indicating that PFK is not as regulating as one may suspect [see Fell's (1997) book for a review of the PFK data]. The problem with this measurement is that the flux control coefficient does not really get at the regulation of glycolytic flux. Traditional thought has placed the regulation of glycolysis at the PFK locus, but the concentration of PFK may be irrelevant since it is the cellular concentrations of activators and inhibitors that really determine the flux through PFK. Changing the total amount of PFK does not change the activity through the PFK locus, just the total potential activity. It is, therefore, conceivable that during the experiments to determine the flux control coefficient of PFK, the concentration of inhibitors and activators that act on PFK were altered enough to allow the cell to maintain the same glycolytic flux. Thus, when performing flux control analysis, one can only determine reliable and meaningful flux control coefficients for enzymes whose activity is not susceptible to inhibition or activation by metabolites external to the pathway. One must be careful when measuring flux control coefficients to ensure that a change in enzyme concentration will bring about a change in enzyme activity.

What happens when the enzymes are only a part of a larger pathway in which the flux is regulated primarily by an enzyme external to the system? This is where the experimenter's knowledge plays a critical role in correctly defining the complete system so that meaningful numbers are obtained. Metabolic control analysis places no restrictions on the size of the pathway so that it can be anything from 2 to 1000 enzymes big. It is important to include relevant reactions but to exclude nonrelevant reactions when defining a system. In this way, meaningful information about the relative contribution of the enzyme under study is obtained. This point is important when considering the truncated linear system of Figure 3.5. In our simulation, the flux was calculated using Eqs. (3.22) and (3.23) with fixed concentrations of the first and last intermediates. The flux control coefficients were obtained by changing the concentration of the enzymes. By doing this, we ensured that the flux was a property of the system (it was defined by the concentration of the first and last intermediates). However, identical results could have been obtained if, instead of fixing the concentration of the first and last intermediates, we fixed the flux. In this case, the flux is external to the system and is under external control. Under these conditions, flux control coefficients are irrelevant because the flux is already set. So what basis in reality do flux control coefficients have? This circular reasoning gets us nowhere, and the test of the method is really in actual cellular measurements with real metabolic pathways. Metabolic control analysis makes no claims about the mechanisms and, indeed, makes no claims about where control lies. It simply asks the question: If one changes a parameter (enzyme concentration, intermediate concentration, etc.), what happens to the flux?

METABOLIC SIMULATION

Metabolic simulation is a wide-ranging topic and one that cannot be covered in any detail by a small section in a chapter. Many studies exist on the subject and the reader is referred to the books listed at the end of this chapter for a thorough examination of the subject. Because of the enormity of the subject, this section will focus on two points raised by the simulation: (1) the interaction between supply and demand on maintaining the concentration of intermediates and (2) the effect of feedback inhibition on pathway kinetics and intermediate concentrations. These two subjects were chosen because they are fundamental to the understanding of pathway interactions and because they provide a link between metabolic control theory and metabolic modeling, albeit on a small scale.

Supply versus Demand

Hofmeyr and Cornish-Bowden (2000) have analyzed the interaction between supply and demand on the concentration of an intermediate for a number of different enzyme systems from an MCA perspective. For a simple system where

$$S \xrightarrow{v_{\text{supply}}} \mathrm{I} \xrightarrow{v_{\text{demand}}} \mathrm{P} \tag{3.50}$$

it is possible to relate the flux control coefficients for supply and demand to the elasticities of the system through the application of MCA. For a two-reaction system, the solution is simple

$$C^J_{\text{supply}} = \frac{\varepsilon_{\mathrm{I}}^{\text{demand}}}{\varepsilon_{\mathrm{I}}^{\text{demand}} - \varepsilon_{\mathrm{I}}^{\text{supply}}} \qquad C^J_{\text{demand}} = \frac{-\varepsilon_{\mathrm{I}}^{\text{supply}}}{\varepsilon_{\mathrm{I}}^{\text{demand}} - \varepsilon_{\mathrm{I}}^{\text{supply}}} \tag{3.51}$$

The concentration control coefficients can also be derived:

$$C^{\mathrm{I}}_{\text{supply}} = -C^{\mathrm{I}}_{\text{demand}} = \frac{1}{\varepsilon_{\mathrm{I}}^{\text{demand}} - \varepsilon_{\mathrm{I}}^{\text{supply}}} \tag{3.52}$$

Equation (3.52) shows that the concentration control coefficient for supply is equal to but opposite in sign to the concentration control coefficient for demand (i.e., the sum of the concentration control coefficients is zero). While this seems trivial, this simple modeling of a two-process system reveals some profound truths about metabolism. First, if the supply reaction for an intermediate is increased or decreased, the demand reactions must be similarly regulated so that the intermediate concentration is unchanged. This is demonstrated in Figure 3.13 where the natural log of the velocity for a process is plotted against the concentration of an intermediate. When supply > demand, the concentration of intermediate increases, and when

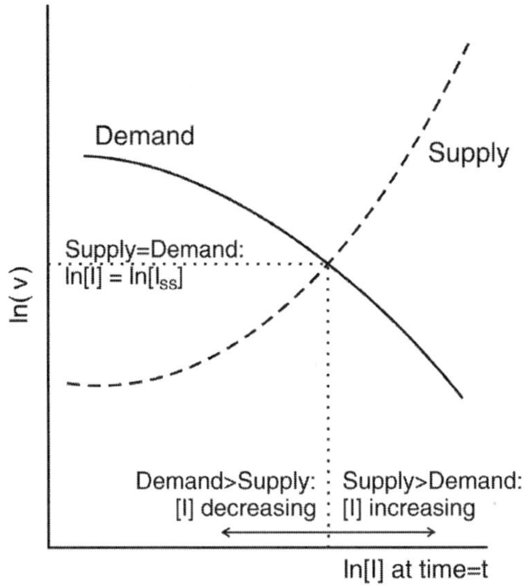

Figure 3.13 Relationship between the concentration of the intermediate at a fixed time and the relative rates of supply and demand reaction blocks. A steady-state condition is only achieved when supply = demand. Otherwise, the concentration of the intermediate decreases (supply < demand) or increases (supply > demand) unless some other factor can control the process.

demand > supply, the concentration of the intermediate decreases. Under both these conditions, there is no steady state. It is only when supply is balanced by demand that a steady-state concentration of intermediate is obtained.

While this conclusion may seem trivial, it is a point that must be emphasized when studying animals that depress their metabolism, as will be discussed in detail in later chapters. It applies, in particular, to animals that successfully survive periods of anoxia by depressing their metabolic rates. Under anoxic conditions, the cell loses the ability to produce ATP from oxidative phosphorylation and is restricted to the very much lower ATP yield of glycolysis—ATP supply is severely reduced under anoxia. Many have argued that regulation of energy-utilizing reactions need only apply in these conditions and that a reduction in cellular energy requirements will suffice to regulate metabolism. Equation (3.51) shows that this is not the case and energy-supplying reactions must also be regulated to maintain homeostasis. Thus, Eqs. (3.51) and (3.52) show that both ATP-producing and ATP-consuming reactions must be similarly decreased during the anoxic episode in order to maintain ATP concentrations at their control levels.

The second interesting conclusion from Eqs. (3.51) and (3.52) is the fact that the concentration control coefficient for supply decreases as the flux control coefficient for supply increases. This means that as a reaction gains control over the flux, it looses control over the concentration of the intermediate. Returning to the example of the anoxic animal, the above analysis shows that, if the ATP-consuming reactions have more control over the flux, the ATP-producing reactions will have more control over the concentration of ATP. Once again, this emphasizes the importance of balancing both supply and demand reactions in response to an external stress to maintain cellular homeostasis and explains much about the regulation of glycolysis during anoxia.

Feedback Inhibition

In the previous section, mathematical analysis of the truncated system of Figure 3.5 revealed some interesting properties of the system. In particular, inhibition of the terminal enzyme had no real effect on the flux through the complete system, although it did result in large increases in the concentrations of some intermediates. This brought about the conclusion that, for a linear system of enzymes, the majority of flux control was exerted by enzymes at the beginning of the pathway. While this is generally true for linear systems of Michaelis–Menten type of enzymes, linear systems, as such, rarely exist in the cell. Most pathways have enzymes that are responsive to inhibitors and activators generated within the pathway or elsewhere that act to influence pathway behavior. Several of the enzymes within the glycolytic pathway, for example, are allosterically affected by metabolites, but the nature of the inhibition/activation differs depending on the cell type and organism. An example of such a system is observed in marine invertebrates where phosphofructokinase, an enzyme at the start of glycolysis, is known to be feedback inhibited by products of glycolysis, such as ATP and PEP. Feedback inhibition occurs when an intermediate at or near the end of the pathway inhibits an enzyme near the start of the pathway. MCA analysis of feedback inhibition shows why this property of PFK is so important to the animal's ability to survive periods of anoxia.

Applying the MCA connectivity theorem to a linear system of enzymes with feedback inhibition:

$$I_0 \overset{v_0}{\rightleftharpoons} I_1 \overset{v_1}{\rightleftharpoons} I_2 \overset{v_2}{\rightleftharpoons} I_3 \overset{v_4}{\rightleftharpoons} I_4 \qquad (3.53)$$

gives the following connectivity relationship for intermediate I_3:

$$C_0^J \varepsilon_{I_3}^0 + C_2^J \varepsilon_{I_2}^2 + C_3^J \varepsilon_{I_3}^3 = 0 \qquad (3.54)$$

Combining Eqs. (3.54) with the two relationships for intermediates 2 and 3 [see Eqs. (3.41) and (3.42)] and the summation theorem [Eq. (3.32)] gives three relationships that can be used to completely resolve the system using the matrix approach.

An inhibitor can act upon an enzyme in three classically defined patterns: competitive, uncompetitive, noncompetitive (as described in Chapter 2), as well as allosterically. In reality, both uncompetitive and noncompetitive are special cases of mixed inhibition so that one really need only consider whether the inhibition is competitive or not. Competitive inhibition is unlikely to produce any significant change in pathway behavior because, as pointed out previously, when the affinity for a substrate is decreased, the system can respond by increasing the concentration of its substrate and maintain flux. Thus, the system effectively "out-competes" the inhibitor by adjusting the concentration of intermediates. Thus, it is only the mixed inhibitors that can affect pathway flux to any degree. Similarly, if the enzyme is allosteric, one need only consider an inhibitor that will lock the enzyme in the lower activity conformation (as described by the Monod–Wyman–Changeux model).

The present example will, therefore, assume that the initial, regulated example is allosteric because we are interested in modeling the effects of feedback inhibition on a purported "rate-controlling" enzyme, such as PFK. The system we are interested in modeling has, in fact, already been considered by Kacser and Burns (1995) and the equations relating the change in flux control coefficients to the inhibition constants have already been worked out. Figure 3.14 shows the results. The simulation provides a fascinating insight into metabolic regulation along a pathway. When the first enzyme is not very responsive to the feedback inhibitor (i.e., it has a low affinity for the inhibitor), the first enzyme exerts a considerable amount of control over the pathway flux. The fact that the first enzyme of a linear pathway tends to have a bigger flux control coefficient has already been discussed (see above). However, if the affinity of the first enzyme for the inhibitor increases, it looses this control to the terminal enzyme of the pathway. This occurs because the pathway becomes more responsive to the concentration of the inhibitor [the substrate for enzyme 3 in Eq. (3.53)], and the concentration of the inhibitor is highly dependent on the activity of the terminal enzyme. At higher inhibitor sensitivities, pathway flux decreases because of the slowing of the first enzyme. This simple system illustrates the interaction that can occur between an enzyme that is responsive to feedback inhibition and a terminal pathway enzyme such as PFK and PK. In many marine animals both PK and PFK are phosphorylated in response to cellular stresses to reduce their affinity for substrate, decrease their overall activity, and make them more responsive to inhibitors. The combination of reducing the PK activity during stress (and/or reducing its affinity for substrate) and increasing the sensitivity of PFK to inhibition by PEP would dramatically slow glycolytic rate as suggested by the response of Figure 3.14. A flux control analysis would probably indicate that virtually all the control of glycolysis rests with PK since it would be responsible for maintaining the concentration of PEP at high levels. So, although the total PFK activity would have virtually no control over the glycolytic flux, its responsiveness to feedback inhibition would be the key to glycolytic regulation under these conditions. This system provides a good example of the interplay between enzymes at both ends of a linear pathway and shows why some enzymes at the start of metabolic pathways may be responsive to feedback inhibitors.

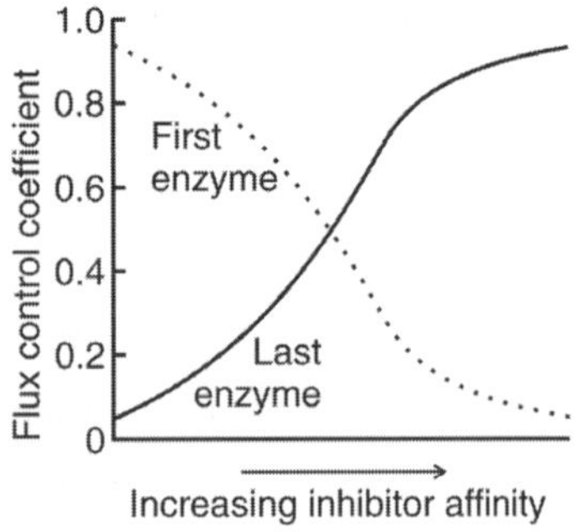

Figure 3.14 Relationship between the sensitivity of the first enzyme of a linear sequence to feedback inhibition and the flux control coefficient. As the first enzyme becomes more susceptible to feedback inhibition, its flux control coefficient decreases while that of the terminal enzyme increases.

FORMATION OF MULTIENZYME COMPLEXES

Initially, enzymologists thought of the cell as a collection of soluble enzymes that were randomly distributed throughout the cytoplasm. As cellular organelles and structures were discovered (the mitochondrion, plasma membrane, microsomal membrane, nucleus, lysosome, golgi apparatus), it became clear that some enzymes were localized to specific areas of the cell. For example, many enzymes of the citric acid cycle are found exclusively within the mitochondria. Later, detailed investigation of purified enzymes revealed the existence of multienzyme complexes. These complexes contain several different enzyme activities that can be co-purified together. They complement one another and often represent sequential reactions. The enzymes of fatty acid synthase, pyruvate dehydrogenase, and cytochrome *c* oxidase are good examples of such multienzyme complexes.

Enzymes that strongly associate during purification are easily identified as existing in large complexes. However, there are many enzymes that do not co-purify but are still believed by some to exist in a complex in the cell. Arguments certainly exist to support this claim. The cell is a crowded milieu with little free water. The average protein content in animal cells varies from 20 to 25% of the total weight. This translates into protein concentrations between 200 and 250mg/mL with minerals, metabolites, and water making up the rest. This high-protein concentration means that the cytoplasm is a highly viscous solution with much of the water in ordered structures situated around proteins and cell organelles. This has implications for the way in which enzymes function *in vivo*. Under these conditions, the random diffusion of substrates and products from enzyme to enzyme, randomly scattered throughout the cytosol, is thought to take too long to support the observed pathway activities. Direct measurement of the diffusion of large proteins in the cytosol shows the effects of this crowding: The proteins diffuse much slower than they do in dilute solutions of ions. In addition, protein–protein self-association is known to be a function of the total protein content with greater association observed at higher total protein concentrations.

With these ideas in mind, enzymologists have begun to question the enzyme organization of the cytosol. Many cellular structures exist within the cytosol (e.g., actin and myosin filaments, subcellular organelles) that can bind enzymes when tested *in vitro*, and one can show that various cytosolic enzymes can bind to each other in addition to binding to cellular structures under *in vitro* conditions. These discoveries led researchers to test whether the enzymes of a single metabolic pathway could be bound together to form a single multienzyme complex. The potential benefits of a multienzyme complex include: (1) enzymes in the complex would work at faster rates because their substrates would be produced in close proximity to their active sites and their products would be quickly removed; (2) enzymes in the complex would respond more quickly to inhibitors and activators (metabolic intermediates would be in close proximity to inhibitor and activator sites); and (3) formation and destruction of the complex could serve as a mechanism to control overall pathway flux. Interpreting the experimental evidence, however, is not simple because of the difficulties encountered in reproducing the cellular conditions *in vitro*. For example, enzyme binding is normally measured with purified components under conditions of low-protein concentration and low ionic strength. Both conditions are opposite to what is found in the cell. These conditions are used because it is extremely difficult to reproduce cellular conditions when performing binding experiments *in vitro*. Nevertheless, several research groups (although not all) support the existence of multienzyme complexes based on evidence obtained from these initial experiments.

The best evidence for the existence of multienzyme complexes is kinetic, and the best kinetic experiments are those examining the enzymes of the glycolytic pathway of smooth muscle. Using whole cells and radioactively labeled glucose, one can demonstrate that the glycolytic pathway of smooth muscle cells is divided between two cellular locations, one that is used to fuel muscle contraction and one that serves the energy needs of plasma membrane transporters and pumps. Glycolytic pathways in both locations use glucose as a substrate, but the glucose comes from different pools so that one can follow individual pathway activity by labeling one of the glucose pools and not the other.

One of the hypothesized advantages of enzyme–enzyme binding is the potential for channeling substrate and product between the active sites of sequential enzymes. It is thought that this would effectively increase the concentration of substrate and/or decrease the concentration of product near the active site. The bound enzyme would, therefore, "see" a higher substrate concentration than one randomly dissolved in the cytoplasm. Examples of specific channeling exist in the literature. In the heart mitochondrion, the existence of the outer mitochondrial membrane allows the creation of an isolated compartment between the inner and outer mitochondrial membrane. This compartment has higher ATP and lower ADP and inorganic phosphate concentrations than those found in the cytosol. This allows the enzyme creatine kinase to have special access to the ATP produced by oxidative phosphorylation to make higher concentrations of creatine phosphate for use in muscle contraction. The outer mitochondrial membrane also provides another well-defined example of substrate channeling. The hexokinase enzyme in mammalian brain appears to bind to the cytosolic side of the outer mitochondrial membrane pore that allows passage of ATP and ADP into and out of the mitochondrion. Binding to this pore is thought to increase hexokinase activity by placing it near a source of high ATP.

It is relatively simple to devise a kinetic test for substrate channeling. One simply has to construct a two-coupled enzyme system as shown in Eq. (3.14). The steady-state concentration of the shared intermediate will reflect the K_m value for the coupling enzyme. It is easier to do this for enzymes that share a single substrate/product pair such as the enzymes that oxidize and reduce NAD. The concentration of NADH is easily followed spectrophotometrically, and the activities of the enzymes can be significantly increased without disturbing the $NADH/NAD^+$ ratio as long as the ratio of enzyme activities is maintained. These tests have shown that many purported cases of enzyme channeling were experimental artefacts. In fact, no kinetic evidence exists to date to support complexes between the enzymes of NADH metabolism or the enzymes of the glycolytic pathway (with the exception of the smooth muscle case described above).

The failure of many experiments to prove the existence of enzyme–enzyme channeling does not mean that

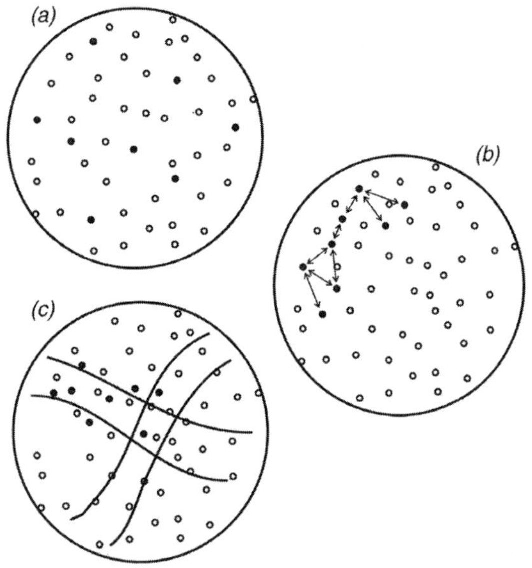

Figure 3.15 Representation of enzyme distribution in a cell showing a (*a*) random, (*b*) "free cytosolic" loose complex formation, and (*c*) loose-complex formation along an actin filament. The model supposes specific interactions exist between the enzymes of a metabolic pathway (black dots). The strength of these interactions is not sufficient for the enzymes to be isolated as a well-defined complex but may help to localize the enzymes within a region of the cytoplasm. If one of the enzymes associates with F-actin with a high affinity, this may help to localize the enzymes around these structures.

enzymes are freely soluble within the cell's cytoplasm. In fact, it is almost certain that they are not as freely soluble as was once thought. What could all the *in vitro* data mean for cellular organization? One idea is that the relatively weak interactions between the enzymes of the glycolytic complex may serve to concentrate the enzymes in a smaller area and reduce the diffusion distance between active sites. In this scenario, weak attractions between sequential enzymes in a pathway may influence their distribution within the cytoplasm so that they are more likely to be closer together than would be predicted by a random distribution (Figure 3.15). This could also serve to increase enzyme activity by reducing the distance that a shared intermediate has to travel before being acted on by the next enzyme in the sequence. Cellular structures may help to provide a framework for these loose associations. For example, it is known that many glycolytic enzymes can bind actin, a protein that exists in filaments through the cell. Actin could, therefore, provide a structure for the formation of a loose complex within the cytoplasm, which may reduce diffusion distances and serve to localize the glycolytic pathway within the cytoplasm. Proving the existence of such associations *in vivo* would be difficult, but theoretical modeling could provide some basis for their existence. This would require the modeling of two randomly diffusing proteins and determining the effect of a weak interaction. The existence of such a complex would provide some reason for the observed interaction between proteins *in vitro*. In addition it may provide some insight into the high glycolytic rates that can be supported in contracting muscle.

REFERENCES

Ainscow, E. K., and Brand, M. D. (1999a). Top-down control analysis of ATP turnover, glycolysis and oxidative phosphorylation in rat hepatocytes. *Eur J Biochem* **263**:671–685.

Ainscow, E. K., and Brand, M. D. (1999b). The responses of rat hepatocytes to glucagon and adrenaline. Application of quantified elasticity analysis. *Eur J Biochem* **265**:1043–1055.

Cornish-Bowden, A. (1995). *Fundamentals of Enzyme Kinetics*, Portland Press, London.

Fell, D. A. (1997). *Understanding the Control of Metabolism*, Portland Press, London.

Fell, D. A., and Sauro, H. M. (1985). Metabolic control and its analysis. Additional relationships between elasticities and control coefficients. *Eur J Biochem* **148**:555–561.

Groen, A. K., and Westerhoff, H. V. (1990). Modern control theories: a consumers' test. In *Control of Metabolic Processes*, Cornish-Bowden, A., and Cárdenas, M. L. (eds.), Plenum, New York, pp. 101–118.

Barman, T. E. (1969). *The Enzyme Handbook*, Springer, New York.

Heinisch, J. (1986). Isolation and characterization of the two structural genes coding for phosphofructokinase in yeast. *Mol Gen Genet* **202**:75–82.

Hofmeyr, J.-H. S., and Cornish-Bowden, A. (2000). Regulating the cellular economy of supply and demand. *FEBS Lett* **476**:47–51.

Kacser, H., and Burns, J. A. (1995). The control of flux: 21 years on. *Biochem Soc Trans* **23**:341–366.

Ricard, J. (1999). *Biological Complexity and the Dynamics of Life Processes*. Elsevier, New York.

Thomas, S., and Fell, D. A. (1998). The role of multiple enzyme activation in metabolic flux control. *Adv Enzyme Regul* **38**:65–85.

Srivastava, D. K., and Bernhard, S. A. (1986). Enzyme–enzyme interactions and the regulation of metabolic reaction pathways. In *Current Topics in Cellular Regulation*, Vol. 28, Academic, New York, pp. 1–68.

4

SIGNAL TRANSDUCTION PATHWAYS AND THE CONTROL OF CELLULAR RESPONSES TO EXTERNAL STIMULI

JUSTIN A. MACDONALD

Cells have developed an inherent ability to react to changes in their environment. The evolution of multicellular organisms has created a requisite need for organisms to spread information among all cells, not just those at the interface with the environment. As a result, cells of the same type can react synchronously to environmental stresses. In all multicellular organisms, elaborate communication networks between cells coordinates the growth, differentiation, and metabolism that occur in diverse tissues and organs. Intercellular communication can progress by direct cell-to-cell contact or by the generation of extracellular signaling molecules. These signaling molecules are synthesized and released by sensing cells and then migrate to other target cells, where they elicit a specific response from those target cells that have receptors for the signal. For most signaling molecules, a complex series of events and cellular machinery, collectively known as signal transduction pathways, links the formation, transmission, and reception of the signal. A major mechanism of signal transmission is protein phosphorylation (or dephosphorylation). It is the purpose of this chapter to describe the major signaling pathways by which signaling molecules exert their effects on cells. The associated cellular machinery that controls protein phosphorylation will also be characterized.

FUNCTION AND STRUCTURE OF SIGNALING PATHWAYS

Intercellular communication relies on specialized cells that perform dual functions: (1) they act as sensors of changing environmental stimuli, and (2) they produce signals (chemical or electrical) that inform other cells of the changing conditions. The specific signal output of a sensing cell (both intensity and duration) is generated in response to a particular stimulus. Ultimately, the signal is registered by a receptor on a target cell and is transmitted and processed further with the help of intracellular signaling networks. Often different environmental stimuli can elicit multiple signals from sensing cells. Target cells have also developed multiple receptors that permit the cells to respond to more than one signal. In addition, target cells possess many different intracellular networks to respond to the large number of potential signals. Each network can respond independently to different signals. Thus, a sophisticated coordination of receptors and intracellular networks permits a single target cell to respond to multiple signals in an efficient manner.

In summary, then, there are several steps linking an environmental stimulus to a molecular/biochemical response in a target cell (Fig. 4.1): (1) signal stimulus, usually in the form of changing environmental conditions, (2) the formation of a signal in a sensing cell in response to an external stimulus, (3) transmission of the signal to the target cell, (4) reception of the signal by the target cell, (5) transduction of the signal across the cell membrane, (6) intracellular transmission of the signal within the target cell, (7) transformation of the signal into a biochemical or electrical response in the target cell, and (8) termination of the signal.

To understand signaling, it is necessary to consider the mechanisms by which signal transfer occurs. The func-

Functional Metabolism: Regulation and Adaptation, edited by Kenneth B. Storey.
ISBN 0-471-41090-X

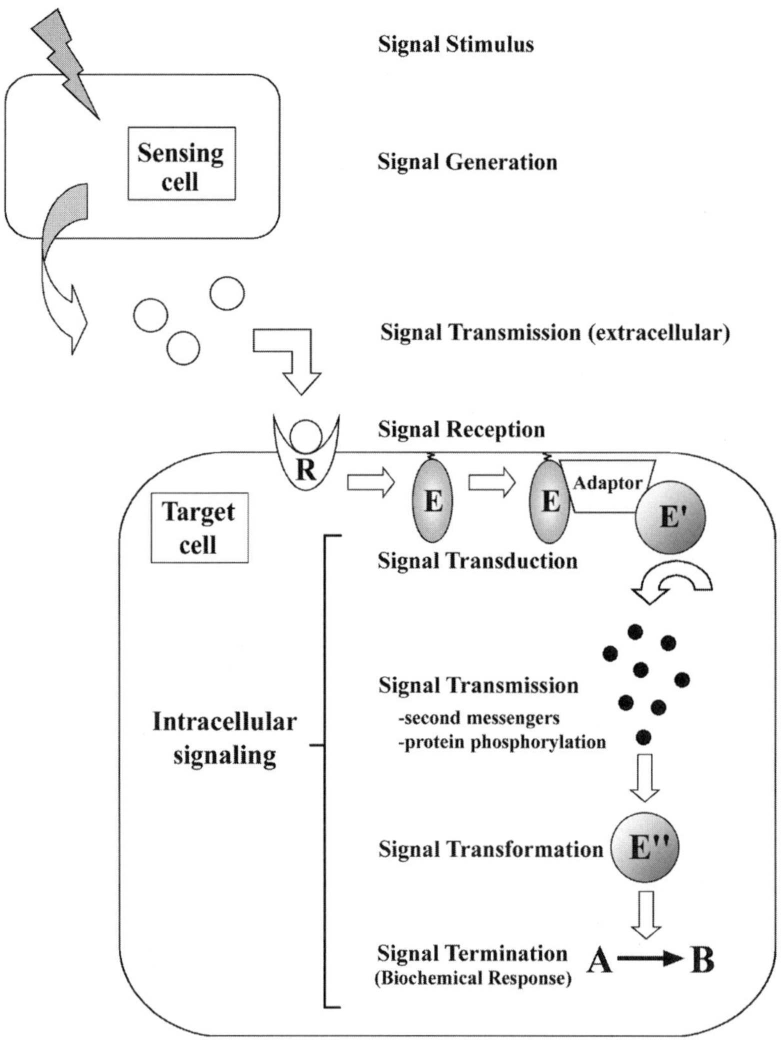

Figure 4.1 The components of intercellular and intracellular communication. The reception of an environmental stimulus triggers a sensing cell to generate a chemical messenger that can be secreted and transported to a target cell. Reception of the chemical signaling molecule by a membrane receptor (R) activates a signal transducing pathway. The signal is relayed to a downstream effector protein (E). Frequently, adaptor proteins coordinate the response of multiple effector proteins (E′). The intracellular transmission of a signal occurs via secondary messengers and protein phosphorylation. Ultimately, a specific biochemical response is generated within the target cell. Signal termination occurs through feedback mechanisms that permit regulation of communication at any of the preceding signaling steps.

tional capacity of multicellular organisms arises from their ability to coordinate the biochemical reactions in the various cells of the organism. In complex biological systems, cellular communication can occur through a number of mechanisms.

- Chemical messengers are distributed both extracellularly in the form of hormones or neurotransmitters (primary messengers) or intracellularly in the form of small molecular weight metabolites (secondary messengers).
- Electrical conduction of impulses by nerve cells allows cells to communicate with other cells at specialized nerve synapses. Electrical signals can be transformed into chemical signals and vice versa.
- Gap junctions are distributed along the lateral surfaces of cells allowing the exchange of small molecules. Gap junctions help to integrate the metabolic activity

of all cells in a common tissue by assuring that they share a common pool of metabolites.

- Intercellular contact via cell surface proteins allows direct communication between cells. Cell surface proteins of one cell interact with specific complementary proteins of another cell. Intracellular protein–protein interactions are also known to facilitate signal transmission within cells.
- Enzymatic control by post-translational modifications (including, but not limited to, protein phosphorylation) helps to coordinate activities of multiple enzymes and pathways within a cell.

Specialized proteins, called receptors, are utilized for the reception of intercellular signals by the target cell. There are two principal routes by which chemical signals can be processed by a target cell. Some signaling molecules are able to pass through the cell membrane to bind and activate receptors located within the cell. Other signaling molecules are membrane-impermeable and, therefore, must exert their effects from outside the cell. Intracellular receptors for membrane-permeable chemical messengers are normally found in the nucleus but can also be located in the cytosol. The membrane-permeable signal molecules include a large family of steroid hormones, such as estrogens, androgens, testosterones, and progesterones. After binding of the hormone, the receptor undergoes a conformational change. The receptor is then able to bind to DNA itself or to proteins that, in turn, interact with DNA. By this process, steroid hormones can directly regulate gene expression. The membrane-impermeable signaling molecules include some hormones, neurotransmitters, and growth factors in higher organisms. Others found in lower organisms include mating factors and pheromones in yeast and folic acid and adenosine cyclic 3′,5′-phosphate (cAMP) in the slime mold, *Dictyostelium discoideum*. Membrane-impermeable signaling molecules bind to specific receptors on the outer surface of the plasma membrane of the target cell. Upon binding of the ligands, these receptors convert the extracellular stimulus to an intracellular molecular form by one of several mechanisms. Hence, the term *signal transduction* (from the Latin *transducere*; to lead across or transfer) was used originally to describe the movement of signals across the cell membrane. The term is now more broadly applied to include all of the cellular architecture involved in the flow of biological information.

For most neurotransmitters, the receptor is a ligand-gated ion channel and the response of the cell is a ligand-induced change in the membrane potential. For most other extracellular signals, a complex series of events links the binding of the signaling molecule to its specific membrane receptor and to its final effects on cellular function. Typically, binding of a ligand results in either the stimulation of an intrinsic enzymatic activity of its receptor or the modulation of a transducing protein. The transducing proteins act to pass signals on from the receptor to downstream proteins, which recruit additional proteins as the next members in the signaling chain. The participating proteins can be enzymes or they can be adapters that recruit other proteins into the signaling chain. Adapters function to target signaling proteins to specific cellular sites and to recruit signaling molecules into multiprotein signaling complexes. The overall chain of events from the activation of the membrane receptor to the ultimate effects on cellular function is referred to as a signal transduction pathway.

The activation of signaling pathway enzymes often leads to the formation of small molecular weight, diffusible chemical messengers called second messengers that facilitate the intracellular transmission of the signal. A number of second messengers have been identified including cyclic-AMP, or cAMP, guanosine 3′,5′-cyclic monophosphate (cyclic-GMP, or cGMP), Ca^{2+}, inositol phosphates, nitric oxide, and certain lipid molecules.

Frequently, the final step in a signal transduction pathway involves the post-translational modification (see Text Box 4.1) of a particular protein that plays a central role in a biological process. A common and important post-translational modification associated with signal transduction in animal systems is the covalent attachment of phosphate onto protein targets (Fig. 4.2). Reversible protein phosphorylation is now known to regulate almost all aspects of cell life. When the phosphorylation state of a protein is altered either by phosphate attachment via a protein kinase or phosphate removal via a protein phospha-

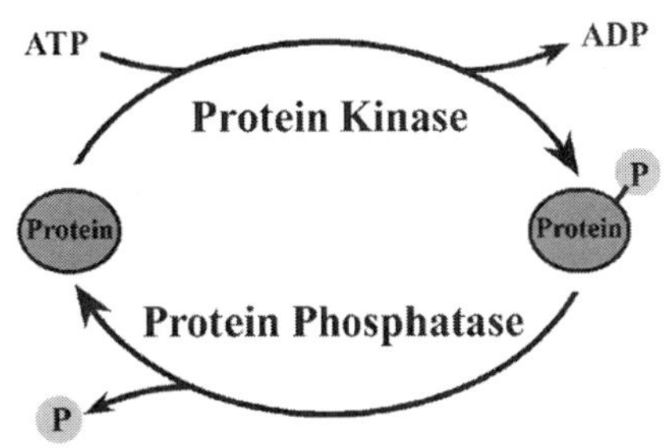

Phosphorylation:	Protein + ATP	⟶	Protein-P + ADP
Dephosphorylation:	Protein-P + H_2O	⟶	Protein + Pi
Net:	ATP + H_2O	⟶	ADP + Pi

Figure 4.2 The protein phosphorylation cycle. Protein kinases catalyze the covalent attachment of phosphate onto a protein substrate whereas a protein phosphatase will only catalyze the removal of covalently attached phosphate from a protein substrate. The net result of the protein phosphorylation cycle is the hydrolysis of adenosine 5′-triphosphate (ATP) to produce adenosine 5′-diphosphate (ADP) and inorganic phosphate (P_i).

TEXT BOX 4.1 POST-TRANSLATIONAL MODIFICATIONS

Connectivity of Biological Studies

Cellular proteins are not invariant products of genes but are subject to a high degree of interdependent processing at the protein level. This critical component of cellular function and regulation includes many kinds of post-translational modifications of proteins. In contrast to the static genome, where all information could in principle be obtained from the DNA of a single cell, the proteome is dynamic and highly dependent not only on the type of cell but also on the state of the cell.

Common Post-translational Modifications

Phosphorylation	Adenylation	Ubiquination
Glycation	Oligomerization	Amidation
Proteolysis	γ-Carboxyglutamic acid formation	Methylation
Acetylation	Hydroxylation	Iodination
Nitrosylation	Glycosylation options	ADP-ribosylation
Lipidation options	*N*-acetylgalactosamine	Oxidation
Myristolation	Galactose	
Palmitoylation	Mannose	
Prenylation	*N*-acetylglucosamine	
GPI (glycophosphatidylinositol)	Sialic acid	

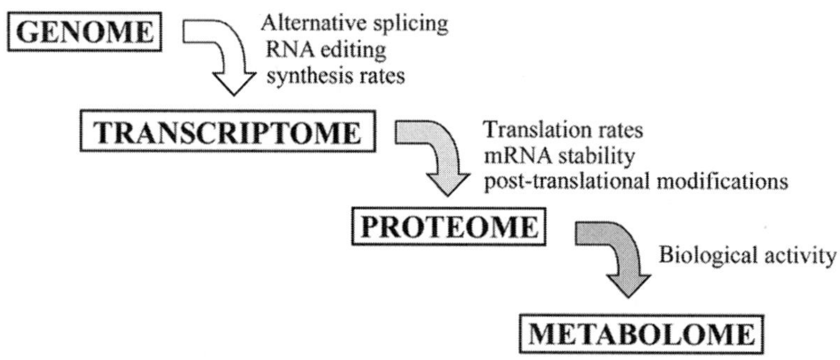

Figure TB4.1 The biological information encoded in the genome is linked through the transcriptome and the proteome to the metabolome. Multiple protein isoforms (proteome) can be generated by RNA processing and post-translational modifications. The mRNA population (transcriptome) can be regulated by message stability and translation efficacy. Post-translational processing can alter a protein's biological activity and change the cellular concentration of metabolic substrates, products and effectors (metabolome).

tase, the biological function of the protein may be enhanced or suppressed. Phosphorylation or dephosphorylation can affect a protein in numerous ways including: (1) increasing or decreasing protein/enzyme activity, (2) enabling a protein to interact with or dissociate from other proteins, (3) allowing a protein to move from one subcellular compartment to another, or (4) marking a protein for destruction. The simplicity, flexibility, and reversibility of this post-translational modification, coupled with the ready availability of adenosine 5′-triphosphate (ATP) as a phosphoryl donor, is presumably why the mechanism has been adopted to regulate so many biological processes. For additional information of protein control via reversible phosphorylation see Chapters 1 and 12.

Protein Kinases and Protein Phosphatases

Protein kinases (and their associated protein phosphatases) are well suited to provide a mechanism of biological control. Protein kinases catalyze protein phosphorylation by transferring the high-energy terminal γ-phosphate group of ATP onto a selected target amino acid, almost always one with a free hydroxyl group, such as serine or threonine. Phosphorylated proteins often have altered conformations (tertiary and quaternary structures; see Chapter 2) because the new phosphate group is more bulky than the hydroxyl group, negatively charged, and more hydrophilic. In the case of enzymes, this can alter activity, usually by affecting the enzyme's affinity for substrate. For nonen-zyme proteins, as indicated above, an altered three-dimensional structure can affect their ability to polymerize or to effectively span cell membranes. Protein phosphoryla-tion/dephosphorylation reactions, like all chemical reactions, are governed by an equilibrium constant [Eq. (2.5)], and their kinetics must obey the Haldane relationship [Eq. (2.33)]. Like many other cellular reactions, the phosphorylation reaction is driven to completion because ATP is the source of high-energy phosphate for the process. The equation can be written as

$$\text{Enzyme} + \text{ATP} \leftrightarrow \text{Enzyme-P} + \text{ADP} \tag{4.1a}$$

with an equilibrium constant given by

$$K_{eq} = [\text{Enzyme-P}][\text{ADP}]/[\text{enzyme}][\text{ATP}] \tag{4.1b}$$

The process is driven far to the right by the relatively large excess of free energy released upon ATP hydrolysis and the relatively (compared to ATP) low free cellular [ADP] (adenosine 5′-diphosphate) concentrations; free ATP is approximately 3 mM while free ADP is only about 60 μM. For this reason, phosphorylation events appear unidirectional, even though they are not. Because enzyme phosphorylation is such an important cellular event, the dephosphorylation process is catalyzed by a separate enzyme. It is common to find cellular mechanisms to regulate both phosphorylation and dephosphorylation. Like phosphorylation, there is an associated equilibrium equation for the dephosphorylation process:

$$\text{Enzyme-P} + H_2O \leftrightarrow \text{Enzyme} + H_2PO_4 \tag{4.1c}$$

with its own equilibrium constant:

$$K_{eq} = [\text{Enzyme}][H_2PO_4]/[\text{Enzyme-P}][H_2O] \tag{4.1d}$$

The overall process for the reaction can, therefore, be written as [combining (4.1a) and (4.1c)]:

$$\text{ATP} + H_2O \leftrightarrow \text{ADP} + H_2PO_4 \tag{4.1e}$$

which, you will note, is simply the equation for the hydrolysis of ATP to ADP and free phosphate. Thus, while phosphorylation events take up some of the energy of ATP hydrolysis, the cycle is complete, and the total energy released only when the phosphorylated enzyme is dephosphorylated. This means that the phosphoenzyme has a higher energy than the dephosphoenzyme and can be thought of as existing in a chemically "activated" state, even though its actual enzyme activity may be lower than that of the dephosphorylated enzyme.

Protein phosphorylation provides a means for rapidly switching the activity of a cellular protein from one state to another (i.e., allows for on–off control in many cases) because the reaction equilibria lie far in the direction of product formation, as indicated above. From a practical point of view, protein kinases catalyze only protein phosphorylation reactions and protein phosphatases only catalyze dephosphorylation reactions. Hence, in theory, the proportions of phosphorylated versus dephosphorylated protein can be varied from very small (approximately 0.02 : 1; i.e., the ratio of free ADP : ATP in the cell) to 1 : 0.02 by changing the relative rates of the protein kinase and phosphatase reactions. However, this degree of change in phosphorylation is very rarely observed in nature. By contrast, it is impossible to achieve this type of control over reversible reactions catalyzed by a single isolated enzyme (not in a metabolic pathway) because altering an enzyme's affinity for substrate will have an equal and opposite effect on its affinity for product by virtue of the Haldane relationship [Eq. (2.33)].

Because the net result of protein phosphorylation is the hydrolysis of ATP (Fig. 4.2), the phosphorylation and dephosphorylation reactions are energetically favorable. However, they must be regulated to prevent wasteful cycling that would drain a cell of its supply of ATP. The ability of a cell to regulate the two reactions independently is undoubtedly a primary reason why this mechanism is used to regulate so many cellular processes.

Another reason why phosphorylation–dephosphorylation cycles are such important control mechanisms is that they possess high sensitivity to signal input. Sensitivity in this case refers to the fact that a small percentage change in the initial input signal produces a much greater percentage change in the final output response. The phosphorylation and dephosphorylation reactions catalyzed by protein kinases and protein phosphatases are said to be zero order with respect to substrate, that is, they will occur at rates that are not dependent on the concentration of the target protein substrate. The reaction rates are independent of the concentration of target protein substrate because the kinase and phosphatase enzymes are working near saturation. The reactions will have zero-order rate constants k_f and k_r, respectively. If the numbers of all target proteins in the cell are considered to be infinitely large and if there is full saturation of kinase and phosphatase enzymes, there are only two possible states: when $k_f > k_r$, all protein targets are phosphorylated and when $k_f < k_r$ all protein targets are unphosphorylated. Thus, a minor change in the environment that influences k_f or k_r can change the phosphorylation state of a target protein completely from one extreme to the other. The extent of the phosphorylation is said to be *ultrasensitive* to changes in signal input. Hence, the signaling sensitivity that results from protein kinases and phosphatases that display zero-order kinetics is referred to as zero-order ultrasensitivity.

One way in which cellular sensitivity is produced is to have protein kinases and protein phosphatases act simultaneously on target proteins with opposing metabolic functions. The best studied example of this is in the control of glycogen breakdown in the muscle and liver. Hormone stimulation by glucagon in the liver or epinephrine in the muscle activates adenylyl cyclase to produce the second messenger, cyclic AMP, that then activates cAMP-dependent protein kinase (PKA). PKA then phosphorylates several targets. PKA phosphorylates and activates glycogen phosphorylase kinase (GPK, the enzyme that activates glycogen phosphorylase) and at the same time phosphorylates and inactivates the protein phosphatase that inactivates phosphorylase kinase. These paired targets of PKA ensure that the PKA-mediated phosphorylation of GPK is not "wasted" due to its immediate dephosphorylation by the phosphatase. This same principle for increasing the sensitivity of response to a signal also applies on a larger scale to control the opposing processes of glycogenolysis versus glycogen synthesis (e.g., glycogen synthase is phosphorylated and inhibited by PKA), an effect that is important because it prevents the rise in intracellular glucose levels during glycogenolysis from triggering a resynthesis of glycogen. On an even larger scale, the same principle of reciprocal control regulates carbohydrate versus lipid catabolism to prevent wasteful recycling of fuels.

Cascades and Cross-talk

A unique organizational feature exhibited by protein kinases and phosphatases is their assembly into a type of linked network known as a signal transduction cascade. These cascades provide greatly increased sensitivity to signals, allow a single signal to be transmitted to stimulate a wide variety of metabolic and gene expression responses, and allow for coordinated cross-talk between pathways when cells are responding to multiple different signals. A key factor of the cascade system is the capability of many protein-phosphorylation systems to switch between an on state and off state.

Signal transduction cascades should not be thought of as linear networks but as a complex weave of interacting pathways. Different signaling pathways communicate with one another in a process commonly known as cross-talk. There are numerous examples of cross-talk between signaling pathways that involve protein phosphorylation. In general, the greater the number of components present in a signal pathway, the more targets there are for cross-talk. Many protein kinases are themselves either positively or negatively regulated by phosphorylation, and the same is true of many protein phosphatases, making them prime targets for cross-talk in pathways using phosphorylation cascades. Signal pathway cross-talk is becoming increasingly important for our understanding of signaling networks.

Signal cascades can branch in both convergent and divergent manners so that the summation of many different inputs into numerous different outputs is possible (Fig. 4.3). Cross-talk can occur between pathways activated by a single receptor (divergent) or pathways activated by different receptors (convergent). In fact, integration of cellular responses elicited by different receptors must occur by cross-talk of signaling pathways. Cross-talk can take place at many levels from the membrane to the nucleus and may involve components that are in common between two pathways. Furthermore, as important as cross-talk is, it is also becoming increasing clear that pathway insulation to prevent cross-talk is an equally important component of metabolic regulation.

An example of convergent pathways that produce the same end effect is illustrated by the response of cells to epinephrine. Some cells have two classes of adrenergic receptors (i.e., receptors that bind epinephrine or norepinephrine): the α-adrenergic receptor and the β-adrenergic receptor. The immediate effect of the binding of epinephrine by the α-adrenergic receptor is to stimulate an elevation of intracellular Ca^{2+} concentration which can activate GPK, among other enzymes. By contrast, β-adrenergic receptors are coupled to adenylate cyclase, so that the immediate effect of the binding of epinephrine to the β-adrenergic receptor is an elevation in cAMP levels. The increase in cAMP leads to activation of PKA which, in turn, can activate GPK. Thus, one hormone can elevate concentrations of two second messengers, Ca^{2+} and cAMP, to activate two signal transduction cascades which converge to evoke the same end response. Other cascades exhibit divergent paths to stimulate cellular responses from a single signal. For example, stimulation of the insulin receptor can activate at least three kinase pathways [PKC, mitogen-activated protein kinase (MAPK), and PKB/Akt] that each then elicit multiple cellular responses (e.g., gene expression, glucose metabolism, glucose transport). Some cascades are both converging and diverging.

Signal pathway cross-talk may also utilize feedback loops that can act in negative or positive manners. Feedback can be introduced at multiple points in a pathway. Typically, protein kinases phosphorylate and activate kinases located immediately downstream in the signal transduction cascade. However, a protein kinase may also phosphorylate the protein kinase that lies upstream of it and, thereby, cause a down-regulation of the pathway. Such a situation is present in the MAPK cascade. For

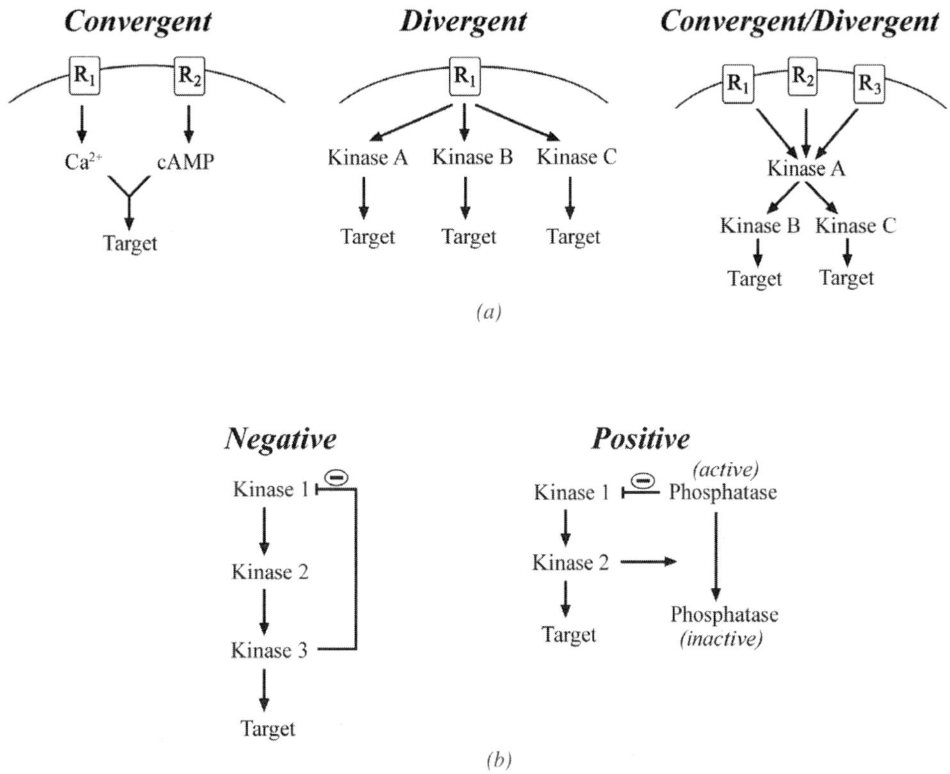

Figure 4.3 Cross-talk and feedback loops in signaling pathways. Examples of three types of cross-talk mechanisms are presented: (a) *Convergent*—different signaling molecules can lead to the activation of multiple signaling pathways (e.g., elevated levels of two second messengers Ca^{2+} and cAMP), which act on a single cellular target. *Divergent*—stimulation of a single receptor can promote the activation of multiple signaling pathways, which act on different cellular targets. *Convergent/divergent*—multiple signals may condense to activate a single signaling protein; this protein may in turn be connected through cross-talk to many additional signaling members and regulate multiple cellular targets. Examples of two types of feedback loops are presented. (b) *Negative*—once activated, downstream signaling members may inhibit the activity of upstream members to promote a decrease in the total flux through the pathway. *Positive*—downstream signaling members may remove inhibitory effects on upstream members, thus increasing the total flux through the pathway.

example, the MAPK, ERK, is activated when it is phosphorylated by MEK, the protein kinase that lies upstream of it. ERK then phosphorylates various downstream targets, but activated ERK also phosphorylates MEK to provide a negative feedback that down-regulates the pathway. Such a system may seem counterproductive but, in fact, allows for very short "pulse" activations of ERK that are needed in certain situations. An example of positive feedback is the situation where a protein kinase may phosphorylate and inactivate the protein phosphatase that is responsible for dephosphorylating the substrate of the kinase. For instance, as described earlier, PKA not only phosphorylates and activates GPK but at the same time phosphorylates and inactivates the protein phosphatase that dephosphorylates GPK. Positive feedback loops ensure that any minor activation of a protein kinase is amplified into a sustained response and help to prevent the futile hydrolysis of ATP that would occur with uncontrolled kinase and phosphatase activities.

One of the major challenges in understanding signaling networks is to elucidate how signaling specificity is achieved when many of the same core signaling pathways are activated by different receptors that ultimately elicit different cellular responses. In principle, signaling specificity can result from the activation of unique signaling pathways by a single receptor but in practise this rarely (never) occurs. More commonly, since cellular responses reflect an integration of outputs from all the pathways activated by a single receptor, signaling specificity can result from the unique combination of signaling pathways activated by the receptor. The exact location of ligand-activated receptors on the cell surface may also dictate which

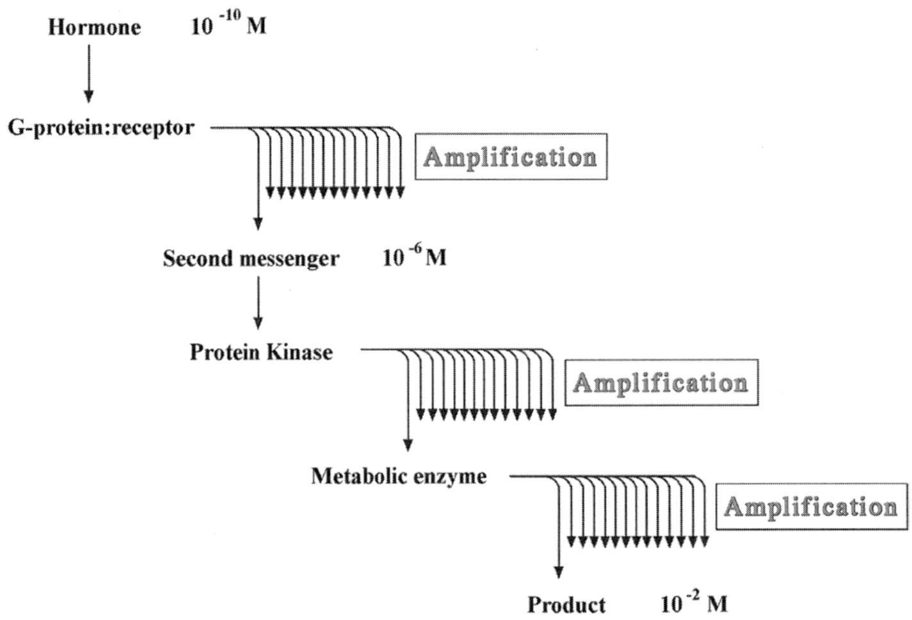

Figure 4.4 Amplification of an extracellular signal by means of a signal transduction cascade. Binding of a single hormone molecule (e.g., epinephrine) can result in the production of millions of product molecules (e.g., glucose). In this example, amplification steps are represented by the synthesis of many second-messenger molecules (e.g., cAMP), the stimulation of many activating enzymes (e.g., glycogen phosphorylase kinase and glycogen phosphorylase), and the production of many product molecules (e.g., glucose). The cascade of amplifying steps can explain the extremely high potency of many hormones.

pathways can be activated; thus spatial separation may be used as another way of generating specificity. Cell type-specific responses to a given extracellular stimulus may also be elicited by the expression of different repertoires of downstream signaling proteins.

Amplification and Signal Thresholds

The cascade of steps linking the binding of a hormone to the final cellular response to that hormone may seem overcomplicated on first inspection. It would be simpler, of course, if a hormone activated its final target (enzyme, gene) directly, thereby saving several steps and sparing cellular energy. However, a cascade system permits the coordination of an entire group of responses (e.g., activation of multiple enzymes in pathways, transcription of multiple genes) by changes in the level of a single messenger molecule. And importantly, the cascade of events provides a huge amplification of an initially tiny signal.

Amplification is distinct from other network parameters such as signal sensitivity and signal threshold. Amplification refers to the situation where the absolute concentration of the final target protein is much higher than that of the input signal (Fig. 4.4). A downstream target of a protein kinase cascade can be present at concentrations several orders of magnitude higher than the hormone that triggers the effect. For example, the concentration of epinephrine needed in the blood to stimulate glycogenolysis and release glucose from the liver can be as low at 10^{-10} M. This hormonal stimulus generates a concentration of more than 10^{-6} M of cAMP in the liver cell. Three additional amplification steps precede the conversion of glycogen to glucose accounting for an additional 10^4 amplification and a final cellular glucose concentration exceeding 10^{-3} M.

One by-product of amplification is the need for signal thresholds. Because signal initiation can be induced with only a few activated receptor molecules per cell, tight control of the on/off switch is required. In many systems, the threshold is as little as a twofold change in the level of input that is needed to initiate or terminate signaling. Signaling thresholds can be different in different cell types permitting organ-specific responses to environmental stimuli.

SIGNAL TRANSMISSION VIA PROTEIN KINASES

Three major types of input are currently known that modulate protein phosphorylation cycles: extracellular signals, cell cycle checkpoints, and environmental or nutritional stress. Each of these types of stimuli can produce intracellular messages that ultimately act by regulating the activity of enzymes that catalyze the phosphorylation of a wide variety of protein substrates, including other signaling proteins, metabolic enzymes, structural proteins, inhibitor proteins, transcription factors, and translation factors. The families of enzymes that catalyze the transfer of a phosphate group from ATP to another protein are called protein kinases. The first known protein kinase was actually GPK, and it was discovered by Krebs and Fischer in 1956 when they determined the mechanism by which nervous stimulation and adrenaline-activated glycogen breakdown. However, in recent years the number of known protein kinases has grown explosively, and novel protein kinase family members continue to appear in the scientific literature. Many of the newly identified protein kinase members were identified using homology-based polymerase chain reaction (PCR) cloning strategies (all protein kinases have sequence similarities). Current estimates suggest the existence of nearly 500 different protein kinase members. It is believed that protein kinases constitute nearly 1% of total cellular protein in a typical mammalian cell.

Protein kinase family members have been identified and characterized throughout the Eukaryota including vertebrate, invertebrates, plants, fungi, and unicellular eukaryotes such as yeast. Phylogenetic mapping has traced the origin of the protein kinase proginator gene to a time prior to the evolutionary separation of the major eukaryotic kingdoms, and many types of protein kinases are highly conserved across phylogeny. Studies in yeast have been instrumental for both the isolation of new protein kinases and the elucidation of their physiological roles. The discovery of protein kinase genes throughout the eukaryotic kingdom has lead to a growing acceptance that protein phosphorylation has been preserved throughout cellular evolution and is instrumental as a primary mechanism of metabolic regulation.

Protein phosphorylation has also been identified in prokaryotic organisms. Protein kinases have been found in bacterial systems that control chemotaxis, nutrient uptake, and gene transcription. Bacterial protein kinases are classified into three general groups: (1) protein histidine kinases, (2) protein Ser/Thr kinases, and (3) phosphotransferases such as those of the phospho*enol*pyruvate-dependent phosphotransferase system involved in sugar uptake.

The most common signal relaying system present in prokaryotic organisms is the two-component system. A prototypical two-component pathway consists of two proteins: a protein histidine kinase (also called a sensor kinase) and a response regulator. The N-terminal section of the histidine kinase functions as an input domain that can detect extracellular signals directly. The C-terminal portion contains a transmitter module that includes a conserved histidine residue. Signal perception regulates autophosphorylation of this histidine residue. The phosphoryl group is then transferred from the histidine of the sensor kinase to a conserved aspartate residue in the receiver domain of the response regulator. The phosphorylation state of the regulator modulates the activity of its output domain. Known output domains include transcription factors or regulators of metabolism. More complex histidine kinase modules contain an additional transmitter protein called a histidine-containing phosphotransfer (HPt) protein. The HPt protein serves as a phosphohistidine intermediate in the transfer of the phosphoryl group from a receiver domain onto another receiver, resulting in a His-Asp-His-Asp phosphorelay. Recently, histidine kinases have also been discovered in plants (see Text Box 4.2).

Histidine kinases are unrelated to the superfamily of protein serine/threonine and tyrosine kinases that contains the overwhelming majority of eukaryotic protein kinases. However, a small but ever-growing number of histidine protein kinases are being identified in eukaryotes as a result of searching gene databases for sequences with high percent identity to the sequences of prokaryote protein kinases. Three of these are prominent mitochondrial protein kinases: pyruvate dehydrogenase kinase, α-ketoglutarate dehydrogenase kinase, and the branched-chain α-ketoacid dehydrogenase kinase (see Chapter 8). Presumably these are derived from the prokaryotic histidine kinases that were present in the ancient prokaryote that became the mitochondrion after endosymbiotic union with the protoeukaryotic cell (see Chapter 20).

Protein kinases use the γ phosphate of ATP to generate several types of phosphate monoesters. In eukaryotes, the alcohol groups on serine or threonine (Ser/Thr protein kinases), or the phenolic groups on tyrosine (Tyr protein kinases) are the most common sites of phosphorylation (Fig. 4.5). However, amino groups on histidine, lysine, or arginine can also be phosphorylated and, in rare occasions, protein carboxyl groups on aspartate or glutamate can also serve as phosphate acceptors. Therefore, many residues in a protein can potentially be phosphorylated.

What parameters define the phosphorylation site of a substrate protein? Not all serine, threonine, or tyrosine residues in a protein are potential phosphorylation targets. The potential for phosphorylation of a specific amino acid residue depends on the primary structure around the residue. By using sequence comparisons of phosphorylation sites as well as defined peptides as substrates, it has been clearly shown that the amino acid sequence in the region surrounding the phosphorylation site is an important determinant of specificity. Different protein kinases exhibit different requirements with respect to the sequences on either side of the residue to be phosphorylated, so that each kinase subfamily has its own consensus sequence

TEXT BOX 4.2 TWO-COMPONENT SIGNAL RELAY SYSTEM AND ETHYLENE SIGNALING IN PLANTS

The two component system is a signaling mechanism used primarily by bacteria to respond to a broad array of environmental signals. *Escherichia coli* alone has at least 30 different pairs of sensor/response regulator proteins. Most recently, the ethylene receptor in the plant *Arabidopsis thaliana* was identified as a two-component histidine kinase homolog.

Small gaseous molecules play important roles in biological signaling in both animal and plant physiology. Nitric oxide is the leading example in mammals. In plants, the premier example of a gaseous signal molecule is the simple hydrocarbon ethylene (C_2H_4). Considered one of the five classic plant hormones, ethylene has long been known to play an important role in fruit ripening (Fig. TB4.2).

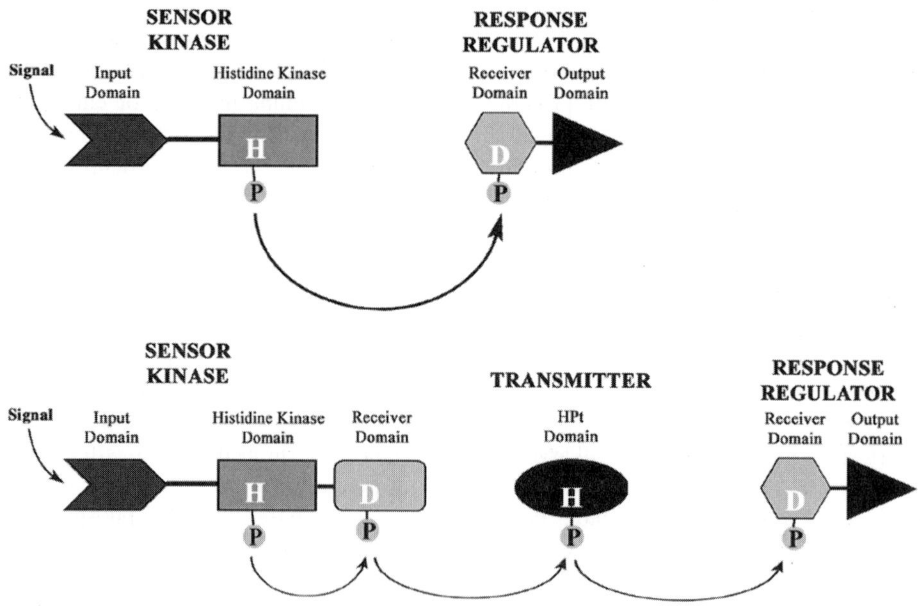

Figure TB4.2 The two-component histidine kinase regulatory system. The simplest two-component system consists of a sensor histidine kinase and a response regulator. The sensor kinase possesses ATP-binding and histidine (H) autophosphorylation regions within the histidine kinase domain. Phosphate transfer is from the sensor kinase to a response regulator. The response regulator also possesses two domains; a receiver domain for aspartate (D) phosphorylation and an output domain which has the ability to bind DNA and activate transcription. In the more complex two-component systems, the sensor kinase may possess an integral receiver domain. In this case, the phosphate group is transferred through a bridge component (transmitter) that carries a histidine-containing phosphotransfer domain (HPt).

for phosphorylation (Table 4.1). A single protein can exhibit from one to several such sites for phosphorylation by an individual protein kinase as well as phosphorylation sites for two or more different kinases so that multiple and/or cooperative phosphorylation is possible. Phosphorylation potential also depends on the overall three-dimensional architecture of the protein substrate (i.e., its tertiary structure). Since phosphorylation is a post-translational event, the protein substrate is in a native, folded state. The accessibility of a particular amino acid residue can influence the probability that it will be phosphorylated.

One exciting outcome of the ongoing genomic sequencing projects is the construction of a comprehensive catalog of the two major subgroups of protein kinases, Ser/Thr kinases and the Tyr kinases, from multiple eukaryotes. The genome of the yeast, *Saccharomyces cerevisiae*,

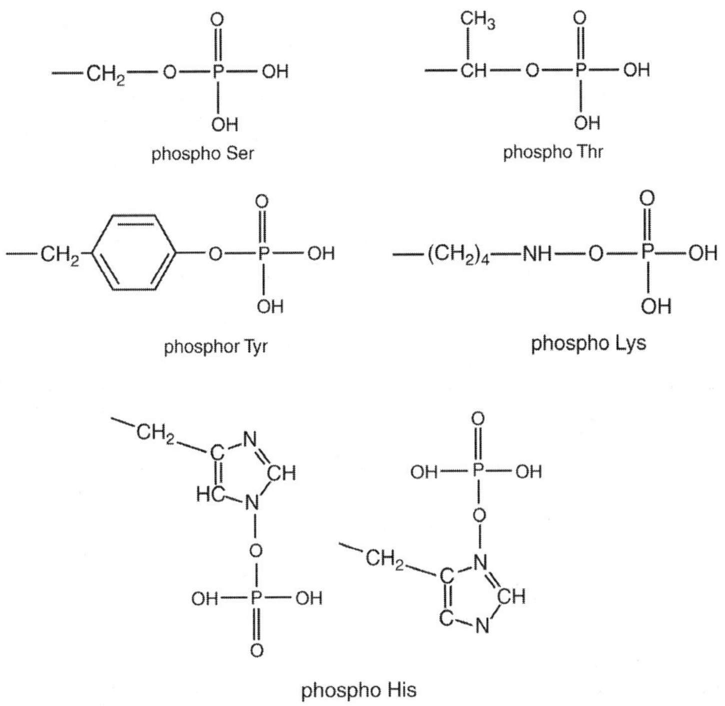

Figure 4.5 Chemical structures of phosphorylated amino acid residues. The sites of phosphorylation are usually the hydroxyl groups of specific threonine (Thr), serine (Ser), or tyrosine (Tyr) residues, but the primary and tertiary amines of lysine (Lys) and histidine (His) residues may also be used. The imidazole side chain of histidine may exist as two tautomers: thus phosphorylation may occur at either of the two N atoms within the ring.

encodes 114 Ser/Thr kinase proteins (but no Tyr kinases) out of 6217 genes; hence, 1.8% of genes are Ser/Thr kinases. The recently completed sequence of the nematode *Caenorhabditis elegans* genome revealed 400 kinase proteins (including 92 Tyr kinases) out of 19,099 genes (2.1% protein kinases). The human genome is predicted to encode more than 1000 protein kinases; currently, ~350 distinct protein kinase gene products are known to exist. Further subdivision of protein kinases into distinct families has been made based on shared structural and functional properties. Similarity in kinase amino acid sequence has proven to be a good indicator of other charac-

TABLE 4.1 Consensus Sequences for Phosphorylation by Protein Kinases

Protein Kinase	Consensus Sequence[a]
cAMP-dependent protein kinase (PKA)	R-X-**S/T**-X, R-R/K-X-**S/T**-X
Protein kinase C (PKC)	X-**S/T**-X-R/K
Myosin light-chain kinase (MLCK)	X-K-K-R-X-X-**S**-X
Cdc2 protein kinase	**S/T**-P-X-R/K
Casein kinase I (CKI)	X-pS-X-X-**S/T**
Casein kinase II (CKII)	X-**S/T**-X-X-E
Phosphorylase kinase (GPK)	K/R-X-X-**S**-V/I
Mitogen-activated protein kinase (MAPK)	P-X-**S/T**-P, X-X-**S/T**-P
Ca^{2+}/calmodulin-dependent protein kinase II (CaMK II)	R-X-X-**S/T**-V/I
Ca^{2+}/calmodulin-dependent protein kinase I (CaMK I)	L-R-X-X-**S/T**-X-X-L

[a]Phosphorylation site is presented in bold type. Single-letter amino acid code: E = glutamic acid, I = isoleucine, K = lysine, L = leucine, P = proline, R = arginine, S = serine, T = threonine, V = valine, X = any amino acid, pS = phosphorylated serine.

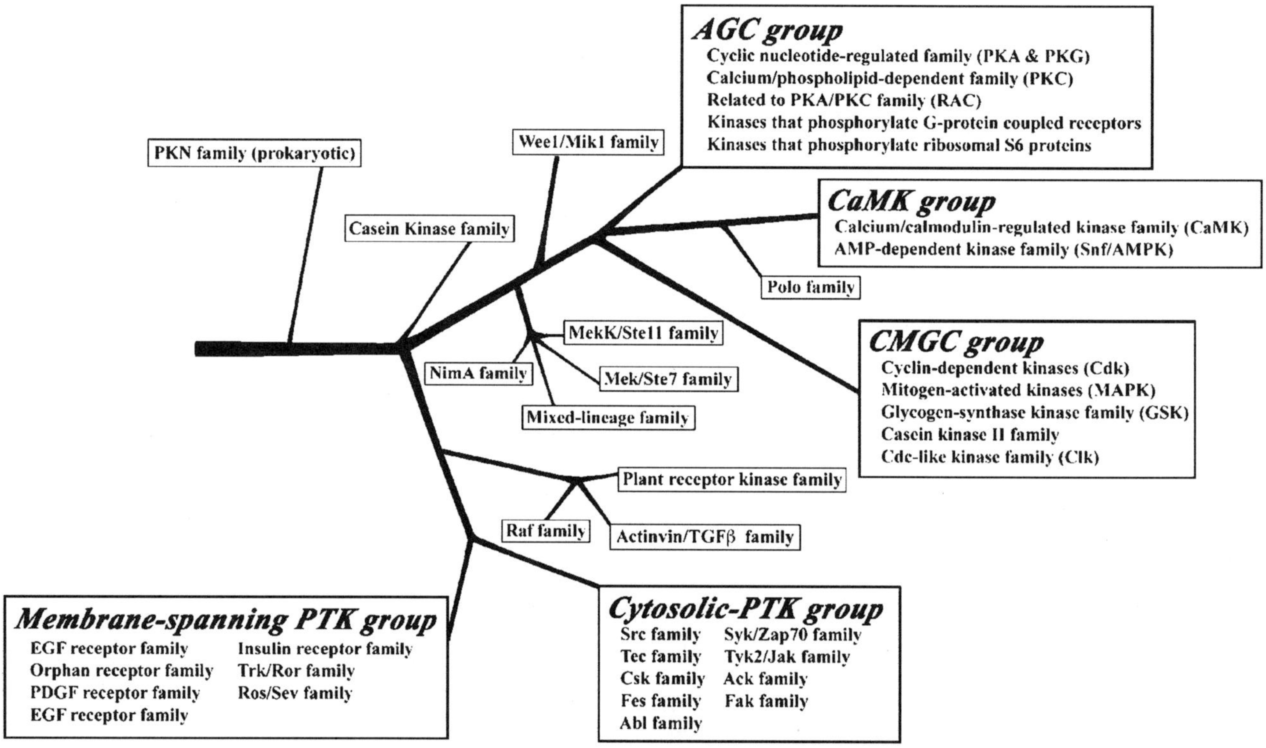

Figure 4.6 Tree diagram showing how the major families of protein kinases are related to one another. Three major groups of serine/threonine protein kinases are presented with the AGC (the cAMP-kinase, cGMP-kinase, and calcium-dependent kinase), CaMK (the calmodulin-dependent kinase group), and CMGC (cyclin, mitogen, glycogen, and casein kinases) groups. Two major groups of protein tyrosine kinases (PTK) are illustrated with the cytosolic and membrane-spanning PTKs. Many families of protein kinases do not belong to one of these five major classifications. Their relation to the major groups of protein kinases are also shown.

teristics held in common by the different members of the family. The protein kinase superfamily of eukaryotes has been subdivided into four large groups of kinases, each consisting of multiple related families (Fig. 4.6):

- The AGC group, which includes among others the cyclic nucleotide protein kinases (PKA and PKG) and the Ca^{2+}/phospholipid-dependent protein kinases (PKC)
- The CaMK group, which includes protein kinases regulated by Ca^{2+}/calmodulin
- The CMGC group, which includes cyclin-dependent protein kinases and the mitogen-activated protein kinases
- The PTK group, which includes protein tyrosine kinases

A large number of kinases do not fit within these four major groups. Many of these "unclassified" kinases are clearly related to one another through structural and functional characteristics but, because of a lack of data, it has been difficult to make broader generalizations that could arrange the families into a larger group. Other miscellaneous kinases do not fall into any major group of eukaryotic kinases. These include histidine kinases, aspartate/glutamate kinases, and a number of protein kinase homologs from deoxyribonucleic acid (DNA) viruses and other prokaryotic organisms.

All enzymes that act as protein kinases share common structural features despite having overall protein sequences that are dissimilar. Many conserved features of the primary structure of protein kinase have been identified through inspection of multiple amino acid sequence alignments. Those regions that are recognized as being invariant or nearly invariant throughout the kinase superfamily (conserved in over 95% of 650 kinase sequences) have been assigned to 12 domains (Table 4.2). These domains are regions that are not interrupted by large amino acid insertions and that contain patterns of conserved residues.

TABLE 4.2 Protein Kinase Domains and Their Role in Phosphate Transfer to Protein Substrates

Domain	Domain Structure	Domain Function
I	β-sheet flap	Clamps and anchors α/β phosphates of ATP
	Hydrogen bonds	Encloses adenine ring of ATP
	Hydrophobic pocket	
II	Salt bridge/ionic bond	Anchors and orients ATP
		Essential for maximum enzyme activity
III	Ionic bond	Stabilizes interactions with α and β phosphates
IV	Hydrophobic pocket	No assigned function
V	Hydrogen bonds	Anchors ATP via adenine ring or ribose sugar
	Hydrophobic pocket	Surrounds adenine ring
	Ionic bond	Binds protein substrate
VIA	α-helix	Support structure
VIB	Contains catalytic residues	"Catalytic loop" facilitates phosphotransfer
	-Asp and Lys	Chelates Mg^{2+} ion
	Ionic bonds	Binds ATP ribose hydroxyl group
	Hydrogen bonds	Binds protein substrate
VII	Hydrogen bonds	Chelates activating Mg^{2+} ion
	Contains "DFG" triplet	Bridges β and γ phosphates of ATP
		Orients γ phosphate for transfer
VIII	"APE" motif	Stabilizes structure of peptide lobe
	Autophosphorylation site	Activation of maximal activity
IX	Ionic bond	Stabilizing catalytic loop
	Hydrophobic pocket	Interacts with protein substrate
X	Poorly conserved	Unclear
XI	Ionic bond	Stabilizes structure of peptide binding lobe

Additional studies of site-directed mutants have shown that each of the 12 kinase domains play essential roles in enzymatic function. These kinase domains in eukaryotic protein kinases impart the catalytic activity. Three separate roles can be attributed to the kinase domains:

1. Binding and orientation of the ATP phosphate donor in a complex with divalent cation (usually Mg^{2+}, sometimes Mn^{2+})
2. Binding and orientation of the protein substrate
3. Transfer of the γ-phosphate from ATP to the acceptor hydroxyl residue (serine, threonine or tyrosine) of the protein substrate

The homologous nature of the kinase domains implies that they all fold into topologically similar three-dimensional base structures. The first three-dimensional structure of a protein kinase that was determined using X-ray crystallography was that of the catalytic subunit of PKA. The three-dimensional (3-D) structure of several other protein kinases is also now known and shows the highly conserved topology of the protein kinase catalytic core. The 3-D model for cAMP-dependent protein kinase and other Ser/Thr kinases is a two-lobed structure with a deep cleft between the lobes, which is recognized as the site of catalysis (Fig. 4.7). One lobe is primarily responsible for binding the protein substrate and initiating phosphotransfer, and the other is primarily responsible for anchoring and orienting the nucleotide. The cleft is flexible and functions as a hinge to bring the two lobes together upon binding of the substrate protein and ATP.

More extensive structural studies completed with kinase–substrate complexes have demonstrated the localization of certain functional segments to parts of the kinase sequence. For example, a glycine-rich loop is involved at the ATP binding site with a consensus sequence Gly-X-Gly-(Phe/Tyr)-Gly-X-Val. Sequence comparisons and mutation experiments have revealed an essential function in the catalysis of phosphate transfer for several conserved amino acid residues (Fig. 4.8). It was originally proposed that an aspartic acid residue present at the active site acted as a catalytic base. However, it is now believed that this negatively charged residue does not accept a proton in any of the reaction steps but, instead, that the hydroxyl hydrogen of serine is shifted to the simultaneously transferred phosphate group of ATP. Another aspartic acid residue is important for the maintenance of

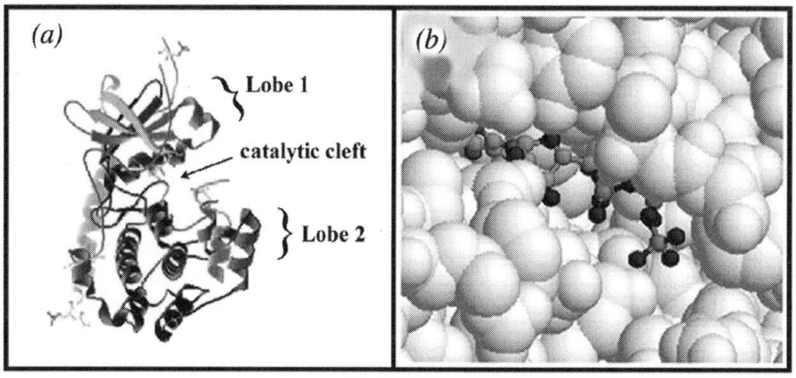

Figure 4.7 (*a*) Schematic drawing of the polypeptide backbone of cAMP-dependent protein kinase (PKA). α Helices are depicted as helical ribbons and β sheets are depicted as arrowed ribbons. The two lobes of the protein and the catalytic cleft are indicated. (*b*) A space-filling model of the ATP binding site within the catalytic cleft of PKA. A molecule of ATP is illustrated with the γ phosphate clearly visible. The nucleotide component of ATP is buried deep within the cleft and is only partially visible. Molecular models of PKA were generated with aid of the Rasmol 3D-image processing program. (See http://www.umass.edu/microbio/rasmol/ and Protein Data Bank http://www.rcsb.org/pdb/index.html.)

the configuration of the active site, and a positively charged lysine residue is required for stabilization of the intermediate reaction state.

Many contact points are involved in the association of the protein substrate with the substrate binding surface of the kinase. Hydrophobic residues of the substrate are associated with their binding equivalents in the hydrophobic pockets of the kinase. Charged residues of the substrate form ionic interactions with the appropriate residues in the kinase and also help dictate the specificity observed in substrate binding. The diversity of contact points helps to explain why certain variations in the sequence adjacent to the phosphorylation site are tolerated. It is not necessary that all possible contacts are formed; indeed, a substrate that only provides some of the contact points can still be bound and phosphorylated, assuming that some of the key amino acids are present. This characteristic results in the indistinct substrate specificity observed for many protein kinases, especially during *in vitro* experiments. Further insight into how kinases are targeted to specific substrates was obtained from the discovery that protein kinases are frequently associated with targeting or anchoring proteins. Thus, substrate specificity is also directed through the regulation of subcellular localization.

It is becoming increasing clear that many protein kinases do not find their physiological substrates by simple diffusion within the cellular mileu. Instead they are frequently directed to particular loci in the vicinity of their substrates by interaction with targeting subunits. In this way, protein kinases (and protein phosphatases) with inherently broad specificities are restricted to particular subcellular locations or organelles and a select subset of protein substrates that are co-localized to that subcellular location. The restriction of protein kinase activity can be achieved either by recruitment of recently activated kinases to targeting subunits or by activation of dormant protein kinases already tethered to targeting subunits. In many cases, multiple signaling proteins are targeted by a single anchoring protein. Different combinations of kinases and phosphatases can be co-localized by a single anchoring protein and targeted to the site of phosphorylation.

KINASE SIGNALING NETWORKS FOR SERINE/ THREONINE PHOSPHORYLATION

The majority of known hormones, neurotransmitters, and other molecules that act in intercellular communication do so by signal transduction pathways that involve G-protein-coupled receptors. The G-protein-coupled receptor family encompass a large family of related proteins and includes, but is in no way limited to, adrenergic, muscarinic, serotonin, light, dopamine, pressure, odorant, adenosine, angiotensin II, and thrombin receptors. The study of G-protein-coupled receptors and their associated regulating proteins has grown into a complex field of its own. These receptors consist of seven transmembrane regions and are, therefore, also referred to as serpentine or 7-TM receptors. The N terminus of a 7-TM is extracellular whereas the C terminus is found in the cytoplasm. With the binding of an extracellular ligand to the N terminus, the 7-TM receptor can interact through its C terminus with a heterotrimeric guanine-nucleotide binding protein (G protein) (see Text Box 4.3 for more information on the structure/func-

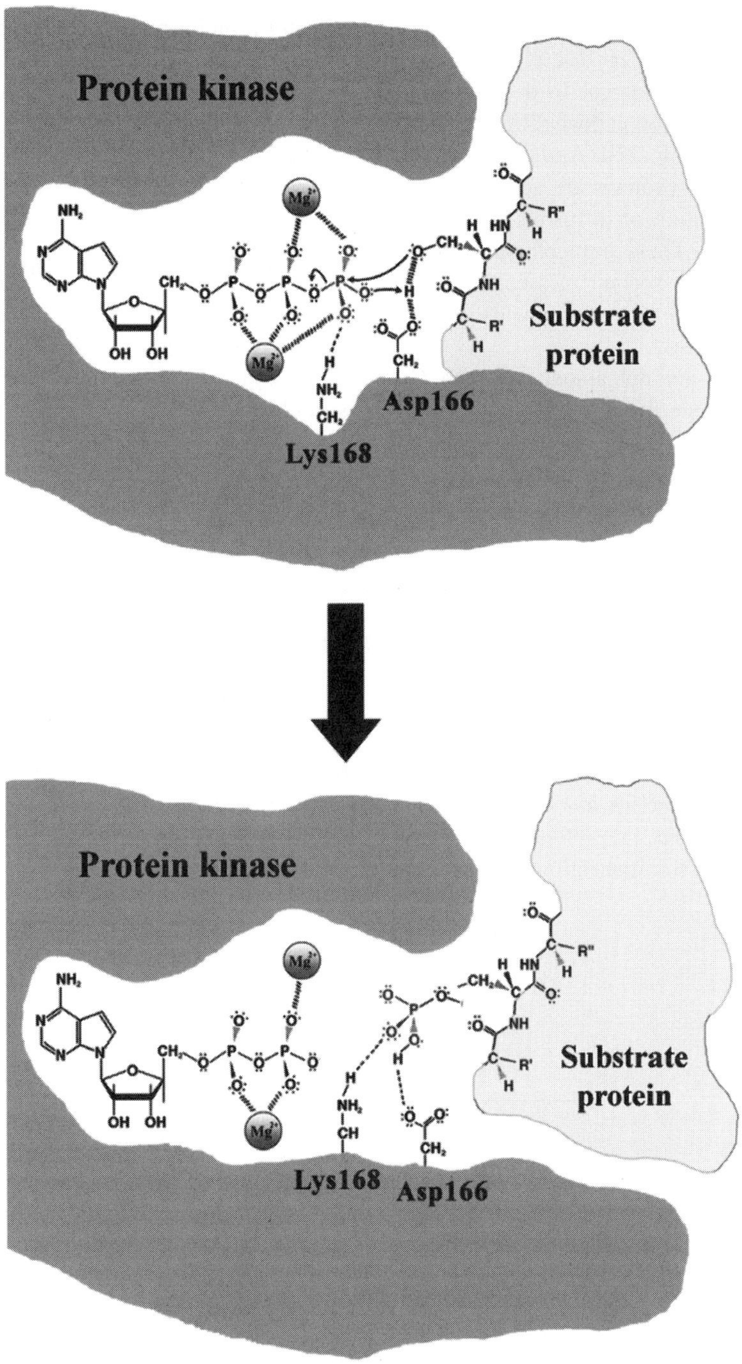

Figure 4.8 Illustration of the protein-kinase-catalyzed mechanism for the phosphoryl transfer from ATP to the serine residue of a protein substrate. The γ-phosphate group of ATP is transferred to the corresponding hydroxyl group of the substrate protein. In the catalytic cleft of the protein kinase, the phosphoester chain of ATP (or corresponding derivative) is complexed by two divalent metal ions (i.e., magnesium). A negatively charged aspartic acid residue (Asp166 in PKA) assists in the simultaneous transfer of the phosphate group from ATP to the substrate protein. A positively charged lysine residue (Lys168 in PKA) helps stabilize the intermediate reaction state.

tion of G proteins). The regulatory G proteins function as switches that can exist in an active or inactive form. In the active form the G protein can signal to downstream components in the signal transduction pathway. In the inactive form, signal transmission is blocked. Among the most important actions of activated G proteins are activation of enzymes that control the formation of diffusible chemical signaling molecules in the cell. These intracellular signaling molecules are referred to as second messengers and include cAMP, cGMP, Ca^{2+}, inositol phosphates, ADP ribose, and certain lipid molecules.

Some membrane receptors are not associated with G proteins. One important class in this group are proteins with intrinsic protein–tyrosine kinase activity. When these receptors are stimulated by an agonist, their tyrosine kinase activity is turned on and they phosphorylate specific effector proteins on particular tyrosine residues. The receptor for insulin, as well as the receptors for some growth factors, are tyrosine kinases. The role of tyrosine kinases and phosphatases in signal transduction will be dealt with in greater detail in Chapter 5.

Second-Messenger-Dependent Protein Kinases

Second-messenger systems allow an entire group of enzymatically catalyzed reactions to be regulated by the level of a single species of molecule. They also allow for the stimulus to be reversed rapidly when second-messenger levels drop. As mentioned earlier, using a cascade of events also provides a huge biomolecular amplification of an initially small signal.

TEXT BOX 4.3 HETEROTRIMERIC G PROTEINS

The identification of G-protein-coupled receptors in the 1970s and elucidation of their role in activating membrane-bound adenylyl cyclase was an important step in understanding how hormone signals were transduced across cell membranes. The superfamily of regulatory G proteins are subdivided by sequence homologies, molecular weight, and subunit structure into three main classes: the heterotrimeric G proteins, the Ras/GTPases, and the translation initiation and elongation factors:

Regulatory GTPase Superfamily

Heterotrimeric G-protein Family
- Gα Gq,α
- Gi,α G12,α
- Gs,α Gβ,γ

Ras Family
- Ras/Rap
- Rho/Rac
- Rab
- Ran
- Arf

Translation Initiation Factor Family
- EF-Tu
- EF-G
- eIF-2

The heterotrimeric G proteins consist of three subunits, α, β, and γ; multiple isoforms of each subunit exist. The α subunit binds the G nucleotide (GDP or GTP) and contains a GTPase activity. The β/γ complex enhances the interaction of the 7-TM receptor with the α subunit. Both the α subunit and the β/γ subunit complexes are able to modulate the activity of effector enzymes. Biochemical analysis of purified components demonstrated that upon ligand binding to the 7-TM receptor, the associated G-protein exchanges bound GDP for GTP and undergoes a conformational change. This action releases the β/γ-subunit complex from the GTP/α-subunit complex and permits the individual complexes to signal downstream effectors. For example, the GTP/α-subunit complex stimulates adenylyl cyclase to produce cAMP. Sustained signaling in response to a stimulus is generally an undesirable event, and signal termination is initiated by a feedback loop when the α subunit (which has GTPase activity) hydrolyzes the bound GTP, either spontaneously or upon interaction with a GTPase-activating protein (GAP) or regulators of G-protein signaling (RGS). GTP hydrolysis causes reassociation of the α subunit with the β/γ subunit to reform the heterotrimer (Fig. TB4.3).

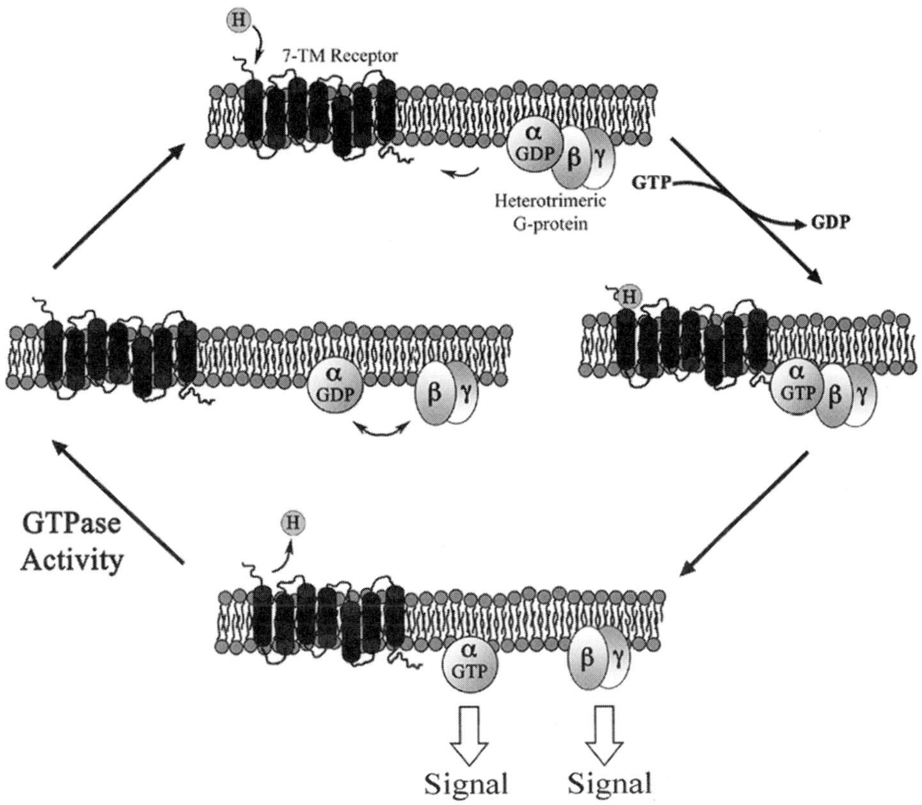

Figure TB4.3 Signal transduction via the G-protein-coupled receptor complex.

The prototypes for second-messenger signaling systems are the cAMP-dependent protein kinase, cGMP-dependent protein kinase, and Ca^{2+}/phospholipid-dependent protein kinase (protein kinase C or PKC) cascades. A common feature of this class of protein kinases is that they are maintained in an inactive state in the absence of second messengers by association with autoinhibitory regions that resemble the phosphorylation sites on substrate proteins. These autoinhibitory regions may contain the actual phosphorylation site sequence or a pseudosubstrate sequence if they do not contain a phosphorylatable residue (e.g., Ser or Thr). The inhibitory region may be on the same subunit as the kinase domain, or it may be present on a separate regulatory subunit.

cAMP-Dependent Protein Kinase (PKA) One of the best understood signal transduction pathways involves the second-messenger cyclic-AMP. Cyclic-AMP is generated following receptor activation by many different hormones and neurotransmitters. Many, if not all, of the components of this signaling cascade (e.g., 7-TM receptor, heterotrimeric G protein, adenylyl cyclase, kinase subunits) have been well-characterized (Fig. 4.9).

As discussed above cAMP-mediated signaling begins with the interaction of an extracellular trigger molecule (typically a hormone or hormone agonist, e.g., adrenaline) with a specific 7-TM receptor. Ligand binding to the receptor triggers a conformational change in the protein that allows the intracellular C-terminal region of the receptor to interact with a heterotrimeric G protein, activating the G protein. The G protein α_s subunit then binds guanosine 5′-triphosphate (GTP), dissociates from the complex, and stimulates adenylate cyclase, an intrinsic membrane protein, to catalyze the formation of cAMP from ATP.

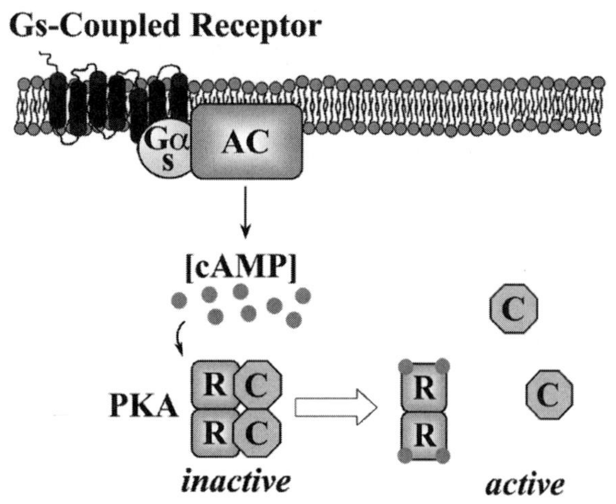

Figure 4.9 The generation of cyclic-AMP (cAMP) and the activation of cAMP-dependent protein kinase (PKA). Hormones or other ligands bind to their respective G-protein-coupled receptor and catalyze the activation of the α subunit of a heterotrimeric G protein ($G_{\alpha}s$). The G protein then stimulates the catalytic activity of adenylate cyclase (AC). Adenylate cyclase produces cAMP from ATP. The cAMP diffuses throughout the cell where it binds to the regulatory subunits (R) of PKA. The binding of cAMP induces a conformational change in the R-subunit structure and causes a dissociation of the catalytic subunits (C). The PKA catalytic subunits, when freed from the inhibitory constraints of the R subunits, are able to phosphorylate various intracellular proteins.

Thus, the extracellular message is transduced through the membrane by the sequential actions of three membrane-confined proteins: a 7-TM receptor, a G protein, and adenylate cyclase.

The formation of cAMP is controlled by the activation of adenylate cyclase. However, another aspect of the regulation of cAMP levels is the degradation of cAMP by a cAMP phosphodiesterase. This enzyme hydrolyzes cAMP by eliminating the ring structure to form 5′-AMP. The synthesis and degradation of cAMP are both subject to complex regulation by multiple hormones and intracellular stimuli. This allows cells to integrate responses to many types of signals from its internal and external environments.

Cyclic-AMP is the second messenger for many hormones and the effects of elevated cAMP differ markedly in various types of cells. However, the effects of cAMP are mediated in a similar manner in all cells via cAMP activation of the cAMP-dependent protein kinase (PKA). The inactive PKA holoenzyme consists of a regulatory subunit dimer that maintains two catalytic subunits in a dormant state (Fig. 4.9). The binding of cAMP to tandem sites on each PKA regulatory subunit removes autoinhibitory contacts in the holoenzyme and releases both of the catalytic subunits. Once released from the regulatory subunit, the catalytic subunits are fully active and free to phosphorylate protein substrates within the cell.

Multiple isoforms of both the catalytic and regulatory subunits of PKA exist. Two regulatory subunits, RI and RII, exhibit distinct cAMP binding affinities and are located at different intracellular sites. RI is predominantly cytoplasmic and is more sensitive to cAMP than RII, allowing dissociation of RI containing tetramers at lower cAMP concentrations. By contrast, the majority of type RII PKA is found associated with subcellular structures. The catalytic subunit, termed PKAc, phosphorylates substrates on serine and threonine residues that are presented in a consensus sequence of Arg-Arg-X-Ser/Thr (where X is any amino acid). Given the frequency of this consensus sequence in the proteome, it is not surprising that overlapping regulatory processes are in place to restrict the action of the kinase. Not only are cellular cAMP levels closely regulated (as described previously) but signal-terminating mechanisms also offer additional compartmentation of the second-messenger pools through desensitization of adenylyl cyclase or localized activation of phosphodiesterases. Furthermore, the regulatory subunits are typically present in excess amounts compared with the catalytic subunits. This overabundance ensures the rapid reformation of the holoenzyme when cAMP levels return to the basal state. A ubiquitous inhibitor protein, PKA-I, also serves to sequester free catalytic subunit as well as to mediate the export of the kinase from the nucleus.

Selectivity in PKA action is achieved by compartmentation of the kinase at different subcellular loci due to interactions with targeting proteins known as the A-kinase anchoring proteins (AKAPs). The first AKAPs were identified in the early 1980s as contaminating proteins that co-purified with the RII regulatory subunit after affinity chromatography on cAMP-Sepharose. AKAPs ranging in size from 15 to 300kDa have subsequently been detected in a variety of tissues with a typical cell expressing 5 to 10 distinct members. Although there is little sequence similarity between individual AKAPs, each anchoring protein contains two types of binding sites: (1) a conserved "anchoring motif" that binds the regulatory subunit of PKA, and (2) a "targeting domain" that directs the subcellular localization of the PKA–AKAP complex through association with a particular type of structural protein, membrane, or subcellular organelle. As a result, PKA is held by the AKAP in an inactive state at various defined subcellular locations, where it is ready to respond to changes in local cAMP concentrations produced by the concerted actions of adenylyl cyclases and phosphodiesterases.

The targeting of the PKA holoenzyme to a discrete intracellular environment is the defining feature of each AKAP. AKAP complexes have been found localized with the plasma membrane, mitochondria, centrosome, microtubular network, and nucleus. Many AKAPs bind multiple signaling enzymes and permit the integration of signals from several distinct second messengers to preferentially control selected phosphorylation events. The best examples of this are the AKAP79, which functions as a signaling scaffold for localizing PKA, PKC, and protein phosphatase 2B (calcineurin) to the postsynaptic densities of neurons, and the AKAP250 (gravin) that targets both PKA and PKC to the plasma membrane and filopodia of cells.

cGMP-Dependent Protein Kinase Cyclic-GMP-dependent protein kinase (PKG) is a serine/threonine-specific kinase that is found in many tissues, although in lower amounts than PKA. cGMP also occurs in cells at concentrations lower than those of cAMP. Analogous to the cAMP system, the levels of intracellular cGMP are regulated at the level of synthesis by soluble or particulate guanylate cyclases and at the level of degradation by phosphodiesterases.

Cyclic-GMP-dependent protein kinase is relatively abundant in lung, cerebellum, and smooth muscle tissue. The enzyme is a dimer of two identical 76-kDa subunits. Unlike PKA, which has separate regulatory and catalytic subunits, the regulatory and catalytic components of PKG reside on the same polypeptide chain. In the absence of cGMP, the regulatory domain inhibits catalytic activity. With cGMP binding (two cGMP binding sites per subunit), a conformational change occurs and inhibition is released. At the enzymatic level there has been considerable debate as to whether PKG and PKA perform the same function in cells. cAMP and cGMP kinases have similar substrate specificities, and most substrates that are phosphorylated by cGMP kinase are phosphorylated equally by cAMP kinase. However, at the physiological level, it appears that in some processes cGMP promotes responses opposite to that of cAMP. For example, activation of the β-adrenergic receptors of cardiac muscle, brain, smooth muscle, and lymphocytes simultaneously produce a rise in the level of cAMP and a drop in the level of cGMP. Conversely, stimulation of the muscarinin acetylcholine receptors in these tissues results in a drop in the level of cAMP and a rise in the level of cGMP.

The production of cGMP is particularly sensitive to Ca^{2+}. In systems where cGMP acts as a second messenger, the simultaneous presence of calcium is an absolute requirement. Guanylate cyclase is inactive at low concentrations of Ca^{2+} and becomes progressively more active with increasing concentrations of free Ca^{2+}. By contrast, adenylyl cyclase requires low concentrations of Ca^{2+} for activation and is inhibited by high concentrations of Ca^{2+}. In view of the difference in the calcium sensitivity of these two enzymes, the relative concentrations of cAMP and cGMP in cells can, in principle, be influenced by intracellular free Ca^{2+}. The greater calcium responsiveness of cGMP synthesis suggests that in some systems Ca^{2+} acts to stimulate the production of cGMP. The merger of Ca^{2+} and cGMP second messengers has been most elegantly displayed in the regulation of vascular smooth muscle contraction via nitric oxide (see Text Box 4.4).

Unlike PKA, which has its own large group of cellular localizing proteins, PKG is not known to associate with a specific family of anchoring proteins. PKG does, however, contain a leucine zipper motif that likely facilitates interactions with other proteins including its putative substrates.

Ca^{2+}-Dependent Protein Kinases Ca^{2+} is a key second messenger involved in many different signaling pathways in eukaryotic cells. A resting cell has a cytosolic Ca^{2+} concentration of roughly 100 nM, a concentration ~20,000-fold lower than that of extracellular Ca^{2+}; and thus cells have an intricate network by which cytoplasmic Ca^{2+}

TEXT BOX 4.4 NITRIC OXIDE AND cGMP-KINASE

The discovery in 1987 of the formation of nitric oxide (NO) in vascular epithelial cells initiated a flurry of new research that ultimately resulted in the awarding of a Nobel Prize in Medicine to its discoverers, R.F. Furchgott, L.J. Ignarro, and F. Murad. A highly reactive gas, NO was initially known as the endothelium-derived relaxing factor (EDRF) prior to its identification. NO is produced from arginine by the enzyme nitric oxide synthase (NOS), the co-product being citrulline (see below). Citrulline and arginine are intermediates of the urea cycle and arginine can be regenerated from citrulline by urea cycle enzymes (Fig. TB4.4).

Nitric oxide has been implicated in a variety of inter- and intracellular signaling processes including vasodilation, immune reaction, and neurotransmission. The predominant pathway of NO function involves stimulation of soluble guanylate cyclase (solGC)

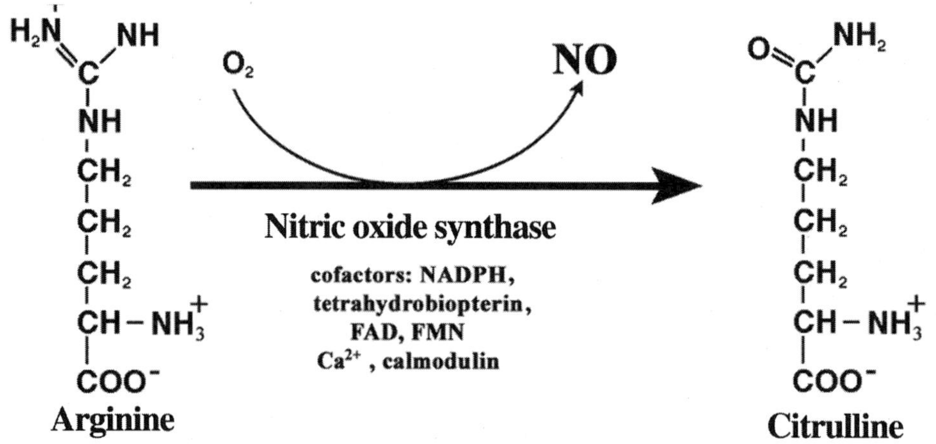

Figure TB4.4 Nitric oxide is a product of nitric oxide synthase.

leading to the formation of cGMP, activation of PKG, and PKG-mediated protein phosphorylation. Other factors such as natriuretic peptides bind to the membrane-associated form of guanylyl cyclase (memGC) and also increase cGMP levels. Nitrovasodilators, such as sodium nitroprusside and other organonitrates, produce NO in the smooth muscle cell, which activates the soluble guanylyl cyclase. Other effects of PKG include alterations in smooth muscle cell gene expression. NO can also elicit a calcium signal via cGMP-dependent ion channels. Although a direct effect of NO on protein kinases or phosphatases has yet to be demonstrated, the fact that NOS is phosphorylated by a number of different protein kinases suggests possible interactions of the NO pathway with a variety of other signaling pathways. NO is a free radical and its actions as one of several reactive nitrogen species are discussed further in Chapter 9 (Fig. TB4.5).

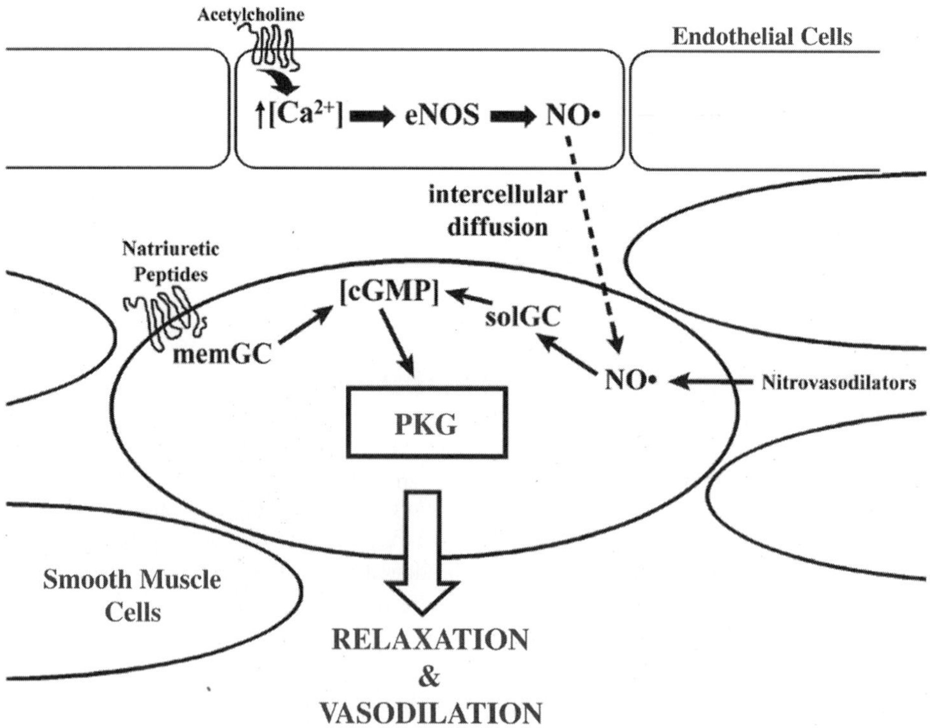

Figure TB4.5 Nitric oxide is a prime mediator of smooth muscle relaxation.

levels are precisely controlled. A few of the many functions regulated by elevated intracellular levels of Ca^{2+} are synaptic transmitter release, muscle contraction, cytoplasmic motility, cell cleavage, oxidative phosphorylation, DNA replication, and various reactions in intermediary metabolism.

Ca^{2+} is extruded from the cell against its electrochemical gradient by two different transport systems present in the plasma membrane. The Na^+/Ca^{2+} exchanger and the Ca^{2+} pump, or Ca^{2+}-ATPase, act in concert to hold the Ca^{2+} concentration in the cytosol at a low level. In the red blood cell, the Ca^{2+} pump is located in the plasma membrane and transports Ca^{2+} ions out of the cell. However, in muscle cells, the Ca^{2+} pump moves Ca^{2+} ions from the cytosol to the interior of the Ca^{2+} storage compartment, the sarcoplasmic reticulum. Increases in cytosolic Ca^{2+} concentrations occur principally by Ca^{2+} influx from the extracellular space through plasma membrane Ca^{2+} channels or via Ca^{2+} release from the internal storage compartments. Various signals stimulate either waves or spikes of increased Ca^{2+} to raise cytosolic concentrations to 1 to $2\,\mu M$ (see also Chapter 8). The signals involved can be either hormonal (ligand-gated Ca^{2+} channels) or electrical (voltage-dependent Ca^{2+} channels). Voltage-dependent and ligand-gated ion channels in the plasma membrane initiate Ca^{2+} entry from extracellular pools; ligand-gated ion channels, such as the inositol trisphosphate receptor present on the ER or SR membranes, can be activated by second-messenger linked G-protein receptors to increase cytosolic Ca^{2+} levels (Fig. 4.10).

There are several mechanisms by which Ca^{2+} performs a regulatory function in the cell. Many enzymes/proteins have specific binding site(s) for Ca^{2+} and their activity is directly dependent on Ca^{2+} binding. Other proteins without enzyme activity also have Ca^{2+}-regulated functions. The latter group plays a role in the Ca^{2+}-dependent activation of enzymes that can activate target proteins. These Ca^{2+} receptors have a high affinity ($K_D < 10^{-6}$) and high selectivity for Ca^{2+} and require an increase in Ca^{2+} concentration for the activation of a target protein.

The most widespread of these Ca^{2+} receptors is calmodulin. Calmodulin is a 148-amino-acid protein (16,680 Da) comprised of four helix–loop–helix protein folding motifs called EF hands. Each of the four EF hands binds one Ca^{2+} ion. The protein is inactive in its native state but binds four calcium ions to form a complex that binds to and activates numerous enzymes and effector proteins. Ca^{2+} binding is a cooperative event so that binding kinetics have a Hill coefficient >1. Binding of one Ca^{2+} ion serves to slightly modify the EF hand conformation, and this subtle change in protein conformation is transduced to the second Ca^{2+}-binding site, giving it an increased binding affinity. Binding of the second Ca^{2+} ion causes the exposure of a hydrophobic pocket that enables calmodulin to recognize the binding domains of target proteins. The Ca^{2+}–calmodulin complex is a signal molecule that is involved in many signal transduction pathways including those involved in cell proliferation, mitosis, muscle contraction, and neuronal signal transduction. Several well-characterized Ser/Thr protein kinases rely on the Ca^{2+}–calmodulin complex. This group includes kinases such as glycogen phosphorylase kinase (GPK), myosin light-chain kinase (MLCK), and the calcium–calmodulin (CaM) protein kinases I, II, III, and IV. CAMK III is also known as eEF-2 kinase because it phosphorylates the eukaryotic elongation factor 2 (eEF-2) of the ribosomal translation machinery. Calcium-dependent kinases are grouped according to whether they are dedicated kinases having a single substrate (GPK, eEF-2 kinase, MLCK) or whether they are multifunctional (CaMK I, II, and IV) with several substrates.

Members of this group of kinases have very similar structures. Like the cyclic nucleotide-dependent protein kinases, the calmodulin-dependent protein kinases are maintained in an inactive state. Each member of the group has a catalytic domain adjacent to a regulatory region that contains an overlapping autoinhibitory domain. The interaction between the autoinhibitory domain and the catalytic domain maintains the kinase in an inactive conformation by preventing binding of the protein substrate or MgATP. Binding of Ca^{2+}–calmodulin alters the conformation of the autoinhibitory domain such that it no longer interferes with substrate binding. There has been great progress recently in the characterization of the kinase members that rely on the Ca^{2+}–calmodulin complex for activity and in the assessment of their physiological roles. Future studies will probably find more members of this group and document additional cellular roles.

CaMKs The multifunctional enzymes, CaM kinases I, II, and IV, are found in the cytoplasm associated with a variety of membrane or cytoskeletal elements or in the nucleus. The enzymes phosphorylate a number of different substrates *in vitro* and show more complicated regulatory features than the dedicated Ca–calmodulin kinases such as MLCK. Studies of the multifunctional CaM kinases has helped determine how Ca^{2+}–calmodulin can bind to a target and relieve an intramolecular autoinhibition that results in activation of the enzyme. In the case of CaMK II this results in autophosphorylation that generates a Ca^{2+}–calmodulin autonomous kinase activity. CaMK IV also can become autonomous of Ca^{2+}–calmodulin, although the precise mechanism responsible for generating this activity state has yet to be defined. On the other hand, CaMK I is always Ca^{2+}/calmodulin-dependent, even when autophosphorylated. Common to all three enzymes is the fact that phosphorylation sensitizes the protein kinase to subsequent Ca^{2+} signals.

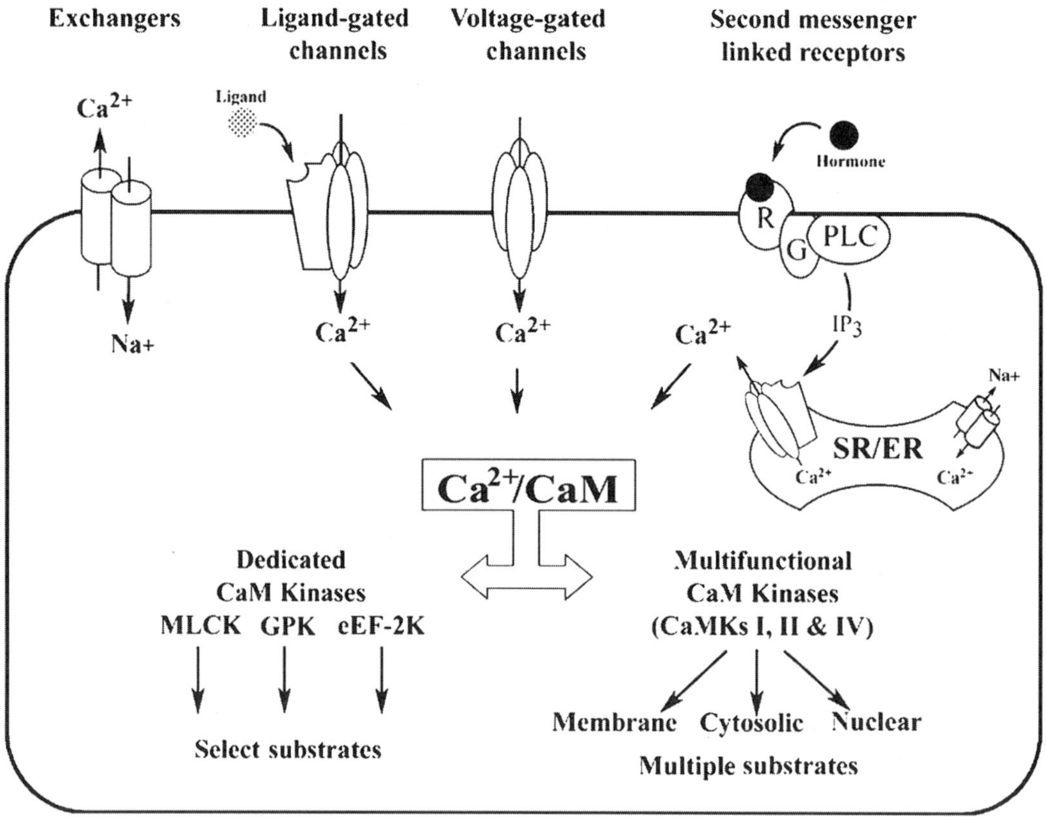

Figure 4.10 Intracellular calcium levels are regulated by diverse mechanisms. Calcium can originate from extracellular or intracellular stores. Ligand-gated channels, voltage-gated channels, and exchangers increase intracellular Ca^{2+} concentrations by transporting calcium ions across the plasma membrane. Second-messenger-linked signal transduction pathways regulate Ca^{2+} release from intracellular sites, the sarcoplasmic reticulum (SR), or endoplasmic reticulum (ER). Hormones bind to their respective receptor (R) and catalyze the activation of a heterotrimeric G protein (G), which in turn stimulates the phospholipase enzyme (PLC) to generate inositol trisphosphate (IP_3). IP_3-activated channels on the SR/ER membrane release calcium into the cytosol. Calcium and the calcium receptor protein (calmodulin; CaM) activate two different groups of Ca^{2+}/CaM-dependent protein kinases. The multifunctional CaM kinases (CaMK I, II, and IV) phosphorylate various cellular targets. The dedicated CaM kinases (myosin light-chain kinase, MLCK; glycogen phosphorylase kinase, GPK; and elongation factor-2 kinase, eEF-2K) phosphorylate single-protein targets.

CAMK II The best characterized of the multifunctional CaM kinases, CaMK II, is present in all cells of vertebrate and invertebrate species. It is a multimeric enzyme composed of 10 to 12 copies of the monomeric catalytic enzyme. Electron microscopy pictures reveal the CaMK II structure to be surprisingly similar to a bicycle wheel—the interaction domains of the enzymes form the spokes of a wheel, and the catalytic domains form the outside (or rim) of the wheel. Multiple isoforms of CaMK II exist; the isoforms that are present in a given cell depend on which of the four different genes (α, β, γ, and δ) is transcribed and which of a considerable number of alternatively spliced forms of each primary transcript is translated. It is also possible to have multiple isoforms in a single cell, and the ratio of the isoforms regulates the heteromultimer formation. Some CaMK II isoforms contain a nuclear localization sequence that directs the enzyme to the nucleus where it has been implicated in the regulation of transcription.

The catalytic domain of each CaMK II monomer is normally present in an autoinhibited state. Activation of a single catalytic subunit of CaMK II requires that the Ca^{2+}/CaM complex bind to it and to an adjacent subunit. This dual interaction allows the catalytic domain of one subunit to phosphorylate a single Thr residue (Thr-286 in CaMK IIα isoform) of the other. Once a subunit becomes phosphorylated, its affinity for Ca^{2+}/CaM is increased by nearly 1000-fold. The high binding affinity prevents Ca^{2+}/CaM from dissociating from the activated CaMK II and preserves the activated state for a longer period of time. Once the Ca^{2+} signal is removed, CaM remains bound to CaMK II for a short period of time, and the enzyme remains in the active state. Most interestingly, the phosphorylated enzyme retains 20 to 80% of the activity of the Ca^{2+}/CaM-activated form after the full dissociation of CaM. The Ca^{2+}/CaM-independent state exists until the removal of the activating phosphate residue, which returns the enzyme to a fully inactive form. This unique mechanism of activation provides CaMK II with a "memory" for transient increases in Ca^{2+} and apparently enables detection and differentiation of repetitive Ca^{2+} pulses. The intensity of CaMK II activation is dependent upon the interval between the occurrence of Ca^{2+} signal pulses. If the Ca^{2+} signaling pulses occur with higher frequencies, a long lasting CaMK II activation is possible since the kinase remains in its fully active state between Ca^{2+} signals. CaMK II has been demonstrated to play an important function in learning and memory, presumably because of its unique ability to respond to Ca^{2+} transients.

The CaMK IIα isoform is expressed exclusively in the brain and is a major component of the postsynaptic membrane (PSM), accounting for almost 1% of total rat cerebral cortex protein. About 20 to 35% of the enzyme exists in soluble form in adult rat forebrain, and the remaining 65 to 80% is particulate and concentrated in the PSM. The physiological function of the enzyme is to increase synaptic strength by phosphorylating ion channels and signaling proteins such as glutamate receptors and *N*-methyl D-aspartate (NMDA) receptors. In addition, CaMK IIα is required for long-lasting changes in synaptic strength such as long-term potentiation, a process that is thought to be involved in learning and memory. These functions of CaMK IIα have been confirmed by knocking out the gene in mice. In addition, the memory capability of transgenic animals is influenced by the overexpression of CaMK II.

The other three isoforms of CaM-kinase II (β, γ, δ) are present in brain but occur in other tissues as well. CaMK II appears to have a role in the mechanism by which an increase in Ca^{2+} in a nerve terminal causes an exocytotic release of neurotransmitter. Synapsin I, an excellent substrate of CaMK II, is found in nerve terminals and binds to the external surface of synaptic vesicles. Bound synapsin prevents vesicles from participating in exocytosis, and phosphorylation of synapsin I causes its dissociation from the vesicles to allow exocytosis to proceed.

CAMK I AND CAMK IV The CaMK I and CaMK IV enzymes are present as monomeric signaling enzymes and are therefore distinct from the multimeric CaMKII. The CaMK I and IV enzymes are closely related and share a common requirement for phosphorylation by an upstream CaMK kinase (CaMKK), in a manner analogous to the mitogen-activated protein kinase cascades (Chapter 5). However, the details of the activating mechanisms of CaMK I and IV differ considerably from the MAP-kinases and from each other. Two CaMKKs, α and β, have been identified in mammals, and these upstream kinases are also Ca^{2+}/CaM binding proteins. Based on an evolving understanding of CaMK I/IV regulation and the molecular cloning of upstream CaMKKs, a Ca^{2+}/calmodulin-dependent protein kinase cascade (CaMK cascade) has been proposed.

CaMK I is widely distributed in various neuronal and nonneuronal tissues, occurring in highest amounts in the cerebral cortex. CaMK IV has a much more restricted distribution; it is most abundant in brain and T lymphocytes but also occurs in testis, ovary, adrenal, skin, and bone marrow. Even though CaMK I and IV are present in virtually all mammalian cells and tissues, virtually nothing is known about their physiologically relevant substrates. Unlike the other CaM kinases, CaMK IV seems to reside in the nucleus of all cells in which it is expressed. This nuclear localization implicates the enzyme in the regulation of gene transcription (see Chapter 5).

The phosphorylation of a single Thr residue in the activation loop is required for both CaMK I and IV to become fully active; however, following this phosphorylation CaMK IV acquires Ca^{2+}/CaM independence, whereas CaMKI remains Ca^{2+}/CaM-dependent. In the absence of Ca^{2+}, CaMK I quickly reverts to the inactive state. In contrast to the relatively simple mechanism by which the activity of CaMK I is regulated, at least three steps are required to activate CaMK IV. First, the Ca^{2+}/CaM complex must bind to CaMK IV; this exposes Thr-196 in its activation loop and allows a low level of protein kinase activity. Second, this low level of protein kinase activity catalyzes the intramolecular autophosphorylation of a number of Ser residues near the N terminus of the protein, which is required to relieve a novel form of autoinhibition. Third, Thr-196 is phosphorylated by a CaMKK that increases protein kinase activity. When activated in this fashion, CaMK IV exhibits an activity that is independent of Ca^{2+}/CaM. Since removal of the Ca^{2+} signal has little effect on its activity, dephosphorylation seems to be the primary mechanism by which CaMK IV is inactivated in the cell. At least in brain extracts and T lymphocytes, CaMK IV and an associated protein phosphatase (PP2A) can be isolated as a complex.

Yeast, nematode, and fungal species contain homologs of CaMKK and CaMK I, an observation that supports the likelihood that CaM kinase cascades are evolutionarily conserved. The roles of these homologs cloned from *S. pombe* and *C. elegans* are not yet known but may be involved in cytoskeletal dynamics or transcriptional regulation. The filamentous fungus *A. nidulans* also displays homologs to CaMKK and CaMK I. In this species the CaMK I gene is essential and required for progression through the nuclear division cycle. Preliminary data suggests that fungi seem to contain a CaMK cascade that functions in both cytoplasm and nucleus. If this proves to be true, then it follows that simpler organisms might contain a single CaMK I/IV-like protein that can act in either the cytoplasm or the nucleus.

CAMK III OR EEF-2 KINASE The eukaryotic elongation factor-2 kinase (eEF-2 kinase), also known as CaMK III, is present in virtually all tissues of vertebrates, as well as various invertebrates. The only known substrate of eEF-2 kinase is eEF-2, a 100-kDa protein that promotes ribosomal translocation, the reaction that moves the ribosome along the messenger ribonucleic acid (mRNA) template during translation. Phosphorylation of eEF-2 by eEF-2 kinase inactivates the protein, and this mechanism is one of several phosphorylation-dependent events that regulate protein synthesis (see Chapter 7 for more details of translational control and Chapter 16 for eEF-2 regulation in mammalian hibernation).

The primary structure of eEF-2 kinase from several eukaryotic organisms was recently determined. Surprisingly, the enzyme does not display any homology to the other calmodulin-dependent protein kinases or to other members of the eukaryotic protein kinase superfamily. However, the putative catalytic domain of eEF-2 kinase appears to be highly similar to the catalytic domain of a recently described myosin heavy-chain kinase (MHCK) from *Dictyostelium*. Thus, eEF-2 kinase, *Dictyostelium* MHCK A and B, and several putative protein kinases from mammalian sources that have not yet been fully characterized may represent a new class of protein kinases. This class seems to be structurally related to the histidine protein kinases of bacterial two-component systems.

Myosin Light-Chain Kinase (MLCK) Myosin light-chain kinase plays a central role in smooth muscle contraction and many nonmuscle motile processes owing to its Ca^{2+}/CaM-dependent phosphorylation of myosin regulatory light chains (MLC). The amount of MLCK in smooth muscle tissues is similar to or greater than that found in skeletal muscle, and both are considerably greater than the amount found in nonmuscle cells and cardiac muscle. Structural analyses and differences in enzyme properties show that there are at least two isoforms of MLCK, a smooth/nonmuscle MLCK and a skeletal muscle MLCK. An increase in cytosolic Ca^{2+} concentration results in the association of Ca^{2+}/CaM with an inactive MLCK. The C-terminal region of MLCK contains an autoregulatory domain that contains the autoinhibitory and CaM-binding domains. The autoinhibitory domain folds back onto the catalytic core, thereby blocking catalytic activity. Following Ca^{2+}/CaM binding, the inactive MLCK is converted to an active form by the displacement of the autoinhibitory region allowing MLCK to phosphorylate MLCs, its only physiological substrate. The activity of smooth muscle MLCK can be modulated by phosphorylation at specific sites that lead to increased K_{CaM} or V_{max} values; however, both nonphosphorylated and phosphorylated MLCKs are tightly regulated by Ca^{2+}/CaM binding, and no evidence exists for physiologically relevant Ca^{2+}/CaM-independent activity.

Glycogen Phosphorylase Kinase (GPK) Glycogen phosphorylase kinase is responsible for catalyzing the phosphorylation of glycogen phosphorylase, thereby activating it and initiating glycogen degradation. The GPK holoenzyme contains four tetramers, each composed of four subunits: two regulatory subunits (α, β), a catalytic subunit (γ), and a calcium binding subunit (δ) that is actually an integral calmodulin. The α and β regulatory subunits are phosphorylated by PKA in response to a hormonal signal.

Glycogen phosphorylase kinase is regulated by changes in intracellular Ca^{2+} levels as well as by hormones that activate the cAMP signal transduction pathway (Fig. 4.11). In the dephosphorylated form, GPK is inactive because the catalytic subunit is inhibited by interaction with the other subunits. An increase in Ca^{2+} partially activates the dephosphorylated enzyme by binding to the δ subunit (calmodulin) and removing the inhibiting subunit interactions. Under conditions of low Ca^{2+}, GPK can also be activated by hormonal signals that raise cAMP levels. PKA fully activates GPK by rapid phosphorylation of the β subunit and by slower phosphorylation of the α subunit. In the phosphorylated form, GPK also requires Ca^{2+}; however, the phosphorylated enzyme is activated at significantly lower Ca^{2+} concentrations than the unphosphorylated enzyme.

An interaction of cAMP and Ca^{2+} signals occurs during glycogenolysis in striated muscle. Activation of GPK can result from both hormonal and neuronal activation of GPK use in different second messengers, but both accelerate the breakdown of glycogen. β-adrenergic stimulation provides immediate activation of GPK at low [Ca^{2+}]. Nerve impulses cause the release of Ca^{2+}, which fuels prolonged muscle contraction. As a result of the elevated [Ca^{2+}], GPK is active even if unphosphorylated.

Ca^{2+}/Phospholipid-Dependent Protein Kinase (PKC) At first glance, the calcium- and phospholipid-dependent second-messenger signaling system appears quite different from the classical cyclic nucleotide pathways that were

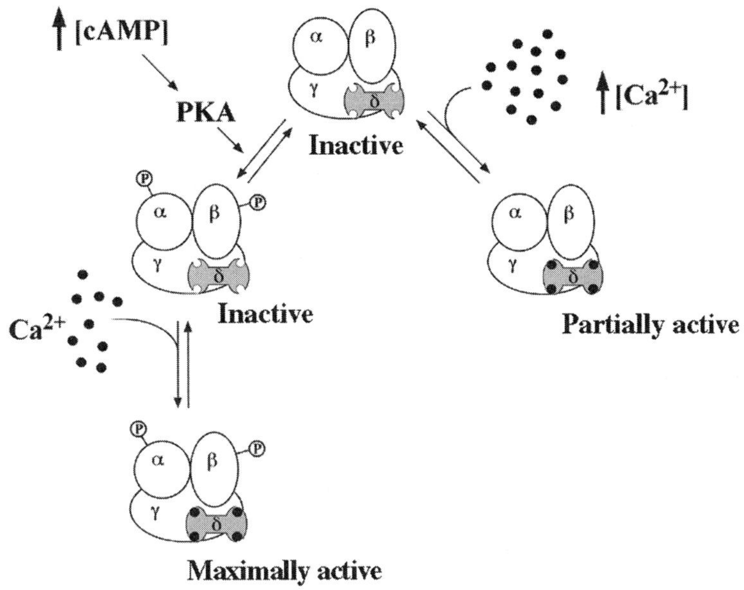

Figure 4.11 Regulation of glycogen phosphorylase kinase by different second-messenger systems in skeletal muscle. Hormonal stimulation of muscle results in increased cAMP production. Activated cAMP-dependent protein kinase (PKA) phosphorylates the α and β subunits of glycogen phosphorylase kinase. In the phosphorylated form, the binding affinity of the δ subunit (calmodulin) for calcium is enhanced. Activation of glycogen phosphorylase kinase is now possible at low calcium concentrations. Nervous stimulation of muscle causes Ca^{2+} channels to open. As the intracellular concentration of Ca^{2+} increases, the δ subunit is saturated with Ca^{2+} and the glycogen phosphorylase kinase is activated. Glycogen degradation begins and ultimately ATP is made available for muscle contraction.

previously described. However, on closer inspection it is apparent that the system of lipid messengers has certain general features that are carried over from the cyclic nucleotide cascades. In both cases, an extracellular signal is transferred through a membrane receptor to a G-protein effector and then to an amplifier enzyme. The amplifier enzyme converts a biologically inert precursor into second messengers that, in turn, activate internal transduction enzymes, namely protein kinases, which are able to modulate cellular enzyme/protein activities.

The main difference between the cyclic nucleotide systems and the calcium- and phospholipid-dependent second messengers is the use of complex phospholipid constituents from the phospholipid bilayer of the cell membrane as the precursors for the production of second messengers. The activated G protein in this case activates a membrane-bound enzyme, phospholipase C (PLC) that catalyzes the hydrolysis of phosphatidylinositol 4,5-bisphosphate (PIP2) into two molecules that act as second messengers, diacylglycerol, or DAG, and inositol trisphosphate, or IP_3. The DAG molecules remain within the lipid bilayer and move by diffusion to interact with a membrane-bound protein kinase termed the calcium- and phospholipid-dependent protein kinase (PKC) (Fig. 4.12). The other product of PIP2 hydrolysis, IP_3, is water-soluble and diffuses away from the membrane through the cytosol to induce the release of Ca^{2+} from the ER or SR. These calcium ions act as messengers to active multiple intracellular functions, including the further stimulation of PKC. Full activation of PKC also depends on the presence of membrane phospholipid phosphatidylserine (PS). PKC in the cell is distributed between a free cytosol pool (inactive) and enzyme bound to the inner surface of the plasma membrane. The enzyme is active only when membrane-bound and is regulated, in part, by changes in the percentage of enzyme that is membrane-bound. DAG and Ca^{2+} promote membrane binding.

The DAG that activates PKC can be produced by PLC (as described above) and also by phospholipase D. The substrate for the PLC route is PIP2, which is produced by phosphorylation of phophatidylinositol, one of the minor membrane phospholipids, by PI kinase to generate phosphatidyl-inositol 4-phosphate (PIP) and then by PIP kinase to form PIP2. PLC is translocated to the membrane upon activation in a signal transduction pathway. The exact mechanism of PLC activation is still unclear, but different PLC isoforms are differentially activated by heterotrimeric G proteins, receptor tyrosine kinases, or by nonreceptor

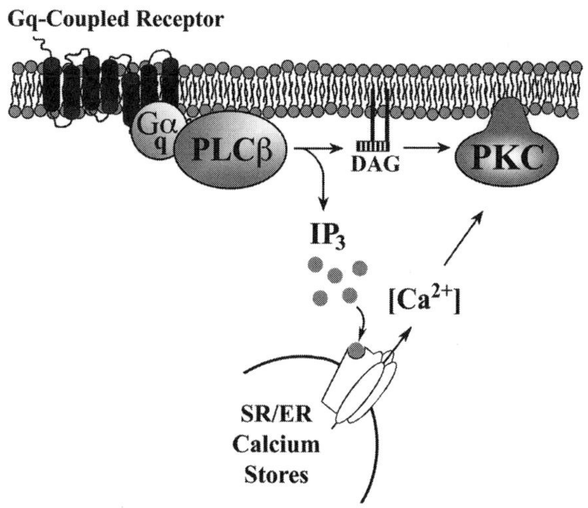

Figure 4.12 The generation of Ca^{2+} as second messenger and the activation of the Ca^{2+}/phospholipid-dependent protein kinase (PKC). Hormones or other ligands bind to their respective G-protein-coupled receptor and catalyze the activation of the α subunit of a heterotrimeric G protein ($G_\alpha q$). The G protein then stimulates the catalytic activity of phospholipase C (PLCβ), which catalyzes the hydrolysis of phosphatidylinositides into inositol trisphosphate (IP_3) and diacylglycerol (DAG). The IP_3 acts as a diffusable signal molecule in the cytosol while the DAG remains anchored in the plasma membrane. IP_3 binds with an IP_3-activated Ca^{2+} pump on the membrane of the sarcoplasmic reticulum (SR) or endoplasmic reticulum (ER). Ca^{2+} ions then move through the channel into the cytoplasm. Cytosolic free Ca^{2+} levels rise from resting levels of 10^{-7} to $>10^{-6}$ M. The PKC enzyme is localized to the plasma membrane by lipid-binding domains; its activity is stimulated when the levels of both Ca^{2+} and DAG increase.

tyrosine kinases in their cytosolic form. The DAG produced from PIP2 in this manner rapidly disappears and supplies a "pulse" stimulation of PKC. A second mechanism of DAG production is via the hydrolysis of phosphatidylcholine (PC), the predominant phospholipid in mammalian cell membranes. PC hydrolysis by phospholipase D produces phosphatidic acid (PA) and choline. PA is then converted into DAG by the action of a phosphomonoesterase. The DAG produced in this manner is persistent, so PKC signaling can be sustained for several hours.

Diacylglycerols generated from either PIP2 or PC activate PKC equally well, but their mechanism of degradation is different. Because the side chains in PC differ from those in PI, different species of DAG are produced. DAGs produced from PIP2 are rapidly converted into PA by the action of a DAG kinase and then recycled into PI. DAGs produced from PC are poorer substrates for DAG kinase and are only slowly degraded.

The PKC occurs in multiple isozymes that have been classified into three groups on the basis of their structure and ability to bind cofactors. The conventional PKC isozymes (α, βI, βII, and γ) are activated by Ca^{2+}, DAG, and phosphatidylserine. The novel PKC isozymes (δ, ε, η, θ, and μ) lack the Ca^{2+}-binding domain and are activated by PS, DAG, and unsaturated fatty acids. The atypical PKC isozymes (ζ, λ, and ι) also lack the Ca^{2+}-binding domain, have some constitutive activity, and are also activated by PS, phosphatidylinositides, and unsaturated fatty acids. All PKC isozymes contain a carboxyl-terminal kinase core and an amino-terminal regulatory segment. The regulatory section contains two important functional modules: an autoinhibitory pseudosubstrate sequence module and a membrane-targeting module. The pseudosubstrate sequence module allosterically regulates access to the substrate binding cavity. The membrane targeting module defines the classes of PKC isozymes. Contained within the membrane targeting module of all PKCs, except the atypical isozymes, is the C1 domain, which permits DAG binding. Because atypical PKCs do not possess a C1 domain, they lack a response to DAG. Conventional PKCs contain a functional C2 domain that binds anionic lipid in a Ca^{2+}-dependent manner. Novel PKCs also contain a C2 domain; however, a modification in the Ca^{2+}-binding pocket in the novel PKCs makes them unresponsive to Ca^{2+}.

Before PKC is able to respond to second messengers, it must first be "primed" by a series of ordered protein phosphorylations. Newly synthesized PKC associates with the membrane. This "juvenile" form of the enzyme is found with the pseudosubstrate segment released from the active site, allowing a full exposure of the activation loop. The recently discovered phosphoinositide-dependent kinase (PDK-1) catalyzes a phosphorylation in the activation loop of the kinase, which acts as an "on switch," promoting the correct alignment of residues in the active site for catalysis and substrate binding. Conventional, novel, and atypical PKCs are all regulated by PDK-1. Following phosphorylation by PDK-1, PKC rapidly autophosphorylates at two sites and undergoes a change in structural conformation that allows the pseudosubstrate to gain access to the active site. This inactive species then dissociates from the membrane to form the inactive cytosolic pool that accounts for the majority of PKC found in unstimulated cells. Elevation of intracellular levels of Ca^{2+} and DAG promote the reassociation of PKC to the membrane through binding by the C1 and C2 domains. Engagement of the enzyme's two membrane binding domains imparts a structural transformation on the protein that releases the pseudosubstrate segment and maximally activates PKC.

Docking and Scaffolding Proteins in PKC Signaling The subcellular location of PKC is critical for its activation, and numerous targeting proteins for PKC have been identified that localize both active and inactivate PKC isozymes. Upon stimulation with hormones that increase intracellular concentrations of DAG and Ca^{2+}, PKCs translocate to the plasma membrane, become activated, and phosphorylate specific substrates. The inactive cytosolic forms of PKC can translocate to different membrane locations where they will be exposed to different pools of substrate proteins once activated. Whereas the attachment of PKC to membranes clearly requires protein–phospholipid interactions, protein–protein interactions seem to facilitate the differential localization of PKC isoforms inside cells. Some members of the AKAP family of PKA anchoring proteins allow a simultaneous association of PKA and PKC at the plasma membrane. Also, several classes of unique PKC targeting proteins have been identified. Substrate-binding proteins (SBPs) bind inactive PKC. The phosphorylation of SBPs by PKC following its activation abolishes the targeting interaction, suggesting that SBPs may associate with the kinase transiently.

Receptors for activated C-kinase (RACKs) are not necessarily substrates for PKC and bind at one or more sites distinct from the substrate-binding pocket of the kinase. RACKs interact only with the activated forms of PKCs, suggesting that PKC binding to RACK occurs after cell stimulation and probably acts to localize the active kinase with a RACK at a specific membrane location. RACK1 was identified and cloned as a protein that bound to the C2 domains of PKC and appears to regulate the subcellular localization of activated PKC. RACK1 also interacts with other C2 domain-containing proteins including PLC-γ1. Moreover, there are isozyme-specific RACKs that presumably anchor individual PKC isozymes close to their physiological substrate(s). Experimental evidence has shown that in cardiac myocytes RACK1 is specific for β_{II}PKC, whereas RACK2 is specific for εPKC. Thus, it appears that the specificity of PKC function may be determined, in part, by the different locations of isozyme-specific RACKs.

PICKs, for proteins that interact with C kinase, are another class of PKC binding protein. These were identified by two hybrid screens in which the catalytic core of the kinase was used as bait. PICK1 binds *in vivo* to a variety of transmembrane proteins, including the AMPA-type glutamate receptor, dopamine transporters, ADP-ribosylation factors, and some receptor tyrosine kinases. Hence, a membrane targeting function for PICKs is implied.

5′-AMP-Dependent Protein Kinase (AMPK) A limited number of protein kinases are regulated by small molecules that are normal cellular metabolites rather than by specifically synthesized second-messenger molecules. Their function appears to have been adapted to provide protection against environmental situations that create nutritional stress. The AMPK cascade does not respond directly to nutritional availability but instead to the cellular stresses that deplete cellular energy levels. It achieves this by sensing the cellular levels of AMP and ATP, and as its name suggests, is activated by AMP. Homologs of the kinase have been isolated from yeast (sucrose nonfermenting-1; SNF1 kinase), mammals (AMPK), and higher plants (SNF1-related kinase-1; SnRK1).

5′-AMP-dependent kinase is a heterotrimer made up of α, β, and γ subunits. Gene analysis in mammals has demonstrated that all three subunits exist as multiple isoforms. The isoforms exhibit subtle differences in tissue distribution, subcellular localization, and regulatory properties. The α subunit of AMPK is a 63-kDa catalytic component. It possesses both catalytic and regulatory domains. The regulatory domains possess autoinhibitory sequences that maintain the kinase in an inactive state by binding to the kinase domain in the absence of AMP. It is believed that AMP activates the kinase by displacing the autoinhibitory domain from the catalytic core. The β subunit is a scaffold on which the other two subunits are assembled. The protein possesses two different binding regions, the KIS (kinase interacting sequence) and ASC (association with SNF1 kinase) domains, which are responsible for interactions with the α and γ subunits, respectively. There is some evidence that the β subunit may also play a role as an adapter protein, directing the AMPK complex to particular substrate proteins. The role of the γ subunit is not as well-defined; however, it is thought that the protein supplies a regulatory role by binding to the α subunit and stabilizing the AMP-bound, active form. An interaction between the γ subunit and the regulatory domain of the α subunit would serve to displace the autoinhibitory region from the catalytic domain, thus maintaining an activated complex.

The AMPK kinase is also a member of a cascade system. AMPK can be activated by phosphorylation by the upstream kinase, AMPKK. This phosphorylation occurs at a site within the activation loop of the α subunit of the kinase. Subsequently, the phosphorylation of AMPK by AMPKK was also found to be stimulated by increases in AMP levels. It is now known that AMP has at least four effects on the AMPK pathway (Fig. 4.13): (1) AMP provides allosteric activation of AMPK, (2) AMP provides allosteric activation of the upstream AMPKK, (3) AMP binds to AMPK and makes it a better substrate for the upstream AMPKK, and (4) AMP binds to phosphorylated AMPK and makes it a worse substrate for inactivating protein phosphatases. The AMPK cascade provides an excellent example of a zero-order, ultrasensitive cascade that responds to the elevation of AMP in cells. Ultrasensitivity in this system is believed to result from at least two factors. First, the primary signaling molecule AMP acts at more than one step in the cascade by activating an upstream kinase, AMPKK, and

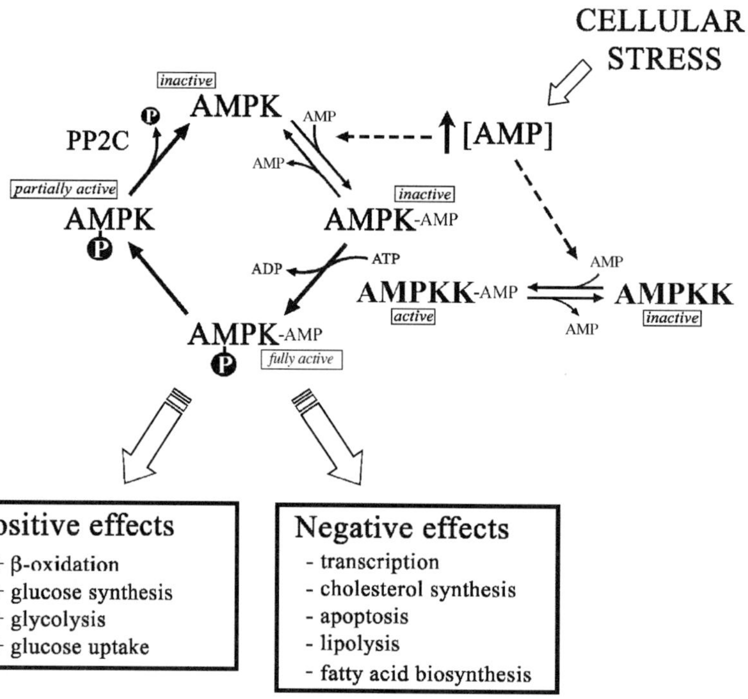

Figure 4.13 Regulation of AMP-dependent protein kinase activity in the cellular response to stress. AMP-activated protein kinase (AMPK) is regulated both by allosteric activation and by phosphorylation by an upstream AMP-activated kinase kinase (AMPKK). Elevation of cellular AMP levels results in an activation of AMPKK activity. AMP binding to AMPK makes the kinase a better substrate for the upstream AMPKK. The activation of AMPK is reversed by protein phosphatase 2C. A major function of the AMP-activated protein kinase pathway is to inactivate key regulatory enzymes of several biosynthetic pathways when the cell is subjected to stress that causes ATP depletion.

by binding to the downstream kinase, AMPK, making it a better substrate for activating phosphorylation by AMPKK and simultaneously making it a poorer substrate for inhibitory dephosphorylation by protein phosphatases, this is, PP2C. And second, under basal conditions, the upstream kinase is saturated with the downstream kinase. More information regarding the ultrasensitive nature of protein kinases is available in Chapter 5.

All of the effects of AMP on AMPK activity are antagonized by high concentrations of ATP. Thus, the AMPK pathway has been proposed to act as a sensor that monitors the cellular energy charge (i.e., the AMP : ATP ratio). When activated, AMPK turns off several key enzymes of ATP-consuming anabolic pathways. The activation of the AMPK pathway is considered to be particularly important during times of cellular stress associated with ATP depletion, such as in cardiovascular ischemia, during metabolic depression in times of environmental challenge (i.e., decreased oxygen availability and heat shock), in endurance exercise in skeletal muscle, in hypoglycemia, and with exposure to metabolic poisons. In yeast the AMPK homolog, SNF1-kinase, is activated in response to glucose deprivation. Activation of SNF1-kinase activity may be the trigger to induce expression of yeast genes that encode proteins required for aerobic metabolism or for catabolism of alternative carbon sources such as sucrose or galactose. The central role AMPK maintains in switching metabolic flux between catabolism and anabolism leads it quite appropriately to be labeled as a system that acts as a cellular "fuel gauge."

The AMP : ATP ratio is known to increase when the oxygen supply is restricted or in the presence of inhibitors of mitochondrial oxidative phosphorylation and so acts as a sensor of the energy state of the cell. In various tissues (liver, adipose, and skeletal muscle), rising levels of AMP activates AMPK, which then phosphorylates and inactivates several key regulatory enzymes in energy-consuming biosynthetic pathways, such as acetyl-CoA carboxylase (fatty acid synthesis), 3-hydroxyl-3-methylglutaryl CoA reductase (sterol biosynthesis), creatine kinase (high-energy metabolite synthesis) and glycogen synthase (glycogen synthesis). AMPK is also thought to mediate the recruitment of glucose transporters (GLUT4) by an unknown mechanism distinct from insulin. The net effect is to shut down energy-expensive anabolic pathways in the cell under situations where ATP availability is limiting.

Apart from suppression of anabolic pathways, AMPK has also been shown to stimulate ATP production under anaerobic conditions in heart. AMPK-mediated activation of 6-phosphofructo-2-kinase (PFK-2) increases the production of fructose-2,6-bisphosphate, a powerful activator of the glycolytic regulatory enzyme, 6-phosphofructo-1-kinase (PFK-1). This phenomenon is superimposed on the well-known allosteric activation of PFK-1 by changes in AMP and ATP concentrations. However, activation of heart PFK-1 and PFK-2 is insufficient to stimulate glycolysis by itself. The stimulation of glucose transport by AMPK would increase the overall substrate supply. Liver and skeletal muscle PFK-2 isozymes do not contain the AMPK phosphorylation site, and therefore PFK-2 activation by AMPK is restricted to tissues containing the heart isozyme. The difference in regulation is intriguing and the reasons for the discrepancy are not yet known.

Cyclin-Dependent Protein Kinases (Cdk) Progression through the cell cycle in mammalian cells requires the coordinated regulation of the activities of several cyclin-dependent protein kinases (Cdks). The principle task of the cell division cycle is to replicate DNA without errors (during S phase) and to segregate the duplicated chromosomal DNA equally into two daughter cells during mitosis (or M phase). In addition to the molecular regulators that drive these processes, a monitoring mechanism ensures that the S phase is completed before mitosis begins, and vice versa. As a result, two gap phases are present in the somatic cell cycle; the G1 gap separates the M and S phases, and the G2 gap separates the S and M phases. These gap phases provide the cell "a moment of reflection" before proceeding on to the next step in the cell division cycle.

The timing of Cdk activation is determined by the association of proteins called cyclins, which are the regulatory subunits of the kinase complex. The kinase activity of all Cdks requires the binding of a cyclin. Each of the phases of the cell cycle is characterized by the expression of a distinct type of cyclin, and fluctuations in cyclin levels represent the primary mechanisms by which Cdk activity is regulated. The *G1* cyclins include cyclins C, D1-3, and E, and their accumulation is rate-limiting for progression from the G1 to S phase. The D-type cyclins assemble with the Cdk4 or Cdk6 kinases, and cyclin E complexes with the Cdk2 kinase. The mitotic or G2 cyclins, which include cyclin A and cyclin B, are involved in the control of G2 to M transition and mitosis. Cyclin A forms complexes with Cdk2 kinase and to a lesser extent Cdc2 kinase, whereas cyclin B is known to form complexes with Cdc2 kinase. Cyclins and Cdks are also regulated by phosphorylation. The assembly of a Cdk with its corresponding cyclin results in a partially active complex. Full activity is achieved following phosphorylation of the Cdk on a conserved Thr residue located within the ATP-binding pocket.

Cyclin-dependent protein kinases activity is predominantly localized in the nucleus. Thus, cyclin/Cdk complexes must not only be assembled and activated at the correct time, but also transported to their sites of action. The nuclear localization of the cyclin/Cdk complexes is mitigated by proteins that possess nuclear localization signals. In the nucleus, the active Cdk/cyclin complexes phosphorylate a number of proteins including a tumor suppressor protein known as the retinoblastoma susceptibility gene product (pRb). pRb controls commitment to progress from the G1 to S phase, at least in part by repressing the activity of the E2F transcription factors known to promote cell proliferation. The biological activity of the pRb protein is inhibited when it is phosphorylated.

FUNCTION AND REGULATION OF PROTEIN PHOSPHATASES

Protein phosphatases are the cellular antagonists of protein kinases and are an indispensable part of signal transduction processes involving protein phosphorylation. Although early studies focused mainly on protein kinases, with dephosphorylation regarded as a constitutively active "housekeeping" activity, recent work clearly indicates phosphatases match their kinase counterparts in specificity, variability, and complexity. The actions of kinases and phosphatases are in a state of dynamic equilibrium, the level of protein phosphorylation at any instant reflecting the relative activities of the protein kinases and phosphatases that catalyze the interconversion processes. Under physiological conditions, phosphate esters of Ser, Thr, and Tyr residues are stable and only show a low rate of spontaneous hydrolysis. Thus, the cell requires enzymes capable of cleaving phosphate esters off proteins in a controlled and regulated manner; this is the role of the protein phosphatases. Nonspecific alkaline and acid phosphatases are also capable of hydrolyzing phosphate esters in cells, but *in vivo* these never act in that mode but instead serve to promote phosphorus scavenging and remobilization within the cell. Protein phosphatases have a damping effect on protein-kinase-mediated signal transduction by diminishing and terminating a signal created by protein phosphorylation. Protein phosphatases can also have a positive effect in signaling pathways since a variety of proteins are inhibited or inactive when phosphorylated and are activated by dephosphorylation.

In keeping with the classification of protein kinases, the protein phosphatases have been divided into two large families on the basis of their substrate specificity. The family of protein-tyrosine phosphatases (PTPases) shows a preference for removing phosphate from tyrosine. The PTPase family includes the dual-specificity phosphatases that are capable of hydrolyzing phosphorylated serine/threonine residues as well as phosphorylated tyrosine. Enzymes in the serine/threonine protein phosphatase family have a preference for removing phosphate from phosphorylated serine and threonine residues. The Ser/Thr protein phosphatases are a large family of structurally related members, but the PTPases are not structurally homologous to the Ser/Thr phosphatases. Recall that, by contrast, all protein kinases (tyrosine and serine/threonine) appear to have been derived from a common ancestoral protein kinase. In the section to follow, the structure and regulation of the major Ser/Thr protein phosphatases will be introduced. The role of protein tyrosine phosphatases, whose intracellular targets include the receptors for growth factors, will be discussed in greater detail in Chapter 5.

Serine–Threonine Protein Phosphatases

Protein phosphatase activity was first discovered as the enzymatic activity that converted the active form of glycogen phosphorylase *a* back to its inactive *b* form. Initially, the nature of serine–threonine protein phosphatases was not known; a large number of studies had isolated protein phosphatase activities, but little effort had been made to establish whether the various activities were related. It was then determined that the majority of the protein phosphatases reported in the literature could be classified into two groups based on simple biochemical characteristics. The enzymes were originally subdivided by enzymatic function into protein phosphatase type-1 (PP-1) or type-2 (PP-2) depending on whether they specifically dephosphorylated the α or the β subunit of GPK and whether they were inhibited by nanomolar concentrations of two small heat- and acid-stable inhibitor proteins, termed inhibitor-1 and inhibitor-2 (Table 4.3). The PP-1 family dephosphorylates the β subunit of phosphorylase kinase whereas the PP-2 family acts on the α subunit. Type-2 phosphatases can be subclassified into three distinct enzymes, PP-2A, PP-2B (also known as calcineurin), and PP-2C based on their dependence on divalent cations. PP-2B activity has an absolute requirement for the Ca^{2+}–calmodulin complex. PP-2C requires Mg^{2+} whereas PP-2A is generally active in the absence of divalent cations. Type-1 and type-2 phosphatases can also be distinguished by their susceptibility to inhibition by okadaic acid, a complex fatty acid produced by marine dinoflagellates. More recently, the classification of phosphatases that dephosphorylate serine and threonine residues has been redefined by gene sequence homologies as members encoded by the PPP and PPM gene families. The PPP family includes the PP-1, PP-2A, and PP-2B phosphatases while the PPM family includes the PP-2C and mitochondrial protein phosphatases.

The three-dimensional structures of the PP-1 catalytic subunit and PP-2B show similarities that have helped to define the amino acid residues important to catalysis in the PPP family. Although the two enzymes show limited amino acid sequence identity, selected amino acid residues at the active site are invariant and participate in substrate recognition and catalytic efficacy. The PP-1 catalytic subunit contains two metal ions at the active site, Fe^{2+} and Zn^{2+}. Dephosphorylation is catalyzed in a single step by a metal-activated water molecule. The single-step S_N2-type reaction mechanism involves a water or hydroxyl molecule acting as a bridge ligand between the two metal ions and positioned directly in line with the P–O scissile bond of the substrate. A conserved histidine residue is in the correct position to act as a general acid, donating a proton to the oxygen atom on the leaving group of the phosphate ester.

Several novel mammalian protein phosphatase members have been identified within the PPP family including PP-4, PP-5, and PP-6. These show minimal sequence similarity to the PP-1 and PP-2 families but are widely distributed in mammalian tissues and, most importantly, possess properties suggesting that they perform distinct cellular functions. PP-4 has been cloned from mammalian and *Drosophila* sources. Located at the centrosome and in the nucleus, PP-4 seems to be a candidate for participation in the initiation of microtubule growth at centrosomes. Immuno-

TABLE 4.3 Biochemical Properties of Some Ser/Thr Protein Phosphatases

	PP-1	PP-2A	PP-2B	PP-2C
Preference for the α or β subunit of GPK	β	α	α	α
Inhibition by inhibitor-1 and inhibitor-2	Yes	No	No	No
Requirement for divalent cations	No	No	Yes (Ca^{2+})	Yes (Mg^{2+})
Requirement for calmodulin	No	No	Yes	No
Inhibition by okadaic acid (K_i)	Yes (20 nM)	Yes (0.2 nM)	Yes (5 μM)	No
Activity toward GPK	High	High	Very low	Very low

localization studies with antibodies to PP-4 show dense levels of the enzyme at the centrosome throughout all stages of mitosis except telophase, the one stage where spindle pole bodies are not present in the dividing cell. Upon reentry into interphase, PP-4 is again found at the centrosome of each daughter cell. PP-5 appears to participate in cell growth with a suggested role in the regulation of ribosomal RNA transcription. Higher levels of PP-5 are found in cells during logarithmic growth than in dividing or serum-deprived cells. PP-5 may also desensitize receptor signals; PP-5-mediated dephosphorylation significantly reduced signaling through glucocorticoid receptors.

Protein Phosphatase 1 Protein phosphatase 1, the prototype for the Ser/Thr phosphatases, has been studied since 1943 as the enzyme responsible for the conversion of glycogen phosphorylase *a* to glycogen phosphorylase *b*. The discovery that the enzyme was a phosphatase came in 1955 at the same time as the discovery of glycogen phosphorylase kinase, findings which marked the beginning of the era of studies of reversible protein phosphorylation as a regulatory mechanism. The number of cellular functions now known to be regulated by PP-1 is extensive.

The enzyme responsible for inactivating glycogen phosphorylase, later confirmed to be the PP-1 catalytic subunit (PP-1c), was isolated with the use of protein denaturants (ethanol and urea) as a 37-kDa protein. However, the treatment of tissue extracts with ethanol generated large increases in the apparent activity of glycogen phosphorylase phosphatase, concomitant with the reduction in the apparent size to that of the free catalytic subunit. These findings were later explained by the identification of an inhibitor protein(s) that was complexed with the catalytic subunit and led to the discovery of the heat-stable PP-1 inhibitor proteins. The existence of protein inhibitors of the PP-1 catalytic subunit (PP-1c) played an important role in unraveling the enzymology of PP-1c from two perspectives: the role of inhibitor proteins in regulating phosphatase activity and the existence of holoenzyme forms. Both of these perspectives are discussed in more detail later in the chapter.

Protein Phosphatase 2A Protein phosphatase 2A (PP-2A) accounts for a major portion of the Ser/Thr phosphatase activity in most cells and tissues. Like the other phosphatases, it was originally considered to be a housekeeping enzyme that passively and nonspecifically antagonized the actions of protein kinases by removing phosphate groups from phosphoproteins. However, it is now known that PP-2A plays a dynamic role in cellular signaling. Its activity is tightly controlled via regulation of its subunit composition and selective targeting to specific subcellular multiprotein signaling modules. Biochemical and genetic studies have demonstrated that PP-2A is an abundant, ubiquitous, and remarkably conserved enzyme. Deletion of the gene that encodes the PP-2A catalytic subunit is lethal in yeast and mice, demonstrating that PP-2A is essential for the normal metabolism of cells. Homologs of PP-2A have been identified in organisms as diverse as algae, fungi, *Drosophila*, higher plants, yeast, and mammals.

The core PP-2A homoenzyme consists of a 36-kDa catalytic subunit (C) that is always associated with a 65-kDa scaffolding subunit (A or PR65) that modulates its enzymatic properties. AC dimers are abundant, but the majority of PP-2A exists as an ABC heterotrimer, the AC dimer associating with one of three isomers of a regulatory subunit (B, B′, B″). The exact function of the B-type subunit is not yet known, but undoubtedly, it influences substrate specificity and subcellular localization. Posttranslational modification of the C subunit, including carboxymethylation and phosphorylation, influence PP-2A subunit composition, catalytic activity, and substrate specificity. PP-2A is involved in a wide variety of cellular processes including intermediary metabolism, muscle contraction, synaptic transmission, RNA splicing, gene expression, and cell cycle progression.

Protein Phosphatase 2B Protein phosphatase 2B (PP-2B) or calcineurin was first identified as a major calmodulin binding protein in brain. Later it was found to have Ser/Thr protein phosphatase activity. PP-2B is a heterodimer of calcineurin A (60kDa) and calcineurin B (19kDa). Calcineurin A is the catalytic subunit and also binds to calmodulin, whereas calcineurin B is the regulatory Ca^{2+}-binding subunit. With four Ca^{2+}-binding loops and 35% sequence identity with calmodulin, calcineurin B is also a member of the EF hand family of Ca^{2+}-binding proteins. The PP-2B catalytic subunit contains a catalytic domain with a long carboxy-terminal extension that inhibits catalytic activity. Interaction of the carboxy-terminal extension with the Ca^{2+}-binding, regulatory subunit relieves the inhibition when it is loaded with Ca^{2+} and explains why the enzyme is dependent upon Ca^{2+} for activity. Interestingly, calcineurin is activated at much lower Ca^{2+} concentrations ($K_d \sim 0.1$ nM) than the CaMKs ($K_d \sim 50$ nM), although the physiological relevance of this has yet to be determined.

Targeting proteins localize PP-2B to specific areas of the cell. In the brain, an inactive form of PP-2B is associated with AKAP79 at the plasma membrane. In T cells, cytosolic PP-2B is found associated with one of its physiological substrates, the transcription factor NFAT. T-cell activation is mediated via activation of PP-2B. This leads to dephosphorylation of NFAT and its translocation into the nucleus, possibly with the phosphatase still attached. In many situations, the direct substrates of PP-2B have not been determined. However, activation of PP-2B dephosphorylates a number of inhibitors of protein phosphatases and protein kinases [e.g., PP-1 inhibitor-1, the RII subunit of PKA, and the dopamine- and cAMP-regulated phosphoprotein of molecular mass 32,000 (DARPP-32)]. It appears that PP-2B may be a key enzyme that controls the activity state of other signaling enzymes with much broader substrate specificities (the reader is referred to Text Box 4.5 for more information regarding the identification of protein phosphatase and kinase targets).

Protein Phosphatase 2C At present, only limited information is available on the function of protein phosphatase 2C (PP-2C). The enzyme was originally identified as a Mg^{2+}-dependent protein phosphatase. PP-2C is a monomer with two major isoforms (α and β). Three α subtypes and two β subtypes have been identified; $\beta 1$ is ubiquitous whereas the $\beta 2$ isoform is only found in brain and heart. PP-2C may play a role in Ca^{2+}-dependent signal transduction in brain by catalyzing the dephosphorylation of CaMK-II.

Studies with *S. cerevisiae* have implicated PP-2C in cellular signaling processes involved in osmoregulation. PP-2C negatively regulates the osmosensing HOG1 (high osmolarity and glycerol) kinase cascade, the functional equivalent of the p38 mitogen-activated protein kinase of mammals. Whether PP-2C regulates the kinase directly or interacts with downstream targets remains to be established. In the plant, *Arabidopsis thaliana*, the *abi1* gene encodes a protein homologous to PP-2C that is essential for the response to the plant hormone, abscisic acid, which among other functions plays a key role coordinating metabolic responses during cold acclimation (see Chapter 17).

Targeting Protein Phosphatases with Anchoring Proteins The involvement of PP-1 and PP-2 family members in so many different cellular events raises the question as to how all of these processes can be regulated independently. This question is even more pertinent when one considers that both PP-1 and PP-2A are expressed at micromolar concentrations in cells and share a 49% sequence homology within their catalytic core (and hence, could possibly catalyze the dephosphorylation of many of the same substrates). Another paradox that has puzzled researchers for some time is how a relatively small number of protein phosphatases could effectively reverse the actions of a much greater number of protein kinases. The human genome codes for ~300 to 350 serine/threonine kinases but only ~40 serine/threonine phosphatases. So how is the specificity and independent regulation of target enzymes maintained if this low number of phosphatases must serve all these kinases? The key to the specificity of PP-1 (and PP-2) catalytic function resides with the interaction of the catalytic subunit with a variety of different regulatory subunits.

TEXT BOX 4.5 MAPPING THE TARGETS OF SIGNALING NETWORKS

One major problem in the field of protein phosphorylation is the difficulty of identifying the true *in vivo* substrates of different protein kinases and phosphatases. Only a few substrates are easily identified (e.g., enzymes with readily detectable effects of phosphorylation on kinetic parameters). Several methods have been tried to find the targets of various kinases and phosphatases. For the identification of protein kinase substrates, the application of specific kinase inhibitors can sometimes be useful, but their use *in vivo* can be unreliable because most of these inhibitors are not truly selective for one kinase. Sometimes, the phosphorylation consensus sequences for newly identified protein kinases are not known; in these cases, degenerate-oriented peptide libraries can be screened to determine the amino acid sequences that provide maximal potential for phosphorylation. With a known phosphorylation consensus sequence, databases can be screened for potential substrates and potential new protein targets identified and then susceptibility to phosphorylation by the kinase concerned can be tested *in vitro*. Both protein kinases and protein phosphatases are directed to their substrates, either by targeting subunits or protein interaction domains on the catalytic subunit itself. Hence, many phosphoproteins also have secondary binding sites on them for these targeting interactions. Therefore, the probability of finding true physiological substrates can be increased by searching protein databases to find those proteins that exhibit both consensus phosphorylation and binding site sequences for the protein kinase of interest.

The identification of physiological substrates for protein phosphatases is equally difficult. Cell permeable phosphatase inhibitors, such as okadaic acid and calyculin A, have been used in some cases to identify phosphatase substrates as those proteins that show increased *in vivo* phosphorylation in the presence of the inhibitors. Genomic deletions of protein phosphatase targeting subunits in yeast have also been very fruitful in the identification of phosphatase substrates.

These regulatory subunits can be divided into two general classes: inhibitory and targeting. Inhibitory subunits potently reduce PP-1c activity toward phosphorylated substrates and can be localized in specific areas of the cell. For example, the nuclear inhibitor of PP-1 (NIPP-1) is found only in the nucleus where it assists in regulating PP-1 activity. Each targeting subunit appears to tailor the catalytic subunit for a specific function by directing it to a particular region of the cell. A key element of the targeting hypothesis is that the cellular activity of PP-1 is only expressed when it is targeted; in other words, specificity is achieved by targeting of PP-1 to a restricted microenvironment. This would rationalize the fact that PP-1 demonstrates relatively nonspecific phosphatase activity toward a variety of different phosphorylated protein substrates. Regulatory subunits have been found that target PP-1 to glycogen particles, myosin, the nucleus, ribosomes, mitochondria, the cytoskeleton, the plasma membrane, and centrosomes, to name a few. Mammalian PP-1 binding proteins that were either isolated by biochemical methods or identified by yeast two hybrid screens include the retinoblastoma gene product (pRb), heat shock protein of 78 kDa (HSP78), p53BP2, splicing factor PSF, ribosomal protein L5, herpes virus γ134.5 protein, human SDS-22 homolog, growth arrest and DNA damage protein of 34 kDa (GADD-34), nuclear inhibitor of PP-1 (NIPP-1), HOX11, spinophilin, p99 PP-1 inhibitor, C-kinase potentiated inhibitor of 17 kDa (CPI-17), myosin protein targeting subunits in smooth and skeletal muscle (MYPT1 and MYPT2), and inhibitor proteins 1, 2, and 3. In addition, several glycogen-binding proteins related to the muscle glycogen targeting subunit, G_M, including G_L, R5, U5, and protein targeted to glycogen (PTG), have been identified. Genetic studies of yeast mutations and the use of the two hybrid screens have revealed over a dozen genes that encode putative PP-1 binding proteins. These include *GAC1*, *REG1*, *REG2*, *SCD5*, *GIP1*, *SHP1*, *GIP2*, and *SDS22* in *S. cerevisiae*. These genes are variously required for control of glycogen metabolism, protein synthesis, glucose repression, meiosis and/or sporulation, and mitotic cell cycle regulation.

Molecular Basis for Multiple PP-1 Binding Proteins The catalytic subunit of PP-1 (PP-1c) has been isolated from many sources as a component of multiprotein complexes. These complexes typically contain PP-1c and a regulatory (targeting or inhibitory) subunit bound in a 1 : 1 ratio. Regulatory proteins also have the potential to simultaneously bind one or two other proteins on additional docking sites. Thus, the PP-1 holoenzyme may have two to four constituent protein members. The 1 : 1 ratio of PP-1c and regulatory subunit implies that the interaction of different targeting subunits with PP-1c is mutually exclusive and that the binding site(s) for different regulatory subunits are probably identical or overlapping. Indeed, analysis of several different regulatory subunits identified a short highly conserved peptide motif with the sequence (R/K)(V/I)XF that was specifically recognized by a binding site on PP-1c. This binding motif is an absolute requirement for PP-1c interaction with its regulatory subunits. Subsequently, the consensus sequence for PP-1c binding motif was divided into one group that contains the motif VXF and one that contains VXW. Both motifs are generally preceded by two to five basic residues before the conserved valine. Interestingly, a number of PP-1c binding sites contain the sequence VSF, which together with the preceding basic residues, forms a PKA phosphorylation site. Phosphorylation at this site disrupts the interaction of the regulatory (or inhibitory) subunit with PP-1c.

The phosphorylation of PP-1c holoenzymes by PKA is now known to be a key regulatory mechanism that disrupts PP-1c localization and action in several different cellular contexts; for example, PP-1c binding to both the nuclear inhibitor protein (NIPP-1) and the muscle glycogen targeting subunit (G_M) is disrupted in this way. While the result of PKA phosphorylation in both of these situations is the disruption of the PP-1c/regulatory protein complex, the net effects on PP-1 activity are quite different. Phosphorylation of the mammalian muscle G_M subunit by PKA (in response to epinephrine) triggers dissociation of the PP-1 catalytic subunit from G subunit, releasing it from the glycogen particle and interrupting its ability to dephosphorylate its glycogen-bound substrates. PKA phosphorylation in this situation results in the down-regulation of PP-1c activity; however, PKA phosphorylation can result in the up-regulation of PP-1c activity, as demonstrated in the next example. Part of the nuclear PP-1 component is present bound to the nuclear inhibitor of PP-1 (NIPP-1). This inactive pool of PP-1 can be activated via phosphorylation of NIPP-1 by PKA which leads to PP-1 dissociation from the inhibitory protein. Free PP-1c is then able to associate with other targeting subunits present in the nucleus and mediate selective nuclear events. The activation of a nuclear PP-1 by PKA might explain why cAMP induces dephosphorylation of certain proteins. In general, the regulation of nuclear phosphorylation is still poorly understood.

A number of additional subunits have been identified that specifically interact with PP-1c to potently inhibit the activity of the enzyme toward all substrates. Two small proteins, inhibitor-1 and inhibitor-2, were originally identified as heat-stable proteins that inhibited PP-1 activity. I-1 is a more effective inhibitor when phosphorylated by PKA. Phosphorylation of I-1 occurs *in vivo* in skeletal muscle after adrenaline injection and also in liver after glucagon administration. The concentration of I-1 is higher than that of PP-1c in several tissues and its phosphorylation state is modulated by hormones that increase or decrease the intracellular concentration of cAMP. Changes in level of I-1 phosphorylation are therefore likely to be ac-companied by alteration of the activity of PP-1. Such an action provides a mechanism for amplifying the effects of cAMP since many substrates for PP-1 are themselves phosphorylated by PKA.

There is no sequence homology between I-1 and I-2, and the inhibition of PP-1c by I-2 is clearly different from that induced by I-1. For example, I-2 is a less potent inhibitor of

PP-1c activity and I-2 inhibition can be reversed upon its degradation. Phosphorylation of I-2 by PKA does not enhance its inhibitory effect. Convincing evidence for the role of I-2 *in vivo* is lacking. The most plausible ideas for I-2 function are as a form of molecular chaperone, perhaps acting as (a) a storage site for PP-1c when it is not bound to a targeting subunit, (b) an inactivator of PP-1c released adventitiously into the cytosol from particulate fractions, or (c) an assistant in the proper conformational folding of newly synthesized PP-1c prior to its delivery to a targeting protein at the appropriate subcellular location.

COORDINATED ACTION OF PROTEIN KINASES AND PROTEIN PHOSPHATASES

Reversible protein phosphorylation controls a huge number of metabolic processes, many of which have been discussed in other chapters. To complete this chapter, the following sections present selected examples of metabolic control by protein kinases and phosphatases, highlighting selected regulatory methods.

Protein Phosphorylation and Regulation of Glycogen Metabolism

Glycogen synthesis and degradation are tightly regulated by hormones and metabolic signals, primarily via modulation of the enzymatic activities of glycogen synthase and glycogen phosphorylase. The primary mechanism by which these enzymes are regulated in response to changing physiological conditions is reversible protein phosphorylation. In the catabolic state (e.g., during fasting or stress), glycogen synthase, the rate-limiting enzyme in glycogen synthesis, is phosphorylated on multiple serine residues by several kinases, including PKA and glycogen synthase kinase 3 (GSK3), resulting in its progressive inactivation. Catabolic signals, namely cAMP and Ca^{2+}, also promote the activation of GPK, which, in turn, phosphorylates glycogen phosphorylase, resulting in activation of the enzyme and a stimulation of glycogenolysis. Under anabolic conditions, the opposite occurs. Insulin stimulates an activation of glycogen biosynthesis by promoting the dephosphorylation of glycogen synthase. Insulin both attenuates the action of kinases such as PKA and GSK3 and activates serine/threonine phosphatases, in particular PP-1. The insulin-dependent dephosphorylation and activation of glycogen synthase is catalyzed by PP-1. PP-1 also contributes to the dephosphorylation and inactivation of glycogen phosphorylase when glycogen synthesis is occurring (Fig. 4.14).

Interestingly, insulin does not activate PP-1 globally, but rather specifically stimulates PP-1 that is associated with the glycogen particle. The catalytic subunit of PP-1 is associated with a glycogen binding protein known as the G subunit. The G subunit is tightly bound to glycogen, and association of PP-1c with the G subunit creates a holoenzyme complex known as PP-1G. PP-1G is a highly active form of the phosphatase that binds to glycogen so that PP-1c can dephosphorylate multiple enzymes associated with the glycogen particle, including the phosphorylated forms of glycogen phosphorylase and GPK. Regulation of the activity of PP-1G can be mediated by phosphorylation of the G subunit at two distinct Ser residues, P1 and P2. Phosphorylation at the P1 site modulates the substrate preference of the PP-1 catalytic subunit by increasing enzymatic activity toward glycogen synthase at the expense of GPK as a substrate. Therefore, phosphorylation of the P1 site has a stimulatory effect on the dephosphorylation of substrates with the overall result being an inhibition of glycogen degradation and a stimulation of glycogen synthesis. By contrast, phosphorylation at the P2 site reduces the affinity of the G subunit for the PP-1 catalytic subunit, leads to the dissociation of the catalytic subunit from glycogen, and contributes to an inhibition of protein phosphatase activity on the glycogen particle. The inactivation of glycogen phosphorylase by PP-1G action is hindered by phosphorylation of the G-subunit at the P2 site.

Hormonal stimulation (e.g., adrenaline) and activation of PKA causes phosphorylation of P2 site of the G subunit. Consequently, the PP-1 catalytic subunit dissociates from the G subunit and is released into the cytosol where it may express activity toward other substrates or, perhaps more likely, is quickly inactivated upon association with the cytosolic inhibitor protein I-2. The catalytic subunit can reassociate with the targeting G subunit only after the phosphorylation at the P2 site is removed. Both PP-2A and PP-2B can complete this function. PP-2B, being a calcium-dependent enzyme, can mediate a stimulatory effect of Ca^{2+} on glucose production by both activating GPK and removing PP-1G inhibition of glycogen phosphorylase. Phosphorylation of the G subunit at the P1 site, but *not* the P2 site, can be elicited by GSK-3 activation following insulin stimulation. Insulin controls glycogen metabolism by influencing the activity of PP-1G, allowing it to dephosphorylate glycogen synthase and thereby activate glycogen synthesis as well as dephosphorylate glycogen phosphorylase to inhibit glycogen degradation.

Protein Phosphorylation and Regulation of Smooth Muscle Contraction

Smooth muscle cells are critical to the normal physiology of many of the organs of the vertebrate body. Smooth muscle cells are the principal component of blood vessels where they regulate vascular tone and play a central role in the pathogenesis of atherosclerosis and other vascular diseases, of the gastrointestinal tract where they regulate

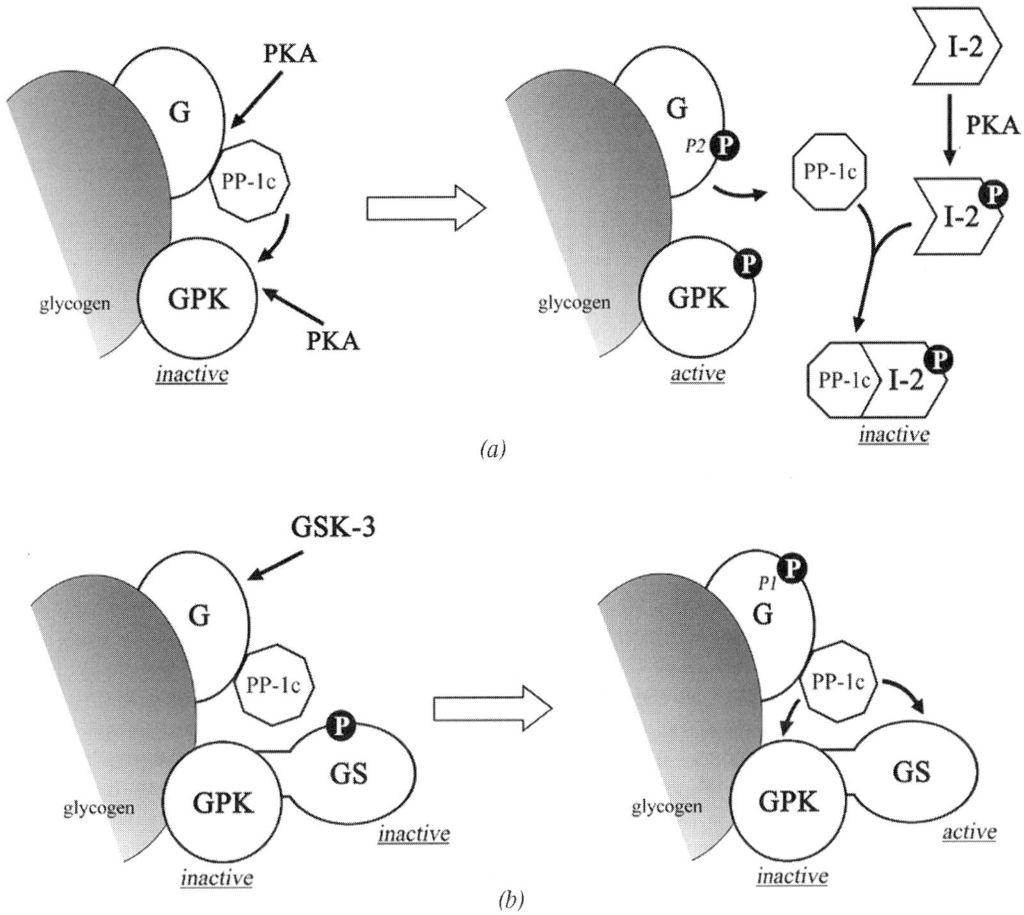

Figure 4.14 Protein phosphatase regulation of glycogen metabolism in muscle. The catalytic subunit of protein phosphatase type I (PP-1c) is localized with the glycogen particle by the glycogen-targeting subunit (G). Under resting conditions, association of PP-1c with the G subunit results in the continuous dephosphorylation and inactivation of glycogen phosphorylase kinase (GPK). (*a*) Upon β-adrenergic stimulation, the cAMP-dependent protein kinase (PKA) can phosphorylate both GPK and the G subunit. Phosphorylation of the G subunit at position 2 (P2) results in the dissociation of PP-1c from the glycogen particle into the cytoplasm. Cytosolic PP-1c is complexed with the phosphatase inhibitor protein type 2 (I2). PP-1c binding with I2 is also regulated by PKA. By sequestering PP-1c away from the glycogen particle in an inactive form, the I2 protein potentiates GPK phosphorylation and promotes sustained glycogen degradation. (*b*) The activity of the glycogen-bound protein phosphatase can also be modulated by insulin stimulation. In this case, glycogen synthase kinase (GSK-3) is activated via the insulin receptor and phosphorylates the G subunit at position 1 (P1). Phosphorylation of the P1 site modulates the substrate preference of PP-1c. The activity of PP-1c toward GPK and glycogen synthase (GS) is increased when complexed with the G protein phosphorylated at P1. Therefore, insulin has a stimulatory effect on glycogen synthesis by initiating the dephosphorylation and activation of GS and the dephosphorylation and inactivation of GPK.

peristalsis of nutritional products through the intestine, of lungs where they contribute to the elastic properties of the alveoli and play a central role in the pathogenesis of asthma, and of other specialized organs including the bladder, uterus, and placenta where they play roles essential to processes specific to these tissues.

Smooth muscle contraction and relaxation are regulated primarily by changes in intracellular calcium levels. Cytosolic Ca^{2+} concentrations increase due to influx through plasma membrane Ca^{2+} channels or via Ca^{2+} release from internal storage compartments. Ca^{2+}–calmodulin complexes form and bind to the inactive form of myosin

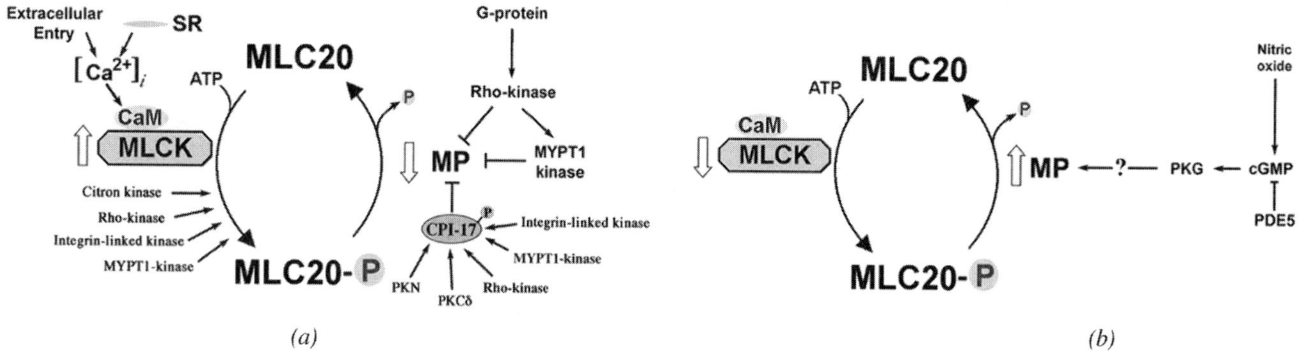

Figure 4.15 Complex signaling pathways regulate the contractile state of smooth muscle cells. (*a*) Smooth muscle contraction can be activated by a calcium-dependent mechanism. Calmodulin (CaM) activation follows increases in intracellular [Ca^{2+}], which originates from extracellular and sarcoplasma reticulum (SR) stores. Contraction results when the myosin regulatory light chain (MLC20) is phosphorylated by a myosin light-chain kinase (MLCK) in a calcium/CaM-dependent manner. However, contraction in the absence of calcium can occur by altering the balance between MLC20 phosphorylation by a number of different calcium-independent kinases (e.g., citron kinase, Rho-activated kinase, integrin-linked kinase, and the MYPT1-kinase) and MLC20 dephosphorylation by myosin phosphatase. Activation of a variety of G-protein-coupled receptors results in the inhibition of MLC20 dephosphorylation. Protein kinases, such as the myosin-associated protein kinase (MYPT1-kinase) and Rho-activated kinase, can phosphorylate and inactivate myosin phosphatase activity. Inhibition of myosin phosphatase also occurs through the phosphorylation and activation of the C-kinase potentiated inhibitor protein (CPI-17) by a number of kinases. (*b*) Smooth muscle relaxation follows a decline in intracellular [Ca^{2+}] as a result of a reduction in MLCK activity. Alternatively, calcium-independent relaxation of smooth muscle occurs in response to changes in cyclic-GMP (cGMP) levels. cGMP-dependent protein kinase (PKG) activates myosin phosphatase activity through an unknown mechanism. Nitric oxide stimulates cGMP production whereas cGMP-phosphodiesterase type 5 (PDE5) catalyzes cGMP degradation.

light-chain kinase (MLCK) to stimulate an activation of the kinase through oligomerization. MLCK can then phosphorylate myosin regulatory light chains (MLC). In smooth and nonmuscle cells, MLC phosphorylation markedly increases actin-activated myosin MgATPase activity and is an essential prerequisite for contraction. In skeletal and cardiac muscle, the actin-activated myosin MgATPase activity is maximal, even in the absence of MLC phosphorylation.

When cytosolic Ca^{2+} concentrations decline, the process undergoes a reversal. Dissociation of calmodulin from MLCK converts the kinase to its inactive monomeric form. MLCK inactivation also allows dephosphorylation of MLC by myosin phosphatase (referred to as MP or SMPP-1M). Originally, it was assumed that SMPP-1M activity was constant and that the only regulation of MLC phosphorylation was via Ca^{2+}–calmodulin control of MLCK. With the development of methods to measure internal free [Ca^{2+}], it was deduced that for a given [Ca^{2+}] the contractile force developed could vary depending on the method of stimulation. This led to other studies that provided strong evidence that MLC phosphorylation could be modulated by agonists without a change in [Ca^{2+}] (Fig. 4.15). These led to the important realization that a change in myosin phosphorylation state was due to a shift in the MLCK : SMPP-1M activity ratio. Either MLCK was activated or SMPP-1M was inhibited. The key discovery was that agonist administration was linked to myosin phosphatase inhibition.

SMPP-1M is a trimer comprising a 110-kDa regulatory myosin targeting subunit (MYPT), the 37-kDa type-1 catalytic subunit (PP-1c), and a 20-kDa protein of uncertain function (M20). There are several isoforms of MYPT and two genes have been identified on human chromosomes 1 and 12. A dominant feature of MYPT structure is a series of ankyrin repeats at the N-terminal end of the molecule; these may assist in the binding of the PP-1c catalytic subunit and the substrate, phosphorylated myosin. In addition, at the N-terminal fringe of the ankyrin motifs is a consensus PP-1c binding motif (KVXF). The M20 protein is known to bind to the C-terminal end of MYPT. The cellular function of the M20 subunit has not been established; however, M20 has recently been reported to possess microtubule binding activity and may play a role in regulating microtubule formation.

It is reasonable to assume that regulation of the activity of myosin phosphatase involves changes in subunit interactions. The focus has been on the MYPT1 subunit as it has already been demonstrated to coordinate the binding of PP-1c and the phosphorylated substrate. Several putative regulatory mechanisms that involve MYPT have been proposed. These include:

1. Arachadonic acid has been proposed to modulate SMPP-1M activity. Arachidonic acid may inhibit SMPP-1M activity by direct interaction with the C-terminal part of MYPT, causing dissociation of the holoenzyme into its subunits, or indirectly by activating a kinase (e.g., PKC) that phosphorylates MYPT, causing inhibition.
2. The activation of a myosin phosphatase inhibitor protein by PKC modulates SMPP-1M activity. C-kinase potentiated inhibitor protein, CPI-17, is a 17-kDa myosin phosphatase inhibitor that becomes a very potent inhibitor of SMPP-1M following phosphorylation by PKC. Phosphorylation of CPI-17 occurs in response to stimulation by agonists such as histamine and phenylephrine and enhances inhibitory potency toward SMPP-1M by over 1000-fold.
3. Developed independently of the arachadonic acid scheme is the inhibition of myosin phosphatase activity resulting from phosphorylation of MYPT. About 80% inhibition of SMPP-1M activity was observed upon phosphorylation of a C-terminal inhibitory site. How the phosphorylation of the C-terminal part of MYPT inhibits PP-1c, which is associated with the N terminus is not yet known, but one possibility is that the phosphorylated residue occupies the active site of PP-1c.

A number of kinases can phosphorylate the MYPT subunit on its C-terminal half, including the Rho-associated kinase (ROK), integrin-linked kinase (ILK), the citron kinase, PKC, PKA, and PKG (Fig. 4.15). The isolated SMPP-1M holoenzyme also contains a kinase that phosphorylates MYPT and inhibits phosphatase activity. This endogenous MYPT-kinase was recently identified and characterized. An important feature of both ROK and the MYPT-kinase is that they directly phosphorylate MLC, increase actin-activated MgATPase activity, and induce contraction of smooth muscle fibers. Therefore, the inhibition can be considered a two-phase process, one phase being the phosphorylation of MYPT and/or CPI-17 with the ensuing inhibition of SMPP-1M activity, and the second phase being a direct phosphorylation of MLC. Obviously, the two processes can act together to amplify a change in the proportion of MLC found in the phosphorylated state. It is possible that SMPP-1M activity may also be activated, for example, following increases in intracellular cAMP and/or cGMP levels. Evidence in support of this is very limited; however, a tentative hypothesis is that phosphorylation of MYPT by PKG alters the cellular distribution of SMPP-1M. This would involve a shuttle of SMPP-1M between a dephosphorylated, membrane-bound, and inhibited state and a phosphorylated, cytoskeletal-associated, and active state.

SUGGESTED READING

Cellular Signaling

Barik, S. (1996). Protein phosphorylation and signal transduction. In *Subcellular Biochemistry*, Vol. 26, *Myo-inositol Phosphates, Phosphoinositides, and Signal Transduction*, Biswas, B. B., and Biswas, S. (eds.). Plenum, New York, pp. 115–164.

Cohen, P. (2001). The role of protein phosphorylation in human health and disease. *Eur J Biochem* **268**:5001–5010.

Conn, P. M., and Means, A. R. (eds.) (2000). *Principles of Molecular Regulation*. Humana, Totowa, NJ.

Downes, C. P., Wolf, C. R., and Lane, D. P. (eds.) (1997). *Cellular Responses to Stress*. Princeton University Press, Princeton, NJ.

Krauss, G. (1999). *Biochemistry of Signal Transduction and Regulation*. Wiley-VCH, Berlag, Weinheim.

Ricard, J. (ed.) (1999). *Biological Complexity and the Dynamics of Life Processes*. Elsevier, New York.

Protein Kinases

Hardie, G., and Hanks, S. (eds.) (1995). *The Protein Kinase Facts Book: Protein-Serine Kinases*. Academic, London.

Liu, W. S., and Heckman, C. A. (1998). The sevenfold way of PKC regulation. *Cell Signal* **10**:529–542.

Steeg, P. S., Palmieri, D., Ouatas, T., and Salerno, M. (2003). Histidine kinases and histidine phosphorylated proteins in mammalian cell biology, signal transduction and cancer. *Cancer Lett* **190**:1–12.

Protein Phosphatases

Bollen, M. (2001). Combinatorial control of protein phosphatase-1. *Trends Biolog Sci* **26**:426–431.

Lee, E. Y. C., Zhang, L., Zhao, S., Wei, Q., Zhang, Z., Qi, Z. Q., and Belmonte, E. R. (1999). Phosphorylase phosphatase: New horizons for an old enzyme. *Front Biosci* **4**:270–285.

Sontag, E. (2001). Protein phosphatase 2A: The Trojan Horse of cellular signaling. *Cell Signal* **13**:7–16.

Wera, S., and Hemmings, B. A. (1995). Serine/threonine protein phosphatases. *Biochem J* **311**:17–29.

5

TYROSINE PHOSPHORYLATION AND THE CONTROL OF CELLULAR INFORMATION

JUSTIN A. MACDONALD

The discovery in the mid-1950s that the enzymatic activity of a protein could be altered by phosphorylation on serine and/or threonine residues was the seminal event that launched the field of signal transduction. This finding was immediately followed by the identification of intracellular secondary messengers [e.g., cyclic adenosine 5′-monophosphate (cAMP)] and the heterotrimeric G-protein-coupled receptors that are responsible for initiating second-messenger synthesis in response to hormone binding (discussed in the previous chapter). Less than a decade later, the first serine/threonine protein kinase was identified, and progress in understanding the mechanisms of serine/threonine phosphorylation has continued to grow at an astounding rate. While research in serine/threonine phosphorylation has revealed many important themes and principles of signal transduction, another equally important aspect of signaling was only uncovered within the last 15 years. A new class of protein phosphorylation, on tyrosine residues, caused the next "revolution" in transmembrane signaling. The present chapter explores the development of the tyrosine phosphorylation research field, addresses recently defined mechanisms of signal propagation, and outlines a tyrosine signaling network that plays an important role in the transfer of cellular information from plasma membrane to nucleus.

The first evidence for the presence of proteins that were phosphorylated on tyrosine residues was obtained from a group of purified viral transforming proteins that were associated with the polyomavirus. These proteins mitigated uncontrolled cell proliferation, and, ultimately, they were identified as a new class of protein kinases that phosphorylated tyrosine. Subsequently, growth factors were the first group of endogenous molecules in eukaryotic systems that were shown to stimulate tyrosine phosphorylation. Growth factors are proteins that are usually present in very low concentrations (nano- or picomolar) and that stimulate cell division or the proliferation of specific cell types, for example, platelet-derived growth factor (PDGF), epidermal growth factor (EGF), transforming growth factor-β (TGFβ), fibroblast growth factor (FGF), and the interleukins. Protein phosphorylation on tyrosine residues was quickly revealed as an important eukaryotic cell signaling mechanism. Environmental stimulators of tyrosine phosphorylation include osmotic stress (both hyper- and hyposmotic), hypoxia, oxygen radicals, hydrogen peroxide, ultraviolet light, heat shock, and various chemicals. Cellular protein phosphotyrosine levels are regulated by the antagonistic activities of the protein tyrosine kinases (PTKs) and the protein tyrosine phosphatases (PTPs).

PROTEIN TYROSINE KINASES

The protein tyrosine kinases are the principal effectors of transmembrane tyrosine signaling. They fall into two subclasses: the receptor protein tyrosine kinases (RPTKs) and the nonreceptor cytoplasmic protein tyrosine kinases (cPTKs). Tyrosine kinases in both subclasses contain a catalytic core domain that is able to phosphorylate tyrosine residues on protein substrates. However, the RPTKs are transmembrane proteins that contain a glycosylated extracellular domain, a single transmembrane sequence, and an intracellular domain that has intrinsic catalytic activity

Functional Metabolism: Regulation and Adaptation, edited by Kenneth B. Storey.
ISBN 0-471-41090-X

that is activated upon ligand binding to the extracellular domain. More is known about RPTKs that are activated by growth factors than those that are activated by environmental stresses (e.g., osmotic stress, hypoxia). Our knowledge of the molecular mechanism of RPTK signaling derives primarily from studies of the PDGF and EGF receptors. Growth factor activation of a RPTK occurs in a series of steps: (1) ligand binding to the extracellular domain; (2) ligand-induced receptor dimerization; (3) mutual transphosphorylation whereby the close proximity of the two catalytic domains in the dimer allows each monomer to phosphorylate tyrosine residues in the activation loop of the other catalytic domain; and (4) autophosphorylation of tyrosines outside the catalytic domain (Fig. 5.1). Once the receptor has phosphorylated itself, it is able to activate the next step in the chain with the help of several adaptor proteins that facilitate binding of effector enzymes to the membrane.

The cPTKs do not contain transmembrane regions and therefore do not fall into the cell surface receptor classification. There are currently at least 33 known vertebrate genes that encode nonreceptor cPTKs. They can be divided into nine protein families: Abl, Fes/Fer, Syk/Zap70, Jak, Tec, Fak, Ack, Src and Csk. Four additional cPTKs (Rlk/Txk, Srm, Rak/Frk, and Brk/Sik) do not appear to belong to any of the defined families. External stimuli can also lead to activation of nonreceptor tyrosine kinases. Many cPTKs form complexes with the intracellular regions of other cell surface receptors (e.g., antigen receptors), which may or may not possess intrinsic cPTK catalytic activity of their own (Fig. 5.2). The best understood of the cPTKs are the members of the Src family. These have a myristilated N terminus that permits their association with the membrane. Src is normally maintained in a phosphorylated, catalytically inactive state and is activated by dephosphorylation of regulatory phosphotyrosine sites to allow Src to form a signaling complex with the cytoplasmic domain of a RPTK.

Early biochemical work emphasized the concepts of metabolic pathways, in which one step leads to a subsequent step in a linear fashion. This was also the dominant paradigm in the early years of signal transduction research because the first major protein kinase cascade that was identified (the cAMP-dependent protein kinase) proceeded in a linear cascade with the amplification of a signal. However, this linear concept has proven inadequate in the description of complex intracellular signaling, particularly with the PTKs. A good example of the failings of the linear transmission model arose during the early investigations into RPTK function. When it was discovered that the receptors for many growth factors were transmembrane PTKs, the search for the next link in the chain of events was initiated. The focus was on finding and identifying the *downstream* proteins that were phosphorylated on tyrosine by the activated ligand receptor. When extracts of growth-factor-stimulated cells were analyzed for phosphotyrosine-containing proteins (using antibodies specific for phosphotyrosine), the most prevalent phosphotyrosine-containing proteins were the RPTKs themselves. This observation was inconsistent with models in which the receptor initiates a signaling cascade by phosphorylating many substrate proteins and required the development of a new model. The key to signal transmission by PTKs turned out to be the creation of binding sites on the receptor. Autophosphorylation

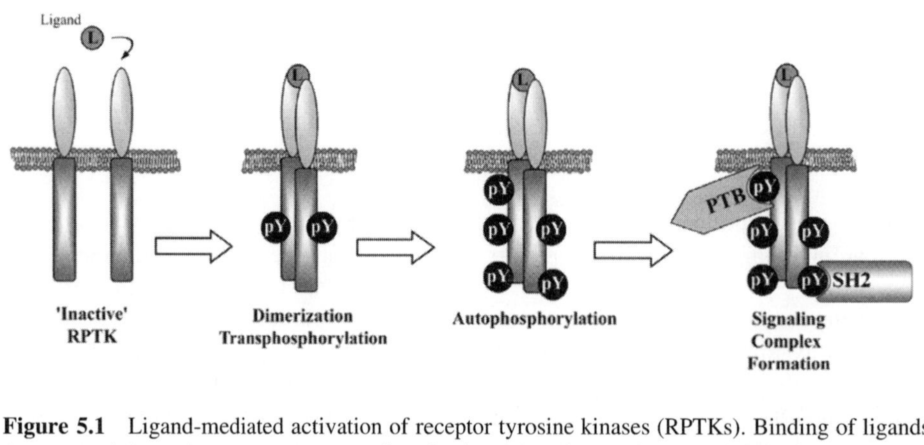

Figure 5.1 Ligand-mediated activation of receptor tyrosine kinases (RPTKs). Binding of ligands (e.g., growth factors) promotes receptor dimerization. Mutual transphosphorylation of tyrosine residues within the catalytic core of the RPTK result in the activation and potentiation of receptor tyrosine kinase activity. The autophosphorylation of tyrosine residues in the noncatalytic, cytoplasmic regions of the RPTK creates docking sites for downstream adaptor proteins. The adaptor proteins bind to phosphorylated tyrosine residues through specialized phosphotyrosine-binding domains such as the Src homology 2 (SH2) and the phosphotyrosine-binding (PTB) domains.

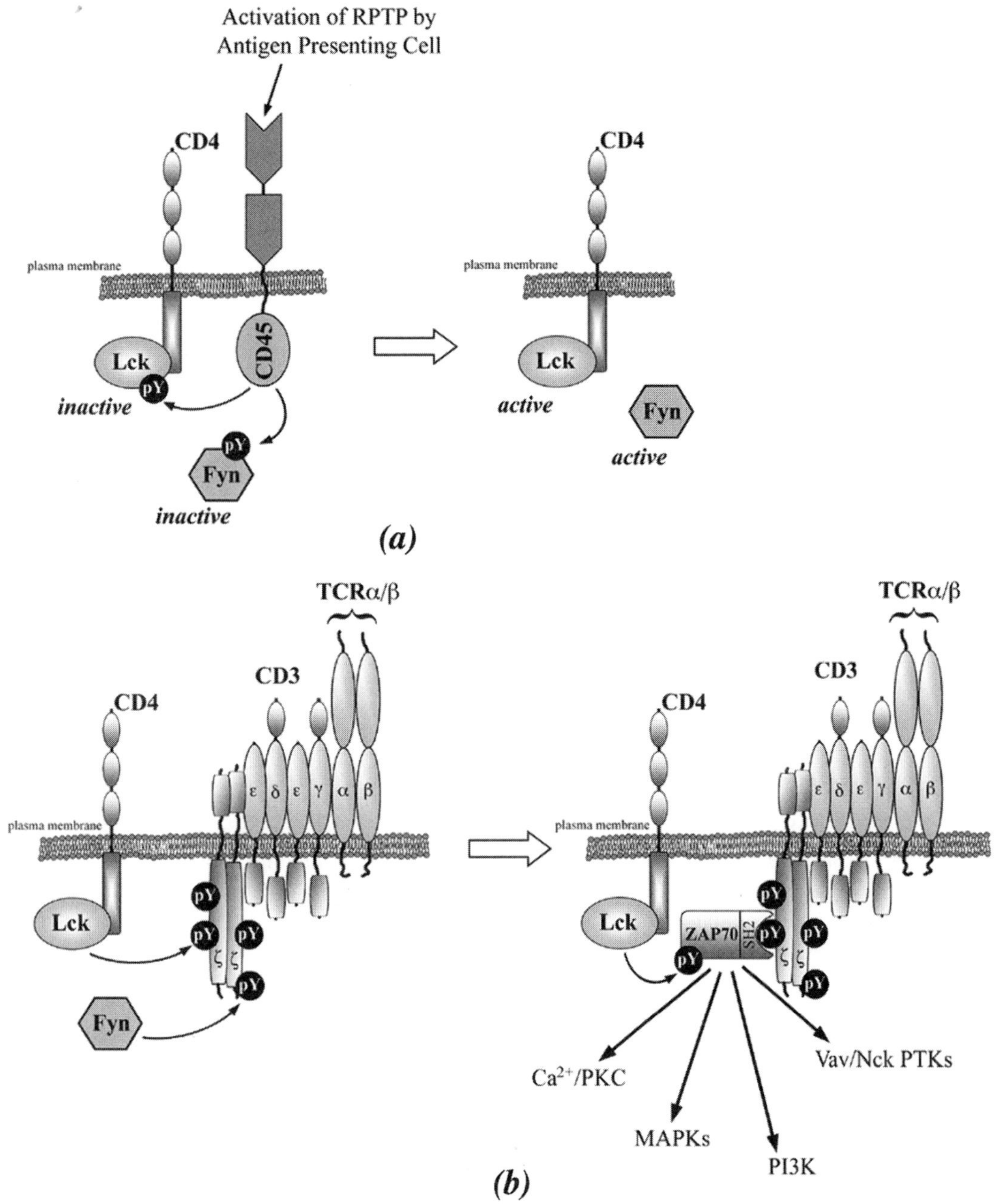

Figure 5.2 Src-family nonreceptor protein tyrosine kinases (PTKs) play crucial roles in initiating T-cell antigen receptor signaling. (a) The initial activation of Src-family kinases, Lck and Fyn, requires the dephosphorylation of an inhibitory phosphotyrosine residue by a receptor protein tyrosine phosphatase (CD45) that is activated through intercellular contact with the antigen-presenting cell. (b) Concurrently, T-cell activation is elicited when foreign peptide presented on the surface of antigen-presenting cells (e.g., dendritic cells, macrophages, and B cells) is recognized by the T-cell antigen receptor (TCR). The TCR is comprised of a ligand binding subunit (the α and β chains) and a signaling subunit (the CD3 δ, ε, and γ chains and the TCRζ chains). Fyn and Lck (present as a multimeric complex with the CD4 receptor) are able to phosphorylate tyrosine residues on the ζ chains of the TCR. Following ζ-chain phosphorylation, ZAP70 (ζ-chain-associated protein kinase) can associate via its SH2 domains with phosphotyrosine residues on the ζ chain of the TCR. After attachment to the ζ chains, ZAP70 is phosphorylated and activated by Lck. This wave of protein tyrosine phosphorylation leads to an increase in intracellular calcium and an activation of downstream signaling pathways including additional PTKs (e.g., Vav and Nck), the calcium/phospholipid-dependent protein kinase (PKC), phosphatidylinositol 3-kinase (PI3K), and the mitogen-activated protein kinases (MAPKs). Ultimately, these signaling pathways activate transcription factors such as the activator protein-1 (AP-1), nuclear factor-$\kappa\beta$ (NF-$\kappa\beta$), or the nuclear factor of activated T cells (NFAT), which lead to the expression of genes that control the T-cell response.

of tyrosine residues within the PTKs created sites for the binding of proteins with high affinity for specific phosphotyrosine-containing peptide sequences. From these studies emerged the idea that signal transduction mechanisms could include an alteration of binding activities in response to a ligand.

Re-evaluation of classical signaling pathways such as the cAMP-dependent signal transmission has subsequently revealed the critical importance of regulated protein–protein interactions in these systems as well (Chapter 4). For example, the details of this relatively simple pathway involve no less than five important protein–protein interactions: (1) the heterotrimeric G protein binds to the liganded (but not the unliganded) receptor, (2) conformational changes induced by receptor binding coupled with guanosine 5′-diphosphate (GDP) release and guanosine 5′-triphosphate (GTP) binding cause the dissociation of the α subunit of the G protein from the β and γ subunits, and from the liganded receptor, (3) the free Gα binds to and activates the adenylyl cyclase, (4) the G$\beta\gamma$ complex binds to and relocalizes the β-adrenergic receptor kinase leading to receptor phosphorylation and desensitization, and (5) cAMP binding causes the dissociation of the active catalytic subunits of cAMP-dependent protein kinase (PKA) from the regulatory subunits, which are themselves tethered by an A kinase binding protein. Much of the current research on signal transduction is concerned with the analysis of protein–protein interactions. These interactions are now thought to be so important to the understanding of signal transduction that one of the first experiments completed on a newly identified protein is a search for interacting partners. The mechanisms of protein–protein interaction and the effects on signal transmission are discussed further in the following sections of this chapter.

Protein-Binding Modules and Specificity

There are a number of recognizable modules that confer binding specificity to signaling proteins (Table 5.1). A prediction of the type(s) of binding interactions expected for a protein can be made on inspection of the amino acid sequence of the protein. Cellular recycling of binding interactions makes sense from an evolutionary point of view in that domains can be shuffled and existing interaction pairs fine-tuned so that specific binding surfaces do not need to arise independently for each pair of interacting proteins. Protein interaction modules that are recognizable by sequence similarity fall into two overlapping classes. The first are those that are independently folding units that confer a characteristic and specialized type of binding interaction. In these modules the most conserved residues are those that are directly involved in binding to ligands. The second broad class of interaction modules are those where the sequence similarity is due to a common folded structure but does not necessarily predict the specific type of binding interaction. These motifs may be repeated many times in the proteins that contain them and assemble with other repeats into higher-order structures.

Three important aspects of protein–protein interactions are specificity, regulation of binding specificity, and flexibility. Specificity, or the affinity for target sites, is a graded phenomenon. Protein–protein interactions with very high specificity tend to have dissociation constants in the range of 10^{-9} M. An example is the association of the inhibitor-2 protein with the protein phosphatase type 1 catalytic subunit. Moderate specificity interactions typically have dissociation constants in the 10^{-8} to 10^{-7} range and include complexes of Src homology 2 (SH2) domains with tyrosine phosphorylated targets. Weak specificity interactions, such as SH3-targeted protein complexes, have affinities in the 10^{-6} to 10^{-5} M range. In general, as the specificity of a protein–protein interaction increases, the number of potential binding proteins will decrease. For this reason, the components of high-specificity protein complexes are usually static, unchanging species. Moderate affinity protein complexes may involve interactions with several different subunits; for example, the binding of the same heterotrimeric G-protein β-subunit with several different α and γ subunits. And low-specificity SH3 domains have the opportunity to associate with proline-rich sites in tens to hundreds of different proteins. Protein domain interactions display surprising flexibility. For instance, single amino acid substitutions can alter the specificity of binding domains. This apparent flexibility may have an evolutionary advantage, in the sense that domain binding specificity might change rapidly, allowing the formation of new signaling connections such as when single-celled organisms became more complex and were faced with increasing environmental challenges. The following subsections give an overview of various binding domains that have key functions in cellular signal transduction.

SH2 Domains Many substrates and adaptor proteins recognized by receptor PTKs contain the discrete SH2 domains. SH2 domains were originally identified as regions of homology between the Src tyrosine kinase and other distantly related kinases. These regions of homology were shown to bind specifically to tyrosine-phosphorylated peptides but not to the corresponding unphosphorylated peptide. Crystal structures of several SH2 domains have been solved together with bound phosphopeptides. These show that the SH2 domain is a compact globular domain of about 100 residues that interacts with proteins that contain a phosphotyrosine residue within a specific sequence. The phosphotyrosine residue is buried in a deep pocket with the residues on the immediate C-terminal side of the phosphotyrosine, making specific contacts with

TABLE 5.1 Protein Binding Modules Implicated in Signal Transduction[a]

	Binding Domain	Core Binding Site Sequence	Overall Size	Domain Containing Protein/Binding Partner
Phosphotyrosine	SH2	PpYXXX	~100	Grb2/EGF receptor PTK
	PTB	NPXpY (NPXpY)	~100–150	IRS-1 docking protein/Insulin receptor
Proline rich	SH3	PXXP	~60	Src PTK/p85 subunit of PI3-Kinase
	WW	PPXY or PPLP	38	Yes-associated protein (YAP)/Yes (Src-like PTK)
Phospho-Ser/Thr	FHA	pTXXD or pTXXX	~65–100	Plant receptorlike kinase (RLK5)/RLK5 associated phosphatase (KAPP)
	14-3-3	XpSX or XpTX	—	Wee1 cell-cycle Y kinase/14-3-3 dimer
Phospholipid	PH	Various	~120	Akt Ser/Thr kinase/PI(3,4)P2
	C1	Various	~50	PKC isoforms/Diacylglycerol or phorbol esters
	C2	Various	~130	PKCβ1 isoform/Calcium
	FYVE	RKHHCR, others?	~60	PI(3)P
	TUBBY	Undefined	~260	TULP1/PI(4,5)P2 and PI(3,4)P2
	PX	Undefined	~120	PI3K and CISK/PI(3)P
Apoptosis/cell survival	DD	Electrostatic	~80–100	Tumor necrosis factor receptor R55/TRADD
	DED	Various	Unknown	Pro-caspase 8/FADD
	CARD	Electrostatic	~90–100	CARDIAK Ser/Thr kinase/Pro-caspase 1
	TRAF	Not defined	~150	TNF-receptor-associated factors (TRAF)/TNF receptors and TRADD adaptors
Miscellaneous	EF-hand	Various	~40	Calmodulin/calcium
	Armadillo	Not defined	~40	β-catenin/APC tumour supressor protein
	Ankyrin	Variable	33	p53 binding protein (53bp2)/p53
	PDZ	S/TXV-COOH	~80–90	*Drosophila* INAD protein/PLC-β and PKC
	TIR	Undefined	~135–160	Toll-like receptors (TLR4)/MyD88 adaptor
	WD	Various	~30	G-protein β subunit/G-protein α, γ subunits
	TPR	W(L/G)YAFAP	~34	p67PHOX/GTP-Rac
	LIM	Undefined	~60	PINCH/Integrin-linked kinase
	Bromo	Acetylated lysine, others?	~110	Snf-2 (yeast AMPK)/Acetylated lysine residues
	Chromo	Not defined	~30–70	HP-1/Histone H3 methylated lysine
	CSD	Hydrophobic residues?	~40–70	HP-1/Ku70
	GEL	Undefined	~120–150	Gelsolin/Calcium and F-actin
Proteolysis	F-Box	Unknown	~42–48	Yeast Cdc4/Skp1 and Rbx1
	Hect	Various	~350	Nedd4/UbcH5
	RING	CXXCXX-XCXXHX	~40–60	Cbl/UbcH7

[a]The number of amino acid residues is given for the "overall size" of the binding domain. The amino acid sequence present in the binding site is presented as the "Core Binding Site Sequence." Single-letter amino acid code: X, any amino acid; C, cysteine; W, tryptophan; H, histidine; G, glycine; Y, tyrosine; A, alanine; F, phenylalanine; D, aspartic acid; N, asparagine; K, lysine; L, leucine; P, proline; R, arginine; S, serine; T, threonine; V, valine; X, any amino acid; pS, phosphorylated serine; pT, phosphorylated threonine; pY, phosphorylated tyrosine; and –COOH, carboxyl terminus.

the surface of the SH2 domain. Since binding is only dependent on a short peptide sequence, the interaction can be mimicked with synthetic peptides, demonstrating that the interaction is largely independent of the larger protein structure.

The physiological importance of SH2 domains became apparent when it was shown that many proteins implicated in signaling contained SH2 domains and that these proteins were often found associated with ligand-activated growth factor receptors. Later, bacterially expressed SH2 domains were shown to bind to tyrosine-phosphorylated proteins, including most importantly, activated RPTKs. Further experiments have shown that these domains serve a general role in signaling in complex eukaryotes by mediating the relocation of SH2-containing proteins in response to changes in tyrosine phosphorylation. The Src SH2 domain is also essential for negative regulation of its activity. Autoinhibition of tyrosine kinase activity occurs when a C-terminal regulatory tyrosine present in the catalytic region is phosphorylated. The SH2 domain is able to bind intramolecularly to this phosphotyrosine on the catalytic domain and creates an inactive conformation of the kinase. Because SH2 domains are absent from unicellular eukaryotes (e.g., yeast), it is believed that SH2 domains and tyrosine kinases are a relatively recent evolutionary innovation developed by multicellular eukaryotes to allow coordinated control by extracellular signals over the responses of multiple cells, tissues, and organs.

PTB Domains Another site that recognizes tyrosine-phosphorylated sites, the phosphotyrosine binding, or PTB, domain, was identified during the analysis of the Shc adaptor protein. Shc also contains an SH2 domain and was known to bind tyrosine-phosphorylated proteins. However, phosphotyrosine binding was not abolished with removal of the SH2 domain, and it became apparent that additional sites consisting of NPXpY motifs were also binding phosphotyrosine. The binding characteristic was eventually mapped to a 160-amino-acid region of Shc with no sequence homology to the SH2 domain. Newly identified proteins containing PTB domains apparently recognize NPXY or related motifs, but binding is not dependent on tyrosine phosphorylation.

SH3 Domains As was the case with the SH2 domains, the SH3 domains were first identified in signaling proteins as a region of homology to the Src kinase. They have subsequently been found in a wide range of proteins including some isolated from yeast. The SH3 domain, which is another compact globular domain of 60 residues, associates in a phosphorylation-independent manner with short proline-rich binding sites in proteins. Most SH3 domains bind to core sites with the consensus PXXP. They are often found in the same proteins as SH2 domains, but this does not indicate any structural or functional similarity in the two domains. SH3-mediated binding has not been shown to be directly regulated. It is likely that in most cases these domains bind constitutively, functioning as a type of intracellular "glue" and not as a switch. Indeed, one of the best characterized roles of an SH3 domain is in a class of proteins called SH2/SH3 adaptors, which serve to mediate the formation of signaling complexes. In *Drosophila melanogaster* and *Caenorhabditis elegans*, the SH2/SH3 adaptor protein Grb2, which contains one SH2 and two SH3 domains, recruits a guanine nucleotide exchange factor (GEF) to the membrane. The GEF–Grb2 complex associates with growth-factor-activated RPTKs through the binding of the Grb2 SH2 domain to phosphotyrosine sites.

WW Domains The WW domain is another small domain used in signaling proteins, which can also interact with proline-containing sequences but with a different sequence specificity than SH3 domains. However, some evidence suggests that WW domains and SH3 domains can potentially bind overlapping sites. The domain name is derived from two conserved tryptophan residues (abbreviated W) spaced 20 to 22 residues apart within the consensus sequence. Two classes of WW domain ligands have been identified, represented by the consensus sequences xPPxY and PPLP. WW and SH3 domains seem to have independently evolved to bind proline-rich hydrophobic ligands with moderate affinity. In addition, a few select WW domains function as phosphoserine- or phosphothreonine-binding modules, suggesting that certain WW domains have evolved an alternate mode of action.

Leucine Zipper Domains The leucine zipper (LZ) domain is a helical structure that forms coiled coils and was originally identified as a highly conserved motif mediating the binding of transcription factors to deoxyribonucleic acid (DNA). The LZ region is defined by a pattern of at least four leucine (L) residues repeated every seventh amino acid that mediates dimerization through the formation of parallel α-helical dimers. An LZ-containing polypeptide must exhibit a hydrophobic surface and adopt a coiled-coil structure to form and stabilize the intrahelical interaction. Usually the zipper is incorporated into a helix–loop–helix conformation called the basic helix–loop–helix–leucine zipper (bHLH/Zip). The LZ motif has been found in a range of important proteins, including those that function as transcription factors in many organisms. Its role in promoting the activation of gene expression through homo- and heterodimerization of transcription factors has been well-documented. However, LZ motifs have also been shown to mediate other protein–protein interactions (e.g., the association of myosin phosphatase with cyclic guanosine 5′-monophosphate (cGMP)–protein kinase, and the oligomerization of phospholamban, the phosphoprotein that regulates Ca^{2+} uptake across the sarcoplasmic reticulum).

PH Domains The plekstrin homology (PH) domain was first identified as a repeated segment in the platelet protein pleckstrin and was subsequently identified by sequence comparison in a number of other proteins. PH domain structures suggest that they ought to bind protein ligands, but despite extensive searches, no convincing protein ligands have been identified. Instead, it appears that PH domains may bind phospholipids. The size of the domain is approximately 100 residues but varies considerably due to insertions in variable loop regions. The three-dimensional structures of a number of PH domains have been solved and reveal that their overall folding patterns are very similar. All PH domain structures currently suggest the presence of a highly polarized electrostatic potential that may favor binding to membranes via the positively charged portion of the domain. Experimentally, several PH domains have been shown to bind to polyphosphorylated inositols and inositol lipids with moderate affinity. For instance, the PH domain of phospholipase-Cδ binds specifically to inositol-1,4,5-triphosphate (IP_3) and phosphatidylinositol 4,5-bisphosphate (PIP_2). Thus, the best guess is that PH domains act as general phospholipid-dependent membrane-anchoring domains and thereby help stabilize the interaction of substrates with the membrane once they are translocated in response to receptor PTK activation.

PDZ Domain The PDZ domain is found in many proteins that localize to specialized submembrane structures such as tight junctions and postsynaptic densities and is thought to perform a scaffolding function in assembling these structures. The name PDZ derives from the first three proteins in which this domain was identified: PSD-95 (a 95-kDa protein involved in signaling at the postsynaptic density), DLG (the *D. melanogaster* Discs Large protein), and ZO-1 (the zonula occludens 1 protein involved in maintenance of epithelial polarity). The domain is found not only in eukaryotes but also in prokaryotes, suggesting it was one of the first protein interaction modules to evolve. The majority of binding sites for PDZ domains are at the extreme C terminus of the target protein. The terminal carboxyl group of the protein ligand is bound by a core sequence that is highly conserved among PDZ domains. PDZ domains are believed to contribute to the assembly of large submembranous complexes. Well-characterized examples of PDZ-mediated clustering include bridging transmembrane receptors to multiple intracellular effector proteins or concentrating transmembrane proteins such as ion channels into specific areas of the cell membrane.

Ankyrin Repeats Ankyrin binding domains (ANK) were first identified as a repeating motif in the membrane matrix protein ankyrin and have subsequently been found by sequence similarities in a wide variety of proteins. To date, ANK domains have been identified in over 1700 different proteins and detected in organisms ranging from viruses to humans. ANK domains mediate protein recognition for a disparate set of proteins, including ion channels/pumps, transcriptional regulators, cytoskeletal organizers, developmental regulators, and cell adhesion molecules. These molecules are present in the nucleus, cytoplasm, plasma membrane, and extracellular milieu. No common theme among the known ANK domain protein targets has been identified. However, the role of ANK domains in mediating protein–protein interactions has been well-documented. The diversity of biological roles for proteins with ANK domains is mirrored in the diversity of unrelated proteins with which they interact. Thus, ANK domains could be viewed as a generalized framework for protein–protein interactions.

WD Repeats This domain is defined by the presence of four or more repeating units containing a conserved core of approximately 40 amino acids that usually ends with tryptophan–aspartic acid (WD). WD repeats are found in a number of eukaryotic proteins that have critical roles in many essential biological functions including adaptor/regulatory modules in signal transduction, pre-mRNA (messenger ribonucleic acid) processing, apoptosis, and cytoskeleton assembly. The β subunit of the heterotrimeric G-protein complex is the prototypic WD-repeat-containing protein. This protein contains chains of WD repeats to form a propeller-like structure with 7 antiparallel β-sheet blades. The only common functional theme of proteins that contain WD domains is to serve as a platform to which other proteins can bind in either stable or reversible interactions.

FHA Domain The FHA domain, or forkhead-associated domain, was originally identified as a conserved region of forkhead-type transcription factors. The FHA domain is 65 to 100 amino acids long. So far, the FHA domain has been identified in more than 200 different proteins and is found primarily in eukaryotic nuclear proteins and in certain prokaryotes, such as mycoplasma bacteria. The FHA domain mediates phosphopeptide interactions with proteins phosphorylated by serine/threonine kinases. An interesting feature of the FHA domain is its apparent specificity for phosphothreonine residues over phosphoserine residues. FHA domains bind to a pTXXD or pTXXX motif. The domain is associated with proteins involved in numerous processes including intracellular signal transduction, transcription, protein transport, DNA repair, and protein degradation.

14-3-3 Proteins Although not technically a binding motif, the 14-3-3 proteins have been implicated as a novel type of chaperone protein that modulates interactions between components of signal-transduction pathways. Homologs of the 14-3-3 proteins have been found in a broad range of eukaryotic organisms (including plants and fungi), but they are apparently absent from prokaryotic systems. The 14-3-3 proteins are 30-kDa polypeptides with 9 closely related family members in mammals. The name 14-3-3 was given to a mammalian brain protein family due to its particular migration pattern on two-dimensional (diethylamino)ethyl (DEAE)–cellulose chromatography and starch gel electrophoresis. The 14-3-3 proteins form homo- and hetero-dimeric cuplike structures that bind to discrete phosphoserine- or phosphothreonine-containing motifs. They are known for their ability to bind multiple cellular protein ligands, with more than 100 binding partners identified to date. The 14-3-3 proteins have been shown to interact with a range of protein kinases, phosphatases, receptors, transcription factors, and other signaling proteins. This implies a role for the 14-3-3 family of proteins in the formation of protein complexes involved in signal transduction. It is now clear that 14-3-3 isoforms are involved in many other cell functions, and this is only one of their many roles. In some instances, 14-3-3 proteins appear to export their binding partners from the nucleus to the cytoplasm in a phosphorylation-dependent manner.

PROTEIN TYROSINE PHOSPHATASES (PTPs)

Until recently, most studies of tyrosyl phosphorylation focused on the PTKs. Characterization of the PTPs proceeded at a slower pace for technical and philosophical reasons. The requirement for suitable phosphorylated substrates that could be dephosphorylated by the enzymes in order to measure PTP activity was a major technological hurdle. Furthermore, PTPs were initially thought to be few in number, to have little substrate specificity, and to be confined to housekeeping roles in the cell. However, it is now clear that the PTPs comprise a large superfamily of related enzymes that are subject to sophisticated modes of regulation and that play critical roles in the control of a wide array of signaling pathways.

All PTPs contain a signature motif, H-X-X-C-X-X-R, that is required for catalytic activity. The cysteinyl residue in the signature motif executes a nucleophilic attack on the phosphate group in the substrate. In addition, an essential aspartyl residue contributes to catalysis by first protonating the phenolate leaving group in the substrate and then serving as a general base to activate a water molecule to promote hydrolysis of the cysteinyl-phosphate intermediate. The structure of the active site of classical PTPs, in particular the depth of the active site cleft, renders it specific for phosphorylated tyrosine residues and it will not cleave phosphoserine or phosphothreonine. Only the phosphorylated tyrosine side chain on a target substrate is of sufficient length to be accessible to the nucleophilic cysteinyl residue; phosphoserine and phosphothreonine residues are too short to be dephosphorylated. Other domains in PTPs are important for activity regulation and cellular localization mechanisms.

The large number of PTPs suggests that each plays a specific role in modulating signaling by PTKs. Moreover, PTPs are highly specific, not only for specific phosphoprotein and even nonprotein (i.e., phospholipid) substrates but also, in some cases, for specific phosphorylation sites within those substrates. Members of the PTP superfamily are categorized as classical PTPs or dual-specificity phosphatases (DSPs). Classical enzymes are further subdivided into the transmembrane, receptorlike (RPTPs), and nontransmembrane forms. Thus, the subcellular localization of particular PTPs contributes to its substrate specificity. Interestingly, in contrast to the general requirement for dimerization to activate RPTKs, dimerization of RPTP domains may serve to inhibit their catalytic activity in some cases.

Examples of DSPs include the mitogen-activated protein (MAP) kinase phosphatases (MKPs) as well as cdc25 family members. MKPs dephosphorylate particular members of the MAP kinase family, whereas cdc25 isoforms dephosphorylate cyclin-dependent protein kinases. Similarity between the classical PTPs and the DSPs is limited to the signature motif. As the name implies, DSPs typically dephosphorylate both phosphotyrosine and phosphoserine/phosphothreonine residues; however, some substrates of selected DSPs are not even proteins but rather phospholipids. Indeed, the best characterized example of a DSP is PTEN (phosphatase and tensin homolog), which has absolute specific activity toward phosphate at the 3 position in the sugar ring of phosphatidylinositol 3,4-bisphosphate (PIP_2) and phosphatidylinositol 3,4,5-trisphosphate (PIP_3).

Protein tyrosine phosphatases can modify RPTK signaling in both negative and positive manners (Fig. 5.3). For example, a positive role for the PTP SHP-2 in mediating signaling from multiple RPTKs has been established, and some PTPs, including CD45, can activate RPTKs during T-cell antigen receptor signaling. Evidence for the opposite function, inhibition of RPTK activities by PTPs, has come from studies where specific PTP genes have been disrupted or deleted. Disruption of the gene encoding PTP-1B in mice results in increased insulin sensitivity and prolonged tyrosine phosphorylation of the insulin receptor. Studies with cell cultures also support this role for PTP-1B as a negative regulator of insulin receptor signaling. Dimerized RPTKs display a resistance to dephosphorylation by PTPs. Whether this resistance is caused by inhibition of PTPs by proteins associated only with dimeric receptor forms or is caused by the dimeric receptor forms being intrinsically poorer substrates for PTPs remains unresolved.

The establishment of PTPs as important regulators of RPTK signaling suggests that the regulation of RPTK-directed PTP activity will be an important mechanism for the control of RPTK activity. The regulation of PTP activity can occur via two main mechanisms: (a) modulation of the specific PTP activity or (b) the direction of PTPs to activated RPTKs by specific recruitment.

Direct Modulation of PTP Activity

Multiple mechanisms for direct modulation of PTP activities exist. These include:

1. *Phosphorylation* The activities of some PTPs are regulated by reversible phosphorylation of selected Ser/Thr residues in the protein with both negative and positive effects, depending on the PTP involved. Other PTPs, including PTP-1B and SHP-1, are activated by Tyr-specific phosphorylation.

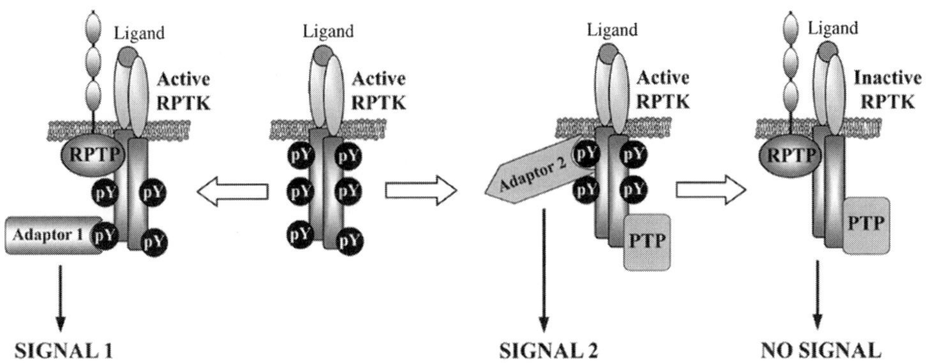

Figure 5.3 Protein tyrosine phosphatases modulate signaling from receptor protein tyrosine kinases. Membrane-linked receptor protein tyrosine phosphatases (RPTP) are able to dephosphorylate phosphotyrosine residues (pY) on receptor protein tyrosine kinases (RTPK). This allows signaling (signal 1) to proceed through adaptor proteins that can bind the remaining phosphotyrosines. Cytosolic protein tyrosine phosphatases (PTP) can catalyze the removal of different phosphotyrosine residues from the active RPTK. This allows different adaptor proteins to bind the RPTK and produce an alternate signal (signal 2). If all phosphotyrosine residues are removed from the RPTK, adaptor proteins are unable to bind, and no signal is generated.

2. *Reversible Oxidation* Reversible oxidation of PTPs can regulate activity. Different oxidants including H_2O_2 can modify the catalytic cysteine residue. Conversion of the cysteine to a sulfenic acid derivative is reversible whereas further oxidation to sulfinic or sulfonic acid derivatives is irreversible. The formation of inter- and intramolecular disulfide bonds with participation of the catalytic cysteine may be involved in activity regulation.

3. *Ligand Interactions* Structural variability of the extracellular domains of various RPTPs suggests that they may have selective ligand interactions. Various RPTP ligands have been identified and shown to regulate the specific activity of RPTPs.

4. *Dimerization* In a manner that is analogous to activation of RPTKs following dimerization, evidence suggests that inhibition of RPTP activity may result from dimerization.

Recruitment of PTPs to RPTKs

Multiple mechanisms also exist for the recruitment of PTPs to RPTKs. These include:

1. *Direct Recruitment to RPTKs* PTPs can be directly recruited to RPTKs via regulated protein–protein interactions (Fig. 5.4). The best understood interaction occurs between phosphotyrosine residues and SH2 domains. For the phosphatases SHP-1 and SHP-2, an N-terminal SH2 domain participates in an intramolecular inhibitory interaction with the PTP catalytic site. Occupation of the SH2 domain by a ligand that contains phosphotyrosine (an activated RPTK) impairs the inhibitory interaction, leading to strong activation of the PTP. Recruitment and activation of PTPs and the subsequent dephosphorylation of the RPTK phosphotyrosine complexed by the SH2 domain then leads to a disruption of the PTP–RPTK complex. For those PTPs that do not contain SH2 domains, additional structural motifs that could mediate direct PTP–RPTK interactions await discovery. It is also possible that the extracellular portions of RPTPs and RPTKs may play a role in directing the intracellular activity of RPTPs.

2. *Indirect Recruitment to RPTKs* The interaction between RPTKs and PTPs may be controlled by the regulated distribution of PTPs to specific membrane compartments. PTP access to substrate RPTKs might occur via binding to docking or adaptor proteins or through the release of PTPs from another cellular location. Regulated proteolyis can also change the cellular localization of PTPs.

The biological output from RPTKs depends ultimately on the combination of signaling pathways that are activated, which is determined in part by the set of SH2-domain proteins recruited to the receptors (Fig. 5.3). The duration, amplitude, and compartmentalization of these signaling complexes also contribute to the eventual output.

FROM MEMBRANE TO NUCLEUS: INTRACELLULAR SIGNAL TRANSMISSION

A group of protein kinases generically termed mitogen-activated protein kinases (MAPKs) transmit signals in response

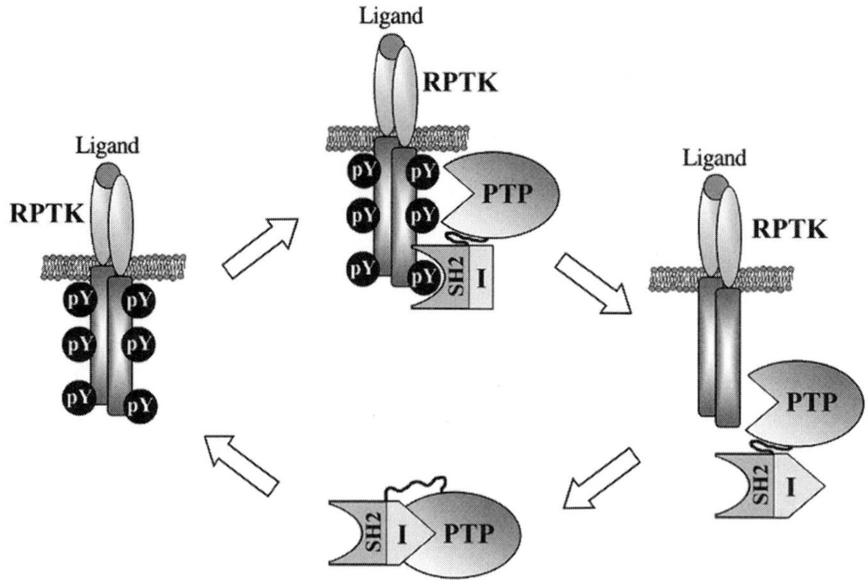

Figure 5.4 Protein tyrosine phosphatases dephosphorylate receptor protein tyrosine kinases. Many protein tyrosine phosphatases (PTP) possess phosphotyrosine binding domains, such as SH2 domains, that permit the association of PTPs with receptor protein tyrosine kinases (RPTK). PTPs can exist in a latent form with catalytic activity blocked by an autoinhibitory domain present within the PTP sequence. Association of a PTP with a phosphotyrosine residue through a SH2 group results in a conformational change in the PTP structure that removes the inhibitory domain (I) from the catalytic site. The PTP is now able to catalyze the removal of phosphotyrosine residues (pY) from the RPTK. Removal of all phosphotyrosine residues results in the dissociation of the PTP–RPTK complex, and the PTP reverts to the inactive state.

to a variety of stimuli including growth factors, mitogens, changes in osmolarity, stress, hormones, ultraviolet (UV) radiation, and chemical treatment. These extracellular signals are linked to cellular responses as diverse as cell growth and proliferation, differentiation, and cell death (apoptosis). Biochemical and genetic evidence has established that activation of the prototypical MAPK occurs via the sequential activation of a series of upstream kinases. The core MAPK module consists of a chain of three protein kinases that conduct a signal from the cell membrane to the inside of the nucleus (Fig. 5.5). Via this chain external events and signals are able to regulate gene expression. The chain begins at the cell membrane where an external signal causes the activation of a receptor kinase, usually a RPTK. The RPTK activates a G protein that, in turn, activates a serine/threonine MAP kinase kinase kinase (MAPKKK). The MAPKKK then phosphorylates a dual specificity MAP kinase kinase (MAPKK), which subsequently phosphorylates a MAP kinase (MAPK). Phosphorylation of the MAPK frequently increases its activity by about 1000-fold (hence, a virtual on–off switch). The MAPK is then able to phosphorylate cytosolic substrates or translocate into the nucleus where it can phosphorylate a variety of nuclear transcription factors.

Three main families of mammalian MAPKs have been studied most extensively. These are: (a) the extracellular signal regulated kinase (ERK) family that are generally activated by mitogens and differentiation-inducing stimuli, (b) the stress-activated protein kinase (SAPK) family, and (c) the $p38^{MAPK}$ family. The latter two are generally activated by stress stimuli such as ischemia/reperfusion, UV radiation, hyperosmolarity, and heat shock.

ERK Pathway

In vertebrates, the best understood MAPK pathway leads to activation of the extracellular signal-regulated kinases, ERK. The ERK pathway is generally, but not exclusively, activated by stimulation of RPTKs. Recent evidence has shown that G-protein-coupled receptors can also stimulate ERK kinase activity. The ERK pathway is used extensively for transcytoplasmic signaling to the nucleus, where the transcription of specific genes is induced through phosphorylation and activation of transcription factors.

An established mechanism of ERK signal transmission is via RPTK dimerization and activation of downstream adaptor proteins. Adaptor proteins are a type of signaling molecule that bind to activated RPTKs via SH2 domains.

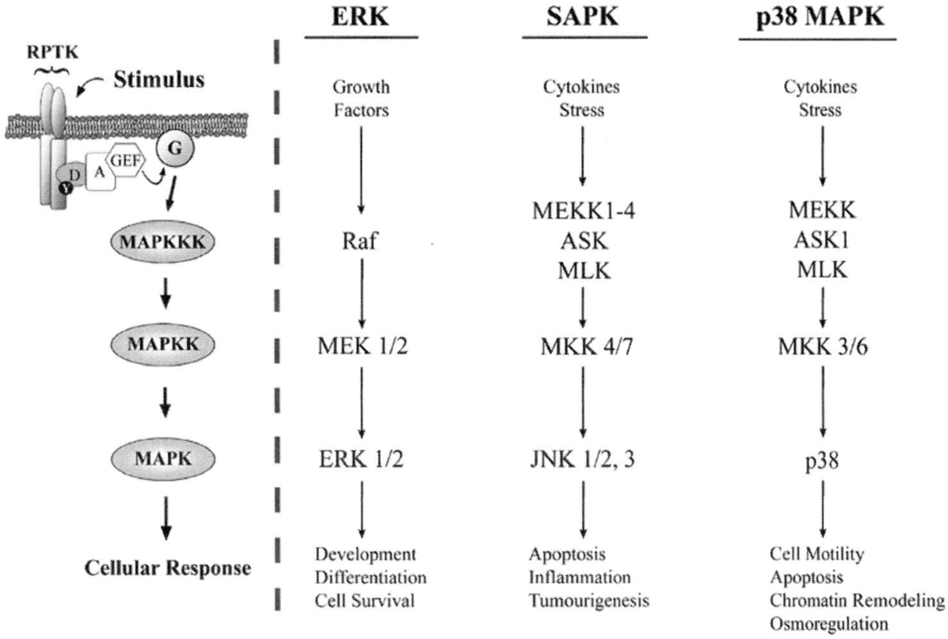

Figure 5.5 Mitogen-activated protein kinase families. The MAPK family includes the extracellular-regulated protein kinases (ERKs), stress-activated protein kinases (SAPKs), and p38 kinase (p38 MAPKs) pathways. The MAPK module consists of a three-tier regulatory cascade of MAPKKK, MAPKK, and MAPK signaling members. The MAPKKKs respond to a variety of extracellular signals, including growth factors and cellular stress. The classical mechanism of MAPK-pathway activation is through receptor protein tyrosine kinase (RPTK) activation. The signal is passed through phosphotyrosine-docking proteins (D) and adaptor proteins (A) to a guanine exchange factor (GEF). The active-GEF facilitates small G-protein (G) activation of a MAPKKK. The MAPKKKs can activate MAPKKs, which in turn can activate MAPKs. Activated MAPKs can control cellular responses by phosphorylating transcription factors, regulatory proteins, and other kinases.

In addition to a SH2 domain, adaptor proteins typically contain one or more interaction domains (PH, PTB, SH3, etc.) permitting an effector protein bound to the second interaction domain to be co-translocated with the adaptor. The co-translocated effector may have intrinsic catalytic activity or be a substrate for RPTK phosphorylation. For example, Sos, a guanine nucleotide exchange factor (GEF) is an effector protein that exchanges GDP for GTP on small G proteins. Some substrates may also have catalytic activity; for instance, SH-PTP2 can be phosphorylated by the PDGF-β receptor and has intrinsic PTPase activity.

One of the best studied adaptor molecules is a growth factor receptor bound protein, Grb2. Grb2 is a 24-kDa protein that essentially consists of little more than a SH2 domain sandwiched between two SH3 domains. Indeed, Grb2 was originally cloned by screening a complementary DNA (cDNA) library with oligonucleotides designed to recognize SH2 domains. Grb2 can be detected in all human tissues, and homologs of Grb2 have been found in *C. elegans* and *D. melanogaster*. Grb2 is normally bound with the GEF, Sos, with the two SH3 domains of Grb2 binding proline-rich regions of Sos. The Grb2–Sos complex can associate with tyrosine-phosphorylated RPTKs, namely EGF and PDGF receptors. The Grb2–Sos complex can bind directly to the EGF receptor but seems to be directed to the PDGF receptor by association with an additional protein. The primary role of Grb2–Sos is the regulation of Ras activity.

Ras is a small GTP-binding protein; it was originally identified as a transforming gene activated in a variety of tumors (it was first identified in viruses that cause rat sarcomas). Like other G proteins, Ras shuttles between an active GTP-bound state and an inactive GDP-bound state. When Grb2 binds with an activated RPTK, it brings the associated Sos to the cytoplasmic face of the cell membrane where Sos can act on Ras to catalyze GDP–GTP exchange. GTP-bound Ras is able to bind to the N terminus of Raf kinase family members, which are MAPKKKs, and brings Raf to the membrane. For more information about the small, monomeric G-protein family, the reader is directed to Text Box 5.1. The binding of Ras with Raf is necessary, but not sufficient,

TEXT BOX 5.1 SMALL G PROTEINS

The G proteins comprise a large family of regulatory GTPases, which include the heterotrimeric G-protein family, the small G-protein or Ras/GTPase family, and the initiation and elongation factor family. The members of the small G-protein family are monomeric proteins with a molecular mass of approximately 20 kDa.

Signal transmission via the small G proteins is inseparably linked with their membrane association. Activated small G proteins pass the signal on to subsequent effector molecules that have enzyme activity. The small G proteins are involved in reaction chains by functioning as molecular switches. The switch function is based on cyclical, unidirectional transitions between an active GTP-bound form and an inactive GDP-bound form. All small G proteins have low intrinsic GTPase activity, and because GDP is bound with high affinity, small G proteins tend to accumulate in the GDP-bound inactive state. The relative activation state of small G proteins can be measured directly as an increase in the GTP/GDP ratio of guanine nucleotides bound to the small G protein.

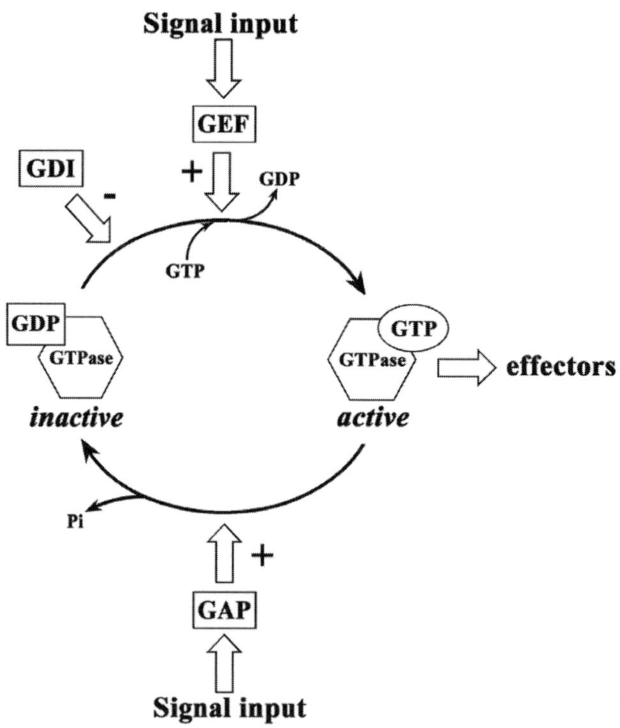

Figure TB5.1 The GTP/GDP ratio of a small G protein is regulated via GTPase activating proteins (GAPs), GTP dissociation inhibitor proteins (GDIs), and GTPase exchange factors (GEFs). The rate of GDP dissociation can be accelerated by GEFs; thus association with a GEF will increase the proportion of small G proteins present in the active, GTP-bound form. The intrinsic GTPase activity of a small G protein may be enhanced by association with a GAP. GDIs inhibit the dissociation of GDP from the G protein and create a cytosolic pool of inactive GDP-bound small G protein.

for Raf activation. Presumably, phosphorylation of Raf occurs once it is translocated to the membrane as a result of its binding to Ras. Activated Raf phosphorylates and activates the cytoplasmic MEK group of MAPKKs, which in turn phosphorylate and activate the ERK MAPKs.

G-protein-coupled receptor activation of MAPK signaling represents the other major strategy for the regulation of cell growth (Fig. 5.6). The events connecting a particular heterotrimeric G-protein-coupled receptor to the MAPK cascade are highly cell-specific and may depend on the receptor type, the type of G protein, and probably the availability, the concentration, and the cellular localization of several signaling proteins with key functions. A classification of the biochemical routes linking G protein receptors to MAPKs has revealed two frequently used pathways: Ras independent and Ras dependent. Ras-dependent pathways involve the transactivation of RPTKs through the activation of Src, a cytosolic PTK. Activation of MAPK signaling by Ras-independent pathway frequently occurs through PKA and PKC.

The regulation of MAPKKKs represents the signal entry point to the MAPK pathways and is accordingly complex. Three MAPKKKs (Raf, Mos, and Mekk) have been identified; all of which can phosphorylate and activate MEK, the MAP kinase kinase. The activation of MAPKKKs has been shown to involve three key phenomena: recruitment to the membrane, typically mediated by an upstream activating protein; oligomerization, often within a multiprotein complex containing additional regulatory components; and phosphorylation by upstream kinases. The mitogenic MAPKKK Raf illustrates all three properties. It is recruited to the membrane by Ras, where it undergoes regulatory phosphorylation, and oligomerization.

MEK was the first example discovered of a dual-specificity kinase, that is, a kinase that can phosphorylate both tyrosine and serine/threonine residues. It activates ERK by phosphorylating both a tyrosine and a serine/threonine residue at a conserved site, the S/T-E-Y phosphorylation site. This site is highly conserved in MAPKs throughout the multicellular eukaryotes, both animal and plant. No other substrates have been identified for MEK, so it may specifically phosphorylate only ERK.

Subsequently, the activated ERK translocates into the nucleus where many of the physiological targets of the ERK signal transduction pathway are located. These include a large and diverse group of transcription factors that are activated by ERK phosphorylation (e.g., Elk-1, c-Myc, and the CCAAT/enhancer binding protein, C/EBPβ). Although transcription factors are important ERK targets, only part of the active ERK pool translocates to the nucleus. Much remains in the cytoplasm and other subcellular compartments. Accordingly, activated ERKs can also phosphorylate cytoplasmic substrates such as cytosolic phospholipase A2 as well as activate other kinases such as the ribosomal S6 kinase (Rsk). ERKs can also regulate gene expression through post-transcriptional mechanisms involving cytoplasmic targets that modulate translational control and mRNA stabilization.

The best-studied MAP kinase homologs are ERK1 (42 kDa) and ERK2 (44 kDa). Studied exhaustively in mammalian systems, this pair of kinases has been reported in various organisms, including yeast, plants, and invertebrates (MAPKs were originally discovered in sea star oocytes). Recently, genes encoding novel MAPKs, such as ERK4, ERK5, and ERK6 have also been cloned from vertebrate sources. These members and the previously identified ERK3 kinase are probably not regulated like the other members of the family.

Several components of the ERK pathway are regulated by phosphatases. The MAPKKs, MAPKs, and possibly Raf are activated by phosphorylation and so the removal of the phosphate moieties results in inactivation. Although *in vitro* experiments have demonstrated that PP-2A will dephosphorylate the activating phosphoserine and phosphothreonine in ERK and MEK, respectively, inactivation of ERKs and MEKs in the cell is believed to be the responsibility of the dual-specificity protein phosphatase, MAPK phosphatase (MKP-1). This phosphatase can dephosphorylate both tyrosine and threonine residues and is very specific for MAPK. Expression of MKP-1 and another homologous phosphatase, PAC-1, is induced by growth factors and other signals that activate MAP kinases so the same signals that activate the kinases may also signal the production of dual-specificity phosphatases that can turn off the kinases. MAPKs may even be involved in transmitting the signal that stimulates phosphatase expression that would provide a feedback mechanism for the regulation of MAPKs.

The kinetics of ERK inactivation differs from cell to cell. PAC-1 and MKP-1 are nuclear-localized proteins and, therefore, are positioned to inactivate active ERKs following their translocation into the nucleus. MKP-related phosphatases may not be important in ERK inactivation in all cell types. Inactivation of ERKs may occur by PP-2A dephosphorylation leaving the phosphotyrosine removal to an unknown PTP. The relative contributions of the different inactivation mechanisms remain to be determined.

Negative feedback is an important mechanism for signal dampening and desensitization. Negative feedback may also be important in MAPK pathways. Activated ERKs can phosphorylate Sos, Raf, MEK, and MKP-1. Phosphorylation of Raf and MEK does not appear to alter their activity, but phosphorylation could alter their subcellular localization or enhance their susceptibility to dephosphorylation by MKP-1. ERK phosphorylates Sos in the C-terminal domain that binds Grb2. This does not affect Grb2 binding but may decrease the affinity of the Grb2–

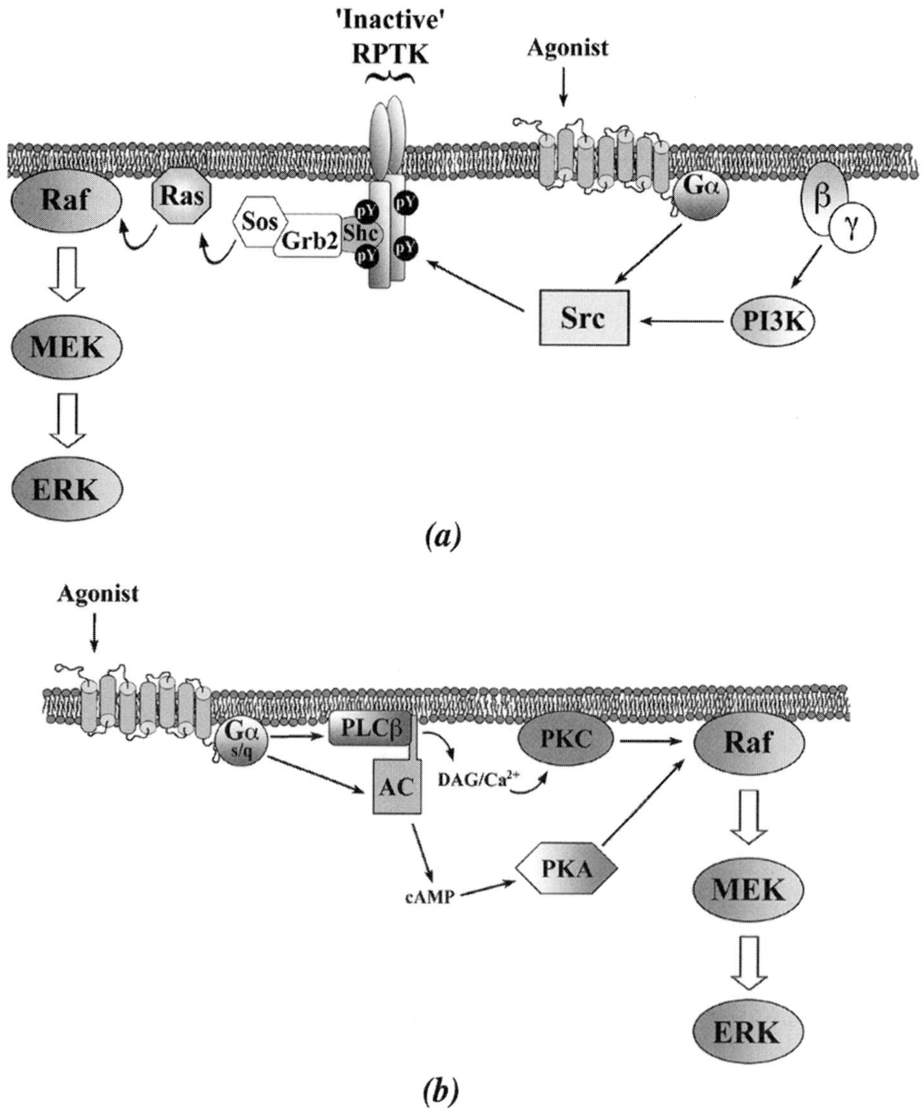

Figure 5.6 The ERK family of mitogen-activated protein kinases can be activated by G-protein-coupled receptors through a Ras-dependent (*a*) or Ras-independent pathway (*b*). Ras-dependent activation of ERK occurs through the "activation" of an inactive receptor protein tyrosine kinase (RPTK). Ligand-bound G-protein-coupled receptor is able to activate Src, a nonreceptor protein tyrosine kinase, directly or through the phosphatidylinosidyl 3-kinase (PI3K) pathway. Src proceeds to phosphorylate tyrosine residues of an inactive RPTK. The signal is then passed from the RPTK through phosphotyrosine-docking proteins (Shc) and adaptor proteins (Grb2) to a guanine exchange factor (Sos). Sos facilitates activation of the small G protein Ras, which in turn can activate Raf (a MAPKKK). Raf can activate MEK, which in turn can activate ERK. Ras-independent activation of ERK occurs through the heterotrimeric G-protein-coupled receptors, which activate downstream accessory proteins such as adenylyl cyclase (AC) and phospholipase C (PLCβ). Second-messenger generation (e.g., cAMP, Ca^{2+}, and diacylglycerol) leads to the activation of Ser/Thr protein kinases such as cAMP-dependent protein kinase (PKA) and the Ca^{2+}/phospholipid-dependent protein kinase (PKC). These second-messenger-dependent protein kinases are able to activate Raf (a MAPKKK).

Sos complex for phosphotyrosine-containing proteins that hold it at the membrane. Activation of the ERK pathway is inhibited by activation of PKA, possibly as a result of direct inhibitory phosphorylation of Raf.

SAPK/JNK Pathway

The stress-activated protein kinases (SAPKs) have also been termed JNK protein kinases because they were first identified by their role as the c-Jun N-terminal kinase that phosphorylated the transcription factor c-jun and regulated its transcriptional activity. The organization, regulation, and function of the JNK/SAPK pathways remain incompletely understood, but it is known that specific JNK/SAPK kinase members play unique functional roles. Work with cell culture systems and investigation of genetic mutants in a number of model organisms has established a variety of distinct physiological roles for individual SAPKs and their associated pathways including embryonic morphogenesis, the regulation of proliferation, and the regulation of the apoptotic response to cellular injury.

With the identification of the individual components of the system, researchers are closer to understanding the mechanisms that determine JNK/SAPK signaling specificity. However, the JNK/SAPK (and p38) cascades are much more complex than the ERK cascades. JNK/SAPKs respond to many stimuli, and a full understanding of the physiological functions and specificities of at least three mammalian SAPK/JNK kinase isoforms, two MAPKKs and 12 upstream MAPKKKs remains elusive. Three isoforms have been identified as the kinases responsible for the phosphorylation of the c-Jun transcription factor; these are called SAPK α, β, and γ or JNK2, 3, and 1, respectively. Two MAPKKs have been identified as upstream activators of the SAPKs, SEK1, and SEK2. The SEKs are dual-specificity kinases that phosphorylate the signature motif T-P-Y in the activation loop of SAPKs. The SEKs are also regulated by a large and diverse array of MAPKKKs.

The c-Jun transcription factor is normally associated with other transcription factors to form the activator protein-1 (AP-1) complex. AP-1 is a heterodimer comprised of bZIP transcription factors, typically c-Jun or JunD, along with members of the fos (usually c-Fos) and activating transcription factor (ATF) family of transcription factors. All bZIP transcription factors contain leucine zippers that enable homo- and heterodimerization, and AP-1 components are organized into Jun-Jun, Jun-Fos, or Jun-ATF dimers. Additional JNK/SAPK targets include other Jun protein family members (JunB, JunD) and the related activating transcription factor-2 (ATF-2), the ternary complex factor (TCF), tumor suppressor p53 protein, Smad3, the nuclear factors of activated T cells (NFAT), and Myc. It was once thought that JNK/SAPK targets were exclusively transcription factors that contrasted with the actions of the ERK and p38 families that phosphorylated substrates outside the nucleus as well. However, at least one cytoplasmic target for JNK has now been identified—a protein involved in mRNA stabilization.

p38 Pathway

Originally described as a 38-kDa protein that underwent Tyr phosphorylation in response to endotoxin treatment and osmotic shock, p38 (the (α isoform) was purified by antiphosphotyrosine immunoaffinity chromatography. cDNA cloning revealed that p38 was a mammalian MAPK homolog that was closely related to the HOG1 kinase, a MAPK found in yeast that is activated by osmotic stress. Like HOG1, p38 is activated by dual threonine and tyrosine phosphorylation at a T-G-Y activation sequence. Four p38 genes have now been identified in mammals: the original p38α isoform, as well as β, δ, and γ isoforms. One of the targets of p38 phosphorylation is another cytoplasmic kinase, MAPK-activated protein kinase 2 (MAPKAP kinase-2), a Ser/Thr kinase, which has been shown to phosphorylate the small 27-kDa heat shock protein (Hsp27). Hsp27 may be important in stabilizing actin filaments during osmotic stress or heat shock. Like the SAPKs, p38 kinase is strongly activated *in vivo* by environmental stresses (a unique role of the p38 kinase pathway is discussed in Text Box 5.2) and inflammatory cytokines and is inconsistently activated by insulin and growth factors. In almost all instances, the same stimuli that recruit the SAPKs also recruit the p38s. One exception is ischemia reperfusion. SAPKs are not activated during ischemia, but rather during reperfusion, whereas the p38s are activated during ischemia and remain active during reperfusion. The basis for this difference is unknown.

The upstream kinases responsible for the activation of the p38 kinase family have been identified as MKK3 and MKK6 and were cloned by degenerate polymerase chain reaction (PCR) using conserved MEK sequences as templates. Both enzymes are highly selective for p38 and do not activate SAPKs or ERKs. MKK3 and MKK6 differ in their substrate selectivity with regard to p38 isoforms, MKK3 preferentially activating p38α and p38β whereas MKK6 can activate isoforms. Similarly, MKK6 is activated by all known p38 activators, whereas MKK3 is more strongly activated by physical and chemical stresses. As with the MEKs, the stress-activated MAPKKs of the p38 pathway are regulated by Ser/Thr phosphorylation. The number, diversity, and complex enzymology of Ser/Thr kinases that act as MAPKKKs upstream of the stress-activated MAPKs is daunting. This complexity is consistent with the multiple different stimuli that recruit the p38

TEXT BOX 5.2 MOLECULAR MECHANISM OF A COPPERTONE TAN

A major environmental stress encountered by humans is solar UV light, which can cause skin inflammation, the induction of proinflammatory cytokines and skin aging, as well as skin cancer, including the highly aggressive and increasingly common malignant melanoma.

The adverse effect of UV light on human skin has led to the development of an effective defense mechanism—the suntan. In response to low levels of UV irradiation epidermal melanocytes increase the production of the pigment melanin in specialized organelles termed melanosomes. The melanosomes are transferred into surrounding keratinocytes where they act to protect against UV-induced DNA damage. Although the tanning response is one of the most obvious manifestations of the effects of UV irradiation on mammalian cells, the signal transduction pathways operating to bring about UV-induced hyperpigmentation are just beginning to be understood.

The *tyrosinase* gene encodes the rate-limiting enzyme for the production of melanin and is absolutely required for pigmentation; the absence of a functional tyrosinase enzyme results in an albino phenotype. Although much of the tanning response comprises a post-translational activation of the melanosome, transcription of the *tyrosinase* gene is UV responsive. The ubiquitous basic helix–loop–helix–leucine zipper transcription factor Usf-1 is required for the UV activation of the *tyrosinase* promoter and is phosphorylated and activated by the stress-responsive p38 kinase.

pathway. The MAPKKKs upstream of SAPK and p38 fall into three broad protein kinase families: the MEK kinases (MEKKs), the mixed lineage kinases (MLKs), and the thousand and one kinases (TAOs).

The p38s (and SAPKs) are the dominant stress-activated Ser/Thr kinases responsible for the activation of the AP-1 transcription factor. Both JNK and p38 are able to phosphorylate the nuclear transcription factor ATF2.

Cross-talk in MAPK Signaling

Cross-talk between distinct MAPK signaling pathways has been demonstrated in experiments where a specific MAPK signaling component has been eliminated (e.g., systems such as the gene knock-out mouse, RNA interference, and siRNA studies). These types of studies have shown that signaling redundancy through cross-talk is integrated into the MAPK modules. Mutation or elimination of specific signaling components often does not result in any change in physiological response (i.e., the terminal kinases are still activated and the spectrum of genes induced upon receptor activation remains constant). These findings may be explained by the existence of more than one means of activating a pathway. Cross-talk within the MAPK modules occurs most commonly between pathways activated by different receptors and can occur at many levels from the plasma membrane to the nucleus (Fig. 5.7). At the cell surface, as discussed previously in the chapter, there is an established link between G-protein-coupled receptors and RPTKs. Cross-talk within a MAPK module is best illustrated with MEKK1. MEKK1 is a MAPKKK and is able to interact with several upstream regulators, including the small G proteins (e.g., Ras), other upstream kinases (e.g., the hematopoietic progenitor kinase, HPK1), and adaptor proteins (e.g., Grb2). Once activated, MEKK1 can go on to stimulate at least three downstream pathways. It can phosphorylate MEK1, which leads to ERK activation, MKK7, which leads to JNK activation, and the I$\kappa\beta$-kinase, which stimulates gene expression via the nuclear factor-$\kappa\beta$. In addition MEKK1 can interact with other downstream adaptor proteins such as the 14-3-3 protein. Given the number of pathways MEKK1 can potentially influence, some form of regulation is necessary to allow selectivity of response. Pathways can be insulated from cross-talk effects through scaffolding proteins (discussed in the next section) that assemble protein kinase signaling cascades. The scaffolding proteins can control the upstream and downstream components with which MEKK1 is able to interact.

SIGNALING THROUGH SPACE AND TIME: THE MAPK CASCADE AS A CELLULAR SWITCH

Multicomponent signaling pathways such as the MAPK cascade are assumed to possess signal amplification. In the three-step, multicomponent MAPK pathway, each kinase can phosphorylate and activate many downstream kinases, resulting in a geometric amplification of the initial signal. A small amount of MAPKKK can bring about the activation of a larger concentration of MAPKK, which can bring about the activation of an even larger concentration of MAPK. One important consequence of the cascade arrangement is the creation of a "switchlike" or ultrasensitive response to an input signal. The individual members of the kinase cascade possess zero-order ultrasensitive kinetics (see Chapter 4 for more on zero-order ultrasensitivity) or sigmoidal responses with Hill coefficients greater than 1. Ultrasensitivity is produced as a result of the distributive mechanism of MAPKK and MAPK activation. The distributive mechanism describes a process in which the two phosphorylation reactions necessary for

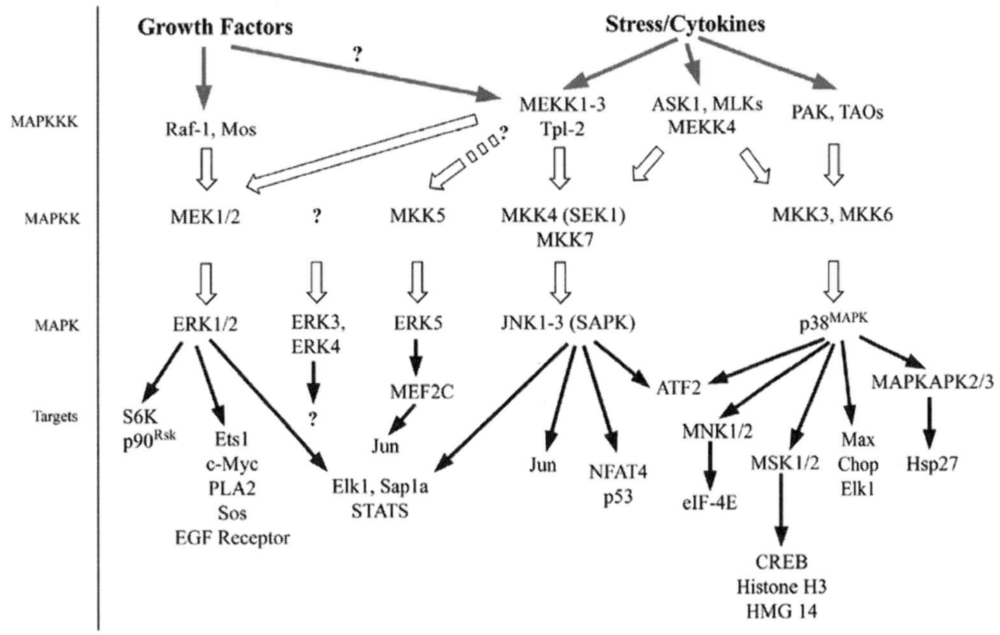

Figure 5.7 Cross-talk between MAPK signaling pathways. Activated MAPKKKs can activate one or several MAPKKs. In contrast, the MAPKKs are relatively specific for a single target MAPK. The MAPKs are able to control a myriad of cellular responses depending on which downstream target is phosphorylated. Included in this group of downstream targets are transcription factors, other kinases, upstream regulators, and other enzymes.

activation of MAPKK and MAPK are separated by full dissociation of the intermediary kinase–substrate complexes.

The responses of the individual levels of a cascade can be combined multiplicatively. If the response of a MAPKK to a MAPKKK has a Hill coefficient of x, and the response of a MAPK to a MAPKK has a Hill coefficient of y, then the apparent Hill coefficient for the response of the MAPK to the MAPKKK will be no greater than $x \cdot y$. As a result, the overall response of a cascade built out of ultrasensitive phosphorylation/dephosphorylation reactions is steeply sigmoidal and approaches an "all or none" character. In contrast, if each individual level of a cascade exhibited a hyperbolic, Michaelis–Menton response to the kinase above it, the overall response of the cascade would also be hyperbolic. This situation has been validated by mathematical modeling and *in vivo* experimentation. The steady-state activity of MEK (a MAPKK) as a function of Mos (a MAPKKK) concentration is approximated by a Hill plot with a Hill coefficient of about 1.7. The steady-state activity of ERK1 (a MAPK) as a function of Mos concentration has a Hill coefficient of 5. This large Hill coefficient means that the first increments of stimulus input produce very little response from the MAPK, but once a threshold is reached, the MAPK responds in a robust fashion. Other important factors regulate the ultrasensitivity of signaling cascades. The presence of positive feedback loops and scaffolding proteins (addressed below) within the MAPK cascade can alter the ultrasensitivity of the system.

ROLE OF SCAFFOLDING PROTEINS

Studies in yeast provided evidence that MAP kinase pathways are assembled from a combination of protein kinases into distinct protein complexes or modules. The minimum MAPK module contains a MAPKKK, a MAPKK, and a MAPK. The components of these modules interact via direct protein–protein interactions and/or are tethered to scaffolding proteins. Importantly, the assembly of distinct MAPK modules appears to allow segregation of MAPK signaling components into units that are responsive to independent stimuli, that obtain appropriate subcellular targeting, that are insulated from similar modules, and that can regulate functionally distinct substrates. Scaffold proteins are thought to contribute to a mechanism that determines signaling specificity by assembling individual components into a complex.

While it is likely that scaffold proteins increase the specificity of signaling by physically separating various signaling complexes, their function may not be limited to this. Researchers theorized that scaffolds would catalyze

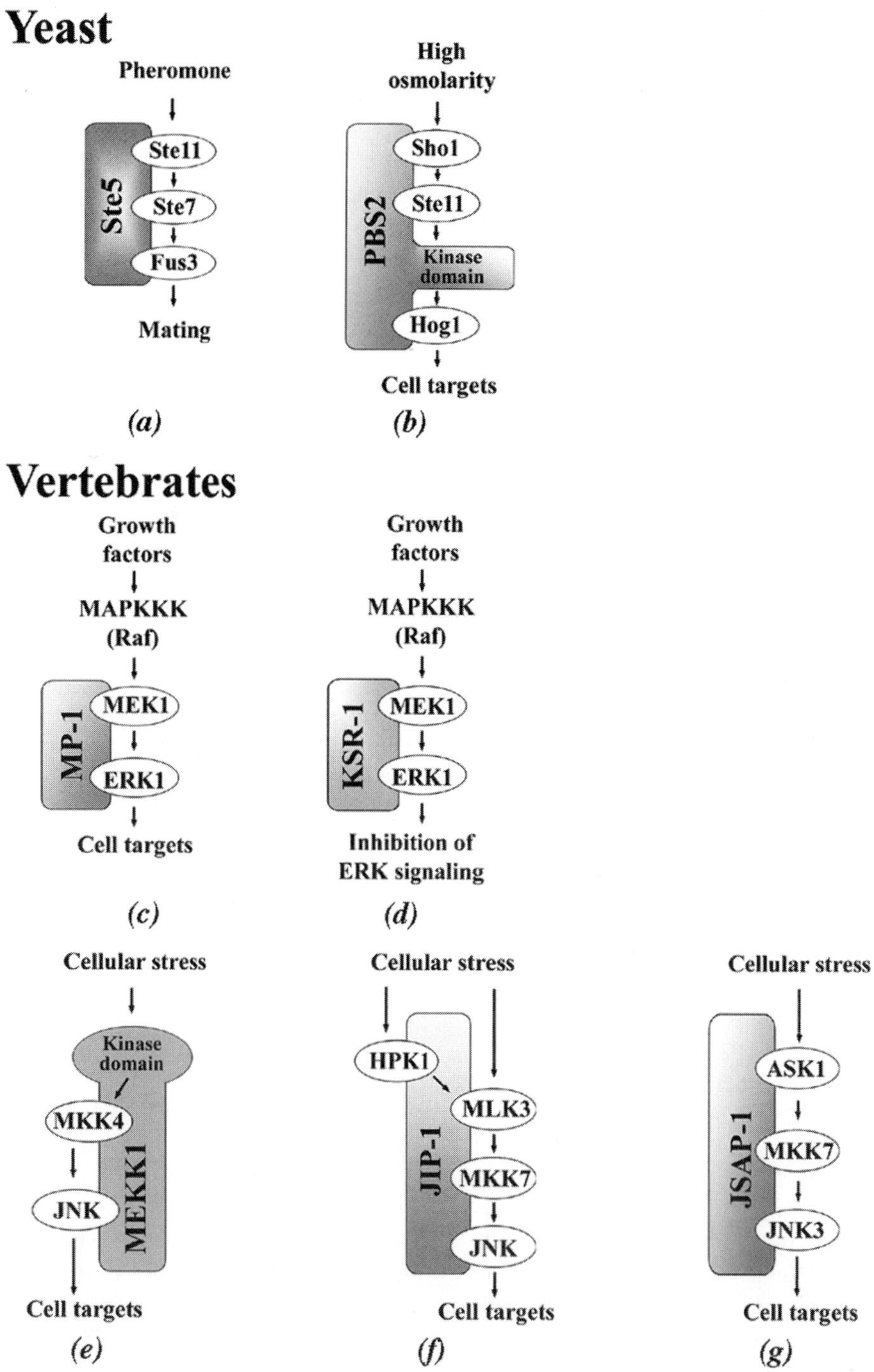

Figure 5.8 Protein scaffolds localize members of the MAPK modules. The yeast Ste5 and Pbs2 scaffolds bind the protein kinase members involved in (*a*) the mating response and (*b*) the high osmolarity reponse signal transduction pathways, respectively. In the mating response, Ste5 is a scaffold protein that binds the members of the MAPK module, Ste11 (a MAPKKK), Ste7 (a MAPKK), and Fus3 (a MAPK). In the osmosensing pathway, Pbs2 acts as both the scaffold protein and the MAPKK; Sho1 is an osmosensor. Ste11 is again the MAPKKK, and Hog1 is the MAPK. These yeast scaffolds use a common MAPKKK member for activation of different downstream MAPKK and MAPK members. In vertebrates, the scaffold proteins (*c*) MP-1 and (*d*) KSR-1 can each bind the same two members, MEK1 and ERK1, of the MAPK module but with different downstream effects. Other scaffolds can bring together all three components of the MAPK module: (*e*) MEKK1 in addition to acting as a MAPKKK, binds MKK4 and JNK; (*f*) JIP-1 binds MLK3 (a MAPKKK), MKK7 (a MAPKK), and JNK (a MAPK) as well as an associate kinase member HPK1; and (*g*) JSAP-1 while binding the same MAPKK as JIP-1 binds different MAPKKK (ASK1) and MAPK (JNK3) members.

the activation of the pathway components by modulating the rate and amplitude of signal transmission, facilitating phosphorylation of successive members of the cascade. Indeed, experimental observations have shown this to be the case; the rate of kinase activation within a scaffold complex is higher than in the absence of the scaffold. The two perspectives on scaffold function are not mutually exclusive: "catalytic" scaffolds also co-localize components and co-localization is a means to enhance mutual interactions and thus signaling efficiency.

It is now commonly believed that the three kinase components of MAPK modules are complexed with a scaffold protein. The prototypical scaffold protein in the MAPK pathway was first identified in yeast (Fig. 5.8). A member of the mating pathway, the protein Ste5 was shown to specifically bind the MAPKKK (Ste11), the MAPKK (Ste7), and a MAPK (Fus3). Because Ste11 participates in at least two other signaling pathways and Ste7 in one other pathway, the formation of a signaling complex with Ste5 specifically connects Ste11 and Ste7 to Fus3 and facilitates activation of this but not other MAPKs. A second example of the organization of MAPK modules by scaffolding proteins in yeast is the high osmolarity response pathway. The osmosensing response pathway includes an osmosensor (Sho1), a MAPKKK (Ste11, the same MAPKKK found in the mating pathway), a MAPKK (Pbs2), and a MAPK (Hog1). Pbs2 performs two functions—it acts as the MAPKK and it acts as the scaffolding protein binding Sho1, Ste11 and Hog1. Thus, the MAPKKK Ste11 is present in the two MAPK modules responsible for the mating response and for osmosensing, but there is no cross-talk between the two modules. The reason for this is that Pbs2 scaffolding of Ste11 selectively regulates its activation by Sho1; likewise, Ste5 scaffolding of Ste11 selectively controls its activation. Scaffolding proteins enable a common MAPKKK to elicit differential activation of downstream targets.

Functionally analogous scaffold proteins have also been identified in mammalian cells. These scaffold proteins include: (a) MP-1 (MEK partner 1), which specifically binds the MAPKK (MEK1) and MAPK (ERK) components of the extracellular signal-regulated kinase pathway; (b) JIP-1 (JNK-interacting protein), which specifically binds the SAPKKK (MLK3), SAPKK (MKK7), and SAPK (JNK1) components of the stress-activated kinase pathway; (c) JSAP1 (JNK/SAPK activating protein 1), which specifically binds the SAPKKK (the apoptosis signal-regulating kinase, ASK1), SAPKK (MKK7), and SAPK (JNK3); and (d) KSR-1 (kinase suppresor of Ras), which specifically binds MEK and ERK. KSR-1 may have a unique physiological role in the regulation of MAP kinase signaling; in contrast to the other MAPK scaffolding proteins, KSR-1 appears to inhibit ligand-induced MAP kinase activation.

Detailed mathematical modeling of the MAPK cascade describing the sequential phosphorylation of the kinases has helped to describe the role of a generic scaffold protein in MAPK pathway signaling. The mathematical models have proven adequate for the prediction of many key control properties of MAPK cascades in various experimental systems. Potential limitations of the scaffold model are discussed in Text Box 5.3. The results have demonstrated that in the presence of optimal scaffold concentrations a substantial increase in the signaling output can be elicited. This is believed to occur because scaffolds can create kinase complexes that are able to channel reac-

TEXT BOX 5.3 CONFLICTING VIEWS ON SCAFFOLD FUNCTION

Does Scaffolding Affect a Pathway's "Switchlike" Behavior?

The current ultrasensitive or switchlike model, which produces the on–off characteristics of the MAPK pathway, requires two critical assumptions: (1) free diffusion of component kinases and (2) activation by dual phosphorylation of MAPKK and MAPK through a distributive process, that is, one in which the two phosphorylation reactions necessary for MAPKK and MAPK activation are separated by full dissociation of the intermediary kinase–substrate complexes. Experimental evidence suggests that ERK activation by MEK does occur through this distributive process rather than by the more efficient processive mechanism (i.e., the two phosphorylation events occur *without* dissociation of the intermediary kinase–substrate complexes). This validates one of the necessary assumptions underlying the ultrasensitive model. However, protein scaffolds actually prevent component diffusion by imposing a fixed stoichiometric relationship between components of a given module. Presumably, by holding the kinase and its substrate in close proximity to each other, the scaffold could convert a distributive reaction to a processive one. On this basis, scaffolds should prevent switchlike control of the MAPK pathway signaling.

Is There Amplification in a Scaffold Complex?

Multicomponent signaling pathways are assumed to function in signal amplification. In a three-step multicomponent MAPK cascade, each kinase can phosphorylate and activate many downstream kinases, resulting in a geometric amplification of the initial signal and an increase in the sensitivity of the response. If the component kinases and scaffold are present at 1 : 1 molar ratios, and a fixed stoichiometry between components of the module is imposed during the activation process, then little or no signal amplification will occur in a multikinase cascade—a violation of a central tenet of signal transduction dogma. So, scaffolds may increase specificity at the expense of signal amplification.

However, sequential interactions can provide amplification without sacrificing specificity if interactions are disrupted upon activation of the downstream kinase. For instance, JNK1 is bound by the N-terminal extension of MKK4 (a MAPKK), which also interacts with the catalytic domain of MEKK1 (a MAPKKK). There is some experimental evidence that the binding of MEKK1 with MKK4 is disrupted after MKK4 activation and prior to JNK1 binding with MKK4. Furthermore, JNK1 binding to MKK4 is also disrupted following JNK1 activation. The result of this process is the creation of an amplification system whereby MAPKKKs and MAPKs are free to phosphorylate more than one substrate protein.

Do Scaffolds Simplify Signaling by Limiting Cross-Talk?

Researchers have theorized that the role of scaffolds is to reduce cross-talk in signaling pathways by locking the various interacting components together in complex. But if the components are locked together, why has this multicomponent system of interacting kinases not been abandoned and replaced by a single kinase molecule? Conclusive *in vivo* evidence has not been found that indicates that a specific scaffold protein actually maintains pathway specificity at the expense of cross-talk. Many kinases display substrate specificity in the absence of scaffold protein, so the kinetic characteristics of the individual kinases may dictate substrate specificity, and hence the amount of cross-talk inherent to the signaling pathway.

Are Scaffolds Inhibitor Proteins in Disguise?

Many scaffold proteins were originally isolated as inhibitory proteins; the JIPs were first known as JNK inhibitor proteins. It is believed that the original inhibition observed for JIP was an overexpression artifact. Artifactual information has also been obtained with MP-1 in cellular overexpression systems; in this situation, an enhancement in signal transduction with MP-1 was only apparent when ERK was co-overexpressed. These studies have left the impression that the true roles of scaffolds in the organization of multicomponent signal cascades may not be obvious to today's researchers without the development of new techniques.

tions by specific enhancement of a catalytic process that would otherwise occur differently in solution. However, if the scaffold concentration is greater than optimal, a significant decrease in signaling can occur. Having a scaffold protein in excess promotes the separation of kinases into nonfunctional complexes, which leads to inhibition of signaling.

The rate of MAPK activation (as well as several other kinetic parameters) during the onset of signaling display nonlinear dependence on the scaffold concentration (Fig. 5.9). That is to say, the kinetic parameters of the pathway all have maximal values, which decrease as the scaffold concentration is either increased or reduced. Not only is there an optimal scaffold concentration for maximal pathway activation, but the optimum does not depend on the binding constants of the kinase–scaffold interactions. The main determinants of the optimal scaffold concentration are the concentrations of the kinases and the character of their mutual interactions. For example, positive cooperation in binding of inactive kinases shifts the position of the optimum scaffold concentration to lower levels. Negative cooperation increases the optimum scaffold concentration and effectively decreases the signaling output from the complex (Fig. 5.10). The sensitivity of pathway activation to scaffold concentration increases with the number of kinases associated with the scaffold. This trait has been verified using experiments in which the binding sites of kinases in a scaffold have been mutated to change or abolish affinity of the kinases for the scaffold.

A scaffold protein can alter the threshold level of signal transmission in a kinase complex. One manner in which to produce an ultrasensitive or switchlike response is to have a distributive rather than processive activation of MAPKK and MAPK (Fig. 5.9). A distributive mechanism results in a sigmoidal signaling input–output curve (i.e., Hill coefficient > 1). Increasing scaffold concentrations gradually diminishes the sigmoidal character of MAPKK and MAPK activation as a function of input signal strength. Attachment of the kinase and substrate to a scaffolding molecule might prevent the dissociation of the intermediary kinase–substrate complexes and result in the interaction becoming processive with signaling losing its threshold properties (i.e., Hill coefficient $= 1$). A processive mechanism of activation describes a process in which the intermediary kinase-substrate complexes do not dissociate between the first and second phosphorylation reaction.

Scaffolds increase the sensitivity and output of signal transduction pathways by reducing the activation threshold and augmenting signal response, respectively. This might be undesirable *in vivo* because the ability of the signaling system to filter out low nonspecific signals in a generally "noisy" environment might be compromised. Thus, the

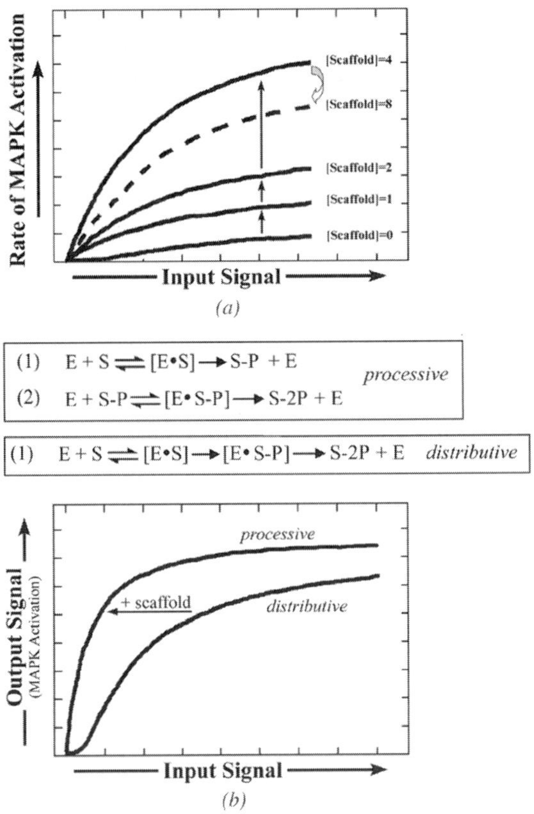

Figure 5.9 Scaffolds affect the signaling parameters of a multicomponent kinase pathway. (*a*) The addition of a scaffolding molecule results in the progressive facilitation of MAPK activation only within a limited range of scaffold concentration. When the amount of scaffold exceeds an optimal value, nonfunctional complexes are formed that impair signaling transmission (dashed line). The amount of input signal corresponds to the activity of an upstream MAPKKK. (*b*) The distributive mechanism of kinase phosphorylation presupposes that the two phosphorylation events required for MAPKK and MAPK activation are separated by full dissociation of the intermediary kinase–substrate complexes. The distributive mechanism results in a sigmoidal signaling curve. Attachment of the kinase and substrate to a scaffolding molecule impedes the dissociation of intermediary kinase–substrate complexes and results in a processive mechanism of phosphorylation.

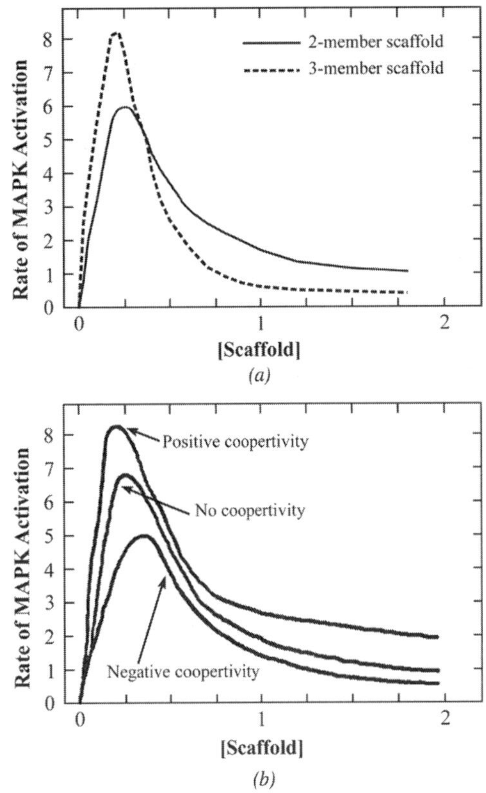

Figure 5.10 Signaling through a kinase–scaffold complex is sensitive to the cooperativity of kinase binding and the number of kinases in the complex. (*a*) The output (i.e., MAPK activation) from a kinase pathway is sensitive to the number of kinases associated with the scaffold protein. It requires considerably more two-member scaffolds than three-member scaffolds to inhibit activity to the same level at high scaffold concentrations. In general, at high scaffold concentrations, the signaling output is inhibited as $1/[\text{Scaffold}]^{m-1}$ (where m is the number of kinases a scaffold binds to assemble a functional complex). (*b*) Cooperativity in kinase binding does not affect the existence of an optimal scaffold concentration but results in a shift of the optimum. No cooperativity, negative, and positive cooperativity are shown. Cooperativity is defined as a change in association constant for binding of a second (inactive) kinase when the first (active) kinase is already bound.

manner in which scaffold concentration is regulated in the absence of and during signaling is extremely important. The scaffold concentration can be altered at a global level by altering the level of gene expression and proteolysis of the scaffold protein. Locally, scaffold amounts can be changed by concentrating the scaffold protein in specific subcellular compartments at the expense of the other parts of the cell or by altering the absolute number of binding sites on the scaffold via protein phosphorylation and/or other posttranslational modifications. The effects of phosphorylation of scaffold proteins in this process are unknown. Global and local regulatory mechanisms are both used *in vivo* to modulate the amount of scaffold. For example, proteolytic degradation of the yeast scaffold Ste5 is reduced when the MAPKKK Ste11 is present in a complex. The Ste5 concentration is also regulated; receptor activation results in translocation of Ste5 scaffold from the cytosol and nucleus to the cell membrane. Ste5 concentration becomes greatest where the concentration of the activated receptor is the highest. Concentration of Ste5 in other subcellular compartments is concomitantly reduced. Therefore, the concentration of the scaffold localized near the activated receptor is significantly higher than the average concentration in the cytosol.

TRANSCRIPTION FACTORS LINKED TO SIGNALING PATHWAYS

Coordinated programs of gene expression are central to complex biological processes. The basic features of gene regulation are highly conserved and apply to both prokaryotes and eukaryotes. The rate of transcription of a gene is controlled by regulatory sequences that surround and extend upstream of the 5′ end of the transcriptional start site (promoter) and can also lie as far as thousands of base pairs (bp) away from the 5′ or 3′ ends of the gene (enhancers) (for review see Chapter 6). Transcription factors bind to short sequence elements within promoters and enhancers and either activate or repress transcription initiation. The function of these factors can be modulated and, indeed, many are targets of signal transduction pathways. In principle, phosphorylation can affect and regulate any or all of the following properties of a transcription factor: nuclear translocation, DNA binding, and transactivation. A general mechanism of regulated translocation has evolved for the rapid and accurate transmission of signals from cell-surface receptors to the nucleus. Protein kinases or transcription factors, present as inactive forms within the cytoplasm, are phosphorylated and then these molecules are translocated into the nucleus.

MAPK Nuclear Targets

Once activated, MAPK family members translocate to the nucleus where they regulate the activity of several transcription factors. Included as targets of all three families of MAPKs are the ternary complex factors (TCFs) that are members of the Ets family of transcription factors. The TCF family members Elk-1, SAP-1, and SAP-2, for example, have roles in regulating the serum response element (SRE). The SRE is located approximately 300 bp upstream of the transcriptional start site in the *fos* promoter and is linked to the serum induction of a number of genes. Serum induction of genes occurs in cultured cell lines when serum (containing growth factors) is added as a growth and nutrient media. The addition of serum resulted in the transcription a number of genes presumably due to the presence of endogenous growth factors. The TCFs contain 3 conserved motifs: an N-terminal section that is necessary to bind DNA, a central region that is the SRF interaction domain, and a C-terminal region that harbors the transactivation domain, which contains several consensus MAPK phosphorylation sites. The transcriptional activity of the TCFs is stimulated by phosphorylation of the C-terminal MAPK sites. TCFs cannot bind the SRE autonomously but require protein–protein interactions with a constitutively bound serum response factor in order to bind the SRE.

Nuclear Factor-κB

Another of the several well-characterized signaling pathways that lead to transcriptional activation, the nuclear factor-κB (NF-κB) pathway is unique in its unusual mechanism of regulation via a ubiquitination and proteosome-mediated pathway. NF-κB regulates the transcription of a large number of genes, particularly those involved in apoptosis (programmed cell death), immunity, and inflammation. Many viruses, including human immunodeficiency virus type 1 (HIV-1) exploit NF-κB to activate their own genes and to stimulate the survival and proliferation of lymphoid cells in which they replicate. Inappropriate regulation of NF-κB is directly involved in a wide range of human disorders, including a variety of cancers, neurodegenerative diseases, and inflammatory conditions (arthritis, asthma, and inflammatory bowel disease).

NF-κB is a transcription factor composed of dimeric complexes of two proteins, p50 (NF-κB1) or p52 (NF-κB2), usually associated with members of the Rel family (p65/RelA, p50/p105, p52/p100, c-Rel, RelB), which have potent transactivation domains. Different combinations of NF-κB/Rel proteins bind distinct κB promoter sites to regulate the transcription of different genes. NF-κB is maintained in an inactive form by sequestration in the cytoplasm through interaction with inhibitory proteins, the IκBs. All IκBs contain multiple ankyrin repeats, and these domains mediate IκB binding to the NF-κB/Rel complex and mask the nuclear localization signal of NF-κB.

The prototypical pathway for NF-κB activation involves its liberation from an inactive complex with an inhibitor protein, IκB (Fig. 5.11). Phosphorylation of IκB by an upstream IκB-kinase triggers its poly-ubiquitinylation; this modification targets IκB for rapid degradation by the 26S proteasome. Proteolytic degradation of IκB immediately precedes and is required for NF-κB nuclear translocation to occur. The degradation of IκB exposes a nuclear localization sequence in the NF-κB subunit resulting in its immediate translocation to the nucleus. This irreversible step in the signaling pathway constitutes a commitment to transcriptional activation. The signal is eventually terminated by export of NF-κB out of the nucleus.

IκBs also play an important role in termination of NF-κB activation. The return of NF-κB to the cytoplasm depends on IκB synthesis, a process that itself requires NF-κB transcriptional activity, as well as IκB-dependent nuclear export. Newly synthesized IκB enters the nucleus and binds NF-κB, thereby enhancing its dissociation from

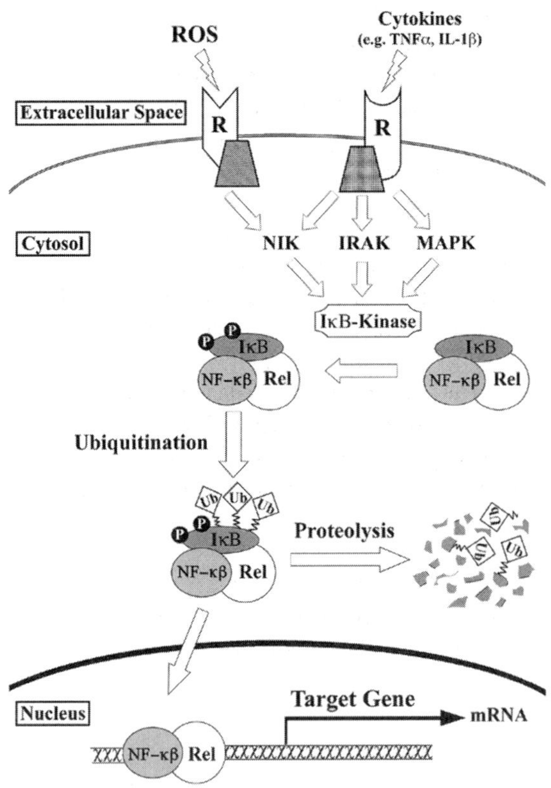

Figure 5.11 Different agents, reactive oxygen species (ROS) and cytokines such as the tumor necrosis factor (TNFα) and interleukin-1β (IL-1β), induce stress in cells and change the expression of many genes by activating the nuclear factor-κB (NF-κB) transcription factor. The activation of NF-κB involves its liberation from an inactive complex with the inhibitor protein, IκB, which resides in the cytosol. Stimulation of the receptor complex (R) leads to the recruitment and activation of a number of different signaling pathways. These kinases, including the mitogen-activated protein kinase (MAPK), the integrin receptor-associated kinase (IRAK), and the NF-κB-inducing kinase (NIK), in turn activate the IκB-kinase. The removal of IκB is controlled by protein phosphorylation on two serine residues (Ser32 and Ser36) by the IκB-kinase. IκB phosphorylation targets the inhibitory protein for ubiquitination (Ub) and subsequent degradation by the 26S proteosome. Upon IκB degradation, the liberated NF-κB/Rel dimer is able to translocate into the nucleus and activate target genes by binding to regulatory elements in enhancers and promoters.

DNA (the affinity of NF-κB for IκB appears to be higher than its affinity to DNA) and causing its removal to the cytoplasm by means of a nuclear export sequence present on IκB.

Much research has focused on the identification of pathways regulating the proteolytic NF-κB activation. MAPKs, and NF-κB-inducing kinase, and interleukin-1 receptor-associated kinase, have been demonstrated to regulate IκB-kinase activity and hence IκB degradation. However, there is increasing evidence for another complex level of NF-κB activation, which involves modulatory phosphorylations of the DNA-binding subunits (e.g., RelA and p50/NF-κB1). These phosphorylations can control several functions of NF-κB, including DNA binding and transactivation properties, as well as interactions between the transcription factor and regulatory proteins. Although their overall impact on NF-κB function has yet to be determined, these phosphorylations will very likely provide a mechanism to fine tune NF-κB function.

Second Messengers and Transcriptional Activation

Recall from the previous chapter that there are a number of intracellular second messengers that are integral to the coordinated control of primary metabolism and cellular physiology. The intracellular level of one second messenger, cAMP, is regulated primarily by adenylate cyclase, which is in turn modulated by various extracellular stimuli mediated by G-protein-coupled receptors. Until recently, it was believed that the effects of cAMP were limited to the metabolic machinery in the cytoplasm. Then it was observed that the phosphorylation state of a number of nuclear proteins was directly linked to the activation or inhibition of the cAMP-dependent signaling pathway by hormone agonists. The cAMP response element (CRE) was subsequently identified as the link between the hormonal stimulation of cells and the modulation of transcriptional activity. A nuclear protein, cAMP response element binding protein (CREB), was soon isolated and found to mediate the stimulation of gene expression in response to factors that elevate intracellular cAMP levels. CREB contains three major structural elements: a DNA binding domain rich in basic amino acids, a leucine zipper responsible for dimerization, and an activating phosphorylation site. Two additional CREB family members were subsequently identified by cDNA library screening: activating transcription factor (ATF-1) and cAMP response element modulator (CREM). The transciptional activity of all three of these proteins is regulated by cAMP and phosphorylation by PKA.

Cyclic AMP response element binding protein, ATF-1, and CREM bind to target DNA sites primarily as homodimers through the leucine zipper region. Although phosphorylation of CREB is required to induce gene expression through CREs, this phosphorylation is not sufficient for gene expression. Far Western blot analysis of nuclear extracts with a ^{32}P-labeled phosphorylated CREB probe revealed a single protein band that associated with phosphorylated CREB. Thus, a CREB binding protein (CBP) was identified and shown to associate with phosphorylated CREB in order to mediate transcriptional induction of cAMP-responsive genes.

Calcium is also a well-established regulator of transcription. Modulation of responses to this ubiquitous second messenger can occur by superposition of coincident Ca^{2+}-independent signals, but there is also growing evidence that the strength, frequency, and location of the Ca^{2+} signal determines specific transcriptional results. The multifunctional Ca^{2+}/calmodulin-dependent protein kinases (CaMKs) may elicit effects on components of transcription complexes, directly connecting Ca^{2+} with changes in gene expression. Their activation is greatly enhanced following phosphorylation catalyzed by upstream kinases in a manner analogous to the MAPK cascades. Based on an evolving understanding of CaMK regulation and cloning of the CaMK kinases (CaMKKs), a CaMK cascade has been proposed.

The transcription factor CREB is one of the potential nuclear targets of the CaMK cascade. The mechanism by which Ca^{2+} induces transcription through the CRE is not fully understood. Both CaMKI and CaMIV can phosphorylate CREB on activating phosphorylation sites and thereby mediate transcription. However, CaMKI is located predominantly in the cytoplasm and whether it directly or indirectly phosphorylates CREB remains in question. By contrast, CaMKIV is present in the nucleus and, therefore, may be a better candidate for activating Ca^{2+}-mediated transcription.

It is possible that the CREB binding protein, CBP, may be an additional target for Ca^{2+}-induced gene transcription. Experiments have demonstrated that CBP is transcriptionally responsive to increases in intracellular Ca^{2+}. The requirement for direct phosphorylation of CBP by CaM kinases, however, is questionable. Although constitutive CaM kinase can induce transcription from fragments of CBP, it is still unclear whether CaM kinases are essential for Ca^{2+} regulation of CBP. CaMKIV has been linked with the regulation of many transcription factors other than CREB and CBP, including AP-1, serum response factor, and ATF-1. ATF-1, like CREB, is a leucine zipper transcription factor that can heterodimerize with other ATF family members as well as with CREB to stimulate transcription through CREs. The regulation of ATF-1 by CaM kinases is remarkably similar to that of CREB. CREB is phosphorylated by CaMKI and CaMKIV to activate ATF-1-dependent transcription. The CaMK cascade may have diverse means to regulate Ca^{2+}-mediated transcription.

COORDINATED CONTROL OF CELLULAR EVENTS: INSULIN SIGNALING

Insulin signaling provides an important example of the action and integration of phosphotyrosine signaling pathways, combining many of the elements of control that we have discussed throughout this chapter. Insulin is an essential peptide hormone that regulates metabolism, growth, and differentiation. Insulin initiates and controls multiple metabolic events at different loci within the cell, including the plasma membrane, cytoplasm, mitochondria, and nucleus (see Chapter 10). These are mediated by complex insulin-stimulated signal transduction pathways.

The biological actions of insulin are initiated when insulin binds to its cell surface receptor. The insulin receptor (IRTK) is a receptor tyrosine kinase but is somewhat unique among RPTKs because it exists in a dimeric form even in the absence of insulin. The insulin receptor has two extracellular α subunits and two transmembrane β subunits that are joined together by disulfide bonds (Fig. 5.12). In the absence of insulin, the α subunits inhibit the tyrosine kinase activity contained in the intracellular portion of the β subunits. When insulin binds to specific regions of the α subunit, a conformational change in the receptor causes activation of the tyrosine kinase domain in the β subunit. The subsequent activation of IRTK by autophosphorylation permits the receptor to interact with other signaling molecules and to phosphorylate tyrosine residues on cellular substrates. Phosphotyrosine motifs on IRTK permit interactions between it and SH2-containing molecules. A myraid of downstream SH2-containing proteins are found associated with the insulin receptor. These include phosphatidylinositol 3-kinase (PI3K), growth factor receptor bound protein (Grb2), SH2-containing phosphatase-2 (SHP-2), GTPase-activating protein (GAP), and phospholipase C-γ (PLC-γ).

Although the autophosphorylated form of IRTK can interact directly with some SH2-containing proteins (e.g., SHP-2, PLC-γ), these interactions are not critical for insulin signaling. Instead, a novel group of substrates for IRTK have been discovered. These are the insulin receptor substrates (IRS-1, 2, 3, and 4), and they perform signal transmission from the IRTK to downstream signaling molecules. Members of the IRS family contain several important domains, including a pleckstrin homology domain and a phosphotyrosine binding (PTB) domain, which are important for mediating interactions with the insulin receptor. The PTB domain binds to specific phosphotyrosine residues of the insulin receptor β subunit. The carboxyl-terminal portion of the IRS proteins contains multiple tyrosine-containing motifs that undergo phosphorylation by the IRTK and serve as docking sites for the SH2-domain-containing proteins that mediate the next steps in the cascade.

Some of the SH2-containing proteins that associate with IRSs have their own enzymatic activity (e.g., SHP-2, PLC-γ), whereas others function as adaptors (e.g., Grb2 and the regulatory p85 subunit of PI3K). These adaptors assist in the formation of specific signaling complexes via simultaneous interactions of multiple SH2 domains on the adaptor protein with both upstream and downstream signal-

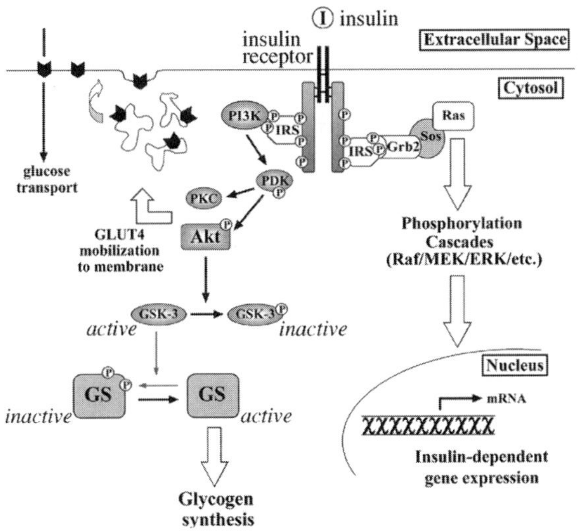

Figure 5.12 Insulin action is regulated by a complex interplay of intracellular signaling pathways. The binding of insulin to its receptor induces autophosphorylation of the receptor on tyrosine residues and thereby stimulates its tyrosine kinase activity toward intracellular substrates such as the insulin receptor substrates (IRS). Tyrosine phosphorylation of IRS leads to the activation of two major signaling pathways, the mitogen-activated protein kinase (MAPK) and the phosphatidylinositol 3-kinase (PI3K) pathways. MAPKs are involved in proliferation and differentiation processes, in particular by regulating insulin-dependent gene expression in the nucleus. The MAPK pathway does not appear to play a significant role in the transmission of the metabolic effects of insulin. In contrast, the PI3K pathway is involved in several of the metabolic effects of the hormone, such as glucose transport and glycogen synthesis. Activation of PI3K generates phosphoinositides thereby recruiting the 3′-phosphoinositide-dependent kinase-1 (PDK-1). In turn, PDK-1 phosphorylates and activates downstream effectors including Akt (also known as PKB) and protein kinase C. Akt and PKC contribute to GLUT-4 translocation. Akt is also directly involved in the regulation of glycogen synthesis. Following its activation by PDK, Akt phosphorylates and inactivates glycogen synthase kinase-3 (GSK-3). This prevents GSK-3 from phosphorylating glycogen synthase (GS) and allows glycogen synthase to remain active in the production of glycogen.

ing molecules. For example, activation of Ras and PI3K, two major effectors of pathways shared by a number of growth factor receptors including the insulin receptor, are activated by SH2-containing adaptor proteins (Fig. 5.12). Grb2 is normally found associated with the guanine nucleotide exchange factor Sos (two SH3 domains of Grb2 bind to proline-rich regions of Sos). Phosphotyrosine motifs on IRS-1 can interact with the SH2 domains of Grb2, bringing the Grb2–Sos complex into contact with Ras. Sos catalyzes the exchange of GTP for GDP on Ras leading to its activation and, in turn, Ras can activate the MAP-kinase pathway and initiate changes in gene transcription. Similarly, the p85 regulatory subunit of PI3K is normally associated with the p110 catalytic subunit. Upon insulin stimulation, phosphotyrosine motifs on IRS proteins interact with SH2 domains on p85 leading to activation of p110 catalytic activity. Signaling downstream of PI3K involves a Ser/Thr kinase cascade that ultimately mediates insulin-stimulated glucose transport as well as activating glycogen synthesis.

Under fasting conditions when insulin levels are low, glucose transport into cells is the rate-limiting step in glucose metabolism. However, when plasma glucose levels are high following a meal, insulin release stimulates glucose transport into muscle and adipose cells (see Chapter 10). This is mediated via the IRTK, which stimulates an increase in the levels of the insulin-responsive glucose transporter (GLUT4) in the plasma membrane. Insulin stimulation directs a redistribution of GLUT4 from a cytoplasmic pool of vesicles to the plasma membrane where they increase glucose import (Fig. 5.12). It appears that PI3K plays a major role in the translocation of GLUT4. Phospholipid products generated by PI3K activate a phosphoinositide-dependent kinase (PDK) by binding to its PH domain. PDK then phosphorylates and activates another Ser/Thr kinase known as Akt (or protein kinase B), which, in turn, stimulates GLUT4 recruitment to the membrane. This action of Akt has been confirmed via overexpression of constitutively active mutants of Akt in adipose cells that show high continuous recruitment of GLUT4 to the plasma membrane. In addition to Akt, other downstream effectors of PI3K signaling are believed to contribute to GLUT4 translocation; the atypical PKCζ and λ isoforms are potential candidates.

In addition to enhancing glucose transport, insulin directs the disposition of glucose within target cells. Insulin can stimulate glucose conversion to glycogen, glucose catabolism to pyruvate and lactate, and, in some cells, fatty acid biosynthesis from glucose. Some of these effects are mediated rapidly by altering target enzyme activity through signaling cascades. For example, following its activation by PI3K, Akt phosphorylates and inactivates glycogen synthase kinase-3 (GSK-3). This prevents GSK-3 from phosphorylating glycogen synthase and allows glycogen synthase to remain active in the production of glycogen. Insulin also modulates the phosphorylation state of key regulatory enzymes so as to diminish the rate of gluconeogenesis and increase the rate of fatty acid synthesis by stimulating the activities of enzymes such as pyruvate dehydrogenase and acetyl CoA carboxylase. Other effects of insulin on catabolic and anabolic flux are mediated more slowly through insulin effects on transcription and translation. Transcript levels of many mRNA species are regulated by insulin. Key examples include glucokinase, pyruvate kinase, phosphoenolpyruvate carboxykinase, and 6-phosphofructo-2-kinase.

CONCLUSION

It is not that surprising that abnormal phosphorylation events should turn out to be a cause or consequence of human disease. About 30% of human proteins contain covalently bound phosphate, and 500 protein kinases and a third that number of protein phosphatases are encoded by the human genome. These regulatory proteins are involved in specific signaling pathways that regulate cell functions as diverse as metabolism, cell cycle progression, cell adhesion, vascular function, differentiation, and angiogenesis. The process of protein phosphorylation and dephosphorylation acts as a molecular switch for modulating these processes and is under strict control in normal cells. A number of human disorders result from mutations in specific protein kinases, phosphatases, and other signaling proteins (Table 5.2).

Because of this, protein kinases and phosphatases have become the dominant targets for the development of therapeutic drugs by academic researchers and the pharmaceutical industry. Several compounds that inhibit the activity of tyrosine kinases are being evaluated as cancer therapeutic agents in clinical trials. Some naturally occurring toxins and pathogens exert their effects by altering the activity of protein kinases and phosphatases. These compounds and their chemical derivatives are being used in the development of therapeutic drugs. Considerations in the development of effective inhibitors include nonspecific actions of compounds, cellular uptake, and tissue specificity. Inhibitors may allow other therapeutic agents additional time to become effective, or they may act synergistically with current treatments. Approaches for regulating kinase and phosphatase gene expression continue to be refined. For example, specific antisense oligonucleotides for inhibiting posttranscriptional processing of messenger RNA, so-called gene therapy, has yielded some success in the clinical arena.

TABLE 5.2 Diseases and Disorders Caused by Abnormal Protein Kinase and Phosphatase Signaling

Disease/Disorder	Kinase/Phosphatase Involved
Breast cancer	ERB2 receptor kinase
Chronic myelogenous leukemia	Abl tyrosine kinase
Chronic myelomonocytic leukemia	PDGF receptor kinase
Ataxia-telangiectasia	ATM kinase
Li-Fraumeni syndrome	Chk2 kinase
Myotonic muscular dystrophy	Myotonin protein kinase
Wolff–Parkinson–White syndrome	AMP-activated protein kinase
Chraniosynostosis	FGF receptor kinase
Papillary renal cancer	Met receptor kinase
Paralytic shellfish poisoning	Protein phosphatase type 1 and 2A
Bubonic plague	*Yersinia* protein tyrosine phosphatase
Non-Hodgkins lymphona	Alk kinase
Diabetes	Insulin receptor kinase
Insulin hypersensitivity	PTP1B tyrosine phosphatase
Leprechaunism	Insulin receptor kinase
X-linked myotubular myopathy	MTM1 tyrosine phosphatase
Organ transplant rejection	PP2B phosphatase
Wolcott–Rallison syndrome	eIF2A kinase

SUGGESTED READING

Tyrosine phosphorylation

Hubbard, S. R., and Till, J. H. (2000). Protein tyrosine kinase structure and function. *Ann Rev Biochemis* **69**:373–398.

Hunter, T. (2000). Signaling-2000 and beyond. *Cell* **100**:113–127.

Li, L., and Dixon, J. E. (2000). Form, function, and regulation of protein tyrosine phosphatases and their involvement in human diseases. *Sem Immunol* **12**:75–84.

Ostman, A., and Bohmer, F. D. (2001). Regulation of receptor tyrosine kinase signaling by protein tyrosine phosphatases. *Trends Cell Biol* **11**:258–266.

Raugei, G., Ramponi, G., and Chiarugi, P. (2002). Low molecular weight protein tyrosine phosphatases: small, but smart. *Cell and Molecular. Life Sci* **59**:941–949.

Schlessinger, J. (2000). Cell signaling by receptor tyrosine kinases. *Cell* **103**:211–225.

van Huijsduijnen, R. H., Bombrun, A., and Swinnen, D. (2002). Selecting protein tyrosine phosphatases as drug targets. *Drug Discovery Today* **7**:1013–1019.

MAPK Pathways

Camps, M., Nichols, A., and Arkinstall, S. (2000). Dual specificity phosphatases: A gene family for control of MAP kinase function. *FASEB J* **14**:6–16.

Johnson, G. L., and Lapadat, R. (2002). Mitogen-activated protein kinase pathways mediated by ERK, JNK, and p38 protein kinases. *Science* **298**:1911–1912.

Schaeffer, H. J., and Weber, M. J. (1999). Mitogen-activated protein kinases: Specific messages from ubiquitous messengers. *Mol Cell Biol* **19**:2435–2444.

Cellular Switches and Scaffold Proteins

Burack, W. R., and Shaw, A. S. (2000). Signal transduction: Hanging on a scaffold. *Curr Opin Cell Biol* **12**:211–216.

Catling, A. D., Eblen, S. T., Schaeffer, H. J., and Weber, M. J. (2001). Scaffold protein regulation of mitogen-activated protein kinase cascade. *Methods Enzymol* **332**:368–387.

Ferrell, J. E. (1999). Building a cellular switch: more lessons from a good egg. *Bioessays* **21**:866–870.

Levchenko, A., Bruck, J., and Sternberg, P. W. (2000). Scaffold proteins may biphasically affect the levels of mitogen-activated protein kinase signaling and reduce its threshold properties. *Proc Natl Acad Sci U S A* **97**:5818–5823.

Pawson, T., and Nash, P. (2003). Assembly of cell regulatory systems through protein interaction domains. *Science* **300**:445–452.

NF-κβ

Karin, M., and Ben-Neriah, Y. (2000). Phosphorylation meets ubiquination: The control of NF-$\kappa\beta$ activity. *Ann Rev Immunol* **18**:621–663.

Janssen-Heininger, Y. M., Poynter, M. E., and Baeuerle, P. A. (2000). Recent advances towards understanding redox mechanisms in the activation of nuclear factor-$\kappa\beta$. *Free Radic Biol Med* **28**:1317–1327.

Smahi, A., Courtois, G., Rabia, S. H., Doffinger, R., Bodemer, C., Munnich, A., Casanova, J. L., and Israel, A. (2002). The NF-$\kappa\beta$ signalling pathway in human diseases: From incontinentia pigmenti to ectodermal dysplasias and immune-deficiency syndromes. *Hum Mol Gen* **11**:2371–2375.

Insulin Signaling

Baumann, C. A., and Saltiel, A. R. (2001). Spatial compartmentalization of signal transduction in insulin action. *Bioessays* **23**:215–222.

Cheng, A., Dube, N., Gu, F., and Tremblay, M. L. (2002). Coordinated action of protein tyrosine phosphatases in insulin signal transduction. *Eur J Biochem* **269**:1050–1059.

Ducluzeau, P. H., Fletcher, L. M., Vidal, H., Laville, M., and Tavare, J. M. (2002). Molecular mechanisms of insulin-stimulated glucose uptake in adipocytes. *Diabetes Metab* **28**:85–92.

Kido, Y., Nakae, J., and Accili, D. (2001). The insulin receptor and its cellular targets. *J Clin Endocrinol Metab* **86**:972–979.

Saltiel, A. R., and Pessin, J. E. (2002). Insulin signaling pathways in time and space. *Trends Cell Biol* **12**:65–71.

6

CONTROL OF TRANSCRIPTION IN EUKARYOTIC CELLS

WILLIAM G. WILLMORE

INTRODUCTION

Metabolic regulation involves the stringent control of the activities of enzymes and functional proteins in response to changes in cellular needs. These changes may include a fluctuation in environmental conditions, nutrient availability, performance demands, advancement to a new stage of development, or a response to foreign materials (e.g., toxins, bacteria, viruses). The checkpoints of metabolism adjust enzyme activities in order to return the cellular milieu to homeostasis or adapt to the new condition.

Regulation of the activities of enzymes and functional proteins can occur at three levels: protein synthesis, protein degradation, and protein modification. Control of synthesis involves the regulation of either transcription (from gene to ribonucleic acid (RNA)) or translation (from RNA to protein). Degradation usually refers to enzyme inactivation by proteolysis but also encompasses some cases where enzymes undergo limited proteolysis to convert them from an inactive to an active form (e.g., fibrinogen to fibrin conversion during blood clotting). Protein modification often takes the form of covalent bonding of groups to proteins such as via phosphorylation, methylation, acetylation, glycosylation, and ubiquitination; these are rapid methods of controlling enzyme activity and are usually employed by cells as a short-term responses to stress. This chapter and the next will deal with enzyme synthesis. The current chapter will explore the control of transcription, whereas Chapter 7 reviews the control of translation. Both chapters will first briefly outline the basal machinery required to fulfill each process and will then focus on the regulation of this machinery in response to various intracellular and extracellular cues. General principles will be emphasized, but as with all rules, there are exceptions and a variety of these will be discussed. Finally, in the current chapter, it is hoped that the reader will gain an appreciation of the modular nature of (a) the nucleic acid elements to which protein factors bind as well as (b) the protein factors themselves. These functional deoxyribonucleic acid (DNA) or protein "modules" are interchangeable and can be linked together in various combinations to perform different roles, such as enhancing transcription of a particular gene in response to a particular extracellular signal. The interchangeable modularity of proteins and DNA elements can be utilized in recombinant DNA technology in order to answer questions in molecular biology.

BASAL TRANSCRIPTION MACHINERY

Transcription is the synthesis of single-stranded RNA from a double-stranded DNA template. It is a multistep process involving many enzymes and proteins, the most important being RNA polymerase. The transcribed RNA product is known as a gene if it encodes (or is translated) into a protein product. Most genes are made up of coding exons with intervening noncoding introns. Introns are later removed to form the mature RNA product composed strictly of exons. The goal of transcription is the unraveling of chromatin DNA and the synthesis of RNA in a regulated fashion.

We will begin with a review of the mechanisms and machinery of nuclear transcription in eukaryotes (transcription of mitochondrial genes differs in some significant ways and is outlined in Text Box 6.1). Nuclear transcription involves preinitiation, initiation, elongation, and termination stages. The preinitiation and initiation stages

Functional Metabolism: Regulation and Adaptation, edited by Kenneth B. Storey.
ISBN 0-471-41090-X

TEXT BOX 6.1 MITOCHONDRIAL TRANSCRIPTION

The mitochondria have their own genome, the mitochondrial DNA (mtDNA), with several copies per mitochondrion. mtDNA is located in the mitochondrial matrix and is inherited through the female germ line. The mitochondrial genome is a closed circular, double-stranded DNA molecule of approximately 16.6kb with "heavy" and "light" strands separating on density gradient centrifugation. Most genes are encoded on the heavy strand, including 2 rRNAs, 14 tRNAs, and 12 polypeptides. The light strand encodes 8 tRNAs and 1 polypeptide. The human mitochondrial genome is very compact; it lacks introns (see Chapter 8) and contains only one regulatory region. This is the apparent result of the transposition of many mitochondrial genes to the nuclear genome over the course of evolution.

Mitochondrial transcription differs from nuclear transcription in a number of ways. First, there are coding differences including: (a) the codon TGA is used for tryptophan in mtDNA, whereas it is a stop codon in nuclear DNA, and (b) AGA or AGG encode stop codons in mtDNA but encode arginine in nuclear DNA. Second, the 22 tRNA species encoded by mtDNA are sufficient to support translation of all mitochondrial protein genes as opposed to the 32 required to translate nuclear-encoded genes. This reduction is achieved through "wobble" of the first anticodon position, allowing the tRNA to recognize all codons of a four-codon family. Third, the heavy or light chains are typically transcribed as single units (unlike the single gene transcription in the nucleus) with a displacement loop (or D-loop) in the heavy strand containing the origin of replication as well as the promoters for mitochondrial transcription. However, mitochondria are not self-supporting entities since transcription and replication, as well as many of the metabolic functions of mitochondria, depend upon nuclear-encoded proteins that are imported into the mitochondria.

Promoters for transcription are present on both the heavy strand (P_H) and the light strand (P_L), facing in the opposite directions. The consensus sequence contained in both is 5′-CANACC(G)CC(A)AAAGAYA-3′ (the transcription initiation site is underlined). Both promoters contain enhancers known to bind to mitochondrial transcription factor A (mtTFA) encoded by the nucleus. The promoters function independently and are recognized by a single subunit, nuclear-encoded, nonselective core RNA polymerase similar to the T7, T3, and Sp6 bacteriophage RNA polymerases. mtTFA contains two high mobility group domains known to be involved in DNA binding. Ten base pairs (one helical turn) must be present between the binding site for mtTFA and the start site of transcription. A second initiation site occurs for heavy strand transcription and may play an important role in rapid translation of mtDNA (such as in rapidly growing cells).

The light strand of mtDNA is transcribed as a single pre-mRNA containing all light strand encoded genes. The heavy strand can be completely or incompletely translated depending upon which site of initiation is utilized. Early termination occurs between the 16S rRNA and the tRNA for leucine. Early termination can also be caused by a protein that binds to the mtDNA at this border site: the mitochondrial transcription termination factor (mTERM). mTERM provides a physical barrier to mitochondrial RNA polymerase resulting in early termination of translation and is known to bend the mtDNA at the point of binding. It is a factor that contains three leucine zippers required for DNA binding. Processing of the primary transcripts is much the same as occurs in the nucleus.

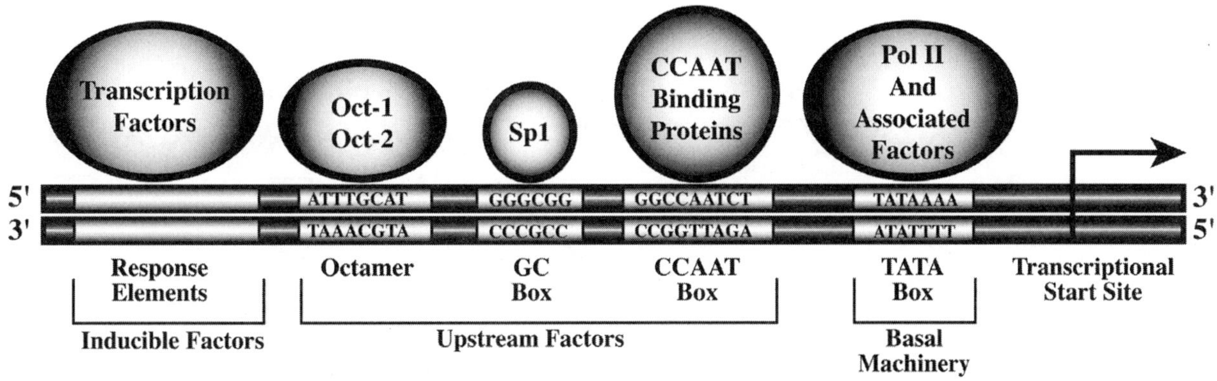

Inducible Factor	Binding Element	Induced By
AP-1	5'-TGAGTCA-3'	Phorbol esters, growth factors, cytokines, neurotransmitters
ATF/CREB	5'-TGACGTCA-3'	Hormones, growth factors, inflammatory cytokines
HIF-1	5'-TACGTGC-3'	Hypoxia
GR	5'-AGAACANNNTGTTCT-3'	Glucocorticoids
NFAT	5'-GGAAA-3'	Calcium
NF-kB	5'-GGGRNYYCC-3'	Viruses, cytokines, immune modulators, inorganic compounds, immune modulators, chemotherapeutics, ER sensors, oxidants
p53	5'-RRRC(A/T)(T/A)GYYY-3'	DNA damage
PPARγ	5'-AGGTCANAGGTCA-3'	Prostaglandins, fibrates, thiazolidinediones, fatty acids
SMAD	5'-GTCTAGAC-3'	Transforming growth factor-β, tumor necrosis factor-β, IL-1β, lipopolysaccharides, interferon-γ
STAT	5'-TTCC(C/G)GGAA-3'	Interferons, interleukins, growth factors, peptide hormones

Figure 6.1 Promoter elements of eukaryotic genes. Promoters contain the binding sites for basal, upstream, and inducible factors. The binding sequences for various factors are show. Abbreviations are AP-1, activator protein 1; ATF/CREB, activating transcription factor/cAMP response element binding protein; HIF-1, hypoxia-inducible factor 1; GR, glucocorticoid receptor; NFAT, nuclear factor of activated T cells; NF-κB, nuclear factor-κB; PPARγ, peroxisome proliferator activated receptor γ; SMAD, sma and mad related proteins; STAT, signal transducers and activators of transcription.

include the binding of various proteins to facilitate the unwinding of chromatin and the interaction of RNA polymerase with the DNA template. These proteins can be divided up into basal, upstream, and enhancer factors. The basal factors are an absolute requirement for transcription, whereas the upstream and enhancer factors may vary extensively in their nature and number present.

The basal factors of transcription include RNA polymerases and their associated proteins. RNA polymerase I, II, and III (pol I, II, and III) catalyze the synthesis of ribosomal RNA (rRNA), message RNA (mRNA), and transfer RNA (tRNA), respectively. Since the focus of this chapter is on the regulation of protein synthesis, only pol II will be mentioned along with its corresponding mechanisms of control. The transcriptional start site is the point on the DNA template where RNA synthesis begins. Basal factors bind to a distinct DNA region upstream of this start site known as the promoter. The "strength" of a promoter refers to how frequently RNA polymerase initiates transcription from it.

In eukaryotic transcription, the promoter contains an element known as the TATA box, a conserved A-T-rich septamer found about 25 base pairs before the transcriptional start site. The consensus sequence for the TATA box is 5′-TATAAAA-3′ (Fig. 6.1). TATA boxes are the binding site of TATA binding protein (TBP) (Fig. 6.2), the first protein to bind to DNA during transcription. TBP is a small protein (~38 kDa) that binds to the minor

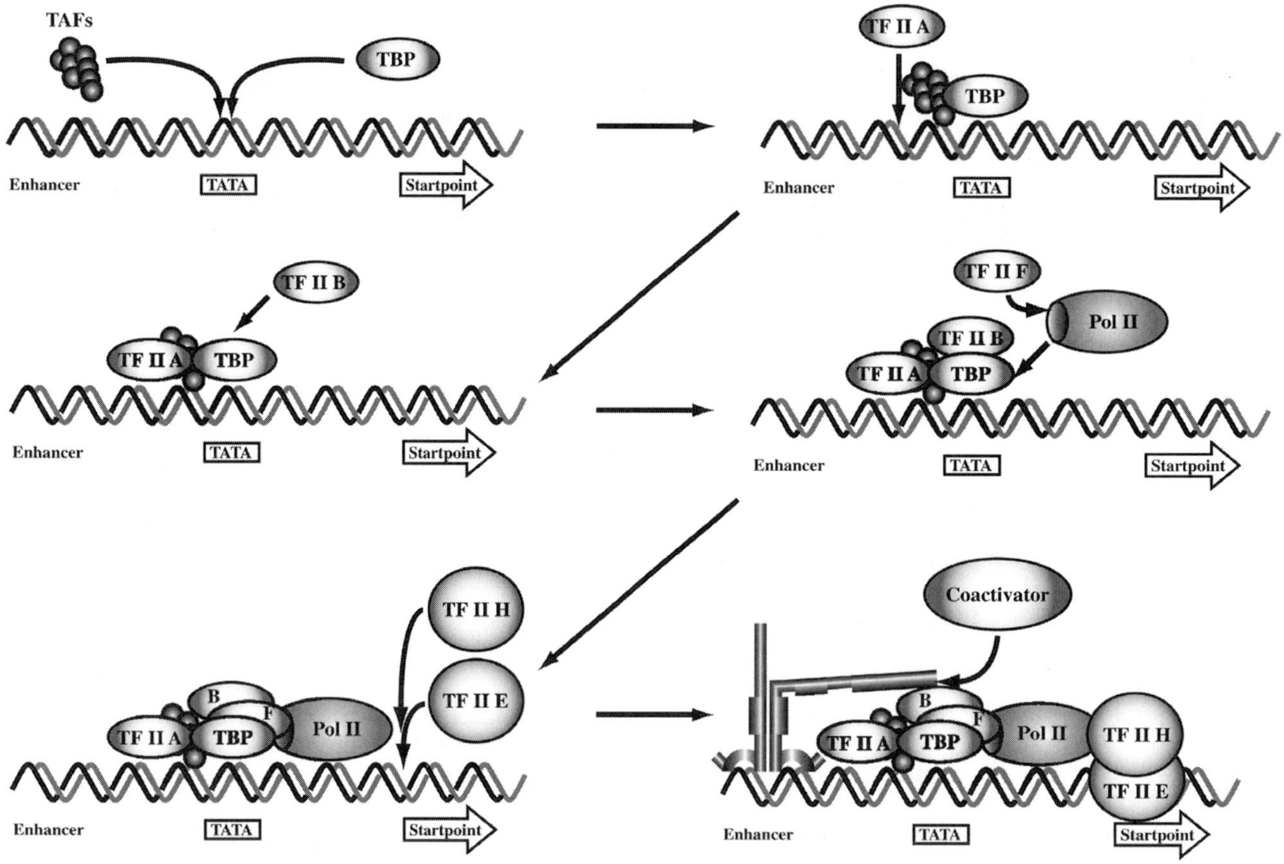

Figure 6.2 Assembly of the basal transcriptional machinery. Binding of TATA-box binding protein (TBP), TBP associated factors (TAFs), and transcription factors (TFIIs) for RNA polymerase II (pol II). The transcriptional start site is shown. The binding of an inducible transcription factor with/without a cooperating coactivator is shown in the last panel.

groove of DNA (most other DNA binding proteins bind to the major groove). This binding causes considerable bending of the DNA helix, facilitating the interaction of other proteins in the basal complex. The TBP protein resembles a saddle, with the inner surface binding to DNA and the outer surface making protein–protein interactions. It contains a 180-amino-acid domain made up of an imperfectly repeated sequence sufficient for binding to DNA. Eleven other proteins, known as TAFs (TBP associated factors), interact with TBP at the TATA box. Together, TBP and its associated TAFs are known as TFIID (or transcription factor D for pol II).

The second factor to bind is TFIIA. TFIIA has two subunits in yeast and three in mammals. It is extremely acidic and binds directly to TBP, stabilizing the binding of TBP to DNA. It also acts as an antirepressor, blocking repressors of transcription that inhibit transcription factor binding or that remove TBP from the DNA.

The third factor to interact with the basal complex is the single subunit protein TFIIB. Like TFIIA, it binds directly to TBP, stabilizing its binding to the TATA box. It has an N-terminal zinc finger domain for DNA binding (see transcription factors below) and a C-terminal direct repeat that makes it proteolytically stable. It recruits pol II through TFIIF and is the most likely target for interaction with regulatory transcription factors.

The TFIIF and pol II form a complex prior to being recruited by the basal apparatus. TFIIF is an $\alpha_2\beta_2$ tetramer (the two subunits sometimes being referred to as RAP74 and RAP30). The larger subunit (RAP74) is an adenosine 5′-triphosphate (ATP)-dependent helicase and the smaller subunit (RAP30) binds to pol II. It collaborates with TFIIB in recruiting pol II. TFIIF may promote transcription elongation and remain associated with pol II during elongation. Eukaryotic pol II is a large protein of 12 subunits. Five of the RNA polymerase subunits (RPBs) are shared by pol I, II, and III. The two largest subunits (L and L′) interact with the promoter DNA. The largest subunit of pol II (RPB1) contains a carboxy-terminal domain (CTD) that is composed of repeats of Tyr-Ser-

Pro-Thr-Ser-Pro-Ser. The serine and threonine residues of this repeat are hyperphosphorylated by TFIIH to initiate transcription.

The next two factors to join are TFIIE and TFIIH. TFIIE is also an $\alpha_2\beta_2$ tetramer. The larger subunit contains a zinc finger domain. TFIIE provides a docking site for the recruitment of TFIIH to the initiation complex and modulates the helicase and kinase activities of TFIIH. TFIIH is composed of nine subunits, two of which contain helicase activities and unwind the DNA template strands at the start site of transcription. This process requires the hydrolysis of ATP. The kinase activity is associated with another subunit of TFIIH. TFIIH is also essential in the transcription-linked repair of DNA damage by the excision-repair pathway. Its binding completes the basal transcription complex and elongation can occur. TBP remains bound to the TATA box during elongation but the other factors dissociate from the complex.

UPSTREAM FACTORS

Upstream factors are conserved DNA sequences recognized by constitutive DNA-binding proteins and located upstream of the transcriptional start site. These proteins affect transcription but may or may not interact with the basal machinery. If they interact with the basal transcription machinery, they are known as transactivating (or transcription) factors. Factors that bind to the basal machinery directly to activate transcription without binding to DNA are known as activating factors. The DNA elements to which upstream factors bind include CCAAT boxes, GC boxes, and octamers (Fig. 6.1).

CCAAT boxes have the consensus sequence of 5′-GGCCAATCT-3′. Inclusion of CCAAT boxes upstream of the promoter increases its strength (i.e., the rate at which the mRNA is transcribed). CCAAT boxes can occur in either forward or reverse orientation, the direction of which does not affect the function. They are recognized by factors of the CCAAT box transcription factor (CTF) family of proteins (CP1, CP2, CP3), all of which are produced by alternative splicing of a single gene. CCAAT boxes are also recognized by albumin CCAAT factor (ACF), nuclear factor 1 (NF1), CCAAT enhancer binding protein (C/EBP), and CCAAT displacement protein (CDP). CDP binds to CCAAT boxes and prevents the other factors from recognizing them. An example of a factor binding to CCAAT boxes is given with the discussion of CCAAT box binding protein (CBP or NF-Y) later in the chapter.

GC boxes have the consensus sequence of 5′-TGGGCGGGGC-3′. They often occur in multiple copies and, like CAAT boxes, can occur in either orientation. They are recognized by specificity factor 1 (Sp1), a monomer of 105 kDa. Sp1 has three zinc fingers at its C terminus and serine/threonine and glutamine-rich regions at its N terminus. This factor, as well as the redox control of its DNA-binding function, will be discussed later in the chapter.

Octamers have the consensus sequence 5′-TTATTTG-CATAA-3′. This sequence is recognized by Oct-1, which is ubiquitous, and Oct-2, which is tissue-specific. Both belong to the POU (Pit-Oct-Unc) family of transcription factors that have a common DNA-binding motif and include the mammalian factor Pit-1, Oct-1, Oct-2, and the nematode protein Unc-86. These factors play critical roles in the progressive development of general body plan, organ commitment, and finally, specific cell types. Oct-1 regulates cell-specific genes by recruiting other transcription factors or coactivators. The DNA-binding POU domain consists of two structurally different but highly conserved subdomains separated by a flexible linker: the POU-specific (POU_S) and the POU-homeodomain (POU_H). This motif partially encircles the DNA to recognize the eight base pairs of the octomer element. Both subdomains bind to DNA on opposite faces of the octomer sequence. POU_S consists of four α helices with the second and third helices forming a structure similar to a helix–turn–helix motif (see DNA-binding domains below). POU_H is similar to other homeodomain DNA-binding proteins and will also be discussed below.

Constitutive factors that bind to upstream elements can be regulated to a certain degree, especially during development. However, it is the inducible factors that bring about the immediate responses to cellular stress. These will be discussed in the next section with specific examples to follow.

INDUCIBLE FACTORS

Upstream factors bind constitutively to their DNA elements. Other DNA-binding proteins, known as inducible factors or transcription factors, bind in response to specific signals. Inducible factors provide the mechanism by which signals from environmental cues are conveyed and result in altered gene expression. Such cues could include a change in a stage of development, a response to a foreign toxin or invading bacterium, cold, heat, light, variation in oxygen or nutrient availability, a change in redox status, or a change in ionic strength of the extracellular milieu. Inducible factors may either increase the rate at which a particular mRNA is transcribed by increasing the rate of initiation by pol II or they may repress transcription by interfering with the binding of pol II. The sequences of DNA recognized by inducible factors are known as enhancer elements. Each inducible factor recognizes an enhancer of a specific sequence. The position of the enhancer elements need not

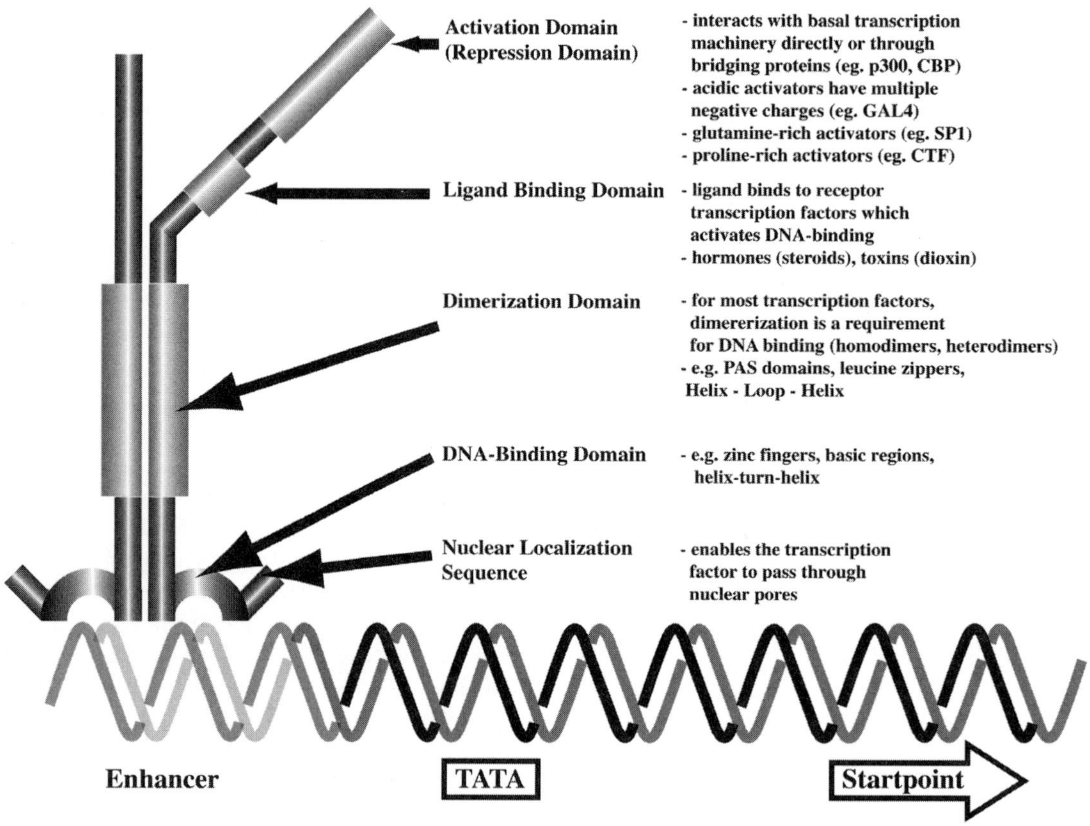

Figure 6.3 The general components of a basic transcription factor. Not all transcription factors have all of the domains shown. All factors contain a nuclear localization signal, a DNA-binding domain, and an activation/repression domain. Only the nuclear receptors contain ligand-binding domains.

be fixed and can occur either upstream or downstream of the promoter. Like upstream factors, enhancer elements can occur in either orientation and often show redundancy of function.

Inducible factors are proteins containing various functional subdomains or motifs (Fig. 6.3). These domains are short amino acid sequences that perform independent functions (protein–protein recognition, DNA sequence recognition, ligand recognition). Thus, the modular nature of transcription factors is revealed. All inducible factors contain a DNA-binding domain. All also contain either an activation or repression domain. The majority also contain dimerization domains that allow them to form heterodimers or homodimers (providing an essential control mechanism for their function; see below). Some transcription factors are activated by the binding of a ligand (e.g., steroids, toxins) and these also have a ligand-binding domain. These subdomains function independently, and hybrid proteins can be created with the DNA-binding domain of one transcription factor fused to the activation domain of a second. The hybrid protein will only recognize the DNA element of the first factor and have the trans-activation characteristics of the second. Thus, the nature of these factors, similar to the DNA elements to which they bind, is modular. The following sections will outline the types of functional domains commonly found in transcription factors.

DNA-Binding Domains (DBD)

The DBDs of transcription factors act as a tether, anchoring the activation or repression domains of these factors close to the basal apparatus. Many transcription factors are classified into families according to the type of DNA-binding domains they contain. DNA-binding domains are often helical regions of the proteins that recognize base pairs exposed in either the major or minor grooves of the double helix of DNA. Electrostatic and hydrogen bonding can occur between amino acid side chains of the DNA-binding domain and the specific DNA base-pair sequence of the enhancer. The DNA elements to which transcription factors bind in response to a particular stimulus are known

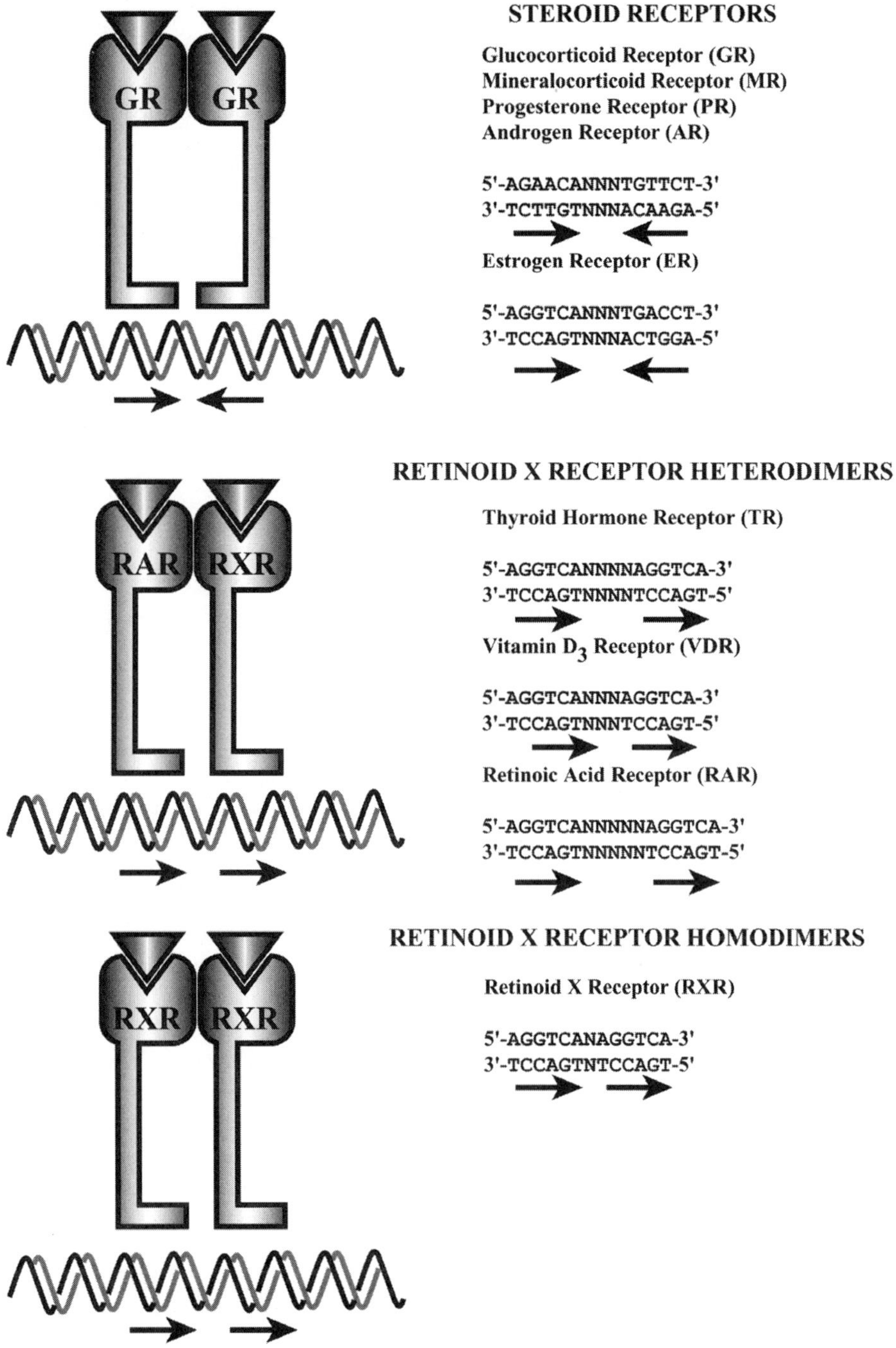

Figure 6.4 Nuclear receptors bind to either palindromic or direct DNA sequence repeats. In many cases, the directions of the repeated sequences determines whether a homo- or a heterodimeric transcription factor binds. The DNA sequences to which several different nuclear receptors bind are shown.

as response elements. For example, transcription factors that respond to glucocorticoids bind to glucocorticoid response elements (GREs), those that respond to hypoxia bind to hypoxic response elements (HREs), and so on. In the case of homo- or heterodimeric transcription factors, DNA half-sites are recognized by each protein (palindromes or direct repeats); the two half-sites together form what is known as an E-box recognition site (Fig. 6.4). The DNA-binding activities of many transcription factors can be measured using electrophoretic mobility shift assays (see Text Box 6.2). Different types of DNA-binding motifs are listed below.

Zinc Fingers Zinc fingers have been found both in proteins that bind to DNA and those that do not (Fig. 6.5). Zinc fingers are regions of the protein, approximately 30 amino acids long, that are folded into a loop around a central zinc atom. The consensus sequence for the zinc finger is Tyr/Phe-X-Cys-X_{2-4}-Cys-X_3-Phe/Tyr-X_5-Leu-X_2-His-X_{3-4}-His. One zinc atom is bound in a tetrahedral formation by two cysteine and two histidine side chains. The loop formed is compact enough to be inserted into the major groove of DNA where it makes two site-specific contacts with DNA per finger. Zinc fingers can occur in tandem repeats (most zinc finger proteins have 2 to 9) for continuous contact with DNA over an extensive length of

TEXT BOX 6.2 STUDYING TRANSCRIPTION FACTORS

Experimentally, almost all functions of transcription factors can be studied. This includes DNA binding, nuclear localization, activation (or repression), and ligand binding. Studies can be performed both *in vivo* and *in vitro* to (a) detect the presence or absence of a particular transcription factor under different conditions or (b) determine if a factor is activated or repressed under various experimental treatments. These techniques can lead to structural and functional studies on transcription factors purified from tissues or cells or expressed in a recombinant system. The DNA-binding functions of transcription factors can be examined using two methods.

Electrophoretic Mobility Shift Assays (EMSAs)

These assays, also called gel-shift assays or gel-retardation assays, rely on the interaction between the transcriptional activator (or repressor) and its target DNA response element (Fig. TB6.1*a*). The principle of the assay is that a protein–DNA complex has a retarded mobility on a nondenaturing polyacrylamide gel as compared with free DNA alone. This retarded mobility depends upon the different size, molecular weight, and charge of the DNA-binding proteins. EMSAs can be performed using transcription factors present in whole cell extracts (cytoplasm and nucleoplasm), nuclear extracts (nucleoplasm only), or that are recombinantly expressed. The specific DNA sequence that the factor binds to is end-labeled with a radioactive phosphate using the enzyme T4 polynucleotide kinase. This enzyme replaces the terminal phosphate of the last nucleotide with a radioactive one. The radiolabeled DNA fragment is then mixed with the cell/nuclear extract and allowed to bind for a short time. The mixture is separated on a nondenaturing polyacrylamide gel. Free DNA probe runs quickly to the bottom of the gel, whereas protein-bound DNA is detected as a specific radioactive band on the gel using autoradiography or a phosphoimager. Furthermore, the identity of the protein bound to the DNA fragment can be determined by adding specific antibodies to different aliquots of the mixture prior to electrophoresis. If the antibody recognizes the transcription factor, then the DNA/transcription factor/antibody complex shows an even greater retardation on the gel that is commonly referred to as a supershift. In some cases, EMSAs can also be conducted using nonradioactive methods, such as by labeling with digoxygenin or biotin.

DNA Footprinting

The second method for determining transcription factor DNA-binding activity is DNA footprinting (Fig. TB6.2). This method also relies on end-labeling of DNA, via the same methods as EMSAs. However, the difference between EMSAs and DNA footprinting is that the DNA-binding sequence for the transcription factor is known in the former method, whereas the latter method is used to determine the sequence of specific binding element. Labeled DNA is mixed with protein extract containing the DNA-binding protein and then subjected to either limited treatment with DNaseI or a chemical that cleaves

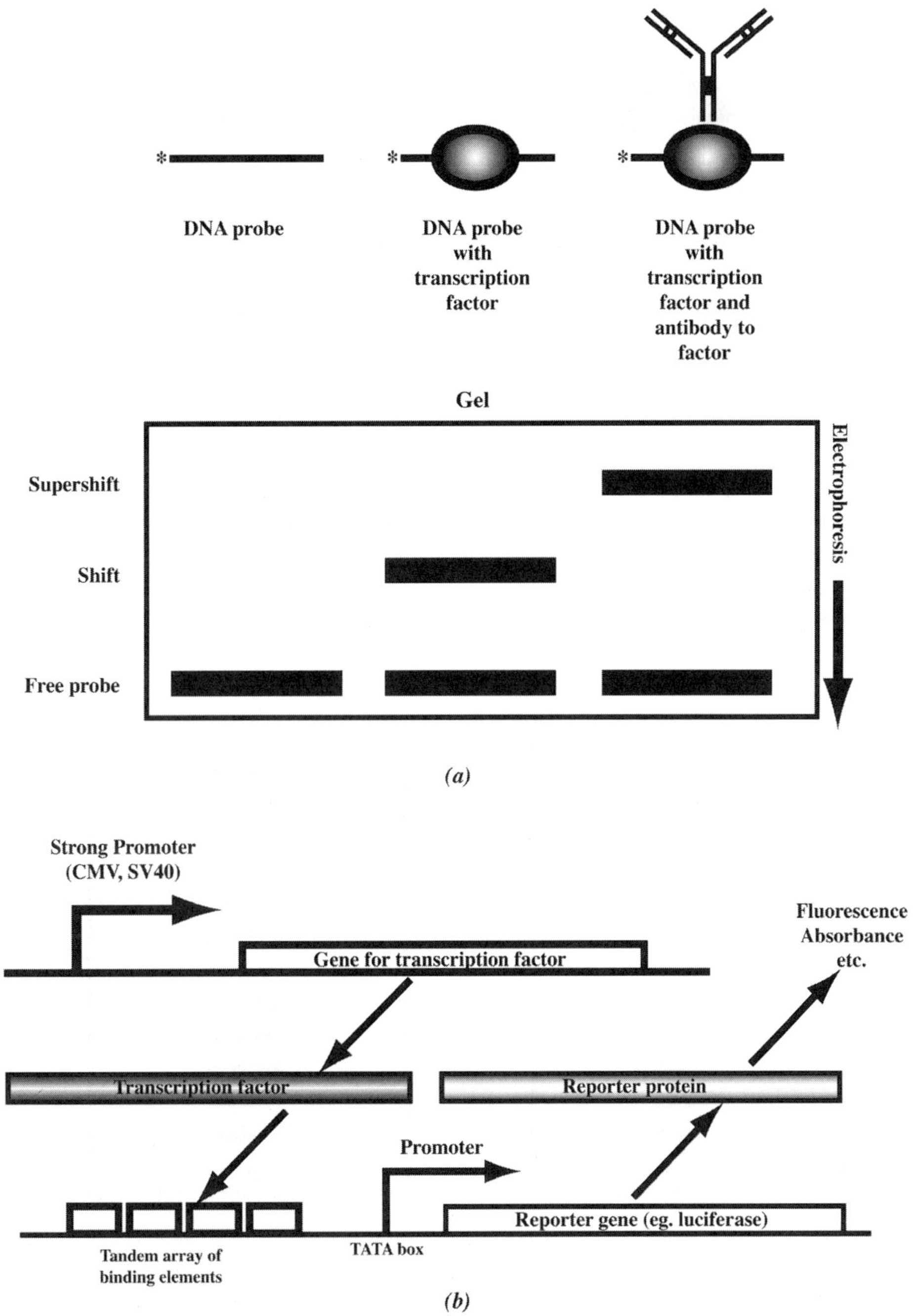

Figure TB6.1 (*a*) Electrophoretic mobility shift assay (EMSA) or gel-shift assay. The binding sequence of the transcription factor is end-labeled with a radioactive phosphate (asterisk). This is mixed with cellular proteins with/without an antibody specific to a subunit of the transcription factor. The mixture is then run on a nondenaturing gel. The binding of protein to labeled DNA results in a band of the slower moving complex (a shift) as its mobility is "retarded" compared to free DNA. If antibody is present, then the DNA is further retarded in what is known as a supershift. (*b*) Reporter assay for transcription factor function. Plasmids containing (a) the gene encoding a transcription factor, and (b) the DNA-binding elements for that factor are cotransfected into cells. Production of transcription factor under the control of a strong promoter results in the binding to the reporter plasmid and the production of a reporter protein (using the endogenous basal transcription machinery) that can be detected by fluorescence or absorbance, depending on the reporter protein chosen.

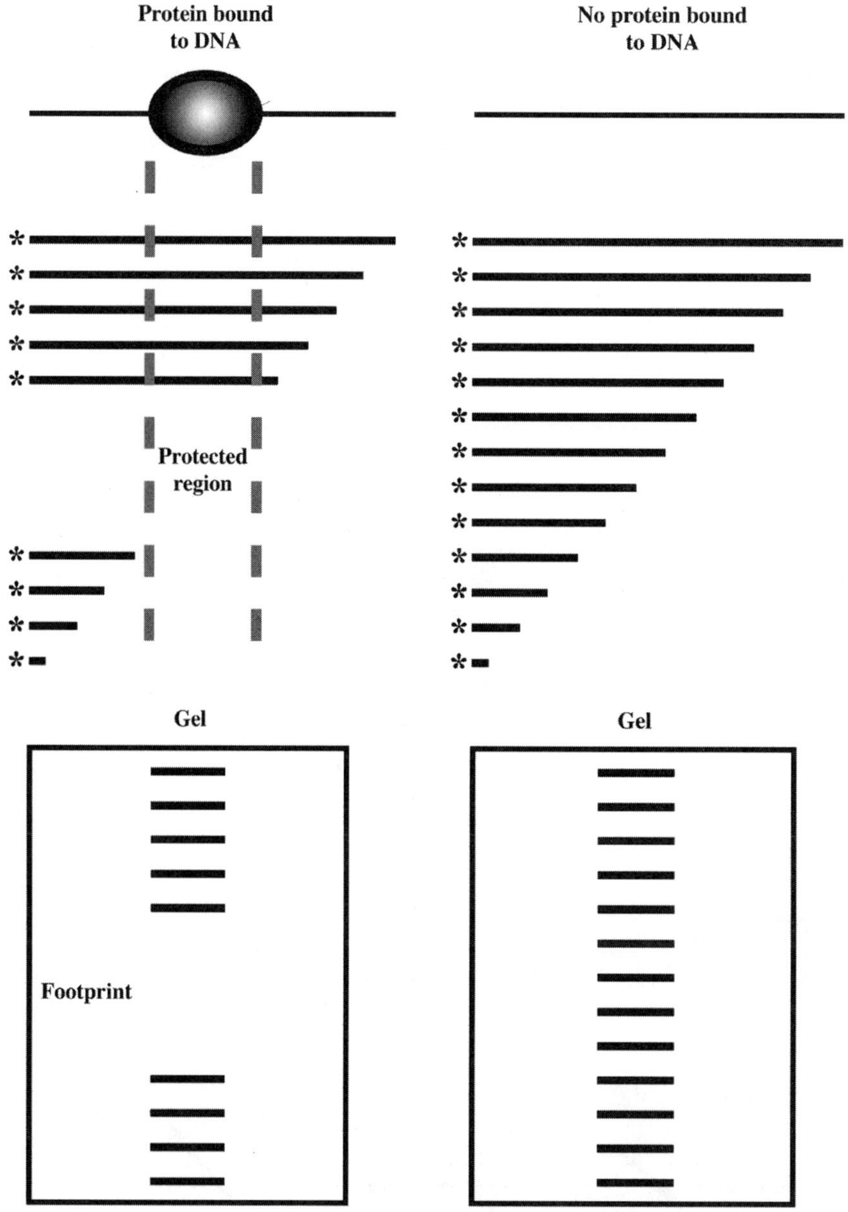

Figure TB6.2 DNA footprinting. The DNA sequence to be tested for protein binding is end-labeled with a radioactive phosphate. It is then either chemically treated or treated with DNase that nicks the DNA at every base. Regions protected by protein bound to the DNA are protected from such attack. Lengths of DNA protected by the protein are missing from the pool of fragmented, labeled DNA. The DNA is then dissociated from proteins and run on a denaturing gel. Blank regions in the banding pattern reveal segments of DNA that are bound by protein. Sequencing reactions using the same uncut DNA can be run alongside of the *footprint* pattern to determine the sequence of DNA protected by the transcription factor.

every phosphodiesterase bond between nucleotides with equal frequency. Chemical cleaving agents usually involve small iron-containing organic molecules. Protein-bound DNA is protected from cleavage by these agents. The treated mixture is then subjected to electrophoresis with subsequent autoradiography. Each sequential band on the gel represents DNA molecules that differ by one nucleotide in length. The labeled fragments containing the protein-binding site will be missing from the gel pattern, leaving a gap where no cleavage is observed. This gap is known as a DNA *footprint*. The sequences of protected fragments are determined by G+A and/or C+T chemical sequencing ladders of the same end-labeled DNA probe run alongside of the footprint.

Cellular Localization

Cellular localization can be conveniently studied in vivo by making fusion proteins between the transcription factor of interest and fluorescent protein. Fusion proteins are created by molecular cloning of transcription factors, in frame, into a fluorescent protein expression plasmid. The fusion of fluorescent protein may be N terminal or C terminal to the transcription factor. Fusion proteins are constructed in such a way as to not disrupt the functions of either of the individual proteins. Fusion of fluorescent protein may be only to a part of the transcription factor that contains the organelle localization sequence (e.g., nuclear, mitochondrial, endoplasmic reticulum, chloroplast) along with other functional domains. Green, yellow, blue, and cyan fluorescent proteins are available. Plasmids containing fusion proteins are transfected into whole cells that are subsequently treated and observed with either confocal or epifluorescent microscopy. Movement to and from the cellular organelle can be determined according to experimental treatment.

Activation and Repression of the Translational Machinery

Activation and/or repression of the basal machinery by transcription factors can be determined *in vivo* using reporter assays (Fig. TB6.1*b*). This assay also involves molecular cloning. However, the DNA elements cloned into plasmids are the DNA-binding sequences (or response elements) of the transcription factors themselves. Response elements are placed in tandem arrays, either in forward or reverse orientation, in front of the promoter of genes of enzymes that can be conveniently measured. The more elements placed in the arrays (or concatamers), the better the chance of transcription factor binding. Reporter enzymes include firefly luciferase or β-galactosidase. Any number of elements can be placed in promoters, and orientation is redundant as transcription factors bind and activate equally well in either orientation (forward or reverse). Cells are transfected with reporter plasmids, treated, lysed, and the enzymatic activities of the lysates are determined. The amount of enzyme produced from the reporter is directly correlated with (a) the binding activity of transcription factor to the cloned elements, and (b) the activation or repression of the basal transcription machinery that binds to the promoter. The fact that DNA-binding elements can be placed in any number in any orientation reveals the modularity of their nature. In organisms, most genes regulated by specific factors will have multiple binding elements for these factors in close proximity to their promoters.

Ligand Binding

Ligand binding to receptors can be measured via the use of radiolabeled ligands. The ratio of bound/free ligand (y axis) can be plotted versus bound ligand (x axis) resulting in a Scatchard plot. Both the total number (from the x intercept of the plot) and the affinity (from the slope of the plot) of the receptors for the labeled ligands can be determined from this graph. As with EMSAs, native (or nondenaturing) conditions must be used.

Modular Nature of Transcription Factors

The modular nature of transcription factors can be shown through the use of fusion proteins (Fig. TB6.3). Since transcription factor domains act independently from one

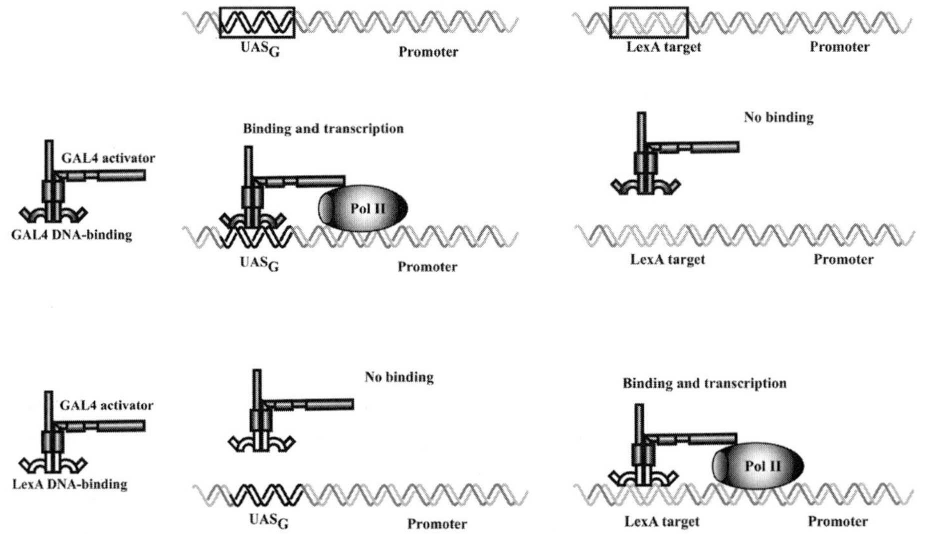

Figure TB6.3 The modularity of transcription factors and the DNA elements to which they bind. The Gal4 transcription factor binds to the UAS$_G$ DNA-binding element and activates transcription from pol II. If the DNA-binding domain of Gal4 is replaced by that of LexA, it can no longer bind to UAS$_G$ but binds to the LexA DNA-binding element. Thus, not only can the domains of transcription factors be switched for others (in the creation of fusion proteins), but the DNA sequences to which they bind are modular and can be exchanged for others or linked together in tandem repeats (during the construction of reporter plasmids).

another, they can be interchanged among factors without affecting each other. For example, the DNA-binding domain of one transcription factor (TF1) can be fused to the activation domain of a second, but differing, transcription factor (TF2). The hybrid protein will only bind to the DNA response element of TF1 (placed close to the promoter) but will only activate in the manner of TF2 (either directly with the basal transcription machinery or through a bridging coactivator). Such hybrid factors can be custom-made to activate specific genes, depending on the response elements upstream from the target genes.

the protein. Examples of proteins that contain zinc fingers include TFIIIA (activates RNA polymerase III), Sp1, the glucocorticoid receptor, and the estrogen receptor.

Helix–Turn–Helices Helix–turn–helices occur in transcription factors that are symmetrical homodimers with each protein recognizing a half-site in the DNA. Helix–turn–helices consist of two α helices joined together by a short stretch of amino acids that form the turn. One α helix (termed the recognition helix) lies in the major groove of DNA where amino acid side chains can hydrogen bond with specific DNA base pairs and the other lies at an angle across the DNA. Examples of helix–turn–helices proteins include the lac, λ, and trp repressor proteins.

A related motif is called the homeodomain (or homeobox) (Fig. 6.6*a*). The original name is derived from its original identification in the *Drosophila* homeotic loci (whose genes determine the final identity of body structures). This domain consists of 60 amino acids that form three α helices separated by two turns. Helix 3 lies in the major groove of DNA, whereas helices 1 and 2 lie at right angles across helix 3 outside of the double helix of DNA. Homeodomains occur in many proteins that regulate embryogenesis in *Drosophila* and are also found in mammalian octomer-binding proteins (Oct-1 and Oct-2).

Basic Regions Basic regions occur in homo- and heterodimeric transcription factors that have either leucine zipper or helix–loop–helix dimerization domains (Fig. 6.6*b*). These regions generally occur at the N terminus of the protein adjacent to the dimerization domain. Dimerization allows the basic regions to be inserted into the major grooves of DNA. Basic regions are approximately 15 amino acids long with at least 6 of the amino acids being

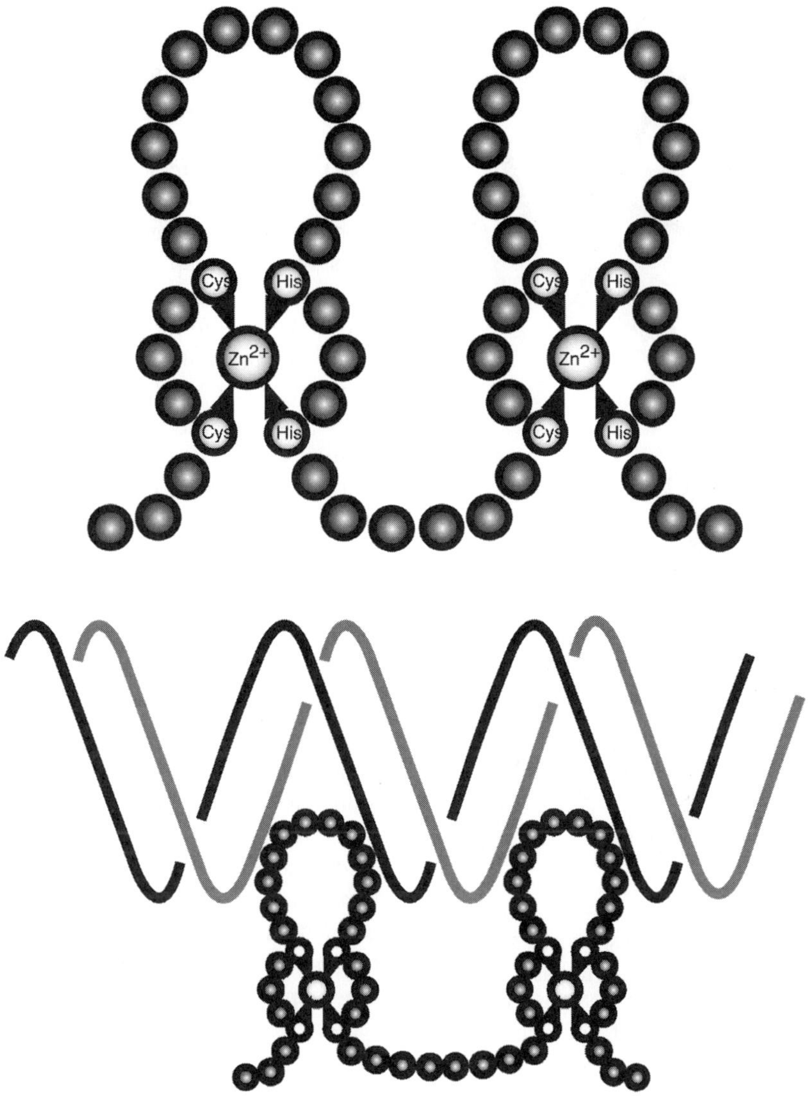

Figure 6.5 Zinc finger DNA-binding motif. The cysteine (Cys) and histidine (His) residues involved in complexing zinc are shown. The bottom panel shows insertion of the zinc fingers into the major grooves of DNA where they make hydrogen bonds and electrostatic interactions with bases in the DNA-binding sequence.

highly conserved in these regions. They form α helices with the basic residues interacting with the DNA. Basic amino acid residues can also interact with phosphates in the DNA backbone. Basic regions that are associated with leucine zippers are termed basic zipper (or bZip) proteins and those associated with helix–loop–helix proteins are termed basic helix–loop–helix (or bHLH) proteins. Examples of bZip proteins include the CAAT enhancer binding protein (C/EBP), cyclic adenosine 5′-monophosphate (cAMP) response element binding protein (CREB), and activator protein-1 (AP-1; Fos and Jun). Examples of bHLH proteins include hypoxia-inducible factor (HIF-1), E12, E47, and MyoD. BETA2 and NeuroD are bHLH proteins that are important in neurodevelopment.

Dimerization Domains (DD)

Dimerization is found in many transcription factors (both homodimers and heterodimers) and is a crucial part of the interaction of these factors with DNA. As mentioned with basic proteins, dimerization domains usually occur adjacent to DNA-binding domains of transcription factors (Fig. 6.3). Single subunits cannot bind to DNA alone; the complexed dimer is required for DNA binding. The reason for this is

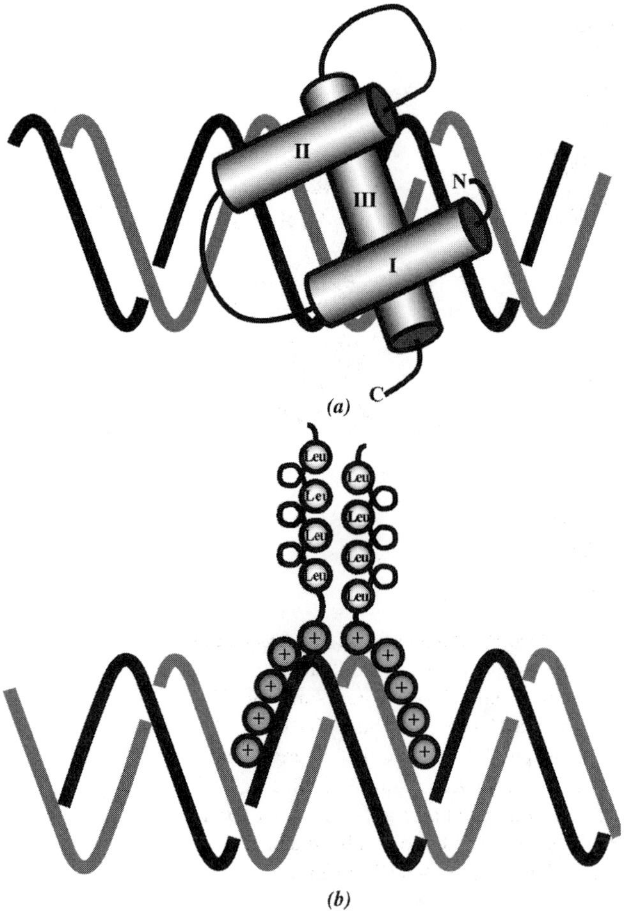

Figure 6.6 (*a*) A homeobox DNA-binding motif. Helix 3 makes contacts with DNA bases in the major groove, whereas helices 1 and 2 lie at right angles to helix 3 outside of the DNA helix. (*b*) A bZIP DNA-binding domain. Basic amino acids lie in the major groove of DNA to form interactions with the response element, whereas the adjacent leucine zipper (shown in green) forms the dimer. Leucine (Leu) residues form a hydrophobic interface for the proteins to interact.

that DNA-binding elements tend to occur as two halves, or half-sites, to which each monomer of the dimer binds. These half-sites contain the same sequence either in parallel with one another (direct repeats) or running in opposite directions (palindromes).

Examples where the DNA half-site determines the dimerization partner can be found in the ligand-binding nuclear receptors (see below). Palindromic DNA-binding sequences can favor homodimers (Fig. 6.4). These include the steroid-binding glucocorticoid, mineralocorticoid, progesterone, androgen, and estrogen receptors. Direct repeats can favor either heterodimers or homodimers. These include the thyroid, vitamin D, retinoic acid, and peroxisome proliferator activated receptors. Exceptions to this rule include the dimeric orphan receptors (retinoid X receptors and hepatic nuclear factor 4).

The types of dimerization domains are numerous and include nonspecific sites of weak binding between proteins. The two most common dimerization domains are helix–loop–helices (HLH) and leucine zippers.

Helix–Loop–Helices The helix–loop–helix motif is a stretch of 40 to 50 amino acids that contains two amphipathic helices separated by a loop of varying length. Proteins of this group form heterodimers and homodimers. Hydrophobic residues on the corresponding faces of the two α helices form the interaction domain between the dimer subunits. Basic DNA-binding regions are located adjacent to the HLH. Dimerization partners that lack the basic regions prevent the transcription factor from binding (Fig. 6.7). Examples of this are the MyoD/E12 and AC-S/da heterodimers. Both heterodimeric transcription factors bind strongly to DNA. However, E12/Id or AC-S/emc heterodimers cannot as both Id and emc lack basic regions adjacent to their HLH regions (see Regulation of Transcription Factors). The binding of a non-DNA-binding partner is a common mechanism of repression of the activity of certain transcription factors.

Leucine Zippers As with the HLH proteins, these proteins form both heterodimers and homodimers. A leucine zipper is an amphipathic α-helix structure approximately 30 amino acids long that contains the hydrophobic amino acid leucine at every seventh residue (Fig. 6.6*b*). The hydrophobic faces of two α helices interact in a parallel orientation to bring the dimer together. While leucine is the most common amino acid at the seventh positions of the zipper, other hydrophobic amino acids sometimes replace it.

Activation Domains (AD)

Activation involves the interaction of transcription factors with the basal transcription machinery resulting in enhanced transcription. The activation domains of transcription factors are distant from the DNA-binding regions of these proteins and may occur more than once in the same protein (Fig. 6.3). Activation domains are thought to interact either with TBP, TFIIB, TFIID, or directly with RNA pol II itself (Fig. 6.2). They may also interact with proteins known to be transcriptional coactivators (see below) that bridge the gap between transcription factor and the basal transcriptional machinery. Transactivation activity of transcription factors can be conveniently measured using reporter assays (see Text Box 6.2).

Activation domains are classified according to their amino acid content: acidic amino acids (aspartate and glutamate), glutamine, and proline. Acidic activation domains are regions of significant negative charge and can form α-helical amphipathic structures. Acidic domains may

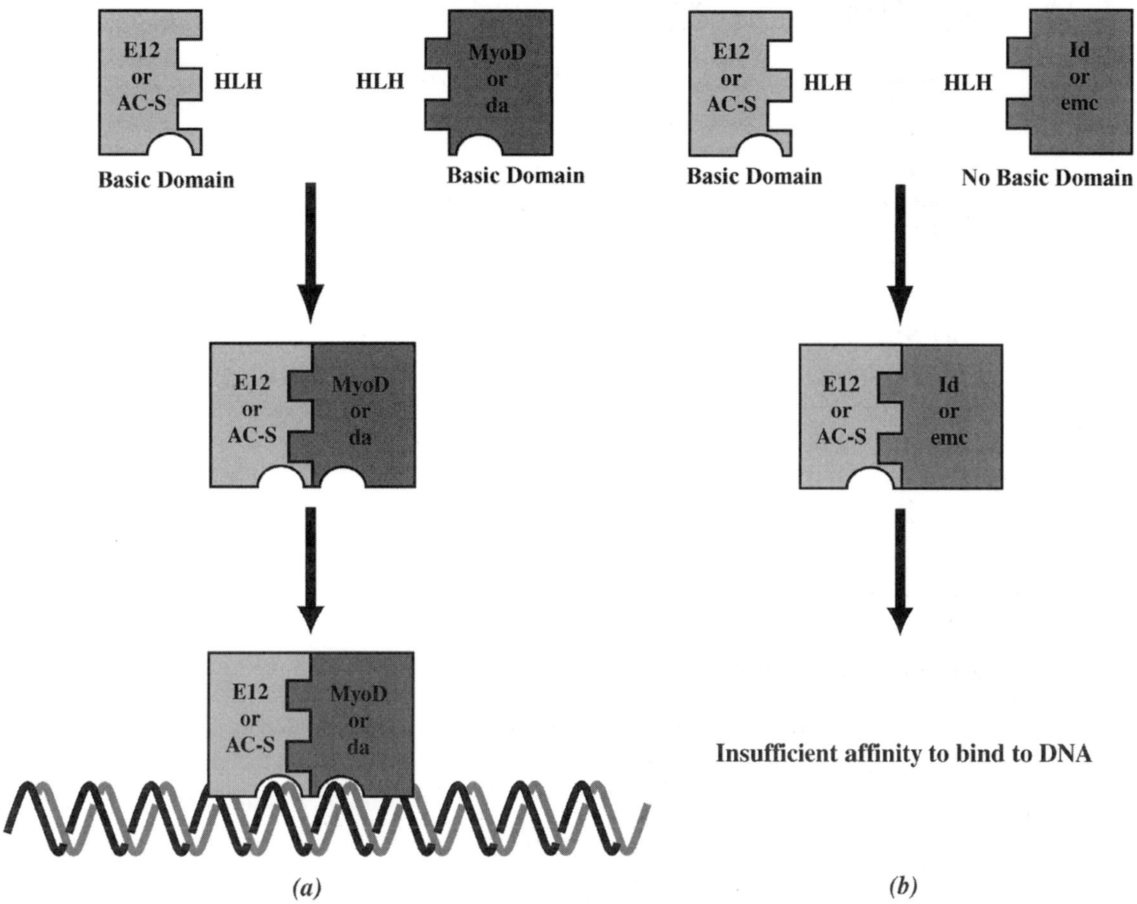

Figure 6.7 Nonfunctional dimerization partners cannot bind to DNA. MyoD/E12 are two bHLH proteins that dimerize to bind to DNA and activate genes involved in muscle differentiation. The inhibitor, Id, a HLH protein that lacks a basic domain, can also dimerize with MyoD or E12. Complexes that include Id, however, cannot bind to DNA and activate gene transcription.

interact with TBP of the basal apparatus, or they may enhance the ability of TFIIB to join the basal initiation complex. There are also some indications that acidic activators can also interact with TFIID. DNA-binding proteins such as GAL4, GCN4, VP16 and the glucocorticoid hormone receptor all contain acidic activation domains. Glutamine-rich transactivation domains interact with the TAFs associated with TBP of the basal transcription machinery. For example, Sp-1 has four glutamine-rich domains involved in activation. CREB contains an N-terminal glutamine-rich transactivation domain, as does Oct-1. Glutamine-rich activation domains stimulate transcription in all eukaryotic cells except yeast. Proline-rich activation domains assist in promoter activation by interacting with histones (in particular histone 3), causing them to dissociate from promoter DNA. CCAAT-binding transcription factor (CTF) is a factor that contains a C-terminal proline-rich activation domain.

Transcriptional coactivators are proteins that can form a bridge between transcription factors and the basal transcription machinery. Coactivators must have certain properties in order to function with transcription factors. First, they must interact with either an activation or a repression domain of the DNA-binding protein. Second, they must not bind to DNA themselves. Third, they must have autonomous activation domains themselves in order to interact with the basal transcription machinery. Coactivators should not enhance basal transcriptional activity on their own. They usually function as multiprotein complexes. Examples of multiprotein coactivators are shown in Table 6.1. Some coactivators (such as PGC-1) interact, not with the activation domain of transcription factors, but with their DNA-binding domains to enhance the likelihood of transcription factor interaction with chromosomal DNA.

Complexes of coactivators perform specific functions. Some coactivators recruit histone acetyltransferases or

TABLE 6.1 Various Coactivators of Transcription and Their Associated Proteins

- *Histone Acetyltransferases*
  - p160 family of coactivators
    - Steroid receptor coactivator 1 (SRC-1), also known as:
      - Nuclear receptor coactivator protein 1 (NcoA-1)
    - SRC-2; also known as:
      - Transcription intermediary factor 2 (TIF2), glucocorticoid receptor interacting protein 1 (GRIP1), NcoA-2
    - SRC-3; also known as:
      - Receptor associated coactivator 3 (RAC3), amplified in breast cancer 1 (AIB1), activator of thyroid and retinoic acid receptors (ACTR), thyroid receptor activator molecule 1 (TRAM-1), CBP interacting protein (pCIP), NcoA-2
  - p300 and CREB binding protein (CBP) transcriptional integrators
    - the p300/CBP-associated factor (pCAF)
- *ATP-Dependent Chromatin Remodeling Complexes*
  - SWItch (SWI)
  - Sucrose non-fermenting (SNF)
  - Brahma-related gene (BRG)
- *Mediator-Like Protein Complexes*
  - Vitamin D_3 receptor interacting protein (DRIP)
  - Thyroid hormone receptor-associated protein (TRAP)
  - Activator-recruited cofactor (ARC)
  - Peroxisome proliferator-activated receptor (PPAR)–binding protein (PBP) complexes
- *Activating signal cointegrators 1 and 2 (ASC-1 and ASC-2)*
- *Peroxisome proliferator-activated receptor–gamma coactivator 1 (PGC-1)*

have histone acetylase activity themselves. Acetylation is reversible and involves the formation of Lys ε-NH_3^+ groups in nucleosomal histones. The result is a diminished electrostatic charge on histones, reducing the affinity of the histone for the negatively charged sugar–phosphate backbone of DNA and causing the DNA to unwind from histones leading to promoter activation. Histones may also be deacetylated by histone deacetylases that reverse DNA unwinding from histones. The ATP-dependent chromatin remodeling complexes become physically and functionally associated with the CTD of RNA pol II. They disrupt nucleosomal arrays in chromatin in an ATP-dependent manner, allowing the binding of TBP and TAFs to the DNA template. Mediator-like protein complexes interact with nuclear receptors in a ligand-dependent manner. Like the ATP-dependent chromatin remodeling complexes, the mediator-like complexes are thought to interact with the CTD of pol II.

Activation domains of transcription factors interact with specific regions of coactivators. Interactions with the coactivator CBP/p300 usually occurs with a cysteine/histidine-rich region of this protein. Transcription factors that have been found to interact with CBP/p300 include nuclear factor-κB (NF-κB) and c-Jun of AP-1. CBP/p300 interacts with TBP and TFIIB of the basal apparatus and interacts with SRC-1 to coactivate nuclear receptors (ligand-inducible transcription factors) involved in steroid metabolism. CBP/p300 also has histone acetylation activity associated with a region of the protein known as the bromodomain. Activation domains of nuclear receptors (NR) (see below) interact with a hydrophobic LXXLL motif (called NR boxes where L is leucine and X is any amino acid) found in the coactivators SRC-1 or glutamate receptor-interacting protein 1 (GRIP1).

Repression Domains (RD)

Although repressors have been less studied than activators, they constitute an important part of gene regulation. Some of the first DNA-binding regulatory proteins to be discovered were repressors. Repressors have the same DNA-binding domains that activators do: basic regions, homeodomains, and zinc fingers. It has been possible to map "repressor domains" on individual repressors, but because of the diversity of their function, few common domains have been characterized.

Repressors may function in a number of different ways. First, they may bind directly to a DNA element recognized by an activator, thereby competing directly with the activator for binding to the same element. This strategy is not very common since many activators are usually present in close proximity to a promoter. Repression by competition would require multiple overlapping DNA-binding sites with activators. Second, repressors may interact directly with an activator (already bound to DNA) and prevent its association with either a coactivator or the basal transcription machinery. This is known as either quenching or masking. Quenching can also occur between two activators that compete for the same DNA-binding site. When this occurs, one activator represses the other under conditions were the former predominates. Third, repressors may bind directly to the basal machinery leading to its silencing, irrespective of the presence or absence of an activator. Examples of repressors that interact directly with the basal transcription machinery include Mot1 and Not (that interact with TBP), Dr1/Drap1 (that interacts with TFIIB), and Srb10/Srb11 (that interacts with pol II and TFIIH).

Repressor proteins contain specific protein–protein interaction domains that are involved in the repression of transcription factor activity. Among these are the broad complex, tramtrack, and bric-a-brac (also known as the POxvirus and zinc finger domain or BTB/POZ) domain, the KRuppel-associated box (KRAB) domain, and the SCAN domain. The SCAN domain is named after the first four proteins in which the domain was discovered: SRE-ZBP, CTfin51, AW-1 (ZNF174), and Number 18 cDNA or ZnF20. BTB/POZ, KRAB, and SCAN domains are present in the amino termini of a large number of Cys2-His2 zinc finger proteins. The zinc fingers of these proteins occur at the carboxyl end. While BTB/POZ occurs in a diversity of organisms ranging from yeast to humans, KRAB and SCAN seem to be vertebrate-specific. These domains may occur in multiples within the same protein and in various combinations (e.g., both KRAB and SCAN domains within a single protein). BTB/POZ domains are responsible for the recruitment of histone deacetylases, resulting in chromatin remodeling and changes in gene expression. KRAB is a 75-amino-acid region containing two predicted amphipathic α helices. KRAB is believed to function by binding a corepressor known as KRAB-associated protein 1 (KAP-1), KRAB-A interacting protein (KRIP-1), or transcriptional intermediary factor 1 β (TIF1β). Once bound to DNA, KRAB recruits corepressors, forming a heterochromatin-like complex and ultimately leading to gene silencing. SCAN domains occur in a wide variety of proteins with or without zinc fingers. Proteins with SCAN domains are involved in a diverse set of biological functions involved in cell development and differentiation. SCAN and BTB/POZ domains are capable of mediating hetero- and homodimerization among proteins within their families. As more proteins are identified with these repressor domains, their role in mediating gene transcription will become increasingly clear.

Like coactivators, corepressors must fill certain criteria. They must interact with the basal transcription machinery. They must not bind to DNA themselves, and they must have repression domains that function autonomously from domains that bind to DNA-binding proteins. Corepressors have intrinsic histone deacetylase activity, thereby tightening the chromosomal structure and preventing transcription. Known corepressors include receptor interacting protein 140 (RIP140), friends of GATA (FOG), C-terminal binding protein (CtBP), transforming growth factor β interacting factor (TGIF), the silencing mediator for retinoid and thyroid receptors (SMRT), and nuclear receptor corepressor (N-CoR).

Ligand-Binding Domains (LBD)

Transcription factors that activate gene expression upon binding of a ligand are known as nuclear receptors. Ligands can include hormones, vitamins, metabolites, and toxins (e.g., dioxin). Major categories of nuclear receptors include steroid receptors, thyroid/retinoid/vitamin D receptors, peroxisome proliferator activated receptors, and orphan (with unidentified ligands) receptors. LBDs, in general, are globular structures composed of 11 to 12 α helices closely folded in a similar manner. The α helices are arranged in a three-layer antiparallel sandwich that becomes more compact upon the ligand binding. Ligands bind deeply in the hydrophobic core of the sandwich. The binding of a ligand results in a conformational change that causes either homodimerization of the nuclear receptor or its association with coactivators or corepressors. In the absence of a binding ligand, some nuclear receptors act as transcriptional repressors. The movement of helix 12 upon ligand binding is mainly responsible for creating a surface for coactivator binding.

Nuclear Localization Signal (NLS)

All activators and repressors must enter the nucleus to bind to their corresponding DNA elements and interact with the basal machinery. The transport of many transcription factors into the nucleus is triggered by changes in differentiation or metabolic state. Therefore, nuclear entry represents a key control point for signal transduction and gene expression. Proteins smaller than 50 kDa can pass freely across the nuclear membrane, whereas larger proteins require a sequence that targets them for active transport into the nucleus. A NLS is a short peptide sequence within proteins that is sufficient to target the protein that contains it to the nucleus. In general, these sequences are arginine- and lysine-rich and can occur as single elements or bipartite repeats of two elements. Fusion of a NLS onto nontargeted proteins will result in accumulation of the fusion protein in the nucleus of the cell.

There are two steps in nuclear localization. The first is energy-independent and involves recognition of the NLS of the target protein and transport to the nuclear pore complex (NPC) (Fig. 6.8). The second is energy-dependent and involves translocation through the pore and into the nucleus. Recognition of the NLS occurs by transport proteins called importins. Importins are composed of an α

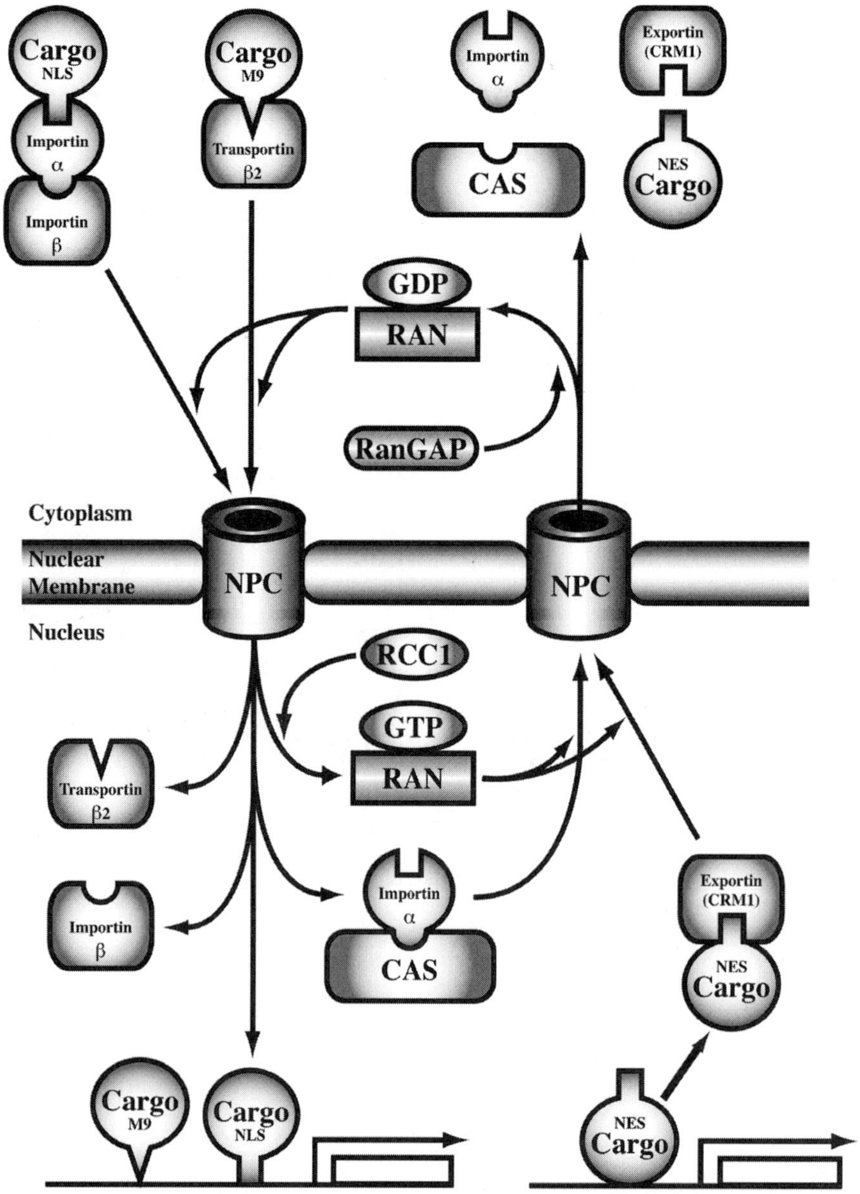

Figure 6.8 Nuclear import and export of proteins. "Cargo" proteins can either be recognized by importin (composed of an α and a β subunit) if they have an NLS or by transportin ($\beta 2$ subunit similar to importin β subunit) if they have an M9 sequence. Transport through the nuclear pore complex (NPC) is assisted by the small GTPase Ran. Ran GTPase-activating protein (RanGAP) generates inactive (GDP-bound) Ran. Once the cargo proteins are imported into the nucleus, they are free to bind to DNA if they are transcription factors. RCC1 facilitates the nucleotide exchange of GDP for GTP once Ran is in the nucleus. Importin α is again exported from the nucleus by cellular apoptosis susceptibility gene (CAS) in the presence of RanGTP. Exportin (CRM1) exports cargo proteins with a NES from the nucleus with the assistance of RanGTP.

and a β subunit. The α subunit recognizes the NLS on the cargo protein. The β subunit binds to the α subunit and is recognized by the GTPase Ran in its GDP bound form. There are numerous α and β isoforms. In general, $\alpha 1\beta 1$ importin is involved in nuclear transport of nuclear-targeted proteins. Import of the target protein through the nuclear pore results in the conversion of GDP to GTP in an ATP-dependent process. Depletion of cellular ATP causes nuclear-targeted proteins to remain in the cytosol. A different NLS, termed M9, is recognized by the importin $\beta 2$

subunit only (also known as transportin), without the involvement of an α subunit.

The NPC is composed of proteins termed nucleoporins that are recognized by importins. These nucleoporins form four rings; a cytoplasmic ring, a central ring, a nucleoplasmic ring, and a joining ring. The NPC also contains eight filaments that extend from the cytoplasmic ring, a central transporter, and a "nuclear basket" formed by eight nucleoplasmic filaments that extend from the nuclear ring to the joining ring. After import into the nucleus, transcription factors are released from the importins and can bind to DNA elements.

Recognition of the NLS of a protein is a regulated process. NLS recognition by importins can be altered by protein phosphorylation in close proximity to the NLS, resulting in either enhanced or inhibited importin recognition of the NLS and nuclear transport. In some cases, the NLS is masked by another protein until a signaling molecule interacts with the masking protein and allows it to come off the NLS, exposing it for interaction with importins. An example of this is the cytosolic inhibition of the nuclear transport of NF-κB (see below) by the inhibitory protein inhibitor of κ-B (I-κB). I-κB masks the NLS of NF-κB, retaining this transcription factor within the cytosol until stress-induced changes to the inhibitory protein cause it to come off, allowing importins to interact with NF-κB.

Nuclear Export Signal (NES)

Nuclear export signals are short sequences sufficient to transport proteins that contain them out of the nucleus into the cytoplasm. NES are leucine-rich peptide sequences. Again, transport involves the GTPase Ran, but in the case of nuclear export, Ran/GTP binds to exportin. Exportin belongs to the importin β family of proteins. During the transport of a protein containing a NES out of the nucleus, GTP is converted to GDP, and the protein is released from Ran and exportin into the cytoplasm (Fig. 6.8).

REGULATION OF TRANSCRIPTION FACTORS

The regulation of transcription factors (that includes activators and repressors) can occur at multiple levels (Fig. 6.9). This allows for flexibility of control under shifting metabolic conditions. Levels of control can include (1) synthesis or degradation of the transcription factor protein, (2) covalent modification or ligand binding to the transcription factor, and (3) binding of inhibitors to the protein to prevent either its localization to the nucleus or its ability to bind to DNA. In the case of heterodimeric transcription factors, this may include a change of partner protein from an inactive partner (no DNA binding) to an active one (DNA binding). In all cases, the active form of the transcription factor is the form that binds to DNA. The following section will review the types of controls exerted on transcription factors and give examples of each.

Synthesis of transcription factors may determine their specificity such that certain factors are only synthesized in certain tissues or only synthesized under certain conditions. A good example is the homeobox proteins (or HOX proteins) that regulate cell differentiation, proliferation, and migration. These are major control factors for the cell cycle, and they also control body pattern formation in *Drosophila*. Some homeobox proteins are oncogenes. Certain homeobox proteins are expressed only during embryogenesis, where they are responsible for angiogenesis and cardiovascular development. Others are only expressed in endothelial cells or in vascular smooth muscle cells.

The availability of a transcription factor may also depend upon its degradation. This form of regulation involves targeted proteolysis of the factor. Proteins are "tagged" for destruction either by posttranslational modification or by the addition of small peptide markers such as ubiquitin or SUMO-1 (also known as sentrin, GMP1, UBL1, PIC1, or Smt3). An example of this is the basic HLH transcription factor hypoxia-inducible factor-1 (HIF-1). This transcription factor is responsible for hypoxia-inducible gene expression (see below and Chapter 15). It is a heterodimer composed of an α and a β subunit. The β subunit is constitutively expressed and is always present. However, the α subunit undergoes an oxidative modification under normal oxygen conditions (21% atmospheric oxygen) that leads to its subsequent ubiquitination and degradation by proteolysis. Under hypoxic conditions, the α subunit is stabilized and is available for dimerization with the β subunit leading to DNA binding of the transcription factor.

Reversible phosphorylation represents another mechanism for regulation of transcription factor function. Phosphorylation by protein kinases or dephosphorylation by protein phosphatases can convert an inactive factor to an active one. For example, phosphorylation of the heat shock transcription factor (HSTF) by members of the mitogen-activated protein kinase (MAPK) family converts it to an active DNA-binding form. By contrast, dephosphorylation of the Jun subunit of AP1 (a heterodimer of subunits Fos and Jun) by casein kinase II (CKII) converts it to its active form.

Ligand binding to receptors can affect their nuclear localization or their DNA binding functions. As mentioned previously, these include the steroid receptors and the dioxin receptors. Examples include the retinoic acid receptor (RAR) and the retinoid X receptor (RXR) (Fig. 6.4). RAR binds the all-trans form of retinoic acid (RA) as a high-affinity ligand whereas RXR has high affinity for

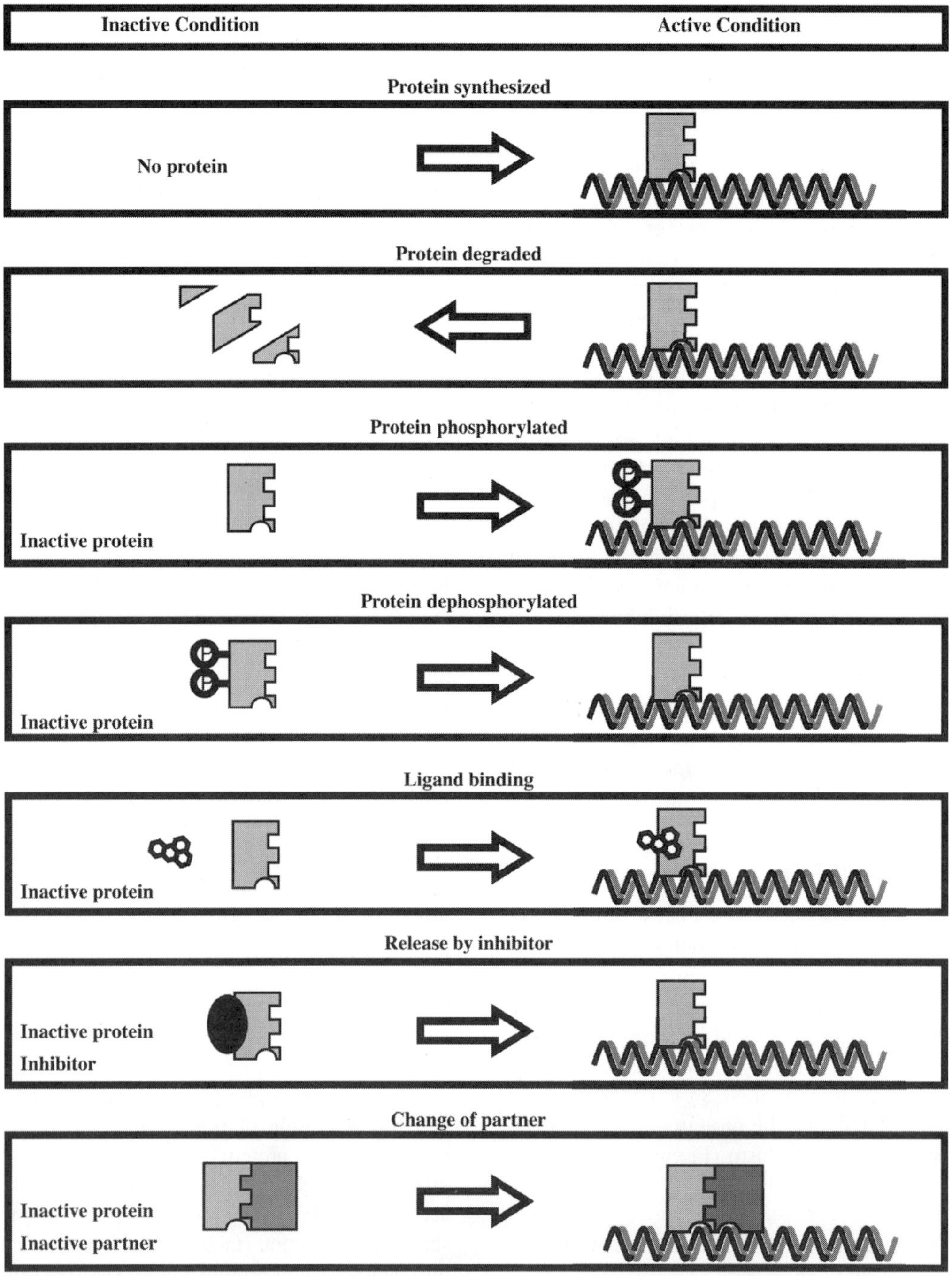

Figure 6.9 Regulation of transcription factor function. Regulation may occur at a number of levels. The overall pool of a particular transcription factor may be either increased or decreased by synthesis (in a particular tissue or cell type) or degradation (in response to external stimuli), respectively. Posttranslational modification of factors (e.g., phosphorylation) can either activate or inactivate a transcription factor. The binding of a ligand (steroid, toxin) or the release of an inhibitory protein may allow factors to translocate to the nucleus and activate transcription. Change from an inactive dimerization partner to an active dimerization partner can allow a transcription factor to bind to DNA and transactivate the basal transcription machinery.

9-cis RA. When all-trans RA is present in high concentrations, RXR forms heterodimers with RAR and binds to the RAR response element to activate genes involved in development and vitamin A metabolism. When 9-cis RA is present in high concentrations, RXR forms homodimers and binds to the RXR response element to up-regulate genes involved in human lipoprotein metabolism. Hence, the type of ligands present determine which dimer forms and which genes are activated.

The binding of an inhibitor to a transcription factor is another way to regulate its availability. An example of this is the heterodimeric transcription factor NF-κB. In nonstimulated cells, an inhibitor protein (I-κB) binds to the heterodimer and retains it in the cytosol, blocking its translocation to the nucleus. Phosphorylation of I-κB dissociates it from the complex, allowing the NF-κB heterodimer to travel to the nucleus and bind to its DNA recognition sequence. Phosphorylated I-κB is released from the complex and rapidly ubiquitinated and degraded by the same protease that degrades HIF-1α. Release of I-κB from the complex occurs during stimulation by a number of inflammatory factors.

Heterodimeric transcription factors may form many partners. In general, both partners of a heterodimer must be functional (i.e., have DNA-binding capability) in order to activate transcription. This is especially true with the basic HLH family of transcription factors. In some cases, the dimerization partner lacks a DNA-binding domain and is nonfunctional (Fig. 6.7). An example of this is the E12 transcription factor. E12 regulates genes involved in muscle differentiation. It is a basic HLH protein that forms a functional dimer with MyoD. MyoD has both an HLH dimerization domain and a basic DNA-binding domain. However, when E12 partners with Id, an HLH protein that lacks a basic region, it can no longer bind to DNA and activate transcription.

EXAMPLES OF TRANSCRIPTION FACTORS INVOLVED IN METABOLIC CONTROL

The list of transcription factors involved in the activation and repression of gene expression is growing on a daily basis. To cover all factors known to date is beyond the scope of this chapter. We will review the principles of metabolic control of transcription by using examples of action and regulation of transcription factors involved in oxygen signaling and redox (reductive/oxidative potential) signaling in the cell. Both signals are related and can be considered stresses when they deviate from their normal concentrations. Variations in intracellular oxygen can include an increase (hyperoxia), a decrease (hypoxia), or a complete absence (anoxia) as compared with normal oxygen concentrations (normoxia). In a similar manner, an increase in prooxidants in a cell causes oxidative stress (see Chapter 12). The two factors are interrelated and changes in the level of oxygen in a cell has a direct effect upon its redox state, hyperoxia and hypoxia being associated with oxidative and reductive conditions in the cell, respectively. The goal of adaptation to either stress is to restore homeostasis; normoxia in the case of oxygen and "normoxidant" (or normal oxidant) in the case of redox.

Changes in oxygen tension or redox may affect the nuclear localization, DNA binding, dimerization, ligand binding, or transactivation/transrepression functions of transcription factors. In most cases, critical sulfur-containing amino acids are involved (cysteine, sometimes methionine). Cysteines in the DNA-binding regions of transcription factors may have hydrogen bonds or electrostatic interactions with bases in the binding elements. Oxidation of cysteines may have effects on the ability to maintain a certain DNA-binding conformation due to metal ion chelation (e.g., zinc fingers). The formation of disulfide bridges can also abolish dimerization, required for the activation of many transcription factors. In the sections below we will discuss transcription factors whose function is altered by oxygen or redox. Examples will be given to reflect the functional roles of transcription factors mentioned under Inducible Factors (DNA binding, transactivation, etc.) as well as the control of such factors as discussed under Regulation of Transcription Factors (synthesis, degradation, inhibitors, ligand binding). In most cases, factors are inactivated by oxidative conditions but, in some cases, they are activated. Finally, examples of redox factors that reduce transcription factors, thereby activating them, will be given.

Factors Repressed by Oxidative Conditions

Hypoxia-Inducible Factor 1 (HIF-1) Hypoxia-inducible factor-1 is a transcription factor that up-regulates genes in response to low oxygen conditions (Fig. 6.10). Genes activated by this factor are of two sorts: (a) those whose protein products function to increase oxygen delivery to tissues through either vasodilation or angiogenesis and (b) those whose protein products increase the capacity for anaerobic ATP output. The former include the genes for erythropoietin, vascular endothelial growth factor, inducible nitric oxide synthase, and heme oxygenase-1, whereas the latter include many of the glycolytic enzymes (glyceraldehyde-3-P dehydrogenase, phosphoglucomutase, aldolase A, lactate dehydrogenase A, phosphoglycerate kinase-1, pyruvate kinase M) as well as the glucose transporter (GLUT-1). HIF-1 is a heterodimeric basic helix–loop–helix protein composed of HIF-1α and the aryl hydrocarbon receptor nuclear translocator (ARNT), also known as the HIF-1β subunit. HIF-1 binds to the consensus sequence

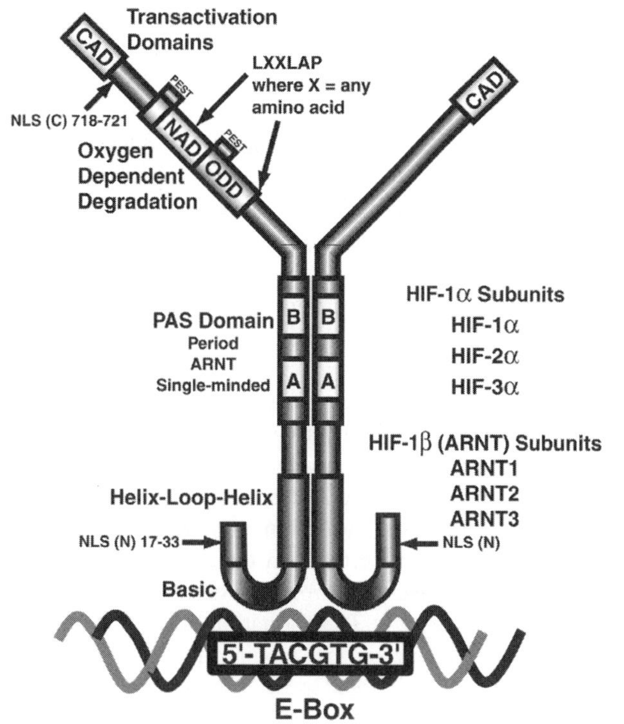

Figure 6.10 The transcription factor HIF-1. HIF-1 is composed of an α subunit and a β subunit (ARNT) of which there are three isoforms each. These isoforms can combine in various ways (e.g., HIF-1α/ARNT2). The ODD domain contains two short amino acid sequences (LXXLAP where X = any amino acid) that are recognized by prolyl hydroxylase-like enzymes and are hydroxylated. This results in HIF-1α oxygen-dependent degradation. The PAS and helix–loop–helix domains are the dimerization sites between the two HIF-1 subunits. The complete heterodimer recognizes and binds to a specific E-box sequence of DNA known as the HRE. The basic regions of the proteins adjacent to the helix–loop–helices are responsible for recognition of this sequence. NAD = N-terminal activation domain; CAD = C-terminal activation domain; ODD = oxygen-dependent degradation domain; PEST = proline, glutamic acid, serine, and threonine-rich sequences; PAS = period/ARNT/single-minded domain; NLS(N) = N-terminal nuclear localization sequence; NLS(C) = C-terminal nuclear localization sequence.

5′-(G/C/T)ACGTGC(G/C/T)-3′ known as the hypoxic response element (HRE). The N-terminal and a C-terminal activation domains for HIF-1 are present only on the subunit. The leucine-rich HIF-1α activation domain interacts with the cysteine/histidine-rich 1 (C/H1) domain of the p300/CBP coactivator. Another domain, present in both HIF-1α as well as ARNT, assists in dimerization. This is known as the PAS domain, named after the three proteins in which the domain was first discovered (period–ARNT–similar). ARNT can heterodimerize with other bHLH PAS proteins including the aryl hydrocarbon receptor (Ahr). Ahr is a ligand-binding protein that responds to toxins; the ligand in this case being dioxin. Ahr/ARNT dimers bind to the Ahr response element present near genes that detoxify polycyclic aromatic hydrocarbons (including the cytochrome p450s).

HIF-1 is a transcription factor that is regulated at multiple levels. The primary mode of regulation is by oxygen tension. HIF-1α is an example of a protein that is regulated by degradation, in this case, degradation triggered by oxygen (Fig. 6.11). HIF-1α contains an oxygen-dependent degradation (ODD) domain that is modified at two key prolines by hydroxylation. The prolines are present in a short amino acid sequence, LXXLAP (where X is any amino acid), of which two occur in the ODD domain. Since hydroxylation is an oxygen-dependent process, the absence of oxygen results in HIF-1α not being hydroxylated and not degraded. ARNT is not hydroxylated and therefore is constitutively present in cells. HIF-1α, however, is hydroxylated in the presence of oxygen by a prolyl hydroxylase-like enzyme. This enzyme contains iron at its catalytic site and can be inactivated by iron chelators. Hydroxylation of HIF-1α leads to its subsequent ubiquitination and degradation by the 26S proteosome under normoxic conditions. However, in the absence of oxygen, HIF-1α is stabilized, becoming available for transportation to the nucleus, dimerization with ARNT, and activation of low oxygen-dependent genes.

Hypoxia-inducible factor 1 (HIF-1) is not only regulated by oxygen at the level of protein stability, but its ability to transactivate gene expression is also regulated by oxygen tension. HIF-1α is also hydroxylated at a critical asparagine in its C-terminal activation domain (CAD) by an asparaginyl hydroxylase. Hydroxylation disrupts HIF-1α ability to interact with p300/CBP, the protein required for bridging the gap between HIF-1 bound to an HRE and the basal transcription machinery. When the CAD of HIF-1α is fused to the GAL4 DBD (see Text Box 6.2), this fusion protein can only transactivate an HRE-controlled reporter in the absence of oxygen. Thus, the regulation of the HIF-1α CAD by oxygen is independent of the regulation of its degradation by oxygen.

Not only does HIF-1 respond to a change in oxygen tension but also to changes in redox. A critical cysteine (C774) in the C-terminal activation domain of HIF-1α is required for the interaction with p300/CBP. Reduction of this cysteine by redox-active proteins, such as thioredoxin and redox factor 1 (Ref-1) (see Chapter 12), enhances this interaction and thus transactivation of the basal machinery by HIF-1α. Thioredoxin and Ref-1 are redox factors that are dithiol hydrogen donors for a variety of target proteins. Critical cysteines in these proteins undergo reversible oxidation–reduction reactions catalyzed by reductases. Along with glutathione, these proteins constitute the major cellular reducing systems.

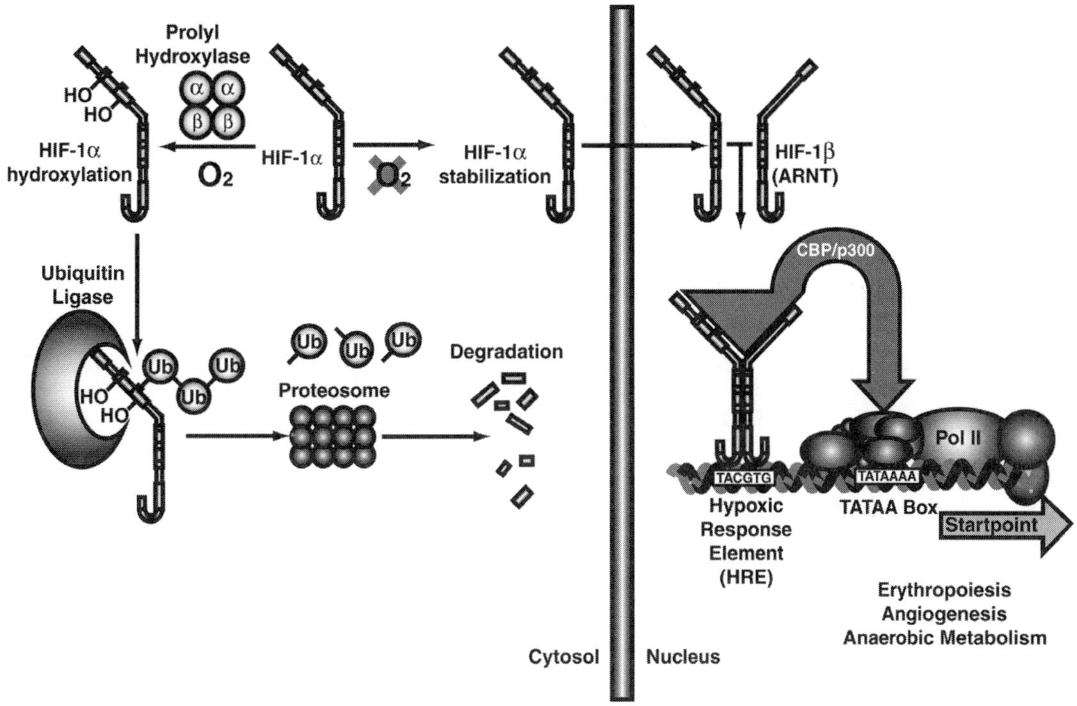

Figure 6.11 Oxygen-dependent degradation of HIF-1α. Oxygen-dependent hydroxylation of HIF-1α by prolyl hydroxylases results in its recognition by ubiquitin ligases. Polyubiquitination of HIF-1α by ubiquitin ligases targets it for proteolytic degradation by the proteosome. In the absence of oxygen (hypoxia), HIF-1 is not hydroxylated and is stable, becoming available for (a) transport to the nuclear compartment, (b) dimerization with ARNT, and (c) binding to HREs and activating gene transcription. Transactivation of the basal transcription machinery is through the coactivator CBP/p300, which interacts with the transactivation domains of HIF-1. Genes that are up-regulated by HIF-1 under low oxygen conditions are those involved in the production of red blood cells (erythropoiesis), the formation of new blood vessels (angiogenesis), and an enhancement of the capacity anaerobic energy production.

To a lesser extent, HIF-1 may also be regulated by protein phosphorylation. This mechanism may regulate HIF-1 in response to growth factor stimulation. Previous studies have shown that the p42/p44 (Erk2/Erk1) mitogen-activated protein kinases (MAPKs) can phosphorylate HIF-1α and enhance the transcriptional activity of HIF-1. However, HIF-1α stabilization under normoxic conditions is independent of MAPK activation. Since growth factors have been shown to recruit the p42/p44/MAPKs via the tyrosine kinase 3 Ras/Raf 3 MEK pathway, this may explain the response of HIF-1 to growth factor stimulation. However, the mechanism by which growth factors stabilize HIF-1α under normoxic conditions still remains unknown.

Activation of the PI3-kinase, Akt1 (also known as protein kinase B), has also been shown to activate HIF-1. Phosphorylation of glycogen synthase kinase 3, which inhibits the translation of HIF-1α protein, by Akt1 may inhibit the function of this kinase and thereby activate HIF-1α translation. Akt1 works via the receptor tyrosine kinase/PI3–kinase/PTEN/Akt kinase pathway. Akt1 is activated in response to various hormones, growth factors, and extracellular matrix components and has been shown to inhibit apoptosis. It has previously been shown that a decrease of HIF-1α in a cell occurs during apoptosis. The effects of HIF-1α on apoptosis may therefore not be a direct effect, but may be a concomitant effect of Akt1 stimulation.

Nuclear Factor-κB A transcription factor, NF-κB is involved in the activation of genes involved in acute phase reactions and the immune response (Fig. 6.12). It is the major transcription factor involved in the up-regulation of stress-induced genes and is activated by factors including the cytokines tumor necrosis factor (TNF) and interleukin-1 (IL-1), phorbol esters, lipopolysaccharides (LPS), double-stranded RNA, protein synthesis inhibitors, ultraviolet (UV) and ionizing radiation, viral transactivator proteins, and reactive oxygen species (ROS). It also activates

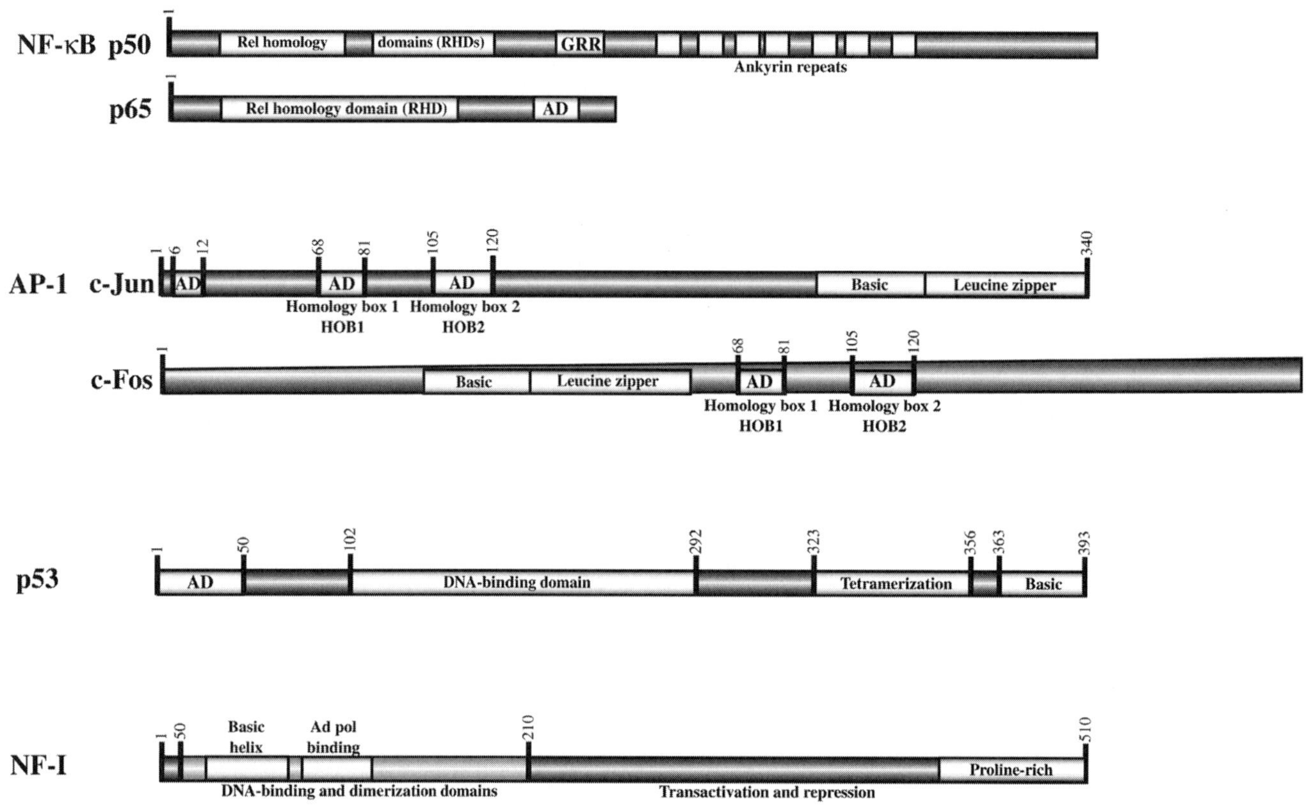

Figure 6.12 The protein domains of NF-κB (p50 and p65 subunits), AP-1 (c-Fos and c-Jun subunits), p53, and NF-I. The Rel homology domains of NF-κB are shown. Ankyrin repeats (association domain of I-κB) in p50 are presented. AP-1 contains homology boxes and a bZip DNA-binding and dimerization domain. p53 contains a region for tetramer formation. GRR, glycine-rich region; AD, activation domain. The adenovirus DNA polymerase interaction domain contained in NF-I is shown.

immunoglobulin κ genes in B lymphocytes. NF-κB is known as the *central mediator* of the human immune response. The NF-κB proteins belong to the Rel family of transcription factors that include five known members: p50(p105), p52(p100), p65 (currently known as RelA), c-Rel, and RelB. The subunits of this family combine to form both heterodimers and homodimers, the most common being NF-κB, which is a p50/p65 heterodimer. The p50 subunits contain only DNA-binding elements, whereas p65 subunits contain both DNA-binding and transactivation domains. The Rel proteins contain a highly conserved Rel homology region (RHR) of about 300 amino acids composed of two immunoglobulin (Ig)-like domains. The RHR contains the dimerization domain, DNA-binding domain, and transactivation domain, as well as the NLS and the site for interaction with its inhibitor protein, I-κB. The DNA-binding portion of the Rel proteins consists of 10 flexible loops extending from the secondary structure of the Ig-like domains. Other proteins that have been found that interact with NF-κB include high mobility group protein (I)Y and Sp1. It can also interact with the coactivator SRC-1. Interestingly, NF-κB up-regulates the transcription of I-κB, thereby providing its own feedback control.

NF-κB is an example of a transcription factor that is regulated by the binding of an inhibitor. Activation of NF-κB by stimulatory factors involves the dissociation of the inhibitory subunit I-κB. I-κB keeps NF-κB sequestered in the cytoplasm, but when I-κB dissociates the NF-κB p50/p65 dimer can travel to the nucleus. Dissociation is triggered by phosphorylation of I-κB by an I-κB kinase (IKK), and the protein is then ubiquitinated and broken down by the 26S proteosome (see Fig. 5.11). I-κB has multiple copies of a 30-amino-acid structure known as an ankyrin repeat. Ankyrin repeats are responsible for most, but not all, of the interaction of I-κB with NF-κB. Once released and translocated to the nucleus, NF-κB binds to a decameric enhancer element (5′-GGGRNYYCC-3′

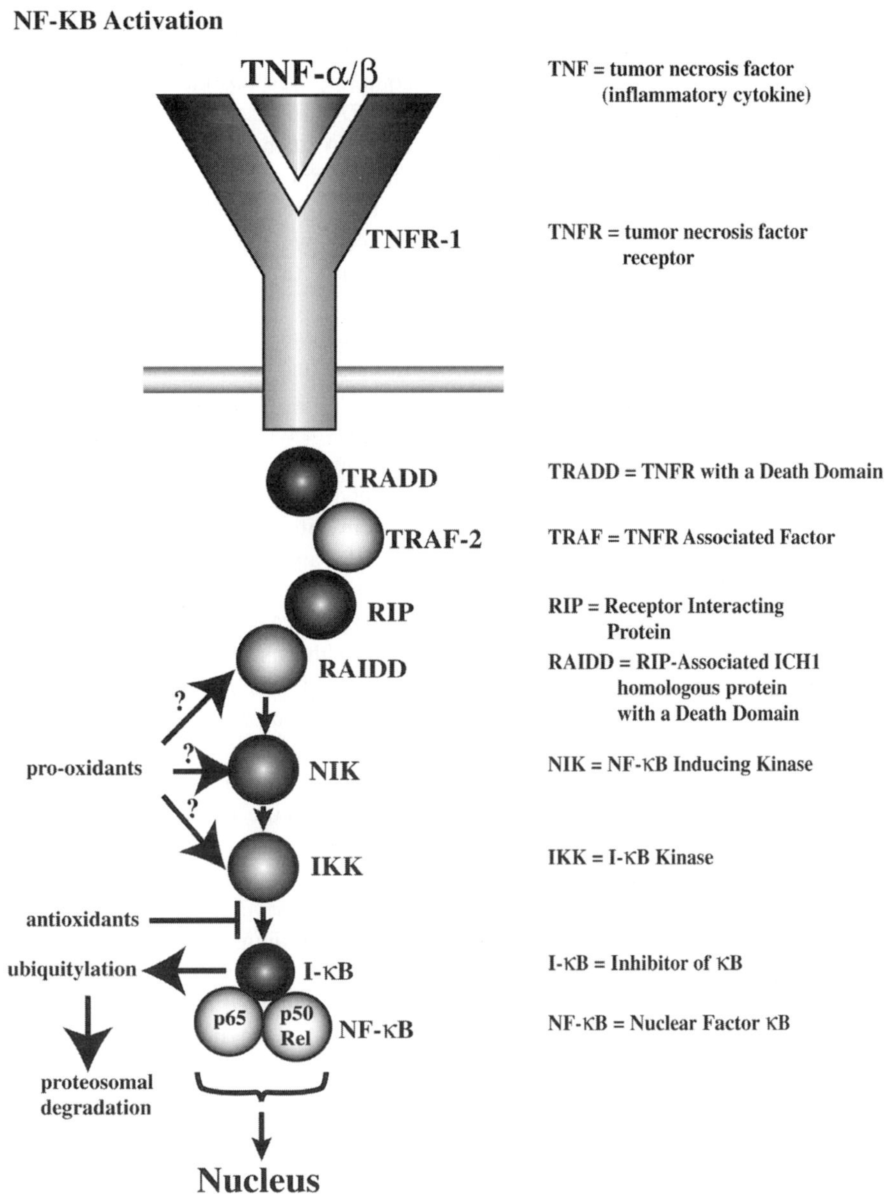

Figure 6.13 The activation pathway of NF-κB by TNF-$\alpha\beta$. The end result of the activation of this pathway is the phosphorylation of I-κB, the inhibitory subunit of NF-κB, and its subsequent release, ubiquitination, and degradation by the proteosome. NF-κB is then free to enter the nucleus to stimulate gene targets. Potential upstream kinase targets of oxidative stress are indicated.

where R = purine, N = any nucleotide, and Y = pyrimidine). It interacts directly with TFIID and TFIIB through an acidic transactivation domain contained on p65.

NF-κB is regulated by redox in two different ways. First, oxidative stress induces NF-κB translocation to the nucleus without the degradation of I-κB. Increased levels of hydrogen peroxide induces phosphorylation of I-κB at Ser32 and Ser36, causing it to dissociate from NF-κB. I-κB is not the direct target of oxidation itself as I-κB cannot be dissociated from NF-κB *in vitro* by direct treatment with hydrogen peroxide. However, it is possible that oxidation activates one of the upstream kinases involved in the phosphorylation of I-κB (Fig. 6.13). Second, a critical cysteine (Cys61) in the DNA-binding domain of the p50 subunit of NF-κB can be oxidized. Oxidation of this cysteine seems to decrease the DNA-binding activity of NF-κB *in vitro* because oxidized glutathione (GSSG) decreases binding, whereas proteins with reducing power such as thioredoxin and Ref-1 increase it. Thioredoxin can associate directly with the p50 subunit of NF-κB. Thus, NF-κB seems to be a factor that is both activated and repressed by oxidative conditions. It remains to be determined which effects play a more important role *in vivo*.

Activator Protein 1 (AP1) Activator protein 1 is a heterodimer of proteins from the Jun (Jun, Jun-B, and Jun-D) and Fos (Fos, Fra-1, Fra-2, and Fos-B) families (Fig. 6.12) (see also Chapter 5). In contrast to NF-κB, AP1 is located entirely in the nucleus. AP1 regulates cell proliferation, differentiation, and transformation and has also been implicated in apoptotic cell death. Genes that are controlled by AP1 include those that encode for metallothionein, collagenase, stromelysin, transforming growth factor β1, interleukin-2, and other cytokine genes. Jun proteins consist of 340 amino acids and contain N-terminal transactivation domains and a C-terminal bZIP domain. Fos proteins consist of 381 amino acids and contain bZIP and transactivation domains in the center of the protein. AP1 heterodimers bind to the consensus sequence 5′-TGAGCTCA-3′ known as the 12-*O*-tetradecanoyl-phorbol-13-acetate (a phorbol ester tumor promoter) response element, or TRE. The activation domains of Jun and Fos are divided into two homology box motifs (HOB1 and HOB2), which cooperate to activate transcription. Jun also contains domains that directly interact with TFIID (TBP, TAF250) and TFIIB. Other activation domains on Fos include an N-terminal and C-terminal activation motif. These activation motifs may interact with TFIID and TFIIB. AP1 can interact with the coactivators SRC-1 and Jun activation domain-binding protein 1 (JAB1).

Like NF-κB, AP1 is a factor that is both induced and inhibited by oxidative stress. This includes both transcriptional and posttranscriptional mechanisms. Transcription of both *c-jun* and *c-fos* is rapidly induced by hydrogen peroxide and other prooxidants. At the posttranslational level, the DNA-binding function of AP1 is reduced by the oxidation of redox-sensitive cysteines. Both cysteines are in the DNA-binding domains of the two subunits: Cys252 in Jun and Cys154 in Fos. Redox factors such as thioredoxin and Ref-1 are able to reduce these critical cysteines and restore the DNA-binding activity of AP-1. The latter protein has been shown to interact directly with Cys252 of the Jun subunit. AP-1 is also strongly activated by hypoxia.

p53 *p53* is a tumor suppressor gene that inhibits tumor cell growth and transformation (Fig. 6.12). It induces apoptosis and is known as the "guardian of the genome" as it blocks cell cycle and gene amplification. The *p53* gene is the most frequently mutated gene associated with human cancers. p53 is a 393-amino-acid nuclear phosphoprotein with an N-terminal transactivation domain, a C-terminal oligomerization domain, and a central DNA-binding domain. It binds to the consensus sequence 5′-RRRC(A/T)(T/A)GYYY-3′ as a tetramer, unlike most other transcription factors that bind to DNA as dimers. A substantial number of genes are up-regulated by p53 response elements. These include *mdm-2* (part of a ubiquitin ligase complex responsible for the degradation of p53), *GADD45* (leads to cell cycle arrest at G1 phase in response to DNA damage), cyclin G, *bax* (promotes cell death through apoptosis), and insulinlike growth factor binding protein 3 gene (IGF-BP3, an antimitogenic protein that is induced by DNA damage). p53 has three main types of domains: (a) N-terminal activation domains (AD1 and AD2) that can be phosphorylated by double-stranded DNA-activated protein kinase (DNA-PK) or casein kinase 1 (CKI) and interact with TFIID (TBP, TAF40, TAF60) as well as TFIIH (p62), (b) a C-terminal domain that includes a tetramerization domain, a basic region, a nonspecific DNA interaction domain, and three NLS that can be phosphorylated by cyclin-dependent kinase (CDK), protein kinase C, and casein kinase 2 (CKII) for NLSI, NLSII, and NLSIII, respectively, and (c) the basic domain at the C terminus contains a central sequence-specific DNA-binding domain and four highly conserved regions that are responsible for making contacts with the major and minor grooves of the p53 response element. The less conserved regions make up a β sandwich that forms a scaffold to support the DNA-binding elements. A proline-rich domain occurs between the activation domain and the specific DNA-binding domain.

p53 is regulated by both hypoxia and redox. Twelve cysteines within the DNA-binding domain can be oxidized, resulting in several p53 conformations that do not function in DNA binding. Cysteine-to-serine mutations show that 4 of the 12 are involved in direct DNA binding. Another three are involved in zinc finger coordination. Thioredoxin and Ref-1 have been shown to restore both the DNA-binding and transactivation functions of p53. Altered p53 tumor suppressor function may provide a link between oxidative stress and carcinogenesis.

Nuclear Factor I (NF-I) Also known as CCAAT transcription factor (CTF), NF-I is a transcription factor involved in the activation of DNA replication and the expression of a large number of cellular and viral genes (Fig. 6.12). NF-I up-regulates the cytochrome P450 1A1 (*CYP1A1*) gene in response to the environmental contaminants, dioxin, and benzo[*a*]pyrene. There are at least four genes (NF-I-A, B, C, X) encoding these factors and different isoforms of proteins arising from the NF-I genes due to alternative mRNA splicing (see next chapter). These isoforms can combine to form homo- and heterodimers that are DNA binding. These ubiquitous activators bind to the palindromic binding sequence 5′-TTGGCNNNNNGCCAA-3′. NF-I can also bind to half-sites (5′-TTGGC-3′) but with reduced affinity. The typical NF-I protein consists of an N-terminal DNA binding and dimerization domain and C-terminal transacti-

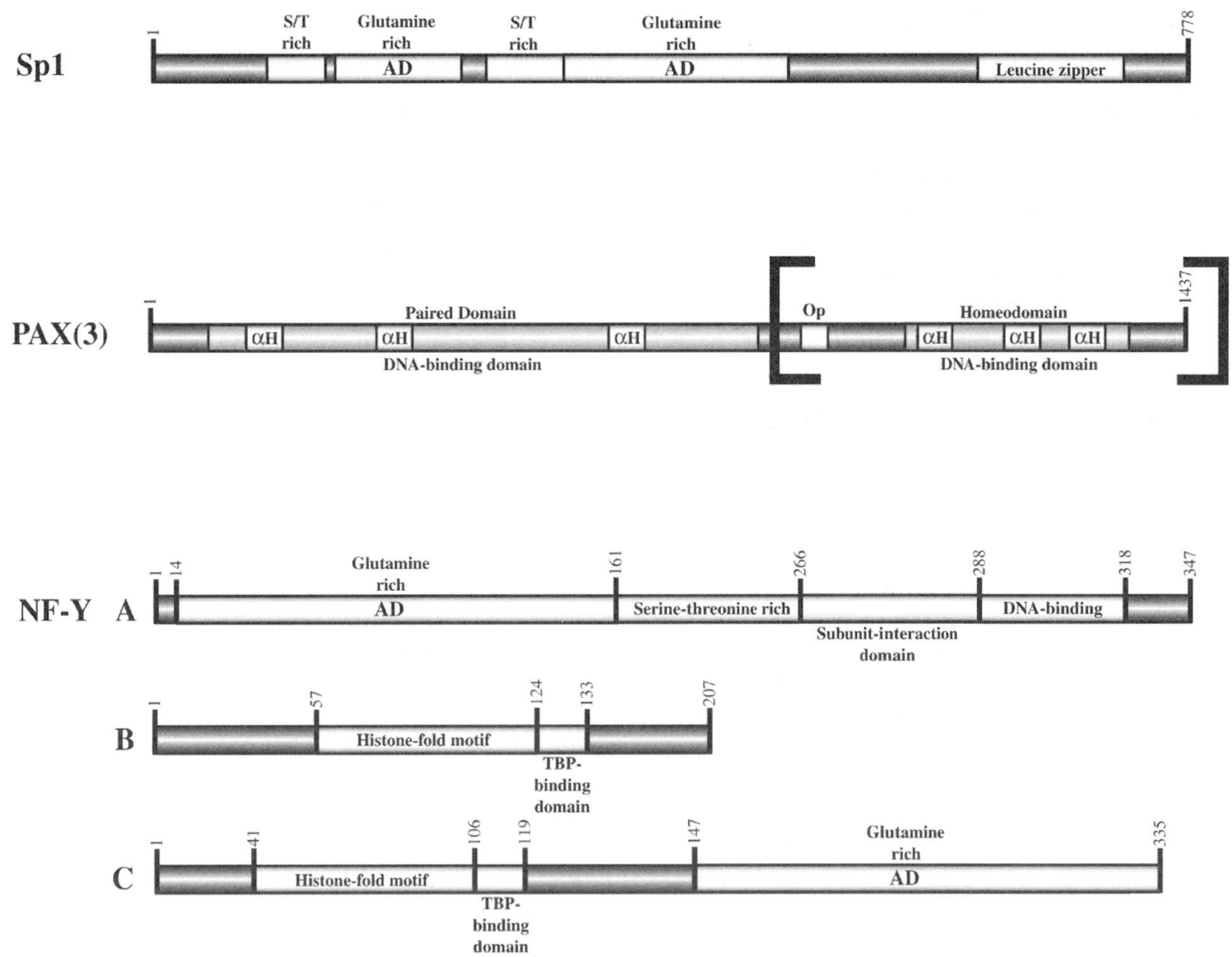

Figure 6.14 The protein domains of NF-I, Sp1, PAX3, and NF-Y (A, B, and C subunits). The glutamine- and serine/threonine-rich transactivation domains of Sp1 are shown. Of the six forms of PAX proteins, PAX3 is the representative form. Other forms of PAX (PAX1, 2, 4, 5, and 6) vary in the domain outlined in brackets. The HFM of NF-Y allow for dimerization. αH, alpha helix; AD, activation domain; TBP, TATA-box binding protein.

vation and/or transrepression domains. The NF-I DNA-binding domain does not have sequence homology with any other known DNA-binding domains in other transcription factors. It is a highly basic α-helical domain. The C-terminus portion of the activation domain of NF-I is proline-rich and interacts with TBP as well as TFIIB. Repression may result from competition for the same DNA-binding sites as other activators. NF-I competes for the same binding site as Sp1 (see below) to inhibit its activation. The mechanism by which the C-terminal repression domains of NF-I work is currently unknown.

Oxidative conditions affect both NF-I DNA binding and transactivation functions. The DNA-binding domain of NF-I contains four cysteines, three of which are required for DNA interaction. Oxidation of these cysteines by hydrogen peroxide treatment or glutathione depletion abolishes DNA binding while treatment with glutaredoxin 1 enhances it. Hydrogen peroxide also represses transactivation from the NF-I transactivation domain when tested with a fusion protein between the Gal4 DNA-binding domain and the NF-I transactivation domain. A critical cysteine in this domain is key to this inactivation.

Specificity Factor 1 (Sp1) Specificity factor 1 is a transcription factor that binds to GC boxes present upstream from a wide range of promoters (Fig. 6.14). Sp1 was first identified as activating transcription from multiple GC boxes in the simian virus 40 (SV40) promoter. It plays a vital role in TATA-less housekeeping gene transcription. It activates the transcription of the genes for aldolase A, pyruvate kinase M2, β-enolase, and dihydrofolate reductase. Sp1 binds to the GC box consensus sequence

5′-TGGGCGGGGC-3′ through three C-terminal zinc-finger motifs (part of the C2H2 zinc finger family of proteins). It contains two glutamine-rich domains in the N-terminal portion of the protein, each associated with a serine/threonine-rich region. The glutamine-rich domains act as strong activation domains and interact with TBP, TAF55, TAF110, and TAF130. The second glutamine-rich region has been shown to be involved in the homomultimerization of this transcription factor. Sp1 is known to be both phosphorylated and glycosylated leading up to its multimerization. Sp1 has overlapping binding sites with other transcription factors, including the zinc finger Kruppel-like factors.

Oxidation affects cysteines in the C2H2 DNA-binding zinc fingers of Sp1. DNA binding of Sp1 is reduced or increased in the presence of oxidizing (such as hydrogen peroxide) or reducing (such as dithiothreitol) agents, respectively. Zinc coordination as well as DNA binding protects Sp1 from oxidative stress. This is particularly important to consider when using reporter plasmids that rely upon promoters and enhancers that contain Sp1 sites (such as simian virus 40 or SV40). Treatments that produce oxidative conditions would have a profound effect on the results using such reporters.

Paired Box (PAX) The paired (PAI and RED) box family containing genes (PAX) is comprised of nine members of transcription factors that function as master regulatory proteins during embryogenesis and organogenesis (Fig. 6.14). They determine cell fate as well as complete structures. They also play a role in oncogenesis as they repress the transcription of the *p53* gene. Repression of transcription by PAX protein may result from squelching as non-DNA-binding forms of PAX can also repress transcription. The PAX proteins are characterized by a conserved DNA-binding motif, a paired domain originally described in the *Drosophila* protein "paired" and an octapeptide motif involved in protein–protein interaction. The 128-amino-acid paired domain consists of two distinct helix–turn–helix subdomains, PAI and RED, both of which bind to DNA. PAX3, PAX4, PAX6, and PAX7 contain a second DNA-binding domain of the paired homeobox type. PAX2, PAX5, and PAX8 contain only remnants of this paired homeobox domain. PAX1 and PAX9 do not contain this second domain. The paired domain is sufficient to mediate sequence-specific DNA binding. The homeobox domain may or may not cooperate with the paired domain in DNA binding. The activation domain of PAX proteins is similar to the Pro-Ser/Thr-rich domains of other transcriptional activators, such as Oct-2 and CTF-1.

The PAI subdomains of the paired domains of PAX5 and PAX8 are redox-sensitive. The formation of a disulfide bridge between two cysteines in this subdomain blocks the DNA binding of these proteins. The RED subdomain is insensitive to redox changes. Ref-1 has been found to control the redox state of the PAI subdomain of PAX5. It remains to be demonstrated whether other PAX proteins are sensitive to oxidation in the same manner.

Nuclear Factor for Y Box Protein (NF-Y) Also termed CCAAT-binding protein (CBP), NF-Y is a heterotrimeric transcription factor, composed of three different subunits, that plays a role in cell senescence (Fig. 6.14). It up-regulates the genes involved in the late G1/S transition of the cell cycle: thymidine kinase, dihydrofolate reductase, and E2F1. NF-Y may recruit histone acetylases [hGCN5 and p300/CBP-associated factor (PCAF)] or deacetylases and therefore may also play a role in chromatin remodeling. It has been shown that NF-YB and NF-YC interact directly with TBP and TAFs (TFIID) *in vitro*. There are three NF-Y subunits, A, B, and C. NF-YA subunits are either 318 or 347 amino acids long and contain a N-terminal glutamine-rich activation domain, a central serine/threonine-rich region, and a C-terminal subunit interaction domain and DNA-binding domain. The DNA-binding domain contains a bipartate NLS and is similar in structure to other CCAAT box-binding proteins. Several factors can recognize CCAAT boxes but most are activated by NF-Y. All three of the NF-Y subunits are required for proper DNA binding. NF-YA can exist in two different forms due to an alternative mRNA splice site, resulting in long and short isoforms of this protein. NF-YB subunits are 207 amino acids long and contain a central histone-fold motif and a TBP-binding domain. NF-YC subunits are 335 amino acids long and contain a N-terminal histone-fold motif (HFM) and TBP-binding domain and a C-terminal glutamine-rich activation domain. The glutamine-rich activation domains are similar to those found in Sp1. The HFMs are similar in sequence to those found in histones H2B and H2A and are composed of three α helices separated by two linker regions. The HFMs allow NF-YB and NF-YC to form a dimer prior to recruiting NF-YA. HMFs may also allow NF-Y to associate with histones themselves. NF-Y binds to the CCAAT box consensus sequence 5′-CTGATTGGYYRR-3′ or 5′-YYRRCCAATCAG-3′ present in the promoters of many constitutive, inducible, and cell-cycle-dependent eukaryotic genes. This conserved motif, known as the Y box, was first identified in the promoter of the major histocompatability complex class II (MHCII) gene. NF-Y has been reported to interact with other transcription factors resulting in the stabilization the DNA-binding potentials of both factors; such partners include Sp1, SREBP1, and RF-X.

Control of NF-Y occurs at multiple levels and involves NF-Y nuclear localization, alternative splicing of NF-YA, and tissue-specific expression of subunits (plants). Nuclear localization of NF-Y increases during the midblastula stage of development in *Xenopus* embryos. The mechanism

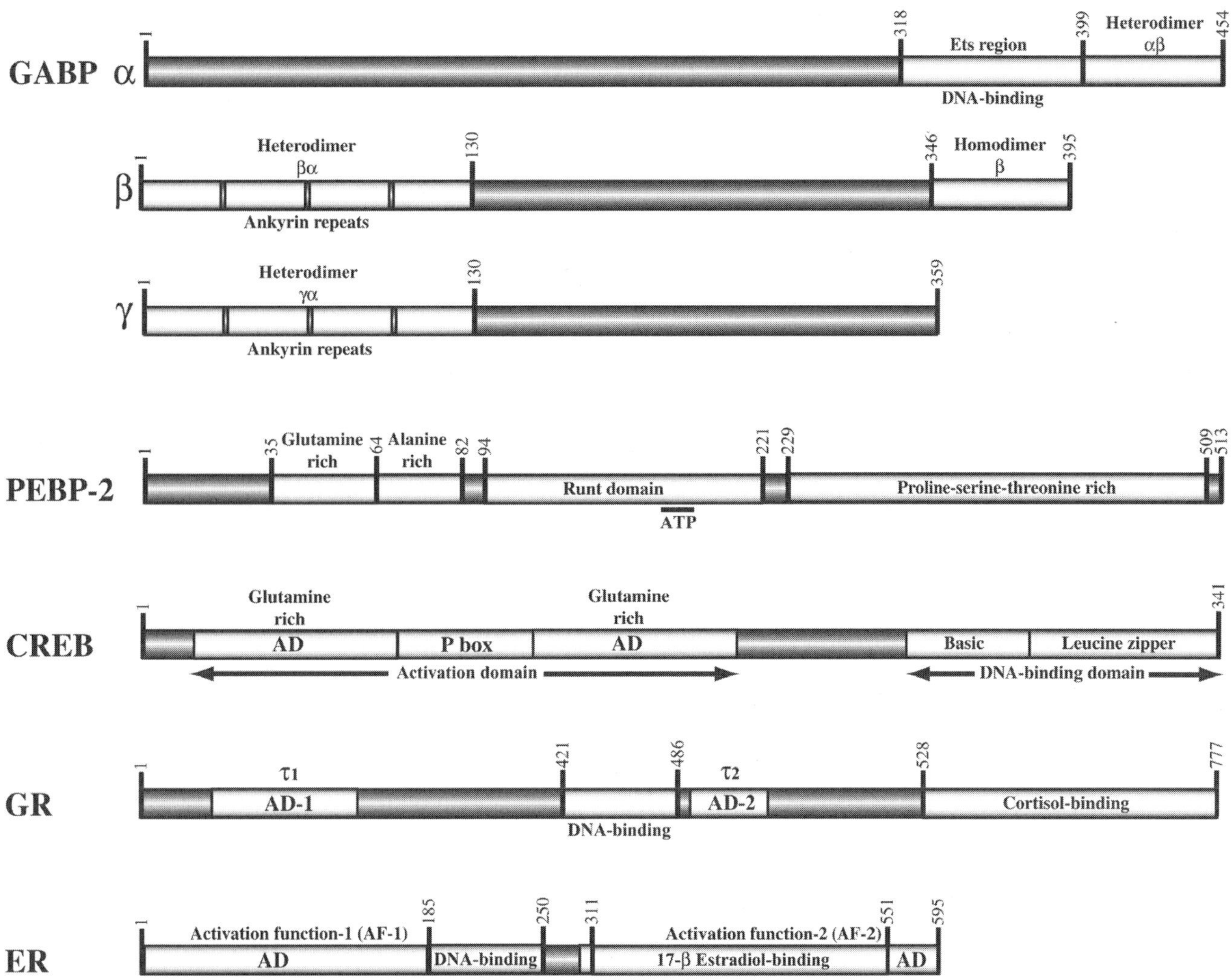

Figure 6.15 The protein domains of GABP (α, β, and γ subunits), PEBP-2, CREB, GR, and ER. The heterodimer and homodimer domains of the GABP subunits are indicated, as well as the characteristic Ets DNA-binding domain of the α subunit. The runt homology domain of PEBP-2 contains an ATP-binding site. The phosphorylation box (P box) of CREB is presented. The steroid (ligand) binding sites of GR and ER are shown in gray. GR contains two (τ_1 and τ_2) activation domains (AD).

of this is not known. The two forms of NF-YA, formed from alternative splicing, are somewhat tissue specific. Both forms are found in all tissues, however, fibroblasts contain predominantly long and lymphocytes contain predominantly short NF-YA. In plants, many different isoforms of the subunits exist. Some isoforms are only found in seedlings, whereas others are found in the mature plant.

Oxidation of two cysteines in the NF-YB subunit, Cys85 and Cys89, abolishes NF-Y assembly and DNA binding. These cysteines occur in the HFM and may affect the initial dimerization of NF-YB with NF-YC. Mutation of these cysteines to serine or the inclusion of Ref-1 restores the assembly and DNA-binding activities of NF-Y.

GA-Binding Protein (GABP) Specific transcription factors are involved in the regulation of metabolism and mitochondrial function (for a description of the regulation of mitochondrial gene expression, see Text Box 6.1). The Ets-related transcription factors are ubiquitous heteromeric transcription factors implicated in the regulation of several (nuclear-encoded) genes involved in mitochondrial energy metabolism including subunits of cytochrome *c* oxidase, ATP synthase, and mitochondrial transcription factor 1 (mtTF1) (Fig. 6.15). Members of the Ets family include nuclear respiratory factor 2 (Nrf-2) or growth-associated binding protein (GABP). The Ets proteins have a characteristic DNA-binding motif and are named after the first protein in which this binding motif was found; the avian erythroblastosis virus E26 oncogene or E twenty six (Ets). GABP, one member of this family, is a strong positive regulator of genes that encode ribosomal proteins. This factor binds to the duplicated Ets regulatory sequence of 5′-GTTCCCGGAAG-3′. There are three known subunits of GABP; α, β, and γ. The α subunit contains a C-terminal DNA-binding domain that is found in all Ets protein family members. This domain is a "winged helix–turn–helix" (also called the *forkhead* motif) and allows the protein to bind to DNA as monomers. The first α helix is separated from the other two by two β sheets. Binding with partner proteins greatly enhances the DNA specificity of the DNA-binding subunit. Two arginines and a tyrosine present in the conserved portions (third α helix) of the Ets domains are responsible for making hydrogen bond contacts with the GGA sequence of the binding motifs present in the major grooves of the DNA helix. Binding causes the DNA to bend and curve around the Ets motif. The α subunit also contains a C-terminal heterodimerization domain for GABPβ. GABPβ and GABPγ contain N-terminal heterodimerization motifs with GABPα. The β subunit contains a C-terminal homodimerization domain made up of a leucine zipper. The heterodimerization domain of the β subunit contains a series of ankyrin repeats whose loops fit into a depression in GABPα resulting in $\alpha\beta$ dimerization.

Oxidative stress inactivates both DNA binding and dimerization of GABP. Sulfhydryl modification of Cys388 and Cys401 inhibits DNA binding of GABPα, whereas modification of Cys421 affects GABPs dimerization with GABPβ. Oxidative inactivation of GABP may provide a link between ROS overproduction by the mitochondria (e.g., during reperfusion after ischemia) and the down-regulation of synthesis of mitochondrial electron transport chain components.

Polyoma Virus Enhancer Binding Protein 2 (PEBP-2) Also known as the core-binding factor (CBF), PEBP-2 is a transcription factor involved in the regulation of lymphoid cell-specific genes (Fig. 6.15). These genes include the mRNAs encoding for polyadenovirus, murine leukemia virus, T-cell receptors (α, β, γ, δ, ε) CD3, myeloperoxidase, neutrophil elastase, granulocyte/macrophage colony-stimulating factor, macrophage colony-stimulating factor receptors, interleukin-3, and granzyme B. It is composed of α and β subunits. Three mammalian genes encode for α subunits (αA, αB, and αC) but only one gene encodes for the β subunit. The α protein has high homology to the *Drosophila* runt protein. The α subunit binds to the consensus DNA sequence 5′-(R/T)ACCRCA-3′. The β subunit does not make contact with DNA but enhances the DNA binding of the α subunit through allosteric interactions. PEBP-2α subunits have very highly conserved N- and C-terminal domains, a central runt homology region, and a C-terminal proline-, serine-, and threonine-rich region (PST region). The runt domains contain a conserved ATP-binding site (GRSGRGKS). PEBP-2αA also contains N-terminal glutamine and alanine stretches just prior to the runt domain. Both the DNA binding and the heterodimerization functions are associated with the runt domain of the α subunits. PEBP-2β subunits only contain a N-terminal runt domain involved in dimerization. Alternative splice forms (see next chapter) of the β subunit exist.

The α subunit of PEBP-2 is a nuclear protein, whereas the β subunit is transported to the nucleus from the cytoplasm. Conserved cysteines, Cys115 and Cys124, present in the runt domain of PEBP-2α can be oxidized. Oxidation affects the ability of PEBP-2α to dimerize with the β subunit. Thioredoxin and Ref-1 act synergistically for activation of the runt domain. Surprisingly, the β subunit further enhances the activity of these proteins and protects the α subunit from oxidation. Oxidative conditions may affect nuclear trafficking of the β subunit to activate transcription of PEBP-2 associated genes.

cAMP Response Element Binding (CREB) Protein The CREB protein is a transcription factor that is regulated by phosphorylation in response to cyclic AMP, Ca^{2+}, growth factors, and inflammatory cytokines (Fig. 6.15). CREB has a role in many physiological systems including long-term memory, circadian rhythms, pituitary function, spermatogenesis, and intermediary metabolism (see Fig. 10.3). Phosphorylation of CREB results in its activation and the recruitment of the coactivator CBP that interacts with the basal transcription machinery. Many kinases can phosphorylate CREB in response to different stimuli. These include cyclic-AMP-dependent protein kinase (PKA), Ca^{2+}/calmodulin-dependent kinase IV, p70 S6 kinase, ribosomal S6 kinase 2, MAPK, and MAPK-activated protein kinase 2 (MAPKAPK-2). CREB contains a N-terminal activation domain consisting of two glutamine-rich domains surrounding a phosphorylation box (P-box or kinase-inducible domain). Phosphorylation of a key serine in this region, Ser133, turns this protein into a powerful activator. It has a bZIP DNA-binding domain at the C terminus that binds to palindromic consensus sequences of 5′-TGACGTCA-3′ (the cAMP-responsive element or CRE). It forms homodimers and heterodimers with other proteins in this class, including cAMP response element modulator (CREM) and activating transcription factor 1 (ATF-1). CREB is the focal point of many diversified signaling pathways involved in adaptation to cellular stress.

Hypoxia results in the phosphorylation of Ser133 and the activation of CREB, whereas oxidative stress reduces this phosphorylation. Pretreatment of cells with hydrogen peroxide decreases the stimulatory effects of nerve growth factor or epidermal growth factor on CREB.

Redox factor 1 (Ref-1, see below) enhances DNA binding of CREB. The Ref-1 promoter also contains a CRE binding site shown to up-regulate Ref-1 in response to oxidative stress. In this case, however, the factor that binds is a c-Jun/ATF-2 heterodimer.

Glucocorticoid Receptor (GR) The GR belongs to the family of ligand-binding (steroid-responsive) nuclear receptors (Fig. 6.15). It is involved in many physiological processes including carbohydrate, protein, and lipid metabolism, and modulation of immune and central nervous system (CNS) responses (see Fig. 10.5). Glucocorticoids promote gluconeogenesis and the formation of glycogen and enhance the degradation of fats and proteins. The GR is also required for the maturation of several organ systems (lung and adrenal gland). The binding of ligand triggers a conformational change that increases the receptor's affinity for DNA. The GR binds as a dimer to a palindromic DNA consensus sequence known as the glucocorticoid response element (GRE) composed of two half-sites (5′-AGAACA-3′) (Fig. 6.4). Each monomer of the dimer recognizes one half-site. The receptors for mineralocorticoids and progesterone also bind to this site. The N-terminal portion of GR contains an immunogenic domain that increases the transcriptional response (termed τ_1). The central DNA-binding domain is composed of two zinc fingers. This domain also contains the main transcriptional activation domain (termed τ_2 or DNA- for the combined region). The C terminus is the steroid-binding domain. This also forms the homodimerization domain of the GR. Surprisingly, deletion of this domain results in a constitutively active DNA-binding receptor. It is suggested that this domain inhibits the DNA-binding domain until the binding of ligand (cortisol). The binding of ligand to GR causes it to translocate to the nucleus. The binding of GR to the GRE causes alteration of the nucleosomal architecture, causing the DNA to unwind from histones. This allows for the binding of other transcription factors and the formation of the preinitiation complex. GR has also been shown to interfere with the activities of CREB, AP1, and NF-κB. In the case of NF-κB, a direct interaction between the two transcription factors has been shown. Therefore, while GR activates transcription of genes involved in the steroid response, it represses transcription of genes controlled by other factors.

Oxidative stress can alter hormonal regulation. The GR is an example of a nuclear receptor that is modified by oxidation. Two cysteines in the DNA-binding domain of the GR can be oxidized by hydrogen peroxide and reduced by dithiothreitol. Oxidation of these cysteines causes both a decrease in ligand-binding activity as well as DNA binding. Oxidation also impairs ligand-dependent and -independent translocation of the GR to the nucleus. This involves the oxidation of Cys481 present in the NLS of the GR. Thioredoxin (see below) associates with the GR DNA-binding domain in the nucleus to restore GR activity under conditions of oxidative stress. Together, this emphasizes that cellular glucocorticoid signaling is modulated by redox state.

Estrogen Receptor (ER) The ER belongs to the same family of steroid-binding nuclear receptors as GR (Fig. 6.15). Estrogen is required for the development of female secondary sex characteristics. The dark side of estrogen is that it is also involved in tumorigenesis in the female reproductive tract and breast. The ER plays a critical role in breast cancer and can be inhibited by an agonist known as tamoxifen. The marker gene for breast cancer, *sP2*, is up-regulated by the ER. There are α and β isoforms of the receptor. Like GR, it has a central DNA-binding domain and a C-terminal ligand-binding domain. The DNA-binding domain contains two zinc fingers. Also like GR, the ER binds as a dimer to a palindromic sequence known as the estrogen response element (ERE) composed of two half-sites (5′-AGGTCA-3′) (Fig. 6.4). The ligand bound is 17-β estradiol. ER has both N-terminal and C-terminal activation domains. These are termed activator functions 1 and 2 (AF-1 and AF-2). AF-1 is ligand-independent and can be phosphorylated by MAPK, whereas AF-2 is ligand-dependent. AF-2 also contains the receptor dimerization domain. ER binds to the coactivators SRC-1/TIF-2, TIF-1, ARA70, and CBP/p300. Both ER and GR interact with heat shock protein 90 (Hsp90). The binding of ligand dissociates this chaperone protein from the receptors, allowing them to dimerize and bind to hormone response elements (HREs; both GREs and EREs). ER can also activate transcription from AP-1 binding sites.

The ER function is also repressed by oxidative stress. Hydrogen peroxide inhibits *sP2* production in breast cancer cells by repressing the ER. Overexpression of thioredoxin or treatment with thiol reducing agents can restore hydrogen-peroxide-treated ER function. Inactivation of ER by oxidative conditions involves cysteines found in the DNA-binding region of the transcription factor. Oxidation of these cysteines induces a conformational change in the ER, resulting in the loss of α-helical structure, DNA binding, and transactivation capabilities.

FACTORS ACTIVATED BY OXIDATIVE CONDITIONS

Insight into transcription factors that are activated by oxidative stress come from studies on *Escherichia coli* and *Salmonella typhimurium*. The SoxRS and OxyR transcription factors are activated by oxidative stress. These proteins are activated at the posttranscriptional level by reactive

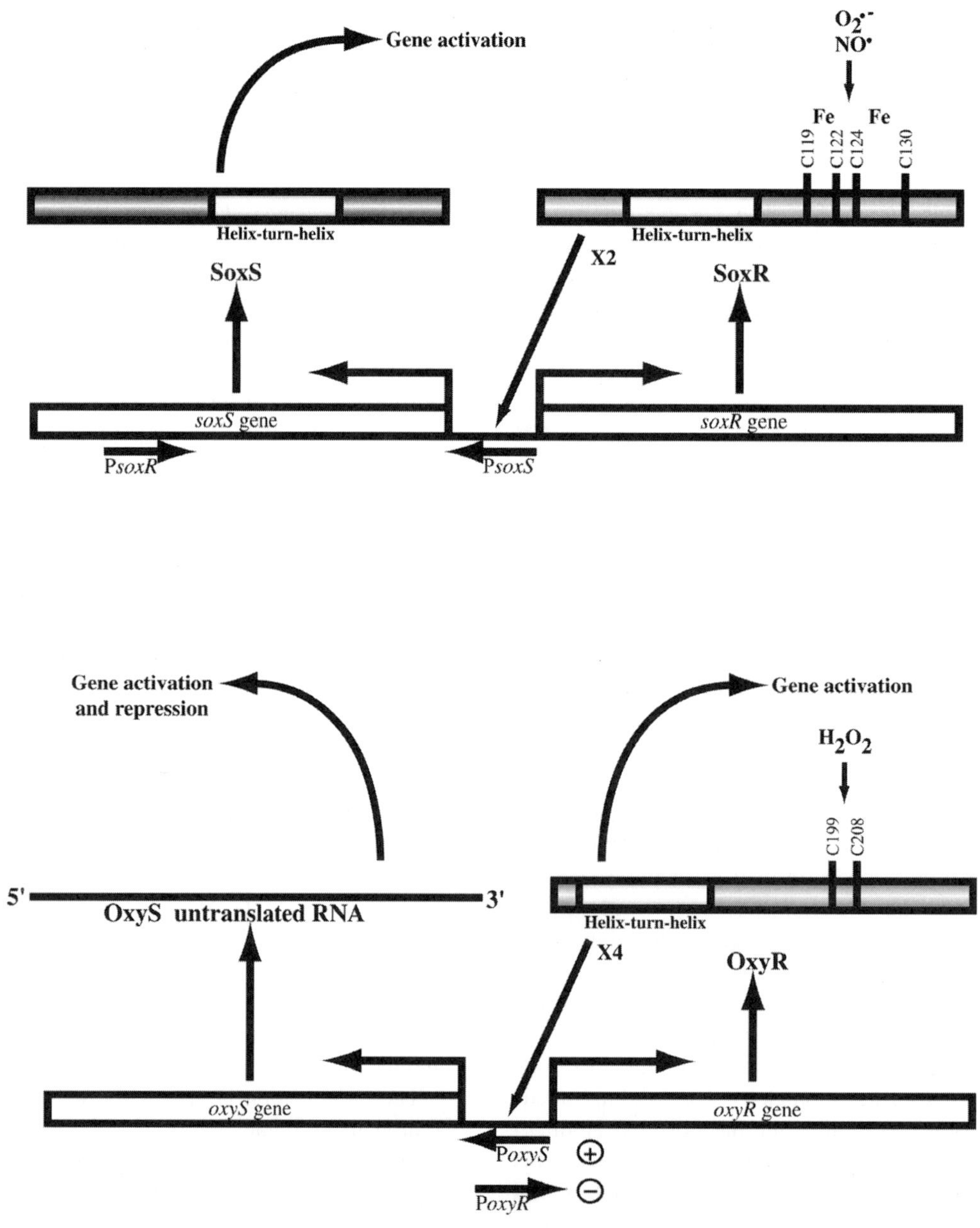

Figure 6.16 Activation of the SoxRS and OxyRS regulons in *E. coli*. The SoxR protein is a [2Fe–2S] cluster protein that responds to superoxide and nitric oxide. Activation involves oxidation of these irons that causes SoxR to dimerize and bind to the promoter (PsoxS) of SoxS. SoxS, in turn, activates genes involved in antioxidant and reducing power defenses. The iron-binding cysteines are shown. Critical cysteines in OxyR are oxidized by hydrogen peroxide, causing the protein to homotetramerize and bind to the promoters of various genes involved in peroxide metabolism as well as glutathione reduction. As well, OxyR up-regulates OxyS, an untranslatable mRNA that is involved in gene activation and repression. OxyS most likely acts as antisense RNA to block assembly of the basal machinery in certain promoters.

oxygen species (ROS). For more information on this topic see Chapter 12.

The SoxRS system is comprised of two proteins, SoxR and SoxS, that are activated in response to superoxide and to redox cycling agents (Fig. 6.16). Genes that are activated by the SoxRS system include manganese-dependent superoxide dismutase, the DNA-repair enzyme endonuclease IV, ferredoxin reductase, and the superoxide-resistant isozymes of fumarase and aconitase. SoxRS activation also up-regulates glucose-6-phosphate dehydrogenase, which produces the reduced nicotinamide adenine dinucleotide phosphate (NADPH) to enhance the reducing power of the cell, as well as the ferric uptake repressor (Fur repressor), which may result in decreased iron uptake into the cell and reduced hydroxyl radical production from free iron. SoxR is a homodimer that contains two active iron–sulfur clusters [2Fe–2S] per protein. When the iron–sulfur clusters are oxidized from $[2Fe{-}2S]^{1+}$ to $[2Fe{-}2S]^{2+}$ by superoxide, SoxR becomes active. SoxR can be reduced to its inactive form *in vitro* by dithionite. Nitric oxide has also been found to activate SoxR. Monothiols promote disassembly of the [2Fe–2S] cluster, whereas dithiols promote assembly. Both SoxS and SoxR contain helix–turn–helix DNA-binding motifs. Activation of SoxR by superoxide, nitric oxide, and other free radicals causes it to bind only to the *soxRS* regulon, the region of the DNA between the *SoxR* and *SoxS* genes that contains both of their promoters. The SoxS protein up-regulates the genes for antioxidant enzymes. SoxS binds to its own promoter to down-regulate its own transcription. The core sequence for the proposed SoxS binding sequence is 5′-ANNGCAY-3′. Activation of the basal transcription machinery involves interaction with the C-terminal domain of the subunit of RNA polymerase II.

The OxyR protein is activated in response to hydrogen peroxide (Fig. 6.16). Genes that are up-regulated by OxyR include catalase, alkyl hydroperoxide reductase, glutaredoxin 1, glutathione reductase, and the Fur repressor. The OxyR protein is a homotetramer. Oxidation by hydrogen peroxide of Cys199 and Cys208 residues activates OxyR-induced transcription. Reduction by glutaredoxin 1 or glutathione inactivates this transcription factor. OxyR has an unusually large helix–turn–helix DNA-binding site that makes contacts at four major grooves on one face of the DNA helix. The consensus DNA-binding sequence of the four repeats is 5′-ATAG-3′. OxyR activates transcription by recruiting pol II to the promoters. As with SoxS, it interacts with the C-terminal domain of the α subunit of pol II. OxyR also induces production of OxyS, an untranslated RNA that acts as both an activator and a repressor of genes in *E. coli*.

REDOX FACTORS

Since the function of most transcription factors is repressed under oxidative conditions (as seen above), it is favorable to keep them in a reduced state. The intracellular milieu of most cells is normally maintained in a reduced state due to high concentrations of intracellular thiols, the main one being glutathione. However, under oxidizing conditions, added reducing power may be required. Additional redox factors can interact specifically with key proteins that are required to be reduced in order to function properly. Redox factors, such as thioredoxin, glutaredoxin, protein disulfide isomerase, and redox factor 1 (Ref-1), interact with transcription factors to maintain them in a reduced state.

Thioredoxin is a family of small proteins that (a) are reduced by thioredoxin reductase in an NADPH-dependent fashion and (b) oxidize other proteins (Fig. 6.17). There are two forms of the protein: thioredoxin 1 is cytosolic and nuclear whereas thioredoxin 2 is mitochondrial. Thioredoxin stimulates cell growth and is an inhibitor of apoptosis. The protein reduces ribonucleotide reductase (required for the conversion of ribonucleotides to deoxyribonucleotides) as well as thioredoxin reductase (converts hydrogen peroxide to water). The two catalytic cysteines, Cys32 and Cys35, are located in the N terminus of all thioredoxins. Thioredoxin 2 has an N-terminal mitochondrial localization sequence (MLS). Thioredoxin forms homodimers as well as heterodimers with other proteins (e.g., Ref-1, p50 subunit of NF-κB). Thioredoxin reduces Ref-1, which, in turn, reduces other proteins. The interaction between p50 and thioredoxin occurs at the critical cysteine (Cys62) found in the DNA-binding region of the NF-κB subunit. A number of stimuli increase thioredoxin 1 expression in cells including hypoxia, lipopolysaccharides, hydrogen peroxide, phorbol esters, viral infection, UV radiation, and oxidative stress. Prooxidants (including hydrogen peroxide) also stimulate the posttranslational nuclear localization of thioredoxin.

A relative of thioredoxin is the protein glutaredoxin (Fig. 6.15). This protein can also reduce ribonucleotide reductase. The reducing power for glutaredoxin, however, comes from reduced glutathione (GSH), a cysteine-containing tripeptide that is abundant in the cell. Glutathione is reduced by NADPH-dependent glutathione reductase. The catalytic cysteines in glutaredoxin are Cys22 and Cys25. Glutaredoxin also regulates the activities of transcription factors such as NF-I, OxyR, and PEBP-2 (see above).

Protein disulfide isomerase (PDI) is a multifunctional protein that serves as a subunit of prolyl 4-hydroxylases and microsomal triglyceride transfer protein (Fig. 6.17). It also serves as a chaperone that assists in the folding of a number of newly synthesized proteins. It binds various peptides within the lumen of the endoplasmic reticulum. It has been reported to bind thyroid hormone, undergo ATP-dependent autophosphorylation, and have dehydroascorbate reductase activities. PDI belongs to the thioredoxin family of proteins and has two thioredoxin-like domains, both containing a CGHC motif: one at the N terminus of the protein and one close to the C terminus. The central portion of the protein contains two redox-inactive thioredoxin folds and the very C terminus of the protein contains a highly acidic region that may be involved in binding Ca^{2+}. A terminal KDEL motif is necessary for the retention of PDI in the endoplasmic reticulum. PDI catalyzes thiol–

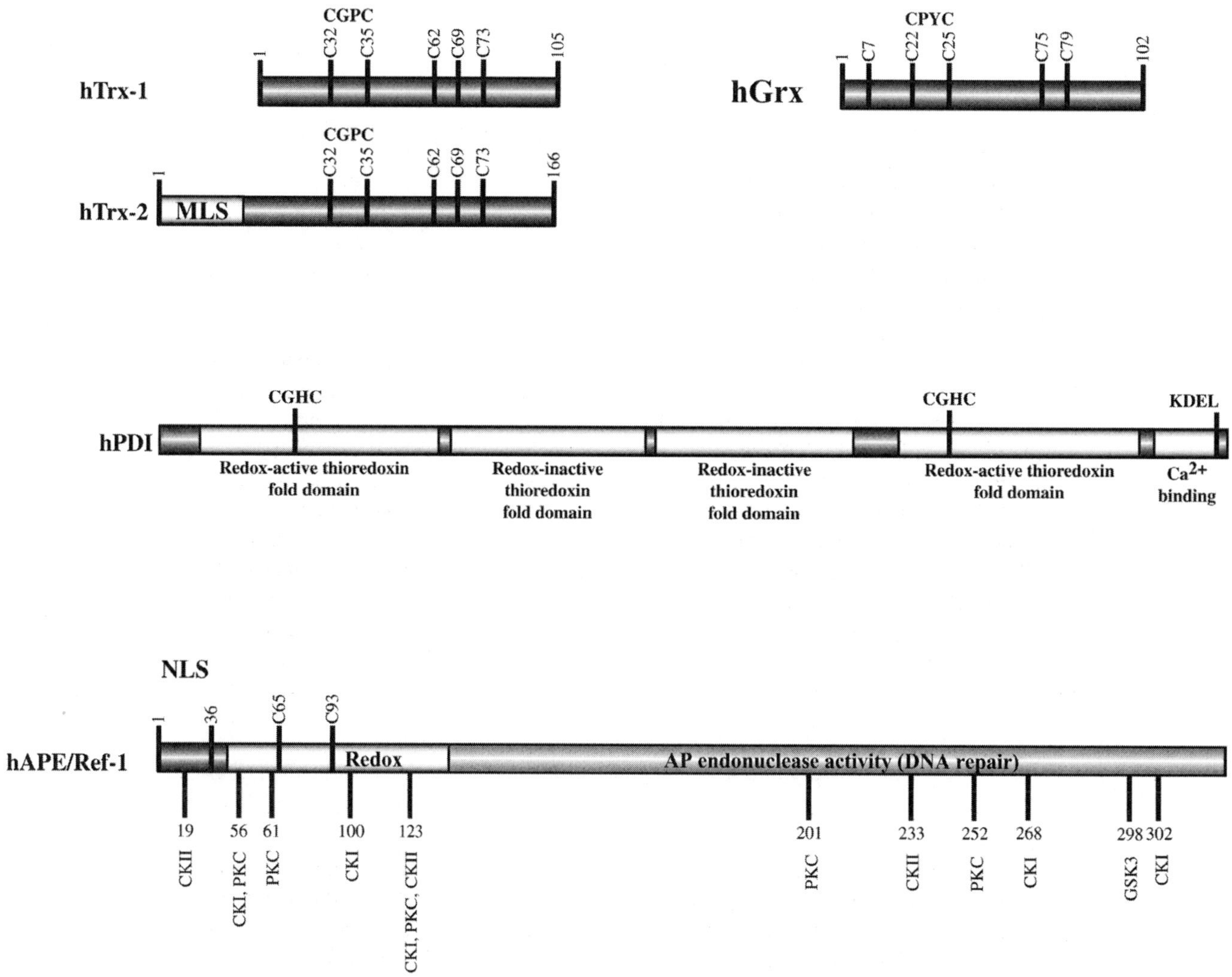

Figure 6.17 Structure of various redox factors. The active cysteines involved in reduction, the CGPC and CPYC sequences for thioredoxin (Trx), and glutaredoxin (Grx) are indicated. Thioredoxin 2 contains a mitochondrial localization sequence (MLS) at its N terminus. The redox-active and -inactive thioredoxin-like domains of protein disulfide isomerase (PDI) are shown. The PDI active cysteine involved in reduction (CGHC) are also shown. A C-terminal KDEL sequence localizes PDI to the endoplasmic reticulum. The regions involved in DNA repair and redox control of Ref-1 are shown. Amino acid residues in Ref-1 involved in redox control (Cys65 and Cys93) as well as the phosphorylation sites by various kinases are marked.

disulfide interchange, both as a homodimer and as subunits of the prolyl 4-hydroxylase tetramer. PDI could potentially maintain localized redox potentials for various transcription factors. PDI function has been found to be important for the DNA-binding functions of such transcription factors as NF-κB and AP-1. PDI also affects the dimerization of E2A basic helix–loop–helix transcription factors and has been found to associate directly (can be copurified) with E2A transcription factors.

Redox factor 1 is also known as apurinic/apyrimidinic endonuclease/redox effector factor 1 (Apel) (Fig. 6.17). It has multifunctional roles, not only in the reduction of various transcription factors (discussed earlier) but also in the base excision repair of oxidatively damaged DNA. The N-terminal portion of the protein is essential for redox control of other proteins. The rest of the protein is involved in DNA repair, containing AP endonuclease activity. The regions involved in redox and DNA repair overlap slightly. Ref-1 is ubiquitous and can autoregulate itself by binding to its own promoter within the genome and inhibiting transcription. The mRNA for Ref-1 is up-regulated by ROS. Interestingly, the inactivation of a CREB binding site in the promoter of Ref-1 abolishes the hydrogen peroxide up-regulation of its gene (see above).

As with thioredoxin, Ref-1 also translocates to the nucleus in response to ROS due to an NLS located between amino acids 1 and 36. The protein can also be posttranscriptionally phosphorylated by CKI and II. Cys65 and Cys93 are critical for redox activity. Thioredoxin can heterodimerize with and reduce Ref-1 through direct association. This association involves Cys32 and Cys35 of thioredoxin. Translocation of thioredoxin from the cytoplasm to the nucleus is essential for formation of the thioredoxin/Ref-1 heterodimer. Along with the factors listed above, Ref-1 has also been shown to stimulate the DNA-binding activity of ATF, CREB, the oncogene *Myb*, and early growth response 1 gene (*Egr-1*). Thus, it provides a unique link between oxidative stress and transcription factor function.

CONCLUSIONS

The regulation of gene expression at the level of transcription is a sophisticated process involving many players guided by diverse stimuli. This chapter has attempted to outline the modularity of such factors and the DNA elements to which they bind. The advent of completely sequenced genomes and recombinant DNA technology has made it possible to engineer factors for specific binding to control the expression of specific genes. Targeted transcription of key genes may overcome certain genetic disorders. This is especially true if there is redundancy among isoforms related to the defective gene product. Overexpression of one isoform would compensate for the nonfunctionality of another. Finally, the modularity of the protein transcription factors as well as the DNA elements to which they bind have allowed molecular biologists to study protein structure and function as well as promoter element function through the use of recombinant DNA technology. By combining different subelements of proteins in the form of fusion proteins or different response elements in the form of reporter constructs, the functionality of these elements can be elucidated.

FURTHER READING

Allen, J. F. (1993). Redox control of transcription: sensors, response regulators, activators and repressors. *FEBS Lett* **332**(3):203–207. *A short but thought-provoking mini-review on redox sensing.*

Collins, T., Stone, J. R., and Williams, A. J. (2001). All in the family: The BTB/POZ, KRAB, and SCAN domains. *Mol Cell Biol* **21**(11):3609–3615. *An excellent outline on repressor domains.*

Evans, A. R., Limp-Foster, M., and Kelley, M. R. (2000). Going APE over ref-1. *Mutat Res* **461**:83–108. *Outlines the role of this redox factor in the control of gene expression.*

Forman, H. J., and Cadenas, E. (eds.) (1997). *Oxidative Stress and Signal Transduction.* Chapman & Hall, New York. *Complete reference for the role of reactive oxygen species in the control of transcription and gene expression.*

Jans, D. A., Chan, C. K., and Huebner, S. (1998). Signals mediating nuclear targeting and their regulation: Application in drug delivery. *Med Res Rev* **18**(4):189–223. *A good review of nuclear import, export, and retention of transcription factors. NLS and NES as well as their regulation are discussed.*

Latchman, D. S. (1997). Transcription factors: An overview. *Int J Biochem Cell Biol* **29**(12):1305–1312. *This introductory article on a whole issue dealing with transcription factors provides an excellent overview of the basic domains of transactivating DNA-binding proteins.*

Mitchell, P. J., and Tjian, R. (1989). Transcriptional regulation in mammalian cells by sequence specific DNA binding proteins. *Science* **245**:371–378. *Although a bit dated, this provides a good basic overview of transcriptional control of gene expression.*

Morel, Y., and Barouki, R. (1999). Repression of gene expression by oxidative stress. *Biochem J* **342**:481–496. *A complete description of the factors known to be modified by oxidative conditions.*

Taanman, J.-W. (1999). The mitochondrial genome: structure, transcription, translation and replication. *Biochim Biophys Acta* **1410**:103–123. *A good review article on transcription and translation in the mitochondria.*

7

TRANSLATIONAL CONTROLS AND PROTEIN SYNTHESIS IN EUKARYOTIC CELLS

WILLIAM G. WILLMORE

INTRODUCTION

Translation is the synthesis of protein from a messenger ribonucleic acid (mRNA) template. As with transcription, translation involves many factors, the most important being the ribosome. The previous chapter dealt with the controls on transcription, most of which generally occur prior to the synthesis of mRNA. The same principle is true for translation in that most of the checkpoints for the regulation of translation occur prior to the synthesis of protein from the mRNA template. These checkpoints ultimately determine the protein that is produced.

The focus of this chapter will be the controls that regulate the basal translational machinery by various factors responding to changes in the intracellular milieu. Control of translation can occur at a number of levels including (1) posttranscriptional RNA processing, (2) mRNA transport from the nucleus to the cytosol, (3) translation by the basal machinery, and (4) mRNA stability and degradation. This chapter will also deal with the general principles of translational control as well as discuss some interesting exceptions to the rules. The reader is also referred to Chapters 15 and 16 for an analysis of translational control in unique animal systems of adaptation to environmental stress (anoxia tolerance hibernation). The modular nature of the mRNA template will be emphasized as well as the factors that bind to it and control translation. We begin with an overview of the basal machinery involved in eukaryotic translation.

BASIC ELEMENTS OF TRANSLATION

Posttranscriptional RNA Processing

The RNA product that is synthesized during transcription has a sequence that is fully complementary to the gene but is quite different from the mature mRNA that is delivered to the ribosome (Fig. 7.1). In brief, unprocessed nuclear mRNA is much longer that the processed cytoplasmic mRNA, is much more unstable, and has greater sequence complexity. This RNA is known as heterogeneous nuclear RNA (hnRNA), and it must be further processed within the nucleus before it leaves to be transcribed in the cytoplasm. The primary transcript contains both exons that are retained in the final gene product and introns that are spliced out. Processing involves the removal of the introns, as well as the formation of a "cap" on the 5′ end of the transcript, methylation, and the addition of a poly(A) tail on the 3′ end of the transcript. This results in a mature mRNA that contains an open reading frame (ORF) (sometimes more than one) that is translated into protein and that is flanked by both 5′ and 3′ untranslated regions (UTRs). The completely processed mRNA is then transported through the nuclear pores to join with ribosomes in the cytoplasm. These steps in RNA processing are explained in greater detail below.

Capping Capping involves the addition of a guanylyl residue to the 5′ end of the hnRNA, just after it is initiated, by the enzyme guanylyl transferase. The enzyme adds a guanosine 5′-triphosphate (GTP) to the 5′ end in reverse

Functional Metabolism: Regulation and Adaptation, edited by Kenneth B. Storey.
ISBN 0-471-41090-X

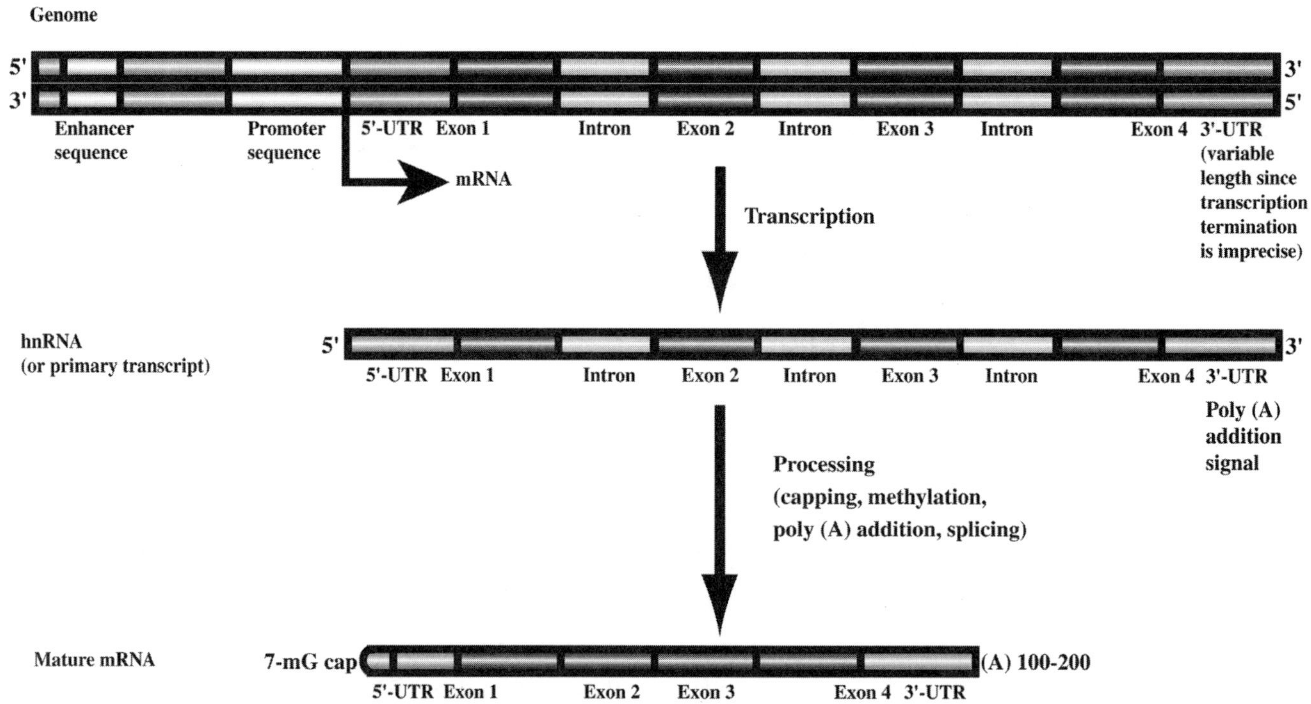

Figure 7.1 mRNA processing from genome to mature mRNA. The primary RNA transcript (hnRNA) is composed of coding exons and noncoding introns. The noncoding 5′- and 3′-untranslated regions (UTRs) are part of the first and final exons (each may be contained in more than one exon), respectively. After the addition of the 5′ cap and polyadenylation of the 3′ end, the introns are spliced out to give the final mature mRNA, which is exported from the nucleus into the cytoplasm for translation.

orientation, removing two phosphate groups from the GTP and one from the 5′ terminal residue of the transcript and leaving a triphosphate bridge between the two. The cap is then methylated at the 7 position of the guanine residue. The resulting 5′ structure is known as a 7-mG cap.

Polyadenylation Gene transcription by RNA polymerase II (pol II) typically continues far past the 3′ end of the final exon of the gene, with the precise point of termination being nonspecific. However, termination of transcription does not occur until after a specific polyadenylation signal sequence (AAUAAA) has been passed. This signal is recognized by a protein known as the cleavage and polyadenylation specificity factor (CPSF; see below) which contains 4 subunits (30, 70, 100, and 160kDa). Addition of the poly(A) tail is made between 10 and 30 nucleotides downstream from this signal where the transcript is cleaved by an endonuclease. Poly(A) is then added posttranscriptionally to the hnRNA (or pre-mRNA) by the enzyme poly(A) polymerase, using adenosine 5′-triphosphate (ATP) as the substrate. The subunits of the specificity factor, the endonuclease, and the poly(A) polymerase all form a complex resulting in tightly controlled polyadenylation. Most eukaryotic mRNAs have 100 to 200 adenine residues added to their poly(A) tail. The exception is histone mRNAs, which lack a poly(A) tail. Capping and polyadenylation of the hnRNA occurs prior to splicing (see below). Addition of long poly(A) tails to eukaryotic mRNAs helps to protect the coding portion of the transcripts from being degraded by RNases.

Splicing A group of six nuclear core proteins (A1, A2, B1, B2, C1, C2) associate with the hnRNA to form a structure known as the heterogeneous nuclear ribonuclearprotein (hnRNP) complex. These are responsible for maintaining the hnRNA in an untangled, accessible conformation. Accessibility is important for the attachment of RNA–protein complexes that are responsible for RNA splicing (see below). Once the hnRNA is capped and polyadenylated, nuclear splicing of the pre-mRNA occurs. Splicing depends upon recognition of very short consensus sequences that are found at the 5′ and 3′ ends of the intron (Fig. 7.2). These short sequences (GU...AG) are very highly conserved. Flanking sequences to these include $G_{100}U_{100}A_{60}A_{74}G_{84}U_{50}$... and ...8YN$C_{78}A_{100}G_{100}$ for the 5′ and 3′ ends of the intron, respectively (where

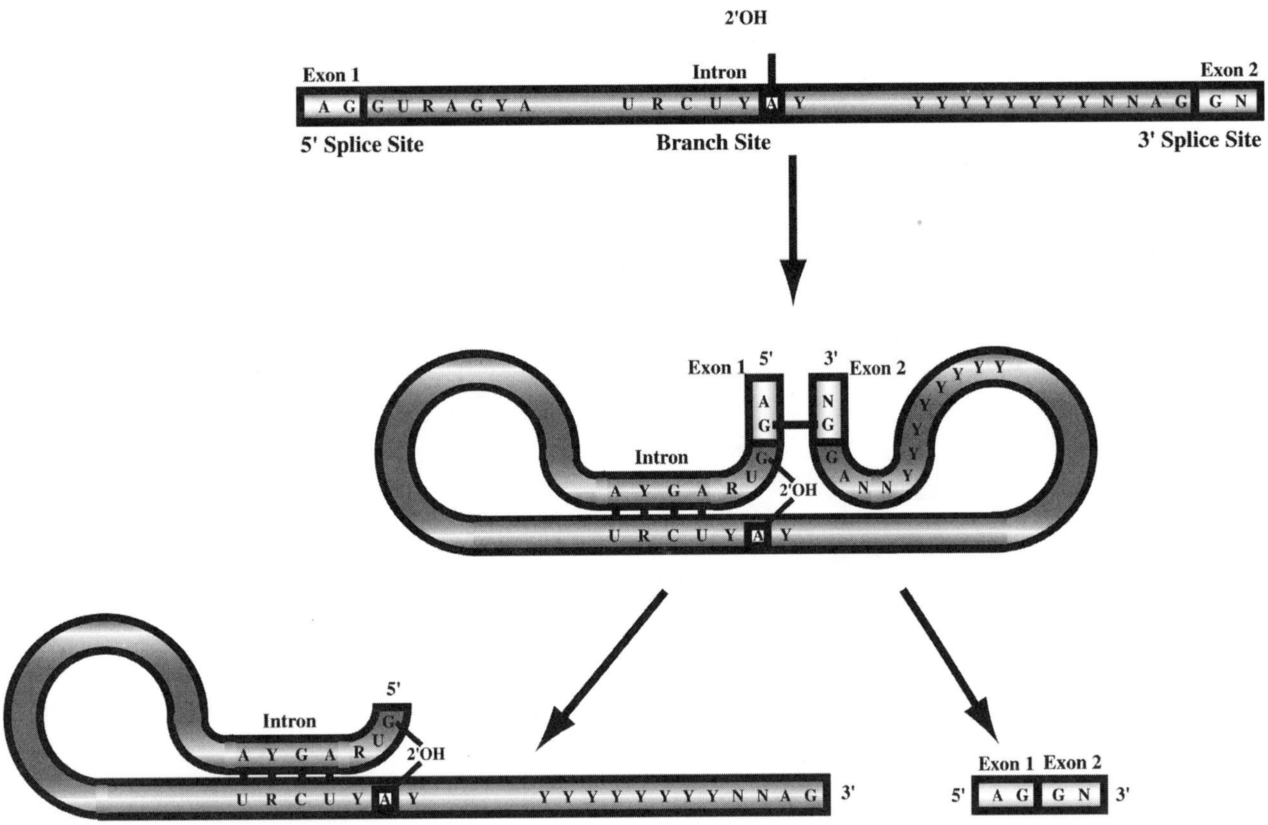

Figure 7.2 Lariat formation in mRNA splicing. Shown is the consensus sequence for the splice sites at the 5′ and 3′ ends of the intron as well as the branch site for attachment of the 5′ end of the intron. The adenine at the invariant branch site is boxed. Y, any pyrimidine; R, any purine; N, any nucleotide. The bottom panel shows the formation of the lariat and the subsequent joining of exons 1 and 2.

Y = pyrimidine and N = any nucleotide; the numbers in subscript designate the percentage frequency of occurrence of each particular base in the consensus sequence). Not only must these short, highly conserved sequences be present in the intron, but a conserved branch site 18 to 40 nucleotides upstream from the 3′ splice site must also be present. This branch site is represented by the sequence $Y_{80}NY_{87}R_{75}A_{100}Y_{95}$ (where R = purine). The branch site is used for attachment of the 5′-phosphate group of the intron's invariant 5′-G (once cleaved) to the 2′-OH at the invariant branch site A to form a 2′,5′-phosphodiester bond. This covalently closed loop is known as the lariat.

Other RNPs then assemble at the hnRNA to catalyze splicing. These include small nuclear ribonucleoprotein particles (snRNPs, "snurps," or "scyrps"). snRNPs consist of a single RNA of 100 to 200 nucleotides in length and about 10 proteins. These snRNPs are classified U1 to U6, depending on which of the small RNAs are present in them (Fig. 7.3). snRNPs assemble at the hnRNA splice sites to form spliceosomes. The U1 snRNP binds first to the 5′ end of the intron to be spliced out. This is followed by U2 snRNP, which binds to the branch point. U4, U5, and U6 snRNPs bind to replace U1 snRNP on the hnRNA. Exon ligation then occurs with the release of U2, U4, U5, and U6 snRNPs. snRNPs are recycled for subsequent nuclear splicing. Each step of this process requires ATP. Alternative nuclear splicing represents a translational control point that will be discussed later on in this chapter.

Features of the mature mRNA are important in translation initiation. The 7-mG cap is recognized by the first of the ribosomal initiation factors to bind to the mRNA. The length and secondary structure present in the 5′-untranslated region (UTR) plays an important role in translational efficiency. Secondary structure in the 5′UTR is an important control point for translation and will be discussed later in the chapter. The 3′UTR and poly(A) tail of the mature mRNA, as well as the proteins that bind to this region, are also important. The mRNA is "scanned" once the complete ribosome binds until the first AUG sequence is located. This is the start codon that is used to initiate transcription.

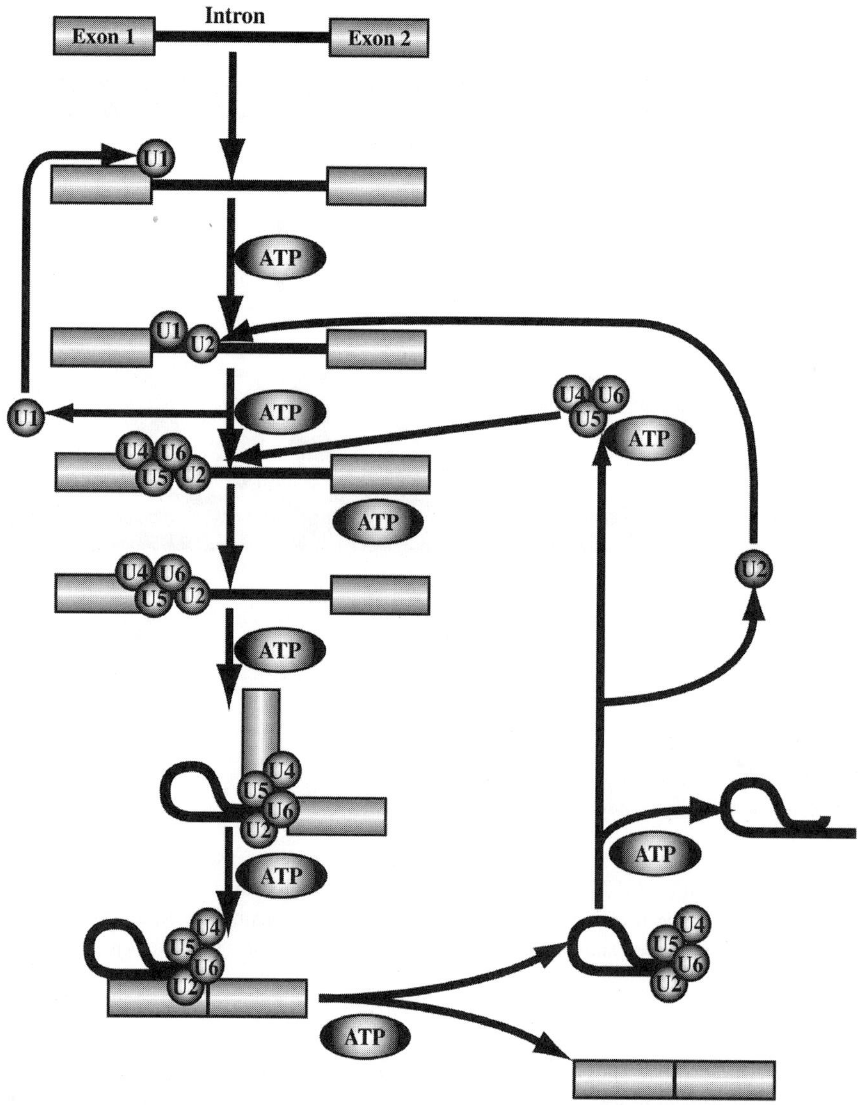

Figure 7.3 Formation of the spliceosome from snRNPs. Ribonucleoprotein complexes U1 to U6 are involved in lariat formation and exon splicing. The intron lariat between exons 1 and 2 is shown as a curled line. Each step of splicing is ATP-dependent.

mRNA Translocation to the Cytosol Translocation of the mature mRNA from the nucleus to the cytosol occurs through the nuclear pore complex (NPC). The process is similar to that described for nuclear export of transcription factors (through their nuclear export sequence, or NES). "Capped" and spliced mature mRNAs with hnRNPs still bound are recognized by a nuclear cap-binding complex (CBC), which binds to the 5′ ends of the mRNAs. The CBC moves through the nuclear pore first and nuclear-restricted hnRNP proteins are removed from the mature mRNA. Most hnRNP proteins lack a NES and are retained within the nucleus. Those that contain a NES follow the mature mRNA into the cytosol from the nucleus. These hnRNP proteins are then replaced by cytosolic mRNA-binding proteins that are involved in mRNA stability and translation, and the hnRNP proteins are transported back into the nucleus. Cytosolic hnRNPs pass back through the NPC through recognition of hnRNP NLS by importins and the GTPase Ran; see the discussion of nuclear localization sequences (NLS) in the previous chapter.

Ribosomes

Once the mature mRNA has reached the cytosol, it is available for translation by ribosomes. Ribosomes are ribonucleoprotein complexes present in the cytosol of all cells

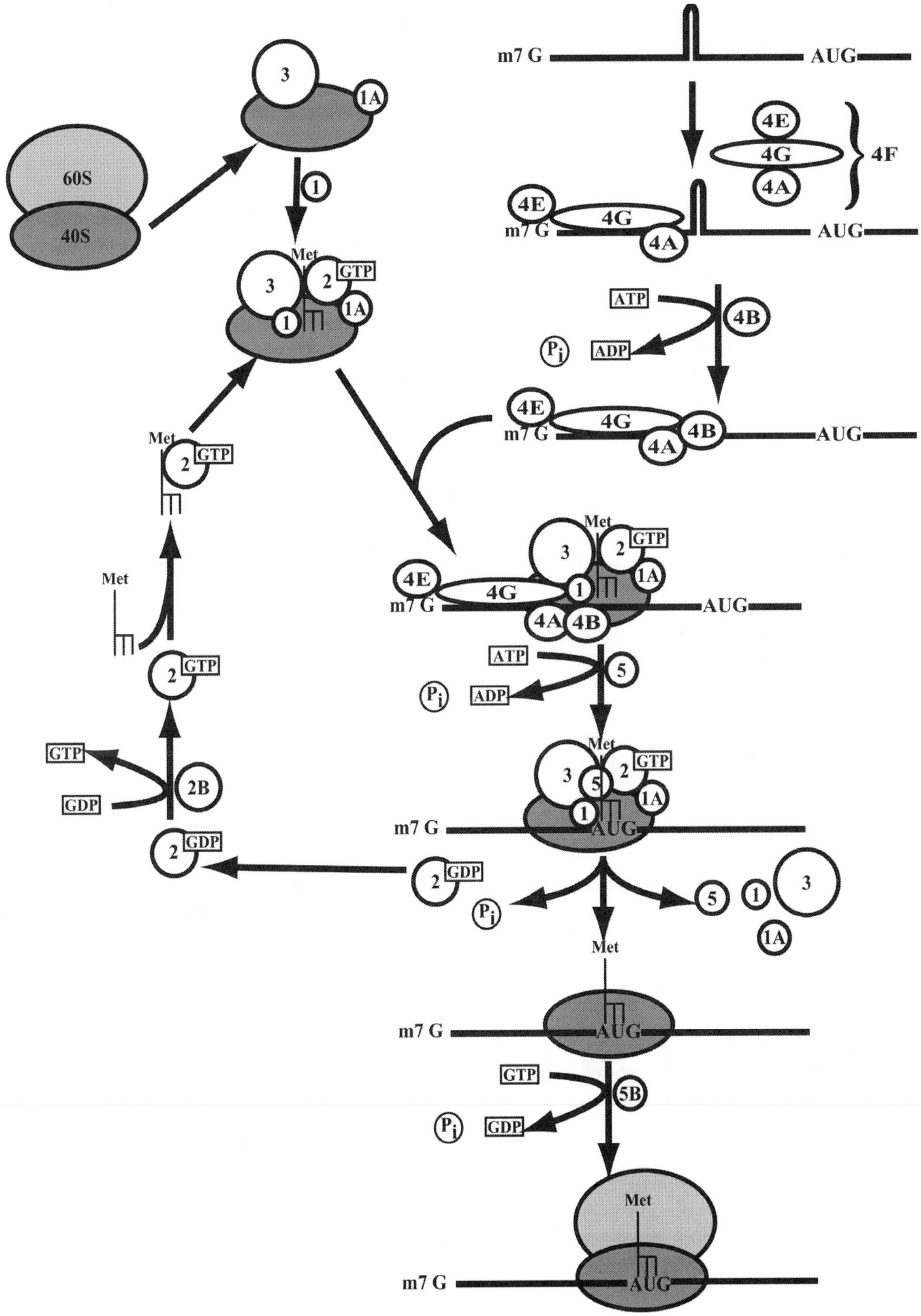

Figure 7.4 The assembly of factors at the mRNA during translation initiation. The 80S ribosome (composed of 60S and 40S subunits) is shown. Eukaryotic initiation factors (eIFs) are indicated with simple numbers or letter/number combinations. The start codon (AUG) as well as the ATP and GTP requiring steps are indicated. See text for details.

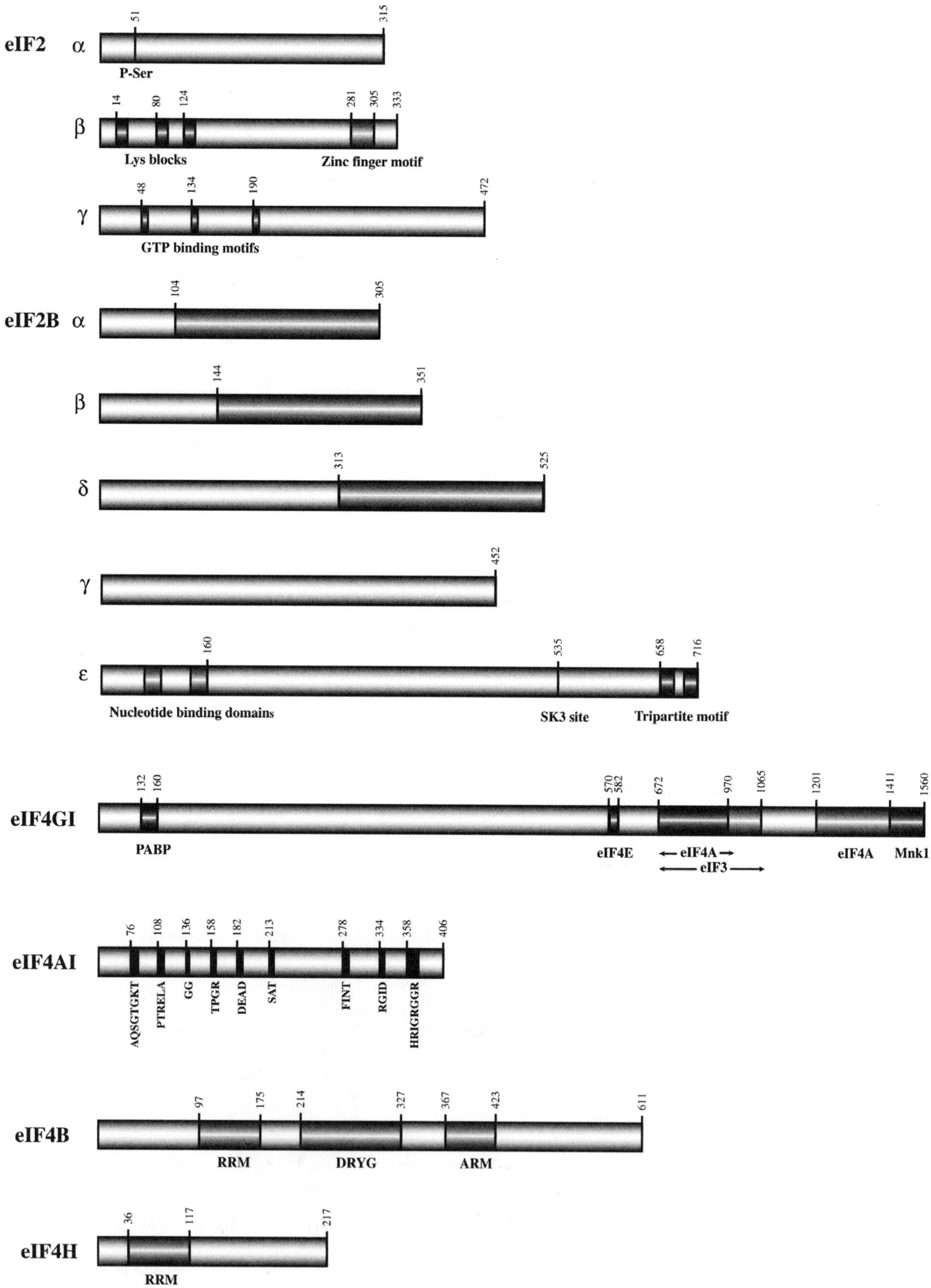

eIF2 α
51
315
P-Ser
β
14
80
124
281
305
333
Lys blocks
Zinc finger motif
γ
48
134
190
472
GTP binding motifs
eIF2B α
104
305
β
144
351
δ
313
525
γ
452
ε
160
535
658
716
Nucleotide binding domains
SK3 site
Tripartite motif
eIF4GI
132
160
570
582
672
970
1065
1201
1411
1560
PABP
eIF4E
eIF4A
eIF3
eIF4A
Mnk1
eIF4AI
76
108
136
158
182
213
278
334
358
406
AQSGTGKT
PTRELA
GG
TPGR
DEAD
SAT
FINT
RGID
HRIGRGGR
eIF4B
97
175
214
327
367
423
611
RRM
DRYG
ARM
eIF4H
36
117
217
RRM

as well as the matrix of mitochondria and the stroma of chloroplasts. The cytosolic eukaryotic 80S ribosome is composed of a small 40S subunit and a large 60S subunit (S = Svedbergs, the sedimentation coefficient used in gradient centrifugation). The 40S subunit is composed of one 18S RNA (1874 bases) and 33 polypeptides. The 60S subunit is composed of one 28S RNA (4718 bases), a 5.8S RNA (160 bases), a 5S RNA (120 bases), and 49 polypeptides. The 80S ribosomes must first dissociate into their 40S and 60S subunits for translation initiation to begin. Dissociation and reassembly on the mRNA translational start site is mediated by eukaryotic initiation factors (eIFs). These are often key targets in the control of translation and their regulation will be discussed later on in the chapter.

Translation Initiation

Each of the initiation factors has a different role to play in assembling and bringing together the ribosome, the mRNA and the initiating Met-tRNA (transfer RNA) (Fig. 7.4). Dissociation of the ribosome is promoted by eIF3 and eIF1A. eIF1A and eIF1 are the smallest (17 to 22 kDa) but the most highly conserved of all of the initiation factors. eIF1 weakly associates with eIF3 and may promote the binding of the 40S ribosome to the initiation codon. This activity is strongly stimulated by eIF1A. It plays a role in AUG codon recognition by Met-tRNA, the binding of Met-tRNA to the 40S ribosome, and ribosomal scanning. It also affects 80S ribosomal dissociation. eIF3 is the largest of the initiation factors (600 kDa) and contains at least 11 different subunits. It has all the same functions as eIF1A as well as interacting with numerous other initiation factors. It likely helps to organize initiation complexes on the 40S ribosomal surface.

A specific tRNA is used to initiate protein synthesis: methionyl-tRNA (Met-tRNA). Met-tRNA binds to eIF2 in the presence of GTP. eIF2 distinguishes the initiator Met-tRNA from other aminoacylated tRNAs by the methionyl residue and the A-U base pair at the end of the acceptor stem. When not bound to the ribosome, eIF2 exists in a complex with guanosine 5′-diphosphate (GDP) and eIF2B, the eIF2 binding protein. Phosphorylation of the GDP to GTP converts inactive eIF2/GDP to active eIF2/GTP. This stimulates dissociation of eIF2-GTP from eIF2B, which then binds Met-tRNA and joins the assembling ribosome. Association of eIF2 with Met-tRNA is a GTP-dependent process and requires eIF2B for catalysis. eIF2 is composed of three subunits; α, β, and γ. The α subunit contains a phosphorylation site at Ser51 and is responsible for AUG start codon recognition (Fig. 7.5). eIF2β has three N-terminal lysine blocks and a C-terminal zinc-finger motif. The lysine blocks have been implicated in eIF2β binding to eIF2B, eIF5, and to mRNA, whereas the zinc-finger motif is involved in initiator codon recognition. eIF2γ contains three GTP-binding motifs. In the presence of eIF5, which acts as a GTPase-activating protein to promote GTP hydrolysis within the 40S initiation complex [consisting of 40S*eIF3*AUG*Met-tRNA(f)*eIF2*GTP], the initial methionine residue is put in place and eIF2-GDP is released and reassociates with eIF2B (Fig. 7.4). eIF2B is composed of 5 subunits; α, β, δ, γ, and ε (Fig. 7.5). The α, β, and δ subunits all have homologous C-terminal domains. The ε subunit contains a N-terminal nucleotide binding domain and a C-terminal tripartite motif. It also contains a glycogen synthase kinase 3 phosphorylation site in the C-terminal portion of the protein. The tripartite motif is involved in binding eIF2β. The association of ε and γ subunits possesses strong guanylate exchange activity. The α, β, and δ subunits form a complex that distinguishes between phosphorylated and nonphosphorylated forms of eIF2α. Phosphorylation of the eIF2α subunit forms a stable complex of phospho-eIF2-GDP-eIF2B that is inactive until released by phosphatase action. Reversible phosphorylation control over eIF2α is one of the most important regulatory mechanisms of translation control (see below and Chapters 15 and 16).

The first factor to bind to the mature mRNA is eIF4F, itself a complex of three factors (eIF4A, eIF4E, eIF4G). The 7-mG cap is recognized by eIF4E. Translation can be inhibited by the binding of eIF4E-binding protein (4E-BP) to eIF4E. In this respect, 4E-BP acts as a translational repressor and is another key regulatory factor in translational control, as will be discussed later in this chapter. eIF4G binds to a convex surface on eIF4E that is opposite to the site that binds the 7-mG cap. This surface also binds 4E-BP. There are two known forms of eIF4G: eIF4GI and eIF4GII. eIF4GI has an N-terminal region that binds poly(A)-binding protein (PABP) (Fig. 7.5).

Figure 7.5 Domain structure of initiation factors eIF2 (α, β, and γ subunits), eIF2B (α, β, δ, γ, and ε subunits), eIF4GI, eIF4AI, eIF4B, and eIF4H. The phosphorylation site at Ser51 of eIF2α and the zinc finger (recognizes the initiator codon) of eIF2β are shown. The lysine blocks involved in both protein and RNA interactions of eIF2β are presented as well as the sites for binding GTP in eIF2γ. Homologous regions of eIF2Bα, β, and δ are represented in gray. The tripartite motif involved in binding to eIF2β and the glycogen synthase kinase 3 (GSK-3) phosphorylation site (labeled SK3 site) are indicated. eIF4GI contains binding sites for PABP, eIF4E, and eIF3. A binding site for Mnk1, which phosphorylates eIF4E, is found at the C terminus of eIF4GI. The motifs shared between eIF4AI and DEAD box proteins are shown. RNA recognition motifs (RRMs) are present in both eIF4B and eIF4H. As well, eIF4B contains aspartate/arginine/tyrosine/glycine (DRYG) and arginine-rich motifs (ARM).

PABP provides another mode of regulatory control over transcription, as will be discussed later. The central portion of eIF4G contains sites for interactions with eIF3, eIF4A, and eIF4E. The C terminus of the protein contains a binding site for the protein kinase Mnk1. Mnk1 binds to eIF4G in response to growth factors, hormones, mitogens, cytokines, or stress and phosphorylates eIF4E, providing yet another input for regulatory control of transcription in response to extracellular stimuli (see below). eIF4A is an ATP-dependent RNA helicase capable of unwinding RNA duplexes (Fig. 7.5). It contains many conserved sequences that places it in the DEAD box family of helicases. Together with eIF4B, large duplexes of RNA can be unwound and prevented from reassociating. Its ATPase activity provides the energy needed for mRNA scanning. ATP binding to eIF4A allows RNA binding to its C-terminal HRIGRXXR motif. Three isoforms of eIF4A exist (I, II, and III), with only I and II being functional. eIF4A may interact with other factors through eIF4F.

The next factors to bind are eIF4B and eIF4H. eIF4B is a homodimer that binds to RNA and promotes the recruitment of ribosomes to the mRNA as well as stimulating the RNA helicase activities of eIF4A and eIF4F (Fig. 7.5). Both eIF4B and eIF4H contain RNA recognition motifs (RRM) in their N termini. eIF4B has an additional aspartate, arginine, tyrosine, and glycine-rich (DRYG) region and an arginine-rich motif (ARM). The RRM (composed of two RNP motifs) and ARM are two nonspecific RNA-binding domains. eIF4B may play a role in RNA–RNA interactions at the start site of transcription. eIF4B is also phosphorylated by two kinases; p70 S6 kinase and PKA. eIF4B binds RNA weakly and stimulates the ATPase activities of eIF4A and eIF4F. eIF4H (Fig. 7.5) is a highly unstable factor. It functions as a monomer and contains the same RRM motif as eIF4B.

The complex of the 40S ribosome with eIF1, eIF1A, eIF2, eIF3, and Met-tRNA then joins the complex of eIF4A, eIF4B, eIF4G, eIF4E, and mRNA to form the complete *40S preinitiation complex* and commence mRNA scanning and AUG recognition (Fig. 7.4). The latter complex forms the docking site on the mRNA for the former. The 40S ribosome preinitiation complex then migrates downstream from the 5′ cap toward the initiation codon. This process consumes ATP. Scanning continues until an RNA–RNA interaction occurs between the initiation codon (AUG) on the mRNA and the anticodon (CAU) on the Met-tRNA. Scanning by the 40S ribosome may be discontinuous and occur in "hops" in a process known as *shunting*. Secondary mRNA structure between the 40S ribosome and the start codon disrupts scanning, and *stalling* of the ribosome over the initiation codon may assist in its recognition. An important step in this recognition is the eIF5-promoted hydrolysis of the GTP bound to eIF2. Conversion of GTP-bound eIF2 to GDP-bound eIF2 results in its weakened affinity for the 40S ribosome and ejection. Dissociation of GDP-bound eIF2 from the complex results in the subsequent rapid dissociation of eIF1, eIF1A, eIF3, and eIF5. This prepares the 40S ribosomal subunit for junction with the 60S ribosomal subunit. When the interaction of codon/anticodon is established, eIF5 activates the GTPase center in eIF2γ by GTPase activating protein (GAP) activity. The C terminal of eIF5 contains a bipartite motif of aromatic and acidic residues, termed AA boxes, that interact with the three lysine blocks of eIFβ2. GDP-bound eIF2 is recycled to GTP-bound eIF2 by eIF2B. GTP-bound eIF2 picks up another Met-tRNA and the cycle continues.

Assembly of the 60S ribosomal subunit with the 40S ribosomal subunit forms the complete functioning ribosome. The complete ribosome is composed of two domains in which there are several active centers. The first domain is the translational domain and holds the A site, the P site, the peptidyl transferase site, and the site for binding the mRNA. The A site (or entry site) is the only center that can be entered by an incoming aminoacyl-tRNA. This site is found at codon "$n+1$" on the mRNA. The P site (or donor site) is occupied by the peptidyl-tRNA, a tRNA to which the elongating polypeptide chain is attached. This site is found at codon n. The peptidyl transferase activity is associated with the 60S ribosomal subunit. The second domain is the exit domain and holds the exit site for the elongating polypeptide chain.

Docking of the 60S ribosomal subunit with the 40S ribosomal subunit requires GTP-bound eIF5B. Although poorly understood, eIF5B may bind to the A site of the 40S and 60S ribosomal subunits, assisting in the proper alignment of the 60S subunit on the 40S complex. It may also ensure placement of the Met-tRNA in the P site of the ribosomal subunits. eIF5B contains three consensus GTP binding motifs. Once the 80S ribosome is formed, initiation is complete and elongation (with subsequent termination) can follow.

Translation Elongation

Elongation in eukaryotes relies upon aminoacylated tRNAs, elongation factors (eEFs), the mRNA/80S ribosome/Met-tRNA complex, and GTP. Of the elongation factors, eEF1 is composed of two proteins; eEF1 and eEF1B (eEF1B contains α, β, and γ subunits). eEF1 is the aminoacyl-tRNA-binding factor and is responsible for bringing in aminoacylated-tRNAs into the A site of the ribosome, hydrolyzing GTP into GDP in the process. eEF1 recycles the GDP to GTP bound to eEF1. Only eEF1Bα and eEF1Bβ contain guanine nucleotide-exchange factor (GEF) domains and facilitate the off-rate for eEF1A-bound GDP. eEF2 is the eukaryotic translocation factor and, using its GTPase activity, moves the ribosome one codon down the mRNA so that the next codon is positioned at the A site. eEF1A and eEF2 both contain well-defined GTP-binding domains to catalyze their GTPase activities. Peptide bonds between amino acids are formed by peptidyl transferase catalytic activity of the complete ribosome. Early experiments suggested that the peptidyl transferase activity of the ribosome was catalyzed by the 23S rRNA (ribosomal RNA). Translocation of the ribosome occurs one codon at a time with the corresponding aminoacylated tRNA (as determined by the anticodon sequence) always coming into the A site. Many ribosomes can translate a single mRNA at the same time. mRNA associated with a series of ribosomes engaged in translation is known as a polyribosome or polysome. The greater the number of ribosomes, the more efficiently a particular mRNA is translated.

Once at termination, or after the stop codon has reached the A site of the ribosome, protein synthesis ceases. Termination or stop codons are recognized by specific proteins known as release factors, which are responsible for polypeptide release from the ribosome. Eukaryotic releasing factor (eRF) is an α_2 dimer. It binds to the ribosomal A site in a GTP-dependent manner and hydrolyses the final peptidyl-tRNA ester bond. GTP is hydrolysed in the process, and release of the peptide, deacetylated tRNA, and ribosome from the mRNA template ensues. Recycling of the ribosome to the start site may occur. When recycling occurs, the mRNA circularizes and enhances recruitment of the 40S ribosomal subunits to the 7-mG cap. The most likely mechanism for this is the association between eIF4G and PABP. This increases the efficiency of translation and will be discussed later in this chapter.

REGULATION OF TRANSLATION

Regulation of translation can be achieved at multiple levels targeting (1) the mRNA, (2) the initiation factors, (3) the elongation factors, and (4) the ribosome. At each level there are multiple possible modes of regulation. The following sections review the mechanisms of translational regulation and provide examples of how controls on translation are integrated with overall cellular responses to stimuli and stresses. A summary of these mechanisms is presented in Table 7.1. Most control mechanisms involve (a) association of a regulatory protein or regulatory RNA with the mRNA, (b) phosphorylation of the basal elements involved in translation, or (c) regulation of mRNA processing. In many cases, regulation results in either activation or repression of assembly of the basal machinery at the start site of translation. This regulation can come from elements proximal to the translational start site or as distal as the poly(A) tail of the mRNA. Distal regulation of translation involves circularization of the mRNA, a theme that is becoming increasingly prevalent in the control of translation.

Regulation at the Level of mRNA

Regulation of mRNA function can involve inherent structures present within the transcript, cis-acting elements that bind to it, or differential processing from pre-mRNA to produce different final end products. In general, regulation derives from *noncoding* elements present in the message. These include the 5′ cap, introns, UTRs, and the poly(A) tail. Even transcription itself can be considered to be regulation by noncoding sequences since the process is regulated by the noncoding promoter and enhancer regions present in the genomic deoxyribonucleic acid (DNA). Regulation at the level of the mRNA transcript can derive from (1) elements present in the primary structure of the mRNA, (2) binding of proteins to the mRNA, (3) alternative splicing of exons present in the pre-mRNA, (4) transport of mRNA from the nucleus to the cytosol, and (5) frameshifting.

Regulation Derived from mRNA Primary Structure Regulation by the primary structure of the mRNA could include elements present in (a) the noncoding 5′ or 3′UTRs of both the pre- and the mature mRNAs or (b) the poly(A) tail of the mature mRNA. These are discussed in detail below.

Control by the 5′UTR Eukaryotic mRNAs have 5′-untranslated regions that are between 20 and 100 nucleotides in length. Genes with 5′UTRs that are shorter than 12 nucleotides cannot be effectively "loaded" onto the ribosome for translation, whereas 5′UTRs that are very long can attach readily to the ribosome and have an increased chance of containing upstream start codons (uAUGs), short upstream open reading frames (uORFs), or secondary RNA structures (stem loops) that control translation. The greater the number of these upstream elements present, the greater the regulation of the gene downstream from them. These elements are particularly common in mRNAs encoding transcription factors, growth factors, and proto-oncogenes, suggesting that their translation is tightly controlled.

UPSTREAM START CODONS (uAUGs) Another mode of translational control is the presence of alternative initiation sites in some mRNA transcripts. These can produce proteins of different lengths that are transcribed from the same mRNA. Indeed, up to 10% of all eukaryotic mRNAs are known to contain uAUGs within their 5′UTRs that do not have stop codons (similar to uORFs without the termination site). Two proteins encoded from different AUGs may be (a) nonoverlapping with an intercis-

TABLE 7.1 Modes of Translational Regulation

Control at the level of the message RNA

Control by the 5′UTR

Short upstream open reading frames (uORFs): Nonoverlapping, overlapping, or in-frame open reading frames using upstream start codons in the 5′UTR that may result in amino-terminal extension of the major protein product and provide additional domains that confer important auxiliary functions to the protein.

Secondary structure and mRNA binding proteins: RNA stem loops to which regulatory proteins bind (RNA response elements) inhibit translation initiation when protein is bound or de-repress translation by removing an inhibitory protein.

Internal ribosomal entry sites (IRES): Sites downstream of the start codon that can bind ribosomes directly and can mediate translation of open reading frames. Initiation of translation in the absence of a functional 5′ cap-binding protein complex eIF4F (eIF4E, eIF4A, eIF4G).

5′-Terminal oligopyrimidine tracts (5′-TOPs): Long repeats of pyrimidines around the transcriptional start site. Bound by cis-regulatory elements. Important in processes like cell proliferation, differentiation, development, and malignant transformation in eukaryotes.

Control by the 3′UTR

Secondary structure and mRNA binding proteins: RNA stem loops to which regulatory proteins bind (RNA response elements) enhance translation initiation through increased mRNA stability or repress translation by interfering with the binding of 5′ cap-binding elements.

Regulatory RNAs: Short RNAs that base-pair and interfere with essential elements required for binding of positive regulators. Act as repressors of translation.

Control at the poly(A) tail

End-to-end regulation by poly(A) binding proteins (PABPs): Proteins that bind to the poly(A) tail and enhance translation by recruiting eIF4G to the mRNA 5′ cap.

Poly(A) binding protein-interacting proteins (PAIPs): Binds to both PABPs and eIF4A to enhance translation.

Changes in poly(A) tail length: Longer poly(A) tails increase translation efficiency and stability of mRNAs in eukaryotes.

Cytoplasmic polyadenylation elements (CPEs) and binding proteins (CPEBs): Sequences that control cytoplasmic polyadenylation (CPE) are located on the 3′UTR close to the polyadenylation signal and are required for polyadenylation and translational activation. However, the proteins that bind to CPEs (CPEBs) can also repress translation by interacting with the protein "maskin," which "hides" the mRNA from the translational apparatus by binding to eIF4E.

Alternative mRNA splicing

Some gene transcripts are spliced in different ways to include or exclude certain exons. This may occur in different tissues of an organism at different developmental stages. The signals for alternative mRNA splicing is a cell-specific protein that interacts with the spliceosome.

mRNA export from the nucleus to the cytosol

Pre-mRNAs associated with snRNPs in spliceosomes are prevented from being transported into the cytosol through the nuclear pore. Export transport proteins associate with both snRNPs and nuclear pore complex (NPC) proteins.

Frameshifting

Deletions or insertions of base pair into a coding sequence that are not multiples of 3 base pair (codons). They change the frame in which triplets are translated into protein.

Control at the level of initiation factors: Phosphorylation of initiation factors (eIF2, eIF3, eIF4E, eIF4G, and eIF4B) occurs under conditions of stress, including heat shock, viral infection, and serum deprivation. Has a role in differentiation.

Control at the level of elongation factors

eEF1A: Phosphorylation by protein kinase C enhances function. Stimulates guanine nucleotide exchange. Also phosphorylated by casein kinase-2.

eEF1Bα: Phosphorylation by protein kinase C enhances function.

eEF1Bβ: Phosphorylation by cell-cycle regulated kinase $p34^{cdc2}$. Possible role in meiosis and development.

eEF1Bγ: Phosphorylation by protein kinase C enhances function.

eEF2: Phosphorylated in response to growth-promoting stimuli, calcium ion fluxes and other agents. Also ADP-ribosylated by diphtheria toxin.

Control at the level of the ribosome: Phosphorylation of ribosomal proteins, in particular S6, which promotes the initiation of translation.

tronic region, having two completely independent ORFs, each with their own stop codon, (b) overlapping, in the same or different reading frames, with the stop codon of the first uORF being within the downstream ORF, and (c) in-frame with the uORF and the downstream ORF using the same stop codon. The control of expression from alternative ORFs is determined by initiation codon selection by the ribosome. uAUGs may be bypassed by the ribosome due to leaky scanning under different stimuli.

Examples of the use of alternative initiation sites can be found in LIP (liver-enriched transcriptional inhibitor protein) and LAP (liver-enriched transcriptional activator protein). LAP (the longer protein) arises from the use of a uORF in-frame with the downstream ORF of LIP (the shorter protein). The alternative initiation site of LAP is found in the 5′UTR of LIP. LIP is expressed due to leaky scanning at the LAP start codon by the ribosome. The leaky scanning occurs at a specific stage of development. Prenatal abundance of LIP gives way to postnatal abundance of LAP with the function of LAP in terminal liver differentiation.

Some mRNA transcripts contain what is called an internal ribosome entry site (IRES). Despite the identification of many cellular and viral IRES elements, no consensus IRES occurs. IRES elements can promote translation in the absence of a functional 5′ cap-binding eIF4F complex (eIF4A, eIF4E, and eIF4G). An IRES provides a means by which a ribosome can bypass blockage in the 5′UTR and also provides a simple way to allow translation under circumstances in which the cap-dependent mechanism is impaired. The presence of an IRES allows a cell to continue synthesis of selected proteins under circumstances of cellular stress when the energy for cap-dependent protein synthesis is limiting (e.g., hypoxia, hibernation). Despite the overall inhibition of cap-dependent protein translation, selected messages can still get translated because of their IRES. In this respect, cap recognition is uncoupled from ribosomal docking. In picornavirus IRES recognition, entry into the IRES also depends upon two factors involved in IRES recognition, the *trans*-acting factors. These are the autoantigen La and the polypyrimidine tract binding protein (PTB). Many of the genes that have an IRES also have uORFs (see below) and are involved in embryonic development [e.g., *Drosophila* Antennapedia (Antp) and Ultrabithorax (Ubx)], differentiation [e.g., platelet-derived growth factor-2 (PDGF2/c-*sis*), transforming growth factor β1 (TGFβ1)], growth and cancer [e.g., insulin-like growth factor II (IGF II), fibroblast growth factor 2 (FGF-2), the proto-oncogene c-*myc*], and stress [vascular endothelial growth factor (VEGF)], as well as eIF4G.

The archetypal example of control by IRES can be found in the PDGF2/c-*sis* mRNA. PDGF is a heterodimeric protein that acts as a powerful mitogen involved in embryogenesis, development, and wound healing. PDGF2 translation is heavily repressed by extensive secondary structure in platelet progenitor cells, the bone marrow megakaryocytes. Upon megakaryocyte differentiation, translation initiation from an IRES in the 5′UTR of PDGF2/c-*sis* becomes activated. The factors involved in this IRES initiation remain unknown but are most likely activated or newly synthesized upon megakaryocyte differentiation.

SHORT UPSTREAM OPEN READING FRAMES (uORFs) During normal translation, the ribosome scans along the mRNA until the first AUG is encountered. However, uAUGs and uORFs present in the 5′UTR can also serve as translation initiation sites. An uORF may (a) terminate prior to the downstream (main) ORF, (b) terminate within the downstream ORF, or (c) be in-frame with the downstream ORF and utilize its stop codon for termination. In general, uORFs tend to be short. Functional uORFs can be found upstream of genes that control cell growth (such as the genes encoding IGF II, PDGF-2, TGFβ, FGF-2, and VEGF) as well as embryogenesis [Antp, Ubx, RARβ2, the proto-oncogenes c-*mos* and c-*myc*, B-cell leukemia/lymphoma-2 (bcl-2), and murine double minute gene-2 (*mdm*-2)]. Others include the genes encoding the β2 adrenergic receptor, ornithine decarboxylase, and thrombopoietin. The functionality of these uORFs can be determined by mutation of the uAUGs. This results in altered expression of the proteins encoded by the main ORF. In some cases, translation of uORFs regulates mRNA stability. Translation of the main ORF in uORF-containing mRNAs occurs by *leaky scanning* and reinitiation. Leaky scanning is the lack of recognition of certain AUGs by the ribosome. This may occur if the AUG is too close to the cap or if it is in a poor context. In vertebrate mRNAs, a consensus "Kozak" sequence [5′-GCCGCC(A/G)CCAUGG-3′] can increase start site efficiency. Reinitiation at the main open reading frame occurs after the uORF peptide has been translated. The ribosome proceeds a short distance from the termination codon of the uORF (from 30 to 40 nucleotides away) and resumes scanning for the second ORF. This discontinuous scanning may be due to ribosomal shunting. In a growing number of cases, it has been shown that the peptide encoded by the uORF contributes to translational inhibition of the main downstream ORF. The peptide most likely interacts with the translational machinery, causing the ribosome to stall, and preventing or delaying termination.

An example of genes controlled by uORF are the α and β forms of CCAAT/enhancer binding proteins (C/EBP). C/EBPs are bZIP transcription factors that are expressed in terminally differentiated hepatocytes and adipocytes. This transcription factor plays a role in cell maturation and differentiation. Overexpression of C/EBPβ causes growth arrest in hepatoma cells. Whereas C/EBP transcripts are found in many cell types to varying degrees, C/EBP protein only occurs in these two cell types. C/EBP translation (both α and β) is inhibited by conserved uORFs found approximately 7 base pairs upstream from the C/EBP ORF. Repression is overcome by mutation that inhibits translation of the uORF. The C/EBP uORF encodes for the peptide MPGEL in human, mouse, and rat. uORF repression of C/EBP translation may involve ribosomal release after uORF termination, resulting in a weaker reinitiation at the C/EBP initiation codon. The small peptide may also stall the translation machinery.

INFLUENCE OF mRNA SECONDARY STRUCTURES The secondary structure of mRNA involves the formation of stem loops, interior loops, and bulge loops. They are created from inverted repeats in RNA sequences that are complementary to one another, known as palindromes. Hairpin loops are single-stranded RNA loops at the ends of double-stranded RNA duplexes. Interior loops occur as single-stranded RNA mismatch regions between two RNA duplexes. Bulges occur when additional sequence of only one strand of RNA mismatches with the sequence of the second strand. The result is a "looping out" of the mismatched strand from one side of the RNA duplex. When secondary structure forms near the 7-mG cap, eIF4A and eIF4B are prevented from binding. When secondary structures form downstream of the cap, they may impose an impenetrable barrier against scanning by the 40S ribosomal preinitiation complex. The 5′UTR RNA duplex formation may also provide a site for RNA-binding proteins that regulate translation (see below).

An example of translational regulation from 5′UTR secondary structure can be found in the mRNA for ornithine decarboxylase (ODC). This is the rate-limiting enzyme involved in the synthesis of the polyamines that are required for cell proliferation. ODC has extensive secondary structure in its 5′UTR. ODC is stimulated by growth factors [e.g., epidermal growth factor (EGF) and PDGF] and insulin. Under mitogen (insulin) stimulation, phosphorylation of 4E-BP occurs. Unphosphorylated 4E-BP binds to eIF4E and prevents it from binding to the cap of the ODC 5′UTR. Upon phosphorylation, 4E-BP releases eIF4E, allowing increased binding of this initiation factor to disrupt the extensive RNA secondary structure of the ODC 5′UTR so that translation of ODC can occur.

Control by the 3′UTR Regulation of translation from the 3′UTR involves the binding of regulatory RNA-binding proteins (both positive and negative regulation) as well as regulatory RNA (negative regulation) to sequences present in the 3′UTR. Proteins that bind to RNA in the 3′UTR usually bind as complexes such that the proteins involved in repression do not always directly bind to RNA.

REGULATORY RNA-BINDING PROTEINS One of the regulatory mechanisms associated with the 3′UTR is the cytoplasmic polyadenylation element-binding protein (CPEB) system that was first described in studies of the regulation of c-*mos*. CPEB can both activate and repress the c-*mos* gene from its 3′UTR. The C-*mos* proto-oncogene encodes a serine–threonine kinase that is critical for the control of vertebrate meiosis and the early embryonic cell cycle. CPEB binds to the 3′UTR of the c-*mos* gene. To repress C-*mos* synthesis, CPEB binds to another protein named Maskin that, in turn, binds to eIF4E. This prevents the interaction of eIF4E with eIF4G; an interaction required to bring the 40S subunit to the RNA of the c-*mos* gene. By contrast, when protein synthesis is activated, CPEB is phosphorylated at Ser174 by the Eg2 kinase. The change in charge causes CPEB to leave its association with eIF4E and associate instead with the CPSF. As mentioned above, CPSF is responsible for adding poly(A) tails to mRNAs. This disrupts the Maskin/eIF4E complex and allows translation of c-*mos* to proceed uninhibited. The CPEB/CPSF complex also enhances poly(A) tail lengthening, a further feature that enhances translation (see below).

REGULATION BY RNA INTERFERENCE Repression of translation can also occur due to RNA interference (RNAi). This occurs when short pieces of RNA form duplexes to inhibit the translational machinery. The best known example of this is *lin-4* repression of the translation of the *lin-14* mRNA in *Caenorhabditis elegans*. *Lin-14* is responsible for heterochrony (proper timing of developmental events) in *C. elegans*. *Lin-4* encodes two short cytoplasmic mRNAs of 22 and 61 nucleotides that bind to the *lin-14* 3′UTR. According to sequence analysis, up to 7 duplexes can be formed between *lin-4* and the *lin-14* 3′UTR. The *lin-4*/*lin-14* hybrids contain "looped-out" regions of mismatched base pairs. Four of these seven looped-out regions contain a mismatched C residue critical for repression. Is it not known how binding causes the repression, but current evidence suggests that inhibition occurs after initiation, possibly during elongation or release of the Lin-14 protein itself. Another possibility is that the *lin-4*/*lin-14* hybrids may bind a currently unknown repressor protein of *lin-14* translation.

REGULATION FROM THE POLY(A) TAIL Another mechanism of translational control involves the poly(A) tail. The two ends of the mRNA may interact or be in close proximity, and this end-to-end regulation can increase or decrease initiation frequency of translation. Circularization of the mRNA enhances translation by facilitating the recycling of ribosomes. The "scaffolding" for mRNA circularization is provided by the cytoplasmic poly(A) binding protein (PABP) (Fig. 7.6). Three different PABP proteins are known: one major isoform and two tissue-specific or inducible forms. The protein partners of PABP include eIF4G, eukaryotic release factor 3 (eRF3), and the PABP interacting proteins (PAIPs 1 and 2). Circularization is maintained by the interaction of PABP and eIF4G. PABP is composed of two parts. It has an N-terminal portion that contains four RRMs and a conserved C-terminal portion (PABC) that contains the protein interaction sites, including the dimerization and nuclear shuttling domains.

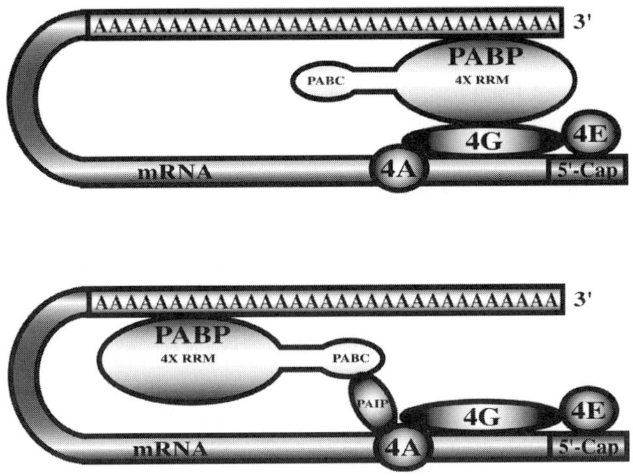

Figure 7.6 Poly(A)-binding protein (PABP) interactions with the basal translation machinery. PABP recognizes poly(A) through four RNA recognition motifs (RRMs) contained in the N-terminal portion of the protein. The opposite face of these motifs interact directly with eIF4G. The poly(A)-binding protein C-terminal domain (PABC) contains sites for interaction with PABP interacting protein (PAIP). PAIP, in turn, interacts with eIF4A. Both interactions enhance recruitment of the basal machinery to the translational start site and enhance the rate of translation.

The PABC is also found associated with the HECT family of ubiquitin ligases. RRMs 1 and 2 anchor PABP to the 3′ end of the mRNA by direct interaction with the poly(A) tail. The α-helical surface opposite to RNA binding then becomes available for interaction with eIF4G. Binding of PABP to poly(A) favors the interaction with eIF4G as the α-helical surface is optimally formed when RRMs 1 and 2 are bound to the poly(A). The PABC binds to PAIPs 1 and 2 as well as to eRF3. PAIP 1 acts as a linker to recruit eIF4A. Because of these interactions, PABP/PAIP 1 can recruit the 40S ribosomal preinitiation complex to the mRNA just as effectively as eIF4E and therefore is cap-independent. Cap-independent docking does not involve eIF4E or eIF4G. mRNAs with and without caps can be translated with equal efficiency via this PABP-dependent mechanism. PABP-dependent translation is favored over cap-dependent when either the cap-binding factors and/or PABP are phosphorylated. Phosphorylation of eIF4G promotes its association with PABP. eIF4B binds to phosphorylated PABP, whereas hypophosphorylated PABP is bound to the eIF4F complex. Phosphorylation of these factors has been found to occur under insulin treatment. PABPs also protect the mRNA from degradation by 5′-3′ endonucleases, a process involving poly(A) tail shortening, decapping, and mRNA cleavage.

Poly(A) length is an important determinant of translation efficiency. The mere presence of a poly(A) tail can stimulate translation. Most mRNAs initially receive poly(A) tails of approximately 250 nucleotides in the nucleus. Upon entering the cytoplasm, poly(A) tails are slowly cleaved. The gain of a poly(A) tail activates transcription whereas the loss inhibits it. Activation is due, in part, to the increased number of PABPs that can bind to longer poly(A) tails. As mentioned above, polyadenylation can occur, not only in the nucleus, but also in the cytoplasm through the action of CPEB. Nuclear polyadenylation requires the AAUAAA signal in the 3′UTR, known as the polyadenylation signal sequence or the nuclear polyadenylation element (NPE). The cytoplasmic polyadenylation element (CPE), also known as the adenylation control element (ACE) is U-rich sequence (with the general structure UUUUUAU) and resides near the NPE. CPEB binds to the CPE and recruits CPSF. CPSF binds to the NPE and, in turn, recruits a cytoplasmic poly(A) polymerase (PAP). PAP is then free to add the cytoplasmic poly(A) tail.

Examples of genes that are regulated by changes in poly(A) length include insulin and the glucose transporter isoform 1 (GLUT-1) in rats. GLUT-1 is involved in the transport of glucose across the blood–brain barrier, a key function since glucose is the principal substrate of the central nervous system. In diabetic rats, the poly(A) tail of GLUT-1 is shortened from ~350 to ~100 nucleotides, a change that is associated with reduced GLUT-1 mRNA stability as assessed by Northern blots. This results in reduced translatability of the GLUT-1 gene. Reduced translatability of mRNAs with shorted poly(A) tails is determined by *in vitro* translation (see Text Box 7.1). Oppositely, treatment with high glucose results in an increase in the insulin poly(A) tail length by about 120 to 140 nucleotides, leading to stimulation of insulin translation followed by insulin action to lower blood glucose levels.

TEXT BOX 7.1 STUDYING TRANSLATION

Several techniques have been useful in the study of translation, each allowing different insights. Among these are:

Gel-Shift Assays and Footprinting

These techniques can also be used to study the interactions of proteins and RNA in much the same way as they are used to study the binding of transcription factors to DNA (see previous chapter). The principles are the same as for DNA gel-shifts, but the template is single-stranded RNA, which is end-labeled in the same manner as DNA. RNA gel-shifts have been used to study the binding of snRNP U1, PABP, and HIV proteins to their respective elements. RNA footprinting involves replacing the DNase used in DNA footprinting with RNase VI. This method is used to determine the RNA sequence that is recognized by the RNA-binding protein.

In Vitro Translation

Translation of specific mRNAs can be performed in a cell-free system. Specific RNAs are cloned into plasmids containing a promoter for bacteriophage RNA polymerase (T7, T3, or Sp6). Cell lysates must be used in order to provide all of the materials (ribosomes, free amino acids, factors, etc.) required for proper translation. For capped mRNA, either wheat germ lysate or reticulocyte lysate are used. For uncapped mRNA, only reticulocyte lysate will result in translated protein. *In vitro* translation can also be used to radiolabel specific proteins in the presence of radioactive amino acids. The amino acids of choice are ^{35}S-labeled sulfur-containing residues (methionine, cysteine). Nonradioactive (biotinylated) amino acids can also be used in this system. The radiolabeled proteins produced are separated by sodium dodecyl sulfate–polyacrylamide gel electrophoresis (SDS–PAGE) and then labeling patterns are assessed by autoradiography or the use of a phosphorimager.

In vitro translation has multiple uses. It can be used to study how the translation of a particular gene occurs. For example, the use of frameshifting, IRES, and alternative splicing can all be determined via this technique. Radiolabeled products that are shorter than the expected size of the full-length protein utilize one of these alternative means of translation. Another use of *in vitro* translation is in the study of protein–protein interactions. The "bait" protein can be tagged for immunoprecipitation, whereas the *in vitro* translated "fish" protein can be radiolabeled. If the immunoprecipitated protein also pulls down the radiolabeled protein, then it can be concluded that the two proteins interact. Other uses of *in vitro* translation include (a) determination of rates of synthesis/degradation of a particular protein, (b) effects of mutation on protein function, and (c) obtaining high expression (large amounts) of a specific protein under the control of a strong promoter.

Gene Knockout

Gene knockout is rapidly becoming a common technique for assessing gene function. The creation of a knockout mouse is a long and tedious process. A quick way of knocking out a gene in cultured cells is through RNA interference (RNAi). In RNAi, the specific gene to be knocked out is placed (as DNA) in a plasmid containing strong promoters to transcribe both sense and antisense RNA in large excess. The plasmid is then placed in cells (a process known as transfection) and both sense and antisense RNA are produced using the cell's transcriptional machinery. Double-stranded RNA is preferred as it is more stable and can readily enter the cell (without being degraded by RNases) as opposed to single-stranded RNA. The antisense RNA hybridizes with the endogenous cytoplasmic mRNA for the specific gene required for knockout, removing that gene from the pool of mRNAs that are translated. Thus, transfected cells will no longer contain the protein for the specific gene of interest. The phenotype of the resulting cells can then be observed over time. Cellular function may be impaired if the protein is essential for survival.

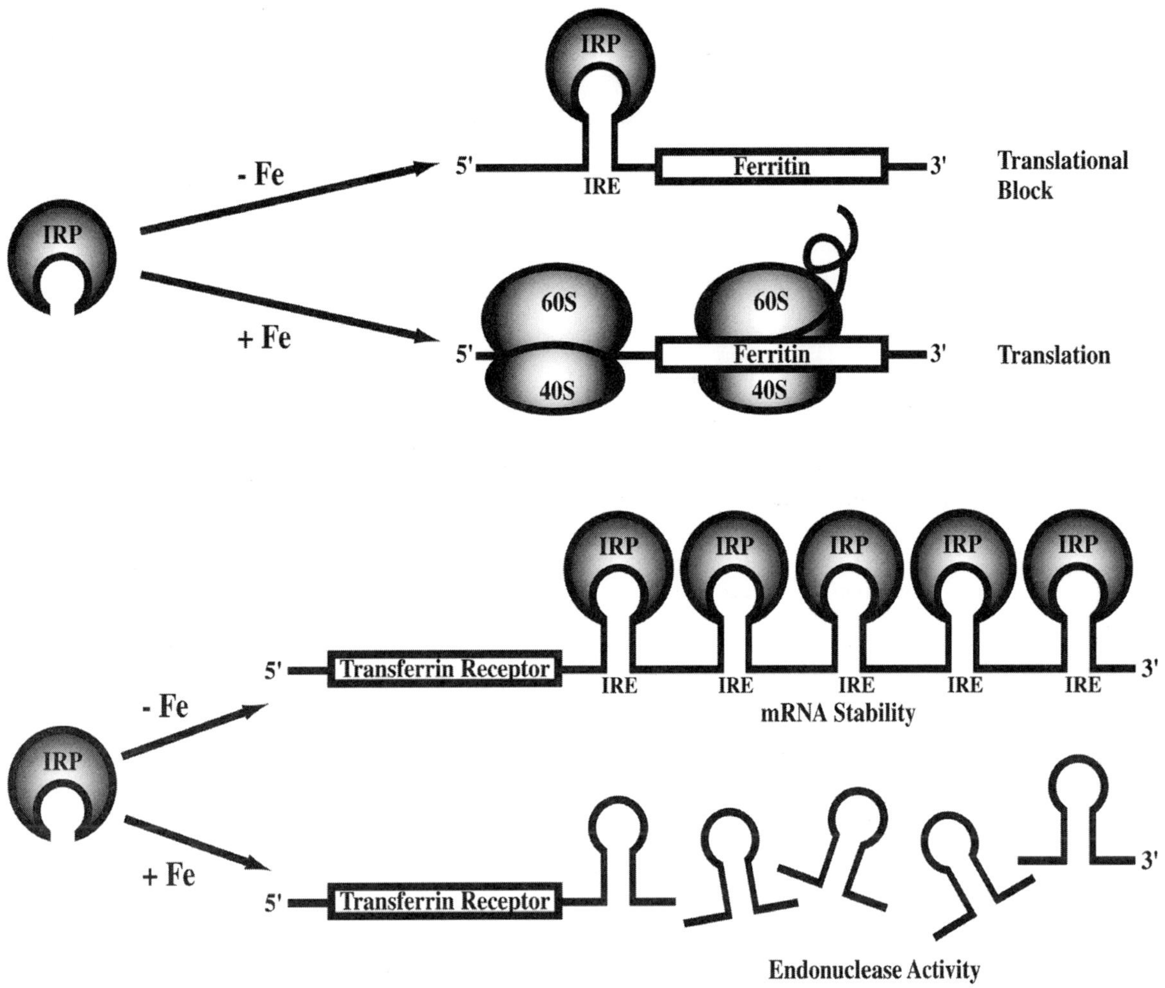

Figure 7.7 Regulation of iron homeostasis by iron regulatory proteins (IRPs). IRP is a dual function protein; it contains an iron–sulfur cluster under high intracellular iron concentrations, but this disassembles in low iron concentrations. When the iron–sulfur cluster is present, the protein has aconitase enzymatic activity. When the iron–sulfur cluster is absent, the cavity formed is available for binding the protein to the iron response elements (IREs) present in the 5′-untranslated region (5′UTR) of the ferritin transcript (one copy) and the 3′UTR of the transferrin receptor transcript (five copies). The binding of IRPs to the IRE of ferritin produces a physical block on ferritin mRNA translation under low iron concentrations that halts the synthesis of the iron storage protein. However, the mRNA of the transferrin receptor is stabilized by the binding of IRPs and is more efficiently translated under low iron conditions. This elevates the content of transferrin receptor to increase the transport of iron into cells under low iron conditions. When IRPs are not bound to the IREs of the transferrin receptor, the transcript is subject to endonuclease cleavage and translation is reduced.

Regulation by RNA Regulatory Elements As mentioned earlier, regulation of mRNA translation can also involve RNA-binding proteins. These proteins are cis-acting elements that act on either the 5′ or the 3′UTR. When proteins bind to the 5′UTR, they usually bind to regulatory elements. When they bind to the 3′UTR, they either bind to regulatory elements or are involved in mRNA stability.

One group of proteins, the iron-regulatory proteins (IRPs), provide an excellent illustration of this mode of control for these are involved in regulation at *both* the 5′ and the 3′UTR (Fig. 7.7). Iron is a key element in cellular growth and metabolism and a major factor in oxidative stress (see Chapter 12). Cellular iron levels are closely regulated by the transferrin receptor, which is responsible for iron uptake, and by ferritin, which stores iron. When iron is present in high concentrations, cellular iron uptake by transferrin is decreased while storage by ferritin is increased. When iron is in low abundance, the reverse occurs. Simultaneous regulation of the translation of both of these genes by iron is achieved, not by the regulation of nuclear transcription but by posttranscriptional control of mRNA translation due to the binding or release of IRPs. One copy of a conserved 28-nucleotide sequence, termed the iron responsive element (IRE), is present in the 5′UTR of the ferritin mRNA, whereas 5 copies are present in the 3′UTR of the transferrin mRNA transcript. The IRE forms an RNA stem loop where the sequence of the loop is almost always CAGUGN (N can be any base but G) (see Text Box 7.2).

Two IRPs (IRP-1 and IRP-2) recognize and bind to the IREs. IRP-1 has a dual function depending upon the iron concentrations present in the cell. When the iron concentration is high, an iron–sulfur cluster (4Fe–4SH) forms in IRP-1, and the protein functions as a cellular aconitase, interconverting citrate and isocitrate in the cytosol. When iron is low, the iron–sulfur cluster disassembles, and the site vacated by iron is available for binding to IREs. In contrast to IRP-1, IRP-2 lacks aconitase activity and is degraded by the proteosome when it is not bound to the IRE when iron is high. IRP-2 is ubiquitinated under iron-replete conditions and this targets it for degradation. Under low iron conditions, the IRPs bind to the IRE in the 5′UTR of ferritin and inhibit translation initiation. Under low iron, IRPs also binds to the IREs present in the transferrin receptor 3′UTR and protect the mRNA from the rate-limiting step of mRNA degradation, endonuclease cleavage. Under high iron conditions, IRPs do not occupy the IREs and that allows polysome formation and increased protein synthesis from the ferritin mRNA and endonuclease cleavage and mRNA degradation of the transferrin receptor mRNA. Other factors influence iron–sulfur cluster assembly, and thus the function, of the IRPs. These include H_2O_2, nitric oxide, hypoxia, and ascorbic acid (vitamin C). Other genes involved in iron regulation also contain IREs in their 5′UTR; these include erythroid 5-aminolevolinate synthase (eALAS), mitochondrial aconitase (m-aconitase), and succinate dehydrogenase iron protein (SDH-Ip). The results of IRP binding and inhibition of the translation of these transcripts include decreased porphyrin synthesis, increased citrate levels, and increased succinate levels for eALAS, m-aconitase, and SDH-Ip, respectively. Reactive oxygen intermediates produced from the later two enzymes decreases upon IRP binding.

TEXT BOX 7.2 MITOCHONDRIAL TRANSLATION

Mitochondria have their own genome and contain their own translation apparatus for producing the proteins encoded by their genes. Mitochondrial translation differs in some respects from cytosolic translation and, in fact, quite closely resembles the prokaryote machinery. Indeed, this is one of the pieces of evidence that points to a prokaryotic origin for the mitochondrion. Mitochondrial ribosomes have a lower sedimentation coefficient (55S) than do cytosolic ribosomes (80S) due to a lower RNA content and smaller sizes for both large (39S) and a small (28S) subunits. However, they contain many more proteins than their cytoplasmic counterparts. Mature mitochondrial mRNA does not have a 7-mG cap and this excludes cap recognition and ribosomal scanning as mechanisms in the translation initiation process. Indeed, the 28S subunit recognizes mRNA in a nonspecific, sequence-independent manner. The only initiation factor known to date is the mitochondrial initiation factor-2 (mtIF-2). This is a GTPase and it facilitates mitochondrial Met-tRNA binding to the 28S ribosomal subunit. Three elongation factors are involved in mitochondrial translation: mtEF-Tu, mtEF-Ts, and mtEF-G. Mammalian mtEF-Tu and mtEF-Ts form a tight complex that cannot be dissociated with GTP alone but can be dissociated in the presence of charged tRNAs and GTP. The process of mitochondrial termination of translation has not been well-studied and little is known about it.

Alternative Splicing Multiple transcripts can be produced from a single gene either by use of alternate promoters or by alternative splicing of the primary transcript. Alternative splicing gives rise to different protein isoforms, each with slightly altered function. Tissue-specific splicing can determine which protein isoform is expressed in a given cell. Exons are "skipped" during the processing of the pre-mRNA to produce the mature mRNA. The mechanism by which alternative splicing occurs is not clearly understood but most likely involves sequence-specific RNA-binding proteins that bind near the splice site and either activate or inhibit splicing. Splicing inhibitors sterically block access of splicing factors (snRNPs in spliceosomes), whereas splicing activators interact with splicing factors to enhance their association with the splice site. Differences in the relative concentrations of hnRNP proteins and splicing factors can influence the selection of alternative splice sites. The concentrations of these factors may differ between cell types. Splicing can either be cis (common) between exons carried on the same RNA molecule or it can be trans (rare) between exons on different RNA molecules. Trans-splicing can only occur

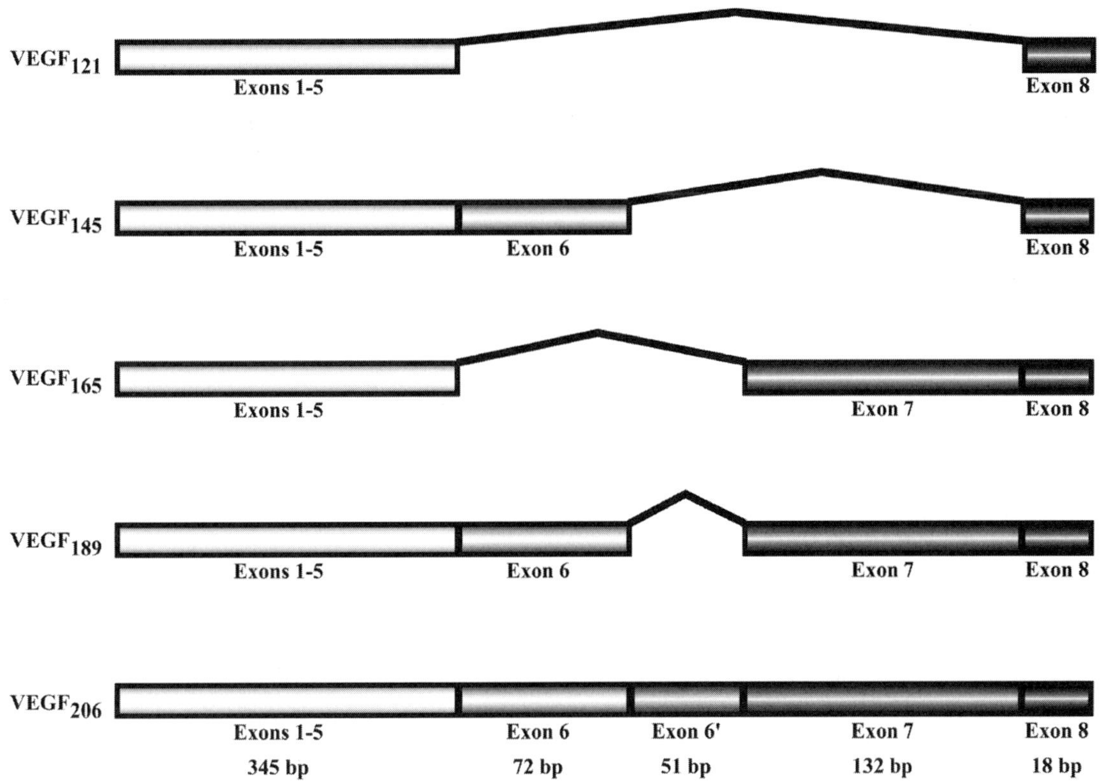

Figure 7.8 Isoforms of vascular endothelial growth factor (VEGF) created by alternative splicing. The exon numbers and lengths are indicated. Alternative splicing is represented by connecting lines between alternative splice sites.

if complementary sequences are present on introns between the trans exons.

Splicing factors interact with snRNPs to direct their activity at the splice site. These factors may be regulated by posttranscriptional modification to either enhance or weaken this association. Serine/arginine-rich (SR) splicing factors (e.g., ASF/SF2) can be phosphorylated by Clk/Sty kinases. They bind, in a sequence-specific manner, to pre-mRNAs. SR proteins have a C terminal rich in SR dipeptides, ideal for extensive phosphorylation. Phosphorylation enhances their interaction with U1 snRNPs, which, in turn, enhances their binding to the splice site.

Vascular endothelial growth factor (VEGF) is a secreted mitogen that is expressed almost exclusively in vascular endothelial cells where it induces microvascular permeability and plays a central role in both angiogenesis (the *de novo* formation of blood vessels from vascular precursor cells) and vasculogenesis. VEGF has many isoforms due to alternative splicing (Fig. 7.8). Many cytokines and growth factors up-regulate VEGF expression or induce VEGF release. VEGF expression is also controlled by hypoxia, the MAPK, PKC, and PKA signal transduction pathways, and by the transcription factors Sp1, AP-1, and AP-2. The binding of a hypoxia-induced stability factor (HuR) increases the half-life of this mRNA. The VEGF transcript also contains an IRES that is advantageous under hypoxic conditions when cap-dependent translation can be inhibited. Six isoforms of human VEGF have been identified: $VEGF_{121}$, $VEGF_{145}$, $VEGF_{165}$, $VEGF_{183}$, $VEGF_{189}$, and $VEGF_{206}$ (the subscript denotes the number of amino acids present in the isoform). $VEGF_{206}$ contains all eight exons of the VEGF gene. Exons 6a + 6b + 7, 6b + 7, 6a + 6b, 6a + 6b, and 6b are missing from $VEGF_{121}$, $VEGF_{145}$, $VEGF_{165}$, $VEGF_{183}$, and $VEGF_{189}$, respectively (* denotes a truncation of exon 6a). $VEGF_{121}$, $VEGF_{165}$, $VEGF_{183}$, and $VEGF_{189}$ have a broad tissue distribution whereas $VEGF_{145}$ and $VEGF_{206}$ are confined to cells of placental origin. $VEGF_{206}$ can be induced in certain cells by incubation with phorbol myristate acetate that stimulates PKC activity. $VEGF_{165}$ is a basic protein (due to the basic residues of exons 6 and 7) and is thus prevented from leaving the cell. VEGF isoforms containing these exons can bind heparin or heparin sulfate, being associated with heparin sulfate proteoglycans in the extracellular matrix and cell surface. In contrast, $VEGF_{121}$, which lacks this region, is mildly acidic and is released freely from cells. $VEGF_{189}$ and $VEGF_{206}$ contain sequence encoded by exon 6 and are completely sequestered in the extracellular matrix.

Another example of tissue-specific alternative splicing is the human mitochondrial ATP synthase γ subunit. The gene for this protein is 23 kilobases long and composed of 10 exons. Exons 1 and 2 contain the N-terminal mitochondrial localization sequence while exons 9 and 10 encode the C-terminal portions of the mature protein. Two isoforms of this protein exist, a heart (H) form and a liver (L) form. The H form is devoid of exon 9 and is found in heart and skeletal muscle. The L form contains exon 9 and is found in brain, liver, and kidney. Both forms are found in skin, intestine, stomach, and aorta. Thus, alternative splicing is an important factor in determining both the tissue specificity as well as the intracellular location of protein isoforms.

Transport of mRNA from Nucleus to Cytosol Another mechanism by which translation can be regulated is to control the transport of the mature mRNA from the nucleus to the cytosol. Transport of incompletely processed mRNAs containing introns would result in defective proteins being translated in the cytosol, with potentially disastrous effects on cellular function. By means that are not fully understood, pre-mRNAs associated with snRNPs in spliceosomes are prevented from being transported into the cytosol. The exception to this is the transport of unprocessed (unspliced) type D retroviral mRNA from the nucleus to the cytosol. Retroviral mRNA contains a constitutive transport element (CTE) that can bind endogenous tip-associated protein (TAP). TAP is an export receptor and contains partially overlapping NLS and NES that allows it to travel freely to and from the nucleus. TAPs accumulate near the nuclear rim and the nucleoplasm and can be transported rapidly out of the nucleus along with their mRNA cargo. The export of intron-containing retroviral mRNAs bypasses the cellular splicing machinery and immune responses to ensure efficient viral replication. The human immunovirus (HIV) relies on a similar pathway except that the CTE is replaced with a sequence (Rev/RRE) to which a viral Rev protein binds and channels the intron-containing HIV mRNA to the export pathway for snRNAs. Thus, viruses have evolved fascinating ways to "hijack" the normal cellular machinery to replicate themselves in cells.

Frameshifting Frameshifting represents another means of regulating translation at the level of the mRNA transcript. Frameshifting happens when a shift from one open reading frame to another occurs during translation (termed *recoding*). This process can be responsive to external stimuli and serve as a regulatory mechanism. The shift may be either one base in the 5′ direction (-1 frameshifting) or in the 3′ direction ($+1$ frameshifting), the latter being less common than the former. A -1 frameshift occurs when the mRNA "slips backwards" relative to both P- and A-site codons on the ribosome, whereas $+1$ frameshifts typically involve either a stop or a rare codon. In some cases the stop codon is overlooked by the ribosome in a process known as *readthrough.* Sequences in mRNAs that predispose ribosomes to shift reading frames are termed *programmed frameshift sites.* A common feature of these sites is the imposition of a translational pause. Translation is slowed down at these sites due to a structural feature of the site. Ribosomes pause at these sites either because of a stable RNA structure (e.g., a pseudoknot of folded, duplexed RNA) or because the sense or nonsense codon in the A site of the ribosome is recognized slowly. A $+1$ frameshift may also occur from a ribosomal "hop" in which the P-site tRNA dissociates from frame 1 of the mRNA while still bound to the ribosome and "lands" one nucleotide downstream in frame 2. This mechanism is known as repairing.

An example of frameshifting as a regulatory mechanism can be found in the translation of an inhibitor of the enzyme ornithine decarboxylase (ODC), know as the antizyme. ODC converts ornithine to putrescine, a component of the polyamine pool. When polyamine levels are high, further synthesis needs to be inhibited, and this is accomplished by antizyme. Antizyme inhibits the import and stimulates the export of polyamines. It also binds to and inhibits ODC, targeting it for proteolytic destruction by the 26S proteosome in much the same way as ubiquitin would. The difference between antizyme and ubiquitin is that only one antizyme molecule is required to present ODC to the proteosome, whereas proteins relying on ubiquitin are usually polyubiquitinated. The synthesis of functional antizyme is promoted by high polyamine levels that stimulate $+1$ frameshifting from the antizyme open reading frame 1 (ORF1) to ORF2. The relatively short ORF1 is overlapped by a much longer ORF2. The functional antizyme contains both the short sequence of ORF1 as well as the complete sequence of ORF2. There is no independent translation of ORF2 and only the $+1$ frameshifting produces a functional antizyme containing this portion (Fig. 7.9). Deletion analysis has identified sequence elements that stimulate frameshifting in the antizyme sequence. These include (1) a 50-nucleotide sequence just 5′ upstream of the frameshift site, (2) a UGA stop codon in ORF1, and (3) a pseudoknot 3 nucleotides downstream of the ORF1 stop codon. The mechanism for antizyme frameshifting may involve repairing.

Regulation of Translation Initiation

Mechanisms of translational control that center on the mRNA transcript (discussed above), as well as mechanisms

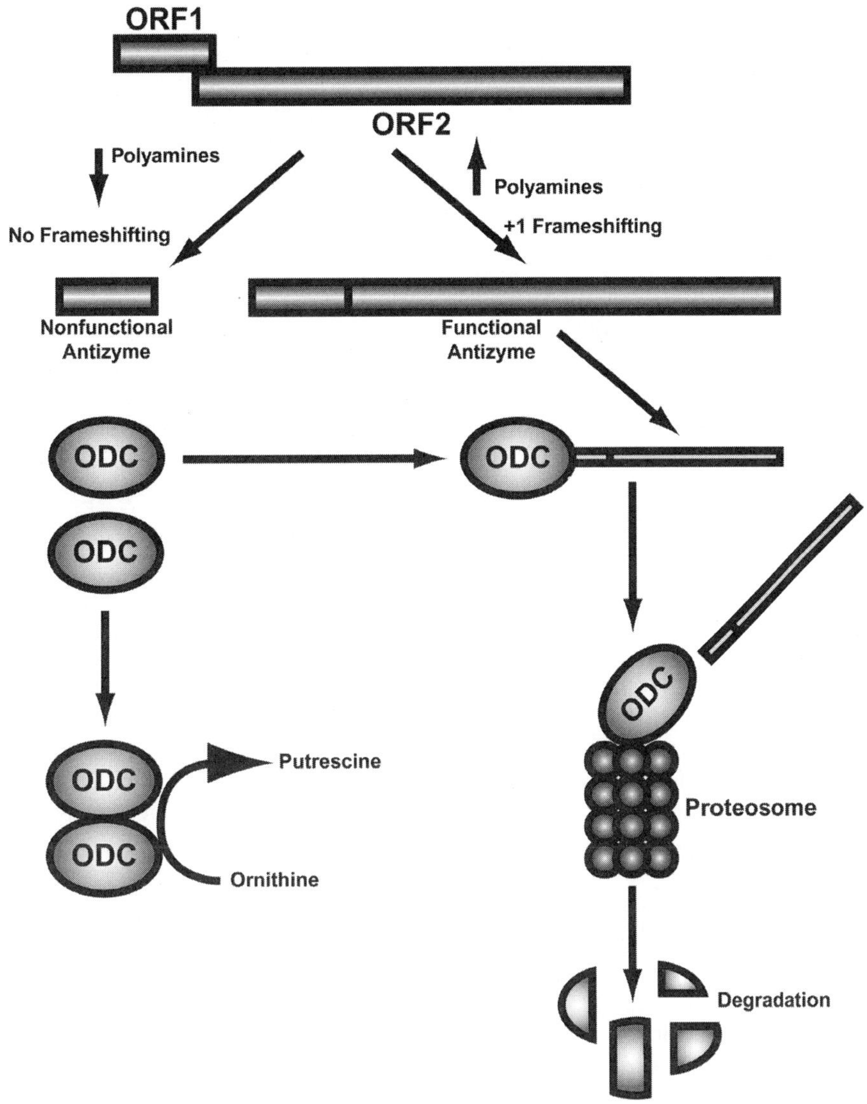

Figure 7.9 Frameshift translation of mammalian antizyme. Increased levels of polyamines cause a +1 frameshift in translation from ORF1 to ORF2 of the antizyme gene resulting in a full-length and functional antizyme protein. The functional antizyme inhibits ornithine decarboxylase (ODC) by binding to the individual subunits of the enzyme and directing them toward proteolytic degradation by the 26S proteosome. In the absence of antizyme, ODC subunits homodimerize to form the functional enzyme that converts ornithine to putrescine.

of transcriptional control (discussed in the previous chapter), provide multiple opportunities for the differential control of the expression of individual genes and their proteins. However, global controls on the overall rate of protein synthesis in cells are also necessary in order to meet challenges that demand either increased protein synthesis (e.g., growth, development, proliferation) or decreased protein synthesis (e.g., starvation, environmental stresses). Protein synthesis is a major energy-consuming process in all cells and must be integrated with both the metabolic demands of the cell and the available energy budget. Control of global protein synthesis is centered on the ribosomes and is implemented primarily via reversible phosphorylation control over the activities of selected ribosomal proteins. Both extra- and intracellular signals regulate the activities of multiple protein kinases and protein phosphatases that, in turn, regulate multiple elements in the basal translational apparatus to produce global changes in protein biosynthesis as well as some specific controls over the translation of certain mRNA types.

Many components of the basal transcription machinery are under reversible phosphorylation control. These include ribosomal proteins (S6), 15 of the initiation factors (eIF2$\alpha\beta$, eIF2B$\delta\varepsilon$, eIF3$\alpha\delta\varepsilon\zeta\eta$, eIF4B, eIF4E, eIF4F$\alpha\gamma$, eIF4G, eIF5), 5 elongation factors (eEF1, eEF1Bα, eEF1Bβ, eEF1Bγ, eEF2), and 8 aminoacyl tRNA synthases (AspRS, GluRS, GlnRS, HisRS, LysRS, MetRS, SerRS, ThrRS). Control may be exerted by direct phosphorylation of the factors or via phosphorylation of binding proteins that interact with these factors. The following section discusses the signals that induce phosphorylation or dephosphorylation, the factors that subject to reversible phosphorylation, the kinases and phosphatases involved, and the effects of reversible phosphorylation on translation.

Phosphorylation of eIF2 results in inhibition of translation. As mentioned previously, eIF2 binds GTP and Met-tRNA and delivers this complex to the 40S ribosomal subunit (Fig. 7.4). The subunit of eIF2 can be phosphorylated at Ser 51 by a number of different kinases in response to a number of different stimuli. These include hemin-regulated inhibitor kinase (HRI) in response to heme deficiency, double-stranded RNA-regulated protein kinase (PKR) in response to viral produced double-stranded RNA, the pancreatic eIF2α kinase or the PKR endoplasmic reticulum-related kinase (PERK) in response to unfolded proteins in the endoplasmic reticulum, and the general control of amino acid biosynthesis kinase 2 (GCN2) in response to serum starvation (the removal from cells in culture of essential nutrients that are normally provided to cultured cells by supplemented blood serum obtained from animals). Phosphorylation of eIF2α greatly increases its affinity for the GTP exchange factor eIF2B. eIF2B is required for eIF2-GDP to recycle to eIF2-GTP, and for the formation of the ternary complex (eIF2-GTP and Met-tRNA). The recycling of eIF2-GDP to eIF2-GTP by eIF2B is inhibited by phosphorylation of eIF2α. Since eIF2α is in much greater abundance than eIF2B, phosphorylation of eIF2α leads to complete sequestering of eIF2B and inhibition of translation.

Phosphorylation of eIF4E at serine 209 enhances translation. eIF4E is the cytoplasmic cap-binding protein. Phosphorylation occurs in response to growth factors, mitogens, cytokines, and hormones, acting via the Ras/Raf/MEK/ERK kinase cascade. Phosphorylation is also stimulated by certain stresses [hyperosmolarity, ultraviolet (UV) irradiation, exposure to lipopolysaccharide, arsenite, or anisomycin], in this case mediated via p38 MAPK. Dephosphorylation is triggered factors including serum deprivation, viral infection, and heat shock; the phosphatase is still unknown. Both phosphorylation pathways converge on Mnk1, a kinase that binds to the C terminus of eIF4G. eIF4G "curls around" so that C terminally bound Mnk1 can phosphorylate N terminally bound eIF4E. Phosphorylation of eIF4E enhances its ability to bind to the mRNA cap. The current model suggests that the phosphate forms a salt bridge with a nearby lysine on the cap-binding site of eIF4E. This effectively "clamps" the cap in eIF4E, allowing translation to ensue.

The translation repressor proteins, 4E-BP1, 2, and 3, are also regulated by phosphorylation. 4E-BPs compete with eIF4G for a common binding site on eIF4B. When 4E-BPs are phosphorylated, their interaction with eIF4E is decreased and translation proceeds. Dephosphorylation of 4E-BPs increase their affinity for eIF4E and translation is inhibited. The influences of many factors on translation are mediated via reversible phosphorylation control of 4E-BPs. Phosphorylation of 4E-BP is stimulated by hormones (e.g., insulin, angiotensin II), growth factors (e.g., EGF, PDGF, NGF, IGF-I, IGF-II), cytokines (e.g., IL-3, GMCSF in combination with steel factor), mitogens (TPA), G-protein-coupled receptor ligands (gastrin, DAMGO), and adenovirus infection. Conversely, serum starvation, amino acid deprivation, picornavirus infection, heat shock, and osmotic shock stimulate the dephosphorylation of 4E-BPs. Six SER/THR phosphorylation sites have been identified in mammalian 4E-BPs, five of which are immediately followed by a proline. The kinases involved include phosphoinositide 3′-OH kinase (PI3K), the SER/THR kinase Akt (also called protein kinase B), and the kinase FKBP12-rapamycin associated protein/mammalian target of rapamycin (FRAP/mTOR). Evidence for PI3K and FRAP/mTOR involvement comes from wortmannin (PI3K inhibitor) and rapamycin (mTOR/p70 S6 kinase inhibitor) inhibition of 4E-BPs phosphorylation, respectively. mTOR is a 290-kDa kinase that coordinates translation in response to mitogenic signals, energy levels, and nutrient sufficiency. During periods of sufficient nutrients, certain amino acids such as leucine, positively regulate mTOR signaling thereby relieving inhibition of translation by 4E-BPs and activating the S6 kinases, which can also regulate translation elongation. During stimulation, the first two (N terminal) of the six phosphorylation sites of 4E-BPs are phosphorylated by FRAP/mTOR. This primes the latter three (C terminal) sites to become phosphorylated (hyperphosphorylation) by a presently unknown kinase, subsequently releasing 4E-BPs from eIF4E and allowing eIF4E to bind to eIF4G and form the active eIF4F complex.

eIF4G activity is also regulated by phosphorylation. eIF4G is the bridging protein between the ribosome and the mRNA. It exists in two forms in the cell: eIF4GI and eIF4GII. eIF4GI is phosphorylated at one site in its N terminus (Ser274) and three sites in its C-terminal hinge region (Ser1108, Ser1148, Ser1192). When eIF4GI is unphosphorylated, it is inactive. One model of phosphorylation has the N terminus of eIF4G folded over the protein and masking the C-terminal hinge region. Serum or mitogen stimulation results in the phosphorylation of the N-terminal site by FRAP/mTOR. This causes a comformational change that results in the unmasking of the C-terminal phosphorylation sites. The C-terminal phosphorylation sites are phosphorylated by an unknown kinase (wortmannin- and rapamycin-insensitive), thereby activating eIF4G. "Unfolding" of eIF4G may open up the sites for association with other factors (PABP, eIF3, eIF4A, eIF4E) resulting in activation.

The PABP interacts with the initiation factors eIF4G and eIF4B (in plants) to promote translation. Phosphorylated PABP exhibits cooperative binding to poly(A) RNA as well as increased binding to eIF4G. The interaction of PABP with eIF4B in plants is decreased by heat shock and during different stages of development. This, however, is not due to dephosphorylation of PABP, but instead to the phosphorylation state of eIF4B. Serum stimulates the interaction between PABP and eIF4G in mammals and *Xenopus laevis*. Rapamycin does not prevent increased association of PABP and eIF4G, which suggests that the mechanisms of association are independent of FRAP/mTOR.

The helicase activity of eIF4F and recruitment of ribosomes to the mRNA are promoted by eIF4B. eIF4B is a phosphoprotein that functions as a dimer and is phosphorylated at Ser406 and Ser422. *In vivo*, eIF4B phosphorylation appears to involve FRAP/mTOR and the S6K1 kinase. However, unlike the 4E-BPs and eIF4G, eIF4B appears to be directly phosphorylated by S6K1 and not by FRAP/mTOR. FRAP/mTOR may phosphorylate S6K1, which, in turn, phosphorylates eIF4B. The effects of phosphorylation on eIF4B activity remain to be determined.

Regulation of Elongation

The regulation of elongation is another way to control global protein synthesis and the expenditure of ATP on this process. Serum and insulin are known to stimulate elongation, whereas anoxia, hypertonicity, and glucagon decrease the rate of elongation. Reversible phosphorylation control has been demonstrated for multiple elongation factors including eEF1A, eEF1B (α, β, γ subunits), and eEF2. eEF1A is phosphorylated at Thr431 by protein kinase Cδ (PKCδ). eEF1Bα is phosphorylated at Ser89, Ser106, and Ser112 by casein kinase-2 (CK-2) whereas EF1B and are phosphorylated at Thr122 and Thr230 by $p34^{cdc2}$ (Cdc2), respectively. Insulin and phorbol esters stimulate phosphorylation of eEF1A and eEF1Bα and γ. Phosphorylation may stimulate guanine nucleotide exchange, and therefore elongation, but this remains to be proven. Phosphorylation of eEF1Bα by CK-2 and eEF1Bβ by Cdc2 has been demonstrated in *Artemia salina* and *X. laevis*, respectively, but in not mammals to date.

The eEF2 is phosphorylated in the GTP-binding domain at Thr56 by eEF2 kinase. eEF2 is the only known substrate for eEF2 kinase, and phosphorylation inhibits eEF2 activity by disrupting its interaction with ribosomes. The kinase has calmodulin-binding activity and is dependent upon Ca^{2+} ions. Activity is stimulated by agents that raise intracellular Ca^{2+} including serum, bradykinin, thrombin, and histamine. Brain ischemia releases glutamate, which also stimulates Ca^{2+} release. β-adrenergic stimulation of PKA can phosphorylate and activate eEF2 kinase, whereas protein phosphatase-2A (PP-2A) can dephosphorylate eEF2 kinase and stimulate elongation. It is not known whether FRAP/mTOR phosphorylates eEF2k.

Regulation at the Level of the Ribosome

Regulation at the level of the ribosome involves the ribosomal protein S6. S6 is phosphorylated by $p70^{S6k}$ kinase (S6k) in response to growth factors and mitogens and phosphorylation increases protein synthesis. S6 lies within the mRNA-binding site of the 40S ribosomal subunit. Phosphorylation of S6 occurs at five serine residues present in the C-terminal cyanogen bromide fragment: Ser235, Ser236, Ser240, Ser244, and Ser247 and enhances the interaction of the ribosome with mRNAs that contain terminal oligopyrimidine tracts (TOPs). TOPs are short ($\sim$8 nucleotides) polypyrimidine tracts that occur at the mRNA transcriptional start site and are known to be transcriptional regulators. They form the core of the translational cis-regulatory element (TLRE). TOPs are found in all mRNAs of all vertebrate ribosomal proteins as well as the mRNAs for the elongation factors eEF1A and eEF2, human PABP, and human hnRNP A1. They mediate the regulation of the translational apparatus at the level of their synthesis. They allow for growth-dependent translational control of the mRNAs in which they are found by causing a shift of these mRNAs from translationally active polysomes in growing cells to translationally inactive RNP particles (subpolysomal fraction) in quiescent cells. This shift occurs between growing and nongrowing states of cells (both in culture and *in vivo*) in response to a wide variety of physiological and pharmacological stimuli. Translation of TOP mRNAs is selectively repressed upon growth arrest at the G_0, S, and M phases of the cell cycle. Thus, they provide an important means for cell growth control by regulating the availability of the ribosomal translation machinery.

Reversible phosphorylation control of ribosomal subunit S6k can be effected by a number of different kinases and phosphatases. S6k is activated by phosphorylation by PDK1, atypical PKCs (λ and ζ), and mTOR. It is inactivated by dephosphorylation by protein phosphatases (2A, 4, and 6). The availability of amino acids stimulates translation by activating mTOR phosphorylation of S6k. mTOR activity on Sk6 is inhibited by rapamycin. Phosphorylation of S6k through atypical PKCs occurs through PI3K and can be inhibited by wortmannin. Posttranslational modification of S6k via phosphorylation represents an important means of immediate control over translation.

CONCLUSIONS

Protein synthesis requires a large investment of energy from the cell. Consequently, the cell's contribution of energy to protein biosynthesis is held in check by tightly regulating the initial steps of transcription and translation. Organisms undergoing metabolic depression must simultaneously down-regulate both protein production and protein degradation. As seen in the previous chapter, multiple controls exist for regulating transcription. Given that this is the case, then one would wonder why translation requires regulation at all. The regulation of translation adds fine control to the expression of genes as proteins. Some systems lack transcriptional control (oocytes, reticulocytes, RNA viruses) and depend upon translational control exclusively. Transcription rates of certain genes can be considerably greater that their translation, keeping the production of the resulting protein curbed. Transcripts can also be accumulated and "held in reserve" by mRNA-binding proteins until the proper cellular signals causes the release of transcripts for translation. The reversible modification of mRNA binding proteins (usually by phosphorylation or dephosphorylation) results in responses that are both rapid and reversible. This means of regulation is particularly important in energy-deprived cells, such as those undergoing stress or responding to stress by metabolic depression. Repression of mRNA translation can also provide a means by which protein production is localized to a specific subcellular region, or a particular stage of development.

The mechanisms by which mRNA translation can be repressed present a potentially important therapeutic tool. With the discovery of natural "antisense" RNAs that control translation (see OxyR, previous chapter), the possibilities for controlling the expression of specific transcripts by gene knockout has become much more straightforward. RNA interference (RNAi) is now commonly used to repress the expression of specific genes *in vivo*. RNAi can be used to silence genes in a number of taxa including plants, invertebrates, and vertebrates. Double-stranded RNA has been successfully introduced into cultured cells of various organisms (including planarians, trypanosomes, zebrafish, *Caenorhabditis elegans*, *Drosophila*, mouse embryos, and most recently, mammalian cells). The approach has now become standard in deciphering the function of a protein of interest. The effects on cell phenotype, once a gene has been knocked out, can be directly observed. The use of RNAi could have broader implications for use in gene therapy. For example, in the field of cancer research, RNAi has been used to inhibit several different target proteins including growth factors, growth factor receptors, proteins involved in cell-cycle regulation, and proteins responsible for the invasive potential of tumor cells. By targeting specific transcripts with double-stranded RNA, the role of these genes in the development of disease can be investigated. Therefore, RNAi provides a highly specific alternative to traditional pharmacological agents in inhibiting translation of disease-related genes.

FURTHER READING

Gray, N. K., and Wickens, M. (1998). Control of translation initiation in animals. *Ann Rev Cell Develop Biol* **14**:399–458. *A complete review of the types of translational control giving specific examples for each.*

Hershey, J. W. B. (1991). Translational control in mammalian cells. *Ann Rev Biochem* **60**:717–755.

Sonenberg, N., Hershey, J. W. B., and Mathews, M. B. (eds.) (2000). *Translational Control of Gene Expression.* Cold Spring Harbor Laboratory Press, Cold Spring Harbor, NY. *A major resource of translational control mechanisms.*

van der Velden, A. W., and Thomas, A. A. M. (1999). The role of the 5′ untranslated region of an mRNA in translation regulation during development. *Int J Biochem Cell Biol* **31**:87–106. *Translational control from the 5′UTR with special emphasis on IRES. Many examples of specific genes are given.*

8

THE MITOCHONDRIA: POWERHOUSE OF THE CELL

KYRA J. COWAN

INTRODUCTION

Mitochondria first attracted molecular attention over a half-century ago. Since that time, many thousands of studies have analyzed all aspects of mitochondrial metabolism and documented the central role of mitochondria in adenosine 5′-triphosphate (ATP) production in eukaryotic cells, giving the mitochondrion its moniker as the *powerhouse of the cell*. The initial sections of this chapter provide an overview of the history, structure/function, and regulation of mitochondrial energy production, including a discussion of metabolic control analysis. Later sections of the chapter discuss selected aspects of mitochondrial metabolism, including protein transport processes, calcium signaling, and the central role of mitochondria in the primary events of programmed cell death (apoptosis). This material only scratches the surface of the functions of this multifaceted organelle and, undoubtedly, studies of the mitochondria will fascinate and inform for many years to come!

Mitochondrial History

In 1913 Otto Warburg documented the role of mitochondria in cellular respiration, but it was not until the 1930s that the urea and tricarboxylic acid (TCA) cycles were outlined. Intact mitochondria were first isolated in the 1940s, with subsequent analyses showing that the mitochondria contain the enzymes of the TCA cycle, fatty acid oxidation, and the respiratory assembly, among others. In the 1950s the mitochondria were determined to be the site for respiration and oxidative phosphorylation, and in the 1960s it was determined that the organelle contained its own deoxyribonucleic acid (DNA). Because of the close structural similarity between mitochondria and prokaryotes, it is believed that mitochondria evolved from symbiotic bacteria living inside primitive cells. The evidence for this includes the shape and size of mitochondria, the sequence of the mitochondrial genome, the size and type of mitochondrial ribosomes, the location of the respiratory assemblies and ATP synthesis complexes, as well as multiple aspects of the boundary membranes such as the composition of the inner membrane and the presence of a second, permeable, porin-expressing outer membrane. This theory, now well accepted, is crucially supported by the fact that mitochondria have their own unique genome, and studies of mitochondrial DNA (mtDNA) clearly support its eubacterial roots. In fact, the closest known ancestors of the mitochondria have been shown to be obligate intracellular parasites, members of the α division of the Proteobacteria. Studies have shown that the evolutionary division begins between the eubacteria *Rickettsia prowazekii* and the mitochondria of the protozoan *Reclinomonas americana.* The transcriptional life of our cells is a truly symbiotic affair—although key subunits of the most important proteins of the respiratory chain are encoded by the mtDNA, most of the proteins that function in the mitochondria are actually encoded in the nucleus. For example, only 3 of the 13 subunits of cytochrome *c* oxidase, the terminal protein complex of the respiratory chain, are encoded by mtDNA (Fig. 8.1).

There are two hypotheses regarding the origin of the eukaryotic cell. The first proposes that there could have been a simultaneous formation of the nucleus and the mitochondria through the fusion of a hydrogen-producing α-Proteobacterium with a hydrogen-requiring, methanogenic Archaebacterium. The second suggests that a series of steps was involved. Initially a fusion of a Proteobacterium and an Archaebacterium could have occurred to form a protoeukaryote lacking mitochondria, and then this was

Functional Metabolism: Regulation and Adaptation, edited by Kenneth B. Storey.
ISBN 0-471-41090-X

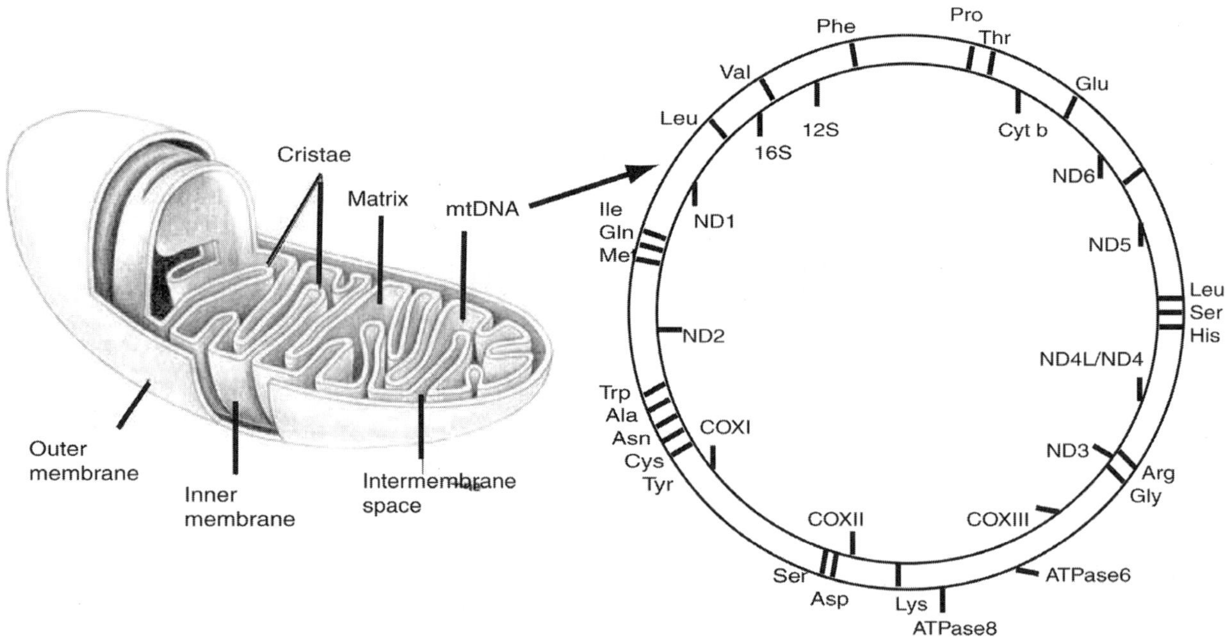

Figure 8.1 Structure of a mitochondrion and map of the human genome showing the genes coding polypeptides and ribosomal RNAs as well as mitochondria-specific tRNAs. The genome consists of only 16,569 base pairs. ND, subunits of NADH-ubiquinone oxidoreductase (complex I); COX, subunits of cytochrome oxidase (complex IV), Cyt *b*, cytochrome *b* (complex III); ATPase 6 and 8, subunits of the F_1-ATP synthase (complex V).

followed by endosymbiosis of an α-Proteobacterium to become the mitochondrion and complete the formation of the eukaryote. However, many questions remain (for more about this see Chapter 20). Scientists continue to search for more ancient mitochondrial genomes than that found in *R. americana*, in order to identify larger and more gene-rich mtDNAs. Studies also continue to define the eukaryotic nuclear genomes of protists, as the genome sequence of *R. prowazekii* clearly confirmed the link between early-division protists and the mitochondrial genome. The coevolution of eukaryotic cells and their microbes is a complex metabolic story, largely as yet untold.

Structure

Mitochondria are found dispersed in the cytosol of all eukaryotes and are responsible for oxygen-dependent ATP synthesis in cells. Recent studies indicate that dispersal is not random, and that mitochondria are strategically placed in sublocales within cells where ATP demand is high. Evidence from electron microscopy as well as some functional studies *in vivo* indicate that mitochondria within a single cell are functionally linked—perhaps even present as ONE functional unit. The organelles are commonly about 0.5 μm in width and 2 μm in length, about the same size as many bacteria. Mitochondria are generally ellipsoid-shaped organelles, with a smooth outer membrane and an invaginated inner membrane that surrounds a dense matrix that contains metabolic enzymes at very high concentrations. Indeed, the matrix *in vivo* is more densely packed with proteins than some protein crystals created in the laboratory—it is essentially an array of *cubically closest packed* protein subunits from membrane to membrane. Virtually no free hydrating water exists within the mitochondria and enzyme-linked reactions and pathways are so crowded that normal rules of diffusion do not apply. In some cases the number of binding sites for small molecular weight molecules [such as reduced nicotinamide adenine dinucleotide (NADH)] are higher than the number of molecules of the metabolite actually present in the mitochondrial matrix. The details of metabolic control, transcription, and translation in such a crowded, highly ordered environment remain to be elucidated.

The outer mitochondrial membrane is very porous, allowing passage of ions and other small molecules up to 5000 Da. This membrane contains very few enzymatically active proteins. The inner membrane is extensively folded into ridges called cristae (Fig. 8.1), the number of invaginations correlating positively with the respiratory activity of the individual cell. The inner membrane has two domains. The first is referred to as the inner boundary domain and is adjacent to the outer membrane, making several close contacts with the outer membrane. The second domain forms the cristae, which are connected to and continuous with the inner boundary membrane. The inner membrane is essentially impermeable to charged or polar molecules but contains multiple specific protein carriers that translocate a variety of molecules including all nucleotides and fuel molecules. New research, utilizing the availability of the complete genome sequence of yeast and humans, indicates that there are many potential mitochondrial membrane transporters present in the genome with sequences that have yet to be characterized.

The space between the outer membrane and the inner membrane is called the intermembrane space, and the space within the inner membrane, inside the folds of the cristae, is called the matrix. It is within the matrix that intermediary metabolism occurs, including fatty acid oxidation and the TCA cycle. Oxidative phosphorylation also takes place inside the matrix but is firmly rooted on the inner mitochondrial membrane. Also found in the matrix is the mitochondrial genome, which codes for selected subunits of inner membrane proteins and for mitochondrial ribosomal ribonucleic acids (rRNAs). Mitochondria have their own ribosomes and other components required for protein translation. Detailed knowledge of the control of mitochondrial transcription (through factors called nuclear respiratory factors (NRF)) and of the control of mitochondrial protein synthesis lags behind our knowledge of the nuclear and cytoplasmic control mechanisms for these processes. During low oxygen stress in lower vertebrates, for example, mitochondrial transcription increases in some tissues at the same time that the majority of nuclear transcription is shut down (see Chapter 15). The control of this process is unclear. Many of the proteins found in the mitochondria are transcribed and translated from nuclear DNA, and the proteins are then imported into mitochondria. Proteins synthesized in the cytoplasm that are fated to end up in mitochondria are translated with inherent targeting signals. These signals allow for transport across the inner membrane by specific membrane translocases.

Mitochondria have their own complement of stress shock proteins, notably the 60- and 10-kDa heat shock proteins. Synthesized in the cytoplasm, they are targeted to the mitochondria where they are involved in chaperone activities. The mitochondrial matrix also has its own complement of protein kinases to regulate the activities of its enzymes. For example, an adenosine cyclic 3′, 5′-phosphate (cAMP)-dependent protein kinase A isozyme is present as well as specific kinases that regulate the key TCA cycle enzymes pyruvate dehydrogenase (PDH) and α-ketoglutarate dehydrogenase (α-KGDH). Interestingly, although the PDH and α-KGDH kinases (as well as the third member of this group, the branched-chain α-ketoacid dehydrogenase kinase) phosphorylate serine residues on their target enzymes, their sequence and structure shows little homology with other vertebrate serine/threonine kinases. Instead, these three mitochondrial kinases are related to the histidine kinases that are found in bacteria, indicating that their origin was with the kinases that were present in the bacterial ancestor of the mitochondrion.

ENERGY METABOLISM

Before discussing the details of the various processes, a quick summary of mitochondrial energy metabolism is useful (Fig. 8.2). The primary fuel for the TCA cycle is acetyl-coenzyme A (acetyl-CoA). This is provided by the importation of pyruvate (the product of glycolysis and of the catabolism of selected amino acids) or from the ß oxidation of fatty acids within the mitochondria. Acetyl-CoA is catabolized by the TCA cycle, releasing $2CO_2$ and generating reducing equivalents in the form of NADH or flavin adenine dinucleotide, reduced from three sites and $FADH_2$ from one site and with one substrate-level phosphorylation forming guanosine 5′-triphosphate (GTP) at the succinyl thiokinase reaction. Reducing equivalents generated in the TCA cycle, as well as those imported from the cytosol via the malate-aspartate or glycerophosphate shuttles, are introduced into the electron transport system and pass through complexes I (or II), III, and IV with the final reduction of oxygen to form water by complex IV (Fig. 8.3). Reducing equivalents from NADH enter the chain at complex I, whereas those from $FADH_2$ enter at complex II so that the oxidation of 1 NADH results in the net formation of ~3 ATP versus ~2 ATP for $FADH_2$. Electron passage through each complex releases free energy with sufficient amounts released at each of complexes I, III, and IV to support ATP synthesis. The chemiosmotic theory of oxidative phosphorylation states that the free energy of electron transport is conserved by pumping hydrogen ions from the mitochondrial matrix into the intermembrane space. A *proton-motive force* is created that includes the electrical potential difference across the membrane ($\Delta\psi$) and a pH gradient. The free energy released on proton translocation back into the mitochondrial matrix drives the phosphorylation of ADP to form adenosine 5′-diphosphate (ADP) to form ATP by complex V (the F_1F_0-ATPase).

Pyruvate Dehydrogenase Complex

During carbohydrate oxidation, pyruvate enters the mitochondrion and undergoes oxidative decarboxylation by the pyruvate dehydrogenase complex to form acetyl-CoA, in the process generating NADH. Acetyl-CoA enters the TCA cycle via conjugation with oxaloacetate to form

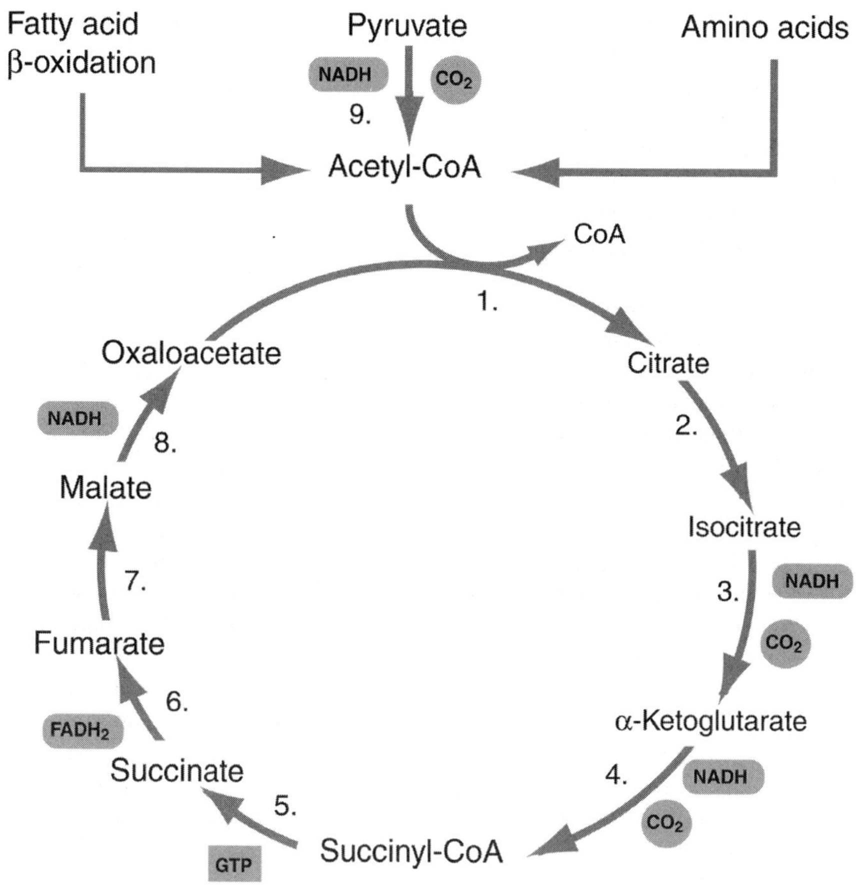

Figure 8.2 The tricarboxylic acid cycle showing inputs of acetyl-CoA from fatty acid, carbohydrate, and amino acid fuel sources and the reactions that yield outputs of reducing equivalents (NADH, $FADH_2$), substrate-level phosphorylation (GTP), or decarboxylation (CO_2). The net yield of one turn of the cycle is $2CO_2$, 1 GTP, 3 NADH + H^+ and 1 $FADH_2$. Enzymes are (1) citrate synthase, (2) aconitase, (3) NAD-dependent isocitrate dehydrogenase, (4) α-ketoglutarate dehydrogenase, (5) succinyl thiokinase, (6) succinate dehydrogenase, (7) fumarase, and (8) malate dehydrogenase. Pyruvate entry into the cycle is via the pyruvate dehydrogenase complex (9).

citrate, and subsequent oxidative decarboxylations at the isocitrate dehydrogenase and α-KGDH reactions coupled with additional oxidations at the succinate dehydrogenase and malate dehydrogenase reactions return the cycle to oxaloacetate while producing $2CO_2$, 1 GTP, 3 NADH + H^+ and 1 $FADH_2$. Fuel selection by the mitochondria varies greatly depending on a variety of factors including species and tissue type, nutritional status, and work/stress load (see Chapters 9, 11, 16). For example, whereas fatty acids are the major oxidative fuel for humans and many other animals, some insect species have an energy metabolism that is based virtually exclusively on the cooxidation of carbohydrates (providing acetyl-CoA) and the amino acid, proline. Proline is catabolized via glutamate as an intermediate and enters the TCA cycle at the α-KGDH reaction to supplement the levels of C4 and C5 intermediates. This pyruvate/proline pair provides the "take-off" fuel for flying insects, the instant power that allows flies and other insects to evade swatting.

Formation of acetyl-CoA from pyruvate [reaction (8.1)] is a critical process for carbohydrate entry into the TCA cycle, and the pyruvate dehydrogenase complex (PDC) is tightly regulated (Fig. 8.4). The complex involves three catalytic enzymes [pyruvate dehydrogenase (PDH), dihydrolipoyl transacetylase and dihydrolipoyl dehydrogenase], each present in multiple copies to make up a final macromolecular complex. Multiple cofactors are involved: thiamine pyrophosphate (TPP), lipoamide, and FAD as catalytic cofactors, and NAD^+ and CoA as stoichiometric cofactors. The PDC also contains PDH kinase and PDH

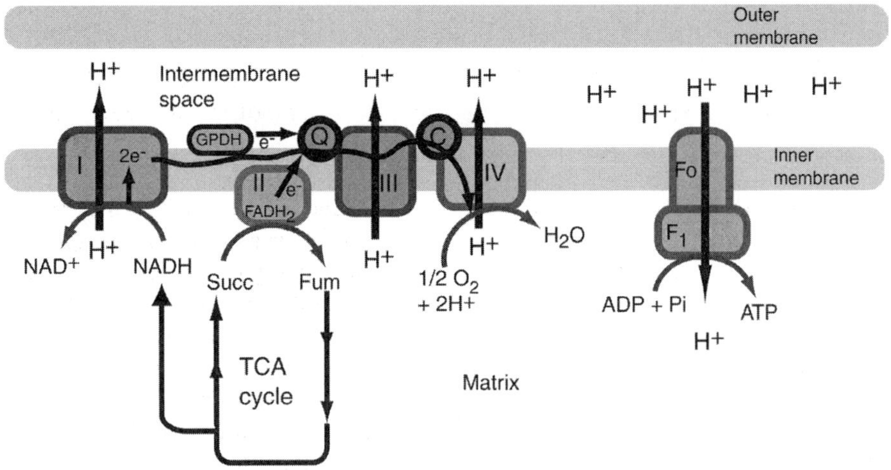

Figure 8.3 Schematic of the electron transport system and oxidative phosphorylation. The NADH and $FADH_2$ produced by the TCA cycle donate electrons to complexes I and II, respectively, and these are then passed via ubiquinone (Q) to complex III and then to cytochrome *c* (C) for transfer to complex IV where the electrons are used to reduce oxygen to H_2O. The glycerol-3-phosphate shuttle carries in electrons from the cytosol, and these are donated to the FAD-containing glycerol-3-phosphate dehydrogenase on the outer face of the inner membrane and also passed on to reduce ubiquinone. Protons are pumped out of the matrix and into the intermembrane space at complexes I, III, and IV, and the electrochemical gradient generated is used to drive ATP synthesis by the F_1F_0-ATPase (complex V).

phosphatase activities that are regulated by signal inputs from cellular metabolism and, via their consequent action in phosphorylating and inhibiting PDH versus dephosphorylating and activating PDH, are critical to the overall regulation of pyruvate oxidation under the influence of hormones (e.g., insulin) and metabolic stresses. The overall reaction is

$$\text{Pyruvate} + \text{NAD}^+ + \text{H}_2\text{O} + \text{CoA} \rightarrow \text{Acetyl-CoA} + \text{NADH} + \text{H}^+ + \text{CO}_2 \qquad (8.1)$$

The reaction steps in the PDC are as follows (Fig. 8.4). First is the decarboxylation of pyruvate by PDH using the coenzyme TPP and creating the reactive intermediate hydroxyethyl thiamine pyrophosphate (HETPP). In the second step, catalyzed by dihydrolipoyl transacetylase, HETPP attacks the lipoamide disulfide, transferring the hydroxyethyl group and oxidizing it to form acetyl lipoic acid. The acyl group is then transferred to coenzyme A by the same enzyme, leaving reduced lipoic acid. The third step oxidizes the reduced lipoic acid using a bound FAD and then transferring reducing equivalents to form NADH.

Cellular control of the PDH complex involves several factors. The enzyme is under allosteric control and is inhibited by high ratios of NADH/NAD, acetyl-CoA/CoA and ATP/ADP. The latter ratio signals the energetic state of the cell and slows pyruvate oxidation when ATP levels are high and/or energy demands low. The acetyl-CoA/CoA ratio rises when lipids are being oxidized as fuels (lipid oxidation always suppresses carbohydrate catabolism) or when biosynthesis from TCA cycle outputs slows (e.g., citrate is exported from the mitochondria carrying acetyl units for fatty acid biosynthesis). The redox ratio NADH/NAD links TCA cycle function with the activity of the electron transport system and is particularly important in halting pyruvate catabolism when oxygen is limiting. Ca^{2+} is also an activator; in muscle, high Ca^{2+} stimulates contraction and activates glycogen phosphorylase to increase glycolytic flux so it is logical that the ion would also activate the PDC, which introduces the glycolytic product, pyruvate, into the TCA cycle.

Reversible phosphorylation provides virtual on–off control of the PDC to gate carbohydrate entry into the TCA cycle in response to hormonal action or environmental stresses. As with glycogen phosphorylase, activity of the PDC is typically expressed as the percentage of total activity present in the active $\underline{a}$ form (PDH$\underline{a}$ is assayed directly; total activity is measured after incubation under conditions that promote enzyme dephosphorylation). The percent PDH$\underline{a}$ is strongly suppressed in many situations, and control at this locus is one of the key events involved in aerobic metabolic rate depression. For example, mammals entering hibernation or snails entering estivation show almost complete inhibition of PDH that helps to

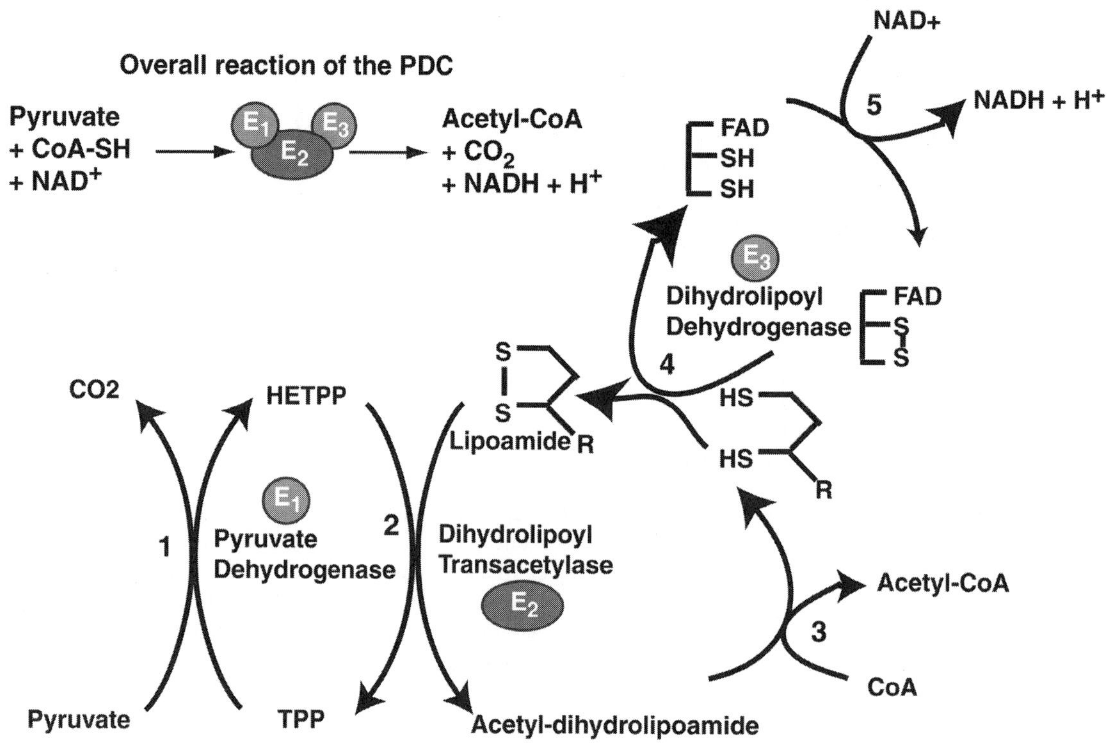

Figure 8.4 Reaction mechanism of the pyruvate dehydrogenase complex catalyzing the conversion of pyruvate into acetyl-CoA. The three enzymes of the complex are shown (pyruvate dehydrogenase, dihydrolipoyl transacetylase, dihydrolipoyl dehydrogenase) as well as five steps in the process: (1) decarboxylation of pyruvate by pyruvate dehydrogenase using the coenzyme TPP creates the reactive intermediate hydroxyethyl thiamine pyrophosphate (HETPP), (2) catalyzed by dihydrolipoyl transacetylase, HETPP attacks the lipoamide disulfide, transferring the hydroxyethyl group and oxidizing it to form acetyl dihydrolipoamide, (3) the acyl group is then transferred to coenzyme A by the same enzyme, leaving reduced lipoic acid, (4) the reduced lipoic acid is oxidized by dihydrolipoyl dehydrogenase using a bound FAD, and (5) reducing equivalents are then transferred to NAD^+ to form NADH.

lower overall metabolic rate by cutting off carbohydrate catabolism. Indeed, increased expression of the PDH kinase gene is one of the highly visible gene expression events as hibernators sink into torpor (see Chapter 16).

The carbon intermediates of the PDC are "channeled" within this three-enzyme complex, and this arrangement—where substrate enters and only a final product emerges—is a model of a phenomenon on metabolic regulation that might be more common than most scientists believe. The PDC is one of a number of tightly bound enzyme complexes in metabolism. Like the PDH, some others are readily isolated as intact complexes (e.g., the ribosomal translation machinery or the 26S proteosome), whereas there is increasingly good evidence that many others are organized in three-dimensional space to facilitate their function *in vivo* but dissociate under *in vitro* extraction conditions (e.g., glycolysis, stress-activated kinase cascades). Recall the advantages for pathway function that derive from complex formation (see Chapter 3). The environment in the mitochondria is just the type of crowded proteinaceous neighborhood where such complexes can be most easily studied. The enzymes involved in β oxidation of fatty acids within the mitochondria form another example of a multienzyme complex where the intermediates are "shielded" within the complex and only a final product (acetyl-CoA) emerges. Furthermore, the TCA cycle itself is likely to be an organized physical entity, linked structurally to the PDC complex and to glutamate dehydrogenase. Experiments on these complexes are difficult to undertake, but some labs continue to pursue the elusive *pathway complex*, which may be the basis of many metabolic functions.

Tricarboxylic Acid Cycle

Otherwise known as the Krebs cycle or the citric acid cycle, the TCA cycle is the common destination of all fuel molecules that are catabolized by the mitochondria: fatty acids, ketones, carbohydrates, and amino acids (Fig. 8.2). Carbon input into the TCA cycle comes from a C2 component (acetyl-CoA) and one or several C4 or C5 inputs (often the carbon skeletons of amino acids) that are processed within the cycle to produce the C4 oxaloacetate that condenses with acetyl-CoA to form citrate. When TCA cycle function is focused primarily on ATP production, the most common fuel input is acetyl-CoA with lesser amounts of C4 and C5 inputs introduced to provide a sufficiently high pool size of oxaloacetate to accept incoming acetyl-CoA units. During exercise in humans and fish, for example, the total pool size of the intermediates of the TCA cycle increases substantially as TCA cycle flux, oxygen consumption, and rate of ATP synthesis increases. The intermediates of the TCA cycle also provide the raw material for a wide variety of biosynthetic pathways; hence, replenishment of intermediates is needed to allow biosynthesis without jeopardizing the energy-producing function of the cycle. Anaplerotic reactions provide this "top up," the most important being the pyruvate carboxylase reaction [reaction (8.2)] which is activated by rising acetyl-CoA concentrations when insufficient C4 moieties are available for incoming C2 units.

$$\text{Pyruvate} + CO_2 + \text{ATP} + H_2O \rightarrow \text{oxaloacetate} + \text{ADP} + P_i \quad (8.2)$$

where P_i is inorganic phosphate. The paragraphs below summarize the enzymatic steps of the TCA cycle and some of the regulatory features of the individual enzymes. The reactions are summarized in Table 8.1.

The first step of the TCA cycle is catalyzed by the citrate synthase (CS):

$$\text{Acetyl-CoA} + \text{oxaloacetate} \rightarrow \text{citrate} + \text{CoA} \quad (8.3)$$

This is a condensation step, forming citryl-CoA, a product that is then hydrolyzed into citrate and CoA. The enzyme shows ordered sequential kinetic behavior; oxaloacetate binds first and stimulates a conformational change in the protein that generates the acetyl-CoA binding site (because of this the enzyme cannot "accidentally" hydrolyze acetyl-CoA). TCA cycle flux can be limited by the supply of both substrates to the CS reaction; acetyl-CoA limitation is generally rectified by increased fuel catabolism, whereas oxaloacetate supplies are elevated by anaplerotic reactions. CS is inhibited by ATP *in vitro*, but the physiological significance of this inhibition has only been shown in the tissues of anoxia-tolerant marine invertebrates.

The next step in the TCA cycle is the isomerization of citrate into isocitrate, carried out by aconitase:

$$\text{Citrate} \leftrightarrow \text{cis-aconitate} \leftrightarrow \text{isocitrate} \quad (8.4)$$

Early studies showed that aconitase appeared to have little or no influence on TCA flux. However, recent studies have show that aconitase is an early target of free radical damage in stressed cells because of the presence of an iron–sulfur cluster in the enzyme. As further discussed in Chapter 12, superoxide radicals inactivate aconitase by reducing Fe^{3+} in the enzyme and causing the release of ferrous iron; such inactivation can have biological significance if it impedes TCA cycle function. Interestingly, the cytoplasmic form of aconitase does "double duty" as an enzyme and as an iron-binding protein that regulates the transcription of the iron-binding proteins ferritin and transferrin as well as mitochondrial aconitase activity, among others (see Text Box 8.1).

The isocitrate produced by aconitase is then involved in the first oxidation step of the TCA cycle, which is catalyzed by NAD-dependent isocitrate dehydrogenase (NAD-IDH)

TABLE 8.1 Enzymes of the Tricarboxylic Acid Cycle

Enzyme	Reaction	$\Delta G^{\circ\prime}$ (kJ/mol)	Cofactors
Citrate synthase	Acetyl-CoA + oxaloacetate + H_2O → citrate + CoA	−32	
Aconitase	Citrate ↔ *cis*-aconitate ↔ isocitrate	13	[4Fe-4S] cluster
Isocitrate dehydrogenase	Isocitrate + NAD^+ → α-ketoglutarate + NADH + CO_2	−21	Mg^{2+} or Mn^{2+}, NAD^+
α-Ketoglutarate dehydrogenase complex	α-ketoglutarate + NAD^+ + CoA → succinyl-CoA + NADH + CO_2	−33	TPP, lipoamide, CoA, FAD, NAD^+
Succinyl-CoA synthase	Succinyl-CoA + GDP + P_i → succinate + GTP + CoA	−2.9	GDP
Succinate dehydrogenase	Succinate + FAD → Fumarate + $FADH_2$	0	FAD, Fe-S clusters
Fumarase	Fumarate + H_2O → L-malate	−3.8	
Malate dehydrogenase	L-Malate + NAD^+ → oxaloacetate + NADH + H^+	29.7	NAD^+
Net reaction	Acetyl-CoA + $2H_2O$ + $3NAD^+$ + FAD + GDP + P_i → $2CO_2$ + 3 NADH + $2H^+$ + $FADH_2$ + CoA + GTP	−49.8	

[reaction (8.5)]. Isocitrate is first oxidized to a ketone (oxalosuccinate) and then decarboxylated to α-ketoglutarate.

$$\text{Isocitrate} + NAD^+ + H_2O \rightarrow \alpha\text{-ketoglutarate} + CO_2 + NADH + H^+ \qquad (8.5)$$

The reaction catalyzes the NAD^+-dependent oxidation of the secondary alcohol, isocitrate, to form oxalosuccinate, which exists transiently while it is decarboxylated to form α-ketoglutarate. Mg^{2+} or Mn^{2+} are required for enzyme activity. NAD-IDH has long been implicated in TCA cycle control. For example, studies with isolated mitochondria in the 1970s convincingly showed that NAD-IDH was rate-limiting to TCA cycle flux in insect flight muscle, whether burning carbohydrate, lipid, or proline fuels. The flight muscle enzyme is activated by ADP and inhibited by Ca^{2+} and hence responds to key changes in muscle metabolism during flight initiation: ADP rises due to ATP hydrolysis by myosin ATPase whereas Ca^{2+} is released from both the sarcoplasmic reticulum and mitochondria to activate muscle contraction. Interestingly, in mammalian systems, Ca^{2+} is an activator of NAD-IDH, and other controls include inhibition by NADH and allosteric activation by ADP. The enzyme in some organisms also undergoes reversible changes in the polymerization state. In some microbes NAD-IDH is subject to reversible phosphorylation (functioning to regulate flow of isocitrate through the glyoxylate bypass), but this does not appear to be the case in animals. Interesting studies of mitochondrial isocitrate metabolism add support for the idea of an enzyme complex (metabolome) of the TCA cycle enzymes. Mitochondria contain both NAD-IDH and an NADP-linked IDH, enzymes that are different in both structure and function (the NADP-linked enzyme is associated with biosynthesis). Careful and complete labeling studies have shown that even though these two enzymes coexist in the mitochondrial matrix, NADP-IDH is <u>not</u> involved in TCA cycle flux. This supports the idea that the TCA cycle must be compartmentalized and that carbon flows entirely through NAD-IDH, even though the activity of the NADP-dependent enzyme is always much higher! Several investigators feel that a complex of <u>all</u> TCA cycle enzymes operates *in vivo* and acts to shorten the diffusion distances of intermediate and coordinates flux through the entire pathway.

The next enzyme in the cycle, α-KGDH, is responsible for catalyzing the oxidative decarboxylation of α-ketoglutarate to succinyl-CoA:

$$\alpha\text{-Ketoglutarate} + NAD^+ + CoA + H_2O \rightarrow \text{succinyl-CoA} + CO_2 + NADH + H^+ \qquad (8.6)$$

This is the second oxidative decarboxylation reaction in the TCA cycle. α-KGDH is an enzyme complex, very similar to the PDC, and consists of the same three types of enzymes; indeed, the same dihydrolipoyl dehydrogenase is in both complexes. The product of both complexes is a CoA ester of the acid that contains one less carbon than

TEXT BOX 8.1 IRON–SULFUR CLUSTERS OF MITOCHONDRIAL ENZYMES

The iron–sulfur cluster-containing proteins of the mitochondria have important roles in catalysis and electron transfer. The four iron atoms in aconitase are bound to four inorganic sulfides and four cysteine sulfur atoms. The cluster participates in the dehydration/hydration reactions of the substrate. Succinate dehydrogenase has a different configuration and contains three types of iron–sulfur clusters: 2Fe–2S (two irons bound to two sulfides), 3Fe–4S, and 4Fe–4S. Other Fe–S-containing proteins in the mitochondria are subunits of complexes I, II, and III. Collectively these proteins are known as nonheme iron proteins or iron–sulfur proteins.

The synthesis of these proteins has been studied in yeast mitochondria. Generally, the components of this synthesis process include Nfs1p, producing sulfur from cysteine to be incorporated into the Fe–S complex. Two other proteins are Isu1p and Isu2p, which are located in the mitochondrial matrix and thought to bind the iron-containing substrate, and catalysis by Nifs1p is thought to create the Fe–S intermediate cluster on the Isu proteins. The addition of reducing reagents results in the release of the product. A final component involved is a protein known as Nfu1p, used for Fe–S protein maturation. Once synthesized the Fe–S cluster is converted into other types of clusters, and then transferred to apoproteins. Addition of reducing equivalents for release of the cluster is thought to be provided by a protein known in yeast as Yah1p (yeast adrenodoxin homolog 1). Its reducing equivalents are thought to come from another yeast protein, Arh1p, an NAD(P)H-dependent adrenodoxin-reductase homolog. Finally, two molecular chaperones found in yeast, Jac1p and Sssq1p (homologs of Hsp40 and Hsp70) are also thought to be involved in Fe–S cluster biosynthesis. They may act to protect the reactive sulfhydryl groups used for Fe–S cluster coordination on the Fe–S apoproteins, or to protect the apoprotein from proteolysis. Much remains to be learned about the biogenesis of Fe–S cluster-containing proteins. Future studies will lead to the elucidation of the mechanisms involved in the mitochondrial synthesis of Fe–S proteins, as well as the mammalian homologs involved in the matrix of the mitochondria.

the substrate. The constituent enzymes of PDC and α-KGDH also share considerable amino acid sequence similarities, and both complexes are subject to similar regulatory controls by metabolites and by mitochondrial protein kinases. Evidence is building that α-KGDH is the undiscovered giant in the control of the TCA cycle. The enzyme is subject to inhibition by its two products, sharing inhibition by NADH with several other enzymes including PDC, CS, and NAD-IDH. Ca^{2+} is an activator that also activates PDC and NAD-IDH. Future studies of the reversible phosphorylation control of this locus will yield new insights into important, hitherto little considered, regulatory parameters regulating the TCA cycle in all cells.

The next reaction is the conversion of succinyl-CoA to succinate by succinyl-CoA synthetase (also called succinate thiokinase) with the energy of the thioester bond used to drive the formation of GTP, a so-called substrate-level phosphorylation:

$$\text{Succinyl-CoA} + P_i + \text{GDP} \rightarrow \text{succinate} + \text{GTP} + \text{CoA} \quad (8.7)$$

This reaction mechanism involves the displacement of the CoA by inorganic phosphate to form succinyl phosphate, followed by transfer of the phosphoryl group to a histidine residue on the protein and the release of succinate, and finally the transfer of the phosphoryl group to guanosine 5′-diphosphate (GDP). Little is known about regulatory controls on this enzyme.

The final reactions of the TCA cycle regenerate oxaloacetate via two dehydrogenase and one hydration reaction. Succinate dehydrogenase (SDH) oxidizes succinate using a bound FAD prosthetic group as the electron acceptor (see Text Box 8.2) and producing fumarate with a double bond between the second and third carbon:

$$\text{Succinate} + \text{FAD} \rightarrow \text{fumarate} + \text{FADH}_2 \quad (8.8)$$

The electrons carried by $FADH_2$ are fed directly into the electron transport chain at complex II, SDH being a membrane-bound enzyme. SDH activity is affected by the composition of the lipids (head groups, degree of unsaturation) of the inner mitochondrial membrane in its vicinity. For example, fish adapted to high versus low environmental temperatures change the lipid composition of their membranes during acclimation in a process called homeoviscous adaptation (see Chapter 16), and this affects SDH activities in their tissues. If the TCA operates *in vivo* as a macromolecular complex, SDH is likely the bound "root" of the group and should be the starting point of studies that look for complex formation.

TEXT BOX 8.2 REDOX COFACTORS OF TCA CYCLE

NAD^+ and FAD are both electron carriers involved in the TCA cycle, but there are several significant differences between the two. NAD^+ is not covalently bound to the dehydrogenases with which it works, and it accepts one hydride ion (H^-, a proton with two electrons) in every dehydrogenase reaction. A second hydrogen atom is released as a proton (H^+). NADH feeds its electrons into the electron transport chain at complex I. Unlike NADH, FAD and FMN (an integral part of complex I, which is the first acceptor of electrons from NADH) are tightly associated with or even covalently bound (as for FAD in succinate dehydrogenase) to their enzyme, and they can accept either two electrons, to form $FADH_2$ or $FMNH_2$, or a single electron, forming a semiquinone. Since the cytochromes of complex III as well as the terminal electron acceptor, oxygen, can only accept electrons one at a time, electron acceptors such as NAD^+ that remove electrons two at a time must pass these on to other carriers that can undergo one-electron redox reactions (i.e., FMN in complex I). Functionally, FAD is typically used as a cofactor in reactions that oxidize alkanes (e.g., succinate) to alkenes (e.g., fumarate), whereas NAD^+ participates in the more exergonic oxidation of alcohols to aldehydes or ketones (see Table 8.1).

In the next reaction, L-malate is formed by the hydration of the fumarate double bond, catalyzed by fumarase:

$$\text{Fumarate} + H_2O \leftrightarrow \text{L-malate} \quad (8.9)$$

Finally, the malate dehydrogenase (MDH) step regenerates oxaloacetate in the third NADH-generating reaction of the cycle:

$$\text{Malate} + NAD^+ + H_2O \leftrightarrow \text{oxaloacetate} + \text{NADH} + H^+ \quad (8.10)$$

Complexes of MDH and CS have been discovered and studied *in vitro*. Since oxaloacetate is present in very low concentrations *in vivo*, complex formation between these two enzymes would serve to channel substrate directly to CS.

Overall, the net reaction of the TCA cycle can be written as follows:

$$\text{Acetyl-CoA} + 3\text{NAD}^+ + \text{FAD} + \text{GDP} + \text{P}_i + 2\text{H}_2\text{O} \rightarrow 2\text{CO}_2 + 3\text{NADH} + \text{FADH}_2 + 1\text{GTP} + 2\text{H}^+ + \text{CoA} \quad (8.11)$$

The direct output of high-energy phosphate carriers (1 GTP per turn of the cycle) is not great, but the energy pay-off of the cycle is contained in the reduced coenzymes generated. Oxidation of these via the electron transport chain with coupling to oxidative phosphorylation leads to a large net ATP synthesis, roughly 2.5 and 1.5 molecules of ATP per molecule of reduced NAD^+ or FAD, respectively (older estimates were 3 and 2).

Control of the TCA Cycle Energy metabolism in the mitochondrion is a very complex system, and it would appear that many regulatory steps are involved. Metabolic control analysis (see Chapter 1) allows for the determination of control coefficients, a measure of the regulatory strength properties of an enzyme in relation to pathway flux. In these analyses, it can be determined how a metabolic event, such as control of the membrane potential ($\Delta\psi$), ATP, and NADH levels, or a metabolic pathway, relies on the activity of the enzyme. Flux control coefficients have been determined in many of the complex systems associated with the mitochondria. In studies conducted by the group of M. Brand in Cambridge (see reading list), metabolic control analysis was applied to incorporate interactions of pathways involved in mitochondrial metabolism with outside pathways. These studies applied the "top-down approach" to metabolic control analysis (MCA). The complexity of interacting pathways was simplified by grouping the reactions as functional 'blocks' and connecting the blocks via primary intermediates.

The key intermediates connected to mitochondrial metabolism are pyruvate, the cytoplasmic NADH/NAD ratio, and the mitochondrial membrane potential. The critical steps of regulation in mitochondrial metabolism were found to be pyruvate and NADH oxidation. In an overall pathway or block, overall control coefficients appear to act as a sum of all the positive and negative control coefficients. Hormone level changes, including those of glucagon and adrenaline, as well as environmental changes also dramatically affect control parameters of all pathways.

The TCA cycle is responsible for providing the electron transport chain with fuel in the form of NADH and $FADH_2$. It is also responsible for providing the cell with substrates for anabolic pathways. It is at these points, where reducing equivalents and substrates for other pathways are produced, that regulation of the cycle takes place. Thus, regulation of the TCA cycle is the regulation of the TCA cycle at its branch points. The following is a list of the regulatory factors involved in mitochondrial metabolism:

1. Availability of substrate, acetyl-CoA, and accumulation of NADH, ATP
2. Abundance of intermediates of the cycle, supplied by glutamate dehydrogenase, pyruvate carboxylase, malic enzyme, and phosphoenolpyruvate (PEP) carboxykinase
3. NADH abundance (NAD/NADH ratio) at PDC, isocitrate dehydrogenase, and α-KGDH
4. ATP availability (ATP/ADP ratio) at PDC, citrate synthase, and isocitrate dehydrogenase
5. The ratio of acetyl-CoA to free CoA; the ratio of acetyl-CoA to succinyl-CoA; and the ratio of citrate to oxaloacetate

BRANCH POINTS: THE TCA CYCLE *IN VIVO*

The TCA cycle is not just used for fuel catabolism; it has multiple interactions with other metabolic pathways in the cell. Outputs from the TCA cycle are very important for biosynthesis. For example, the use of carbohydrates for fatty acid biosynthesis is linked via the TCA cycle intermediate, citrate. Acetyl-CoA introduced from the PDC is used to form citrate, and then citrate is exported from the mitochondria to provide the source of acetyl-CoA in the cytosol for fatty acid and triglyceride synthesis as well as for the synthesis of some amino acids. Alpha-ketoglutarate and oxaloacetate are also outlets to amino acid synthesis, and succinyl-CoA is involved in heme synthesis. Multiple anaplerotic reactions also feed into the TCA cycle, and the carbon skeletons of a variety of amino acids feed into C5 and C4 intermediates of the cycle for catabolism. Some interactions of the TCA cycle with other pathways are discussed below.

Glyoxylate Cycle

The glyoxylate cycle represents a mechanism by which acetyl-CoA, instead of being catabolized to CO_2, can be used for the biosynthesis of 4-carbon compounds (Fig. 8.5). The cycle begins with the condensation of oxaloacetate and acetyl-CoA to form citrate by CS and then aconitase action to produce isocitrate. Isocitrate is then cleaved by the first of two unique enzymes of the cycle, isocitrate lyase, to form glyoxylate and succinate. Succinate can be used for biosynthetic purposes. Glyoxylate then condenses with a second molecule of acetyl-CoA to form malate, via the enzyme malate synthase. Malate is then

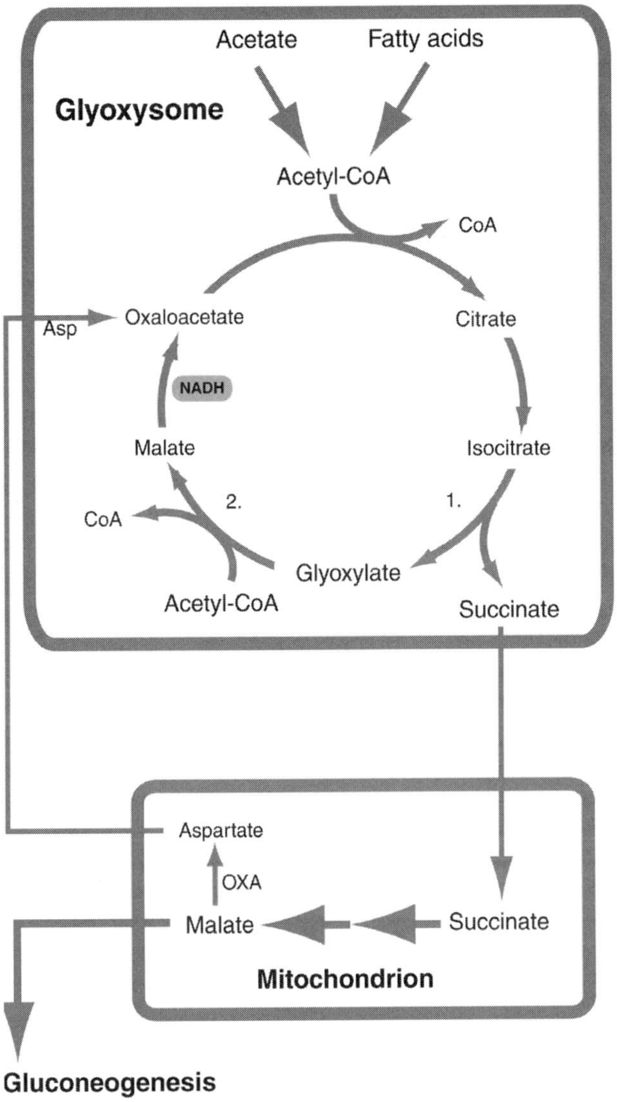

Figure 8.5 The glyoxylate cycle showing interactions between the glyoxysome and the mitochondrion. The action of the glyoxylate cycle allows organisms to use two-carbon acetyl-CoA units, derived from lipids or from acetate, as building blocks for the synthesis of a four-carbon intermediate (succinate) that can then be used for sugar synthesis via gluconeogenesis or for other biosynthetic reactions. The two unique reactions of the cycle are (1) isocitrate lyase and (2) malate synthase. Anaplerotic input to the cycle comes at the level of oxaloacetate that is transferred from the mitochondria as aspartate. The glyoxylate cycle is found in only a few invertebrate animals but is quite common in plants and some microorganisms.

reduced to oxaloacetate to complete the cycle. The net reaction of the glyoxylate cycle is as follows:

$$2\ \text{Acetyl-CoA} + \text{NAD}^+ + 2\text{H}_2\text{O} \rightarrow \text{succinate} + 2\text{CoA} + \text{NADH} + \text{H}^+ \quad (8.12)$$

The glyoxylate cycle is found in plants and some bacteria. The presence of the cycle allows organisms to make use of acetate as a fuel or, in plants, allows the lipid stores of germinating seeds to be used for the biosynthesis of the carbohydrates needed for shoot and root growth before the plant acquires the ability to produce glucose via photosynthesis. In plants, the cycle is sequestered into a unique organelle, the glyoxysome, a modified peroxisome. The glyoxysome shares intermediates with the mitochondria; mitochondrial oxaloacetate can be exported in the form of aspartate and taken up to provide oxaloacetate in the glyoxysome, whereas succinate can be exported back to the mitochondria to enter gluconeogenic reactions. The very different fates of isocitrate in the two cycles means that reciprocal regulation of the two pathways is important, particularly in bacteria where the cycles are not sequestered into separate organelles. Hence, in bacterial systems, reversible phosphorylation control over isocitrate dehydrogenase of the TCA cycle determines the relative use of isocitrate by the two pathways in response to the energy needs of the cell.

The possible presence of the glyoxylate cycle in animal cells has always been controversial. Some studies have reported finding isocitrate lyase and malate synthase in the fat body of insects and in some marine invertebrates, but with careful controls these were shown to be artifacts. However, the two enzymes have been found in several nematode species that produce acetate as an anaerobic end product. In *Caenorhabditis elegans* an unusual situation occurs: a bifunctional 106-kD protein catalyzes both isocitrate lyase and malate synthase activities. Levels of messenger RNA (mRNA) and protein for the bifunctional enzyme were high in early larvae that had never been fed and also rose when later stage larvae were shifted from a fed to fasted state. It has been suggested that glyoxylate cycle enzymes would be highly beneficial for hibernating mammals that rely almost exclusively on lipid catabolism to support long months of torpor. Indeed, one report indicated their presence in hibernating black bears but it was later discredited; the enzymes are also absent from other hibernating species such as ground squirrels and little brown bats. It appears that hibernators can meet their carbohydrate requirements from other sources, in particular gluconeogenesis from glycerol (released from triglyceride hydrolysis). The absence of a glyoxylate cycle in humans and most other animals means that carbohydrate conversion to lipids is one-way, and this is part of the reason that glucose metabolism and lipogenesis is so tightly regulated (see Chapter 9).

Urea Cycle

The urea cycle is another pathway with important reaction steps that takes place partly within the mitochondria and has an intimate link with the TCA cycle. The cycle is the means of disposing of ammonium ion, NH_4^+, which is formed from the breakdown of amino acids. High NH_4^+ levels are toxic to the nervous system, making excretion a necessity. Aquatic organisms eliminate NH_4^+ directly in urine or across gills and other epithelia. Terrestrial organisms cannot afford the high water loss associated with ammonia excretion and so convert ammonium ion to less toxic forms: urea that can be excreted with much lower amounts of water or uric acid that is secreted with minimal water as a virtual paste. Ureotelic vertebrates synthesize urea in the liver. Urea is effectively the condensation product of $2NH_4^+$ and $1CO_2$ and is produced in an overall reaction as follows.

$$2NH_4^+ + HCO_3^- + 3ATP + 2H_2O \rightarrow \text{urea} + 2ADP + 2P_i + AMP + PP_i + 6H^+ \quad (8.13)$$

In sharks and other elasmobranch fishes, urea is maintained in very high concentrations (typically over 300 mM) in order to provide osmotic balance with seawater. Estivating anurans (frogs and toads) also build up high levels of urea during long-term estivation (dormancy) in arid environments to retard water loss from their bodies (see Chapter 14).

The urea cycle was elucidated in 1932 by Hans Krebs and Kurt Henseleit, well before Krebs worked out the TCA cycle. The cycle spans mitochondrial and cytoplasmic compartments with the first two reactions of the cycle contained within the mitochondria and remainder being cytosolic (see Figure 9.17). When amino acids are metabolized in various tissues of the body, the amino groups are frequently transferred to alanine or glutamine that are then transported to the liver. Transamination reactions transfer the amino group of alanine (or other amino acids) to glutamate and then the actions of the mitochondrial matrix enzymes, glutaminase and glutamate dehydrogenase, release the NH_4^+ within the matrix where it is immediately condensed with HCO_3^- to form carbamoyl phosphate by the ATP-dependent carbamoyl phosphate synthetase (CPS) reaction. The carbamoyl group is then transferred to ornithine to form L-citrulline by ornithine transcarbamoylase, resulting in PP_i release. Citrulline is transported out of the mitochondria to the cytosol where it reacts with aspartate to form argininosuccinate. This is a condensation reaction catalyzed by argininosuccinate synthethase. It occurs in an ATP-dependent fashion, resulting in the formation of AMP and pyrophosphate (PP_i). PP_i is hydrolyzed to two inorganic phosphates, so effectively two ATP are required for this step. Argininosuccinate lyase then is responsible for the cleavage of argininosuccinate into arginine and fumarate. The fumarate is then transferred back to the mitochondria, entering the TCA cycle where it is converted to malate and then to oxaloacetate that can be converted to aspartate again. The arginine is hydrolyzed to ornithine and urea, a reaction catalyzed by arginase in the cytosol. Details of the cycle and its role in human amino acid metabolism are discussed in more detail in Chapter 9.

In a fascinating aside, whole animal studies indicate that marmots and bears can recycle their urea after it is made, instead of excreting it during their winter hibernation. The mechanism of this recycling and any changes that would be required to the regulation of urea synthesis in liver mitochondria is largely unexplored.

Fatty Acid Oxidation

Fatty acids represent a highly concentrated fuel source, and their oxidation provides more than twice the energy release as the same weight of carbohydrate or amino acids. In animal cells, fatty acid oxidation takes place primarily in the mitochondria via a series of reactions collectively called ß oxidation that sequentially cleave two-carbon fragments off the long-chain fatty acid. The product, acetyl-CoA, is then passed to the TCA cycle.

The first step in fatty acid utilization is the priming or activation of the fatty acids, forming CoA derivatives in an ATP-dependent manner (Mg^{2+} is a cofactor, as for all enzymes that hydrolyze ATP) [reaction (8.14)]. This is catalyzed by a family of acyl-CoA ligases (specific for short-, medium- or long-chain fatty acids), located either on the endoplasmic reticulum or on the outer mitochondrial membrane.

$$RCOO^- + ATP + CoA \rightarrow \text{acyl-CoA} + AMP + PP_i \quad (8.14)$$

Acyl-CoA moieties cannot traverse the inner mitochondrial membrane and so, before entry, are converted to a carnitine derivative by carnitine palmitoyl transferase 1 (CPT1), an outer membrane enzyme:

$$\text{Acyl-CoA} + \text{carnitine} \leftrightarrow \text{acyl-carnitine} + \text{CoA} \quad (8.15)$$

Fatty acyl-carnitine moieties then cross the inner mitochondrial membrane via a carrier protein, and on the other side carnitine palmitoyl transferase II reconverts the fatty acid to a CoA derivative. Current evidence indicates that primary control over long-chain fatty acid oxidation *in vivo* is exerted at the level of fatty acid entry into the mitochondrion by regulating the enzyme CPT1. CPT1 activity is controlled by reversible phosphorylation and by metabolite inhibition via malonyl CoA (see Chapter 9 for more information on the control of lipid oxidation in humans). CPT1 phosphorylation is regulated in a tissue-specific manner by both hormone activity and in response to environmental stress. Lipolysis is also regulated by altered gene expression via the action of the peroxisome proliferator activating receptors (PPARs), transcription factors that bind to promoter sites on DNA and up-regulate the synthesis of CPT1, cytoplasmic fatty acid binding proteins, and enzymes of β oxidation. For example, up-regulation of genes involved in fatty acid oxidation is crucial to the survival of hibernating mammals that slim down dramatically from their obese prehibernation state while they sleep away the winter months—giving hope to drug companies that this elixir can be bottled and sold (see Chapter 16).

Once inside the mitochondrial matrix, the four enzymatic steps of β oxidation remove successive 2-carbon units from the fatty acyl-CoAs. The first step is a dehydrogenation reaction catalyzed by fatty acyl-CoA dehydrogenase and yielding an α,β-*trans*-enoyl-CoA and $FADH_2$; electrons from $FADH_2$ are funneled into the electron transport chain. Next is the hydration of the double bond to form L-ß-hydroxyacyl-CoA via enoyl-CoA hydratase. In the third step, dehydrogenation of the hydroxyl group ß-hydroxyacyl-CoA dehydrogenase forms β-ketoacyl-CoA and generates NADH that is also transferred to the electron transport chain. In the final step, thiolase mediates the attack of the nucleophilic thiol sulfur of CoA on the β-ketoacyl-CoA, cleaving off an acetyl-CoA moiety and regenerating a shorter acyl-CoA that undergoes another turn through the cycle. For fatty acids with an even number of carbon chains, the process continues until the final 4-carbon acetoacetyl-CoA is cleaved into 2 acetyl-CoA. The process of fatty acid oxidation, for example, the complete oxidation of palmitoyl-CoA, yields a considerable amount of energy: 108 ATP along with $16CO_2$ and $123H_2O$ plus CoA. The $\Delta G^{\circ\prime}$ for the net reaction is -2340 kcal/mol.

Additional reactions of fatty acid metabolism occur in the mitochondria, including the oxidation of unsaturated fatty acids and of odd-chain fatty acids (as found in plants and some marine organisms) and the formation of ketone bodies (mainly in the liver). Oxidation of unsaturated fatty acids, such as oleic or linoleic acid, which contain cis double bonds, require the activities of three additional enzymes. For example, oleoyl-CoA can cycle through the β-oxidation reaction process three times, until the remaining Δ^3-*cis*-dodecenoyl-CoA has to be converted into its trans isomer, via enoyl-CoA-isomerase, bypassing acyl-CoA-dehydrogenase. The Δ^3-*trans*-dodecenoyl-CoA can then be the substrate for enoyl-CoA-hydrase, and the oxidation process continues as per usual. Polyunsaturated fatty acids such as linoleic acid require an additional enzyme, 2,4-dienoyl-CoA reductase. After the beginning of the ß-oxidation process, whereby the cis bond is isomerized into the trans configuration followed by further oxidation, the new substrate, Δ^4-*cis*-enoyl-CoA, is subjected to catalysis by acyl-CoA dehydrogenase to form Δ^2-*trans*,Δ^4-*cis*-dienoyl-CoA. This product cannot be a substrate for enoyl-CoA hydrase, so here 2,4-dienoyl-CoA reductase converts the product to Δ^3-*trans*-enoyl-CoA, then enoyl-CoA isomerase converts it to the Δ^2-trans isomer. After this conversion, β oxidation can begin again.

Odd-number fatty acid chains are processed by ß oxidation in the usual manner until a final three-carbon product, propionyl-CoA, is formed. Propionyl-CoA is then converted into succinyl-CoA in three enzymatic steps: (1) propionyl-CoA carboxylase using biotin as a cofactor synthesizes the four-carbon D-methylmalonyl-CoA, (2) isomerization via methylmalonyl-CoA racemase forms L-methylmalonyl-CoA, and (3) methylmalonyl-CoA mutase using the cofactor cobalamin catalyzes the formation of succinyl-CoA. Succinyl-CoA then enters the TCA cycle where it can have anaplerotic or gluconeogenic fates. The same reactions provide the means for using propionate as a metabolic fuel, which is particularly important for ruminant animals because propionate is a major product of the bacterial fermentation of plant foodstuffs in the rumen. Cows, for example, exhibit extremely high rates of gluconeogenesis from propionate. Furthermore, similar reactions catalyzing the reverse production of propionate from succinyl-CoA are important to ATP generation under anoxic conditions in a variety of invertebrate species (see Chapter 15). In these animals, the TCA cycle is run in reverse when oxygen is lacking with substrate inputs at oxaloacetate catabolized through to succinyl-CoA and then diverted into the reactions of propionate synthesis. The advantages for these anoxia-tolerant animals are twofold: (1) two sites of substrate-level ATP generation occur at the fumarate reductase and methylmalonyl-CoA decarboxylase reaction, and (2) propionate is a volatile end product that is easily excreted.

Ketogenesis

The intramitochondrial formation of ketone bodies is closely linked with fatty acid oxidation. When acetyl-CoA levels are high, synthesis begins with a reversal of the thiolase reaction to convert two acetyl-CoAs to one acetoacetyl-CoA. This product then reacts with a third acetyl-CoA to form ß-hydroxy-ß-methylglutaryl-CoA (HMG-CoA) catalyzed by HMG-CoA synthase. HMG-CoA lyase then cleaves the product to form acetoacetate and acetyl-CoA. Acetoacetate can be processed further to ß-hydroxybutyrate by ß-hydroxybutyrate dehydrogenase located on the inner mitochondrial membrane or to acetone via acetoacetate decarboxylase. Acetoacetate and ß-hydroxybutyrate are the main ketone products that are circulated to other organs, but the smell of small amounts of volatile acetone, released in the breath, is the sure sign of a ketogenic state and a well-known diagnostic in diabetes detection. In mammals, ketone bodies are synthesized in the liver and, as discussed in Chapter 9, they represent water-soluble metabolites of fatty acids that can be easily transported to other organs for use as aerobic substrates. They are particularly important substrates during prolonged fasting because they can augment the fuel supply for organs such as brain that normally oxidize only carbohydrate and cannot use fatty acids as fuels. For this reason, they are also important substrates in states of long-term dormancy where lipids are the primary fuel supply such as mammalian hibernation (see Chapter 16).

Amino Acid Degradation as Fuel for TCA Cycle

The products of the catabolism of amino acids include pyruvate, acetyl-CoA, and carboxylic acid intermediates of the TCA cycle; these can be oxidized or used for gluconeogenesis or ketogenesis (see Chapter 9 for more information on human amino acid catabolism). The catabolism of four amino acids (proline, glutamine, arginine, and histidine) converges in the synthesis of glutamate, and carbon is fed into the TCA cycle as α-ketoglutarate by the glutamate dehydrogenase reaction that releases the NH_4^+ for urea biosynthesis. Carbon skeletons from several other amino acids feed in at the level of succinyl-CoA, often with propionyl-CoA or acetoacetyl-CoA as intermediaries. Phenylalanine and tyrosine give rise to fumarate, whereas aspartate and asparagine are converted to oxaloacetate. The three branched-chain amino acids (leucine, isoleucine, and valine) are oxidized primarily in muscle; after the amino group is removed by a transaminase, they undergo oxidative decarboxylation and conversion to CoA derivatives by the branched-chain α-keto acid dehydrogenase complex. This enzyme is analogous to the PDC and α-KGDH and is also regulated by reversible phosphorylation, being phosphorylated and inactivated when dietary availability of branched-chain amino acids is low.

Amino acids are not a major oxidative fuel source in mammals and are "spared" even during fasting in favor of the oxidation of fats and carbohydrates. This is not the case in all parts of the animal kingdom; for example, various species of fish have a metabolism that preferentially oxidizes amino acids, excreting large amounts of ammonia into the surrounding water. Several species of beetles also use proline as their primary fuel source for migratory flights of many hundreds of miles—the purest burn of amino acids known in nature.

RESPIRATORY CHAIN

The oxidative reactions of the TCA cycle result in the transfer of electrons to NAD^+ and FAD, creating NADH and $FADH_2$ [reactions (18.16) and (18.17)]. To maintain TCA cycle function the oxidized coenzymes must be continuously replenished and within the mitochondria. This occurs via transfer of the electrons to the electron transport system.

$$NADH + H^+ + \tfrac{1}{2}O_2 \rightarrow NAD^+ + H_2O \quad \Delta G^{\circ\prime} = -52.6\,\text{kcal/mol} \tag{8.16}$$

$$FADH_2 + \tfrac{1}{2}O_2 \rightarrow FAD + H_2O \quad \Delta G^{\circ\prime} = -43.4\,\text{kcal/mol} \tag{8.17}$$

Electrons are shuttled through a series of enzyme complexes in the respiratory chain and, key to energy generation, is the concomitant export of protons from the matrix into the intermembrane space and their subsequent reentry while driving ATP synthesis. The redox couples involved in electron transport are summarized in Table 8.2.

The electron carriers involved in the transfer of electrons to O_2, the heme-containing cytochromes, are localized in the inner membrane of the mitochondria. There are three different groups of cytochromes, distinguished as *a*, *b*, and *c* based on their different subgroups on the porphyrin ring and binding sites of the porphyrin on the protein. There are two subtypes of cytochromes per group (cytochromes *c*, *c*1, cytochromes b_L, b_H, cytochromes *a*, a_3) (Fig. 8.3). Overall, the electron transport chain consists of four large enzyme complexes, responsible for the transfer of reducing equivalents to O_2. These multisubunit complexes of proteins are:

TABLE 8.2 Redox Couples

Succinate + CO_2 + $2H^+$ + $2e^-$ ↔ α-ketoglutarate + H_2O	−0.67
Glycerate-3-phosphate + $2H^+$ + $2e^-$ ↔ glyceraldehyde-3-P + H_2O	−0.55
α-Ketoglutarate + CO_2 + $2H^+$ ↔ isocitrate	−0.38
NAD^+ + H^+ + $2e^-$ ↔ NADH	−0.32
FMN + $2H^+$ + $2e^-$ ↔ $FMNH_2$	−0.22
FAD + $2H^+$ + $2e^-$ ↔ $FADH_2$	−0.22
Pyruvate + $2H^+$ + $2e^-$ ↔ lactate	−0.19
Oxaloacetate + $2H^+$ + $2e^-$ ↔ malate	−0.17
Fumarate + $2H^+$ + $2e^-$ ↔ succinate	−0.03
Cytochrome b_L (Fe^{3+}) + e^- ↔ Cytochrome b_L (Fe^{2+})	−0.03
UQ + H^+ + $2e^-$ ↔ UQH	0.03
Cytochrome b_H (Fe^{3+}) + e^- ↔ Cytochrome b_H (Fe^{2+})	0.05
UQ + $2H^+$ + $2e^-$ ↔ UQH_2	0.11
UQH + H^+ + e^- ↔ UQH_2	0.19
Rieske Fe-S (Fe^{3+}) + e^- ↔ Fe-S (Fe^{2+})	0.28
Cytochrome c_1 (Fe^{3+}) + e^- ↔ cytochrome c_1 (Fe^{2+})	0.23
Cytochrome *c* (Fe^{3+}) + e^- ↔ cytochrome *c* (Fe^{2+})	0.24
Cytochrome *a* (Fe^{3+}) + e^- ↔ cytochrome *a* (Fe^{2+})	0.28
Cytochrome a_3 (Fe^{3+}) + e^- ↔ cytochrome a_3 (Fe^{2+})	0.35
O_2 + $4H^+$ + $4e^-$ ↔ $2H_2O$	0.82

Units are V for volts.

Complex I: The NADH dehydrogenase complex that uses NADH to reduce ubiquinone (UQ) to form UQH_2

Complex II: The succinate dehydrogenase complex that uses $FADH_2$ to reduce UQ to form UQH_2

Complex III: The cytochrome *bc*1 complex that transfers electrons from UQH_2 to cytochrome *c*

Complex IV: The cytochrome oxidase complex that transfers electrons from cytochrome *c* to O_2 to form H_2O

Complex I The initial complex of the electron transport chain (ETC) oxidizes the NADH received from the TCA cycle, the PDC, and ß oxidation. Within this large complex of 42 subunits, the 2 electrons from a hydride ion are passed first from NADH to FMN, which is tightly bound to the complex. $FMNH_2$ then passes electrons through several iron–sulfur clusters on different iron–sulfur proteins and ultimately to UQ to form UQH_2. Accompanying this is an export of 4 protons into the intermembrane space and the uptake of other 2 protons from the matrix. A net of 1 proton is translocated into the intermembrane space per electron transferred.

Intramitochondrial protein kinases can regulate complex I. Protein kinase A phosphorylates 2 proteins of the complex (out of 43 proteins in total) and causes an increase in NADH cytochrome *c* reductase and a decrease in superoxide production. Oppositely, activation of pyruvate dehydrogenase kinase phosphorylates a different component of complex I and causes a sharp increase in superoxide production from this complex (see Chapter 12 for more about superoxide production by the electron transport chain).

Complex II Complex II is the succinate dehydrogenase complex. This complex is not involved in proton translocation; its job is simply to transfer electrons via $FADH_2$ to UQ. Succinate dehydrogenase is the only TCA cycle enzyme that is embedded in the mitochondrial inner membrane. The two protons that are obtained from the oxidation of succinate to fumarate are transferred to FAD. $FADH_2$ then passes the electrons through iron–sulfur clusters and finally to UQ to form UQH_2.

Complex III Complex III is also known as cytochrome *bc*1. It consists of two *b*-type cytochromes, b_L and b_H (heme-containing polypeptides); a *c*-type cytochrome, *c*1 (iron–sulfur containing polypeptide); plus four to six other subunits. Electrons flow through complex III from UQH_2 to cytochrome *c*. The process is complicated but designed to accommodate the switch between the two-electron carrier UQ and the *b* and *c* cytochromes that are one-electron carriers. The net result is the oxidation of UQH_2 and the formation of two molecules of reduced cytochrome *c*. The reduction of UQ and the oxidation of its redox partner on complex III is again accompanied by the net vectoral transport of protons from the matrix to the intermembrane space.

Complex IV Complex IV is also known as the cytochrome oxidase complex. This complex is responsible for the overall transfer of electrons from cytochrome *c* to oxygen, and therefore the tetravalent reduction of O_2 to H_2O. In the structure of cytochrome oxidase, there are three copper atoms, as well as hemes that are part of cytochromes *a* and *a*3. CuB is one of the copper atoms and is located near to the iron heme of cytochrome *a*3. Cytochrome *a* is closely but not as closely associated with the other two copper atoms, denoted CuA. In the reactions that take place over this complex, four electrons are added to reduce O_2 to H_2O. Although the transfer of one electron to O_2 to produce $O_2^{\cdot-}$ is thermodynamically unfavored, complex IV overrides this problem through the binding of CuB and the iron group of cytochrome *a*3 to O_2 and then transferring two electrons per reaction.

Cytochrome *c* oxidase has been shown to have a "slip" of proton pumping when the proton-motive force is very high or if there is a chemical modification of subunit III, such that H^+/e^- stoichiometry is reduced. Cytochrome *c* oxidase is phosphorylated on subunits II (and/or III) and Vb by protein kinase A (Fig. 8.6). The effect of phosphorylation on the bovine heart enzyme is to turn on the allosteric inhibition of the enzyme by high ATP so that respiration is controlled by the ATP/ADP ratio, keeping $\Delta\psi$ low and the efficiency of energy transduction high. A Ca^{2+}-dependent protein phosphatase dephosphorylates cytochrome *c* oxidase (COX), resulting in a loss of ATP inhibition, a slip of proton pumping, increasing the pumping (decreased H^+/e^- stoichiometry) and the rate of respiration, and ATP synthesis but decreasing efficiency of energy transduction. Recent research has also shown the importance of nitric oxide in the control of cytochrome *c* oxidase and respiration (see Text Box 8.3).

The ATP synthesis and electron transport in the mitochondria are tightly coupled reactions under most conditions, although it was only in 1961 that it came to be understood how these two distinct processes could be coupled. The chemiosmotic hypothesis, developed by Peter Mitchell, states that the free energy of electron transport is conserved by pumping protons from the matrix into the intermembrane space to create an electrochemical H^+ gradient across the inner mitochondrial membrane. The electrochemical potential of the gradient is exploited to synthesize ATP during the controlled reentry of protons into the matrix via the F_1,F_0-ATPase, also known as ATP synthase or complex V, in the inner mitochondrial membrane. As indicated above, electron flow through com-

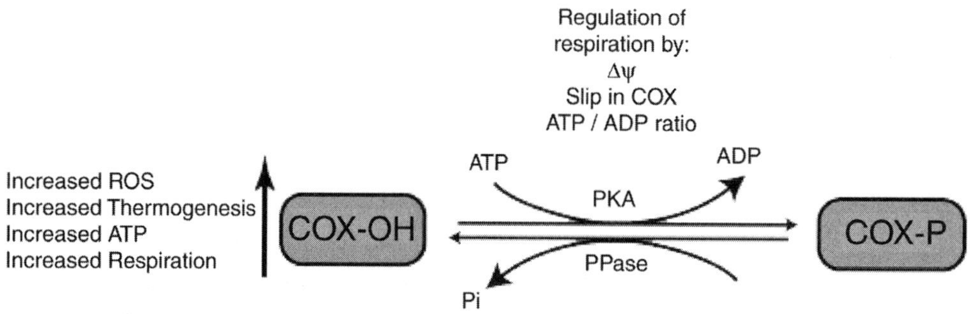

Figure 8.6 Regulation of metabolism through cAMP-dependent protein kinase phosphorylation of cytochrome *c* oxidase (COX). Here, phosphorylation increases the slip in proton pumping in COX at high ψ (proton-motive force). Increasing mitochondrial Ca^{2+} concentrations promote COX dephosphorylation (COX-OH), whereas increased mitochondrial cAMP stimulates phosphorylation (COX-P).

plexes I, III, and IV of the electron transport chain each results in proton extrusion into the intermembrane space.

Complex V The synthesis of ATP during oxidative phosphorylation occurs at complex V, the F_1F_0-ATP synthase. The F_0 component of the complex is the proton channel that is imbedded in the inner membrane and is composed of three different types of polypeptide chains (*a*, *b*, *c*). The F_1 component is the ATP synthase that is located on the matrix side of the inner mitochondrial membrane and is composed of five types of polypeptides (α_3 β_3 γ δ ε).

TEXT BOX 8.3 CONTROL OF RESPIRATION BY NITRIC OXIDE

Another regulator of the respiratory chain in the mitochondria is nitric oxide. Cellular nitric oxide is produced by three different isoforms of NO synthase (eNOS, nNOS, and iNOS) and is a cell permeable, gaseous information molecule. NO inhibits cytochrome *c* oxidase (complex IV) and, with extensive exposure to NO, or in the presence of a more reactive form of NO (NO can react with superoxide to form peroxynitrite, OONO-), it can inhibit respiration altogether. Peroxynitrite can damage or inhibit complexes I, II, IV, and V, aconitase, creatine kinase, mitochondrial membrane potential, mitochondrial DNA, and superoxide dismutase. At low levels of NO, cytochrome *c* oxidase is bound to NO through its oxygen-binding sites. NO is a competitive inhibitor of oxygen consumption, 90% of which occurs at cytochrome *c* oxidase. For more information on NO and OONO- see Chapters 4 and 12.

The overall reaction for ATP synthase is as follows:

$$ADP^{3-} + P_i^{2-} + H^+ \leftrightarrow ATP^{4-} + H_2O$$

This reaction occurs readily even in the absence of any proton-motive force. Once the nucleotides are bound to the β subunits, the reaction occurs quickly, and the equilibrium constant for this reaction is very close to 1. The proton-motive force does not affect the equilibrium constant for this reaction very much. In fact, it is believed that the proton-motive force is required for an alteration in the conformation of the three β subunits of F_1. The F_1 complex has three nonequivalent adenine nucleotide binding sites, one for each pair of α and β subunits. At any given moment, one of these sites is in a conformation that binds ATP tightly, one that binds ADP loosely, and the third is unoccupied in the "open" state. The proton-motive force causes rotation of the central shaft (the γ subunit) contacting each $\alpha\beta$ pair in sequence and altering their conformation: The subunit in the tight conformation switches to the open conformation, and in doing so its ATP dissociates, the loose conformation switches to tight, which promotes condensation of the bound ADP with P_i form ATP, and the open conformation switches to loose and loosely binds ATP and P_i.

Metabolic Control Analysis

Metabolic control analysis (MCA, see Chapter 1) has been applied to assess the regulation of mitochondrial oxidative phosphorylation. For this, the process was divided into four functional subsets: pyruvate oxidation, NADH oxidation, mitochondrial proton leak, and phosphorylation. The results showed several interesting phenomena. Not unexpectedly, oxidative phosphorylation (including the proton leak) exerted negative control over pyruvate oxidation. ATP production by the mitochondria showed significant control over ATP consumption. Overall, the use of ATP in cells depends largely on its supply by oxidative phosphorylation. Control analysis studies indicate that ATP depletion controls respiration rate and that mitochondrial membrane potential has little effect in comparison to changes in ATP levels. Increasing membrane potential is effectively opposed by a drop in the supply of ATP.

The application of MCA has also helped in deciphering the effects of mutations in mtDNA that cause diseases. For example, COX deficiency results in a mitochondrial myopathy, a muscle disorder. Recent studies have shown that there are threshold effects, namely, COX has been shown to have a very low excess capacity, and therefore very small changes in overall COX activity have a significant effect on energy metabolism. Flux control coefficients for COX are relatively high, and there is a strong control by COX over respiration.

Uncoupling for Thermogenesis

The link between electron transport and ATP synthesis is not obligatory in all cells under all conditions. The two can be uncoupled by chemicals such as 2,4-dinitrophenol and by the action of a naturally occurring uncoupling protein (UCP) that acts as a protonophore to allow H^+ reentry into the matrix by a route that is not linked with the F_1F_0-ATPase (Fig. 8.7). UCPs were first discovered as the mechanism underlying nonshivering thermogenesis by brown adipose tissue, a specialized thermogenic tissue containing extremely high numbers of mitochondria as well as large lipid reserves. High rates of fatty acid oxidation and oxygen consumption occur; but, as a result of "short-circuiting" in the presence of the UCP, the energy that would normally be used to drive ATP synthesis is dissipated instead as heat. Brown adipose is found in hibernat-

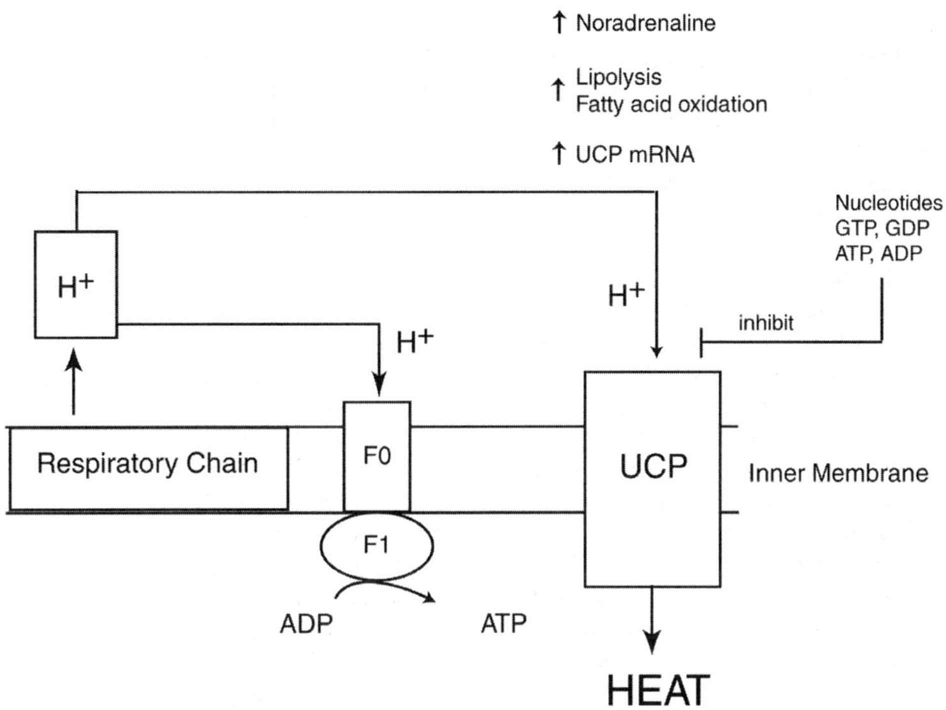

Figure 8.7 In the presence of uncoupling proteins, proton reentry from the intermembrane space into the mitochondrial matrix is diverted away from the F_1F_0-ATP synthase so that the proton-motive force is no longer used to drive ATP synthesis. Energy is released instead as heat. Isoform 1 of UCP that occurs in thermogenic tissues acts as a protonophore to facilitate proton reentry into the matrix. Its molecular mechanism is described in detail in Chapter 16. Thermogenesis in brown adipose tissue is triggered by noradrenaline signals that activate lipolysis and fatty acid oxidation and also up-regulate the expression of the UCP1 gene.

At least seven different proteins are in the TOM complex. Tom40 (the number designates molecular weight) forms the actual channel, and Tom20 and Tom70 are the initial receptors for cleavable preproteins and carrier preproteins, respectively. Preproteins with a presequence are guided by a chain of binding sites on Tom20, Tom22, and Tom5, then through the Tom40 channel to then interact with the intermembrane space (IMS) domain of Tom22.

ing mammals (except bears) and in rodents throughout their life, but in most other mammals (including humans) it is present for just a short time after birth. The regulation of thermogenesis by brown adipose tissue will be considered in greater detail in the section on mammalian hibernation in Chapter 16. UCP1, the brown-adipose-specific isoform, has been extensively characterized and four other isoforms, UCP2 through 5, are now known. Although UCP1 clearly functions in thermogenesis, the others probably do not. Exciting new research on the ubiquitous UCP2 shows that it has a role in protecting mitochondria from damage by reactive oxygen species (ROS). Targeted disruption of the UCP2 gene reduced the ability of cells to protect against ROS, whereas mice that overexpressed UCP2 showed diminished damage and better recovery after experimental stroke. In cultured neurons UCP2 reduced cell death and inhibited caspase-3 activation induced by oxygen and glucose deprivation (mimicking the substrate starvation of stroke). UCP2 shifted the release of ROS from the mitochondrial matrix to the extramitochondrial space. In cultured cardiomyocytes, UCP2 overexpression similarly suppressed markers of cell death, attenuated Ca^{2+} overload, and ROS production and prevented the catastrophic loss of mitochondrial inner membrane potential (induced by H_2O_2), which is an early event in cell death.

In some species of plants, thermogenesis results not from uncoupling of ATP synthesis from the electron transport chain but from the passage of electrons through an alternative respiratory chain. When needed, an alternative NADH dehydrogenase passes electrons directly to UQ, bypassing complex I and its proton pumping, and then complexes III and IV are bypassed by a UQH_2 oxidase that transfers electrons directly to oxygen. Energy that could have been trapped as ATP is all released as heat, and the plants (e.g., skunk cabbage) use this heat to grow and flower before the snow fully melts and to produce and disperse foul-smelling volatile chemicals that attract insect pollinators.

ATP–ADP TRANSLOCASE RELEASE RATIO

A key aspect of mitochondrial function, the provision of ATP for the cell, requires the transport of ADP and ATP across the inner mitochondrial membrane. This is accomplished by the ADP–ATP translocase (AAT) (also known as the adenine nucleotide carrier), an enzyme that allows for the coupled transport of ADP and ATP to and from the mitochondrial matrix, respectively. AAT is a dimer containing an adenine nucleotide binding site, which alternately faces the cytosol and the inner mitochondrial matrix. Depending on the membrane potential, this dimer "everts" from matrix to cytosolic sides of the membrane or the reverse, carrying bound ATP or ADP, respectively. If the membrane potential is high, the translocase is more likely to be involved in eversion from the matrix to the cytosol, as bound ATP has one more negative charge than ADP. Eversion without the presence of a bound nucleotide is negligible, and both nucleotides bind to the translocase in the absence of Mg^{2+}. Overall, the translocase does not operate without the coupling of transporting ATP out of the matrix and ADP in. As the translocase functions, the net membrane potential decreases, and thus this exchange of nucleotides across the membrane is not energy efficient. However, it is an essential reaction for the continuance of oxidative phosphorylation. The transporter is tightly regulated *in vivo* and can be regulated at the gene level by the actions of several hormones. Species that can survive anoxia or freezing show a stress-induced up-regulation of both AAT mRNA as well as up-regulation of other mitochondrial transporters (see Chapter 17).

MITOCHONDRIAL IMPORT OF PROTEINS

The mitochondrial genome encodes 2 rRNAs, 22 tRNAs (transport RNAs) and 13 polypeptides. The polypeptides are all subunits of respiratory chain components: 7 subunits of complex I, 1 of complex II, 3 of complex IV, and 2 of ATP synthase (complex V) (Fig. 8.1). These polypeptides are synthesized on mitochondrial ribosomes using mitochondrial tRNAs. However, more than 98% of the proteins in mitochondria are actually nuclear-encoded, are synthesized as preproteins on cytoplasmic ribosomes, and then are targeted to the mitochondria by recognition of N-terminal targeting sequences. Cytoplasmic import factors transport the newly synthesized preproteins to the outer membrane of the mitochondria and then transporter systems translocate them across the outer membranes and inner via TOM (translocase of the outer membrane) and TIM (translocase of the inner membrane), respectively (Fig. 8.8). Proteins are directed to one of four targeted regions within mitochondria: the outer membrane, the intermembrane space, the inner membrane, and the matrix. Mechanisms of protein import into mitochondria appear to be tightly conserved in evolution. To date, much of the information comes from studies on yeast, *Saccharomyces cerevisiae*, and considerable work remains to be done to confirm the mechanisms in vertebrate cells and determine if significant differences occur, particularly in regulatory mechanisms.

Many preproteins are synthesized with cleavable presequences that are recognized by receptor subunits of the TOM and TIM, but others (e.g., the family of metabolite carriers of the inner membrane) have targeting sequences imbedded within the protein sequence. Preproteins are recognized by mitochondrial import stimulation factors (MSFs) that target the preproteins by binding to their pre-

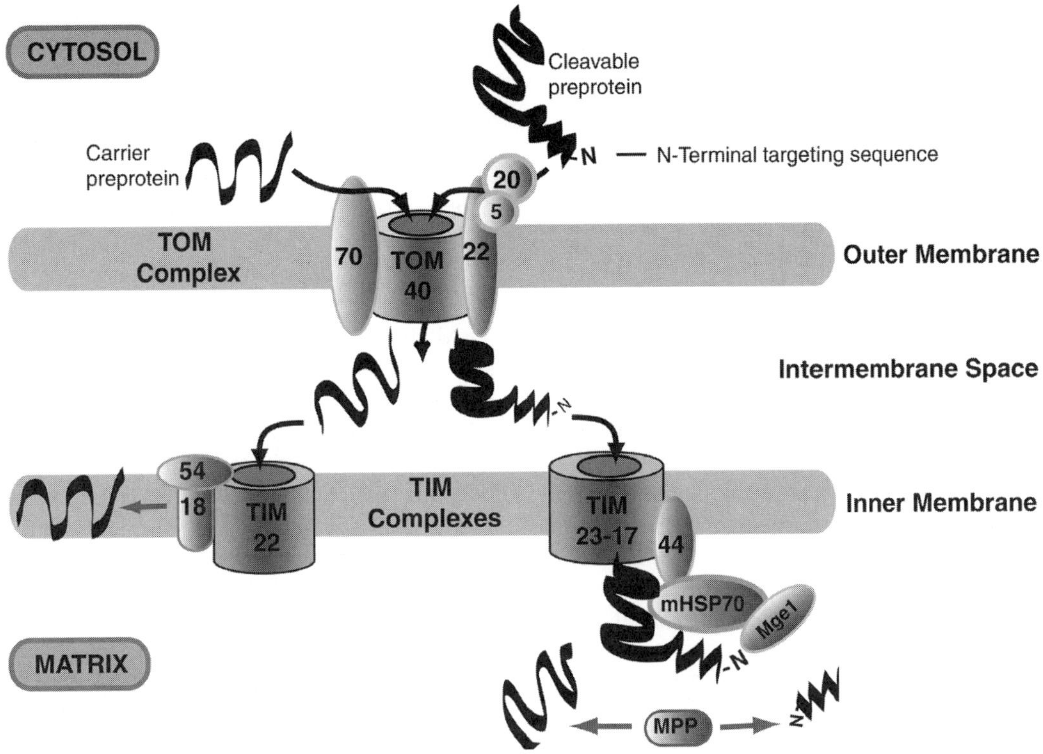

Figure 8.8 Protein import systems of the outer and inner mitochondrial membranes: TOM (translocase of the outer membrane) and TIM (translocase of the inner membrane). The TOM pore mediates transfer of all proteins into the intermembrane space. Preproteins with positively charged N-terminal cleavable targeting sequences that are destined for the mitochondrial matrix are received by the receptor Tom20 and passed via Tom22 (numbers indicate the molecular mass of the protein) into the pore, which consists mostly of Tom40 along with Tom5, 6, and 7. On the other side of the membrane they are received by the intermembrane space domain of Tom22. Proteins are then passed to the TIM23 complex consisting of Tim23 and 17 and through to the matrix with the help of Tim44, the mitochondrial heat shock protein (mHsp70) and mitochondrial GrpE (Mge1), a nucleotide-exchange factor, which aids mHsp70 in the hydrolysis of ATP. Both the mitochondrial membrane potential, $\Delta\psi$ and ATP hydrolysis, are required for import of the positively charged preproteins into the matrix. Once in the matrix, the N-terminal import signal is cleaved by the processing protease, MPP. The import of inner membrane proteins destined to be part of inner membrane carrier complexes again uses the TOM complex but proteins are recognized by the Tom70 receptor. Once in the intermembrane space the proteins are aided into their positions in the inner membrane by the TIM22 complex, consisting of Tim22, 54, and 18 with Tim9, 10, and 12.

sequences and directing them to the Tom receptor where they are released from the MSFs in an ATP-dependent manner. At least seven different proteins are in the TOM complex. Tom40 (the number designates molecular weight) forms the actual channel and Tom20 and Tom70 are the initial receptors for cleavable preproteins and carrier preproteins, respectively. Preproteins with a presequence are guided by a chain of binding sites on Tom20, Tom22, and Tom5, then through the Tom40 channel to then interact with the intermembrane space (IMS) domain of Tom22. Several positively charged amino acid residues on the presequence interact with negatively charged residues on both the TOM and TIM proteins to aid binding and transport. Tom6 and Tom7 modulate assembly and dissociation of the translocase.

Once in the intermembrane space, there are multiple fates for the preproteins. Matrix-targeted preproteins are translocated across the inner membrane by the Tim23–Tim17 channel complex aided by the Tim44 protein on the inner side of the membrane and the matrix mitochondrial (mt) Hsp70, which reversibly binds to the Tim. Mge1 (mitochondrial GrpE, a nucleotide-exchange factor located in the mitochondria) aids mtHsp70 in the hydrolysis of ATP. Tim11 is involved in the import and sorting. Both the mitochondrial membrane potential ($\Delta\psi$) and ATP hydrolysis are required for import of the positively charged preproteins into the matrix. Once in the matrix, the amino-terminal import signal is removed by the mitochondrial processing protease, MPP.

Membrane spanning proteins, destined for the inner membrane, do not have presequences, and these are inserted into the inner membrane with the help of the Tim22 complex, consisting of Tim22, Tim54, and Tim18, assisted by the proteins Tim9, Tim10, and Tim12. Proteins encoded by the mitochondrial genome are aided in moving into their positions in the inner membrane by the proteins Oxa1 and Pnt1.

Intermembrane space proteins, which also do not have presequences as preproteins, are also translocated across the outer membrane but cannot pass through the inner membrane. Proteins targeted to the outer membrane also do not have classic presequences. These proteins have internal targeting sequences that target them to the Tom40 channel, but once inserted into the channel they are stopped by a "stop-transfer" mechanism and released into the membrane with the help of the Tom7 protein.

MITOCHONDRIAL OXYGEN SENSING

The search for the cellular oxygen sensor has been a sort of biochemical Holy Grail for many decades. Given the enormous importance of oxygen for cellular ATP generation by aerobic organisms as well as the fact that oxygen is a substrate in dozens of other cellular reactions and that ROS are now proving to be key elements in intracellular signal transduction, it would not be surprising to find that multiple oxygen sensing mechanisms exist in cells. Although the field continues to search for (and may have found) the oxygen sensor, it may also be true that multiple types of sensors exist in order to transduce low oxygen signals for different purposes or that multiple, seemingly unrelated sensors are in fact responding to a master sensor. At different times, various studies have championed different mechanisms for cellular oxygen sensing, including the involvement of heme proteins (the first suggestion), oxygen-sensitive ion channels, mixed-function NAD(P)H oxidases, mitochondrial proteins, mitochondrially generated ROS, and oxygen-sensitive transcription factors including hypoxia-inducible factor CHIF-1α and nuclear transcription factor kappa-B (NF-κB).

Some ion channels are able to alter their conductance based on PO_2 levels; several oxygen-sensitive K^+ channels have been identified, and recent studies have revealed an oxygen-sensitive Ca^{2+} channel in neuronal tissue. It is not yet certain if ion channels themselves are the oxygen sensors, but most recent evidence suggests that effects are mediated by membrane proteins containing heme groups or redox-sensitive sites such as NADPH oxidase. The mechanism of O_2 sensing is tied up in the responsiveness of these channels to oxidizing and reducing agents. Activity changes are regulated by changes in the redox state of thiol residues. The interplay of hypoxia in cells, free radical levels, and the state of thiol residues are complex and controversial. Hypoxia in many cases actually causes a counterintuitive increase in free radical concentrations, complicating the control mechanisms of thiol sensitive sensors under these conditions.

Alternately, it might be that the reduced nicotinamide adenine dinucleotide phosphate (NADPH) oxidase system, with its core catalytic complex of gp91phox and p22phox subunits (together making the flavo-cytochrome b558) plus its regulatory complex (including subunits p47phox, p67phox, and as well GTPase proteins Rac-1 or Rac-2) is a key part of the sensing system. The enzyme converts oxygen to superoxide, accompanied by NADPH oxidation to $NADP^+$. This complex is expressed particularly in those cells that are extremely O_2 sensitive. Oxygen sensing may be integrated into the mechanisms that generate superoxide, where under normoxic conditions, the NADPH oxidase would remain in the activated state, and under conditions of reduced available oxygen levels, there would be a decrease in electron transport, a loss in the production of superoxide, and the cytosol redox state would be more reduced. The generation of ROS relative to available oxygen might also result in the reduction of activity of ion channels (e.g., the potassium channels). Recent studies on knockout mice that were lacking one of the catalytic subunits, gp91phox, do not support this hypothesis. These animals could respond to hypoxic conditions normally. In other words, even though these transgenic animals had fewer ROS generated in their lungs, the ion channel response to hypoxia was normal and not changed as compared to nontransgenic mice. It is likely that there is a more complicated scenario at hand in terms of vasoconstriction in the lungs during hypoxia.

Heme protein involvement in oxygen sensing was first detected in the response of erythropoietin (EPO) to hypoxia. EPO is the hormone that stimulates the proliferation and differentiation of erythroid cells to enhance the production of red blood cells under hypoxia (see Chapter 18). The fact that the effects of high versus low oxygen on EPO expression could be mimicked by known modulators of oxygen interaction with heme proteins (Co^{2+}, Ni^{2+}, carbon monoxide) provided strong evidence of heme protein participation as an intermediary in the process. The transactivator of the EPO gene is HIF-1, which is now known to activate the expression of at least 30 genes encoding enzymes or proteins that aid cellular responses to low oxygen (see Chapters 6 and 15). During hypoxia, HIF-1α subunits are stabilized, allowing them to dimerize with HIF-1ß subunits and activate gene expression. Oxygen sensitivity is derived from oxygen-dependent hydroxylation of two proline residues on HIF-1α by prolyl-4-hydroxylase. When oxygen is plentiful, these are hydroxylated, and this sets the stage for very rapid HIF-1α degradation whereas when oxygen in low HIF-1α remains unmodified and is free to dimerize with HIF-1ß.

Being the major consumers of oxygen, it has long been supposed that the mitochondria must also have oxygen sensing mechanisms. Recent studies indicate that they do but that it is not specifically oxygen that is sensed but rather the consequences of variation in oxygen on the production of ROS by the respiratory chain. So, whereas superoxide formation by complex I or III was considered for many years to be a "physiological defect" of the electron transport chain, it is now becoming obvious that mitochondrially-generated ROS have major roles in signaling pathways that mediate cellular responses to changes in oxygen tension (for more information about ROS production by the mitochondria, see Text Box 13.3). Hypoxia-induced increases in ROS generation have now been demonstrated in several cell–tissue systems. Studies with cardiomyocytes have been particularly instructive because physiological function (contraction), oxygen consumption, and ROS production can all be assessed and correlated. These cells show a graded response to hypoxia, decreasing contraction and decreasing oxygen consumption in response to declining oxygen availability (but reversible upon reoxygenation). Using a fluorescent dye (2,7-dichlorofluorescein) that reacts with ROS, graded increases in fluorescence were seen during hypoxia, indicating increased ROS production. Addition of antioxidants abolished these increases as well as the hypoxic inhibition of contraction, whereas superfusion of the cells with H_2O_2 mimicked the effects of hypoxia. A reduced maximum velocity of cytochrome oxidase was also identified as a response to hypoxia in this system, and, when azide was used to experimentally inhibit cytochrome oxidase under normoxic conditions, the hypoxic responses were seen—graded ROS increase, graded decrease in contraction. Hence, there is evidence that sensing may be linked with a reduction in cytochrome oxidase activity, whereas signaling emanates from the effect that this has on superoxide production at earlier points in the respiratory chain. New studies have also shown that ROS generated by mitochondria influence the accumulation of HIF-1α, which provides a link from mitochondria oxygen sensing to hypoxia-induced gene expression.

CALCIUM SIGNALING AND THE MITOCHONDRIA

Calcium uptake in the mitochondria has fascinated researchers over the last four decades. Originally it was thought that the mitochondria were the main storage organelles for calcium in cells other than cardiac and skeletal muscle cells. However, in the 1980s it determined that the endoplasmic reticulum (ER) (and some of its subcompartments) was in fact the primary site of calcium storage in the cell (i.e., analogous to the role of the sarcoplasmic reticulum in Ca^{2+} sequestering in muscle) and that calcium flux in the cytosol had little dependence on the mitochondria. Nevertheless, there is a broad and complex assembly of mitochondrial calcium transporters and channels, and the activities of key enzymes of mitochondrial energy metabolism including the PDC, NAD-IDH, and α-KGDH are activated by calcium. Several cellular-level phenomena are now known to be crucially dependent on mitochondrial import and export of calcium.

Ca^{2+} in the cytosol is typically held at very low levels (around 100 nM), and bursts of Ca^{2+} release into the cytoplasm activate a variety of metabolic processes. Calcium influx into the cytoplasm can come from the extracellular medium in response to stimuli including membrane depolarization, stretch, extracellular agonists, and intracellular messengers via many types of ion channels such as the voltage-operated channels that function in nerve and muscle tissues. The sodium–calcium exchanger and the plasma membrane Ca^{2+}-ATPase extrude Ca^{2+} again. Calcium is also released into the cytoplasm from internal stores in the ER or sarcoplasmic reticulum (SR) via IP_3-gated channels or ryanodine receptors. Calcium is returned to the ER/SR via the action of the sarco(endo)plasmic Ca^{2+}-ATPase. Whenever Ca^{2+} is released into the cytoplasm, its levels also rise in mitochondria. Calcium enters the mitochondria electrophoretically via a Ca^{2+} uniporter and exits again predominantly via a Na^+/Ca^{2+} exchanger but also by sodium-independent pathways, including a H^+/Ca^{2+} exchanger or through permeability transition pores (Fig. 8.9).

Calcium accumulation in the mitochondria can be prevented by chemically collapsing the proton gradient using an uncoupler, or by altering the environmental conditions for the cell, such as exposing the cell to anoxia or impairing mitochondrial respiration. During hypoxia, with cytosolic Ca^{2+} levels elevated and mitochondrial Na^+ levels high, the mitochondria can also become loaded with calcium due to a reversal of the Na^+–Ca^{2+} exchanger. Elevated calcium levels in the mitochondria have a significant impact on metabolic pathways in the mitochondria. As noted above, three of the key regulatory enzymes are activated by Ca^{2+} so that stimuli that increase mitochondrial Ca^{2+} levels automatically increase TCA cycle flux, which elevates NADH production and stimulates mitochondrial respiration, increasing $\Delta\psi$, and driving increased rates of ATP production. In the case of pyruvate dehydrogenase, one of the key regulatory features is control by a calcium-dependent phosphatase, which dephosphorylates the enzyme and converts it to its active form. Hence,

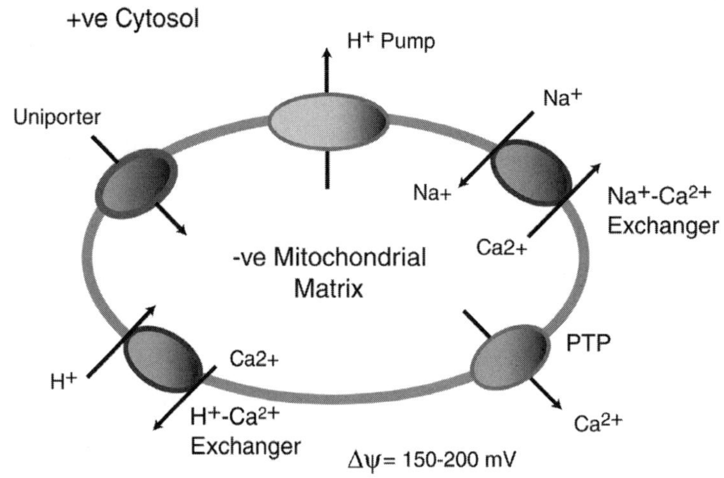

Figure 8.9 Mitochondrial calcium import and export. The membrane potential of the mitochondria is believed to be negative relative to the cytosol. Rate-limiting to the uptake of calcium into the mitochondria is H^+ pumping. Calcium influx into the cytoplasm occurs electrophoretically through a uniporter, and exits again via a Na^+/Ca^{2+} exchanger or by sodium-independent pathways, including an H^+/Ca^{2+} exchanger or through permeability transition pores (PTP). Whenever Ca^{2+} is released into the cytoplasm, its levels also rise in the mitochondria.

calcium levels in the mitochondria have a direct role in modifying the rate of ATP production in the cell, and this coordinates ATP availability with the increased rates of various Ca^{2+}-dependent processes that are activated by elevated Ca^{2+} in the cytosol (e.g., muscle contraction).

The uptake of calcium into the mitochondria results in the depolarization of the mitochondrial membrane and is electrogenic. This process has an effect on cellular signaling in both excitable and nonexcitable cells. Generally, it has been shown that, in excitable cells such as neurons and at neuromuscular junctions, cytosolic calcium levels rise dramatically with calcium uptake through voltage-gated calcium channels, and that recovery from this uptake is biphasic, with one extremely rapid phase (e.g., microseconds) of recovery time followed by a prolonged phase that takes several minutes. Inhibition of calcium uptake in the mitochondria results in a decrease in the rate of the rapid recovery event and abolishes the secondary phase of the biphasic recovery. Thus, it is believed that the mitochondria, by responding to elevated cytosolic calcium with a rapid Ca^{2+} uptake into these organelles, allows a more rapid recovery of cytosolic functions from the Ca^{2+} stimulus.

In nonexcitable cells, Ca^{2+} levels in the cytosol are also affected by mitochondrial Ca^{2+} pools. In this case, Ca^{2+} release from the ER in response to a stimulus (e.g., a hormone) involves the calcium-permeable channels. The creation of a cytosolic calcium wave coincides with a depolarization of the mitochondrial membrane, and this could be associated with Ca^{2+} influx into the mitochondria as the calcium wave occurs. If mitochondrial uptake of calcium is inhibited, then the wave of cytosolic calcium is increased twofold. Thus, the influx of calcium into the mitochondria appears to lower local calcium levels, acting as a buffer for the propagation of calcium waves through the cytosol following stimulation.

Dramatic changes in intramitochondrial calcium are seen under conditions of ischemia/reperfusion in oxygen-sensitive organisms. During ischemia (interruption of blood flow) a drop in respiratory rate is accompanied by mitochondrial swelling. This affects the activity of calcium transporters. Heart studies have shown that pretreatment with Ca^{2+}-channel blockers significantly improves recovery times during reperfusion; in these studies, mitochondrial Ca^{2+} concentrations were decreased, as was left ventricular stiffness, and compliance rose. In deep hypoxia and anoxia, the mitochondrial formation of superoxide, hydrogen peroxide, and peroxynitrite results in dramatically increased release of calcium from the mitochondria. This release of calcium can result in the activation of calcium-dependent proteins in the cytosol, including proteases and nucleases, increasing the potential for cell death.

The most dramatic conditions of altered mitochondrial calcium involve those of oxidative stress. Under extreme conditions, the mitochondria alter their roles as important regulatory sites for cytosolic calcium levels, and become players in the initial events resulting in cell death. Under conditions of anoxia or ischemia, prevention of mitochondrial calcium uptake can in fact be protective to cells such as neurons. In neurons, research involving manipulation of mitochondrial free calcium during conditions of ischemia and reperfusion has shown that matrix calcium levels themselves influence cellular injury.

AGING, DISEASE, AND THE MITOCHONDRIA

Mitochondrial functions, both genetic and bioenergetic, are compromised over time and many studies indicate that mitochondrial insufficiencies play a key role in cellular aging. The damage caused by oxidative stress is a key culprit as mitochondria are a primary target for ROS attack, particularly because an electron leak from ubisemiquinone ($QH^{\cdot}$) is the greatest source of cellular ROS production. Indeed, the actual levels of $QH^{\cdot}$ in the cell are a good measurement for the potential for ROS generation (see Chapter 12 for more about ROS damage). Oxidative damage to mtDNA is particularly critical and is even more significant than damage to nuclear DNA because this damage strongly magnifies and accumulates over time.

During the mitochondrial aging process, various functions of the organelle decrease dramatically over time, including electron transport chain activity, energy output, and transcriptional activity and fidelity. Multiple deletion mutations accumulate over time, and many of these mutations are the same as those that occur in mitochondrial genetic diseases. These mitochondrial DNA mutations accumulate the most in cells that are postmitotic and oxidatively active. In aged cells, there is a dramatic rise in both mitochondrial superoxide and hydrogen peroxide, which are normally detoxified in the presence of the enzyme superoxide dismutase. Mutations in the 13 polypeptides that are coded by mtDNA (Fig. 8.1) significantly affect respiratory chain function, and inhibition of the electron transport chain is at least partially involved in the increased amount of superoxide produced. Deletion mutations in mtDNA occur in much greater numbers in patients suffering from Alzheimer's disease, Huntington's disease, and chronic ischemic heart disease, and these mutations are all hypothesized to occur with accompanying ROS accumulation. Mitochondrial defects also cause or are implicated strongly in a variety of other diseases, including cancer. The interplay of free radicals, electron transport chain activity, and DNA transcription clearly holds the key to many disease puzzles.

Mitochondrial DNA mutates at a very high rate and, among others things, affects the generation of ROS, the production of ATP for the cell, and activation of programmed cell death (apoptosis). Mitochondrial DNA mutations are at the root of Leber's hereditary optic neuropathy (LHON) and LHON dystonia, which are the result of the same mutation in a subunit of the NADH dehydrogenase (Table 8.3). LHON, in comparison to dystonia, presents itself later in life in patients and is represented by a lower percentage of the mutant DNA. The mutation in LHON causes a drop in complex I activity, causing patients to suffer from sudden blindness in midlife, as the optic nerve dies. In dystonia, patients develop mental retardation, speech impairment, slowed growth and movement disorders, with some degeneration of the basal ganglia.

Leigh's syndrome, another mitochondrial disease, results from mutations in the gene for subunit 6 of the mitochondrial F_1F_0-ATP synthase, impairing mitochondrial ATP synthesis. This is a heteroplasmic mutation, and, when the mutation occurs in the mitochondrial population at low levels, it causes neuropathy, ataxia, and retinitis pigmentosa (a condition called NARP). When the mutation presents itself at high levels, it is called Leigh's syndrome, and symptoms can occur that are ultimately fatal, including delay in development, regression, ophthalmoplegia, optic atrophy, ataxia, hypotonia, and spasticity.

Mitochondrion-associated diseases are typically degenerative and are inheritable genetic diseases. They arise frequently from a genetic defect in a mitochondrially encoded protein causing protein malfunction, but defects in nuclear-encoded mitochondrial proteins are also known. Disease can also alter the number of mitochondria and/or their morphology. Generally, it has been found that the primary mitochondrial functions that are affected and that manifest themselves in the form of disease are the generation of ROSs, the production of ATP for the cell, and programmed cell death (apoptosis). Mutations of mtDNA are inherited through the maternal cell line, whereas mutations that occur in nuclear genes that encode mitochondrial proteins can show Mendelian inheritance; phenotypes are often wide-ranging, diverse, and overlapping. All mitochondrial diseases tend to be late onset and progressive, and thus inheritance and age are both key factors in the severity of the disease.

TABLE 8.3 Human Diseases Resulting from Mutations of Mitochondrial Genes

Disease	Subunit/Protein/ Complex
LHON (Leber's hereditary optic neuropathy)	ND1 & ND4 & ND6/complex I
LHON dystonia	ND6/complex I
Leigh's disease	ATPase 6
MERRF (myoclonic epilepsy with ragged-red fibers)	ATPase 8
MELAS (mitochondrial encephalomyopathy, lactic acidosis, and strokelike episodes)	ND1/complex I
NARP (neuropathy, ataxia, and retinitis pigmentosa)	ATPase 6

Generally speaking, defects at virtually any locus in the short mitochondrial genome have disastrous consequences for the organism. Examples of mutation diseases include MERRF (myoclonic epilepsy with ragged red fibers) and MELAS (myopathy, encephalopathy, lactic acidosis and strokelike episodes), both of which involve point mutations in transfer RNAs that result in a reduction in mitochondrial protein synthesis and in complex I and IV activities. Kearns–Sayre syndrome and progressive external ophthamoplegia (PEO) both involve large-scale deletions. Late-onset Alzheimer's disease can occur correlated with a specific protein synthesis mutation. For most of these diseases, the complete pathway mechanisms that ultimately lead to cellular malfunctioning are not yet worked out. However, studies attempting to repair the effects of these diseases are ongoing. For example, in order to repair the defect in mitochondrial proton gradients and restore calcium homoeostasis in diseases such as in NARP or MERRF, one approach has been to inhibit the mitochondrial efflux of calcium. By emphasizing the role of ion homeostasis and by maintaining mitochondrial calcium balance, the ATP and calcium response was salvaged in the cell and cell death prevented.

Mutations in nuclear genes that encode mitochondrial proteins are also deleterious and usually result in an inhibition of oxidative phosphorylation and/or destabilization of mitochondrial DNA. A mutation in an 18-kDa structural protein of complex I results in mental retardation in children, hypotonia, and some basal ganglia degeneration. A mutation in frataxin, which is targeted to the inner mitochondrial membrane and is involved in iron transport, results in Friedrich's ataxia, and the symptoms include hypertrophic cardiomyopathy, peripheral neuropathy, and cerebellar ataxia as a result of iron overload in the mitochondrial matrix, faulty production of iron–sulfur proteins and increased iron-mediated ROS production (see Chapter 13 for more information).

APOPTOSIS

Apoptosis, or programmed cell death, has been the focus of thousands of studies in recent years because of an increasing realization of the role that it plays in the development and differentiation of tissues and organs, in the responses to viruses, environmental stresses, or wounding of tissues, and to multiple disease states including cancer.

Generally speaking there are two types of cell death: apoptosis and necrosis. Necrosis refers to the death of the cell whereby the integrity of the cellular organelles is lost; organelles break up, cytoplasm is contaminated with breakdown products, and the plasma membrane itself ruptures after cell swelling, radiating cellular damage products to surrounding cells and tissues. By contrast, apoptosis is a regulated process of cell death; damage is contained and managed and an apoptotic cell does not seed necrotic damage to its neighbors. However, the plasma membrane of the apoptotic cell does bleb, flipping phosphatidylserine moieties from the inner leaflet to the outer surface. In addition, the chromatin condenses and the cytosol decreases in volume. The process of apoptosis involves the activation of caspases and endonucleases and the degradation of DNA into fragments of delimited sizes. These discrete fragments of cleaved DNA can be diagnostic of the whole process; apoptotic cells lysates, when run on acrylamide gels and stained for DNA, show a distinct DNA, "laddering."

The model in Figure 8.10 represents the involvement of the mitochondria in nonapoptotic pathways (necrosis) that lead to cell death. As a result of certain cell death stimuli, nonapoptotic pathways involving osmotic disequilibrium within the mitochondria lead to expansion of the matrix and rupturing of the mitochondrial membrane. Once the outer membrane is ruptured, the electrochemical gradient ($\Delta\psi$) across the inner membrane is lost, oxidative phosphorylation fails, and there is an overproduction of ROS.

Apoptotic destruction is different. Central to apoptosis are several families of proteins that have been highly conserved throughout evolution. These proteins include the Bcl-2 family of proteins, APAF-1 (apoptotic protease activating factor 1), and the caspase family of proteases. The

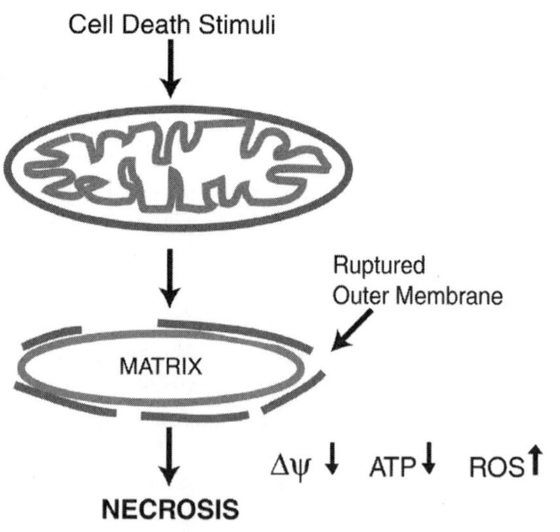

Figure 8.10 Involvement of mitochondria in cell death via necrosis. Cell death stimuli lead to an osmotic disequilibrium within the mitochondria. This causes an expansion of the matrix and rupturing of the outer mitochondrial membrane. Once the outer membrane is ruptured, the electrochemical gradient ($\Delta\psi$) across the inner membrane is lost, and there is an overproduction of ROS and a drop in ATP production.

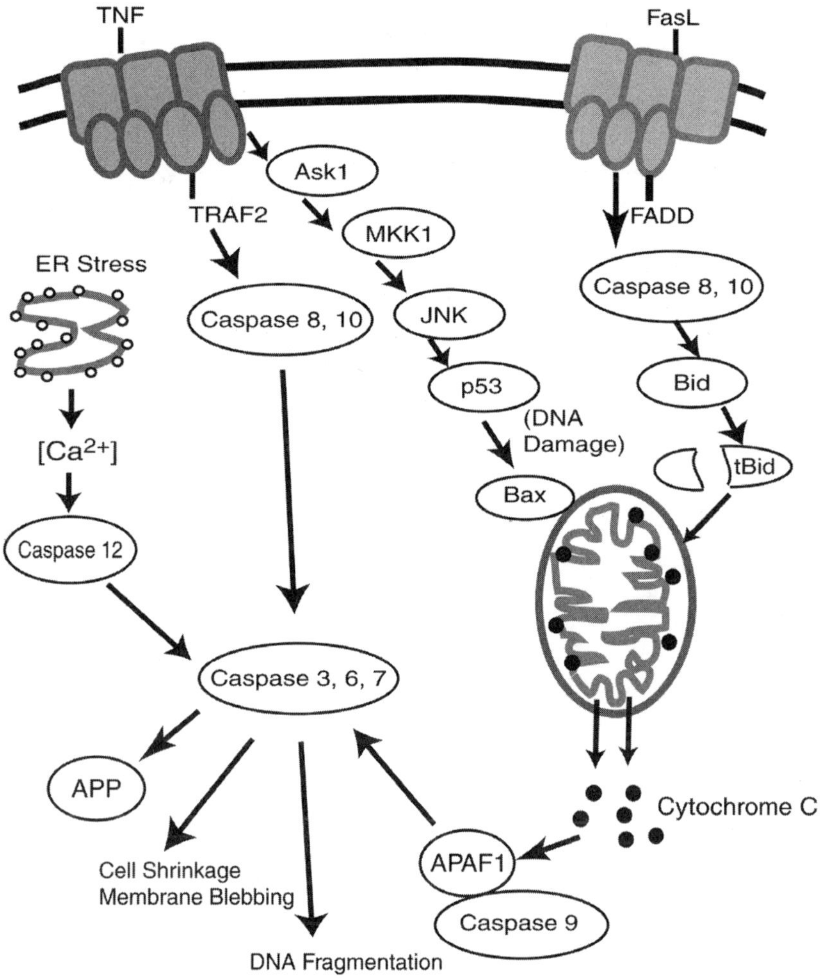

Figure 8.11 General overview of programmed cell death via apoptosis showing the involvement of mitochondria. Cell death is characterized by shrinkage of the cell, membrane blebbing, and fragmentation of cellular DNA. Regulators of apoptosis include the caspases, and once this family of proteases is activated, both nuclear and cytoskeletal proteins are cleaved and cell death is induced. As shown here, the mitochondria are key to the initiation of cell death. Cytochrome *c* release from the mitochondria is involved in the activation of caspase 9. Some programmed cell death initiators or stimuli include proteins FasL and TNF, DNA damage, and ER stress. FasL and TNF cause the activation of caspases 8 and 10. DNA damage causes caspase 9 activation. ER stress (along with Ca^{2+}) activates caspase 12.

latter group, the caspases (meaning cysteinyl aspartate-specific proteases) are initially translated as pro-proteins that are "dormant" until activated by proteolytic cleavage. Once caspases are activated, nuclear and cytoskeletal proteins are cleaved and cell death is induced. The release of cytochrome *c* from the mitochondrial matrix, coupled to the activation of caspase 9, is a pivotal step in apoptosis. Programmed cell death initiators or stimuli include the protein FasL, DNA damage, ER stress, and tumor necrosis factor (TNF). FasL and TNF lead to the activation of caspases 8 and 10. DNA damage causes the activation of caspase 9 and ER stress along with calcium release results in activation of caspase 12 (Fig. 8.11).

It is believed that cell death involves the mitochondria in multiple ways. First, there is a release of mitochondrial proteins, including cytochrome *c*, which is responsible for the activation of the caspases. Newly released cytochrome *c* then comes together with APAF-1 and procaspase 9, resulting in the formation of the so-called apoptosome, which causes activation of caspase 9 and subsequent apoptosis. Cytochrome *c* involvement in this process can be inhibited by Bcl-2 on the surface of the mitochondria. Interestingly, the release of cytochrome *c* may have two outcomes, depending on cell type: necrosis if the cell type in question expresses excessive amounts of caspase inhibitors, or apoptosis if it does not. Other proteins released from the mitochondria that are involved in inducing apoptosis include procaspase 3, apoptosis-inducing factor (AIF), and Smac (second mitochondrial activator of caspases). The latter, which is released by mitochondria when cytochrome *c* is released, plays a role in preventing the actions of inhibitors of caspase activity, or IAPs (inhibitors of apoptosis), and thus facilitates the action of the cell death signaling pathway.

Figure 8.12 shows the involvement of the mitochondria in the activation of caspases, resulting in subsequent apoptosis. The release of cytochrome *c* into the cytosol occurs through a variety of stimuli, including calcium, oxidants, and Bax, a Bcl-2 family protein. Active APAF-1, which is activated by binding to cytochrome *c*, binds to procaspase 9, initiating the proteolytic signaling pathway. Cell death stimuli trigger proapoptotic members of the Bcl-2 protein family, such as Bax and Bak, resulting in the opening of channels or pores, releasing cytochrome *c* and other proteins.

The mitochondria undergo morphological changes during the progression of apoptosis, including some increased density in the matrix and a reduction in the size. In addition, mitochondria alter their distribution in the cell; for example, they have been found to cluster around the outside of the nucleus during some of the early events of apoptosis. It is not known, however, whether these morphological or distribution changes play a role in the signaling events of apoptosis that involve the mitochondria.

One of the key caspases involved in programmed cell death, caspase 3, is activated by the released mitochondrial cytochrome *c*. Cytochrome *c* enters the cytosol from the mitochondrion, binds to APAF-1, and then in the presence of ATP or dATP, this complex activates procaspase 9. Caspase 9 then catalyzes the cleavage of procaspase 3 into active caspase 3.

Cytochrome *c* release was found to rely on a group of proteins, the Bcl-2 family, comprised of both antiapoptotic and apoptosis-stimulating proteins. Fifteen Bcl-2 protein family members have so far been identified in mammals. Each contains at least one of four conserved peptide regions, known as the Bcl-2 homology domains or BH (BH1 to BH4) (Fig. 8.13). These domains are largely α helical and regulate each other by the formation of homo- or heterodimers. Bcl-2 and Bcl-xL are involved in blocking apoptosis; both of these proteins have at least three BH domains. Bax, Bak, and BH3-only proteins such as Bid and Bad trigger the apoptotic signaling pathways—Bax and Bak both contain BH1, BH2, and BH3 domains and resemble Bcl-2 in structure. BH3-only proteins require this domain for binding to other Bcl-2 family members

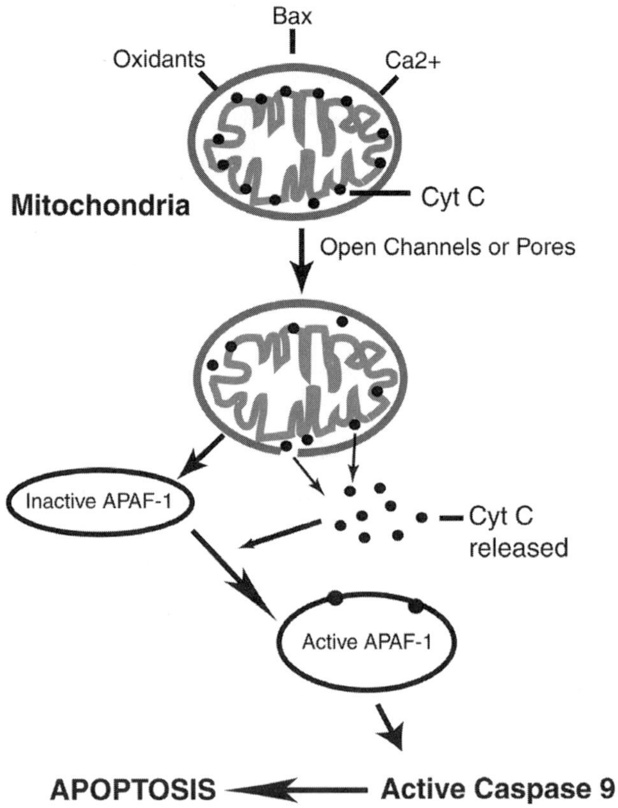

Figure 8.12 Triggers for programmed cell death showing the involvement of the mitochondria in the activation of caspases and subsequent apoptosis. The release of cytochrome *c* and other proteins into the cytosol occurs under a variety of stimuli, including Ca^{2+}, oxidants, and proapoptotic members of the Bcl-2 protein family, such as Bax and Bak, resulting in the opening of channels or opening of pores. The activation of Apaf-1 occurs through the binding of cytochrome *c* to inactive Apaf-1. Active Apaf-1 can then bind to procaspase 9, leading to further caspase activation and apoptosis.

and for their involvement in the apoptotic signaling pathway. Bcl-xL actually has a structure resembling that of a protein that creates pores in the membranes of cells—diphtheria toxin. It is thought that the Bcl-2 proteins that are antiapoptotic actually protect the mitochondria from breakdown, whereas the apoptotic Bcl-2 family members create the pores in the outer mitochondrial membrane to release cytochrome *c* into the cytosol.

The release of mitochondrial proteins is controlled by the type and amounts of the several members of the Bcl-2 family of proteins; when the antiapoptotic proteins Bcl-2 or Bcl-xL are overexpressed in many different cell types, cytochrome *c* release from the mitochondria is prevented, as is subsequent caspase activation and cell death. Additionally, the expression of the apoptotic signaling protein Bax results in the release of cytochrome *c* in the

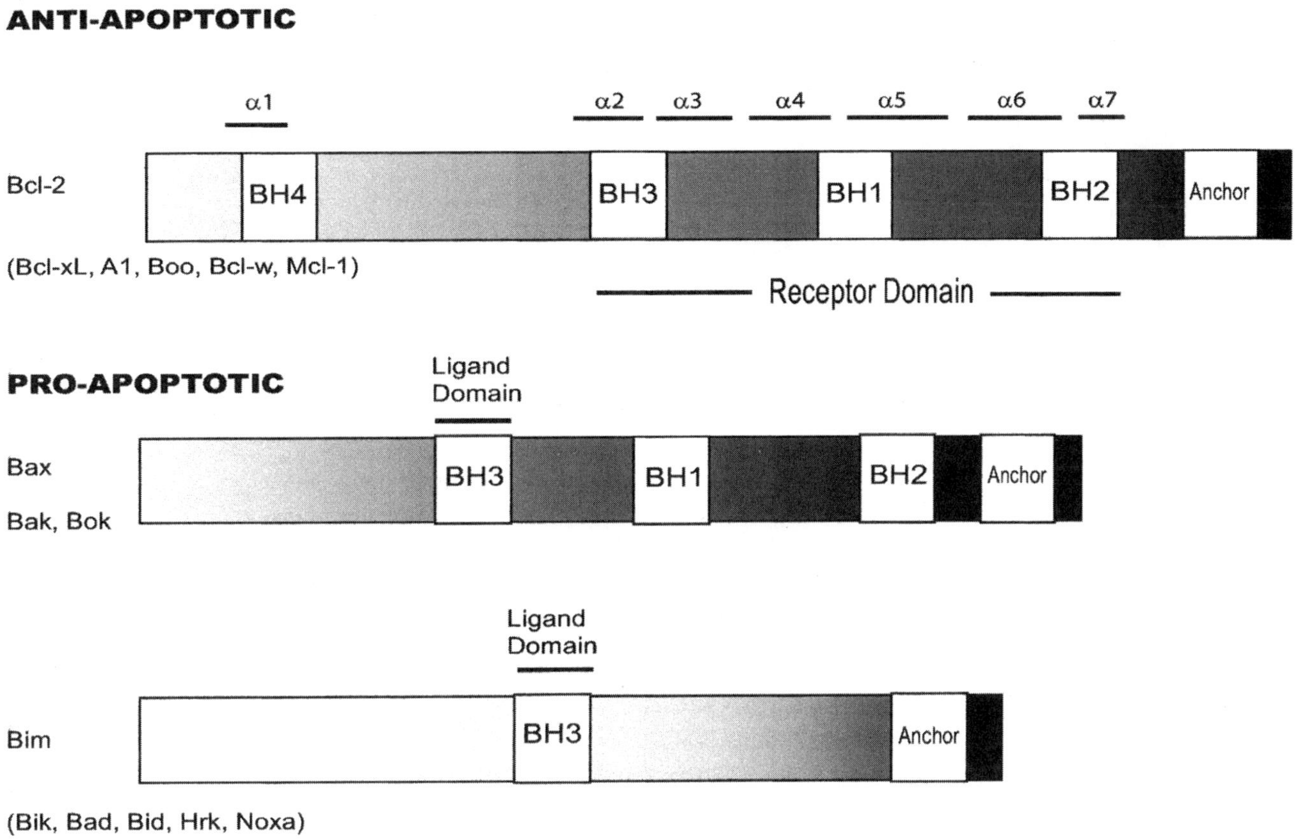

Figure 8.13 Antiapoptotic proteins of the Bcl-2 type each containing at least one of four conserved peptide regions, the Bcl-2 homology domains or BH (BH1 to BH4). Bcl-2 and Bcl-xL inhibit apoptosis; both of these proteins have at least three BH domains. Proapoptotic proteins such as Bax, Bak, and BH3-only proteins such as Bid and Bad trigger the apoptotic signaling pathways. Bax and Bak both contain BH1, BH2, and BH3 domains. Involvement of Bax in apoptosis requires translocalization to the mitochondrial membrane. Bax has a hydrophobic domain (ANCHOR) to anchor the protein into the outer mitochondrial membrane. The α-helices are indicated which can form hydrophobic cores in the three-dimensional structure.

absence of any other cell death signals. This Bcl-2 protein family member can cause the release of cytochrome *c* in the presence of a caspase inhibitor, indicating that caspases are not involved in the release of cytochrome *c* from the mitochondria under stimuli that cause cell demise.

The location of the Bcl-2 family members has been gradually determined, and they are not all bound to the mitochondrial membrane. In fact, only Bcl-2 and Bcl-xL have so far been found to be membrane-bound. Other Bcl-2 family members, such as Bid, Bad, and Bim, have been found to localize to the mitochondrial membrane from the cytosol only during the apoptotic process. Once there, responding to signaling from the cytosol, they can regulate the activities of the membrane-bound Bcl-2 proteins (and the release of cytochrome *c*) by protein–protein interactions. The signals that trigger the binding of these regulatory proteins include cleavage of Bid by caspase 8 and dephosphorylation of Bad. However, the three-dimensional structure and regulation of the protein complexes have yet to be elucidated. Figure 8.14 shows the involvement of a proapoptotic protein, Bax, which has a hydrophobic domain to anchor the protein into the outer mitochondrial membrane, in apoptosis. Other Bax family members include Bak, Bok, Bik, Bad, and Bid; all contain BH3 domains although most do not have the BH2 or BH1 domains. First, following a death stimulus, Bax translocates to the mitochondrial outer membrane. Once there, the conformation of Bax is altered to reveal its N terminus, triggering oligomerization and insertion into the mitochondrial membrane. The presence of this new integral membrane protein then allows for cytochrome *c* release. It has been found that both Bid and Bcl-xL regulate Bax activity by binding to it and preventing changes to Bax.

The Bax family is diverse and regulation of these proteins is much more complex than is shown in Figure 8.14. They can be activated by multiple stimuli, including changes in cellular pH, reversible phosphorylation

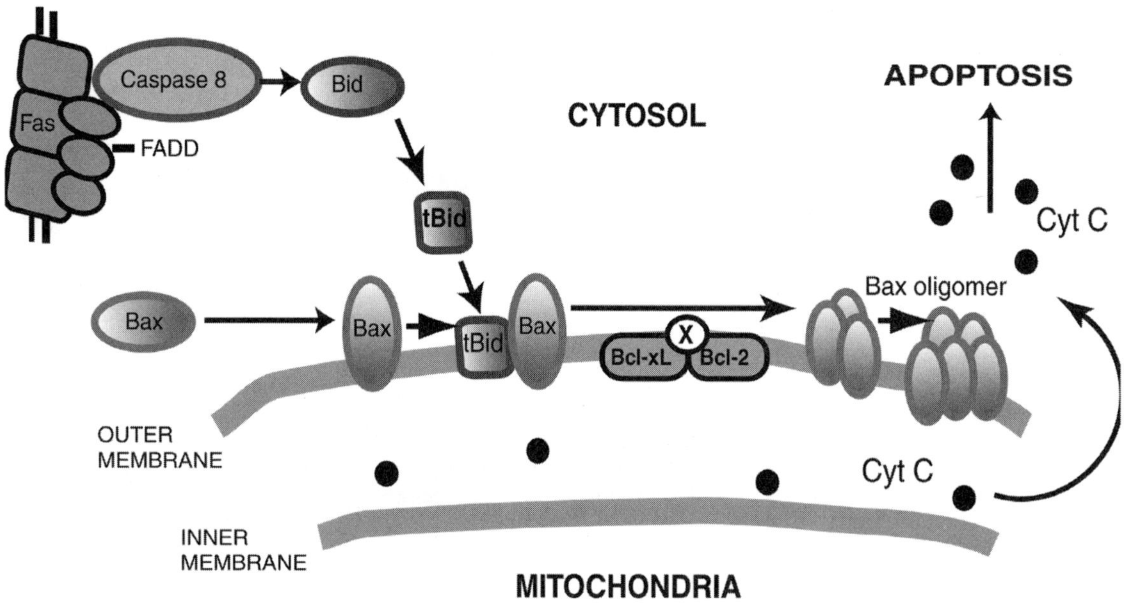

Figure 8.14 Fate of Bax during apoptosis. Both Bcl-2 and Bcl-xL are involved in regulating Bax activity by binding to it and preventing changes to Bax. This figure shows the fate of Bax following a death stimulus, which includes Bax translocation to the mitochondrial outer membrane, change in the conformation of Bax, oligomerization, and insertion into the mitochondrial membrane, resulting ultimately in cytochrome *c* release. As a result of a death stimulus, Bid is cleaved at its amino terminus by caspase 8, and truncated Bid (tBid) can then insert inside the mitochondrial membrane; this is necessary for cytochrome *c* release (this is prevented by Bcl-xL and Bcl-2 in the mitochondria). Circled X = inhibition.

(activation by phosphatases, inhibition by kinases), and activation of the activity of caspases by cleavage. An increase in cellular pH has been shown to promote a conformational change in Bax to its active form and therefore make it more susceptible to caspase activity. Bad, a proapoptotic protein, can be phosphorylated by protein kinase A, protein kinase B (Akt), or $p90^{RSK}$ allowing Bad to bind to the 14-3-3 protein, sequestering this protein away from the mitochondria (Fig. 8.15). A phosphatase activity prevents this binding, allowing Bad to bind to Bcl-2 through BH3, preventing Bcl-2 activity. Bid has been shown to be activated by the proteolytic activity of caspase 8 (as also shown in Fig. 8.14). Bid does not have a hydrophobic sequence, but through proteolytic processing its BH3 domain is exposed to bind to Bax and Bcl-2, inhibiting Bcl-2 and activating Bax.

Two mechanisms have been proposed for the involvement of the Bcl-2 proteins in the release of cytochrome *c* from the mitochondrial inner membrane: (1) formation of cytochrome *c* conducting channels by the Bcl-2 proteins, and (2) rupture of the mitochondrial outer membrane through changes regulated by the Bcl-2 proteins. The foundation for the first theory is that the Bcl-2 family members, especially Bcl-2 and Bax, can form ion channels in vesicles and bilayers, similar to pores triggered by bacterial toxin. Thus, it is believed that the proapoptotic members of the Bcl-2 protein family form channels on the outer mitochondrial membrane, which are big enough to allow for the passage of mitochondrial proteins such as cytochrome *c*. Release of cytochrome *c* itself from the inner membrane possibly involves a two-step process: (i) removal of cytochrome *c* from its membrane-anchoring protein cardiolipin through lipid peroxidation to form soluble cytochrome *c*, and (ii) release of cytochrome *c* through the outer membrane. (See Text Box 8.4.)

The second theory, involving the rupture of the outer mitochondrial membrane, could involve the inhibition of the exchange of mitochondrial ATP for cytosolic ADP during apoptosis. Here, it is thought that the voltage-dependent anion channel (VDAC), located in the outer membrane, and the adenylate translocator (ANT), located in the inner membrane, are impaired, resulting in hyperpolarization of the inner membrane due to the fact that F_1F_0-ATPase activity is stopped and H^+ relocalization into the matrix is inhibited. Hyperpolarization, or the increase in mitochondrial membrane potential, results in the swelling of the matrix, and it is thought that the Bcl-2 family of proteins is somehow involved in regulating one of these steps.

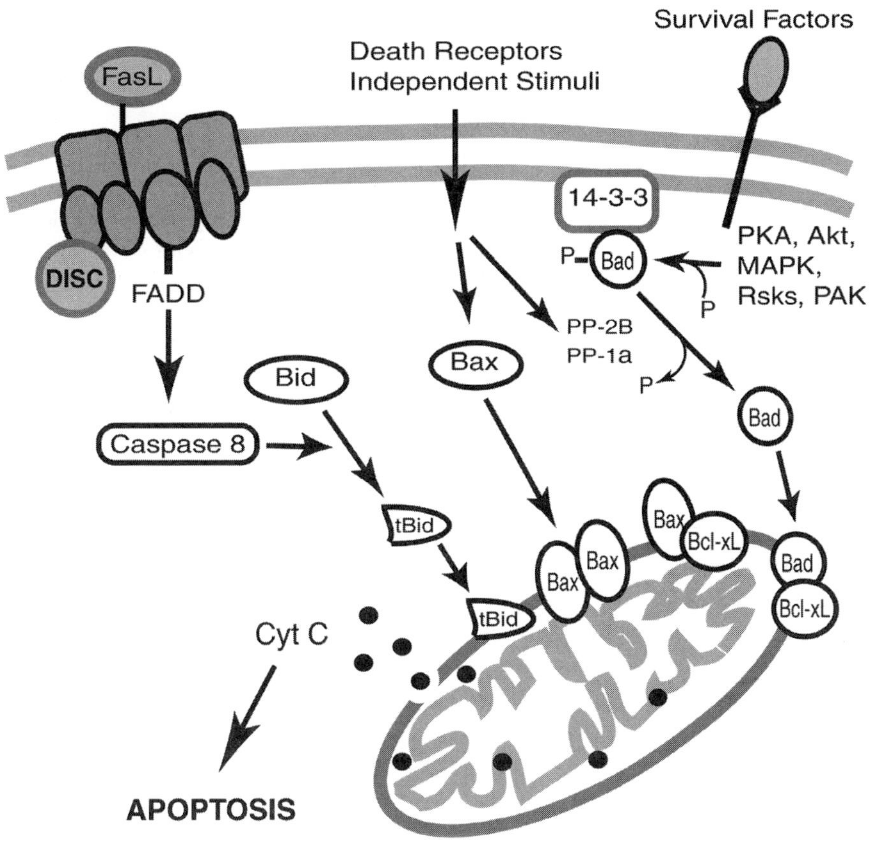

Figure 8.15 A cell death signal, such as Fas ligand (FasL), binds to its receptor, Fas. Trimerization of the receptor allows formation of the death-inducing signaling complex, DISC, which involves the association of the cytoplasmic region of Fas, FADD (an adaptor protein, Fas-associating protein with death domain) and procaspase 8. Activated caspase 8 then cleaves Bid, ultimately causing cytochrome *c* release and cell death. Caspase 8 can also directly activate downstream caspases. Death receptors and independent stimuli can also trigger translocation of Bax and Bad to the mitochondria. In cells that are not dying, Bad is phosphorylated by survival factor signaling pathways, which include PKA, Akt, $p90^{RSK}$, mitogen-activated protein kinases (MAPKs) and PAK (p21-activated kinase). Bad is then sequestered to the cytosol by binding to protein 14-3-3. During apoptosis, Bad is dephosphorylated by calcineurin (PP-2B) or protein phosphatase 1α and translocates to the mitochondria to bind to Bcl-xL, displacing Bcl-xL from Bcl-xL-Bax heterodimers, and therefore Bcl-xL inhibition of Bax is prevented.

An alternate possibility is that the permeability transition pore (PTP) is the key element in the process.

The PTP consists of an association of several proteins: cyclophilin D, a matrix protein, VDAC, and ANT. As the matrix has a high concentration of solutes, opening of the PTP, which occurs due to the rise in Ca^{2+} concentrations associated with apoptosis, results in a decrease in P_i, adenine nucleotides, ROS and changes in pH, and a general nonselective exposure of the inner membrane to larger molecules up to 1.5 kDa. The H^+ gradient dissipates, and the respiratory chain is uncoupled. The mitochondrial membrane potential and the chemical balance between the matrix and the cytosol are disrupted, leading to a swelling of the matrix space. The cristae of the inner membrane actually have a much larger surface area than the outer membrane, and therefore, the swelling of the matrix leads to outer membrane rupture. The pore can also be affected by alterations in Ca^{2+} levels in the cytosol and oxidants, causing the release of apoptosis-inducing caspases. This second theory for the rupture of the mitochondrial membrane is still under debate, and the events outlined may prove to be a result of cytochrome *c* release, as opposed to the cause. However, support for the theory comes from the experiments that showed that inhibitors of ANT,

TEXT BOX 8.4 DEBATE ON MITOCHONDRIAL PARTICIPATION IN APOPTOSIS

Some scientists disagree with the idea that mitochondria are central to apoptosis. They cite the fact that studies with *in vitro* cell models provide a nonphysiological environment that can be unreliable, as well as the fact that some studies have shown that mitochondria are not involved in apoptosis. These dissenting studies have mostly been conducted using whole tissue or whole animal systems and caspase inhibitors. Each of these studies suggests that caspase activation precedes any of the events that occur at the mitochondria.

Nevertheless, other studies conducted on whole mammal systems, using these very caspase inhibitors, have shown that, upon exposure to certain conditions such as UV light, these inhibitors do in fact prevent the caspase signaling pathways from occurring. However, cytochrome *c* release from the mitochondria still occurs, and apoptosis of the cell takes place. This provides support for the importance of the role of the mitochondria in the onset of cellular apoptosis.

including cyclosporine and bongkrekic acid, also inhibit PTP opening and prevent apoptosis during *in vitro* experiments. This indicates an important role for the PTP.

CONCLUSION

Even though the physical details of the electron transport chain, oxidative phosphorylation, and the metabolic pathways inside mitochondria are now very well known, there will continue to be interest in mitochondrial studies. New mechanisms that impact on the regulation of energy metabolism continue to be uncovered and require exploration such as the regulatory role(s) of ROS and the role of crowding and protein–protein interactions in channeling metabolic flux in mitochondria. The critical role of mitochondria in apoptosis, their ability to produce ROS as both damage and signaling products, their role in oxygen sensing and in Ca^{2+} signaling, and the involvement of mitochondria and/or the mitochondria genome of various diseases will provide decades of further studies. New directions in linking neurodegenerative disease studies, and aging in general, to mitochondrial dysfunction indicate that the mitochondria are not only responsible for crucial aspects of cell survival in an aerobic environment, but they are also the source of production of damaging ROS, and are key players in the onset of programmed cell death.

TEXT BOX 8.5 APOPTOSIS—A COMPARATIVE VIEW

The basic features of apoptosis are now well-known and are highly conserved across phylogeny. A general view of apoptosis is a process for ridding a body of damaged and dying cells without stimulating the inflammatory reactions that accompany necrosis. But apoptosis is also a highly important part of differentiation and senescence in the normal life of many organisms. In plants, for example, apoptosis is an important way of recovering nutrients from structures that are no longer needed. For example, once a flower has been pollinated, apoptosis of petals and stigma allow nutrients to be recovered from these organs that are no longer needed. Similarly, when seeds are ripe, the softening of the fruit tissues or splitting open of seed pods is accomplished via apoptosis with recovery of nutrients from the cells. On a megascale, many marine cephalopods such as squid mate only once in their life. Immediately after mating (or after egg-laying in females) both sexes undergo massive cell and organ destruction—millions of animals activating death pathways simultaneously.

Multiple situations must also exist where a stress that causes apoptosis in one species has no effect on another. The molecular mechanisms that underlie these diverse responses have received little attention. For example, freezing of mammalian tissue (such as during cryosurgery) turns on death pathways not only in the directly affected cells but also in a penumbra of cells surrounding the damage site. However, freeze-tolerant animals, such as frogs and turtles that naturally endure weeks of freezing during the winter (see Chapter 17), show no apoptotic (or necrotic) response by their cells to freezing; indeed, there is initial evidence of increased levels of antiapoptotic proteins in their organs. A similar situation occurs in anoxia-tolerant species (see Chapter 15). Oxygen deprivation that would rapidly stimulate apoptosis in mammalian cells produces no signs of cell injury or apoptotic activity in many kinds of anoxia-tolerant species including lower vertebrates such as turtles, or invertebrate snails and mussels—species that routinely survive days and even months without oxygen. How does an organism know what is a survivable stress and what is not? What are the adaptive changes that prevent an apoptosis in response to a stress that should otherwise be damaging or lethal?

SUGGESTED READING

Brand, M. D., and Ainscow, E. K. (1999). Regulation of energy metabolism in hepatocytes. In *Technological and Medical Implications of Metabolic Control Analysis*, Cornish-Bowden, A. J., and Cardenas, M. L. (eds.), Kluwer Academic Publishers, New York, pp. 131–138.

Chandel, N. S., and Schumacker, P. T. (2000). Cellular oxygen sensing by mitochondria: Old questions, new insight. *J Appl Physiol* **88**:1880–1889. *This review gives a thorough overview of the insights into deciphering the oxygen-sensing mechanism in the mitochondria.*

Duchen, M. (2000). Mitochondria and calcium: From cell signaling to cell death. *J Physiol* **529**:57–68. *Very detailed analysis of mitochondrial calcium signaling, specifically mitochondrial calcium uptake and its impact on cell survival.*

Green, D. R., and Reed, J. C. (1998). Mitochondria and apoptosis. *Science* **281**:1309–1312. *An excellent summary of the involvement of the mitochondria in apoptosis.*

Neupert, W. (1997). Protein import into the mitochondria. *Ann Rev Biochem* **66**:863–917. *A very thorough, though not very recent, analysis of the import of protein into the mitochondria.*

Salvioli, S., Bonafe, M., Capri, M., Monti, D., and Franceschi, C. (2001). Mitochondria, aging, and longevity—a new perspective. *FEBS Lett* **492**:9–13. *A short review that provides a general overview of the involvement of the mitochondria in apoptosis, as well as a viewpoint not provided in the text.*

Science (1999), Vol. 283, pp. 1475–1497. *Special section on mitochondria. Includes the following three reviews, each giving a solid discussion of current understanding about different aspects of mitochondria.*

Gray, M. W., Burger, G., and Lang, B. F. (1999). Mitochondrial evolution. *Science* **283**:1476–1481.

Saraste, M. (1999). Oxidative phosphorylation at the fin de siecle. *Science* **283**:1488–1493.

Wallace, D. C. (1999). Mitochondrial diseases in man and mouse. *Science* **283**:1482–1488.

9

HUMAN ENERGY METABOLISM IN HEALTH AND DISEASE

Steven C. Greenway

INTRODUCTION

The word *metabolism* is derived from the Greek word for *change* and encompasses all the processes that convert nutrients into energy [adenosine 5′-triphosphate (ATP)] and synthesize or degrade biological molecules. Metabolism consists of enzymatic reactions organized into pathways. The catabolic or energy-yielding pathways are those responsible for the oxidative degradation of complex nutrient molecules (carbohydrate, lipid, or protein) obtained either from feeding or from cellular reserves. The breakdown of these complex molecules results in the formation of simpler molecules. These reactions are usually exergonic; some energy is released as heat but most is recaptured as ATP. Protons released during catabolism are captured by oxidized nicotinamide adenosine dinucleotide ($NADP^+$) for use in reductive biosynthetic reactions or by oxidized nicotinamide adenosine dinucleotide (NAD^+) for use in feeding the mitochondrial proton pump that powers ATP synthesis. Conversely, anabolic pathways are those that are responsible for the synthesis of varied and complex biomolecules from simpler precursors. These endergonic reactions require an input of energy in the form of ATP to drive them and utilize reducing power supplied by reduced NADP (NADPH) and reduced NAD (NADH).

This chapter will briefly review the fundamentals of human carbohydrate, lipid, and protein metabolism supplemented with recent developments in our understanding of these processes. Due to the nature of scientific research and the questions being asked, most of these new developments have been uncovered in animal models or *in vitro* in cultured cells. Whether the processes discovered in these model systems remain true *in vivo* in human physiology still remains to be seen in many cases. Various human diseases are included as examples of what happens when the pathways of energy metabolism are disrupted.

HUMAN CARBOHYDRATE METABOLISM

Glucose is arguably the most important fuel for human metabolism. Glucose has a central role in metabolism due to the fact that many tissues (e.g., brain, red blood cells, exercising skeletal muscle) require an uninterrupted flow of glucose to supply their energy needs. The typical adult human requires 190 g of glucose per day, 150 g used by the brain and 40 g for other tissues! Except in conditions of starvation (discussed later), glucose is the sole source of energy for the brain and in the resting state provides fuel to all other tissues. Glucose is a flexible metabolite that can be converted to other forms of fuel, including lipids (e.g., fatty acids, cholesterol, steroid hormones), amino acids, and nucleic acids. Except for vitamins, essential amino acids, and essential fatty acids, all metabolites needed in the human body can be synthesized from glucose. The concentration of glucose in the blood is maintained within a narrow range (~5 mM). This homeostasis is of critical importance. If blood glucose levels fall too low (hypoglycemia), then seizures, coma, and even death can occur. Conversely, if blood glucose levels are elevated (hyperglycemia) for long periods of time, such as in poorly controlled diabetes mellitus, the consequences can

Functional Metabolism: Regulation and Adaptation, edited by Kenneth B. Storey.
ISBN 0-471-41090-X

include blindness, renal failure, cardiac and peripheral vascular disease, and neuropathy.

In addition to the diet, glucose is also available from body glycogen stores (primarily found in liver and skeletal muscle) or can be synthesized via gluconeogenesis from metabolites, including amino acids, lactate, and glycerol. The balance between glucose utilization, glucose storage, and glucose synthesis depends on the hormonal and nutritional status of the individual. The fate of glucose also depends on the individual tissue concerned, for example, as described next for liver, muscle, and brain.

The liver has a central role in glucose metabolism and can direct the sugar into multiple uses including: (a) catabolism via glycolysis and the tricarboxylic acid (TCA) cycle to produce ATP, (b) storage as glycogen (glycogenesis), (c) utilization as a carbon precursor for the biosynthesis of metabolites (e.g., fatty acids, nucleotides, many others), and (d) generation of NADPH reducing power via the pentose phosphate pathway for use in biosynthesis and antioxidant defense (5 to 10% of dietary glucose is processed via this pathway). The liver also plays a central role in both glycogenolysis (the breakdown of glycogen) and gluconeogenesis (the synthesis of glucose from noncarbohydrate precursors), both of which contribute to the supply of blood glucose for use by other tissues. Venous drainage from the gut and pancreas passes through the hepatic portal vein before entry into the general circulation. Therefore, after a meal the liver is bathed in nutrients and insulin. The liver retains more than 50% of the glucose presented to it by the portal system and also takes up amino acids, lipids, and other nutrients, thereby "smoothing out" potentially wide fluctuations in blood nutrient levels that would otherwise be presented to peripheral tissues.

Skeletal or cardiac muscle can also use glucose for energy metabolism directly or store it as glycogen. However, in contrast to liver, glycogenolysis in muscle provides glucose only for endogenous cellular use; glucose is not released into the bloodstream. A unique property of glucose as a tissue substrate is that it can still be metabolized (via glycolysis to lactate) in the absence of oxygen. This anaerobic pathway is very important for ATP generation in exercising muscle (particularly high intensity exercise by white muscle; see Chapter 11), is critical for the survival of tissues that do not possess mitochondria such as red blood cells, and prolongs survival of all tissues when they are oxygen-deprived.

The brain receives glucose from the bloodstream and oxidizes glucose to supply its energy needs. Brain has minimal glycogen stores and it does not contribute any glucose to the rest of the body. The absolute dependence of brain energy metabolism on glucose is one of the reasons why glucose homeostasis is defended so vigorously. Under extreme conditions, such as starvation, the brain is able to use ketone bodies (products of lipid oxidation) as an alternative source of energy but still requires glucose, which is typically provided from gluconeogenesis in the liver when body glycogen reserves are minimal.

In the typical North American diet 50% of calories come from the consumption of starch, sucrose, and lactose. Digestion breaks down these complex carbohydrates to the simple monosaccharides glucose, galactose (see Text box 9.1), and fructose, which are then absorbed and transported via the bloodstream to the tissues where they are metabolized. Glucose enters a cell via specific membrane-bound transport proteins that permit the facilitated uptake of glucose. Five different isoforms of the glucose transporter (GLUT) have been found, each with a tissue-specific distribution. After glucose enters the cell, it is rapidly phosphorylated to form glucose-6-phosphate (G6P) by the enzyme hexokinase.

TEXT BOX 9.1 GALACTOSEMIA

Galactose sugar is introduced into glycolysis via the actions of three enzymes, galactokinase [reaction (TB 9.1)], galactose-1-phosphate uridyltransferase [reaction (TB 9.2)], and phosphoglucomutase [reaction (TB 9.3)]:

$$\text{Galactose} + \text{ATP} \rightarrow \text{galactose-1-P} + \text{ADP} \quad \text{(TB 9.1)}$$

$$\text{Galactose-1-P} + \text{UDP–glucose} \rightarrow \text{glucose-1-P} + \text{UDP–galactose} \quad \text{(TB 9.2)}$$

$$\text{Glucose-1-P} \rightarrow \text{glucose-6-P} \quad \text{(TB 9.3)}$$

Galactosemia is an autosomal recessive disorder caused by galactose-1-phosphate uridyltransferase deficiency with an incidence of 1 in 60,000 live births and an early onset of symptoms. The newborn infant normally receives up to 20% of its caloric intake as lactose, a disaccharide formed from glucose and galactose. Without the transferase enzyme, the infant is unable to metabolize galactose-1-phosphate, the accumulation of which results in injury to kidney, liver, and brain cells. Symptoms of the disease in infants include jaundice, an enlarged liver, vomiting, hypoglycemia, convulsions, lethargy or irritability, and feeding difficulties with poor weight gain. If diagnosis is delayed, damage to liver (cirrhosis) and brain (mental retardation) becomes progressively more severe and irreversible. Elimination of galactose from the diet reverses growth failure and renal and hepatic dysfunction.

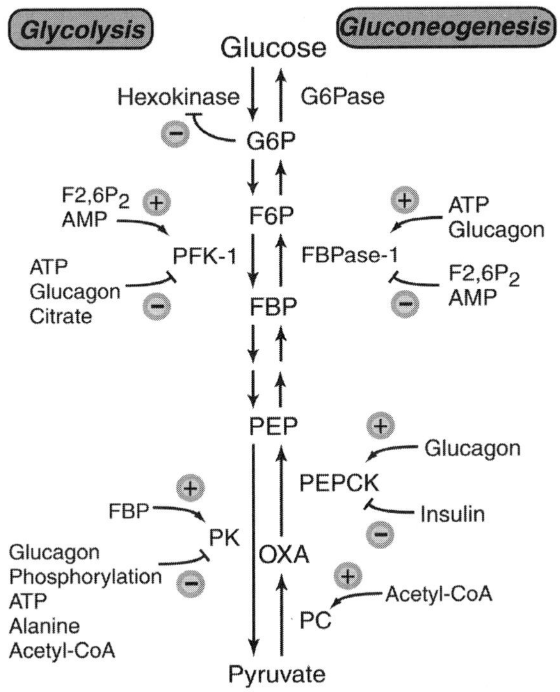

Figure 9.1 The regulation of glycolysis and gluconeogenesis in the liver. The conversion of glucose to pyruvate (glycolysis, left side of figure) begins with the phosphorylation of glucose by hexokinase to produce glucose-6-phosphate (G6P), which can feedback inhibit hexokinase to decrease flux through glycolysis. The next major regulatory step in glycolysis is the reaction catalyzed by phosphofructokinase-1 (PFK-1), the conversion of fructose-6-phosphate (F6P) to fructose-1,6-bisphosphate (FBP). PFK-1 is activated by fructose-2,6-bisphosphate ($F2,6P_2$) and AMP and inhibited by ATP, glucagon, and citrate. FBP provides feedforward activation of pyruvate kinase (PK), and PK activity is inhibited by glucagon and protein phosphorylation as well as ATP, alanine, and acetyl-CoA. The opposing reactions of gluconeogenesis (right side of figure), which include three reactions that bypass the irreversible, exergonic reactions of glycolysis, begin with the conversion of pyruvate to oxaloacetate (OXA) by pyruvate carboxylase (PC), whose activity is enhanced by acetyl-CoA. OXA is then converted to phosphoenolpyruvate (PEP) by PEP carboxykinase (PEPCK), whose expression is regulated by insulin and glucagon. Fructose-1,6-bisphosphatase (FBPase-1), which bypasses PFK-1, is inhibited by $F2,6P_2$ and AMP (the converse of PFK-1) and activated by ATP and glucagon. Finally, glucose-6-phosphatase (G6Pase) is responsible for the dephosphorylation of G6P to glucose, which is then released into the blood for uptake by peripheral tissues.

Glycolysis

The central pathway for G6P catabolism is glycolysis, whereby one molecule of G6P is catabolized to two molecules of pyruvate with the net production of two ATP molecules (Fig. 9.1). All cells of the human body contain the

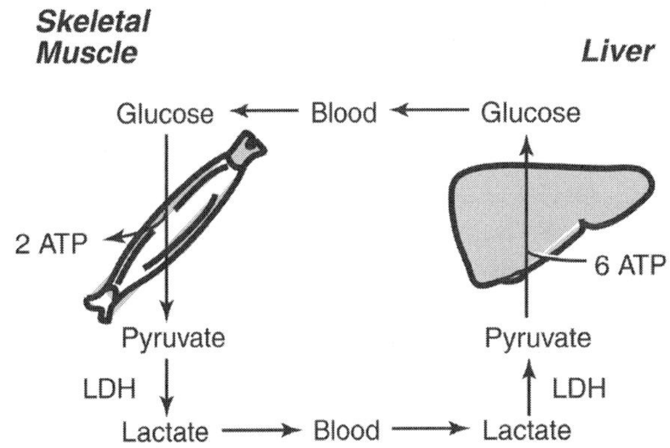

Figure 9.2 The Cori cycle. Lactate, produced by anaerobic glycolysis in exercising skeletal muscle, is released into the bloodstream where it is taken up by the liver and converted to pyruvate by lactate dehydrogenase (LDH). This pyruvate is then used by the liver as a substrate for gluconeogenesis. Glucose produced by the liver can be released again into the bloodstream for use by other tissues. Oppositely directed LDH actions in liver and muscle are achieved by the presence of different LDH isoforms in the two organs as well as a low $NADH/NAD^+$ ratio in liver versus a high ratio in vigorously exercising muscle.

glycolytic pathway, but the fate of the pyruvate produced depends on the specific cell type and the availability of oxygen. Most tissues possess the mitochondrial pyruvate dehydrogenase complex (PDC), which catalyzes the conversion of pyruvate to acetyl-CoA to provide acetyl-CoA as a precursor for biosynthesis or as a substrate for mitochondrial oxidative metabolism. However, some cells (e.g., red blood cells; see Chapter 18) lack mitochondria and generate their ATP solely from glycolysis, whereas in others the ATP demand can at times exceed the aerobic capacity and necessitate additional ATP generation via anaerobic glycolysis (e.g., exercising white skeletal muscle; see Chapter 11). In such cases, anaerobic glycolysis ends with the conversion of pyruvate to lactate by lactate dehydrogenase (LDH), a reaction that regenerates the NAD^+ that is needed by the glyceraldehyde-3-phosphate dehydrogenase reaction. The lactate end product is frequently exported to be used as an aerobic fuel by organs such as the heart or reconverted to glucose via gluconeogenesis in the liver in a process known as the Cori cycle (Fig. 9.2).

Gluconeogenesis

In the resting state in humans glycogenolysis accounts for 70 to 80% of hepatic glucose output with gluconeogenesis accounting for the remainder. However, gluconeogenesis assumes much greater importance in situations of prolonged fasting when there is no intake of carbohydrate and glycogen stores have been depleted. Liver is the

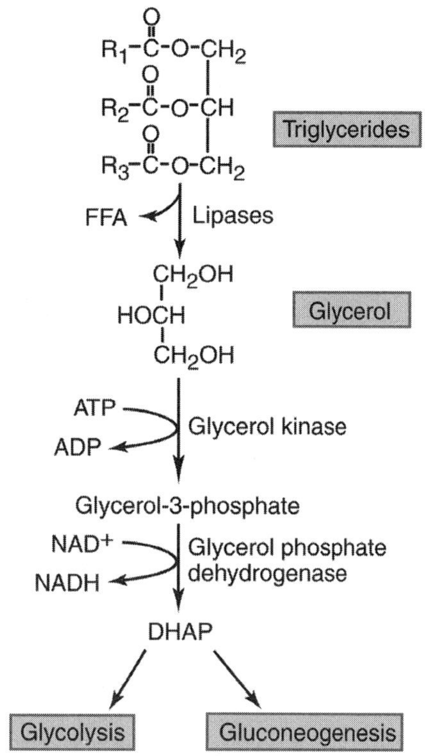

Figure 9.3 Glycerol metabolism. The digestion of triglycerides by lipases produces free fatty acids (FFA) and glycerol. The enzyme glycerol kinase phosphorylates glycerol to produce glycerol-3-phosphate, which is then oxidized by glycerol phosphate dehydrogenase to produce dihydroxyacetone phosphate that can be reintroduced into the reactions of glycolysis or gluconeogenesis.

primary site of gluconeogenesis, although the kidney contributes ~10% of all the glucose produced in this manner (with an increased fraction during starvation). The primary substrates for gluconeogenesis are pyruvate, lactate, glycerol, and amino acids. The rate of gluconeogenesis increases in response to a high-protein diet and has an important role in the utilization of excess amino acids consumed during a meal.

The synthesis of glucose from pyruvate or lactate is essentially a reversal of glycolysis, making use of most of the reactions of glycolysis but bypassing two key irreversible reactions, pyruvate kinase (PK) and 6-phosphofructo-1-kinase (PFK-1) (Fig. 9.1). Liver and kidney also possess the enzyme, glycerol kinase, which catalyzes the phosphorylation of glycerol (released during triglyceride hydrolysis) and allows glycerol-3-phosphate to reenter the pool of glycolytic triose phosphates (Fig. 9.3). The availability of lactate as a gluconeogenic substrate increases substantially after periods of muscle exercise, whereas the contribution of glycerol to gluconeogenesis becomes more important when lipolysis is accelerated such as during fasting or in diabetes. However, overall, amino acids, in particular alanine and glutamine, are the major substrates for gluconeogenesis in liver. These are derived both from diet and as products of muscle protein and amino acid catabolism. Gluconeogenesis is discussed in more detail under Human Protein Metabolism later in this chapter.

Glycogen Metabolism

Glycogen is the storage polymer for carbohydrate in animals. The largest depots are found in liver and skeletal muscle. Up to 10% of liver mass can be glycogen, and, although glycogen represents only about 1% of skeletal muscle mass, the much greater amount of skeletal muscle than liver in the body means that total body reserves of glycogen are about twice as high in muscle as in liver. Glycogen is a polymer with branch points occurring approximately every fourth glucose molecule along the linear chains. This branched structure permits the storage of many glucose molecules as a compact glycogen granule. For glycogen synthesis, or glycogenesis, G6P is converted to glucose-1-phosphate (G1P) by phosphoglucomutase (PGM) and then to uridine diphosphate (UDP)–glucose by glucose-1-phosphate uridylyl transferase. Glycogen synthase (GS) catalyzes the addition of a single UDP–glucose molecule onto the glycogen polymer (Fig. 9.4). Another enzyme, 1,4-α-glucan branching enzyme, is responsible for the formation of branch points. In humans, approximately one-third of the G6P used for glycogen synthesis in the liver is derived from gluconeogenesis, whereas muscle cells directly utilize glucose taken up from the blood for glycogen synthesis.

Although large amounts of glucose are stored as glycogen in the human liver, this resource is depleted after only a 24-h fast. When blood glucose levels decrease during fasting, the breakdown of hepatic glycogen stores via glycogenolysis is the first line of defense to prevent hypoglycemia. Glycogen degradation is accomplished by glycogen phosphorylase (GP) and a debranching enzyme (Fig. 9.5). GP phosphorylates individual glucose molecules to produce G1P, which is promptly converted to G6P by PGM. In the liver most of this G6P is dephosphorylated by G6Pase to provide glucose for export. By contrast, in most other organs, including muscle, the G6P derived from glycogenolysis is catabolized within the cell.

Regulation of Carbohydrate Metabolism

The phosphorylation of glucose after it has entered the cell is catalyzed by hexokinase. Multiple isozymic forms of hexokinase are known that differ in their kinetic properties, regulatory mechanisms, and tissue location. Hexokinase I dominates in skeletal muscle and other peripheral tissues,

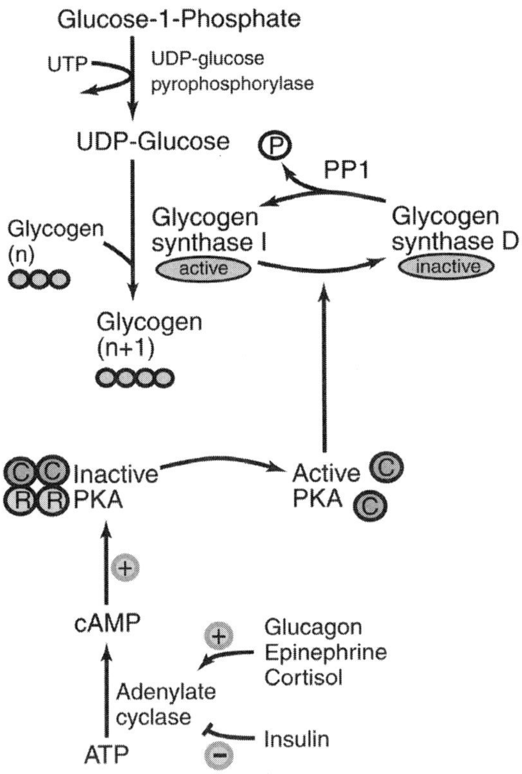

Figure 9.4 Regulation of glycogen synthase (GS). G6P is converted to glucose-1-phosphate (G1P) by phosphoglucomutase and used by uridine diphosphate (UDP)–glucose pyrophosphorylase in the presence of uridine triphosphate (UTP) to synthesize UDP–glucose. GS then adds the glucosyl unit from UDP–glucose to the C-4 hydroxyl group at a nonreducing end of a glycogen strand. GS exists in two forms; an active, dephosphorylated G6P-independent GS*I* and a less active, phosphorylated, G6P-dependent GS*D*. Interconversion between these two forms is accomplished by phosphoprotein phosphatase 1 (PP1) and cyclic-AMP (cAMP)-dependent protein kinase A (PKA). The enzyme adenylate cyclase (AC) is activated by glucagon, epinephrine, and cortisol and inhibited by insulin. Activation of AC increases the production of cAMP; binding of cAMP to the inactive PKA holoenzyme leads to the dissociation of the regulatory subunits (R) from the catalytic subunits (C), which are then free to phosphorylate and inactivate GS.

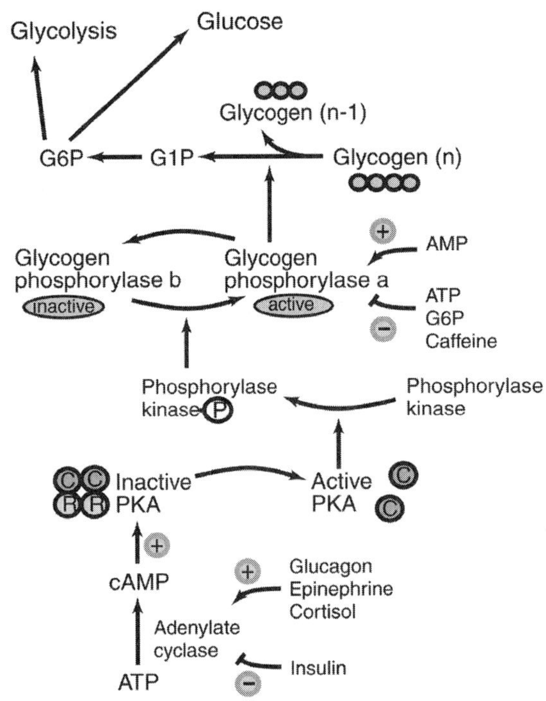

Figure 9.5 Regulation of glycogen phosphorylase (GP). The cleavage of glucose units from the nonreducing ends of glycogen molecules is catalyzed by GP. This releases G1P that is converted to G6P and then fed into multiple fates including glycolysis or G6Pase-mediated dephosphorylation to produce glucose for export. Like GS, GP is regulated by reversible phosphorylation that interconverts the enzyme between active phosphorylase *a* (GP*a*) and the less active phosphorylase *b* (GP*b*); these reactions are accomplished by phosphorylase kinase and PP1. GP is also subject to allosteric regulation, being activated by AMP and inhibited by ATP, G6P, and caffeine. Phosphorylase kinase is also regulated by reversible phosphorylation, being converted to its active form by PKA under positive stimulation from glucagon, epinephrine, or cortisol.

its activity coordinated with the activity of the insulin-responsive glucose transporter, GLUT4. Since hexokinase I is inhibited by its product, G6P, a balance between glucose uptake and glucose phosphorylation is maintained.

By contrast, hexokinase IV (also known as glucokinase) is the dominant form in liver. Glucokinase is not inhibited by G6P and in conjunction with the GLUT2 glucose transporter ensures rapid uptake of glucose into the liver when blood sugar levels are high. Glucokinase is also found in pancreatic β cells, but transcription of the glucokinase gene and processing of the primary transcript differs between the liver and the pancreas, resulting in two glucokinase isoforms. This divergence allows for specific tissue responses to hormones and/or nutrients. For example, in liver, but not in the pancreatic β cell, insulin stimulates transcription of the glucokinase gene and glucagon inhibits it. In the β cell, but not in liver, expression of the glucokinase gene is responsive to glucose. Hence, glucokinase expression is regulated hormonally by the insulin/glucagon ratio in liver, whereas blood glucose levels regulate pancreatic β-cell glucokinase expression.

In liver, the glucokinase reaction is opposed by the G6Pase reaction. This enzyme utilizes G6P produced from either gluconeogenesis or glycogenolysis and converts it to glucose for export into the circulation. Futile cycling of glucose between the glucokinase and G6Pase reactions is prevented by spatial separation of the two enzymes, regulatory controls on enzyme activities, and hormonal controls on gene expression. Thus, insulin stimulates glucokinase expression but inhibits expression of the G6Pase gene; hence, the insulin response after a meal favors the uptake, phosphorylation, and storage of glucose.

The first committed step of the glycolytic pathway is the reaction catalyzed by PFK-1, and this reaction is subject to regulation by several allosteric effectors (Fig. 9.1). PFK-1 activity is activated by adenosine 5′-monophosphate (AMP), inorganic phosphate (P_i), ammonium ion (NH_4^+), and fructose-2,6-bisphosphate ($F26P_2$) and inhibited by ATP, hydrogen ions (H^+), and citrate. The gluconeogenic counterpart of PFK-1 is fructose-1,6-bisphosphatase (FBPase-1), which is also subject to allosteric regulation (Fig. 9.1). Activators of PFK-1 (AMP, $F26P_2$) are inhibitors of FBPase-1, whereas ATP, which inhibits PFK-1, activates FBPase-1. ATP, AMP, and P_i all signal the energy status of the cell, H^+ signals cellular pH, citrate signals the amount substrate available for the TCA cycle, and $F26P_2$ signals the insulin/glucagon ratio. During hypoxia or during strenuous exercise, hydrogen ions and lactate are produced and the hydrogen ions inhibit PFK-1 and glycolysis to protect against further lactate buildup. Citrate signals the cell that substrate for the TCA cycle is plentiful, and, by inhibiting PFK-1, it slows glycolysis and further spares glucose, particularly under conditions where lipids are being used as the primary fuel.

In liver, $F26P_2$ is a major regulatory molecule that exerts major control over the relative rates of carbon flux through glycolysis versus gluconeogenesis. Levels of $F26P_2$ are regulated by a bifunctional enzyme that has both 6-phosphofructo-2-kinase (PFK-2) activity and fructose-2,6-bisphosphatase (FBPase-2) activity (Fig. 9.6). The enzyme is regulated by insulin and glucagon. After a meal, when blood glucose increases, insulin levels increase and activate protein phosphatase 2A (PP2A), which dephosphorylates the bifunctional enzyme resulting in increased PFK-2 activity and increased levels of $F26P_2$, which stimulates glycolysis. Conversely, during fasting, insulin levels decrease and glucagon levels rise, stimulating the synthesis of cyclic-AMP (cAMP) and the activation of cAMP-dependent protein kinase (PKA). PKA phosphorylates the bifunctional enzyme causing increased FBPase-2 activity and decreased $F26P_2$ levels that inhibits PFK and favors FBPase-1 activity and gluconeogenesis. Nutrients and hormones also regulate transcription of the PFK-2/FBPase-2 gene with insulin increasing transcription and glucagon inhibiting transcription.

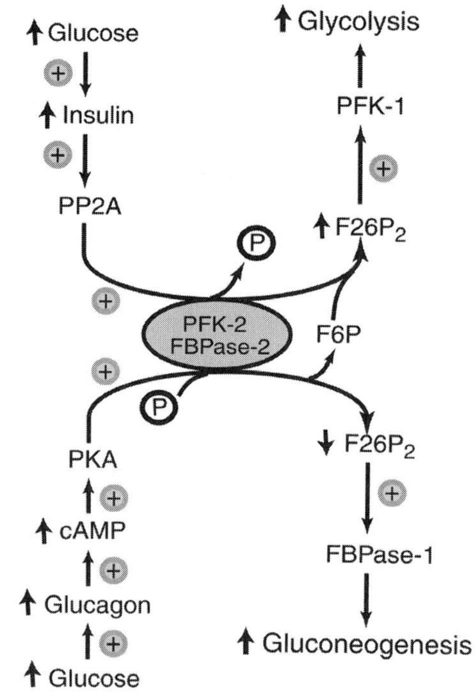

Figure 9.6 Fructose-2,6-bisphosphate ($F26P_2$) metabolism. Levels of the allosteric effector $F26P_2$ are regulated by a bifunctional enzyme that has both 6-phosphofructo-2-kinase (PFK-2) activity and fructose-2,6-bisphosphatase (FBPase-2) activity. Increased blood glucose, for example, after a meal, stimulates increased insulin secretion that activates protein phosphatase 2A (PP2A) to dephosphorylate the bifunctional enzyme and increase the PFK-2 activity; this stimulates glycolysis (top half of figure). Conversely, during fasting, insulin levels decrease and glucagon levels rise, stimulating the synthesis of cAMP and the activation of PKA. PKA phosphorylates the bifunctional enzyme to increase FBPase-2 activity and reduce $F26P_2$ levels; this FBPase-1 activity and gluconeogenesis (bottom half of figure).

Another important regulatory locus for glycolytic versus gluconeogenic control involves the pyruvate kinase (PK) and phosphoenolpyruvate carboxykinase (PEPCK) reactions (Fig. 9.1). Four isoforms of PK have been characterized: L is the major form in liver and is also present in kidney and small intestine, L′ or R is in red blood cells, M1 is the dominant form in muscle, heart, and brain, and M2 is found in fetal tissues and most other adult tissues. Insulin, glucose, fructose, and glycerol all increase expression of the L isoform in liver. PK catalyzes the conversion of phosphoenolpyruvate (PEP) to pyruvate; it is typically inhibited by ATP (signaling a high energy state) and alanine (signaling plentiful gluconeogenic substrate) and is activated by feedforward effects of the product of PFK-1, $F16P_2$. In liver, a glucagon-mediated increase in cAMP stimulates PK phosphorylation by PKA to inhibit the enzyme and slow glycolysis. Insulin has the reverse effect. The phosphorylated form of PK is not only less active but is also much more sensitive to the effects of allosteric inhibitors.

Phosphoenolpyruvate carboxy kinase is an important regulatory step in gluconeogenesis, and together with the pyruvate carboxylase (PC) reaction, circumvents the PK reaction. PEPCK is primarily regulated by changes in enzyme concentration as a result of changes in gene expression in response to diet and hormonal signals. Fasting or glucagon elevate levels of cAMP which raises the transcription rate of PEPCK as well as the stability of the messenger ribonucleic acid (mRNA) product. Conversely, insulin and high blood glucose both decrease transcription of the PEPCK gene and reduce the stability of its mRNA product. The catecholamines, epinephrine and norepinephrine, when released, can stimulate gluconeogenesis by activating PC. Catecholamines do not play an essential role in maintaining the level of blood glucose during fasting, but increased secretion of catecholamines compensates and prevents hypoglycemia if glucagon secretion is deficient. Cortisol activates gluconeogenic enzymes and increases transfer of free amino acids to the liver. However, unlike glucagon and catecholamines, the stimulatory effect of cortisol takes several hours to occur. Growth hormone increases hepatic glucose production by altering substrate availability and promoting enzyme induction. Insulin also indirectly decreases gluconeogenesis by inhibiting the peripheral release of gluconeogenic precursors. Glucagon increases hepatic alanine uptake and increases the proportion of alanine that is converted to glucose. Glucagon also has a lipolytic effect in the liver to increase available free fatty acids for catabolism and release glycerol for gluconeogenesis.

Between meals and during normal periods of fasting (i.e., overnight), the maintenance of blood glucose levels depends almost entirely on the breakdown of glycogen stores. Glycogenolysis can be induced by glucagon and the catecholamines, epinephrine and norepinephrine. The dominant stimulus for glycogen breakdown in liver is via the glucagon-mediated rise in intracellular cAMP. The rise in cAMP activates PKA, which phosphorylates the inactive *b* form of GP to convert it to the active *a* form. Simultaneously, glycogen synthase is phosphorylated and inactivated, thereby avoiding simultaneous glycogen breakdown and formation. By contrast, insulin is responsible for stimulating glycogen synthesis in liver and muscle from the glucose that was absorbed after a meal. Insulin inhibits glycogenolysis and directs G6P toward glycogen biosynthesis by increasing the activity of GS and decreasing the activity of GP.

Glycogen synthase is interconverted between the active, G6P-independent form (or I form) or the inactive, G6P-dependent form (or D form) by phosphorylation (Fig. 9.4). Various protein kinases can phosphorylate and inactivate GS including glycogen phosphorylase kinase (GPK), Ca^{2+}/calmodulin-dependent kinase, protein kinase C, glycogen synthase kinase-3, and casein kinases I and II. In liver, both glucagon and epinephrine (acting through β-adrenergic receptors) can stimulate this PKA-mediated process, whereas in skeletal muscle epinephrine stimulation of PKA triggers GS inactivation. GS is dephosphorylated to return it to the active I form by protein phosphatase 1 (PP1). PP1 is subject to inhibition by a protein called inhibitor 1, which is itself phosphorylated and activated by PKA. Thus, when glucagon levels rise the increase in cAMP that activates PKA both phosphorylates and inactivates GS and at the same time prevents reactivation of GS by phosphorylating inhibitor 1 in order to suppress PP1 activity.

Glycogen phosphorylase is converted to its active *a* form by phosphorylation and is also subject to allosteric activation by AMP and inhibition by ATP (Fig. 9.5). PKA catalyzes the phosphorylation of GPK, which in turn phosphorylates GP. GPK can also be stimulated by Ca^{2+}-mediated phosphorylation via CaMK, a mechanism that is particularly important in exercising muscle. AMP allosteric activation of GP*b* is also important in muscle and helps to mediate the rapid increase in glycogenolysis that is needed in working muscle because one of the first metabolic changes to occur when muscle contraction begins is a decrease in ATP, which is rapidly translated into a multifold rise in AMP. In the liver glucagon is the primary stimulus that initiates glycogenolysis, and hepatic GP is sensitive to glucose more than to AMP and ATP. Glucose binds to hepatic GP and induces a conformational change so that GP*a* becomes a better substrate for protein phosphatase and is dephosphorylated to the inactive *b* form.

Glucose concentration, independent of regulatory hormones, also affects glucose metabolism. The effect of glucose can occur through transcription factors as described later in this chapter, and metabolites generated by the oxidation of glucose can also have important regulatory effects, for example, G6P. Other mediators, for example, adenosine, prostaglandins, and cytokines (tumor necrosis factor-α, interleukin-1β, interleukin-6), may also have effects on glucose metabolism, but, since these chemical messengers can also affect the "classical" regulatory hormones it is difficult to determine the significance of their effects *in vivo*.

Transcriptional Regulation of Carbohydrate Metabolism

The effects of hormones on enzyme regulation in energy metabolism have been known for some time. However, in recent years the emphasis has shifted to analyze hormonal control of gene expression. This section will summarize some recent advances in understanding the intracellular signaling mechanisms involved in the regulation of genes of carbohydrate metabolism.

The production of glucose by the liver, during fasting, for example, is important in maintaining stable levels of blood glucose, and this process is tightly regulated by hormonal signals. The mechanism of this hormonal regulation was recently shown to involve PGC-1, originally named because of its action as a PPAR-gamma coactivator (PPAR being the peroxisome proliferator-activated receptor, a transcription factor) that participates in the PPAR-mediated expression of genes involved in fatty acid metabolism. PGC-1 belongs to a class of molecules known as coactivators that are components of the transcription factor complexes that control gene transcription.

During fasting, levels of insulin drop and glucagon and glucocorticoid levels increase. Glucagon stimulates an increase in cAMP levels and PKA activity. One of the targets that is phosphorylated by PKA is the cyclic AMP response element binding (CREB) protein, a transcription factor (Fig. 9.7). CREB then binds to the cAMP response element (CRE) that is upstream of the PGC-1 gene (among others) and induces its expression. Elevated PGC-1 protein in turn potentiates the transcription of various gluconeogenic genes including PEPCK, G6Pase, FBPase-1, and PC. PGC-1 also appears to interact with the hepatic nuclear factor-4α (HNF-4α) and the glucocorticoid receptor at the glucocorticoid response element (GRE). Thus, PGC-1 promotes cooperativity between the cAMP and glucocorticoid signaling pathways during gluconeogenesis. Of note, PGC-1 transcription is inhibited by insulin and this explains the ability of insulin to suppress gluconeogenesis by blocking the expression of enzymes such as PEPCK.

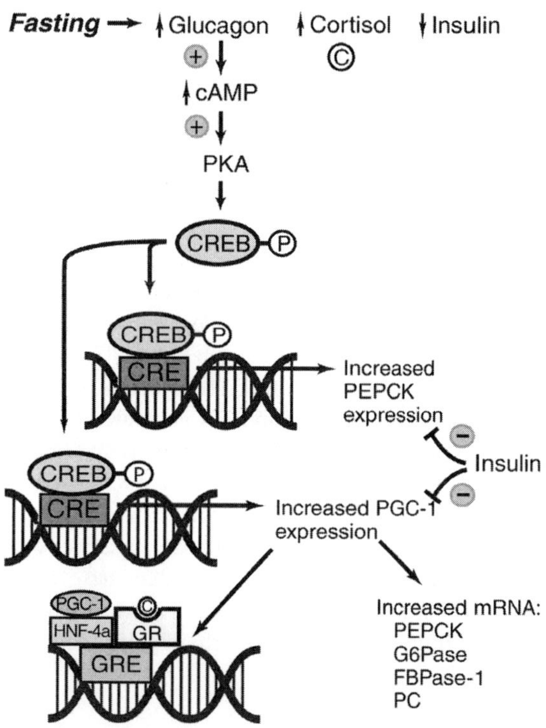

Figure 9.7 The role of PGC-1 in transcriptional regulation of carbohydrate metabolism. Fasting leads to increased levels of glucagon and cortisol (C) and decreased levels of insulin. This change in hormone balance activates adenylate cyclase, increases cAMP production, activates PKA, and increases the phosphorylation of the transcription factor, CREB. Binding of phosphorylated CREB to its response element (CRE) leads to increased expression of the genes for PEPCK and PGC-1. The up-regulation of these genes is counteracted by an increase in insulin. Increased PGC-1 stimulates increased transcription of various gluconeogenic genes including PEPCK, G6Pase, FBPase-1, and PC. PGC-1 also interacts with hepatic nuclear factor-4α (HNF-4α) and the glucocorticoid receptor (GR) at the glucocorticoid response element (GRE).

Pyruvate Dehydrogenase

The conversion of pyruvate to acetyl-CoA is carried out by the pyruvate dehydrogenase complex (PDC), which is comprised of three different enzymes: E1 is pyruvate dehydrogenase, E2 is dihydrolipoamide transacetylase, and E3 is dihydrolipoamide dehydrogenase. The acetyl-CoA produced by PDC in the mitochondria can have both catabolic (oxidation via the TCA cycle) and anabolic fates (e.g., fatty acid synthesis, isoprenoid/cholesterol/steroid synthesis). PDC links glycolysis and the TCA cycle in tissues such as cardiac muscle, skeletal muscle, and brain, which have a high demand for oxidative carbohydrate catabolism, and also provides carbon substrate for fatty acid synthesis in adipose tissue and liver (Fig. 9.8). Since the fates of the committed carbon are so divergent, and since the conversion of pyruvate to acetyl-CoA is irrevocable in animals (although plants and microbes can reconvert acetyl-CoA to four-carbon intermediates using the glyoxylate cycle), PDC is tightly regulated. For example, strong inhibition of PDC in most organs during fasting or starvation suppresses both the anabolic and catabolic uses of carbohydrate, helps to shift most organs to lipid-based energy metabolism, and spares liver carbohydrate reserves for a single function in the supply of glucose to organs (e.g., brain) that are completely dependent on it. As discussed in Chapter 16, control over PDC is also a key element of metabolic rate depression in hibernating mammals. Inactivation of PDC also helps to promote gluconeogenesis by limiting the oxidation of gluconeogenic substrates such as alanine, pyruvate and lactate. Conversely, during times of plenty, the conversion of glucose/pyruvate to fatty acids allows the storage of fuel reserves in the most efficient form.

Pyruvate dehydrogenase complex is subject to multiple regulatory controls. Its end products, acetyl-CoA and NADH, are inhibitors as is ATP, whereas AMP activates the enzyme. The enzyme is also controlled by reversible phosphorylation of serine residues on E1. Phosphorylation by pyruvate dehydrogenase kinase (PDK) inactivates PDC, whereas pyruvate dehydrogenase phosphatase returns PDC to its active form. PDK is an integral component of the PDC complex and is also sensitive to

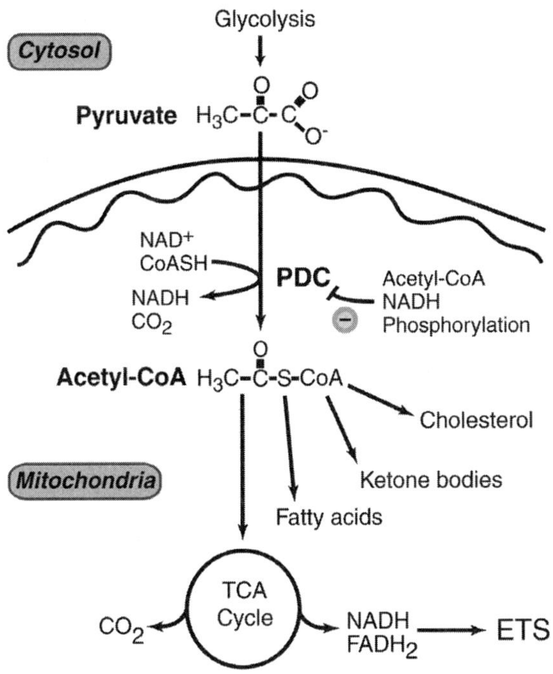

Figure 9.8 Regulation of the pyruvate dehydrogenase complex (PDC). Pyruvate produced by glycolysis is converted to acetyl-CoA by the mitochondrial PDC. This reaction is inhibited by acetyl-CoA, NADH, and phosphorylation of PDC by the PDKs. Acetyl-CoA produced by PDC can be used for a variety of metabolic processes, including cholesterol, ketone, or fatty acid synthesis or may be oxidized by the TCA cycle.

allosteric activation by acetyl-CoA and NADH, whereas the phosphatase is activated by Mg^{2+} and Ca^{2+}, high Mg^{2+} being a reflection of a reduced ATP/ADP ratio.

Four isoforms of PDK have been identified (PDK1–4). They coexist with two PDC phosphatases that belong to the protein phosphatase 2C (PP2C) family. Covalent modification of PDC by phosphorylation/dephosphorylation is believed to be the most important regulatory mechanism in both the short- and long-term. Short-term regulation is well-characterized with PDC phosphatases being activated by Ca^{2+} (e.g., in working muscle) and PDKs being inhibited by pyruvate and activated by acetyl-CoA and NADH. Expression of different PDK isoforms is important in regulating PDC activity in the longer term and under different conditions (e.g., starvation, hibernation). For example, in rats, starvation and diabetes increase the amounts of PDK4 mRNA and protein in cardiac and skeletal muscle, whereas in liver PDK2 increases substantially with a lesser increase in PDK4. These changes in PDK expression are reversed with refeeding. By contrast, PDK activities are unchanged in brain during starvation, consistent with the brain's continuous reliance on glucose oxidation for energy production. The increase in PDK4 expression is mediated by PPARs and glucocorticoids and the decrease is mediated by insulin.

HUMAN LIPID METABOLISM

Fat, or adipose tissue, is a loose connective tissue composed of lipid-filled cells (adipocytes) supported by a network of blood vessels, collagen fibers, fibroblasts, and immune cells. Brown adipose tissue is found in newborn human infants and mammals such as rodents where it plays an important metabolic role in thermogenesis due to its high concentration of mitochondria. White adipose tissue is the primary type of adipose tissue in humans and is able to store an almost limitless amount of triglycerides.

Although there are multiple forms and roles for lipids in human metabolism, it is the triglycerides that have a central role in human energy metabolism. Approximately 85% of the stored energy reserves of an average 70-kg man are in the form of triglycerides, primarily in adipose tissue. Compared with carbohydrate and protein, fat is the most efficient storage form for fuels, when catabolized releasing about 9 kcal/g versus 4 kcal/g for carbohydrate and protein. Fat also requires less water for storage, only about 1 g of water per gram of fat compared to 4 g of water per gram of carbohydrate or protein.

After ingestion of a meal, dietary fats are mixed with bile salts in the small intestine and dispersed into mixed micelles of bile salts and triglycerides. Micelle formation provides access for interaction with pancreatic lipase that cleaves the long-chain triglycerides into diacylglycerols, 2-monoacylglycerols (fatty acid moiety attached at position 2 on the glycerol backbone), long-chain fatty acids, and glycerol (Fig. 9.9). These reform into tiny micelles that are taken up into intestinal epithelial cells. There they are reconverted into triglycerides and packaged with cholesterol, phospholipids, and lipoproteins into chylomicrons that are released into the lymph and then into the blood to be delivered to tissues. Within the capillaries of selected organs, the cell surface enzyme, lipoprotein lipase (LPL), hydrolyzes the triglycerides to release fatty acids and glycerol that are taken up by cells. In tissues such as muscle, the fatty acids are oxidized as fuels whereas in adipocytes, they are rapidly reconverted to triglycerides and stored. Adipose tissue lacks glycerol kinase and cannot use the glycerol produced by LPL so the glycerol travels back to the liver where it is recycled. After being digested by LPL, 85 to 90% of triglycerides are removed from the chylomicrons and the resulting chylomicron remnants are cleared from the plasma by the liver. Unlike long chain fatty acids, medium- and short-chain fatty acids are able to pass directly into the portal vein for transport to the liver as free fatty acids.

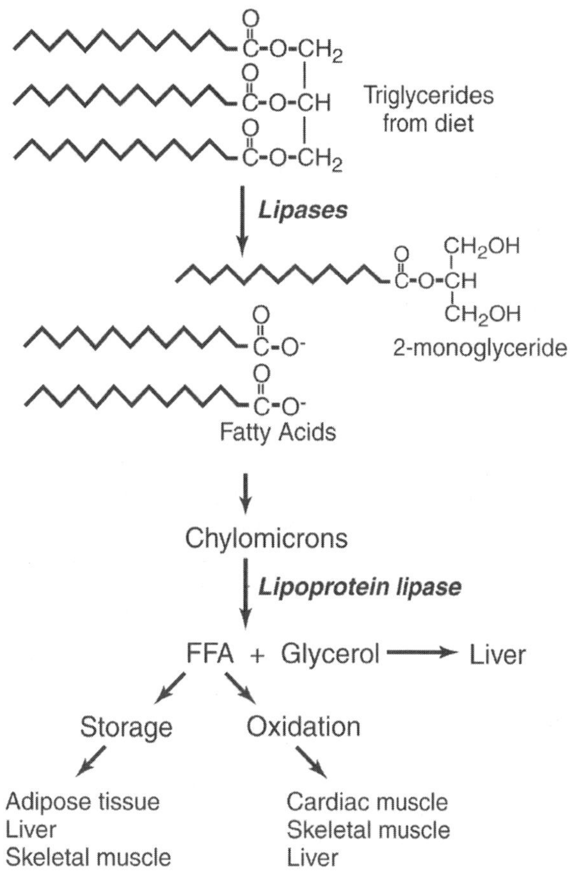

Figure 9.9 Triglyceride metabolism. Triglycerides consumed in the diet are hydrolyzed by pancreatic lipases to FFAs and 2-monoglycerides. These molecules are absorbed through the intestinal wall and triglycerides are subsequently regenerated and packaged into chylomicrons for transport in the blood. Digestion of the packaged triglycerides by lipoprotein lipase (LPL) produces FFA and glycerol. Glycerol is returned to the liver where it is used for gluconeogenesis. FFAs have a more diverse fate and can be either stored in adipose tissue, heart, or skeletal muscle or oxidized by liver, cardiac, and skeletal muscle depending on the body's energy balance.

Whereas chylomicrons provide the primary means of transport of dietary lipids, other lipoproteins transport endogenous lipids. Very low density lipoproteins (VLDL) are synthesized by the liver and function to transport triglycerides that are synthesized by liver to adipose tissue and muscle. Lipase action gradually converts VLDL to intermediate-density lipoproteins (IDL) and then low-density lipoproteins (LDL) that are removed from circulation by endocytosis after interacting with the LDL receptor. Defects in the metabolism of these lipoproteins leads to dyslipoproteinemias; those involving LDL metabolism are often associated with premature atherosclerosis and coronary artery disease.

Lipogenesis

Fat synthesis, or lipogenesis, encompasses the processes of fatty acid synthesis and then triglyceride synthesis, and takes place in both liver and adipose tissue. When a meal rich in carbohydrates and low in fats is consumed, the excess carbohydrate must be metabolized. A small amount can be used immediately as a substrate but most is processed to restore glycogen reserves, fuel the biosynthesis of many molecules, or converted to triglycerides. The conversion of carbohydrate to fat is known as *de novo* lipogenesis (Fig. 9.10). *De novo* lipogenesis is not the pathway of first resort in response to an intake of surplus carbohydrate. The most important responses to increased carbohydrate intake are increased glycogen storage and increased whole body carbohydrate oxidation. It has long been recognized that animals reduce carbohydrate oxidation in times of energy starvation, to preserve lean body mass (by reducing the need for gluconeogenesis from amino acids). The converse also seems to apply since carbohydrate oxidation is increased in times of carbohydrate surplus. Dietary carbohydrates in excess of energy requirements are stored in modestly expandable tissue glycogen stores during the daytime and is then released from the liver for oxidation during the overnight fast. Hence, the main daily cycle consists of daytime carbohydrate storage and night-time carbohydrate release, not daytime *de novo* lipogenesis followed by night-time fat oxidation. Furthermore, the high-fat content (40%) of the modern Western diet may suppress *de novo* lipogenesis. However, *de novo* lipogenesis has a significant role in certain situations, for example, when a high-carbohydrate diet is consumed, during fetal development, in premature infants, and may contribute to the high blood levels of triglycerides seen in alcoholic liver disease.

In humans, triglyceride synthesis from carbohydrate takes place in both the liver and adipose tissue. Pyruvate, formed via glycolysis, enters the mitochondria from the cytosol, where it is decarboxylated to acetyl-CoA by PDC. Citrate synthase then catalyzes the condensation of acetyl-CoA with oxaloacetate to form citrate which, during fatty acid synthesis, is transported back into the cytosol and reconverted to acetyl-CoA and oxaloacetate. The first committed step in fatty acid biosynthesis is the formation of malonyl-CoA from acetyl-CoA and bicarbonate by the enzyme acetyl-CoA carboxylase (ACC). Both acetyl-CoA and malonyl-CoA are then activated by binding with acyl carrier protein (ACP), and then molecules of acetyl-ACP and malonyl-ACP are introduced into the fatty acid synthase (FAS) complex of seven enzymatic activities. The two precursors are joined, with the release of CO_2 from malonyl-CoA, and processed through two NADPH-dependent reduction steps and one dehydration step to form butyryl-ACP. Subsequent cycles import more

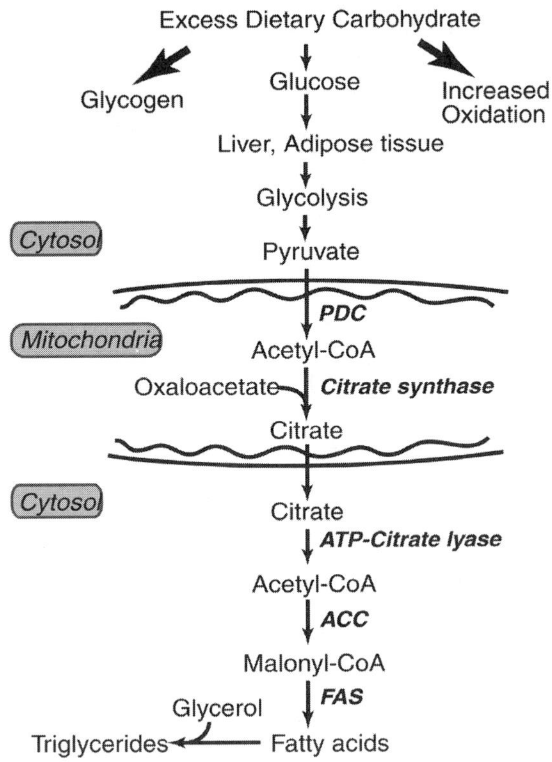

Figure 9.10 *De novo* lipogenesis. The predominant fates of excess carbohydrate are storage as glycogen or increased oxidation. However, excess glucose can also be used to form triglycerides. Excess glucose is taken up by either the liver or adipose tissue and converted to pyruvate by glycolysis, then to acetyl-CoA by PDC, and then complexed with oxaloacetate to form citrate. Citrate is exported from the mitochondria and cleaved by ATP-citrate lyase to produce oxaloacetate and acetyl-CoA, the latter becoming the substrate for cytosolic fatty acid synthesis. ACC converts acetyl-CoA to malonyl-CoA in an irreversible reaction that represents the first committed step in fatty acid synthesis and is the only reaction of the process that occurs outside of the multienzyme fatty acid synthase (FAS) complex. The acetyl and malonyl groups are transferred to acyl carrier protein (ACP) by acetyl transferase and malonyl transferase, respectively, and then acetyl-ACP and malonyl-ACP are complexed by β-ketoacyl synthase (also known as acyl-malonyl ACP condensing enzyme) followed by three steps (reduction, dehydration and reduction) to produce a 4-carbon butyryl-ACP molecule. In the next cycle of the process, butyryl-ACP condenses with another malonyl-ACP to create a six-carbon saturated acyl-ACP molecule. Each cycle adds a further 2-carbon unit until a 16-carbon palmitoyl-ACP is produced, and then this is hydrolyzed to release palmitate. Palmitate may then undergo desaturation and/or elongation reactions to form other fatty acids, and, in liver or adipose tissue, these newly formed fatty acids are esterified to glycerol-3-phosphate to form triglycerides ready for transport or storage.

malonyl-ACP units until a final 16-carbon palmitoyl-ACP is made and then palmitate is released. Palmitate can then undergo desaturation and/or elongation reactions to form other fatty acids. In liver, newly formed fatty acids are esterified to glycerol-3-phosphate and incorporated into a VLDL molecule for transport to adipose tissue for storage. Adipose tissue can also synthesize fatty acids and triglycerides directly from glucose via these same reactions.

Regulation of Lipogenesis

Ingestion of a high-carbohydrate meal induces key lipogenic enzymes in the liver, including those involved in fatty acid synthesis (e.g., ACC, FAS), triglyceride synthesis, NADPH-generating enzymes, and selected glycolytic enzymes. Elevated blood glucose stimulates lipogenesis both directly as a substrate and as an inducer of the expression of lipogenic genes and indirectly via its effects on insulin/glucagon release. To date, the role of glucose versus insulin in the induction process is not completely clear. By contrast, glucagon or epinephrine inhibits lipogenesis in liver and growth hormone (GH) dramatically reduces lipogenesis in adipose tissue. GH may interfere with insulin signaling at the postreceptor level or affect the phosphorylation state of transcription factors and thereby decrease the expression of lipogenic enzymes.

Acetyl-CoA carboxylase links fatty acid and carbohydrate metabolism through the shared intermediate acetyl-CoA. Furthermore, the ACC reaction is irreversible and gates a pathway that is very expensive for the cell (costing one ATP for each malonyl-CoA made and two NADPH for each two-carbon addition to the growing fatty acid). Hence, regulation of ACC is extremely important, and it is accomplished by multiple mechanisms (Fig. 9.11). Diet and hormones affect ACC activity. A carbohydrate-rich, low-fat diet stimulates the expression and activity of ACC, whereas starvation and diabetes reduce ACC activity by repressing the expression of the ACC gene or by increasing the phosphorylation state of the ACC protein (or both). Glucagon effects on liver include phosphorylation of ACC, mediated via PKA, whereas insulin stimulates dephosphorylation. Insulin and glucose also stimulate ACC gene expression, apparently via the ChoRF and sterol regulatory element bending proteins (SREBP-1) transcription factors. ACC is also controlled by allosteric effectors (activation by citrate, inhibition by long-chain fatty acyl-CoA) and reversible polymerization. Levels of citrate increase when there is abundant carbohydrate substrate yet low demand for carbohydrate oxidation, conditions that favor the channeling of acetyl-CoA out of the TCA cycle and into fatty acid synthesis. Long-chain fatty acyl-CoA provides both feedback inhibition of synthesis and a means of inhibiting synthesis under starved conditions when fatty acid catabolism is activated. Eukaryotic ACC is a dimer but is only active when it polymerizes into a filamentous form. Citrate promotes polymerization and long-chain fatty

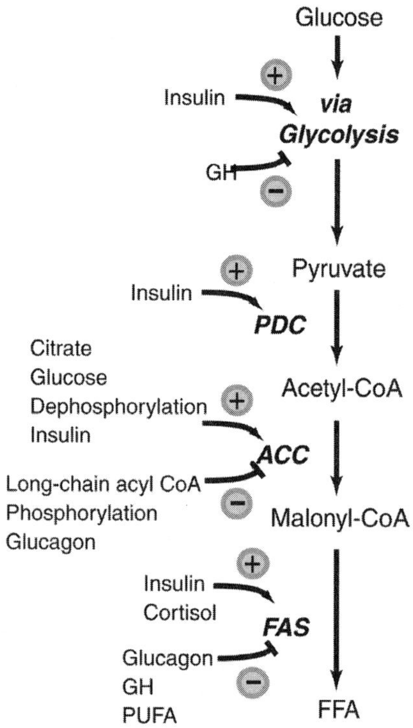

Figure 9.11 Regulation of lipogenesis. Lipogenesis is regulated at several steps. The initial passage of glucose through glycolysis is regulated by insulin and growth hormone (GH). Insulin also stimulates the activity of the PDC. ACC catalyzes the first committed step of fatty acid synthesis, and this irreversible reaction is subject to strict regulation. Citrate and glucose allosterically activate ACC, whereas long-chain acyl-CoAs inhibit ACC. Insulin and glucagon exert opposing effects on ACC, and protein phosphorylation plays an important role in regulating ACC activity. FAS is also subject to regulation by hormones and is inhibited by polyunsaturated fatty acids (PUFA).

acyl-CoAs promote depolymerization. Protein phosphorylation inactivates the enzyme via two mechanisms: (a) causing polymer dissociation and (b) changing sensitivity to allosteric effectors, the phosphoenzyme being also less sensitive to activator and more sensitive to inhibitors. Eight different serine phosphorylation sites occur on the liver ACC and six different protein kinases are known to phosphorylate it. However, it appears that PKA and the AMP-dependent protein kinase (AMPK) are the most physiologically relevant kinases.

AMP-Dependent Protein Kinase The AMPK is a ubiquitous serine/threonine kinase composed of three subunits, the catalytic α subunit and the β and γ regulatory subunits. AMPK has been described as a cellular fuel gauge that acts to restore the cellular energy balance. AMPK acts as a sensor of the cellular AMP : ATP ratio and when cellular levels of AMP rise, activation of AMPK switches off a range of energy-expensive cellular processes including fatty acid biosynthesis (via control of ACC), sterol biosynthesis (via phosphorylation of HMG-CoA reductase), and hormone-sensitive lipase (HSL; reducing lipid export from adipocytes and opposing PKA-mediated lipolysis) (Fig. 9.12). Decreasing lipolysis, a catabolic process, may seem paradoxical in a low-energy state, but it is hypothesized that inhibition of HSL prevents the futile, and expensive, resynthesis of triglycerides that will occur if free fatty acids are not immediately released from the adipocyte. (See Text Box 9.2.)

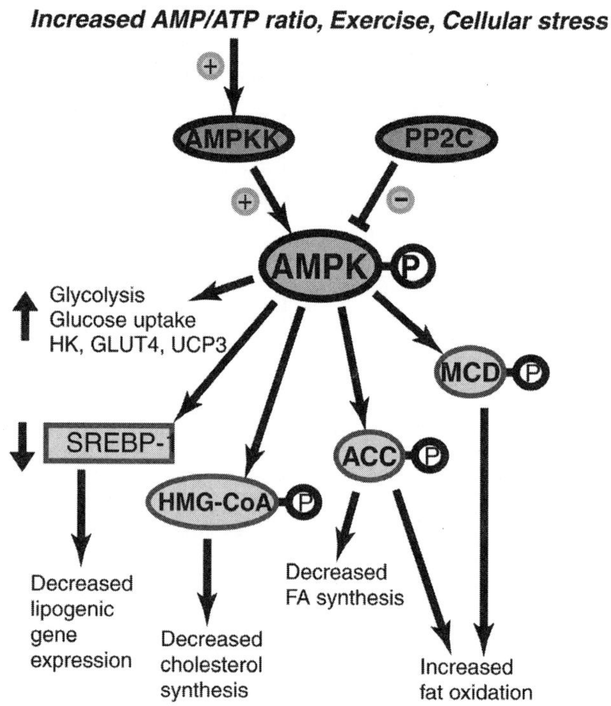

Figure 9.12 Regulation and effects of AMP-activated protein kinase (AMPK). An increase in cellular AMP relative to ATP as a result of exercise or cellular stress (e.g., hypoglycemia, ischemia, hypoxia, heat shock) leads to the activation of AMPK kinase (AMPKK), which phosphorylates and activates AMPK. This action is opposed by protein phosphatase 2C (PP2C). Active AMPK has multiple effects that decrease cellular energy expenditure and increase carbohydrate and lipid catabolism. Inhibition of SREBP-1 leads to decreased expression of the genes for lipogenic enzymes (e.g., ACC, FAS, L-PK, S_{14}) and together with the phosphorylation of ACC decreases fatty acid (FA) synthesis. Phosphorylation of ACC also decreases malonyl-CoA synthesis and, with the phosphorylation of malonyl-CoA decarboxylase (MCD), which increases the decarboxylation of malonyl-CoA, reduces the cellular concentration of malonyl-CoA, and releases inhibition of fat oxidation. Phosphorylation and inhibition of HMG-CoA reductase (HMG-CoA) decreases cholesterol synthesis. Effects of AMPK on carbohydrate metabolism occur mainly in skeletal and cardiac muscle and include increased enzyme expression, increased uptake of blood glucose, and increased glycolysis.

TEXT BOX 9.2 THE WOLFF–PARKINSON–WHITE SYNDROME

The Wolff–Parkinson–White (WPW) syndrome affects 1 to 3 persons per 1000. Affected persons are born with an abnormal electrical pathway between the atria and ventricles in addition to the normal conduction pathway. The presence of this abnormal pathway can lead to life-threatening ventricular arrhythmias. Recent analysis has shown that the underlying gene mutation occurs in the gene that encodes the $\gamma 2$ regulatory subunit of AMPK, *PRKAG2*. It appears that nonphysiologic activation of AMPK by mutations in its subunits increases glucose uptake (by stimulating translocation of GLUT4 to the plasma membrane) and increased hexokinase activity, which leads to glycogen accumulation. It is this excess glycogen accumulation in conductive tissues that causes electrophysiologic abnormalities, although the mechanism of how glycogen accumulation causes abnormal conduction remains unknown.

The AMP-dependent protein kinase does not appear to be regulated by extracellular signals such as hormones but is phosphorylated and activated by an upstream kinase, AMPK kinase (AMPKK), which is also stimulated by AMP and may also be sensitive to palmitoyl-CoA. AMPKK activity is increased by exercise as well as cellular stresses (heat shock, hypoglycemia, ischemia, and hypoxia). AMPK is dephosphorylated and inactivated by PP2C. AMPK promotes increased fatty acid oxidation via two distinct but interrelated pathways. Phosphorylation of ACC by AMPK leads to decreased malonyl-CoA synthesis, and concurrent phosphorylation and activation of malonyl-CoA decarboxylase (MCD) increases decarboxylation of malonyl-CoA. Both pathways lead to decreased levels of malonyl-CoA, which favors increased fatty acid oxidation. AMPK also affects carbohydrate metabolism. Activated AMPK increases glucose transport into cardiac and skeletal muscle cells and in skeletal muscle increases expression of HK, GLUT4, and several mitochondrial proteins including uncoupling protein 3 (UCP3). AMPK has also been shown to phosphorylate PFK-2 to increase production of fructose-2,6-bisphosphate, a key activator of PFK-1 and glycolysis. Other effects of AMPK that have been noted include phosphorylation of creatine kinase (likely to prevent reversal of the reaction and conversion of ATP to ADP) and nitric oxide synthase (to increase production of nitric oxide, which results in vasodilation and increased blood flow to ischemic tissues).

The drug metformin is used for the treatment of type 2 diabetes mellitus to decrease hyperglycemia. Metformin also reduces the concentration of certain lipids in the bloodstream. Although it has been used for many years, the actual mechanism of metformin action is only now being elucidated. Metformin appears to act by stimulating the phosphorylation of AMPK. The beneficial effects of metformin on plasma lipids appears to arise from activation of AMPK, which leads to decreased ACC and MCD activities, decreased lipogenesis, decreased cholesterol synthesis, and increased fatty acid oxidation. Furthermore there is decreased expression of the transcription factor SREBP-1, which decreases the expression of lipogenic enzymes including ACC, FAS, L-PK, and S_{14} (see later section). Decreased levels of SREBP-1 may improve insulin resistance and a reduction in liver fat content may ameliorate hepatic insulin resistance. In addition to effects on lipid metabolism metformin, via AMPK, decreases hepatic glucose production (possibly through decreased expression of PEPCK and G6Pase) and increases skeletal muscle glucose uptake.

Fatty Acid Synthase Fatty acid synthesis is also controlled by regulation of FAS (Fig. 9.11). Activity is high in the fed state when dietary carbohydrate is high and low during fasting or when dietary fat levels are high, especially levels of polyunsaturated fatty acids (PUFAs). No allosteric or reversible phosphorylation controls on FAS have been identified, but regulation occurs at the transcriptional and posttranscriptional levels. FAS protein levels are largely determined by mRNA abundance, which is, in turn, determined by the rate of gene transcription. Carbohydrate induces hepatic transcription of the enzymes of the FAS complex under both insulin and glucocorticoid stimulation. Expression is suppressed by PUFAs and sterols and by the administration of glucagon or GH.

Transcriptional Regulation of Lipogenesis

Transcriptional regulation of lipogenesis is mediated by the SREBPs, which are transcription factors that regulate the expression of genes connected with cholesterol and fatty acid metabolism (Fig. 9.13). SREBPs are members of a family of deoxyribonucleic acid (DNA)-binding proteins known as the basic/helix–loop–helix/leucine zipper family that function as dimers to activate multiple genes. Three forms of SREBP have been identified. SREBP-1a and SREBP-1c derive from a single gene by alternative promoter usage, whereas SREBP-2 is the product of a separate gene. SREBP-1a is expressed primarily in cell lines, and it appears that SREBP-1c is more physiologically relevant. SREBP-1c is primarily involved in the activation of genes involved in fatty acid biosynthesis, whereas SREBP-2 is more selective for genes involved in cholesterol homeostasis. SREBPs recognize the sterol response element (SRE, a 10-base-pair sequence) in the promoter

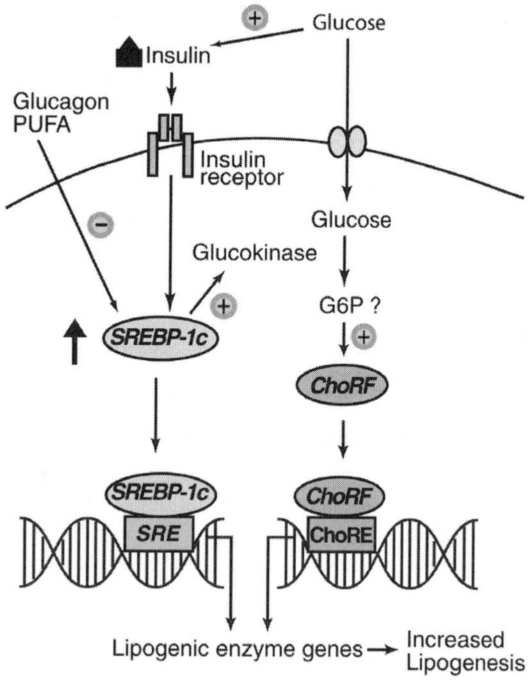

Figure 9.13 Transcriptional regulation of lipogenesis. Elevated blood glucose levels stimulate insulin release, and activation of the insulin receptor leads to the induction of SREBP-1c mRNA and protein. SREBP-1c binds to its response element (SRE) in the promoter region of many lipogenic enzyme genes where it functions to stimulate transcription. Up-regulation of SREBP-1c is opposed by glucagon and polyunsaturated fatty acids (PUFA). Increased uptake of glucose, and increased glycolysis in the hepatocyte is facilitated by the SREBP-1c activation of glucokinase, and high glucose leads to increased expression of genes that contain a carbohydrate response element (ChoRE). The identity of the transcription factor (ChoRF) that activates the ChoRE is still not known.

region of genes encoding a variety of lipogenesis-related proteins including the LDL receptor and enzymes of cholesterol and fatty acid biosynthesis. SREBP synthesis is increased by glucose and insulin but decreased by PUFAs. SREBPs seem to be required for adipogenesis (proliferation of adipose tissue) but are not strongly adipogenic by themselves. The role of SREBPs in adipogenesis is less clear than their role in lipogenesis.

Elevated blood glucose levels following a high-carbohydrate meal promote secretion of insulin from the pancreatic β cell. Insulin acts on the hepatocyte insulin receptor and leads to the induction of SREBP-1c mRNA and protein. SREBP-1c binds to the promoter region of many lipogenic enzyme genes where it stimulates transcription. Simultaneously, the high glucose levels feed an elevated glycolytic rate in the hepatocyte. This is mediated in part by the presence of the high K_m glucokinase, which allows G6P production to increase proportionately with postprandial glucose levels. Because glucokinase expression is induced by insulin, probably through SREBP-1c, the two pathways are coordinated. Increased glucose in hepatocytes also leads to increased expression of genes that contain a carbohydrate response element (ChoRE), first identified in the L-PK gene and now also known to occur in the ACC and FAS genes as well as in a protein, S_{14}, which has a regulatory role in stimulating lipogenesis when carbohydrate levels are high. The nature of the transcription factor (ChoRF) that activates the ChoRE is still not known, but the presence of two copies of the consensus sequence for ChoRF binding in the 5′ untranslated region of genes suggests that it is a dimer but distinct from SREBP. It appears to mediate gene up-regulation responses in response to high glucose, and the rapid response time of gene expression to high glucose suggests probable activation of ChoRF by phosphorylation or metabolite activation, rather than by transcriptional up-regulation of the ChoRF protein.

Hence, genes involved in carbohydrate utilization for lipogenesis can be activated by two mechanisms, via insulin stimulation of SREBP or via glucose stimulation of ChoRF. Some genes respond only to SREBP (e.g., glucokinase), others have only ChoRE binding sites (e.g., L-PK), and others bind both transcription factors (e.g., FAS, S_{14}). Why this is remains to be determined. However, it is likely that this system of dual control provides finer regulation of the levels of individual gene products and allows for multiple regulatory inputs by multiple effectors of the lipogenic process. For example, PUFAs inhibit the expression of lipogenic enzymes apparently by inhibiting SREBP whereas glucagon inhibition of these genes appears to be mediated via ChoRF. Another possibility is that since SREBPs also up-regulate PPARγ, which is involved in adipocyte differentiation, the presence of a second signal (glucose) ensures that when SREBPs are up-regulated for another purpose (e.g., adipocyte differentiation rather than lipogenesis) gene products not needed in this process are not wastefully transcribed.

Lipolysis

Adipose tissue is the main storage depot for lipid fuel reserves (triglycerides) in humans. When fatty acids are required for fuel, hormone-sensitive lipase (HSL) in adipocytes hydrolyzes triglycerides to yield three free fatty acids (FFA) and one glycerol molecule (Fig. 9.14). Once released into the bloodstream the FFAs bind to albumin for transport to other tissues where they diffuse into cells, their rate of uptake being mainly concentration-driven. Fatty acids are key fuels for the body during fasting or starvation, are the preferred substrate for cardiac muscle, and support aerobic exercise by skeletal muscle (see Chapter 11). The

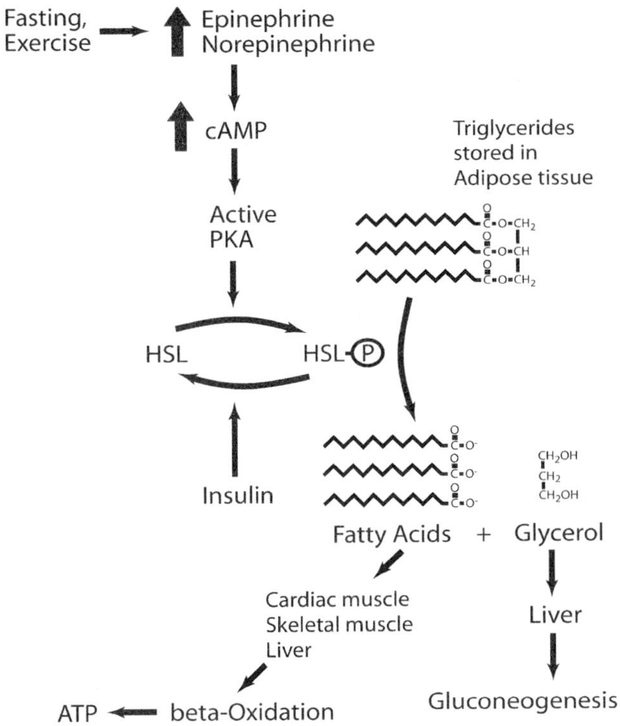

Figure 9.14 Regulation of lipolysis. Under conditions where ATP generation from lipolysis is required (e.g., fasting, prolonged exercise), epinephrine and norepinephrine stimulate adenylate cyclase to raise cAMP levels and activate PKA. PKA phosphorylates and activates hormone-sensitive lipase (HSL), an action opposed by insulin. Active HSL hydrolyzes stored triglycerides to release FFA and glycerol. FFA circulate to cardiac and skeletal muscle and to liver, where they are oxidized to produce ATP. Glycerol is taken up by the liver and used as a substrate for gluconeogenesis.

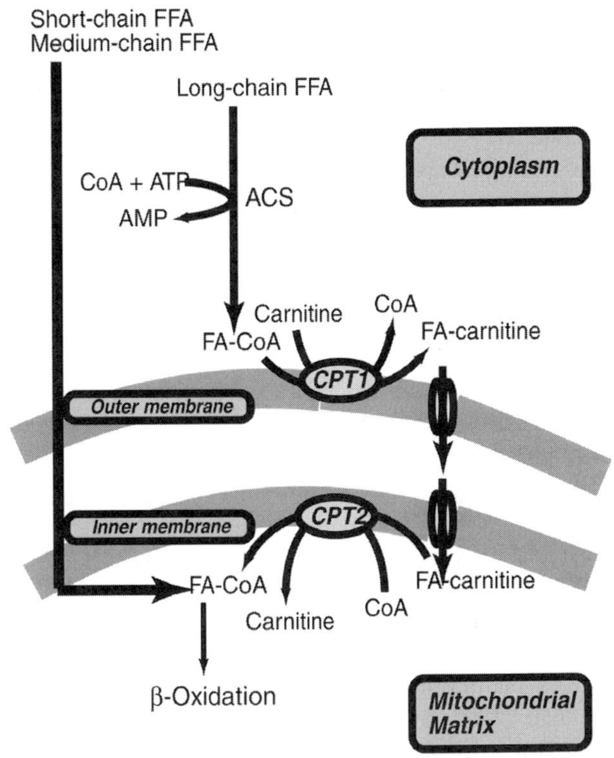

Figure 9.15 Long-chain fatty acid–carnitine shuttle. Whereas short- and medium-chain FFA are able to freely diffuse into the mitochondria, long-chain FFA cannot cross the inner mitochondrial membrane. Carnitine palmitoyltransferase-I (CPT1), located on the outer mitochondrial membrane, catalyzes the transfer of the long-chain fatty acyl group from CoA to carnitine. The fatty acyl-carnitine then crosses the mitochondrial outer and inner membranes. Once in the mitochondrial matrix, carnitine palmitoyltransferase-II (CPT2) regenerates the fatty acyl-CoA that is then delivered to the reactions of β oxidation.

glycerol produced from triglyceride hydrolysis is mainly taken up by liver, providing a gluconeogenic substrate.

The rate of FFA oxidation is also proportional to its concentration in the blood, there being little or no storage of FFAs (or reconversion back to triglycerides) in most tissues. Inside the cell, FFAs bind to fatty acid binding proteins and are transported to the mitochondria. Fatty acid oxidation has three main components: (1) a transport system to move long-chain fatty acids into the mitochondria, (2) the intramitochondrial β-oxidation pathway that cleaves the long-chain fatty acid to produce acetyl-CoA, and (3) the catabolism of acetyl-CoA via the TCA cycle coupled with the electron transport system and oxidative phosphorylation to produce ATP. Unlike carbohydrates, FFAs cannot be used as ATP-generating fuels in the absence of oxygen.

The majority of FFAs used as fuel are long-chain acids (10 to 20 carbons), so our discussion will focus on them (Fig. 9.15). FFA catabolism begins with the formation of fatty acyl-CoA derivatives at the outer mitochondrial membrane. These cannot pass through the inner membrane space, so a carrier system is used. The enzyme carnitine palmitoyltransferase-I (CPT1), located on the outer mitochondrial membrane, catalyzes the transfer of the long-chain fatty acyl groups from CoA to carnitine. The fatty acyl-carnitine molecule then crosses the inner mitochondrial membrane via a specific transporter in exchange for free carnitine that is transported out. Once in the mitochondrial matrix, carnitine palmitoyltransferase-II (CPT2) on the inner aspect of the inner membrane regenerates the fatty acyl-CoA that becomes the substrate for β oxidation. Both CPT1 and CPT2 are enriched at contact sites between the outer membrane and the peripheral inner membrane, suggesting that long-chain acyl-carnitine formation and uptake is facilitated by the submitochondrial localization of these two proteins. The contact sites are also the loci for extensive trafficking of other compounds (e.g., proteins, phospholipids) between extra- and intramitochondrial compartments as well as the binding sites for enzymes such as hexokinase and creatine kinase on the cytoplasmic face and for the interaction of mitochondria with the cytoskeleton.

β oxidation is catalyzed by four enzymatic steps that result in the successive cleavage of two-carbon acetyl-CoA fragments off the fatty acid. Acetyl-CoA is then oxidized by the TCA cycle. The complete oxidation of one palmitoyl-CoA (16-carbon) to CO_2 and H_2O yields 129 mol of ATP.

Regulation of Lipolysis

Fatty acid release from storage triglycerides in adipose tissue is regulated via reversible phosphorylation control of the enzyme HSL. During fasting or exercise, HSL is activated via PKA-mediated phosphorylation. Depending on the physiological state, the initiating hormone can be glucagon, epinephrine or ß-corticotropin that stimulates cAMP production and PKA activation (see Chapter 4 for details of the signal transduction pathway). Growth hormone also stimulates HSL. In the fed state, when insulin levels are high, HSL is dephosphorylated and inactive, thereby preventing fat mobilization. Indeed, lipolytic inhibition is indicated by a strong decrease in the net glycerol output from adipose tissue. Insulin-mediated inhibition of HSL, along with insulin-mediated stimulation of lipogenesis, shifts the metabolic profile of adipose tissue into a net uptake and storage of fuel under fed conditions.

Fatty acid oxidation by other tissues is regulated primarily via control over CPT1, so that once fatty acyl-CoAs enter the mitochondrial matrix, they are rapidly catabolized via the β-oxidation pathway, which has little or no regulatory control. However, other sites of regulation occur in the TCA cycle including a form of feedback inhibition from high acetyl-CoA levels, which leads to the synthesis of malonyl-CoA. Malonyl-CoA in muscle cells is now believed to constitute a fuel sensor that is involved in cross-talk between glucose and fatty acid metabolism. When glucose is the primary oxidative fuel, for example, after eating, PDC is active and malonyl-CoA levels rise because of the abundance of citrate, which serves as a precursor of cytosolic acetyl-CoA and as an activator of ACC. Under these conditions fatty acid oxidation is suppressed by the inhibitory action of malonyl-CoA on CPT1. Newly synthesized fatty acids, synthesized from dietary glucose, are converted to triacylglycerols for storage rather than being catabolized. This ensures that fatty acid synthesis and oxidation do not occur simultaneously. However, under catabolic conditions, such as during fasting or in uncontrolled diabetes, malonyl-CoA concentrations fall and release the inhibition of CPT1, thereby activating fatty acid oxidation. The reduction in malonyl-CoA is mediated in part by glucagon. Decreased flux through glycolysis and reduced pyruvate availability for citrate synthesis leads to reduced availability of acetyl-CoA for malonyl-CoA production for fatty acid biosynthesis.

Peroxisome Proliferator-Activated Receptors The expression of lipolytic genes has recently been shown to be under major control by the PPARs. These were first cloned as the nuclear receptors that mediate the effects of synthetic compounds called peroxisome proliferators on gene transcription. This role is now considered relatively minor compared to the central roles of these receptors in human energy metabolism. PPARs are ligand-dependent transcription factors that form a subfamily of the nuclear hormone receptor superfamily, which includes others such as the glucocorticoid or thyroid hormone receptors. They bind to a specific PPAR response element in the N-terminal region of target genes. PPARs bind to their response element as a heterodimer with a retinoid X receptor (RXR). Upon binding, a conformational change opens a binding cleft for the recruitment of other coactivators, such as PGC-1. The search for natural ligands of PPARs has so far identified PUFAs and selected prostaglandins, whereas two important classes of drugs, the fibrates and the thiazoldinediones, are good synthetic ligands. Three PPAR isoforms (α,β,γ) have been identified to date.

Expressed highly in brown adipose tissue and liver, PPARα is also found in kidney, heart, and skeletal muscle. The target genes of PPARα are involved in β oxidation of fatty acids in mitochondria and peroxisomes and ω oxidation in microsomes, including CPT1. The results of PPARα stimulation include increasing fatty acid uptake and fatty acid oxidation and decreasing lipoprotein assembly and transport. During fasting, fatty acids are oxidized to acetyl-CoA and a portion are also converted to ketone bodies by the liver. Both of these processes are strongly stimulated by PPARα, the expression of which is elevated during fasting. Fatty acids are ligands for PPARs, so it is possible that the large amounts of fatty acids liberated from adipose tissue can stimulate their own metabolism by activating PPARα. Experiments with PPARα-null mice show that PPARα is important in the hepatic response to fasting. In the fed state the phenotype of PPARα-null is quite mild (because of suppression of hepatic fatty acid oxidation), but when fasting these mice suffer from a defect in fatty acid oxidation and ketogenesis resulting in elevated plasma FFA, hypoketonemia, hypothermia, and hypoglycemia. The hypoglycemia emphasizes the important interplay between fatty acid and glucose metabolism in energy homeostasis. Feeding PPARα-null mice a high-fat diet led to a huge accumulation of lipid in the liver and since fatty acid uptake and triglyceride synthesis are unaffected, this accumulation must result from a defect in fatty acid oxidation. The phenotype of PPARα-null mice resembles the phenotype of humans who suffer from a genetic defect of one of the enzymes of fatty acid oxidation, for example medium-chain acyl-CoA dehydrogenase (MCAD) deficiency, which is the most common fatty acid oxidation defect in humans. These patients suffer from severe fasting-induced hypoglycemia, often have a fatty liver and show a plasma metabolite profile that is similar to the profile observed for the PPARα-null mice. PPARα has a number of other functions including lowering triglycerides, raising HDL levels, and acting as an antiinflammatory agent. The antihyperlipidemic fibrates (used to lower triglycerides and increase HDL in dyslipidemic patients) are ligands for PPARα and are thought to mediate their lipid-lowering effects by agonism of PPARα. Less is known about PPARβ, but it may be involved in increasing HDL levels. PPARβ is found in many tissues but the highest expression is in the gut, kidney, and heart.

Involved in adipocyte differentiation and lipid metabolism, PPARγ is mainly expressed in adipose tissue and to a lesser extent colon, immune system, retina, and skeletal muscle. Overexpression of PPARγ in fibroblasts causes them to mature to white adipocytes. Genes regulated by PPARγ include, among others, the adipocyte isoform of fatty acid binding protein, LPL, fatty acid transport protein, acyl-CoA synthase, PDK4, PEPCK, and leptin. PPARγ also up-regulates UCP1, an uncoupling protein found in the mitochondria of brown adipose tissue, which is an important regulator of oxidative phosphorylation and adaptive thermogenesis (see Chapter 16 for its role in mammalian hibernation). PPARγ is also a receptor for the thiazolidinedione (TZD) class of insulin-sensitizing drugs for type 2 diabetes. In clinical studies TZDs have been associated with weight gain possibly owing to stimulation of adipogenesis. PPARγ expression is increased by insu-lin and SREBP-1, suggesting a lipogenic effect in addition to its adipogenic effect. PPARγ cooperates with the CCAAT/enhancer-binding protein (C/EBP) and SREBP transcription factors in controlling adipocyte differentiation.

Increased storage of triglycerides in adipose results in increased production of the hormone, leptin (see Chapter 10), by the tissue as part of a feedback mechanism to limit further food intake and weight gain. Consistent with the role of PPARγ in promoting lipogenesis, leptin gene expression in adipose tissue is under negative control by PPARγ. PPARγ may attenuate the increase in leptin expression during feeding to limit wasteful lipolysis and fatty acid oxidation processes, which are stimulated by leptin, and decreased PPARγ expression may lead to increased leptin levels and, as a result, lower food intake and weight gain.

Many contradictions are present with PPARγ: (1) How can a factor predominantly expressed in adipose tissue improve insulin sensitization and glucose utilization in muscle, the primary site of glucose disposal? (2) Since obesity is a strong risk factor for type 2 diabetes, how does activation of a nuclear receptor involved in adipogenesis have antidiabetic effects? (3) PPARγ is essential for adipocyte development and the formation of fat depots, but on the other hand PPARγ agonists are powerful drugs that are used clinically to treat insulin resistance and hyperglycemia associated with type 2 diabetes. Although homozygous PPARγ-null mice $-/-$ were not viable, mice heterozygous for PPAR$\gamma-/+$ showed normal glucose tolerance and enhanced insulin sensitivity, the same phenotype induced by PPARγ agonists. This paradox suggests that PPARγ activators might not act purely as agonists, and the effect of PPAR modulators may be gene-context-specific. However, PPARγ is a direct link between fatty acid concentrations and the regulation of gene transcription in fat cells.

Ketogenesis

When a person is well-nourished, acetyl-CoA formed during the β oxidation of fatty acids is oxidized to CO_2 and H_2O in the TCA cycle of the liver. However, in the catabolic state seen during prolonged fasting or starvation, the liver converts a large portion of this acetyl-CoA into ketone bodies, primarily acetoacetate and β-hydroxybutyrate (Fig. 9.16). Ketone bodies are water-soluble metabolites of fatty acids that are readily transported to other organs for use as aerobic substrates. They are particularly important because they are the only fuel other than glucose that the brain can use for energy production. During prolonged fasting, liver glycogen reserves are depleted and gluconeogenesis can be limiting, so ketones become a major fuel for the brain. The liver produces ketone bodies, but it cannot catabolize them because it lacks the mitochondrial enzyme, succinyl CoA:3-ketoacid CoA transferase, required for the activation of acetoacetate to form acetoacetyl-CoA. Therefore, ketone bodies leave the liver to be taken up by peripheral tissues, where they are oxidized for energy. The net result of ketone body oxidation is 125 mol of ATP per mole of original palmitate. The rate of formation of ketone bodies is directly proportional to the rate of fatty acid oxidation. Thus, when the rate of mobilization of fatty acids from adipose tissue is high, hepatic fatty acid oxidation and ketone body production are also high.

HUMAN PROTEIN METABOLISM

The dietary intake of protein is essential for life. When protein intake is restricted or insufficient in a child, growth is limited. The loss of protein is also associated with increased mortality in many human diseases (see Cachexia later in this chapter). Body protein serves a

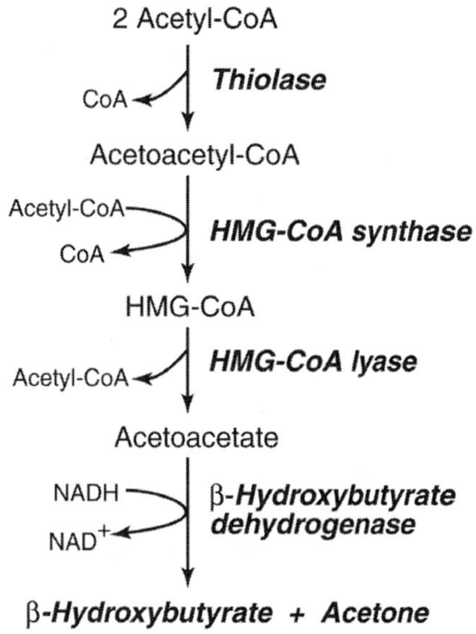

Figure 9.16 Ketogenesis. The ketone bodies, acetoacetate, acetone, and β-hydroxybutyrate, are produced in the liver from acetyl-CoA. The first reaction, catalyzed by thiolase, condenses two molecules of acetyl-CoA. A third molecule of acetyl-CoA is used to produce β-hydroxy-β-methylglutaryl-CoA (HMG-CoA), which is then converted to acetoacetate and acetyl-CoA by HMG-CoA lyase. Acetoacetate is then reduced by β-hydroxybutyrate dehydrogenase to produce β-hydroxybutyrate and acetone.

variety of functions, but almost half of all the body's protein is found in just four proteins: collagen, actin, myosin, and hemoglobin. Furthermore, $\sim$40% of total body protein is located in skeletal muscle. In terms of energy metabolism, skeletal muscle represents a substantial reserve of amino acids that can be oxidized as a fuel or used for gluconeogenesis but, unlike the use of glycogen or lipid fuel reserves, muscle protein is not simply a storage form for energy. Loss of muscle protein represents loss of functional muscle mass, reducing muscle strength and endurance, and hence, human metabolism is designed to minimize this loss in all but extreme cases (see the later discussion of human energy metabolism in fasting and starvation). Proteins are continually synthesized and degraded, but, for an individual in protein balance, the amount of protein being degraded is equal to the protein intake. Entry into the body's pool of amino acid is via the ingestion and breakdown of dietary proteins or the *de novo* synthesis of amino acids. Removal of amino acids from the body's pool occurs through protein synthesis or excretion, the latter usually occurring as oxidation of the carbohydrate moiety to CO_2 and disposal of the nitrogen moiety as ammonium ion or urea. A high rate of protein turnover is required so that we can adapt quickly to changes that affect our bodies such as environmental stress, infection, or injury. However, this adaptability has a cost and protein turnover, including the degradation and the synthesis of new protein, that uses about 15% of basal energy expenditure.

Dietary protein is digested by gastric (pepsin) and pancreatic proteases, the latter including both endopeptidases (trypsin, chymotrypsin, elastase) and exopeptidases (aminopeptidase, carboxypeptidase). These break the peptide bonds in the stomach and small intestine to produce a mixture of free amino acids (30%) and oligopeptides (70%) with lengths of 2 to 6 amino acids. Oligopeptides may be digested further or transported into cells where they are further broken down. Free amino acids are also transported into intestinal cells and then pass to the liver via the portal vein. Amino acids may be transported across the intestinal cell membrane via active Na^+-dependent transport (primarily), facilitated diffusion, or transport linked to the γ-glutamyl cycle. Cotransport of Na^+ and amino acid from the outside of the cell to the inside is driven by the lower intracellular concentration of Na^+. This allows cells to concentrate amino acids from the intestinal lumen. A relatively large free amino acid pool is maintained in the blood to supply tissues with a continuous supply of individual amino acids for the synthesis of proteins, neurotransmitters, and other nitrogen-containing compounds.

Amino Acid Metabolism, the Urea Cycle and Gluconeogenesis

Dietary amino acids are transported to various tissues for metabolism; this occurs primarily in liver but also in intestine, skeletal muscle, and kidney where selected amino acids have very specific roles. In liver, the fate of ingested amino acids is either (a) protein biosynthesis (liver produces a wide variety of secreted proteins including clotting factors and albumin) or (b) gluconeogenesis plus urea synthesis. Previously, we have stressed the importance of gluconeogenesis during fasting, but it is also the major pathway for amino acid catabolism after eating a meal.

The major metabolic process involved in the metabolism of dietary amino acids is their oxidation to glucose and CO_2. Intestine processes three amino acids in particular; nearly all of the dietary glutamate and aspartate and approximately two-thirds of the glutamine are catabolized within the intestinal mucosal cells. These three amino acids are oxidized to produce end products, including alanine, ammonium ion, and lactate that are released into the bloodstream. Glutamine derived from muscle is also an important fuel for the intestine, but glucose is not a major oxidative fuel. Muscle is the primary site for the catabolism of branched-chain amino acids (leucine, isoleucine, and valine); these are oxidized to form glutamine and alanine and then released into the blood for transport

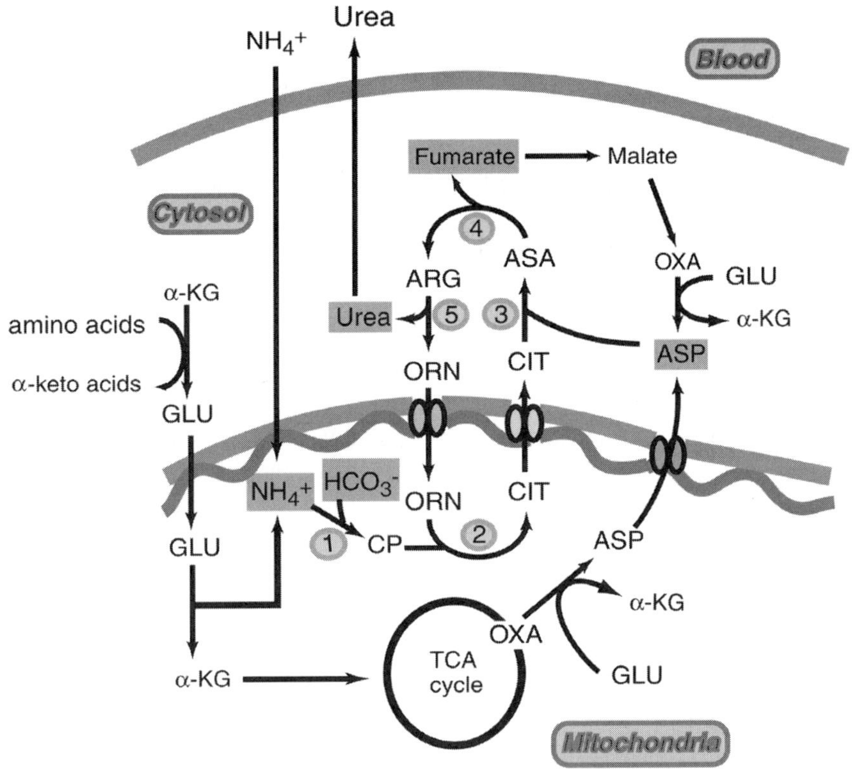

Figure 9.17 The urea cycle. Amino groups removed from a variety of amino acids are transferred to glutamate. The glutamate dehydrogenase reaction releases NH_4^+ in the mitochondrial matrix and some NH_4^+ is also directly taken up from the plasma. The urea cycle begins with two mitochondrial reactions: (1) the ATP-dependent condensation of NH_4^+ and CO_2 via carbamoyl phosphate synthetase to form carbamoyl phosphate (CP) and (2) condensation of CP with ornithine (ORN) to form citrulline (CIT), catalyzed by ornithine transcarbamoylase. CIT is exported from the mitochondria and combines with ASP to form arginosuccinate (ASA) via the cytoplasmic arginosuccinate synthetase reaction (3). ASA is hydrolyzed to form fumarate and arginine (ARG) by argininosuccinate lyase (4), and then arginase (5) completes the cycle by cleaving ARG to form urea and regenerating ORN, which reenters the mitochondria. The net result of the urea cycle is the production of one molecule of urea from two ammonium ions and one bicarbonate. Urea is exported from the liver and delivered to the kidney for excretion. Fumarate can be recycled back into the TCA cycle or can be used for gluconeogenesis.

to the liver. The kidney converts glutamine (released from muscle) and glycine to bicarbonate, which is important for maintenance of whole-body acid–base balance, and serine, which is transported to the liver.

Amino acid metabolism in the liver is unique in that it is the only organ in the body that possesses all of the enzymes of the urea cycle and, therefore, has the capacity to detoxify ammonium ion by synthesizing urea. In liver, amino groups from many amino acids (particularly alanine) are transferred in transaminase reactions to glutamate. The resulting keto acids are fed into oxidative or gluconeogenic reactions, whereas the glutamate feeds excess amino groups into the reactions of the urea cycle for disposal of the nitrogen (Fig. 9.17). The action of glutamate dehydrogenase in the mitochondrial matrix releases NH_4^+ (some NH_4^+ is also directly taken up from the plasma). Two mitochondrial reactions begin the synthesis of urea. First, NH_4^+ is joined with bicarbonate to form carbamoyl phosphate by the ATP-dependent carbamoyl phosphate synthetase (CPS) reaction. Carbamoyl phosphate then condenses with ornithine to form citrulline, catalyzed by ornithine transcarbamoylase (OTC). Citrulline is exported from the mitochondria and in the cytoplasm arginosuccinate synthetase (AS) joins citrulline and aspartate to form arginosuccinate; aspartate can be derived from either mitochondrial or cytoplasmic pools. Argininosuccinate is hydrolyzed to form fumarate and arginine by argininosuccinate lyase (AL), and then arginase completes the cycle by cleaving arginine to form urea and regenerating ornithine, which reenters the mitochondria. Urea is exported from the liver and delivered to the kidney for excretion, whereas the other output of the cycle, fumarate, can be recycled to aspartate or drawn off into gluconeogenic reactions.

The importance of the urea cycle for maintaining ammonium ion at low levels is emphasized by the consequences of several rare genetic or inborn errors of metabolism, including deficiencies in any of the five urea cycle enzymes, which cause hyperammonemia. Ammonium is highly toxic to the central nervous system, and clinical symptoms and signs of hyperammonemia include vomiting and neurologic abnormalities such as ataxia, confusion, convulsions, and agitation or irritability that may alternate with periods of lethargy and may progress to coma. Sodium benzoate increases ammonium clearance from the body and, unless the patient has an arginase deficiency, the administration of arginine helps to treat a urea cycle defect by supplying ornithine and *N*-acetylglutamate. To decrease ammonium production by intestinal bacteria an oral antibiotic and a laxative are also given.

The interactions of the TCA cycle, urea cycle, and enzymes of gluconeogenesis metabolize dietary amino acids and the products of peripheral tissue amino acid metabolism (predominantly alanine and also glutamine, serine, and ammonium ion) to glucose, urea, and CO_2. This converts a group of specialized fuels, the amino acids, which for the most part cannot be rapidly oxidized by most organs, into glucose, the universal fuel. It allows the majority of the energy supplied as dietary protein to be made available to all tissues of the body without the need for each tissue to synthesize the complex array of enzymes required for amino acid catabolism.

Regulation of Protein Metabolism

Insulin and growth hormone are the predominant anabolic hormones as far as protein metabolism is concerned. Insulin can simultaneously stimulate protein synthesis (in the presence of amino acids) and block protein breakdown. Growth hormone is associated with the long-term regulation of growth and levels are highest in children, especially during growth spurts. Growth hormone effects are mediated in part by the insulin-like growth factors (IGF-I and IGF-II) that are synthesized by the liver. Testosterone also stimulates protein synthesis and synthetic testosterone-like compounds, anabolic steroids, promote the retention of body protein.

Glucagon, cortisol, and epinephrine are protein catabolic hormones that stimulate a loss of body protein and inhibit muscle protein synthesis. Glucagon, glucocorticoids, and thyroid hormone increase the rates of amino acid catabolism as well as hepatic amino acid uptake, gluconeogenesis, and ureagenesis. The stimulation of the urea cycle by cortisol and glucagon makes sense when we remember the close linkage between the urea cycle and gluconeogenesis. Elevated glucagon interferes with insulin's ability to inhibit protein degradation and promotes gluconeogenesis from amino acids and lactate. Cortisol decreases protein synthesis and increases protein degradation in muscle while also stimulating the expression of liver enzymes involved in amino acid oxidation and gluconeogenesis.

Transcriptional Regulation of Protein Metabolism

The C/EBP transcription factor family are basic-leucine zipper (bZIP) factors that are critical for the gene expression of urea cycle enzymes. C/EBPα-deficient mice show reduced mRNA levels for CPS, AS, AL, and arginase as well as decreased protein amounts of CPS, OTC, AL, and argninase. These mice have a phenotype of severe liver disease with multiple biochemical abnormalities including coagulopathy, hypoglycemia, hypoalbuminemia, hyperbilirubinemia, and hyperammonemia. C/EBPα-deficient mice also show reduced mRNA levels for GS, albumin, and PEPCK.

Urea cycle genes are induced by glucagon and glucocorticoids and repressed by insulin. Another member of the C/EBP family, C/EBPβ, is necessary for glucocorticoids and glucagon to exert their effects *in vitro*, but *in vivo*, the loss of C/EBPβ in genetically engineered mice is compensated for an unknown mechanism (possibly C/EBPα) and enzyme levels are unaffected.

Another important regulator of the urea cycle is PPARα, which by inhibiting mRNA expression of enzymes involved in amino acid catabolism has an effect opposite to that of C/EBPα. In mice, PPARα influences the expression of numerous genes whose protein products are involved in transamination, deamination, the urea cycle (all five enzymes), oxidation of α-keto acids, amino acid interconversions, and the synthesis of amino-acid-derived products. PPARα activation appears to lead to an overall decrease in amino acid degradation, and the increase in plasma urea in PPARα-null mice supports this conclusion.

The effects of PPARα on the expression of genes involved in amino acid metabolism may occur via a negative response element or PPARα may stimulate the expression of another transcription factor that serves as a transcriptional repressor. Since fatty acids are ligands for PPARα, the suppressive effect of PPARα on urea cycle enzymes may provide a potential explanation for the inhibitory effect of fatty acids on ureagenesis and ammonia detoxification. During prolonged fasting, fatty acid oxidation becomes the major source of energy for the liver, an effect mediated by PPARα. At the same time the relative contribution of amino acid metabolism to hepatic ATP production, which is dominant in the fed state, declines. The reciprocal relationship between fatty acid oxidation and nitrogen metabolism is indicated by concentrations of plasma ketones and urea; plasma ketones increase during fasting whereas plasma urea decreases. The simultaneous increase in ketone body concentration and decrease in urea concentration during fasting in mice is actually due to the action of a single factor, PPARα, which balances the needs of the two pathways by altering the expression of genes involved. The fatty acid–amino acid balance may be similar to the fatty acid–glucose cycle proposed (Text Box 9.3), with fatty acids inhibiting amino acid utilization in a situation comparable to their inhibition of glucose.

HUMAN ENERGY METABOLISM IN THE FED STATE

Ingestion of a meal leads to an increase in plasma insulin levels and a decrease in circulating catabolic hormones. In the liver, insulin stimulates increased glucose uptake. After conversion to G6P, carbon is preferentially stored as glycogen (insulin activates GS) or, if carbohydrate intake is high enough, G6P is used to synthesize fatty acids and triglycerides. Lipogenesis is promoted by insulin (mediated by SREBP-1c) and glucose stimulation of the expression of lipogenic genes. Triglycerides are packaged into VLDLs and released into the circulation for re-uptake and storage by adipocytes. In the absorptive period the liver is also presented with high levels of amino acids that are processed in several ways: (a) re-release into the blood to supply other tissues, (b) increased hepatic protein synthesis, (c) gluconeogenesis, or (d) use as a substrate for mitochondrial oxidation and ATP production.

After feeding there is also an increase in fatty acid biosynthesis and increased triglyceride storage in adipose tissue. This is accomplished by an insulin-mediated increase in the synthesis and release of LPL to break down circulating triglycerides and enhance their importation into adipocytes. Insulin also stimulates SREBP-1c and PPARγ to increase expression of genes involved in glucose and fatty acid uptake, fatty acid synthesis, and fatty acid conversion to triglycerides. As triglyceride levels rise, production of the adipocyte hormone leptin is stimulated as part of a feedback mechanism to limit further food intake and weight gain (notably, leptin is under negative control by PPARγ).

Feeding also suppresses lipolysis. Again, this is mainly insulin-mediated, in particular via insulin-stimulated dephosphorylation of HSL. With a reduced rate of lipolysis in adipose tissue, the concentration of circulating free fatty acids is also lowered and, hence, their availability to other tissues as a metabolic fuel is reduced. In liver the active synthesis of fatty acids in the fed state, stimulated by insulin effects on ACC, causes a concomitant suppression of lipolysis due to the inhibitory effects of malonyl-CoA on CPT1. This ensures that β oxidation and fatty acid synthesis do not occur simultaneously in a futile cycle.

In skeletal muscle in the fed state insulin stimulates the uptake of nutrients. Insulin stimulates recruitment of the GLUT4 glucose transporters into the plasma membrane and increased glucose uptake is used to replenish muscle glycogen stores. Amino acid uptake is also increased,

TEXT BOX 9.3 RANDLE HYPOTHESIS

The metabolism of fatty acids is reciprocally linked with the metabolism of carbohydrates; in muscle, for example, if the oxidation of fatty acids decreases, the oxidation of glucose or glycogen must increase to meet the energy requirements of the cell. In 1963, P.J. Randle attempted to explain the relationship between glucose and fatty acids. Essentially, his theory states that the availability of free fatty acids dictates the rate of fat oxidation in muscle and that fatty acid oxidation directly inhibits glycogen and glucose metabolism. This is known as the glucose–fatty acid cycle. Randle observed that increased fatty acid flux led to elevated acetyl-CoA concentrations in mitochondria and high acetyl-CoA caused both product inhibition of pyruvate dehydrogenase activity (thereby slowing carbohydrate entry into the TCA cycle) and elevated citrate concentrations. High citrate allosterically inhibited PFK-1 to provide further inhibition of carbohydrate catabolism, and PFK-1 inhibition, in turn, led to an accumulation of G6P, which caused feedback inhibition of further glucose phosphorylation by hexokinase.

Although these allosteric-type controls remain important, the inhibitory effect of fatty acids on glucose metabolism may also be exerted at other levels. Exposure to free fatty acids may inhibit insulin action in muscle by acting downstream of the insulin receptor. Evidence for this includes observations that elevated plasma free fatty acids inhibited the insulin-dependent tyrosine phosphorylation of insulin receptor substrate-1 and its association with phosphatidylinositol 3-kinase. The diminished activation of this pathway might result from a direct interaction of fatty acids or metabolites with components of the phosphorylation pathway. Supporting this is a strong correlation between accumulation of intramuscular lipids and insulin resistance. The PPAR transcription factors are another example where free fatty acids may act as a ligand to exert an effect on metabolism. Thus, the cell has different mechanisms for sensing the concentrations of metabolites that are just as important in regulating cellular and body homeostasis as receptors for hormones.

and, although the oxidation of amino acids is also increased in the immediate postprandial period, there is a net positive balance of protein. In the fed state, fatty acids are not an important fuel for skeletal muscle. In brain, glucose is the exclusive fuel in the fed state since the brain does not possess significant stores of triglycerides, and free fatty acids do not cross the blood–brain barrier efficiently.

HUMAN ENERGY METABOLISM DURING FASTING AND STARVATION

We all fast for short periods of time when we sleep and humans sometimes endure longer fasts. Many mammals fast for months at a time; for example, hibernating mammals may not eat for 6 to 9 months (see Chapter 16). Simple starvation can be defined as inadequate intake of protein and/or calories. Starvation can be short-term (fasting) or long-term (chronic protein-energy undernutrition). Starvation is most often caused by lack of food but, in developed countries, undernutrition is most often related to medical causes (e.g., decreased appetite or anorexia due to illness, drugs, dementia, or other psychiatric conditions). Starvation not only causes an imbalance in energy homeostasis, it elicits a systemic orchestrated response consisting of metabolic changes that are the reciprocal of those seen in the fed state. In the absence of food, plasma levels of glucose, amino acids, and triglycerides decrease; insulin secretion falls and glucagon release increases. The body switches to a regulated mobilization and catabolism of stored reserves of glycogen, triglycerides, and protein, and the priorities of metabolism become: (a) maintaining an adequate blood glucose level to sustain the energy requirements of the brain and other glucose-dependent tissues, (b) mobilizing fatty acids from adipose tissue and ketone bodies from liver to supply fuel to all other tissues, and (c) minimizing protein catabolism so as to limit the loss of lean mass (see Text Box 9.4).

In a normal 70-kg man metabolic fuels available at the start of a fast are 0.2 kg of glycogen, 15 kg of fat, and 6 kg of protein. Carbohydrate is a quantitatively insignificant fuel reserve, which is why prolonged starvation causes the body to shift from a metabolism based on carbohydrate to a lipid economy. After 24 h of fasting, the use of glucose as a fuel has decreased with only 15% of liver glycogen stores remaining, and the rates of hepatic glucose production and whole-body glucose oxidation have decreased. The relative contribution of gluconeogenesis to hepatic glucose production does increase as the rate of hepatic glycogenolysis declines. Gluconeogenesis persists at a reduced rate using lactate, glycerol from hydrolyzed triglycerides, and amino acids derived from muscle breakdown, all to help maintain glucose supply for the brain. As fasting continues, the conversion of glutamine to glucose in the kidney represents almost 50% of total glucose production.

During short-term starvation (1 to 14 days of fasting), the decline in plasma insulin, increase in plasma epinephrine, and increase in lipolytic sensitivity to catecholamines stimulate adipose tissue lipolysis. The increase in fatty acid delivery to the liver, in conjunction with an increase in the

TEXT BOX 9.4 MALNUTRITION

Improper or inadequate food intake or inadequate absorption of nutrients can all lead to malnutrition. Chronic malnutrition usually involves deficits in more than a single nutrient and has many of the same characteristics as starvation but develops over a longer time period. Growth failure, impaired protein synthesis, and reduced immunity to disease are all sequelae of chronic malnutrition. Clinical syndromes of malnutrition include kwashiorkor, associated with edema resulting from low protein intake, and marasmus in which protein and calorie intakes are deficient and there is no edema.

The term *marasmus* refers to the simple starvation, or protein–calorie malnutrition, which develops when energy intake fails to meet metabolic demand. It can result from an inadequate caloric intake due to insufficient diet, to improper feeding habits, or to metabolic or congenital abnormalities. Protein malnutrition can result from insufficient intake of good quality protein, impaired protein absorption (e.g., chronic diarrhea), abnormal protein losses (e.g., in the urine or due to infection, hemorrhage, or burns), or the failure of protein synthesis (e.g., in chronic liver failure). Marasmus is recognized by ketosis and decreased muscle and fat mass. Normal serum albumin levels persist until very late stages when hypoalbuminemia and tissue edema result from compromised hepatic function. Treatment involves providing adequate calories and protein, accompanied by fluid and micronutrient support.

Kwashiorkor, the most serious and common form of malnutrition in the world today, results from a severe deficiency of protein in the diet and an inadequate caloric intake that usually begins after breast feeding stops and continues to about 5 years of age. In early stages, protein malnutrition is subtle, but children are usually lethargic or irritable. Advanced illness is characterized by inadequate growth, loss of muscle tissue, edema, and fluid retention, and increased susceptibility to infections. Hypoalbuminemia is typically present and ketosis typically absent.

ratio of plasma glucagon–insulin concentration, enhances the production of ketone bodies by the liver. A maximal rate of ketogenesis is reached by 3 days of starvation, and plasma ketone body concentration is increased 75-fold by 7 days to provide 70% of the brain's energy needs. The use of ketone bodies by the brain greatly diminishes glucose requirements and spares the need for muscle protein degradation to provide fuel for gluconeogenesis. If early protein breakdown rates were to continue throughout starvation, a potentially lethal amount of muscle protein would be catabolized in less than 3 weeks. In contrast to the depletion of fat or glycogen stores, depletion of protein stores has functional consequences because proteins are important structural components of cells, particularly of muscle. Only about one-third of available protein can be catabolized as a fuel supply without compromising vital functions.

During long-term starvation (14 to 60 days of fasting), the body relies almost entirely on lipids stored in adipose tissue for its fuel, which provides more than 90% of daily energy requirements. Muscle protein breakdown decreases to less than 30 g/day, causing a marked decrease in urea nitrogen production and excretion. The decrease in osmotic load diminishes urine volume to 200 mL/day, thereby limiting fluid requirements. Total glucose production decreases to approximately 75 g/day, providing fuel for glycolytic tissues (40 g/day) and the brain (35 g/day) while maintaining constant plasma glucose concentration.

Simultaneously with alterations in substrate utilization, the body takes further protective measures to maintain energy balance and a loss of body mass for as long as possible. Responses to a starvation diet typically include a decrease in physical activity and a decrease in resting metabolic rate. Resting metabolic rate decreases by about 15% after 7 days of starvation, and starvation can lower resting metabolic rate by as much as 40%. Energy is conserved by a decrease in physical activity caused by fatigue, increased conversion of active thyroid hormone to its inactive form, and suppressed sympathetic nervous system activity.

Four mechanisms are involved in switching the metabolism of the liver between well-fed and starved states: (1) substrate availability, (2) negative and positive allosteric effectors, (3) reversible protein phosphorylation of key enzymes, and (4) changes in the amounts of key enzymes. The last is a long-term adaptive mechanism that results from alterations in gene expression and protein degradation. Starvation causes metabolic reprogramming characterized by a transcriptional shift toward the promotion of fatty acid oxidation, protein breakdown, and gluconeogenesis. Hepatic glycogen is used rapidly during starvation, and tissues that require glucose are dependent on hepatic gluconeogenesis primarily derived from lactate, glycerol, and alanine. In addition hepatic synthesis of glucose is closely linked with synthesis of urea.

Hormones provide the signals that initiate and maintain the metabolic programs that allow us to survive prolonged periods without food (the period of survival without water, however, is much, much shorter). The signals to begin digesting adipose tissue triglycerides and skeletal muscle proteins are a fall in the plasma level of insulin and an increase in the concentration of glucagon caused by limited intake of dietary carbohydrate and, therefore. a low plasma glucose concentration. Hypoinsulinemia facilitates mobilization of free fatty acids, starvation ketosis, and catabolism of skeletal muscle protein.

TEXT BOX 9.5 THE REFEEDING SYNDROME

The refeeding syndrome is a potentially lethal condition seen in individuals who are undergoing refeeding after prolonged malnourishment, for example, in individuals with marasmus or kwashiorkor, cancer, chronic alcoholism, or anorexia nervosa. The clinical features of this syndrome include severe electrolyte and fluid imbalances and metabolic derangements including abnormal glucose metabolism, hypophosphatemia (low phosphate), hypomagnesemia (low magnesium), hypokalemia (low potassium), and vitamin deficiencies such as thiamine/vitamin B_1. Due to chronic malnutrition, the body has come to rely on lipid oxidation as its primary fuel rather than carbohydrates. The ingestion of glucose during refeeding causes insulin release, which results in increased protein synthesis and the increased cellular uptake of glucose, phosphate, potassium, magnesium, water, and thiamine. The resulting electrolyte imbalances can cause neurological complications such as delirium or seizures and potentially fatal cardiac arrhythmias.

CACHEXIA

Cachexia and starvation are the two major paradigms of malnutrition. Cachexia, which means *poor condition* in Greek, can be defined as the accelerated loss of skeletal muscle associated with a chronic inflammatory response. Some of the diseases that lead to a chronic inflammatory response include infections, cancer, acquired immune deficiency syndrome (AIDS), congestive heart failure, rheumatoid arthritis, tuberculosis, chronic obstructive pulmonary disease, cystic fibrosis, Crohn's disease, and chronic renal failure. Cachexia involves weight loss, particularly from skeletal muscle and adipose tissue, anemia, electrolyte and water abnormalities, and anorexia.

Muscle wasting can be either slow and gradual as is seen in conditions such as immobilization (disuse atrophy), heart or respiratory failure, hypothyroidism, or chronic renal failure; in these cases, wasting results from decreased protein synthesis without an increase in the rate of protein degradation. However, muscle loss can also be extremely rapid, as seen in diseases such as sepsis, cancer, and trauma where elevated rates of both protein synthesis and proteolysis occur. This malnutrition occurs despite adequate feeding and is thought to be secondary to cytokine production.

In the nutritionally deprived human, insufficient intake of nutrients, especially amino acids, with secondary changes in hormonal balance is the primary reason for loss of muscle mass. Starvation is characterized by substantial fat loss but less severe loss of lean tissue and responds quickly to complete nutritional support. By contrast, in the critically ill patient with severe injury or infection, an increase in the rate of protein catabolism caused by a systemic inflammatory response is responsible for a specific decline in muscle mass and refeeding does not reverse the macronutrient changes (see Text Box 9.5).

The acute-phase response is the body's immediate response to localize and limit injury following tissue damage. It consists of stereotyped, coordinated adaptations ranging from behavioral to physiologic. One of these is the hepatic synthesis of large quantities of proteins, including binding proteins, protease inhibitors, complement factors, fibrinogen, and many others. The acute-phase response has nutritional implications. It is energy-intensive with high rates of hepatic protein synthesis that require large quantities of essential amino acids. The need for essential amino acids drives the catabolism of skeletal muscle protein under the stimulation of stress hormones—glucagon, cortisol, and epinephrine. The survival value is obvious. An injured individual has an impaired ability to obtain exogenous protein by hunting/gathering, and skeletal muscle, which makes up $\sim$40% of the body weight in men and $\sim$33% in women, contains the body's largest pool of protein. Mobilization of this amino acid reserve is an effective adaptation over the short term because skeletal muscle mass can be replaced rapidly after recovery.

However, problems develop when the process is chronic because the depletion of skeletal muscle mass contributes increasingly to morbidity and mortality. The acute-phase response includes coordinated adaptations of intermediary metabolism that differ from those of starvation. One important difference is that, whereas energy reserves are conserved in starvation by suppressing metabolic rate, in disease or trauma the stress response actually stimulates a hypermetabolic state that accelerates the rate of muscle protein loss above normal. It was originally believed that hypermetabolism was the direct cause of weight loss in cachexia but it is now known to be insufficient by itself. As part of the inflammatory response, which is the body's attempt to control and repair damage to itself, cytokines are released (see Text Box 9.6). Cytokines regulate the acute-phase response and cachexia. Proinflammatory cytokines are protein mediators that are secreted from immunocompetent cells and other cells that mediate the acute-phase response. Their mode of action is predominantly local—paracrine or autocrine. However, cytokines can also produce systemic effects.

TEXT BOX 9.6 INFLAMMATION AND CYTOKINES

Inflammation is a complex reaction that occurs in vascularized connective tissue involving blood and circulating cells (e.g., neutrophils, monocytes, eosinophils, lymphocytes, basophils, and platelets), blood vessels, and cellular and extracellular constituents of connective tissue. Inflammation is fundamentally a protective response that attempts to rid the organism of both the initial cause of cell injury (e.g., microbes and toxins) and the consequences of injury (e.g., necrotic cells and tissues) and is closely intertwined with the process of repair. Inflammation attempts to destroy or isolate the injurious agent while initiating cellular mechanisms to heal and reconstitute damaged tissue. Repair begins during the early phases of inflammation but reaches completion usually after the injury has been neutralized.

Inflammation is divided into acute and chronic patterns with acute inflammation lasting for minutes to a few days and characterized by the exudation of fluid and plasma proteins (edema) and the emigration of leukocytes, predominantly neutrophils. Chronic inflammation is of longer duration and is associated with the presence of lymphocytes and macrophages, the proliferation of blood vessels, fibrosis, and tissue necrosis. The vascular and cellular responses of both acute and chronic inflammation are stimulated by chemical factors (e.g., proinflammatory cytokines) derived from plasma or cells and triggered by the inflammatory stimulus. Such mediators amplify the inflammatory response and influence its evolution. Inflammation is terminated when the injurious stimulus is removed and the mediators are either dissipated or inhibited (e.g., by anti-inflammatory cytokines). Without inflammation, infection would be unchecked and wounds would never heal. However, inflammation may be potentially harmful since it may cause local complications (e.g., formation of scar tissue) or may have systemic, life-threatening effects (e.g., septic shock, anaphylaxis) if the inflammatory response exceeds the body's biological controls.

Cytokines are small proteins produced by many cell types (principally activated lymphocytes and macrophages, but also endothelium, epithelium, and connective tissue cells) that modulate the function of other cell types. Cytokines are produced during immune and inflammatory responses, and secretion of these mediators is transient and closely regulated. Cytokines mediate their effects by binding to specific receptors on target cells, and the expression of cytokine receptors can be regulated by a variety of exogenous and endogenous signals. Cytokine activities include regulation of lymphocyte function, activation of inflammatory cells, natural immunity, and stimulation of hematopoiesis.

Interleukin (IL)-1 and tumor necrosis factor (TNF)-α are two of the major cytokines that drive inflammation and are produced by activated macrophages stimulated by cellular insults (e.g., toxins, physical injury, immune complexes). Their most important actions are their effects on endothelium (endothelial activation), leukocytes, and fibroblasts and induction of the systemic acute-phase reactions. In endothelium they induce the synthesis of endothelial adhesion molecules and chemical mediators (including other cytokines, chemokines, growth factors, eicosanoids, and nitric oxide), promote the production of enzymes associated with matrix remodeling; and increase the surface thrombogenicity of the endothelium. TNF-α also causes aggregation and priming of neutrophils, leading to augmented responses of these cells to other mediators and the release of proteolytic enzymes from mesenchymal cells, thus contributing to tissue damage. IL-1 and TNF-α (as well as IL-6) also induce the systemic acute-phase responses associated with infection or injury, including fever, loss of appetite, the production of slow-wave sleep, the release of neutrophils into the circulation, the release of adrenocorticotropic hormone (ACTH) and corticosteroids, and, particularly, with regard to TNF-α, the hemodynamic effects of septic shock (hypotension, decreased vascular resistance, increased heart rate, and acidosis).

Anti-inflammatory cytokines, such as IL-4, IL-10, and transforming growth factor (TGF)-β, block this process or at least suppress the intensity of the cascade. A balance between the effects of proinflammatory and anti-inflammatory cytokines is thought to determine the outcome of disease, whether in the short-term or long-term.

Of the various cellular proteolytic pathways, the ATP-dependent ubiquitin–proteasome pathway has the dominant role in the regulation of protein turnover. Proteins to be degraded are marked by covalent binding to a small protein, ubiquitin. They are then transported to a barrel-shaped complex, the 26S proteosome, within which proteolysis occurs. The proteosome structure is designed to avoid nonspecific protein digestion. The ubiquitin–proteosome pathway is the mediator of protein degradation in cachexia. Under normal circumstances the activity of this pathway is stimulated by glucocorticoids and thyroid hormone and is inhibited by insulin. However, in chronic disease several proinflammatory cytokines, including tumor necrosis factor-α (TNF-α) and interleukin (IL)-1, stimulate the production of ubiquitin mRNA and lead to an enhancement of muscle proteolysis.

Depletion of lean body mass in cancer cachexia is a major factor responsible for the reduced survival of cancer patients. However, whereas muscle protein synthesis is depressed, synthesis of secretory proteins such as acute-phase reactants by the liver is actually increased so that there may be no change in total body protein synthesis. In cancer patients muscle protein synthesis accounted for only 8% of total body protein synthesis versus 53% for healthy control subjects. A proteolysis-inducing factor (PIF) has recently been isolated from muscle and shown to be a sulfated glycoprotein of molecular weight (MW) 24 kDa. This factor induces protein degradation by up-regulating the ubiquitin-proteasome proteolytic pathway. PIF does not affect appetite but does activate the transcription factors NF-κB and STAT3, resulting in increased production of IL-6, IL-8, and C-reactive protein. In addition to its effects on muscle, PIF may contribute to cachexia by stimulating cytokine and acute-phase protein production.

Although anorexia is commonly associated with cachexia, it is unlikely that weight loss arises primarily from the reduction in food intake. Levels of circulating proinflammatory cytokines (e.g., IL-1β, IL-6, and TNF-α) are elevated in cancer and appear to act through central receptors in the hypothalamus to affect appetite, specifically the melanocortin receptors (MC3R and MC4R), which are receptors for α-melanocyte-stimulating hormone (α-MSH), and activation of these receptors induces anorexia (Chapter 10). It has recently been shown that blockade of the hypothalamic melanocortin receptors (MC3R and MC4R) improved food intake and reversed weight loss in rats with prostate cancer. Blocking these receptors in rats with prostate cancer increased energy balance enough to replenish nutrient stores and change hormonal patterns from those found in a fasting state to those found in the fed state despite ongoing cachexia. However, it was not known if the weight gain was due to fluid, fat, or lean tissue mass, and it was not known if blockade of

M3R and M4R decreased the sensation of illness and food aversion or truly increased appetite. These results may provide the pathway for the anorexia that accompanies cancer and cachexia and may lead to therapeutic targets to prevent this anorexia. However, the mechanism for the cachexia that accompanies cancer still remains unknown. Although it is still unknown how diseases such as cancer or AIDS induce unopposed weight loss, it is hypothesized that the cytokines synthesized and released during illness, either systemically or specifically in the brain, activate the hypothalamic catabolic pathways and interfere with the normal response to weight loss.

Studies searching for a treatment for cachexia have repeatedly shown that hypercaloric feeding is unable to increase lean mass and especially skeletal muscle mass. In these studies the weight gain was almost entirely due to fat accumulation. The supply of extra amino acids, even in the presence of sufficient fuel (fat or glucose) to meet the energy needs of the patient, simply resulted in a stimulation of amino acid oxidation with the production of urea and the diversion of the carbon skeletons into more fat synthesis. The reason for this is that the increase in protein degradation in cachexia is greater in magnitude than the possible increase in protein synthesis during hypercaloric feeding. Pharmacologic therapies that have attempted to treat this condition have included appetite stimulants (e.g., megesterol acetate, tetrahydrocannabinol), growth hormone, anabolic steroids, cytokine inhibitors, and anti-inflammatory agents. None of these have proven successful, and ultimately the strongest predictor of outcome is the severity of the underlying illness, and the best chance for treatment, at this point, is correction of the underlying disease.

METABOLIC CHANGES IN CANCER

In the early 1920s, Otto Warburg observed that tumor cells produced an unusual amount of acid and, even in the presence of oxygen, they preferentially obtained energy by converting glucose to lactic acid by anaerobic glycolysis. This is now known as the *Warburg effect*. The rate of glucose uptake by a tumor is typically an order of magnitude greater than that of a normal tissue and correlates with tumor aggressiveness and prognosis. Whereas normal tissues maintain a balance between cellular proliferation and vascularization so that adequate oxygen supply to all cells can be ensured, in solid human tumors this balance is typically lost with the result that large regions within the tumor become hypoxic. The rapidly dividing tumor cells quickly grow beyond the maximum distance from microcapillaries, which is needed for effective oxygen diffusion to cells. Chronic hypoxia occurs in any tumor mass that is greater than 100 to 200 μm away from a functional blood supply. Therefore, angiogenesis is crucial for tumor survival. However, even with angiogenesis, hypoxia is still widespread in solid tumors since cancer cells are much more prolific than the recruited endothelial cells that form the new blood vessels.

This hypoxia stimulates multiple metabolic responses, typically mediated via the production of the hypoxia-inducible transcription factor, HIF-1 (HIF-1 action as a transcription factor and its role in other situations of hypoxia stress are discussed in Chapters 6 and 14, respectively). HIF-1 induces the expression of multiple genes, typically in two categories: (a) those that improve the capacity for cellular anaerobic metabolism (e.g., glucose transporters, glycolytic enzymes) and (b) those that enhance oxygen delivery (e.g., increase red blood cell production, stimulate capillary growth). HIF-1 is a heterodimer composed of the subunits HIF-1α and HIF-1β, both belonging to the helix–loop–helix PAS family of transcription factors. HIF-1 interacts with coactivators and binds to the hypoxia-response element (HRE) of genes to induce their expression.

In addition to hypoxia, changes in glucose concentration also activate the genes for various glycolytic enzymes via the carbohydrate-response element (ChoRE), which matches the consensus binding-site sequences for MYC and HIF-1. Another transcription factor, USF2, also regulates glycolytic enzyme gene expression in response to changes in cellular glucose concentrations. Transgenic animals overexpressing c-*Myc* in the liver have increased glycolytic enzymes in the liver and also overproduce lactate. Furthermore, the gene for LDH-A, which participates in normal anaerobic glycolysis, and is frequently increased in human cancers, was recently identified as a c-*Myc*-responsive target. Thus, activation of the HIF-1, USF, and *Myc* pathways appears to be an adaptive response by cancer cells to both hypoxia and hypoglycemia. Although both HIF-1 and c-*Myc* can up-regulate the expression of glycolytic genes, other target genes differ between the two. HIF-1 induces glycolytic gene expression in hypoxic cells that also undergo growth arrest. By contrast, c-*Myc* stimulates glucose uptake, glycolysis, and overall metabolism as well as activating the cell cycle machinery for cell proliferation. Therefore, the widespread deregulation of the c-*Myc* gene in human cancers may be a major contributor to the enhanced tumor glycolysis and may confer a growth advantage to tumor cells deprived of oxygen.

The high metabolic rate of the growing tumor and the presence of multiple areas throughout the tumor that are hypoxic means that tumors produce large amounts of lactic acid and have regions that are acidic. This lactate is converted back into glucose in the liver; hence, the Cori cycle functions between tumor and liver (Fig. 9.2). Indeed, Cori cycle activity has been found to increase from 20% in normal subjects to 50% of glucose turnover in cachetic cancer patients and accounts for the disposal of 60% of the lactate produced by the tumor. There is also increased glucose synthesis from alanine and glycerol. Gluconeogenesis uses six ATP molecules for lactate to glucose conversion compared with the two ATP that are gained from glucose catabolism to lactate. Hence, extensive Cori cycle activity is energy-inefficient for the host, and this futile cycle may be responsible, at least in part, for the increased energy expenditure reported in patients with some forms of cancer.

The tumor cells in undervascularized regions either adapt to the hypoxic and acidotic stress or die. Adaptation to a low-oxygen environment has serious consequences. Hypoxia can promote a higher mutation rate and select for a more metastatic and malignant phenotype. In addition to hypoxia, acidosis may also up-regulate genes in tumors, and acidotic and/or hypoxic conditions also lead to an increased mutation frequency that may be due to a decrease in nucleotide excision repair. There also appears to be a role for hypoxia/acidity in induction of invasion and/or metastasis. Furthermore, hypoxia/acidity also confers therapeutic resistance to tumors; for example, hypoxic tumors are more resistant to radiation therapy, and low pH and hypoxia confer resistance to chemotherapy.

SUGGESTED FURTHER READING

Champe, P. C., and Harvey, R. A. (1994). *Biochemistry (Lippincott's Illustrated Reviews)*. J. B. Lippincott, Philadelphia. *This textbook provides a review of clinically relevant biochemistry.*

Choy, E. H. S., and Panayi, G. S. (2001). Cytokine pathways and joint inflammation in rheumatoid arthritis. *N Eng J Med* **344**(12):907–916.

Dang, C. V., and Semenza, G. L. (1999). Oncogenic alterations of metabolism. *Trends Biochem Sci* **24**(2):68–72. *The metabolism of tumours and the importance of hypoxia.*

Gollob, M. H., Green, M. S., Tang, A. S., Gollob, T., Karibe, A., Hassan, A., Ahmad, F., Lozado, R., Shah, G., Fananapazir, L., Bachinski, L., and Roberts, R. (2001). Identification of a gene responsible for familial Wolff-Parkinson-White syndrome. *N Eng J Med* **344**:1823–1831. *Describes one method for discovering new genes underlying disease and hypothesizes how an enzyme important in metabolism could affect the electrical pathways of the heart.*

Horton, J. D., Goldstein, J. L., and Brown, M. S. (2002). SREBPs: Activators of the complex program of cholesterol and fatty acid synthesis in the liver. *J Clin Invest.* **109**(9):1125–1131. *A review of how the transcription factors known as sterol regulatory element-binding proteins regulate mammalian lipid homeostasis.*

Kersten, S., Desvergne, B., and Wahli, W. (2000). Roles of PPARs in health and disease. *Nature* **405**:421–424. *A review of the involvement of the peroxisome proliferator-activated receptors in human diseases.*

Luster, A. D. (1998). Chemokines: Chemotactic cytokines that mediate inflammation. *N Eng J Med* **338**(7):436–445. *These two articles discuss cytokines and chemokines (chemotactic cytokines) and their roles in inflammation and human disease.*

Marks, D.B., Marks, A. D., and Smith, C. M. (1996). *Basic Medical Biochemistry: A Clinical Approach.* Williams & Wilkins, Baltimore. *This textbook provides a review of clinically relevant biochemistry.*

Randle, P. J., (1998). Regulatory interactions between lipids and carbohydrates: the glucose fatty acid cycle after 35 years. *Diabetes Metab Rev* **14**(4):263–283.

Semenza, G. L. (2001). HIF-1, O2 and the 3 PHDs: How animal cells signal hypoxia to the nucleus. *Cell* **107**:1–3. *A review of the role of the nuclear transcription factor HIF-1 and the cellular response to hypoxia.*

Stipanuk, M. H. (ed.). (2000). *Biochemical and Physiological Aspects of Human Nutrition.* W. B., Saunders, Philadelphia. *Comprehensive textbook of human nutrition as well as metabolism.*

Tisdale, M. J. (2002). Cachexia in cancer patients. *Nature Rev Cancer* **2**:862–871. *Dysregulation of appetite and metabolism in cancer and mechanisms of cachexia.*

10

HORMONES IN HUMAN METABOLISM AND DISEASE

STEVEN C. GREENWAY

HORMONAL REGULATION OF HUMAN METABOLISM

Introduction

The endocrine system, along with the nervous system, receives and disseminates information about the body's state (nutrition, development, stress, etc.) and coordinates the actions of distant and distinct cells and tissues to regulate energy metabolism. The endocrine system employs internal chemical messengers, hormones, that are produced and released by specific types of cells. Most hormones are proteins, peptides, or amino acid derivatives, whereas others are formed from cholesterol or its precursors. Hormones either act locally or are transported through the bloodstream to bind to a distant cellular receptor in order to regulate the function of another tissue. The hormone receptors and the intracellular signal transduction cascade(s) responsible for transmitting the hormonal signal into a cellular action are as varied as the hormones themselves. The endocrine system is tightly regulated, and once the stimulus for secretion stops the hormone is rapidly removed. Once secretion halts, the duration of a hormone's effect is determined by several factors: (a) the extent of hormone binding to plasma transport proteins, (b) the rate of its degradation (e.g., in the liver), and (c) the rate of its excretion.

Hormones involved in metabolism are responsible for regulating fuel utilization, including the storage of any fuel in excess of current requirements and the appropriate mobilization of those stored fuels on demand. Some hormones such as thyroid hormone and epinephrine also have a direct influence on basal metabolic rate and, thus, affect how quickly nutrients are consumed. Metabolic hormones maintain blood glucose levels within a constant, narrow range despite wide fluctuations in glucose intake by regulating glucose, lipid, and protein metabolism. Insulin and glucagon are the paramount hormonal regulators of anabolism and catabolism, respectively. There is no other counterpart for insulin, but four other hormones apart from glucagon exhibit catabolic activity: epinephrine, norepinephrine, cortisol, and growth hormone. Together these hormones and glucagon act to oppose the actions of insulin and protect the body against hypoglycemia. The body defends against a falling blood glucose concentration by first decreasing insulin secretion and then by releasing glucagon. When glucagon secretion is deficient, recovery from hypoglycemia is impaired. Epinephrine is the third line of defense against a falling blood glucose, but the role of epinephrine in normal glucose homeostasis is generally critical only when glucagon is deficient. Hypoglycemia develops or progresses when both glucagon and epinephrine are deficient and insulin is present. Growth hormone and cortisol, both of which tend to increase plasma glucose concentrations after several hours, are involved in the defense against prolonged hypoglycemia. The maintenance of blood glucose homeostasis is arguably one of the most important regulatory tasks of the body. Failure of glucose homeostasis leads to either hypoglycemia or hyperglycemia, both of which have damaging consequences. Another recently discovered hormone that plays a key role in fuel metabolism is leptin. This hormone is synthesized and secreted by adipocytes in proportion to the amount of lipid stored. It acts as a signal of body energy stores and is increased in obesity but decreases during fasting or weight loss. This chapter will focus on several of the metabolic hormones involved in fuel metabolism as well as provide an analysis of the consequences of a

Functional Metabolism: Regulation and Adaptation, edited by Kenneth B. Storey.
ISBN 0-471-41090-X

disruption in body energy metabolism and homeostasis in two increasingly common human diseases, obesity and diabetes mellitus.

Insulin

Insulin is the major anabolic hormone of the body. It promotes the utilization of dietary glucose, inhibits the mobilization of other fuels (e.g., inhibition of lipolysis), promotes the storage of glucose as glycogen or fat, and suppresses glucose synthesis by the liver. Overall, insulin acts to reduce blood glucose levels and reestablish glucose homeostasis after the ingestion of a meal. Insulin also increases cellular amino acid uptake, protein synthesis, and growth.

Insulin is a polypeptide hormone synthesized and secreted by the β cells of the islets of Langerhans in the pancreas. It is initially produced as a preprohormone that is converted in the rough endoplasmic reticulum to proinsulin (a single polypeptide folded so that disulfide bonds can form between residues on the C- and N-terminal portions). Proinsulin is cleaved in the Golgi complex to remove a connecting sequence (C peptide) from the center of the molecule to produce mature insulin, which consists of two polypeptide chains (A and B) linked by the disulfide bonds.

A variety of signals generated by the ingestion of food help "prime" the pancreatic β cells to release insulin. These include hormones associated with digestion such as cholecystokinin, secretin, gastrin, and gastric inhibitory peptide, as well as glucagon and glucagon-like polypeptide I and some amino acids (especially arginine). However, actual release of insulin is controlled by the levels of blood glucose (Fig. 10.1). Glucose enters the β cells via the glucose transporter protein (GLUT2), where it is phosphorylated by glucokinase and subsequently oxidized with the generation of adenosine 5′-triphosphate (ATP). The increase in ATP closes ATP-sensitive potassium channels resulting in a depolarization of the β-cell plasma membrane. The hypoglycemic effect of the sulfonylureas, which are widely used in the treatment of type 2 diabetes, is due to their inhibition of these ATP-sensitive potassium channels, triggering the same effect as high ATP. Depolarization activates a voltage-dependent calcium channel that elevates intracellular calcium levels and stimulates insulin-containing vesicles to fuse with the plasma membrane and release insulin. As insulin is secreted, new insulin molecules are synthesized so that secretion is maintained until blood glucose levels fall. Insulin levels in the blood rise to their highest 30–45 min after ingesting a meal, and levels return to baseline approximately 2 h later. Insulin is rapidly removed from the circulation (half-life of 6 min) and degraded by the liver and to a lesser extent by kidney and skeletal muscle.

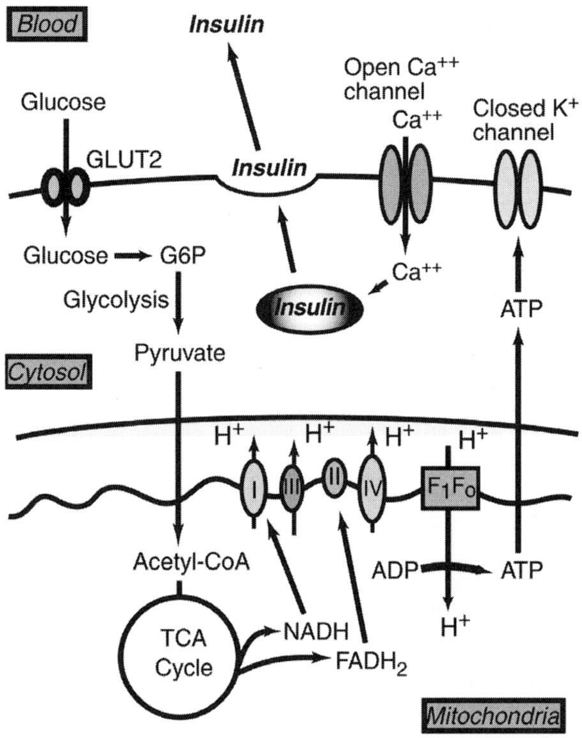

Figure 10.1 Mechanism of insulin secretion. Glucose enters the pancreatic β cell via the glucose transporter GLUT2, where it is phosphorylated by glucokinase and subsequently metabolized by glycolysis and the TCA cycle to produce NADH and $FADH_2$, which donate electrons to the electron transport chain. Protons (H^+) are pumped out by complexes I, III, and IV creating a proton electrochemical gradient. When protons reenter the mitochondria through ATP synthase, ATP is generated from ADP. ATP is pumped out of the mitochondria in exchange for ADP by the adenine nucleotide carrier (not shown). An increase in the cytosolic ATP : ADP ratio inhibits the ATP-sensitive potassium channel (K^+) leading to depolarization of the cell, which opens voltage-gated Ca^{2+} channels. The increase in the intracellular Ca^{2+} concentration stimulates insulin-containing vesicles to fuse with the plasma membrane and release insulin.

Circulating insulin binds to a specific receptor on target tissues. The insulin receptor is encoded by a gene located on human chromosome 19 and is composed of two α subunits and two β subunits linked by disulfide bonds. Humans lacking insulin receptors are severely growth-retarded at birth, which indicates the importance of insulin as a fetal growth factor in humans. The insulin receptor is a receptor tyrosine kinase, but, unlike other receptors of this class that are activated through a process of ligand-induced dimerization, the insulin receptor exists as a preformed dimer, and the binding of its ligand, insulin, results in a conformational change. Insulin binding to the extracellular α subunit of the receptor brings the two α subunits closer together, and this enables ATP to bind to the intracellular domain of the β

subunit. ATP binding activates the β subunit, a tyrosine kinase, to phosphorylate itself, and autophosphorylation increases the kinase activity of the insulin receptor by three orders of magnitude. This then enables the insulin receptor to phosphorylate a variety of intracellular proteins that transduce the insulin signal into metabolic actions (Fig. 10.2). Members of the family of insulin receptor substrate (IRS) proteins, of which four have been identified to date (IRS1–4), appear to be mandatory for insulin-stimulated glucose uptake (mediated by the insulin-stimulated translocation of GLUT4 glucose transporters from the cytoplasm to the plasma membrane), but phosphorylation of other insulin receptor targets (e.g., Gab-1, c-Cbl, and Shc) are also required to facilitate glucose uptake. Tyrosine phosphorylation by the insulin receptor increases the affinity with which these proteins bind other signaling molecules, and the formation of these protein–protein complexes activates various signaling pathways.

Tyrosine phosphorylation of the IRS proteins induces their binding to the Src-homology 2 (SH2) domain of p85, the regulatory subunit of phosphatidylinositol-3-kinase (PI3K), which, in turn, leads to an interaction with the catalytic domain of PI3K, p110, at the plasma membrane and stimulates increased conversion of phosphatidylinositol-4,5-bisphosphate (PIP2) to phosphatidylinositol-3,4,5-trisphosphate (PIP3). Blockade of PI3K activation and PIP3 synthesis has been shown to abolish the ability of insulin to stimulate glucose transport in muscle and adipose tissue, and expression of constitutively active p110 mutants leads to elevated rates of glucose transport. PIP3 acts as an intracellular messenger, causing activation of phosphatidylinositol-dependent kinases, changes in intracellular trafficking, and growth stimulation. PIP3 generated in response to insulin stimulates the phosphoinositide-dependent kinase (PDK) that phosphorylates and activates two classes of serine/threonine kinases, Akt (also known as protein kinase B) and the atypical protein kinase C (PKC) isoforms ζ and λ. The atypical PKCs may play a role in insulin-controlled glucose transport and can activate the mitogen-activated protein (MAP) kinase pathway and the transcription factor, nuclear factor κB (NFκB), leading to increased expression of selected genes. Despite the general agreement that PI3K activity is necessary for insulin-stimulated glucose uptake, additional signals are required. Both Akt and PKC are important in glucose uptake, but the missing links between these enzymes and GLUT4 translocation remain unknown. Phosphorylated Akt can also translocate to the nucleus to activate gene expression and has the ability to phosphorylate proteins that regulate lipid synthesis, glycogen synthesis, cell survival, and protein synthesis. Targets of Akt include glycogen synthase kinase 3 (GSK3), which is phosphorylated and inhibited by Akt. Active GSK3 phosphorylates and inhibits glycogen synthase (GS), but inhibition of

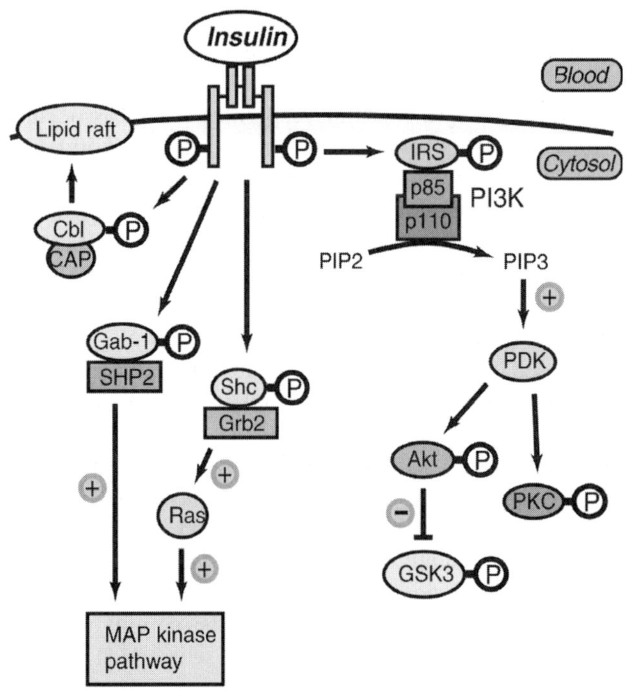

Figure 10.2 Insulin receptor and activated intracellular pathways. Binding of insulin to its receptor stimulates the tyrosine kinase activity of the insulin receptor, which leads to the autophosphorylation of the receptor in addition to the tyrosine phosphorylation of several intracellular substrates, including Cbl, Gab1, Shc, and the IRS family. Phosphorylation of Cbl is facilitated by the adaptor protein CAP, and this complex subsequently translocates to the lipid raft subdomain of the plasma membrane. Phosphorylation of Gab1 allows it to interact with the tyrosine phosphatase SHP2, which leads to activation of the MAP kinase pathway. This pathway is also activated when Shc is phosphorylated by the insulin receptor. Phosphorylated Shc interacts with the adaptor protein Grb2 causing activation of Ras, which is an upstream activator of the MAP kinase family. The IRS family has four members that can interact with a number of intracellular proteins but most notably bind to p85, the regulatory subunit of PI3K. The interaction of phosphorylated IRS and p85 allows the catalytic subunit of PI3K, p110, to catalyze the conversion of PIP2 to PIP3. PIP3 binds to both PDK and Akt, co-localizing them to the plasma membrane and allowing PDK to activate Akt. Akt in turn phosphorylates and inhibits GSK3, resulting in the dephosphorylation of substrates of GSK3, which leads to increased glycogen and protein synthesis. PDK also phosphorylates the atypical protein kinase C isoforms ζ and λ (PKC).

GSK3 allows the dephosphorylation and activation of GS, which contributes to the insulin-induced stimulation of glycogen synthesis. For additional discussion of the insulin receptor and the signal transduction pathway that it triggers see Chapter 5.

The PI3K-mediated pathway may not be sufficient by itself to transduce the effects of insulin and, indeed, several signaling pathways are activated in parallel by insulin. One of these involves the proto-oncogene c-*Cbl* and the Cbl-activating protein (CAP). Tyrosine phosphorylation of c-*Cbl* by the insulin receptor leads to its association with CAP via its SH3 domain. Following phosphorylation, the c-Cbl/CAP complex translocates to a lipid raft or caveolae and subsequent signal transduction may contribute to GLUT4 translocation. Lipid rafts are dynamic assemblies of proteins and lipids that float freely within the liquid-disordered bilayer of cellular membranes but can also cluster to form larger, ordered platforms. Rafts are composed of sphingolipids and cholesterol in the outer exoplasmic leaflet and are connected to phospholipids and cholesterol in the inner cytoplasmic leaflet of the lipid bilayer. These assemblies are fluid but more ordered and tightly packed than the surrounding bilayer. Two further targets for the insulin receptor are Gab-1 and Shc. When the insulin receptor phosphorylates Gab-1, it interacts with the SH2 domain-containing tyrosine phosphatase SHP2, whereas phosphorylation of Shc permits its interaction with the adaptor protein Grb2, which leads to activation of the proto-oncogene *Ras*. Both of these pathways lead to activation of the MAP kinase cascade, a pathway known to be involved in many cellular events (especially gene expression) but whose role in the action of insulin remains largely unknown.

As the primary director of anabolism, insulin has a multitude of effects on carbohydrate, lipid, and protein metabolism in numerous tissues. The primary targets for insulin are skeletal muscle, cardiac muscle, adipose tissue, and liver. Overall, insulin inhibits catabolic reactions in favor of anabolic reactions that act to store biomolecules.

Dietary carbohydrate stimulates insulin release, a response that serves to limit the rise in blood glucose during the absorption of foodstuffs. The increase in plasma insulin promotes glucose uptake by insulin-sensitive tissues and leads to glucose oxidization for energy production, glucose storage as glycogen (in liver and muscle), or glucose utilization for biosynthesis, particularly for *de novo* lipogenesis. Glucose uptake is the rate-limiting step in glucose utilization and storage. Insulin stimulates the transport of glucose into myocytes and adipocytes by stimulating the translocation of a specific glucose transporter isoform, GLUT4, from the cytosol to the cell surface. This action of insulin causes a 10- to 40-fold increase in cellular glucose uptake. Upon entering the cell, glucose is rapidly phosphorylated by hexokinase, whose expression is also increased by insulin, and then glucose-6-phosphate (G6P) is either processed via glycolysis to various fates or stored as glycogen. This latter is mediated by insulin activation of glycogen synthetase (GS). G6P and UDP–glucose allosterically activate GS so that the glucose that insulin has driven into the cell can be efficiently stored as glycogen. Insulin inhibits the protein kinases, including PKA and GSK3, that can phosphorylate and inactivate GS but also activates protein phosphatase 1 (PP1), which dephosphorylates and activates GS. Activation of PP1 also results in dephosphorylation and inactivation of glycogen phosphorylase (GP), thereby inhibiting glycogenolysis when glycogen biosynthesis is active. Efficient utilization of ingested glucose is ensured by the inhibition of glucose-6-phosphatase (G6Pase) and the inactivation of GP whenever dietary glucose is available. The combined inhibition of these two enzymes along with activation of GS accounts for the rapid repletion of glycogen stores after a meal.

In addition to controlling the activities of metabolic enzymes via reversible protein phosphorylation, insulin also regulates the expression of a number of genes that encode liver enzymes. For example, insulin affects the expression of hepatic gluconeogenic and glycogenic enzymes by both transcriptional and posttranslational mechanisms. Insulin decreases transcription of the genes encoding gluconeogenic enzymes such as phosphoenolpyruvate carboxykinase (PEPCK), fructose-1,6-bisphosphatase, and G6Pase and increases transcription of glycolytic enzymes such as glucokinase and pyruvate kinase and lipogenic enzymes such as fatty acid synthase (FAS) and acetyl-CoA carboxylase (ACC). Recent studies suggest that many of these insulin-mediated changes in gene transcription may be mediated by an increase in levels of the sterol-response element binding protein (SREBP) transcription factor, whose messenger ribonucleic acid (mRNA) levels are elevated by insulin (see Chapter 5).

With regards to lipid metabolism, insulin increases lipoprotein lipase (LPL) activity that promotes digestion of very low density lipoproteins (VLDL) and chylomicrons in the bloodstream so that fatty acids are taken up and stored as triglyceride (TG) in adipose tissue. Insulin further facilitates TG synthesis by increasing glucose uptake into adipocytes where it can have two fates, conversion to glycerol-3-P to provide the "backbone" for TG synthesis or use as a substrate for fatty acid biosynthesis. In addition to promoting the storage of glucose as TG in adipose tissue, insulin also profoundly inhibits lipolysis in adipocytes, primarily through inhibition of hormone-sensitive lipase (HSL). This action inhibits the release of nonesterified fatty acids (NEFA) from adipose tissue, thereby lowering plasma NEFA levels and releasing NEFA-mediated inhibition of glycolysis and pyruvate oxidation (as well as the inhibitory effect of NEFA on insulin signaling). Most, if not all, of these insulin-dependent changes in enzyme activities are mediated by protein dephosphorylation reactions, due to a combination of protein kinase inhibition and phosphatase activation. As with carbohydrate metabolism, the reactions involved in the storage of macromolecules are favored, and the reactions of macromolecule catabolism are efficiently inhibited so that futile cycles of storage and breakdown are avoided. Overall, with regards to carbohydrate and lipid metabolism, the rise in serum glucose and insulin that occurs after a meal, in combination with a reduced concentration of circulating NEFA, results in an increase in the portion of the body's metabolic energy that is derived from carbohydrate oxidation and a decrease in that derived from fat oxidation.

Insulin also has an anabolic effect on protein metabolism by stimulating amino acid uptake and protein synthesis and by inhibiting protein breakdown. Muscle wasting occurs in insulin-deficient diabetic patients, and the reversal of wasting with insulin treatment is a clear demonstration of the anabolic effect of insulin. Studies with type 1 diabetic (insulin-dependent) patients have shown that during insulin deprivation both whole-body protein breakdown and synthesis were elevated, but net whole-body protein loss occurred because protein breakdown exceeded protein synthesis. When insulin is given, skeletal muscle protein synthesis seems to be insensitive to insulin in humans, but the decrease in whole-body protein breakdown is largely the result of the decrease in skeletal muscle protein breakdown. However, this may also be related to the lack of a sufficient amino acid supply for increased protein synthesis.

Glucagon

Of the counterregulatory hormones that act in opposition to insulin, glucagon has the primary role and is, therefore, the major hormone that regulates fuel mobilization and catabolism. Insulin and glucagon secretion are normally smoothly and oppositely regulated to ensure that blood glucose is maintained within a narrow range that avoids the perils of either hypo- or hyperglycemia. Glucagon is a polypeptide hormone synthesized as a prohormone in the α cells of the islets of Langerhans in the pancreas. Preproglucagon is synthesized on the rough endoplasmic reticulum and converted to proglucagon, which is proteolytically cleaved to produce glucagon. It is rapidly metabolized by liver and kidneys with a plasma half-life of 3 to 5 min.

Glucagon release is regulated primarily by glucose and insulin. Hypoglycemia inhibits insulin release and stimulates glucagon secretion, whereas hyperglycemia has an inhibitory effect on glucagon secretion that is independent of insulin but with a mechanism that is poorly understood. Insulin can also inhibit glucagon release directly since venous blood flow from the β cells passes by the α cells where stimulation of the α-cell insulin receptor down-regulates glucagon production by an unknown mechanism. Likewise, glucagon can inhibit insulin secretion directly via its receptor on the β cell that likely acts in a cyclic adenosine 3′,5′-monophosphate (cAMP)-dependent mechanism. In general, insulin release becomes negligible or absent when plasma glucose levels are reduced to 80 to 85 mg/dL (4.4 to 4.7 mM). Above this level plasma insulin increases in near linear fashion as a function of plasma glucose concentration. Glucagon release begins to occur when plasma glucose concentration falls to approximately 50 mg/dL (2.7 mM), and maximal suppression of glucagon release occurs at a plasma glucose of 150 mg/dL (8.3 mM). In response to falling plasma glucose levels, glucagon is secreted from α cells into the hepatic portal circulation and is believed to act exclusively on the liver under physiological conditions. Glucagon activates liver glycogenolysis, and to some extent gluconeogenesis, and increases hepatic glucose production within minutes. This increase is transient and glucose production returns toward basal rates over about 90 min. The effect is transient because of glucose-induced insulin secretion and the autoregulatory effect of hyperglycemia, although other factors may be involved. Because glucagon, unlike epinephrine, does not mobilize gluconeogenic precursors, glucagon alone has relatively little impact on gluconeogenesis.

The primary stimulus for the release of glucagon from the pancreatic α cell is low blood glucose, but glucagon is also secreted in response to amino acids and epinephrine. Epinephrine acts to increase glucagon secretion in times of stress regardless of serum glucose level so that both hormones can stimulate liver glycogenolysis and ensure a maximal output of glucose to fuel the "fight or flight" reflex.

The mechanism of action for glucagon is better understood than that of insulin (Fig. 10.3). Once released into the bloodstream glucagon binds to cell surface receptors on target cells that are coupled to adenylate cyclase (AC) by G proteins (see Chapter 4 for more information on G-protein-mediated signaling pathways). A stimulatory guanine-nucleotide-binding protein (G_s) activates AC and is opposed by an inhibitory protein (G_i). It is the interaction between these two G-protein subunits, the stimulatory and the inhibitory, that determines AC activity. AC produces cAMP, which triggers the inactive tetrameric form of cAMP-dependent protein kinase (PKA) to dissociate and release the active catalytic subunits. PKA catalytic subunits phosphorylate key enzymes to either activate (e.g., GP) or inactivate (e.g., GS) them, with the net result being increased catabolism. For example, the effects on GP and GS stimulate glycogen breakdown with a concomitant inhibition of glycogen synthesis.

Like insulin, glucagon has effects on all forms of energy metabolism. Glucagon elevates blood sugar by stimulating the breakdown of liver glycogen (but not muscle glycogen). Glucagon also stimulates the hepatic oxidation of fatty acids and the formation of ketone bodies from acetyl-CoA, but the lipolytic effect of glucagon on human adipose tissue is minimal. Just as glucagon and insulin oppose each other in the reversible phosphorylation of metabolic enzymes, they also have opposing effects on

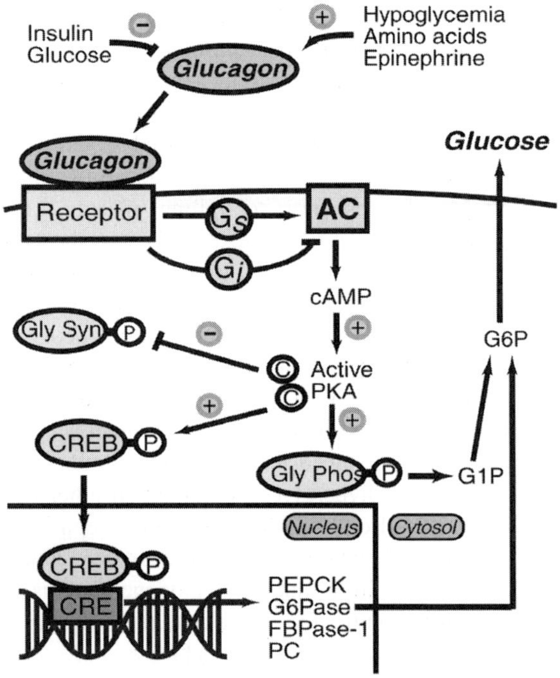

Figure 10.3 Mechanism of glucagon action. Glucagon release from the pancreatic α cells is repressed by insulin and glucose but stimulated by low blood sugar (hypoglycemia), amino acids, and epinephrine. After its release glucagon binds to its cell surface receptor, which is coupled to adenylate cyclase (AC) by G proteins. AC is stimulated via a stimulatory guanine-nucleotide-binding protein (G_s) (and opposed by an inhibitory protein G_i). Active AC produces cAMP, which activates PKA by stimulating the dissociation of the inactive tetramer to release the active PKA catalytic subunits. PKA phosphorylates multiple protein targets with effects that include inactivation glycogen synthetase (Gly Syn) to inhibit glycogen synthesis and activation of phosphorylase kinase, which in turn phosphorylates and activates glycogen phosphorylase (Gly Phos) to stimulate glycogen breakdown and glucose release into the bloodstream. The phosphorylation of the transcription factor, CREB, increases the transcription of enzymes involved in gluconeogenesis (PEPCK, G6Pase, FBPase-1, PC).

gene transcription. For example, elevation of hepatic cAMP levels under glucagon stimulation increases PEPCK gene expression (supporting gluconeogenesis) during hypoglycemia (via the actions of CREB and PGC-1 transcription factors: see Chapter 4), whereas, as noted earlier, insulin inhibits expression of gluconeogenic genes.

Epinephrine and Norepinephrine

The catecholamines, epinephrine, norepinephrine, and dopamine, are the hormones secreted by the sympathetic nervous system. Epinephrine is produced by the medulla of the adrenal gland and released into the systemic circulation. Norepinephrine is found in the peripheral nerve terminals adjacent to target cells within innervated tissue. Dopamine acts primarily as a neurotransmitter in the central nervous system and will not be discussed here. Norepinephrine and epinephrine (formed by the methylation of norepinephrine by the enzyme phenylethanolamine-*N*-methyltransferase) are secreted in response to stress, including pain, hemorrhage, exercise, hypoxia, and hypoglycemia. Catecholamines act widely in the body affecting many cardiovascular (e.g., the fight or flight response) and metabolic processes and have an important role in preventing hypoglycemia. Catecholamine release is stimulated by the release of acetylcholine from preganglionic nerves, which depolarizes the receptor cell causing exocytosis of the catecholamine-containing granules. Once released, catecholamines are rapidly removed from the circulation (half-life of 2 min) by several mechanisms including reuptake by secretory cells, uptake by effector cells, or degradation by the liver. The amount of endogenous catecholamines released at peripheral nerve endings and the plasma concentration of epinephrine are the main determinants of the physiologic responses to activation of the sympathetic nervous system.

Epinephrine and norepinephrine have metabolic effects directed toward mobilization of fuels from their storage sites for oxidation by cells in order to meet the increased energy requirements of acute and chronic stress. They also simultaneously suppress insulin secretion. These hormones act directly on liver, adipose tissue, skeletal muscle, and the α and β cells of the pancreas to influence fuel metabolism. The actions of epinephrine are both direct and indirect and are mediated through both α-adrenergic and β-adrenergic receptors (Fig. 10.4). The α- and β-adrenergic receptors are transmembrane receptors that interact with G proteins (see Chapter 4). The response to epinephrine depends on the particular receptors and G proteins expressed by each cell type. For example, binding of epinephrine to α_1-adrenergic receptors leads to activation of phospholipase C (PLC) via the G protein G_q, whereas if epinephrine binds to the α_2-adrenergic receptor the G protein, G_i, is released and able to inhibit adenylyl cyclase to decrease the formation of cAMP. By contrast, binding of epinephrine to a β-adrenergic receptor stimulates G_s to activate adenylyl cyclase and increase the amount of intracellular cAMP.

Epinephrine acts independently of changes in other hormones or substrates to increase hepatic glycogenolysis and gluconeogenesis. In humans, the hepatic effect is mediated predominantly through β_2-adrenergic mechanisms, although a small direct α-adrenergic stimulation of hepatic glucose production has been reported. Epinephrine also increases the concentration of available gluconeogenic precursors (e.g., lactate, alanine, and glycerol) to produce an increase in glucose production. In addition to increasing

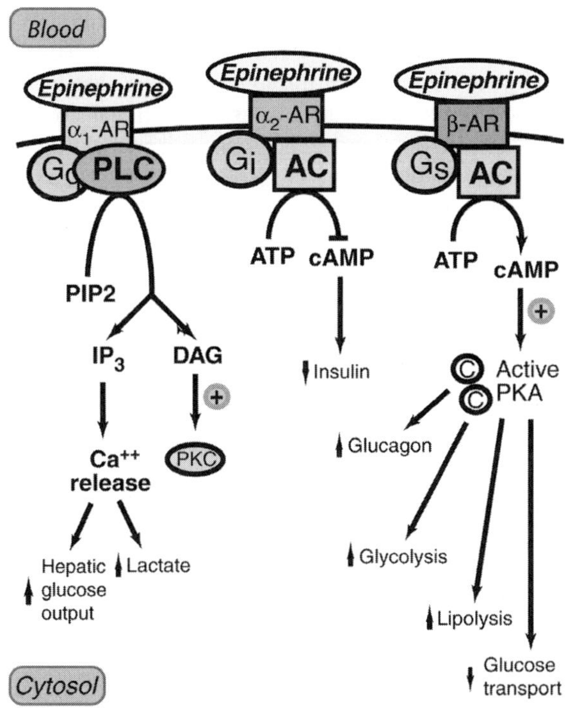

Figure 10.4 Catecholamine receptors. Epinephrine and norepinephrine bind to three distinct adrenergic receptors whose tissue distribution determines the cellular effects of epinephrine or norepinephrine stimulation on that tissue. Activation of the α_1-adrenergic receptor leads to activation of phospholipase C (PLC) via the G-protein G_q resulting in production of inositol triphosphate (IP_3) and diacylglycerol (DAG) from PIP2. IP_3 stimulates the release of Ca^{2+} from the endoplasmic reticulum, which stimulates the liver to increase glucose output and increases lactate production by skeletal muscle. Binding of epinephrine to the α_2-adrenergic receptor releases the G-protein G_i, which inhibits adenylyl cyclase (AC) to decrease the formation of cAMP. Conversely, binding of epinephrine to a β-adrenergic receptor stimulates G_s to activate AC, which increases the amount of intracellular cAMP resulting in dissociation of the active catalytic subunit from the regulatory subunit of PKA. Active PKA has multiple effects, including increasing levels of glucagon, which leads to increased glycogenolysis and increased gluconeogenesis. Glycolysis is also activated in skeletal muscle, which increases the availability of lactate and alanine, important precursors for gluconeogenesis in liver and kidney. PKA increases lipolysis in adipose tissue, which mobilizes glycerol (another gluconeogenic precursor) and fatty acids. The net effect of epinephrine release is an increase in blood glucose concentration.

blood glucose, α_2-adrenergic limitation of insulin secretion is an important indirect hyperglycemic action of epinephrine. β-adrenergic stimulation of glucagon secretion also occurs, but its contribution to the hyperglycemic effect of epinephrine appears to be minor under physiologic conditions. Epinephrine also stimulates the breakdown of adipose cell lipids by activating HSL to convert TG into free fatty acids and glycerol.

Norepinephrine exerts hyperglycemic actions by mechanisms assumed to be similar to those of epinephrine, except that norepinephrine is released primarily from the terminals of sympathetic postganglionic neurons. These terminals are adjacent to adrenergic receptors on target cells within the innervated tissues. Electrical stimulation of hepatic sympathetic nerves decreases glycogen content, increases glucose release, and causes hyperglycemia in animals and in humans. By contrast, parasympathetic stimulation increases the hepatic glycogen content and decreases hepatic glucose release.

Epinephrine and norepinephrine share some effects but also have some different effects on glucose metabolism. Epinephrine increases hepatic glucose production and inhibits insulin-induced glucose uptake in skeletal muscle and adipose tissues, leading to a rapid increase in plasma glucose concentration. By contrast, norepinephrine contributes less to hepatic glucose production but increases glucose uptake and its utilization in skeletal muscle and adipose tissues by the potent β_3-adrenergic receptor that is independent of insulin action. Although the relative importance of the effects of norepinephrine and epinephrine on fat metabolism are still to be resolved, activation of the sympathetic nervous system increases lipolysis and decreases TG-rich lipoprotein accumulation in white adipose tissue, again mediated by the β_3-adrenergic receptor, leading to decreased fat stores.

Central neural mechanisms regulate appetite and food intake, but the sympathetic nervous system should function properly to compensate for increased energy intake by increasing energy expenditure. Mice with a genetic disruption of the specific neurotransmitter receptors involved in sympathetic effects are usually obese and have impaired glucose tolerance. Dysfunction of the sympathetic nervous system should therefore be considered, in addition to food intake, as a primary cause of central nervous system (CNS)-mediated obesity (see later section).

Cortisol

Cortisol is the primary glucocorticoid in humans. This hormone is essential for life since it maintains critical processes during times of stress and limits the reactions of inflammation (see Text Box 10.1). Most of its effects are permissive, meaning that cortisol is not responsible for the initiation of metabolic processes but is necessary for their full expression. Cortisol is important in the stress response and blood levels of cortisol increase in response to virtually any type of stimulus that threatens, or is perceived to threaten, body homeostasis. Cortisol alters metabolic, endocrine, nervous, cardiovascular, and immunologic systems to promote survival under stress. The

TEXT BOX 10.1 HORMONES AND THE STRESS RESPONSE

The stress response can be defined as a reaction of the nervous and endocrine systems that allows an organism to adapt to perturbations of homeostasis resulting from either changes in the external environment (e.g., temperature change, being attacked by a bear) or in the internal milieu (e.g., injury, disease, malnutrition). Stress induces a systemic reaction to tissue injury and/or infection known as the acute-phase response that comprises changes in behavior, body temperature, and production and release of cytokines and hormones. The acute-phase response is considered to be a crucial step in reestablishing homeostasis and involves both induction and suppression of liver proteins. Some of these proteins are necessary for tissue repair and clearance of cell debris or endotoxins as well as other essential homeostatic functions. They are induced by a synergistic action of glucocorticoids and cytokines such as IL-1, IL-6, and TNF-α. Catecholamines and cortisol, released in response to stress, suppress the inflammatory and immune responses and prevent them from damaging healthy tissue in their zeal to repair damaged tissue. However, suppression of the immune system can also have negative effects, especially during chronic stress. Therefore, hormones such as GH, IGF-I, TH, and prolactin may act as immunostimulatory mediators to counteract the actions of the immunosuppressive hormones. As a result of their immunoenhancing effects, the cumulative and potentially immunosuppressive effects of stress may be avoided.

hypothalamic–pituitary–adrenal (HPA) axis functions to respond to a variety of internal and external stresses that challenge homeostasis. The system comprises the neuronal pathways linked to release of catecholamines from the adrenal medulla in the fight or flight response (see previous section) and the control of cortisol production by the adrenal cortex.

Basal secretion of cortisol is necessary for the normal function of most tissues and even small deviations from normal circulating levels of these steroids produce a wide variety of biochemical and physiological changes. Adrenocorticotropic hormone (ACTH) is the principal hormone stimulating cortisol synthesis and secretion. ACTH is synthesized within the anterior pituitary as part of a much larger precursor, pro-opiomelanocortin (POMC). Pituitary ACTH release is stimulated primarily by corticotropin-releasing hormone (CRH) and to a lesser extent by arginine vasopressin (AVP). CRH is synthesized in neurons within the paraventricular nucleus of the hypothalamus. When released, CRH binds to specific receptors to stimulate POMC gene transcription through a process that includes the activation of adenylate cyclase. Additional control is provided through an endogenous circadian rhythm, stress, and feedback inhibition by cortisol itself.

The proinflammatory cytokines, notably interleukin-1 (IL-1), interleukin-6 (IL-6), and tumor necrosis factor α (TNF-α), also increase ACTH secretion, which allows the HPA axis to respond to an inflammatory stimulus. Physical stresses (e.g., fever, surgery, burn injury, hypoglycemia, hypotension, exercise) increase ACTH and cortisol secretion through central actions mediated by CRH and AVP. Food ingestion is a further stimulus to ACTH secretion, and ACTH is secreted in a pulsatile fashion with a circadian rhythm so that levels are highest on waking and decline throughout the day, reaching their lowest values in the evening. An important aspect of CRH and ACTH secretion is the negative feedback control exerted by glucocorticoids. Cortisol inhibits POMC gene transcription in the anterior pituitary and CRH and AVP mRNA synthesis and secretion in the hypothalamus. This negative feedback allows cortisol to play a key regulatory role in the basal control of HPA axis activity and in the termination of a stress response by acting on the hypothalamus and other brain structures to ultimately inhibit ACTH release. This limits the amount of time that tissues are exposed to the catabolic, lipogenic, antireproductive, and immunosuppressive effects of cortisol.

Once released over 90% of circulating cortisol is bound, predominantly to the α_2-globulin cortisol-binding globulin (CBG), and the circulating half-life of cortisol varies between 70 and 120 min. Free cortisol enters target cells by passive diffusion and binds to the intracellular glucocorticoid receptor (GR), which belongs to the thyroid/steroid hormone receptor superfamily of transcription factors (Fig. 10.5; see also Chapter 6). The inactive GR is complexed to heat shock proteins (HSP 90 and HSP 70) in the cytosol, but after binding of cortisol the GR–cortisol complex dissociates from HSP and translocates to the nucleus where homodimers of GR bind to the glucocorticoid response element (GRE) to activate gene transcription. The activated receptors can also inhibit the actions of other transcription factors by direct protein–protein interactions and can affect mRNA stability. In keeping with the diverse array of actions of cortisol, many hundreds of glucocorticoid responsive genes have been identified. However, two important mechanisms whereby cortisol invokes an anti-inflammatory effect involve interactions with two important proinflammatory transcription factors, activator protein 1 (AP-1) and NFκB. AP-1 is composed of c-fos and c-jun subunits and GR binding to c-jun represses AP-1 activation. In a similar fashion the binding of activated GR to NFκB prevents initiation of an inflammatory process. Further-

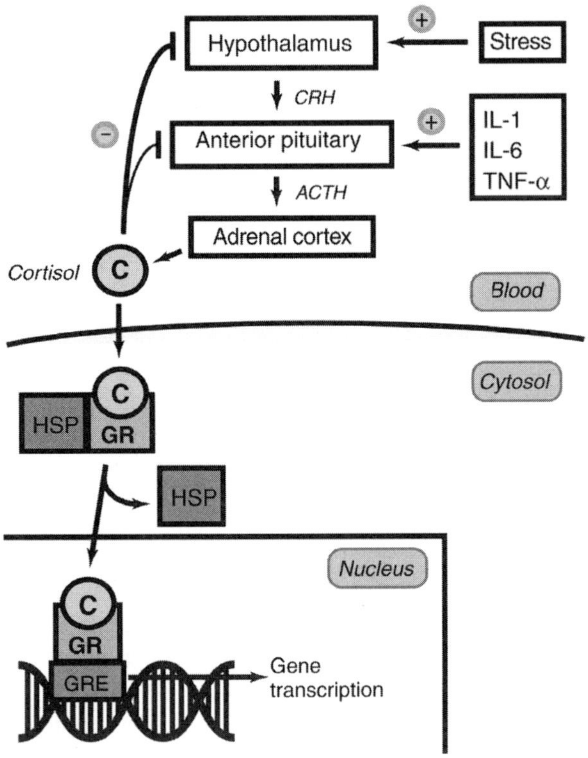

Figure 10.5 Hypothalamic–pituitary–adrenal axis and cortisol receptor. In response to stress (e.g., fever, surgery, burn, hypoglycemia) CRH released from the hypothalamus or proinflammatory cytokines (IL-1, IL-6, TNF-α) stimulate production of ACTH, which acts on the adrenal cortex to release cortisol (C). Cortisol provides negative feedback to decrease further CRH and ACTH release. Cortisol diffuses through the plasma membrane of target cells and binds to the glucocorticoid receptor (GR), which is complexed with heat shock proteins (HSP) in the inactive state. Binding of cortisol activates the GR, which dissociates from HSP and enters the nucleus, where it either interferes with the binding of proinflammatory transcription factors (see text) or binds to the glucocorticoid response element (GRE), resulting in the transcription of glucocorticoid-responsive genes.

more, glucocorticoids, via the GR, induce expression of the inhibitory protein IκB, which binds and inactivates NFκB, and both GR and NFκB compete for the limited availability of coactivators that include cyclic adenosine monophosphate response element binding protein (CREB) binding protein and steroid receptor coactivator 1.

As one of the counterregulatory or glucose-elevating hormones, cortisol increases blood glucose through actions on glycogen, protein, and lipid metabolism. However, unlike glucagon and epinephrine, the hyperglycemic effects of cortisol do not appear for several hours. In the liver, cortisol stimulates glycogen deposition by increasing GS gene expression and inhibiting GP. Hepatic glucose output increases via the gene transcription of key enzymes involved in gluconeogenesis, principally G6Pase and PEPCK. In peripheral tissues (muscle, fat), cortisol inhibits glucose uptake and utilization. In adipose tissue, lipolysis is activated and results in the release of NEFA into the circulation. An increase in total circulating cholesterol and TGs is observed, but high-density lipoprotein (HDL) cholesterol levels fall. Glucocorticoids also have a permissive effect on other hormones, including catecholamines and glucagon. The net effect is to cause insulin resistance and an increase in blood glucose concentrations at the expense of protein and lipid catabolism.

In response to chronic stress, cortisol acts to make fuel available so that when epinephrine is released in the acute situation the organism can fight or flee. Cortisol increases the concentration of fatty acids in the blood and the released glycerol is available for gluconeogenesis. The major catabolic effect of cortisol is to facilitate the conversion of protein in muscles and connective tissue into glucose and glycogen. To do this, cortisol stimulates the breakdown of muscle protein to increase plasma amino acid levels, which can then be used for gluconeogenesis. Chronically, the effects of excess cortisol on adipose tissue are more complex, at least in humans, in whom the deposition of visceral or central adipose tissue is stimulated. In addition to inducing insulin resistance in muscle tissue, cortisol causes catabolic changes in muscle, skin, and connective tissue. Cortisol also affects the expression of other hormones. Although glucocorticoids stimulate growth hormone (GH) gene transcription *in vitro*, glucocorticoids in excess inhibit linear skeletal growth, probably as a result of catabolic effects on connective tissue, muscle, and bone and inhibition of the effects of IGF-I. Cortisol also suppresses the thyroid axis, probably through a direct action on thyroid-stimulating hormone (TSH) secretion. In addition, cortisol inhibits 5′-deiodinase activity that mediates the conversion of thyroxine to the active hormone, triiodothyronine. Cortisol also acts centrally to inhibit gonadotropin-releasing hormone pulsatility and release of luteinizing hormone (LH) and follicle-stimulating hormone (FSH).

In addition to metabolic and hormonal effects, cortisol suppresses inflammation and prevents overactivity of the immune system. Cortisol inhibits production of cytokines and other mediators of inflammation but also stimulates the acute-phase response and influences T- and B-cell function. Glucocorticoids affect immune function, and the immune system affects cortisol release, which suggests that immune cells limit their own activity by stimulating the secretion of cortisol. Glucocorticoids are important drugs in the treatment of chronic inflammation and autoimmune disease and both exogenous and endogenous glucocorticoids (i.e., cortisol) act to limit the immune/inflammatory response so that the deleterious effects of a hyperactive immune system, for example, autoimmune disease, are avoided.

In conditions of chronic glucocorticoid excess, such as Cushing syndrome, there is significant muscle wasting and weakness, progressive loss of protein, atrophy and weakness of muscles, thinning of skin, hyperglycemia, and decreased sensitivity to insulin. Bone formation is reduced and less calcium is absorbed and more is excreted into the urine, leading to osteopenia or osteoporosis. Cortisol also stimulates appetite and the laying down of additional fat in the central areas. With an excess of cortisol, the extremities lose fat and muscle, whereas the trunk and face become fatter. The mechanism for this fat redistribution is unknown.

A deficiency of cortisol (hypoadrenalism) can either be primary (glucocorticoid deficiency due to adrenal disease) or secondary (ACTH deficiency). An example of primary hypoadrenalism is Addison disease, which occurs rarely (incidence of 0.8 per 100,000) usually secondary to autoimmune destruction of the adrenal glands. There is increased skin pigmentation (possibly due to prolonged, increased stimulation of the melanocortin-2 receptor by ACTH), and clinical features are related to the rate of onset and severity of adrenal deficiency. In many cases, the disease has an insidious onset and a diagnosis is made only when the patient presents with an acute crisis during an intercurrent illness. Acute adrenal insufficiency, also known as an Addisonian crisis, is a medical emergency manifesting as hypotension and acute circulatory failure. Alternatively, the patient may present with vague features of chronic adrenal insufficiency (e.g., weakness, fatigue, weight loss, nausea, intermittent vomiting, abdominal pain, diarrhea or constipation, general malaise, muscle cramps, arthralgia, and symptoms suggestive of postural hypotension). These features regress upon treatment with replacement corticosteroids.

Growth Hormone

Growth hormone is considered to be one of the hormones that acts in opposition to insulin as a counterregulatory and catabolic signal, but it has complex effects on energy metabolism in addition to its effects on other aspects of human metabolism. GH release is stimulated by growth hormone-releasing hormone (GHRH) and inhibited by somatostatin, both of which are secreted by the hypothalamus. GH is normally secreted in a pulsatile fashion by the anterior pituitary with surges occurring several times during the day and at night during sleep. In addition to control by the hypothalamus, GH secretion responds to a variety of hormonal and metabolic signals. GH release is stimulated by deep sleep, norepinephrine (via α-adrenergic pathways), gherlin, sex steroids (estrogen, testosterone) stress, and fasting. Insulin-induced hypoglycemia and high-protein meals also stimulate GH secretion. Conversely, GH release is inhibited by hyperglycemia, glucocorticoids, and high concentrations of free fatty acids in addition to obesity, leptin, insulin-like growth factor-I (IGF-I), norepinephrine (by β-adrenergic pathways), and thyroid hormone.

Growth hormone is a single-chain polypeptide hormone consisting of 191 amino acids but circulating GH molecules comprise several heterogeneous forms. A 22-kD peptide is the major physiologic form, accounting for 75% of pituitary GH secretion, and alternative splicing of the GH gene produces a 20-kD molecule that accounts for about 10% of pituitary GH production. Once released into the bloodstream, GH binds to circulating GH-binding proteins. There are two forms of binding proteins, a 20-kD low-affinity binding protein, unrelated to the GH receptor, and a 60-kD high-affinity binding protein, which corresponds to the extracellular domain of the hepatic GH receptor and binds half of the circulating 22-kD GH form. The GH-binding proteins function to minimize acute oscillations in blood GH levels associated with pulsatile pituitary GH secretion and prolong the plasma half-life of GH by decreasing its clearance by the kidneys. The high-affinity binding protein (BP) also acts as a competitive inhibitor to prevent GH from binding to its cell surface receptor.

The liver contains abundant GH receptors, and several peripheral tissues also express modest amounts of receptor, including muscle and fat. The GH receptor is composed of 620 amino acids, weighs 70 kD, and consists of an extracellular ligand-binding domain, a single membrane-spanning domain, and a cytoplasmic signaling component. GH binding to its receptor causes a cascade of cellular events (Fig. 10.6). First, dimerization of the GH receptor occurs, followed by the activation of janus kinase 2 (JAK2), which is linked to intracellular signal transduction and transcription (STAT) of proteins. Phosphorylated STAT proteins are directly translocated to the cell nucleus, where they elicit GH-specific target gene expression by binding to nuclear deoxyribonucleic acid (DNA). GH also induces numerous other intracellular signaling pathways including c-fos, IRS-1 phosphorylation, MAP kinase, and protein kinase C. How these seemingly overlapping pathways converge to integrate the cellular effects of GH is at present unclear. IGF-I, a critical growth factor induced by GH, is probably responsible for most of the growth-promoting activities of GH and also directly regulates GH receptor function. Intracellular GH signaling is inhibited by the suppressor of cytokine signaling (SOCS) proteins, which disrupt the JAK/STAT pathway and thus disrupt GH action. In transgenic mice with deletion of SOCS-2, gigantism develops, presumably because of unrestrained GH action. Tyrosine phosphatases, such as SHP-2, might also contribute to inhibiting GH receptor signaling by dephosphorylating tyrosines in the GH receptor and/or JAK2.

Growth hormone has important effects on growth as well as energy metabolism. During puberty and the pubertal

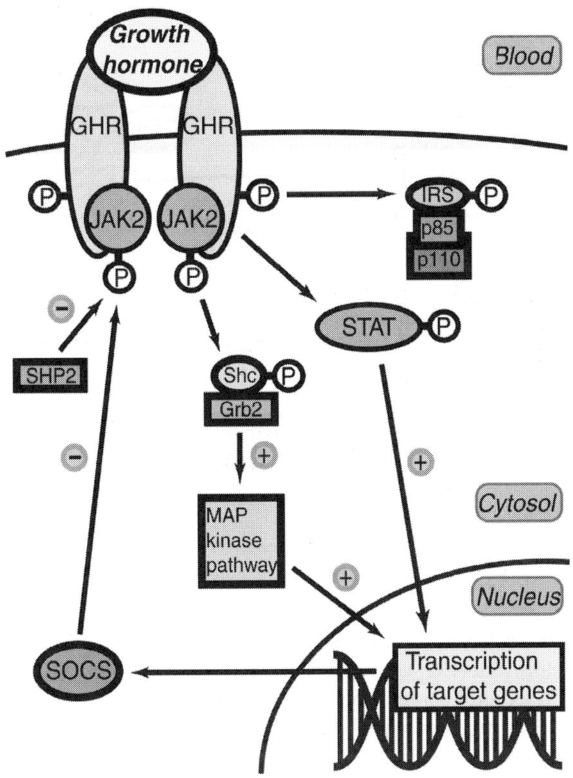

Figure 10.6 Growth hormone receptor. Binding of growth hormone (GH) results in dimerization of two GH receptors (GHR), which increases the affinity of each receptor for JAK2. The two receptor-associated JAK2 molecules are in close proximity, so that each JAK2 can phosphorylate the activating tyrosine of the other JAK2 molecule, thereby activating it. Activated JAK2 then phosphorylates itself and the cytoplasmic domain of the GHR. These phosphotyrosines within the GHR and JAK2 form binding sites for signaling proteins. JAK2 phosphorylates IRS proteins, which are thought to lead to activation of PI3K. GH activation of PI3K via IRS protein might be important for GH stimulation of glucose transport. JAK2 phosphorylates Shc, leading to activation of the MAP kinase pathway. JAK2 also phosphorylates STAT transcription factors, which then translocate to the nucleus. Both the MAP kinase pathway and STAT proteins are important for GH regulation of gene transcription. GH-induced expression of SOCS proteins inhibits further GHR signaling by decreasing the activity of JAK2. Tyrosine phosphatases, such as SHP-2, might also contribute to inhibiting GH receptor signaling by dephosphorylating tyrosines in the GHR and/or JAK2.

growth spurt, there is an increase in the frequency and amplitude of nightly GH surges accompanied by an increase in serum IGF-I. GH continues to be secreted in adulthood after growth cessation, implying important metabolic functions of GH in adult life. Endogenous GH concentrations antagonize insulin action at both hepatic and peripheral sites, resulting in decreased glucose utilization and increased lipolysis. GH also is anabolic, resulting in urinary nitrogen retention, decreased plasma urea levels, and increased muscle mass. GH increases the availability of fatty acids and glycerol, which are oxidized for energy, and indirectly decreases the oxidation of glucose and amino acids because GH-stimulated lipolysis, in part by activation of HSL, elevates the serum free fatty acids (FFAs) that are the preferred substrates of muscle. GH also increases the sensitivity of adipocytes to the lipolytic action of catecholamines and decreases adipose sensitivity to the lipogenic actions of insulin, thereby reducing TG synthesis in adipocytes. When plasma insulin levels are low, for example, during fasting, GH enhances FFA oxidation to acetyl-CoA, which, in combination with increased production of FFAs from adipose tissue, enhances the production of ketone bodies. As a result of lipolysis, the increased amount of glycerol reaching the liver provides substrate for gluconeogenesis and stimulates liver glycogen synthesis. GH suppresses liver glucose catabolism and decreases glucose uptake by fat and muscle cells. Through these actions, and by decreasing sensitivity to insulin, GH causes an increase in blood glucose levels.

Growth hormone also increases the transport of amino acids into muscle cells, which provides the substrate for protein synthesis, and increases muscle RNA and DNA synthesis. This results in a decrease in blood amino acid concentrations and a decrease in blood urea nitrogen, resulting in a positive nitrogen balance. GH also has an effect on bone, stimulating increased bone mass and epiphyseal growth. The anabolic effect of GH replacement in GH-deficient children and adults is well-established. Studies show an increase in muscle mass and a decrease in fat mass. In addition to the increase in muscle mass, an improvement in muscle function has also been reported that included both muscle strength and exercise capacity. GH, through the IGFs, also stimulates the growth and calcification of cartilage. Protein synthesis in general is increased by GH mainly as a result of decreased amino acid oxidation. The complex metabolic effects of GH can be viewed as a means of increasing protein synthesis and favoring fat as the main source of calories so as to spare amino acids for protein synthesis. At the same time, utilization of glucose is also suppressed. GH-deficient children have decreased fasting glucose levels, decreased insulin secretion, and increased insulin sensitivity with increased glucose utilization and blunted hepatic glucose release. GH replacement increases fasting glucose and insulin levels and restores hepatic glucose production.

Increased protein turnover and negative nitrogen balance are also characteristic features of critical illness (see section on Cachexia in Chapter 9). The loss of protein from functioning muscle can lead to a compromise of essential organs and may slow wound healing and tissue repair. This negative nitrogen balance is believed to arise from a resistance to GH and decreased production of IGF-I.

However, in a randomized controlled trial, high doses of GH given to critically ill patients in an intensive care unit, for example, postsurgical or posttrauma patients, was associated with higher mortality. The reason for this increased mortality is unknown, but patients treated with GH died from organ failure, septic shock, or uncontrolled infection, which suggests that GH may be involved somehow in regulation of the immune system. Thus, GH has a variety of metabolic actions some of which are independent of IGF activity but, in general, anabolic actions of GH are mediated via the IGF peptides.

Insulin-like Growth Factors

Insulin-like growth factors (IGFs) participate in the growth and function of almost every organ in the body. Two IGFs have been identified to date, IGF-I and IGF-II. These closely related proteins, with a structure similar to that of proinsulin (insulin with the C peptide still in place), are synthesized mainly by the liver (80%). GH stimulates IGF production and release, and the newly secreted IGF-I feeds back to the pituitary gland to inhibit GH release. Whereas both IGFs are essential for embryonic development, the effects of IGF-I predominate after birth and the physiologic role of IGF-II in the adult is unknown. IGFs are able to bind to the insulin receptor but two classes of IGF receptors have been described. The mitogenic and metabolic actions of both IGF-I and IGF-II appear to be mediated solely through the IGF-I receptor (IGF-IR) with the role of the IGF-II receptor (IGF-IIR) still unclear.

The IGF-IR has a high degree of homology with the insulin receptor (IR) and is also composed of two α subunits and two β subunits. This receptor binds both IGF-I and IGF-II with high affinity, but the affinity for insulin is generally 100-fold less, thereby explaining the relatively weak mitogenic effect of insulin. With the binding of its ligand, IGF-IR undergoes autophosphorylation and is then able to phosphorylate tyrosine residues on other intracellular proteins, such as IRS-1 and IRS-2 (Fig. 10.7). Various proteins can associate with phosphorylated IRS-1, such as PI3K, the phosphotyrosine phosphatase Syp2, the oncogenic protein Nck, and Grb2. Src-homology domain–containing protein (Shc) is phosphorylated by IGF-IR, which allows it to interact with Grb2 and activate Ras, and the MAP kinase cascade, which leads to the activation of transcription factors and gene transcription. Thus, phosphorylation of IRS-1 by either IGF-IR or IR activates multiple signaling cascades that ultimately influence nuclear transcription and gene expression. It is, presumably, at this level that the IGF peptides exert their mitogenic and anabolic actions. Given that insulin and IGF peptides activate similar, if not identical, signaling pathways through their own specific receptors, it is unclear how the cell distinguishes between these overlapping ligands. Whether this merely reflects the relative levels of receptors or whether divergent downstream pathways exist for insulin and IGF action remain questions for future investigation.

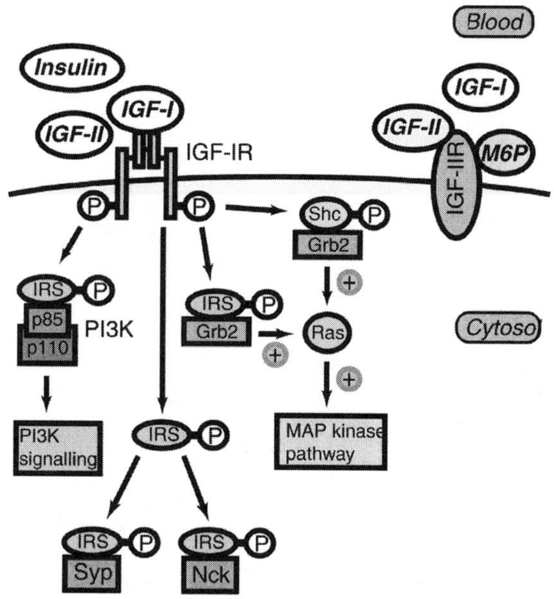

Figure 10.7 Insulin-like growth factor receptors. The IGF-I receptor (IGF-IR) has a high degree of homology with the insulin receptor being composed of two α subunits and two β subunits. This receptor binds both IGF-I and IGF-II with high affinity but insulin binds only weakly. With the binding of its ligand IGF-IR undergoes autophosphorylation on tyrosine residues and is then able to phosphorylate tyrosine residues on other intracellular proteins, such as IRS-1 and IRS-2 (IRS). Various proteins can associate with phosphorylated IRS-1, such as PI3K (p85 and p110), the phosphotyrosine phosphatase Syp2, the oncogenic protein Nck, and Grb2. Shc is phosphorylated by IGF-IR, which allows it to interact with Grb2 and activate Ras and the MAP kinase cascade, which leads to the activation of transcription factors and gene transcription. The IGF-II receptor (IGF-IIR) bears no structural homology with either the insulin receptor or IGF-IR and does not contain tyrosine kinase activity or any other recognized signal transduction mechanism. IGF-IIR binds IGF-II and M6P-containing lysosomal enzymes with high affinity, IGF-I binds with lower affinity, and insulin does not bind at all. The precise role of IGF-IIR is unclear.

The IGF-IIR bears no structural homology with either IR or IGF-IR and is identical to the mannose-6-phosphate (M6P) receptor, but the physiological significance of this receptor is unknown (Fig. 10.7). IGF-IIR does not contain an intrinsic tyrosine kinase domain or any other recognized signal transduction mechanism. IGF-IIR binds only IGF-II and M6P-containing lysosomal enzymes with high affinity. IGF-I binds with lower affinity, and insulin does not bind at all. Binding of IGF-II to its receptor seems to be a mechanism for clearing IGF-II from the blood rather than initiating a signaling cascade or IGF-II may affect the sorting of lysosomal enzymes. Since loss of IGF-IIR in mice is associated with macrosomia and fetal death, this receptor may act as a growth inhibitory component of the IGF system.

Evidence exists to suggest that the IR and IGF-IR share a series of biological actions, while each also maintains others that are unique. The most extreme demonstration of different biological functions for the two receptors comes from studies of the receptor-deficient mice. In mice with a general knockout of the IR gene, death occurs shortly after birth due to severe ketoacidosis and metabolic dysregulation. By contrast, whole-body knockout of IGF-IR leads to pronounced growth retardation and early postnatal death due to respiratory insufficiency and failure to thrive.

In addition to control by GH, the IGFs are also regulated by IGF-binding proteins (IGF-BPs). After secretion, IGF binds to an IGF-BP in the plasma for transport to its receptor. There are six IGF-BPs that specifically bind IGFs, but not insulin, with IGF-BP3 binding over 95% of the circulating IGF. IGF-BPs are involved in regulating the effects of IGFs, and the IGF-BPs can either inhibit or enhance the effect of IGFs depending on the isoform and the tissue type. IGF-BPs have four major functions: (a) they act as transport proteins and control the removal of IGFs from the blood, (b) they prolong the half-life of IGF, (c) they localize IGF to specific tissues and cells, and (d) they regulate IGF binding to its receptor and indirectly control the effects of IGF. IGF-BPs can also have direct effects on cellular functions and at least one IGF-BP is a ligand for a specific cell surface receptor. The importance of the IGF-BPs was recently documented in a study where starved rats were given either IGF-I alone or IGF-I + IGF-BP3. Only the IGF/IGF-BP mixture reversed the catabolic state caused by starvation and stimulated protein synthesis in muscle while maintaining normal concentrations of amino acids and glucose in the blood. Thus, muscle protein synthesis can continue, even in the state of severe catabolism caused by starvation, so long as IGF-I reaches the muscle. Specific proteases (found in blood, semen, cerebrospinal fluid, and urine) degrade the IGF-BPs, and an increase in the concentration of these proteases has been reported in patients with cancer and muscle wasting. Therefore, these proteases may also modify IGF effects by decreasing levels of their transporter protein.

In vitro, IGFs cause progression in the cell cycle, stimulate cell proliferation, inhibit cell death, induce cellular differentiation, and can stimulate hormone secretion from many cell types. *In vivo*, receptors for IGF-I are widespread and IGF-I effects can be identified in many cell types. The best characterized action of the IGFs are their stimulatory actions on cell proliferation, but IGFs can also inhibit cell death and induce differentiation. IGF-I is responsible for mediating most of the anabolic actions of GH, including cell growth and differentiation, and GH and IGF-I have comparable anabolic and growth effects. However, the two hormones have opposing effects on glucose and FFA metabolism. IGF-I has insulin-like effects, including decreasing blood glucose, inhibiting lipolysis, and decreasing protein breakdown. *In vivo*, IGF-I causes hypoglycemia but is much less potent than insulin at reducing FFA levels. IGF-I seems to affect human skeletal muscle protein metabolism mainly by increasing protein synthesis. Fasting, as well as various conditions of protein-calorie malnutrition [marasmus, kwashiorkor, anorexia nervosa, celiac disease, acquired immune deficiency syndrome (AIDS), inflammatory bowel disease], causes resistance to GH and impaired signaling from the GH receptor, which results in decreased IGF-I synthesis and lower circulating concentrations of IGF-I. In catabolic states, such as severe sepsis or trauma, the serum IGF-I concentration is low and there is resistance to GH. The decrease in IGF-I parallels changes in nitrogen balance, so decreased IGF-I levels may result in decreased protein synthesis or increased protein catabolism. In general, the magnitude of IGF-I reduction parallels the severity of the nutritional insult. This decline in IGF-I levels is not due to reduced GH stimulation since humans have an increase in GH during dietary restriction. Therefore, it appears that in undernourished individuals IGF-I production is insensitive to GH. Reduction in hepatic GH-binding capacity could be one mechanism responsible for the fasting-induced decline in IGF-I.

Mice that are IGF-I deficient are very small at birth (40% of normal) and also show growth arrest after birth. Thus, IGF-I seems to be required for growth at all stages of development. GH deficiency results in normal growth *in utero* but growth arrest after birth. This seems to indicate that IGF-I is not regulated by GH until after birth and that IGF-I is likely under genetic control rather than endocrine control before birth.

Interestingly, some tumors, including muscle cancers (e.g., rhabdomyosarcoma, leiomyosarcoma) and Wilm's tumor in the kidney, secrete large amounts of a prohormone form of IGF-II ("big IGF-II"), which stimulates glucose uptake by the tumor and other tissues such as muscle and fat. Hepatic glucose production and insulin secretion are decreased by hypoglycemia and by the direct inhibitory effects of big IGF-II on the pancreatic β cells. In addition, big IGF-II inhibits the secretion of GH, which decreases the synthesis and release of IGF-I and IGF-BP3. Decreased levels of IGF-BP3 enhance the effects of circulating IGF-II.

Thyroid Hormone

Thyroid hormone (TH) has effects on almost all tissues of the body. Overall, TH can be viewed as a tissue growth factor that acts by stimulating protein synthesis and degradation. The positive influences of TH on normal body growth are derived largely from stimulation of protein synthesis, but large excesses of TH lead to accelerated protein catabolism. TH also affects most aspects of glucose metabolism and has multiple effects on lipid metabolism. TH influences nearly all major metabolic pathways. The most obvious and well-known action is the increase in basal energy expenditure (metabolic rate). TH does not intervene in the minute-to-minute regulation of metabolism, unlike insulin and glucagon, but rather modulates the activities of metabolic pathways on a medium- or long-term basis, either by direct action or by modifying the activities of other regulatory hormones such as insulin, glucagon, and catecholamines.

The thyroid gland releases two hormones, thyroxine (T_4) and the more potent triiodothyronine (T_3). T_3, with a 15-fold higher binding affinity for the thyroid receptor (TR) compared to T_4, is the active thyroid hormone responsible for the activation of TH-dependent genes. T_4 serves primarily as a prohormone. The synthesis and release of TH is tightly regulated by a negative feedback system involving the hypothalamus, pituitary, and thyroid gland. Thyrotropin-releasing hormone (TRH) binds to receptors on cells in the anterior pituitary, which secrete thyroid-stimulating hormone (TSH). TSH is the primary regulator of TH production. Binding of TSH to its receptor in the thyroid causes an increase in intracellular cAMP and an activation of PKA leading to T_3 and T_4 synthesis. In hyperthyroid Graves' disease circulating autoantibodies are directed against the TSH receptor and behave as thyroid-stimulating antibodies. These antibodies act as TSH agonists and compete with TSH for binding to its receptor and activation of adenylate cyclase. Both TRH and TSH release are inhibited by increased amounts of T_3 and T_4.

Dietary iodine is required for TH synthesis, and the inclusion of iodine in table salt has been a major public health advance in the prevention of hypothyroidism resulting from iodine deficiency. The majority of the TH released into the circulation is T_4 with only 20% of the daily production of T_3 being derived from direct secretion by the thyroid and 80% from peripheral T_4 conversion. Almost all T_4 is bound to carrier proteins such as thyroxine-binding globulin (TBG) and albumin, and these serum transport proteins serve as a reservoir of hormone that can be released when needed. Only 0.03% of total T_4 and 0.3% of total T_3 in the blood are in the unbound state, but it is the unbound fraction that has biological activity. Unbound, free TH is the only form capable of diffusing across the target cell membrane to interact with the nuclear TR to initiate metabolic effects, and free TH is also the form responsible for feedback regulation and that undergoes degradation. The bound hormone acts merely as a reservoir. The concentration of free TH determines the metabolic state and it is this concentration that is defended by homeostatic mechanisms.

The most important pathway for T_4 metabolism is its outer ring (5′) monodeiodination to produce active T_3. This reaction is catalyzed by type 1 and type 2 deiodinases (D_1 and D_2). D_1 is highly expressed in human liver and kidney, but D_2 is more widely distributed and is found in skeletal and cardiac muscle, the central nervous system, skin, and the pituitary gland. D_1 is usually located on the plasma membrane, whereas D_2 is found in the endoplasmic reticulum, and this may explain why the T_3 produced by D_1-catalyzed reactions readily enters the plasma, whereas that generated by D_2 enters the nucleus. Inner ring deiodination, catalyzed primarily by type 3 deiodinase (D_3), inactivates T_4 and T_3. These deiodination reactions can be considered physiologically activating and inactivating pathways that control T_3 concentrations in peripheral tissues.

Thyroid hormone enters cells both by passive diffusion (like cortisol) and by an energy-requiring membrane carrier. As mentioned earlier, T_3 has a 15-fold greater affinity for the TR than T_4, and therefore the majority of the major biological effects of TH are mediated by T_3. The primary effects of TH are exerted at the level of gene transcription via the TR found inside the nucleus of target cells (similar to glucocorticoid receptors; see Fig. 10.5), and these TRs are found in virtually all tissues. The subsequent binding of the activated TR to the thyroid-response element (TRE) is the next step in regulating gene expression. TRs may form complexes with one another or with retinoic acid receptors (RXRs), another family of nuclear receptors, when binding to the TRE. In the absence of T_3, an unliganded TR binds to the TRE and represses gene transcription. However, in the presence of T_3 ligand, the TR that is heterodimerized with the RXR undergoes a conformational change resulting in gene activation (Fig. 10.8). TR binding may also be influenced by TR phosphorylation, but the effects that this has on gene transcription and which kinases are responsible is still unknown. The interaction of TRs with other intranuclear proteins, including repressors and coactivators, may also have important effects on TR gene regulation. Again, the effect of these other proteins *in vivo* still remains unknown at this time. Although TH is a key regulator of metabolism, few target genes have been identified to date. Binding of T_3 to the TR leads to increased formation of specific mRNAs and proteins, such as those coding for growth hormone, malic enzyme, myosin heavy-chain α, and the calcium pump of the sarcoplasmic reticulum. T_3 suppresses transcription of other genes such as the genes coding for TSH subunits.

An estimate of an individual's basal metabolic rate (BMR) is made by measuring oxygen consumption while fasting and at rest. Since energy expenditure is related to functioning tissue mass, oxygen consumption is then usually related to body surface area. Calculated in this way, basal oxygen consumption (resting energy expenditure) is higher in men than in women and declines rapidly

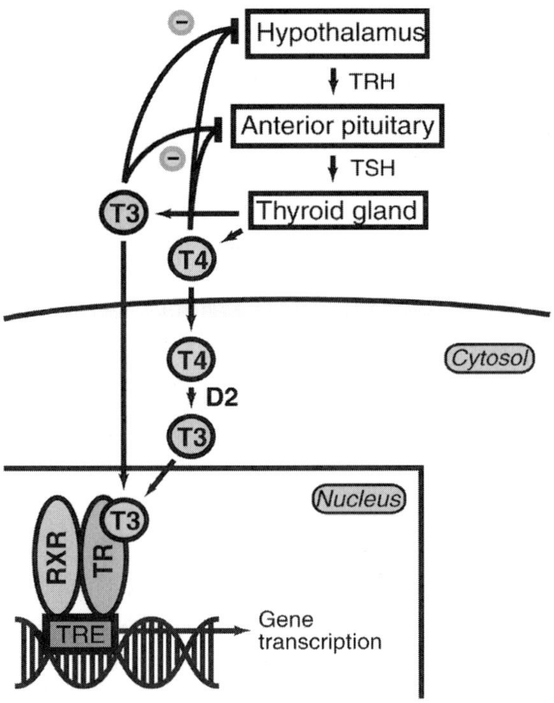

Figure 10.8 Thyroid hormone receptor. In response to metabolic and hormonal signals, the hypothalamus produces TRH, which causes the pituitary to secrete TSH, which stimulates the release of T_4 and T_3 from the thyroid gland. Circulating free T_3 or, more commonly, T_3 produced by the deiodination of T_4 (catalyzed by type 2 deiodinase, D_2) then enters the target cell nucleus. The binding of T_3 to the TR bound as a heterodimer with a retinoid X receptor (RXR) at the thyroid hormone response element (TRE) leads to the transcription of TH-responsive genes.

from infancy to the third decade of life and more slowly thereafter. Under basal conditions, approximately 25% of oxygen consumption is due to energy expenditure in visceral organs (e.g., liver, kidney, heart), 10% in the brain, 10% in respiratory activity, and the remainder in skeletal muscle. A major action of T_3 is to regulate cellular oxygen consumption, and it has been estimated that approximately 40% of the body's resting oxygen consumption is regulated by TH via changes in cellular enzymes, electron transport, and protein synthesis. Abnormal, usually elevated, values for BMR are seen in patients with burns and with systemic diseases (e.g., febrile illnesses, pheochromocytoma, malignancies, seizures, anxiety). In healthy, euthyroid individuals BMR ranges from −15 to +5% when compared to normal values. TH increases energy expenditure and heat production (i.e., increases BMR) as manifested by weight loss, increased caloric requirement, and heat intolerance; and, therefore, in hyperthyroid patients values for BMR may reach +25 to +50%. Conversely, BMR values for hypothyroid patients are commonly −20% and may be as low as −40%. Excess TH affects metabolism profoundly (as indicated by the marked increase in BMR), and hypermetabolism, involving all major fuel sources, is a hallmark of hyperthyroidism. Human hyperthyroidism is a catabolic state accompanied by multiple metabolic abnormalities, with increased energy expenditure and excessive mobilization and utilization of metabolic substrates.

With regards to carbohydrate metabolism, studies in humans have clearly shown that TH increases glucose production and utilization. However, the effect of TH is dependent on concentration as well as systemic conditions. TH increases glucose uptake by the intestine, adipose tissue, liver, and muscle. In both muscle and liver cells TH enhances the stimulatory effect of epinephrine on glycogen breakdown and glycolysis, and TH increases gluconeogenesis in the liver. Whereas low doses of TH enhance glycogen synthesis in the presence of insulin, large doses stimulate glycogenolysis.

All aspects of lipid metabolism, including synthesis, mobilization, and degradation, are influenced by TH, but degradation is affected more than synthesis. TH stimulates an increase in lipolysis and increased concentrations of FFA and glycerol in the blood reflect this. However, the mechanism of TH action leading to the increased lipolysis and lipid oxidation characteristic of hyperthyroidism are poorly understood. TH may sensitize the adipocyte to the lipolytic action of epinephrine by increasing the number of β_2 adrenoceptors and increasing cellular levels of cAMP (by decreasing phosphodiesterase activity) or may increase HSL activity. Increased lipolysis leads to an increase in blood levels of FFAs and in thyrotoxic states elevation of circulating FFAs constitutes the major metabolic fuel with the oxidation of lipids responsible for more than 60% of the resting energy expenditure in hyperthyroid patients. Increased lipolysis is also accompanied by increased ketone production by the liver.

Simultaneously with enhanced lipolysis and lipid oxidation, TH increases lipogenesis. TH accomplishes this increase in lipogenesis by increasing transcription of acetyl-CoA carboxylase, FAS, malic enzyme, ATP citrate lyase, and L-pyruvate kinase and may increase levels of insulin and glucose to modify the metabolic and hormonal environment of the liver. Simultaneous fatty acid oxidation and synthesis may represent a futile cycle of metabolites, which contributes to increased energy expenditure and, hence, an elevated metabolic rate. Fatty acids derived from adipose tissue are the primary source of substrate for TH-induced calorigenesis, and the early increase in lipogenesis simply serves to maintain fat stores. The increase in lipogenesis and lipolysis assures maintenance of adipose pools at operationally satisfactory levels, and the difference between fat consumption and the rate of lipogenesis must be maintained by increased food consumption and, to a variable extent, diminished body growth.

Thyroid hormone also appears to stimulate cholesterol synthesis, but it is thought that the low plasma cholesterol level of hyperthyroid patients is due to an increased clearance rate. Similar to effects on lipolysis and lipogenesis, TH effects on cholesterol metabolism occur through the expression of genes [e.g., low-density lipoprotein (LDL) receptor, cholesterol ester hydrolase, and cholesterol acyltransferase]. Patients with hypothyroidism usually have a marked increase in serum cholesterol that resolves promptly with TH replacement therapy.

The effect of TH on protein metabolism is characterized by a profound stimulatory effect on the synthesis of numerous cytosolic, mitochondrial, and secretory (e.g., albumin, hormones) proteins. TH enhances protein synthesis and muscle growth via stimulation of gene expression. TH is essential for growth but also has a profound catabolic effect on skeletal muscle under both hypo- and hyperthyroid conditions. Thus, both conditions are characterized by muscle weakness caused by a decrease in muscle efficiency rather than a loss of muscle mass.

The brain also affects energy expenditure by means of the hypothalamic–pituitary–thyroid axis. It is clear that increases or decreases in TH are associated with parallel changes in energy expenditure and that relatively small changes in TH produce significant effects. TH levels have been found to drop during starvation, an effect that is dependent upon falling leptin levels and mediated by decreased expression of hypothalamic TRH. Hence, reduced TH levels may contribute to the starvation-induced decrease in thermogenesis that aids conservation of body fuel reserves. The mechanism by which TH stimulates thermogenesis is not established, but it seems to be due to multiple effects on various aspects of energy metabolism such as substrate cycling, ion cycling, and mitochondrial proton leaks. Mitochondrial gene expression is reduced in hypothyroid animals and stimulated upon administration of TH. Selected nuclear genes that encode mitochondrial proteins have TREs, indicating that this hormone is working directly on these genes through TH receptors. In addition to these effects on nuclear genes there have been reports that TH and its receptor can translocate into mitochondria to affect transcription patterns of genes on the mitochondrial genome.

The importance of TH in growth and development is emphasized by the effects of hypothyroidism in children. TH has major effects on brain development both *in utero* and during neonatal development, and neonatal hypothyroidism due to genetic causes or iodine deficiency can cause mental retardation and neurological defects. This developmental impairment is reversible if T_4 is supplied, and this is the reason why all babies are screened at birth for hypothyroidism. However, few TH-regulated genes have been found in the brain. In bone, TH is critical for normal growth and development and hypothyroidism in children causes short stature.

Leptin

In 1994 it was demonstrated that mutation of a single gene was responsible for the greatly increased appetite and weight gain seen in an obese mouse model. Furthermore, administration of the protein product of this gene to these obese mice reversed their body weight, food intake, and serum insulin and caused leanness in normal mice. The product of this gene is a novel hormone that was named leptin (derived from the Greek *leptos*, meaning thin). The human homolog of this gene was found, and subsequent work has shed light on the central role of leptin in the regulation of body weight. However, studies on leptin have shown that it is far from being the cure for obesity, and that body weight regulation is a tightly integrated complex system that defies easy manipulation.

Leptin is a 146-amino-acid cytokine-like peptide that is expressed primarily in white adipose tissue. Leptin is synthesized and secreted by fat cells in proportion to the amount of lipid stored in the body and acts as a signal of body energy stores. In situations where there is increased body fat, such as in obesity, there is increased leptin synthesis and secretion that acts to reduce food intake. Conversely, a decrease in body fat, for example, during fasting or weight loss, causes a decrease in leptin synthesis and secretion, which initiates a complex response that includes behavioral (hunger and food seeking) and biochemical (efficient metabolism, suppression of reproduction, linear growth, and thyroid hormone levels) changes that favor survival during periods of limited fuel availability. Reintroduction of food rapidly increases leptin levels. It is this fluctuation in leptin levels that is believed to maintain body weight within a relatively narrow range. Interestingly, a block on leptin production appears to be part of the mechanism that lets many small mammals accumulate huge fat reserves that increase their body mass by as much as 50% prior to winter hibernation (see Chapter 16).

Leptin has complex effects on the storage and metabolism of fats and carbohydrates. Leptin exerts its effects both through the central nervous system (discussed in the section on Obesity) and also by direct action on peripheral tissues via leptin receptor signaling. The leptin receptor has a docking site for janus kinases (JAK), a family of tyrosine kinases, that phosphorylate members of the signal transduction and transcription (STAT) family of intracellular proteins (similar to the growth hormone receptor; see Fig. 10.6). The STAT proteins then translocate to the nucleus where they stimulate transcription of specific target genes. The leptin receptor gene is expressed in the choroid plexus, hypothalamus, and many peripheral tissues, including pancreatic cells and adipose tissue. Changes in leptin concentration have effects on many other organ systems, including reproduction, the immune system, and bone formation, which indicates that leptin is an important means by which changes in nutrition affect physiology.

Leptin is primarily synthesized by adipocytes and has significant metabolic effects on this tissue. Leptin acts directly on adipocytes to alter the transcription of genes involved in lipogenesis, lipolysis, and energy metabolism and may even trigger cell death. Leptin also regulates lipid homeostasis in other tissues, including liver, muscle, and pancreas, by up-regulating the expression of genes that encode oxidative enzymes, stimulating FFA oxidation and suppressing TG synthesis. However, suppression of TG accumulation in nonadipose tissue, such as muscle and liver, may contribute to insulin resistance.

Recent findings suggest that leptin induces a novel form of lipolysis in adipocytes. Lipolysis that occurs during fasting results in an increase in both glycerol and FFA release from adipocytes. These FFAs are then converted by the liver to ketone bodies (see previous chapter). However, leptin increases lipolysis in adipocytes without causing an increase in FFA release or ketone body formation. Leptin reduces expression of acetyl-CoA carboxylase (ACC1) and fatty acid synthase (FAS) and increases carnitine palmitoyltransferase-I (CPT1) and acyl-CoA oxidase (ACO) expression, possibly via peroxisome proliferator-activated receptor α (PPARα), which is also increased by leptin through an unknown mechanism. The net effect of this leptin stimulation in the adipocyte is a reduction in fatty acid synthesis and an increase in the mitochondrial uptake and oxidation of fatty acids. Leptin-induced glycerol release is unaccompanied by FFA release, which explains why fat loss due to leptin overexpression does not result in increased blood levels of FFA and ketones. Leptin causes FFA to be oxidized within the adipocytes rather than exported to the liver for oxidation to ketoacids.

The interactions between leptin and other metabolic hormones are beginning to be understood. Glucocorticoid effects on leptin synthesis and secretion have been demonstrated in rodents and humans, both *in vivo* and *in vitro*, and cortisol appears to counteract the central effects of leptin and reverse the decrease in food intake and body weight caused by leptin. The sympathetic nervous system plays an important role both in the regulation of energy expenditure and in adipose tissue lipolysis. Catecholamines and β-adrenoceptor agonists have been shown to inhibit leptin production both *in vitro* and *in vivo*. Insulin and glucocorticoids act directly on adipocytes to increase leptin synthesis and secretion and may function as long-term regulators of leptin expression. Growth hormone and thyroid hormone may also affect leptin levels, but the mechanism, effects, and significance of these interactions are not yet known. (see Text Box 10.2)

TEXT BOX 10.2 LEPTIN AND CANNABINOIDS

Cannabinoids are found naturally in the body. These endocannabinoids, such as anandamide, bind to cannabinoid receptor proteins and act to increase appetite. Not surprisingly, plant cannabinoids such as the active ingredient in cannabis, Δ^9-tetrahydrocannabinol (THC), also bind to these receptors creating the well-known effect of marijuana smoking on appetite ("getting the munchies"). In fact, THC is frequently used to help patients, such as those with AIDS, from losing too much weight due to the cachexia associated with their illness. Recently, it was shown that leptin, which suppresses the activity of other appetite-stimulating peptides (NPY and AGRP), also decreases levels of endocannabinoids. This seems to add another branch of complexity to the leptin system.

HORMONAL DYSREGULATION AND HUMAN DISEASE

Obesity

Obesity is a disease in which the accumulation of body fat adversely affects health. The detrimental effects of obesity are demonstrated by its association with the development of type 2 diabetes mellitus, coronary heart disease, an increased incidence of certain forms of cancer (e.g., endometrial, breast, prostate, and colon cancers), respiratory complications (e.g., obstructive sleep apnea), and osteoarthritis of both large and small joints. Presentation of the adverse effects of obesity tends to be delayed and may not appear for as long as 10 years. However, the risk of death increases and life expectancy decreases with increasing weight.

Body fat is usually estimated by the body mass index (BMI), which assumes that differences in weight for persons of the same height is due to fat mass. The BMI is calculated by dividing the weight in kilograms by the square of the height in meters. Individuals with a BMI of 25.0 to 29.9 are overweight and are obese if the BMI is greater than 30.0. Although there are limitations to the use of the BMI, there is a linear relationship between BMI and the incidence of several chronic conditions caused by excess fat, including type 2 diabetes, hypertension, coronary heart disease, and gallstones when the BMI is less than 30, and risks for all these conditions are greatly increased when the BMI is greater than 30.

As discussed earlier, the hormone leptin has profound effects on appetite and body weight regulation, and part of the effect of leptin is mediated centrally in the hypothalamus by the leptin-regulated central melanocortin circuit

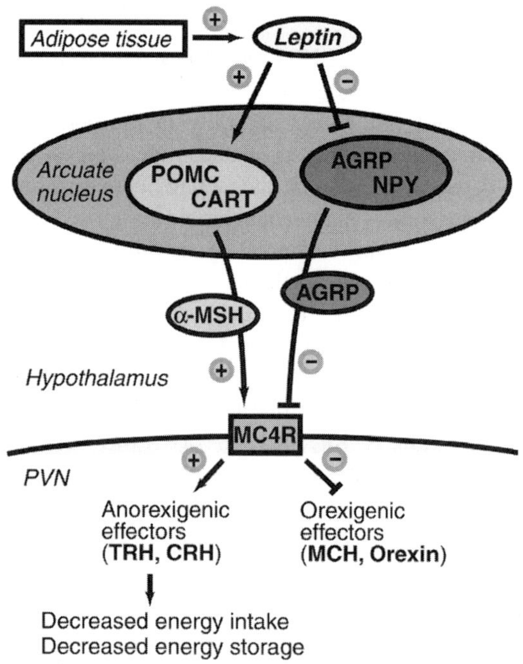

Figure 10.9 Leptin-regulated central melanocortin pathway. An increase in body fat leads to an increase in leptin production by adipose tissue. Leptin acts on neurons in the arcuate nucleus of the hypothalamus in the brain to stimulate the expression of α-MSH, derived from POMC, and CART. Leptin also decreases expression of the orexigenic peptides NPY and AGRP. α-MSH acts as an agonist for MC4R found in the paraventricular hypothalamic nucleus (PVN) of the hypothalamus to activate anorexigenic neural pathways (leading to synthesis and release of CRH and TRH) and inhibit orexigenic pathways and the neuropeptides (MCH, orexin) they produce. Stimulation of MC4R results in decreased energy intake and decreased energy storage, which will decrease the size of adipose tissue deposits and leptin production. AGRP acts as an MC4R antagonist to decrease signaling through this receptor and stimulate appetite and food intake.

(Fig. 10.9). This is the best described and most clinically relevant circuit of weight regulation, but it is not the only system involved. Serum leptin concentration is dependent both on the size of the adipose tissue mass and on the fed/fasted state of the individual; that is, fasting decreases leptin concentration without marked changes in body lipid content. Leptin released from adipocytes crosses the blood–brain barrier to act on its receptor, which is expressed in at least two different groups of neurons in the hypothalamus. These neurons are found within the arcuate nucleus, which has been described as the master control center for both short- and long-term weight regulation. One group of neurons expresses neuropeptide Y (NPY) and agouti-related protein (AGRP), two peptides that increase food intake (orexigenic agents). Another group of neurons expresses the cocaine- and amphetamine-related transcript (CART) and pro-opiomelanocortin (POMC), the precursor of the α-melanocyte stimulating hormone (α-MSH). These peptides, POMC, α-MSH, and CART, decrease food intake (anorexigenic agents). AGRP and α-MSH are antagonistic ligands for a common receptor, the melanocortin 4 receptor (MC4R), which is expressed primarily in the brain. Activation of MC4R by α-MSH reduces food intake, whereas binding of AGRP to MC4R increases feeding. Leptin acts to suppress the activity of the NPY/AGRP-expressing neurons and stimulates activity of the POMC/CART-expressing neurons. Thus, when leptin is present, it stimulates the release of neuropeptides that act to reduce food intake. When leptin is absent, for example, during fasting, the neurons expressing peptides that stimulate food intake are maximally active. In an individual at his or her stable body weight, the two pathways are balanced. The leptin-regulated central melanocortin pathway links peripheral signals (e.g., leptin) and areas of the brain known to control appetite (i.e., the arcuate nucleus of the hypothalamus), but it is still not known how these melanocortin signals produce downstream effects on appetite or energy expenditure. One model involves leptin-regulated nerve terminals projecting onto neurons within the paraventricular hypothalamic nucleus (PVN), which regulates pituitary hormone secretion. For example, the cells that produce TRH are found in the PVN and, therefore, leptin could regulate expression of the metabolic hormones, discussed earlier in this chapter, to exert its effects on food intake and energy utilization.

The neural circuits that regulate appetite and food intake are also influenced by metabolic fuels and hormones. Neurons in the hypothalamus that respond to changes in glucose levels (low brain glucose promotes feeding) may be the same as, or functionally linked to, those neurons that respond to leptin and express the peptides discussed above. The flux of glucose in the pentose phosphate pathway, which has been suggested to be a cellular sensor of energy availability, is also linked to the stimulation of leptin gene expression and secretion. A role for lipid mediators in metabolic sensing may be suggested by the recent observation that inhibitors of FAS potently inhibit food intake through actions in the brain. Insulin also has an important role in the control of energy homeostasis by the central nervous system. Circulating insulin levels reflect the amount of energy stored as fat in the body, and low insulin levels during weight loss act to increase activity of anabolic neural pathways and decrease activity of catabolic pathways (see Text Box 10.3).

Although mutations that result in leptin deficiency or leptin resistance lead to massive obesity in both rodents and humans, leptin has limits in controlling obesity. As fat mass increases, further increases in leptin have only a limited ability to suppress food intake and prevent obesity. It appears that the antiobesity role of leptin has

TEXT BOX 10.3 GASTRIC PEPTIDES: GHERLIN AND PYY

Whereas leptin acts as a satiety signal in the long-term to indicate overall body energy status, the hormone gherlin is important in regulating satiety in the short-term (e.g., after eating a meal). Gherlin was initially described as an inducer of GH release, but more recently its importance as an orexigen (appetite-stimulant) has been appreciated. Gherlin is released by the stomach and levels rise just before meals (as well as during food restriction or starvation) and fall rapidly after meals. Injecting gherlin into human volunteers had a powerful effect in determining the amount of food they subsequently ate. Overproduction of gherlin can lead to obesity as has been shown in people with Prader-Willi syndrome (characterized by hyperphagia and weight gain leading to severe obesity) who have extremely high levels of gherlin. The premeal increase in gherlin may be important in triggering the desire to eat, and gherlin may also contribute to hunger and possibly other adaptations that accompany fasting or starvation. Gherlin antagonists may lead to weight loss and agonists may be of benefit in cachectic patients.

Another hormone involved in weight regulation is the peptide YY_{3-36} (PYY), which is released postprandially by the gut in proportion to the caloric content of a meal. The peptide is an agonist of the NPY Y2 receptor expressed on NPY neurons in the arcuate nucleus. Injection of PYY decreases NPY secretion, increases POMC neuronal activity, and reduces food intake in rodents and humans.

been limited through evolutionary pressure to allow fat storage in times of plenty. The limited efficacy of leptin in inducing weight loss in obese patients may be related to the development of resistance to the action of leptin after prolonged exposure to increased plasma leptin concentrations. There are various mechanisms that have been described to account for leptin resistance, including the fact that leptin crosses the blood–brain barrier by a saturable mechanism and that prolonged stimulation of the JAK-STAT pathway activates inhibitory components such as the suppressor of cytokine signaling 3 (SOCS-3) that limits the effect of agonists. Obesity is often accompanied by elevated levels of blood cortisol due to conversion of inactive cortisone to active cortisol in the adipocytes, which may also contribute to decreased central responsiveness to leptin. Leptin insensitivity, which is observed in most obese rodents, is associated with decreased sympathetic outflow, leading to obesity and impaired glucose tolerance. Pima Indians have a high frequency of missense mutations of the β_3-adrenergic receptor gene (β3-AR) (*Trp64 Arg*). Those with the mutation develop early-onset type 2 diabetes and tend to have a low metabolic rate. This mutation is also associated with abdominal obesity and resistance to insulin, an increased capacity for gain weight, a reduced ability to lose weight, high BMI, and lower lipolytic activities. These findings indicate that secondary decreased sympathetic effects, including a secondary perturbation of leptin and β3-AR signaling, could be a cause of human obesity and type 2 diabetes. Thus, the sympathetic nervous system has an important role in the regulation of glucose and fat metabolism. Dysfunction of the sympathetic nervous system regulation of organs could predispose to obesity and type 2 diabetes (see later section on Diabetes Mellitus).

Treatment with exogenous leptin is clearly effective in individuals or rodents with leptin deficiency with dramatic declines in appetite, body weight, and food intake. However, leptin deficiency and leptin receptor defects in humans are rare, and the majority of obese humans are leptin-resistant rather than leptin-deficient. This resistance to leptin action may explain why treatment with exogenous leptin had only modest effects on appetite and body weight with higher doses of leptin resulting in greater loss of fat mass.

Body weight is determined by genetic, environmental, and social factors that affect energy intake and expenditure. The contribution of genetic factors to the pathophysiology of obesity are presently estimated to be between 25 and 40%. Mutations in leptin and its receptor as well as mutations in the genes for POMC and the enzyme that processes the POMC protein to α-MSH lead to a phenotype of severe, usually early-onset obesity. Recently, mutations in MC4R have been implicated as strongly contributing to hyperphagia-induced morbid obesity. Deleting the gene for MC4R causes obesity in mice, and heterozygous mice have moderate obesity as well. Furthermore, 4 to 5% of severe human obesity appears to be due to a mutation in this gene with most affected humans having a single mutant allele, which causes haploinsufficiency, rather than a dominant-negative mechanism. This suggests that this pathway is required for normal energy homeostasis and is extremely tightly regulated.

Although mutations in single genes leading to obesity have been described, in general, human obesity is polygenic, and the genes responsible for obesity have been found to be the same genes that regulate body weight. "Thrifty" genes are thought to conserve fuel reserves during periods of famine, but they constitute a risk for developing obesity and related conditions (e.g., type 2 diabetes mellitus) when energy is abundant. Throughout evolution there has been a balance between thrifty genes or factors that conserve fuel/energy and "antithrifty" genes or energy-mobilizing factors. This balance allowed survival during periods of food scarcity or deprivation without the negative consequences of excess adipose tissue in times when food was plentiful. The ability to store energy as fat in adipose tissue is an important mechanism for individual survival and reproductive capacity. For example, an obese human weighing 250 pounds has enough energy stores to endure a fast of approximately 150 days! However, in the twenty-first century, easy access to energy-dense foods and decreased physical activity have made these genes maladaptive.

Genetics determine our body weight and our susceptibility to weight gain, and this state is then maintained physiologically with deviations in body weight, in either direction, triggering a potent counter response that resists change. Measurement of autonomic nervous system activity in people losing and gaining weight suggests that body weight is a regulated variable, similar to temperature, pulse, and blood pressure. Furthermore, an obese person who has lost weight actually has a lower BMR than a slender, never-obese person of the same weight. The decline in leptin levels, stimulated by dieting or weight loss, may contribute to the decrease in BMR that accompanies weight loss. The decrease in leptin levels leads to energy conservation by decreasing TH-induced thermogenesis and gonadotrophin secretion while at the same time increasing secretion of glucocorticoids that mobilize energy stores. Thus, body weight appears to be regulated by feedback mechanisms that react to decreases in energy intake with an increase in appetite-stimulating mechanisms and a decrease in energy expenditure.

Although genetics has an important role in body fat regulation, the prevalence of obesity is best explained by behavioral and environmental changes that have resulted from technological advances. The epidemic of obesity in the Western world in the twenty-first century is largely a result of the increased availability of food, the modern composition of food (e.g., high sugar and high fat content), and the generally reduced amount of physical exercise in our daily lives (see Text Box 10.4).

TEXT BOX 10.4 THE METABOLIC SYNDROME X

A new syndrome comprised of five associated diseases has recently been described. The metabolic syndrome X is the label applied to individuals with: (1) obesity, (2) insulin resistance and hyperinsulinemia, (3) dyslipidemia, (4) impaired glucose tolerance and/or type 2 diabetes, and (5) hypertension. Central or abdominal obesity is the type associated with this syndrome, but hyperinsulinemia and insulin resistance are suggested to be the causal features of this syndrome. Insulin resistance represents impaired insulin signaling, resulting from a defect in the postreceptor signal transduction pathway that is normally stimulated by insulin binding to its receptor. However, no signal or feature of insulin action has been found to be causal. Insulin-resistant individuals can remain glucose tolerant if the pancreas compensates for this defect by secreting large amounts of insulin, but type 2 diabetes develops when insulin-resistant persons cannot sustain this state of compensatory hyperinsulinemia. Dyslipidemia consists of high serum triglycerides and low HDL-cholesterol. Hypertension is the most independent of the features associated with this syndrome. The question remains as to whether this is one disease or simply a cluster of many aging or maturity-related risk factors because causal links and underlying mechanisms are not known. Many of the associations of this metabolic syndrome seem to be consequences of obesity, and the metabolic syndrome X may be the negative effects of obesity in a subset of genetically unfortunate individuals.

Diabetes Mellitus

In diabetes there is inappropriate regulation of carbohydrate and lipid metabolism by insulin. This leads to chronic hyperglycemia that causes multiple complications including atherosclerosis and coronary artery disease, kidney failure, blindness, and nontraumatic limb amputation. The two major forms of diabetes, types 1 and 2, have similar complications but result from entirely different disease processes. Type 1 diabetes, which represents 5 to 10% of all cases, is characterized by an absolute dependence on exogenous insulin and is thought to develop due to the autoimmune destruction of the insulin-producing β cells of the pancreas. Type 2 diabetes is much more common and is characterized by a progressive resistance to the effects of insulin that eventually overwhelms the ability of the β cell to compensate. Type 2 diabetes arises from a heterogeneous etiology, with secondary effects from environmental influences.

Insulin resistance is the failure of tissues to respond appropriately to the normal, circulating concentration of insulin. It develops due to a combination of factors including obesity, aging, a sedentary lifestyle, and a genetic predisposition. Insulin resistance can arise from defects in insulin signal transduction, changes in the expression of proteins or genes that are targets of insulin action, cross-talk from other hormonal systems, or metabolic abnormalities. Mutations in the insulin receptor are not that common in the general population, and this suggests that postreceptor defects are largely responsible for the failure of insulin action in diabetes.

The first detectable manifestation of insulin resistance is inadequate glucose uptake into skeletal muscle. Peripheral insulin resistance leads to an elevated blood glucose level and, in response, the β cells of the pancreas increase basal and postprandial insulin secretion. However, at some point the β cells are no longer able to compensate and fail to respond appropriately to an elevated blood glucose. This leads to the loss of glucose homeostasis and impaired glucose tolerance that can progress to diabetes.

In addition to hyperglycemia patients with type 2 diabetes have elevated levels of FFAs and TGs in the blood. They also demonstrate excessive deposition of fat in various tissues, including the muscle bed. This disruption of lipid homeostasis may have an important role in the development of insulin resistance. At some point early in the development of type 2 diabetes, adipocytes become resistant to the antilipolytic effects of insulin, by an as yet unknown mechanism. This resistance leads to an increase in levels of plasma FFAs that progressively worsens as β-cell function deteriorates and insulin secretion decreases.

High levels of circulating FFAs can have a multitude of detrimental effects on the liver. High levels of FFAs impair hepatic uptake of insulin leading to further hyperinsulinemia. This increases the expression of SREBP-1c and leads to overexpression of lipogenic enzymes. Malonyl-CoA concentrations rise inhibiting CPT1 and leading to fatty acid esterification, increased liver fat content, VLDL secretion, and hypertriglyceridemia. If intracellular long-chain acyl-CoA levels rise sufficiently, they offset the inhibitory effect of malonyl-CoA on CPT1, with the result that both fatty acid esterification and oxidation will be enhanced. Increased fatty acid oxidation in liver promotes gluconeogenesis from glycerol, both by maintaining pyruvate carboxylase in an active state and by stimulating up-regulation of key gluconeogenic enzymes (e.g., PEPCK, G6Pase). This sets the scene for inappropriately high rates of glucose production by the liver. Increased hepatic glucose output is primarily responsible for the fasting hyperglycemia seen in diabetes and is a major component of postprandial hyperglycemia.

Other negative effects of high levels of FFAs include interference with insulin action in muscle and liver due to inhibition of the insulin-dependent tyrosine phosphorylation of IRS1 and its subsequent downstream interactions. FFAs are thought to stimulate insulin release, and a possible mechanism for this has recently been described. The G-protein-coupled receptor GPR40, which is abundantly expressed in the pancreas, functions as a receptor for long-chain fatty acids and, by stimulating GPR40, long-chain fatty acids amplify glucose-stimulated insulin secretion from the pancreas.

Free fatty acids also up-regulate the expression of the uncoupling protein (UCP2) gene in β cells. Uncoupling proteins dissipate the proton gradient across the inner mitochondrial membrane that is otherwise used to drive ATP formation from adenosine 5′-diphosphate (ADP) via the ATP synthase reaction. In this way substrate oxidation can be uncoupled from ATP synthesis, allowing the energy to be released as heat instead (see Chapter 16 for the roles of UCPs in mammalian thermogenesis during hibernation). Expression of the UCP2 gene is increased and glucose-stimulated insulin secretion is impaired in pancreatic islet cells of rats fed a high-fat diet. The increase in UCP2 caused by FFA may protect against reactive oxygen species produced when the rate of fatty acid oxidation and subsequent mitochondrial respiration is high. However, it has recently been shown in mice that this increase in UCP2 expression may have significant negative repercussions. Increased expression of UCP2 in pancreatic β cells leads to decreased mitochondrial coupling that decreases the ratio of ATP to ADP and thereby decreases insulin secretion that leads eventually to the development of type 2 diabetes. Mice made obese by mutation in the leptin gene have higher levels of UCP2 in the pancreatic islet cells, but mice engineered with a deficiency of UCP2 in the pancreatic islet cells showed increased ATP levels and enhanced insulin secretion.

Since peripheral insulin resistance is often the first demonstrable change detected in states of impaired glucose tolerance, investigators have attempted to understand this abnormality by studying insulin-stimulated glucose metabolism in muscle, the organ that mediates most of this effect. Knockout mice lacking the muscle insulin receptor manifested moderate insulin resistance and a mild metabolic disorder but did not develop either type 2 diabetes or any major metabolic changes. The lack of overt diabetes in mice with insulin receptor defects in muscle could be explained by the normal function of the insulin receptor in other important tissues (e.g. β cells, liver, adipose tissue), but insulin-stimulated glucose uptake into specific muscle subgroups was not totally abolished, suggesting that insulin signaling in muscle may also be mediated through another pathway (e.g., via the IGF-IR). By contrast, muscle-specific disruption of GLUT4 results in severe insulin resistance and, in some mice, overt diabetes. Elimination of the insulin receptor gene in pancreatic β cells produced diabetic mice, and, when the insulin receptor was eliminated in brain, mice demonstrated increased food intake and obesity and associated insulin resistance. This observation, although only in mice, suggests a coordinated interaction between tissues in the regulation of glucose homeostasis that is more complex than previously anticipated.

Other studies engineered a dominant-negative mutant of the IGF-IR that was mutated to abolish ATP binding within

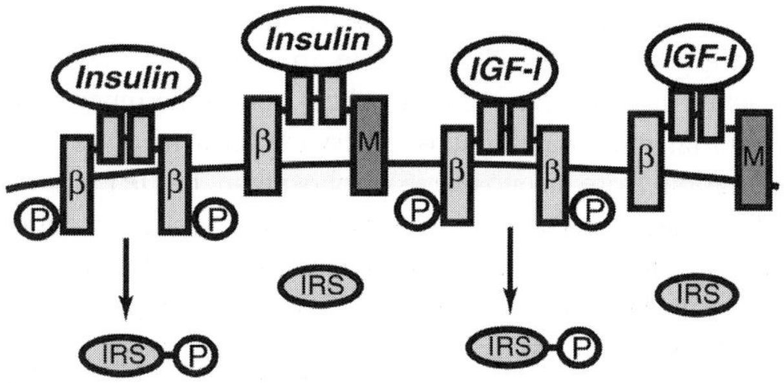

Figure 10.10 Mutation of the IGF-I receptor affects insulin and IGF-I signaling. Expression of a dominant-negative mutant of the IGF-I receptor (IGF-IR) resulted in the formation of nonfunctional hybrid receptors between the mutant IGF-IR and endogenous IGF-IRs and insulin receptors (IR). When two functional β subunits (β) combine, the receptor is functional and is able to autophosphorylate in order to initiate the signaling pathway that results in phosphorylation of IRS-1. However, when the mutated (M) β subunit combines with another mutated β receptor (not shown) or with a normal β subunit, then a nonfunctional hybrid is formed. Failure of this nonfunctional receptor to autophosphorylate results in functional inactivation of the hybrid receptors and markedly reduces insulin-induced glucose uptake into muscle since IRS-1 is not phosphorylated.

the β subunit and expressed under the control of the muscle creatine kinase promoter to ensure its expression only in skeletal muscle. Expression of this mutant resulted in a marked loss of function of both the IGF-IR and the insulin receptor (IR) as reflected in a marked decrease in glucose uptake upon stimulation with IGF-I or insulin. Furthermore, these mice were insulin-resistant and rapidly developed overt diabetes. This dominant-negative effect may be mediated by the formation of nonfunctional hybrid receptors between the mutant IGF-IR and endogenous IGF-IRs and IRs (Fig. 10.10). Failure of the mutant hemireceptor to autophosphorylate prevents phosphorylation of the endogenous hemireceptor and, thereby, results in functional inactivation of the hybrid receptors and markedly reduced insulin-induced glucose uptake into muscle. Increased expression of IR/IGF-IR hybrids has been reported in the skeletal muscle of patients with severe insulin resistance. The phenotype of the mice with nonfunctional skeletal muscle IGF-IR is similar to mice lacking skeletal muscle GLUT4 (severe insulin resistance, fasting hyperglycemia, glucose intolerance), and because GLUT4 is a downstream target of the IR and the IGF-IR, these studies suggest that disruption of both the receptors (IGF-IR and IR) or of the signaling pathway mediating the increase in plasma membrane GLUT4 content in muscle can lead to a diabetic phenotype. By contrast, disruption of the IR signaling pathway in muscle can be functionally compensated for by the IGF-IR.

Free fatty acids or disruptions in insulin signaling may indeed be responsible for insulin resistance, but research is revealing that molecules secreted by adipocytes also have a role. These molecules include tumor necrosis factor α (TNF-α), leptin, and two newly discovered molecules, the forkhead transcription factor (FOXC2) and the hormone resistin.

The TNF-α is expressed in normal adipocytes, is overexpressed in adipocytes from obese people, and can cause insulin resistance through effects on insulin-mediated cellular signaling pathways. The mechanism of action is unknown, but TNF-α may act by reducing the kinase activity of the insulin receptor, possibly by increased serine phosphorylation of the receptor. Rodents treated to reduce circulating levels of TNF-α demonstrated increased insulin receptor kinase activity and increased insulin sensitivity. TNF-α also stimulates catabolism in the adipocyte by blocking lipid synthesis and LPL expression while activating lipolysis.

Speculation regarding the primary defect that leads to the metabolic derangements that characterize type 2 diabetes has examined the possible role of leptin in the pathogenesis of diabetes. Rodents that lack either leptin or the leptin receptor have low sympathetic activity and go on to develop obesity and symptoms of type 2 diabetes. Treatment of normal animals with leptin produces increased sympathetic tone and decreases TG stores. Defective leptin signaling and low sympathetic activity decreases the oxidation of FFAs in skeletal muscle, which leads to increased intramyocellular lipid concentrations causing decreased glucose uptake and insulin resistance. To compensate for the decreased insulin sensitivity in muscle, the pancreas increases insulin secretion to maintain normal glucose tolerance. Chronic hyperinsulinemia further diminishes fatty acid oxidation, raising the concentrations of fatty acyl-CoA and TG, which interferes with the pathways associated with normal glucose uptake. Elevated concentrations of insulin also increase the output of VLDL by the liver. Circulating VLDL will deposit in the muscle and adipose tissue, which in the early stages of the disease will increase fat deposition. If, on top of this, hyperphagia is added (secondary to defective leptin signaling), the situation deteriorates. Eventually, the ability of the adipose tissue to reesterify FFAs is overwhelmed, increasing their concentration in plasma. Excessive FFA will enhance glucose production by the liver, precipitating impaired glucose tolerance.

The transcription factor FOXC2 appears to be a key regulator of adipocyte metabolism. Overexpression of FOXC2 in mice resulted in systemic changes to glucose, insulin, and lipid metabolism that acted to prevent the consequences of caloric overload (i.e., increased adipose tissue and weight gain). The intraabdominal white adipose tissue of mice overexpressing FOXC2 acquired a histology similar to that of brown adipose tissue (i.e., greater numbers of mitochondria) and intrascapular brown adipose tissue became hypertrophic (note that brown adipose is a thermogenic tissue found in rodents and other small mammals, as well as neonates of many mammals including humans). Furthermore, the brown adipose tissue-specific uncoupling protein (UCP1) was up-regulated in abdominal white adipose tissue and genes important for insulin action, adipocyte differentiation and metabolism, sensitivity to adrenergic stimuli, and intracellular signaling were up-regulated in both white and brown adipose tissue. Mice overexpressing FOXC2 also seemed to have an increased sensitivity to the β-adrenergic/cAMP/PKA pathway and compared to control animals had less body fat and demonstrated slower weight gain when fed a high-fat diet. FOXC2-induced up-regulation of HSL in white adipose tissue could provide an increased flow of FFA to the mitochondria where enhanced UCP1 levels would dissipate the energy released from lipid oxidation as heat. The enlarged brown adipose tissue may also produce a factor that suppresses appetite and prevents the FOXC2 transgenic mice from eating more to compensate for their loss of adipose tissue. FOXC2-mediated changes appear to act to protect against obesity and suggest that FOXC2 is an antithrifty gene that defends against obesity and is an important regulator of adipocyte metabolism with the ability to alter the metabolic efficiency of adipose tissue in response to dietary changes. The induction of FOXC2 will determine how efficiently excess calories are converted to TG with pronounced expression of FOXC2 preventing an increase in adipose tissue mass. Decreased expression of FOXC2 or insensitivity to its actions could result in obesity, insulin resistance, and type 2 diabetes.

A novel peptide hormone named resistin has recently been discovered. It is secreted only by adipocytes and expression is decreased by a class of antidiabetes drugs known as thiazoladinediones. These drugs are known to decrease insulin resistance, but their mechanism of action has been unknown. However, it is known that the thiazoladinediones are mediated via the peroxisome proliferator-activated receptor γ (PPAR-γ), and this nuclear receptor protein apparently switches off the expression of resistin. This has been confirmed in studies with PPAR-γ agonists that have been shown to enhance insulin sensitivity by decreasing resistin expression. Hence, it is hypothesized that decreased expression of resistin is involved in reducing insulin resistance. Moreover, resistin is overexpressed in several rodent models of obesity (both genetic and diet-induced), and thiazoladinediones reduce the amount of resistin in the rodent bloodstream. Thus, resistin appears to be a link between diabetes and obesity and provides an explanation for how increased adiposity leads to insulin resistance. This hypothesis is supported by studies in adipocytes, where neutralization with resistin antiserum-enhanced insulin-stimulated glucose uptake, and insulin action was blunted by recombinant resistin. Administration of resistin to mice impaired glucose tolerance without reducing insulin levels and decreased sensitivity to the effects of insulin. Neutralization of resistin reduced the hyperglycemia of obese, insulin-resistant mice, at least in part, by improving sensitivity to insulin. These data suggest that resistin is a unique hormone whose effects on glucose metabolism are antagonistic to those of insulin.

Diabetic ketoacidosis (DKA) is an acute complication of type 1 diabetes mellitus. It can occur in undiagnosed type 1 diabetes or in type 1 diabetics suffering from infection or other stressors (e.g., trauma, myocardial infaction, surgery, or poor compliance with insulin dosing). It can also develop in type 2 diabetics under severe stress such as sepsis, trauma, and major surgery. The underlying mechanism of this disorder is an acute insulin deficiency. This deficiency leads to elevation of counterregulatory hormones such as glucagon, catecholamines, cortisol, and growth hormone that cause a rapid mobilization of fuel stores in muscle and fat. There is increased flux of amino acids to the liver for conversion to glucose and increased flux of fatty acids, also to the liver, for conversion to ketone bodies (acetoacetate and β-hydroxybutyrate) resulting in amounts of gluconeogenesis and ketogenesis that are inappropriate. The low insulin–glucagon ratio also promotes increased ketone production. However, in the absence of insulin, peripheral utilization of glucose and ketones is reduced. This combination of increased production and decreased utilization leads to accumulation of glucose and ketones in the blood. Hyperglycemia has significant effects that contribute to the metabolic derrangements characteristic of this disorder. Once the concentration of glucose in the blood exceeds the renal threshold, glucose "spills" over into the urine, and this causes an osmotic diuresis that depletes the vascular volume. The ketoacidosis may also cause vomiting, which further exacerbates dehydration. Dehydration reduces renal blood flow, which impairs renal glucose excretion and further worsens urine hyperosmolality. Renal hydrogen ion excretion is also impaired, and this worsens the metabolic acidosis that occurs from the accumulation of ketones. Prolonged acidosis can compromise cardiac output and reduce vascular tone, which can lead to cardiovascular collapse with generation of lactic acid, which adds to the metabolic acidosis. The progressive increase in blood osmolality and acidity leads to a decrease in the level of consciousness, progressing to coma and death unless intervention, in the form of exogenous insulin, occurs. Thus, a decrease in the availability of insulin leads to metabolic derrangements that contribute to the severity of the illness in a vicious cycle.

A hyperglycemic, hyperosmotic, dehydrated state can also develop from insulin deficiency but, unlike DKA, there is no ketosis. This syndrome occurs in older patients with type 2 diabetes, and the onset of symptoms is usually more gradual. A partial or relative insulin deficiency may initiate the syndrome by reducing glucose utilization by muscle, fat, and the liver while at the same time inducing hyperglucagonemia and increasing hepatic glucose output. The end result is hyperglycemia that leads to glycosuria and osmotic diuresis with obligatory water loss (as in DKA). However, it is believed that the presence of even small amounts of insulin is sufficient to prevent the development of ketosis by inhibiting lipolysis in the adipose stores. Thus, even though a low insulin–glucagon ratio promotes ketogenesis in the liver, the limited availability of precursor free fatty acids from the periphery restricts the rate at which ketones are formed.

SUGGESTED READING

Ahima, R. S., and Flier, J. S. (2000). Leptin. *Ann Rev Physiol* **62**:413–437. *Everything you wanted to know about leptin but were afraid to ask.*

Griffin, J. E., and Ojeda, S. R. (eds.) (2000). *Textbook of Endocrine Physiology*. Oxford University Press, Oxford, *Basic science and clinical information on all known human hormones.*

Larsen, P. R., Kronenberg, H. M., Melmed, S., Polonsky, K. S. (eds.) (2003). *Williams Textbook of Endocrinology*, 10th ed. W.B. Saunders, St. Louis. *A comprehensive textbook of endocrinology that provides an in-depth discussion of all known hormones.*

Lewis, G. F., Carpentier, A., Adeli, K., and Giacca, A. (2002). Disordered fat storage and mobilization in the pathogenesis of insulin resistance and type 2 diabetes. *Endo Rev* **23**:201–229. *A review of our current understanding regarding the pathogenesis of type 2 diabetes.*

Nature Insight (2000). Obesity. *Nature* **404**:631–677. *A special issue devoted to the topic of obesity that discusses epidemiology, etiology, energy expenditure, and metabolism and treatment options.*

Saltiel, A. (2001). New perspectives into the molecular pathogenesis and treatment of type 2 diabetes. *Cell* **104**:517–529. *A review of new insights into the pathophysiology and genetics of type 2 diabetes and new therapeutic interventions.*

Spiegelman, B. M., and Flier, J. S. (2001). Obesity and the regulation of energy balance. *Cell* **104**:531–543. *A review of the current state of knowledge regarding the pathogenesis of human obesity.*

11

SKELETAL MUSCLE METABOLISM AND PLASTICITY

DENIS R. JOANISSE

INTRODUCTION

The main function of skeletal muscle is to perform mechanical work at the expense of potential chemical energy, usually for postural support and movement. This activity is a defining feature of higher animal life as we know it. However, skeletal muscle is also involved in many other physiological processes. For example, it is a key player in thermogenesis; and, as a metabolic organ, it is important in fuel and energy homeostasis. This is not surprising when one considers that skeletal muscle is the largest single tissue in the body, typically representing 40 to 50% of the total mass of a lean adult human. Mammalian skeletal muscles also display an impressive and wide spectrum of functional and metabolic characteristics. The rate of force production, fatigue resistance, and the nature of energy metabolism are but some of the features that can vary between muscles. Most impressive is the capacity of a given muscle to modify its functional characteristics to meet the demands of the organism. This skeletal muscle plasticity is observed in response to many stimuli, such as exercise training and detraining, growth and differentiation factors, hormones, and nerve impulses. Most animal and human studies have focused on peripheral skeletal muscle when researching the impact of these factors, for example, those of the legs and arms. This is due to the fact that these muscles are the ones that are mainly recruited during exercise, as well as being those most easily accessible for recovery (biopsy) or analysis. Whether all skeletal muscles respond similarly to all stimuli remains an unanswered question.

It is important to remember that skeletal muscle does not act alone. Important communication and coadaptation at the whole organism level must take place to optimize muscle function. For example, rates of oxygen and extramuscular fuel supply must be closely matched to muscle demand. Although not extensively covered in this chapter, it is important to keep this in mind when considering muscle function and adaptation.

SKELETAL MUSCLE PLASTICITY

Dynamic State of Muscle Fibers

Skeletal muscle cells are remarkable for their ability to dramatically change their composition and metabolic capacity as a result of changing functional demands. Thus, marathon runners and sedentary individuals have skeletal muscles with very different capacities for aerobic metabolism. As is usually observed in adaptation, most morphological and metabolic changes in skeletal muscle require repeated use or prolonged inactivity. In fact, the nature and extent of the observed adaptations are dependent on the duration and the type of the stimulus, or on the nature and length of its absence. These adaptations result in a match in the functional capacity of muscle with the actual needs of the organism, this equipoise being the most energy and resource efficient. Thus, skeletal muscle adaptation follows the principle of symmorphosis, which states that structural components of biological systems are quantitatively matched to the functional demands.

In general, aerobic or endurance training improves the capacity of the cardiovascular system to deliver oxygen to muscle as well as the capacity of the muscle for aerobic energy production. The latter occurs in part via an increased capacity for lipid catabolism. Strength or anaerobic training, on the other hand, increases muscle strength as

Functional Metabolism: Regulation and Adaptation, edited by Kenneth B. Storey.
ISBN 0-471-41090-X

well as its resistance to fatigue during high-intensity work. This type of training also favors a greater capacity for the use of carbohydrates for energy production, without improving the capacity for lipid use. A high-intensity intermittent type of training, which recruits both aerobic and anaerobic metabolism, results in increased strength as well as an improved capacity for aerobic metabolism. Chronic activity leads to these changes by the modulation of protein synthesis at pretranscriptional and transcriptional levels, as well as posttranslational modifications. Detraining, or a period of insufficient training stimulus, will lead to a partial or total loss of the adaptations produced by previous training. These effects will vary with the length of the detraining period as well as the level of activity during this period. For example, the maintenance of a minimal level of physical activity or the use of therapeutic electrical stimulators can significantly decrease the rate of muscle atrophy and metabolic changes.

Muscle Fiber Types

The classification of muscle into red and white types is ancient. Red muscles have fibers containing more capillaries, myoglobin, and mitochondria, making them better equipped for aerobic activities than white muscle. Thus, red muscle is better suited to long-term activities such as postural maintenance, whereas white muscle is considered to have a "flight or fight" role in the production of high velocities of contraction and rapid energy production based largely on anaerobic metabolism. The basis for the obvious color difference between the two muscles lies in the myoglobin content within the myofibers. In the presence of oxygen, this muscle-specific hemoglobin-like protein, which is involved in intramyocellular oxygen transport and storage, imparts a red color to muscle. Since myoglobin content is linked to the aerobic capacity of the cell, it is not surprising that red muscle is often referred to as aerobic muscle and white muscle as glycolytic muscle. Today, the myofibers that make up these muscles are usually classified based on the combination of myosin heavy-chain isoforms they contain (I, IIa, and IIx in postnatal humans); type I and IIA fibers are usually equated with red muscle, and type IIX fibers with white muscle (note that lowercase *a* and *x* are used for myosin isoforms, uppercase *A* and *X* designate fiber types).

Within a given muscle, however, not all fibers are identical. For example, human soleus muscle is typically made up of about 80% type I fibers with very few if any type IIX fibers, whereas the triceps shows a greater variety of fibers typically containing about 37% type I, 37% IIA, and 26% type IIX fibers. In fact, when one also considers the highly variable composition in contractile and metabolic proteins and the different physiological properties of each cell within a given muscle, it is likely that there are no two identical cells within a given muscle. Furthermore, the function of any individual muscle at any given moment is a reflection of the number, type, and properties of the fibers that are recruited at that time.

TABLE 11.1 Muscle Fiber Type Classification Systems and Characteristics of the Motor Units Related to Them[a]

Classification of fibers by			
MHC isoform	Type I	Type IIa	Type IIx (or IIb)
Contraction speed	ST	FTa	FTb
Contraction speed and metabolism	SO	FOG	FG
Color (myosin content)	Red	Red	White
Motor Unit			
Classification	S	FR	FF
Fibers/motoneuron	10–200	300–3000	300–3000
Size of motoneuron	Small	Large	Large
Nerve impulse conduction	Slow	Fast	Fast
Contraction time	110 ms	50 ms	50 ms
Motor unit strength	Weak	Strong	Strong

[a]Abbreviations: ST, slow type; FT, fast type; SO, slow oxidative; FOG, fast oxidative glycolytic; FG, fast glycolytic; S, slow; FR, fast fatigue resistant; FF, fast fatiguable.

Different classification schemes based on structural, metabolic, or functional properties of myofibers have been used in the last century, and these are summarized in Table 11.1. Classification of fiber types by myosin heavy-chain (MHC) composition is common. Muscle fiber typing by MHC composition is usually achieved in one of two ways. Histological staining of myosin–adenosinetriphosphatase (ATPase) activity can be done following preincubation at different pH values (Fig. 11.1); this takes advantage of the differential pH lability of this activity among the MHC isoforms. Second, immunohistochemical determination with MHC isoforms can be accomplished using type-specific antibodies. Various electrophoretic approaches are also available to measure the relative content of MHC isoforms from single-fiber or tissue homogenates.

Links between skeletal muscle fiber type distribution and high levels of athletic performance are well-established. In general, endurance athletes have a very high proportion of type I fibers in their leg muscles (commonly >75% in quadriceps), whereas the opposite is true of sprinters or other athletes in high-intensity short-duration events where type II fibers predominate (typically >80% in quadriceps). Although genetics play a role in the determination of an individual's skeletal muscle fiber type profile (accounting for ~45% of the variation in a population), it is now widely accepted that the proportion of muscle fiber types can be changed with physical training, and that this will

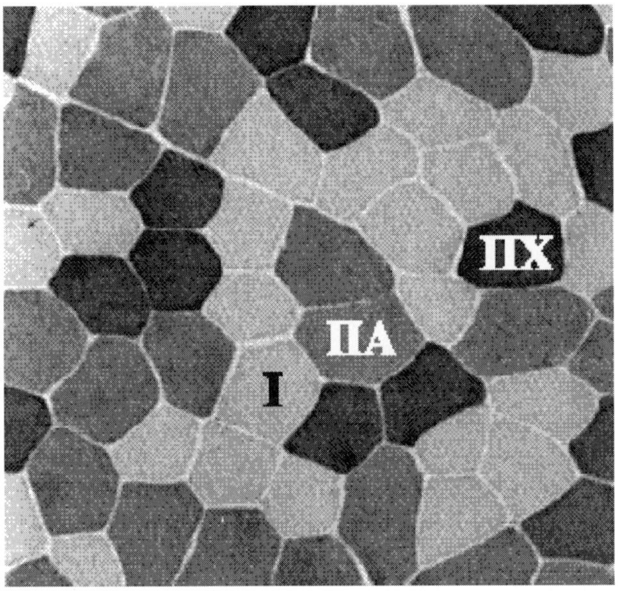

Figure 11.1 Histochemical determination of skeletal muscle fiber types (types I, IIA, IIX) by myosin ATPase activity staining following preincubation at pH 9.4. Cross-sectional view of human vastus lateralis (part of the quadriceps group of muscles).

have an important impact on the functional and metabolic capacity of the whole muscle. This is most easily observed when the chronic contractile activity of a muscle group is changed significantly, such as during training or detraining. Single-fiber analysis has shown that more than one myosin heavy-chain isoform can exist within a single myocyte, for example, a combination of IIa and IIx or I and IIa. This gives rise to a series of structurally and functionally intermediate fibers (Table 11.2). Since no single fiber has been shown to contain type I and IIx MHC isoforms at the same time, it is believed that the transformation of muscle fibers follows a progressive pattern, that is, I ↔ IIA ↔ IIX.

Of course, changes to MHC isoforms are not sufficient to transform a muscle fiber from one type to another. In fact, many other differences occur. These include, but are not limited to, differences in the isoform expression of other myofilament and sarcoplasmic reticulum proteins (e.g., myosin light chains, tropomyosin, troponin C, I, and T, sarcoplasmic reticulum ATPase; Table 11.3), structural and organelle differences (e.g., the Z line of the sarcomere is thicker and mitochondrial volume density is increased in type I fibers), as well as the expression of enzymes of energy metabolism [e.g., elevated glucose transporter (GLUT4) expression in type I fibers, higher glycolytic activities in type II fibers]. A summary of some of the differences between fiber types is given in Table 11.4.

The nature of the training (or detraining) stimulus plays a determining role in changes to fiber type proportions. Endurance activities mainly recruit type I fibers, which are well-equipped for aerobic metabolism. Endurance training, which leads to an increase in the muscle's capacity to maintain activity levels for a long period of time, does so in part by increasing the proportion of type I and/or IIA fibers (usually at the expense of IIX fibers). On the other hand, strength or sprint training, based on explosive efforts that depend mainly on anaerobic metabolism for adenosine 5′-triphosphate (ATP) production, are generally associated with a decreased content of type I fibers in favor of an increase in type II fibers that are better equipped for this

TABLE 11.2 Myosin Heavy-Chain (MHC) Composition of Major Human Muscle Fiber Types

Fiber Type	MHC Isoform
I	Iß
IC	I > IIa
IIC	IIa > I
IIAC	IIa ≈ I
IIA	IIa
IIAX (IIAB)[a]	IIa ≈ IIx (IIb)[a]
IIX (IIB)[a]	IIx (IIb)[a]

[a] Both the B and X nomenclature are found in the literature and are interchangeable when used with respect to human skeletal muscle. However, the properties of the human-type IIb MHC isoform most closely match those of the rat IIx isoform, and many now consider the IIX terminology more appropriate. Note that IIB fibers from rodent skeletal muscle are not the equivalent of IIB fibers in humans.

TABLE 11.3 Main Isoforms of Human Thin and Thick Filament Proteins from Type I and II Muscle Fibers

Protein	Type I Fiber	Type II Fiber
Myosin heavy chain	MHC Iß	MHC IIa
		MHC IIx
Myosin light chain		
Regulatory	MLC 2s	MLC 2f
Alkaline (essential)	MLC 1sa	MLC 1f
	MLC 1sb	MLC 3f
Tropomyosin	Tm α slow/cardiac	Tm α fast
	Tm ß	Tm ß
Troponin C	TnC slow/cardiac	TnC fast
Troponin I	TnI slow	TnI fast
Troponin T	TnT 1s	TnT 1f
	TnT 2s	TnT 2f
		TnT 3f
		TnT 4f

Source: Adapted from Gardiner (2001).

TABLE 11.4 Ultrastructural and Cytochemical Characteristics of Human Muscle Fibers

		Fiber Type		
	Characteristic	I	IIA	IIX
Physiology	Contraction velocity	Slow	Fast	Fast
	Myosin ATPase	Slow	Fast	Fast
	Contraction time (duration)	Long	Short	Short
	Ca^{2+} reuptake	Slow	Fast	Fast
	Fatigue resistance	+++	++	+
	Recruitment order	Early	Mid	Late
Morphology	Color	Red	Red	White
	Fiber diameter	+	+++	+++
	Capillary density	+++	++	+
	Mitochondrial density	+++	+++	+
	Z-line thickness	+++	++	+
	Sarcoplasmic reticulum volume	+	+++	+++
	T-tubules volume	+	+++	+++
	Motor end-plate area	+	+++	+++
Enzyme activities	Myosin ATPase	+	+++	+++
	Hexokinase	+++	++	+
	Glycogen phosphorylase	+	++	+++
	Glycolytic enzymes	+	++	+++
	Oxidative enzymes	+++	++	+
	Myoglobin	+++	++	+
Metabolites	Glycogen	++	+++	+++
	Triglycerides	+++	++	+

type of metabolism. Changes to muscle fiber type proportions occur as soon as a few weeks following the start of relatively high-volume training.

Changes to muscle fibers resulting from increased activity can easily be lost during periods of relatively reduced activity or inactivity. It is during prolonged immobilization of limbs (such as with a cast) that the greatest changes in muscle fiber type proportions have been observed in humans, usually most pronounced in the endurance-trained. In these individuals, decreases of 25% in the proportion of type I fibers are regularly noted following relatively short periods (~6 weeks) of immobilization. These losses are reversible with retraining.

Despite these observations, the correlations between athletic performance and fiber type distribution are modest (but significant). Other factors, such as cardiovascular performance and skeletal muscle metabolic capacity, also contribute to performance.

Muscle Morphology: Fiber Size and Capillary Density

Skeletal muscle hypertrophy or atrophy following changes in activity levels are well-known phenomena. Under normal circumstances, changes to muscle mass following training or detraining in humans do not result from changes to the number of myofibers (hyperplasia or hypoplasia). Instead, increases in the volume of existing myofibers (cell hypertrophy or atrophy) are mainly responsible for the observed changes. For example, comparisons between sedentary and strength-trained individuals reveal about twofold increases in myofiber diameter (~5000 μm^2 vs. ~10,000 μm^2). The fiber size differences tend to be much smaller between sedentary and endurance-trained individuals; the latter often display increased type I fiber diameter (indeed, type I fibers can become >20% larger than type II fibers in highly trained endurance athletes).

Since an increased cross-sectional muscle area is correlated with increased muscle strength and power, it is not surprising that chronic overloading or high-intensity activity results in increased myofiber diameter. Other quantitative changes within the myofibers occur with hypertrophy that also help to increase the strength of contraction. In endurance-trained individuals, ~80% of the cell volume is occupied by the myofibrils. In weightlifters, this increases to ~85%. Combined with the larger fibers in the latter, the total surface of myofibrils within a given myofiber is significantly greater, allowing the production of greater strength during contraction. Such an increase in myofiber size does not contribute to enhanced performance in endurance activities.

Detraining or immobilization has severe consequences on muscle mass. Most commonly observed on limb muscles following casting or bed rest, losses of up to 50% of initial volume can occur within 6 to 8 weeks. These changes are intimately linked to changes to myofiber size. Losses are less severe with bed rest than with complete immobilization, typical leg muscle volume loss after 15 to 20 weeks of bed rest ranging between 20 and 30%. The speed at which muscle atrophy progresses is particularly disconcerting when one considers that most of the loss occurs in the first few weeks of immobilization.

The number of capillary contacts with a muscle fiber is correlated with the capacity of the fiber for aerobic metabolism. Thus, type I fibers have a greater capillarization that type IIA fibers, which are, in turn, better irrigated that type IIX fibers (Table 11.5). Furthermore, the number of capillaries that irrigate muscle fibers can be modified in response to training and detraining. Chronic aerobic activity, such as that seen during endurance training, will lead to an increase in the number of capillaries surrounding the fibers. Thus, type I and IIA fibers from highly trained endurance athletes (these athletes have few if any type IIX fibers) have more capillaries than sedentary or moderately active individuals. In extreme cases of endurance activity, such as international-level endurance cyclists, the number of capillaries surrounding some fiber types (notably the highly aerobic type I) can reach 10 or more. Significant increases in capillary numbers can occur within the first few weeks of training, and this increase in muscle perfusion allows better gas, substrate, and product exchange between muscle cells and the circulation, thus favoring aerobic energy production and a greater capacity to maintain repeated contractions over the long term.

In contrast to endurance training, programs geared toward the development of muscle mass and strength, focusing on high-intensity anaerobic activity, typically do not increase the number of capillaries surrounding muscle

TABLE 11.5 Typical Capillary Contacts with Muscle Fibers in Human Vastus Lateralis (Part of the Quadriceps Muscle Group)[a]

	Sedentary	Active	Endurance Trained
Type I	4.5	5.5	6.5
Type IIA	4	5	6
Type IIX	3.5	4	ND

[a]ND; not determined, as endurance trained athletes have few if any type IIX fibers.

cells. Combined with the myofiber hypertrophy that occurs with these training programs, the end result can in fact be a net decrease in the irrigation of muscle fibers (when measured as capillaries per fiber area) when compared to sedentary individuals. For example, following a 20-week strength training program in young women, capillary density decreased from 522 to 492 capillaries/mm^2 of muscle. A consequence of such an adaptation is that muscle becomes less well equipped for the extraction of circulating substrates and oxygen and, thus, is less resistant to fatigue during long-term activity. Of course, individuals that pursue such a training program are not looking to improve their performance in endurance activities, but to increase their explosiveness.

Detraining also has an important impact on muscle capillarization. In cross-country skiers that participated in a 1500-km Swedish expedition carrying an ~25-kg backpack, triceps brachii biopsies clearly revealed that skeletal muscle capillarization could be increased following training (~15% after 3 weeks and 40% after 8 weeks), but also that these changes almost completely disappeared within 33 weeks following the end of the expedition. Even though this and other studies have clearly established the plasticity of the vasculature surrounding muscle fibers, the angiogenic factors responsible for these changes are still a matter of investigation.

Metabolic Profile

There is no doubt that the metabolic profile of muscle can be changed with training. Changes to the metabolic machinery occur according to the type and intensity of training. In general, endurance (aerobic) training significantly increases the capacity of muscle for oxidative metabolism, by increasing the mitochondrial machinery. This type of training also increases the capacity and dependence of the aerobic machinery on lipid oxidation at a given activity level. On the other hand, strength or sprint training leads to increased anaerobic and glycolytic capacities of muscle, in order to provide the high ATP flux required for these types of exercises. The specificity of these adaptations optimizes the capacity of muscle to the real needs of the organism. Further details on the adaptation of these systems follow in appropriate sections below.

Since ATP production is central to contraction, it is not surprising that much work in adaptation to training focuses on the machinery of energy metabolism. However, other cellular processes are important and essential in the adaptation of muscle cells. One of the next great challenges in muscle biochemistry is the elucidation of the mechanisms and signals responsible for the adaptive changes of muscle.

FUELS AND THEIR CONTROL

Energy Use and Production in Muscle

The range of metabolic rates observed in muscle is unmatched by any other tissue, increasing by as much as 200-fold in working muscle. From a resting level of about 2 mL kg^{-1} min^{-1}, muscle oxygen consumption can account over 90% of the body's aerobic metabolism during exercise. Thus, skeletal muscle shows a remarkable ability to respond to the functional demands imposed upon it by acutely changing conditions.

All muscular activity is dependent on the availability of substrates for ATP production. Muscle cells can tap circulating (blood glucose or lipids) or intracellular (glycogen, lipids, phosphagens) reserves. Which of these reserves is used depends on the nature of the activity performed. High-intensity activities require a very high rate of ATP production that can only be satisfied from the exploitation of intracellular phosphagens and/or glycogen; however, these two resources are depleted after a short time, leading to muscle fatigue. During work of lower intensity, muscle cells extract a higher proportion of their substrates from the circulation and catabolize it aerobically. Still, intramuscular glycogen remains the preferred carbohydrate for muscle ATP production during contraction. Under normal conditions, the use of protein is a minor process in exercising skeletal muscle and will not be discussed here. Only during prolonged starvation does protein become an important fuel for muscle oxidation.

Fuel Availability and Use

Levels of ATP within muscle are finite. During high-intensity work, ATP supplies can only support contraction for a few seconds before they are exhausted (Table 11.6). In fact, calculations reveal that a marathoner needs well over 20 kg of ATP to complete a race (some calculations suggest as much as 60 kg!). Furthermore, cellular ATP levels must not be permitted to decrease below about 30% of their initial values because ATP is required for myofibers to return to their initial resting state (e.g., the return of Ca^{2+} into the sarcoplasmic reticulum). Clearly, an efficient system must be in place to replenish ATP in order to sustain the high rates of ATP turnover required for prolonged muscle work. Oxidative phosphorylation, glycolysis, the creatine kinase reaction, and the adenylate kinase reaction are the primary sources of ATP for working skeletal muscle.

The major fuels used by skeletal muscle are the phosphagens [phosphocreatine (PCr) in vertebrates, phosphoarginine, or others in invertebrate muscles], glucose, glycogen, lactate, triglycerides, and free fatty acids. Glucose and glycogen can be catabolized aerobically (using oxidative phosphorylation) or anaerobically (substrate-level phosphorylation). Lipids and lactate can only be used for ATP production in the presence of oxygen. As seen above, different muscle fiber types are better equipped for aerobic or anaerobic ATP production and, consequently, are preferentially used under these conditions. In a fairly simplistic, but nonetheless accurate view, substrates are chosen so that the rate of ATP production that their catabolism provides (their power for ATP production) is closely matched to the rate of ATP use for the particular activity being performed. During light exercise, fuels are degraded to CO_2 and H_2O via aerobic processes, whereas activity of increasing intensity relies more on anaerobic ATP production. Although oxidative phosphorylation is the most efficient mode for ATP production (in terms of the absolute extraction of potential energy from substrates and the number of ATP produced per substrate unit), it is a lengthy and relatively slow process. Generally, ATP production processes with the highest power output (the greatest amount of ATP that can be produced in a given time) rely on fuels found in the smallest quantities, while those processes with the highest capacity (the highest total quantity of ATP that can be produced) are only able to sustain power outputs of low or moderate intensity.

TABLE 11.6 Maximum Rates of ATP Production Achieved According to Choice of Substrate and Nature of Metabolism Used

Substrate for ATP Production	Maximum Rate of ATP Synthesis	Concentration (per Dry Weight of Muscle)	Estimated Maximal Duration at Maximal Activity
ATP	12 mmol/kg/s[a]	25 mmol/kg	Few seconds
PCr	9 mmol/kg/s	100 mmol/kg	6–10 s
CHO (muscle glycogen, anaerobically to lactate)	5 mmol/kg/s	500 mmol/kg (glycogen)	Not limiting
CHO (muscle glycogen, aerobically)	3 mmol/kg/s	500 mmol/kg (glycogen)	90–120 min
Free fatty acids (circulating)	1.5 mmol/kg/s	Practically unlimited	Not limiting

[a]Note that for ATP the rate given is the maximum rate of consumption. CHO, carbohydrate; PCr, phosphocreatine.

TABLE 11.7 Main Energy Reserves Available to Skeletal Muscle[a]

		Grams	Kilocalories	Minutes of Marathon Possible (at ~20 kcal/min)
Carbohydrate	Liver glycogen	75	308	~15
	Skeletal muscle glycogen	350	1,435	~72
	Blood-borne glucose	15	62	~3
	Total	440	1,805	
Lipids	Subcutaneous	7,800	70,980	~3,550
	Intramuscular	161	1,465	~73
	Total	7,961	72,445	

[a]These are typical values for a reference man (70 kg, nonobese) in the postabsorptive state of a normal diet.

Anaerobic metabolism, which provides a very high rate of ATP production during periods of intense activity, is largely provided for by intramyocellular substrate reserves. Substrates for aerobic metabolism, on the other hand, can be supplied from intramyocellular stores or from the circulation. Triglycerides and carbohydrates are the main fuels for aerobic metabolism and are stored inside and outside the muscle. Intramyocellular stores include glycogen granules and lipid droplets, and these are strategically located near mitochondria for easy access during aerobic metabolism. Circulating fuels originate mainly from liver gluconeogenesis and glycogenolysis and adipose tissue lipolysis. Although adipose tissue lipid stores represent the vast majority (over 90%) of aerobic fuel reserves in healthy individuals (Table 11.7), their use during physical activity is limited by the low rate of ATP production that can be achieved from their oxidation. In prolonged low-level activity or during fasting, however, they present a formidable energy reserve.

Aerobic metabolism is, of course, dependent on the availability of oxygen to working muscle. Maximal oxygen entry into the muscle cell is assured by an important decrease in partial pressure across the capillary–sarcoplasm barrier (about 20 torr). Inside the muscle cell (which typically has a partial pressure of O_2 about 10-fold lower than within the circulation), the oxygen transporter myoglobin is about 50% saturated even in working muscle, a level sufficient for supporting maximal rates of oxidative phosphorylation. Only under conditions of extreme hypoxia does muscle oxygenation decrease significantly. There is still debate as to whether this indicates that oxygen is or is not limiting to aerobic metabolism during high intensity work; for example, it is possible that even with "sufficient" myoglobin saturation, the diffusion into mitochondria and to the cytochromes could be reduced. In either case, the aerobic system cannot supply ATP at the rate necessary to satisfy the needs of the muscle at high work intensities, which must instead turn to anaerobic pathways.

Phosphagens: Phosphocreatine and ATP Although PCr is present in low levels in other tissues, muscle is unique in using PCr as a quantitatively important substrate for rapid ATP replenishment. The transfer of a phosphate group from the energy-rich bond between the nitrogen of the guanido group and the phosphate residue of creatine phosphate to adenosine 5′-diphosphate (ADP) allows the rapid production of ATP, a reaction catalyzed by the enzyme, creatine kinase, in a process sometimes known as the anaerobic alactic system (Fig. 11.2).

The creatine phosphate reaction is reversible, and under conditions of low ATP demand, PCr is rapidly replenished. The use of PCr for ATP production is also advantageous

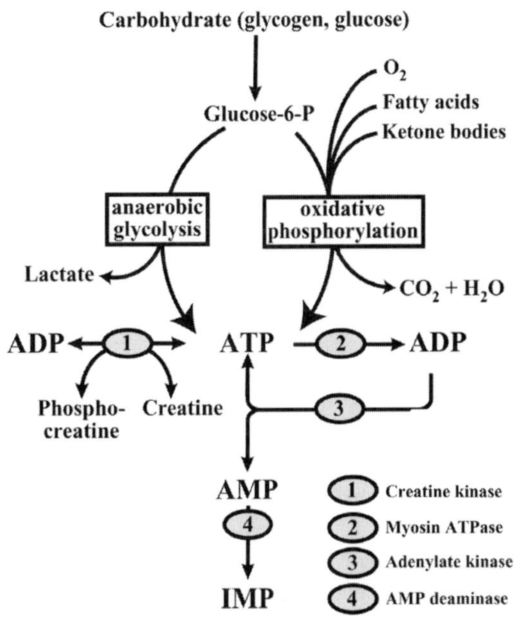

Figure 11.2 Overview of the major metabolic pathways of ATP consumption and production in skeletal muscle.

during periods of intense activity since the reaction consumes a proton (H^+), thus reducing the myocellular acidification associated with anaerobic glycolysis. The PCr system is designed for efficient delivery of ATP to sites of ATP hydrolysis during activity; creatine kinase is located near actin and myosin (near the site of actomyosin ATPase) (see Chapter 14 for a discussion of binding to myofibrils), at the sarcoplasmic reticulum (where ATP is used for Ca^{2+} reuptake) as well as near the sarcolemma (where ATP is used to fuel the Na^+–K^+ pump). A further pool of the enzyme is located on the inner mitochondrial membrane where it serves to regenerate PCr from ATP produced by oxidative phosphorylation.

The rate of PCr depletion increases in proportion to the relative intensity of exercise, and a greater than 90% depletion of PCr is common during high-intensity exercise. Although the creatine kinase reaction is very fast and can sustain the rate of cellular ATP generation needed during high-intensity exercise, the PCr pool is limited in size and its hydrolysis can only maintain ATP levels for a few seconds during such work. Thus, the phosphagen system has low capacity for ATP production. However, the half-time for recovery of PCr reserves during rest periods is typically ~30 s, making PCr a rapidly renewable energy source during intermittent, high-intensity work.

Phosphagens are found at higher concentrations in type II fibers than in type I fibers. In type I fibers, PCr content is ~75 mmol/kg dry weight, whereas in type II fibers this approaches ~85 mmol/kg dry weight. A higher concentration of phosphagen in type II fibers is advantageous given the relative importance of these fibers in high-intensity activities where a high flux of ATP production is required. Endurance training has little or no effect on muscle PCr levels, but this type of training does reduce the depletion of phosphagen during submaximal work because of a better exploitation of aerobic metabolism. However, high-intensity or sprint training leads to an increase in muscle phosphagen stores. For example, 4 to 5 months of strength training can increase PCr concentration by 5% and ATP levels by 20%. These changes are sufficient to permit the maintenance of a very high intensity effort for a few more seconds.

Glycogen and Glucose: Glycogenolysis and Glycolysis Extracellular glucose and intracellular glycogen serve as important substrates for skeletal muscle glycolysis during exercise. Glycogen is the preferred carbohydrate for muscle ATP production. At moderate exercise activities (e.g., 60% of the $V_{O_2,max}$), glycogen accounts for about 85% of all carbohydrates oxidized at the onset of activity. Over time the reliance on glycogen diminishes somewhat, in response to diminishing muscle glycogen stores and increased glucose import into myofibers. Still, after 120 min of continuous moderate activity, glycogen still accounts for about 70% of all carbohydrates oxidized by muscle. Reliance on glycogen as the carbohydrate fuel of choice increases with increasing effort (as does a shift toward anaerobic catabolism), and during high-intensity exercise little or no exogenous glucose contributes to muscle ATP production. A reflection of its important role in fueling muscle glycolysis, glycogen is the most important intramyocellular fuel reserve, usually accounting for about 1% of the total tissue mass (Table 11.7).

Glycogen is synthesized from glucose in the muscle at rest and mobilized when muscle work begins. Diet plays an important role in determining the content of whole-body glycogen storage: Skeletal muscle and liver glycogen content can increase or decrease by more than twofold with high- and low-carbohydrate diets, respectively. Glycogen phosphorylase catalyzes the phosphorolysis of glycogen to form glucose-1-phosphate, which is then fed into the reactions of glycolysis. Patients suffering from McArdle's syndrome (type V glycogen storage disease) lack skeletal muscle glycogen phosphorylase and thus cannot access glycogen as a muscle fuel. The absence of any significant accumulation of blood lactate in these patients during activity reinforces the idea that glycogen, and not glucose taken up from the blood, is the preferred form of carbohydrate fuel for muscle work. Upon entry into the glycolytic pathway, the catabolism of hexose phosphates can proceed either aerobically using oxidative phosphorylation or anaerobically with lactate accumulating. The choice depends on the energy needs of the muscle and the availability of oxygen. Glycogen breakdown in muscle is under hormonal control, with epinephrine activating phosphorylase kinase in a cyclic adenosine 5′-monophosphate (cAMP)-dependent pathway. Ca^{2+} released from the sarcoplasmic reticulum during contraction further activates skeletal muscle phosphorylase. Glycolysis can be activated very quickly in muscle, reaching its maximum rate in less than 5 s during intense exercise. An overview of skeletal muscle glycogen metabolism is given in Figure 11.3.

A key step in the fate of carbon units from glycogenolysis or glycolysis occurs at the level of pyruvate. Pyruvate can enter the mitochondria via the pyruvate dehydrogenase complex and be completely broken down by oxidation to CO_2 and H_2O, a process that yields 38 mol of ATP per mole of oxidized glucose-1-P released from glycogen (or 36 mol if the starting material is glucose). By contrast, the yield of ATP from anaerobic glycogenolysis ending in lactate is much lower; only 3 mol ATP are produced per mole of glucose-1-P (or 2 mol starting from glucose). Thus, anaerobic glycolysis has only a modest capacity for ATP production; the ratio of moles ATP produced per mole substrate consumed is greater than that of PCr hydrolysis but far less than that of oxidative phosphorylation. Despite the increased capacity for ATP production of aerobic carbohydrate catabolism, the rate of ATP generation (the number

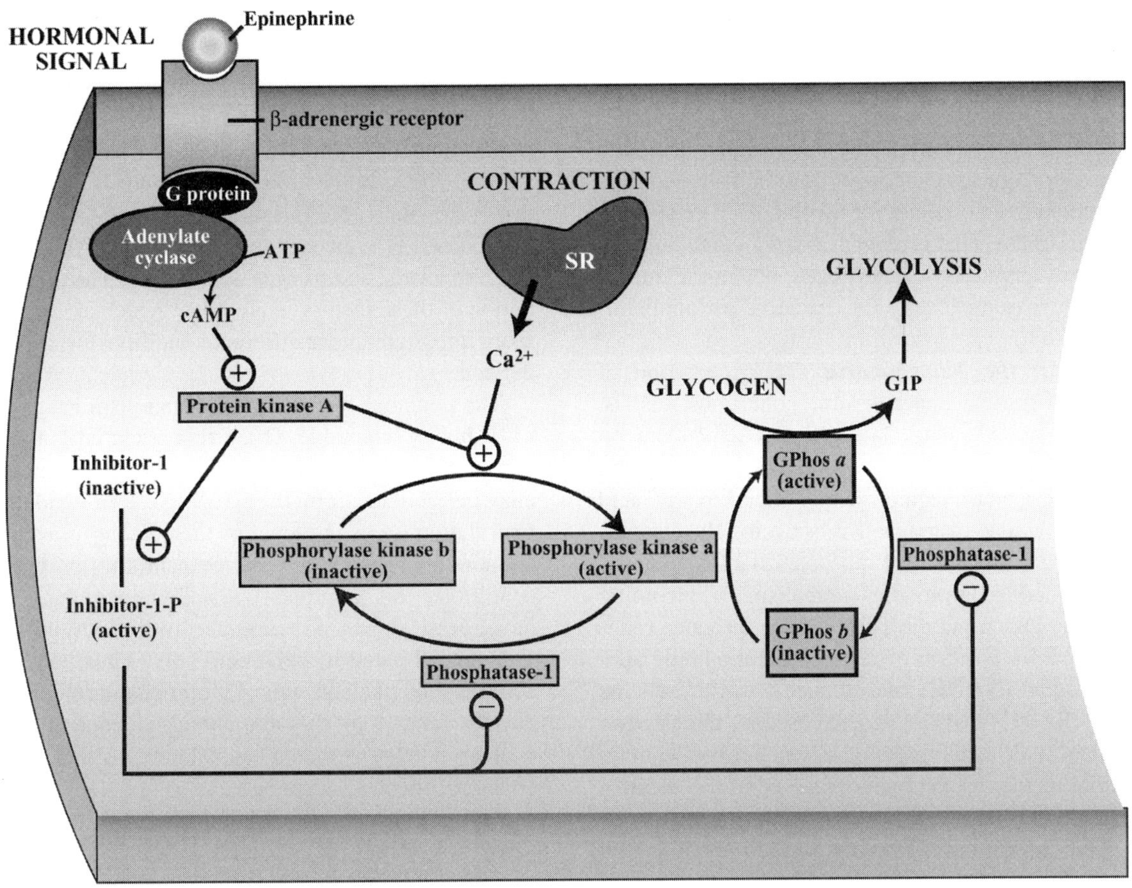

Figure 11.3 Overview of glycogen phosphorylase regulation in skeletal muscle. G1P, glucose-1-phosphate; GPhos, glycogen phosphorylase; SR, sarcoplasmic reticulum; +, activation; −, inhibition.

of ATP that can be produced in a given period of time) is much reduced when compared to anaerobic glycogenolysis and glycolysis. *In vivo*, the capacity for glycolytic ATP production is limited by both the availability of substrate (glycogen) and the production of factors that are inhibitory to muscle activity (fatigue).

Muscle glycogen use is linked to exercise intensity and to the fiber types recruited. During low-intensity work, glycogen reserves in type I fibers are used but at a relatively low rate. During higher intensity work, along with the activation of motor units that contain type II fibers, an increasing dependence on intracellular fuels for ATP production leads to an increased rate of glycogen depletion in all fiber types.

In a given individual, the glycogen content of type II fibers tends to be somewhat higher than that of type I fibers. Endurance or strength/sprint training results in an increase in myofiber glycogen content to a similar degree in all fiber types. This increased glycogen content can be explained in part by increased glycogen synthase maximal activity or activation state (percent of enzyme in the active form), which also permits a more rapid recovery of muscle glycogen levels after the end of contraction.

The effects of endurance training on the glycolytic potential of muscle, often measured as the maximum activity (V_{max}) of phosphofructokinase (PFK) (the enzyme that gates hexose-phosphate entry into the triose-phosphate portion of glycolysis), are less pronounced than the changes seen in muscle oxidative potential (discussed below). In general, the glycolytic potential remains unchanged or can be slightly decreased following many weeks of endurance training. This results in part from the improved capacity of muscle to use lipid fuels following this type of training. On the other hand, training involving high-intensity repeated exercises can significantly increase the expression of skeletal muscle glycolytic machinery. Intimately linked with glycolysis, the capacity of skeletal muscle to import glucose from the blood can also be altered with training. Levels of the muscle glucose transporter, GLUT4 (see below), can increase with endurance training and contribute to the increased capacity for uptake of circulating glucose as an oxidative fuel that is observed in trained individuals.

Following endurance training, muscle also becomes more economical with respect to intramyocellular glycogen use; this results from an increase in muscle mitochondria content, increasing the capacity for aerobic ATP generation (notably from lipids). This *glycogen-sparing effect* helps to prolong activity by delaying glycogen depletion and is linked to an increased capacity for oxidative metabolism.

Oxidative Fuels: the Mitochondria The generation of ATP in skeletal muscle under aerobic conditions, that is, in the presence of adequate O_2 supply, is dependent on the mitochondrial processes of fatty acid β oxidation, the pyruvate dehydrogenase complex, the tricarboxylic acid cycle (TCA cycle; also called the Krebs cycle), the electron transport chain, and oxidative phosphorylation. Briefly, acetyl-CoA generated from the catabolism of incoming substrates is processed via the TCA cycle to produce reducing equivalents in the form of reduced nicotinamide adenine dinucleotide (NADH) and reduced flavin adenine dinucleotide ($FADH_2$), which are then used by the electron transport chain to drive the proton gradient across the inner mitochondrial membrane. The energy stored in this electrochemical gradient is used to drive the oxidative phosphorylation of ADP to form ATP using oxygen as the final electron and H^+ acceptor. The primary substrates for oxidative phosphorylation in skeletal muscle are carbohydrates (blood glucose or myofiber glycogen) and fatty acids (intramyocellular or circulating).

The glycolytic end product, pyruvate, enters the mitochondria via the pyruvate dehydrogenase complex (PDC), which catalyses its conversion to acetyl-CoA. Long-chain fatty acids are imported into mitochondria via the carnitine shuttle [which includes carnitine palmitoyl transferase (CPT) activity], and acetyl-CoA units are sequentially cleaved from fatty acids by the reactions of the β-oxidation pathway. The acetyl-CoA produced by both fatty acid and carbohydrate entry into the Krebs cycle is an important regulator of their mitochondrial import: it is inhibitor of PDC (decreasing pyruvate uptake), and following its transport to the sarcoplasm it can be converted to malonyl-CoA, a potent inhibitor of CPT.

The capacity for oxidative phosphorylation is determined from the number and size of mitochondria (better represented as the volume density of mitochondria within the cell) as well as the capacity for oxygen transport and storage within the myofiber that is determined by the intracellular levels of myoglobin. Respiratory control refers to the closely regulated changes in ATP synthesis and hydrolysis that occur under aerobic conditions, thus maintaining ATP homeostasis. Mitochondria are key players in this regulation, as is the supply of substrate for mitochondrial import and oxidation. Mitochondrial respiration is controlled by a number of factors. Interestingly, it is possible that the controlling elements might differ according to the state of the myofiber as well as its origin. Major contributors to the control of oxidative phosphorylation, as determined from studies with isolated mitochondria, are the concentrations of free ADP, inorganic phosphate (P_i), the redox potential ([NADH]/[NAD^+]), calcium levels, and oxygen availability. The relative contribution of these factors *in vivo* (i.e., which are quantitatively more important under different conditions) is, as yet, poorly defined.

The potential for oxidative metabolism has been shown to be highly adaptable. Often this potential is measured as the maximum activities of a number of mitochondrial enzymes, either from the TCA cycle or oxidative phosphorylation, or as the level of the content of a single mitochondrial protein such as cytochrome *c*. Under most conditions this approach is valid since these measures have been shown to correlate well with mitochondrial volume estimated morphometrically. Thus, changes to the activities of skeletal muscle mitochondrial enzymes are closely related to the mitochondrial content, which has been used to argue that the actual composition of organelles changes little with training.

Training status clearly impacts on the potential for aerobic metabolism measured in this fashion. The mitochondrial volume density of skeletal muscle increases in proportion to the volume of training that invokes aerobic metabolism (Table 11.8). Succinate dehydrogenase and citrate synthase activities can be two- or threefold higher in skeletal muscle from endurance-trained individuals when compared to activities in sedentary individuals. Endurance training of sedentary individuals typically increases skeletal muscle oxidative potential in the order of 30 to 40%. Encouraging for those wishing to start a training program is the observation that a low volume of training (e.g., 20 min of cycling per day) can induce an important increase in the oxidative potential of skeletal muscle. It is also worth noting that a training program

TABLE 11.8 Mitochondrial Content of Human Vastus Lateralis According to Training Status

Subjects	Mitochondrial Content (% of Fiber Surface)
Sedentary	3.5–5.7
Active	4.3–6.6
Long-distance runners	8.1
Professional long-distance cyclists	11.4

Source: From H. Hoppeler (1986). *Int J Sports Med* 7:187–204.

that includes mainly high-intensity intermittent exercises is also very efficient in increasing the oxidative potential of skeletal muscle. The capacity for ATP and PCr regeneration during rest periods in this type of exercise are in fact intimately related to the aerobic potential; thus, an increase in the latter is beneficial in this type of training.

It is interesting to note that the mitochondria within a myofiber are not all the same. Two populations have been identified. By localization and function, skeletal muscle mitochondria are classified as either subsarcolemmal (SS) or intermyofibrillar (IMF). Subsarcolemmal mitochondria, which account for about 10 to 15% of the total mitochondrial volume, are located near the sarcolemma and have lower enzyme activities and respiratory rates then IMF mitochondria, which are embedded among the myofibrils. Functionally, SS mitochondria are believed to supply ATP for membrane or nuclear functions, while IMF mitochondria would provide ATP primarily for the contractile apparatus. It is interesting to note that during endurance (aerobic) training the greatest increase in numbers occurs among the SS mitochondrial population. This, and their localization near the sarcolemma, suggests that this population may be involved in satisfying ATP requirements for local events such as transmembrane transport (e.g., of circulating fatty acids) or increasing ATP supply for Na,K–ATPase activity. In reality, the physical separation between the two populations is not so clear. Strong evidence indicates that skeletal muscle mitochondria are not distinct organelles but more closely resemble a continuous membrane system, termed the mitochondrial reticulum. In this regard, the SS and IMF populations would share a common structure but would have different regional membrane and protein composition in response to different local needs and influences.

Mitochondrial biogenesis is well-known to occur following endurance training and in response to other factors, such as thyroid hormones. Mitochondrial content can be increased by 50 to 100% in as little as ~6 weeks of training in some models, resulting in a greater capacity for oxygen consumption per gram of tissue. Improved endurance performance during this time outpaces the relatively small training-induced changes in maximal whole-body oxygen consumption and is more closely matched to the changes in skeletal muscle aerobic capacity. Contractile activity from voluntary movements or from chronic electrical stimulation have been shown to promote mitochondrial biogenesis by mechanisms that are independent of other physiological conditions that increase mitochondrial content, such as thyroid hormone treatment. The stimuli for biogenesis appear to originate within the contracting cells themselves. For example, fibers that are not recruited during exercise training do not display adaptation of their mitochondrial content. Although the precise modulators of mitochondrial biogenesis remain an area of intense study, calcium signaling and disturbances to ATP turnover (both resulting from contractile activity) are believed to be key initiators, and a number of downstream kinases, phosphatases, and nuclear factors have been linked to these. In order to maintain increased mitochondrial content (and associated increases in aerobic performance and capacity for lipid oxidation), regular muscle activity must be performed since mitochondrial proteins typically have a half-life of about 1 week.

A high potential for aerobic metabolism is not uniquely the result of exercise training. Hereditary factors appear to play an important role, and this explains, in part, why elite athletes have such a high potential for aerobic metabolism. For example, the vastus lateralis muscle from high-level endurance cyclists has a citrate synthase activity (considered an excellent marker of the potential for aerobic activity) that is at least twice as high as that from sedentary individuals. Neither high-volume exercise training nor electrical stimulation protocols (up to 8 h per day) have been able to elevate the citrate synthase activity of sedentary humans to these high activities inherent in elite athletes.

Adaptations to the capacity for lipid catabolism are also linked to those of mitochondrial density since β oxidation occurs within these organelles. Endurance training also results in increased activities of CPT (the limiting activity in the import of long-chain fatty acids into mitochondria) and hydroxyacyl-CoA dehydrogenase (HADH; one of the four enzymes of mitochondrial fatty acid β oxidation), enhancing the capacity for increased utilization of lipids for ATP production during submaximal exercise. Substrate availability also increases following endurance training, through increased intramyocellular lipid levels as well as an increased capacity for lipid uptake from the circulation. For example, 6 weeks of ergocycle training at 70% of $V_{O_2,max}$ resulted in 22, 90, and 144% increases in intramyocellular lipid reserves in type I, IIA, and IIX fibers, respectively. Lipoprotein lipase (LPL) activity, considered by some as limiting the entry of circulating triglyceride lipids into myofibers, also increases following endurance training. The effects of high-intensity training on myofiber lipid content and uptake are still unclear.

The impact of detraining or immobilization on the oxidative potential of muscle are also quite consistent. Decreases of 30 to 40% in mitochondrial enzyme activities and volume density are typically observed following a relatively long period (up to 20 weeks) of inactivity. In general, the decrease in the oxidative potential with inactivity occurs more rapidly than the gains that are observed with exercise training.

Myoglobin serves as the oxygen carrier within muscle cells, in an analogous fashion to hemoglobin in the circulation. Muscle fibers that have a rich potential for aerobic metabolism and a greater interaction with the circulation due to high capillarization (type I and to a lesser extent type IIA fibers) also have a greater myoglobin content. Endurance training can increase muscle myoglobin content by up to 80%.

Despite parallel changes to skeletal muscle enzyme activities, substrate use, and the actual capacity for work following training and detraining, other changes are necessary to improve whole-body aerobic performance. In fact, it is difficult to quantify the contribution of the skeletal muscle metabolic machinery to performance. Maximal oxygen consumption at effort ($V_{O_2,max}$) appears to depend more on changes to cardiovascular function following training, the link with the mitochondrial machinery being less clear. This has led to the suggestion by some researchers that the aerobic capacity of muscle is of little importance for endurance performance. This is not a satisfactory interpretation from a biological design or intuitive sense, and recent evidence shows that under some conditions muscle oxygen uptake does become limiting. It is more likely that cardiovascular and muscle systems work in concert to optimize the exploitation of resources. For example, in collaboration with a greater vascularization of myofibers, an increased skeletal muscle aerobic capacity would help to preserve intramyocellular glycogen by increasing the proportion of ATP production that is generated from the oxidation of lipids or circulating glucose, thus delaying the appearance of muscle fatigue.

Adenylate Kinase and AMP Deaminase During high-intensity activity, adenylate kinase and AMP deaminase work together to supplement ATP production. Adenylate kinase (also known as myokinase or adenylate kinase) catalyzes the following reaction:

$$2\ \mathrm{ADP} \leftrightarrow \mathrm{ATP} + \mathrm{AMP}$$

and provides another short-term solution for ATP production in muscle (Fig. 11.2). This highly reversible reaction (with an equilibrium constant near 1) minimizes the buildup of ADP during high rates of ATP hydrolysis and recovers some energy potential in ADP by converting 50% of accumulated ADP back to ATP. The AMP produced by this reaction is then deaminated to form inosine monophosphate (IMP) by AMP deaminase:

$$\mathrm{AMP} + \mathrm{H_2O} \rightarrow \mathrm{IMP} + \mathrm{NH_3}$$

Catabolism of AMP helps to favor continued ATP production by adenylate kinase, which in turn aids the maintenance of a high [ATP]/[ADP] ratio and prevents product inhibition by ADP of ATPase reactions.

Following high-intensity activity, observed decreases in total myocellular adenylates can reach over 20%, and this is accounted for by the activity of AMP deaminase with an equivalent increase in IMP. The ammonia produced is protonated to form ammonium ion (NH_4^+) by reaction with H^+, which helps to lessen the net acid accumulation during anaerobic work. The ammonium ion can be removed to the circulation for further metabolism in the liver or kidney, but NH_4^+ is also an effective activator of muscle glycolysis, stimulating the enzyme PFK to promote glycolytic ATP production. The adenylate kinase/AMP deaminase pair of reactions is generally called into play during high-intensity activity. Hence, it is not surprising that the activities of these enzymes are higher in type II (fast twitch, glycolytic) muscle fibers.

Of course, muscle does not act alone in the regulation of energy production. The integration of muscle function with that of the circulatory system, whether it be for the supply of oxygen or other substrates (glucose, triglycerides), or for the removal of inhibitory or damaging products of metabolism (lactate, CO_2, NH_4^+, etc.), is an essential element. In fact, when considering adaptation to the changing demands on muscle (whether it be training, detraining, immobilization, etc.), it has become clear that a combination of central (neural), cardiovascular, as well as muscle-specific changes are involved.

CONTROL AND INTEGRATION OF FUEL USE

As the intensity of muscle activity increases, skeletal muscle becomes more reliant on intracellular fuels for ATP production. A shift toward anaerobic intracellular glycogen use is seen, at the expense of lipids and circulating glucose. As the duration of an activity increases (particularly for periods greater than 1 h), the contribution of carbohydrate oxidation is decreased in favor of lipid oxidation, while at the same time the proportion of oxidized carbohydrates originating from circulating glucose increases. All of these changes occur without important changes in blood glucose levels. How muscle cells regulate the use of these substrates for ATP production has received considerable attention. Important regulatory mechanisms have been described at the level of substrate uptake (including O_2), the regulation of key enzymes or pathways of aerobic and anaerobic metabolism, and the impact of muscle fatigue.

Muscle energy production is also complicated by the availability of both intramyocellular and extramuscular substrates. Storage of substrates in tissues other than muscle (e.g., adipose, liver) offers a convenient means of overcoming the limitations of storage space within myofibers, but accessing these substrates is not without difficulties. The liberation of substrates from these distant tissues, their transport within the circulatory system, and their final uptake by contracting muscle cells are complex events that require precise regulation. Consider the obligatory movement across cell membranes (generally requiring specific transporters) and the transport of hydrophobic lipids within aqueous media, only two of the many facets of the complicated use of exogenous fuels. By comparison, transport problems for intramyocellular fuels are fewer, especially when one considers that most of these are stored as lipid droplets or glycogen granules near or immediately adjacent to mitochondria. A few potential key players and mechanisms have emerged in our understanding of the integration of skeletal muscle energy metabolism, and some are discussed here.

ADP in Energy Metabolism

The activity of myosin ATPase during contraction releases ADP, which plays a central role in driving ATP production. Three key processes are influenced by increased ADP levels. First, the creatine kinase reaction is pushed toward ATP formation. Second, ADP serves as a substrate and allosteric activator of enzymes of glycolysis. Finally, mitochondrial respiration (active, state 3) is activated by rising ADP levels.

In myofibers containing increased mitochondrial volume densities (e.g., following endurance training, or in type I fibers compared to type II fibers), a greater sensitivity of mitochondrial respiration to ADP is observed. Essentially, a smaller increase in the concentration of ADP (from myosin ATPase activity) is required to attain the same level of total mitochondrial oxygen consumption within the cell. The ADP-dependent activation of respiration of each mitochondrion (or mitochondrial surface) is activated to a lesser degree in the presence of lower ADP concentrations (i.e., the kinetics do not change), but the sum of the respiration of all mitochondria (or of the total mitochondrial volume) results in a greater overall mitochondrial respiration within the cell. Thus we observe a greater reliance in type I fibers or in endurance trained individuals on mitochondrial respiration for ATP production at a given rate of ADP production, with a PCr sparing effect as well as an attenuation of glycolysis and lactate formation. In type II fibers or in untrained individuals, the relatively lower sensitivity of mitochondrial respiration to ADP results in a proportionally lower increase in total mitochondrial respiration rate during exercise, with ATP needs made up instead by increased glycolytic activity with lactate formation, as well as ATP generation from PCr hydrolysis and the adenylate kinase reaction. Increasing work intensity, with its associated increase in the rate of ADP production, leads to the activation of all of these pathways in all fiber types.

Carbohydrate Metabolism and Glucose Uptake

Carbohydrates are key fuels for muscle energy production. Their levels can also be limiting to exercise duration. Thus, the depletion of intramyocellular glycogen is a key factor in muscle fatigue during low intensity long-term exercise, such as long-distance running. Although intracellular stores of glycogen are significant (in fact, most of the body's glycogen is stored in muscle; see Table 11.7), these supplies are nonetheless finite and must be preserved to allow for very long term exercise. Thus an increased capacity for uptake of circulating glucose is linked with enhanced performance as a result of endurance training.

Glycogen Metabolism in Muscle The synthesis and breakdown of glycogen must be tightly coordinated to prevent futile cycling and the waste of energy-rich phosphates. The key players in glycogen metabolism are the enyzmes, glycogen phosphorylase and glycogen synthase, determining the rate of glycogen breakdown and synthesis, respectively. These enzymes function on the surface of glycogen particles, which in skeletal muscle are mainly found associated with the sarcoplasmic reticulum with smaller amounts distributed in the cytoplasm and near the sarcolemma. (See Text Box 11.1.) Central to the regulation of these enzymes is posttranslational modification by reversible phosphorylation catalyzed by protein kinases and protein phosphatases. Phosphorylation activates glycogen phosphorylase and inactivates glycogen synthase. Thus, the same stimulus regulates both activities but in opposite directions. In periods of rest or during recovery following exercise, glycogen synthase is active and directs carbon units toward glycogen production for later use. When exercise begins, however, mechanisms are in place that inhibit glycogen synthase while activating phosphorylase to provide the glucose-1-phosphate substrate for glycolysis. This control is refined so that the rate glycogen breakdown increases with the need for glycolysis as a provider of ATP, such as during increasing exercise intensity.

Glycogen metabolism in skeletal muscle is under hormonal control, and the key hormone responsible for the activation of glycogenolysis during exercise is epinephrine. Although glucagon also plays a role in regulating glycogenolysis in liver, it is not a factor in skeletal muscle. Epinephrine is released into the circulation during exercise or in the anticipation of exercise, as well as during periods of stress (the fight or flight response). Levels of epinephrine also increase with increasing intensity of the exercise. Epinephrine promotes glycogen breakdown via a cAMP-mediated pathway (Fig. 11.3). Binding of cAMP to the regulatory subunit of protein kinase A induces the dissociation of the regulatory subunits from the catalytic subunits, which are then active. Protein kinase A catalytic subunits directly phosphorylate glycogen synthase (leading to its inactivation) and phosphorylase kinase *b* to convert it to its active *a* form. In turn, phosphorylase kinase *a* phosphorylates the inactive glycogen phosphorylase *b* form to convert it to its active *a* form. Inactive phosphorylase kinase *b* can

TEXT BOX 11.1 MUSCLE FATIGUE—WHAT IS THE KEY PLAYER?

The production of lactate has long been associated with the rapid decline in contractile function during high-intensity exercise, known as muscle fatigue. Thus, it was suggested early on that lactate might inhibit contraction. However, lactate ions themselves have little effect on muscle contraction. Instead, acidosis (reduced pH) from the production of H^+ by the dissociation of lactate (a strong acid) is considered a classic cause of skeletal muscle fatigue during high-intensity work (intracellular pH can decrease by ~0.5 pH units during intense activity). Recent work has challenged this view, as acidosis appears to have little effect on mammalian muscle contraction at physiological temperatures. Also, the temporal correlation between fatigue and acidosis is not always present, the force of contraction sometimes recovering more rapidly than pH following activity. Instead, inorganic phosphate (P_i), produced mainly as a result of PCr depletion in working muscle, might be the single most important contributor to muscle fatigue during high-intensity exercise.

Elevated P_i can influence muscle function in a number of ways. These include a direct effect on myofilaments by reducing cross-bridge force and Ca^{2+} sensitivity, as well effects on Ca^{2+} release and sequestration by the sarcoplasmic reticulum by increasing the open probability of the sarcoplasmic reticulum Ca^{2+} release channels and slowing sarcoplasmic reticulum Ca^{2+} pumps. Recent work points to another, potentially most important, mechanism for P_i-mediated muscle fatigue. Evidence suggests that P_i can move from the myoplasm into the sarcoplasmic reticulum. During intense activity, sarcoplasmic reticulum P_i levels have been calculated to reach levels that could precipitate Ca^{2+} in the sarcoplasmic reticulum, forming calcium phosphate. The consequent reduced availability of Ca^{2+} for release during subsequent contraction could directly impact on performance.

It must be stressed that other factors also contribute to muscle fatigue, especially during maintained continuous maximal contraction where the propagation of membrane action potentials is believed to become a key player. Also, fatigue following lengthy, relatively low intensity types of exercise (e.g., marathon running) is primarily related to factors such as dehydration and the depletion of intramuscular glycogen stores. Although glycogen-depleted muscle cells do display a failure in sarcoplasmic reticulum Ca^{2+} release, it does not appear dependent on P_i accumulation.

also be activated by binding four calcium ions to a special regulatory subunit, and this is important in the further activation of glycogen breakdown following the release of calcium from the sarcoplasmic reticulum during contraction. Dephosphorylation of glycogen phosphorylase proceeds through the action of protein phosphatase-1.

In its inactive *b* form, glycogen phosphorylase is under positive allosteric control from AMP and IMP and negative control from ATP and glucose-6-phosphate (G6P). The physiological impact of AMP and IMP is believed to be limited only to those conditions in which these effectors increase significantly in muscle cells, such as extreme intensity exercise (where AMP levels can triple). Glycogen phosphorylase *a* and *b* forms have high K_m values (27 mM) for one of the substrates, inorganic phosphate (P_i). P_i levels increase noticeably during contraction, especially as the result of PCr mobilization during high-intensity work; intracellular P_i levels can range from about 1 mM at rest to over 20 mM following heavy exercise. Elevated AMP concentration decreases the K_m for P_i of both the *a* and *b* forms of glycogen phosphorylase. Overall, then, glycogen breakdown in skeletal muscle proceeds maximally under conditions of decreased energy charge (high AMP), high P_i (from PCr mobilization), high cytoplasmic calcium (released from the sarcoplasmic reticulum), and high epinephrine levels. These are all hallmarks of increasing exercise intensity.

Glycogenesis is regulated mainly at the level of glycogen synthase. The unphosphorylated active form is known as the *I* form because its activity is independent of G6P concentration. The phosphorylated inactive *D* form is dependent on elevated G6P concentrations for allosteric activation. At least 10 sites can be phosphorylated on glycogen synthase, leading to a gradation in its activity and its sensitivity to G6P. Phosphorylation of glycogen synthase can be carried out by a number of protein kinases; notably, in skeletal muscle these include glycogen synthase kinase, protein kinase A, phosphorylase kinase, and a calcium-dependent kinase. As with glycogen phosphorylase, protein phosphatase-1 is responsible for the dephosphorylation of glycogen synthase. Insulin stimulates glycogen synthase activity, and glycogen formation, by enhancing protein phosphatase-1 activity and by inactivation of glycogen synthase kinase via a protein kinase B (or Akt) mediated phosphorylation cascade. Glycogen concentration itself can also inhibit glycogen synthase to place an upper limit on the amount of glycogen stored in skeletal muscle.

Regulation of Glycolysis in Muscle Glycolysis, mostly from glycogen-derived glucosyl units, plays a major role in skeletal muscle ATP production during periods of limited oxygen availability, during changes in exercise intensity from the resting to the active state (by virtue of its rapid activation), and during aerobic activity of medium to relatively high intensity (generally above about 50% of the $V_{O_2,max}$ in humans).

The control of glycolytic flux is tightly regulated in muscle. This control arises in part from the availability of glycogen, glucose, G6P, and ADP. Following ATP hydrolysis (e.g., during contraction), cytoplasmic ADP levels increase, enhancing ADP availability for the substrate-level phosphorylation reactions of glycolysis. However, substrate availability is a relatively minor contributor to the overall control of glycolytic flux, except in extreme conditions such as glycogen depletion following extremely long periods of activity. Instead, the modulation of key enzymatic steps in the pathway is usually considered to be the major mechanism of glycolytic regulation. In skeletal muscle, these steps include glycogen phosphorylase (for glycogen phosphorolysis—discussed above) and hexokinase (HK; for glucose entry into the pathway).

An important control on the HK reaction is feedback inhibition by elevated G6P levels. This leads to a decrease in glucose uptake into the cell and utilization by glycolysis. This situation can occur when glycogenolysis is proceeding at a sufficient rate to satisfy the demands of glycolysis or when downstream reactions of glycolysis are slowed down.

As in other tissues, skeletal muscle PFK plays a particularly important role in the control of glycolytic flux. A number of mechanisms are involved in PFK regulation. These include changes to enzyme polymerization, reversible phosphorylation, as well as allosteric effects. Both negative and positive effectors of skeletal muscle PFK activity have been identified, usually from *in vitro* studies. Negative effectors of skeletal muscle PFK include ATP, citrate, and H^+. In addition to serving as a substrate for the PFK reaction, ATP can bind to a negative allosteric site to inhibit the enzyme. Hydrogen ions promote the binding of ATP to this allosteric site. Citrate is released into the cytosol from mitochondria under conditions of rising levels of Krebs cycle intermediates and serves to integrate the rates of glycolysis and the Krebs cycle to slow the rate of glycolysis (and the consumption of carbohydrate reserves) to match mitochondrial demand for fuel. Citrate acts by promoting the disassembly of the more active tetrameric form of PFK (the M4 isoform is the main one in muscle) to a less active dimeric form. As will be seen below, citrate may also play an important role in the regulation of lipid versus carbohydrate use.

Positive effectors of skeletal muscle PFK activity include fructose-1,6-bisphosphate ($F1,6P_2$), ADP, AMP, P_i, and NH_4^+. The enzyme has independent sites for allosteric binding by $F1,6P_2$, P_i, and NH_4^+ whereas ADP and AMP act similarly, sharing a single allosteric binding site. During high-intensity work, ATP levels only decrease a little due to buffering by PCr, but this small decrease produces a proportionally larger percentage rise in ADP and a strong percentage increase in AMP due to the activity of adenylate kinase. P_i levels also increase in proportion to the decline in PCr, and NH_4^+ concentrations increase following the deamination of AMP. Together, these serve as important signals to the myocyte to increase flux through glycolysis by activating PFK. Although another metabolite, fructose-2,6-bisphosphate, is a major regulator of PFK activity in liver, it does not play a significant role in the regulation of skeletal muscle PFK activity.

Partitioning of Glucose Metabolism Glucose taken up by muscle is partitioned between glycogen synthesis and glycolysis. Glycolytic products (lactate, pyruvate, alanine) are either released into the circulation or used as substrates by the mitochondria, being introduced via PDC into the TCA cycle (alanine and lactate are first converted to pyruvate by the action of alanine aminotransferase and lactate dehydrogenase, respectively). PDC activity is believed to be rate-limiting for glucose oxidation, and under normal circumstances glycolytic flux exceeds the rate of pyruvate oxidation leading to the baseline levels of circulating lactate, pyruvate, and alanine.

Glycogen synthesis is regulated both at the level of glucose uptake (GLUT4; see below) and by glycogen synthase activity. The rate-limiting nature of glucose transport on glycogen synthesis explains the postexercise carbohydrate loading strategy used by many athletes involved in endurance events. By ingesting carbohydrate as soon as possible following activity, the increased capacity for glucose transport (i.e., GLUT4 levels at the membrane) that is stimulated by both exercise (insulin-independent) and feeding (insulin-dependent) result in the highest possible rate of glycogen resynthesis.

Glycolytic products from anaerobic glycolysis (of which lactate is quantitatively the most important) can have multiple fates: (a) oxidized by muscle mitochondria during the recovery phase, especially in red fibers, (b) exported into the blood for use as an oxidative substrate by other organs (e.g., heart, brain), or (c) recycled into glucose by hepatic gluconeogenesis (known as the Cori cycle) and resupplied to the muscle. Lactate can itself enter mitochondria via the mitochondrial monocarboxylate transporter MCT1, and lactate dehydrogenase within the organelle initiates its catabolism. Myofibers with a high mitochondrial content, such as type I fibers, are preferentially used for the oxidation of lactate originating within the same fibers or from neighboring glycolytic fibers.

Glucose Uptake: The GLUTs Glucose, given its hydrophilic nature, does not freely cross cell membranes. Instead, the task of glucose uptake from the circulation is given to a family of glucose transporters, named GLUT1 through GLUT7. Each isoform has a different tissue distribution as well as unique kinetic parameters. In skeletal muscle, GLUT1, 4, and 5 are present. GLUT1 resides in the sarco-

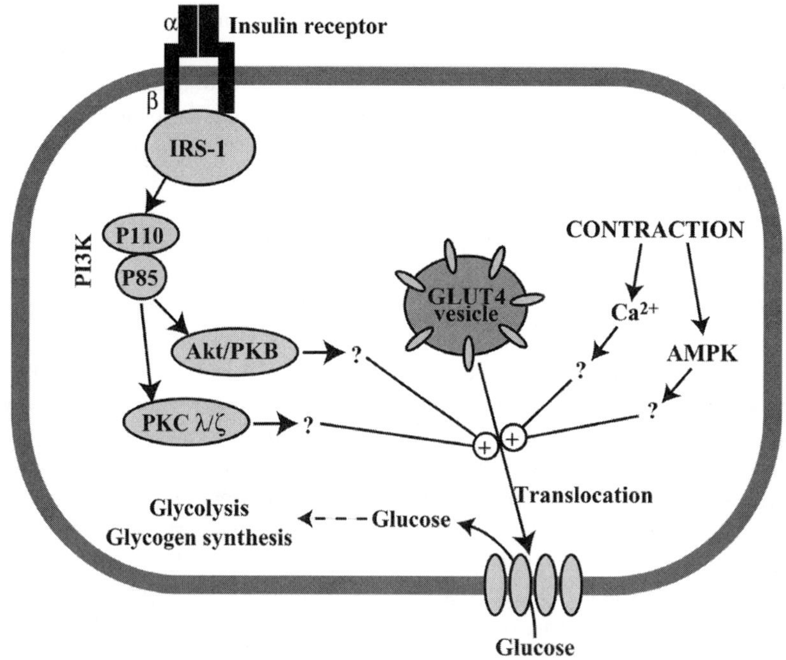

Figure 11.4 Insulin- and contraction-induced translocation of GLUT4 from intracellular vesicles to the sarcoplasm. Although key upstream effectors of GLUT4 have been identified, the downstream events remain unclear. Akt/PKB, serine/threonine protein kinase B or Akt; AMPK, AMP-activated protein kinase; GLUT4, glucose transporter isoform 4; IRS-1, insulin receptor substrate 1; PI3K, phosphoinositide 3-kinase; PKC λ/ζ, protein kinase C lambda/zeta isoform; α and β, alpha and beta subunits of the insulin receptor; +, activation.

lemma independently of stimulation or activity and is believed to be responsible for the basal glucose uptake of skeletal muscle. GLUT5 also appears to reside permanently in the sarcolemma and is believed to be involved in fructose transport (fructose can be metabolized to lactate or serve as a substrate for glycogenesis). GLUT4 is the most abundant ($\sim$15 to 20 times more than GLUT1) and important of the glucose transporters in skeletal muscle and is normally the limiting factor in glucose uptake during muscle work. GLUT4 is unique among the transporters in this family in that, upon stimulation, it can translocate from an intracellular storage site to the sarcolemma and the transverse tubules. Translocation is stimulated by contraction or insulin and increases the V_{max} of glucose transport into the cell (Fig. 11.4). This reaction is extremely important for glucose homeostasis since skeletal muscle accounts for 70 to 80% of insulin-stimulated glucose uptake (see Chapters 9 and 10). Although direct effects on intrinsic GLUT activity remain a possibility, most of the insulin-dependent increase in glucose uptake appears to stem from an increase in the numbers of GLUT4 transporters integrated into the membrane. GLUT4 expression is also related to the metabolic nature of the myofiber, higher levels found in oxidative fibers than in glycolytic fibers.

Endurance training increases the capacity for glucose transport by increasing the content of skeletal muscle GLUT4. However, this same training decreases the utilization of blood glucose during submaximal activity in favor of lipid uptake and oxidation, resulting at least in part from a decreased translocation of GLUT4 following training. This is an important adaptation in order to avoid fatigue during activities of long duration. Increased GLUT4 content under these conditions might serve a particularly important role in glycogen resynthesis during periods of recovery.

Insulin-Dependent and Independent Recruitment of GLUT4 Insulin causes the translocation of GLUT4 from the microsomal fraction to the plasma membrane (Fig. 11.4). Insulin binding to its receptor (IR) stimulates the tyrosine kinase activity of the β subunit leading to phosphorylation of intracellular proteins, notably the insulin receptor substrates (IRS1 and IRS2). These in turn activate phosphatidylinositol-3-kinase (PI3K) and the production of the phosphoinositides PIP2 and PIP3 as second messengers. The downstream activation of Akt/PKB and atypical protein kinase C (PKC λ and ζ) ultimately leads to the translocation of intracellular GLUT4-containing vesicles to the cell membrane and an increase in glucose transport. The precise chain of events between kinase activation and translocation of GLUT4 vesicles remains unclear.

Numerous studies have shown that contraction also stimulates glucose uptake by skeletal muscle but in an insulin-independent manner. Thus, whereas the effect of insulin is inhibited by wortmannin (an inhibitor of the PI3K signaling pathway), the contraction-induced increase in glucose uptake is not. Furthermore, the enhanced glucose uptake resulting from insulin and contraction are generally reported to be additive. In both cases, increased glucose uptake results from the translocation of GLUT4 from vesicles within the sarcoplasm to the sarcolemma of the myofibers. The signaling events leading to GLUT4 translocation during contraction are still unclear. Calcium release from the sarcoplasmic reticulum appears to be a necessary first step, but downstream events remain elusive. The mitogen-activated protein kinase signaling cascade, although activated during contraction, does not appear to be necessary for increased glucose transport. However, some studies suggest that glucose uptake by skeletal muscle is mediated in part by a cascade that involves the AMP-activated protein kinase (AMPK; see the AMPK section below for details).

Upon entry into myocytes, glucose is rapidly phosphorylated to glucose-6-phosphate by hexokinase to make it available for metabolism. Under normal physiological conditions very little free glucose is found in the cell, suggesting that the rate-limiting step in the utilization of extracellular glucose as a muscle fuel is, in fact, glucose transport, although this point is still debated.

Fatty Acid Transport and Entry into Muscle Cells

Long-chain free fatty acids (FFA) are an important energy source for muscle metabolism and are stored as triacylglycerols in extramuscular tissues (adipocytes and liver) as well as in myocytes. Their contribution to total oxidative metabolism depends on numerous factors, including diet, training status, and the duration and intensity of the activity. Adipose tissue stores represent quantitatively the most important whole-body fuel reserve (Table 11.7), but their access to muscle mitochondria is nontrivial. A complex series of steps are involved in FFA use, including their release into the circulation from adipose tissue or from the digestive system, their transport to muscle, their uptake into muscle cells, and finally their transport within myocytes to mitochondria and their uptake by these organelles. The elucidation of the factors responsible for fatty acid import into skeletal muscle and the partitioning of lipid use remain very active areas of research.

The low solubility of free fatty acids in aqueous solutions requires them to be bound to soluble proteins for their transport through plasma and the sarcoplasm, a system analogous to that of oxygen transport by hemoglobin and myoglobin. In plasma the main fatty acid carrier is albumin, which supports total plasma FFA concentrations as high as 2 mmol/L. Within myofibers, cytosolic fatty acid binding proteins (FABPc) are believed to be the main carriers of intracellular lipids, although lateral diffusion (the transport of fatty acids along intracellular membranes linking the sarcolemma to other cellular locales) might also be involved. FABPs also play a role in enhancing cellular uptake of circulating fatty acids and in promoting fatty acid metabolism.

The mechanisms of fatty acid transport across the sarcolemma have yet to be clarified. Generally, both simple and facilitated diffusion are believed to be involved. Evidence showing saturation kinetics for fatty acid transport across myocyte membranes suggests the presence of a protein carrier, but the clear demonstration of a fatty acid transporter remains elusive. Complicating matters, other work shows that diffusion can take place even in the absence of protein. Potentially important skeletal muscle fatty acid transporters include the plasma membrane fatty acid binding protein (FABPm), the fatty acid translocase (FAT), and the fatty acid transport protein (FATP). Upon entry into myocytes FFA can either be shunted toward storage (as triglycerides) or be activated for oxidation in mitochondria via fatty acyl-CoA formation (a reaction catalyzed by acyl-CoA synthetase).

The precise contribution of intracellular, mostly triglycerides (TG), versus circulating lipids to aerobic ATP production under different conditions in muscle is also unclear. The nature and intensity of exercise have an impact on the choice of the lipid pool, and training status and gender influence intracellular lipid use during activity of a similar intensity. Intramuscular triglycerides are believed to play a key role in ATP production. At low intensities, more than 50% of oxidized lipids can be from intramyocellular pools. Type I fibers store approximately twice as much triglyceride as type II fibers, which is not surprising given their higher capacity for aerobic metabolism and their preferential recruitment during low-intensity activities. The importance of circulating lipids as fuels is highlighted by a significant increase in plasma FFA concentration during long-term exercise. Whereas circulating glucose levels remain constant (largely the result of increased release from the liver), plasma FFA concentrations can more than double after 60 to 90 min of moderate intensity endurance exercise (about 60% $V_{O_2,\max}$). This increase occurs via activation of adipocyte lipolysis through increased sympathetic nervous stimulation as well as decreased blood insulin levels. During exercise of higher intensity, FFA release and oxidation decrease. Generally, intramuscular TG can account up to half of the fat oxidized during aerobic exercise, with most of the remainder derived from circulating FFAs. A third source of lipid fuels, plasma triglycerides, are accessible to skeletal muscle cells following cleavage by lipoprotein lipase (LPL) but are believed to contribute no more than ~10% of the lipid fuel for exercise.

Interplay between Lipid and Carbohydrate Metabolism

Physical activity is associated with important increases in both carbohydrate and fatty acid oxidation. The relative rate of use of these fuels varies with the intensity of the activity performed, with an ever-increasing dependence on carbohydrates for aerobic ATP production with increasing exercise intensity. At rest, lipids serve as the main fuel for aerobic metabolism, as shown by the relatively low respiratory quotient (typically ~0.82 for the whole body in humans and usually lower for muscle itself). During low-intensity exercise (e.g., 30 to 40% $V_{O_2,max}$), lipids remain the major oxidative substrate although the absolute rates of both lipid and carbohydrate oxidation increase. At exercise intensities of around 50% of the V_{O_2},max, the contribution of carbohydrates to total fuel oxidation becomes greater than that of lipids. This moderate level of activity is also associated with the peak rate of FFA release from the adipose tissue, whereas during high-intensity exercise, the rates of adipose tissue FFA release are actually similar to those seen at rest. This differs from the pattern of glucose release from the liver, which continues to increase with increasing exercise intensity. During very high intensity work (>90% $V_{O_2,max}$) carbohydrate oxidation continues to increase, whereas the rate of fatty acid oxidation may actually decrease. Intramyocellular glycogen utilization increases exponentially with increasing exercise intensity and remains a dominant source of carbohydrate at moderate to high intensities. How do myofibers control lipid versus carbohydrate use? This question has and continues to generate much research and speculation. A few major regulatory factors have been identified and are discussed below.

Randle's Glucose–Fatty Acid Cycle As first set out by Sir Philip Randle in 1963, the glucose–fatty acid cycle (also known as the Randle cycle) is a homeostatic model that proposes that the balance between glucose catabolism and fat oxidation is determined by the intracellular availability of fatty acids for oxidation. In this model, acutely increased fatty acid availability and oxidation leads to an inhibition of carbohydrate uptake and oxidation. Mechanistically, inhibition of glycolysis and glucose uptake are mediated by products of lipid oxidation. In summary, fatty acids taken up from the circulation enter the mitochondria and undergo β oxidation to produce acetyl-CoA that then enters the Krebs cycle. Increased mitochondrial concentrations of acetyl-CoA and NADH suppress the activity of the PDC to slow pyruvate entry into the Krebs cycle. Elevated acetyl-CoA also leads to increased citrate concentrations, and citrate released into the sarcoplasm inhibits PFK. PFK inhibition results in a rise in G6P levels that feeds back to inhibit the activity of hexokinase and glucose uptake by myocytes, as well as inhibiting glycogenolysis (Fig. 11.5).

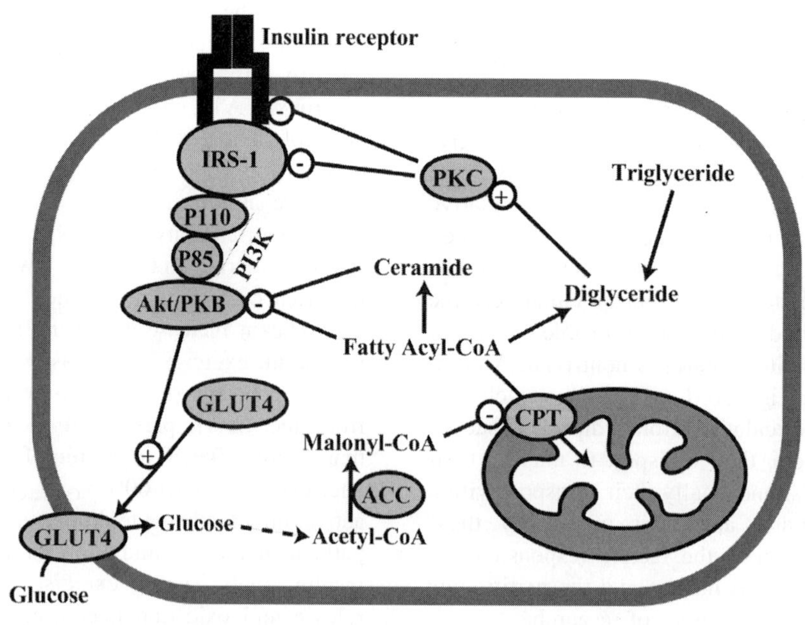

Figure 11.5 Interaction between lipid and insulin signaling in skeletal muscle. ACC, acteyl-CoA carboxylase; CPT, carnitine palmitoyl transferase; +, activation; −, inhibition. Refer to the legend of Figure 11.4 for other abbreviations.

Considerable evidence shows that the Randle glucose–fatty acid cycle operates in humans. At rest, lipid infusion or increased plasma FFA (e.g., from an overnight fast) do result in decreased insulin-stimulated glucose uptake and oxidation at the whole-body level. Studies under well-controlled conditions (such as euglycemic-hyperinsulinemic clamps) also show that maintaining or increasing circulating FFA concentrations inhibits insulin-stimulated glucose uptake by skeletal muscle. However, the partitioning of glucose metabolism between glucose oxidation and glycogen synthesis in skeletal muscle under these conditions is not precisely what is predicted by the mechanisms of the glucose–fatty acid cycle. While a decrease in glucose oxidation associated with PDC inhibition is seen as predicted, a greater decrease in glycogen synthesis is observed as a result of decreased glycogen synthase activity. That is not predicted by the model. In at least some tissues palmitoyl-CoA appears to inhibit glycogen synthase, but whether this occurs in skeletal muscle is unclear.

Thus, other mechanisms beyond the glucose–fatty acid cycle operate in the control of lipid versus carbohydrate metabolism. This is also seen in the response of skeletal muscle lipid oxidation to exercise. During low- to moderate-intensity exercise (up to perhaps about 50% of $V_{O_2,max}$), skeletal muscle glycogen utilization and carbohydrate oxidation tend to decrease (relatively) when plasma FFA concentrations are increased (although the absolute rate of oxidation of all substrates increases). With increasing intensities of exercise, the sensitivity of carbohydrate metabolism to the inhibitory effects of lipids decreases and carbohydrates become the predominant fuels for muscle ATP generation.

Recent models suggest that acutely increased circulating FFA could inhibit glucose oxidation by modulating GLUT4 translocation and/or activity, or by reducing the levels of allosteric activators of carbohydrate metabolism (e.g., P_i and AMP). Together, these would lead to decreased concentrations of pyruvate, which would decrease the contribution of carbohydrate to acetyl-CoA production and oxidative metabolism, the effect that is observed at low exercise intensity. But what happens at high exercise intensities? Increasing levels of other effectors such as AMP, which rise with increasing exercise intensity, may play a key role in the control of glycolysis and fat oxidation. They mediate the shift toward carbohydrate oxidation via increased AMP kinase activity, acetyl-CoA carboxylase inactivation, and malonyl-CoA accumulation.

AMP-Activated Protein Kinase A member of a metabolite-sensing protein kinase family, 5′-AMP-activated protein kinase (AMPK) is found in all eukaryotes. Roles for this kinase range from the regulation of energy metabolism through to transcriptional control. Environmental changes and changing energy demand can modulate AMPK activity, as is observed following exercise, ischemia or hypoxia, and nutrient starvation.

Active as a heterotrimer AMPK consists of α (catalytic), β, and γ subunits. The α subunit contains the kinase domain and binds ATP. The $\alpha 1$ isoform is widely distributed (liver, pancreas, adipose tissue, skeletal muscle), whereas the $\alpha 2$ subunit is predominantly expressed in skeletal muscle and appears to be responsible for most of the effects of AMPK on muscle energy metabolism. In human skeletal muscle, the stable activation of the $\alpha 2$ subunit by phosphorylation following exercise has been demonstrated.

The phosphorylation cascade involving AMPK has been called the *cellular fuel gage*. In particular, the kinase appears to play a key role in the regulation of carbohydrate and fat metabolism. In addition to limiting further ATP depletion by shutting down anabolic pathways (in particular fatty acid and cholesterol synthesis) under conditions of high-energy consumption (i.e., sensed by alterations in cellular energy charge), it also initiates changes that help maintain cellular ATP levels (i.e., increased substrate catabolism).

Regulation of AMPK Activity Processes that activate AMPK activity, such as AMP production during exercise, are finely tuned to the rate of energy use within the cell, with increased rates of ATP utilization leading to AMPK activation. The cellular ratio of AMP to ATP is believed to be an important and precise regulatory mechanism of AMPK activity since ATP is a strong antagonist of AMP effects on the enzyme. AMPK is also allosterically inhibited by physiological concentrations of PCr (or an increased PCr : Cr ratio): decreasing levels of PCr (or a decreased PCr : Cr ratio) during high-intensity exercise would provide important deinhibition of AMPK.

Elevated AMP (or decreased ATP : AMP ratios) remains the focal point of AMPK regulation. AMP is a strong regulator of AMPK activity via both allosteric and posttranslational mechanisms that include: (1) allosteric activation of AMPK by AMP; (2) binding of AMP to AMPK makes it a better substrate for AMPK kinase (AMPKK), the upstream kinase that regulates AMPK via protein phosphorylation; (3) AMP binding to AMPK decreases its susceptibility to protein phosphatases (in particular protein phosphatase-2C); and (4) AMPKK is allosterically activated by AMP. A 200-fold increase in AMPK activity can be achieved by the combined effect of these mechanisms. For more information about the AMPK signal transduction pathway, see Chapter 4.

Targets of AMPK Activity Of particular interest to skeletal muscle are the roles of AMPK in the regulation of creatine kinase activity, glucose metabolism, and free fatty acid metabolism. Metabolic processes that might be related to AMPK activation include increased glycogen concentration, increased GLUT4 content, and increased activities of hexokinase and of mitochondrial enzymes.

Creatine kinase (CK) is a reversible enzyme that mediates the synthesis and the catabolism of phosphagen reserves in skeletal muscle. At rest, the levels of ATP and PCr are high and in equilibrium. Following the onset of muscle activity, the contraction-induced decrease in ATP pulls the CK reaction, producing ATP and Cr from PCr. But as PCr stores are depleted, the equilibrium of the reaction will shift toward PCr synthesis from ATP and Cr. Inhibiting PCr synthesis under these conditions of elevated cellular ATP demand will help to preserve ATP for other more pressing cellular needs (e.g., to maintain contraction). AMPK appears to be a key player in the regulation of this process since it has been found to phosphorylate and inhibit creatine kinase activity. Briefly, following the onset of activity, AMPK will become activated by a decreasing PCr : Cr ratio, as well as increasing AMP levels (or a decreasing ATP : AMP ratio; this is especially true during very intense work). As the intensity of the activity increases, so do the levels of these AMPK activators, leading to the maximal inhibition of CK after PCr depletion. The end result is the maximal availability of ATP for contraction-related energy-consuming processes, and not for PCr resynthesis.

Exercise and electrical stimulation increase skeletal muscle AMPK activity and glucose transport, as does chemical activation of AMPK. In studies where AMPK is inactivated, skeletal muscle glucose uptake is decreased by up to ~30%. The AMPK-mediated activation of glucose uptake in working muscle is most noticeable in type II fibers. In myofibers with higher mitochondrial contents (type I), the higher capacity for lipid oxidation at a given work intensity and well as the increased sensitivity of mitochondrial respiration to small changes in ADP mean that ADP levels do not rise as high during muscle work as in type II fibers. This, in turn, means that AMP production via the adenylate kinase reaction is lower, producing a lesser activation of AMPK and a lower recruitment of GLUT4 to the sarcolemma. This same mechanism would account for the training-induced decrease in GLUT4 recruitment in type I fibers, as a consequence of the training-induced increase in mitochondrial content. Increased glucose transport in muscle promotes both glycogen-sparing during activity (due to increased glucose availability) as well as glycogen resynthesis during recovery. Hence, elevated levels of GLUT4 in the plasma membrane may persist for some time following the end of contraction and thereby continue to promote high glucose uptake. This is one of the benefits of physical activity in people that suffer from diabetes.

At the transcriptional level, chronic AMPK activation (stimulated with the drug 5-aminoimidazole-4-carboxamide-1-β-D-ribofuranoside) has been shown to increase the expression and activity of a number of mitochondrial enzymes and proteins (including citrate synthase, succinate dehydrogenase, and cytochrome *c*). Although not all mitochondrial elements are increased in response to such treatment, these data strongly suggest that AMPK plays a role in mitochondrial biogenesis.

Malonyl-CoA and Fuel Selection Malonyl-CoA is believed to be a key player in the control of lipid oxidation in skeletal muscle and in other tissues (see also Chapter 9). It is a potent allosteric inhibitor of skeletal muscle CPT1, the rate-limiting activity for long-chain fatty acid entry into mitochondria. Increased malonyl-CoA levels have been observed in periods of reduced skeletal muscle lipid oxidation, such as during insulin infusion. During exercise, decreased levels of malonyl-CoA are seen with a concurrent increase in fatty acid oxidation.

In skeletal muscle, malonyl-CoA is synthesized by the ß isoform of acetyl-CoA carboxylase (ACC_β, also called ACC_2). Unlike the ACC_α (ACC_1) isoform found in lipogenic tissues such as liver and adipose tissue, ACC_β is insulin- and glucagon-insensitive and is not phosphorylated by cyclic-AMP-dependent protein kinase. Instead, AMPK-mediated phosphorylation of ACC_β decreases its activity by increasing the K_a for citrate activation and increasing the Michaelis constant for the ACC_β substrates ATP, acetyl-CoA, and bicarbonate. This ultimately leads to a decrease in malonyl-CoA levels and to a stimulation of fatty acid import and oxidation within mitochondria. Ischemia, vigorous exercise, prolonged exercise, and electrical stimulation of muscle have been shown to increase skeletal muscle AMPK activity with an accompanying inhibition of ACC_β and an increase in fatty acid oxidation.

Much work remains to be done to clearly define the precise mechanisms of substrate choice during exercise of changing intensity. The relative contribution of fatty acids and carbohydrates at different exercise intensities appear to depend on the action of other common metabolic modulators. During activity, important increases in both cytosolic AMP and citrate levels are seen. The interplay of their opposite effects on lipid oxidation and glycolysis are believed to be the key factors in substrate selection (Fig. 11.6). AMP effects appear to dominate during high-intensity efforts, leading to a relative decrease in lipid oxidation and increase in glycolysis. Other allosteric activators such as NH_4^+ probably also play a role by stimulating glycolytic flux, thus promoting carbohydrate entry into mitochondria and ultimately leading to increased sarcoplasmic malonyl-CoA concentrations and inhibition of CPT and lipid oxidation.

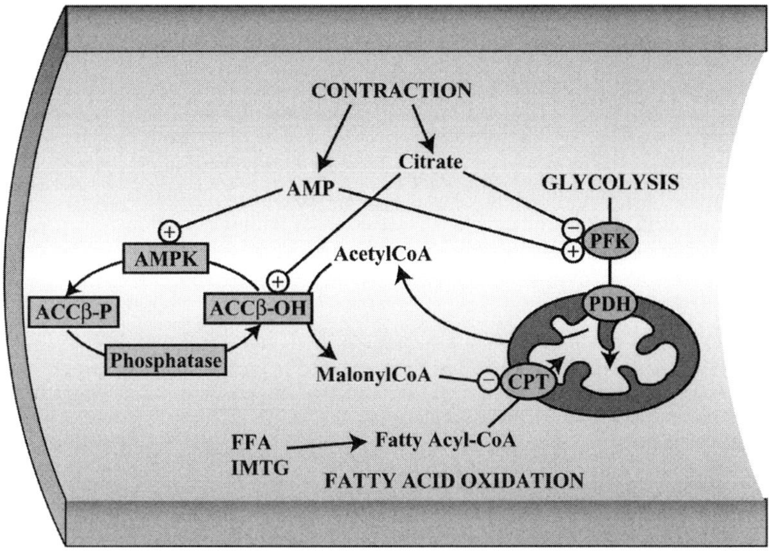

Figure 11.6 Dual mechanism for the regulation of skeletal muscle glycolysis and fatty acid oxidation. Under conditions of high glucose availability or increased Krebs cycle intermediate levels, citrate is released into the cytoplasm and inhibits both glycolysis (via PFK inhibition) and lipid oxidation (by activating ACC_β, leading to malonyl-CoA accumulation and CPT inhibition). During moderate muscle activity, increased AMP levels activate PFK and deinhibit CPT by promoting the phosphorylation and inactivation of ACC_β via AMPK. During high-intensity exercise both cytosolic citrate and AMP are increased, but the effects mediated by AMP appear to predominate and glycolysis becomes dominant; this may also be mediated through the accumulation of other activators of carbohydrate catabolism such as NH_4^+. ACCß-P and ACCß-OH, phosphorylated and dephosphorylated forms of acetyl-CoA carboxylase; CPT, carnitine palmitoyl transferase; FFA, free fatty acids; IMTG, intramyocellular triglycerides; PDH, pyruvate dehydrogenase; PFK, phosphofructokinase; +, activation; −, inhibition.

Reactive Oxygen Species and Nitric Oxide: Role in Skeletal Muscle Function

Over the past decade, the physiological significance of reactive oxygen species (ROS) and nitric oxide (NO) has become a major area of interest in studies of metabolic control and function (for more on this subject see Chapter 12). In skeletal muscle both are known to impact on contraction as well as regulate substrate uptake and metabolism.

Nitric Oxide in Skeletal Muscle Production of nitric oxide (NO; actually a generic term that encompasses NO and a variety of its related molecules such as *S*-nitrosothiols, metal NO complexes, and peroxynitrite, collectively called reactive nitrogen species, RNS) increases during activity in skeletal muscle. NO plays a key role in stimulating the dilation of arterioles to increase blood flow to skeletal muscle during activity so that the delivery of substrates and oxygen can be increased. In addition, NO impacts on other skeletal muscle functions including excitation-contraction coupling, myocyte differentiation, respiration, and glucose homeostasis.

Nitric oxide, a gas, is formed by nitric oxide synthase (NOS) using arginine as a substrate:

$$\text{L-arginine} + \text{NADPH} + O_2 \rightarrow \text{L-citrulline} + {}^{\bullet}\text{NO} + \text{NADP}^+$$

Nitric oxide has a very short half-life (3 to 4 s) in tissues. It reacts quickly with oxygen and superoxide, in the latter case forming peroxynitrite ($ONOO^-$). Given its short half-life, the regulation of signaling occurs at the level of NO biosynthesis. The NOS enzymes are very complex and are dependent on a number of redox cofactors (NADPH, FAD, FMN, heme, and tetrahydrobiopterin) to complete the five-electron oxidation of an amidine nitrogen of arginine. Three isoforms of these cytosolic enzymes are found, encoded by different genes, and are named from the order of their discovery and their tissue distribution (Table 11.9). Many recent publications favor the numerical designation, as it has become clear that most NOSs have a wide tissue distribution.

TABLE 11.9 Nitric Oxide Synthase Isoform Characteristics

Name	Localization	Chronic Regulation of Activity	Activators and Modifications
NOS1 nNOS (*neuronal*)	Sarcolemma Enriched at the neuromuscular junction; NH_2-terminal PDZ domain binds to α_1-syntrophin, a dystrophin associated protein	Increased with crush injury, muscle activity, aging, cytokines Decreases with denervation	Calcium- and calmodulin-dependent activation
NOS2 iNOS (*inducible*)	Diffuse (myofibrils, membrane; localization may depend on condition)	Low levels in normal muscle Increased in chronic heart failure, autoimmune pathologies, cytokines	Ca^{2+} independent
NOS3 eNOS (*endothelial*)	Co-localizes with mitochondrial markers Co- and posttranslational fatty acylation (myristoylation and palmitoylation) may be involved in targeting	Increased with chronic exercise (electrical stimulation), cytokines	Calcium- and calmodulin-dependent activation

Mammalian skeletal muscle is known to express all major NOS isoforms, including a muscle-specific splice variant of the neuronal isoform (NOS1) (known as nNOSμ). Found in mature heart and skeletal muscle, this tissue-specific isoform results from alternative splicing of nNOS pre-RNA (ribonucleic acid) leading to a 34-amino-acid insert. NOS isoform location and expression within skeletal muscles is influenced by a variety of factors including species, age, developmental stage, activity level, muscle fiber-type distribution, and exposure to cytokines and growth factors. The subcellular localization of NOS is isoform-specific in adult skeletal muscle (Table 11.9).

The majority of skeletal muscle NOS activity is attributed to nNOS. In rodent muscle, NOS activity is proportional to type II fiber content, whereas in humans the activity in type I and II fibers is more evenly distributed. The increase in NOS activity with acute contraction or electrical stimulation is consistent with calcium activation of nNOS via a calmodulin-dependent mechanism.

Decreased force production, the inhibition of excitation-contraction coupling, decreased mitochondrial respiration, and increased glucose uptake are all linked to NO production and action. Decreased force appears to arise from a direct inhibitory effect of NO on actomyosin ATPase activity, leading to reduced cross-bridge cycling. Ryanodine receptor-calcium release channels and guanylate cyclase activities are also modulated by NO. The ryanodine receptor is a key player in the release of calcium from the sarcoplasmic reticulum during excitation-contraction coupling. This is the process by which signals at the cell surface are coupled to the release of calcium from the sarcoplasmic reticulum and ultimately lead to actin–myosin interaction and fiber contraction. *S*-nitrosylation of the ryanodine receptor at a single cysteine has been shown to increase its activity in a reversible fashion. Guanylate cyclase in skeletal muscle is implicated in the inhibition of force production as well as promoting glucose uptake and utilization. NO has been shown to bind to the heme moiety of the soluble guanylate cyclase leading to its activation and the increased production of cGMP (Fig. 11.7). cGMP analogs, phosphodiesterase inhibitors (which prevent cGMP breakdown), and NO donors that increase cGMP levels all inhibit excitation-contraction coupling in skeletal muscle. The link between NO and cGMP has also been

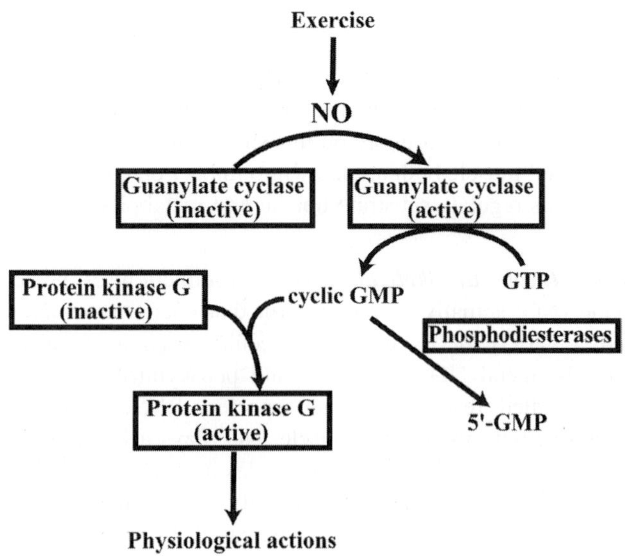

Figure 11.7 Nitric oxide (NO)-mediated activation of the guanosine 3′-5′-cyclic monophosphate (cGMP) pathway.

established *in vivo*, since *mdx* mice (a model of Duchenne muscular dystrophy lacking dystrophin) that are deficient in nNOS fail to show any increase in their cGMP levels following electrical stimulation. The functional link is further reinforced by the observation that both nNOS and cGMP-dependent kinase are found at high levels at the neuromuscular end plate.

The force-decreasing properties of NO could also result from its indirect effects on other systems. Recent evidence strongly suggests that ROS, and perhaps superoxide in particular, are obligatory for optimal force production and the promotion of excitation-contraction coupling in skeletal muscle. Thus, both twitch and tetanic forces developed in submaximal contraction are increased by superoxide, and the addition of superoxide dismutase decreases force development. Since NO is a potent superoxide scavenger (producing peroxynitrite), the removal of superoxide by NO could suffice to decrease force during contraction.

Other cellular targets that have been shown to be inhibited by NO include cytochrome-*c* oxidase, creatine kinase (inhibited by *S*-nitrosoglutathione, which is produced following NOS activation), sarcoplasmic reticulum Ca^{2+}–ATPase, as well as glyceraldehyde-3-phosphate dehydrogenase, complex 1 of the respiratory chain, and aconitase. Therefore, mitochondrial respiration and calcium homeostasis are inhibited by NO production in skeletal muscle.

Redox Modulation of Skeletal Muscle Glucose Transport The redox state of myocytes is of primary importance in the regulation of a number of skeletal muscle functions, including carbohydrate metabolism. Beyond direct effects on enzymes of carbohydrate metabolism (e.g., glycolysis), recent work has outlined the particular importance that redox state plays in glucose transport into skeletal muscle cells, a rate-limiting step in carbohydrate metabolism. Mediators of this redox modulation notably include ROS and RNS (sometimes collectively known as reactive oxygen/nitrogen species, RONS).

The precise RONS species that are responsible for increased glucose transport have yet to be clearly delineated. This is due partly to the interrelationship between many of the compounds and the possibility that their reaction products could, in fact, be responsible for their effect (e.g., peroxynitrite, formed by the reaction of superoxide and nitric oxide). Still, numerous studies have now demonstrated the impact of RONS on glucose transport. For example, hydrogen peroxide (H_2O_2) is an insulin-mimetic agent, stimulating glucose transport into skeletal muscle and increasing exercise-induced glucose transport. H_2O_2 levels are increased after exercise and, therefore, could serve as an endogenous signal for increasing glucose transport. Also, incubation with nitric oxide donors such as sodium nitroprusside (SNP) has been shown to increase glucose transport in skeletal muscle.

To date, the precise involvement of different GLUT isoforms in the RONS-mediated increase in glucose transport remains unclear and is an active area of research. However, both an increase in GLUT1 protein content and an increase in GLUT4 translocation to the cell membrane have been invoked depending on the stimulus used and muscle preparation. The signaling pathways that mediate the RONS effect on glucose transport are also areas of intense research. The monomeric G protein $p21^{ras}$ as well as cGMP have been evoked in RONS-mediated glucose transport. Whether one or both are the key player *in vivo* remains to be determined.

PERSPECTIVES

The complexity of the interaction of skeletal muscle with the nervous system and other tissues has only begun to be understood. At present, the intracellular pathways that transduce the physiological stimuli into signals that impart phenotypic change in skeletal muscle are largely unknown. Even the mechanisms that govern normal muscle function and maintenance are obscure, although the synergistic effects of factors such as neural signaling, growth factors, cytokines, and hormones undoubtedly play important roles. Complicating the matter, evidence of the heterogeneity of the response of individuals to changing functional demands, at the level of whole-body function or muscle-specific characteristics, is ever-increasing. Why a similar training program prescribed to two otherwise comparable individuals does not elicit the same qualitative or quantitative response is a question that we cannot answer at present. Gene and gene–environment interactions, key players in biological diversity, certainly contribute to these observed differences. Yet we have very little information on the precise factors or mechanisms involved in these processes. The next great leap in our understanding of skeletal muscle function and adaptation will include the elucidation of the mediating factors responsible for the changes we observe.

REFERENCES

Reviews

Allen, D. G., and Westerblad, H. (2001). Role of phosphate and calcium stores in muscle fatigue. *J Physiol* **536**:657–665. *An interesting discussion of the mechanisms responsible for, and the presentation of, new models related to muscle fatigue.*

Balon, T. W., and Yerneni, K. K. (2001). Redox regulation of skeletal muscle glucose transport. *Med Sci Sports Exercise* **33**:382–385. *An overview of the importance of redox regulation in relation to skeletal muscle glucose transport and carbohydrate metabolism. Discussion of the role of reactive* *oxygen and nitrogen species, as well as the putative signaling cascades involved.*

Hood, D. A. (2001). Invited review: Contractile activity-induced mitochondrial biogenesis in skeletal muscle. *J Appl Physiol* **90**:1137–1157. *A thorough discussion of the events and players involved in contraction-induced mitochondrial biogenesis, including signals that activate the process, the genes involved and their regulation, and the coordinated expression of nuclear and mitochondrial genes.*

Kelley, D. E., and Mandarino, L. J. (2000). Fuel selection in human skeletal muscle in insulin resistance: A reexamination. *Diabetes* **49**:677–683. *A discussion of the Randle cycle and why it may not explain insulin resistance in skeletal muscle from patients with type 2 diabetes.*

Kemp, B. E., Mitchelhill, K. I., Stapleton, D., Michell, B. J., Chen, Z. P., and Witters, L. A. (1999). Dealing with energy demand: The AMP-activated protein kinase. *Trends Biochem Sci* **24**:22–25. *An excellent overview of the structure, function, and regulation of AMP-activated protein kinase.*

Reid, M. B. (2001). Invited review: Redox modulation of skeletal muscle contraction: What we know and what we don't. *J Appl Physiol* **90**:724–731. *An interesting discussion of the impact of redox state on muscle contraction. Especially provocative are the discussions on the controversial physiological relevance of redox modulation and the source of regulatory species for this process.*

Stamler, J. S., and Meissner, G. (2001). Physiology of nitric oxide in skeletal muscle. *Physiol Rev* **81**:209–237. *A very thorough summary of everything nitric oxide: Production, regulation, physiological impact.*

Westerblad, H., Allen, D. G., and Lannergren, J. (2002). Muscle fatigue: Lactic acid or inorganic phosphate the major cause? *News Physiol Sci* **17**:17–21. *An interesting discussion of why lactic acid accumulation and acidosis do not seem to explain fatigue in muscle, and why other factors such as inorganic phosphate accumulation from creatine phosphate breakdown might.*

Textbooks

Gardiner, P. F. (2001). *Neuromuscular Aspects of Physical Activity.* Human Kinetics, Champaign, IL. *A very comprehensive and up-to-date overview of the neuroscience, kinesiology, molecular biology and physiology related to muscle physiology. From muscle fibers to responses to physical activity.*

Hargreaves, M. (ed.). (1995). *Exercise Metabolism.* Human Kinetics, Champaign, IL. *An excellent source for information on metabolic processes during exercise. Although perhaps dated in some areas, the fundamental principles presented here remain relevant.*

Hargreaves, M., and Thompson, M. (eds). (1999). *Biochemistry of Exercise X.* Human Kinetics, Champaign, IL. *Part of an ongoing series of books on the biochemistry of systems involved in exercise. From signaling to adaptation in skeletal muscle, the cardiovascular system, and the nervous system, these reviews provide up-to-date information on many topics of interest to the field.*

Houston, M. E. (2001). *Biochemistry Primer for Exercise Science*, 2nd ed. Human Kinetics, Champaign, IL. *An extremely well written and accessible textbook on the essentials of biochemistry, with an emphasis on examples taken from the exercise world.*

Radak, Z. (ed.). (2000). *Free Radicals in Exercise and Aging.* Human Kinetics, Champaign, IL. *Considered a key work in the field, this book examines in great detail the sources of radicals, their impact on cellular processes, and their relationship to aging and exercise.*

12

OXYGEN IN BIOLOGY AND BIOCHEMISTRY: ROLE OF FREE RADICALS

MARCELO HERMES-LIMA

12.1 FREE RADICALS AND OXIDIZING AGENTS IN BIOLOGY

A *free radical* is defined as any atom or molecule that contains unpaired electrons and has independent existence (hence, the term *free*). A *dot* is used to represent a free radical species. The abstraction or gain of one electron by a nonradical molecule may (or may not) convert it to a radical species. Free radicals can have positive, negative, or neutral charges [reactions (12.1.1) and (12.1.2)].

$$\text{A} \rightarrow \text{minus one electron} \rightarrow \text{A}^{+\bullet} \qquad (12.1.1)$$

$$\text{B} \rightarrow \text{plus one electron} \rightarrow \text{B}^{-\bullet} \qquad (12.1.2)$$

For most of the nineteenth century, radicals were regarded as parts of molecules and without free existence. In the late nineteenth century, however, it was believed that even if free radicals did exist, their very nature would preclude the possibility of isolating them. Free radicals would have such a fleetingly short existence that the available analytical methods of the period could not detect (or isolate) them. However, the solution-phase studies of Moses Gomberg proved the "free" existence of the triphenylmethyl radical in 1900, which opened the field for much more interesting discoveries. In the mid-twentieth century, the invention of free-radical-based polymerization reactions became very important for the industrial world. Two good examples are the polymerization reactions of styrene and butadiene for the production of synthetic rubber and the polymerization of various vinyl monomers for the production of many plastics.

In the late 1930s Leonor Michaelis proposed that *all* oxidation reactions involving organic molecules would be mediated by free radicals. This *radical* and incorrect prediction prompted interest in the role of free radicals in oxidative biological processes. Up to the early 1970s, there was still very little evidence that free radicals could affect living organisms. Thirty years later, however, the evidence is just too great to be covered in a single book chapter. The following sections will focus on the principles of free radical chemistry and highlight various roles of free radicals in biology and medicine, with selected sections on some key historical events in the study of free radicals in metabolism. Further information connecting free radicals with selected human pathologies are discussed in Chapter 13.

12.1.1 Free Radical Reactions in Biology

In cells, one-electron abstraction of molecules can yield sulfur-, oxygen-, carbon-, and nitrogen-centered free radicals. For example, removal of one electron (and a proton, H^+) from a –SH group of a protein by a radical species ($R^\bullet$) yields a sulfur-centered free radical [reaction (12.1.3)]. If $R^\bullet$ had just one unpaired electron, the reaction would convert it to a nonradical species. Another example is the one-electron removal from bis-allylic C–H bonds of polyunsaturated fatty acids (PUFAs) that yields a carbon-centered free radical [reaction (12.1.4)]. This reaction can

Functional Metabolism: Regulation and Adaptation, edited by Kenneth B. Storey.
ISBN 0-471-41090-X

initiate lipid peroxidation damage of biological membranes (see Section 12.3).

$$R^{\bullet} + \text{protein-SH} \rightarrow \text{one-electron transfer to } R^{\bullet} \rightarrow \text{protein-S}^{\bullet} + R^{-} + H^{+} \quad (12.1.3)$$

$$R^{\bullet} + \text{(PUFA)-CH(bis-allylic C–H bond)} \rightarrow \text{(PUFA)-C}^{\bullet} + RH \quad (12.1.4)$$

High-energy splitting of nonradical molecules also yields free radicals. A classic reaction is the radiation-induced homolysis of water [reaction (12.1.5)], yielding hydroxyl radical ($^{\bullet}OH$) and a hydrogen atom ($H^{\bullet}$), the simplest kind of free radical. A free radical can also add to a nonradical molecule, yielding a free radical product ($A^{\bullet} + B \rightarrow A\text{-}B^{\bullet}$) or a nonradical product ($A^{\bullet} + B \rightarrow A\text{-}B$); a good example in this case is the hydroxylation of aromatic compounds by $^{\bullet}OH$ [e.g., reaction (12.1.6) shows the hydroxylation of phenylalanine]:

$$H_2O + \text{energy} \rightarrow \text{activated } H_2O \rightarrow H^{\bullet} + {}^{\bullet}OH \quad (12.1.5)$$

$${}^{\bullet}OH + \text{phenylalanine} \rightarrow \textit{ortho-}, \textit{para-}, \text{ or } \textit{meta-}\text{tyrosine} \quad (12.1.6)$$

Two radicals can also react with each other to yield a nonradical product ($A^{\bullet} + C^{\bullet} \rightarrow A\text{-}C$). When the two reactants are the same free radical species with one unpaired electron each, then a dismutation reaction may take place; one example is the dismutation of the carbon-centered ascorbate radical (ascorbyl$^{\bullet}$), yielding ascorbate (vitamin C) and dehydroascorbate [DHA; reaction (12.1.7)].

$$2 \text{ ascorbyl}^{\bullet} \rightarrow \text{ascorbate (reduced product)} + \text{DHA (oxidized product)} \quad (12.1.7)$$

Radical molecules, with one or more unpaired electrons, can be attracted to a magnetic field; this means that most radicals have paramagnetic characteristics. Application of the correct electromagnetic energy causes absorption of that energy by the unpaired electrons and an absorption spectrum is obtained (usually unique for each radical molecule), generally in the microwave region. This is the basic principle of the well-known technique of electron spin resonance (ESR; also known as electron paramagnetic resonance, EPR), which has been used for the study of biologically important radicals since the 1970s. An example of an ESR signal of a radical molecule (ascorbyl$^{\bullet}$) is depicted in Figure 12.1 (see also Chapter 13 for an example of ESR usefulness in experimental medicine).

Due to the instability of many radicals (which preclude their direct detection), *spin trapping* techniques are used to study them. This is based on the reaction of radicals with nitroso molecules (R-NO), giving stable nitroxide radicals. Each nitroxide radical has a characteristic ESR spectrum, which can be properly studied. The most commonly used nitroso compounds for biochemical studies are 5,5-dimethyl-1-pyrroline-*N*-oxide (DMPO) and α-phenyl-*N*-*tert*-butylnitrone (PBN) (see Section 12.1.2.5). These spin traps are also useful for *in vivo* studies because the nitroxide radical products can be quantified, for example, in plasma samples.

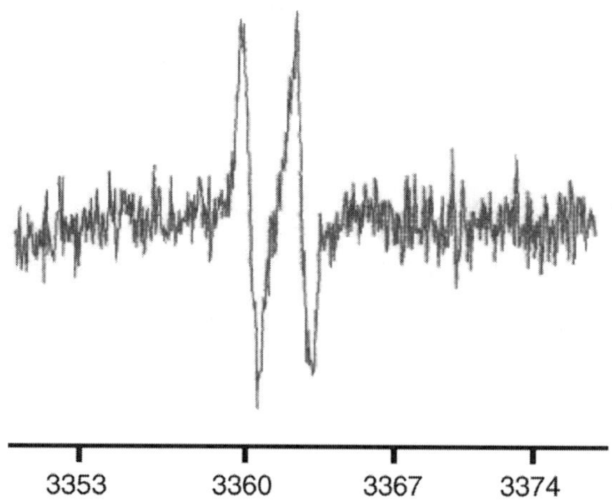

Figure 12.1 A typical two-lined ESR spectra of ascorbyl radical. The spectral data was obtained in an associated Brazilian laboratory after incubation of 5 μM Fe(III)–EDTA with 1 mM ascorbate, in neutral pH. Values on the *x*-axis represent the magnetic field (in Gauss).

Note: All illustrations in this chapter were prepared by Guilherme Pintarelli, University of Brasilia Medical School.

In the following subsections, we will discuss the biological role of oxygen radicals and related species, including nitrogen radicals and other reactive oxygen/nitrogen nonradical species (Table 12.1). The literature often calls

TABLE 12.1 Reactive Oxygen and Nitrogen Species (ROS and RNS)

Radicals	Nonradicals
Superoxide (O_2^{-})	Hydrogen peroxide (H_2O_2)
Hydroperoxyl ($HOO^{\bullet}$)	Alkyl hydroperoxides (LOOH)
Peroxyl ($LOO^{\bullet}$)	Singlet oxygen ($^1O_2$)
Alkoxyl ($LO^{\bullet}$)	Ozone (O_3)
Hydroxyl ($^{\bullet}OH$)	Hypochlorous acid (HOCl)
Nitric oxide (nitrogen monoxide; $^{\bullet}NO$)	Peroxynitrite ($ONOO^{-}$)
Nitrogen dioxide ($^{\bullet}NO_2$)	

these species *reactive oxygen species* (ROS), a term that often includes nitric oxide, peroxynitrite, and other reactive nitrogen species (RNS). For the present discussion, we have decided to differentiate between ROS and RNS for didactic purposes (e.g., nitric oxide is considered an RNS in this chapter). The biological roles of sulfur- and carbon-centered radicals are also discussed in Sections 12.2 and 12.3.

12.1.2 Reactive Oxygen Species

Oxygen gas, which is also known as dioxygen or diatomic oxygen, is a free radical species. In the ground state it contains two unpaired electrons located in different π-antibonding orbitals; these electrons have parallel spins ($\uparrow\uparrow$). The spin restriction rule makes it difficult for oxygen, in the absence of a catalyst, to receive a pair of electrons with parallel spins when oxidizing a molecule. Thus, oxygen must receive one electron at a time.

Several biologically relevant donors (Table 12.2) are able to induce one-electron reduction of oxygen, yielding another radical species: the superoxide ion radical ($O_2^{-\bullet}$). Superoxide is usually represented in the literature without the dot (O_2^-), and we will do the same. Further one-electron reduction reactions yield hydrogen peroxide (H_2O_2, a nonradical species), hydroxyl radical ($^\bullet OH$), and finally hydroxy ion (represented below as OH^-, *which is not the same as* $^\bullet OH$):

$$O_2 \underset{e^-}{\rightarrow} O_2^- + 2H^+ \underset{e^-}{\rightarrow} H_2O_2 \underset{e^-}{\rightarrow} OH^- + {}^\bullet OH \underset{e^-}{\rightarrow} OH^- \qquad (12.1.8)$$

TABLE 12.2 Endogenous Sources of Superoxide and Hydrogen Peroxide

Superoxide

- Autoxidation of small molecules[b] (e.g., adrenalin, noradrenalin, cysteine, reduced pteridines, L-DOPA,[b] dopamine,[b] 5-aminolevulinic acid[b])
- Autoxidation of aqueous Fe^{2+} and certain ferrous iron complexes (e.g., Fe^{2+}–citrate, Fe^{2+}–ATP, Fe^{2+}–ADP)
- Autoxidation of oxyhemoglobin, oxymyoglobin, and *Escherichia coli* flavohemoglobin
- Mitochondrial electron chain (at the ubiquinone and NADH-dehydrogenase sites)[b]
- Cytochrome P450 system of endoplasmic reticulum
- Electron chain of chloroplast (at photosystems I and ferredoxin sites)
- NADPH oxidase of phagocytes and some endothelial cells[b]
- Xanthine oxidase,[b] aldehyde oxidase, and fungi galactose oxidase
- Cyclooxygenases and lipoxygenases[c]

Hydrogen Peroxide

- Dismutation of O_2^- and/or $HOO^\bullet$ (spontaneous or catalyzed by SOD)
- Microbial glucose oxidase and glycoate oxidase from plant peroxisomes
- Other peroxisomal oxidases (e.g., amino acid oxidase, urate oxidase)
- Amine oxidases (e.g., monoamine oxidase, diamine oxidase, lysyl oxidase)

[a]These reactions are usually catalyzed by transition metals.
[b]See Chapter 13 for more details.
[c]See article by Kühn and Borchert (*Free Radic Biol Med* **33**:154–172, 2002) for more details.

12.1.2.1 Singlet Oxygen and Ozone Ground state oxygen can also be converted to singlet oxygen ($^1O_2$) by interaction with triplet-excited molecules, such as excited-protoporphyrin IX. Singlet oxygen is a relatively long-lived (microseconds) nonradical species with outer electrons in antiparallel spins ($\uparrow\downarrow$). Due to the lack of spin restriction $^1O_2$ has high oxidizing power and is able to attack membrane PUFAs, amino acid residues in proteins, deoxiribonucleic acid (DNA), and carotenoids. Skin damage in some types of porphyrias is attributed to the oxidizing effects of $^1O_2$ (see Text Box 13.3). Measurement of light emission from the decay of $^1O_2$ (e.g., at 1270 nm in the infrared region) is often used to quantify $^1O_2$.

Ozone is also a nonradical triatomic species. In the high atmosphere, ozone is Earth's protector against ultraviolet (UV) radiation. However, it is also formed in urban smog and is extremely reactive and a lung poison; notably, some large cities like São Paulo, Brazil, have ozone meters along with normal traffic signs! The $E^{0\prime}$ value of ozone makes it a powerful oxidizing agent that can react with proteins, DNA, PUFAs (see Section 12.3.1), and with small antioxidants, such as vitamin C and uric acid [reaction (12.1.9)]:

$$O_3 + \text{target} \rightarrow O_3^{-\bullet} + \text{oxidized target} \quad (E^{0\prime} = +0.89\ \text{V}) \qquad (12.1.9)$$

Ozone is also highly toxic to crops and forests. Interestingly, transgenic tobacco plants overexpressing superoxide dismutase, an antioxidant enzyme (see Section 12.2.1), in chloroplasts have high resistance to O_3.

12.1.2.2 Superoxide and Hydrogen Peroxide About 1 to 4% of all oxygen consumed by vertebrates (and possibly by higher invertebrates) produces O_2^- radicals. Mitochondria are the main site of O_2^- production in cells, by means of the "leaky" electron transport system (see details in

Chapter 13); other sources are listed in Table 12.2. However, there is concern that the estimates of mitochondrial O_2^- generation (mostly done *in vitro* with isolated mitochondria) are about 10-fold higher than the amount of O_2^- formed *in vivo*. In any case, even if "only" 0.1 to 0.4% of consumed oxygen produces O_2^-, it is still an enormous production of free radicals.

Superoxide is a good example of a free radical species that can act as either an oxidizing [it can oxidize the reduced form of nicotinamide adenine dinucleotide (NADH) to the oxidized form of NAD (NAD^+)] or reducing agent, the later action being much more relevant in biology. Superoxide is able to reduce Fe^{3+} bound to citrate [reaction (12.1.10)], cytochrome C, ferritin (an iron storage protein; see Text Box 12.1), and enzymes containing iron–sulfur clusters (such as aconitase), but not iron bound to transferrin (a plasma iron transporting protein; see Text Box 12.1). The rates of reaction of O_2^- with lipids, proteins, and DNA are too slow to have biological significance:

$$Fe^{3+} - \text{citrate} + O_2^- \rightarrow Fe^{2+} - \text{citrate} + O_2 \quad (12.1.10)$$

The protonated form of superoxide, hydroperoxyl radical ($HOO^\bullet$; also represented in the literature as $HO_2^\bullet$), has a lower reduction potential than O_2^- [reactions (12.1.11a) and (12.1.11b)] and is able to abstract hydrogen from PUFAs. The pK_a value of $HOO^\bullet$ is 4.8 and the acid microenvironment near biological membranes favors its formation:

$$O_2 + e^- \rightarrow O_2^- + e^- + 2H^+ \rightarrow H_2O_2$$
$$(E^{0\prime} = -0.33 \text{ and } +0.94\,\text{V, respectively}) \quad (12.1.11a)$$

$$O_2 + e^- + H^+ \rightarrow HOO^\bullet + e^- + H^+ \rightarrow H_2O_2$$
$$(E^{0\prime} = -0.46 \text{ and } +1.06\,\text{V, respectively}) \quad (12.1.11b)$$

Moreover, the reaction of O_2^- with "free Fe^{3+}" yields a perferryl intermediate [reaction (12.1.12)], which may react with PUFAs and induce lipid peroxidation. Superoxide also reacts with nitric oxide at a very high rate and produce peroxynitrite (see Section 12.1.3):

$$Fe^{3+} + O_2^- \leftrightarrow [Fe^{2+} - O_2 \leftrightarrow Fe^{3+} - O_2^-] \leftrightarrow Fe^{2+} + O_2 \quad (12.1.12)$$

One electron reduction of either O_2^- or $HOO^\bullet$ yields H_2O_2 [reactions (12.1.11a) and (12.1.11b)]. Dismutation of superoxide [reaction (12.1.13)], either spontaneous or catalyzed by the enzyme superoxide dismutase (SOD; see

TEXT BOX 12.1 IRON IN HEALTH AND DISEASE*

Iron is a relatively abundant element on our planet and is vital for almost all known organisms. In humans, iron is present at 35 to 50 mg/kg of body mass. It plays a role in important cellular processes such as DNA synthesis, electron transport, and oxygen transport. Iron is a component of many cellular proteins and enzymes, including hemoglobin (containing 67% of the body iron), myoglobin, catalase, peroxidase, cytochromes, and nitric oxide synthase. It also influences its own homeostasis by regulating the expression of iron-metabolism-related proteins in response to changes in intracellular iron concentration. The wide usefulness of iron in metabolism is due to its redox chemistry: the capacity for donating or accepting electrons in the ferric (Fe^{+3}) or ferrous (Fe^{+2}) oxidation states (iron ions are free radicals since they contain unpaired electrons in the 3*d* orbital). However, this capability can also be harmful because it promotes the generation of oxygen radicals via the Fenton reaction [see reaction (12.1.16)] or via autoxidation reactions. Because iron is a crucial element in multiple cellular processes, a closely regulated balance must be maintained between iron uptake, transport, utilization, and storage.

Iron is taken up from the diet in the form of both heme and nonheme iron. Heme iron is internalized via a heme-receptor process at the cell surface of absorptive enterocytes. Within the cell, heme degradation occurs via a heme oxygenase, releasing inorganic iron to be captured by the ferritin storage protein or transported across the basolateral membrane into the plasma. In the human intestine, the absorption of nonheme inorganic iron is more efficient as the ferrous iron. A mucosal ferrireductase enhances the availability of ferrous iron, which is absorbed at the apical enterocyte membrane by a divalent cation transporter, termed Nramp2. Ferric iron also can be internalized by a membrane complex called paraferritin with subsequent reduction intracellularly.

Iron that is diverted to storage is complexed with ferritin, a 450-kDa protein that can store up to 4500 atoms of iron. The synthesis of ferritin responds to intracellular iron concentration, high iron concentration inducing ferritin expression whereas low iron concentration inhibits expression. This iron-mediated feedback of ferritin expression is regulated, not at the transcriptional level but by control over mRNA translation, as are many

*By Ricardo G. Oliveira, Egle M. Siqueira, and Marcelo Hermes-Lima, Universidade de Brasília, Brazil.

other proteins involved in iron metabolism (e.g., transferrin receptor, Nramp2, ferritin, and 5-aminolevulinic acid synthase). Two elements are crucial to this process, namely iron response elements (IRE) and iron regulatory proteins (IRP).

Iron response elements are stem-loop structures, located in the 5′ (e.g., ferritin) or 3′ (e.g., transferrin receptor) untranslationed regions of mRNA transcripts encoding proteins of iron metabolism. These structures serve as a binding site for IRP. Depending on the location of the IRE, at the 5′ or 3′ region, the binding complex IRE–IRP represses or enhances translation. In iron-deficiency status, the binding of IRP to the IRE in the 5′ region of ferritin mRNA blocks its translation. On the other hand, the binding of IRP to the IRE in the 3′ region of the transferrin receptor mRNA stabilizes the mRNA and enhances its translation to increase the number of transferrin receptors at the cell surface. In high intracellular iron status, IRP shows an aconitase activity (an enzyme of the tricarboxylic acid cycle) and does not bind to the IRE of ferritin mRNA; hence, translation of ferritin mRNA is enhanced. By contrast, transferrin receptor mRNA is probably degraded by an endonuclease, due to the lack of formation of an IRP–IRE complex.

Iron exits enterocytes via an iron transporter called ferroportin located at the basolateral membrane. The metal is transferred to the plasma protein, transferrin, an 80-kDa glycoprotein. The protein has two specific binding sites, occurring as monoferric or diferric transferrin, and its affinity for iron is pH dependent with low affinity at acidic pH. Ferrous iron exported into the plasma undergoes a ferroxidase reaction via the copper-containing oxidase ceruloplasmin. Uptake of iron from the plasma into other tissues is mediated through the binding of the transferrin–iron complex to the transferrin receptor, a dimeric glycoprotein of 80 kDa on the plasma membrane, followed by endocytosis. In the cytosol, the endosomal pH is reduced by a proton pump-mediated influx of H^+. Low pH causes iron to be released from the transferrin (still bound to its receptor) followed by export into the cytoplasm by the Nramp2 transporter where inorganic iron is again stored into ferritin or associated with a diverse group of ligands including organic anions (e.g., phosphates and carboxylates) and polypeptides.

A large number of diseases are accompanied by iron overload due to errors in iron metabolism. One such condition is hemochromatosis, an autosomal recessive iron overload disease where a mutation in the Hfe protein (the hemochromatosis protein that exerts an inhibitory effect on the transferrin receptor, causing increase in iron absorption) results in an accumulation of iron, mainly in the liver. Tissue iron accumulation can lead to morphological alterations, such as cirrhosis, probably due to the iron-mediated generation of ROS [for more details see Miret et al. (2003) *Annu. Rev. Nutr.* **23**:283–301]. Iron overload also occurs in other diseases such as β-thalassemia where a defect in the production of hemoglobin β-chains results in mild to severe anemia and chronic iron deposition in liver and heart. Other diseases result from defects in iron transport (e.g., aceruloplasminemia and atransferrinemia) and from secondary iron disorders, such as chronic inflammation-associated anemia. Iron is also involved in neurodegenerative pathologies, including Alzheimer's and Parkinson's diseases (see Chapter 13).

To avoid the deleterious effects of iron overload, drugs are used to reduce intracellular iron content. These drugs are called chelators (from the Greek *khele*, pincers) and they bind to metal ions to form, in some cases, an inert complex. An ideal iron chelator has an easy oral absorption, low production costs, a high and selective affinity for iron, low toxicity, the ability to cross membranes, and a high antioxidant capacity to inhibit metal-mediated oxyradical formation.

Deferoxamine is an iron chelator that is used clinically in the treatment of β-thalassemia and in some hemochromatosis patients. However, it has some disadvantages such as poor oral absorption, high costs, and a short half-life. Other compounds under evaluation for possible use in the treatment of iron overload diseases include L1 (1,2-dimethyl-3-hydroxypyrid-4-one; still highly controversial), certain plant polyphenols with metal chelating activity (see Text Box 12.4), and pyridoxal isonicotinoyl hydrazone (PIH). PIH was developed in 1970s by Prem Ponka (McGill University, Montreal) and is currently studied by our laboratory in Brazil. It is well-absorbed orally, economical, and effective in inhibiting *in vitro* lipid peroxidation, oxidative DNA damage, ascorbate oxidation, and 2-deoxyribose degradation through the formation of an inert complex of PIH with Fe^{3+} that does not catalyze oxyradical formation (see *Biochim Biophys Acta* **1620**:15–24, 2003). Further studies are needed to uncover other drugs (including new PIH analogs) that meet the requirements for an ideal iron chelator.

Section 12.2.1), is a main source of H_2O_2 *in vivo*. It is postulated that the fast reaction of O_2^- with $HOO^\bullet$ is a main source *in vivo* of noncatalyzed H_2O_2 formation [reaction (12.1.14); rate constant of $8 \times 10^7\ M^{-1}\ s^{-1}$]. Moreover, the speed of the dismutation reaction increases with the decrease in pH. The "overall" rate of spontaneous dismutation of superoxide is considered to be $4.5 \times 10^5\ M^{-1}\ s^{-1}$. Superoxide can also oxidize ascorbate, yielding H_2O_2 at a rate constant of about $3 \times 10^5\ M^{-1}\ s^{-1}$ [reaction (12.1.15)]:

$$2O_2^- + 2H^+ \rightarrow H_2O_2 + O_2 \quad (12.1.13)$$

$$O_2^- + HOO^\bullet + H^+ \rightarrow H_2O_2 + O_2 \quad (12.1.14)$$

$$O_2^- + \text{ascorbate} \rightarrow \text{ascorbyl}^\bullet + H_2O_2 \quad (12.1.15)$$

Hydrogen peroxide is a relatively poor oxidizing agent, having an $E^{0\prime}$ value (+0.32 V for the pair $H_2O_2/^\bullet OH$) lower than ozone, $HOO^\bullet$, or superoxide. For example, H_2O_2 cannot abstract hydrogen from PUFAs. What makes H_2O_2 a potentially dangerous species is its ability to easily cross biological membranes and its high stability. Thus, H_2O_2 produced in mitochondria (which actually arises from O_2^- formation that is followed by dismutation into H_2O_2), for example, could move throughout the rest of the cell to produce effects in multiple other compartments such as in the nuclei. However, in the intracellular environment, the presence of many enzymatic and nonenzymatic processes of H_2O_2 decomposition (e.g., transition metals, selected enzymes, and small molecules) keeps H_2O_2 at a low concentration, ranging from 10^{-9} to 10^{-8} M (notably, the intracellular O_2^- concentration is even lower, at around 10^{-11} to 10^{-10} M).

12.1.2.3 Hydroxyl Radical Many of the oxidizing effects of H_2O_2 on DNA, lipids, and proteins that were observed in the past are now known to have been caused by the interaction of H_2O_2 with transition metals, mainly $Fe^{2+}$and Cu^+, yielding $^\bullet OH$ radicals [reaction (12.1.16) shows the *Fenton reaction*] or other highly reactive oxometallic species, such as ferryl ($Fe^{4+}{=}O$). One electron reduction of H_2O_2 splits the O–O bond, forming $^\bullet OH$ and OH^- [see reactions (12.1.8)]. Some small organic peroxides can also undergo Fenton reactions. Hydroxyl radical can also be formed by radiation-induced homolysis of water [see reaction (12.1.5)] or H_2O_2 (reaction 12.1.17), and by the reaction of hypochlorous acid with O_2^- [reaction (12.1.18)].

$$H_2O_2 + Fe^{2+}(\text{or } Cu^+) \rightarrow Fe^{3+}(\text{or } Cu^{2+}) + OH^- + {}^\bullet OH \quad (12.1.16)$$

$$H_2O_2 + \text{energy} \rightarrow 2{}^\bullet OH \quad (12.1.17)$$

$$HOCl + O_2^- \rightarrow Cl^- + O_2 + {}^\bullet OH \quad (12.1.18)$$

The oxidizing power of a mixture of H_2O_2 with Fe^{2+} salts (but not with Fe^{3+} salts) on tartaric acid was first reported in the form of a note by 22-year old Henry J.H. Fenton, who was still a Cambridge undergraduate student at that time (*Chemical News* **33**:190, 1876). Fenton never mentioned the formation of $^\bullet OH$ as the intermediate species in tartaric acid oxidation, even in later publications (see also: Fenton 1984), but nonetheless, the reaction was later named after him. The concept of $^\bullet OH$ radical production appeared only in the early 1930s, soon after Fenton's death. German chemists Haber and Weiss (*Proc R Soc Lond [A]* **147**:332–351, 1934) proposed that $^\bullet OH$ radicals are formed from the reaction of superoxide with H_2O_2 in the presence of iron. Rediscovery of these reactions, in connection with a biological role occurred only in the late 1970s.

It is widely accepted that the Fenton reaction may be the most relevant source of $^\bullet OH$ formation in biology. The reaction of H_2O_2 with aqueous Fe^{2+} is very slow (rate constant of $10^2\ M^{-1}\ s^{-1}$) and is considered of limited biological significance by many current authors. However, iron bound to low-molecular-weight compounds such as adenosine 5′-triphosphate (ATP) or citrate increases the reaction rate by two orders of magnitude. On the other hand, the Fenton reaction with copper salts proceeds at a rate constant of $5 \times 10^3\ M^{-1}\ s^{-1}$.

For the Fenton reaction to continue, Fe^{3+} must be converted back to Fe^{2+} by reducing agents such as O_2^- [see reactions (12.1.10) and (12.1.19b)] and ascorbate. The scheme below shows the so-called *Haber–Weiss reactions*:

$$H_2O_2 + Fe^{2+} \rightarrow Fe^{3+} + OH^- + {}^\bullet OH \quad (12.1.19a)$$

$$O_2^- + Fe^{3+} \rightarrow Fe^{2+} + O_2 \quad (12.1.19b)$$

$$\overline{H_2O_2 + O_2^- \rightarrow O_2 + OH^- + {}^\bullet OH} \quad (12.1.19c)$$

Hydroxyl radical is one of the most reactive species known. It reacts with almost all kind of molecules with reaction rate constants ranging from 10^7 to $10^{10}\ M^{-1}\ s^{-1}$; indeed, about half of these rate constants are higher than $2 \times 10^9\ M^{-1}\ s^{-1}$. This is so fast that the reaction is controlled by the diffusion rate of $^\bullet OH$ (and the presence or absence of a specific target near the site of $^\bullet OH$ formation!). The $E^{0\prime}$ value for $^\bullet OH$ (+2.31 V) is also the highest among the biologically relevant free radicals (see Table 12.3).

Hydroxyl radicals are involved in the initiation of lipid peroxidation in biological membranes as well as oxidation and damage to proteins, nuclear DNA, and mitochondrial DNA (see Sections 12.3 to 12.5). Carbohydrates and ribonucleic acid (RNA) are also relevant targets of $^\bullet OH$. The $^\bullet OH$ radical participates in the etiology or worsening of multiple human diseases and pathological/stress conditions including cancer, neurological disorders, chronic inflammatory diseases, atherosclerosis, myocardial infarction, iron-overload diseases, and even muscle fatigue in extenuating exercise (see Chapter 13). Moreover, $^\bullet OH$ is also involved in the natural aging process by causing cumulative damage to DNA and proteins in cells.

TABLE 12.3 Standard Reduction Potentials ($E^{0\prime}$) of Selected Substances[a]

Half-cell	Volts	Half-cell	Volts
H_2O/hydrated electron (e_{aq}^-)	−2.87	Fe^{3+}/Fe^{2+} (aqueous)	0.11
$CO_2/CO_2^{\bullet -}$	−1.80	Fe^{3+}–EDTA/Fe^{2+}–EDTA	0.12
O_2, H^+/$HOO^\bullet$	−0.46[b]	$CoQH^\bullet$, H^+/$CoQH_2$ (ubiquinol)	0.20
Fe^{3+}–deferoxamine/Fe^{2+}–deferoxamine	−0.45	Ascorbyl$^\bullet$, H^+/ascorbate	0.28
Fe^{3+}–transferrin/Fe^{2+}–transferrin	−0.50[c]	H_2O_2, H^+/$^\bullet OH$, H_2O	0.32
$O_2/O_2^{-\bullet}$	−0.33,[b] −0.16[d]	Trolox$^\bullet$, H^+/Trolox	0.48
NAD^+, H^+/ NADH	−0.32	α-Toc$^\bullet$, H^+/α-TocH	0.50
Lipoic acid, $2H^+$/DHLA	−0.32[e]	Urate$^\bullet$, H^+/urate	0.59
GSSG, $2H^+$/2GSH	−0.24[e]	PUFA$^\bullet$/PUFA	0.60
Fe^{3+}–ferritin/Fe^{2+}–ferritin	−0.19	$O_3/O_3^{-\bullet}$	0.89
DHA/ascorbyl$^\bullet$	−0.17	$O_2^{-\bullet}$, 2 H^+/H_2O_2	0.94
CoQ (ubiquinone), H^+/$CoQH^\bullet$	−0.04	$ROO^\bullet$, H^+/ROOH	0.77–1.44[f]
		Fe^{3+}-(*o*-phen)$_3$/Fe^{2+}-(*o*-phen)$_3$	1.15
		$HOO^\bullet$, H^+/H_2O_2	1.06
		$^\bullet OH$, H^+/H_2O	2.31

[a]Standard reduction potentials at pH0 and 25°C (E^0) were recalculated for pH 7.0; these values are then $E^{0\prime}$. Basic instructions: Electrons tend to flow from the half-cell of a lower $E^{0\prime}$ to the half-cell of a higher $E^{0\prime}$. For example, it is predicted that the following reaction is possible: DHLA + GSSG → 2 GSH + lipoic acid (electrons flow from DHLA to GSSG). This is so because the $E^{0\prime}$ value of the pair lipoic acid/DHLA is −0.32 V and the $E^{0\prime}$ value for GSSG/2 GSH is −0.24 V.
[b]Calculated at 1 atm of oxygen.
[c]Currently accepted value.
[d]Calculated using an aqueous oxygen concentration of 1 M.
[e]Two-electron reduction.
[f]$E^{0\prime}$ value depends on the type of peroxyl radical.
Source: Data from Garry R. Buettner (1993). *Arch Biochem Biophys* **300**:535–543.

12.1.2.4 Reactive Oxygen Species Formed from PUFAs The process of lipid peroxidation (see Section 12.3 for details) results in the formation of three important reactive oxygen intermediates from PUFAs: alkyl hydroperoxides (LOOH), alkyl peroxyl radicals ($LOO^\bullet$), and alkoxyl radicals ($LO^\bullet$).

As in the case of H_2O_2, alkyl hydroperoxides are not radical species, but they are unstable in the presence of transition metals. Their decomposition causes the formation of an array of different molecules, including aldehydes and alkanes. Some antioxidant enzymes are involved in the conversion of alkyl hydroperoxides into their respective alcohols; these include glutathione peroxidases, glutathione *S*-transferases, and alkyl peroxidases (see Section 12.2 for details).

Alkyl peroxyl radicals (and alkoxyl radicals) are very reactive species and also participate in the process of propagation of lipid peroxidation. The $E^{0\prime}$ values for the various $LOO^\bullet$ species vary between +0.77 and +1.44 V, which is the range of the $E^{0\prime}$ values for $HOO^\bullet$, ozone, and superoxide. Vitamin E is a very important biological reducing agent for alkyl peroxyl radicals (converting $LOO^\bullet$ to LOOH); however, it reacts very slowly with $LO^\bullet$ species (see Sections 12.2.2.3 and 12.3.4). Other biologically important organic peroxides (and peroxyl radicals as well) not derived from PUFAs include those produced from thymine, cholesterol, and certain amino acid residues.

12.1.2.5 Quantitative Determination of O_2^-, H_2O_2, and $^\bullet OH$ Determination of ROS in simple chemical systems or in complex biological preparations can be done in various ways. Some techniques are supposed to be specific for certain species, such as $^\bullet OH$, but there is wide disagreement about this. Assays based on ESR techniques are of great value for the detection of O_2^- and $^\bullet OH$ and will be discussed later in this section.

Superoxide radicals, when generated in high quantities from pulse radiolysis or from KO_2 salt, and in simple chemical/biochemical systems, can be detected by UV absorption at 245 nm (whereas $HOO^\bullet$ absorbs at 225 nm). Several molecules can also be used to indirectly quantify O_2^- spectrophotometrically, including cytochrome C (the ferric form is reduced by O_2^-, measured at 550 nm), acetylated cytochrome C (a way to avoid interference from cytochrome oxidase), adrenalin (oxidized by O_2^- to form adrenochrome), and nitroblue tetrazolium (NBT; reduced by O_2^- to the blue-colored formazan). These are the traditional or popular reagents used. A problem with the use of NBT and adrenalin is that both can produce O_2^- when they are subjected to ROS attack. Ferric-cytochrome C (as well as acetylated cytochrome C) can also be reduced by other molecules that might be present in the tissue extracts, such as ascorbate. Thus, addition of purified CuZn–SOD to the reaction mixture is a necessary control to determine whether reduction of the target molecule is really due to O_2^-. Quantification of light emission from luminol (after reaction with O_2^-) is also a popular technique, but other ROS can also cause light emission from luminol. Formation of formazan precipitates can also be useful for histochemical determination of O_2^-.

Determination of H_2O_2 can be done by various ways. At millimolar concentrations, H_2O_2 can be directly measured at 240 nm or by reaction with $KMnO_4$. In simple chemical systems, it can also be measured in micromolar concentrations by an oxygen electrode after the addition of catalase (which decomposes H_2O_2 to H_2O and O_2). Spectrophotometric or fluorometric methods employing horseradish peroxidase (HRP) plus a substrate that can be oxidized by H_2O_2 (such as scopoletin) are also very common.

In another method, 2′,7′-dichlorofluorescin diacetate can be enzymatically deacetylated to 2′,7′-dichlorofluorescin (DFCH) and then oxidatively converted to the fluorescent compound 2′,7′-dichlorofluorescein DFC; (excitation at 488 nm, emission at 525 nm). Conversion of DFCH to DFC can be done by HRP plus H_2O_2 or by direct reaction with strong oxidants such as peroxynitrite (see below), HOCl, or $^{\bullet}OH$. DFC imaging techniques became popular in the late 1990s for overall determination of oxidative stress in cell culture.

Determination of $^{\bullet}OH$ can be made through hydroxylation reactions, such as the fluorometric determination of benzoate hydroxylation (good for *in vitro* assays) or high-pressure ligand chromatography (HPLC) quantification of salicylate hydroxylation. The latter is often used for *in vivo* studies, where hydroxylation products (2,3- and 2,5-dihydroxybenzoate and catechol) can be detected in blood or urine. Ortho-tyrosine (as well as the *meta-* and *para-*isomers) is a product of $^{\bullet}OH$ attack on phenylalanine residues; *o*-tyrosine is a "classic" biomarker of protein oxidation under oxidative stress conditions (see Section 12.4).

The reaction of $^{\bullet}OH$ reaction with dimethyl sulfoxide (DMSO) (forming methanesulfinic acid and $^{\bullet}CH_3$) and 2-deoxyribose (producing malondialdehyde) has been used for the development of useful colorimetric techniques. However, other ROS (such as ferryl and other $^{\bullet}OH$-like species) are also capable of oxidizing DMSO and 2-deoxyribose; and, therefore, these assays cannot be viewed as specific for $^{\bullet}OH$ detection. Despite this, the 2-deoxyribose assay is still the most popular spectrophotometric assay for $^{\bullet}OH$ (and for the study of antioxidants that react with $^{\bullet}OH$) due to its simplicity and low cost. The product of the reaction, malondialdehyde, forms a pink adduct with thiobarbituric acid (TBA) that is measurable at 532 nm.

Determination of $^{\bullet}OH$ by quantifying the hydroxylation of the nitroso compound DMPO (at 10 to 100 mM) in ESR experiments is a very popular approach. The DMPO–OH radical molecule, with a typical four-lined ESR signal, is reasonably stable for several minutes, making it good for quantification of $^{\bullet}OH$ formation, mostly for *in vitro* studies. However, other reactive species may also produce DMPO–OH and thus the method cannot be considered as specific for $^{\bullet}OH$ detection (although mechanisms exist to confirm $^{\bullet}OH$ involvement). Superoxide radicals also react with DMPO to form a DMPO–OOH adduct, also with a well-defined ESR signal. However, the DMPO–OOH adduct decays quickly (within 1 to 2 min) to DMPO–OH, which needs to be taken into account when interpreting results. Another nitroso compound that is frequently used for $^{\bullet}OH$ quantification is PBN. The low toxicity of PBN makes it useful for *in vivo* studies.

12.1.3 Reactive Nitrogen Species

Nitric oxide is a gas (and a free radical species) that was identified in the 1980s as the endothelium-derived relaxing factor (EDRF). The relaxation of vascular smooth muscle cells by $^{\bullet}NO$, which is produced in nearby endothelial cells, was the very first function of $^{\bullet}NO$ to be discovered. Nitric oxide interacts with the heme prosthetic group of the soluble guanylate cyclase, prompting cyclic guanosine 5′-monophosphate (cGMP) formation and activating cGMP-dependent ion channels and kinases. Other functions of $^{\bullet}NO$ are discussed elsewhere in this book (see Section 12.6 and Chapters 4 and 6).

Nitric oxide synthase (NOS) catalyzes the formation of $^{\bullet}NO$. There are three classic homodimeric isoforms of NOS, named after the tissues/conditions where they were characterized and cloned: neuronal NOS (nNOS, or NOS1), inducible NOS (iNOS, or NOS2), and endothelial NOS (eNOS, or NOS3). These isozymes use the same cosubstrates—oxygen and NADPH (reduced nicotinamide adenine dinucleotide phosphate) [reaction (12.1.20)]—and the same cofactors [flavin adenine dinucleotide (FAD), tetrahydrobiopterin, heme, and calcium/calmodulin] but differ with respect to their mode of regulation and tissue expression. NOS2 has transcriptional and post-transcriptional inducibility and regulation of messenger RNA (mRNA) levels by cytokines; cytokines also mediate the mRNA expression of NOS1 and NOS3. Moreover, very recent evidence has indicated the existance of a fourth type of NOS, the mitochondrial one (mtNOS;

see Lacza et al. *Free Radic Biol Med* **35**:1217–1228, 2003).

$$\text{L-arginine} + \text{NADPH} + O_2 \rightarrow \text{L-citrulline} + {}^{\bullet}\text{NO} + \text{NADP}^+ \quad (12.1.20)$$

Nitric oxide contains an unpaired electron in the last orbital, making it a free radical molecule. It reacts relatively slowly with O_2 producing the orange-brown gas nitrogen dioxide (${}^{\bullet}NO_2$), a very reactive radical species [reaction (12.1.21)]. Further reactions of ${}^{\bullet}NO$ with ${}^{\bullet}NO_2$ will eventually produce nitrite (NO_2^-), which is the major decomposition product of ${}^{\bullet}NO$ [reactions (12.1.22) and (12.1.23)]. The calculated half-life of ${}^{\bullet}NO$ due to its reaction with O_2 is about 10 min and 15 h at 1 and 0.01 μM, respectively (monitoring of ${}^{\bullet}NO$ can be done with specific electrodes). The facts that ${}^{\bullet}NO$ does not have net charge and is high permeable across biomembranes makes it a good signal transduction molecule.

$$2{}^{\bullet}NO + O_2 \rightarrow 2{}^{\bullet}NO_2 \quad (12.1.21)$$

$${}^{\bullet}NO + {}^{\bullet}NO_2 \leftrightarrow N_2O_3 \quad (12.1.22)$$

$$N_2O_3 + H_2O \rightarrow 2NO_2^- + 2H^+ \quad (12.1.23)$$

$${}^{\bullet}NO + O_2^- \rightarrow ONOO^- \quad (12.1.24)$$

Even though ${}^{\bullet}NO$ is quite stable under certain conditions *in vitro*, it disappears within seconds *in vivo*. Nitric oxide reacts very quickly with the heme group of hemoglobin (see Chapter 18); it also reacts with O_2^- at near diffusion rates ($6.7 \times 10^9\ M^{-1}\ s^{-1}$), yielding peroxynitrite [$O{=}NOO^-$, represented usually as $ONOO^-$; reaction (12.1.24)]. It is noteworthy that ${}^{\bullet}NO$ controls the intra- and the extracellular concentrations of O_2^- (by forming peroxynitrite) and vice versa. The antioxidant enzyme superoxide dismutase (see Section 12.2.1.2) plays a key role in extending the physiological lifetime of ${}^{\bullet}NO$ by limiting the amount of O_2^- available to react with ${}^{\bullet}NO$.

Peroxynitrite is a highly stable and toxic nonradical anion. Molar concentrations at pH 13 can be stored for days in the refrigerator. Moreover, peroxynitrite-mediated oxidations can be mediated by both ground state peroxynitrous acid (ONOOH) and by the activated form of the acid (ONOOH*). (*Note*: The asterisk in ONOOH* means that the molecule contains electrons in the activated state rather than in the ground state.) The base ($ONOO^-$) is also a powerful nucleophile. It has been estimated that about 20% of peroxynitrite is in the acid form at pH 7.4; ONOOH (and/or ONOOH*) is highly reactive, with an $E^{0\prime}$ value (+2.10 V) approaching that of ${}^{\bullet}OH$. Peroxynitrite (and/or the acid form) is able to oxidize most biological molecules including DNA, RNA, proteins, and lipids.

Most of the damaging effects that were attributed to ${}^{\bullet}NO$ in the early 1990s are now attributed to peroxynitrite and/or ${}^{\bullet}NO_2$. Important biological markers of RNS action include 3-nitrotyrosine (a nitration product of tyrosine; see Section 12.4) and 8-nitroguanine (nitration product of the DNA base guanine; see Section 12.5).

Nitrosothiols (or thionitrates) are other important nonradical RNS. They are found in plasma at about 1 μM concentrations (mostly as *S*-nitrosoalbumin and nitrosohemoglobin). For example, the reaction of the radical form of glutathione (a thiyl radical; $RS^{\bullet}$) with ${}^{\bullet}NO$ yields *S*-nitrosoglutathione, a molecule with signaling properties. It is assumed that N_2O_3 or peroxynitrite mediate the formation of nitrosothiols.

$${}^{\bullet}NO + RS^{\bullet} \rightarrow RSNO \quad (12.1.25)$$

Some synthetic organic nitrates such as linsidomine (SIN-1) and the nitrosothiol *S*-nitroso-*N*-acetylpenicillamine (SNAP) are import ${}^{\bullet}NO$ donors. These molecules are often used in experiments to study ${}^{\bullet}NO$ effects. In the case of SIN-1, it is now clear that peroxynitrite is also formed in the reaction media due to autoxidation of SIN-1 followed by O_2^- formation and reaction with ${}^{\bullet}NO$. Thus, addition of SOD to the reaction media may enhance the yield of ${}^{\bullet}NO$.

12.2 ANTIOXIDANT DEFENSES

All aerobic living forms have defenses against ROS. Even various obligate anaerobe bacteria, which are not usually exposed to oxygen, have endogenous defenses against oxygen toxicity. Oxygen accumulation in Earth's primitive atmosphere and bodies of water may have provided a strong selection pressure for organisms with capacities to utilize oxygen in their energy metabolism as well as organisms that were capable of handling the toxic by-products of oxygen-based metabolism. On the other hand (as proposed by Denham Harman, University of Nebraska; *Proc Natl Acad Sci USA* **78**:7124–7128, 1981), too much protection by antioxidants on the ancient Earth would not have been favorable for evolution because it would have prevented the oxidative alterations to DNA that led to mutations and eventually to more highly developed life-forms.

Currently, existing life-forms are equipped with a very complex system for the control of free radical damage to cell constituents. The key importance of antioxidant defenses and their ancient origin is emphasized by the fact that some of the enzymatic defense systems are incredibly conserved, with strong conservation of amino acid sequences across wide phylogenetic ranges, from bacteria and humans.

The antioxidant defense systems of living forms can be divided into four subclasses:

1. Primary antioxidant defenses, of enzymatic or nonenzymatic nature, that deal directly with oxygen reactive species.
2. Auxiliary defenses that support the function of the primary antioxidant system (e.g., by recycling or synthesizing substrates of antioxidant enzymes).
3. Metal-complexing proteins/enzymes (e.g., ferritin, transferrin, ceruloplasmin, metallothionein) and low-molecular-weight compounds that prevent or minimize the participation of iron or copper (and other heavy metals) in free radical generation (see Text Box 12.1).
4. Enzymatic repair systems that repair biomolecules damaged by ROS and RNS. This last group of defenses consists mostly of a large array of enzymes that repair oxidized DNA (see Section 12.5). Moreover, a few newly discovered enzymes also provide some degree of repair to oxidized proteins (see Sections 12.2.1.5 and 12.4.2).

Most nonenzymatic antioxidant defenses are not synthesized endogenously by animals but must be provided by their diet. The plant and bacterial world has provided an incredible number of different substances (of dietary relevance or not) with potential antioxidant activity *in vivo* (plant phenols and carotenoids are the main groups of compounds), even though many substances also display pro-oxidant activity *in vitro*. The role of endogenous/dietary antioxidants in biology and medicine will be discussed in Section 12.2.2.

Several metals and other nonmetal elements (such as selenium) can also be considered as part of the antioxidant defense system. They are constitutive cofactors of several antioxidant enzymes (see Section 12.2.1) and of numerous other metabolic enzymes that keep the organisms alive. For example, in humans, copper is essential for ceruloplasmin and for a class of superoxide dismutase (CuZn–SOD; zinc is also crucial for this enzyme). Iron and manganese are, respectively, constituents of catalase and of the mitochondrial form of SOD (Mn–SOD) whereas selenium is crucial for several antioxidant enzymes. Iron and copper, in particular, are both essential and detrimental (as catalysts of ROS formation) for life (see Text Box 12.1).

12.2.1 Enzymatic Antioxidants, Glutathione, and Thioredoxin

The main enzymatic defenses against ROS include superoxide dismutase (several SOD forms exist), catalase, and selenium-dependent glutathione peroxidase (four GPx forms exist). SOD catalyzes the dismutation of O_2^- into O_2 and H_2O_2 [reaction (12.2.1)]. Catalase has a major role in the decomposition of H_2O_2 [reaction (12.2.2)], as does selenium-dependent GPx, which catalyzes the decomposition of H_2O_2 and also of organic hydroperoxides using the tripeptide glutathione in its reduced form (GSH) as a cosubstrate [reactions (12.2.3) and (12.2.4)]. Current evidence indicates that selenium-dependent GPx and another small protein thioredoxin (Trx, Section 12.2.1.5) are able to decompose peroxynitrite ($ONOO^-$).

$$2O_2^- + 2H^+ \rightarrow H_2O_2 + O_2 \quad (12.2.1)$$

$$2H_2O_2 \rightarrow H_2O + O_2 \quad (12.2.2)$$

$$H_2O_2 + 2GSH \rightarrow GSSG + 2H_2O \quad (12.2.3)$$

$$L\text{–}OOH + 2GSH \rightarrow GSSG + L\text{–}OH + H_2O \quad (12.2.4)$$

Other peroxidases have relevant actions in cellular H_2O_2 detoxification, including ascorbate peroxidase (a hemeprotein of 28 to 34 kDa from plants, particularly abundant in chloroplasts) and cytochrome c peroxidase (CcP; from bacteria, fungi, and some trematode worms) [reactions (12.2.5) and (12.2.6)]. Plants do not have GPx. Glutathione *S*-transferases (GSTs) catalyze the conjugation of GSH with xenobiotics [reaction (12.2.7)], including aldehydes produced during lipid peroxidation (see Section 12.3), and they also display selenium-independent GPx activity toward organic hydroperoxides [reaction (12.2.4)]. Alkyl peroxidases [reaction (12.2.8)] and thioredoxin peroxidases (see Section 12.2.3.5) also participate in the defense against organic peroxides.

$$H_2O_2 + \text{ascorbate} \rightarrow \text{dehydroascorbate} + 2H_2O \quad (12.2.5)$$

$$H_2O_2 + Fe^{2+}\text{–cythocrome } c \rightarrow Fe^{3+}\text{–cythocrome } c + 2H_2O \quad (12.2.6)$$

$$GSH + \text{xenobiotic} \rightarrow GS - \text{xenobiotic} \quad (12.2.7)$$

$$L\text{–}OOH + NAD(P)H + H^+ \rightarrow L\text{–}OH + H_2O + NAD(P)^+ \quad (12.2.8)$$

Several auxiliary enzymes are also involved in antioxidant defense. Glutathione reductase (a FAD-containing enzyme widely distributed in animals, plants, and microorganisms) functions to recycle glutathione, converting the oxidized form of glutathione (glutathione disulfide; GSSG) back to GSH using the reducing power of NADPH [reaction (12.2.9)]. Hexose monophosphate shunt enzymes including glucose-6-phosphate dehydrogenase [G6PDH; reaction (12.2.10)] and 6-phosphogluconate dehydrogenase are major suppliers of the necessary NAPDH.

$$GSSG + NADPH + H^+ \rightarrow 2GSH + NADP^+ \quad (12.2.9)$$

$$NADP^+ + \text{glucose-6-P} \rightarrow NADPH + H^+ + \text{6-phosphogluconate} \quad (12.2.10)$$

Ascorbate recycling is catalyzed by NAD(P)H-dependent ascorbate free radical reductase, a flavoenzyme present in plants, algae, fungi, and animals (this enzyme is highly active in kidney and liver of many vertebrates) [reaction (12.2.11)] and by dehydroascorbate (DHA) reductase. DHA reductase is a 23 to 28-kDa monomeric plant enzyme present in both chloroplasts and cytosol that uses the reducing power of GSH [reaction (12.2.12)].

$$NAD(P)H + H^+ + \text{ascorbyl}^\bullet \rightarrow \text{ascorbate} + NAD(P)^+ \quad (12.2.11)$$

$$2GSH + DHA \rightarrow \text{ascorbate} + GSSG \quad (12.2.12)$$

In addition, other enzymes function in the biosynthesis of GSH and in the recycling of sulfhydryl protein mixed disulfides, the latter by the thioredoxin system. These auxiliary defenses will be discussed in Section 12.2.1.5.

Moreover, the isoforms of heme oxygenase, HO-1, has been recently demonstrated to be of key relevance for antioxidant protection in cells. Heme oxygenase is the rate-limiting enzyme in heme catabolism leading to the generation of biliverdin, carbon monoxide, and free iron. HO-1 is a stress-responsive protein induced by various oxidant agents. An increase in HO-1 activity in cells generally occurs together with an increase in ferritin levels (an iron storage protein, see Text Box 12.1), which is stimulated by iron release from heme degradation.

The enzymatic antioxidant system works in a concerted way (Fig. 12.2). If, for example, there is an acute increase in H_2O_2 formation in a mammalian cell, GPx and/or catalase will decompose the peroxide to lower H_2O_2 to the previous steady-state concentration. However, if overformation of H_2O_2 is a chronic condition, where a higher steady-state concentration of H_2O_2 is established, then cells will increase the biosynthesis of H_2O_2-decomposing enzymes (GPx and/or catalase) to increase the capacity for H_2O_2 decomposition. Cells can also increase the biosynthesis of GSH, which can act as an antioxidant on its own in addition to its role as a cosubstrate of GPx and GST. These defense actions prevent the accumulation of oxidative damage to cell components, the consequences of which are highly variable in different cell types. For example, increased H_2O_2 formation can mediate $^\bullet OH$ formation (by metal-catalyzed reactions), which may

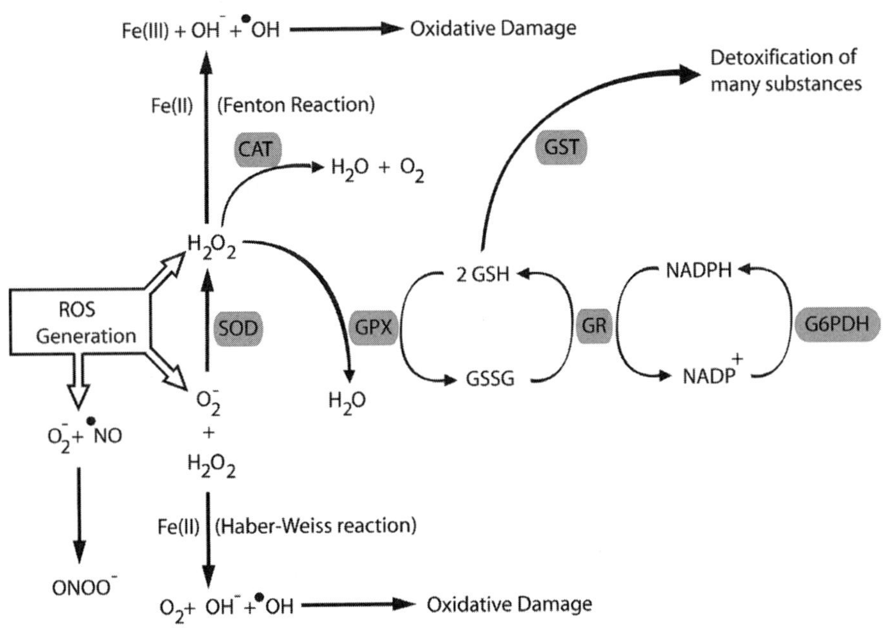

Figure 12.2 Enzymatic antioxidant defenses work in concert to protect cells against reactive oxygen species. The abbreviations SOD, CAT, GST, GR, GPX, and G6PDH represent the enzymes superoxide dismutase, catalase, glutathione *S*-transferase, glutathione reductase, glutathione peroxidase and glucose-6-phosphate dehydrogenase, respectively.

damage many kinds of biomolecules. Moreover, increased O_2^- formation (or exposure to O_2^--generating compounds) might prompt a response of the different SOD isoforms to control superoxide-mediated Fe^{3+} conversion to Fe^{2+} or peroxynitrite formation (upon reaction of O_2^- and $^{\bullet}NO$; Section 12.1) [see reactions (12.2.13)]:

$$O_2^- \rightarrow \text{dismutation to } H_2O_2 \qquad (12.2.13a)$$

$$O_2^- \rightarrow \text{reaction with } Fe^{3+} \text{ or } {}^{\bullet}NO \qquad (12.2.13b)$$

Furthermore, formation of O_2^- and H_2O_2 (and also certain lipid hydroperoxides) can occur as a consequence of signal transduction pathways (see Section 12.6). In this case, enzymatic antioxidants are necessary to ensure that the redox signal is not maintained indefinitely.

In the following subsections we will focus on some key antioxidant enzymes (SOD, catalase, GPx, GST), as well as the glutathione and thioredoxin systems.

12.2.1.1 Superoxide Dismutase The superoxide dismutase family contains four different enzymes with important structural and distribution differences. The most striking difference among the family members is the metal bound to them: There are two forms of copper–zinc SOD, a manganese SOD and an iron SOD. A nickel-containing SOD was recently identified in the bacterium *Streptomyces coelicolor*, but description of this enzyme is outside of the scope of this text.

CuZn–SOD is a 32-kDa dimeric protein that was initially described as being restricted to the cytoplasm of eukaryotes (see Text Box 12.2). However, recently it has also been detected in lysosomes, peroxisomes, nuclei, and the mitochondrial intermembrane space. Virtually all multicellular organisms contain CuZn–SOD in all their tissues; the enzyme is also present in various species of bacteria, located in the periplasmic space.

CuZn–SOD accelerates spontaneous O_2^- dismutation at pH 7.0 by several orders of magnitude at neutral pH (from $4.5 \times 10^5\ M^{-1}\ s^{-1}$ without an enzymatic catalyst to $1.6 \times 10^9\ M^{-1}\ s^{-1}$ with the enzyme), allowing the organisms to effectively control the intracellular steady-state levels of O_2^-. The copper ions in CuZn–SOD work in the dismutation reaction by means of alternate oxidation and reduction [reactions (12.2.14a) and (12.2.14b)]. Each subunit contains one copper ion, and the metal interacts with the imidazole rings of four histidine residues. The zinc ion is important for the maintenance of the structure of the enzyme and is located close to the active site. Most of the enzyme surface is negatively charged except for the active site and a nearby "track" that is positively charged, attracting O_2^-. Inhibitors of CuZn–SOD activity include cyanide and diethyldithiocarbamate, a compound

TEXT BOX 12.2 DISCOVERY OF SOD AND ITS FUNCTIONS

CuZn–SOD was first isolated from bovine erythrocytes in the late 1930s by T. Mann and D. Keilin. The crystalline copper-containing protein with molecular weight of 34 kDa was called *hemocuprein*. However, no enzymatic activity could be detected. Two decades later it was purified from human erythrocytes and baptized *erythrocuprein*. Copper proteins of similar size and copper content were isolated from other tissues, including human brain and bovine liver but, still, no apparent enzymatic function was observed. It was suggested that these copper proteins served as metal stores.

In the same year that men landed on the moon, Prof. Irving Fridovich and his predoctoral trainee Joe McCord, from Duke University (North Carolina), published a paper in *The Journal of Biological Chemistry* (**244**:6049–6055, 1969) that was the starting point for a revolution in biology and medicine. The authors reported that "erythrocuprein" (renamed superoxide dismutase) was able to catalyze the conversion of O_2^- into H_2O_2 and O_2 [see reaction (TB12.1)]. Activity of SOD was also observed in several other organs, including heart, brain, and skeletal muscle.

In those days, the only known biological source of O_2^- was the xanthine oxidase reaction, but there was literature showing that O_2^- could be generated in oxygenated aqueous solutions under X-ray or γ-ray irradiation. The study of O_2^- was still relegated to "exotic" radiation chemists—with very little biological connection. The interesting thing was that it was well-known in the years following the Manhattan Project that certain ionizing radiation sources (including those of natural origin) could have tremendous damaging effects on the organisms but only a few scientists had connected irradiation with oxygen-mediated toxicity! (See Section 12.5.1.)

In Fridovich's words (from p. 6055 of the 1969 paper), the "abundance of superoxide dismutase activity in the variety of animal tissues assayed suggests that the enzyme might play a significant, even vital, role in protecting the organism against the damaging effects of the superoxide radical." This was the basis for the "superoxide theory of oxygen toxicity," proposed by Fridovich and McCord a few years later. In the 1970s it become evident that organisms could produce O_2^- by various sources, including mitochondria, activated leukocytes, and autoxidation of several biomolecules. The superoxide theory of oxygen toxicity postulated that O_2^- was the intermediate of oxygen-mediated toxicity that could oxidize several import biological targets and was linked to the etiology of certain human disorders.

In the 1980s the importance of O_2^- was downgraded to be "just" an intermediate in the process of the metal-catalyzed formation of $^{\bullet}OH$ radical when it was realized that $^{\bullet}OH$ is the species that is really responsible for free-radical-induced toxicity. Indeed, $^{\bullet}OH$ is many orders of magnitude more reactive than O_2^-. Many studies were published describing SOD as a producer of H_2O_2 for Fenton reactions [see reactions (TB12.1) and (TB12.2)], which would make SOD a prooxidant! Others have recognized that SOD is biologically important because it keeps iron in the Fe^{3+} state [by removing O_2^- from solution; see reaction (TB12.3)], thus indirectly inhibiting Fenton-induced $^{\bullet}OH$ formation.

$$2O_2^- + 2H^+ \rightarrow H_2O_2 + O_2 \quad \text{(SOD speeds up this reaction)} \quad \text{(TB12.1)}$$

$$H_2O_2 + Fe^{2+} \rightarrow Fe^{3+} + OH^- + {}^{\bullet}OH \quad \text{(Fenton reaction)} \quad \text{(TB12.2)}$$

$$O_2^- + Fe^{3+} \rightarrow Fe^{2+} + O_2 \quad \text{(SOD inhibits this reaction)} \quad \text{(TB12.3)}$$

Interesting scientific battles occurred in the 1980s, with Fridovich defending the fundamental role of SOD and O_2^- in living organisms. More scientific efforts in the mid-1990s proved that O_2^- plays key roles in biology, including in the formation of $ONOO^-$ (via reaction with nitric oxide, see Section 12.1.3), inactivation of relevant enzymes with [Fe–S] clusters such as aconitase (with removal of ferrous ions, that participate in $^{\bullet}OH$ formation), and participation in cellular redox-signaling mechanisms and signal transduction pathways (see Section 12.6). So, in the end, Fridovich was right, and he is currently considered the father of the oxyradical science world. For more information about Fridovich's personal life and science achievements, read his recent review entitled "With the help of giants" (*Ann Rev Biochem* **72**:1–18, 2003).

that removes copper from the enzyme.

$$\text{SOD–Cu}^{2+} + O_2^- \rightarrow O_2 + \text{SOD–Cu}^+ \quad \text{(oxidation of } O_2^-) \quad (12.2.14a)$$

$$\text{SOD–Cu}^{+} + O_2^- \rightarrow O_2^{2-} + \text{SOD–Cu}^{2+} \quad \text{(reduction of } O_2^-) \quad (12.2.14b)$$

A different type of CuZn–SOD is the extracellular SOD (EC-SOD), a tetrameric glycoprotein with a molecular mass of 135 kDa. There are several isoforms of EC-SOD, and they seem to be bound to cell surfaces, especially in the lung and blood vessels. The biological function of EC-SOD appears to be connected with the extracellular control of O_2^- interaction with $^{\bullet}NO$ (which forms $ONOO^-$).

Mn–SOD is widely distributed in bacteria, yeast, plants, and animals. It is a 40-kDa enzyme with four subunits in mammals (two or four subunits in other species) that contains a Mn^{3+} ion in the catalytic site. Mn–SOD is insensitive to cyanide and is located in the mitochondria. The activity of Mn–SOD in mammals is about 1 to 10% of the total SOD activity (CuZn–SOD plus Mn–SOD). The enzyme plays a pivotal role in safeguarding mitochondria from oxidative stress as well as in regulating intramitochondrial O_2^- concentration (and consequently H_2O_2 concentration). Indeed, animals lacking Mn–SOD ("Mn–SOD knockouts") are incapable of surviving without antioxidant supplementation (see Text Box 12.3).

TEXT BOX 12.3 DETERMINING SOD ACTIVITY AND UNDERSTANDING ITS *IN VIVO* ROLE

There are many assays for the determination of SOD activity, and it is often difficult to compare activities between studies when different techniques are used. The basis of most SOD assays is the determination of the capacity of the sample to inhibit by 50% the reaction of O_2^- (generated in a variety of ways) with a detector molecule. The most commonly used SOD assay is based on O_2^- formation by xanthine plus xanthine oxidase and detection by determining the rate of reduction of Fe^{3+}–cytochrome *c* (Cc).

Other generators of O_2^- include Mn–EDTA/mercaptoethanol and oxygen, illuminated flavins plus oxygen, direct addition of KO_2 powder (the rate of KO_2 disappearance is determined spectrophotometrically), and autoxidation of adrenalin and pyrogallol (quantified colorimetrically). Other detectors of O_2^- include luminol (measurement of light emission), nitroblue tetrazolium (NBT), adrenalin, and hydroxylamine (quantified colorimetrically or by O_2 uptake). Nondenaturating gel electrophoresis has also been used to detect SOD activity in the presence of riboflavin–NBT. This causes formation of blue formazan in the gel, except in the areas corresponding to the SOD bands. The distinction between Mn–SOD and CuZn–SOD in tissue homogenates (or in column fractions during purification) is usually based on the inhibition of CuZn–SOD (but not Mn–SOD) by cyanide or H_2O_2 in parallel assays.

Overexpression of Mn–SOD or CuZn–SOD in transgenic mice (insertion of the human Mn– or CuZn–SOD gene in mice is the strategy of choice) and SOD-defective animals are good models to evaluate the function of SOD in biology. Transgenic mice with elevated levels of CuZn–SOD are more resistant to O_2 toxicity (and free-radical-generating compounds) than controls. Transgenic mice with increased lung Mn–SOD activity are more resistant to lung damage caused by 95% oxygen than control mice.

Knockout CuZn–SOD mice strains develop normally and show no overt motor deficits by 6 months in age. However, mice deficient in CuZn–SOD exhibit marked vulnerability to motor neuron loss after oxidative axonal injury. At the other extreme, homozygous mutant mice lacking Mn–SOD die within the first 10 days of life with a dilated cardiomyopathy, accumulation of lipids in liver and skeletal muscle, metabolic acidosis, and loss of activity of mitochondrial enzymes in several organs. Thus, the lack of control over O_2^- concentration in the mitochondrial matrix by Mn–SOD has drastic consequences. These observations suggest that the components of the antioxidant system might reorganize themselves to cope with the loss of CuZn–SOD, but not with the loss of Mn–SOD, when no exogenous stress is applied to the animals. Under oxidative stress conditions the lack of CuZn–SOD may become a problem for the organism.

The fourth form of SOD is an iron-containing enzyme (Fe–SOD) that was first described in the cell matrix of *Escherichia coli* (as was Mn–SOD). This bacterial enzyme has two or four subunits, with an amino acid sequence similar to Mn–SOD. Some bacterial species contain both Mn–SOD and Fe–SOD, and others contain just one form. Interestingly, recent reports have detected Fe–SOD in chloroplasts of higher plants. However, Fe–SOD has never been observed in the animal kingdom.

The activities of Mn–SOD and CuZn–SOD (see Text Box 12.3 for methods of SOD quantification) are highly regulated by the redox state of cells. Under oxidative stress conditions, it is common to detect increased cellular activities of both enzymes (see Section 12.6 for redox regulation of SODs). However, there are several genetic/pathological conditions that cause a decrease in SOD activities, resulting in disrupted redox balance and causing oxidative stress. One example is found in progeria, a premature aging disease. Human skin fibroblasts from affected individuals have decreased levels of Mn–SOD (but unchanged CuZn–SOD activity) as well as reduced catalase and GPx activities. The diminished capacity for H_2O_2 and O_2^- removal that ensues decreased ability of progeria cells to minimize oxidative damage may be a key factor in the disease. Indeed progeria cells have increased levels of carbonyl proteins, which are products of oxidative damage to proteins (see Section 12.4.3).

12.2.1.2 Catalase The earliest observations of catalase action date to the early nineteenth century, when the French chemist Thénard observed bubbling of H_2O_2 solutions (i.e., release of O_2) when these were added to animal tissue slices. The isolation of catalases (from tobacco, yeast, and blood) was first accomplished at the beginning of the twentieth century. It was initially imagined that H_2O_2 decomposition was a universal property of enzymes and, hence, the enzyme was named catalase because of its ability to *catalyze* H_2O_2 decomposition. In the 1930s the heme prosthetic group of liver catalase was identified.

Mammalian catalase is a homotetrameric protein of about 240 kDa that contains one heme group and one NADPH molecule per subunit. The enzyme is localized primarily in peroxisomes, but smaller amounts are found in mammalian heart mitochondria; both of these organelles are responsible for intense H_2O_2 production. Catalase is also present in most organisms, including vertebrates, invertebrates (see Table 12.4 for examples), plants, fungi, and bacteria. In vertebrates, catalase is present in all tissues with particularly high activity in erythrocytes, liver, kidney, and adipose tissue (200 to 1500 U/mg protein in humans); by contrast, catalase activity in brain is very low (5–20 U/mg protein in humans). Catalase is exceptionally active in insects where its activity may compensate for the absence of selenium-dependent GPx activity in many species. Some parasitic trematode worms, such as the liver fluke, have no catalase and it is believed that cytochrome *c* peroxidase (CcP, found in the intermembrane space of mitochondria) replaces its function in this case.

Bacteria have two types of catalase or hydroperoxidases, HPI and HPII, that differ in structure and kinetic properties. HPI from *E. coli* has four 80-kDa subunits, contains two molecules of protoheme IX, and displays catalase activity ($2H_2O_2 \rightarrow O_2 + 2H_2O$) and a broad peroxidase activity ($AH_2 + H_2O_2 \rightarrow A + 2H_2O$; AH_2 is any substrate that is oxidized). HPII is a homohexameric protein that contains one prosthetic heme *d* group per 93-kDa subunit and displays only catalase activity. HPI is believed to be more important than HPII for H_2O_2 resistance. Moreover, studies in the mid-1980s of the adaptive responses of *E. coli* toward H_2O_2 exposure via the induction of HPI (but not SOD) opened the field to studies of the genetic regulation of the expression of antioxidant enzymes (this is further explored on Section 12.6.1).

The assay of catalase activity is usually done spectrophotometrically by following the rate of disappearance of millimolar amounts of H_2O_2 at 240 nm. Other assays are

TABLE 12.4 Hepatic Activities of Antioxidant Enzymes and Levels of Glutathione (GSH + GSSG) in Several Animals Species

	SOD (U/mg)	Catalase (U/mg)	GST (U/mg)	GR (mU/mg)	cGPx (mU/mg)	GSH-eq (μmol/gww)
Rat	50–90	350–400	0.4–0.5	25–35	600–1500	7–9
Human	40–90[b]	1200–1500	NA	30–50	20–50	4.0–5.0
Turtles[c] and snakes[d]	10–50	70–200	0.6[d] and 2.0[c]	10–30	150–300	1.0–1.2[d] and 3.0[c]
Toads[e] and frogs[f]	15–50	100–600[f] and 1300[e]	0.7–1.5	5–20	40–150	0.6–1.7
Goldfish[g]	40–50	100–150	0.5–0.6	25–30	300–600	2.5–3.0
L. littorea[h] and land snails[i]	25[h] and 50–130[i]	20[h] and 180–210[i]	0.4–1.2	10–40	5–15	0.3[h] and 1.6–2.8[i]
Insect larvae[j]	60–150	40–150	0.05–0.40	0.2–0.4	0[k]	0.9–1.1

[a]This table shows the approximate range of enzyme activities (expressed per milligram protein) and GSH levels [per gram wet weight (gww)], taken from published values for mean ± SEM. Total SOD activity was reported, using the Mn–EDTA/mercaptoethanol/NADH assay [see Paoletti et al. (1986). *Anal Biochem* **154**:536–541].
[b]We converted an average activity (from the literature) of 107,000 units/gww determined by the KO_2 assay in SOD units (per mg/protein) of Paoletti's assay.
[c]Freshwater red-eared slider *Trachemys scripta elegans*.
[d]Garter snake *Thamnophis sirtalis*.
[e]Spadefoot toad *Scaphiopus couchii*.
[f]Frogs *Rana pipiens* and *R. sylvatica*.
[g]Goldfish *Carassius auratus*.
[h]Marine gastropod *Littorina littorea*.
[i]Land snails *Helix aspersa* and *Otala lactea* [see Ramos-Vasconcelos and Hermes-Lima (2003) *J Exp Biol* **206**:675–685].
[j]Larvae of the fly *Eurosta solidaginis* and larvae of the moth *Epiblema scudderiana* [see Joanisse and Storey (1998). *Insect Biochem Mol Biol* **28**:23–30].
[k]Selenium-dependent GPX (cGPx) activity was not detectable, however, selenium-independent GPX activity (GST-Px) was 2.0 to 5.0 mU/mg protein.
NA, Information not available.
Source: Most data were adapted from Marcelo Hermes-Lima et al. (2001). Antioxidant defenses and animal adaptation to oxygen availability during environmental stress. In *Cell and Molecular Responses to Stress*, Vol. 2, K. B. Storey and J. M. Storey (eds.). Elsevier Science, Amsterdam, pp. 263–287.

based on the quantification of O_2 evolution from H_2O_2 decomposition. The catalytic cycle of mammalian catalase involves the reaction of Fe^{3+}-catalase with one H_2O_2 molecule, forming compound I, which contains iron in a formal valency state of Fe^{5+} (it is probably Fe^{4+} bound to a porphyrin π-cation radical). Then, compound I receives two electrons from another H_2O_2 molecule, yielding H_2O and O_2. Interestingly, compound I was the first enzyme–substrate (ES) complex to be detected spectroscopically, confirming the validity of the [ES] hypothesis of Michaelis and Menten (see Chapter 2).

Catalase can be inhibited by aminotriazole, which works by "locking in" compound I, thus arresting the catalytic cycle of the enzyme. Administered to animals or plants, this inhibitor has been a very useful tool for understanding the biological role of catalase. Catalase is also sensitive to cyanide and azide. Most studies suggest that catalase is most effective in dealing with oxidative stress when the intracellular concentrations of H_2O_2 are highly elevated. Small increases in H_2O_2 seem to be better controlled by selenium-dependent GPx (see Section 12.2.1.3).

A lack of catalase can, in most cases, be overcome by a reorganization of other components of the antioxidant apparatus, as long as no further oxidative challenge occurs. For example, we observed in our laboratory in Brazil that administration of aminotriazole (at 1 g/kg) to goldfish causes a transient 80 to 90% reduction in liver catalase activity after 12 h (which lasts for several hours before the fish starts to synthesize more catalase) without increases in the levels of products of lipid peroxidation or protein oxidation in liver. Bacteria and yeast with deleted catalase genes have near-normal growth but are more sensitive to H_2O_2 exposure. Humans with acatalasemia (or Takahara's disease), a very rare condition caused by a mutation of the gene encoding catalase, have normal lives and normal life spans, except for increased incidence of mouth ulceration. This ulceration is caused by H_2O_2-producing *Streptococcus* bacteria normally found in the gums. Since the cells of acatalasemic subjects lack catalase activity, they are more sensitive to the effects of H_2O_2 or H_2O_2-generating compounds.

12.2.1.3 Glutathione Peroxidase Glutathione peroxidases (GPx) are four different selenoenzymes that have similar functions. These enzymes are typical of the animal kingdom, and some of them have been found in the last few years in plants and yeast. More recently, a gene with homology to GPx was shown to contribute to the antioxidant defense on Streptococcus.

12.2.1.3.1 Classical GPx The "classical" form of GPx (cGPx) was first identified in erythrocytes in 1957 and has the capacity to prevent oxidative breakdown of hemoglobin. In 1973 the German group of L. Flohé identified cGPx as a selenium-dependent enzyme; this was the first selenoprotein identified in mammals. cGPx is a homotetramer, with one selenium atom per subunit and ranges in size from 76 to 105 kDa depending on the animal species. Selenium participates in the catalytic cycle of cGPx and is present as a selenocysteine residue at the active site. Interestingly, selenocysteine is coded by the codon UGA, which is usually a stop codon.

Classical GPx catalyses the decomposition of H_2O_2 and some organic hydroperoxides, such as fatty acid hydroperoxide (LOOH) and the artificial substrates cumene hydroperoxide and *tert*-butyl hydroperoxide, using GSH as a cosubstrate [see reactions (12.2.3) and (12.2.4)]. The enzyme cannot reduce membrane-bound lipid hydroperoxides unless they are removed from the phospholipids by phospholipase A2. Very recent evidence has shown that cGPx is able to decompose $ONOO^-$ to nitrite (NO_2) at high rates using the reducing power of GSH [reaction (12.2.15)]. This is a highly important enzymatic activity for minimizing the cellular toxicity of $ONOO^-$:

$$ONOO^- + 2GSH \rightarrow GSSG + H_2O + NO_2 \qquad (12.2.15)$$

Organs with high metabolic rate, such as liver, lung, and kidney, produce more H_2O_2 by their mitochondria (see Section 12.1) and have higher activities of cGPx than do other tissues. Brain, on the other hand (which also has high metabolic rate), has exceptionally low levels of cGPx. cGPx activity has been observed in all vertebrates (fish and rodents have specially high cGPx activities) and in many invertebrates including earthworms, land snails, and crustaceans (see Table 12.4 for hepatic cGPx activity in several animal species). Many insects do not have cGPx, but instead they display selenium-independent GPx activity, which is due to GST isoforms with peroxidase activity (see Section 12.2.1.4).

Depriving animals of dietary selenium causes a severe reduction in cGPx activity (other GPxs are also affected by selenium deprivation, however, with different time-dependent responses) and is associated with a series of muscular and/or cardiac problems, depending on the animal species. Severe selenium deficiency in humans causes degenerative cardiac disorders, typical symptoms of Keshan disease. This disease was first reported in the 1930s in a Chinese area with selenium-poor soil, causing selenium deficiency (blood selenium below 0.3 to 0.4 μM, as compared to 1 to 3 μM for controls). Keshan disease is treated with small doses of sodium selenite (Na_2SeO_3; high quantities of Na_2SeO_3 present in many "orthomolecular antioxidant formulas" can be very toxic!).

Several studies have suggested that cGPx is not merely complementary to catalase but has a critical role in the inhibition/prevention of cellular oxidative challenge. A classical study of the mid-1980s by Pamela E. Starke and John L. Farber (from Hahnemann University School of Medicine, Philadelphia, Pennsylvania) demonstrated that cell cultures where catalase was inhibited (by aminotriazole) were much less affected by oxidants (e.g., paraquat or H_2O_2) than cells with impaired *in situ* function of cGPx (achieved via inhibition of glutathione reductase activity by 1,3-bis(2-chlorethyl)-1-nitrosourea (BCNU), which impairs the recycling of GSH). At low doses of H_2O_2 or paraquat (a generator of O_2^- and, consequently, of H_2O_2), the cells that were incubated with 1,3-bis(2-chlorethyl)-1-nitrosourea (BCNC) showed higher mortality than those incubated with aminotriazole. On the other hand, at high doses of oxidants the cell mortality was similar in either BCNU- or aminotriazole-treated cells (see Starke and Farber, 1985).

The *in vivo* role of cGPx in defense against oxidative stress was better understood when researchers from Ithaca, New York, studied cGPx knockout mice strains—that is, mice completely lacking cGPx activity [Cheng et al. (1998) *J Nutr* **128**:1070–1076]. The animals had normal development up to 15 months [and normal phospholipid hydroperoxide GPx (PHGPx) activity, another form of GPx that decomposes phospholipid hydroperoxides; see Section 12.2.1.3.2] and then were injected with paraquat. The knockout mice were highly sensitive to paraquat administration (at 50 mg/kg), showing mortality rates 10 times higher than controls and increased (4-fold) lipid peroxidation in liver. Moreover, selenium-deprived mice (which causes severe reduction in cGPx activity but has no effect on PHGPx) were also highly affected by paraquat, with mortality rates similar to knockout mice with normal selenium contents. In addition, mice overexpressing cGPx (1.6 and 2.6 times higher activities in kidney and lung, respectively) had mortality rates only one-tenth those of controls when administered paraquat at 125 mg/kg. Moreover, other studies showed that neuronal cells from cGPx knockout mice are more sensitive to H_2O_2 than control cells. These results indicated that under nonstress conditions, other antioxidant defenses can substitute for a lack of cGPx. However, when there is an overproduction/exposure to H_2O_2, cGPx is of fundamental relevance for safeguarding cell homeostasis.

12.2.1.3.2 Other Glutathione Peroxidases and Their Biological Functions In 1982, a form of GPx capable of reducing phospholipid peroxides in pig liver was identified in Fulvio Ursine's laboratory in Italy and was named phospholipid hydroperoxide glutathione peroxidase (PHGPx) [Ursini et al. (1982) *Biochim Biophys Acta* **710**:197–211]. This monomeric enzyme also reduces hydroperoxide groups of cholesterol esters (Ch-OOH; formed upon oxidation of cholesterol; see Section 13.4), thymine (formed upon oxidation of DNA; see Section 12.5), and lipoproteins. The small size and hydrophobic surface of PHGPx has been implicated in its ability to react with complex membrane lipids. PHGPx is present in most tissues from mammals, but its activity is much lower than cGPx (except in testis). Although PHGPx has very low activity compared to cGPx (4 to 20 mU/mg vs. 600 to 2000 mU/mg protein in rat liver, respectively), several studies have demonstrated that it is much more efficient in reducing membrane lipid hydroperoxides than cGPx plus phospholipase A2.

In the late 1980s American and Japanese scientists purified and identified a plasmatic tetrameric form of selenium-GPx (plasma GPx or pGPx). This is a small glycoprotein (about 23 kDa) that is kinetically and structurally different from the cGPx that is present in erythrocytes. pGPx is also capable of reducing phospholipid hydroperoxides. The discovery of pGPx occurred because of its very different ability to respond to selenium reintroduction (following severe selenium deficiency) than cGPx. Plasma GPx activity returned to normal 2 to 4 weeks after selenium reintroduction whereas 3 to 4 months was required before the erythrocyte enzyme activity returned to normal. This is because circulating red cells do not synthesize new proteins, and thus recovery depends on the turnover of red blood cells (which live for about 120 days; see Chapter 18). Once selenium is again available, red blood cell precursors will synthesize cGPx, and activity will be restored to the circulating pool of erythrocytes. Interestingly, pGPx seems to use either thioredoxin or glutaredoxin (see Section 12.2.1.5) as its physiological thiol substrates. The main source of pGPx synthesis is the kidney, which exports the enzyme into the plasma. Other organs also exhibit mRNA for pGPx, including liver, pancreas, muscle, and placenta (that releases pGPx into the amniotic fluid).

The fourth form of GPx is the gastrointestinal-GPx (GI-GPx) isoform. This enzyme was characterized in 1993, and it is a tetrameric selenoprotein that functions to degrade soluble hydroperoxides. It is restricted to the epithelial lining of the gastrointestinal track (in rats, these cells have no cGPx), possibly being the first line of defense against ingested hydroperoxides. Indeed, rats fed with hydroperoxide-rich diets have higher GI-GPx activity than controls. Moreover, mouse strains that are more sensitive to 1,2-dimethylhydrazine (DMH)-induced colon cancer have less GI-GPx activity. This suggests that GI-GPx may defend the organism against the carcinogenic effects of hydroperoxides in the gastrointestinal track.

The function of GPx enzymes as merely antioxidant defenses (a very important function) has recently being challenged. For example, pGPx has been linked with the regulation of extracellular hydroperoxide concentration in signal transduction pathways (see Section 12.6). PHGPx has been also implicated in modulating the function of lipid hydroperoxides as activators of lipoxygenase and cyclooxygenase, mediators in inflammation, and as signal molecules in apoptosis and receptor-mediated signal transduction.

The activity of GPx is usually quantified using a coupled assay with glutathione reductase (1 U/mL), NADPH (0.2 mM), GSH (5 mM), and H_2O_2 (added last, usually at 0.5 mM), with oxidation of NADPH followed at 340 nm. This assay gives the activity of the "selenium-dependent GPx," or simply Se–GPx (which is mostly cGPx). Several researchers use cumene hydroperoxide to measure "GPx activity" in their studies. However, they are in fact measuring the sum of Se–GPx and selenium-independent GPx activity (or GST–Px, an activity of GST), which cannot decompose H_2O_2. The percentage of GST–Px activity in relation to the total GPx in mammalian livers ranges from 30 to 85%; in insects it can even be 100%.

12.2.1.4 Glutathione* S*-transferases Glutathione *S*-transferases constitute a large family of multifunctional enzymes involved in GSH conjugation to xenobiotics and aldehydic products of lipid peroxidation (such as 4-hydroxyalkenals; see Section 12.3.3.2). Conjugates formed in hepatic cells are often excreted into bile using ATP-dependent pumps; they can also be degraded and acetylated to form *N*-acetylcysteine conjugates (mercapturic acids), which are excreted in the urine. Excretion of a mercapturic acid was first isolated in 1879 in the urine of dogs that had been fed bromobenzene. Only in the late 1950s was it discovered that mercapturates derive from GSH conjugation. GSTs are also very important for the biogenesis of prostaglandins and leukotrienes. GSTs are widely distributed in all life-forms including bacteria, fungi, plants, and animals. All animals investigated to date have GST homo- or heterodimers that are active in the cytosol or nucleus. Mammalian liver is particularly rich in GST activity; 5 to 10% of its cytosolic protein is GST!

Structural and functional studies on vertebrate GSTs have divided the several isoenzymes in four major classes: alpha (α), mu (μ), pi (π), and theta (θ) (two minor classes are sigma and kappa). These classes are based upon N-terminal amino acid sequences, substrate specificities, sensitivity to inhibitors, isoelectric point, and immunological analysis. Eric Boyland and co-workers from England discovered the existence of multiple forms of GST in the 1960s. Most GSTs have molecular masses of 40 to 55 kDa. Measurement of "total GST activity" is based on the reaction of GSH with 1-chloro-2,4-dinitrobenzene (CDNB), forming *S*-(2,4-dinitrophenyl)-glutathione, which is quantified at 340 nm. This assay, however, does not measure the θ class of GSTs because of their lack of reactivity toward CDNB.

Glutathione *S*-transferases also display selenium-independent GPx activity, designated as GST–Px, which is measured using the cumene hydroperoxide assay (see Section 12.2.1.3). Endogenous substrates of GST–Px include fatty acid hydroperoxides, Ch-OOH, and thymine hydroperoxide (formed upon oxidation of thymine), but not H_2O_2. The GST–Px activity is differentially distributed among the GST classes, being highest for GSTθ, moderate for GSTα, and low for the other isoforms.

Elegant studies in the late 1970s from Helmut Sies' laboratory in Germany showed that GST–Px activity is relevant in the detoxification of organic hydroperoxides in mammals. They studied GSSG efflux from rat liver under oxidative stress. When GSSG accumulates in the liver (due to cGPx-mediated oxidation of GSH to GSSG), it is pumped out into the circulation. The authors infused H_2O_2 into isolated livers from rats that had been fed a selenium-deficient diet (which severely depletes cGPx activity) and detected no GSSG efflux. However, significant GSSG efflux was observed from these livers when they infused *tert*-butylhydroperoxide (*t*-BOOH), which can be metabolized by GST–Px [reaction (12.2.16)].

$$t\text{-BOOH} + 2\text{GSH} \rightarrow t\text{-BOH} + H_2O + \text{GSSG} \quad (12.2.16)$$

Glutathione *S*-transferases are very important in cancer resistance because many endogenously produced aldehydes and many P450-activated xenobiotic drugs can damage DNA (see Section 12.5). For example, a GSTα found in mice catalyzes the detoxification of *exo*-8,9-epoxide (formed by P450 activation of aflatoxin B_1, a potent hepatocarcinogen found in badly stored peanuts), but rats do not express an enzyme with corresponding activity. This probably accounts for the much greater sensitivity to aflatoxin B_1-induced hepatocarcinogenesis of rats compared with mice. Induction of GSTs has also been linked with resistance of cancer cells toward chemotherapy because many anticancer drugs (or their P450-activated forms) are GST substrates. Tremendous efforts have been made to develop rationally designed GST inhibitors for adjuvant treatment during chemotherapy. One candidate for clinical use is the diuretic drug, ethacrynic acid, a good inhibitor of most GST isoforms.

The study of GST isoforms of insects is also relevant for crop sciences. This is because many crop-plague insects have potent enzymatic activities that can deactivate insecticides via GSH conjugation. Herbivorous insects have very high natural GST activities due to the presence of toxic compounds in the plants they feed on (notably, these toxic compounds are the plant's defensive insecticides).

Moreover, it has been recently demonstrated in Germany that manipulation of the expression of a single GST can modulate the response to oxidative stress in *Caenorhabditis elegans*. This study was the first to show a direct protective effect of GST *in vivo* (Leiers et al. *Free Radic Biol Med* **34**:1405–1415, 2003).

12.2.1.5 Glutathione and Thioredoxin Systems Glutathione is widely distributed in plants, animals, fungi, and bacteria in intracellular concentrations ranging from 0.1 to 10 mM. It also occurs in extracellular fluids and plasma at much lower levels. This antioxidant tripeptide is present primarily in its reduced from, GSH (γ-GluCysGly; see Fig. 12.3), with the ratio between GSH concentration and the disulfide oxidized form (GSSG) varying from 5 to 150, in a species- and tissue-specific manner. The GSH/GSSG ratio also changes during aging, oxidative stress conditions, and disease and is considered a relevant marker of oxidative stress (e.g., a diminution in the GSH/GSSG ratio is correlated with increased production of hydroperoxides). Cells are able to export either GSH (liver is a major exporter; see below) or GSSG into the plasma. Under oxidative stress conditions, the export of GSSG (see Section 12.2.1.4) helps cells to maintain a negative reduction potential.

Glutathione plays several roles in biology. It is classically viewed as a substrate for GPx- and GST-catalyzed reactions. GSH has also nonenzymatic antioxidant activity and can react with various carbon-centered radicals, singlet oxygen, $^{\bullet}OH$, and reactive nitrogen species to yield the GSH–thiyl radical ($GS^{\bullet}$) and other products. One product of the reaction of GSH with nitric oxide is *S*-nitrosoglutathione (GSNO), which may help to prevent $ONOO^-$ formation by neutralizing nitric oxide. Glutathione is also involved in the cellular uptake of amino acids (see below), the metabolism of ascorbic acid (see introduction to Section 12.2), leukotriene synthesis, glutaredoxin synthesis (see below), regulation of the activities of many enzymes with critical thiol groups, and in the mechanisms of protein folding and protection against oxidation of protein –SH groups. GSH is also involved in the mechanism of redox signaling (see Section 12.6).

Reduced glutathione (GSH) is synthesized in vertebrate cells by a two-step process involving γ-glutamylcysteine (γ-GluCys) synthetase and GSH synthetase [reactions (12.2.17) and (12.2.18), respectively]. GSH is a competitive inhibitor of γ-GluCys synthetase, providing feedback control of GSH formation. Intracellular levels of GSH are also controlled by secretion into plasma, as well as by enzymatic oxidation to GSSG, or conjugation with GST substrates (that are pumped out of the cells). γ-GluCys synthetase can be inhibited by buthionine sulfoximine

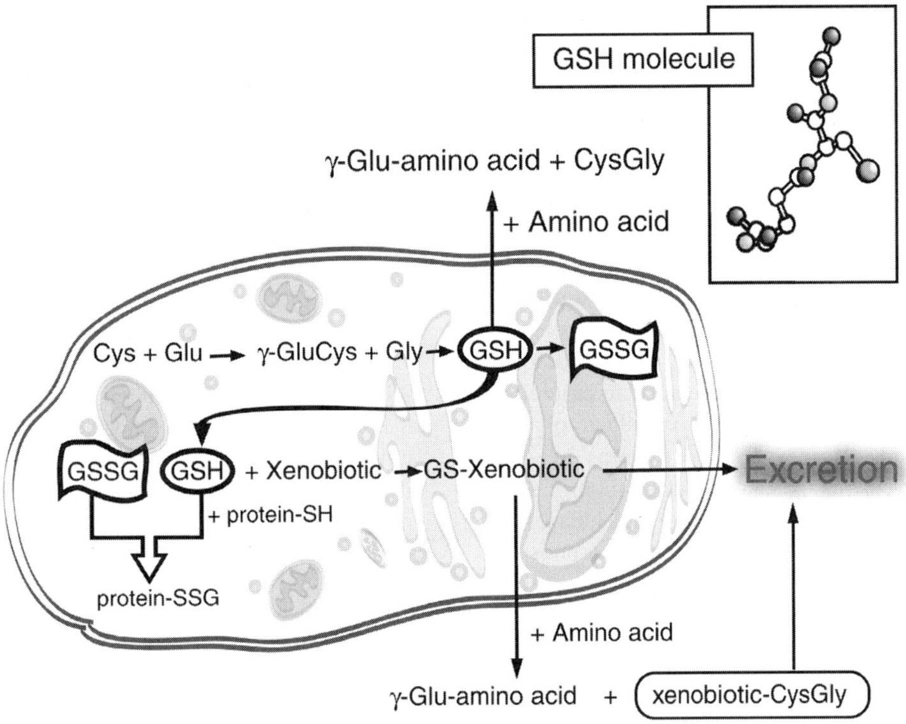

Figure 12.3 Glutathione metabolism in liver cells. The figure shows the biosynthesis of GSH and the fates of GSH and GSSG.

(BSO), which is a widely used tool for the study of the effects of GSH depletion.

$$\text{Glu} + \text{Cys} + \text{ATP} \rightarrow \gamma\text{-GluCys} + \text{ADP} + \text{P}_i \quad \text{(catalyzed by}\gamma\text{-GluCys synthetase)} \qquad (12.2.17)$$

$$\gamma\text{-GluCys} + \text{Gly} + \text{ATP} \rightarrow \text{GSH} + \text{ADP} + \text{P}_i \quad \text{(catalyzed by GSH synthetase)} \qquad (12.2.18)$$

Once GSH is pumped out of the cells, the ectoenzyme γ-glutamyl-transpeptidase (γGT, with the catalytic site facing the extracellular milieu) can catalyze the extracellular reaction of GSH with amino acids, yielding γ-glutamyl-amino acids and cysteinyl-glycine (CysGly) [see reaction (12.2.19) and Fig. 12.3]. γGT also catalyzes the reaction of amino acids with GSH conjugates (e.g., GSH-xenobiotic adducts produced by GST and then pumped out), yielding γ-glutamyl-amino acids and a CysGly conjugate (xenobiotic-CysGly) (Fig. 12.3). Adducts of xenobiotics with either GSH or CysGly can be converted to mercapturic acids (see Section 12.2.1.4) and then excreted in the urine.

To recycle the substrates for GSH synthesis, CysGly is hydrolyzed to the respective amino acids by ecto- or endopeptidases. Moreover, γ-glutamyl-amino acids can be converted to 5-oxoproline (plus an amino acid), which is used for the ATP-dependent synthesis of glutamic acid [Glu; reactions (12.2.20)]. Figure 12.3 summaries the various fates of GSH in cells.

$$\text{GSH} + \text{amino acid} \rightarrow \gamma\text{-glutamyl-amino acid} + \text{CysGly} \qquad (12.2.19)$$

$$\gamma\text{-glutamyl-amino acid} \rightarrow \text{5-oxoproline} + \text{amino acid} \rightarrow \text{Glu} \qquad (12.2.20)$$

There are a large number of assay methods for GSH and GSSG. These include HPLC-based determinations with spectrofluorometric or electrochemical detection as well as spectrophotometric assays. The most commonly used assay measures the reaction of GSH with 5,5′-dithiobis-2-nitrobenzoic acid (DTNB) in the presence of NAPH and glutathione reductase (GR), which produces a yellow product quantified at 412 nm. This assay actually measures total glutathione equivalents (GSH + 2GSSG; since both forms are present in biological samples) because GSSG is converted to GSH by GR plus NAPH. GSSG content alone is then determined in parallel assays after derivatization of GSH with 2-vinylpyridine. The content of "GSH itself" is then determined by subtraction.

Plants have not only GSH but also two analogs, γ-GluCysAla (homoglutathione) and γ-GluCysSer, which have similar antioxidant functions. Some parasitic protozoa such as *Trypanosoma cruzi* (the agent of Chagas disease) lack GSH and use the peptide trypanothione (which is glutathione covalently bonded to the polyamine spermidine) as a scavenger of H_2O_2. This protozoa also contains thioredoxin but not catalase nor GPx.

A reasonable amount of glutathione forms mixed disulfides with proteins (see below) and, therefore, is not "free" in cells. Accumulated GSSG can react with –SH groups of many proteins (a process called protein-*S*-thiolation) and cause their inactivation [reaction (12.2.21a)]. Protein-*S*-thiolation may result from attack of $GS^\bullet$. Moreover, protein-mixed disulfides are more susceptible to proteolytic degradation. GSSG can also induce crosslinkage of –SH groups at the surface of enzymes and proteins [reaction (12.2.21b)]. This crosslinkage causes inactivation of some enzymes (e.g., phosphofructokinase, pyruvate kinase) but activation/stabilization of others (e.g., glucose-6-phosphatase, fructose 1,6-bisphosphatase). Hence, glutathione-mediated protein thiolation is a mechanism of posttranslational regulation of metabolic enzymes. As discussed in Chapter 1, disulfide–dithiol enzyme interconversion is a particularly important mechanism of enzyme control in photosynthetic plants.

$$\text{GSSG} + \text{protein-SH} \rightarrow \text{protein-SSG} + \text{GSH} \tag{12.2.21a}$$

$$\text{GSSG} + \text{protein-(SH)}_2 \rightarrow \text{protein-SS} + 2\text{GSH} \tag{12.2.21b}$$

Auxiliary defenses such as thiol–disulfide oxidoreductases can remove the glutathione group bridged to proteins as well as re-reduce thiol–disulfides in proteins. These oxidoreductases include thioltransferase (or glutaredoxin, an 11-kDa polypeptide) and the 12-kDa polypeptide thioredoxin (Trx; which contains two adjacent –SH groups). Both are present in the cells of most organisms. In mammalians, the two main thioredoxins are Trx-1, the cytosolic and nuclear form, and Trx-2, a mitochondrial form. Thioredoxin in the oxidized form (Trx-SS) is recycled to the reduced form, Trx-$(SH)_2$, by a FAD-containing selenoenzyme thioredoxin reductase (TrxR), which uses NADPH as its cosubstrate [reaction (12.2.22)].

$$\text{Trx-SS} + \text{NADPH} + \text{H}^+ \rightarrow \text{Trx-(SH)}_2 + \text{NADP}^+ \quad \text{(catalyzed by TrxR)} \tag{12.2.22}$$

Thioredoxin can also reduce oxidized thiol groups in proteins and is a physiological substrate for the plasmatic form of GPx (pGPx; see Section 12.2.1.3.2). It also supplies the reducing equivalents to ribonucleotide reductase and thioredoxin peroxidase (TPx), an enzyme that decomposes hydroperoxides using Trx as a cosubstrate (TPx used to be called thiol-specific antioxidant up to the early 1990s). TPx is particularly important for microorganisms and helmintic parasites that lack GPx and/or catalase activity. Mammals seem to have at least two TPx isoforms: One is cytosolic and one mitochondrial. An interesting example of a physiological function of TPx is found in the adrenal gland. This gland produces high levels of H_2O_2 as a result of active sterol biosynthesis and displays very high constitutive levels of mitochondrial TPx.

Thioredoxin is present in cells at much lower levels than GSH (about 1 to 15 μM in bacteria and bovine tissues) and, at these low levels, may not act itself as a physiological antioxidant. However, it is highly important as a substrate of TPx (and possibly of pGPx). The levels of Trx can increase under certain stress/pathological conditions to provide an adaptive response. For example, increased levels of Trx-1 are found in many human tumors, associated with aggressive tumor growth. High levels of plasma Trx were also observed in hepatitis C virus-infected patients with hepatocellular carcinoma or chronic hepatitis (42 to 44 ng/mL; compared with 25 to 35 ng/mL for controls).

12.2.2 Low-Molecular-Weight Antioxidants

Animals produce several nonenzymatic low-molecular-weight metabolites that have physiologically relevant antioxidant actions. These include glutathione (and its analogs, discussed in Section 12.2.1.5), ascorbate (for humans, guinea pigs, and few other vertebrates, ascorbate is a vitamin; see Section 12.2.2.3), melanin (a product of tyrosine oxidation and polymerization, protects the skin against UV radiation), melatonin, and uric acid (see Section 12.2.2.1 for the last two molecules).

Other metabolites have *in vitro* antioxidant action, but their physiological role is still under debate. This list includes pyruvate (reacts *in vitro* with H_2O_2, $ONOO^-$, and HOCl), lactate (reacts *in vitro* with O_2^- and $^\bullet OH$ but cannot block lipid peroxidation; lactate might be relevant as an antioxidant in muscle during exercise), estrogens (protects LDL particles and membranes against *in vitro* peroxidation at concentrations that are 10^4 to 10^6 higher than physiological levels; therapeutic use as antioxidant is questionable), lipoic acid, coenzyme Q, and bilirubin (see Section 12.2.2.1 for the last three molecules).

Dietary antioxidants include vitamin E (Section 12.2.2.3), carotenoids (such as β carotene, a precursor of vitamin A; Section 12.2.2.4), and plant polyphenols (these are discussed in Text Box 12.4). Figure 12.4 depicts several nonenzymatic antioxidants.

Methods for the *in vivo* and *in vitro* assay of low-molecular-weight antioxidants are too numerous to be dis-

TEXT BOX 12.4 PLANT POLYPHENOLS AND HUMAN HEALTH*

Phenolic compounds are ubiquitous in plants and are abundant in fruits, vegetables, seeds, bark of trees, coffee, wine, and tea. They are produced by plant secondary metabolism from glucose that is processed to produce two main intermediates, shikimate and acetate, which are then used to generate secondary metabolites. Shikimate originates from aromatic amino acids (tryptophan, phenylalanine, tyrosine), precursors of aromatic compounds. Some secondary metabolites derive from shikimate or acetate, while some are produced by a combination of these intermediates or their derivatives, for example, flavonoids, anthraquinones, and tannins. In plants, phenolics work as structural components, and have antioxidant, antimicrobial, and antiviral activity. Many flavonoids possess attractive colors and participate in the ecology of plants by making the flowers and fruits attractive to birds and bees. Some are excellent scavengers of free radicals and metal ions. More than 4000 plant phenolics have been described. The classes of plant phenolics found in the human diet are *flavanols* (including catechin, epicatechin, and epicatechin gallate; black and green teas and red wine are rich in flavanols), *flavanones* (such as hesperidin, found in citrus fruits), *flavonols* [including kaempferol, myricetin, rutin, and quercetin (see Fig. 12.4); broccoli, onion, berries, and red wine are flavonol-rich], *flavones*, *anthocyanidins* (such as cyanidin, found in berries), and *phenylpropanoids* (including caffeic and chlorogenic acids; a fresh coffee is a souce of these). Genistein and daidzein are the main representantives of *isoflavones*; soy is the main source of these phenolics.

The beneficial effects of plant phenols (including polyphenols) have been investigated extensively in cells and tissues *in vitro* and in animal models, documenting their antiradical and antioxidant properties and their participation in prevention and therapy of several diseases. Because their relevance to human health has not been fully examined by direct measurements in humans, it is important to be able to extrapolate bioactivities observed *in vitro*, into a physiological scenario *in vivo*. In this regard, some findings are of interest. For example, black tea polyphenols decrease LDL cholesterol and LDL oxidation in rats and increase HDL content. Therefore, they can be beneficial for the treatment of atherosclerotic disorders, hypertension, and type-2 diabetes. Other studies have reported enhanced protection against oxidative stress afforded to red blood cells by polyphenols; for example, plant phenolics from blueberries ameliorate the experimentally induced formation of ROS in red blood cells using an *in vitro* model following dietary consumption by 6-month-old male rats. Polyphenols may provide protection against ROS in various cell systems. They also exhibit cytostatic and cytotoxic effects in tumorigenesis, some being able to inhibit invasion by highly metastatic cells. Polyphenols also inhibit the activity of an array of enzymes including lipoxygenases, cyclooxygenases, monooxygenases, collagenases, xanthine oxidase, protein kinases, and also increase the activity of antioxidant enzymes, such as SOD. Therefore, they may ameliorate inflammatory responses, modulate cell signaling, and improve the antioxidant status of cells.

Besides all the potential beneficial effects of polyphenols such as flavonoids and catechin, it is important to emphasize that their biological and pharmacological effects may depend upon their behavior as either antioxidants or prooxidants in the cell environment. These opposite actions may be related to their redox activity and position and the number of hydroxyls on the molecule. Flavonoids can either inhibit or enhance the formation of hydroxyl radical ($^{\bullet}OH$) by Fenton-like reactions, depending upon the redox activity of the flavonoid and the nature of the metal ion participating in $^{\bullet}OH$ formation. Most polyphenols (including flavonoids) are metal chelators and can either favor or inhibit the participation of the metal (either iron or copper) in $^{\bullet}OH$ formation. For example, we have shown that polyphenol tannic acid chelates Fe^{2+} and prevents its reaction with H_2O_2 (see *Biochim Biophys Acta* **1472**:142–152, 1999); in Brazil we are currently studing the relation between antioxidant activity of caffeic acid and biphenol ellagic acid (see Fig. 12.4) and their ability to chelate iron. Moreover, many polyphenols act as potent $^{\bullet}OH$ scavengers. Studies to elucidate the activity–structure relationships of these compounds are extremely important to distinguish the beneficial versus detrimental effects of polyphenols. Collectively, the available data suggest that polyphenols can improve human health, preventing carcinogenesis and cardiovascular diseases. For more information see reviews from Bravo (*Nutr Rev* **56**:317–333, 1998) and from Lambert and Yang (*Mut Res* **523**:201–208, 2003). We also recomend you to read the interesting interview given by Professor B.N. Ames entitled "Ames agrees with mom's advice: Eat your fruits and vegetables" (*JAMA*, **273**:1077–1078, 1995).

*By Élida G. Campos and Marcelo Hermes-Lima, Universidade de Brasília.

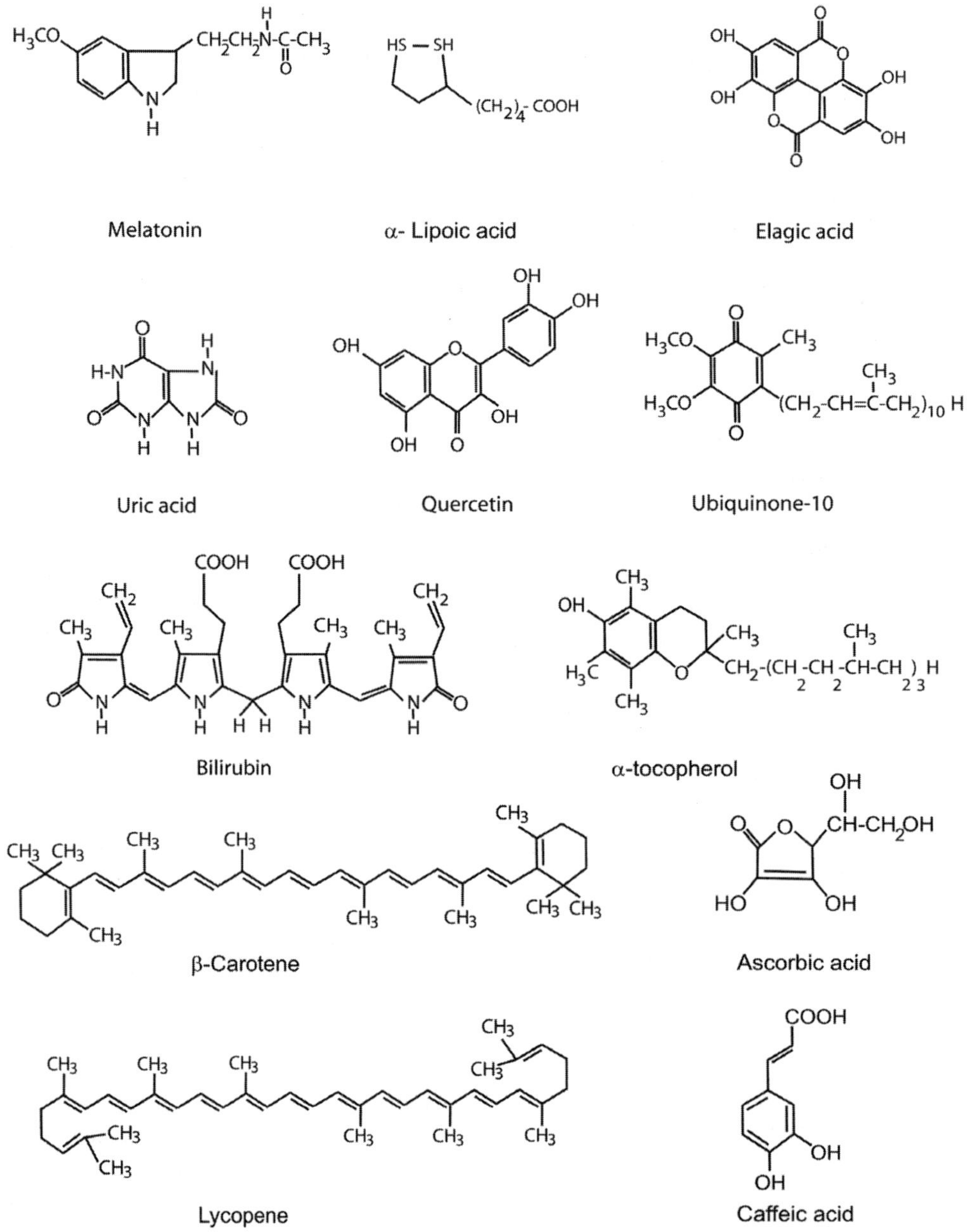

Figure 12.4 Selected nonenzymatic antioxidants of low molecular weight.

cussed in this chapter but we recommend *The Handbook of Antioxidants* (Cadenas and Packer, 1996) for this purpose. However, assessment of the antioxidant capacity of molecules is often based on two main factors. First, the standard reduction potentials (at pH 7.0; giving $E^{0\prime}$) may predict whether a redox reaction between two molecules is possible or forbidden. For example, a comparison of the $E^{0\prime}$ of half-cells (Table 12.3) indicates that dehydrolipoic acid (DHLA) could easily reduce GSSG to GSH. However, $E^{0\prime}$ values can be very different under physiological concentrations (and temperatures) and thus must be corrected by the Nernst equation to give an "effective" reduction potential. In these cases, reactions that may have appeared "impossible" become possible. One example is the autoxidation of ubisemiquinone [$CoQH^{\bullet}$, see reaction (12.2.24)], which does happen *in vivo*. Second, redox reactions that are thermodynamically possible can, in some cases, be kinetically irrelevant. In other words, they are too slow for quick biological responses. Accurate theoretical predictions of the putative actions of antioxidants or prooxidants can be subjected to so many other factors (e.g., effect of co-adjuvant metals, binding of the reagents to membranes and/or proteins) that it is often better just to assess the reaction experimentally *in vitro* and then, as appropriate, follow up by seeking *in vivo* evidence of a particular antioxidant or prooxidant action.

12.2.2.1 Selected Endogenous Molecules with Antioxidant Activity Melatonin (Fig. 12.4) is produced from serotonin in the pineal gland and seems to regulate circadian rhythms. It is able to inhibit lipid peroxidation *in vitro* and neutralizes $^{\bullet}OH$, singlet oxygen, nitric oxide, and HOCl. However, in order to work as an endogenous antioxidant, it would have to be present in concentrations that far exceed its physiological concentrations. There are, however, new indications that melatonin has a function in redox-regulation mechanisms (e.g., stimulating increased activity of some antioxidant enzymes), immunomodulation, cellular growth, and bone differentiation. Moreover, melatonin has been used experimentally as a protective agent against a wide variety of processes and compounds that damage tissues via free radical mechanisms.

Lipoic acid is a thiol-prosthetic group in α-ketoacid dehydrogenase enzyme complexes in the mitochondria. Lipoate synthase, which catalyzes the formation of two C–S bonds from octanoic acid, is responsible for endogenous lipoic acid formation. Although the metabolic relevance of lipoic acid has been known for 40 years, it was only recently recognized as a potential antioxidant metabolite. It is able to react with $ROO^{\bullet}$, HOCl, $^{\bullet}OH$, and $ONOO^-$, and it is also a copper and iron chelating agent. Dehydrolipoic acid (DHLA, the reduced form of lipoic acid; see Fig. 12.4) is a powerful reducing agent (Table 12.3) and can recycle GSSG to GSH *in vitro* (see footnote *a* to Table 12.3), ascorbyl radical to ascorbate [reaction (12.2.23)], and the free radical form of vitamin E (see Section 12.2.2.3) to vitamin E. However, the endogenous levels of lipoic acid and DHLA are very low in organisms, suggesting that they may not work *in vivo* as either a free radical scavenger or as reducing agent (in the case of DHLA).

$$\text{DHLA} + \text{ascorbyl}^{\bullet} \rightarrow \text{lipoic acid} + \text{ascorbate} \quad (12.2.23)$$

On the other hand, the use of exogenous lipoic acid for treatment of a number of liver conditions linked to oxidative stress, including alcohol-induced damage, metal intoxication, and CCl_4 poisoning, has been successful in many cases. Lipoic acid effectively protects the brain against oxidative stress associated with ischemia and reperfusion and also reverses some of the effects provoked by vitamin E deficiency in mice. These observations point the way toward the clinical use of lipoic acid as an antioxidant.

Coenzyme Q (or ubiquinone; see Fig. 12.4) has been known since 1957 as a key component of the mitochondrial respiratory chain. Although most studies have focused on ubiquinone-10, several other ubiquinones occur in nature, located in most cell membranes and in lipoproteins. One-electron oxidation of ubiquinol ($CoQH_2$) yields the radical species ubisemiquinone ($CoQH^{\bullet}$), which can be sequentially oxidized to ubiquinone (CoQ) (see $E^{0\prime}$ values in Table 12.3). These reactions are fully reversible providing, for example, a continuous flow of electrons in and out of the respiratory chain. Dismutation of ubisemiquinone ($2CoQH^{\bullet} \rightarrow CoQ$ and $CoQH_2$) is also a thermodynamically favorable reaction that seems to take place *in vivo*. Moreover, ubisemiquinone is considered a highly relevant source of mitochondrial O_2^- formation [reaction (12.2.24) is thermodynamically favorable when calculating the "effective" $\Delta E^{0\prime}$ value, which takes in consideration the *in vivo* concentration of reactants and products; see Schafer and Buettner, *Free Radic Biol Med* **30**:1191–1212, 2001]. Ubiquinone can also be reduced to ubiquinol by the FAD-containing enzyme DT-diaphorase, which helps to maintain a low cellular $CoQ/CoQH_2$ ratio.

$$CoQH^{\bullet} + O_2 + H^+ \rightarrow O_2^- + CoQ \quad (12.2.24)$$

$$CoQH_2 + LOO^{\bullet} \rightarrow CoQH^{\bullet} + LOOH \quad (12.2.25)$$

On the other hand, ubiquinol efficiently protects not only membrane phospholipids from lipid peroxidation [reaction (12.2.25)] but also mitochondrial DNA and membrane proteins from free-radical-induced oxidative damage. In LDL undergoing lipid peroxidation, ubiquinol seems to be relevant for the recycling of α-tocopherol [see reaction (12.2.31)]. The importance of ubiquinol as an *in vivo* antioxidant is still a matter of debate, even though there are indications of positive effects of coenzyme Q therapy in patients with several heart pathologies.

Uric acid (Fig. 12.4; or urate in neutral pH) is produced by the oxidation of xanthine catalyzed by xanthine oxidase or xanthine dehydrogenase. Too great a production of urate leads to gout, an inflammatory joint disease caused by the precipitation of urate crystals (most nonprimates are able to convert urate to allantoin and urea). Urate concentration in human plasma is in the range of 50 to 900 μM, being high enough to react with ozone, $^{\bullet}OH$, $ONOO^-$, and peroxyl radicals. Upon oxidation, urate is converted to the resonance-stabilized urate radical (urate$^{\bullet}$), which can either be recycled back to urate by ascorbate or converted (nonenzymatically) to allantoin [reactions (12.2.26) and (12.2.27)]. An indirect proof that urate may function *in vivo* as an antioxidant is the fact that allantoin is produced in patients with diseases associated with oxidative stress, such as Wilson disease (see Chapter 13) and hemochromatosis (see Text Box 12.1). Interestingly, researchers from the Federal University of Rio de Janeiro, Brazil, have proved that urate is the most important nonenzymatic antioxidant defense in a blood-sucking insect (they produce large amounts of urate without ever developing gout!) [Souza et al. (1997) *Free Radic Biol Med* **22**:209–214].

$$\text{Urate} + {}^{\bullet}\text{OH} \rightarrow \text{urate}^{\bullet} + H_2O \quad (12.2.26)$$

$$\text{Allantoin} \leftarrow\leftarrow \text{urate}^{\bullet} + \text{ascorbate} \rightarrow \text{urate} + \text{ascorbyl}^{\bullet} \quad (12.2.27)$$

Bilirubin, another end product of a metabolic pathway (see Fig. 12.4), is an insoluble molecule produced by mammals from heme degradation. It has powerful antioxidant activity *in vitro* against peroxyl radicals and singlet oxygen. However, unlike uric acid, an antioxidant role for bilirubin *in vivo* is still unproven.

12.2.2.2 Vitamin C Water-soluble ascorbic acid (Fig. 12.4) is produced from glucose metabolism by plants and most animals. The enzyme-catalyzed reaction of gulono-γ-lactone (synthesized from glucose) plus oxygen yields ascorbate and H_2O_2. However, various mammalian species, such as guinea pigs and all primates, lack the ability to produce ascorbic acid, which is essential for several metabolic reactions, including noradrenalin biosynthesis, hydroxylation of proline and lysine for pro-collagen formation, and iron absorption in the gut (by reducing Fe^{3+} to Fe^{2+}). A deficiency of vitamin C intake in humans causes the "sailor's disease," scurvy, a disease associated with osteoporosis, hemorrhaging, and anemia. The Dutch doctor Ronsseus first reported the need for the consumption of oranges for scurvy prevention in sailors in the midsixteenth century. The levels of ascorbate in adult human plasma range from 30 to 100 μM, which can be maintained by daily absorption of 60 to 100 mg ascorbate. In 1970 Nobel laureate Linus Pauling popularized the use of megadoses of vitamin C, one gram per day, for preventative use against the common cold (a still unproven issue) and in the 1990s as an antioxidant supplement.

Apart from the metabolic functions of ascorbate, it is a potent nonenzymatic antioxidant defense, able to recycle vitamin E (at least *in vitro*; see Section 12.3.4) and to scavenge peroxynitrite, ozone, singlet oxygen, O_2^{-}, $HOO^{\bullet}$, and ${}^{\bullet}OH$. Considerable evidence shows that ascorbate inhibits lipid peroxidation and oxyradical-mediated damage to proteins and DNA *in vitro*. On the other hand, ascorbate can also be considered a prooxidant due to its ability to recycle iron from the ferric to the ferrous state, thus providing a Fenton reagent for ${}^{\bullet}OH$ formation. Indeed, several *in vitro* experiments designed for metal-catalyzed ${}^{\bullet}OH$ formation are initiated by addition of ascorbate (0.5 to 3 mM) to the reaction medium.

One-electron oxidation of ascorbic yields the ascorbyl radical, which is converted by another one-electron oxidation to dehydroascorbate (DHA); nonenzymatic degradation of DHA yields oxalate and other products [reactions (12.2.28)]. Recycling of ascorbate in animals and plants is enzyme-catalyzed [see reactions (12.2.11) and (12.2.12)]. It is noteworthy that the ascorbyl radical has low reactivity toward biomolecules. Ascorbyl radical can be easily detected in human plasma samples by ESR techniques (see Section 12.1 and Fig. 12.1). There is evidence that plasma of iron-overload patients (see Text Box 12.1) have decreased concentrations of ascorbate, whereas the ESR signal for ascorbyl radical is increased.

$$\text{Ascorbate} \underset{(-e^-)}{\rightarrow} \text{ascorbyl}^{\bullet} \underset{(-e^-)}{\rightarrow} \text{DHA} \rightarrow \begin{matrix}\text{oxalate and}\\ \text{other products}\end{matrix} \quad (12.2.28)$$

According to the British researchers Barry Halliwell and John M.C. Gutteridge (1999; see pages 203 to 205), evidence that points to an unequivocal antioxidant role for ascorbate in mammals is still missing. For example, mutant rat strains that do not synthesize vitamin C (ODS rats) have higher levels of lipid peroxidation products, but, interestingly, ascorbate supplements did not reduce lipid peroxidation in these rats. Moreover, even though the consumption of fruits and vegetables (main sources of dietary vitamin C) can be associated with decreased risk of cardiovascular diseases and of several types of cancer—specially of the gastrointestinal tract—the actual beneficial effect of vitamin C itself is still uncertain. The results from many epidemiological studies or double-blind placebo-controlled studies show either beneficial effects or no effects, and they depend on the specific disease, genetic factors, and age and lifestyle (exercise, smoking habits, use of antioxidant or vitamin supplements) of the subjects. Furthermore, data using biomarkers of oxidative damage to DNA bases (cancer is associated with DNA oxidative damage; see Section 12.5) have given no clear evidence that vitamin C supplements can decrease the levels of oxidative DNA damage *in vivo*, except perhaps in subjects with very low vitamin C intakes.

12.2.2.3 Vitamin E Vitamin E is composed of eight naturally occurring lipid-soluble compounds, α-, β-, γ-, and δ-tocopherols (see Fig. 12.4) and α-, β-, γ-, and δ-tocotrienols. All of these function as antioxidants in membranes to prevent the propagation of lipid peroxidation (see Section 12.3.4 for details). In animals, the most active component of the "vitamin E family" is α-tocopherol, which is the form present in highest concentrations in cell membranes and in the plasma as part of lipoproteins. The levels of total tocopherol in human plasma range between 18 and 28 μM, with important variations among different populations.

Vitamin E was first detected as a fat-soluble factor essential for rodent reproduction in the early 1920s and was isolated from wheat germ oil in 1936. It was later shown to be essential in the diet of all animals (which classifies it as a vitamin) and as capable of preventing hemolytic anemia in premature babies. Sources of vitamin E in the human diet include vegetable oils, nuts, margarines, wheat germ, and green leafy vegetables.

The reaction of α-tocopherol (α-TocH) with peroxyl PUFAs [$LOO^\bullet$, or other radicals, $R^\bullet$; see reaction (12.2.29)] yields α-tocopheryl radical (α-Toc$^\bullet$) and LOOH. α-Tocopheryl radicals can be nonenzymatically recycled back to α-tocopherol by other antioxidants, such as ascorbate, coenzyme Q [reactions (12.2.30) and (12.2.31)], and carotenoids (see Section 12.2.2.4). The continuous recycling of α-tocopherol (and other tocopherols) seems to explain why the concentration vitamin E is so low in biological membranes. For example, there is one α-tocopherol molecule per 2000 molecules of mitochondrial phospholipid and 6 to 10 molecules of α-tocopherol per human LDL particle.

$$\alpha\text{-TocH} + R^\bullet \rightarrow \alpha\text{-Toc}^\bullet + RH \qquad (12.2.29)$$

$$\alpha\text{-Toc}^\bullet + \text{ascorbate} \rightarrow \text{ascorbyl}^\bullet + \alpha\text{-TocH} \qquad (12.2.30)$$

$$\alpha\text{-Toc}^\bullet + CoQH_2 \rightarrow CoQH^\bullet + \alpha - \text{TocH} \qquad (12.2.31)$$

Trolox, a soluble synthetic vitamin E analog, is also a powerful antioxidant that is effective against lipid peroxidation. Note that the $E^{0\prime}$ value for the Trolox$^\bullet$/Trolox pair is almost the same as the value for α-Toc$^\bullet$/α-TocH (Table 12.3).

Vitamin E deficiency in mammals causes higher susceptibility to lipid peroxidation when animals are exposed to oxidizing agents. α-Tocopherol may indirectly inhibit protein and DNA injury as well because (i) products of lipid peroxidation can damage protein residues (see Section 12.4) and DNA bases and (ii) peroxidation of membranes can prompt calcium-dependent oxidative stress (see Section 12.3), leading to DNA and protein oxidation.

Vitamin E contents in humans (either normal dietary contents or levels achieved with vitamin E supplements) have been correlated with the incidence of cardiovascular diseases and various types of cancer showing either beneficial effects or no effects. It is not easy to isolate the effect of vitamin E itself from the effects of various other dietary antioxidants and many other confounding factors. Possibly, only individuals severely deprived of vitamin E (and with very bad diet habits!), which are more prone to be at risk for cardiovascular diseases, have a better response to vitamin E supplementation (see also Section 13.4).

Vitamin E, mostly α-tocopherol, has other nonantioxidant functions in organisms. It participates in the structure of membranes and in signal transduction pathways. For example, α-tocopherol is an inhibitor of protein kinase C. These observations indicate that health-benefits roles of vitamin E may not be solely attributable to the antioxidant actions of vitamin E.

12.2.2.4 Carotenoids*   In the 1930s it was discovered that vitamin A originates in the gut and liver from carotenoids. Although ~600 carotenoids have been described, only about 50 have provitamin A activity. The major carotenoids in human plasma are α- and β-carotene, lycopene (see Fig. 12.4), cryptoxanthin, and lutein; these are almost exclusively found associated with lipoproteins. In terms of the total body pool of carotenoids, major storage organs include the liver and adipose tissue. Excellent dietary sources are dairy products, yellow and green vegetables, fish, eggs, and organ meats; tomato is very rich in lycopene.

The antioxidant potential of carotenoid molecules was first described in the early 1930s by Monaghan and Schmitt (*J Biol Chem* **96**:387–395, 1932) while studying a lipid peroxidation process. In the late 1960s, Foote and Denny (*J Am Chem Soc* **90**:6233–6235, 1968) showed the quenching of singlet oxygen ($^1O_2$) by carotenoids. The antioxidant behavior of carotenoids is dependent on their structures and on the nature of the oxidizing species. Most information on their antioxidant potential has been obtained from *in vitro* experiments that test individual carotenoids in organic solvents with one species of ROS. The interaction of carotenoids with $^1O_2$ occurs by a transfer of the excitation energy to the carotenoid or by a chemical quenching of $^1O_2$; the latter results in the irreversible destruction of the carotenoid molecule. Carotenoids may interact with oxygen radicals by three main ways: electron transfer, hydrogen abstraction, and addition of a radical species [reactions (12.2.32) to (12.2.34) for the case of $ROO^\bullet$; carotenoids are abbreviated as CAR].

$$ROO^\bullet + CAR \rightarrow ROO^- + CAR^{\bullet +} \qquad (12.2.32)$$

$$ROO^\bullet + CAR \rightarrow ROOH + CAR^\bullet \qquad (12.2.33)$$

$$ROO^\bullet + CAR \rightarrow (ROO\text{-}CAR)^\bullet \qquad (12.2.34)$$

In biological systems carotenoids rarely occur as free molecules but are associated with proteins or lipoproteins, and their distribution is very heterogeneous in tissues. The microenvironment where the carotenoid molecule is located may have a profound effect on its antioxidant properties. β-Carotene and lycopene, for example, lie parallel

*Section 12.2.2.4 was prepared in collaboration with Ph.D. student Sandra Arruda, from Universidade de Brasília.

to the membrane surface, within the hydrophobic core, whereas the dihydroxy carotenoids such as zeaxanthin span the membrane entirely. β-Cryptoxanthin and zeaxanthin have greater protective effects against peroxyl radicals in the aqueous phase of liposomal membranes as compared to β-carotene and lycopene, even though β-carotene and zeaxanthin have the same conjugated C=C chain length. The predicted collision rates of β-carotene with peroxyl radicals within the membrane would be very low and, in this case, these molecules are expected to be poor antioxidants. Therefore, differences in the antioxidant actions of various carotenoids can be attributed to differences in their location within the lipid bilayer.

Carotenoids can function synergistically with α-tocopherol and vitamin C to provide an effective barrier against oxidation. An integrated mechanism for the interaction of vitamins C and E with β-carotene has been proposed, where the carotenoid molecule recycles vitamin E [reaction (12.2.35)] and the resulting carotenoid radical is, in turn, "regenerated" by vitamin C [reaction (12.2.36)].

$$\text{CAR} + \alpha\text{-Toc}^{\bullet} \rightarrow \alpha\text{-TocH} + \text{CAR}^{\bullet+} \quad (12.2.35)$$

$$\text{CAR}^{\bullet+} + \text{ascorbate} + \text{H}^{+} \rightarrow \text{CAR} + \text{ascorbyl}^{\bullet} \quad (12.2.36)$$

The synergistic protection afforded by carotenoids and other co-antioxidants is dependent on the balance between all the components. In humans that smoke, a diminished plasma vitamin C concentration, compared with nonsmokers, leads to an accumulation of β-carotene radical due to an inefficient "regeneration" process. The antioxidant behavior of carotenoids is, in part, dependent upon the partial oxygen pressure; at low pO_2 carotenoid molecules act *in vitro* as a chain-breaking antioxidant. At a high pO_2 carotenoid radicals may react with oxygen to produce a carotenoid peroxyl radical, which may function as a prooxidant.

Epidemiological studies show good correlation between declining incidence of cardiovascular diseases, such as ischemic heart disease, and an increase in dietary intake and plasma concentration of carotenoids and vitamin A. Plasma concentrations of β-carotene above 0.4 to 0.5 μmol/L and of vitamin A in the range of 2.2 to 2.8 μmol/L have been proposed to reduce the risk of such diseases. However, those studies do not exclude some lifestyle choices (as discussed above for the case of vitamin E) that may reduce the risk of cardiovascular diseases.

Human studies with rigorous control of different variables may provide a better view of the role of carotenoids in public health. The double-blind placebo-controlled studies ATBC (Alpha-Tocopherol Beta-Carotene; see *N Engl J Med* **330**:1029–1035, 1994) and US CARET (β-carotene and retinol efficacy trial; *N Engl J Med* **334**:1150–1155, 1996) tested whether vitamin E and/or β-carotene supplementation could reduce the incidence of cancers and cardiovascular diseases in former smokers or workers exposed to asbestos. Curiously, analysis of both trials found a higher incidence of fatal coronary heart disease and lung cancer in β-carotene-supplemented groups. On the other hand, experimental studies in rats subjected to ischemia-reperfusion stress (see Section 13.5) showed a significant decrease in plasma and liver retinol, suggesting that carotenoids may act against oxidant injury in infarction.

Much more work is required to reconcile the vastly different results from intervention trials and epidemiological and experimental studies regarding the *in vivo* role of carotenoids. In any case, it is a good precaution for smokers (who will not quit this bad habit!) to avoid β-carotene supplements.

12.3 LIPID PEROXIDATION

Most of today's knowledge about the chemical aspects of lipid peroxidation in biological systems was borrowed from the oil and food industry. It has long been known that rancidity of oils and meat is caused by lipid peroxidation. The first determination of the relevance of oxygen to oil viscosity during storage was done in the early nineteenth century (see Section 12.3.2). However, it was not until the 1960s that we became aware that lipid peroxidation could be a relevant event in biology and medicine.

12.3.1 Introduction

The process of lipid peroxidation is considered to be a major cause of cell injury and death. Many pathological conditions and human diseases are connected with lipid peroxidation, as will be discussed later in this chapter. Basically, lipid peroxidation is a chain reaction, in most cases catalyzed by transition metals, where strong oxidants cause the breakdown of membrane phospholipids that contain polyunsaturated fatty acids (PUFAs). Lipid peroxidation damage to biological membranes can have several levels of severity depending on the nature and concentration of the oxidant, ranging from local reductions in membrane fluidity to full disruption of bilayer integrity. LDL particles are also major targets of lipid peroxidation.

Lipid peroxidation of endoplasmic or sarcoplasmic reticulum may cause uncontrolled Ca^{2+} flux into the cytoplasm. This can be a disastrous event for cellular metabolic processes since elevated Ca^{2+} can activate/deactivate enzymes at inappropriate times. Moreover, Ca^{2+}-dependent proteases, phospholipases, and endonucleases can undergo uncontrolled activation, causing degradation of enzymes, membranes, and DNA. Calcium also stimulates nitric oxide synthase, which will prompt $ONOO^-$ formation from nitric oxide and O_2^- (see Section 12.1.3). Furthermore, peroxidation of mitochondrial membranes alters the functioning of the respiratory chain, which could lead to insufficient ATP production and exhaustion of cell energy. Peroxidative damage to mitochondria, a process that is also mediated by Ca^{2+}, causes swelling, membrane disruption, and a loss of the selective permeability of the inner membrane.

Peroxidation of the plasma membrane can affect membrane permeability (the membrane may even be ruptured) and the proper functioning of hormone receptors and membrane proteins involved in signal transduction pathways. Moreover, cell contents, from small molecules to proteins, can be released into the extracellular fluids as a consequence of membrane rupture (see Text Box 12.5 for the effect of H_2O_2 in the induction of cellular lipid peroxidation).

12.3.2 Initiation and Propagation Phases of Lipid Peroxidation

The initiation of lipid peroxidation involves abstraction of a hydrogen atom of a methylene group of a PUFA molecule (LH), caused by the oxidizing species and forming a carbon-centered lipid radical ($L^\bullet$) [reaction (12.3.1); see also Fig. 12.5]. Methylene-bridged double bonds of common PUFA species [18 : 2(ω-6) linoleic, 18 : 3(ω-3) α-linolenic, 20 : 4(ω-6) arachidonic, 20 : 5(ω-3) eicosapentaenoic, and 22 : 6(ω-3) docosahexaenoic acids] are much more easily oxidized than are monounsaturated fatty acids [such as 16 : 1(ω-7) palmitoleic and 18 : 1(ω-9) oleic acids]. This is due to the weaker nature of the bis-allylic C–H bonds (bond dissociation energy (BDE) of about 315 kJ/mol) compared to monoallylic C–H bonds (BDE of about 370 kJ/mol). Moreover, the higher the number of double bonds of a PUFA, the greater the susceptibility of the fatty acid molecule to undergo peroxidation. Thus, 18 : 2(ω-6) and 22 : 6(ω-3) have the lowest and the highest susceptibilities, respectively, for peroxidation among common PUFA species.

Thermodynamics is useful to explain why bis-allylic hydrogen is highly oxidizable (the $E^{0\prime}$ of the $PUFA^\bullet$/PUFA couple is +0.60 V; see Table 12.3) as compared with aliphatic hydrogen from saturated fatty acids ($E^{0\prime} = +1.9$ V). By thermodynamic rules, several oxidizing agents can abstract hydrogen from PUFAs. Observe in Table 12.3 that oxygen and H_2O_2 cannot react with PUFAs due to their half-cell $E^{0\prime}$ values (−0.33 V and +0.32 V for the couples O_2/O_2^- and $H_2O_2/^\bullet OH$, respectively). Superoxide radical cannot initiate peroxidation, even though its $E^{0\prime}$ value (+0.94 V for the couple O_2^-/H_2O_2) would allow it. This is because of the charged nature of superoxide, making it very unlikely to enter the hydrophopic interior of the membrane bilayer. However, the protonated form of superoxide ($HOO^\bullet$; $E^{0\prime}$ value of +1.06 V for the couple $HOO^\bullet/H_2O_2$) is able to abstract hydrogen from PUFAs; the rate constant of the reaction is about $2 \times 10^3\ M^{-1}\ s^{-1}$.

Hydroxyl radical is to be considered the most important species involved in the initiation of lipid peroxidation. The rate constant for the reaction of $^\bullet OH$ with artificial membranes under very controlled conditions is about $5 \times 10^8\ M^{-1}\ s^{-1}$. However, there is major controversy in the literature about the nature of the $^\bullet OH$ sources and the locations and velocities of $^\bullet OH$ formation when considering the biological/pathological relevance of $^\bullet OH$-mediated initiation of lipid peroxidation. Hydroxyl radicals also have an extremely short half-life (a few nanoseconds), and they react with many other cell targets before reaching the interior of a membrane bilayer. Several authors have considered that peroxynitrite, $^\bullet NO_2$, perferryl species ($Fe^{3+}{-}O_2^-$), ferryl species ($Fe^{4+}{=}O$), and even the radical form of vitamin E (see Section 12.3.3) are of importance for the initiation reactions of lipid peroxidation. Ozone can also react with PUFAs generating ozonides, which decomposes to cytotoxic aldehydes; thus, ozone is not an initiator of lipid peroxidation. Iron ions may also participate in the initiation step of peroxidation by directly promoting the formation of perferryl and ferryl species, as well as catalyzing the formation of $^\bullet OH$ radicals. Copper ions are also biologically active in promoting $^\bullet OH$ formation and lipid peroxidation, especially in studies of copper-mediated peroxidation of LDL particles (see discussion in Chapter 13 about atherosclerosis).

After a carbon-centered lipid radical forms, it tends to stabilize by means of electron resonance forming a conjugated diene [reaction (12.3.2); see Fig. 12.5]. Conjugated dienes can be measured at 234 nm and studies with artificial bilayer preparations, microsomes, LDL particles, or cell cultures under free radical attack have shown a marked time-dependent increase in absorbance at 234 nm. Thus, measurement of the kinetics of conjugated diene formation under different experimental protocols is an important tool in biochemistry, with clinical and biological applications. This assay allows for the measurement of the effect of agents that increase (oxidants) or decrease (antioxidants) the rate of lipid peroxidation reactions. The assay is also relevant for studies of the peroxidability of different membrane compositions. Preformed conjugated dienes in animal and plant tissues can also be quantified using a second-derivative spectrum analysis after lipid extraction in organic solvents. This technique has been subject to much criticism due to the fact that it is not possible to determine the absolute amount of conjugated diene in the samples. However, when comparing results with

TEXT BOX 12.5 CELL DAMAGE CAUSED BY H_2O_2-MEDIATED LIPID PEROXIDATION

As mentioned in Section 12.3.1, lipid peroxidation can cause major injury to cells and tissues. In the early 1990s, David Janero and co-workers from Ciba Pharmaceuticals, New Jersey, provided an excellent example of the damaging effects of H_2O_2 in cultures of neonatal-rat cardiomyocytes (see *J Cell Physiol* **149**:347–364, 1991). They observed that H_2O_2 caused dose-dependent lipid peroxidation (by measuring TBARS, conjugated dienes and endogenous vitamin E contents). The illustration (Fig. TB12.1) shows the concentrations of conjugated dienes and TBARS as a function of the period of exposure of cardiomyocytes to 250 μM H_2O_2. The peaks of concentration of conjugated dienes and TBARS occurred at 30 and 45 min, respectively; similar results were observed with 50 or 500 μM H_2O_2. The offset in peak concentration occurs because conjugated dienes are one of the first products of the "lipid peroxidation pathway" (see Fig. 12.5), whereas TBARS are termination products and therefore take longer to accumulate. Moreover, both are produced and decomposed; conjugated diene levels diminish due to the ongoing chain of peroxidation reactions whereas malondialdehyde and other aldehyde products that can react with TBA (see Text Box 12.7) are metabolized by aldehyde dehydrogenases. In addition, vitamin E concentration in cardiomyocytes decreased very rapidly upon H_2O_2 exposure, indicating that it was used as a chain-breaking antioxidant.

The lethal effect of H_2O_2 on cardiomyocytes was attested by the fall in ATP content. The lag period preceding ATP loss (see Fig. TB12.1) is indicative of a process that is induced by lipid peroxidation. Sarcolemmal rupture of the cardiomyocytes mediated by H_2O_2 was demonstrated by leakage of the enzyme lactate dehydrogenase into the medium. Addition of Trolox or iron chelators (*o*-phenanthroline and deferoxamine) to the cell cultures before addition of H_2O_2 (at 500 μM) prevented lipid peroxidation and sarcolemmal disruption. These results indicate that lipid peroxidation is mediated by iron-catalyzed reactions. H_2O_2 added to the cultures reacts with intracellular iron and produces ROS (via Fenton reactions) that initiates/propagates the peroxidation process, leading to membrane disruption and cell death.

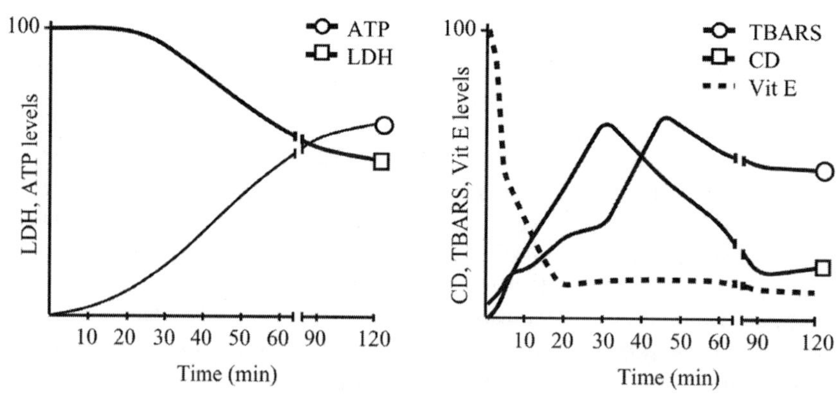

Figure TB12.1 Oxidative stress in cultured cardiomyocytes induced by H_2O_2. The figure on the left shows the decline in cellular ATP levels and the augmentation of extracellular lactate dehydrogenase (LDH) levels (caused by leakage from cells) after challenge by 0.25 mM H_2O_2. The figure on the right shows the corresponding changes in α-tocopherol (Vit E) and lipid peroxidation products: conjugated dienes (CD) and thiobarbituric acid reactive substances (TBARS). Data modified from Janero et al. (1991).

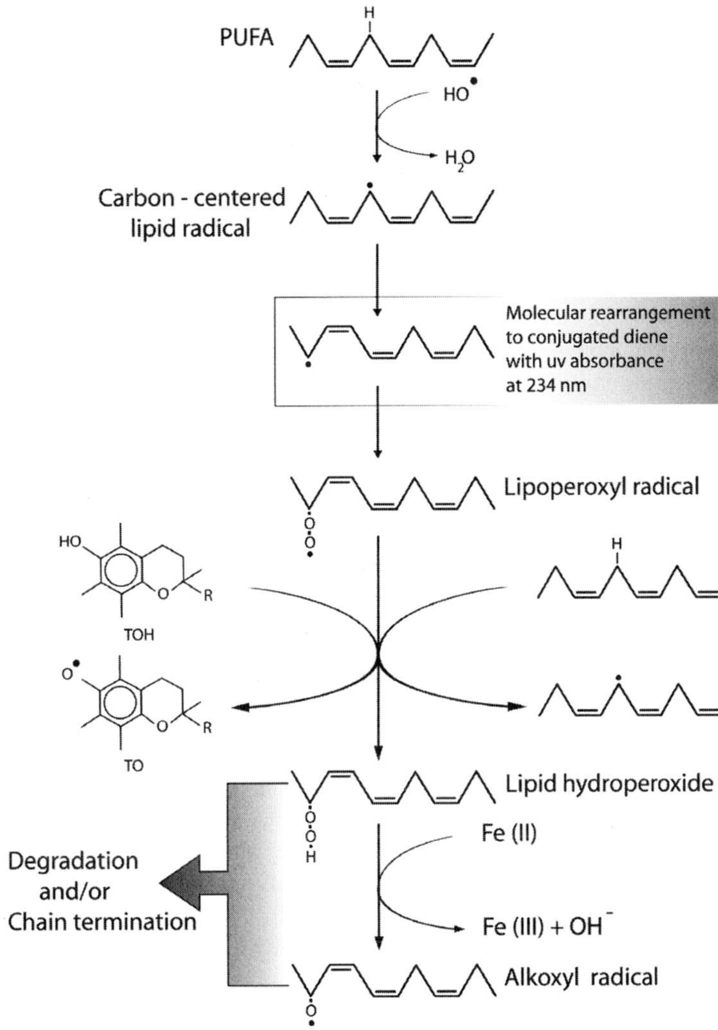

Figure 12.5 The basic scheme of lipid peroxidation reactions. A lipid molecule reacts with an oxidizing agent (•OH in this case) resulting in a carbon-centered lipid radical. Upon reaction with oxygen, it is converted in a lipoperoxyl radical, which may react (i) with another PUFA (producing a chain reaction, since a new carbon-centered lipid radical is produced) or (ii) with α-tocopherol (TOH), which is converted in tocopheryl (TO), producing in either case a lipid hydroperoxide. The later is then converted to an alkoxyl lipid radical upon reaction with iron. Lipid hydroperoxides and alkoxyl radicals undergo numerous routes for degradation and chain termination.

proper controls, relative values of conjugated dienes (percent increases or decreases) have been well-accepted in the literature.

Initiation reactions:

$$LH + R^{\bullet} \rightarrow L^{\bullet} + RH \quad (12.3.1)$$

$$L^{\bullet} \rightarrow \text{conjugated diene } L^{\bullet} \quad (12.3.2)$$

$$O_2 + \text{conjugated diene } L^{\bullet} \rightarrow LOO^{\bullet} \quad (12.3.3)$$

The next step in the *peroxidation pathway* is the reaction of a PUFA-conjugated diene (a carbon-centered $L^{\bullet}$ species) with molecular oxygen to yield a peroxyl radical ($LOO^{\bullet}$) [reaction (12.3.3); see Fig. 12.5] that feeds the propagation step of peroxidation. $L^{\bullet}$ can also undergo reaction with other $L^{\bullet}$ isomers, but this is only quantitatively relevant under very low O_2 concentrations. ESR methods of spin-trapping (see Section 12.1), mainly using nitroso compounds such as DMPO and PBN, have proven the formation of $LOO^{\bullet}$ in biological membranes. Interestingly, it has been calculated that, in LDL particles, a single $L^{\bullet}$ can initiate the oxidation of 30 to 50 other PUFA molecules.

Propagation reactions:

$$LOO^\bullet + LH \rightarrow L^\bullet + LOOH \quad (12.3.4)$$

$$Fe^{2+} + LOOH \rightarrow Fe^{3+} + LO^\bullet + OH^- \quad (12.3.5)$$

$$Fe^{3+} + LOOH \rightarrow Fe^{2+} + LOO^\bullet + H^+ \quad (12.3.6)$$

Propagation of lipid peroxidation takes place by the reaction of $LOO^\bullet$ with a methylene group on a nonoxidized PUFA molecule to yield a lipid hydroperoxide (LOOH; see Text Box 12.6) and another PUFA carbon-centered radical ($L^\bullet$) [reaction (12.3.4)]. The new carbon-centered radical can react with O_2 to start a chain reaction (Fig. 12.5). The O–O bond of $LOO^\bullet$ or LOOH species can also undergo cyclization and/or fragmentation, leading to termination reactions (see Section 12.3.3).

The most common (indirect) method for assessing the peroxidation reaction is O_2 uptake (using a classical Clark-type electrode). Since this method is very simple [it determines reaction (12.3.3)], many studies have taken advantage of its usefulness for determining the effects of oxidants and antioxidants under *in vitro* conditions.

The oldest assessment of peroxidation reactions was done early in the nineteenth century by de Saussure who determined air absorption (using a mercury manometer) into walnut oil during long-term storage (see Halliwell and Gutteridge, 1999). Over the course of 8 months of observation very little air was absorbed, only 3 times the volume of oil (actually oxygen was being consumed and incorporated into the oil). After that, within 10 days, a fast reaction occurred and air absorption was then 60 times the oil volume; the oil viscosity also increased incredibly, producing an "evil smell" in de Saussure's words. Finally, the rate of air absorption decreased gradually over the following 3 months. What de Saussure had observed was the 3 main phases of lipid peroxidation: *initiation*, *propagation*, and *termination*. The kinetics of both conjugated diene formation and O_2 consumption in biomembrane preparations in modern experiments also display a slow phase (lag phase, or initiation) followed by a very rapid phase (log phase, or propagation) and a termination phase (Fig. 12.6).

TEXT BOX 12.6 LIPID HYDROPEROXIDE DETERMINATION

Various methods have been used to determine LOOH in biological samples. Iodometric determinations of LOOH were mostly employed in the 1970s and 1980s. These are based on the reaction of iodide (I^-) with LOOH, yielding L-OH (an alcohol) and iodine (I_2), the latter quantified with sodium thiosulfate. This reaction is not very useful for tissue samples because many other biological oxidizing agents can catalyze I_2 formation. Determination of LOOH by a glutathione peroxidase (cGPx) assay that measures the yield of NADPH oxidation at 340 nm ($LOOH + 2GSH + NADPH \rightarrow L\text{-}OH + H_2O + GSSG + NADP^+$), was developed in the 1980s. However, this assay needs highly purified cGPx (an expensive enzyme) and cannot measure LOOH from membranes unless these are first disrupted using purified phospholipase. Determination of LOOH by oxidation of a ferrous complex with xylenol orange at acid pH has been used for membrane preparations. The resulting Fe^{3+}–xylenol orange complex is measured at 580 nm (the reaction involves propagation mechanisms that are still not fully understood).

In the mid 1990s, Hermes-Lima and co-workers adapted the xylenol orange assay for use with tissue extracts to compare control versus stressed conditions in biological and clinical studies (see *Free Radic Biol Med* **19**:271–280, 1995). For example, it is well-known that ischemia/reperfusion of mammalian organs induces an increase in lipid peroxidation products, including increased levels of conjugated dienes (see Chapter 13). The ventricular levels of iron–xylenol orange-reactive LOOH increased by about 1.6-fold after 20 min of coronary occlusion, followed by 10 min of reperfusion. The corresponding increase in ventricular TBARS (a classical assay for peroxidation; see Text Box 12.7) was about 1.5-fold in these experiments.

12.3.2.1 Metals and Lipid Peroxidation Propagation Iron and copper ions play a pivotal role in the process of lipid peroxidation, mostly by means of LOOH decomposition (metals are also involved in the initiation of lipid peroxidation, by catalyzing the formation of ROS). Fe^{2+} and Fe^{3+} [bound to ligands such as ethylenediaminetetraacetic acid (EDTA), citrate, or adenosine diphosphate (ADP)] may react with LOOH, producing either peroxyl ($LOO^\bullet$) or alkoxyl species ($LO^\bullet$), which feed the propagation process [reactions (12.3.5) and (12.3.6)].

For example, Fe^{2+}-citrate or Fe^{2+}–ADP (but not Fe^{2+} alone) can induce severe damage *in vitro* to mitochondrial functions, such as loss of transmembrane electrochemical potential. The severity of the damage is correlated with the increase in the initial rate of O_2 uptake (caused by lipid peroxidation) of the mitochondrial membranes and the formation of aldehydic by-products of peroxidation (see Section 12.3.3.2).

Certain iron chelators (see Text Box 12.1) such as deferoxamine and pyridoxal isonicotinoyl hydrazone (PIH) are

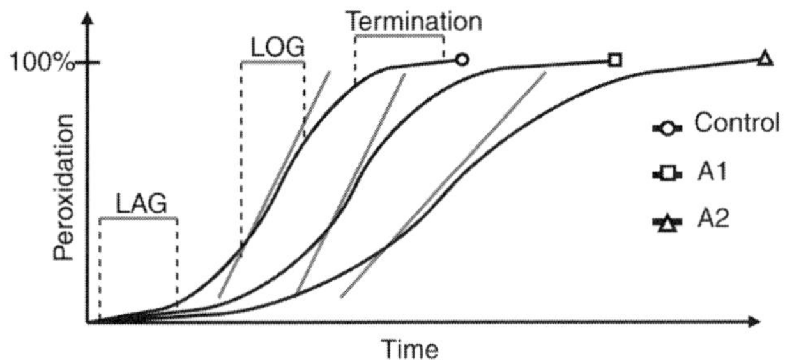

Figure 12.6 A typical time course of a lipid peroxidation reaction (the *y* axis represents product formation, such as TBARS or conjugated dienes), showing the three phases of the process: initiation (lag phase), propagation (log phase), and termination. It also shows the effect of two types of antioxidants (A1 and A2). Both augment the period of the initiation phase, but just one of them (A2) affects the log phase of the reaction.

able to inhibit lipid peroxidation in biomembranes by forming a complex with Fe^{3+} [see reaction (12.3.7) for the case of PIH]. The resulting complex is not redox active and thus cannot receive electrons from LOOH [or very low rates of reaction (12.3.8) may occur]. By functionally arresting $LOO^\bullet$ formation, these chelators may stop the propagation of the peroxidation process and lead to a diminished formation of by-products of lipid peroxidation, such as aldehydes.

$$Fe^{2+}\text{-citrate} + 2PIH \rightarrow Fe^{2+}\text{–}PIH_2 + \text{citrate} \rightarrow Fe^{3+}\text{–}PIH_2 \quad (12.3.7)$$

$$Fe^{3+}\text{–}PIH_2 + LOOH \rightarrow Fe^{2+}\text{–}PIH_2 + LOO^\bullet + H^+ \quad (12.3.8)$$

In the 1980s, American researcher Steven D. Aust proposed that iron-catalyzed lipid peroxidation depends on an optimum 1 : 1 ratio of Fe^{2+} to Fe^{3+} at the beginning of the process [Minotti and Aust (1987) *J Biol Chem* **262**:1098–1104]. Aust suggested that a $Fe^{2+}\text{–}Fe^{3+}\text{–}O_2$ complex is involved in the initiation of the peroxidation reaction. This view has permeated the literature for over two decades, even though the proposed oxidant was never isolated. Many authors (including myself) do not believe this complex exists.

Lipid peroxidation mediated by iron and copper ions has been associated with the etiology of several human diseases. Metal-mediated peroxidation is also considered a "worsening event" in many diseases and pathological process. One of the best-studied cases is the liver cirrhotic condition caused by hemochromatosis, a genetic disease with high prevalence in North America that is caused by a defect in the iron absorption mechanism that leads to ferritin–iron accumulation in liver and heart (see Text Box 12.1). This continuous load of hepatic iron increases the levels of redox-available "free iron," which is iron bound to low-molecular-weight molecules (such as citrate, ATP, ADP, and phosphate). Free iron may then initiate/propagate peroxidation reactions, inflicting membrane damage and causing the formation of toxic peroxidation termination products (such as aldehydes, see section below), which can react with DNA and proteins.

In addition, several neurodegenerative disorders, such as Parkinson's disease and Alzheimer's disease, are connected to increased iron levels in certain brain areas, leading to lipid peroxidation and neuronal dysfunction (see Chapter 13). Copper-mediated peroxidation of LDL particles has been also associated with the etiology of plaque formation in atherosclerosis. The determination of termination products of lipid peroxidation has been very useful for the study of lipid peroxidation in human diseases.

12.3.3 Termination of Peroxidation Reaction and Its Consequences

Termination of the peroxidation chain is caused by (i) reaction between two lipid radical species [reaction (12.3.9)], (ii) decomposition of LOOH, $LO^\bullet$, and $LOO^\bullet$ species to form many by-products (including hydrocarbon gases, F_2-isoprostanes, and aldehydes) [reaction (12.3.10)], or (iii) the presence of lipid-soluble phenolic antioxidant compounds such as α-tocopherol or butylated hydroxytoluene (BHT, an artificial antioxidant used extensively in the food industry) [reaction (12.3.11)]. Antioxidant molecules (AH) react with $LOO^\bullet$ yielding LOOH and a phenolic radical that is resonance-stabilized ($A^\bullet$). For example, tocopheryl radical is formed upon reaction of α-tocopherol with $LOO^\bullet$ (Fig. 12.5; see Section 12.3.4 for further discussions about chain-break antioxidants).

$$LO^{\bullet} + LO^{\bullet} \rightarrow LOOL \qquad (12.3.9)$$

$$LOOH \rightarrow\rightarrow\rightarrow \text{aldehydes, alkanes, and isoprostanes (mostly metal-mediated reactions)} \qquad (12.3.10)$$

$$AH + LOO^{\bullet} \rightarrow A^{\bullet} + LOOH \qquad (12.3.11)$$

The fate of the termination products of lipid peroxidation is also important for their effects in the cells and for analytical biochemistry. We shall next discuss some relevant classes of peroxidation by-products.

12.3.3.1 Hydrocarbon Gases Volatile hydrocarbon gases, mainly ethane and pentane, are produced by β-scission reactions of $LO^{\bullet}$ species in metal-catalyzed reactions. Although formation of hydrocarbons from lipid peroxidation constitutes only a minor degradation pathway, they are of special interest for noninvasive studies in laboratory animals and in humans because these volatile gasses can be collected from exhaled breath for analysis by gas-liquid chromatography.

Alkane formation is dependent on the nature of the PUFA. Oxidation of omega-3, omega-4, omega-6, and omega-7 PUFAs enriches the yield of ethane, propane, pentane, and hexane, respectively. Because omega-3 and omega-6 PUFAs are the most abundant PUFAs in biological membranes, ethane and propane are routinely quantified by researchers. Ethane production as a marker of lipid peroxidation has been especially well-documented and validated in studies with rats.

Several studies have correlated increased formation of breath alkanes in humans or laboratory animals under stress conditions (e.g., hyperoxia, extenuating exercise, smoking, liver transplantation), natural aging, and in patients suffering from several pathologies [e.g., rheumatoid arthritis, cystic fibrosis, acute myocardial infarction, multiple sclerosis, schizophrenia, acquired immunodeficiency syndrome (AIDS), Alzheimer's disease]. For example, increased breath pentane has been found in patients with alcoholic hepatitis (1.7 ± 0.7 pmol/mL) and alcoholic cirrhosis (3.4 ± 0.7 pmol/mL) as compared with controls (0.9 ± 0.1 pmol/mL). Hepatic lipid peroxidation is well-known to be elevated in alcoholic cirrhosis. Determination of breath pentane in nonsmoking patients suffering from Crohn's disease, which is an acute intestinal inflammation with intense influx of neutrophils, indicated significantly higher levels than in control subjects (11.6 ± 1.7 vs. 5.8 ± 0.5 pmol/kg/min). Control smokers had breath pentane levels (9.4 ± 1.4 pmol/kg/min) that were similar with those found in patients with Crohn's disease.

12.3.3.2 Aldehydic Products Aldehydes are the most relevant class of products of lipid peroxidation in quantitative terms. The carbonyls formed during lipid peroxidation are composed of several subclasses including *n*-alkenals, 2-alkenals, 2,4-alkadienals, alkatrienals, hydroxyalkenals, hydroxylperoxy-alkenals, α-dicarbonyls, alkanes, alkenes, and (un)saturated ketones. Major carbonyls produced during peroxidation of PUFAs are malondialdehyde (MDA), hexanal, 4-hydroxy-2,3-*trans*-nonenal (4-HNE), 4-hydroxy-2,3-*trans*-hexenal, and 2-propenal (acrolein) (Fig. 12.7).

Several methods have been used for the determination of soluble carbonyls. The most popular one is the formation of thiobarbituric acid reactive substances (TBARS) (see Text Box 12.7). Others include derivatization with dinitrophenylhydrazine (DNPH) followed by separation with HPLC and gas-chromatography coupled with electrochemical determination (CG–EC). These methods have been applied to blood, tissue samples, and urine. For example, rats under treatment with classical peroxidation-inducing chemicals, such as Fe(III)-NTA (nitrilotriacetic acid) or CCl_4, show increased levels of soluble carbonyls in the urine. TBARS (and MDA determination) has also been widely used for studies in animal comparative biochemistry, plant physiology, and food science. Various stress conditions are associated with increased TBARS/MDA (either in specific tissues or determined in the urine or plasma; many of these examples are discussed further in Chapter 13) including hyperoxia, extenuating exercise, reperfusion injury in rodent organs, burn injuries, AIDS, cystic fibrosis, acute myocardial infarction, Alzheimer's disease, Parkinson's disease, rheumatoid arthritis, and diabetes.

The aldehydic products of lipid peroxidation are of a highly toxic nature. The dialdehydes can crosslink and aggregate membrane proteins. For example, valine residues of hemoglobin are very reactive with MDA, forming a stable product with a half-life of several days. Moreover,

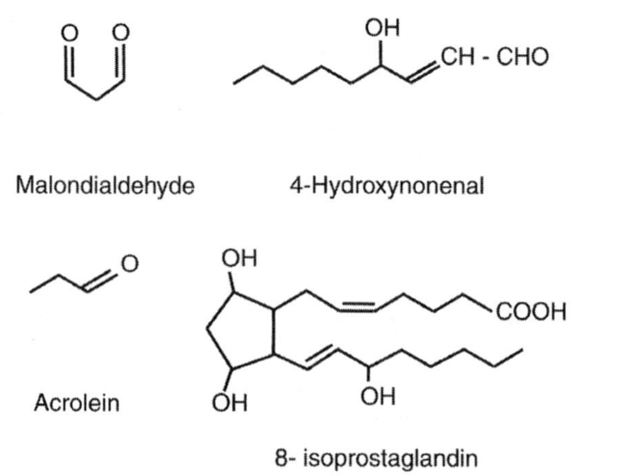

Figure 12.7 Selected end products of lipid peroxidation.

TEXT BOX 12.7 THE TBARS ASSAY

Malondialdehyde (MDA) is one of the end products of lipid peroxidation. It reacts with thiobarbituric acid (TBA) under acid pH and elevated temperature to form a pink fluorescent MDA:TBA (1:2 ratio) adduct. Hundreds of studies have used the TBA test (also called thiobarbituric-acid-reactive substances assay or TBARS assay) to quantify MDA formation. However, this assay is neither specific for MDA (other aldehydes formed during peroxidation are also reactive) nor is MDA the main aldehyde product of lipid peroxidation. Thus, the use of the TBARS assay for determination of MDA content in biological samples (using calibration curves with pure MDA) gives an overestimation of the true concentration of this molecule. Moreover, the yield of MDA in peroxidation reactions is dependent on the membrane composition and the nature of the oxidant insult. Since MDA is just one of many products of lipid peroxidation, determination of MDA by HPLC (or by methods of chromatography coupled with mass spectrometry) may underestimate the true extent of the lipid peroxidation process under study. In addition, MDA is not a metabolically inert product of peroxidation; several aldehyde dehydrogenases are capable of converting MDA (and other aldehyde products of peroxidation) to their corresponding alcohols.

Even though measurements of TBARS have been under severe attack from many researchers (a ballot was proposed in the 1995 International Free Radical Meeting for a ban of the TBARS assay!), it is still considered a relevant biomarker of lipid peroxidation in tissue sample preparations, especially for comparative purposes. However, to make conclusions about a specific sample, other methods of measuring lipid peroxidation should be used concomitantly.

accumulation of aldehyde-modified soluble enzymes and membrane proteins compromises normal metabolic functions and may cause cell dysfunction and death. This seems to be a relevant factor in neuronal dysfunction in Alzheimer's disease (see Section 13.2).

Aldehydes also react with DNA, mainly with guanine bases, forming stable adducts. Levels of the fluorescent products (excitation at 315 to 390 nm, emission at 420 to 460 nm) of the interaction of DNA with aldehydes correlate positively with the disruption of DNA physical-chemical properties and the loss of template activity for RNA polymerase. Recently, gas chromatography–mass spectrometry (GC–MS) has been used to identify adducts of aldehydes with DNA. For example, DNA from disease-free human liver was found to contain 5400 MDA–deoxyguanosine adducts per cell. Several studies have shown that DNA modifications caused by lipid peroxidation products have cytotoxic, mutagenic, and carcinogenic consequences (see Section 12.5).

12.3.3.3 Chemiluminescence The reaction of two alkoxyl PUFA radicals yields an alcohol (L-OH) in the ground state and an excited-state ketone [L=O*; see reaction (12.3.12)], which can emit light as it decays to the ground state. Peroxyl radicals can undergo self-reactions yielding singlet oxygen ($^1O_2$) and a carbonyl product that may or not be in the excited state (triplet carbonyl). The determination of light emission from these Russell-type reactions has been an important tool for lipid peroxidation research. Recent studies that have followed chemiluminescence in isolated LDL particles under peroxidation stress have shown good correlation with more traditional methods for quantifying lipid peroxidation, such as measurement of conjugated dienes. Chemiluminescence can also be useful for probing the oxidation resistance of membrane preparations (e.g., caused by lipid-soluble antioxidants under evaluation) without the need of time-consuming extraction procedures.

$$LO^\bullet + LO^\bullet \rightarrow L{=}O^* + L\text{-}OH \qquad (12.3.12)$$

Low-level chemiluminescence can be detected in cells, tissue samples, and isolated organs with the use of highly sensitive photodetectors. The use of chemical enhancers of light emission, such as lucigenin and luminol, can also be helpful. The technique has the advantage of being a noninvasive method for continuous determination of lipid peroxidation. For example, chemiluminescence by using lucigenin was applied for studies of oxidative stress associated with ischemia and reperfusion in isolated rat heart. Lucigenin (at 10^{-5} M) was used in the perfusates to enhance the luminescence signal. Chemiluminescence decreased from 219 ± 11 counts per second (cps) at baseline to 149 ± 9 cps during a 12-min ischemia but markedly increased to a peak of 476 ± 36 cps during 3 to 5 min of reperfusion before decreasing again to near control values over the next 15 to 20 min. Addition of SOD to the perfusate (2000 U/min) caused a significant reduction in chemiluminescence during reperfusion, suggesting that O_2^- radicals participate is the complex mechanism of lipid peroxidation. The studies from the early 1990s demonstrated that free radical generation and lipid peroxidation occurs during the reperfusion phase of the ischemia/reperfusion cardiac stress (see Section 13.5).

12.3.3.4 Isoprostanes A group of prostaglandin-like compounds, known as F_2-isoprostanes, are produced by a free-radical-catalyzed mechanism during peroxidation of arachidonic acid. Their generation involves the formation of peroxyl radical isomers of arachidonic acid, which undergo endocyclization to form prostaglandin-like compounds with subsequent reduction to F_2-isoprostane-like molecules. F_2-isoprostanes are released in the free form from esterified arachidonic acid undergoing peroxidation possibly by the action of phospholipases. The F_2-isoprostane, 8-isoprostaglandin (8-EPI-PGF_{2a}; see Fig. 12.7) was identified some years ago as a product of *in vivo* peroxidation in rats induced by CCl_4.

Determination of F_2-isoprostanes requires sophisticated and expensive methods such as GC–MS. Simpler immunodetection assays for 8-EPI-PGF_{2a} in tissue samples, plasma, and urine have been used since the early 1990s, but there is still controversy over the reliability of this assay for quantification of absolute levels of 8-EPI-PGF_{2a}. In addition, recent studies have shown that isoprostanes are not merely markers of lipid peroxidation. They can also invoke biological and pathological responses in several cell types, especially in lung cells.

As an example of F_2-isoprostane determination in a stress condition, it has been observed increased plasma concentrations of isoprostanes associated with cigarette smoking. Plasma levels of free and esterified F2-isoprostanes were significantly higher in smokers (0.24 ± 0.15 and 0.57 ± 0.22 nM, respectively) than in nonsmokers (0.10 ± 0.02 and 0.35 ± 0.06 nM). Smoking had no short-term effects on the circulating levels of F2-isoprostanes. However, the levels of free and esterified F2-isoprostanes fell significantly (by 38 and 25%, respectively) after 2 weeks of abstinence from smoking. In another study, 8-EPI-PGF_{2a} (measured in the esterified form by a GC–MS technique) was found to be increased in the plasma of non-insulin-dependent diabetes mellitus (NIDDM) patients (0.93 ± 0.07 nM, $n = 39$) when compared to control subjects (0.28 ± 0.04 nM, $n = 15$).

Interestingly, both cigarette smoking and NIDDM have been linked with oxidative stress when biomarkers of DNA oxidative damage (see Section 12.5), protein oxidation (see Section 12.4), and other markers of lipid peroxidation have been assessed.

12.3.4 Antioxidants That Function Against Lipid Peroxidation

Many lipid- and water-soluble molecules provide defenses against lipid peroxidation. These include vitamin E, carotenoids, plant polyphenols, bilirubin, lipoic acid, coenzyme Q, melatonin, uric acid, and various synthetic antioxidants, such as probucol (an antiatherogenic drug; see Section 13.4), BHT, and Trolox (a water-soluble analog of α-tocopherol). Ascorbate can be either a prooxidant by recycling metals to the reduced state, or a quencher of free radicals that could initiate lipid peroxidation. In the following discussions we will focus on α-tocopherol.

As discussed earlier in this chapter (Section 12.2.3), all components of vitamin E are able to function as antiperoxidative agents. In the case of α-tocopherol, it works as an antioxidant of peroxidation reactions by donating hydrogen to lipid hydroperoxyl radicals ($LOO^\bullet$), resulting in tocopheryl radical and LOOH [reaction (12.3.13), which has a rate constant 10,000-fold higher than the reaction of $LOO^\bullet$ with lipids]. Other antioxidants also react with $LOO^\bullet$, including other vitamin E components, coenzyme Q, uric acid, bilirubin, BHT, and probucol. The hydrophobic tail of α-tocopherol anchors the molecule in the membrane, positioning the phenolic –OH groups for reaction with $R{-}OO^\bullet$ groups (from oxidized phospholipids) at the hydrophilic surface of the membrane. The propagation of *in vitro* peroxidation reactions of biomembranes will only occur when α-tocopherol is depleted; thus, it is called a chain-breaking antioxidant. The increase in α-tocopherol concentration in membranes under properoxidative conditions will further elongate the "lag phase" of lipid peroxidation, but usually not changing the "log phase" (see Fig. 12.6).

$$\alpha\text{-TocH} + LOO^\bullet \rightarrow LOOH + \alpha\text{-Toc}^\bullet \quad (\text{rate constant} = 10^6\ M^{-1}\ s^{-1}) \qquad (12.3.13)$$

$$\alpha\text{-Toc}^\bullet + AH(\text{antioxidant}) \rightarrow A^\bullet + \alpha\text{-TocH} \qquad (12.3.14)$$

$$\alpha\text{-Toc}^\bullet + LH \rightarrow L^\bullet + \alpha\text{-TocH} \qquad (12.3.15)$$

Tocopheryl radical (α-$Toc^\bullet$) can be stabilized by resonance; it can be "recycled" to α-tocopherol by water-soluble ascorbate or lipid-soluble coenzyme Q [reaction (12.3.14)]. Accumulation of α-$Toc^\bullet$ may lead to initiation of lipid peroxidation [reaction (12.3.15)], which explains the prooxidant effect of α-tocopherol. Tocopherol-mediated lipid peroxidation (TMP) is currently a subject of many studies and conflicting conclusions. One can argue that the low rate constants of TMP reactions (about $0.1\ M^{-1}\ s^{-1}$) make them insignificant under *in vivo* conditions, especially considering that the steady-state levels of α-$Toc^\bullet$ are very low. Moreover, several antioxidants may reduce very quickly α-$Toc^\bullet$ to α-tocopherol (rate constants ranging from 10^4 to $10^6\ M^{-1}\ s^{-1}$).

12.4 PROTEIN OXIDATION

Awareness by the scientific community of the clinical and biological relevance of protein oxidation started relatively later in the short history of free radical research. The boom of studies on this subject started only in the early 1990s. Most investigations have focused on oxidative damage to specific purified enzymes (or structural proteins) or on studies in cells/organelles and tissue samples, where overall protein oxidation is determined and linked to general processes of oxidative stress. As in the case of lipid peroxidation, many diseases and pathological (or even physiological) processes have been connected with protein oxidation (see below).

Unlike the situation of DNA damage, there are few cellular mechanisms involved in providing protection against protein oxidation or acting to repair damaged proteins (see Section 12.4.2). Instead, damaged proteins are targeted for degradation (often via ATP-dependent ubiquitination) by endogenous proteases, including cathepsin c, calpain, trypsin, and the proteasomes. The recycling of oxidized proteins by means of proteolysis is the way that cells prevent the accumulation of malfunctioning proteins. Nonoxidized amino acids are released to be reused for biosynthesis of new proteins. It is interesting that protease biosynthesis is up-regulated under oxidative stress conditions.

Proteins can be damaged by various free-radical-mediated mechanisms including irradiation, products of lipid peroxidation and sugar oxidation (mostly aldehydes, forming carbonyl adducts with proteins), nitric-oxide-derived metabolites, and metal-catalyzed oxidation systems. Other nonradical mechanisms are also of great importance such as cold and heat denaturation, pressure-induced denaturation, and nonenzymatic glycation of proteins (by Amadori chemistry). Glycated proteins are also subjected to oxidation that leads to advanced glycation end products (AGE), which are of major importance in aging and Alzheimer's disease. Glycation of many relevant proteins including the lens proteins (causing cataract) and hemoglobin is also associated with diabetes.

TABLE 12.5 Oxidation of Amino Acid Residues and Some of Their Products

Phenylalanine → 3-nitrophenylanaline,[a] *ortho*-, *para*-, and *meta*-tyrosine
Tyrosine → 3-nitrotyrosine,[a] dityrosine, dihydroxyphenylalanine (DOPA)
Histidine → 2-oxo-histidine
Tryptophan → 6-nitrotryptophan,[a] hydroxytryptophan, kynurenine, formylkynurenine
Methionine → methionine sulfoxide, methionine sulfone
Cysteine → disulfite derivatives
Leucine and valine → hydroxy derivatives
Threonine → 2-amino-3-ketobutyric acid
Lysine and arginine → carbonyl derivatives
Proline → carbonyl derivatives, 2-pyrrolidone

[a]Products of reactive nitrogen species (RNS) attack.
Source: Data adapted from Stadtman and Levine (2000).

Superoxide radicals inactivate selected enzymes including aconitase (by reducing Fe^{3+} in the enzyme molecule and causing the release of ferrous ions) and other [Fe–S] cluster-containing dehydratases. Inactivation of aconitase is of tremendous biological significance because it is a key enzyme for the normal functioning of the Krebs cycle. Alkyl peroxides (mostly LOOH) and H_2O_2, in general, are not capable of inflicting relevant protein damage. Various reports of H_2O_2-induced *in vitro* enzyme inactivation must be carefully examined because the presence of contaminating transition metals in the solutions can promote $^{\bullet}OH$ formation via Fenton-like reactions.

$$O_2^{-} + Fe^{3+}\text{–ligand} \rightarrow Fe^{2+} + \text{ligand} + O_2 \qquad (12.4.1)$$

$$Fe^{2+} + \text{protein} \rightarrow Fe^{2+}\text{– protein} \qquad (12.4.2)$$

$$Fe^{2+}\text{– protein} + H_2O_2 \rightarrow Fe^{3+}\text{– protein} + OH^{-} + {}^{\bullet}OH\,\text{(site-specific formation)} \qquad (12.4.3)$$

$$Fe^{3+}\text{– protein} + {}^{\bullet}OH \rightarrow \text{oxidized protein} + Fe^{3+} \qquad (12.4.4)$$

Hydroxyl radicals formed by various types of irradiation can oxidize virtually all amino acid residues of proteins. Irradiation studies with purified proteins have led to the identification of most of the products of oxidation of amino acid residues (see Table 12.5). Moreover, peptide bond cleavage can also result from attack by ROS on glutamyl, aspartyl, and prolyl residues. The most important source of protein oxidation seems to be meditated by the binding of low amounts of iron or copper to the protein structures. This prompts the formation of radical species, upon reaction with H_2O_2 or alkyl peroxides, and induction of site-specific protein damage [reactions (12.4.1) to (12.4.4)]. Metal-catalyzed oxidation is involved in the damage to most enzymes, membrane receptors, and structural proteins.

12.4.1 *In Vitro* Studies on Protein Oxidation

Many studies were conducted through the 1990s on oxidative damage to purified proteins. Typically, the focus was on the relationship between enzyme activity and the concentration of the oxidizing agent, whereas enzyme structural changes were also assessed by fluorescence studies, electrophoresis, and immunoblotting. Levels of oxidized products of amino acid residues, such as carbonyls, oxidized –SH groups, and dityrosine were also quantified.

An interesting example of the study of protein inactivation by oxidation was conducted several years ago in Earl Stadtman's laboratory (National Institutes of Health, Maryland). It assessed the effect of micromolar concentrations of Fe^{2+} and H_2O_2 on the loss of activity by glucose-6-phosphate dehydrogenase (G6PDH) purified from *Leuconostoc mesenteroides* (Szweda and Stadtman, *J Biol Chem* **267**:3096–3100, 1992). Enzyme inactivation was correlated with the formation of one carbonyl group (see Section 12.4.3) per enzyme subunit, showing that inactivation is the result of site-specific oxidative modification. The results suggested that Fe^{2+} binds to the glucose-6-phosphate binding site and that interaction of the enzyme-bound Fe^{2+} with H_2O_2 leads to the oxidative modification of amino acid residues that are essential for enzyme activity. H_2O_2 alone was unable to trigger a loss of enzyme activity. This example is of special relevance because G6PDH is key to NADPH production for antioxidant defense (see Section 12.2).

Another interesting study determined the inactivation of purified RNase and lysozyme by means of irradiation or copper-catalyzed oxidation. Using fluorometric methods, increased formation of *ortho*-tyrosine and dityrosine in both enzymes under oxidative attack was observed.

TEXT BOX 12.8 DETERMINATION OF PROTEIN THIOLS

Reduced thiols from soluble or membrane proteins are widely measured by the Ellman procedure, which was developed in the 1950s. This reaction with 5,5′,dithiobis-2(2-nitrobenzoic acid) (DTNB) yields *p*-nitrothiophenol anions that absorb at 412 nm (extinction coefficient = 13.6 mM^{-1} cm^{-1}). A note of caution for experimental design is the fact that the *p*-nitrothiophenol anion is oxidized/decomposed by ROS. Thus, addition of catalase to the assays before inclusion of DTNB can help to prevent severe underestimation of protein –SH groups in studies with purified enzymes or cell extracts under oxidative stress. For example, ROS induce the loss of 30 to 45% of –SH groups within 30 min in purified rabbit liver metallothionein, a protein that has cysteine as one-third of its amino acid content. These determinations would be jeopardized if catalase was not used in the analytical procedure (Suzuki et al., *Free Radic Biol Med* **9**:479–484, 1990).

12.4.2 Oxidation of Sulfur-Containing Residues and Tyrosine Residues

Cysteine and methionine residues are highly susceptible to oxidation by ROS. Cysteine residues are converted to disulfides and methionine residues are converted to methionine sulfoxide (MeSOX) residues. Virtually all biological systems contain thiol–disulfide oxidoreductases (see Section 12.2.1.5) and MeSOX reductases, which can reconvert the oxidized residues into the original forms. These are the only oxidative modifications to proteins that can be repaired. Moreover, it has been proposed that the cycle of oxidation/re-reduction of methionine residues functions as a "sink" for free radical attack on proteins (notably, the oxidation of exposed methionine residues generally has little effect on enzyme activities). Thus, methionine residues can be considered oxyradical scavengers. It is of relevance that the activity of MeSOX reductase is decreased in brains of Alzheimer's patients, and this may be connected with the increased damage to the methionine-containing amyloid β-peptide (Aβ) in this disease (see Section 13.2).

Assessment of the loss of protein –SH groups (by means of oxidation) is also a relevant indicator of oxidative stress affecting proteins. Several reactive species such as *t*-butyl hydroperoxide, menadione, $ONOO^-$ and $^{\bullet}OH$ are capable of oxidizing protein –SH groups. The oxidative loss of activity by many enzymes (in studies with purified enzymes) correlates with the oxidation of sulfhydryl groups. Protein thiols are also easily measurable in tissue samples (see Text Box 12.8) and constitute a relevant marker of oxidative stress.

Reactive nitrogen species (RNS) are also capable of reacting with several amino acid residues, including tyrosine, cysteine, tryptophan, and methionine. Particularly interesting is the reaction of nitrogen dioxide ($^{\bullet}NO_2$) and peroxynitrite with tyrosine residues, yielding 3-nitrotyrosine and dityrosine (also formed from $^{\bullet}OH$ attack), which can alter enzyme activities. Determination of dityrosine is done by fluorescence detection (excitation 284 nm, emission 410 nm) in purified enzymes after acid hydrolysis. Formation of 3-nitrotyrosine can be measured at 428 nm (extinction coefficient = 4.2 mM^{-1} cm^{-1}) from purified enzymes such as bovine serum albumin. Moreover, detection of 3-nitrotyrosine in tissue samples (e.g., in brain of Alzheimer's patients) by means of GC–MS or Western blotting is considered a relevant biomarker for *in vivo* damage induced by RNS.

12.4.3 Carbonyl Protein

The carbonyl derivatives of proteins are the most relevant products of free radical attack on proteins. Protein carbonyl can also be formed from the reaction of aldehydes (formed from lipid peroxidation and sugar oxidation) with proteins. The presence of carbonyl proteins in cell and tissue samples has become a widely accepted biomarker of oxidative stress. This has led to the development of several highly sensitive methods for their determination. The quantification of protein carbonyl groups is done by derivatization with 2,4-dinitrophenyl-hydrazine (DNPH), acid precipitation, and measurement of absorbance at 350 to 380 nm (extinction coefficient = 22 mM^{-1} cm^{-1}). Other assays are based on immunohistochemical detection of carbonyl proteins and HPLC separation of carbonyl protein adducts with DNPH.

Protein carbonyl in control samples from many different sources averages about 2 nmol/mg protein, which is approximately 0.1 mol of carbonyl per mole of protein. If this is not proven by future research to be an overestimation, it means that about 10% of cellular proteins have a carbonyl group in their structures. This may represent the steady-state between damaged protein formation and damaged protein degradation. A two- to threefold increase in carbonyl protein content under oxidizing conditions could create a high percentage of dysfunctional proteins that may be a major contributor to cell damage and death due to oxidative stress.

Increased levels of carbonyl protein are present (in tissue samples) in many stress conditions and pathologies associated with oxidative stress including smoking, extenuating exercise, hyperoxia (in laboratory animals), ischemia/reperfusion of mammalian organs, rheumatoid arthritis, cystic fibrosis, Alzheimer's disease, Parkinson's disease, alcoholic liver disease, cataractogenesis, and diabetes (several of these conditions/diseases are further discussed in Chapter 13). Aging is also related to carbonyl protein accumulation (see Text Box 12.9). However, not all cases have clear-cut "radical explanations." For example, an interesting study was done in the 1990s with postmortem brain tissue from patients with Parkinson's disease (see Chapter 13) and age-matched controls. In brain areas associated with Parkinson's disease, such as the substantia nigra and caudate nucleus, there were significant increases in protein carbonyl levels. These results are of apparent clinical relevance since increased lipid peroxidation and increased levels of the DNA oxidation product 8-OH-dGua (see Section 12.5) have been previously observed in the substantia nigra of Parkinson's disease patients. However, increased carbonyl levels were also found in areas of the brain that may not be affected in Parkinson's disease. This may suggest that oxidative stress is related to the therapeutic treatment of Parkinson's disease with L-DOPA (3,4-dihydroxyphenylalanine), which can exert prooxidant properties *in vitro* (see Chapter 13).

12.5 DNA OXIDATIVE DAMAGE

Damage to DNA by ionizing radiation and oxygen radicals (as well as DNA repair) has been a field of intense investigation for the last 40 years. It is an impossible task to cover all the chemical, analytical, biological, and clinical aspects of it in a single textbook chapter. Thus, an overview of several aspects of cellular oxidative stress connected to DNA damage is presented below.

12.5.1 DNA Damage: Introductory Remarks

Determinations of DNA damage were part of the first studies of the biological effects of free radicals. These studies began in the post–World War II atomic era, when researchers became interested in understanding the biochemical basis of the toxicity of ionizing radiation. Rebeca Gerschman from Argentina was the first to theorize, in a 1953 lecture at the University of Rochester, that ionizing radiation damage and oxygen toxicity were due to free radical formation. The proposal was published one year latter in the journal, *Science*, in collaboration with Daniel L. Gilbert (a medical student at that time but currently at the National Institutes of Health in the United States) and co-workers. Later they reported that high oxygen pressures change the viscosity of DNA solutions. In 1958, they also correlated their *in vitro* data with studies of mouse mortality induced by high oxygen tensions. The overall idea was that the univalent reduction of oxygen, caused by ionizing radiation or metabolic processes (see Scheme 12.1), would cause the formation of reactive intermediates that could damage cell components such as DNA. They were right. But in those days oxygen was still considered only a "good boy," essential for aerobic life. Only with the discovery of SOD (see Section 12.1 and Text Box 12.2) and the establishment of the role of ROS in biology in the early 1970s was more attention given to the link between DNA damage by ionizing radiation and oxygen.

In the early 1980s, studies of DNA damage by ionizing radiation were "replaced" by the studies of the effects of H_2O_2, since H_2O_2 is formed from the radiolysis of water. Furthermore, H_2O_2 is simpler and safer to work with.

$$O_2 + e- \rightarrow O_2^- + e- \rightarrow O_2^= + e- \rightarrow O^= + O^-$$
$$O^- + e- \rightarrow O^=$$

Scheme 12.1 Scheme of univalent reduction of oxygen exactly as proposed by Gerschman and co-workers [Oxygen poisoning and X-irradiation: A mechanism in common. *Science* **119**:623–626 (1954)], showing the formation of hydrogen peroxide (shown as $O_2^=$), superoxide radical (O_2^-), and hydroxyl radical (shown as O^-); OH^- was represented as $O^=$. Compare this 1950s scheme with the currently accepted one [reactions (12.1.8)]. The poor acceptance of women in medical sciences in the 1950s contributed to the low impact of Gerschman's radically different ideas (see Gilbert, *Ann NY Acad Sci* **899**: 1–14, 2000).

TEXT BOX 12.9 CARBONYL PROTEIN AND AGING

Natural aging is associated with a number of changes in biochemical parameters, including an increase in carbonyl protein levels in many animals species. The carbonyl content of protein in cultured human fibroblasts increases exponentially as a function of the age of the fibroblast donor. There is also a clear relationship between the carbonyl content of different tissues (human, rat, fly) and the fraction of the life span of the species (Fig. TB12.2). Moreover, dietary caloric restriction increases the average life expectancy (but not the maximum life span) of several species, and this correlates with reduced amounts of carbonyl protein, compared with age-matched controls.

Why would carbonyl protein increase with aging? There are several factors. (1) With aging, there is a decrease in the levels of several low-molecular-weight antioxidants that can quench free radicals and prevent protein oxidation. (2) Proteolytic activity decreases during aging, contributing to accumulation of damaged proteins. (3) Some damaged proteins with intra- or intermolecular crosslinks between a carbonyl group and a lysine residue are resistant to proteolysis and may also inhibit the activity of the 20s proteasome. That would further contribute to accumulation of damaged proteins, being particularly important in long-lived nondividing neuronal cells. (4) Aging may also increase the rate of mitochondrial generation of O_2^-, which could set up a situation for increased protein oxidation. These ideas are consistent with the observations of increased carbonyl protein levels in the frontal and occipital poles of human brain, eye lens proteins, and skeletal muscle as a function of age.

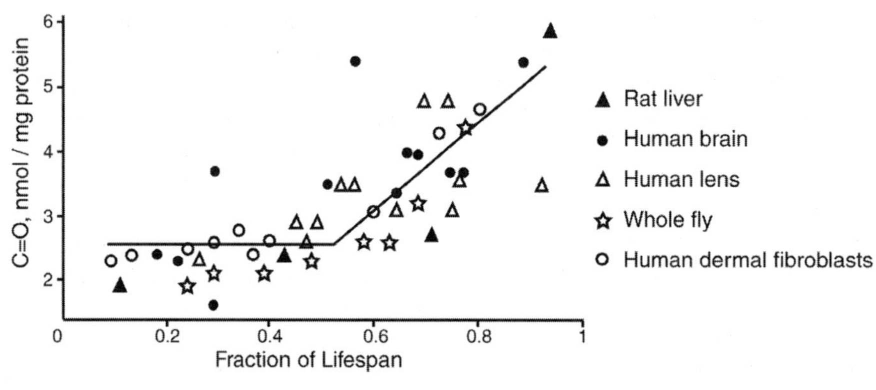

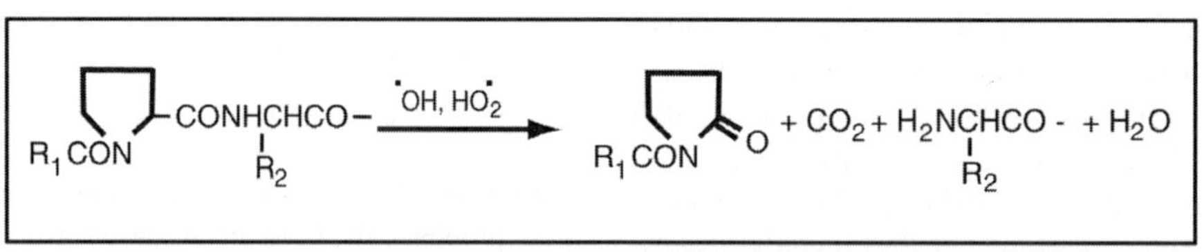

An example of a carbonyl product formation (2-pyrrolidone).

Figure TB12.2 Carbonyl protein versus aging. The figure was obtained (with permission by Dr. Stadtman) from Stadtman and Levine (2000). As an example of carbonyl formation, it shows the oxidation of proline residues leading to 2-pyrrolidone formation and peptide bond cleavage.

Even researchers from radiation chemistry/biochemistry laboratories started to study the toxic effects of H_2O_2 on cells and DNA, and they made major advances in elucidating the chemistry/biochemistry of DNA fragmentation, modification, and repair.

12.5.2 DNA as Target of Reactive Oxygen Species

Damage to DNA has been observed in many cell types, from invertebrates to mammals, and in response to multiple different forms of oxidative stress. Oxidative damage to DNA participates in many stress conditions and diseases, especially in carcinogenesis and aging (see Section 12.5.5). Moreover, DNA fragmentation is related to apoptosis, which is a crucial anticancer physiological mechanism of organisms. Exposure to ozone, ionizing radiation, chemical/metal-catalyzed generation of $^{\bullet}OH$ radicals, hyperbaric oxygen, metabolic by-products of xenobiotic compounds (e.g., aldehydes, H_2O_2), iron-overload, and nitrogen reactive species all induce DNA strand breaks and/or base modifications (see Section 12.5.3). It is also interesting to note that the quantification of oxidative damage to DNA has been widely developed as a probe to test the potentially damaging (and potentially mutagenic) effects of various chemicals, natural compounds, and metals. These studies use different DNA sources including plasmid DNA, purified mammalian nuclear DNA, isolated mitochondria (for studies of mtDNA), and cells under conditions of oxidative stress.

Superoxide radicals and H_2O_2 are not directly involved in DNA damage. However, their interaction with transition metals promotes DNA strand breaks and base modifications. Almost two decades ago Rogerio Meneghini's group at the University of São Paulo, Brazil, proposed that H_2O_2 induces DNA damage (measured as DNA fragmentation by alkaline digestion techniques; see Section 12.5.4) in cultured cells indirectly, by means of its interaction with intracellular transition metals, mainly iron (MelloFilho et al. *Biochem J* **218**:273–275, 1984). They also proposed that iron ions bound to the DNA structure would react with H_2O_2 (which crosses biological membranes easily) causing Fenton-type reactions, site-specific $^{\bullet}OH$ generation, and thus DNA damage. This mechanism was actually much like the one proposed years later for metal-catalyzed protein oxidation (see Section 12.4).

It is known from *in vitro* studies that iron and copper bind tightly to DNA at neutral pH values. Copper appears to bind preferentially to GC-rich residues in DNA. However, there are still doubts about whether iron or copper binds to DNA *in vivo*. Indirect evidence points to this in the case of iron, including the presence of a P-ATPase that transports Fe^{3+} into the nucleus, as observed by Meneghini's research group. New evidence suggests that superoxide mediates the removal of iron (as Fe^{2+}) from protein-bound iron–sulfur clusters (and many of these clusters are in the nuclei). This Fe^{2+} may bind to DNA, react with H_2O_2, and lead to $^{\bullet}OH$-mediated site-specific DNA damage (see Fig. 12.8).

Until the mid-1990s, an alternative proposal made by S. Orrenius in Sweden (see *Trends Pharmacol Sci* **10**:281–285, 1989) for the mechanism of DNA damage was well accepted by the scientific community. Orrenius's ideas were that oxidative stress would cause DNA damage (as fragmentation, but not base oxidative modification) indirectly. Free radicals would induce lipid peroxidation and rupture of the endoplasmic reticulum, which could release Ca^{2+} into the cytoplasm and activate Ca^{2+}-dependent endonucleases. This idea began to lose favor when studies in the 1990s showed that membrane-permeable iron chelators (such as 1,10-phenanthroline, which prevents Fe^{2+} reaction with H_2O_2) protected DNA from oxidative damage and cell killing under conditions of oxidative stress. On the other hand, the role of calcium in the processes of cellular redox-regulation and oxidative stress cannot be underestimated. Moreover, Ca^{2+} activation of endonucleases, caused by loss of cellular calcium homeostasis, has been recognized recently as a relevant component of the apoptotic processes.

It is also relevant to mention the literature on polyphenols and phenolic acids (see Text Box 12.4) that are present in red wines, vegetable oils, and many dietary vegetables which include compounds such as quercetin, caffeic acid, murin, tannic acid, elagic acid, and kaempferol. These may act as either antioxidants or prooxidants, but studies have shown that their action is mostly related to the capacity to chelate transition metals and/or to intercalate in the DNA structure.

12.5.3 Oxidative Damage (and Repair) to DNA Bases

Hydroxyl radicals can damage all DNA bases. They can react with thymine and cytosine yielding many products including cytosine glycol and thymine glycols, as shown in Table 12.6 (see also Fig. 12.9). Thymine–hydroperoxide is also detected after free radical attack on DNA. Reaction of $^{\bullet}OH$ with guanine and adenine residues at C4, C5, and C8 positions leads to hydroxylated and deaminated products, such as 8-hydroxyguanine (8-OH-Gua), 8-hydroxyadenine (8-OH-Ade), 2,6-diamino-4-hydroxy-5-formamidopyrimidine (FapyGua), and other modified bases with an open ring. Pyrimidine dimers are also relevant products of UV damage to DNA (mediated by $^{\bullet}OH$ formation), resulting from crosslinking of adjacent pyrimidines. Singlet oxygen also damages DNA bases, generating mostly guanine-derived products, including 8-OH-Gua and FapyGua.

Reactive nitrogen species (RNS) also produce base modifications, resulting in 8-nitroguanine (this is not

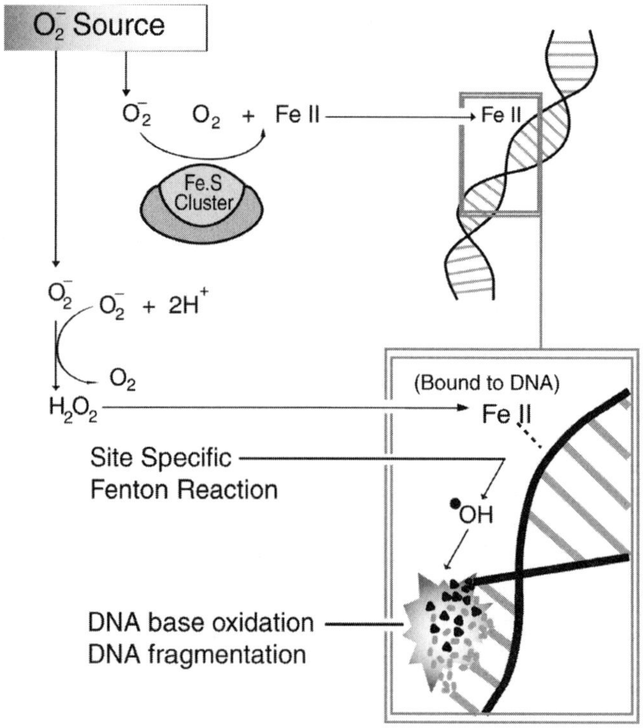

Figure 12.8 The role of iron in DNA oxidative damage. Superoxide radicals formed from cellular sources (such as mitochondria and endoplasmic reticulum) undergo (i) dismutation into H_2O_2 and (ii) cause Fe(II) release from [Fe-S]-containing proteins. When Fe(II) bound to DNA reacts with H_2O_2, it produces site-specific Fenton reaction (and $^{\bullet}$OH formation), leading to DNA oxidative damage.

TABLE 12.6 Selected Products of DNA Base Oxidation/Modification by ROS and RNS

Base	Product
Guanine	8-Hydroxyguanine (8-OH-Gua)
	2,6-Diamino-4-hydroxy-5-formamidopyrimidine (FapyGua)
	8-Nitroguanine[a]
	Xanthine[a]
Adenine	8-Hydroxyadenine (8-OH-Ade)
	4,6-Diamino-5-formamidopyrimidine (FapyAde)
	Hypoxanthine[a]
Thymine	Thymine glycol (*cis*- and *trans*-)
	5-Hydroxymethyluracil
	5-Hydroxy-5-methylhydantoin
	5,6-Dihydrothymine
	5-Hydroxy-6-hydrothymine
Cytosine	Cytosine glycol
	5-Hydroxyhydantoin
	5-Hydroxycytosine
	5-Hydroxyuracil
	5,6-Dihydroxyuracil
	Uracil[a]

[a]Products of RNS attack.

formed by $^{\bullet}$OH attack), 8-OH-Gua, FapyGua, thymine glycol, and 5-hydroxy-methyluracil. Formation of 8-OH-Gua (and other "typical" $^{\bullet}$OH-oxidation products such as FapyGua) as products of the direct action of $ONOO^-$ on DNA has been a matter of debate because they might also be formed as a result of DNA attack by decomposition products of ONOOH, such as $^{\bullet}$OH.

In general, mispaired, oxidized, and deaminated bases (or the nucleotides with the damaged bases) are removed from DNA by repair enzymes. Nonspecific endonucleases can remove a stretch of DNA containing an oxidized base, whereas DNA glycosylases remove one specific oxidized base. Methods that quantify these excised bases and nucleotides in cell extracts, intercellular fluids, blood, and urine are widely used to assess oxidative damage to DNA. For example, measurement of 8-OH-Gua in systems under oxidative stress provides an evaluation of the balance between the rates of DNA damage and repair. If there is an increase in 8-OH-Gua in blood or urine samples from animals under a certain stress/pathological condition, this could reflect an increase in DNA damage and/or a decrease in repair. Other methods for evaluating oxidative damage to DNA involve purification and digestion of the DNA (enzyme-catalyzed or by acid hydrolysis),

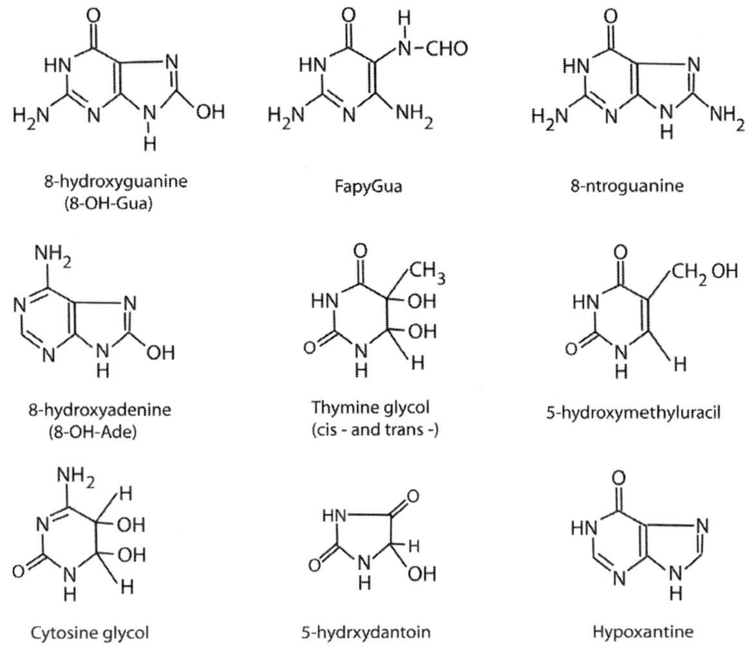

Figure 12.9 Selected oxidized/modified DNA bases.

so that the actual amounts of oxidized nucleosides or bases can be quantified.

8-Hydroxydeoxyguanosine (8-OH-dGua) and its free base 8-OH-Gua (also called 8-oxo-Gua) are the products most frequently measured as indicators of oxidative DNA damage. This product constitutes about 5% of all DNA adducts. Quantification is typically by one of two methods: (1) HPLC separation coupled with electrochemical detection (HPLC-EC) after bases are removed from DNA by enzymatic hydrolysis, or (2) gas chromatography–mass spectrometry (GC–MS) analysis after DNA is hydrolyzed in weak acid under heating.

The validity of the two methods is highly debated because both over- and underestimation of results can occur. HLPC analysis usually gives concentrations of 8-OH-dGua that are several-fold higher than results obtained from GC–MS. The GC–MS method for 8-OH-Gua quantification is also subject to overestimation due to 8-OH-Gua generation as a result of the chemical hydrolysis of 8-OH-dGua during the assay or from RNA free radical oxidation. Thus, RNA-free samples are needed for determinations of DNA oxidation products. Baseline levels of 8-OH-dGua in rodent livers (rat, mouse, and hamster, measured by HPLC-EC) were described in the mid-1990s as being in the range of 1.5 to 50 molecules of 8-OH-dGua per 10^6 DNA bases, which is equivalent to a range of 0.6 to 20 mol of 8-OH-dGua per 10^5 guanines. The current lowest baseline estimates of the ratio 8-OH-dGua/guanine in rat hepatocytes and HeLa (human transformed epithelial) cells are around 0.5×10^{-5}. Even though the battle for the correct measurement of baseline levels of 8-OH-dGua (and other modified bases) continues, comparative measurements are still of high relevance for clinical and biological studies. For more detail on the measurement of DNA oxidation see the recent trial published at *Free Radic Biol Med* **34**:1098–1099, 2003).

Levels of 8-OH-dGua in urine and tissue samples are significantly increased in humans and laboratory animals under conditions of oxidative stress (see Text Box 12.10). Examples include the presence of augmented 8-OH-dGua levels in patients suffering from systemic lupus erythematosus (in lymphocytes) and amyotrophic lateral sclerosis (in the spinal cord), cystic fibrosis, rheumatoid arthritis, and diabetes (in urine samples), as well as in postmortem brain analyses of patients with Parkinson's and Alzheimer's diseases. Currently, 8-OH-dGua is the most popular biomarker of free radical damage to DNA.

12.5.4 Other Types of DNA Alterations, Including Strand Breaks and Fragmentation

Oxygen and nitrogen reactive species also cause various other alterations to DNA structure. Crosslinks between DNA bases and amino acid residues of proteins are one example. Loss of purine bases, leaving apurinic sites, is another consequence of free radical attack on DNA. Cyto-

TEXT BOX 12.10 DETERMINATION OF 8-OH-dGua IN SELECTED STUDIES

Japanese scientists reported in 1990 that injecting rats with ferric nitrilotriacetate (Fe–NTA), a known renal carcinogen, induces a significant increase in the levels of 8-OH-dGua and lipid peroxides in kidney. The dose-dependent effect of Fe–NTA in 8-OH-dGua formation was parallel with the appearance of nephrotoxic responses in terms of serum biochemical and histopathological changes. The researchers also determined the levels of 8-OH-dGua in several organs of aging rodents (comparing 6- to 30-month-old rats). They observed increased 8-OH-dGua levels in liver and kidney of aged animals.

Studies from a research group in Singapore revealed that the levels of 8-OH-dGua in the sperm DNA of smokers are significantly higher than in the sperm DNA of nonsmokers (6.2 ± 1.7 vs. 3.9 ± 1.3 mol of 8-OH-dGua per 10^5 mol of dGua). Moreover, it has been found that infertile men have about twice the amount of 8-OH-dGua in sperm DNA as do fertile men (the results were also correlated with conventional seminal parameters). Thus, oxidative damage to sperm DNA might be important in the etiology of male infertility.

Genomic damage was also correlated with several autoimmune diseases (associated with inflammation; see Chapter 13) by researchers from England. Levels of 8-OH-dGua, determined by HPLC, were 6.8 ± 0.8 mol of 8-OH-dGua per 10^5 mol of dGua in blood lymphocytes from healthy subjects. However, the levels of 8-OH-dGua were significantly increased in lymphocytes of patients suffering from rheumatoid arthritis (9.8 ± 1.6), systemic lupus erythematosus (13.7 ± 2.8), vasculitis (10 ± 3.2), and Behcet's disease (9.2 ± 1.9). However, 8-OH-dGua levels did not correlate with disease duration, disease severity, or age.

Systemic DNA lesion was studied in non-insulin-dependent diabetes mellitus (NIDDM) by assessments of urinary 8-OH-dGua. Researchers from Finland collected urine samples (24-h collection) from 81 NIDDM patients 9 years after the initial diagnosis and of 100 control subjects matched for age and gender. The total urinary excretion of 8-OH-dGua was markedly higher in NIDDM patients than in the controls (68.2 ± 39.4 vs. 49.6 ± 37.7 μg). Moreover, glycosylated hemoglobin (the usual clinical test for diabetes) at high levels was associated with elevated concentrations of urinary 8-OH-dGua. Interestingly, NIDDM patients also have increased plasma levels of biomarkers of lipid peroxidation, such as 8-EPI-PGF_{2a} (see Section 12.3.2.4).

sine can be deaminated and produce uracil (this can have mutagenic consequences, see Section 12.5.5). In addition, deoxyribose (the sugar of the DNA structure) is fragmented by $^{\bullet}OH$, yielding an array of products, including peroxyl radicals and carbonyls (MDA is one product of DNA damage). Damage to deoxyribose may cause DNA strand breaks.

Studies on double-strand breaks to DNA are of special relevance for understanding oxidative stress processes. Many analytical methods have been employed for this purpose. However, some of them are problematic because they actually increase the number of nicks due to manipulations and, thereby, overestimate DNA fragmentation. Thus, DNA fragmentation assays are mainly used to quantify relative levels of damage in control versus experimental samples.

Many assays use gel electrophoresis to produce a "footprint" of the nuclear DNA fragmentation pattern. The fluorescent dye ethidium bromide is used to stain the DNA. The anticancer drug bleomycin is well-known for its effects on DNA fragmentation. It forms a complex with the DNA structure and, in the presence of transition metals, causes oxygen-dependent single- and double-stranded lesions to DNA. Different oxidizing agents can produce unique DNA footprints; the pattern may depend on the base composition of the DNA. For example, Japanese researchers observed in the late 1990s that cupric nitriloacetate, a potent carcinogen, causes severe nuclear DNA fragmentation to human promyelocytic HL-60 cells in culture, as observed by gel electrophoresis. They also observed increased 8-OH-dGua formation caused by cupric nitriloacetate. The effect was dose-dependent and was associated with apoptosis of the HL-60 cells under stress.

The DNA plasmids are often used for simple model studies of the effects of oxidants and antioxidants. Intact plasmids are supercoiled (SC) whereas plasmids containing single-strand breaks will form open-circular (OP) forms; double strands will cause fragmentation. SC and OP forms migrate differently on agarose gel electrophoresis and damage can be quantified by measuring the loss of the SC band and/or the increase in the OP band intensity. Metal-catalyzed $^{\bullet}OH$ formation, singlet O_2, and $ONOO^-$ are able to cause plasmid DNA fragmentation and damage; antioxidants, such as catalase, and certain metal chelators can prevent DNA strand-break formation.

12.5.4.1 Alkaline Digestion or Unwinding of DNA for Determination of DNA Fragmentation Other classical assays to determine DNA fragmentation are based on alkaline digestion of the DNA followed by enzymatic incorporation of radioactively labeled nucleotides. Enzyme-catalyzed incorporation of tritiated thymidine is often used. The fragments are then separated (by HPLC or regular chromatographic columns) and the radioactivity quantified. The use of cesium salt gradients to separate fragmented DNA labeled with H^3-thymidine (without alkaline digestion) is another method to analyze DNA damage. This class of assays (which can be used for cells or tissue samples) highly overestimates the actual number of DNA nicks, but, in relative terms, oxidatively damaged DNA will have more fragments than controls.

It was observed two decades ago that incubation of isolated rat liver with ferrous ions causes a dose-dependent fragmentation of DNA, measured by alkaline digestion, as well as lipid peroxidation. Interestingly, in this study DNA damage was inhibited by chain-break antioxidants (BHT and vitamin E; see Section 12.3). This study by Thomas Shires (*Biochem J* **205**:321–329, 1982) was one of the first to show that products of lipid peroxidation contribute to DNA fragmentation.

Another method for analyzing DNA damage uses alkaline exposure of DNA at low temperature to produce unwinding of intact or nicked double helix; unwinding is faster in DNA with strand breaks. A fluorescent probe that intercalates with DNA (e.g., ethidium bromide) is then introduced and used to determine the extent of DNA damage. Using this assay (fluorometric analysis of DNA unwinding, FADU assay), it has been observed that 100 μM $ONOO^-$ causes DNA strand breaks in isolated rat thymocytes (a process inhibited by Trolox) and oxidation of cellular thiols.

12.5.4.2 TUNEL and Comet Assays to Quantify DNA ***Damage*** The enzymatic labeling techniques used to detect and quantify DNA fragmentation are based on the addition of fluorescent bases (or modified bases, detected using antibodies) to free OH groups created by strand breaks. The currently popular TUNEL assay (TUNEL stands for TdT-mediated X-dUTP nick end-labeling; TdT = terminal deoxynucleotidyl transferase) is an example of this kind of methodology. These techniques can be applied to cell cultures provided that the cells are made permeable for the entry of the enzymes and substrates.

Another popular technique of the late 1990s is the comet assay. This is based on cell electrophoresis; after exposure to an oxidizing agent (e.g., H_2O_2 or hyperbaric oxygen), membranes are removed by treatment with a detergent, followed by staining of the DNA. Fragmented DNA migrates more quickly to the anode than does intact DNA and so a "tail" is produced, very similar to that of a comet in the sky; the bigger the comet tail the greater the DNA fragmentation. The presence of oxidized DNA bases can also be detected by exposing cell lysates to endonuclease III, which cuts the DNA at sites of oxidized pyrimidines.

12.5.5 Biological Effects of DNA Damage

The most studied consequences of oxidative DNA damage are the mutagenic and carcinogenic effects (Fig. 12.10). Aging, cigarette smoking, and several human diseases have been also connected to DNA damage (see end of Section 12.5.3 and Text Box 12.10). Damage to mitochondrial DNA (mtDNA) is also a key factor in several human degenerative pathologies.

Organisms have a finite capacity to deal with continuous insults to DNA bases (e.g., oxidation, nitration, deamination) using their endogenous repair mechanisms. Excessive oxyradical generation (caused by many cellular processes and by a huge number of substances) can overwhelm the capacity of repair enzymes to correct DNA structure. Misreading of altered bases can then lead to mutations during DNA replication. For example, *C:G → T:A transversion mutations* (see Fig. 12.11) may arise from misreading of 8-OH-Gua during DNA replication because 8-OH-Gua can also pair with adenine. Likewise, 8-OH-Ade can

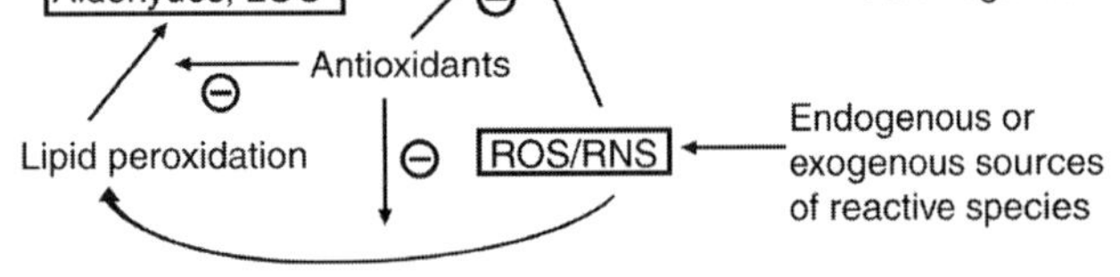

Figure 12.10 Connection between DNA oxidative damage and DNA repair with basic biological effects. The (−) symbol indicates inhibitory actions of antioxidants.

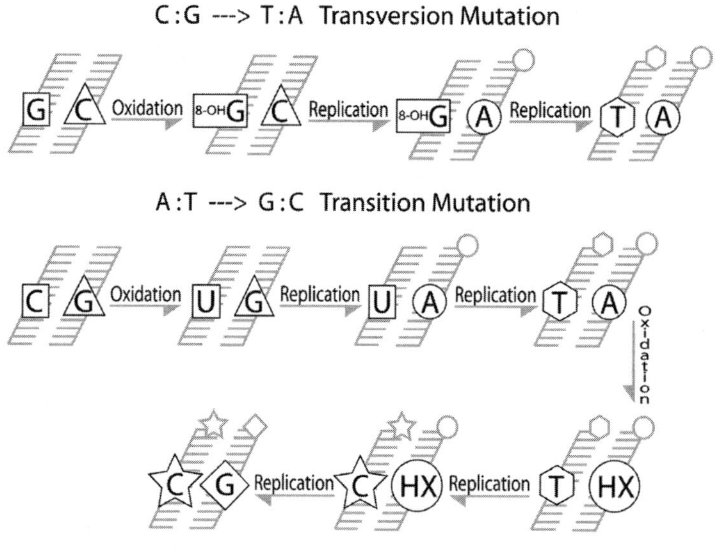

Figure 12.11 Two types of DNA mutations. After every DNA replication, the figure shows a new DNA strand marked with a symbol. U and HX represent uracil and hypoxanthine, which are oxidative deamination products of cytosine and adenine, respectively, upon RNS attack.

mispair with guanine, causing another transversion mutation. Moreover, oxidative deamination of cytosine to uracil (which pairs with adenine) may induce *A:T → G:C transition mutations* because oxidative deamination of adenine to hypoxanthine (caused by $ONOO^-$ attack on adenine) will form a hypoxanthine–cytosine pair during new replication (Fig. 12.11).

Other effects not directly related to mutations include the fact that ring-fragmented bases can block DNA replication and that DNA–protein crosslinks interfere with chromatin unfolding, DNA repair, replication, and transcription. In addition, DNA lesions (induced by ROS or RNS) may cause activation of the chromatin-bound enzyme poly(ADP-ribose) polymerase (PARP), which is responsible for ADP-ribosylation of proteins involved in DNA repair. Excessive activation of PARP may deplete the cellular NAD^+ pool, challenge the normal regulation/function of metabolism, and interfere severely with ATP synthesis.

Damage to mtDNA also has severe consequences to cells. Mitochondria have DNA repair mechanisms that are much less efficient than the nuclear DNA repair systems. The physical proximity of mtDNA to the intramitochondrial sites of O_2^- and H_2O_2 formation, as well as the absence of histones on mtDNA, make it a highly susceptible target for oxidative damage. Moreover, the intramitochondrial Fe^{2+} pool, which can be released from aconitase by O_2^-, could prompt Fenton reactions and $^{\bullet}OH$-mediated mtDNA oxidative injury. Damage to mtDNA may compromise cellular bioenergetics since several proteins (or protein subunits) involved in ATP synthesis are encoded by mtDNA. For example, it has been observed that mtDNA from rat hepatoma cells are susceptible to iron-mediated strand breaks, in contrast to mtDNA from normal hepatocytes. The much higher levels of Mn–SOD (the mitochondrial form of SOD; see Section 12.1.1) and coenzyme Q in hepatocytes, in comparison with hepatoma cells, may explain the absence of mtDNA strand breaks in normal cells. Indeed, Mn–SOD is an unique form of cell defense against oxidative stress; the absence of this enzyme in knockout mice strains has devastating effects and causes very high mortality (see Text Box 12.3).

12.6 FREE RADICALS AND SIGNAL TRANSDUCTION

A few more key points about reactive oxygen/nitrogen species remain to be discussed. First, conditions of increased free radical formation (that do not cause immediate cell death) stimulate cells to raise the activities of antioxidant enzymes and the synthesis of low-molecular-weight antioxidants. This has been well-described in multiple tissues and species under many different conditions of oxidative stress. However, the signaling mechanisms involved in these antioxidant responses were a mystery until the late 1980s, when Bruce Demple's team from Harvard University started to reveal some secrets of these mechanisms in bacteria [see paper by Greenberg and Demple (*EMBO J* **7**:2611–2617, 1988) for the effects of H_2D_2]. They observed that H_2O_2 (or O_2^- and $^{\bullet}NO$, in later publications) can regulate the expression of genes that code enzymes directly or indirectly linked to bacterial survival under an ROS-enriched environment. Even though a wealth of information has been amassed on the mechanisms of redox regulation in bacteria (see Section 12.6.1), much has still to be learned about the components, regulation, and function of comparable systems in higher organisms.

Second, if ROS (and RNS, such as $^{\bullet}NO$) are able to regulate the production of antioxidant defenses, they may also be involved in the regulation of many other cellular processes, directly or indirectly linked with the redox state of cells. Currently, it is known that over 100 genes are activated in response to oxidative stress in mammalian cells. For example, changes in the cellular redox state modulate transcriptional activation of metallothionein and collagen genes, posttranscriptional control of ferritin mRNA, and activation of transcription factors including Myb and Egr-1. Several proto-oncogenes are transcriptionally activated by increased cellular oxidation. Exposure of normal or transformed cells to UV radiation or H_2O_2 stimulates the expression of *jun*-B, *jun*-D, c-*fos*, and *fos*-B. Results obtained over the last 10 years in many laboratories have proven that ROS and RNS can be considered relevant second messengers in signal transduction in many molecular cascades. However, redox-regulation of vertebrate cells, leading to cell response and adaptation to numerous stimuli, including oxidative stress, is a very complex and intricate process and its full comprehension is still in its infancy. One interesting example addressed in Section 12.6.3 is the involvement of ROS in the process of oxygen sensing, that is, in the ability of cells/organisms to respond to changes in oxygen tensions.

12.6.1 Bacterial Regulons *sox*RS and *oxy*R

Escherichia coli has independent multigene responses to two kinds of oxidative stress: elevated H_2O_2 triggers the *oxy*R regulon, and excess O_2^- (and $^{\bullet}NO$) triggers the *sox*RS regulon. Singlet oxygen has also been recently implicated in *soxRS* activation in *E. coli*. These regulons coordinate the induction of numerous promoters related to protective responses. Several similar mechanisms exist in other bacterial species, including Mycobacteria.

Gene activation by *sox*RS occurs in two stages: (1) constitutive SoxR protein is activated by a redox signal (see below for the mechanism of SoxR activation), which strongly activates the transcription of *soxS* mRNA; and (2) SoxS protein is synthesized and induces at least 15 genes [see reactions (12.6.1) to (12.6.3)]. The products of these genes include endonuclease IV (initiates repair of oxidatively damaged DNA; see Section 12.5.3), Mn–SOD (see Section 12.2.1.1), G6PDH (to supply NADPH reducing power for antioxidant enzymes; see Section 12.2.1), and superoxide-insensitive fumarase C (a metabolic enzyme). These gene products are all directly or indirectly involved in adaptive responses to O_2^--induced oxidative stress. Other gene products that are triggered by the *sox*RS regulon are related to antibiotic resistance and efflux pumps. Thus, *sox*RS contributes to clinical antibiotic resistance, in part, by counteracting compounds that induce oxidative stress:

$$O_2^- + \text{SoxR–[2Fe–2S]}^{2+}\text{(inactive)} \rightarrow O_2 + \text{SoxR–[2Fe–2S]}^{3+}\text{(active)} \quad (12.6.1)$$

Active SoxR → triggers production of *soxS* mRNA and SoxS protein (12.6.2)

SoxS protein → activates transcription of various genes (Mn–SOD, G6PDH, etc.) (12.6.3)

The SoxR protein of *E. coli* is a homodimer containing two [2Fe–2S] clusters per monomer. One-electron oxidation of [2Fe–2S] clusters by O_2^- or $^{\bullet}NO$ causes the activation of the SoxR protein (by means of an allosteric/torsional event) as a transcription factor. It is interesting to note that O_2^- usually causes damage to Fe–S clusters of enzymes, but not to SoxR. The role of $^{\bullet}NO$ in the activation of bacterial SoxR has a direct relevance for bacterial defense against the destructive products of phagocytes. As pointed out by Bruce Demple, several bacteria have "evolved to sense and respond to a major cytotoxic weapon of the immune system."

In addition to the effects of $^{\bullet}NO$ and O_2^-, excess H_2O_2 causes peroxide resistance in *E. coli*. This resistance is dependent on the *oxy*R gene, which governs a series of genes that constitute the *oxy*R regulon. The OxyR protein activates the synthesis of gene products with clear involvement in antioxidant defense; these include glutathione reductase, glutaredoxin, NADPH-dependent alkyl hydroperoxidase, and catalase (HPI; see Section 12.2.1.2). OxyR also activates the synthesis of the gene products of *fur* (encoding an iron-binding repressor of iron transport) and *dps* (encoding a DNA- and iron-binding protein that protects DNA from oxidative attack). This gene product is an iron-binding repressor of iron transport, which prevents loading of iron when excess of H_2O_2 is present (this is a good strategy against Fenton-mediated $^{\bullet}OH$ formation). Activation of the transcription factor is dependent on the H_2O_2-mediated oxidation of two specific –SH groups per monomer in its tetrameric structure. Following oxidative stress, deactivation of OxyR seems to be mediated by the GSH–glutaredoxin system, reducing OxyR disulfide bonds.

12.6.2 Oxidants and Cellular Redox Signaling

Extracellular signaling molecules such as insulin, growth factors, and cytokines induce changes in cell metabolism via complex mechanisms that involve transmission of the signal from the plasma membrane to the nucleus, where gene expression is altered. The first step of a signaling cascade generally involves the activation of receptors that either have protein kinase activity or activate protein kinases (and/or phospholipases) in the cytoplasm. The signal is eventually transmitted to the nucleus, where it activates the transcription factors that regulate gene expression.

Stimulation of cells by ROS causes signal transduction via the same signaling pathways as those triggered by growth factors or cytokines (in some instances ROS bypass steps of the signaling chains). For example, oxidants activate receptor-type protein tyrosine kinases (RTK), non-receptor-type protein tyrosine kinases (PTKs), and downstream signaling components including mitogen-activated protein (MAP) kinases, protein kinase C (PKC), phospholipase C-γ, and calcium flux in cells.

12.6.2.1 ROS, Phosphorylation, and Signal Transduction As pointed out by Robert Floyd's group, from the Oklahoma Medical Research Foundation, "the common paradigm in all redox-sensitive signal transduction pathways is the presence of the intermediate protein kinases which are activated by phosphorylation of specific regulatory domains." H_2O_2 and other oxidants mediate these activations/phosphorylations (Hensley et al. *Free Radic Biol Med* **28**:1456–1462, 2000).

For example, the binding of insulin to its receptor tyrosine kinase (RTK) increases the receptor's enzymatic activity and results in the phosphorylation of both the receptor (autophosphorylation) and the insulin receptor substrate-1 (IRS-1) (see Chapter 5). It has been observed that H_2O_2 has insulin-mimetic effects, such as the stimulation of metabolism, via the activation of the insulin receptor and the phosphorylation of IRS-1. Moreover, both epidermal growth factor (EGF) and platelet-derived growth factor (PDGF) stimulate cell growth via activation of their respective receptors, and phosphorylation of the EGF and PDGF receptors is also induced by exogenous treatment with H_2O_2, stimulating downstream signaling systems.

It is interesting that H_2O_2 not only activates early components of signaling cascades, but H_2O_2 is also produced endogenously upon the binding of growth factors to their cellular receptors, causing activation of RTKs. The stimulation of formation of H_2O_2 then effects downstream components of signal transduction cascades. The first clear-cut evidence implicating H_2O_2 as an endogenous messenger was obtained in the mid-1990s by Toren Finkel's group, from the National Institutes of Health, Maryland (Sunderesan et al. *Science* **270**:296–299, 1995). They observed that stimulation of rat vascular smooth muscle cells by PDGF transiently increased the intracellular concentration of H_2O_2. The response of vascular smooth muscle cells to PDGF, which includes tyrosine phosphorylation, MAP kinase stimulation, DNA synthesis, and chemotaxis, was inhibited when the growth-factor-stimulated rise in H_2O_2 concentration was blocked (by increasing the intracellular concentration of catalase or by the addition of the chemical antioxidant *N*-acetylcysteine).

Sue G. Rhee and research associates, from the National Institutes of Health, obtained similar results (see Bae et al. *J Biol Chem* **272**:217–221, 1997) working with human epidermoid carcinoma cells stimulated with EGF. They observed that growth-factor-stimulated H_2O_2 production is a key event for tyrosine phosphorylation. Rhee also suggested that inhibition of protein tyrosine phosphatase (PTP) activity by H_2O_2 is required for EGF-induced protein tyrosine phosphorylation to be manifested (see Section 12.6.2.1.1). Indeed, PTPs contain a nucleophilic cysteine in the active site, which can be oxidized by H_2O_2; GSH restores the enzyme to the active form.

The enzymatic source(s) of the endogenous H_2O_2 that is produced during stimulation by growth factors (or cytokines) remained elusive until very recently. Activation of cyclooxygenases or lipoxygenases (see Table 12.2) by growth factors has been suggested as relevant sources of superoxide and H_2O_2 generation. NADPH oxidase (similar to the enzyme found in phagocytes; see Section 13.3) is currently considered the main primary source of ROS involved in signaling in mammalian cells such as smooth muscle cells, chondrocytes, and kidney epithelium.

12.6.2.1.1 Phosphatases and ROS Although oxidants stimulate many PTKs (such as Lck, Fyn, Src, Syk, and Lyn) and RTKs, it seems that oxidants do not act directly on them. Rather, it seems clear that inhibition of PTPs by oxidants is the crucial event in the activation/phosphorylation of the EGF-receptor, Lck, and Fyn. Thus, blockage of phosphatase action (by the use of inhibitors) allows maximal signal output through the protein kinase cascade. Reactivation of PTPs seems to be mediated by thioredoxin plus GSH (see Section 12.2.1.5) and leads to dephosphorylation of intermediate protein kinases and transcription factors (see below), thereby terminating the redox-sensitive signal.

On the other hand, chronic situations of oxidative stress (and ROS generation) may overwhelm the normal capacity of cells to turn-off signal transduction cascades, which may be harmful. Pharmacological agents that maintain phosphatase (PTPs) activity during an oxidative challenge would be expected to antagonize the redox signaling process. Indeed, there is a great deal of evidence showing the antagonizing

Growth factor binding to receptor → RTK autophosphorylation
→ phosphorylation of Grb-2/Sos → SHC–Grb-2–Sos complex formation (activation by H_2O_2)
→ activation of G-protein Ras
→ Raf-1 or Mekk activation (via phosphorylation) (*)
→ MKK/MEK activation (via phosphorylation) (activation by H_2O_2) (**)
→ ERK or JNK activation (via phosphorylation) (***)
→ TCF formation (and phosphorylation by JNK or ERK)
→ TCF–SRF complex formation at the serum response element (SRE) of c-*fos* → c-*fos* gene expression

Note: Looking at the MAP kinase pathway, some stages are activated by H_2O_2 and others phosphorylation: MAPKKK (*) → MAPKK (**) → MAPK (***)

Scheme 12.2 JNK/ERK pathways and ROS.

effects of exogenous antioxidants (or overexpression of antioxidant enzymes) on upstream signal transduction pathways.

12.6.2.1.2 MAP Kinases One of the most studied families of signal transduction pathways is the MAP kinase family (see Chapter 5). Activated RTKs stimulate the MAP kinase cascade, mainly via the activation of Ras, a small guanosine 5′-triphosphate (GTP) binding protein. Four MAP kinase subfamilies have been identified to date: extracellular signal-regulated kinases (ERK), c-Jun N-terminal kinase/stress-activated protein kinases (JNK/SAPK), p38 MAP kinases, and Big MAP kinases. All these pathways contain redox-sensitive sites and several of the MAP kinases are directly activated by H_2O_2.

In the ERK pathway, the activation of growth factor(s) receptors results in tyrosine autophosphorylation and prompts tyrosine phosphorylation of Grb-2/Sos, which binds to an adapter protein the SH-containing (SHC) protein. The resulting SHC–Grb-2–Sos complex activates the G-protein Ras (Ras-GDP is converted to Ras-GTP) (see Text Box 5.1 on page 136). The formation of the SHC–Grb-2–Sos complex is also stimulated by H_2O_2. Activated Ras stimulates the phosphorylation of Raf-1 (Raf-1 is a MAP kinase kinase) by means of PTKs and serine/threonine kinases and PKC. Once activated, Raf-1 phosphorylates MKK/MEK (MEK1, but not MEK2, is stimulated by H_2O_2 treatment), which then activates ERK by means of phosphorylation (see Scheme 12.2 and Fig. 5.5 on page 135, for other Map kinase pathways). Oxidative stress also activates protein phosphatases which can dephosphorylate MAP kinases. This might represent a feedback control of ROS activation of MAP kinase.

Members of the ERK, JNK, and p38 MAP kinase subfamilies phosphorylate the transcription factors Elk-1 and SAP-1, which then associate with other nuclear proteins to form the ternary complex factor (TCF). TCF associates with another transcription factor, the serum response factor SRF, which finally activates the expression of genes such as c-*fos* (Scheme 12.2). The c-*fos* gene is one of the best-studied early response genes, thus the study of its redox induction has been useful in elucidating the mechanisms of gene expression by ROS.

12.6.2.1.3 Protein Kinase C (PKC) and Phospholipase Cγ (PLCγ) The PKC is activated by diacylglycerol (produced from receptor-mediated hydrolysis of inositol phospholipids) or phorbol 12-myristate 13-acetate (PMA). PKC is also regulated by means of phosphorylation, lipid cofactors, and calcium binding. PKC can activate transcription factors nuclear factor κB (NF-κB) and activator protein 1 (AP-1) (see Section 12.6.2.2), which are subject to redox-regulation.

Treatment of cells with H_2O_2 and redox-cycling quinones leads to stimulation of PKC activity. Moreover, Japanese scientists recently observed that low levels of H_2O_2 induce tyrosine phosphorylation of several PKC isoforms (independent of receptor-coupled hydrolysis of inositol phospholipids), prompting enhanced activity of the phosphorylated PKC. Menadione, which generates O_2^-, also induced tyrosine phosphorylation as well as activation of PKC isoforms.

Furthermore, PKCs contain unique structural features that are susceptible to oxidative modification. The N-terminal regulatory domain contains zinc-binding, cysteine-rich motifs that are readily oxidized by H_2O_2. When oxidized, the autoinhibitory function of the regulatory domain is compromised and, consequently, cellular PKC remains active. The C-terminal catalytic domain contains several reactive cysteines that are targets for various chemopreventive antioxidants such as selenocompounds, polyphenolic agents, and vitamin E analogs. Modification of these cysteines decreases cellular PKC activity. Thus, PKC may be subjected to a complex redox-regulation, which affects its role in tumor promotion and the control of cell growth. This makes PKC an interesting target for pharmacological manipulation.

Intracellular calcium homeostasis is also regulated by the redox state of cells. Hydrogen peroxide activates PLCγ phosphorylation, turning on the production of inositol 1,4,5-trisphosphate (Ins-1,4,5-P_3). Moreover, the Ins-1,4,5-P_3 receptor/calcium channel in the endoplasmic reticulum is activated (causing a rise in cytoplasmic calcium) not only by the binding of Ins-1,4,5-P_3 but also by the direct action of oxidants on the Ins-1,4,5-P_3 receptor. Recent evidence indicates that the redox state of cells directly modulates the activities of Ca^{2+}-ATPases and the $Ca^{2+}-Na^+$ exchanger.

12.6.2.2 Transcription Factors and ROS When upstream signaling pathways finally reach the nucleus, they stimulate changes in gene expression mediated by the activation of several transcription factors, such as NF-κB, AP-1, the tumor suppressor p53, and the pro-oncogene product Myb (see Chapter 5). The action of several transcription factors is redox sensitive. For example, the activity of AP-1 (which is formed by the oncogene products Jun and Fos) is inhibited by oxidation of conserved cysteine residues located in its DNA-binding site domain. Reducing agents (such as the protein Ref-1) restore the DNA-binding activity of AP-1. Moreover, cysteine residues in a reduced state are also essential for the DNA-binding activity of NF-κB. Thioredoxin and Ref-1 can sustain the transcriptional activity of NF-κB (see Scheme 12.3). For more information about redox regulation of transcription factors, see Chapter 6.

The earliest evidence that showed direct effect of oxidants on a protein involved in signal transduction appeared in the early 1990s in studies on NF-κB. NF-κB is a multiprotein complex that activates the transcription of a variety of genes involved in the early defense reactions of higher organisms. In nonstimulated cells, NF-κB resides in the cytoplasm in an inactive complex with the inhibitor IκB. Pathogenic stimuli cause the release of the IκB subunit (in the case of IκBα, it is hyperphosphorylated and then degraded by the 26S proteasome). This allows NF-κB to enter the nucleus, bind to DNA control elements and induce the synthesis of specific mRNAs (see Chapter 6). Since the mid-1980s it has been known that activation of NF-κB is triggered by a variety of agents including the cytokines interleukin-1 and tumor necrosis factor (TNF), viruses, endotoxins, phorbol esters, UV light, and ionizing radiation. In 1991 German scientists reported that low concentrations of H_2O_2 also activate NF-κB by causing the phosphorylation and release of IκB; this is prevented by various antioxidants (Schreck et al. *EMBO J* **10**:2247–2258, 1991).

Activation of NF-κB is currently considered a relevant marker of oxidative stress in several pathological situations, including HIV-1 (human immunodeficiency virus) infection, atherosclerosis, reperfusion stress, and Alzheimer's disease. NF-κB is also a factor involved in the transcriptional activation of genes encoding γ-GluCys synthetase (see Section 12.2.1.5) and Mn–SOD.

12.6.3 Oxygen Sensing and ROS

Scientists have been curious for decades about how major metabolic pathways are regulated at the gene level by changes in oxygen tension, including responses to oxygen limitation by hypoxia-sensitive species and adaptive responses to hypoxia/anoxia by anoxia-tolerant species. It has been postulated from experiments with rat carotid body preparations that a heme-containing oxidase (similar to the NADPH oxidase of phagocytes) might work as an oxygen-sensing protein. This oxidase could produce O_2^- (that is dismutated to H_2O_2) in proportion to different oxygen tensions. H_2O_2 production is PO_2-dependent, being highest under normoxia and lowest under hypoxia.

The erythropoietin (EPO) gene is one of the well-known genes whose expression is enhanced under hypoxia (see Chapter 6). It has been demonstrated in Hep G2 cells that EPO expression is under the control of PO_2-dependent H_2O_2 production. In hypoxia, when H_2O_2 production decreases, full expression of the EPO gene occurs. The role of H_2O_2 as the signaling molecule in the oxygen response has been also substantiated in studies with other genes that are induced by hypoxia including aldolase A, phosphoenolpyruvate carboxykinase (PEPCK), glucokinase (see Text Box 12.11), and tyrosine hydroxylase.

Exogenous sources of ROS and/or ligand-stimulated ROS formation

⇓

Activation of RTKs or PTKs (e.g., by ROS-mediated inhibition of PTPs)

⇓

Activation (mediated or not by ROS) of downstream pathways
(MAP kinases, PKC, PLCγ, calcium signaling, etc.)

⇓

Activation of transcription factors (NF-κB, AP-1, p53, etc.)

⇓

Expression of stress-response gene products

Scheme 12.3 ROS and signal transduction.

TEXT BOX 12.11 CARBOHYDRATE METABOLISM, OXYGEN SENSING, AND ROS

In liver, the periportal and perivenous hepatocytes are subjected to different oxygen tensions (70 and 35 mmHg, respectively), causing a hepatic *metabolic zonation.* Hepatocytes from the periportal zone have more gluconeogenic capacity than perivenous hepatocytes, which are better equipped for glycolysis. Hydrogen peroxide formation is higher in the periportal zone (maximum production of 1.2×10^{-9} M) than the perivenous zone (0.8×10^{-9} M).

When analyzing cultured rat hepatocytes, Helmut Acker and co-workers (from Germany; see *News Physiol Sci* **15**:202–208, 2000) recently observed that glucagon-dependent PEP carboxykinase (PEPCK) mRNA production is reduced under low oxygen tensions. However, addition of H_2O_2 (50 μM) to the cell cultures under low oxygen tensions restored maximal PEPCK mRNA production. On the other hand, insulin-mediated production of glucokinase mRNA transcripts is stimulated under low oxygen tensions but inhibited by H_2O_2. These results indicate that ROS (produced under different oxygen tensions) have the ability to regulate both gluconeogenic and glycolytic pathways, at least in rat liver.

The authors also provided evidence that H_2O_2 is only an intermediate in Fenton-mediated $^{\bullet}OH$ radical formation and that $^{\bullet}OH$ is the chemical species more closely linked with the regulation of PEPCK and glucokinase genes. Possibly, $^{\bullet}OH$ may oxidize –SH groups in certain candidate transcriptional factors, thus shifting the balance between reduced and oxidized states.

The main transcription factor involved in the oxygen-sensing mechanisms is the hypoxia inducible factor 1 (HIF-1) (see Chapter 6), whose activity seems to be redox-regulated. HIF-1 is a heterodimer protein complex that activates transcription through binding to specific hypoxia-responsive sequences present in genes that are activated by hypoxia, including those for glycolytic enzymes, growth factors, and vasoactive peptides. Hypoxia induces HIF-1 complex formation by stabilizing the HIF-1α subunit, which under normoxic conditions is degraded by an ubiquitin–proteasome system. The possibility of either ROS-induced HIF-1 activation (under hypoxia) or ROS-induced HIF-1 deactivation (under normoxia) is still a matter of intense debate (see Michiels et al. *Free Radic Biol Med* **33**:1231–1242, 2002).

SELECTED REFERENCES

Abuja, P. M., and Albertini, R. (2001). Methods for monitoring oxidative stress, lipid peroxidation and oxidation resistance of lipoproteins. *Clin Chim Acta* **306**:1–17. *This article presents an update of assays for the determination of lipid peroxidation and their biomedical application.*

Ahmad, S. (ed.) (1995). *Oxidative Stress and Antioxidant Defenses in Biology.* Chapman & Hall, New York. *This book provides a wealth of information about antioxidant enzymes and oxidative stress in many microorganisms, plants, invertebrates, and vertebrates.*

Allen, R. G., and Tresini, M. (2000). Oxidative stress and gene regulation. *Free Radic Biol Med* **28**:463–499. *This review provides a huge tabular summary of the many redox effects on gene expression and signaling pathways that were available through the end of 1999.*

Brigelius-Flohé, R. (1999). Tissue-specific functions of individual glutathione peroxidases. *Free Radic Biol Med* **27**:951–965. *Interesting review on the history of glutathione peroxidases and their role in health and disease.*

Cadenas, E., and Packer, L. (eds.) (1996). *Handbook of Antioxidants.* Marcel Dekker, New York. *This book gives a comprehensive overview of the chemical, biological, and clinical aspects of antioxidant molecules. The most relevant antioxidant systems in biology are reviewed individually, including metabolism, molecular action, and physiological relevance, as well as health issues, genetics, and potential therapeutic applications.*

de Zwart, L. L., Meerman, J. H. N., Commandeur, J. N. M., and Vermeulen, N. E. P. (1999). Biomarkers of free radical damage. Applications in experimental animals and in humans. *Free Radic Biol Med* **26**:202–226. *This article reviews the usefulness of current techniques in lipid peroxidation and DNA damage for the study of oxidative stress in animal models and in humans.*

Fenton, H. J. H. (1984). The oxidation of tartaric acid in presence of iron. *J Chem Soc Proc* **10**:157–158. *This is the first full description of the oxidation of tartaric acid by hydrogen peroxidase and* Fe^{2+}.

Halliwell, B., and Gutteridge, J. M. C. (1999). *Free Radicals in Biology and Medicine*, (3rd ed.). Clarendon Press, Oxford. *The students from my laboratory call this 900-page book the "bible." It is indeed the most complete source of information on free radicals, antioxidants, and oxidative stress ever published.*

Janero, D. R., Hreniuk, D., and Sharif, H. M. (1991). Hydrogen peroxide-induced oxidative stress to the mammalian heart-muscle cell (cardiomyocyte): Lethal peroxidative membrane injury. *J Cell Physiol* **149**:347–364. *If you are going to work with the effect of oxidants in cells, the reading of this classic article is a must. A brief description of the article's contents is covered in Text Box 12.5.*

Josephy, P. D. (1997). *Molecular Toxicology*, Oxford University Press, New York. *An excellent book on xenobiotic biotransformation linked with mutagenesis and oxygen toxicology, while also providing a historical perspective.*

Koppenol, W. H. (1993). The centennial of the Fenton reaction. *Free Radic Biol Med* **15**:645–651. *This is a brief biography of the life and Cambridge career of Professor Henry J.H. Fenton, with special emphasis on the Fenton reaction. A list of Fenton's 80 publications (from 1876 to 1920) is also provided.*

Sen, C. K., Sies, H., and Baeurle, P. A. (eds.) (2000). *Antioxidant and Redox Regulation of Genes*. Academic, San Diego. *This book examines how oxidants and antioxidants may regulate, at the biochemical and molecular level, a variety of physiological and pathological processes.*

Stadtman, E. R., and Levine, R. L. (2000). Protein oxidation. *Ann NY Acad Sci* **899**:191–208. *This is a brief and updated discussion of free-radical-mediated protein oxidation and the role of protein oxidation in animal aging and pathological processes.*

Starke, P. E., and Farber, J. L. (1985). Endogenous defenses against the cytotoxicity of hydrogen peroxide in cultured rat hepatocytes. *J Biol Chem* **260**:86–92. *This classic study shows beautifully how antioxidant enzymes work against oxidative stress in cells. The authors also make good use of enzyme inhibitors. This is one of the readings I recommend for students interested in oxidative stress.*

Wallace, S. S. (2002). Biological consequences of free radical-damaged DNA bases. *Free Radic Biol Med* **33**:1–14. *Interesting and up-to-date review on DNA oxidation methods of quantification and implication towards mutagenesis.*

13

OXIDATIVE STRESS AND MEDICAL SCIENCES

MARCELO HERMES-LIMA

13.1 OXIDATIVE STRESS IN BIOLOGY AND DISEASE

Helmut Sies, from Dusseldorf, Germany, originally proposed the concept of oxidative stress in the mid-1980s. He defined oxidative stress as "an imbalance between oxidants and antioxidants in favor of the oxidants, potentially leading to damage" (Sies, 2000, p. 102). Under normal circumstances there is an equilibrium between free radical generation (and/or their concentration) and their removal by antioxidants (including the ability of cells to repair oxidative damage). Many pathological conditions and diseases are associated with either an increase in the generation of free radicals or a decrease in antioxidant capacity; sometimes with both conditions. The balance between oxidants and antioxidants lies at the border between health and many human diseases and pathological conditions, including neuronal diseases, inflammation, atherosclerosis, reperfusion injury (these four conditions/diseases are addressed below), iron overload (see Text Box 12.1), carcinogenesis (see Section 12.5), porphyria (see Text Box 13.1) and diabetes (see comments in Chapter 12).

The concept of oxidative stress is not only applicable to humans, laboratory animals, or mammalian cells. A huge number of studies have focused on invertebrate and lower vertebrate free radical toxicology, as well as on plants, fungi, and bacteria. Increased free radical formation prompts an adaptive cell response by mobilizing or augmenting the biosynthesis of endogenous antioxidants. However, under certain circumstances, the antioxidant apparatus is overwhelmed by excess free radical formation. This leads to an oxidative stress condition.

Moreover, oxidative stress is not only the result of a pathologic state. There are several conditions that cause physiological oxidative stress, that is, situations where oxidative damage is part of a natural life cycle of an organism (in some cases, life cannot go on without stress!). Examples are muscle exercise, fertilization, bacterial killing by phagocytes (see Section 13.3), natural aging (addressed in Chapter 12), and arousal from hibernation or dormancy in small rodents and a variety of other animals. Furthermore, as addressed in Chapter 12, reactive oxygen species (ROS) and reactive nitrogen species (RNS) are also of key relevance in signal transduction pathways, in both normal and pathological conditions.

13.2 FREE RADICALS AND NEURONAL DISORDERS

Many studies have correlated oxidative stress with a number of neuronal disorders, including brain damage in Wilson's disease, Friedreich's ataxia, amyotrophic lateral sclerosis (ALS), Parkinson's disease, Alzheimer's disease, and brain ischemia (see Section 13.6.4 for general discussions on ischemia). In all these cases there is evidence for increased levels of biomarkers of free radical damage, provoked by oxygen and nitrogen reactive species. However, each case differs as to how free radicals are generated and how/where they affect the normal cell metabolic homeostasis. Mitochondrial dysfunction may also be linked to neurodegenerative diseases through a variety of different pathways, including increased free-radical generation, impaired calcium buffering, release of apoptotic mitochondrial factors (e.g., cytochrome *C* and apoptosis inducing

Functional Metabolism: Regulation and Adaptation, edited by Kenneth B. Storey.
ISBN 0-471-41090-X

TEXT BOX 13.1 ABOUT PORPHYRINS, MADNESS, WEREWOLVES, AND FREE RADICALS[†]

Acute intermittent porphyria (AIP) and porphyria erythropoietica are two inborn disorders of human metabolism where reactive oxygen species are implicated: oxygen radicals in the former disease and singlet oxygen in the latter. Both diseases prevail in Northern Europeans and result from deficient biosynthesis of enzymes of the heme biosynthetic pathway, porphobilinogen deaminase and protoporphyrin-IX ferrochelatase, respectively. Consequently, 5-aminolevulinic acid (5-ALA) plus porphobilinogen (PBG), two heme precursors, accumulate in tissues of AIP patients and are excreted in their urine. The PBG present in urine gives it a wine color, hence the term "porphyria" (derived from *porphurus*, Greek for *purple*). A high amount of protoporphyrin IX (PP-IX) is deposited in the skin of porphyria erythropoietica patients.

Beginning in the late 1980s, results from Bechara's research group have shown that 5-ALA undergoes iron-catalyzed oxidation by molecular oxygen to yield 4,5-dioxovaleric acid (DOVA), ammonium ions, and H_2O_2. This reaction is propagated by O_2^-, and $^\bullet OH$ is co-produced via the iron-catalyzed Haber–Weiss reaction. The first evidence connecting 5-ALA with cellular oxidative stress was the observation that ROS generated from 5-ALA aerobic oxidation caused injury to isolated liver mitochondria [see Hermes-Lima et al. (1991). *Biochim Biophys Acta* **1056**:57–63).

$$\underset{\textbf{5-ALA}}{H_2C(NH_2)CO(CH_2)_2CO_2^-} + O_2 \rightarrow \underset{\textbf{DOVA}}{HCOCO(CH_2)_2CO_2^-} + NH_4^+ + H_2O_2$$

5-ALA oxidation also induces *in vitro* and *in vivo* oxidative damage to red muscle and liver mitochondria, to liver DNA, and to γ-butyric acid (GABA) receptors of cortical synaptic membranes. 5-ALA may also be intimately involved in iron metabolism since, *in vitro*, it releases iron from ferritin and activates the iron regulatory protein (IRP-1). Also, in liver biopsy samples of AIP patients, the elevated 5-ALA is correlated with iron deposits in hepatocytes. These effects of 5-ALA may produce some of the typical symptoms of AIP syndrome: muscular weakness, primary liver cancer, and neurological dysfunctions (behavior alterations, hallucinations, intense abdominal pain). Interestingly, two famous bloody wars took place in the reign of symptomatic AIP carriers—the American War of Independence during the reign of King George III, the Mad King (1738–1820) of Britain, and World War I, led by the last German Emperor Wilhelm II (1859–1941).

In individuals with porphyria erythropoietica, dermal accumulation of PP-IX causes dramatic cutaneous reddening, inflammation, and blistering in areas exposed to the sun, as well as facial hirsutism. The biochemistry behind these manifestations is attributed to PP-IX-photosensitized formation of singlet oxygen ($^1O_2$), a highly oxidant form of molecular oxygen (see Section 12.1). Porphyrins as well as chlorophyll, rose Bengal, methylene blue, and chlorpromazine act as photoreceptors in this process. Photon absorption by these "dyes" excites them to the singlet (fluorescent) state, which is followed by intersystem crossing to the triplet (phosphorescent) state and energy transfer to colliding oxygen molecules. The acceptor, ground state triplet oxygen, is then promoted to the very electrophilic singlet state. Biological damage driven by the triad of oxygen, dye, and light is called *photodynamic action*. This same principle can be harnessed in methods to kill viruses, fungi, and tumor cells (photodynamic therapy) or to design insecticides and herbicides.

[†]Etelvino J.H. Bechara, University of São Paulo, Brazil.

PP-IX (ground state singlet) + $h\nu \rightarrow$ PP-IX^{1*} (excited singlet) $\rightarrow$ PP-IX^{3*} (excited triplet)

PP-IX^{3*} + O_2 (ground state triplet) $\rightarrow$ PP-IX + $^1O_2$ (excited singlet)

$^1O_2$ + biomolecule targets $\rightarrow$ chemical and biological damage

People carrying porphyria erythropoietica develop nocturnal habits to avoid sunlight, and a reddish and hairy semblance are thought to be responsible for the creation, in Medieval times, of werewolf and vampire myths. Ironically, in folklore, vampires can be frightened by garlic, a vegetable that is exceptionally rich in thiols, which can offer protection from peroxidation reactions. Perhaps, the immortal vampires don't want to be "cured"!

factor, AIF), and the induction of mitochondrial permeability transition (opening of a calcium-dependent "megapore" recognized years ago as a major player in mitochondrial damage). This can lead to both apoptotic and necrotic cell death.

It is relevant to point out that several discrepancies appear when comparing specific results on neuronal oxidative stress from different laboratories. This may be due to factors such as differences in the length of postmortem intervals before removal of brain samples, or different techniques for sample processing and analysis.

13.2.1 Oxidative Stress in Wilson's Disease, Friedreich's Ataxia, and ALS

13.2.1.1 Wilson's Disease An autosomal-recessive disorder of copper metabolism Wilson's disease results from the absence or dysfunction of a copper-transporting P-type adenosinetriphosphatase (ATPase) that leads to impaired biliary copper excretion and disturbed synthesis of holoceruloplasmin (ceruloplasmin is a blood protein involved in copper storage and detoxification). In addition, copper overload occurs in several organs, including liver and brain. Oxidative stress caused by increased copper levels arises due to Fenton chemistry (see Section 12.1.2) and leads to dysfunction of mitochondrial energetics and cell damage. Plasma copper concentration is also increased in Wilson's disease, affecting the levels of plasma antioxidants. Copper chelation therapy by means of *d*-penicillamine or triethylenetetramine administration is the current form of clinical management of Wilson's disease [other chelators have been proposed, including pyridoxal isonicotinoyl hydrazone (PIH); see Text Box 12.1] and antioxidant-based therapy has been proposed as a complementary form of treatment.

13.2.1.2 Friedreich's Ataxia The most common form of autosomal recessive spinocerebellar ataxia, with an incidence of 1 : 50,000 in the European population, Friedreich's ataxia is often associated with a cardiomyopathy. Friedreich's ataxia derives from a defect in frataxin, a protein involved in regulating mitochondrial iron metabolism. This leads to problems in Fe–S cluster formation in aconitase (a Krebs cycle enzyme) and mitochondrial complex I, as well as faulty cellular bioenergetics, mitochondrial iron overload, and increased free-radical-mediated damage. A recent study has shown significant improvement in muscular and cardiac levels of adenosine 5′-triphosphate (ATP) in patients treated with both coenzyme Q and vitamin E for 6 months.

13.2.1.3 Amyotrophic Lateral Sclerosis (ALS) Neurodegeneration in ALS is characterized by the specific loss of central and peripheral motor neurons. There are several different mutations associated with familial ALS, but about 20% of the cases have a defect in the gene encoding CuZn–SOD, located on chromosome 21. Several mutated forms of CuZn–SOD have been described and, in these cases, neuronal degeneration correlates with loss of CuZn–SOD activity. Increased carbonyl protein, nitrotyrosine and 8-OH-dGua, biomarkers of oxyradical and $ONOO^-$ attack (see Sections 12.4 and 12.5), have been observed in ALS spinal cords. Moreover, there are mitochondrial abnormalities in liver and skeletal muscle biopsies, as well as a decrease in cytochrome oxidase activity in motor neurons, that could be associated with oxidative damage.

The partial loss of CuZn–SOD activity in ALS was formerly thought to be due to increased neuronal oxidative stress caused by a general decrease of antioxidant capacity. This view began to be challenged when CuZn–SOD-deficient mice were found to suffer no neuronal abnormalities, at least when they are young (see Text Box 12.3). However, there is new evidence that several mutant CuZn–SOD isoforms in ALS induce free radical generation due to a peroxidase-like activity (as well as reduced "normal" SOD activity). Thus, this could partly explain the origin of increased free radical formation in ALS. Indeed, recent experiments with mice expressing a

CuZn–SOD mutant of ALS show motor neuron degeneration and mitochondrial damage.

13.2.2 Parkinson's Disease (PD) and Oxidative Stress

Parkinson's disease is a chronic, progressive disorder of the central nervous system that belongs to a group of conditions called motor system disorders. PD is the direct result of the loss of cells in the substantia nigra, an area of the brain (the name of this region is due to the presence of neuromelanin, a black pigment). Usually, loss of 50 to 75% of the substantia nigra's dopaminergic cells triggers PD symptoms.

Parkinson's disease has been known since ancient times. An English doctor, James Parkinson, first described it extensively in 1817. The thoroughness of his analysis is such that researchers and clinicians are still urged to read his original notes on the condition. At least one million people in the United States are estimated to have PD. Many of them, perhaps half, are thought to be undiagnosed (for more information visit www.michaeljfox.org).

13.2.2.1 ***Involvement of Free Radicals in PD Etiology*** Much evidence links PD with oxidative stress. Damage to the brain's substantia nigra has been correlated with iron accumulation, increased lipid peroxidation and deoxyribonucleic acid (DNA) oxidation (as determined by 8-OH-dGua), and depletion of GSH. Increased carbonyl protein and advanced glycation end products (AGE, see Section 12.4) were also observed recently in the substantia nigra of PD patients. Elevated iron levels may cause increased free radical formation, which possibly accounts for the decrease in antioxidant capacity and increase in oxidative damage to neuronal cell components. However, the increase in iron concentration in the substantia nigra has been connected recently to the advanced stages of the disease, suggesting that this phenomenon may be a secondary, rather than a primary initiating event. In any case, the loss of iron homeostasis in PD and in other neurodegenerative disorders is a relevant event for neuronal dysfunction. The autoxidation of iron–neuromelanin complexes is proposed to be an important source of ROS in advancing PD.

Other evidence has indicated that L-DOPA (3,4-dihydroxyphenylalanine; a precursor of dopamine) therapy, which is beneficial at the onset of PD, may ultimately be a further cause of oxidative stress and worsening of PD. This might happen because dopamine is a well-known H_2O_2 generator via monoamino oxidase B [MAO-B; see reaction (13.1)]. In the presence of transition metals, H_2O_2 will lead to $^{\bullet}OH$ formation [reactions (13.2)]. L-DOPA and dopamine also react with $O_2{}^-$ and peroxyl radicals, in the presence of iron, and initiate chains of events leading to oxidative stress. It is possible that the abnormal intracellular iron distribution in substantia nigra cells, when they are still active in dopamine production, could set up a condition for increased free radical formation and oxidative stress. Later events would be cell dysfunction and death, and an overall decrease in dopamine formation and secretion by substantia nigra.

$$\text{Dopamine} + O_2 + H_2O \rightarrow \text{L-DOPA} + H_2O_2 + NH_3 \qquad (13.1)$$

$$H_2O_2 + \text{iron} \rightarrow {}^{\bullet}OH \rightarrow \text{cell damage} \qquad (13.2)$$

If the oxidative stress hypothesis proves to be right, it is possible that antioxidant-based therapies and the use of effective iron chelators could be of relevance as alternative forms of PD management. Indeed, a different treatment uses deprenyl (a specific inhibitor of MAO-B), which inhibits MAO-B-mediated H_2O_2 formation without effects on dopamine biosynthesis.

13.2.2.2 ***Toxic Agents, Oxidative Stress, and PD*** Another hypothesis that links PD with oxidative stress is related to free radical generation by xenobiotic drug metabolism, which could promote (chronically or acutely) oxidative damage to substantia nigra's neurons. One example is 1-methyl-4-phenyl-1,2,3,6-tetrahydropyridine (MPTP), a heroin substitute of the 1970s, which causes severe substantia nigra destruction and PD symptoms. Moreover, MPP^+, a MPTP metabolite, causes inhibition of mitochondrial ATP synthesis, loss of mitochondrial calcium homeostasis, and increased formation of O_2^- radical (and consequently $ONOO^-$ generation, via reaction with nitric oxide; see Section 12.1.3). Iron and nitric oxide are also involved in the toxic effects of MPTP. Interestingly, many neuroscientists use MPTP to produce animal models of PD.

Thus, several environmental toxins and pesticides could be involved in neuronal damage, which could also target substantia nigra, and produce PD symptoms. Proving these links is a complex process because they not only involve free radical/neuronal biochemistry but also epidemiological surveys correlating PD and toxin exposure. As pointed out by Halliwell and Gutteridge (1999), a currently studied candidate is *n*-hexane, widely used in modern society, which is metabolized to aldehydic neurotoxins (remember that many aldehydes can promote alterations in DNA and proteins; see Section 12.3.2.2 in Chapter 12). Rotenone, a pesticide and a classic inhibitor of mitochondrial electron transport, was recently shown to provoke PD symptoms in rats.

13.2.3. Alzheimer's Disease: The Free Radical Connection

Alzheimer's disease (AD) is a neurodegenerative disorder characterized by severe loss of memory and cognitive functions that affects approximately 4 million persons in the United States alone, with a prevalence of 3 and 18.7% in persons 65 to 74 and 75 to 84 years old, respectively. Between one-third and one-half of the population over 85 years is affected by AD. With an aging society, it is expected that 9 million people will be affected by AD in the United States by the year 2040.

The German doctor Alois Alzheimer made the first clinical and pathological description of an AD case in 1907 when examining a 51-year-old woman suffering from dementia. Upon autopsy, Alzheimer observed massive cerebral cortical neuron loss, the presence of silver-positive neurofibrillary tangles, and alterations subsequently known as senile plaques. All these observations are typical of postmortem analysis of AD brains.

13.2.3.1 Involvement of Oxidative Stress in AD Etiology The search for the molecular etiology of AD has been intense and has included a search for genetic defects, altered amyloid precursor protein (APP) processing (inducing accumulation of amyloid β precipitates; see Text Box 13.2), glutamate neurotoxicity, trace metal toxicity,

TEXT BOX 13.2 ALZHEIMER'S DISEASE, Aβ, AND OXIDATIVE STRESS

The most common explanation for the onset of AD is the precipitation of amyloid beta peptides ($A\beta$), leading to neurotoxic fibril formation. Several types of extracellular $A\beta$ are formed from the processing of the amyloid precursor protein (APP). Numerous experimental efforts have been made to understand and inhibit fibril formation. Several researchers have also connected $A\beta$ deposits with the oxidative stress hypothesis of AD. In the mid-1990s D. Allan Butterfield, from the University of Kentucky, proposed an interesting explanation that $A\beta$ itself, independently of its precipitation, may be directly responsible for the free-radical formation [reaction (TB13.1)] and consequent damage to neuronal membrane systems, leading to lipid peroxidation and oxidation of membrane proteins. The presence of increased iron levels may further contribute to the propagation of the peroxidation process, causing release of toxic aldehydes and lipid hydroperoxides to the cytoplasm.

Oxidative damage to the neuronal plasma membrane may lead to a loss of ionic homeostasis (increase in calcium influx and inhibition of Ca^{2+} – ATPase-mediated calcium efflux) and an increase in intracellular calcium. This event causes calcium-dependent activation of nitric oxide synthase (bringing about $ONOO^-$ formation upon reaction of nitric oxide with O_2^-) and uncontrolled activation of calcium-dependent proteases (such as calpain). The loss of calcium homeostasis also contributes to mitochondrial dysfunction, which may increase the normal rates of mitochondrial O_2^- production (leading eventually to $^\bullet OH$ and/or $ONOO^-$ formation). This chain of events could explain the process of neuronal cell dysfunction and death in AD brain [reaction (TB13.2)]. Butterfield's proposal is based on investigations of the oxidizing effects of $A\beta$ peptides in membrane preparations, isolated enzymes, and cell cultures. It was also observed that certain $A\beta$ peptides generate ESR-detectable ROS in simple *in vitro* incubations [see Varadarajan et al. (2000), *J Struct Biol* **130**:184–208].

$$A\beta + O_2 \rightarrow\rightarrow \text{oxidized-}A\beta + \text{ROS} \tag{TB13.1}$$

(Possibly a process mediated by transition metals)

$$\begin{aligned}&\text{Plasma membrane} + \text{ROS}\\ &\quad\rightarrow \text{damage to membrane protein and lipids}\\ &\quad\rightarrow \text{loss of } Ca^{2+} \text{ homeostasis}\\ &\quad\rightarrow \text{mitochondrial dysfunction} \rightarrow \text{necrosis}\end{aligned} \tag{TB13.2}$$

The key question to ask of this oxidative stress hypothesis is: Why is only a certain percentage of aged persons subjected to $A\beta$-induced toxicity? This leads to other questions: Are there $A\beta$ isoforms (mutant or not) that are more prone to free radical generation? Are there genetic and/or environmental factors contributing to differential neuronal cell vulnerability to $A\beta$ toxicity? Is it possible that if we live long enough, the natural processes connected with aging (including increased free radical brain damage) would make us all AD patients?

George Perry and Mark A. Smith, from Case Western Reserve University, Ohio, proposed a totally different view for the role of the senile plaques (composed mostly of amyloid-β deposition) in AD. They observed, for example, that the percentage of amyloid burden (i.e., the area of $A\beta$ deposition, obtained from autopsy of 22 AD cases) is inversely proportional to the neuronal levels of RNA oxidative damage, measured as cytoplasmic 8-OH-Gua. The authors proposed that $A\beta$ deposition may be a compensatory response against oxidative stress. In other words, $A\beta$ deposition could be regarded as a protective response of the brain. This is also supported by the fact that many aged and even middle-aged individuals (with intact cognition) often show extensive Aβ deposits. According to Perry and co-workers, when talking about the current pharmacological approach to AD, "we are playing a dangerous game in focusing efforts on the removal of amyloid-β, which could quite likely have the opposite effect to that promised: of a return to cognition." [Perry et al. (2000). *Lancet* **355**(9205):757, 2000].

In the radical approach to AD by Perry and Smith, aging could be at the center of a system of causes leading to a dysfunctional and damaged brain. Amyloid would be just one of these causes (or even a consequence!). Others include: (a) increased sensitivity to oxidative stress that may be the result of alterations in membrane constituencies, reductions in the synthesis of endogenous antioxidants, and increases in membrane lipid peroxidation and redox active iron; (b) increases in inflammation (a process that also produces ROS and RNS) possibly mitigating the effects of anti-inflammatory agents; and (c) regional alterations in lipid membrane composition, which could have consequences in the membrane fluidity and function of ion pumps and channels (also causing loss of calcium homeostasis). Thus, aging could influence one or more of the causes of AD, predisposing people for AD as they age. For a more complete view of these radical ideas see Joseph et al. [*Neurobiol Aging* **22**:131–146 (2001)] and Perry et al. (*Comp Biochem Physiol C* **133**:507–513 (2002)].

deficit of mitochondrial ATP synthesis, free-radical-induced oxidative stress, and neuronal cell death. The evidence that links AD with free-radical-mediated cell damage derives from increased levels of biomarkers of oxidative stress [thiobarbituric acid reactive substances (TBARS), 4-hydroxynonenal, isoprostanes, carbonyl protein, AGE, nitrotyrosine, 8-OH-dGua, and DNA strand breaks; see Sections 12.3 to 12.5] in brain areas that are typically affected by AD, as compared with age-matched controls. Both mitochrondrial DNA (mtDNA) and nuclear DNA are affected in AD. Decreased levels of several polyunsaturated fatty acids (PUFAs), such as arachidonic and docosahexaenoic acids (which are highly susceptible to lipid peroxidation), are also found in AD brain.

Moreover, glutathione *S*-transferase (GST) activity is reduced in AD brains, suggesting that the capacity to deal with aldehydes and lipid hydroperoxides (via GST-Px activity; see Section 12.2.1.4) is compromised. Mitochondrial energy metabolism is also affected in AD, with reduction in the activity of various components of the respiratory chain, such as cytochrome oxidase (COX activity is decreased by 16 to 38% in the cerebral cortex of AD subjects).

Increased levels of iron were observed in different regions of AD brain (amygdala, hippocampus, and olfactory pathway), in comparison with age-matched controls. The regions where iron was elevated are the ones that show severe degenerative changes in AD. Elevated levels of iron could increase the potential for free radical generation and neuron damage. Increased aluminum ions have also been found in AD brain, even though there is not a consensus among neuroscientists and epidemiologists as to their relevance. Aluminum ions can accelerate lipid peroxidation reactions and thus could be an additional player in the oxidative stress process.

13.2.3.2 Formation of Free Radicals in AD brain What would cause increased free radical formation in AD? The increase in iron concentration could be just part of the story. A great number of studies have focused on the role of amyloid β peptide (Aβ) in AD etiology. It is a soluble component of the plasma and cerebrospinal fluid, derived from APP processing. It has also been recognized that Aβ is the major protein constituent of the senile plaques, present in virtually all AD cases. Genetic studies of early-onset familial AD (FAD; which is a rare disease) offer some evidence for the central role of Aβ in the pathogenesis of AD. Several FAD mutations have been found in the APP gene, which is encoded by chromosome 21. Individuals with Down syndrome also have Aβ deposits, possibly due to overexpression of APP (these persons have three copies of chromosome 21).

There is considerable amount of evidence linking free radical production by Aβ (either soluble or in the fibril precipitated form) and the development of oxidative stress and neuronal damage in AD (see Text Box 13.2). On the other hand, acute inflammatory response (a common phenomena in AD brain), which could be caused by fibrillar Aβ, prompts respiratory bursts of activated phagocytes (see Section 13.3) and generation of nitric oxide and O_2^{-}. This leads to $^{\bullet}OH$ radical formation via Fenton chemistry, as well as $ONOO^{-}$ formation. The generation of ROS and RNS may trigger oxidative damage to neuronal membranes, loss of calcium homeostasis, mitochondrial degeneration, and cell dysfunction.

It is likely that AD is associated with multiple etiologies and pathogenic mechanisms (see Text Box 13.2). However, there is solid evidence showing that oxidative stress is an important part of the mechanism of neurodegeneration in AD brain. If so, the use of brain-accessible antioxidants and nontoxic iron chelators could be useful as therapeutic strategies. Indeed clinical trials with vitamin E administration in AD patients resulted in some attenuation of disease development.

13.3 INFLAMMATION AND OXIDATIVE STRESS: AN OUTLOOK

It has long been known that phagocytes are key components of the immune defense system. They are used to kill potentially pathogenic microorganisms such as bacteria and fungi using free radicals as weapons. However, under certain conditions, phagocytes may produce too many free radicals and "unintentionally" cause oxidative stress and damage to nontarget cells. Many disorders have been connected with free radical production by phagocytes.

13.3.1 "Manufacture" of Free Radicals by Phagocytes

Formation of free radicals by activated phagocytes is associated with a sharp and transitory increase in oxygen uptake well known as a "respiratory burst." The respiratory burst was first observed in the 1930s and was associated with increased mitochondrial respiration. Two decades later, it was shown that mitochondria were not directly involved with the respiratory burst because the phenomenon occurred in the presence of cyanide. In 1961, it was shown that activated phagocytes produced H_2O_2 during the respiratory burst, which could be involved in bacterial killing. Subsequently, in the early 1970s (after the discovery of SOD) Bernard Babior and co-workers, from La Jolla, California, demonstrated that activated phagocytes produce mainly O_2^-; hydrogen peroxide formation was just the result of O_2^- dismutation.

Babior also demonstrated that O_2^- formation during the respiratory burst was catalyzed by the FAD-containing enzyme reduced nicotinamide adenine dinucleotide phosphate (NADPH) oxidase [see reaction (13.3)]. Under resting conditions in phagocytes the enzyme is in a dissociated form that includes both membrane-bound subunits (such as the flavo-cytochrome b_{558}), which are present in vesicles, and cytosolic subunits including p47 and p67. During phagocyte activation, p47 is phosphorylated and this triggers the association of the membrane-bound subunits with the cytosolic ones. The next step is the fusion of the vesicle containing the assembled enzyme with the membrane of the phagocytic vesicles that have engulfed bacterial pathogens, followed by O_2^- production. Patients with chronic granulomatosus disease have phagocytes with dysfunctional NADPH-oxidase and cannot produce O_2^-. These patients are highly susceptible to infections caused by bacteria and fungi.

Formation of O_2^- and H_2O_2 in the phagocytic vesicles triggers $^{\bullet}OH$ production due to iron-catalyzed Haber–Weiss reactions, and this is part of the weaponry to kill the engulfed bacteria [reaction (13.4)]. Other oxidants are also key players in the bactericide action of phagocytes. Nitric oxide is formed by the induction of nitric oxide synthase (NOS) when phagocytes are activated. The reaction of O_2^- with nitric oxide leads to $ONOO^-$ formation, whereas decomposition of $ONOO^-$ yields $^{\bullet}OH$ and nitrogen dioxide [$^{\bullet}NO_2$, also very reactive; reaction (13.5)]. Finally, the heme-containing enzyme myeloperoxidase, which is secreted into phagocytic vesicles, catalyzes the reaction of H_2O_2 with chlorine ions to produce highly reactive hypochlorous acid (HOCl) or the unprotonated form OCl^- [see reaction (13.6)]. Br^- and I^- are also substrates of myeloperoxidase but of much less physiological significance. Moreover, nonenzymatic reactions of HOCl with H_2O_2 or with ammonia yield, respectively, singlet oxygen [$^1O_2$; see reaction (13.7)] and chloramine (NH_2Cl), which is an oxidizing agent more toxic than HOCl:

$$2O_2 + NADPH \rightarrow NADPH^+ + H^+ + 2O_2^- \quad \text{(catalyzed by NADPH-oxidase)} \qquad (13.3)$$

$$2O_2^- + 2H^+ \rightarrow H_2O_2 \,\text{(plus iron)} \rightarrow {}^{\bullet}OH \qquad (13.4)$$

$$^{\bullet}NO + O_2^- \rightarrow ONOO^- \,(\text{plus } H^+) \rightarrow ONOOH \rightarrow {}^{\bullet}OH + {}^{\bullet}NO_2 \qquad (13.5)$$

$$H_2O_2 + Cl^- \rightarrow H_2O + OCl^- \quad \text{(catalyzed by myeloperoxidase)} \qquad (13.6)$$

$$HOCl + H_2O_2 \rightarrow H_2O + Cl^- + O_2^{\,1} + H^+ \qquad (13.7)$$

In summary, an array of oxidizing agents are produced by activated phagocytes, including O_2^-, H_2O_2, $^{\bullet}OH$, $^{\bullet}NO$, $ONOO^-$, $^{\bullet}NO_2$, HOCl (or OCl^-), $^1O_2$, and NH_2Cl. These are the oxidizing agents that should kill bacteria and pathogenic fungi.

13.3.2 Oxidative Stress Induced by Phagocytes

When too many oxidants are produced by phagocytes during inflammatory events (not necessarily related to microbial infection), they can instead have damaging effects. For example, chronic inflammatory processes are linked with an increased risk of carcinogenesis. Several laboratories found that the products of activated neutrophils can induce mutations in the DNA of nearby cells; such mutations can give rise to carcinogenic processes. Generation of several oxidants by inflammatory cells has been linked to oxidation of low-density lipoproteins (LDL) particles and arterial damage in atherosclerosis (see Section 13.4). Oxidative stress in postischemic processes, for example, in heart, has also been linked with increased free radical formation by neutrophils (see Section 13.5). Moreover, the pathophysiology of rheumatoid arthritis (RA), acute respiratory distress syndrome (ARDS), emphysema, and cystic fibrosis is also connected to phagocyte-induced free radical formation. In all these cases, antioxidant-based therapy could be a relevant alternative experimental procedure.

*13.3.2.1 **Rheumatoid Arthritis (RA)*** This disease is characterized by chronic inflammation of the synovial tissue, affecting up to 3% of the population in most countries. The increased presence of activated macrophages and neutrophils in the synovial fluid suggests that oxidants formed by those cells could partly account for damage to the joint tissues. Indeed, neutrophils from the synovial fluids of RA patients have augmented *ex vivo* O_2^- production. Moreover, levels of carbonyl protein, nitrotyrosine, and products of lipid peroxidation are increased in the synovial fluid of RA patients, as well as levels of urinary 8-OH-dGua and breath pentane (indicators of systemic increases in DNA oxidation and lipid peroxidation, respectively; see Chapter 12). Furthermore, "free iron" levels are augmented in the synovial fluids of RA patients, creating an environment for increased $^{\bullet}OH$ formation.

*13.3.2.2 **Acute Respiratory Distress Syndrome (ARDS)*** A clinical condition characterized by leakage of fluid into the alveoli, mostly caused by shock, ARDS results in damage to the pulmonary endothelium. There is evidence to suggest that oxidants produced by activated neutrophils could be partly responsible for the damage. Indeed, in patients with ARDS, the lungs are full of neutrophils. Phenomenological data support the role of oxidative stress in ARDS, including increased plasma levels of 4-hydroxynonenal and carbonyl proteins and decreased levels of GSH (and increased GSSG; as discussed in Chapter 12, this is evidence for peroxide-mediated oxidative stress) in the alveolar lung fluid.

*13.3.2.3 **Emphysema*** In emphysema there is excessive release of elastase (and other proteases) by active neutrophils and inactivation of α1-antiproteinase, a protein that inhibits protease activity. The greatly increased activity of alveolar phagocytes in smokers could cause oxidative inactivation of α1-antiproteinase, which could make the tissue more susceptible to proteolytic damage. Indeed, alveolar fluids from smokers have higher elastase activity and lower α1-antiproteinase activity when compared to controls. A critical methionine residue of α1-antiproteinase is a relevant target for oxidation. This is in agreement with the observation of high levels of methionine sulfoxide (see Section 12.5) in alveolar fluids from smokers.

*13.3.2.4 **Cystic Fibrosis*** An an inherited disease, affecting 1 per 2500 Caucasians, cystic fibrosis is caused by a defect in a single protein involved in chlorine transport across epithelial surfaces. The defect has consequences that affect many organs but is seen most dramatically in lung, where it results in severe accumulation of mucus and chronic lung infection. Production of free radicals by activated phagocytes seems to participate in lung tissue destruction. Indeed, neutrophil counts are increased in the lung. Moreover, patients with cystic fibrosis have (i) an augmented plasma concentration of TBARS and dityrosine, (ii) increased carbonyl protein, nitrotyrosine, and free iron in the sputum, (iii) decreased levels of antioxidants in the plasma (ascorbate and β-carotene) and sputum (GSH), and (iv) increased breath pentane levels and augmented urinary excretion of 8-OH-dGua. Furthermore, similar to emphysema, α1-antiproteinase activity is decreased in the lung of cystic fibrosis patients.

13.3.3. More on Microbial Killing: A Radical New Hypothesis

Anthony W. Segal and coworkers, from the University College London, published in March 2002 a new hypothesis for the mechanism of bacterial killing by neutrophils (see Reeves et al. *Nature* **416**: 291–297, 2002) that may change totally our view about this matter. They observed that mice deficient in neutrophil-granule proteases but normal in respect of O_2^- formation are unable to resist staphylococcal and candidal infections. They proposed that O_2^- production in the phagocytic vacuole (leading to other ROS) of activated neutrophils is not the main factor for bacterial killing. The highly increased anionic charge in the vacuoles (caused by massive O_2^- formation) is compensated for by a pH-dependent surge of K^+ ions that cross the vacuole membrane. In their words "the consequent rise in ionic strength engenders the release of cationic granule proteins including elastase and cathepsin G." Under these conditions the proteases are activated, causing bacterial destruction.

13.4 ATHEROSCLEROSIS AND FREE RADICALS: TO BE OR NOT BE?

Cardiovascular diseases (CVD) are the major source of morbidity and mortality in the Western word, claiming nearly 1 million lives per year in the United States alone of an estimated 60 million Americans that have CVD. The principal manifestations of CVD are heart attack and stroke, which represent the clinical end result of a systemic vascular process known as atherosclerosis. This is usually considered a disease of aging. However, excess LDL cholesterol, diabetes, hypertension, and cigarette smoking can provoke premature atherosclerosis.

Thrombosis and lipids have been known to be involved in atherosclerosis since the nineteenth century, based on autopsy studies. This condition is manifested in three stages known as early, developing, and mature lesions. Early lesions are characterized by the presence of nodular areas of lipid deposition, the "fatty streaks," composed of cholesterol-filled cells, as well as smooth muscle cells.

13.4.1 Free Radicals, LDL Oxidation, and Atherogenesis

The free radical hypothesis for atherosclerosis begins with the oxidation of LDL particles, forming oxidized LDL particles (ox-LDL). Some evidence indicates that binding of LDL particles to proteoglycans of the arterial wall precedes LDL modification. The accumulation of ox-LDL particles stimulates the production of proinflammatory cytokines from local arterial cells and leukocytes. Various cytokines and growth factors can activate mononuclear phagocytes that accumulate within the nascent atheroma to produce further cytokines and growth factors that amplify the process. Other protein factors, induced by ox-LDL particles, augment the expression of scavenger receptors on the surface of macrophages, which promote ox-LDL endocytosis (this is actually considered a defense mechanism!) and the further transformation of lipid-filled macrophages to foam cells, the hallmark of the early atheroma. Moreover, primary formation of O_2^- and $^{\bullet}NO$ by inflammatory cells (see Section 13.3) may prompt $^{\bullet}OH$ and $ONOO^-$ generation, which induces a "second round" of oxidative stress within the early lesion phase. Foam cells eventually die (possibly by apoptosis), and extracellular lipids may accumulate in the lesions, which are at this point forming fibrous dome-shaped plaques (the *developing lesions*). Later events include calcification and physical disruption of the plaque leading to thrombosis (the *mature lesions*).

The big questions for the hypothesis that oxidative stress participates in atherosclerosis are: Is oxidation of LDL particles *in vivo* indeed a relevant process of early lesions? Which oxidizing agents are involved in LDL modifications? Can antioxidants both inhibit LDL oxidation and minimize atherogenesis?

13.4.2 What Is to Blame for Primary LDL Oxidation?

There is very little doubt that ox-LDL accumulates in human atheromas. However, the precise mechanisms that trigger LDL oxidation are still a matter of intense debate. Aldehyde conjugation in lysine residues has been immunologically detected in lipoproteins (mostly in ApoB, which makes about 95% of the apoprotein content in LDL) in early lesions. These modifications in ApoB make LDL particles recognizable by scavenger receptors in macrophages. As previously discussed (see Section 12.3), aldehydes can be formed as end products of peroxidation of phospholipids, which are present at the LDL surface (the molar ratio PUFA : LDL is about 1300, so there is plenty of oxidizable substrate in LDLs), and react with ApoB. Moreover, biomarkers of lipid peroxidation and protein oxidation, such as isoprostanes, MDA, carbonyl protein, and di-tyrosine, are increased in early lesions.

13.4.2.1 Metal-Catalyzed LDL Oxidation It has been long proposed that oxidation of LDL could be initiated *in vivo* by metal-catalyzed processes. Copper ions are the most active metals in inducing LDL peroxidation *in vitro*, and the reaction is inhibited by chain-breaking antioxidants such as vitamin E, BHT, and probucol (a cholesterol-lowering drug). However, plasma copper is highly regulated and plasma "free copper" concentrations may not be high enough to prompt LDL peroxidation or ApoB oxidation. "Catalytic" iron and copper have been detected in advanced human atheromas, but this may not account for LDL oxidation in early lesions. Moreover, O_2^- can also be formed from NADPH oxidase activity of phagocytes (see Section 13.3) and induce metal-catalyzed $^{\bullet}OH$ formation and LDL oxidation. The problem with this hypothesis is that the initial oxidation of LDL starts before accumulation of macrophages and other phagocytes. Thus, LDL oxidation might be initiated by oxidizing agents formed from other sources; vascular/endothelial cells seem to be good candidates.

In agreement with the putative role of transition metals in the oxidation of LDL components, the research group of Jukka T. Salonen recently showed a positive correlation between serum ferritin concentration (as a marker of body iron) and plasma levels of cholesterol oxidation products (oxysterols) in 669 Finnish men (see Tuomainen et al. *Free Radic Biol Med* **35**: 922–928, 2003).

13.4.2.2 Nitric Oxide and LDL Oxidation There is some new evidence showing NADPH oxidase activity and O_2^- formation in vascular cells. Thus, the formation of too much O_2^- may prompt $ONOO^-$ formation [$ONOO^-$ degradation also yields $^{\bullet}OH$ and $^{\bullet}NO_2$ radicals; reaction (13.5)] due to the reaction of O_2^- with endothelium-generated nitric oxide ($^{\bullet}NO$). This could promote LDL peroxidation and oxidation/nitration of tyrosine residues in ApoB. Indeed, nitrotyrosine is accumulated in early lesions.

On the other hand, endothelium-generated $^{\bullet}NO$ is inhibited by atherogenesis and by lipid peroxidation, which causes decreased $^{\bullet}NO$-mediated control of arterial relaxation and vascular homeostasis. This can be an escalating step toward arterial dysfunction. Probucol, vitamin E, and high levels of ascorbate (as well as L-arginine, a precursor of $^{\bullet}NO$ formation) seem to increase $^{\bullet}NO$ bioactivity in vascular endothelium in animal models of atherosclerosis. The use of competitive inhibitors of nitric oxide synthase (such as L-NAME) in rodents can also induce atheroscletotic lesions.

13.4.2.3 Effect of Other Agents on LDL Oxidation Hypochlorous acid may also participate in LDL modification. Phagocyte myeloperoxidase produces the highly oxidizing agent HOCl (see Section 13.3.1), which can attack many protein targets including ApoB. There is evidence for increased 3-chlorotyrosine, a marker of HOCl attack in protein, in early lesions. Another type of LDL modification that may be atherogenic and proinflammatory is glycation, followed by formation of advanced glycation end products (AGE). This may be especially relevant in diabetic patients.

Oxidation of cholesterol molecules during LDL peroxidation yields several products, including cholesterol hydroperoxides (Ch-OOH) and epoxy-cholestanols, which are toxic to arterial cells. However, the actual *in vivo* role of this process in atherogenesis is still uncertain.

13.4.3 Vitamin E and Atherosclerosis

Several epidemiological studies were designed to verify the role of vitamin E in CVD, but they produced conflicting results. Recently, the double-blind placebo-controlled trials HOPE (with 9541 patients at high risk for myocardial infarction) and GISSI (with 11,324 patients with myocardial infarction) showed that intake of vitamin E supplements for several years had no positive effects in men and women in relation to CVD. On the other hand, the well-known CHAOS study determined that vitamin E supplements reduced the incidence of nonfatal heart attacks in 2000 British patients with established coronary artery disease by 77%. Prospective cohort studies published in 1993, the Nurses' Health Study (with 87,000 female subjects) and the Health Professionals' Follow-up Study (39,000 male subjects) verified an inverse association between coronary artery disease and intake of vitamin E. (Keaney Jr. 200; listed in the Selected References).

There is no clear explanation for the discrepancies among different studies. It is possible, as proposed in 2000 by John Keaney, Jr., from Boston University School of Medicine, that lipid peroxidation plays a modest role in atherogenesis in humans. This radical proposal would explain discrepancies in the effects of vitamin E (dosages of 0.1 to 5 g/kg diet) in many animal studies, where it can inhibit LDL oxidation without affecting the outcome of atherogenesis. On the other hand, probucol usually inhibits both LDL peroxidation and the development of early lesions in animal models.

Another explanation for the disagreements in the epidemiological studies is that in well-nourished populations most individuals are already receiving the maximum benefit of antioxidants through a healthy diet. So, only individuals at "heightened oxidative stress" would benefit from "extra" vitamin E intake. However, this explanation does not account for the positive effects of vitamin E intake detected in the 1993 cohort studies.

13.5 ROLE OF FREE RADICALS IN ISCHEMIA AND REPERFUSION

Ischemia is defined as an arrest of blood flow to a tissue or organs. Of major importance is the disruption of oxygen delivery, but during ischemia substrates are no longer delivered by the blood and waste products cannot be removed. When the interruption of blood flow is prolonged, it can have devastating effects on metabolism and organ functioning. Well-known examples of ischemic damage are myocardial infarction and stunning, stroke, and organ removal for transplantation (see Chapter 12). Certain organs are more susceptible to oxygen-rich blood deprivation than others. For example, a few minutes of ischemia causes irreversible brain damage in mammals, whereas skeletal muscle can recover after much longer episodes.

In the case of heart, coronary occlusion resulting from atherosclerotic plaques or vasospasm results in severe reduction in myocardial blood flow, leading to myocardial cell injury or necrosis. About 1.5 million Americans develop myocardial infarction per year, leading to nearly 200,000 deaths. The severity of the myocardial damage is proportional to the period under ischemia, bringing about diminished and/or fatal cardiac function. Treatment of acute myocardial ischemia, with the aim of restoring blood flow, involves the use of antithrombolytic agents or coronary balloon angioplasty. Although the reestablishment of perfusion is a necessary intervention, and reduces infarct size and overall mortality, it causes a chain of events that produces additional myocardial cell dysfunction, a phenomenon termed *reperfusion injury*. Depending on the length of the ischemic period, reperfusion can be even more damaging than ischemia itself. Interestingly, however, reperfusion injuries were not seen in laboratory animals if deoxygenated perfusate was used (even though such perfusion does not restore organ function!). This "oxygen paradox" remained a mystery to medicine until the free radical ideas appeared in the 1970s.

In the case of brain, stroke is a major cause of serious long-term disability, with about 600,000 Americans suffering a new or recurrent stroke per year. The annual direct and indirect cost of stroke in the United States is beyond $40 billion. Ischemic brain injury is caused not only by a regional incomplete ischemia (ischemic stroke), but also by the consequences of a transient cardiac arrest, leading to global brain ischemia. The high ATP turnover demanded by neuronal tissues and normally supplied by the aerobic oxidation of glucose leads to a very rapid energy depletion when both oxygen and glucose delivery by the blood is disrupted. This is associated with membrane depolarization (loss of Na^+ and K^+ intracellular/extracellular gradients needed for propagation of action potentials) and disruption of calcium homeostasis (loss of calcium gradients), leading to increased cytoplasmic calcium and calcium-activated phospholipases. The restoration of oxygen-rich blood flow may restore aerobic bioenergetics but also brings about neuronal reperfusion injury.

13.5.1 Postischemic Free Radical Generation

It is now recognized that reperfusion injury is associated with the overgeneration of ROS (and RNS as well) result-

TEXT BOX 13.3 MITOCHONDRIAL ROS FORMATION IN PHYSIOLOGICAL AND PATHOLOGICAL CONDITIONS*

Among the many physiological sources of ROS (and RNS), such as eicosanoid metabolism, nitric oxide production, cytochrome P-450 systems, and phagocytosis processes, the mitochondrial respiratory chain is considered to be the most important. In this organelle, it has been estimated that 1 to 4% of the oxygen consumed is deviated into ROS production. However, very recent determinations indicate that under approximate physiological conditions this number may be only about 0.1%.

Early experiments by Alberto Boveris (from Argentina) and Britton Chance (from the United States) suggested that complex I from isolated mitochondria generates H_2O_2 during reoxidation of the complex's flavin mononucleotide (see *Biochem J* **134**:707–716, 1973). In addition, they observed that H_2O_2 production increased with increasing oxygen tension. Superoxide is actually the primary species formed by mitochondria, and it is then dismutated to H_2O_2 either spontaneously or via a SOD-catalyzed reaction. A second proposed site for O_2^- production during mitochondrial respiration is the ubiquinone–cytochrome *bc*1 segment of complex III, which transfers electrons from ubiquinol to cytochrome *c*. Electron transfer within the ubiquinone–cytochrome *bc*1 complex involves a ubisemiquinone radical intermediate [see reaction (12.2.24)], which has been shown to be reductant for O_2^- generation in complex III [see Turrens et al. (1985). *Arch Biochem Biophys* **237**:408–415]. It has also been proposed that other flavin-containing enzymes and iron–sulfur centers in the respiratory chain are sites of ROS production. Superoxide production from mitochondria is regulated by the metabolic state, pO_2, and ADP availability and the rate of ATP formation. Nitric oxide also modulates mitochondrial O_2^- formation by regulating electron flow in the respiratory chain.

The phenomenon of ischemia reperfusion in mammalian organs (see Section 13.5) is associated with a considerable rise in postischemic ROS production. This can be explained by the relatively high oxygen tension during reoxygenation interacting with a fully reduced respiratory chain, leading to increased O_2^- and H_2O_2 formation. Moreover, there is also evidence that iron, and perhaps copper, is released as a free metal during the ischemic phase. Thus, increased postischemic H_2O_2 formation, in the presence of augmented free iron levels, leads to Fenton-mediated $^{\bullet}OH$ formation and oxidative damage to mitochondria and other cell constituents. Oxidative damage to mitochondrial proteins (such as those of the electron chain, F_0F_1-ATPase, aconitase, and the adenine nucleotide translocase), leading to defective ATP formation, has also been linked to neurodegenerative disorders such as Parkinson's and Huntington's diseases.

For many years, mitochondrial ROS formation was considered to be a "physiological defect" in the normal process of electron transfer that results in oxygen reduction to water. This was because mitochondrial ROS formation has been found in all species tested including yeast, plants, invertebrates, and vertebrates. Alterations in aerobic metabolic activity has also been correlated with mitochondrial ROS formation; organisms with high metabolic rates usually produce more mitochondrial ROS than organisms with low metabolic rates. A balance between mitochondrial ROS formation, antioxidant capacity, and membrane phospholipid composition (particularly the unsaturation index) has also been linked with aging.

However, since the mid-1990s, mitochondrial ROS production has been considered to be a relevant source of O_2^- and H_2O_2 for physiological redox signaling pathways and also for apoptotic processes. The release of cytochrome *c* from mitochondria is a primary step in apoptosis, and this release may create a more reduced state of the respiratory electron chain, prompting increased mitochondrial O_2^-/H_2O_2 formation, which is one of the many steps toward apoptotic cell death. Elegant experiments from a University of São Paulo research team showed that isolated mitochondria from yeast mutants that lacked cytochrome *c* show higher H_2O_2 production than wild types. Addition of exogenous cytochrome *c* to isolated mitoplasts significantly diminished H_2O_2 formation [see Barros et al. (2003), *Free Radic Biol Med* **35**:179–188]. It is possible that the "antioxidant" action of cytochrome *c* addition is due to a decrease in electron accumulation at earlier steps of the respiratory chain, thus minimizing O_2^-/H_2O_2 formation from those sites.

For more information on mitochondrial ROS formation (and their effects) we recommend the recent reviews by Enrique Cadenas and Kelvin J.A. Davies (*Free Radic Biol Med* **29**:222–230, 2000) and by Gustavo Barja (*Free Radic Biol Med* **33**:1167–1172, 2002), from Spain.

*Ricardo G. Oliveira and Marcelo Hermes-Lima, University of Brasilia.

TEXT BOX 13.4 POSTISCHEMIC FREE RADICAL DETECTION BY ESR

An elegant demonstration of free radical overproduction during reperfusion of ischemic heart was made in the late 1980s by Garlick and co-workers (*Circ Res* **61**: 757–760). By analyzing the reaction of free radicals with the spin-trap PBN (added to the perfusion solutions), the authors showed no increase in the basal levels of ESR signals of PBN adducts (see information on ESR techniques in Section 12.1) during 15 min of ischemia in isolated rat heart. However, ESR signals increased dramatically within 4 min of reperfusion before decreasing again within several minutes. This study clearly demonstrated that a burst of free radical production takes place during early reperfusion. Interestingly, reperfusion using deoxygenated solutions caused no increase in ESR signal. Many other studies have demonstrated that an increase in lipid peroxidation occurs immediately after reperfusion of ischemic organs.

Direct demonstration of free radical production just after reperfusion was first published in 1987 by Jay Zweier and co-workers, from The Johns Hopkins Medical Institutions (Baltimore, Maryland) (*Proc Natl Acad Sci USA* **84**:1404–1407, 1987) from analysis of ESR signals from isolated rabbit hearts that were deep-frozen under three conditions: control, ischemic (10 min), and postischemic (10 s following 10 min ischemia). The ESR analysis was made in samples that were pulverized under liquid nitrogen (at −196°C such a low temperature was used to "freeze" the radical molecules). In samples taken at 10 s following reperfusion, they observed highly increased signals as compared to control and ischemic samples. The ESR

ing in free-radical-mediated damage. Mitochondria are a main cellular site of free radical generation under normal conditions. It is estimated that 1 to 4% of all oxygen consumed by mitochondria is converted into reactive oxygen species. This is caused by "electron leak" at the respiratory chain, primarily leading to O_2^- generation. The greater the reducing state of the respiratory chain, the greater the availability of electrons that can leak from the chain, and the greater the production of O_2^- radicals. This process is also dependent on the oxygen concentration and tension. During ischemic events, the respiratory chain is in a reduced state because there is little or no oxygen that can be converted to H_2O by cytochrome oxidase. During reperfusion, the quick influx of oxygen to ischemic tissues causes overgeneration of ROS (see Text Box 13.3) and overall damage to cell constituents. Mitochondria from several organs/tissues (including brain and heart) seem to be the main site of ROS generation at the onset of reperfusion [see Text Box 13.4 on electron spin resonance (ESR) studies of free radical detection in postischemic heart].

The mitochondrial burst of ROS production during reperfusion can overwhelm existing cellular antioxidant defenses and cause damage to macromolecules including DNA, proteins, and membrane lipids. Moreover, postischemic peroxidation of endoplasmic reticulum causes a further increase in cytoplasmic calcium concentration (as discussed above for the brain, partial loss of calcium homeostasis takes place during ischemia due to disruption of aerobic ATP supply) that can lead to uncontrolled activation of phospholipases and proteases. Calcium activation of nitric oxide synthase (NOS) may also provoke increased formation of $^{\bullet}NO$ and consequently of $ONOO^-$. In mammalian systems undergoing reoxygenation or reperfusion, these free-radical-induced events can lead to severe cell damage, apoptosis and organ failure.

Other important sources of ROS and/or RNS formation in postischemic tissues are the enzyme xanthine oxidase (producing O_2^- radicals, which can be converted to other reactive species) and NADPH oxidase in activated phagocytes (see Section 13.3). There is evidence that allopurinol, an inhibitor of xanthine oxidase, improves functional recovery of the stunned myocardium and reduces infarct size in dogs. Even though this enzyme may be responsible for ROS generation in the canine heart (as well as in mammalian gut and liver, other organs that can be subjected to ischemia and reperfusion), its activity is just too low in human heart to be a relevant player in postischemic oxidative stress. The role of xanthine oxidase in reperfusion injury was originally proposed for intestinal ischemia in the mid-1980s. During ischemia, the breakdown of ATP (due to hypoxic conditions) causes accumulation of hypoxanthine and xanthine, which are substrates of xanthine oxidase. When oxygen is reintroduced to the system, xanthine oxidase may produce higher concentrations of O_2^- and, consequently, of $^{\bullet}OH$ radicals.

There is evidence that activated neutrophils do contribute to ROS/RNS generation and necrosis in postischemic myocardium (but possibly not in myocardial stunning, which is not accompanied by neutrophil accumulation). Moreover, the vascular endothelium can also be a relevant source of ROS/RNS in events of ischemia and reperfusion. As stated by David Lefer and Neil Granger (see Selected References), from Louisiana State University, the vascular endothelium can adopt an inflammatory phenotype that promotes the recruitment and activation of leukocytes in postischemic tissue. This may promote a second and

TEXT BOX 13.5 NATURE'S FACULTATIVE ANAEROBES—HOW DO THEY DEAL WITH REOXYGENATION-INDUCED OXIDATIVE STRESS?*

Although alien to mammalian life, many invertebrates and cold-blooded vertebrates are excellent facultative anaerobes that experience natural situations where oxygen availability to their tissues varies widely. Many can live without oxygen for days, weeks, or even months at a time. As discussed in Chapter 15, this includes various freshwater turtles that hibernate underwater, some fish that experience seasonal oxygen depletion of the waters in which they live (including cold temperate waters or in warm ponds in the Amazon jungle), and many gill-breathing intertidal marine mollusks that must endure oxygen deprivation during each aerial exposure at low tide. Other organisms experience not just anoxia but also ischemia. With each freezing exposure, freeze-tolerant animals endure extended periods of ischemia due to the freezing of blood plasma and all other extracellular body fluids (see Chapter 17). Hundreds of species of insects, many intertidal marine invertebrates, and various amphibians and reptiles living in seasonally cold climates have developed freeze tolerance. These anoxia- and ischemia-tolerant animals express a variety of biochemical adaptations that sustain their survival during oxygen deprivation, one of these being antioxidant protection.

In the early 1990s, at Carleton University, Ottawa, we hypothesized that hypoxia/anoxia-tolerant species should also have effective antioxidant defenses to deal with the oxidative stress that might occur with the reintroduction of oxygen into their tissues. After days, weeks, or months without oxygen, its sudden reintroduction should create a danger of ROS overgeneration and oxidative damage to cells, much like the reperfusion stress experienced by mammalian organs. The big difference was that these animals readily survive the stress of oxygen reperfusion, and, hence, they constitute good model systems for trying to assess the factors that prevent or minimize reperfusion stress.

We observed that in several cases, the activity of antioxidant enzymes (usually catalase, SOD, or selenium-dependent GPx) and levels of GSH were increased in the organs of anoxia- or freeze-tolerant species while under oxygen deprivation or freezing exposure. That was the case in (i) red-sided garter snakes *Thamnophis sirtalis parietalis* and wood frogs *Rana sylvatica* during freezing (at $-2.5°C$ for 5 and 24 h, respectively), (ii) goldfish *Carassius auratus*, garter snakes, leopard frogs *Rana pipiens*, freshwater snails *Biomphalaria tenagophila*, and marine snails *Littorina littorea* under anoxia (for 8, 10, 30, 24, and 144 h, respectively), and (iii) leopard frogs under dehydration stress at 5°C (50% loss of body water after 92 h, a condition that imposes great restriction on blood flow to internal organs). We also observed that red-eared turtles *Trachemys scripta elegans* increase the activity of alkyl hydroperoxidase reductase (an enzyme that decomposes lipid hydroperoxides) and GSH-synthetase in some organs after 20 h of underwater anoxia. Moreover, the activity of tissue antioxidant enzymes (and levels of GSH) was generally higher in the freeze-tolerant wood frog when compared to freeze-intolerant frogs; anoxic-tolerant turtles also showed high constitutive tissue levels of antioxidant enzymes and GSH when compared to nontolerant vertebrates. Similar observations were made in the mid-1990s by Margaret Rice (New York University) when comparing ascorbate concentration in the brain of anoxic-tolerant and nontolerant vertebrates: The anoxic-tolerant species showed higher ascorbate levels. [Rice et al. (1995) *J Neurochem* **64**:1790–1799].

These data on increased antioxidant content during anoxia/freezing exposure [see for example: Hermes-Lima and Storey (1993) *Am J Physiol* **265**: R646–R652; Lushchak et al. (2001) *Am J Physiol* **280**: R100–R107] could be considered an apparent paradox. Why would antioxidant defenses be improved (apparently by enzyme synthesis) in tissues under conditions of low or zero oxygen when ATP availability for protein synthesis is at a premium? Under these situations there would be too little ROS formation to turn on signal transduction mechanisms that normally lead to the biosynthesis of antioxidant enzymes and GSH. We still do not know the answers to this paradox, but we have a reasonable physiological explanation. It appears that anoxia/freeze-tolerant species prepare for the eventual return of oxygen and the oxidative stress that it will cause by up-grading their antioxidant defenses while under low oxygen conditions. This prepares organs with enhanced protection against potential ROS overgeneration that can occur immediately when oxygen-rich blood flow resumes (i.e., during reoxygenation after anoxia, thawing after freezing, or rehydration after dehydration). Indeed, measurement of lipid peroxidation products during these recovery processes generally show very little sign of the production of damage products as a result of oxidative stress. Thus, although there is still much to be learned, it appears that well-developed antioxidant defenses are an integral part of natural anoxia and ischemia tolerance. For a more

*Marcelo Hermes-Lima, Janet M. Storey, and Kenneth B. Storey.

comprehensive view of the role of ROS and antioxidants in anoxic/hypoxic-tolerant species, read Hermes-Lima and Zenteno-Savín (2002) listed in the Selected References.

more prolonged phase of free radical generation (the first phase being the free radical burst right after reoxygenation) and consequently of oxidative damage.

13.5.2 Antioxidants versus Reperfusion Injury

Several studies have shown that addition of antioxidants to perfusion solutions can have beneficial effects against reperfusion injury in heart, kidney, and brain. This is a relevant strategy because endogenous antioxidant capacity can be overwhelmed by oxyradical overgeneration during reoxygenation. Moreover, in some cases, the activities of certain endogenous antioxidant enzymes are actually decreased during ischemia in mammalian organs. On the other hand, there are many invertebrates and cold-blooded vertebrates that are highly tolerant of extended anoxia exposure (Chapter 15) and reoxygenation. Interestingly, these typically display high antioxidant defenses, either constitutively or induced by anoxia exposure (see Text Box 13.5).

Addition of antioxidants after the onset of reperfusion is, however, without much effect. This is because overproduction of free radicals has already occurred. Several antioxidants are capable of reducing postischemic oxidative stress in heart, brain, and kidney; examples are SOD, polyethylene glycol bound SOD (PEG-SOD has longer plasma half-live than free SOD), SC-52608 (a low-molecular-weight SOD mimetic), U74006F (a chain break antioxidant), the oxyradical scavengers dimethylurea and *N*-2-mercaptopropionyl glycine, and the iron chelator deferoxamine. The efficacy of deferoxamine demonstrates the involvement of transition metal ions in free radical production during reperfusion, possibly via Fenton reactions. Delocalization of iron during ischemia in several organs, such as kidney (the cause is still under debate), seems to be of relevance for iron-mediated oxidative stress and cell damage during reperfusion.

Many studies that have administered exogenous antioxidants for myocardial infarction have had disappointing results. Similarly, not all studies of antioxidant effects on brain ischemia/reperfusion have shown neuroprotection. One reason for the variability may be the ability (or not) of the antioxidants to gain access to the sites of free radical production during reperfusion. On the other hand, the use of exogenous antioxidants and allopurinol has been beneficial in improving the efficiency of experimental organ transplants, especially kidney, after cold (0 to 5°C), hypothermic (5 to 30°C), or warm ischemic storage (37°C) (see Chapter 19). Cold or hypothermic storage are known to prolong kidney metabolic viability by delaying ATP depletion and the loss of calcium homeostasis.

SELECTED REFERENCES

Babior, B. M. (2000). Phagocytes and oxidative stress. *Am J. Med* **109**:33–44. *This is an interesting and easy-to-read review for biology undergraduates and medical professionals on the role of free radicals in inflammation, written by the scientist who opened this field in the 1970s.*

Blake, D., and Winyard, P. G. (Eds.) (1995): *Immunopharmacology of Free Radical Species*. Academic, London. *The role of free radicals in several human pathological conditions is discussed in this book.*

Dröge, W. (2002). Free radicals in the physiological control of cell function. *Physiol Rev* **82**: 47–95. Excellent review on the role of free radicals in cell physiology, human pathologies and natural aging.

Halliwell, B., and Gutteridge, J. M. C. (1999). *Free Radicals in Biology and Medicine*, 3rd ed. Clarendon, Oxford. *This incredible 900-page book offers a wealth of information linking free radicals and oxidative stress with a number of human pathologies.*

Hermes-Lima, M., and Zenteno-Savín, T. (2002). Animal response to drastic changes in oxygen availability and physiological oxidative stress. *Comp Biochem Physiol C* **133**:537–556. *This is an almost encyclopedic review on the role of antioxidants as part of the biochemical adaptations of animals to withstand some extremes of the environment.*

Keaney, Jr., J. F. (2000). Atherosclerosis: From lesion formation to plaque activation and endothelial dysfunction. *Mol Aspects Med* **21**:99–166. *This is an excellent review article on the biochemical and epidemiological facts of atherosclerosis.*

Lefer, D. J., and Grander, N. (2000). Oxidative stress and cardiac disease. *Am J Med* **109**:315–323. *This is another easy-to-read review for biology undergraduates and medical professionals on the role of free radicals in heart diseases, with a special focus on ischemia and reperfusion stress.*

Markesbery, W. R. (1997). Oxidative stress hypothesis in Alzheimer's disease. *Free Radic Biol Med* **123**:134–147. *This article offers "all-you-wanted-to-know" about ROS-induced oxidative damage in Alzheimer's disease, updated to December 1996.*

Sies, H. (Ed) (1985). *Oxidative Stress*. Academic, London. *This book is listed for historical reasons. Helmut Sies defines for the first time the concept of oxidative stress (see pages 1 to 8 of his book).*

Sies, H. (1986). The biochemistry of oxidative stress. *Angew Chem Int Ed* **25**: 1058–1071. This is the initial major article dealing with the biochemistry of oxidative stress. In the words of Dr. Sies, it is "suitable for teaching, may be even still now".

Sies, H. (2000): *Encyclopedia of Stress*, Vol. 3. Academic, San Diego, pp. 102–105.

14

BIOCHEMICAL ADAPTATION

KENNETH B. STOREY

> Living creatures press up against all barriers; they fill every possible niche all the world over.... We see life persistent and intrusive—spreading everywhere, insinuating itself, adapting itself, resisting everything, defying everything, surviving everything!
>
> —John Arthur Thomson, 1920

All organisms strive to remain alive, to maintain homeostasis, and to optimize growth and reproductive potential in the face of continual challenges from both their internal and external environments. The concept of *biochemical unity* (Chapter 1) tells us that all organisms have the same basic biochemical components—the same building blocks (macromolecules, fuels, substrates, cofactors), and the same basic metabolic pathways, membrane structure, and transcription and translation machinery. Yet metabolism is called upon to adjust to innumerable stresses, major and minor, on timescales that range from instantaneous to evolutionary. That organisms can do this is obvious from the fact that life on Earth has radiated into every conceivable environment. Living organisms are found in the frigid Antarctic and boiling hot springs, in the ocean depths and at the tops of mountains. Life abounds in anoxic sulfurous mud flats, in hypersaline lakes, and in the driest deserts. All this is possible because of biochemical adaptation—changes to the structure, function, regulation, and integration of biological molecules and metabolic processes. By means of biochemical adaptation, organisms can maintain both *metabolic control* (adjusting the output of a metabolic pathway in response to an external signal) and *metabolic regulation* (maintaining metabolic parameters relatively constant over time, despite fluctuations in external conditions). To review these concepts see Chapter 1.

Biochemical adaptation is necessary for two main reasons: (1) all biological molecules and all biochemical reactions are directly susceptible to perturbation by multiple environmental parameters—for example, temperature, pressure, pH, ionic strength, solute concentrations, water availability, radiation, and attack by free radicals, and (2) all cells and organisms, in order to remain viable, must maintain an adequate level of energy turnover through an adequate supply of the energy currencies of the cell, primarily adenosine triphosphate (ATP), which is used to drive thermodynamically unfavorable reactions, and reduced nicotinamide adenine dinucleotide phosphate (NADPH), which is used for reductive biosynthetic reactions. Environmental stresses can disrupt the physical properties of biological molecules, the activities and regulatory properties of functional proteins (e.g., enzymes, transporter proteins), and the integrated functioning of the biochemical reactions in individual pathways, individuals cells, and even intact organisms. These concepts and many other aspects of biochemical adaptation and physiological evolution are treated more fully by Hochachka and Somero (2002), and this excellent resource is highly recommended to the reader.

The vast majority of organisms on Earth are highly susceptible to multiple changes in their physical environment. For example, most organisms are ectotherms (cold-blooded) and, hence, a change in environmental temperature means a change in body temperature. A change in temperature dictates an automatic change in the rates of virtually all metabolic reactions, as well as conformational changes to the structure of proteins and changes in the fluidity of membranes. Living within our mammalian bodies, humans do not often fully appreciate the environmental stresses that rule the lives of most other organisms on Earth. This is because a large evolutionary effort has been

Functional Metabolism: Regulation and Adaptation, edited by Kenneth B. Storey.
ISBN 0-471-41090-X

put into the creation of a near-constant internal environment for our cells and organs. The temperature of our core organs rarely deviates from $\sim$37°C, oxygen supply is rarely interrupted, our skin has good resistance to water loss, and our bodies rigorously regulate cellular conditions of pH, osmolality, and ionic strength. This relative freedom from environmental constraints comes at a high cost—a high rate of food consumption is needed to support the metabolic rates of mammals that are about four- to sevenfold higher than those of equivalently sized reptiles. However, a constant internal environment and high body temperature provide key advantages for mammals and birds, including an ability to remain active under adverse environmental or seasonal conditions as well as support for advanced brain functions. This latter has brought humans even greater freedom from environmental constraints for it has given us the capacity to substitute man-made technology for metabolic adaptation. Other organisms are not so lucky.

ENVIRONMENTAL STRESSES

We will begin with a brief outline of some of the effects of environmental parameters on biological molecules and metabolic functions. This is a selective list, focusing only on those stresses that are utilized in the examples of biochemical adaptation discussed in this chapter and the next three.

Oxygen

Oxygen availability has two main actions in metabolism—essentially, one good and one bad!

Oxygen the Good Oxygen is a substrate of many biological reactions. Most famously, it is the terminal electron acceptor of the cytochrome *c* oxidase reaction of the mitochondrial respiratory chain, but oxygen is also a substrate in about 200 other cellular reactions. Because of the huge difference in ATP output from oxidative phosphorylation compared with fermentative reactions, oxygen-dependent ATP synthesis is utilized by most multicellular organisms. Many organisms, including humans, are highly sensitive to reduced oxygen levels with variable abilities to endure low oxygen (hypoxia) and little capacity to survive full oxygen depletion (anoxia). Other organisms are good facultative anaerobes that can switch easily between aerobic and anoxic lifestyles, whereas others function as obligate anaerobes (mainly various microbes and some animals such as intestinal parasites). Biochemical adaptations for dealing with oxygen availability will be dealt with in greater detail in Chapter 15.

Oxygen the Bad Reactive oxygen species (ROS) (e.g., superoxide, hydrogen peroxide, hydroxyl radical) are byproducts of the electron transport chain and products of selected enzymatic and nonenzymatic reactions in cells. It has been estimated, for example, that 1 to 4% of all O_2 consumed by vertebrates is "lost" in the production of superoxide radicals, chiefly from a "leaky" electron transport chain. ROS cause damage to many kinds of cellular macromolecules, and all cells must have effective antioxidant defenses to quench their production and repair or catabolize damage products. Changes in oxygen availability as well as other stresses (e.g., iron overload, radiation, xenobiotics) increase ROS production and necessitate adaptive responses by antioxidant defenses. Metabolic adaptations for dealing with ROS are dealt with in Chapter 12.

Temperature

The vast majority of organisms on Earth are ectotherms whose body temperature always closely matches ambient temperature. Temperature change affects biological molecules in multiple ways.

1. Temperature affects the rates of all chemical and biochemical reactions. In most cases, reaction rates double for every 10°C increase in temperature; this is designated as a temperature quotient or $Q_{10} = 2$. However, some metabolic reactions have Q_{10} values <2, whereas others range up to 3 or 4, and so differential effects of temperature change on the rates of cellular reactions can disrupt metabolic homeostasis.

2. Temperature alters the strength of the weak bonds that are critical to the conformation of macromolecules and to the conformational changes that occur during ligand binding and catalysis in enzymes. A decrease in temperature increases the stability of hydrogen bonds, van der Waals interactions, and electrostatic bonds but decreases the stability of hydrophobic bonds, and vice versa for a temperature increase. Because each protein/enzyme differs in the number and type of weak bonds that are important to its conformation, ligand binding, or other properties, a change in temperature affects each protein/enzyme slightly differently with the potential to disrupt the integrated functioning of multicomponent pathways. For example, a temperature increase may increase the K_m for the substrate of one enzyme while decreasing the K_m of another.

3. Both high and low temperatures can denature proteins. Cold denaturation is often reversible, but heat can irreversibly denature proteins at temperatures not too far about the normal biological range. For example, exposure to 55 to 60°C denatures many types of proteins from mesothermic organisms, including humans, but many thermophilic microorganisms live happily near 100°C.

4. Temperature affects the flexibility of fatty acids and phospholipids and changes the fluidity of biological membranes.

5. Below 0°C biological water can freeze, and organisms must take steps to either avoid freezing (e.g., with antifreezes) or endure freezing (e.g., cryopreservation). Biochemical adaptations for dealing with life below 0°C will be dealt with in greater detail in Chapters 16 and 17.

Water and Ions

Water is the universal solvent of all biological reactions and a substrate or product of many, its bulk is a key structural element in the bodies of all organisms, and the physical properties of water (e.g., melting point, surface tension, weak bonding) impact on many biochemical and physiological parameters. The structures and functions of most biological macromolecules are sensitive to changes in the type and concentration of inorganic ions as are selected vital functions, such as the maintenance of membrane potential difference. Stresses caused by variation in water and ion contents can have multiple effects including:

1. Variation in organismal water content necessarily changes ion concentrations and thereby affects biochemical reactions and metabolic functions, all of which have both high and low limits on the tolerable concentrations of Na^+, K^+, Ca^{2+}, and other ions.
2. Cell volume changes due to hyper- or hypo-osmotic stresses can cause physical damage to cells and destroy the integrity of the cell membrane. Shrinkage beyond a critical minimum can cause a breakdown of membrane bilayer structure due to excessive compression whereas excessive volume increase can burst cells.
3. All biological macromolecules are coated by a "shell" of vicinal water that contributes to their conformation and, in cases of extreme desiccation or freezing, the loss of vicinal water may cause denaturation of macromolecules.

MECHANISMS OF METABOLIC REGULATION

Previous chapters have discussed multiple mechanisms of metabolic regulation, and all of these are employed in biochemical adaptation. Table 14.1 lists many of these mechanisms, and the remainder of this chapter illustrates the principles of how these mechanisms can be used in biochemical adaptation with selected examples from studies of animal adaptation to environmental stress. Central to biochemical adaptation is the regulation of enzymes, and most of the discussion below relates to enzymatic adaptation, although the mechanisms are typically applicable to most proteins. In the field of biochemical adaptation, the regulation of central enzymes of intermediary energy metabolism (particularly those of glycolysis) has received by far the most attention because of the critical role that the maintenance of energy supply has in the adaptation to any environmental stress. It is these enzymes that we will also use as our primary examples.

TABLE 14.1 Some Mechanisms of Biochemical Adaptation

Enzyme/Protein Level
Changes in protein or enzyme amount
Changes in isoform/isozyme type and properties
Changes in the concentrations of substrates and effectors of enzymes
Changes in the kinetic and regulatory properties of enzymes
Covalent modification of enzymes and proteins
Protein–protein binding and changes in subcellular location
Influence of low-molecular-weight stabilizers
Transcriptional and Translational Level
Evolution of novel protein or enzyme types
Changes in response elements and transcription factors
Control over protein translation

ENZYME ADAPTATION

Enzymes are the catalysts of cells and most of the "business" in cells is conducted by enzymes. Central to biochemical adaptation is the control of enzyme function, and multiple mechanisms have been designed to tailor enzymes for optimal function in the cells/tissues/organisms in which they reside and to provide an appropriate range of responses to deal with metabolic demands and environmental stresses. Adjustments to enzymes in response to environmental stress can make use of any and all of the mechanisms of metabolic regulation that have been discussed in previous chapters. The discussion here will review many of these mechanisms, summarized in Table 14.1, and highlight instructive examples of biochemical adaptation.

Changes in Enzyme or Protein Amount

The amount of each enzyme or protein in a cell is a primary determinant of the capabilities of different cells, tissues, and species, and changes in enzyme/protein amount provides the coarse control of metabolism that is a key part of cellular response to many external signals (hormones, environmental stress, etc.). On an evolutionary timescale, differences in

TABLE 14.2 Activities of Muscle Enzymes[a]

	Rat Red Fibers	Rat White Fibers	Hummingbird Red Muscle	Tuna White Muscle
Glycogen phosphorylase	21	29	31	106
Hexokinase	1.3	0.7	9.2	0.8
Pyruvate kinase	246	405	672	1295
Lactate dehydrogenase	555	849	230	5492
Citrate synthase	19.5	5.5	343	2

[a]Selected glycolytic enzymes as well as citrate synthase from the tricarboxylic acid cycle are compared in the red and white muscle fibers of the vastus lateralis muscle of rats, the red pectoral muscle of hummingbirds, and the white muscle of tuna. All data are units per gram wet weight.

the amounts of selected enzymes (typically quantified as enzyme maximal activity per gram wet weight) helps to define the metabolic capacities of different cell/tissue types and the differences in these capacities between homologous cells/organs of different species.

Skeletal Muscle: A Study in Evolutionary Design The above concepts can be illustrated using the constitutive activities of selected enzymes of carbohydrate catabolism in skeletal muscles of different species. Recall from Chapter 11 that red muscle (slow twitch, type I) is designed for endurance work and primarily generates ATP from the aerobic catabolism of substrates and mitochondrial oxidative phosphorylation, whereas white muscle (fast twitch, type II) is designed for high-intensity burst work and relies mainly on ATP generated from creatine phosphate hydrolysis and anaerobic glycolysis (ending in lactate accumulation).

Data for the vastus lateralis muscle in rats show typical differences between the red and white fibers of mammalian skeletal muscle (Table 14.2). White fibers clearly have the higher glycolytic capacity with a greater capacity for glycogen degradation via glycogen phosphorylase (GP) and higher activities of the terminal enzymes of glycolysis, pyruvate kinase (PK), and lactate dehydrogenase (LDH) (Fig. 14.1). By contrast, red fibers have a higher activity of hexokinase (HK) and a greater HK : GP ratio (0.062 in red vs. 0.024 in white), which indicates a greater reliance on blood-borne substrates (glucose) as fuel. Red muscle also has a very much higher citrate synthase (CS) activity, which correlates with its greater capacity for ATP generation via mitochondrial oxygen-dependent metabolism.

The comparison between species in Table 14.2 shows the extremes that can be achieved by evolution when optimizing muscle enzyme activities for the very high power outputs needed for hovering flight by hummingbirds or for high-speed burst swimming by predatory tuna. Hummingbird red muscle still shows a well-developed glycolytic capacity, but the extremely high CS activity (which is matched by other mitochondrial enzymes) shows the tremendous capacity of the muscle for oxidative metabolism

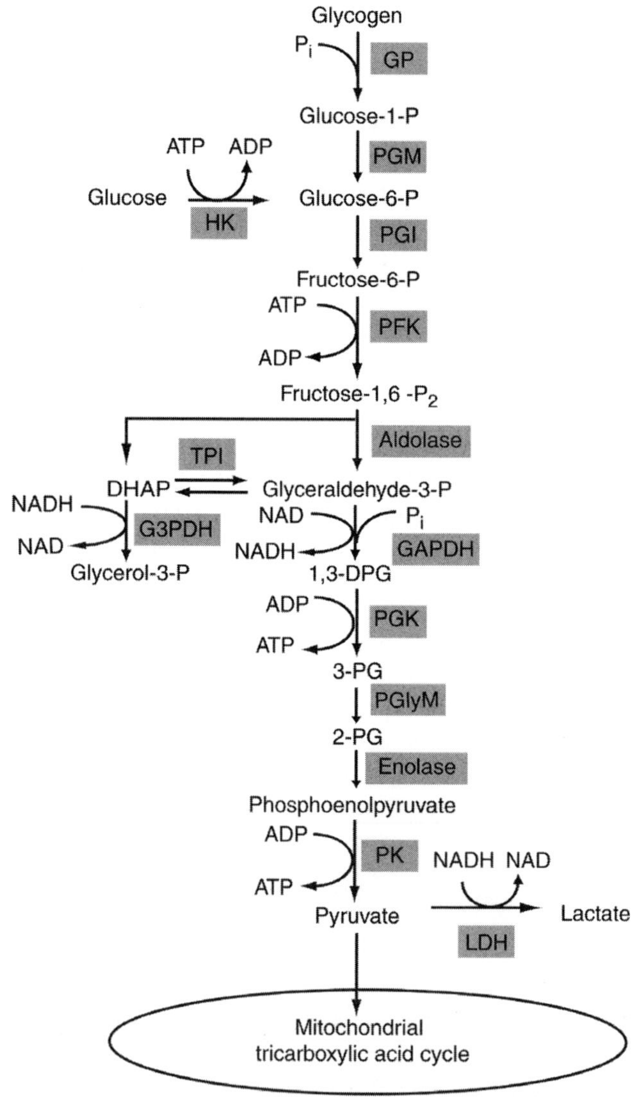

Figure 14.1 The pathway of glycolysis. Enzyme abbreviations are: GP, glycogen phosphorylase; HK, hexokinase; PGM, phosphoglucomutase; PGI, phosphoglucoisomerase; PFK, 6-phosphofructo-1-kinase; GAPDH, glyceraldehyde-3-phosphate dehydrogenase; TPI, triosephosphate isomerase; PGK, phosphoglycerate kinase; PGlyM, phosphoglyceromutase; PK, pyruvate kinase; LDH, lactate dehydrogenase.

and correlates with a mitochondrial density in hummingbird muscle that is the highest found in any vertebrate muscle. By contrast, the swimming muscles of tuna contain the highest LDH activities ever measured and provide the muscle with the ability to generate enormous bursts of power output fueled by anaerobic glycolysis. The extremely low HK : GP ratio (0.0075) also indicates a reliance on huge reserves of glycogen in tuna muscle for substrate support of anaerobic glycolysis.

Stress-Induced Changes in Enzyme Amount Changes in the amount of an enzyme in a cell redefines the maximum capacity of its catalyzed reaction and, when the enzyme involved has a high flux control coefficient (see Chapter 1 for metabolic control theory), also exerts a major influence on overall metabolic flux through the pathway in which it resides. Numerous signals or stresses can stimulate changes in the amount of an enzyme or protein in a cell. This coarse control typically occurs over a relatively long time frame for it involves adjustments to protein synthesis (transcription and/or translation) or protein degradation and differs from the immediate or short-term responses to the same stress that can be achieved via fine controls on the activity of individual enzyme molecules (discussed in later sections). Examples of coarse control over enzyme amount include the effects of exercise training or electrical stimulation in elevating the activities of glycolytic and/or tricarboxylic acid cycle enzymes in muscle (see Chapter 11) and the actions of hormones in stimulating changes in the enzymatic profiles of tissues (see Chapter 10). Environmental stresses can similarly induce changes in enzyme activities in order to overcome the perturbing effect of the stress or to restore homeostasis when the stress is prolonged. For example, in hypoxia-sensitive species such as humans a low oxygen signal, acting via the hypoxia-inducible transcription factor (HIF), stimulates increased synthesis of multiple glycolytic enzymes as well as plasma membrane glucose transporters. This raises the capacity for anaerobic ATP synthesis when ATP production by oxidative phosphorylation is limited by low O_2 availability (see Chapters 6 and 15 for more on HIF function and hypoxia adaptation, respectively).

The process of cold acclimation in fish provides another good example of stress-induced changes in enzyme amount. When fish are transferred from warm to cold water, multiple changes are made to readjust metabolic functions for the new temperature. Several involve coarse controls on enzyme amount with two of the prominent changes being: (1) a rapid increase in the synthesis of desaturase enzymes that functions to restore fluidity to cold-rigidified membranes by increasing the number of unsaturated phospholipids in membranes, and (2) an increase in the activities of multiple mitochondrial enzymes as part of a general proliferation of mitochondria in the cold. For example, trout acclimated to 4 versus 18°C showed 40% higher activities of citrate synthase and cytochrome *c* oxidase in red muscle of cold-acclimated fish and 70% higher activities in white muscle. This adjustment occurs because the combination of low animal metabolic rate and higher dissolved oxygen content in cold water permits aquatic ectotherms to rely more heavily on aerobic ATP production to support muscle work at colder temperatures and, hence, mitochondrial enzymatic capacity responds accordingly.

Many organisms also show seasonal adjustments to enzyme activities that track predictable changes in environmental conditions. Such adjustments are frequently triggered via photoperiod cues that modulate production of one or more hormones that, in turn, alter gene expression and protein synthesis. In cold-hardy insects, for example, the activities of glycogen phosphorylase and polyol dehydrogenase rise sharply during the early autumn to support a massive biosynthetic effort that converts as much as 20% of the insect's body weight into the antifreeze agent, glycerol, to provide winter cryoprotection (see Chapter 17).

Changes in Enzyme and Protein Properties

Changes to the properties of proteins, particularly enzymes, is a major instrument of biochemical adaptation. Alterations to enzymes can include changes in substrate/ligand affinity, in the types of effectors to which they are sensitive, in their affinity for activators and inhibitors, and in the influence of physical parameters (temperature, pH, ion concentrations) on enzyme/protein properties. As a result, homologous enzymes or proteins in different species may show considerably different kinetic and regulatory properties that tailor the enzyme/protein to the particular metabolic needs of the cell/tissue/organism or allow it to function optimally under the range of environmental conditions that the organism normally experiences. Evolutionary time has also produced isoforms of proteins (encoded on different genes yet catalyzing the same reaction) and alloforms (protein variants of a single gene that can have slightly different properties). Isoforms may differ in their tissue distribution (e.g., liver versus muscle isoforms), subcellular location (e.g., cytoplasmic versus mitochondrial isoforms), expression patterns at different developmental stages, functional roles, and regulatory mechanisms and can play critical roles in defining the metabolic capabilities of different tissues.

Enzyme Properties Chapters 1 and 2 outlined enzyme kinetics and multiple mechanisms of enzyme regulation including modulation of substrate availability, the actions of activators and inhibitors, and the effects of changes in parameters such as pH, temperature, and ionic strength. One of the most powerful and plastic mechanisms of biochemical adaptation is the tailoring of enzyme kinetic and

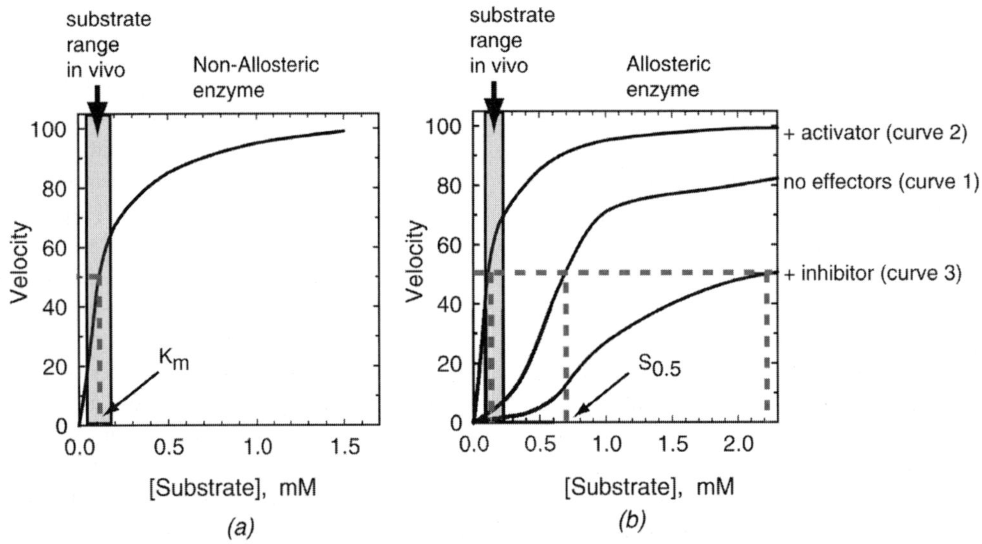

Figure 14.2 Relationship between substrate concentration and enzyme velocity for (a) nonallosteric and (b) allosteric enzymes. The nonallosteric enzyme shows hyperbolic substrate saturation kinetics, in this case with a K_m value (substrate concentration that produces half-maximal velocity) of 0.12 mM. The allosteric enzyme shows a strongly sigmoidal relationship between [S] and velocity in the absence of effectors (curve 1), in this case with a $S_{0.5}$ value (equivalent to K_m for a sigmoidal relationship) of 0.7 mM. Addition of an activator shifts the V vs. [S] curve to the left, creates a hyperbolic relationship and lowers $S_{0.5}$ to 0.12 mM. Addition of an inhibitor causes a right shift in the V vs. [S] curve and raises $S_{0.5}$ to 2.2 mM. The shaded bar shows the range of cellular substrate concentration *in vivo*.

regulatory properties to adjust enzyme function to meet the metabolic demands placed on cells, tissues, and organisms. Changes in the kinetic and regulatory properties of enzymes, integrated with the cellular environment in which the enzymes function, can produce a range of adaptive responses that can create both compensatory adjustments to allow metabolic functions to continue under stress conditions or opportunistic adjustments that alter metabolism under new environmental conditions.

Figure 14.2 shows a schematic of some of the common patterns of enzyme function. Many enzymes show hyperbolic substrate saturation curves and function under near-equilibrium conditions with very low control coefficients (Fig. 14.2*a*). They are often present in high amounts in tissues (relative to the pacemaker enzymes of their pathway) and act primarily as throughput conduits in pathway flux. In glycolysis, enzymes such as phosphoglucoisomerase, phosphoglyceromutase, aldolase, enolase, and lactate dehydrogenase function in this manner (Fig. 14.1). Studies of such enzymes have frequently demonstrated that their K_m values (substrate concentration producing half-maximal velocity) are well-matched with the concentration range of their substrates *in vivo*.

Compensatory adjustments to the K_m values of such enzymes have been demonstrated in a number of instances, and these presumably help to fine-tune enzyme function, particularly in the face of environmental stress. Temperature change is one such stress that affects the activities, conformations, and ligand binding properties of all enzymes, and adjustments to enzyme properties are often seen in the responses of ectothermic animals to temperature change. For instance, on an evolutionary timescale enzymes can be modified to optimize their function for the mean thermal environment in which they must operate. For example, extensive studies of LDH [reaction (14.1)] from multiple species of fish ranging from the Antarctic to the tropics have shown that an optimal K_m value for pyruvate of about 0.2 mM is invariably found when the enzyme is assayed at temperatures within the normal environmental range for the species, be it 0 to 2°C for Antarctic fish or 30 to 35°C for tropical fish. However, within as little as 5 to 10°C above the normal range, the K_m value rises rapidly (i.e., enzyme affinity for pyruvate declines); for example, the K_m pyruvate of Antarctic fish LDH was ~0.4 mM when assayed at 15°C. The molecular basis of this phenomenon is small changes to the amino acid sequences of LDH in different species that readjust the ratios of hydrophobic and hydrophilic bonds within the protein for optimal function within a given thermal window. For example, LDH from Antarctic fish, has several more glycine residues in key areas of the protein than does the enzyme from temperate fish, and these increase the flexibility of the cold-adapted orthologs [refer to Hochachka and Somero (2002) for an expanded treatment].

$$\text{Pyruvate} + \text{NADH} + \text{H}^+ \longleftrightarrow \text{Lactate} + \text{NAD}^+ + \text{H}_2\text{O} \quad (14.1)$$

Allosteric enzymes offer many more opportunities for adaptive control (Fig. 14.2*b*). Substrate binding is modulated in one of several ways. First, most such enzymes are multimeric and show homotropic allosterism, meaning that substrate binding to one subunit alters the conformation of other subunits to positively enhance substrate binding to them. This results in a sigmoidal relationship between enzyme activity and substrate concentration that makes the enzyme highly sensitive to small changes in substrate (see Chapter 1). These enzymes are also typically sensitive to regulation by multiple nonsubstrate effector molecules. Effectors bind at allosteric sites away from the active site and, by doing so, change enzyme conformation and alter enzyme–substrate binding properties. Activators alter enzyme conformation to increase substrate affinity, creating a left shift in the velocity (V) versus substrate concentration [S] curve whereas inhibitors shift the V versus [S] curve to the right to greatly reduce substrate affinity (Fig. 14.2*b*). Binding by allosteric activators can also modify enzyme sensitivity to inhibitors and vice versa. Such heterotropic allosterism offers multiple opportunities for influencing enzyme rate by metabolites other than substrates, products, and cofactors. Feedforward activation can come via allosteric effects by products of earlier reactions in the pathway whereas feedback inhibition can result from sensitivity to the ultimate end product of a pathway. Sensitivity to the energy status of the cell can be imparted via the presence of allosteric sites for ATP, adenosine diphosphate (ADP), or adenosine monophosphate (AMP) and metabolites from competing pathways can exert effects. With the multiple opportunities for allosteric regulation of such enzymes (often combined with additional control of the enzyme by reversible phosphorylation; see below), it is not surprising that allosteric enzymes are often the pacemaker enzymes of pathways. The ability to modify allosteric regulatory properties of enzymes also provides a huge range of opportunities for biochemical adaptation. Enzyme control in different species/tissues can be modified by deleting or adding regulation by different allosteric effectors, by changing enzyme sensitivity to one or more effectors, and by modifying enzyme/effector interactions with respect to parameters such as pH and temperature. For example, later in this chapter we will discuss the effects of widely differing sensitivities to L-alanine inhibition by pyruvate kinases in the regulation of glycolysis in different species. In the next section, the interactions of the enzyme 6-phospho-fructo-1-kinase [PFK; reaction (14.2)] with its effectors and pH is emphasized, whereas in Chapters 16 and 17 several examples are given of the interactions between temperature and enzyme properties in the initiation and regulation of various biochemical adaptations that support cold-hardiness.

$$\text{Fructose-6-P} + \text{ATP} \rightarrow \text{fructose-1,6-P}_2 + \text{ADP} \quad (14.2)$$

The PFK of animal tissues is an excellent example of an allosteric enzyme and is frequently considered to be a pacemaker of glycolysis. PFK is the first committed step of glycolysis, its ATP-utilizing reaction gating the entry of hexose phosphates into the triose phosphate portion of glycolysis (Fig. 14.1). The reaction is functionally irreversible *in vivo* and during gluconeogenesis is bypassed by the fructose-1,6-bisphosphatase reaction. PFK is regulated in a multitude of ways: by allosteric effectors, by reversible phosphorylation, via reversible polymerization of the active tetramer, and in muscles via binding interactions with myofibrillar components. We will return to PFK many times in this chapter to provide examples of different principles of metabolic regulation.

The regulation of PFK by substrates, allosteric effectors, and parameters such as pH and temperature is complex and provides multiple opportunities for adaptive control of enzyme activity. In general, PFK shows a sigmoidal relationship between fructose-6-phosphate (F6P) substrate concentration and enzyme velocity and, in the absence of effectors, and the $S_{0.5}$ value for F6P is far above the physiological concentrations of F6P. ATP is the second substrate of PFK (K_m ~0.05 mM) but is also a powerful inhibitor of the enzyme at higher concentrations that are closer to physiological ATP levels (typically 2 to 8 mM). Hence, under normal cellular levels of F6P and ATP, net PFK activity would be extremely low, and the enzyme is highly dependent on the influence of powerful activators to produce the major changes in PFK activity that are needed when the demand for glycolytic flux is high. These activators include AMP, which is a key signal of low energy status in cells. Powerful AMP effects on PFK are a major factor in the activation of glycolytic ATP output under situations where cellular energy supply is stressed (e.g., muscle work, hypoxia). PFK is also activated by inorganic phosphate (the net product of creatine phosphate conversion to ATP followed by ATP hydrolysis) and by NH_4^+ and inosine monophosphate (IMP) (the products of AMP deaminase action on AMP), all of which also signal high energy demand, particularly in working muscle. Glycolysis also has another role in the supply of carbohydrate for biosynthetic reactions and a second powerful activator, fructose-2,6-bisphosphate (F2,6P_2), mediates

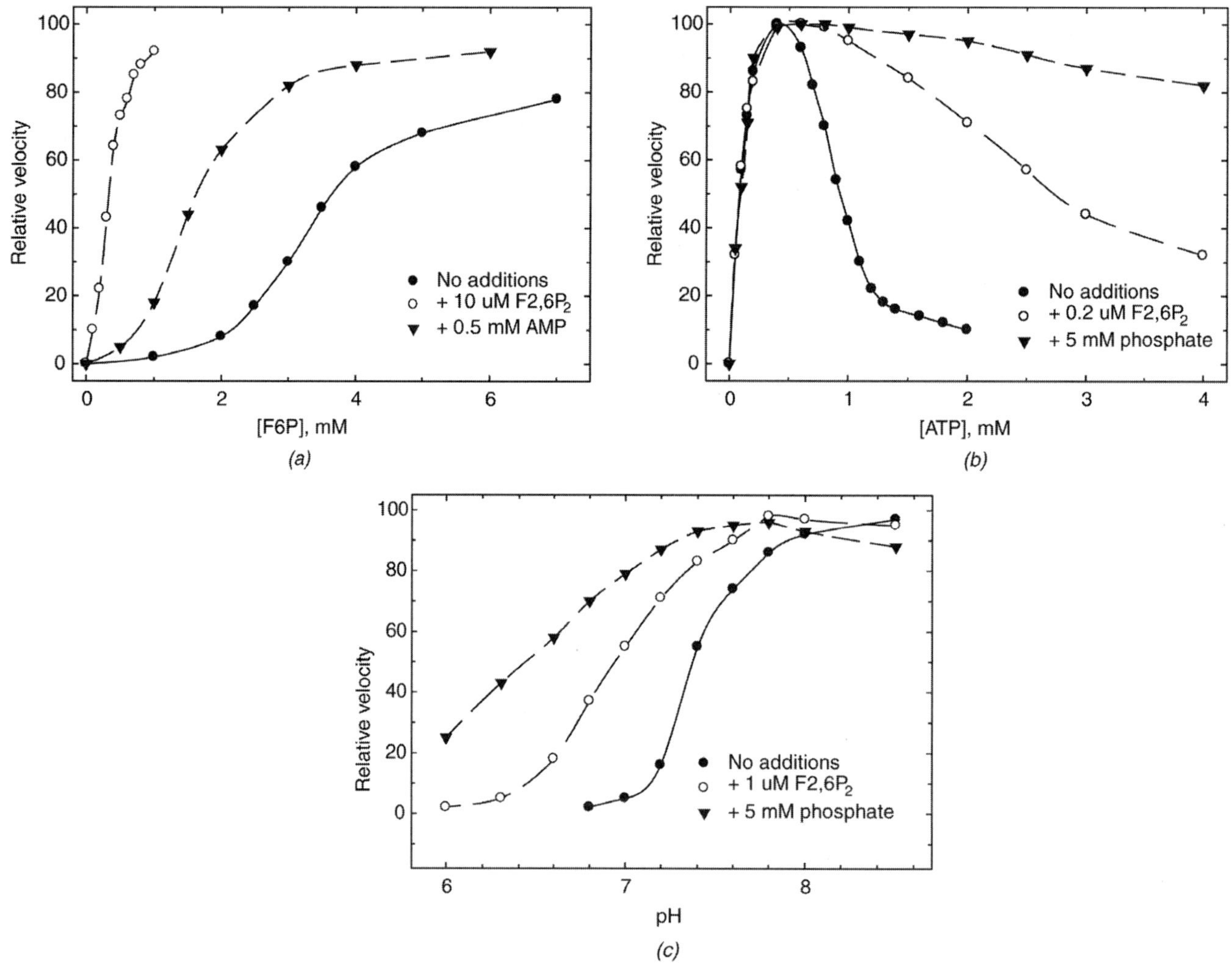

Figure 14.3 Effect of allosteric activators on properties of trout white muscle phosphofructokinase under simulated metabolic conditions in resting muscle: (*a*) fructose-6-phosphate substrate saturation curves at pH 7.2 with 7.5 mM ATP as cosubstrate, (*b*) ATP substrate saturation curves at pH 7.2 with 0.2 mM F6P as cosubstrate, and (*c*) pH profile. [Data compiled from J. Y. Su and K. B. Storey (1994). *Int. J. Biochem* **26**:519–528.]

PFK activity in this role. The enzyme is also affected by inhibitors that slow glycolytic rate when energy levels are high (ATP inhibition) or when biosynthetic demands decrease and cause a buildup of glycolytic (phosphoenolpyruvate) or tricarboxylic acid cycle (citrate) intermediates. Furthermore, PFK is highly sensitive to changes in pH within the physiological range with acidification reducing affinity for F6P, increasing ATP substrate inhibition, and reducing the effects of activators.

The PFK from the white (type II) muscle of trout provides an excellent example of the complexities of PFK regulation. In the absence of effectors, with a physiological level of ATP cosubstrate (7.5 mM), and an assay pH of 7.2 (the pH typical of resting muscle), PFK shows very low affinity for F6P; the $S_{0.5}$ value was 3.9 mM with a Hill coefficient, n_H, of 4 indicating high sigmoidicity (remember that $n_H = 1$ for a hyperbolic relationship; Chapter 1) (Fig. 14.3*a*, Table 14.3). When compared with the *in vivo* concentration of F6P in fish muscle, which is <0.20 mM, it is obvious that enzyme activity under these conditions would be very low. Furthermore, ATP substrate inhibition of PFK is high under resting conditions in muscle with an ATP I_{50} value (concentration producing 50% inhibition) of 0.9 mM when measured at physiological F6P levels (Fig. 14.3b, Table 14.3). However, in the muscle at rest low levels of activators such as $F2,6P_2$ can profoundly change PFK kinetics to regulate low rates of carbohydrate use for biosynthesis. As little as 0.2 μM $F2,6P_2$ strongly activates the enzyme and in the presence of 10 μM $F2,6P_2$ the $S_{0.5}$ for F6P was lowered from 3.9 to 0.2 mM

TABLE 14.3 Effects of Metabolite Modulators on the $S_{0.5}$ and n_H for Fructose-6-Phosphate of Trout Muscle PFK under ATP Concentrations and pH Values Representing "Resting" (ATP = 7.5 mM, pH 7.2) and "Exercised" (ATP = 3.5 mM, pH 6.6) Cellular Conditions[a]

	Resting Conditions (Modulator)	$S_{0.5}$ (mM)	n_H	Exercised conditions (Modulator)	$S_{0.5}$ (mM)	n_H
Control (no additions)	—	3.4	4.1	—	18.5	4.8
+ AMP	0.05 mM	1.3	3.5	0.50 mM	2.2	1.8
+ Inorganic phosphate	25 mM	0.2	1.0	50 mM	0.5	1.8
+ NH_4Cl	1 mM	0.9	4.0	5 mM	4.3	2.5
+ F2,6P_2	10 μM	0.2	1.8	10 μM	1.0	2.0
+ All 4 modulators	As above	0.07	1.4	As above	0.15	1.0

[a]The modulator concentrations chosen represent those found *in vivo* in resting versus exercised trout muscle.
Source: Data modified from J. Y. Su and K. B. Storey (1994). *Int J Biochem* **26**:519–528.

(Table 14.3) whereas the I_{50} for ATP was increased from 0.9 to 19 mM.

High-speed swimming demands a huge increase in power output from working muscle (10- to 100-fold), and in white muscle this is supplied by the rapid depletion of creatine phosphate reserves and by an immediate activation of glycolysis to produce ATP from the catabolism of glycogen to lactate. During muscle work, intracellular pH falls by about 0.6 units. Obviously, PFK must be activated during exercise, but how is this achieved? PFK activity falls rapidly at pH values below 7 (Fig. 14.3*c*) and the enzyme's kinetic parameters are negatively affected by reduced pH. For example, when measured at pH 6.6, the values for $S_{0.5}$ F6P and I_{50} ATP were 26.3 and 0.2 mM (compare these with the values for pH 7.2 assays above). Here is where activators that signal low energy levels play a huge role. Activators including inorganic phosphate and AMP lessen enzyme sensitivity to low pH and reduce the inhibitory effects of ATP (Figs. 14.3*b* and 14.3*c*). For example, the I_{50} value for ATP rose by 10-fold in the presence of 5 mM inorganic phosphate; notably, inorganic phosphate levels can increase *in vivo* by at least 20 mM during intense muscle exercise as the result of creatine phosphate hydrolysis. Coupled with a fall in cellular ATP levels of about 50% during exercise, net PFK inhibition by ATP is greatly reduced in working muscle. Enzyme affinity for F6P is also vastly improved by the actions of allosteric activators. Table 14.3 shows the influence of four activators on the $S_{0.5}$ for F6P and the n_H of F6P binding comparing both individual effects and the additive effects of multiple activators and utilizing conditions of effector concentrations and pH that mimic either "resting" or "exercised" muscle conditions. All activators individually lower both $S_{0.5}$ and n_H but, more importantly, strong additive effects of activators occur so that $S_{0.5}$ for F6P can be lowered from 18.5 mM in the absence of effectors to 0.15 mM in the presence of the four activators. This 123-fold change lowers the $S_{0.5}$ of the enzyme to a value that closely matches in vivo F6P concentrations (0.16 mM) in working muscle.

ISOFORM/ISOZYME PROPERTIES

A wide range of functional demands on an enzyme or protein cannot always be met by a single enzyme/protein form, and in many cases, isozymes/isoforms have evolved that may differ in kinetic properties, sensitivity to effectors, susceptibility to covalent modification, binding interactions with other macromolecules, subcellular location, tissue distribution, and so forth. The five isoforms of the vertebrate plasma membrane glucose transporter (GLUT) provide an instructive example of differential isoform distribution and function. The primary mode of glucose movement across cell membranes is facilitated diffusion mediated by GLUT proteins that move the sugar down its concentration gradient. Multiple GLUT isoforms are known, each with properties geared for different functions. GLUT1 is the passive glucose transporter of erythrocytes and many other tissues and is probably the "ancestral" form. It facilitates one-way uptake of glucose from the plasma into cells. GLUT2 is found chiefly in liver, and unlike the other GLUT2 isoforms functions effectively in two-way glucose transport into or out of hepatocytes. This capacity is crucial for its central role in liver, the organ that maintains plasma glucose homeostasis under both fed and starved conditions. GLUT3 is expressed in brain and nervous tissue and appears to act in concert with GLUT1 to ensure that glucose, the primary substrate for oxidative metabolism in the brain, is always in adequate supply. Muscles and adipose tissue have another isoform, GLUT4. This isoform mediates the insulin-dependent uptake of glucose into cells under fed conditions to provide sugar for energy metabolism and the synthesis of fuel reserves (glycogen in muscle, triglycerides in adipose). The action of insulin is to stimulate a translocation of inactive transporters from intracellular vesicles to the cell surface and thereby increase transport capacity. Finally, GLUT5 has a unique role in the small intestine, functioning in the uptake of a variety of dietary hexoses, particularly fructose. By diversifying into many different isoforms, a wide variety of specialized needs and organ-specific functions can be met.

Among enzymes, differences in the kinetic and regulatory properties of isozymes are critical to their function and to the control of the pathway in the tissue in which they reside. A good example of differential isozyme properties is the muscle and liver isozymes of pyruvate kinase (PK):

$$\text{Phosphoenolpyruvate} + \text{ADP} \rightarrow \text{pyruvate} + \text{ATP} \quad (14.3)$$

In skeletal muscle PK is typically a high activity enzyme (Table 14.2) and, as the second ATP-producing reaction of glycolysis (Fig. 14.1), PK has a critical role to play in the high rates of ATP output demanded by working muscle, particularly in white muscle. In liver, by contrast, PK is present in lower activities and is closely regulated. Glycolysis in liver plays a substantial role in providing carbohydrates for biosynthesis when glucose is plentiful, but the pathway must be shut down under either of two conditions: (a) when glycogenolysis must be directed into the synthesis and export of glucose, such as under starved conditions, and (b) when high levels of gluconeogenic substrates need to be converted to glucose. Differences in PK properties between skeletal muscle and liver isozymes emphasize these organ-specific roles. Table 14.4 shows how this is done.

Muscle PK is generally a nonallosteric enzyme that shows a velocity (V) versus substrate concentration (S) relationship that is hyperbolic (i.e., Hill coefficient=1). The enzyme shows a high maximal activity (Table 14.2) and a low control coefficient that supports its role in "high-throughput" glycolysis in working muscle. The enzyme is designed with a high affinity (low K_m) for its carbohydrate substrate, phosphoenolpyruvate (PEP), the K_m value typically falling well within the range of *in vitro* PEP concentrations. These properties of skeletal muscle PK are much the same in many different phylogenetic groups, as can be seen by comparing mammalian and squid muscle PK in Table 14.4.

By contrast, the liver isozyme of PK in mammals is an allosteric enzyme that shows a sigmoidal relationship between PEP concentration and enzyme velocity (Hill coefficient is ~1.5). Substrate affinity for PEP by liver PK is lower than that of the muscle enzyme [in Table 14.4 compare muscle PK with the values for dephosphorylated (DeP) liver PK] with K_m values that are typically higher than the normal range of PEP concentrations *in vivo.* The high K_m keeps enzyme activity low except when the activator, fructose-1,6-bisphosphate (FBP), is present. The enzyme is also inhibited by L-alanine. The two effectors shift the V versus [S] curve to the left or right, respectively, and alter the shape of the curve. A rise in FBP, the product of the PFK reaction (an earlier step in glycolysis) (Fig. 14.4), signals an activation of glycolysis, and feedforward activation of PK coordinates the rates of these two loci. On the other hand, a rise in alanine signals the presence of amino acids that can fuel gluconeogenesis and, under those circumstances, PK needs to be inhibited in order to allow pyruvate reconversion to glucose. Table 14.4 also shows how phosphorylation of the PK protein alters the properties of liver PK, and this is dealt with in the next section.

A dramatic example of the adaptation of the properties of PK isozymes is illustrated by the data for whelk muscle PK in Table 14.4. Whelks are large marine snails that have a high tolerance for oxygen deprivation. Adaptation of PK properties plays a key role in anoxia tolerance, as will be discussed in Chapter 15, but here it is instructive to see how very much the properties of PK can be altered when there is a specific need. In contrast to muscle PK from mammals or squid (another mollusk), whelk muscle PK is both an allosteric enzyme and susceptible to reversible phos-

TABLE 14.4 Kinetic and Regulatory Properties of Pyruvate Kinases from Mammalian and Mollusk Sources[a]

	Mammalian Muscle	Squid Muscle	Mammalian Liver		Whelk Muscle		Whelk "Liver"	
			DeP	P	DeP	P	DeP	P
K_m PEP (mM)	0.02	0.15	0.3	0.8	0.07	0.85	0.38	1.10
K_m ADP (mM)	0.30	0.40	0.25	0.25	0.27	0.25	0.21	0.22
K_a Fructose 1,6-P_2 (μM)	N.E.	N.E.	0.06	0.13	0.05	1.3	0.16	0.48
K_i L-alanine (mM)	N.E.	N.E.	0.70	0.35	24.5	0.05	3.9	0.48

[a]The liver and whelk enzymes are interconverted by reversible phosphorylation between dephosphorylated (DeP) and phosphorylated (P) forms. The "liver" of whelks is an organ called the hepatopancreas. The DeP form of whelk PK is the form present in aerobic tissues whereas the enzyme is converted to the P form under anoxic conditions (see Chapter 15). N.E. = no effect.

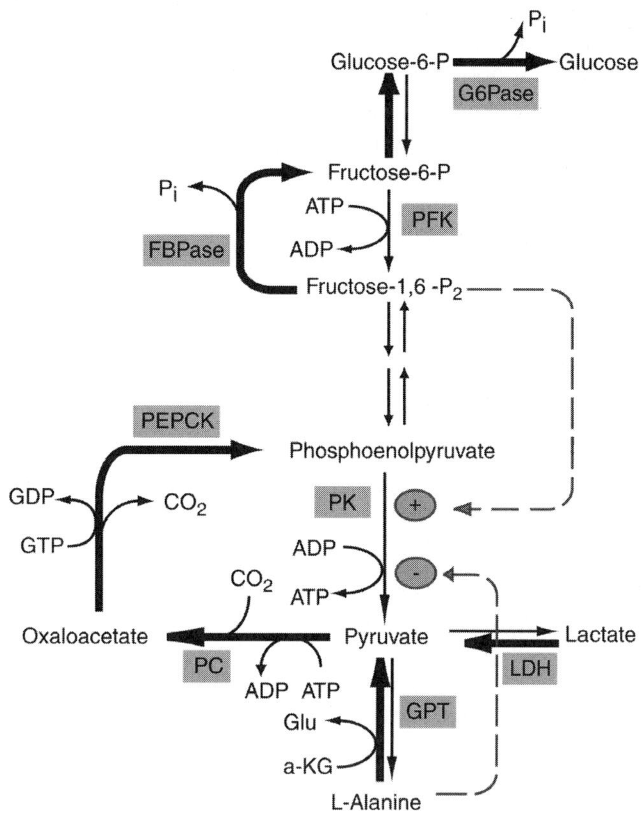

Figure 14.4 Gluconeogenesis using lactate or alanine as substrates, showing the reactions that bypass the phosphofructokinase (PFK) and pyruvate kinase (PK) loci of glycolysis. Other abbreviations are: G6Pase, glucose-6-phosphatase; PEPCK, phosphoenolpyruvate carboxykinase; PC, pyruvate carboxylase, GPT, glutamate-pyruvate transaminase; LDH, lactate dehydrogenase; Glu, glutamate; α-KG, alpha-ketoglutarate. Two modes of allosteric control of PK are also shown (gray dashed lines): feedforward activation by fructose-1,6-P_2 and inhibition by L-alanine.

phorylation, much like the mammalian liver enzyme. In its dephosphorylated form it probably does not function very differently from the mammalian or squid muscle PK; its low K_m PEP would facilitate its participation in muscle glycolysis during exercise and, although inhibited by alanine, the very high K_i value means that alanine effects would be of limited consequence *in vivo.* However, once the enzyme is phosphorylated, it becomes highly sensitive to alanine inhibition, and this is a critical factor in PK control in anoxia-tolerant species. PK from the "liver" (in mollusks, its an organ called the hepatopancreas) of whelks also shows this enhanced sensitivity to L-alanine in the phosphorylated form, but in other properties it is not unlike the mammalian liver enzyme.

Thus, we can see several principles of enzyme adaptation from this examination of PK properties: (1) isozymes can be designed with very different kinetic properties that suit their organ-specific functions, (2) enzymes that perform similar "jobs" display similar kinetic properties across wide phylogenetic groups, (3) selected kinetic properties (such as the K_m PEP) are highly malleable, whereas others (such as the K_m ADP) are not, and (4) enzyme sensitivity to one or more allosteric effectors can be enhanced or reduced to place enzymes under different regulatory controls in different circumstances.

Enzyme Control via Reversible Protein Phosphorylation

Cellular proteins can undergo posttranslation modification via multiple kinds of covalent modification (see Chapter 1). Of these, the mechanism of greatest importance to metabolic regulation in animal cells is reversible protein phosphorylation. Protein kinases transfer the γ-phosphate group from ATP onto an amino acid residue in a covalent attachment that is stable unless it is cleaved by a protein phosphatase (for review see Chapter 4). The first protein kinase to be appreciated was cyclic 3′,5′-adenosine monophosphate (cAMP)-dependent protein kinase (known as PKA), that phosphorylates serine or threonine (Ser/Thr) residues on a wide variety of enzymes, notably enzymes of glycogen metabolism that were the first targets of PKA research. In recent years, the number of known protein kinases has grown explosively. Indeed, in 2002, an electronic analysis of sequences in the human genome revealed 448 sequences that contained the critical residues essential for kinase function. The major groups of human protein kinases are listed in Table 14.5, and the roles and regulation of protein kinases in signal transduction pathways are discussed in much greater detail in Chapter 4. Here, we are primarily concerned with the effects of phosphorylation and dephosphorylation on the control of metabolic enzymes and other functional proteins.

The covalent attachment of one or more negatively charged phosphate groups onto a protein changes its conformation and its charge. The difference in charge between phosphorylated and dephosphorylated protein forms is often exploited for the separation or purification of the two forms via techniques such as ion exchange chromatography or isoelectric focusing (Fig. 14.5). These separation techniques, as well as selected key differences in kinetic behavior, are also used for judging the effects of signals or stresses on the interconversion of the enzyme/protein between high and low phosphate forms. A change in phosphorylation state can affect a protein in one of several ways, including (a) on–off control over activity, (b) changes to kinetic properties, (c) changes in enzyme binding to other proteins, and (d) other effects.

TABLE 14.5 Major Subfamilies of Protein Kinases Encoded on the Human Genome, Classified by Kinase Catalytic Domain

Subfamily Name	Number of Members	Representatives in the Group
AGC	59	Protein kinases A, G, C
CAMK	81	Calcium/calmodulin-dependent kinases
CMGC	72	Cyclin-dependent kinases, mitogen-activated protein kinases, glycogen synthase kinase, casein kinase 2
TK	94	Receptor and nonreceptor tyrosine kinases
CK1	12	Casein kinase 1

Source: From Krupa and Srinivasan (2002).

On–Off Control over Activity Many enzymes are converted between active and inactive states by reversible phosphorylation. The classic example of this is glycogen phosphorylase (GP) [reaction (4.4)], which cleaves glu-

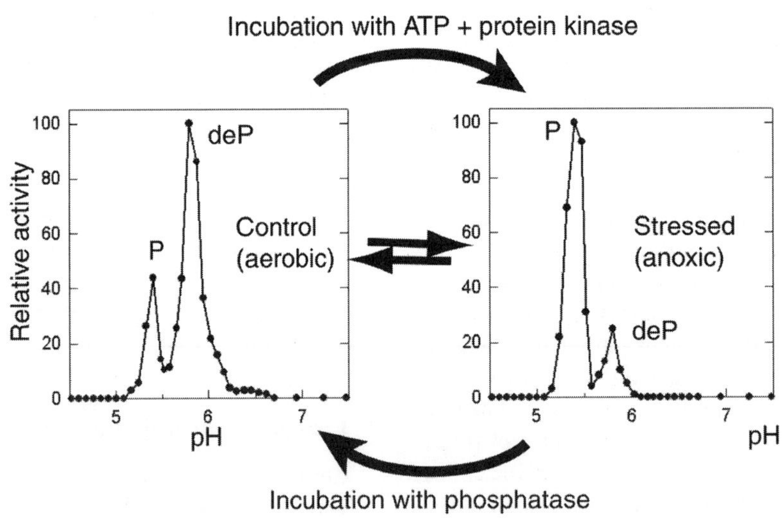

Figure 14.5 Assessment of stress effects on enzyme phosphorylation state. Tissue extracts from control and stressed animals are subjected to isoelectrofocusing, in this case producing two peaks of enzyme activity eluting with isoelectric points of 5.4 and 5.8. Environmental stress (anoxia exposure in this case) changes the relative amount of activity in each peak. To determine which is the phosphorylated form, tissue extracts are pretreated *in vitro* with either (1) protein kinase + ATP, which shifts activity into the peak at pI 5.4 and identifies it as containing the phosphoenzyme (P), or (2) phosphatase incubation, which shifts activity into the pI 5.8 peak and identifies it as the dephosphorylated (deP) enzyme. The figure is derived from the effects of anoxia exposure on pyruvate kinase in muscle of a marine mollusk; the deP form predominates under aerobic conditions and is converted to the P form under anoxic conditions (see Chapter 15).

cose-1-P units off the glycogen polymer:

$$\text{Glycogen}_{(n)} + \text{P}_i \rightarrow \text{glucose-1-P} + \text{glycogen}_{(n-1)} \quad (14.4)$$

Phosphorylation of GP converts the inactive *b* form to the active *a* form. Simultaneously, phosphorylation has the opposite effect on glycogen synthase, inactivating that enzyme and thereby providing reciprocal control over glycogen synthesis versus breakdown. Although GP*b* can have activity *in vitro* when presented with high concentrations of AMP as an allosteric activator, it is unlikely that the *b* form contributes in any significant way to glycogen breakdown *in vivo*. Hence, phosphorylation of GP by phosphorylase kinase and its reversal by protein phosphatase 1 (PP-1) provides effective on–off control of the enzyme.

On–off control via reversible phosphorylation allows cells to maintain a high enzymatic potential that can be unleashed virtually instantaneously whenever it is needed. It provides the speed of response that longer-acting coarse controls (protein synthesis or degradation) cannot and a range of response (fold-activation) that is virtually impossible to achieve with other means of enzyme regulation such as changes in the concentrations of enzyme substrates or effectors. For example, when a fly or bee takes off, the percentage of GP in the active *a* form in flight muscle can rise from <20% to 80 to 100% within 15 s (triggered via Ca^{2+}/calmodulin activation of phosphorylase kinase) (see Fig. 1.6 and 4.11) in order to support the 20- to 100-fold increase in metabolic rate that is needed to power flight (dipteran and hymenopteran insects power flight with carbohydrate fuels).

Reversible phosphorylation providing on–off control is also found at many other key loci in cellular metabolism. For example, another critical locus in carbohydrate catabolism is regulated in this way. The pyruvate dehydrogenase complex (PDC) [reaction (14.5)] gates carbohydrate entry into the oxidative reactions of the tricarboxylic acid cycle:

$$\begin{aligned}\text{Pyruvate} + \text{CoA} + \text{NAD}^+ &\rightarrow \text{acetyl-CoA} \\ &+ \text{CO}_2 + \text{NADH} + \text{H}^+ \quad (14.5)\end{aligned}$$

In this case, phosphorylation turns *off* the PDC, and control at this locus is important in regulating the relative use of carbohydrate versus lipid as aerobic fuels for the mitochondria. Indeed, phosphorylation-mediated inactivation of the PDC is a key element in the "carbohydrate sparing" that occurs during starvation or torpor (see discussion of hibernation in Chapter 16). It promotes the switch to lipid fuels by most organs in order to conserve carbohydrates for organs such as brain that have few other alternatives. It is not surprising, then, that the initiating enzyme in the mobilization of lipid reserves, triglyceride lipase (also called hormone-sensitive lipase), is oppositely affected by phosphorylation and turned *on*.

On–off control via reversible phosphorylation is also extremely important in the regulation of signal transduction cascades. Numerous protein kinases are themselves controlled in this manner by upstream kinases. For example, the activities of mitogen-activated protein kinases (MAPKs) increase about 1000-fold when they are phosphorylated. Hence, the dephosphorylated forms are effectively inactive *in vivo*. This sort of on–off control when applied to multiple enzymes in a signal transduction cascade is also highly effective in producing a huge amplification of a signal (see Chapter 4). The exploration of kinase cascades and many other proteins that are regulated via reversible phosphorylation (e.g., ribosomal initiation factors and elongation factors) has been greatly aided by a recent technological advance, the development of phospho-specific peptide antibodies (see Text Box 14.1).

Phosphorylation-Induced Changes to Kinetic Properties Reversible phosphorylation provides on–off control for some enzymes, but for others it provides the means of making major adjustments to enzyme kinetic and regulatory properties, often changing the sensitivity of the enzyme to selected allosteric effectors and frequently facilitating large changes in pathway flux.

Table 14.4 illustrates this for the PK reaction. Compare the properties of the dephosphorylated and phosphorylated forms of mammalian liver PK. It can be seen that phosphorylation of the PK protein reduces enzyme affinity for substrate ($S_{0.5}$ PEP increases by 2.7-fold), reduces enzyme sensitivity to the activator fructose-1,6-P_2 (K_a increases by 2.2-fold), and increases enzyme sensitivity to inhibition by L-alanine (K_i decreases by 50%). Phosphorylation also changes the relationship between velocity and PEP substrate concentration, increasing the sigmoidicity of the relationship (the Hill coefficient increases). These changes are illustrated graphically in Figure 14.6. The relationship between [S] and *V* for the dephoshorylated and phosphorylated enzyme forms is reminiscent of the relationships seen in Figure 14.2 for an allosteric enzyme in the absence versus presence of an activator, but the key difference is that phosphorylation is a stable modification that is independent of changes in metabolite effector concentrations and permanent until reversed by the action of a phosphatase. All of the effects of phosphorylation on PK properties produce an enzyme that is less active *in vivo* because the phosphoenzyme has reduced substrate affinity, is less affected by activators, and is more strongly affected by inhibitors.

In addition, the reduced activity of the phosphoenzyme is frequently compounded *in vivo* by changes in the cellular levels of effectors. For example, in mammalian liver, PK is phosphorylated when it is necessary to turn off glycolysis and activate gluconeogenesis instead. Under these

TEXT BOX 14.1 PHOSPHO-SPECIFIC PEPTIDE ANTIBODIES

The development of phospho-specific peptide antibodies has created a minirevolution in the study of biochemical regulation, particularly for evaluating signal transduction pathways. The activity states of many proteins/enzymes are controlled in a virtual on–off manner by reversible phosphorylation so changes in the relative amount of phosphoprotein provides an excellent indicator of relative protein activity and of protein response to a stimulus or stress. Protein activity state is assessed using two kinds of antibodies, one that is raised against the whole protein and one that is raised against a short peptide that contains the phosphorylated amino acid residue (e.g., serine, threonine, tyrosine). The latter reacts only with the phosphoprotein and not with the dephosphorylated form. As Figure TB14.1 shows, a stress typically elicits a change in the relative amount of the phosphoprotein with little or no change in total protein content. An excellent source for further information on this topic is the website of Cell Signaling Technology, Inc., at www.cellsignal.com.

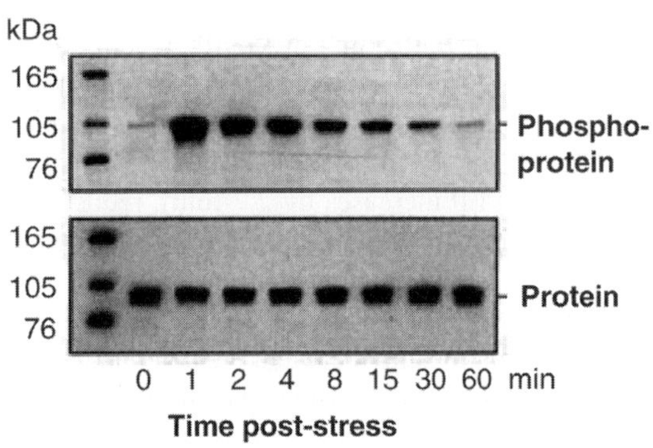

Figure TB14.1 Western blots showing the time course of activation of a 105-kDa protein by protein phosphorylation. Upper panel shows stress-induced changes in the amount of phosphoprotein determined with a phospho-specific polyclonal antibody (activation is maximal within 1 min and the declines to return to control levels by 60 min). Lower panel shows that total protein remains constant.

conditions, the concentration of L-alanine in liver cells is elevated because lactate and alanine, produced by working muscle, are two major substrates for gluconeogenesis in the liver. A severalfold increase in alanine concentrations, combined with the conversion of PK to its phosphorylated form that is strongly inhibited by alanine, create a powerful net suppression of PK activity *in vivo* so that carbon flow can be directed through the gluconeogenic reactions

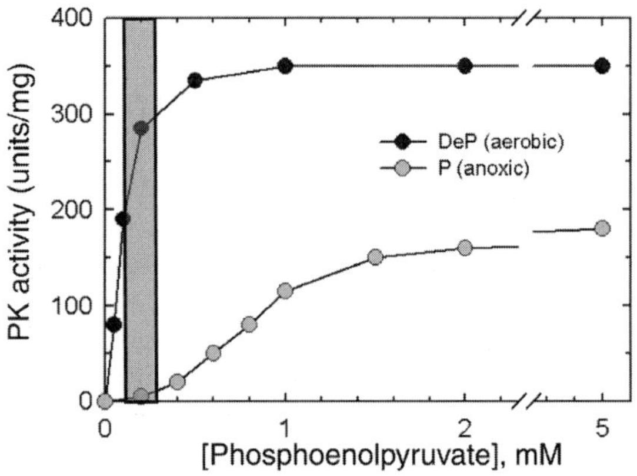

Figure 14.6 Relationship between pyruvate kinase (PK) activity and phosphoenolpyruvate (PEP) substrate concentration for the dephosphorylated (DeP) and phosphorylated (P) forms of the enzyme isolated from muscle of a marine mollusk exposed to aerobic versus anoxic conditions. The vertical bar shows the *in vivo* range of PEP substrate concentrations, 0.15 to 0.30 mM, in the muscle and indicates that the anoxic enzyme form would be virtually inactive *in vivo*. The calculated $S_{0.5}$ values (substrate concentration giving half-maximal activity) are 0.07 mM for the aerobic form and 0.85 mM for the anoxic form.

that bypass PK, namely pyruvate carboxylase and PEP carboxykinase (Fig. 14.4).

Phosphorylation-Dependent Changes in Binding to Other Macromolecules The conformational changes induced by protein phosphorylation or dephosphorylation often alter the ability of a protein to bind to another subcellular component. An excellent early example of this effect was provided during studies of mammalian skeletal muscle PFK. Muscle exercise stimulated PFK phosphorylation, but the kinetics of the purified phospho-PFK differed very little from those of the dephosphorylated form. This was puzzling until it was determined that the primary consequence of the phosphorylation of muscle PFK was to increase PFK binding to F-actin. This positioned this regulatory enzyme of glycolysis (the ATP-producing pathway) into proximity with the myofibrils (the ATP-utilizing motor) to potentially increase the efficiency of ATP turnover during exercise. A number of other muscle enzymes also associate with myofibrils, as will be discussed in a later section.

Reversible protein phosphorylation is also involved in hundreds of other binding interactions. For example, many transcription factors are phosphorylated via protein kinases (often by the MAPKs) to produce active factors that are then capable of binding to deoxyribonucleic acid (DNA) (see Chapter 6). Protein kinases and protein phosphatases are frequently compartmentalized to selected subcellular locations via binding interactions with targeting proteins. Phosphorylation of targeting proteins can either

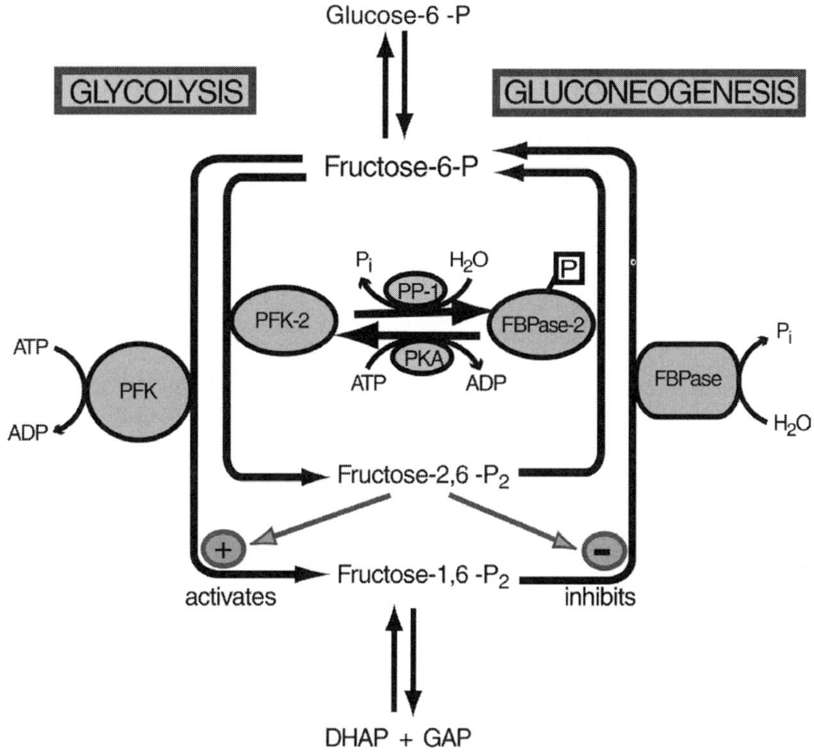

Figure 14.7 Reciprocal control of glycolysis versus gluconeogenesis via the allosteric effects of fructose-2,6-bisphosphate on phosphofructokinase (PFK) and fructose-1,6-bisphosphatase (FBPase). Phospofructokinase-2 (PFK-2), the enzyme that produces fructose-2,6-P_2, is bifunctional and synthesizes fructose-2,6-P_2 in its dephosphorylated form; but after phosphorylation by protein kinase A (PKA) it acts as a fructose-2,6-bisphosphatase to hydrolyze fructose-6-P.

disrupt or promote such binding. For example, PP-1 is targeted to the glycogen particle by its interaction with a glycogen-binding (G) targeting subunit. When glycogenolysis is activated under adrenergic stimulation (PKA mediated), phosphorylation of the G subunit by PKA causes PP-1 to dissociate from the glycogen particle and prevents the phosphatase from dephosphorylating GP at the same time that PKA is stimulating its phosphorylation (see Figure 4.14).

Other Enzyme-Specific Effects An excellent example of this is the case of the enzyme 6-phosphofructo-2-kinase (PFK-2). Only discovered in the early 1980s, PFK-2 and its product, fructose-2,6-bisphosphate (F2,6P_2), caused a minirevolution in our understanding of the control of glycolysis by identifying F2,6P_2 as an extremely potent effector of glycolytic versus gluconeogenic flux due to its reciprocal control over PFK and fructose-1,6-bisphosphatase (FBPase) [reaction (14.6)] (Fig. 14.7). This unique monomeric enzyme is bifunctional and capable of catalyzing either the synthesis or breakdown of F2,6P_2 depending on the phosphorylation state of the protein.

$$\text{Fructose-1, 6-P}_2 + \text{H}_2\text{O} \rightarrow \text{fructose-6-P} + \text{P}_\text{i} \quad (14.6)$$

When dephosphorylated, the enzyme produces F2,6P_2, which is a powerful activator of PFK and a strong inhibitor of FBPase, the gluconeogenic enzyme that bypasses the PFK reaction. Phosphorylation of the enzyme by PKA causes a reversal of the reaction, causing the enzyme to function as a fructose-2,6-bisphosphatase (FBPase-2) and catabolize F2,6P_2. In mammalian liver it has been concluded that high F2,6P_2 mediates a "fed" signal that allosterically activates PFK to allow carbohydrate to be processed via glycolysis into various biosynthetic outputs. Under "starved" or stressed conditions, however, liver biosynthesis is inhibited and glycogenolysis (activated by glucagon or epinephrine stimulation of PKA) is directed into the synthesis of glucose for export. Inhibition of PFK-2 by phosphorylation stops F2,6P_2 synthesis so levels of the activator drop and, combined with the direct effects of phosphorylation of the kinetic properties of liver PFK (discussed above), this leads to a strong inhibition of the PFK locus that promotes glucose export. Under the same conditions, gluconeogenesis is typically activated to provide glucose output from this second source, and the phosphorylation-mediated promotion of the FBPase-2 reaction drains remaining

$F2,6P_2$ and releases FBPase from inhibition so that it can participate in gluconeogenesis (Fig. 14.7).

Biochemical Adaptation via Reversible Phosphorylation Given the wide array of actions and consequences that protein phosphorylation can cause, it is obvious that there are innumerable opportunities for using reversible phosphorylation as a mechanism of biochemical adaptation. Many specific examples are discussed in the next three chapters, but some of the principles for the adaptive use of reversible phosphorylation include:

1. *Susceptibility to Phosphorylation* An enzyme/protein may be susceptible to phosphorylation in some species but not in others, and the development of reversible phosphorylation control can provide a mechanism for exerting strong control on an enzyme under unusual or novel stress conditions. For example, skeletal muscle pyruvate kinase (PK) in mammals is typically not a phosphoenzyme. However, whelk muscle PK is a phosphoenzyme, and anoxia-induced phosphorylation produces major changes in enzyme properties that strongly suppress enzyme activity (Table 14.4). Another example provides a twist on this story. PK in the skeletal muscle of most amphibians follows the typical vertebrate pattern, but in species that estivate (estivation is a seasonal torpor induced by arid conditions), such as spadefoot toads, skeletal muscle PK has become a phosphoprotein. In this case, when toads enter estivation, the amount of dephosphorylated PK rises substantially and dephospho-PK shows kinetic differences (higher K_m for PEP, higher K_a for $F1,6P_2$) that would suggest that it is the less active form *in vivo.* Conversion of skeletal muscle PK to a less active form during estivation is one component of a general suppression of metabolic rate during estivation, many other aspects of which are also regulated by reversible phosphorylation of enzymes. It is interesting, however, that it is the dephosphorylated form of toad muscle PK that is the less active form, whereas in whelk muscle and vertebrate liver it is the phosphorylated form of PK that is less active. Hence, there are no fixed rules about whether phosphorylation of a given protein must activate or inactivate an enzyme.

2. *Consequences of Phosphorylation for Enzyme Kinetic Properties* Adaptive change in the quantitative effect of phosphorylation on enzyme properties can be critical for altering enzyme and/or pathway function in response to an imposed stress. Table 14.4 shows that the effects of phosphorylation on mammalian liver PK and molluskan muscle PK are qualitatively the same: In both cases, the phosphoenzyme shows a higher K_m for PEP, a higher K_a for $F1,6P_2$, and a lower K_i for L-alanine. However, the quantitative effects of phosphorylation on enzyme properties are hugely different for each of these parameters. Most notably, a two-fold difference in the K_i for alanine between phosphorylated and dephosphorylated forms of mammalian liver PK has been magnified to a 490-fold difference between the two enzyme forms in the mollusk. As will be discussed more fully in Chapter 15, the kinetic differences between the two PK forms in mollusk muscle provide a virtual on–off control over PK *in vivo* that is critical to redirecting glycolytic carbon flow into an alternative pathway of fermentative ATP production in anoxia-tolerant marine mollusks.

3. *Susceptibility to Different Protein Kinases* Enzymes may be phosphorylated by more than one kind of protein kinase, each with different consequences for enzyme function. Each protein kinase mediates enzyme response to a different signal so alone or in combination, enzyme phosphorylation by multiple protein kinases can create a wide range of enzyme responses to different stresses. Continuing with the example of PK, mammalian liver PK is well-known to be phosphorylated by PKA, which contributes to glycolytic inhibition when gluconeogenesis is needed. In marine mollusks, however, PK is susceptible to phosphorylation by protein kinases A, G, and C, but the major changes in PK kinetic properties that are induced by anoxia exposure (such as seen for whelk muscle in Table 14.4) are mimicked only by the actions of cyclic-3′,5′guanosine monophosphate (cGMP)-dependent protein kinase (PKG) on the protein.

Enzyme Control via Reversible Binding to Subcellular Macromolecules

The early history of enzymology explored the kinetic and regulatory properties of purified enzymes to characterize each individual enzyme catalyst and lay the framework for our understanding of how enzymes catalyze reactions and how they are influenced by allosteric and other effectors. The subsequent identification of posttranslational modification of enzymes provided the key to understanding how changes to the activity states of enzymes could be made in a manner that was independent of changes in cellular substrate and effector concentrations and provided the mechanism by which external stimuli could make large and stable changes to the activities and properties of selected enzymes. Coincident with the explosion of research on the role of posttranslational modification in enzyme control has been interest in a third mode of regulation, that of the spatial or three-dimensional control of metabolism in cells. This mode of control also provides multiple opportunities for biochemical adaptation, various examples of which will be discussed in this and the next chapter. Below, we will first briefly explore three issues:

1. Protein–protein binding interactions between enzymes and their regulatory proteins

2. Targeting of enzymes to specific subcellular sites via enzyme binding interactions with structural macromolecules
3. Formation of multienzyme complexes

Protein–Protein Regulatory Interactions Binding interactions between two (or more) proteins are now well-known as mechanisms of metabolic regulation. Binding interactions frequently regulate enzyme activity. The classic example is that of PKA. In its inactive form, PKA is a tetramer made up of two regulatory and two catalytic subunits. However, the binding of cAMP to the regulatory subunits causes their dissociation from the catalytic subunits, which are then free to phosphorylate target proteins. Numerous other signal transduction enzymes also have binding interactions that affect their activity (Chapter 4). For example, PP-1 activity is regulated by one of two inhibitor proteins that bind to it. The transcription factor nuclear factor kappa B (NFκB) shows another principle in its interactions with its inhibitor protein, IκB. Not only does the interaction with IκB hold NFκB in an inactive form, it also keeps the protein in the cytoplasm. Stimuli that activate NFκB do so by causing the phosphorylation and ubiquitination of IκB, causing it to dissociate from NFκB and allowing NFκB to migrate into the nucleus where it activates the transcription of various target genes.

Targeting of Enzymes to Subcellular Locations Protein–protein binding interactions are also important in targeting enzymes to very specific subcellular locations. Again, this is an extremely important mechanism for controlling enzymes involved in signal transduction (Chapter 4) and obviously presents many opportunities for adaptive change. Interaction with a targeting protein is well-known in the regulation of both PKA and PP-1 as well as the MAPKs. The regulatory subunits of PKA bind with A-kinase anchoring proteins (AKAPs) that feature a conserved "anchoring motif" to bind the regulatory subunit of PKA and a "targeting domain" that interacts with a particular subcellular component. Multiple AKAP proteins are known that interact with different structural proteins, membranes, or cellular organelles. As a result, PKA can be positioned at multiple subcellular sites where it is ready to respond to changes in local cAMP concentrations.

A similar system of targeting proteins is also key to PP-1 regulation. Indeed, this mechanism provides the answer to a paradox that has puzzled researchers for some time [the reader is referred to Cohen (2002) for an expanded treatment]. The human genome codes for ~300–350 serine/threonine kinases but only ~40 serine/threonine phosphatases. So how is the specificity and independent regulation of target enzymes maintained if this low number of phosphatases must serve all these kinases? The answer is with regulatory subunits. More than 50 regulatory subunits have been identified that can interact with the catalytic subunit of PP-1, and the formation of each of these complexes confers distinct phosphoprotein substrate specificities, restricted subcellular locations, and diverse regulation to PP-1. Regulatory subunits have been found that target PP-1 to glycogen particles, myosin, the nucleus, ribosomes, mitochondria, the cytoskeleton, and centrosomes, to name a few. In each location the regulatory subunit positions the enzyme near to its potential phosphoprotein targets. For example, association of PP-1 with the glycogen binding (G) subunit targets the enzyme to the glycogen particle where it can dephosphorylate active GP and halt glycogenolysis. Phosphorylation of the mammalian muscle G subunit by PKA (in response to epinephrine) triggers dissociation of the PP-1 catalytic subunit from G subunit, releasing it from the glycogen particle and interrupting its ability to dephosphorylate its three glycogen-bound substrates: GP, phosphorylase kinase, and glycogen synthase (see Figure 4.11).

Many metabolic enzymes and functional proteins are also localized to specific sites in the cell via protein–protein binding interactions, typically with structural macromolecules. For example, as mentioned previously, the GLUT4 glucose transporter is distributed between the plasma membrane and intracellular vesicular compartments. One of the actions of insulin stimulation on muscle cells is to translocate GLUT4-containing vesicles to the plasma membrane by stimulating a release of these vesicles from intracellular tethering sites and allowing them to move to the plasma membrane where docking and fusion of the vesicles is mediated via interaction with specific proteins.

Several other enzyme binding interactions are also well-known. Hexokinase isozymes I and II bind to mitochondria, whereas type III has a perinuclear localization. Type I hexokinase binds to mitochondria through an interaction with porin, the protein that forms channels through which metabolites traverse the outer mitochondrial membrane. Experiments have suggested that in this location, hexokinase selectively uses intramitochondrially generated ATP and that this provides a way of coordinating glucose entry into glycolysis with cellular ATP demands. Kinetic properties of the type I form (inhibition by G6P that is relieved by inorganic phosphate) also fit the enzyme for a catabolic function that is sensitive to the energy needs of the cell. By contrast, types II and III appear to serve anabolic roles such as providing G6P for glycogen synthesis or for the pentose phosphate pathway.

In skeletal muscle there is strong evidence of the association of enzymes involved in ATP production with elements of the ATP-utilizing myofibrillar contractile apparatus and that this association is key to muscle function. For example, the enzyme creatine kinase [CK; reaction (14.7)], which delivers ATP from creatine phosphate reserves, is localized in muscle cells at key subcellular sites where ATP demand is high during muscle work: near the contractile machinery (the site of actomyosin ATPase), at the sarcoplasmic reticulum (where ATP is used for Ca^{2+} reuptake), and near the sarcolemma (where ATP is used to fuel the Na^+-K^+ pump). A further pool of the enzyme is located on the inner mitochondrial membrane where it serves to regenerate PCr from ATP produced by oxidative phosphorylation. The MM-dimer of CK specifically interacts with the sarcomeric M-line, and a recent study showed that binding was localized to two pairs of lysine residues that are present in M-CK but are not found in the B-CK isozyme that is widely distributed in other tissues. The principle here is that addition or removal of protein–protein binding capabilities can fulfill organ-specific needs for the placement of selected enzymes or isozymes at specific subcellular locations.

Another enzyme involved in muscle adenylate metabolism, AMP deaminase [reaction (14.8)], is also regulated in this way. AMP deaminase binds to myosin, where it is positioned to help optimize the supply of ATP to the myosin ATPase. AMP deaminase drains the pool of AMP that forms as a result of the adenylate kinase reaction [AK; reaction (14.9)]:

$$\text{Phosphocreatine} + \text{ADP} \rightarrow \text{creatine} + \text{ATP} \tag{14.7}$$

$$\text{AMP} \rightarrow \text{IMP} + \text{NH}_4^+ \tag{14.8}$$

$$\text{ADP} \longleftrightarrow \text{ATP} + \text{AMP} \tag{14.9}$$

The AK reaction helps to increase ATP supply to myosin ATPase if the rate of ATP production from muscle phosphagen pools, glycolysis, or mitochondrial oxidative phosphorylation is less than optimal and ADP begins to accumulate. The AK reaction scavenges ADP and converts it to ATP and AMP. Removal of AMP from the adenylate pool via the AMP deaminase reaction helps to keep the AK reaction running in the direction of ATP synthesis, and, in addition, the production of ammonium ion by this reaction has two effects: (a) It uses up protons that would otherwise further increase the acidification of the working muscle cell, and (b) NH_4^+ is a strong activator of PFK and provides an additional signal to stimulate PFK activity and glycolytic rate when ATP supply is limiting. Note that PFK also responds to two other signals of energy limitation: AMP is a strong activator and declining ATP reduces ATP substrate inhibition of the enzyme. The importance of AMP deaminase binding to myosin has been documented in multiple studies that have shown an increase in the amount of enzyme bound to myosin under physiological conditions where muscle ATP demand is increased (exercise) or where ATP supply is compromised (hypoxia). For example, a 90-min exposure of fish to hypoxia (low oxygen) increased the percentage of bound AMP deaminase in muscle from ~20 to 50%.

There is also good evidence that some of the glycolytic enzymes have binding interactions with myofilaments that are important to the control of glycolysis in working muscle. Recall that the most prominent effect of protein phosphorylation on skeletal muscle PFK was to increase its affinity for F-actin, an action that would bring this regulatory enzyme of glycolysis into a close association with the myofibrils. The physiological relevance of PFK binding is supported by data that show that (a) PFK retains significant binding to subcellular particulate matter under the ionic strength and metabolite concentrations found *in vivo* (various other glycolytic enzymes bind at low ionic strength but are solubilized at physiological ion concentrations), (b) PFK binding increased at lower pH values, which would enhance binding under the increasingly acidic conditions of working muscle, and (c) PFK binding to F-actin has a net activating effect due to positive effects on enzyme properties (higher affinity for F6P substrate, greater sensitivity to AMP and inorganic phosphate as activators, less sensitivity to ATP and citrate as inhibitors). By contrast, the kinetic effects of enzyme-F-actin binding interactions were all inhibitory when aldolase, glyceraldehyde-3-phosphate dehydrogenase (GAPDH), PK, and LDH were assessed. New studies also show that PFK, CK, and AK are targeted to the I-band and M-band regions of the sarcomere through association with a protein called DRAL/FHL-2, an adapter protein that, in turn, interacts with the elastic filament protein, titin. Another study has also shown that the presence of PFK, and to a lesser extent aldolase (but not PK or phosphoglycerate kinase), is necessary to achieve MM-CK binding to the myofibrillar M-line, which further demonstrates the importance of ordered enzyme binding to muscle structural components for the supply of ATP to the contractile apparatus.

Other advances in understanding the role of enzyme binding interactions in muscle function have come from studies with *Drosophila* flight muscle. In fruit fly muscle the enzyme glycerol-3-phosphate dehydrogenase (G3PDH) isozyme 1 (the muscle-specific isozyme) was found localized at Z-disk and M-lines and co-localized with it (and dependent on G3PDH-1 presence) were GAPDH, aldolase, triosephosphate isomerase, phosphoglycerate kinase, and phosphoglyceromutase. In mutant flies that were null for G3PDH-1, none of the other enzymes localized to either location. When G3PDH-3 (the isozyme of nonmuscle tissues) was substituted in transgenic flies, there was similarly no enzyme localization at either the Z-disks and M-lines; this isozyme lacks the C-terminal tripeptide (Glu-Asn-Leu) that is characteristic of G3PDH-1. Furthermore, these transgenic flies could not fly. The conclusion from this research is that the presence of a full complement of glycolytic enzyme activities is not enough to support muscle work. These enzymes must also be co-localized appropriately to create the correct structural organization of the glycolytic enzymes in relationship to the ATP-utilizing muscle fibers.

Support for the physiological importance of glycolytic enzyme binding to subcellular structural elements has also come from multiple studies that documented changes in the percentages of bound versus free enzyme in response to experimental states (e.g., muscle exercise, hypoxia) that change the demand for ATP production. These studies used several different methods of cell disruption to preserve (theoretically) glycolytic enzyme binding interactions with particulate matter. A commonly used method, for example, is extraction in an isomotic sucrose solution that should not disrupt ionic binding interactions between proteins and allow the separation of enzymes into soluble (in the supernatant) versus bound (in the pellet) fractions. (Note: After centrifugation, enzyme in the pellet is subsequently released for assay by a second extraction in a dilute buffer.) Other methods used high pressure or ultracentrifugation of minced tissue pieces to retrieve free enzymes in extruded cellular "juice" while leaving behind bound enzymes. In quantitative terms the results from these techniques were often very different but, qualitatively, there was general agreement across multiple species and tissues that changes in glycolytic rate were positively correlated with changes in the percentages of bound glycolytic enzymes. Figure 14.8 shows two such studies. Exhaustive swimming exercise by trout (strongly increasing glycolytic rate) increased the amounts of several glycolytic enzymes that were associated with the particulate fraction of the cell whereas anoxia exposure of whelk heart (reducing glycolytic rate due to metabolic rate depression) reduced the content of bound enzymes. Note that in both cases changes in the amount of bound PFK occurred. Increased enzyme binding to the particulate fraction under conditions of increased cellular ATP demand would theoretically place glycolytic enzymes into close association with the ATP-utilizing myofibrillar apparatus. Thus, these sorts of studies produced strong evidence for the physiological importance of glycolytic enzyme distribution between bound and free states as a means of metabolic regulation and biochemical adaptation.

However, difficulties arose when attempts were made to reconcile these data with the kinetic and binding behavior of glycolytic enzymes *in vitro*. Several studies evaluated glycolytic enzyme binding with F-actin and, except for PFK (described above), the results were generally unimpressive. F-actin interactions with several glycolytic enzymes were shown to be ionic in nature with good binding demonstrated under the low ionic strength and low protein conditions of an *in vitro* assay but with binding virtually eliminated when ionic strength, metabolite levels, or protein concentrations were raised to mimic cytoplasmic conditions. Furthermore, several enzymes were kinetically inhibited when bound; for example, aldolase binding to F-actin was competitively inhibited by the aldolase substrate, $F1,6P_2$,

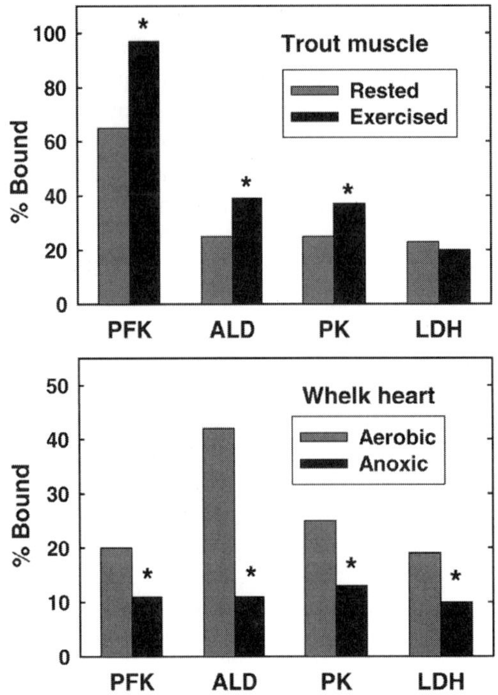

Figure 14.8 Effect of stresses on the binding of glycolytic enzymes to the subcellular particulate fraction. Fresh tissues were gently homogenized in a sucrose solution and centrifuged to separate free (supernatant) from bound (pellet) enzyme fractions; the pellet was then reextracted in dilute buffer to release bound enzymes for assay. Metabolic situations that increase glycolytic rate (e.g., muscle exercise in trout) typically increase the percentage of bound enzyme, whereas situations that decrease metabolic rate (e.g., anoxia-induced metabolic rate depression in whelks) are accompanied by reduced enzyme binding. *, Significantly different from the corresponding control state, $P < 0.05$.

so that there is no way that aldolase could both catalyze its reaction and remain bound to F-actin. PK and LDH binding to particulate matter were similarly disrupted by substrates (PEP, NADH).

Multienzyme Complexes The classic view of metabolism within a cell was that the vast majority of enzymes were freely soluble entities within the cytoplasm or within the lumen of various organelles. Compartmentation of selected enzymes/pathways into organelles was accepted as was the presence of a number of enzymes as integral membrane proteins, but the vast majority of metabolic enzymes, as well as their substrates, were considered to be randomly distributed in a cytoplasmic "soup." Pathway flux depended upon diffusion of substrates from one enzyme to the next. This view arose because a vast majority of enzymes, no matter how gentle the cell disruption or homogenization technique, always appeared in the soluble cytoplasmic fraction. This view began to change from the 1960s onward with two types of advances: (1) the development of cell disruption techniques that apparently preserved delicate binding interactions and (2) histochemical techniques that showed uneven distribution of enzymes in the cytoplasm and preferential association of certain enzymes near structural elements in the cell. It is now known that the cytoplasm is very heterogeneous and that multiple mechanisms exist for creating subcellular associations of enzymes with other macromolecules in order to localize selected reactions to specific subcellular locations (see above). The reader is also referred to Chapter 3 for detailed treatments of the consequences of enzyme complex formation for enzyme activity and pathway flux.

Some types of tightly bound multienzyme complexes are unquestionably integral to selected cell functions. For example, the ribosome is clearly a complex of multiple enzymes and structural proteins that synthesizes new proteins just as the proteasome is a multicomponent digestive complex that degrades proteins; multienzyme complexes are also involved in DNA synthesis (see Chapters 6 and 7). As we will also see in Chapters 15 and 16, the association versus dissociation of multienzyme complexes such as the ribosome is a major mechanism of cellular response to stress; for example, the regulated disaggregation of polysomes is a key element in the strong suppression of protein synthesis under stress conditions that limit cellular ATP availability. Fatty acid synthase, pyruvate dehydrogenase, and cytochrome *c* oxidase are also good examples of metabolic "enzymes" that are, in fact, multienzyme complexes. For example, the pyruvate dehydrogenase complex is an association of three enzymes that carry out the reduction, decarboxylation, and thioester formation steps of the net reaction.

But what about enzymes in pathways that are traditionally thought of as being soluble, such as glycolysis. The idea of a *glycolytic complex* involving an association of all of the enzymes of glycolysis together has resurfaced in the scientific literature a number of times. Such a complex could serve to channel substrate from one enzyme to the next through the entire pathway to facilitate efficient carbohydrate catabolism and ATP output. If also anchored close to cellular sites with high ATP demand (e.g., myofilaments), a highly efficient coupling of ATP production and ATP use could be created that could, for example, power high-intensity white muscle work. During the 1980s and early 1990s several lines of evidence were developed to support the existence of a glycolytic complex, including (a) positive correlations between glycolytic rate and the percentage of bound glycolytic enzymes under multiple metabolic states that either increased or decreased ATP demand (Fig. 14.8) and (b) glycolytic binding interactions *in vitro* with structural proteins such as F-actin. However, selected glycolytic enzymes could be bound to structural elements without necessarily creating a full glycolytic complex, and as discussed above, several enzymes were actually inhibited when bound to F-actin. Indeed, mathematical modeling using the properties of free and bound enzymes and considering the percentages of each glycolytic enzyme that could be bound to F-actin under physiological ionic strength showed that a functional glycolytic complex that could channel substrate from glucose input to lactate output was unlikely. Hence, the idea of a glycolytic complex has been out of favor for the last several years. What is interesting, however, is that new studies are reopening interest in this idea. The data from *Drosophila* that show that flies cannot fly if G3PDH and its associated enzymes are not localized to specific areas of the contractile apparatus and the studies showing that PFK, CK, and AK bind to the DRAL/FHL-2 protein that associates with titin both support the idea that (a) glycolytic enzymes form binding associations with myofibrillar proteins that benefit ATP supply during contraction, and (b) a glycolytic complex is probably not built around enzyme associations with F-actin. Instead, the model seems to be that selected enzymes (e.g., G3PDH, PFK) interact with specific anchoring proteins and that these enzymes in turn can associate with other enzymes, potentially ultimately building an entire glycolytic complex. Hence, an older idea (glycolytic complex formation) seems poised for a come-back but one in which there will likely be many more players and much more complexity, including multiple myofibrillar proteins that can act as binding sites, multiple anchoring or targeting proteins, various modulating proteins, and also highly specific sequences in which different enzymes can bind to the complex. Studies over the next few years should be intriguing.

BIOCHEMICAL ADAPTATION AND MACROMOLECULAR PROTECTANTS

In most of the cases of enzymatic regulation discussed above, the purpose of biochemical adaptation is to make selective changes to the regulation of one or more individual enzymes or proteins in order to make very specific changes to metabolism or compensate for very specific effects of stress on individual enzymes/proteins. However, environmental stress can also have generalized consequences for many cell functions and/or macromolecules. In these situations, it is inefficient to consider modifying each protein to optimize it for stress resistance and more efficient to utilize one of several types of protectants or stabilizers for "global" stress relief. Metabolite protectants are reviewed briefly below and those involved in cold adaptation are discussed in greater detail in Chapter 17.

A variety of low-molecular-weight metabolites are used as protectants in the face of environmental stresses such as dehydration, freezing, or salinity change. These have two main functions: (1) cell volume regulation and (2) stabilization of protein conformation or membrane bilayer structure. Metabolite protectants generally fall into two groups, carbohydrates and nitrogenous compounds. Carbohydrates include monosaccharides (e.g., glucose), disaccharides (e.g., trehalose), and a variety of polyhydric alcohols (e.g., glycerol, sorbitol, and others). Nitrogenous small molecules include urea, trimethylamines, betaine, and multiple amino acids including alanine, glycine, and proline. The same metabolite can be used in different contexts across phylogeny. For example, high glycerol levels provide antifreeze protection to cold-hardy insects (see Chapter 17), minimize cell water loss for algae placed in hyperosmotic media, and stabilize macromolecular structures in organisms that endure extreme desiccation (e.g., brine shrimp cysts). Glycerol is also well known to experimental biochemists as an excellent stabilizer of purified proteins and is widely used in medical science for the cryopreservation of cells and tissues.

Cells strive to maintain homeostasis, which includes maintaining an optimal cell volume and holding the ionic strength of body fluids within strict limits. Beyond these limits irreparable harm can occur; for example, mammals show nervous system dysfunction when plasma sodium rises by 30 to 60 mM above normal. Stresses that cause excessive water loss from cells (e.g., desiccation, freezing, hyperosmotic stress) create multiple problems because: (a) water is a key structural element in all cells, particularly for organisms without bones or exoskeletons, (b) water loss (or water gain) puts structural strain on membranes that can lead to the breakdown of the bilayer structure, (c) multiple cell functions (e.g., nerve signal conductance, cell sensitivity to external stimuli), and enzymatic activities require both overall ionic strength and the concentrations of individual ions to be maintained within narrow limits, and (d) maintenance of a surrounding vicinal shell of water is critical for native protein conformation and situations of extreme dehydration cause protein denaturation.

The majority of metabolite protectants have a primary physiological role in cell volume regulation and a secondary role, when cell water loss is unavoidably extreme, in the physical protection of macromolecules. The function in cell volume regulation is due to the colligative effects of high concentrations of low-molecular-weight metabolites in retarding water loss from cells. Physical stabilization of proteins typically arises because of the preferential exclusion of protectants from contact with the surface of proteins in aqueous solutions; this stabilizes the native conformation of proteins surrounded by their "shells" of water (see Chapter 17 for more about the action of cryoprotectants on both proteins and membranes).

Certain metabolites such as polyhydric alcohols and neutral amino acids are favored as protectants for several reasons. High solubility in the aqueous cytoplasm is a must in order to achieve the high concentrations (0.5 to 2 M) that are typically needed to have a substantial colligative effect in retarding cell water loss. But such high concentrations of many solutes (e.g., inorganic ions) would have hugely negative effects on protein conformation and enzymatic activities. Hence, useful protectants must be *compatible solutes*, which means that their presence in high concentrations must have little or no effect on the functions of cellular enzymes and pathways. That this is true has been shown for many different enzymes but is illustrated in Table 14.6 for the enzyme 6-phosphogluconate dehydrogenase, the second enzyme in the pentose phosphate pathway. In present case, the Michaelis constant (K_m) for substrate of the enzyme was assessed at two temperatures and in the presence/absence of glycerol, sorbitol, and KCl. At 22°C, the addition of high levels of glycerol (500 mM), sorbitol (250 mM), or both polyols had little effect on the K_m, but high KCl greatly reduced enzyme affinity for substrate by increasing K_m by sevenfold. However, when glycerol and sorbitol were added to assays containing high KCl, the negative effects of KCl on K_m were largely overridden. The same relationship was seen when the enzyme was assayed at 5°C, but, additionally, the low-temperature-stimulated rise in the K_m value under control conditions (no additions) was also partially reversed by the presence of polyols.

Nitrogenous metabolites are also frequently used as protectants. All animals maintain an intracellular ionic strength (made up mainly of K^+ but also Na^+, Cl^-, Ca^{2+}, and other ions) of 200 to 300 mM with other metabolites and proteins adding about another 200 mM. However, marine animals

TABLE 14.6 Effects of Glycerol and Sorbitol on the Michaelis Constant (K_m) for Substrate of 6-Phosphogluconate Dehydrogenase from a Cold-Hardy Insect: Reversal of Negative Effects of Low Temperature and High Salt[a]

	K_m 6-Phosphogluconate (μM)	
	Assay at 22°C	Assay at 5°C
Control (no additions)	64	119
+ Glycerol	53	60
+ Sorbitol	52	52
+ Glycerol and sorbitol	38	64
+ KCl	451	471
+ Glycerol, sorbitol and KCl	119	232

[a]When present, the concentrations of additions were 500 mM glycerol, 250 mM sorbitol (both similar to natural winter levels in the insect), and 500 mM KCl. Compared with the control K_m at 22°C, a temperature decrease to 5°C doubles the K_m whereas addition of KCl raises K_m by ~sevenfold. In the presence of polyols, the temperature effect is reversed and the KCl effect is greatly reduced.

have to reconcile this with a seawater ionic strength of ~1000 mM (mostly Na^+ and Cl^-). Marine invertebrates make up the difference in osmolality with high intracellular pools of amino acids, betaines, and other nitrogenous metabolites. These pools are raised or lowered as needed to allow animals to compensate for increases or decreases in seawater salinity. Again, the nitrogenous metabolites used are typically compatible solutes that do not disrupt enzyme kinetic or regulatory properties. However, one interesting situation deserves comment. Elasmobranch fishes (sharks, rays) use high concentrations of urea and trimethylamines/betaines in their body fluids to balance their internal osmolality with that of seawater. Urea is well-known to biochemists as a protein denaturant, and so researchers wondered how shark enzymes would function in the presence of constant high urea, about 400 mM in animals adapted to full-strength seawater. Indeed, the kinetic properties of a number of shark enzymes were negatively affected by high urea when assayed *in vitro*. However, these same enzymes responded positively to the addition of trimethylamines/betaines and, when added together in the physiological ratio (approximately 2 : 1), the negative effects of urea on enzyme properties were fully counteracted by the positive effects of trimethylamines/betaines. Hence, sharks have evolved a balanced system of multiple osmolytes that, as a whole, is compatible with unperturbed enzyme function although individual components can perturb some enzymes.

Urea is also commonly accumulated by terrestrial organisms that undergo desiccation stress. For example, urea accumulation is key to water retention during estivation (torpor) in toads and frogs that live in arid environments. These animals are only active briefly during a short summer rainy season, but then they retreat underground to estivate for 9 to 10 months of the year. As the desert soil dries out, the animals respond to increasing desiccation stress by breaking down muscle protein and accumulating urea, the nitrogenous end product of protein catabolism in their body fluids (see Chapters 8 and 9 for urea cycle function). Urea levels can reach 300 mM in plasma and tissues of spadefoot toads, but unlike the elasmobranch situation no counteracting solutes accumulate. However, in this case, no counteracting solute system is needed because natural levels of urea in estivating toads actually have little or no effect on estivator enzymes. Figure 14.9 shows the effect of 200 mM KCl on four toad enzymes. In all cases, enzyme activity was inhibited including a 90% reduction in the activity of glutamate dehydrogenase, a key enzyme of amino acid catabolism. However, urea inhibited only one enzyme. The comparable situation was documented for many other metabolic enzymes from liver and muscle of spadefoot toads. Because toads can lose about half of their total body water during estivation, intracellular K^+ and Cl^- concentrations could rise during estivation to a level that can have serious inhibitory effects on cellular enzymes. The accumulation of urea in estivator tissues helps to prevent this. Acting as a colligative osmolyte, the accumulation of high amounts of urea retards water loss from cells and from the organism as a whole. This, in turn, minimizes the increase in intracellular inorganic ion

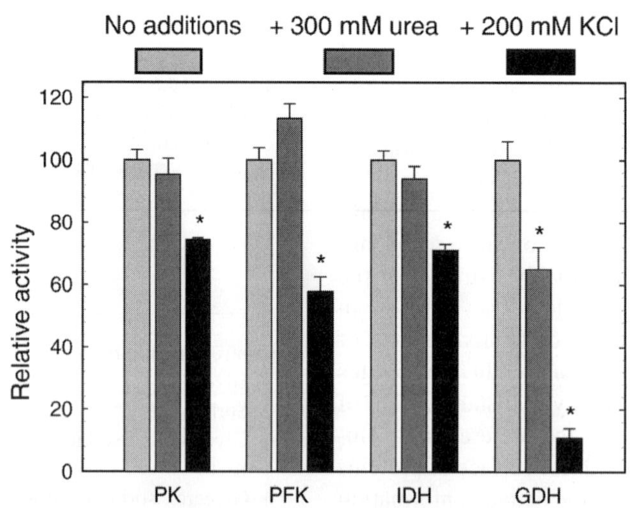

Figure 14.9 Effect of the addition of 300 mM urea or 200 mM KCl on the maximal activities of enzymes from spadefoot toad skeletal muscle. Enzymes are: PK, pyruvate kinase; PFK, phosphofructokinase; IDH, NAD-dependent isocitrate dehydrogenase; GDH, glutamate dehydrogenase. Data are expressed relative to control activities (no additions) and are means ± SEM, $n = 3\text{-}4$. *, Significantly different from the corresponding control, $P < 0.01$. See Storey (2002) in the suggested reading list.

concentrations and prevents ion concentrations from rising into a range that would inhibit enzymatic activity.

ADAPTIVE CONTROL OF TRANSCRIPTION

Organisms maintain an astonishing array of genes; many are constitutively expressed either in all or selected cell types, but many others are differentially expressed only in certain situations such as during growth, differentiation, and sexual maturation or in response to external stimuli or stressors including pathogens, changes in food availability, and changes in environmental parameters. The ability to detect and analyze stress-induced changes in gene expression has improved immensely in recent years and has led a revolution in the way that researchers can approach problems in biochemical adaptation. Previously, the main approach to a problem in biochemical adaptation was to identify some outstanding metabolic feature that was present in stress-tolerant organisms but not stress-intolerant forms and then work "from the bottom-up" to determine how that feature functioned and how it was regulated at both a protein and gene level. Examples from the field of cold-hardiness illustrate this point (see Chapter 17 for greater expansion). One of the first discoveries in the field of insect cold-hardiness was that animals produced huge quantities of glycerol (reaching ~2 M in concentration or as much as 20% of the animal's body weight) that protected cells for survival at −40°C or lower. This initial discovery has led to thousands of research studies that have explored the distribution of glycerol and other cryoprotectants in hundreds of species, the regulation of cryoprotectant synthesis/degradation, the actions of cryoprotectants in protecting cells and proteins, and the applied use of these metabolites in medical organ cryopreservation. Similarly, a measured difference (hysteresis) between the freezing and melting points of the blood plasma of cold-water marine fish led to the discovery of antifreeze proteins (AFPs) and spawned another huge field that has explored AFPs in multiple ways: physiological function, molecular binding interactions with ice, and, most recently, the control of gene expression and the identification of novel mechanisms of gene evolution (see below).

However, the recent explosive growth in gene screening technologies, including new complementary DNA (cDNA) arrays (Text Box 14.2) has now made it possible to explore biochemical adaptation "from the top-down" by providing the means to identify those genes that are up-regulated in response to a stress and then analyze the corresponding response of protein levels and determine the action of the protein in contributing to stress tolerance. This approach has been particularly useful in two situations: (a) identifying gene/protein changes that are important but do not have a major metabolic "footprint" that would allow them to be

TEXT BOX 14.2 NEW FRONTIERS IN BIOCHEMICAL ADAPTATION

A revolution is underway in biology and medicine supported by two major advances: (1) the sequencing of the genomes of multiple species and (2) the development of cDNA microarrays. These have been combined to provide the ability to simultaneously assess the responses to an experimental parameter by thousands of genes and provide a broad-based and comprehensive analysis of cell/organ responses to stimuli (e.g., hormones, nutrition, development, aging), environmental stresses, or disease. For decades, the experimental sequence for studying organismal response to stress has been (a) identify a perturbed or novel physiological function, (b) seek an underlying protein/enzyme adjustment, (c) characterize the protein/enzyme adaptation, and (d) where appropriate, analyze gene structure and/or transcriptional control. Gene screening technology inverts this sequence and provides researchers with a second way of tracking down biochemical adaptations based on the identification of genes that are strongly up- or down-regulated by an imposed stress. Very importantly, array screening is "opening the eyes" of researchers to the coordinated involvement of many aspects of cellular metabolism in the response to an external stimulus. Major advances provided through gene screening can include: (1) discovery of genes/proteins and their cell functions/pathways that were never previously suspected as being part of an adaptive response, (2) analysis of the responses to a stress by functionally related groups of genes/proteins (e.g., mitochondrially encoded genes, genes for transmembrane transporters), and (3) identification of responses by multiple genes that are under control by a specific signal transduction pathway or that share a common response element.

MICROARRAY TECHNOLOGY

State-of-the-art glass slide microarrays can have up to 31,500 nonredundant cDNAs bound to them, most of which are identified genes, whereas others are expressed sequence tags representing genes of, as yet, unknown function. Figure TB14.2 outlines the basic procedure of array screening showing the following steps: (1) mRNA is isolated from control versus experimental tissues, (2) reverse transcriptase is used to create first-strand cDNA from the mRNA template in the presence of one of two fluorescent dyes (either Cy3 or Cy5) conjugated to dCTP, (3) equal amounts of control and experimental cDNA samples are mixed and applied to the microarray for hybridization, (4) slides are washed to remove nonspecific binding and centrifuged to dry, (5) slides are placed in a microarray reader and scanned twice to provide fluorescent intensity profiles for each dye (excitation and emission wavelengths are, respectively, 550 and 570 nm for Cy3 and 649 and 670 nm for Cy5), and (6) false coloring is applied to each intensity profile, and then images are superimposed and typically show differences in signal intensity as follows: red (experimental signal > control), yellow (experimental = control), or green (experimental < control). A ratio experimental : control (E : C) transcript levels of about 2 : 1 or greater is generally considered to be significantly up-regulated. A color image of a hybridized array and a more detailed technical explanation and animated tutorial are available from the Microarray Centre (www.microarrays.ca).

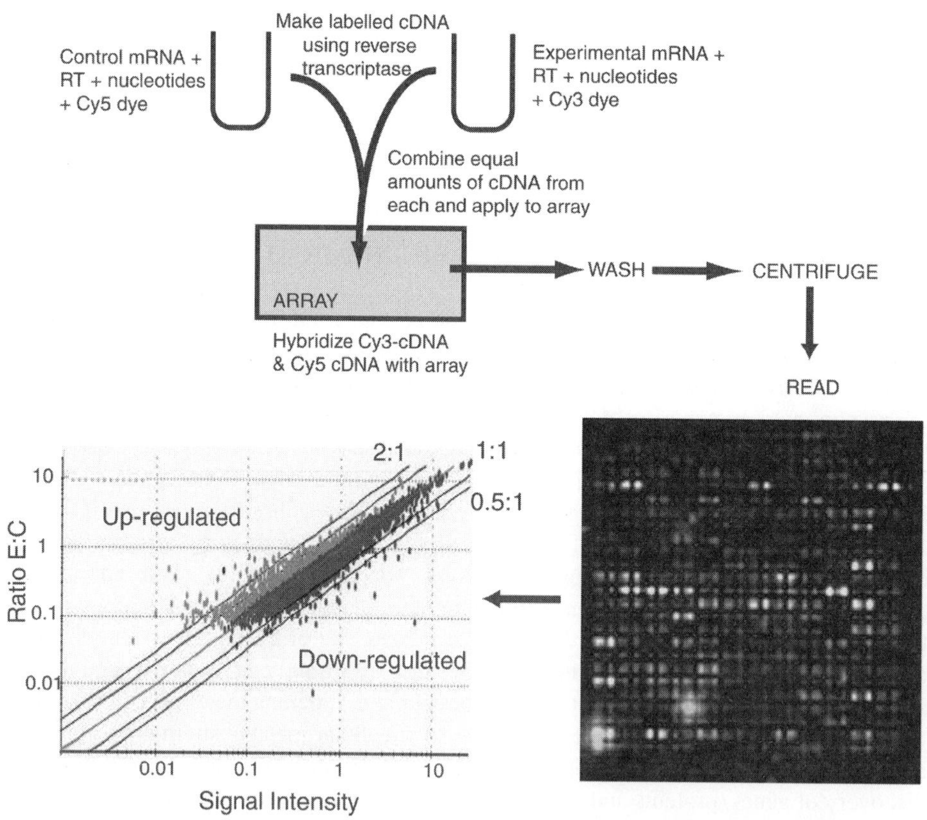

Scatter plot showing Ratio Experimental:Control versus intensity of control. Genes with an E:C ratio of 2:1 or greater (upper left) are putatively up-regulated

Scan FLUORESCENCE at 2 wavelengths, colorize & superimpose control & experimental images

Figure TB14.2

HOMOLOGOUS VERSUS HETEROLOGOUS SCREENING

Whereas homologous screening is obviously best (e.g., human tissue samples assessed with human cDNA arrays), comparative biochemists can make good use of heterologous screening to assess differential gene expression in unusual animal systems. For example, we have used human arrays to assess gene expression changes in hibernating ground squirrels and bats (see Chapter 16). The gene sequence differences between species do limit the degree of cross-hybridization but still allow analysis of thousands of genes. For example, we found 85 to 90% cross-reaction when cDNA prepared from ground squirrel tissues was screened on a human 19,000-gene cDNA array but only 18.4% cross-reaction when cDNA from a marine snail was used. The low cross-reaction of snail cDNA is disappointing, but the analysis still revealed over 300 genes that were putatively up-regulated in response to anoxia stress (including protein phosphatases and kinases, mitogen-activated protein kinase-interacting factors, translation factors, antioxidant enzymes, and nuclear receptors), providing a plethora of new "leads" in the search for the regulatory mech-anisms underlying anoxia tolerance (see Chapter 15).

found through a bottom-up approach, and (b) finding novel proteins that were never before implicated in stress tolerance.

Figure 14.10 shows the application of this top-down approach in the discovery and assessment of a novel gene called *liver 16* or *li16*, which contributes to freezing survival by the wood frog, *Rana sylvatica.* The wood frog is one of only a handful of amphibians and reptiles that naturally endures whole-body freezing during winter hibernation (see Chapter 17). The search for freeze-responsive genes began with the construction of a cDNA library made from messenger ribonucleic acid (mRNA) isolated from liver of frozen frogs. The library was screened using cDNA probes prepared from mRNA of control frogs (held at 5°C) and frozen frogs (24 h at −2.5°C) and a freeze-responsive clone was isolated. Nucleotide sequencing provided a partial sequence, and then this was extended using the technique of 5′RACE (rapid amplification of cDNA ends) to provide the full open reading frame that encoded a 12.8-kDa protein. Both nucleotide and putative protein sequences were compared with all others stored in GenBank but had no matches. This suggested that the *li16* gene encodes a protein with a unique function in vertebrate freeze tolerance. The technique of Northern blotting was used to confirm that the *li16* gene was up-regulated during freezing. Samples of mRNA from liver of control frogs and frogs frozen for 4, 8, or 24 h were separated on formaldehyde gels, transferred to nylon membranes, and then hybridized with *li16* probe, the results showing a mean 3.6-fold increase in *li16* transcript levels after 24 h. A stress-induced elevation of transcripts is most likely due to increased gene transcription but can also arise from post-transcriptional controls (e.g., decreased rate of transcript degradation). Hence, the nuclear runoff assay was used to directly test the rate of transcript synthesis. Freshly isolated nuclei from both control and frozen frogs were incubated with ^{32}P-uridine triphosphate to label elongating transcripts and after 30 min the levels of radiolabeled *li16* transcripts were assessed by dot blots using *li16* probe. As Figure 14.10 shows, nuclei from frozen frogs produced substantially more *li16* transcripts, levels being 2.4-fold higher than in control frogs; this confirmed increased gene transcription during freezing (however, transcript levels of a constitutively expressed gene were unchanged). Finally, since stress-induced up-regulation of a gene is of little or no functional consequence if it is not translated into protein, Li16 protein levels were examined via Western blotting. Li16 protein was quantified using antibodies raised against the C-terminal decapeptide of the protein. Western blots revealed the presence of an appropriately sized 13- to 14-kDa protein in wood frog liver, and the content of Li16 protein increased progressively during freezing to a final 2.4-fold higher than control levels. The function of the Li16 protein awaits further experimentation but, to date, some facts are helping to narrow its possible functions: (a) The protein has no signal sequence and is therefore not exported from hepatocytes, (b) expression is limited to liver, heart, and gut, (c) the protein is also induced by anoxia exposure, which suggests a possible role in aiding ischemia resistance during freezing, and (d) transcript levels in tissue slices respond to stimulation by cGMP but do not respond to cAMP.

Changes in the amounts of selected proteins in cells is a major mechanism of biochemical adaptation, and regulation of gene transcription and protein translation are integral to biochemical adaptation. Multiple opportunities for transcriptional control exist (summarized in Chapter 6), and these can all be utilized to achieve biochemical adaptation.

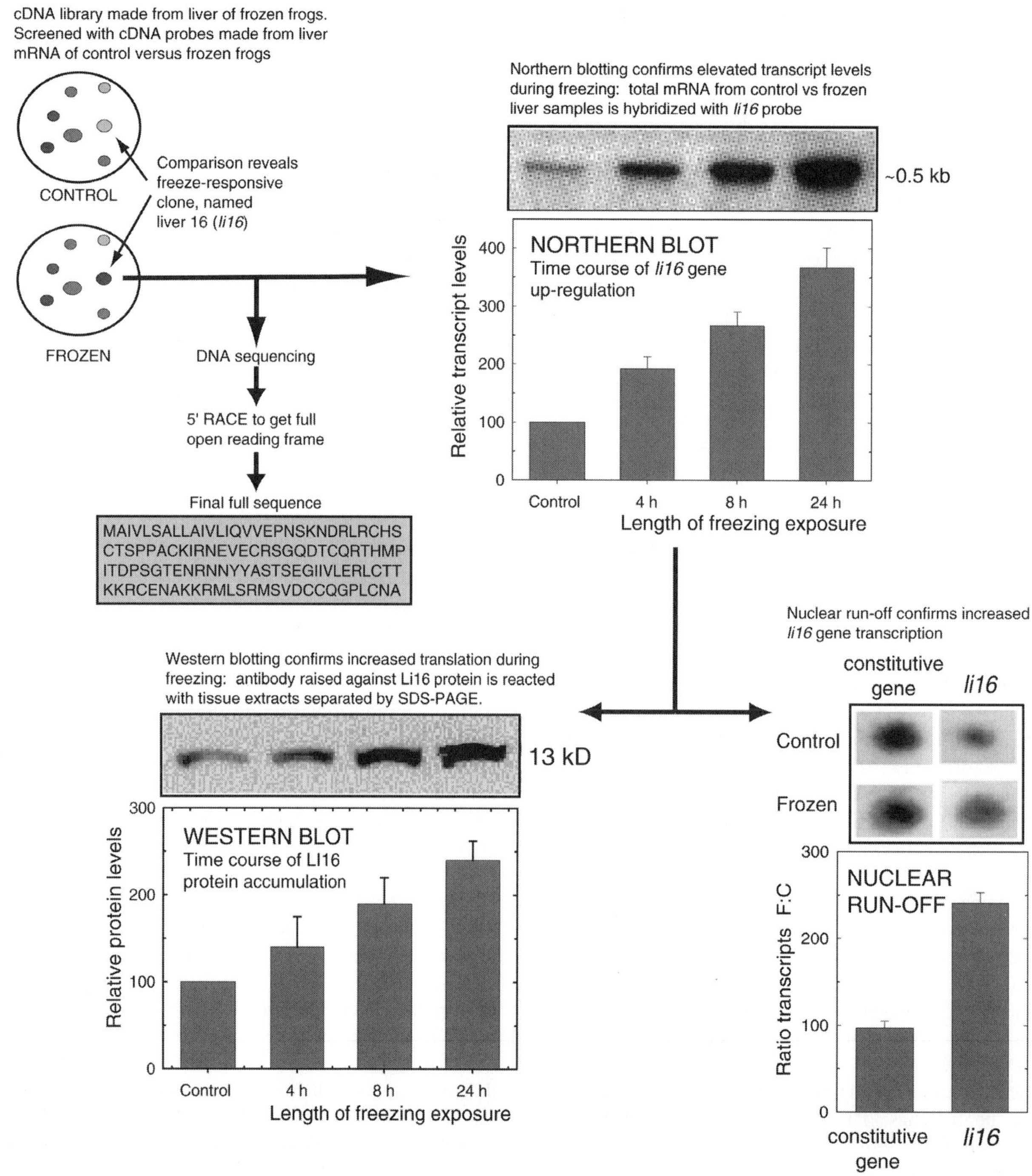

Figure 14.10 Methods for discovery and analysis of stress-induced genes highlighting the analysis of a freeze-induced novel gene, *li16*, from liver of the wood frog, *Rana sylvatica*. Techniques are explained in the text. [Example data are excerpted from J. D. McNally, S. Wu, C. M. Sturgeon, and K. B. Storey (2002). *FASEB J* **16**:902–904.]

These include:

1. Evolution of new genes encoding new protein types that address key adaptive needs.

2. Duplication of genes to insert multiple copies in the genome to greatly increase the number of mRNA transcripts that are produced in response to a signal.

3. Modification of existing or elaboration of new response elements in the 5′ untranslated region of a gene so that transcription of an individual gene can respond to a new transcription factor or to new combinations of factors.

4. Elaboration of new transcription factors or modification of the stimuli that activate transcription factor synthesis to provide a new or modified transcription factor response to a signal. Modification of transcription factor response (in either type or amount) would typically alter the responses of all target genes under its control to provide a modified yet coordinated response by multiple genes to a signal.

5. Modification of one or more elements of the signal transduction pathway leading from cell surface to gene expression that may include: the cell surface receptor, a second messenger metabolite, and one or more protein kinases often linked in a cascade that ultimatly regulate transcription factor synthesis. Multiple components in the pathway allow for multiple regulatory inputs that could permit:

   a. One signal to be spread out to initiate multiple effects on different aspects of cell metabolism such as up-regulation of selected genes, stimulation of fuel metabolism, and modulation of membrane ion conductance
   b. Diverse signals to activate the same coordinated set of responses or subsets of the responses
   c. Signal response to be modulated by inputs from other extracellular or intracellular sources

Two examples drawn from the field of animal cold-hardiness provide excellent illustrations of the principles of biochemical adaptation at the translational level. The first shows how the elaboration of new transcription factors (mechanism 4 above) can be utilized to bring preexisting gene responses to dehydration under the control of a cold trigger, whereas the second shows how new genes can be constructed and proliferated (mechanisms 1 and 2 above) to fulfill a requirement for an antifreeze protein.

Modification of Transcription Factors and Response Elements

Many plants and animals that live in temperate, polar, and alpine regions of Earth have developed freeze tolerance, the ability to endure the accumulation of ice in extracellular fluid spaces while defending the liquid state of the cytoplasm (see Chapter 17). Several of the adaptations that aid freeze tolerance are concerned with cell volume regulation, including mechanisms that defend a critical minimum cell volume against cell water loss into growing extracellular ice crystals, and the proliferation of stabilizers that help to preserve the structural integrity of membranes when cells shrink. In plants, these same mechanisms are found worldwide in drought-resistant plants (to help cells resist evaporative water loss), and this gave rise to the idea that adaptations for freeze tolerance grew out of preexisting strategies of desiccation resistance. How was this accomplished? The answer lies in gene regulation.

Genes involved in desiccation resistance in plants all contain a drought-responsive element (DRE) in their 5′ promoter region that binds a DRE-binding (DREB) transcription factor that is produced in response to a dehydration signal. When researchers looked at cold-responsive genes, they identified a cold-repeat element (CRT) in the 5′ promoter of cold-responsive genes that bound a CRT binding factor (CBF). Interestingly, when the DRE and the CRT sequences were finally compared in the model freeze-tolerant plant, *Arabidopsis thaliana*, they were found to be identical. Hence, the same set of cell volume regulatory genes could be turned on by either the DREB or CBF transcription factors (Fig. 14.11). However, although the DREB transcription factor is widely distributed, only cold-hardy plant species possess the CBF transcription factor. Transcription and synthesis of CBF is stimulated by cold or freezing exposure through a signal transduction pathway that has not yet been worked out. Thus, through the development a single new transcription factor (CBF), plants were able to exert coordinated control over a diverse group of genes whose products aid drought resistance and place the expression of these genes instead under the control of a cold signal. For more about this topic, see Chapter 17.

Elaboration of New Genes and Proteins

As much as possible, metabolism is conservative, making use of existing genes and proteins and modifying them as little as possible to adapt to new situations. For example, as little as one amino acid substitution in the sequence of the enzyme LDH can differentiate the pressure-sensitive enzyme of shallow-water fish from the pressure-insensitive form displayed by deep sea fish.

In some cases, however, a truly novel protein is needed for a unique new function and then some very creative solutions are found. Two instructive examples of how new genes and proteins are created comes from recent research on the origin of antifreeze proteins in coldwater marine fishes. The freezing point of vertebrate plasma is about $-0.5^{\circ}C$, whereas that of seawater is $-1.86^{\circ}C$. Nonetheless, polar fish live year-round in water that is colder than the freezing point of their blood; in temperate regions, surface-dwelling and inshore fish also show the same anomaly during the winter months. In these situations, body fluids are prevented from freezing by the presence of antifreeze proteins (AFPs) that bind to microscopic ice crystals and prevent them from growing (AFP action is discussed in greater detail in Chapter 17). In evolutionary terms, the

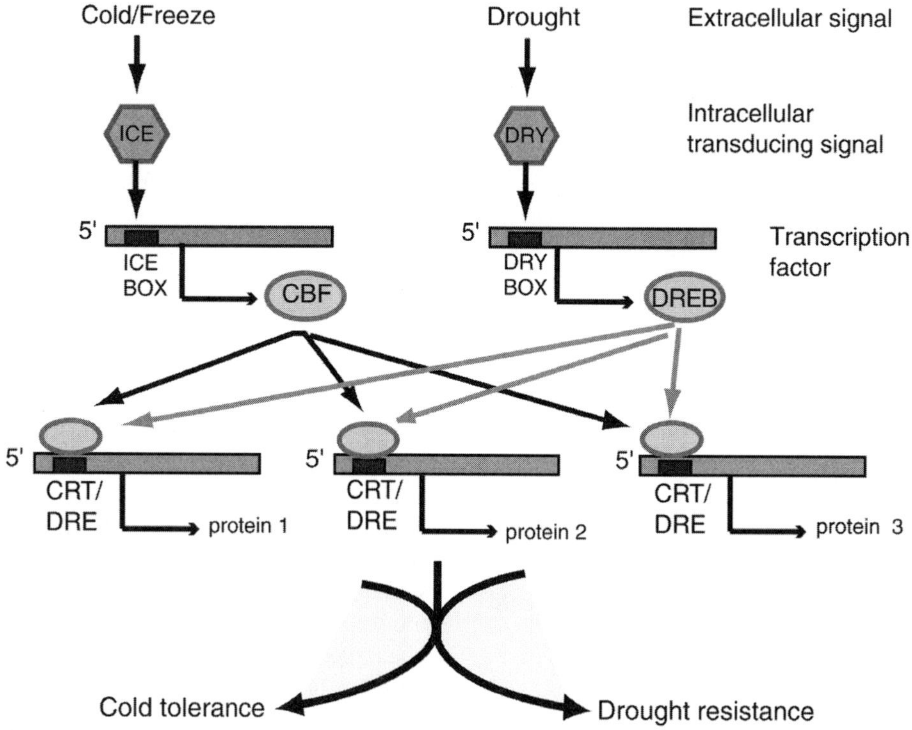

Figure 14.11 Schematic illustrating how a group of plant genes that aid drought resistance were put to a new use in freeze tolerance through the elaboration of a new transcription factor that allowed the genes to respond to a second environmental signal, cold. Genes involved in drought resistance contain a drought-responsive element (DRE) with a sequence TACCGACAT in their 5′ promoter region. This responds to the DREB transcription factor that is produced in response to a "DRY" signal. The same response element occurs in genes that respond to cold stimuli where it is known as the cold-repeat (CRT) element. Through the elaboration of a single new transcription factor, CBF (CRT binding factor), these multiple drought-responsive genes were made responsive to cold. CBF responds to an "ICE" (meaning "inducer of CBP expression") signal that interacts with a cold-regulatory element, the "ICE-box," present in the promoter region of the CBF. [Redrawn from M. F. Thomashow (1999). *Ann Rev Plant Physiol Plant Mol Biol* **50**:571–599. Used with permission.]

pressure to develop antifreezes was rapid and recent. The need for them arose only 10 to 14 million years ago when sea level glaciation began to recur after an interruption of >200 million years. How are new genes produced?

Five main types of AFPs are known, each in different families of fish, and two of these provide intriguing stories of gene adaptation. The cysteine-rich type II AFPs illustrate the evolution of an existing gene/protein into a new function. Type II AFPs are found in sea raven, smelt, and herring, species that diverged long before the glaciation that stimulated the need for AFPs. Nonetheless, both structural and sequence analysis has shown that the AFPs of all three species were derived from the Ca^{2+}-dependent (C-type) lectins, proteins that play a role in immunity by recognizing surface carbohydrates on pathogens. AFPs from all three fish show up to 30% sequence identity with the carbohydrate recognition domains (CRDs) of C-type lectins. Apparently, the ability of the CRDs to recognize and bind to galactose or other sugars was modified to create an AFP that recognizes and binds to ice. Interestingly, in two of the three species, the AFP retained one of the two calcium-binding domains of the C-type lectin and is Ca^{2+}-dependent, whereas the sea raven AFP lost both domains and is Ca^{2+}-independent.

Interestingly, plant AFPs also appear to be derived from older protein types. Carrot AFP is leucine rich and shows high sequence similarity with polygalacturonase inhibitor proteins that are secreted by plants as antifungal agents. AFPs in winter rye also show derivation from pathogenesis-related secretory proteins, the progenitor proteins apparently being ß(1-3)endoglucanase and chitinase. As in fish, therefore, plant AFPs appear to have arisen quite recently on the evolutionary timescale through modification of preexisting protein types.

The antifreeze glycoproteins (AFGPs) of Antarctic notothenioids and Arctic cod illustrate a different strategy

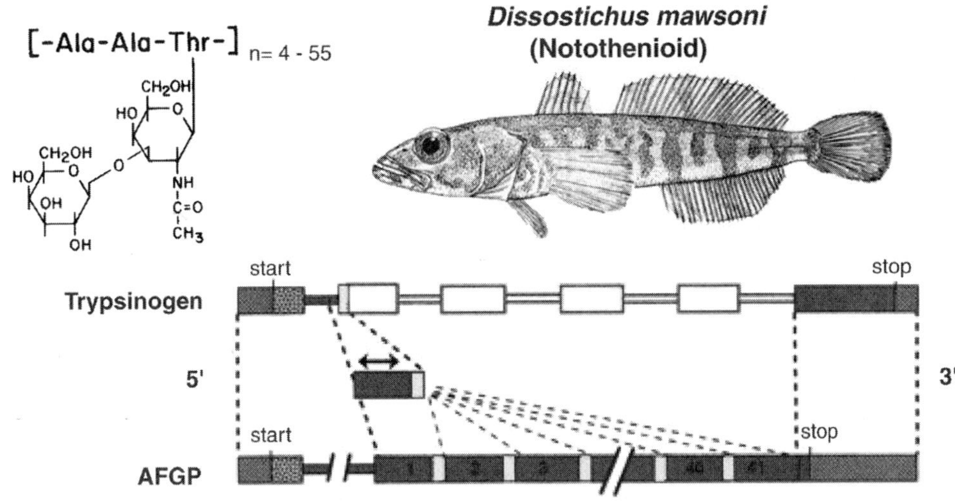

Figure 14.12 Evolution of the antifreeze glycoprotein (AFGP) gene from the pancreatic trypsinogen gene in the Antarctic notothenioid fish *Dissostichus mawsoni.* Upper left panel shows the structure of the glycotripeptide repeat with the disaccharide galactose-*N*-acetylgalactosamine attached to the threonine residue; different sizes of AFGPs repeat the tripeptide from 4 (2.6 kDa) to 55 (34 kDa) times. The lower panel shows the region of the trypsinogen gene from which the 9 nucleotide segment coding for the tripeptide was taken; this region spans the boundary between the first intron (thin boxes) and the second exon (thick boxes). The segment was iteratively repeated to produce tandem repeats in the AFGP gene (in this case 41) with retention of the 5′ and 3′ regions; white segments between the AFGP coding regions represent proteolytic cleavage sites. The 5′ signal sequence is key for secreting the AFGP into the digestive tract where the antifreeze coats ingested ice crystals. [The lower panel is reproduced from J. M. Logsdon and W. F. Doolittle (1997). *Proc Nat Acad Sci U.S.A.* **94**:3485–3487. Used with permission.]

of gene/protein evolution. AFGPs consist of a family of isoforms of different sizes (2.7 to 34 kDa) that are all composed primarily of a simple glycotripeptide repeat (Thr-Ala/Pro-Ala)$_n$ with galactose-*N*-acetylgalactosamine attached to each Thr (Fig. 14.12). How this odd structure came to be was unknown until researchers began to address the structure of the notothenioid AFGP genes. Each gene encodes a large polyprotein precursor with multiple AFGP molecules; for example, the first gene that was characterized coded for a polyprotein of 46 AFGP molecules (44 of isoform 8 and 2 of isoform 7) linked in direct tandem by a 3-residue spacer that was the posttranslational cleavage site. Notably, the creation of a direct tandem linkage is a highly efficient way of "quickly" constructing a new gene under intense selective pressure and arranging for the synthesis of copious quantities of antifreeze from a single mRNA. The clue to the origin of this strange gene came when the flanking sequence downstream from the 3′ stop codon was analyzed. A search in GenBank showed ~80% homology with the C terminus of a fish trypsinogen, the precursor of the gut protease, trypsin. Then, when the AFGP and trypsinogen genes from notothenioid pancreas were isolated and sequenced, it became clear that the former clearly evolved from the latter. The 5′ and 3′ ends of the trypsinogen gene were conserved in the AFGP gene (94 and 96% identical, respectively) with the 5′ end contributing the secretory signal sequence needed to allow AFGP secretion into extracellular spaces (e.g., blood plasma, lumen of the gut). Studies showed that the AFGP coding region evolved by repeated duplication and rearrangement of a 9-bp (base pair) segment crossing the first exon–intron boundary of the trypsinogen gene. Hence, unlike the type II AFPs, the nototheniid AFGP did not arise by divergence of a preexisting protein. Instead, a new gene was created from a very small DNA segment of the trypsinogen gene, part of which was noncoding. Interestingly, the genome of the nototheniids actually contains a "historical" record of the events of AFGP evolution. Searches have revealed a gene that encodes for a protease with a Thr-Ala-Ala element, a gene that encodes for both a protease and an antifreeze, and other genes that encode for just antifreezes.

A surprising twist occurred, however, when researchers began to characterize the AFGP genes of northern cod. Although the protein sequences of nototheniid and cod AFGPs are virtually identical, the gene for cod AFGP bears no resemblance at all to the trypsinogen gene and clearly has a very different evolutionary origin. Furthermore, the AFGP genes from the two types of fish use different codons for the amino acids of the functional tripeptide and are constructed with very different intron–exon organization and spacer sequences so that very different processing of polyprotein precursors must occur. Therefore, evolution of the cod and nototheniid AFGP genes were independent events, and they represent excellent examples of convergent evolution where the same protein product arose from two different genomic origins.

ADAPTIVE CONTROL OF TRANSLATION

Other opportunities for adaptive responses at the gene/protein level exist via controls on translation as well as controls on protein degradation. Our treatment here will be brief because general aspects of translational control are discussed in Chapter 7, whereas Chapters 15 and 16 provide multiple illustrations of the adaptive regulation of protein translation that occurs during metabolic rate depression in anoxia-tolerant and hibernating animals. The little that is known about adaptation responses of proteolytic mechanisms is also discussed in Chapter 15.

In most cases, an increase in the rate of transcription of a gene automatically leads to a comparable increase in the rate of translation of the mRNA transcript and a proportional rise in the levels of the protein. Because of this and because of the relatively short half-life of most mRNA transcripts, changes in the number of mRNA transcripts of a given gene are typically good indicators of both a change in the rate of gene transcription and a change in the rate of synthesis of the encoded protein. Hence, techniques that quantify mRNA levels in control versus stressed states (e.g., Northern blotting, reverse transcription-polymerase chain reaction, cDNA array screening) are widely used with good accuracy to predict whether a stress has altered the rate of transcription and translation of a given gene and its protein (Fig. TB14.2).

However, there are several opportunities for differential controls that unlink transcription and translation. A temporal separation can occur resulting, for example, in mRNA up-regulation in response to a stress but a delay in protein synthesis until the recovery period after the stress is removed (e.g., see Chapter 16 for the example of the organic cation transporter regulation in mammalian hibernation). In other cases mRNA transcripts may be stabilized and held constitutively in an untranslatable form by the binding of specific regulatory proteins that are only released in response to a specific signal/stress (e.g., see Chapter 15 for the example of ferritin regulation).

Multiple modes of translational control can contribute to biochemical adaptation. These include global controls that increase or decrease the overall rate of translation and specific controls that affect the translation of individual types of mRNA transcripts. Among the mechanisms that will be discussed in the next chapters are:

1. Changes in the proportion of ribosomes in translationally active polysome versus translationally silent monosome fractions and the differential distribution of mRNA species between these two pools.
2. Differential transcript stabilization by multiple mechanisms including transcript binding to poly A binding proteins.
3. Regulation of mRNA entry, ribosome assembly, and peptide elongation by reversible protein phosphorylation control over the function of ribosomal initiation and elongation factors.
4. Mechanisms for circumventing the normal mode of transcription initiation (m^7G-capped-dependent initiation) by the presence of an internal ribosome entry site (IRES) in the mRNA transcripts of selected genes. The presence of an IRES allows the continued transcription of certain mRNA species under situations where the transcription of the majority of transcripts is blocked.

The present chapter has described many of the types of mechanisms that can be used for biochemical adaptation. The following three chapters explore the use of these mechanisms in greater detail in dealing with specific situations: oxygen limitation (Chapter 15), winter survival of small mammals by hibernation (Chapter 16), and survival at subzero temperatures by enduring the freezing of body fluids (Chapter 17).

SUGGESTED READING

General Principles of Biochemical Adaptation

Hochachka, P. W., and Somero, G. N. (2002). *Biochemical Adaptation: Mechanism and Process in Physiological Evolution.* Oxford University Press, Oxford, UK. *Principles of biochemical adaptation; easy to read and illuminating.*

Price, N. C., and Stevens, L. (2000). *Fundamentals of Enzymology.* Oxford University Press, Oxford, UK. *Basic principles of enzyme function and regulation, control of pathways and multienzyme complexes, clinical aspects.*

Storey, K. B. (2002). Life in the slow lane: Molecular mechanisms of estivation. *Comp Biochem Physiol A* **133**:733–754. *Principles of metabolic regulation that allow organisms to enter a hypometabolic or dormant state including enzyme responses, reversible phosphorylation control, signal transduction, and role of urea as an enzyme protectant during desiccation. An integrated example of how the mechanisms of biochemical adaptation discussed in this chapter are put together to achieve long-term survival of organisms under stress.*

Storey, K. B., and Brooks, S. P. J. (1995). The basis of enzymatic adaptation. In *Principles of Medical Biology*, Vol. 4, Bittar, E., and Bittar, N. (eds.) JAI Press, Greenwich, CT, pp. 147–169. *Review of multiple mechanisms of enzyme control in cells.*

Storey, K. B., and Storey, J. M. (Eds.) (2000–2002). *Cell and Molecular Responses to Stress*. Three volume series: Vol. 1, *Environmental Stressors and Gene Responses* (2000). Vol. 2, *Protein Adaptations and Signal Transduction* (2001). Vol. 3, *Sensing, Signaling and Cell Adaptation* (2002). Elsevier, Amsterdam. *Multivolume, multiauthor series focused on mechanisms of metabolic regulation and biochemical adaptation.*

Willmer, P., Stone, G., and Johnston, I. (2000). *Environmental Physiology of Animals*. Blackwell Science, Oxford, UK. *Comprehensive textbook dealing with physiological adaptations to many of the same environmental stresses as are dealt with at the biochemical level in the present chapter and the subsequent three chapters.*

Specific Topics

Cohen, P. T. W. (2002). Protein phosphatase 1—targeted in many directions. *J Cell Sci* **115**:241–256. *Overview of protein phosphatases.*

Fletcher, G. L., Hew, C. L., and Davies, P. L. (2001). Antifreeze proteins of teleost fish. *Ann Rev of Physiol* **63**:359–390. *Review of the structure, function, regulation, and evolutionary origin of fish antifreeze proteins.*

Krupa, A., and Srinivasan, N. (2002). The repertoire of protein kinases encoded in the draft version of the human genome: Atypical variations and uncommon domain combinations. *Genome Biol* **3**(12): research0066.1–0066.14. *Overview of the families of mammalian protein kinases.*

Sullivan, D. T., MacIntyre, R., Fuda, N., Fiori, J., Barrilla, J., and Ramizel, L. (2003). Analysis of glycolytic enzyme co-localization in *Drosophila* flight muscle. *J Exp Biol* **206**:2031–2038. *Current ideas about and latest techniques for investigating enzyme association with the muscle myosin motor.*

15

OXYGEN LIMITATION AND METABOLIC RATE DEPRESSION

KENNETH B. STOREY AND JANET M. STOREY

INTRODUCTION

Oxygen is crucial to the lives of most multicellular organisms on Earth. As humans, we are critically aware that an interruption of oxygen supply to our most oxygen-sensitive organ, the brain, for more than about 5 min can cause irreparable damage or death. Many other organisms also live highly oxygen-dependent lives. For example, the power requirements of high-speed hovering flight by hummingbirds and many insects are so great that without oxygen flight is impossible. Indeed, the flight muscles of bees and flies, which power flight with carbohydrate oxidation, have become so oxygen-dependent that lactate dehydrogenase has been deleted so that there is actually no way for these muscles to generate useful amounts of adenosine 5′-triphosphate (ATP) from anaerobic glycolysis. However, despite our anthrocentric view that oxygen is crucial to life, many other organisms actually survive equally well in the presence or absence of oxygen. These facultative anaerobes often live in environments where oxygen can become depleted either intermittently or at regular intervals. Many others are obligate anaerobes that never use oxygen; most of these are microbes but a variety of parasites (trematodes, nematodes, cestodes) that live in the gastrointestinal tract of other animals also lead anaerobic lives. In this chapter we will explore some principles of metabolic regulation as they apply to oxygen limitation. First, we will assess the reasons why inadequate oxygen supply causes metabolic injury to oxygen-sensitive species (such as ourselves) and examine the mechanisms that are used in attempts to regain adequate tissue oxygenation or to enhance glycolytic ATP production. Second, we will examine the biochemical mechanisms that allow facultative anaerobes to switch between oxygen-dependent and oxygen-independent life, with a particular emphasis on metabolic rate depression as the key survival strategy for enduring interruptions of oxygen supply.

The discussion in this chapter will be largely limited to what is termed *environmental* hypoxia (low oxygen) or anoxia (no oxygen), meaning limitation of oxygen availability to the organism. This can arise from a variety of circumstances, such as a move to a much higher altitude, breath-hold diving by lung-breathing animals, aerial exposure of gill-breathing animals, and oxygen depletion in ice-covered ponds or in polluted waters. Environmental anoxia is frequently differentiated from *functional* anoxia, which refers to situations where high-intensity work by muscles (e.g., sprint running) outstrips the capacity of muscle oxidative metabolism to produce ATP (see Text Box 15.1 and Chapter 11).

In the previous chapter environmental stresses were said to affect metabolism in two ways, by direct perturbation of the structure/function of biological molecules and biochemical reactions or by jeopardizing the energy currencies of the cell. Two main actions of oxygen were also listed—a "good" role in cellular energy metabolism and a "bad" role in the damage done by reactive oxygen species to macromolecules. The bad role of oxygen can be categorized as a direct perturbation by a stressor on biological molecules, and Chapters 12 and 13 discuss the damage to cellular macromolecules by reactive oxygen species (ROS) as well as the antioxidant defenses that mitigate this damage. Note, however, that the bad role of ROS is being seriously reevaluated at the present time, and

Functional Metabolism: Regulation and Adaptation, edited by Kenneth B. Storey.
ISBN 0-471-41090-X

TEXT BOX 15.1 FUNCTIONAL HYPOXIA OR ANOXIA: HIGH-INTENSITY MUSCLE WORK

Low-level exercise or even sustained medium- to high-intensity exercise (e.g., marathon running, hovering flight, cruiser swimming) can almost always be supported by ATP generation from mitochondrial oxidative phosphorylation using lipids or carbohydrates as the fuel (depending on the species). However, situations of high-intensity "burst" exercise (e.g., sprint running, the sudden leap of a frog) demand an instantaneous power output by contracting muscle fibers that outstrips the capacity of aerobic metabolism to supply ATP. Such high power demands are dealt with in two oxygen-independent ways: (1) a rapid mobilization of muscle phosphagen reserves provides an instant supply of ATP, and (2) glycolysis is rapidly activated to generate ATP from the catabolism of endogenous glycogen with the accumulation of a glycolytic end product (Fig. 15.1) (also see Chapter 11 on muscle metabolism). In vertebrate animals, the phosphagen is creatine phosphate (supplied by the creatine kinase reaction) and the glycolytic end product is lactate, produced by the reduction of pyruvate by lactate dehydrogenase. Both mechanisms supply high rates of ATP output for very short times (generally just a few seconds) but with a significant cost to oxygen-based metabolism that must replenish creatine phosphate and glycogen reserves during the recovery period after burst exercise (this is called an oxygen debt).

Some invertebrate animals use the same phosphagen and anaerobic end product to support burst muscle work, but a number of alternatives also exist. For example, in addition to creatine phosphate, there are seven other phosphagens (and corresponding phosphagen kinases) among invertebrates. The phosphagen kinase reactions show differences in thermodynamic poise, and the phosphagens differ in some physical properties including intrinsic diffusivity. Overall, the big "winners" in evolutionary terms are creatine phosphate and arginine phosphate, the latter being widespread in the "higher" invertebrate groups (insects, crustaceans, and other arthropods as well as mollusks). Alternatives to lactate also occur widely among invertebrates (outside of the Arthropoda); multiple other pyruvate oxidoreductases are used that couple the reductive condensation of pyruvate with an amino acid to form an end product called an imino acid or an "opine" (the alpha-amino nitrogen is linked with the C2 keto group of pyruvate) (Fig. 15.1). The most notable alternative system occurs in working muscles of mollusks. Here, arginine phosphate is the phosphagen and the glycolytic end product is octopine (named because it was first found in *Octopus*). The enzymes involved are arginine kinase (ArgK) and octopine dehydrogenase (ODH):

$$\text{Phosphoarginine} + \text{ADP} \rightarrow \text{arginine} + \text{ATP} \quad (\text{ArgK})$$

$$\text{Pyruvate} + \text{arginine} + \text{NADH} + \text{H}^+ \rightarrow \text{octopine} + \text{NAD}^+ + \text{H}_2\text{O} \quad (\text{ODH})$$

So, in a high-speed chase between a sperm whale and its prey, a giant squid, the whale will deplete creatine phosphate and accumulate lactate in its working muscle, whereas the jet propulsion swimming of the squid catabolizes arginine phosphate and accumulates octopine. Table TB15.1 shows an example of the effects of exhaustive exercise and hypoxia exposure on metabolite levels in the mantle muscle of a cephalopod mollusk. Both stresses deplete muscle glycogen, but exercise causes a greater mobilization of phosphagen pools (recall from Chapter 11 that phosphagen is mobilized first during burst muscle work) and ATP. Octopine accumulates as the glycolytic end product, and the total arginine pool (phosphagen + free arginine + octopine) is maintained constant.

The production of octopine actually appears to have some advantages over lactate as an end product including: (a) by consuming arginine, the product of phosphagen hydrolysis, octopine synthesis can help to "pull" the

TABLE TB15.1 Effects of Exhaustive Swimming Exercise or Severe Hypoxia on Muscle Metabolites in a Cephalopod Mollusk, the Cuttlefish *Sepia officinalis* [a]

	Control	Exercised	Hypoxic
Glycogen	23.7	4.5	2.2
Pyruvate	0.1	0.8	0.2
ATP	8.7	2.2	4.5
Arginine phosphate	33.6	3.8	18.8
Arginine	29.6	45.3	37.5
Octopine	0.2	8.6	3.6
Sum Arg-P + Arg + octopine	63.4	57.7	59.9

All values in μmol/g wet mass. [a]Both glycogen and phosphagen are substrates for exercise and hypoxic metabolism in cuttlefish, but, although glycogen depletion is similar in both exercise and hypoxia, exercise causes a greater depletion of ATP and arginine phosphate and a greater accumulation of octopine. Note that the sum of arginine phosphate, arginine, and octopine remains constant. For exercise animals were stimulated to jet repeatedly until exhausted. For hypoxia animals were given 30 min exposure to seawater that had been bubbled with nitrogen gas to lower P_{O_2} to 10 mm Hg from a normal value of 130 mm Hg.

Source: Data compiled from K. B. Storey and J. M. Storey (1979). *J Comp Physiol* **131**:311–319.

ArgK reaction toward ATP synthesis, (b) octopine accumulation causes less cellular acidification than does lactate, and (c) by combining the products of phosphagen hydrolysis and glycolysis into a single end product, cellular osmolality is not perturbed. This latter is quite important for the cells of marine invertebrates because these animals maintain isoosmotic balance with seawater; hence, a net accumulation of osmotically active particles (such as if lactate accumulated) would result in cell swelling due to water influx.

A number of other opines occur as glycolytic end products in different species including alanopine, strombine, lysopine, nopaline, and tauropine formed from the reductive condensation of pyruvate with alanine, glycine, lysine, proline, and taurine, respectively. These are almost exclusively found in marine organisms (although a few freshwater animals such as clams have low activities of one or more of these enzymes together with lactate dehydrogenase). However, a more universal replacement of lactate as a glycolytic end product has not occurred. This seems to be because marine invertebrates have large pools of the necessary amino acid substrates within their cells; these amino acids and other nitrogenous compounds provide about half of the osmotic balance needed to defend cell volume against the high ion concentrations in seawater. Freshwater and terrestrial animals lack these amino acid pools that would also provide the cosubstrates for opine synthesis.

whereas ROS clearly cause much damage to cellular macromolecules, many new studies are now showing that they also have key roles in intracellular signaling. In its good role, oxygen primarily affects energy currencies of the cell. When oxygen availability drops below a critical value, cellular energy supply can be compromised because the rate at which ATP is generated from oxidative phosphorylation in the mitochondria falls below the rate at which the many ATP-utilizing cellular reactions are consuming ATP. If this imbalance continues unabated, ATP availability for metabolism soon becomes limiting, metabolic damage accrues, and viability can be compromised.

There are two ways that metabolic regulation can be applied to deal with the effects of low oxygen availability on cellular energetics. The first is to increase the production of ATP by oxygen-independent mechanisms, primarily utilizing the glycolytic pathway, in order to meet the demands of ATP-consuming cellular reactions. The second is to reduce the cell's need for ATP production by suppressing the rate of ATP use by ATP-consuming cellular reactions. Virtually all organisms have some capacity, however short term, for increasing ATP output by oxygen-independent means (a capacity that is highly elevated in white muscles to deal with functional anoxia; see Text Box 15.1), but only facultative anaerobes have developed the aggressive metabolic controls necessary to strongly suppress metabolic rate to a level that can be sustained indefinitely by anaerobic ATP generation.

OXYGEN LIMITATION—THE PROBLEM

All cells strive to maintain homeostasis, a balance between the rate of ATP production by central catabolic pathways and the rate of ATP utilization by innumerable cellular processes. Oxygen-dependent ATP production by oxidative phosphorylation in the mitochondria makes the most efficient use of metabolic fuels, extracting their full energetic potential while fully "burning" fuels to CO_2 and H_2O. Not surprisingly, then, most organisms on Earth have taken up an aerobic lifestyle, but, in doing so, they can place themselves in jeopardy if oxygen availability is compromised. If oxygen is cut off, cells have two immediate energy supply problems: (1) lipid fuels become useless, as do most amino acids, because their catabolic pathways include no oxygen-independent substrate-level phosphorylation reactions by which ATP can be made and (2) phosphagen depletion plus greatly increased glycolytic flux rarely, if ever, meets the unmodified ATP demands of cellular metabolism for more than a few seconds or minutes.

The problem is particularly acute for organs such as brain that consume tremendous amounts of energy; for example, the human brain represents about 2% of our body weight but consumes about 25% of our energy budget. An interruption of blood supply (ischemia) to mouse brain causes an almost instantaneous increase in glycolytic rate of 4- to 7-fold, but this only partially compensates for the 18-fold decrease in ATP yield per glucose catabolized (a net of 2 ATP is produced per 1 glucose catabolized to 2 lactate versus 36 ATP if glucose is instead catabolized to CO_2 and H_2O). Brain is further compromised in its attempt at compensation because, with blood flow cut off, the brain is also deprived of its primary fuel source, blood glucose, and has only small reserves of endogenous carbohydrate (glycogen) to fuel glycolysis. Hence, within seconds, cellular ATP levels begin to fall, and within 5 min of oxygen deprivation as much as 90% of the ATP is depleted in mammalian brain. A high proportion of brain energy budget is committed to supporting ion-motive ATPases in their role in maintaining membrane potential difference; for example, the sodium–potassium ATPase alone is responsible for 5 to 40% of cellular ATP turnover, depending on cell type. When only 50 to 65% of brain ATP is lost, membrane depolarization occurs, and this sets off multiple negative consequences. Depolarization results in a rapid uptake of Na^+ (when ATP is limiting, Na^+ influx through ion channels is unopposed by oppositely directed pumps) and water that is followed by an influx of Ca^{2+} through voltage-gated Ca^{2+} channels. The collapse of the sodium gradient causes the sodium–glutamate cotransporters to eject glutamate into the extracellular space where this neurotransmitter triggers a range of events, including activation of excitatory neurotransmission by stimulating *N*-methyl-D-aspartate (NMDA) receptors that are responsible for a significant part of the Ca^{2+} influx. Transient elevation of cytosolic Ca^{2+} is a critical signaling mechanism in the implementation of many normal cell functions, but sustained high Ca^{2+} triggers a range of pathological changes, including the activation of phospholipases and proteases that can lead to damage and death of neurons and all other cells (see also Chapter 8).

If the restoration of oxygen-rich blood flow occurs soon enough, these injuries caused by low cellular bioenergetics can be reversible, but the restoration of blood flow to previously ischemic tissues also stimulates a whole new set of injuries. Reperfusion injury, such as occurs after heart attack or stroke, was discussed in detail in Chapter 13 and is caused by a burst of ROS generation, chiefly superoxide radicals, from the highly reduced electron transport chain when oxygen is suddenly reintroduced. ROS-mediated damage arising from the bad role of oxygen can result in direct damage by free radicals to macromolecules, including deoxyribonucleic acid (DNA), proteins, and membrane lipids. Indirect damage can also arise such as from an inability of sarco- or endoplasmic reticulum membranes that are damaged by peroxidation to properly resequester Ca^{2+}, thereby exacerbating the Ca^{2+}-mediated damage caused by low energetics in the anoxic/ischemic phase. Hence, anoxia-tolerant animals need to have adaptations that deal with both the stresses on cellular energetics under anoxia and the potential for ROS-mediated damage during the return to aerobic life.

OXYGEN LIMITATION—RESPONSES IN OXYGEN-SENSITIVE SYSTEMS

Many organisms, including humans, are highly dependent upon oxygen and have little or no ability to survive if deprived of oxygen. If oxygen availability to tissues drops below a critical level, then a suite of compensatory responses, both physiological and biochemical, are activated with two goals: (1) to improve oxygen delivery to tissues and (2) to increase glycolytic ATP production to compensate for the reduced ATP output from oxidative phosphorylation. Both goals are addressed with immediate (seconds to minutes) and longer term adjustments (hours to days). Consider, for example, the case of an Acapulco resident taking a trip to Mexico City. Upon stepping off the airplane at an elevation of 2300 m, the traveler is deposited in an atmosphere that contains 30% less oxygen than at his sea-level home and immediate compensation for the reduced oxygen availability begins. Physiological responses include an immediate increase in ventilation rate and heart rate, an increase in hemoglobin unloading of oxygen, and a release of stored red blood cells from the spleen. These events attempt to increase oxygen uptake into the blood, oxygen-carrying capacity, and oxygen delivery to organs. The immediate biochemical event is an activation of oxygen-independent routes of ATP synthesis via an increase in glycolytic rate in all tissues as well as the consumption of creatine phosphate reserves in those tissues (e.g., muscles) that have significant reserves of phosphagen.

Control of Glycolysis

The activation of glycolysis in response to low oxygen stress has been extensively researched; in particular, this was initially driven by the desire to understand the molecular basis of the Pasteur effect (see Text Box 15.2). Initial work during the 1960s focused on the allosteric regulation

---

TEXT BOX 15.2 THE PASTEUR EFFECT

In 1861 Louis Pasteur published the results of experiments on yeast that showed that under anoxic conditions yeast consumed more sugar per unit mass than under aerobic conditions. His conclusion was that "oxygen inhibits fermentation," but in recent years the Pasteur effect is usually discussed as the corollary of the original "anoxia increases fermentation." In other words, in the presence of oxygen, yeast use mitochondrial oxidative phosphorylation to gain the maximum ATP yield from substrate catabolism; but, when oxygen is lacking, the rate of fermentation rises dramatically and ethanol is produced as the terminal product of glycolysis. The primary constraint on the Pasteur effect is substrate availability. Per mole of glucose consumed, aerobic metabolism produces 18 times more ATP than does fermentation. Hence, to sustain unaltered cellular ATP demands in anoxia, the rate of glucose utilization by glycolysis would have to increase by 18-fold. The virtually unlimited substrate availability that this requires is possible in some circumstances, for example, for yeast growing in a sugary mash of malted barley and hops! Excretion of end products (in this case, ethanol) also prevents a toxic buildup of intracellular waste (and yields a nice beer). However, for most multicellular organisms with fixed reserves of carbohydrate fuel in their tissues, the Pasteur effect is only a short-term solution to anoxic insult and fails as a long-term solution.

---

of enzymes and particularly on the control of 6-phosphofructo-1-kinase (PFK-1) as being central and rate-limiting to overall glycolytic rate [reaction (15.1)] (Fig. 15.1):

$$\text{Fructose-6-P} + \text{ATP} \rightarrow \text{fructose-1,6-P}_2 + \text{ADP} \quad (15.1)$$

The activity of PFK-1 is highly sensitive to substrate inhibition by high levels of ATP and to allosteric activation by

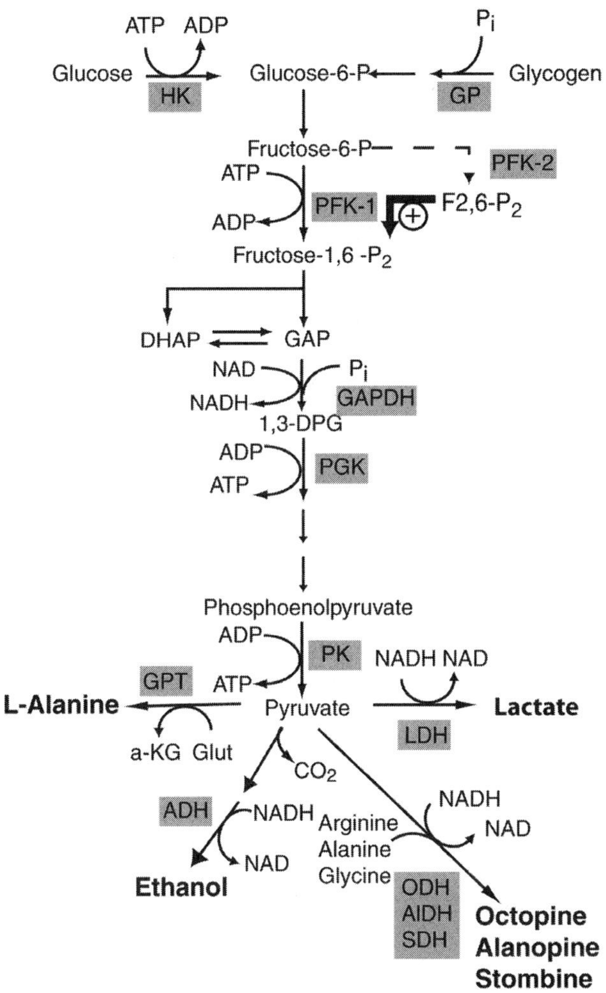

Figure 15.1 Anaerobic glycolysis showing inputs from glucose or glycogen and a range of possible end products derived from pyruvate. Also shown is the production of fructose-2,6-bisphosphate ($F2,6P_2$) and its activating effect on PFK-1. Enzymes are: HK, hexokinase; GP, glycogen phosphorylase; PFK-1, 6-phosphofructo-1-kinase; PFK-2, 6-phosphofructo-2-kinase; GAPDH, glyceraldehyde-3-phosphate dehydrogenase; PGK, phosphoglycerate kinase; PK, pyruvate kinase; LDH, lactate dehydrogenase; GPT, glutamate-pyruvate transaminase; ADH, alcohol dehydrogenase; ODH, AlDH, and SDH, octopine, alanopine, and strombine dehydrogenases, respectively. Abbreviated metabolites are: GAP, glyceraldehyde-3-phosphate; 1,3-DPG, 1,3-diphosphoglycerate; α-KG, α-ketoglutarate; Glut, glutamate.

adenosine 5′-monophosphate (AMP). When ATP demand outstrips ATP availability in cells, ATP levels fall, and because of the near equilibrium of the adenylate kinase reaction (2 ADP $\rightarrow$ ATP + AMP) and the relative levels of ATP, adenosine 5′-diphosphate (ADP), and AMP in cells, a small percentage decrease in ATP concentration can cause an immediate severalfold rise in AMP levels. For example, recall from Chapter 1 that a 10% decrease in [ATP] in insect flight muscle during takeoff caused an immediate 2.5-fold increase in [AMP]. Allosteric control of PFK-1 was long regarded as the basis of glycolytic activation in the Pasteur effect. Control comes from strong allosteric activation of PFK-1 by AMP as well as ADP, inorganic phosphate (which rises when ATP and creatine phosphate hydrolysis is high), and NH_4^+ (produced from AMP deamination to inosine monophosphate (IMP), particularly in working muscles), all of which rise under hypoxic/anoxic conditions, as well as relief from inhibition by ATP and citrate, both of which decline in hypoxia/anoxia. When glucose is the substrate, increased glucose uptake (by GLUT transporters) and increased glucose phosphorylation by hexokinase are also regulatory factors in the Pasteur effect. Hexokinase control comes from activation by inorganic phosphate, inhibition by the hexokinase product, glucose-6-phosphate, and reversible binding of hexokinase to the mitochondrial membrane. In the 1980s the central role of PFK-1 in glycolytic rate control was further cemented with the discovery of fructose-2,6-bisphosphate ($F2,6P_2$), another powerful activator of PFK-1 (see Chapter 14 and Fig. 14.7). $F2,6P_2$, synthesized by 6-phosphofructo-2-kinase (PFK-2), rises during anoxia in yeast and by activating PFK-1 facilitates increased use of carbohydrate as a fermentative substrate. In vertebrates, $F2,6P_2$ involvement in activating glycolysis in hypoxia/anoxia is very much organ-specific; it occurs in heart but not in skeletal muscle or liver. In liver, $F2,6P_2$ actually falls under hypoxia/anoxia and contributes to the PFK-1 inhibition that is necessary to divert glycogenolysis into glucose export in order to feed the substrate needs of other organs.

In the late 1980s yet another factor in glycolytic rate control was discovered, the AMP-activated protein kinase (AMPK) (see Chapter 4). Ultimately, this enzyme now appears to be a major regulator of the low-energy signal in cells, and many effects of ATP depletion are now known to be mediated through this pathway rather than via the direct effects of adenine nucleotides on enzymes. Increased [AMP] in cells directly activates AMPK and also activates its upstream kinase (AMPK kinase) such that AMPK is strongly activated under cellular situations that either inhibit ATP production (hypoxia, glucose limitation, uncouplers, and other stresses) or stimulate ATP consumption (exercise) (see Fig. 14.13). Active AMPK switches on catabolic pathways and switches off anabolic ones (e.g., 3-hydroxy-3-methylglutamyl coenzyme A (CoA) reductase and acetyl-CoA carboxylase, key enzymes in steroid and fatty acid biosynthesis, respectively). In ischemic heart, AMPK plays a key role in glycolytic activation with two main effects: (1) an activation of PFK-2 to elevate $F2,6P_2$ levels and stimulate PFK-1 and (2) stimulation of the translocation of GLUT4 to the plasma membrane. Hence, AMPK both improves glucose uptake into the ischemic heart and enhances PFK-1 activity by stimulating production of its potent allosteric activator, $F2,6P_2$. In skeletal muscle, AMPK does not affect $F2,6P_2$ production but increases the expression of GLUT4, hexokinase, and several mitochondrial enzymes, changes that are also features of endurance training.

Hypoxia-Induced Gene Expression

The response to hypoxia/ischemia stress by oxygen-sensitive systems includes not just acute activation of glycolytic rate but also gene expression responses that elevate the overall glycolytic capacity of cells and organs in preparation for possible prolonged hypoxia exposure. These gene expression responses are coordinated by the hypoxia-inducible transcription factor (HIF-1). The structure, regulation and action of HIF-1 is discussed in detail in Chapter 6 (see also Fig. 6.11). Recall that under oxygenated conditions the HIF-1α subunit is modified by an oxygen-dependent prolyl hydroxylase enzyme that targets the subunit for rapid degradation by the proteasome. When oxygen availability is low, this modification does not occur and HIF-1α is stabilized, forms a dimer with the HIF-1ß subunit, migrates to the nucleus, and activates the transcription of a suite of genes. Gene responses that improve glycolytic capacity include increased amounts of multiple enzymes in the glycolytic pathway (e.g., hexokinase, glyceraldehyde-3-P dehydrogenase, aldolase A and C, lactate dehydrogenase A, phosphoglycerate kinase-1, pyruvate kinase M) as well as increased levels of glucose transporter isoforms 1 and 3 to increase the capacity of cells to import glucose across the plasma membrane. Other genes that are up-regulated by HIF-1 include those whose products have angiogenic or vasodilatory functions such as erythropoietin that stimulates red blood cell synthesis, vascular endothelial growth factor that stimulates capillary growth, inducible nitric oxide synthase, and heme oxygenase.

Overall, then, the response to low oxygen by oxygen-sensitive organisms such as humans is a compensatory one. Mechanisms are activated that increase oxygen delivery to cells and that enhance ATP production by oxygen-independent routes. If the hypoxic stress is not too severe, acclimation will occur over time, and when the oxygen-carrying capacity of blood is sufficiently elevated, acute responses such as hyperventilation and high rates of glycolysis are abated. Many organisms are much more hypoxia tolerant than humans. They are prepared with better constitutive capacities, both physiological and biochemical, for dealing with hypoxic excursions and inducible responses are triggered at much lower pO_2 values (see Text Box 15.3). Our treatment of hypoxia tolerance here has been very brief, but this field is a huge one and the reader is referred to the suggested reading list for more information on the subject.

FACULTATIVE ANAEROBIOSIS

Life on Earth began in an anoxic environment, and glycolysis, the fundamental pathway of energy production from substrate fermentation, is still present in all organisms (see Chapter 20). However, although glycolysis remains integral to energy generation under low-oxygen conditions, the limitations of glycolysis as a general means of ATP supply are immediately obvious when compared with the advantages of oxygen-based fuel catabolism. These include the following:

1. *Substrate Options* Aerobic metabolism can make use of carbohydrate, lipid, and protein as oxidative fuels for mitochondrial pathways, whereas under anaerobic conditions organisms are restricted to the use of carbohydrates and a few amino acids as fermentative fuels. For many organisms, this means that their major stored fuel reserves (lipids) are useless if oxygen supply fails.
2. *Energy Yield* Glycolysis yields a net of only 2 mol ATP/mol glucose catabolized to lactate (or 3 mol ATP/mol glucose-1-phosphate cleaved from glycogen) compared with 36 (or 38) mol ATP produced if sugar is fully catabolized to CO_2 and H_2O by mitochondria. Hence, the normal rate of ATP use by aerobic cells can only be sustained under anoxia by a massive increase in the rate of fermentative glycolysis.
3. *Efficiency of Substrate Use* By rearranging the above statements, it is obvious that to generate equal amounts of ATP, anaerobic glycolysis (ending in lactate) would use 18 times more glucose than would be needed for aerobic ATP production from carbohydrate oxidation. Hence, anaerobic glycolysis can be viewed as a wasteful use of substrate, something that few organisms can normally afford.
4. *End Products* The incomplete catabolism of carbohydrates via anaerobic glycolysis results in the accumulation of waste products, generally lactate in animals and ethanol in plants, that if not excreted or "neutralized" in some manner can cause serious toxicity. By contrast, the volatile end product of mitochondrial respiration, CO_2, is readily excreted or exhaled.

TEXT BOX 15.3 HYPOXIA TOLERANCE

A distinction can be made between anoxia tolerance, the ability to survive without oxygen, and hypoxia tolerance, the ability to endure low-oxygen conditions. Many organisms can endure extended periods of low oxygen but cannot survive total oxygen lack. Hypoxia tolerance varies considerably between species and between organs and developmental stages within a given species. For example, no mammal is capable of sustained life without oxygen (anoxia), but many have a much greater tolerance for hypoxia than humans do. These include burrowing mammals (e.g., moles) whose tunnels are not always well-ventilated, alpine mammals that live at high altitudes, breath-hold divers such as whales and seals, and hibernating species such as ground squirrels that take only a few breaths per hour while in torpor (read more about hibernation in Chapter 15). Fetal and neonatal mammals are also much more hypoxia-tolerant than adults of the same species. In fetal mammals this is because oxygen must be derived from the maternal circulation and, indeed, oxygen tension in fetal brain is less than half the normal value for adults. The birth process also frequently imposes hypoxic and/or ischemic episodes.

However, although functioning under hypoxic conditions, such animals are generally not oxygen-limited. For example, the fact that hibernating mammals sustain torpor for many months by the slow oxidation of their huge adipose reserves shows that they cannot be oxygen-limited because lipids can only be catabolized aerobically. Similarly, the vast majority of dives by marine mammals are fully aerobic. These animals have made multiple physiological adjustments that enhance their "on-board" reserves of oxygen, including extremely high concentrations of the oxygen storage protein (myoglobin) in muscle, a high red blood cell count (hematocrit), and a huge reserve of red blood cells in the spleen. Selective vasoconstriction during the dive imposes hypoxic and ischemic conditions on those organs that can withstand the stress (e.g., kidney, intestine) and thereby enhances the oxygen available to sensitive organs (e.g., brain). Only in dives of unpredictably long durations is strong metabolic rate depression and a reliance on the high glycolytic capacity of diver organs needed to deal with low-oxygen stress. For more information on hypoxia tolerance, especially about mammalian diving and high altitude adaptations, consult P. W. Hochachka and G. N. Somero. *Biochemical Adaptation: Mechanism and Process in Physiological Evolution*, Oxford University Press, Oxford, 2002.

Despite the obvious limitations of a reliance on anaerobic glycolysis for ATP synthesis, there are numerous situations where this is necessary. As discussed in Text Box 15.1, intense muscle work (functional anoxia) is one of them. Another is the specialized needs of selected cell types. For example, to maximally pack each cell with hemoglobin, red blood cells have lost both mitochondria and nuclei. Their only option for ATP synthesis is the fermentation of glucose; fortunately, glucose is readily available to them from the surrounding plasma. Cells of the retina also function without oxygen because the infiltration of capillaries to bring oxygen to the retina would disrupt the visual field. Finally, many organisms experience environmental anoxia as part of their normal lives. They live in environments where oxygen can become depleted on either short-term or seasonal time scales or have lifestyles that cause them to be cut off from oxygen delivery intermittently. When oxygen supplies fail, these organisms simply switch to fermentative pathways and employ multiple intriguing biochemical adaptations to ensure their long-term survival without oxygen. Two of the most commonly studied animal models of facultative anaerobiosis are freshwater turtles that can live submerged underwater for many months over the winter (Text Box 15.4) and intertidal marine mollusks that are deprived of oxygen at each low tide (Text Box 15.5).

Mechanisms of Long-Term Anoxia Survival

The limitations of anaerobic glycolysis as an efficient method of ATP generation (listed above) also provide us with a basic list of factors that should be addressed to optimize anaerobic survival: (a) access to large supplies of fermentable fuels, (b) supplementing the basic glycolytic pathway with other reactions that increase ATP yield per glucose catabolized, (c) minimizing cytotoxicity of end products by enhancing buffering capacity or making products that can be excreted easily, (d) meeting the ATP demands of the anoxic cell by either greatly increasing glycolytic ATP output or greatly decreasing rates of ATP use in anoxia, and (e) maintaining or upgrading antioxidant defenses to deal with potential oxidative stress during the transition from anoxic to aerobic life. Each of these are discussed below.

Fermentable Fuels For most organisms, anaerobic metabolism basically means carbohydrate catabolism via glycolysis using polysaccharide reserves (glycogen in animals, starch in plants) or free sugars as substrates. A few selected amino acids can also be used in reactions that provide "substrate-level" phosphorylation of ADP, but lipid fuels, because they all feed into acetyl-CoA for catabolism in the tricarboxylic acid (TCA) cycle, can only be used to produce ATP by oxidative phosphorylation

TEXT BOX 15.4 CHAMPION OF VERTEBRATE ANAEROBIOSIS—THE FRESHWATER TURTLE

Many turtle species spend much of their lives underwater, diving for food or to evade predators. Winter survival for many freshwater turtles is also ensured by underwater hibernation that provides an escape from freezing temperatures. Aerobic metabolism can easily support short-term dives (although perhaps with some muscle lactate accumulation if swimming is vigorous), but for long-term hibernation, other strategies are needed to allow these lung-breathing animals to survive. Some species of turtles (mostly soft-shelled varieties) have solved the problem by using extrapulmonary mechanisms of oxygen uptake; oxygen is extracted from the water across the epithelium lining the throat or the cloaca. In cold water when metabolic rate is low, this strategy is sufficient to sustain aerobic metabolism. Other turtles have perfected facultative anaerobiosis. Turtles belonging to the *Trachemys* and *Chrysemys* genera, which include the common red-eared slider (*T. scripta elegans*) of the pet store trade and the painted turtles (*C. picta*) that are a familiar sight in ponds and rivers across North America, can live submerged in cold water for 3 to 4 months without oxygen. The secrets of their survival include strong metabolic rate depression to ~10% of the aerobic rate at the same temperature, large stores of glycogen fuel loaded into all organs, and the ability to both buffer and store lactic acid in their bony shell. The extreme anoxia tolerance of turtle brain and heart has been appreciated by medical science, and these are widely used as models for understanding the molecular events that impart anoxia and ischemia resistance to organs.

TEXT BOX 15.5 INVERTEBRATE FACULTATIVE ANAEROBES—MOLLUSKS OF THE SEASHORE

Anoxia tolerance is critical for the survival of many kinds of marine invertebrates and has been particularly well-studied in mollusks, including bivalves (e.g., mussels, clams, oysters) and gastropods (e.g., periwinkles, whelks). For these animals, natural oxygen deprivation can result from a variety of factors: (1) gill-breathing species that live in the intertidal zone (especially sessile creatures such as mussels) are deprived of oxygen when the waters retreat with every low tide, (2) burrowing and benthic mollusks can encounter anoxic bottom sediments, (3) high silt or toxin levels in the water as well as predator harassment can force shell valve closure, leading to substantial periods of "self-imposed" anoxia, and (4) animals in small tide pools can be oxygen-limited when animal and plant respiration depletes the oxygen supply in the water. The anoxia tolerance of the intertidal species found at high latitudes (e.g., periwinkles, blue mussels) is also an important contributor to freezing survival by these animals for, when extracellular fluids freeze, cells must rely on anaerobic energy production to remain viable until they thaw again (see Chapter 17). The remarkable anoxia tolerance of these marine shellfish (that includes a profound anoxia-induced metabolic rate depression) accounts for the long shelf-life of fresh shellfish in seafood markets. Anaerobiosis in mollusks has several components including large tissue reserves of glycogen and fermentable amino acids, modified pathways of fuel catabolism that increase the ATP yield compared with glycolysis alone, production of volatile end products that are easily excreted, and strong metabolic rate depression.

in the mitochondria. For microorganisms, unlimited supplies of fermentable sugars are sometimes available from the extracellular medium, and indeed, work by Louis Pasteur studying yeast provided the first analysis of fuel use by organisms growing under aerobic versus anaerobic conditions (see Text Box 15.2). Among multicellular animals, however, carbohydrate catabolism under anoxia typically draws upon tissue glycogen reserves, which are generally much higher in organs of anoxia-tolerant species compared with intolerant species. Furthermore, in cases where anoxia exposure is seasonal (e.g., turtles that hibernate underwater; see Text Box 15.4), glycogen reserves are built up to high levels during prehibernation feeding. In vertebrates, glycogen reserves are particularly high in the liver, and this organ is the main supplier of carbohydrate, in the form of blood glucose, to other organs during anoxia. In many invertebrates, however, there is less centralization of glycogen stores and most organs retain their own supplies of fermentable fuels. Indeed, this sort of self-sufficiency has made the tissues of bivalve mollusks (e.g., gill, mantle, adductor muscle) excellent and widely used models for *in vitro* studies of anaerobic metabolism. Figure 15.2 shows a typical pattern of glycogen consumption and end-product accumulation in the gill of the marine clam, *Mercenaria mercenaria*, over the course of 4 days of anoxia exposure.

Some amino acids can also be used as anaerobic fuels. Aspartate is the most notable one, but asparagine, glutamate, and glutamine can also be used. Anoxia-tolerant marine invertebrates have been particularly successful in

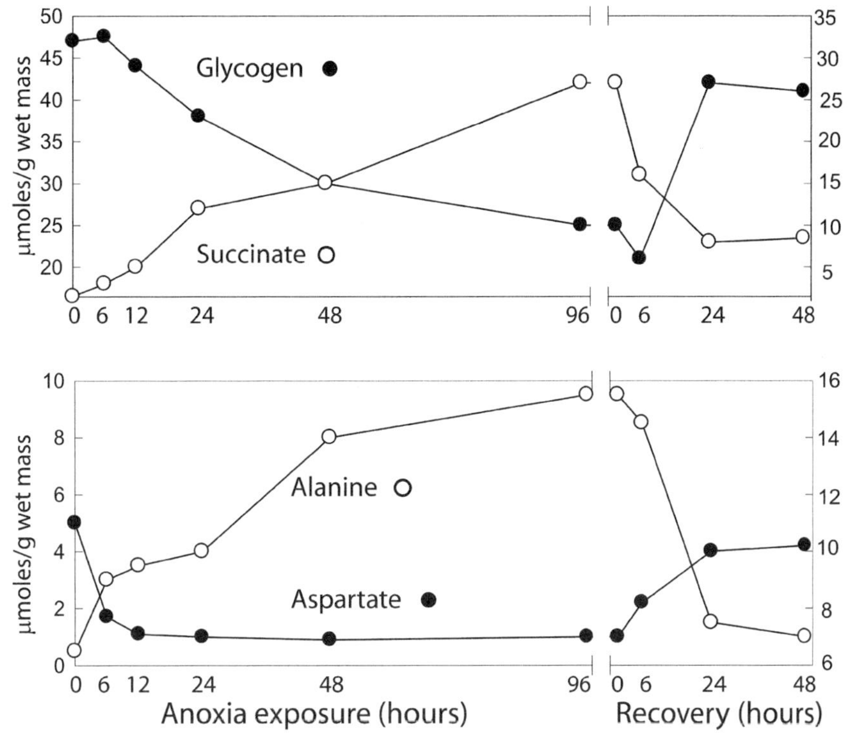

Figure 15.2 Metabolism in gill of the marine clam, *Mercenaria mercenaria*, over 4 days of anoxia exposure and 2 days of aerobic recovery. Glycogen and aspartate are utilized as substrates, whereas alanine and succinate accumulate as products. Note the rapid initial catabolism of aspartate, whereas glycogen fuels long-term anaerobiosis. All metabolites are largely restored to control levels within 24 h of aerobic recovery. [Data compiled from S. A. Korycan and K. B. Storey (1983). *Can J Zool* **61**:2674–2681.]

integrating the anaerobic catabolism of glycogen and aspartate (Fig. 15.3). Aspartate is rapidly mobilized early in anoxia exposure (Fig. 15.2 shows that aspartate pools are largely depleted within the first 6 h in *M. mercenaria*), whereas glycogen takes over as the sole fuel for longer term anaerobiosis. However, for intertidal invertebrates that are cut off from oxygen during twice-daily aerial exposures at low-tide, 6 h often represents the extent of a natural anaerobic excursion, and hence the pathways of coupled fermentation of glycogen and aspartate (Text Box 15.5, Fig. 15.3, and discussed below) may be the ones that support most normal aerial exposures at low tide. Aspartate is catabolized by reactions that reverse the segment of the TCA cycle running from oxaloacetate to succinate with ATP produced in a substrate-level phosphorylation by the fumarate reductase reaction (Fig. 15.3).

Improved Energy Yield and Alternative End Products For most animals, the end product of anaerobic carbohydrate catabolism via glycolysis is lactate (in plants it is ethanol). The action of lactate dehydrogenase (LDH) [reaction (15.2)] in catalyzing the reduced nicotinamide adenine dinucleotide (NADH)-dependent reduction of pyruvate to form lactate regenerates the NAD that is needed by the glyceraldehyde-3-phosphate dehydrogenase (GAPDH) reaction and allows glycolysis to continue without a net depletion of NAD cosubstrate (Fig. 15.1). However, anaerobic glycolysis with lactate as the end product has two limitations: (a) a low yield of ATP per glucose catabolized and (b) significant cellular acidification. Several alternative end products to lactate have appeared in anoxia-tolerant species that provide enhanced ATP yield per glucose catabolized and/or a reduced acid buildup. These include several kinds of "opines," ethanol, alanine, succinate, propionate, and acetate (Fig. 15.1):

$$\text{Pyruvate} + \text{NADH} + \text{H}^+ \rightarrow \text{lactate} + \text{NAD}^+ + \text{H}_2\text{O} \qquad (15.2)$$

$$\text{Pyruvate} + \text{amino acid} + \text{NADH} + \text{H}^+ \rightarrow \text{opine} + \text{NAD}^+ + \text{H}_2\text{O} \qquad (15.3)$$

Alternative pyruvate oxidoreductases, the so-called opine dehydrogenases [reaction (15.3)], are quite widespread

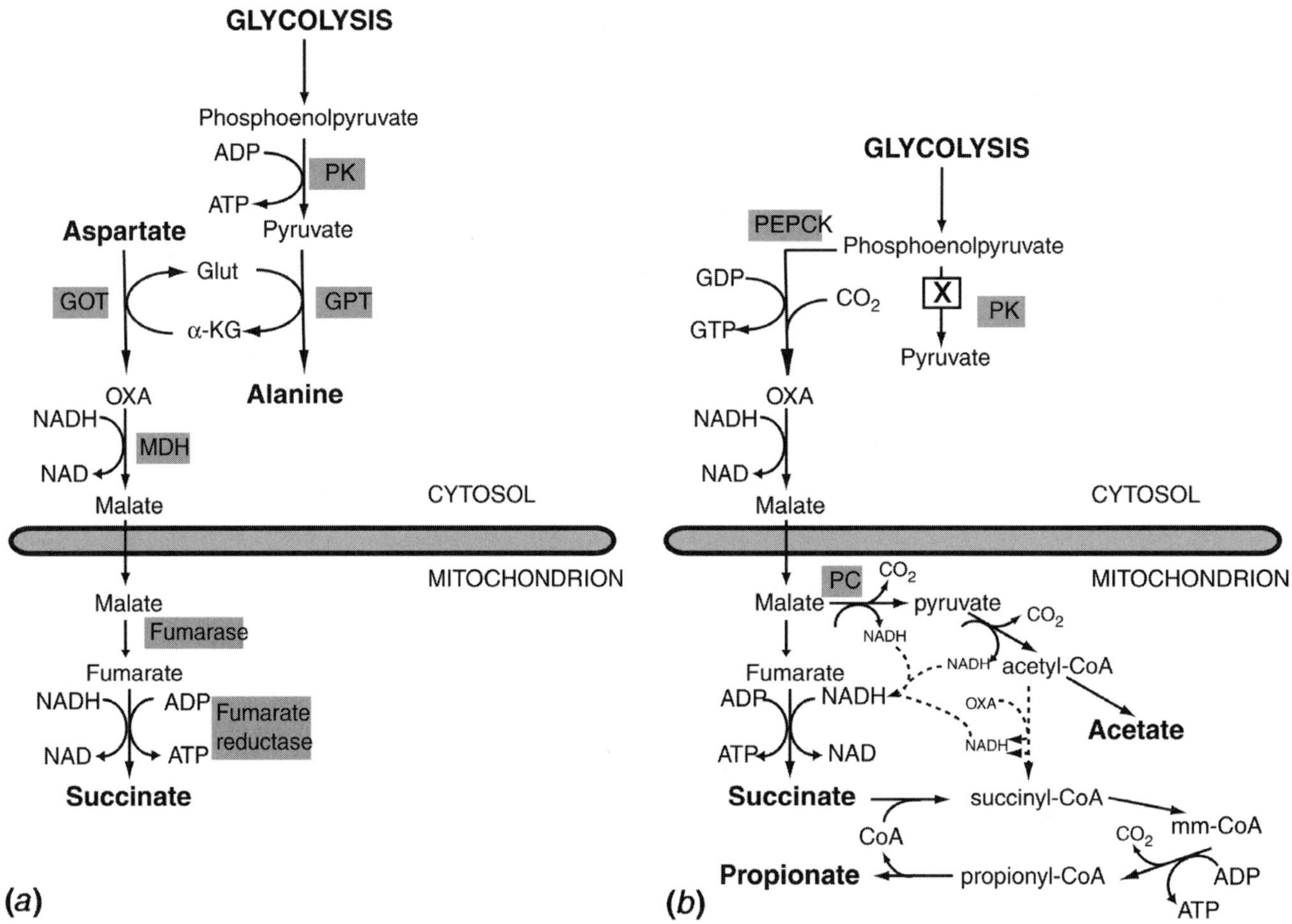

Figure 15.3 Patterns of anaerobic metabolism in marine molluscs. (*a*) Early in an anoxic excursion, the pyruvate produced by glycolysis is converted to alanine in a reaction that is coupled with aspartate conversion to oxaloacetate (OXA); the enzymes involved are glutamate–pyruvate transaminase (GPT) and glutamate–oxaloacetate transaminase (GOT). OXA reduction to malate by malate dehydrogenase (MDH) regenerates cytoplasmic NAD that would otherwise have been the function of lactate dehydrogenase. In the mitochondria, malate is converted to succinate with ATP generated at the fumarate reductase reaction. (*b*) Later in anoxia when aspartate pools are depleted, glycolytic carbon is shunted directly into the reactions of succinate synthesis. Inhibition of the pyruvate kinase (PK) reaction promotes the conversion of phosphoenolpyruvate (PEP) to OXA via the enzyme PEP carboxykinase. The intramitochondrial generation of NADH needed for succinate synthesis is still somewhat controversial but appears to involve malate dismutation with a portion of the malate used to generate NADH via malate conversion to acetyl-CoA using the enzymes pyruvate carboxylase and pyruvate dehydrogenase. The acetyl-CoA could then be processed via a low rate of forward functioning of the tricarboxylic acid cycle, generating more NADH, or acetyl-CoA could be converted to acetate. Production of the volatile end products, acetate and propionate, occurs only in prolonged anaerobiosis. The reactions linking succinate to propionate involve methylmalonyl-CoA (mm-CoA) as an intermediate and ATP generation at the propionyl-CoA carboxylase reaction.

among marine invertebrates where they may supplement or completely replace LDH. The structure of all opines is functionally equivalent to that of octopine (see Text Box 15.1), and the main amino acids used by these reactions are arginine, alanine, glycine, serine, lysine, proline, and taurine. There is no energetic advantage to producing these end products compared with lactate, but in the case of octopine or alanopine, their synthesis incorporates one of the other end products of anaerobic metabolism. For octopine this is the arginine released from phosphagen hydrolysis (see Text Box 15.1), and for alanopine it is the alanine that is a major product of anaerobic glycolysis (see below). However, in general, the opine end products more commonly accumulate as products of muscle work (functional anoxia) than of environmental anoxia. For example, Figure 15.2 shows a net alanine accumulation by *M. mercenaria* gill of about 8 μmol/g wet mass over 4 days of anoxia exposure, but over the same time course we found that the net increase in alanopine was only about 0.5 μmol/g.

For many marine invertebrates (particularly bivalve or gastropod mollusks), a coupled fermentation of glycogen and aspartate occurs as the initial response to anoxia. Glycogen is catabolized to the level of pyruvate, and aspartate is converted to succinate (Fig. 15.3*a*). The pathways are linked in two ways: (a) amino group transfer—glutamate–oxaloacetate transaminase removes the amino group from aspartate to produce oxaloacetate and glutamate–pyruvate transaminase adds the amino group to pyruvate to produce alanine as the glycolytic product—and (b) cytoplasmic redox balance—the NADH generated by the GAPDH reaction of glycolysis is regenerated by the malate dehydrogenase reaction (replacing LDH). Aspartate levels are high in tissues of marine mollusks, and this paired catabolism of glycogen and aspartate can support several hours of anaerobiosis with alanine and succinate typically accumulating in a 1 : 1 ratio. Figure 15.2 illustrates this pattern for *M. mercenaria* gill and shows that aspartate reserves are depleted after about the first 6 h. Alanine production supports the same ATP output as does lactate synthesis, but the energetic advantage of the coupled system comes from succinate production because ATP is generated at the fumarate reductase reaction (Fig. 15.3). Hence, the coupled conversion of 1 mol glucose plus 2 mol aspartate into 2 mol alanine and 2 mol succinate produces a net of 4 mol ATP/mol glucose catabolized, compared with the 2 mol that would result if lactate was the sole product.

When anaerobiosis is prolonged and aspartate pools are depleted, the amino acids are "cut out" of the scheme and carbon from glycolysis is directed straight into the reactions of succinate synthesis. The energy yield of direct fermentation of glucose to succinate is also 4 mol ATP/mol glucose. In this scheme, glycolysis proceeds as normal to the level of phosphoenolpyruvate (PEP), but then, instead of continuing into the pyruvate kinase (PK) reaction, carbon is diverted into the PEP carboxykinase (PEPCK) reaction that yields oxaloacetate plus guanosine 5′-triphosphate (GTP) (or ATP) (Fig. 15.3*b*). This switch is regulated by the anoxia-induced phosphorylation of PK that strongly inhibits PK in all tissues (this is discussed more later). Note that in humans and most other animals PEPCK is typically a key player in gluconeogenesis, the coupled actions of pyruvate carboxylase and PEPCK circumventing the PK reaction to reconvert pyruvate to PEP. In many anoxia-tolerant invertebrates, however, the reverse reaction is emphasized under anoxia, and the enzyme converts the three-carbon PEP into the four-carbon oxaloacetate that can be fed into the pathway of succinate synthesis.

The energy yield per mole of glucose catabolized can be further raised to 6 mol ATP if succinate is further catabolized to produce propionate, a volatile fatty acid. A need to maintain redox balance in the anaerobically functioning mitochondria typically requires the dismutation of malate to produce both propionate and acetate. This scheme is found in many marine invertebrates (but typically only when anoxia is prolonged) and is also the basis for normal carbohydrate fermentation in parasitic helminths that live in the anoxic lumen of the intestines of other animals. Interestingly, the novel carbon metabolism of parasitic helminths was first suggested by the seemingly odd observation that CO_2 fixation was a required part of anaerobic glycogen catabolism in these animals; CO_2 incorporation was later found to occur at the PEPCK reaction.

End Product Excretion and Minimizing Cytotoxicity As is evident from the comparison of the net ATP yield from glucose fermentation to lactate (2 ATP) versus glucose oxidation to CO_2 and H_2O (36 ATP), the incomplete fermentation of glucose leaves a lot of potential energy trapped in the end product. Synthesis of selected alternative products from glucose such as succinate or propionate can extract considerably more energy and often have the bonus of creating less acidity and/or being easy to excrete (e.g., propionate and acetate are volatile). Anaerobic metabolism always results in tissue acidification, and low pH can have negative consequences for many cellular enzymes (recall from Chapter 1 that all enzymes are sensitive to pH change and normally show distinct pH optima). Hence, the synthesis of end products that enhance ATP yield, are less acidifying, or can be readily excreted are all bonuses for facultative anaerobes, and many species have adopted one or more of these options to minimize cytotoxicity during long-term anaerobiosis.

Before we proceed further, it is useful to digress briefly to note that, strictly speaking, it is not the production of "lactic acid" that causes cellular acidification in situations where ATP production is derived from glycolysis (e.g., muscle exercise, anoxia). Protons are actually produced as the products of ATPase action [ATP is hydrolyzed to ADP, inorganic phosphate (P_i), and H^+], and so it is the ATP-consuming reactions that actually generate H^+. Under aerobic conditions this proton output is balanced by proton consumption during ATP synthesis by oxidative phosphorylation, and net cellular pH is unchanged. However, when glycolysis is the only source of ATP synthesis, this balance breaks down because the glycolytic pathway is not a net consumer of protons. Hence, ATP turnover in anoxia (glycolytic ATP production plus cellular ATP consumption by ATPases) results in the net production of H^+. For example, 1 mol glucose converted to 2 mol lactate and producing 2 mol ATP results in a net production of 2 mol H^+ when the ATP is hydrolyzed. Rearranged, this means that 1 mol ATP can be turned over for every mole of H^+ produced. When glucose is fermented to succinate, however, the net proton output is less because 2 mol ATP can be turned over per mole of H^+ produced, and for propionate this rises to 3 mol ATP turned over per mole of H^+ accumulated. Hence, there are clear advantages to succinate or propionate production for anaerobes: their synthesis yields more ATP per mole of glucose consumed and their production results in a lower net acid buildup as compared with lactate production. However, a slow acidification still occurs in all anaerobic systems, and so additional choices must be made to (a) elevate buffering capacity and/or (b) detoxify anaerobic end products by excreting them. Different organisms have made different choices.

Turtles have optimized buffering. Freshwater turtles of the *Trachemys* and *Chrysemys* genera are the champion facultative anaerobes of the vertebrate world (see Text Box 15.4). Anaerobiosis for them is simply glucose/glycogen catabolism to lactate, a high-acid buildup option. However, they turn a peculiarity of their morphology (their large shell) into a metabolic advantage in two ways. First, calcium and magnesium carbonates are released from the shell, the carbonate providing plasma buffering of H^+. Second, lactate is taken up and stored by shell and bone; for example, amounts of 136 and 164 mmol/kg wet mass were found in these compartments in 125-day anoxic turtles compared with plasma lactate levels of 155 mM (lactate is <2 mM in controls). Nearly half of the total body lactate that was accumulated during anoxia was sequestered and buffered within shell and bone, and combined with the buffering provided by carbonate released from shell and bone (determined from the rise in plasma Mg^{2+} and Ca^{2+}), nearly 75% of the total lactic acid buffering in anoxia could be attributed to this source. Hence, shell and bone are crucial to long-term anoxia survival for turtles. Furthermore, because this strategy requires no end-product excretion, turtles end their anaerobic excursions with a plentiful supply of lactate that can be consumed as an aerobic fuel or returned to glycogen stores when air breathing is again possible.

Species of anoxia-tolerant fish have optimized the strategy of end-product excretion. Goldfish and crucian carp show the highest anoxia tolerance among fish species and use this capacity to allow them to survive in small pools that can become oxygen-depleted, particularly when ice-locked in the winter. Lactate is still the product of anaerobic metabolism in most organs, but it is exported into the blood and delivered to skeletal muscles. There the LDH reaction is reversed and the pyruvate is converted in two steps to ethanol and CO_2. The enzymatic reactions involved are pyruvate decarboxylase [reaction (15.4)] and alcohol dehydrogenase [reaction (15.5)]:

$$\text{Pyruvate} \rightarrow \text{acetaldehyde} + CO_2 \tag{15.4}$$

$$\text{Acetaldehyde} + \text{NADH} + H^+ \rightarrow \text{ethanol} + \text{NAD}^+ \tag{15.5}$$

The pyruvate decarboxylase seems to be a novel function of the pyruvate dehydrogenase complex (and therefore the reaction may occur mostly in red muscle), and the conversion to ethanol is catalyzed by the unusual presence of very high activities of alcohol dehydrogenase in the skeletal musculature of these fish. Both ethanol and CO_2 are then excreted by the gills. Clearly, this scheme is wasteful in terms of the carbon reserves of the animal, but these fish still have the option of continuing to feed in anoxic waters, which is unlike the situation of most other facultative anaerobes that are either hibernating (e.g., turtles) or withdrawn into closed shells (e.g., intertidal mollusks). Notably, intestinal parasites that excrete propionate and acetate are also "wasteful" of the carbon reserve and energy potential of these products, but they can afford this because their hosts supply them with a virtually constant supply of sugars from digesting foodstuffs.

Most anoxia-tolerant marine invertebrates have discarded lactate as an anaerobic end product and, furthermore, they have combined multiple strategies in order to design highly efficient systems of fermentative metabolism. They use fermentative pathways that optimize energy output by coupling glycogen and aspartate catabolism and utilize alternative pathways of fermentation that produce extra ATP in substrate-level phosphorylations (Fig. 15.3). They have also optimized the synthesis of a range of low-acid products (e.g., alanine, succinate, propionate, acetate) that includes volatile products (propionate, acetate). Notably, the volatile products accumulate in large amounts only when anaerobiosis is prolonged to several days so that under normal tidal cycles of aerial exposure there would be very little net carbon loss from tissues because the main products of short-term anaerobiosis are alanine and succinate. Finally, mollusks mobilize calcium carbonate from their shell to enhance buffering capacity during anoxia.

Meeting ATP Demand The ATP output of fermentative pathways is always much lower than that of oxidative metabolism, and hence, there are only two options for reestablishing a balance between ATP consumption and ATP production in anoxia. One is to raise anaerobic ATP production up to a level that can meet the unaltered demands of ATP-utilizing reactions. This can be accomplished over the short term by greatly accelerating the rate of glycolysis but often fails very quickly (human brain energetics are irreversibly compromised after <5 min of anoxia). Some organisms can sustain high rates of ATP production during long-term anaerobiosis in cases where substrate supply is virtually unlimited (e.g., microorganisms growing in a sugary broth; see Text Box 15.2). However, multicellular organisms would quickly and wastefully deplete their internal reserves of fermentable fuels by using this strategy, and therefore, long-term anoxia survival by most facultative anaerobes has come to depend on strong metabolic rate depression. By reducing net ATP use in anoxia to <10% of the corresponding aerobic rate, organisms can meet ATP demand using fermentative pathways of ATP production and gain a 10-fold or more extension of the time that their fixed internal carbohydrate reserves can fuel anaerobic survival. By reducing ATP turnover in anoxia, animals also reduce the extent of acidification during long-term anaerobiosis as well as the net buildup of end products (that need to be excreted or that must be restored to glycogen pools via ATP-expensive gluconeogenic reactions when oxygen returns). Anoxic excursions are often of unpredictable duration, and so the ability to ration carbohydrate reserves and sustain homeostasis for the longest possible time is critical. Metabolic rate depression is quantitatively the most important of the biochemical adaptations that support anaerobiosis and is discussed in detail below.

Antioxidant Defenses Although it may seem contradictory, all anoxia-tolerant animals that we have examined have very well developed antioxidant defenses, both enzymatic and metabolite, for dealing with the generation of ROS. This is true not only of anoxia-tolerant vertebrates and invertebrates but also of freeze-tolerant animals and hibernating mammals and, therefore, high antioxidant defenses may be generally associated with situations where organisms experience very wide variation in oxygen availability. Under ischemic or hypoxic situations, the electron carriers of the mitochondrial respiratory chain become reduced, and when oxygen is reintroduced, an immediate reoxidation of these carriers takes place and results in a transient overproduction (or burst) of ROS. This burst of ROS production is well known to be the basis of postischemic reperfusion injuries in mammalian organs recovering from an ischemic event. Anoxia-tolerant species appear to avoid ROS-triggered metabolic injury by sustaining high constitutive levels of antioxidant defenses and supplementing this, in some cases, with anoxia-induced increases in the activities of antioxidant enzymes or elevated glutathione pools. This subject is discussed in greater detail in Text Box 13.5.

METABOLIC RATE DEPRESSION

Metabolic rate depression (MRD) is a widespread response to many types of stress and is found throughout phylogeny (see Text Box 15.6). The biochemical mechanisms of MRD received the greatest initial study from researchers studying anoxia tolerance, and hence the subject is introduced here but the phenomenon will be explored further in the next chapter when we discuss mammalian hibernation. As mentioned earlier, there are two main solutions to situations that disrupt an organism's capacity to produce ATP at a rate that adequately supplies the ATP demands of metabolism. One is compensation, which works well in the short term, and the other is conservation. If ATP production is decreased due to low oxygen availability, then organisms turn down their rate of ATP consumption until a new rate of ATP turnover is established where ATP production again equals consumption. MRD is a key component of anoxia survival for all anoxia-tolerant organisms for it allows metabolic rate to be lowered to a level that can be supported by the ATP output from fermentative pathways alone. For example, turtles submerged under water suppress their anoxic metabolic rate to only 10 to 20% of their comparable resting metabolic rate when breathing air at the same temperature. In marine mollusks, anoxic metabolic rate is reduced to only 2 to 10% of the aerobic value.

How is metabolic rate depression achieved? Several extrinsic factors contribute to energy savings. For example, organisms in hypometabolic states typically show few voluntary movements by skeletal muscles, most do not eat so the costs of digestion and absorption are saved, and heart beat, breathing, and kidney filtration rates are all greatly reduced. Metabolic rate is also suppressed by accompanying factors that are typically present, including low pO_2, elevated pCO_2, low pH, and sometimes lower temperature. For instance, many ectothermic organisms voluntarily seek cooler temperatures when challenged by hypoxia and thereby use a decrease in body temperature to help reduce tissue demands for oxygen. Indeed, a hypoxia–hypothermia connection is also suspected as part of the mechanism by which hibernating mammals lower their metabolic rate and body temperature as they sink into torpor (see Text Box 16.5). However, intrinsic mechanisms within cells appear to account for at least half of the whole animal energy savings in the hypometabolic state. These induce stable suppression of the rates of multiple metabolic processes to provide both a net reduction in metabolic rate and selective targeting of specific cell functions that are unneeded in the hypometabolic state.

Metabolic suppression is not necessarily applied uniformly to all processes within a cell or to all organs within an organism. Priorities exist between organs; for example, the suppression of protein synthesis in different

TEXT BOX 15.6 METABOLIC RATE DEPRESSION—TORPOR, DORMANCY, HIBERNATION, DIAPAUSE, ESTIVATION, AND MORE

The ability to strongly suppress metabolic rate and sink into a hypometabolic state is a lifesaver for many organisms and is found in virtually all phylogenetic lineages. Metabolic rate depression (MRD) is, in fact, one of the most powerful defenses that animals have developed to deal with environmental stress of many kinds. The ability to sink into a dormant or torpid state allows the organism to effectively "wait out" the stress, extending by 10- or 20-fold or more the length of time that the animals can live off internal reserves of stored fuels (chiefly glycogen or triglycerides). MRD is widespread in nature as a way of ensuring long-term survival when environmental conditions are incompatible with normal life (e.g., too hot, too cold, too dry, no oxygen). It is an integral component of torpor, hibernation, estivation, diapause, dormancy, anaerobiosis, and cryptobiosis.

As discussed at length in this chapter, MRD is key to anaerobic survival, but it is also widely used in aerobic circumstances. Nightly torpor, during which metabolic rate drops by 20 to 30%, preserves just enough body fuel to allow many small birds and mammals to live to see the next day. Winter hibernation allows many mammals to "sleep" through the winter. Hibernation can last as long as 9 months in Arctic or alpine environments, and the profound MRD of the hibernator produces energy savings of as much as 90% compared with the costs of maintaining a high and constant body temperature throughout the winter (see Chapter 16). Life in arid regions of Earth is aided by estivation, an aerobic dormancy induced by dry and/or hot conditions; a 70 to 90% reduction in metabolic rate allows lungfish, frogs, toads, and snails, among others, to survive for many months until the next rainy season. Diapause arrests the developmental cycle of many insects and other invertebrates to allow them to "wait out" excessively hot, cold, or dry conditions or to synchronize the transformation of a whole population to the next developmental stage (e.g., adult emergence). Extremes of hypometabolism are found in cryptobiosis where a virtual ametabolic state in many seeds, spores, cysts, emrbyos, and eggs produces a life extension that can stretch to years, decades, or even centuries (for more information see Text Box 17.1).

organs of ground squirrels during hibernation ranged from 0 to 85%, the zero value belonging to brown adipose tissue that seems to need to maintain full biosynthetic potential to support its central role in thermogenesis (see Chapter 16). Priorities also exist within cells as to which metabolic activities are maintained in the hypometabolic state. Table 15.1 illustrates this with the fractional use of ATP by different metabolic processes in liver cells (hepatocytes) of turtles under aerobic versus anoxic conditions. In normoxic cells, five main ATP-consuming processes were identified with ion pumping by the Na^+K^+–ATPase and protein synthesis using the largest portions of cellular ATP. Under anoxic conditions, however, total ATP turnover in liver cells fell by 94%, and each of the ATP-consuming processes was differently affected. Na^+K^+–ATPase activity decreased by 75%, whereas protein synthesis was suppressed by 93% and gluconeogenesis was undetectable in anoxic cells. As a result of this reordering of metabolic priorities, the sodium/potassium pump became the dominant energy sink in anoxic hepatocytes, consuming 62% of total ATP turnover. These priorities appear to derive from the ATP sensitivity of different energy-consuming reactions; in isolated thymocytes the pathways of macromolecular biosynthesis were shown to be most sensitive to energy supply, followed by sodium cycling and then calcium cycling across the plasma membrane, and finally the mitochondrial proton leak was least sensitive to ATP.

The molecular mechanisms of MRD are a subject of much study at present, and while the full picture is not yet available, the critical importance of certain mechanisms, such as reversible protein phosphorylation, is known and is discussed below. In general, mechanisms of MRD need to be readily reversible so that the constitutive metabolic machinery of cells can be retained in readiness to respond rapidly when organisms arouse from the hypometabolic state. Hence, hypometabolism is not associated

TABLE 15.1 Fractional Use of Cellular ATP Turnover by Different Cellular Activities in Turtle Hepatocytes Incubated under Aerobic versus Anaerobic Conditions

	Normoxia	Anoxia	Suppression in Anoxia (%)
Na^+K^+–ATPase	28.5	62.3	75
Protein synthesis	36.1	20.8	93
Protein degradation	16.6	9.1	94
Urea synthesis	3.0	7.8	70
Gluconeogenesis	17.0	0	100

Source: Data reworked from P. W. Hochachka, et al. (1996). *Proc Natl Acad Sci USA* **93**:9493–9498.

with a major loss of metabolic capacity by cells, but instead reversible controls are applied to coordinately suppress the rates of all metabolic processes while leaving intact the potential to rapidly return to the normal state.

Reversible Protein Phosphorylation

The most powerful and widespread mechanism of MRD is control over the activities of enzymes and functional proteins via reversible protein phosphorylation. As discussed in Chapter 14, reversible phosphorylation can affect enzyme function in multiple ways, including on–off control over activity, marked changes in enzyme properties, or altering enzyme-binding interactions with partner proteins.

Control of Anaerobic Glycolysis in Marine Mollusks The first indication that reversible protein phosphorylation played a role in MRD came from studies of the control of fuel catabolism in anoxia-tolerant marine mollusks. Anoxia-induced phosphorylation of PK proved to be key to the control of the PEP branch point and the partitioning of glycolytic carbon between the aerobic (PEP is fed into the Krebs cycle) and anoxic (PEP is fed into the reactions of succinate synthesis) routes of carbohydrate catabolism. By strongly inhibiting PK, PEP can be rerouted into the PEPCK reaction to produce oxaloacetate that is then fed into succinate synthesis (Fig. 15.3*b*). Aerobic and anoxic variants of PK were found in anoxia-tolerant marine mollusks, and these were separable on ion exchange chromatography and showed stable differences in kinetic properties (see Figs. 14.5 and Table 14.4). As Table 14.4 showed, our studies of PK from both muscle and hepatopancreas of whelks (*Busycon canaliculatum*) showed that the anoxic form of PK has a much lower affinity for PEP as a substrate, is much less sensitive to fructose-1,6-bisphosphate ($F1,6P_2$) as an activator, and is much more susceptible to inhibition by the anaerobic end product alanine. We traced the differences between the aerobic and anoxic forms of PK to their contents of covalently bound phosphate. Aerobic PK is the low-phosphate form and anoxic PK is the high-phosphate form, and the two are interconvertible via treatments with protein kinases versus protein phosphatases. Anoxia-induced phosphorylation of PK coupled with anoxia-induced changes in the concentrations of substrates and effectors of PK *in vivo* (particularly the large accumulation of alanine that occurs in virtually all marine mollusks under anoxia; Fig. 15.2), produces a situation where PK activity is virtually shut off under anoxic conditions to allow effective rerouting of glycolytic carbon via PEPCK into the pathway of succinate synthesis. Anoxia-induced phosphorylation of PK has now been documented in many different kinds of marine invertebrates and appears to be a universal mechanism for controlling the PEP branch point in order to make the switch from aerobic to anoxic routes of carbohydrate catabolism. Interestingly, however, it is not always the phosphorylated enzyme that is the low-activity, anoxic form. In the periwinkle, *Littorina littorea* (a small intertidal gastropod mollusk), we found that the opposite situation occurred; anoxia exposure stimulated a dephosphorylation of PK and the low-phosphate enzyme form shows reduced substrate affinity and enhanced inhibition by alanine compared with the high-phosphate aerobic enzyme form.

Subsequently, we found that anoxia-induced phosphorylation regulated not just PK but also other loci in glycolysis in anoxia-tolerant marine mollusks. This changed our view of the importance of reversible phosphorylation control from that of a mechanism that controlled a single branch point to a mechanism that coordinated glycolytic rate. Ultimately, we broadened this view even more and know now that the mechanism is a widespread and general mechanism for the coordinated suppression of all aspects of metabolism in hypoxia/anoxia-tolerant organisms. Our studies of glycolytic rate control showed that, in addition to PK control, reversible phosphorylation regulated glycogen phosphorylase (GP), PFK-1, and PFK-2 under anoxia. Figure 15.4 shows the coordinated effects of anoxia-induced protein phosphorylation on these four enzymes in whelk gill. In this case, the activities of GP*a* (the active enzyme) and PFK-2 fell over 20 h of anoxia exposure to 48 and 41% of aerobic values, whereas PK maximal activity was suppressed to just 16%. Suppression of PFK-2 activity had a major impact on $F2,6P_2$ levels, which fell to just 2% of aerobic values, a change that would then have major consequences for PFK-1 activity. PFK-1 phosphorylation state was assessed by changes in one of its kinetic parameters, the I_{50} value for PEP, which was reduced during anoxia to 20% of its aerobic value. Similarly, the I_{50} value for alanine of PK fell to just 4% of the aerobic value in anoxia. It is important to notice that the time course of all these changes is very similar, showing the greatest change over the first 4 h of anoxia exposure. This emphasizes the coordinated nature of the glycolytic suppression response to anoxia.

Having confirmed the role of reversible enzyme phosphorylation as a key mechanism of anoxia-induced MRD in marine mollusks, the question became whether this was a universal mechanism of MRD. Multiple studies have now confirmed this in many ways with demonstrations that (1) glycolytic enzymes are also targets of anoxia-induced phosphorylation in anoxia-tolerant vertebrate animals (e.g., goldfish, turtles) (see below); (2) reversible phosphorylation control of glycolytic enzyme activities also occurs in situations of aerobic MRD, such as during estivation in terrestrial snails and toads (see Text Box 15.7) or during hibernation in mammals (see Chapter 17); and (3) reversible phosphorylation regulates

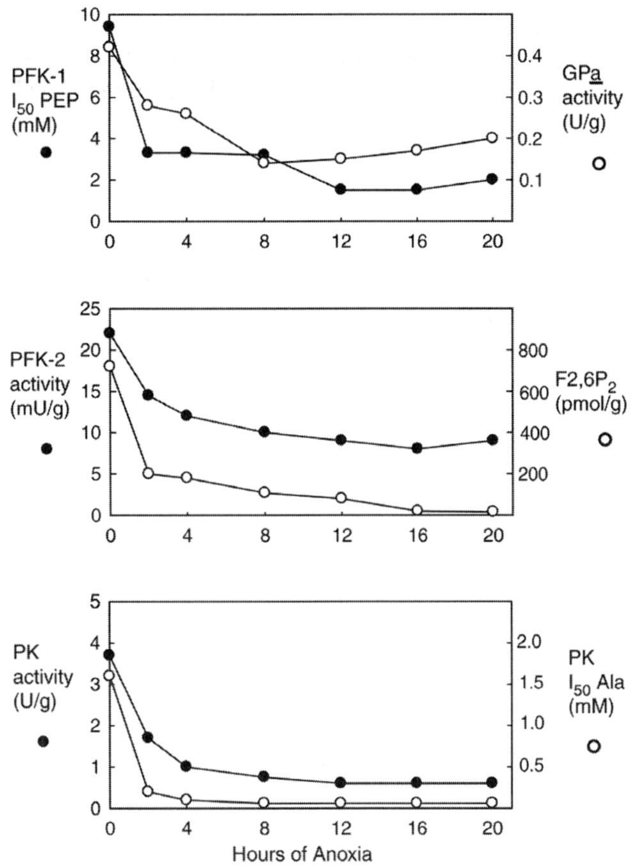

Figure 15.4 Coordinated changes in the activities and properties of glycolytic enzymes and fructose-2,6-bisphosphate ($F2,6P_2$) levels in gill of the whelk, *Busycon canaliculatum*, over the course of 20 h of anoxia exposure. Shown are (*a*) the activity of the active *a* form of glycogen phosphorylase (GP*a*) and the I_{50} value for phosphoenolpyruvate (PEP) of 6-phosphofructo-1-kinase (PFK-1), (*b*) the activity of 6-phosphofructo-2-kinase (PFK-2) and levels of its product, $F2,6P_2$, (*c*) the activity of pyruvate kinase (PK) and its I_{50} value for L-alanine. Activities are in units (or milliunits) per gram wet mass and concentrations are in millimolar or pmol/g wet mass. [Modified from K. B. Storey (1993). Molecular mechanisms of metabolic arrest in mollusks. In P. W. Hochachka, P. L. Lutz, T. J. Sick, M. Rosenthal, and G. van den Thillart, (eds.). *Surviving Hypoxia: Mechanisms of Control and Adaptation*. CRC, Boca Raton, FL, pp. 253–269.]

multiple other cell functions during hypometabolism, including suppression of pyruvate dehydrogenase activity (regulating pyruvate entry into the TCA cycle), the activities of ion-motive ATPases, and inhibition of the ribosomal translation machinery.

Control of Carbohydrate Metabolism in Anoxia-Tolerant Vertebrates Anoxia-induced reversible phosphorylation supplies coordinated control over the activities of

TEXT BOX 15.7 ESTIVATION

Estivation is a state of aerobic torpor. The underlying trigger for estivation is typically arid conditions, often accompanied by a lack of food availability and high environmental temperatures. Metabolic rate is typically reduced by 70 to 90%, and water conservation strategies are used including burrowing underground, apnoic breathing patterns, formation of cocoons or other physical barriers to water loss, and elevation of body fluid osmolality (see discussion of urea accumulation in Chapter 14). The physiology and biochemistry of estivation has been most extensively studied in two groups: pulmonate land snails and anuran amphibians (such as the spadefoot toad of the American Southwest). The first demonstration that the biochemical mechanisms of metabolic arrest that are used by anoxia-tolerant animals also underlie MRD in the aerobic state of estivation occurred during studies of land snail estivation. These studies confirmed that reversible protein phosphorylation that produced stable changes in the activity states of enzymes was a general principle of MRD in both aerobic and anoxic systems.

Land snails entering estivation in response to food and water deprivation showed stable changes in the properties of glycolytic enzymes (GP, PFK-1, PK) and strong suppression of $F2,6P_2$ levels in their tissues—responses that frequently parallel the patterns seen during anoxia in marine molluscs (see Fig. 15.4). For example, PK from foot muscle of estivating *Otala lactea* showed reduced affinity for PEP substrate (a 50% increase in K_m), increased sensitivity to inhibition by L-alanine and ATP (I_{50} values decreased by 60 and 40%, respectively), and a increase in isoelectric point from pH 5.85 to pH 6.2 as compared with control snails. Significantly, these same changes in PK properties were also achieved when land snails were given anoxia exposure under a nitrogen gas atmosphere. Treatments *in vitro* with protein kinases or phosphatases showed that, as in anoxia-tolerant mollusks, reversible phosphorylation of glycolytic enzymes was responsible for estivation-induced changes in enzyme function and a cGMP-dependent protein kinase was again implicated. However, in estivating snails an additional enzyme target of control over carbohydrate metabolism was found—pyruvate dehydrogenase (PDH). The percentage of PDH present in the active, dephosphorylated form decreased from ~98% in control snails to 60% over the first 1 to 1.5 days of estivation but rebounded within 1 h when snails were aroused by spraying with water. Control over PDH allows a coordinated suppression of both glycolytic and mitochondrial carbohydrate catabolism in the estivating states. The importance of

PDH control is discussed more extensively in Chapter 16 where it is a key element of MRD in mammalian hibernation.

Estivation in vertebrates also uses these same principles of metabolic control. Isoelectric focusing of skeletal muscle extracts from spadefoot toads, *Scaphiopus couchii*, revealed the presence of two forms of PK and PFK whose proportions changed during estivation and that were interconvertible by reversible phosphorylation. Interestingly, however, in spadefoot toad muscle the effect of estivation was to increase the proportion of the dephosphorylated enzymes, and kinetic analysis revealed that, contrary to the situation in mollusks, the dephosphorylated enzymes were, in this case, the less active enzyme forms. For muscle PFK this fits well with the known controls on vertebrate muscle PFK. Phosphorylation of vertebrate muscle PFK occurs during exercise and the phospho-enzyme shows increased binding to myofibrils in active muscle that helps to localize glycolytic ATP production near the sites of ATP use by the myofibrillar ATPase (see Chapter 14). Thus, the increased content of the low-phosphate form of PFK in muscle of estivating toads is consistent with a reduced glycolytic rate accompanying the overall MRD of the estivating state.

The changes in phosphorylation state of both PK and PFK in toad organs correlated well with the suppression of the activities of both protein kinases A and C during estivation. As in other situations of MRD, the metabolic potential of organs is largely retained in estivation with relatively few changes seen when the maximum activities of a wide variety of metabolic enzymes were surveyed. However, one general exception to this was the response of enzymes of antioxidant defense, which were generally lower in estivating animals than in aroused toads. This is consistent with the idea of reduced oxidative stress under the low metabolic rate of the estivating state and the need to restore defenses against ROS when metabolic rate rises by severalfold when animals arouse from torpor.

For more information on estivation consult K. B. Storey. Life in the slow lane: molecular mechanisms of estivation. *Comp Biochem Physiol A* **133**:733–754 (2002).

glycolytic enzymes in tissues of anoxia-tolerant vertebrates. Regulation of liver enzymes is particularly prominent, and this is related to the role of liver as the central store of carbohydrate fuel in the vertebrate body (see Chapter 9). Apart from liver, skeletal muscle is the only organ with major glycogen reserves, but muscle does not export glucose (its glucose transporter is for import only). The main way that muscle can export fuel is as end products, lactate or alanine, that are typically products of muscle work. These are significant aerobic fuels for organs such as heart, but they are useless under anoxic conditions. Hence, glucose supply by liver is key to the anoxia survival of the whole animal, particularly for the survival of brain, which has only low levels of endogenous glycogen. Not surprisingly, then, the percentage of GP present as the active *a* form in turtle liver increased from 14 to 26% after 5 h of anoxia exposure, but note that this increase is not large for an enzyme that is supplying glucose fuel to all other organs. Furthermore, GP activity was generally unaffected in other organs, which is consistent with the idea of strong MRD in anoxia that lowers ATP demand to a rate that can be fueled by low rates of carbohydrate fermentation. In recent years the application of metabolic control analysis to multiple systems has repeatedly highlighted the importance of substrate supply, rather than ATP consumption, in regulating metabolic rate, so this suggests that control by liver over substrate supply to the whole body may be a critical part of MRD.

Table 15.2 shows that anoxia exposure of goldfish triggers stable and coordinated changes to the properties of three liver glycolytic enzymes, very similar to the situation seen in marine mollusks. Glycogen phosphorylase activity is reduced by two mechanisms: decreases in the total amount of enzyme and in the percentage of enzyme in the active *a* form. In combination, this lowers the activity of GP*a* in anoxia to just 50% of the aerobic value (note

TABLE 15.2 Effect of 24 h in N_2 Bubbled Water on Properties of Glycolytic Enzymes in Goldfish Liver[a]

	Aerobic	Anoxic
Glycogen phosphorylase, U/g	3.6	2.5
% *a*,	80	55
Fructose-2,6-P_2, nmol/g	7.77	0.75
6-Phosphofructo-1-kinase		
$S_{0.5}$ fructose-6-P, mM	1.56	2.33
K_a fructose-2,6-P_2, μM	0.17	0.08
K_a AMP, μM	0.29	0.37
I_{50} ATP, mM	1.43	1.25
Isoelectric point	3.86	4.30
Pyruvate kinase		
$S_{0.5}$ PEP, mM	0.45	0.80
K_a fructose-1,6-P_2, μM	0.20	0.30
I_{50} L-alanine, mM	25.1	11.6
Isoelectric point	3.87	4.30

[a]All values are the means of 3–8 independent determinations and all anoxic values shown are significantly different from the corresponding aerobic values.

Source: Data are compiled from K. B. Storey (1987). *Physiol Zool* **60**:601–607 and M. S. Rahman and K. B. Storey (1988). *J Comp Physiol* **157**:813–820.

that, unlike the situation in turtle liver mentioned above, GP*a* activity was very high in liver of control fish). Anoxia exposure stimulated multiple changes in the properties of PFK-1, including reduced affinity for its substrate, fructose-6-P, and altered sensitivities to both activators and inhibitors. The K_a value for F2,6P_2 decreased by 50% in anoxia, which indicates that the enzyme is more sensitive to this activator, but liver F2,6P_2 content dropped 10-fold in anoxia so, overall, the influence of F2,6P_2 on PFK-1 is much reduced in anoxia. This makes sense because in liver F2,6P_2 acts as an anabolic signal to promote carbohydrate use for biosynthetic purposes, and such activity needs to be suppressed under anoxic conditions. Stable modification of PK properties was also seen with reduced substrate affinity, decreased sensitivity to activation by F1,6P_2, and increased sensitivity to alanine inhibition, all changes that would suppress PK activity in anoxia. Furthermore, both PFK-1 and PK show distinct changes in their isoelectric points under anoxia; this indicates a change in the net charge of the proteins and is frequently diagnostic of protein phosphorylation. Other organs of goldfish show selective changes in many of these parameters in anoxia, consistent with organ-specific metabolism in anoxia. We have also documented stable changes to the properties of GP, PFK-1, and PK as well as F2,6P_2 levels in response to anoxic submergence in the organs of freshwater turtles.

Signaling Mechanisms Inducing Events of Hypometabolism The clearly defined differences in kinetic parameters between phosphorylated and dephosphorylated enzyme forms (e.g., the 25-fold difference in the PK I_{50} for alanine seen in Fig. 15.4) provides a useful tool that can be exploited to explore the regulation of anoxia-induced enzyme phosphorylation. The effects of hormone second messengers and other treatments in mimicking anoxia-induced effects on mollusk enzymes have been explored *in vitro* with isolated tissues or tissue extracts. For both PK and PFK-1 such treatments were highly consistent in showing that anoxia-induced effects on the kinetic properties of the enzymes in marine mollusks were mimicked by treatments with cyclic 3′,5′-guanosine monophosphate (cGMP). For example, when extracts of radular retractor muscle from aerobic whelks were incubated with Mg–ATP plus the second messengers of protein kinases A [cyclic 3′,5′-adenosine monophosphate (cAMP)], C (Ca^{2+}-+ phorbol 12-myristate 13-acetate), or G (cGMP) and then PK kinetics were evaluated, the K_m value for PEP and the I_{50} value for alanine changed significantly only in response to cGMP. Treatment with cGMP raised the K_m from 0.05 mM (control) to 0.23 mM and lowered the I_{50} from 10 to 2.5 mM, replicating the effect of anoxia on the enzyme. This evidence and similar results for other species have clearly linked the events of anoxia-induced MRD in marine mollusks to control by the protein kinase G (PKG) mediated signal transduction pathway. However, despite this documented link, to date there has been little analysis of this protein kinase in facultative anaerobes. We have documented the presence of a PKG in marine whelks, *B. canaliculatum*, and shown that it stimulates the phosphorylation of PK. Furthermore, we also found a second-messenger-independent protein kinase that specifically phosphorylates PK but not PFK-1 or GP. This specific PK kinase is probably regulated by PKG, the cascade system of control allowing powerful regulation of the PEP branch point in anoxia. A serine/threonine PK phosphatase of the type 2C group reverses the effects of PK kinase. In other systems, the PKG signal transduction pathway is triggered by nitric oxide (NO). Although NO has not yet been explored as a possible regulator of aerobic–anaerobic transitions in anoxia-tolerant mollusks, this is clearly an area with exciting new research possibilities.

Another possible regulatory factor in MRD is acidosis. A reduction in cellular pH values accompanies all arrested states, in both anaerobic systems and situations of aerobic metabolic arrest such as hibernation and estivation. However, whether pH change triggers or controls MRD is questionable. Under anoxic conditions, the slow development of acidosis over the long term does not fit well with the rapid and early changes in enzyme phosphorylation patterns that trigger MRD or with the fact that anoxia-tolerant species clearly take steps to minimize the extent of tissue acidosis. Nonetheless, as we will see several times in the discussion of metabolic arrest in this and the next chapter, cellular acidosis clearly creates a metabolic context that facilitates various events of metabolic arrest. Associated with acidosis and likely the more important signaling influence on MRD is hypercapnia (high CO_2). Respiratory acidosis due to CO_2 retention as a result of breath-hold (apnoic) breathing patterns is a feature of most arrested states. The influence of high CO_2 has been strikingly illustrated with land snails, where the simple elevation of CO_2 in the air (pCO_2 raised to 65 mmHg) resulted in a decrease in the oxygen consumption of active snails by 50% within 1 h. However, as soon as CO_2 was removed, metabolic rate rebounded. Hypercapnia also stimulated a similar reduction in oxygen consumption by anoxia-tolerant marine worms. CO_2 retention is now a well known event during entry into many situations of MRD, and it is rapidly reversed when hypometabolism ends. For example, hyperventilation to clear accumulated CO_2 is one of the first events when hibernating mammals begin the arousal process. Signaling and regulatory events linking CO_2 and organ enzymatic responses during dormancy have received very little exploration to date, but some of our results with whelk muscle PK (Table 15.3) suggest that this could become a useful area of study.

TABLE 15.3 Effect of Anoxia, pH Change, and CO_2 on Pyruvate Kinase Kinetics During *in vitro* Incubations of Whelk Radular Retractor Muscle[a]

	Tissue pH	K_m PEP (mM)	I_{50} Alanine (mM)
Aerated, control	7.64	0.12	18.3
Aerated, pH 5.5	6.39	0.29	7.8
100% N_2	6.97	0.54	3.20
95 : 5% $N_2 : CO_2$	6.64	1.80	1.00
95 : 5% $N_2 : CO_2$, pH 9.5	7.35	1.62	0.70

[a]Incubations were in artificial seawater; in two cases, seawater pH was altered (pH 5.5, pH 9.5) in order to lower or raise tissue pH. Lowering tissue pH in aerated tissues had minimal effect on PK properties compared with the effects of anoxia (N_2 gas bubbling). The presence of CO_2 in anoxic incubations stimulated much greater changes in PK properties that were not altered when tissue pH was held artificially high.

Source: Data are reworked from S. P. J. Brooks and K. B. Storey (1989). *J Exp Biol* **145**:31–43.

These studies used *in vitro* tissue incubations and showed that the anoxia-induced increase in the K_m value for PEP and decrease in the I_{50} value for alanine of PK were much more pronounced when anoxia exposure was the result of bubbling seawater with a N_2–CO_2 (95 : 5) mixture rather than with 100% N_2 alone. Furthermore, these changes to enzyme properties still occurred even if tissue pH was kept artificially high in anoxia, indicating that it is not the acidifying effect of CO_2 that stimulated the enzyme modification.

Control of Membrane Transport and Related Functions

Membrane Ion-Motive ATPases and Ion Channels ATP consumption by ion-motive ATPases is one of the biggest energy expenditures in all cells as well as one of the most crucial. Not only does the failure of transmembrane ion gradients impair normal cell functions (e.g., signal transmission, neural conductivity, muscle contactility) but also membrane depolarization can trigger cascades of degenerative events, as explained earlier. It is not surprising, therefore, that anoxia-tolerant animals strongly reduce ATP expenditures on transmembrane ion movements in anoxia and coordinate a net suppression of both influx and efflux of ions by a process that has been called "channel arrest." This phenomenon has been particularly well studied in the brain of anoxia-tolerant turtles, and the summary here will pertain mostly to those animals.

Ion gradients across the plasma membrane (Na^+ and Ca^{2+} high outside, K^+ high inside cells) are maintained by ATP-dependent ions pumps that move ions against their concentration gradients versus facilitated movements of ions down their concentration gradients through ion channels. To reduce ATP expenditure without dissipating ion gradients, ion channel conductance is suppressed in anoxia. Studies have shown that K^+ leakage is significantly lower in brains of anoxic turtles compared with aerobic controls and that Ca^{2+} channel activity decreases progressively with length of anoxia. Closure of conductance channels is also accompanied by a decrease in sodium channel abundance in plasma membranes and silencing of NMDA receptors. NMDA receptors are a subfamily of glutamate receptors; they are high-flux, ligand-gated cation channels that are highly permeable to Ca^{2+} and a major source of Ca^{2+} entry into anoxic/ischemic brain. Strong inhibition of Ca^{2+} influx into turtle cells in anoxia is particularly crucial because Ca^{2+} levels in cerebrospinal fluid can rise six-fold within 10 days during anoxic submergence of turtles due to the mobilization of shell calcium carbonate to enhance plasma buffering of accumulating lactate. Activity of the Na^+K^+–ATPase ion pump in turtle brain was also reduced by 30 to 35% during anoxia but was rapidly reversible upon reoxygenation. The mechanism of Na^+K^+–ATPase suppression is likely to be reversible protein phosphorylation, which we have shown underlies the inhibition of this ion pump in hibernating mammals (see Chapter 16). Silencing of NMDA receptors has been linked with three mechanisms operating on different time scales: (a) short term—within 8 min of anoxia exposure NMDAR activity (Ca^{2+} influx and open probability) in turtle brain was reduced by 50 to 60% mediated by activation of protein phosphatases 1 or 2A; (b) intermediate term—by 2 h of anoxia a rise in intracellular Ca^{2+} of about 35% would trigger Ca^{2+}/calmodulin-mediated suppression; and (c) long term—reversible removal of NMDARs from the plasma membrane occurred when anoxia exposures exceeded 3 days.

Studies of anoxia-tolerant animals and of ischemia–reperfusion in hypoxia-sensitive mammals both agree on the importance of adenosine as a signaling molecule that triggers adaptations to protect tissues from injury. Adenosine is released by hypoxic or ischemic tissues and stimulates multiple protective effects. In the ischemic mammalian heart, for example, these include a reduction in myocardial oxygen demand through negative effects on the rate and force of contraction, promotion of ATP production from glycolysis, a reduction in oxygen free radical release, a stimulation of vasodilation, and an inhibition of neutrophil and platelet aggregation so that blood vessels do not become blocked during the interruption of blood flow. Adenosine is synthesized from AMP by the enzyme 5′-nucleotidase, and adenosine levels rise quickly under anoxia or ischemia stress and stimulate one of four subtypes of G-protein-coupled adenosine receptors. Signaling via the A1 adenosine receptor is particularly important and contributes to preconditioning, the phenomenon whereby short initial periods of ischemia greatly reduce the metabolic damage that accrues from a subsequent extended ischemic challenge. The importance of adenosine to MRD was confirmed in studies with brain of anoxia-tolerant turtles. Adenosine in turtle brain peaked first after 2 to 3 h of anoxia exposure and then subsequent pulses of adenosine release occurred with longer anoxia exposures. Interestingly, adenosine (but not any of several other putative neurotransmitters) increased three- to four-fold under anoxia and hypercapnia in the anoxia-tolerant marine worm, *Sipunculus nudus*, and adenosine infusion into coelomic fluid of normocapnic worms suppressed oxygen consumption. These data suggest that adenosine plays a role in metabolic rate depression across phylogeny.

One of the known effects of adenosine action in turtle brain is a suppression of the activities of ATP-dependent ion channels, mediated by the A1 receptor. Suppression of K^+ leakage has been documented and, in particular, adenosine caused a strong suppression of excitatory neurotransmission in turtle brain by reducing Ca^{2+} entry into cells via NMDAR by more than 50%. The positive effects of adenosine during preconditioning are mediated intracellularly by at least two signal transduction pathways (in heart): protein kinase C (PKC) and the p38 mitogen-activated protein kinase (MAPK). In our studies of turtle liver and brain we found PKC activation during anoxia in both organs; the percentage of membrane-bound (active) PKC rose from 21 to 45% within 1 h of anoxia exposure in turtle liver, whereas in hindbrain the percent bound was 42% in controls and 69% in turtles after 5 h of anoxic submergence. Interestingly, PKC was suppressed in another part of the brain (the cerebrum) after 5 h of anoxia, a result that may indicate differential responses or differential timing of the development of anoxia-protective mechanisms in different parts of the brain. Hence, studies of the downstream effects of PKC and p38 signaling in anoxic organs are warranted. One of the intracellular end results of adenosine signaling (and of preconditioning) appears to be opening of the mitochondrial ATP-dependent K^+ (K_{ATP}) channel, which may have one of several protective consequences.

Adenosine also stimulates NO release in the vascular endothelium, and NO has multiple effects in ischemic/anoxic tissue, including stimulation of vasodilation, activation of PKC, and stimulation of K_{ATP} channel opening. Recall that NO can act via the cGMP signal transduction pathway and that cGMP has been shown to stimulate multiple adjustments by anaerobic energy metabolism in anoxia-tolerant mollusks. Hence, we now have good evidence of the types of signaling molecules, receptors, signal transduction pathways, and at least some of the subcellular targets that are involved in anoxia tolerance. What is needed now is a comprehensive characterization of all of these events within a single anoxia-tolerant animal model system in order to confirm or refute the importance of the many molecular mechanisms of receptor signaling and signal transduction that have been implicated from studies of both anoxia-sensitive and anoxia-tolerant systems.

Mitochondrial Functions The mitochondrion is known as the powerhouse of the cell because its primary function is the generation of ATP from the oxygen-dependent catabolism of fuels. But what happens to mitochondria under anoxic conditions? Clearly the electron transport system becomes reduced, TCA cycle activity is halted, and ATP synthesis stops, but during anaerobiosis the organelles must be maintained in a viable state in order that aerobic metabolism can be rapidly resumed and apoptosis avoided (see Chapter 8 for the role of mitochondria in apoptosis). Similar to the situation with the plasma membrane, key functions of the mitochondria are dependent on the maintenance of ion gradients across membranes, creating a membrane potential difference. In particular, maintenance of the proton-motive force is key to mitochondrial survival.

Under aerobic conditions, the proton-motive force is established by proton pumping out of the matrix into the intermembrane space via complexes I, III, and IV of the respiratory chain. Proton return is via the F_1F_0–ATPase and drives ATP synthesis. Under anoxic conditions, the respiratory chain becomes reduced and can no longer pump H^+. To prevent the collapse of the proton-motive force, mitochondrial "treason" occurs; the organelles switch roles from being major ATP producers to being potential major users of cellular ATP. The occurs because the F_1F_0–ATPase reverses its function and becomes an ATP-dependent proton pump that takes over the role of H^+ extrusion from the matrix. The ATP needed to fuel this function is imported from the cytosol by an accompanying reversal of the adenine nucleotide translocator. Clearly, then, it would be advantageous for organisms to limit this mitochondrial consumption of ATP in anoxia. Recent studies have considered two possibilities: (1) direct inhibition of F_1F_0–ATPase activity and (2) reduced proton conductance by the inner membrane. The data favors direct inhibition of the F_1F_0–ATPase in anoxia. In studies using intact mitochondria from skeletal muscle of anoxia-tolerant frogs, the calculated rate of ATP consumption by the F_1F_0–ATPase in anoxia was only ~4% of the enzyme's corresponding rate of ATP production under aerobic conditions. Even so, this amount of ATP consumption by the F_1F_0–ATPase could still consume ~9% of the total ATP turnover in anoxic frog muscle, and together with the ~75% of anoxic ATP turnover that is devoted to Na^+K^+–ATPase activity, it is obvious that anoxic tissues spend a huge proportion of their energy budget on maintaining ionic homeostasis. The mechanism of F_1F_0–ATPase inhibition in anoxia-tolerant species is still being investigated but could involve an F_1–ATPase inhibitory subunit (IF_1) that has been identified in a number of species and is known to bind to the ATPase under low pH and nonenergizing conditions to inhibit ATP hydrolysis.

Another consideration for systems under oxygen limitation (e.g., severe hypoxia) is the mitochondrial proton leak, protons that leak back into the matrix without passing through the F_1F_0–ATPase and driving ATP synthesis. The futile cycling of protons that results partially uncouples respiration, reducing the efficiency of energy conservation. About 20% of mammalian standard metabolic rate has been attributed to such mitochondrial proton cycling. In situations of low oxygen availability (hypoxia) such a wasteful consumption of oxygen seems counterintuitive, especially in hypoxia-tolerant systems that use MRD to suppress ATP expenditure when oxygen is limiting. So, what happens to proton leak in hypoxia? Studies with muscle mitochondria from aerobic versus deeply hypoxic frogs indicated an ~50% reduction in proton leak under hypoxic conditions (based on state 4 respiration rates). Furthermore, the reduction in proton leak seemed to be caused by a decrease in the activity of the electron transport chain and not by a change in inner membrane proton conductance. The data suggest that proton cycling is reduced in parallel with the reduction in metabolic rate and that this results from a decrease in the rate of substrate oxidation and a decrease in the size of the proton-motive force in hypoxia. Indeed, under low oxygen pressures such as actually occur *in vivo* in the intracellular microenvironment of the mitochondria in normoxia (estimated to be 0.3 to 0.4 kPa), recent studies have shown that proton leak and uncoupled respiration are actually very low and that phosphorylation efficiency is high. The implication is that the high-proton-leak rates measured in most *in vitro* studies of isolated mitochondria (that use air-saturated incubation medium at ~20 kPa) may actually be a response of the organelles to hyperoxia that may contribute to minimizing oxidative stress under unusually high oxygen concentrations.

Control of Protein Synthesis

Protein synthesis consumes a substantial portion of cellular ATP turnover under aerobic conditions (Table 15.1), using about 5 ATP equivalents per peptide bond formed. Synthesis is well known to be sensitive to the availability of ATP and amino acids, and suppression of protein synthesis appears to be a proactive response to multiple stress situations, such as starvation and hypoxia in mammals. Inhibition of protein synthesis provides substantial energy savings to cells under stress. It is not surprising, therefore, that multiple studies have confirmed that protein synthesis is suppressed in an organ-specific manner as part of MRD in stress-tolerant animals, not only as part of anoxia tolerance but also in hibernation, estivation, and diapause. For example, in marine snails (*L. littorea*) ^{3}H-leucine incorporation into protein by hepatopancreas in anoxic snails was just 50% of the aerobic value. Furthermore, the inhibition of protein synthesis occurred very quickly, dropping within the first 30 min of anoxia exposure and remaining low for the remainder of a 48-h excursion. The proactive nature of this response is illustrated by the fact that ATP limitation for protein synthesis would not a factor within the first 30 min under the nitrogen gas atmosphere because ATP levels in tissues of anoxic snails only begin a slow decline after many hours of anoxia exposure. Protein synthesis inhibition also occurs under anoxic conditions in vertebrates; recall from Table 15.1 that the rate of protein synthesis was reduced by 93% in turtle hepatocytes under anoxia whereas a 95% inhibition was measured in liver of anoxic goldfish. Recall also that the fractional use of cellular ATP by protein synthesis is disproportionally reduced in anoxia; protein synthesis consumed 36.1% of cellular ATP turnover in aerobic turtle hepatocytes but only 21% in anoxic hepatocytes Similarly, when killifish embryos enter diapause, the fractional use of ATP by protein synthesis drops from 36% to negligible. Strong inhibition of protein synthesis in a hypometabolic state makes intuitive sense. The biosynthesis of new proteins is something of a "luxury item" that can be forfeited under stress in favor of the maintenance of more critical cellular functions such as sustaining central ATP-producing systems, the ability to sense and respond to stimuli, and osmotic and ionic balance.

Inhibition of protein synthesis during hypometabolism could be regulated in two ways: (1) by reduced messenger ribonucleic acid (mRNA) substrate availability and (2) by inhibition of the ribosomal translational machinery. Substrate availability is an influence on any metabolic process, but, in general, neither total mRNA content nor the specific mRNA transcript levels of most constitutively expressed genes are altered in hypometabolic states; this has been confirmed in studies with disparate systems, including anoxia survival in mollusks and brine shrimp and hibernation in mammals (Chapter 16). For example, when we used complementary DNA (cDNA) array screening to assess anoxia-responsive gene expression in *L. littorea*, 89% of genes assessed showed no change in transcript levels in anoxic snails, approximately 10% showed putative up-regulation, and only 0.6% showed suppressed mRNA levels in anoxia.

Instead, reversible control over the rate of protein synthesis in anoxia is vested primarily in the regulation of the ribosomal translation machinery. Two mechanisms are particularly important: (1) reversible phosphorylation of selected proteins of the translational machinery and (2) the physical state of ribosome assembly. Ribosomes are large multimeric complexes of proteins; individual subunits have structural, enzymatic, and regulatory roles in protein synthesis. The regulation of translation is discussed in detail in Chapter 7, so the current discussion will focus on what is known about the targeted suppression of translation during anoxia-induced MRD. As noted above, the levels of most mRNA transcripts are maintained in arrested states, but some are up-regulated to provide new proteins with specific functions for survival under stress (see the later section on anoxia-induced gene expression). Therefore, control of translation in anoxia must accomplish three goals: (1) strong suppression of the overall rate of protein synthesis in cells, (2) stabilization and/or storage of most existing mRNA transcripts so that they are not degraded over long periods of hypometabolism and are available for translation when normal metabolic functions are reestablished, and (3) provide for the selected biosynthesis of a few stress-specific proteins. How is this accomplished?

Control of Translation by Reversible Protein Phosphorylation Both translation initiation and polypeptide elongation are inhibited as parts of MRD. Both have been traced to reversible phosphorylation control over specific ribosomal subunits (for a discussion of the regulation of elongation, see Chapter 16). A key element in translation initiation is the eukaryotic initiation factor 2 (eIF2), which introduces initiator methionyl–transfer RNA (tRNA) into the 40S ribosomal subunit (Fig. 15.5) (also see Chapter 7). Phosphorylation of the alpha subunit of eIF2 (eIF2α) blocks this function because phospho-eIF2α is an inhibitor of the guanine nucleotide exchange factor eIF2B and prevents the recycling of eIF2α between successive rounds of peptide synthesis. This mechanism of global protein synthesis inhibition is well known in all eukaryotic systems and is a response to stresses, including viral infection, heat shock, nutrient deprivation, induction of apoptosis, and hypoxia/ischemia. For example, when protein synthesis in rat brain was measured during recovery after a 10-min ischemic episode, net synthesis was reduced by 85%, and this corresponded with ~23% of eIF2α in the phosphorylated state compared with ~1% under normoxia. We found a similar response in marine snails under anoxia. Western blotting using antibodies that detected both total eIF2α and the phosphopeptide showed a 15-fold elevation of phospho-eIF2α in hepatopancreas of anoxic snails with no change in total eIF2α content (Fig. 15.5) (refer also to Text Box 14.1 for an explanation of the methodology used). This correlated with a 50% decrease in total protein biosynthesis in anoxia. However, this situation was rapidly reversed during aerobic recovery and, within 1 h after oxygen was reintroduced, phospho-eIF2α content was fully depleted.

To date, studies of the inhibition of translation initiation during metabolic arrest have focused mainly on the regulation of eIF2α, but protein synthesis inhibition during ischemia or starvation in mammals also involves controls on other initiation factors. These may also be targets of reversible suppression during MRD. For example, the eukaryotic initiation factor 5 (eIF5) is also regulated by reversible protein phosphorylation; it acts as a GTPase-activating protein to promote GTP hydrolysis within the 40S initiation complex [composed of 40S*eIF3*AUG*Met-tRNA(f)*eIF2*GTP] (see Fig. 6.4). Both ischemia and amino acid starvation also regulate eukaryotic initiation factor 4. Subunit eIF4E and its binding protein (4E-BP1) are both dephosphorylated under stress in rat brain (concurrent with eIF2α phosphorylation) with dephosphorylation of 4E-BP1 allowing it to bind to eIF4E and inactivate it. By contrast, subunit eIF4G shows proteolytic fragmentation after ischemia/reperfusion.

The involvement of protein phosphorylation in the control of translation inhibition during anoxia-induced MRD obviously implicates the actions of one or more protein kinases in carrying out this process and suggests the mechanism by which protein synthesis inhibition can be coordinated with the suppression of other metabolic functions in anoxia. However, it is yet known which protein kinases mediate the response to anoxia. Several protein kinases can phosphorylate eIF2α in mammals, representing multiple signals (e.g., ischemia, amino acid availability); these are discussed in Chapter 7.

Translation inhibition during anoxia may also involve suppression of the activities of the phosphatases that dephosphorylate the ribosomal phosphoproteins. Both protein phosphatase (PP) 1 and 2A can dephosphorylate phospho-eIF2α, and PP-1 seems to have this role *in vivo*. Studies of mammalian ischemia/reperfusion are inconclusive as to whether phosphatase control is an important factor in translation inhibition, but studies of phosphatase responses during natural MRD have shown a consistent reduction of PP-1 and PP-2A activities in tissues during anaerobiosis as well in estivation and freezing.

Ribosome Aggregation State The activity state of the protein-synthesizing machinery in a cell/tissue can generally be inferred from the state of ribosomal assembly. Active translation occurs on polysomes (aggregates of ribosomes moving along a strand of mRNA), whereas monosomes are translationally silent. Polysome dissociation has long been recognized as a cellular response to stress; for example, hypoxia, starvation, and diabetes all trigger polysome dissociation in rat tissues. Recently this mechanism has been confirmed as an integral feature of the suppression of protein biosynthesis during MRD in both anoxia-tolerant and hibernating systems (see Chapter 16).

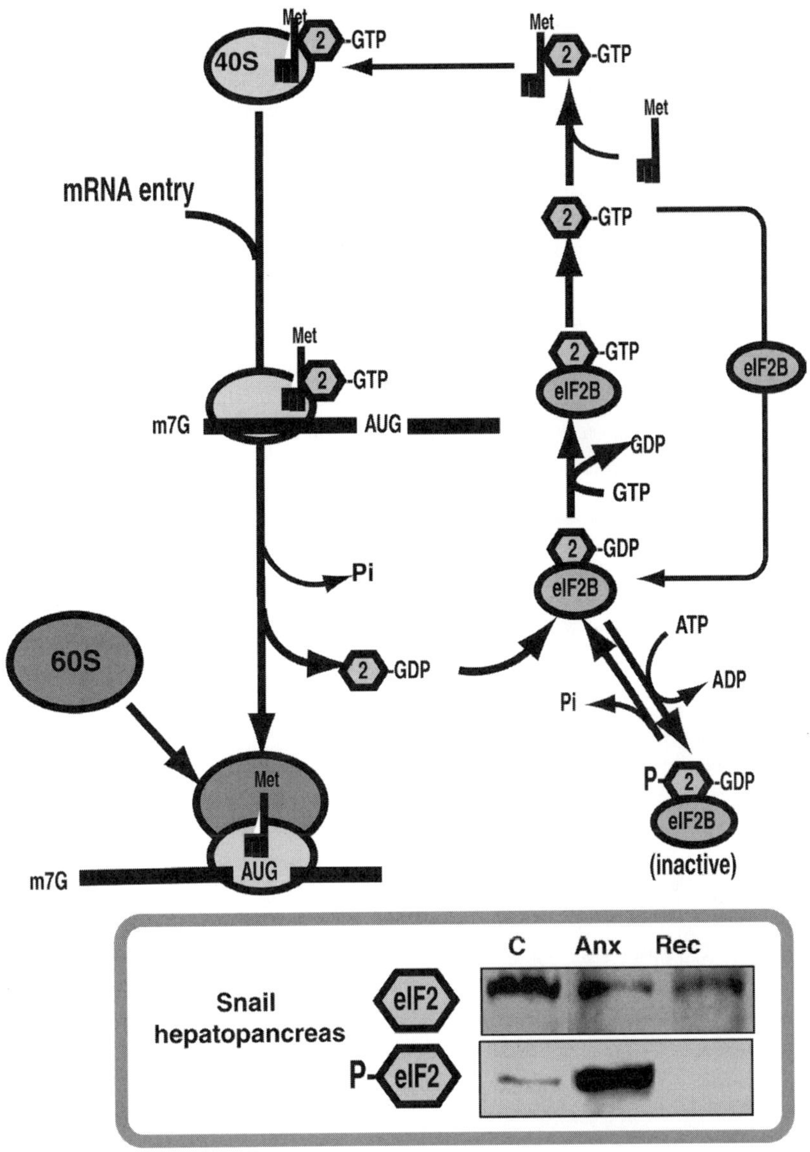

Figure 15.5 Reversible phosphorylation control of the eukaryotic initiation factor 2 (eIF2). eIF2 delivers the initiating methionine tRNA to the 40S ribosomal subunit. Phosphorylation of the α subunit of eIF2 inhibits translation because phospho-eIF2α acts as a dominant inhibitor of the guanine nucleotide exchange factor, eIF2B, and prevents the recycling of eIF2α between successive rounds of peptide synthesis. Inset shows the strong increase in the amount of phospho-eIF2α content, as determined by Western blots, in hepatopancreas from anoxic (Anx) marine snails, *Littorina littorea*, compared with aerobic controls (C). However, after 1 h of aerobic recovery (Rec) this is fully reversed and no phospho-eIF2α remains.

To assess the state of ribosomal assembly, tissue extracts are separated on a sucrose gradient. Polysomes migrate to the denser fractions, whereas monosomes are found in the less dense fractions. Figure 15.6 shows an example of the effects of anoxia exposure on the polysome profile in extracts of hepatopancreas from the marine snail, *L. littorea.* Under aerobic conditions, most ribosomes are present in the higher density polysome fractions (fractions 5 to 11 in the figure); RNA analysis via absorbance measurements at 254 nm (shown as bars) or ethidium bromide staining (not shown) both confirmed high RNA content in these fractions. Samples from each fraction were then separated on agarose gels and blotted to nylon membranes and Northern blotting was used to detect

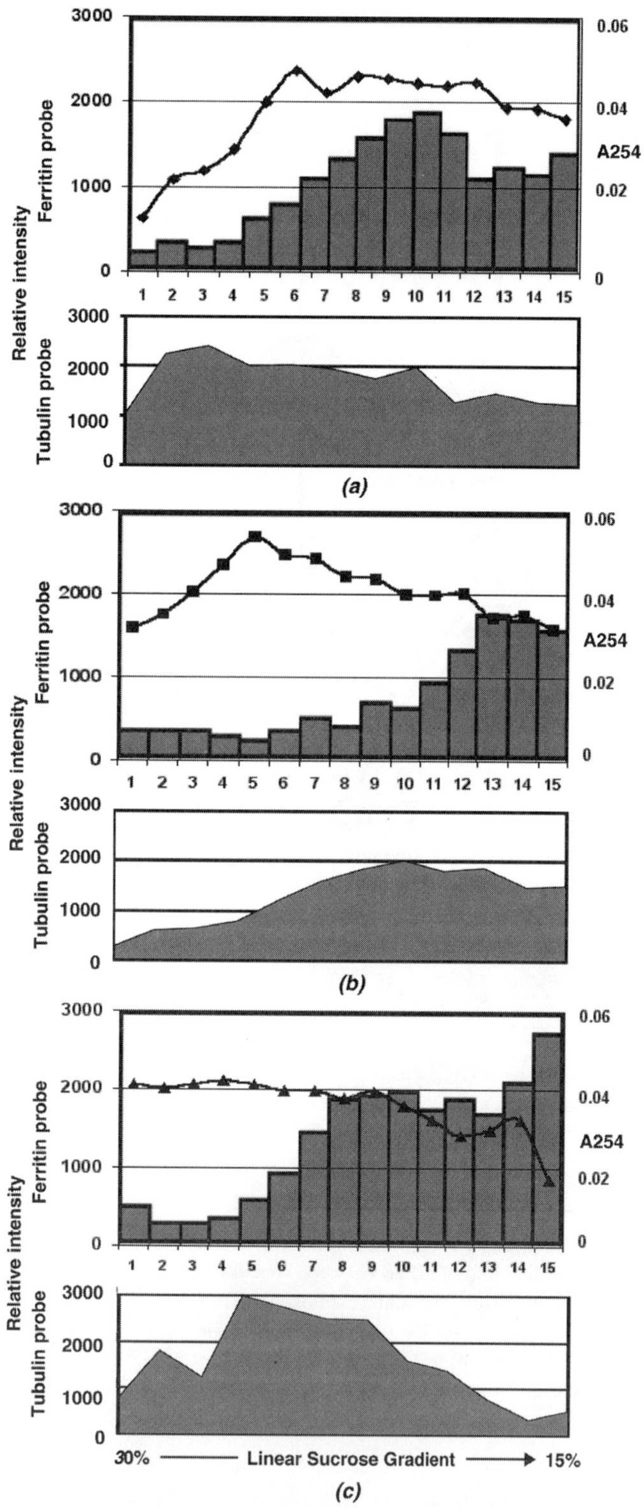

Figure 15.6 Polysome profiles of *L. littorea* hepatopancreas extracts prepared from snails under the following conditions: (*a*) aerobic, (*b*) 72-h anoxic, and (*c*) 6-h aerobic recovery after 72-h anoxia. Postmitochondrial supernatants were centrifuged on 15 to 30% continuous sucrose density gradients. Fractions were collected with high sucrose (30%) in fraction 1 and decreasing to 15% in fraction 15. In the upper panels bars show A_{254}, representing the relative amount of total RNA (mostly ribosomal RNA) in each fraction. Line graphs show the distribution of mRNA transcripts of ferritin heavy chain, a gene that is up-regulated in anoxia. Lower panels show the corresponding distribution of transcripts of a constitutively active gene, α-tubulin. To assess both ferritin and α-tubulin, mRNA was isolated from each fraction of the sucrose gradient, separated on agarose gels, blotted onto nylon membranes, and then hybridized with ^{32}P-labeled cDNA probe for each gene. Transcript levels were determined from the scanned intensity of bands on these Northern blots. [From K. Larade and K. B. Storey (2004) *J Exp Biol* **207**:1353–1360.]

mRNA transcripts of a constitutively expressed gene, α-tubulin, and an anoxia-induced gene, ferritin. Under aerobic conditions both α-tubulin (shown in the lower panels) and ferritin (line graph in upper panels) mRNA was found in the polysome fractions (Fig. 15.6*a*). This confirms a state of active translation under aerobic conditions. However, when anoxia was imposed by a nitrogen gas atmosphere, the peak of ribosomal RNA shifts to a lower density (fractions 12 and higher), indicating dissociation of polysomes. This began within 24 h of anoxia exposure and, as shown in Figure 15.6*b*, by 72 h there is little evidence of any polysomes remaining. Tubulin mRNA also shifted into the monosome fractions, which indicates that the mRNA pool of constitutively active genes is generally conserved in the hypometabolic state but not translated (a conclusion also reached from other evidence). However, transcripts of ferritin, a gene up-regulated by anoxia in *L. littorea* (see later section on gene expression), stayed in the high-density fractions in association with the few remaining polysomes. This suggests that ferritin transcripts are actively translated in the snail hepatopancreas under anoxia. Within 6 h of the shift back to aerobic conditions, Fig. 15.6*c* shows that the situation was normalized for renewed aerobic protein synthesis: Polysome reassembly had recurred, tubulin mRNA had returned to the polysome fraction, and ferritin mRNA remained with the polysomes.

Message Selection for Translation During Hypometabolism Although entry into a hypometabolic state is not the time for extensive ATP expenditure on the synthesis of new proteins, nonetheless all systems of natural MRD that have been examined (e.g., anoxia-tolerant organisms, hibernation, estivation, freezing) show up-regulation of selected gene transcripts and increased synthesis of selected proteins (see also Chapters 16 and 17). Transcript up-regulation does not always mean that levels of the protein also rise in the hypometabolic state; sometimes protein synthesis is delayed until the recovery period (see Chapter 16). However, various messages are translated during hypometabolism despite the general inhibition of ribosomal translation by the mechanisms discussed above. One way that this is accomplished is via message selection. As mentioned earlier, subunit G of initiation factor 4 (eIF4G) undergoes proteolytic fragmentation in response to stress (e.g., ischemia/reperfusion, amino acid starvation) in mammalian systems and may be similarly affected in hypometabolic systems (see Fig. 7.4). Fragmentation changes the types of mRNAs that can be translated because intact eIF4G is needed to allow eIF4E-bound m^7G-capped mRNAs (the vast majority of cellular mRNAs) to bind to the small ribosomal subunit. When intact eIF4G is missing and eIF2α is phosphorylated and inactive, message selection changes dramatically. Only those messages that contain an internal ribosome entry site (IRES) can be translated because an IRES allows the message to bypass the normal mechanism of binding to the small ribosome subunit (see Chapter 7). For example, during amino acid starvation an up-regulation of selected gene transcripts containing an IRES occurs despite a general inhibition of the cap-dependent translational apparatus. In almost all cases, entry into a hypometabolic state is, in essence, a form of starvation since animals typically do not have access to food (consider shell valve closure in mollusks in response to aerial exposure or underground hibernation or estivation), and they typically switch to a dependence on internal fuel reserves. Therefore, it is likely that the protein synthesis inhibition response of MRD grew out of the preexisting mechanisms that regulate suppression of nonessential processes during starvation or other forms of stress. The IRES mechanism of mRNA message selection could then be employed to accomplish the specific up-regulation of selected genes that have protective functions during natural hypometabolism.

One gene that is well-known to use an IRES is the hypoxia-inducible factor-1 (HIF-1). As discussed earlier, HIF-1 mediates various gene responses to hypoxia in oxygen-sensitive organisms, and its activity is primarily regulated by the availability of the HIF-1α subunit. As such, HIF-1α must be translated under ATP-limited hypoxic conditions even if overall protein synthesis is inhibited. When the distribution of HIF-1α mRNA was assessed on a polysome profile, it was found that HIF-1α transcripts remained in the polysome fraction during hypoxia, unlike the majority of mRNA types. Sequence analysis showed that HIF-1α contains a long and guanosine-cytosine-rich 5′-untranslated region (5′UTR) that is typical of an IRES. Vascular endothelial growth factor (VEGF), one of the genes stimulated by HIF-1, also contains an IRES so this confirms the importance of an IRES for the selective up-regulation and translation of stress-responsive genes. These results suggest that a common characteristic of genes that are up-regulated during MRD and whose transcripts remain associated with polysomes for translation under stress conditions may be the presence of an IRES in the 5′UTR. To date, however, this has not been confirmed.

Protein Degradation

Given that protein synthesis is strongly suppressed in hypometabolic states but, upon arousal, animals show no significant deficit of cellular proteins, it is obvious that protein degradation must also be suppressed during hypometabolism. Indeed, Table 15.1 shows that this is the case for turtle hepatocytes when challenged with anoxia. Coordinated inhibition of the synthesis versus degradation of proteins results in a new lower net rate of protein turnover in hypometabolic states and effectively increases protein longevity. For example, the half-life of cytochrome oxidase increased 77-fold in anoxic encysted embryos of brine shrimp compared with normally developing embryos. Not only is suppressed protein turnover important for ATP savings as part of MRD but also the protein "life extension" that results minimizes the accumulation of nitrogenous end products of proteolysis during hypometabolism and, hence, the costs of processing, storing, and/or excreting wastes. Note, then, that Table 15.1 also shows that the rate of urea production drops by 70% in anoxic hepatocytes as compared with aerobic controls.

To date, very little is known about the mechanisms of proteolytic suppression during MRD. Ubiquitin-dependent proteolysis is a major mode of cellular proteolysis; conjugation with a polymer of ubiquitin tags proteins for degradation by the 26S proteasome (see Fig. 6.11), whereas monoubiquitinated proteins are targeted for endocytosis and degradation in lysosomes. Brine shrimp exposed to anoxic conditions showed ubiquitin conjugate levels that were just 7% of normoxic values, and this suggested a block on proteolysis under anoxia at the level of ubiquitin conjugation. Ubiquitination resumed in embryos when aerobic conditions were reestablished and cellular pH and ATP levels had normalized again. Our data from anoxia-tolerant turtles might also support control at the level of protein ubiquitination because we found that the activity of the multicatalytic proteinase complex (a key part of the 26S proteosome) did not change in liver of anoxic versus aerobic turtles, suggesting that inhibitory control over protein degradation was located elsewhere. However, in hibernating mammals, levels of ubiquitin-conjugated protein rose two- to threefold during torpor, suggesting that the inhibitory block in this case might be on the proteolysis machinery rather than the ubiquitin-tagging system. It is not yet known whether one or both of these mechanisms represent the common mode of proteolysis control during anaerobiosis or other modes of hypometabolism.

ANOXIA-RESPONSIVE GENE EXPRESSION

Control over the coordinated expression of subsets of genes is an excellent mechanism of biochemical adaptation and is employed in many instances. As discussed earlier, the mammalian response to hypoxia is a coordinated up-regulation of the genes that improve oxygen delivery or glycolytic capacity of tissues, mediated by the HIF-1 transcription factor. The compensation response differs from the conservation strategies that are used by hypoxia/anoxia-tolerant species to lower the oxygen demands of tissues (e.g., MRD) and extend the time that organisms can live using endogenous reserves of fermentable fuels. The exploration of gene expression responses to anoxia by facultative anaerobes is still at an early stage but, not surprisingly, the anoxia-responsive genes identified to date are not the same as those that are up-regulated in hypoxia-sensitive organisms. For example, we found that anoxia exposure does not stimulate an increase in the activities of glycolytic enzymes in tissues of freshwater turtles or marine mollusks (oysters or littorines); the glycolytic capacities of tissues of anoxia-tolerant animals are apparently always optimized to allow smooth transitions between aerobic and anaerobic states. Indeed, until very recently, HIF-1 had not been found in anoxia-tolerant species, and it is still not known whether HIF-1 or an unknown transcription factor mediates anoxia-induced gene expression in these organisms. Logically, the involvement of a different transcription factor would make sense in order to allow anoxia-tolerant species to separate HIF-1-mediated angiogenic or erythropoietic responses that animals could need for a variety of purposes from anoxia-induced gene responses that stimulate and regulate long-term hypometabolism. To date, exploration of gene expression responses to anoxia in anoxia-tolerant species has focused on two main groups: freshwater turtles and marine mollusks.

Gene Expression Responses to Anoxia in Turtle Organs

Submergence in nitrogen-bubbled water is widely used to study both metabolic and gene expression responses to anoxia in turtles; in cold water, turtles can endure many weeks of this treatment, which mimics the conditions of underwater winter hibernation. Our first analysis of anoxia-responsive genes in turtles came from screening of a cDNA library prepared from heart of anoxia-exposed turtles, *Trachemys scripta elegans*. This revealed the anoxic up-regulation of four genes, and very surprisingly, three of these were encoded on the mitochondrial genome. One was homologous with the mitochondrial WANCY (tryptophan, alanine, asparagine, cysteine, and tyrosine) tRNA gene cluster, another was identified as the gene (*Cox1*) for cytochrome C oxidase subunit 1 (COI), and the third was *Nad5* encoding subunit 5 of NADH-ubiquinone oxidoreductase (ND5). When we performed comparable screening of a cDNA library made from brain of anoxic turtles, mitochondrially encoded genes were again up-regulated, this time represented by subunit 4 of ND as well as cytochrome *b* (CYTb). ND4/5, CTYb, and CO1 are components of complexes I, III, and IV, respectively, of the electron transport chain. All of these complexes are large polymers; for example, complex I can contain as many as 42 polypeptide subunits whereas complex IV has 13 dissimilar subunits. Six of the subunits of complex I and 3 of complex IV are encoded on the mitochondrial genome, whereas the rest are nuclear (see Chapter 8). Northern blots revealed a 3- to 5-fold increase in *Nad5* and *Cox1* transcript levels within 1 h of anoxic submergence in heart, whereas *cytb* and *Nad4* rose by about 6- and 12-fold in liver over the same time. All four transcripts showed differential up-regulation in other organs as well and transcript levels were reduced again within 1 to 5 h after a return to aerobic conditions. The significance of the up-regulation of these mitochondrially encoded proteins in response to anoxia remains unknown, but we are now finding that the phenomenon is widespread during entry into hypometabolism, occurring also during hibernation and freeze tolerance (see Chapters 16 and 17). Hence, there is growing evidence that mitochondrial gene expression as well as specific controls on mitochondrial enzymes/proteins is an integral part of anoxia tolerance. Recall the discussion earlier in the chapter about the regulation of the F_1F_0–ATP synthase in anoxia, and in Chapter 16 we outline reversible phosphorylation control of ND (complex I) by pyruvate dehydrogenase kinase (inhibiting) and mitochondrial protein kinase A (activating). ND is also now recognized as the primary site of superoxide production by the electron transport chain, and this superoxide produced by ND is known to have important signaling roles in cells. Thus, the combination of new information on the regulation of mitochondrial enzymes and new discoveries about mitochondrial gene up-regulation during transitions to hypometabolic states show that the mitochondria have important regulatory roles in the control of cellular metabolic rate via mechanisms that we are just only beginning to appreciate.

Gene Expression Responses to Anoxia in Marine Snails

Recent studies of anoxia-induced gene expression in our laboratory have focused on the marine gastropod, *L. littorina*, and have documented the anoxia responsiveness of genes including ribosomal protein L26, ferritin heavy chain, metallothionein, cytochrome *c* oxidase subunit 2 (*Cox2*), and several novel genes. *Cox2* joins the list of mitochondrially encoded genes that are up-regulated under stresses, including anoxia, that initiate MRD. One of the novel genes, named *kvn*, showed a six-fold up-regulation of transcript levels in *L. littorea* hepatopancreas during anoxia exposure that was reversed within 1 h of aerobic recovery. The gene codes for a protein of 99 amino acids with a predicted molecular weight of 12 kDa. A 15-residue hydrophobic signal sequence at the N terminal, as well as a lack of any known retention signals or subcellular localization motifs, suggests that the protein is excreted into the hemolymph. Active translation during anoxia was inferred from the high proportion of *kvn* transcripts in the polysome fraction. Ongoing studies are seeking its function.

Transcript levels of the gene encoding the ribosomal protein L26 rose three- to fivefold in *L. littorea* tissues over the course of 2 to 4 days of anoxia exposure but fell again within 1 h of aerobic recovery. This pattern suggests a functional need for enhanced L26 protein in anoxia, and indeed, nuclear run-off assays (explained in Chapter 14) confirmed that the change in L26 transcript levels was the result of increased transcription of the L26 gene, not an alternative form of regulation such as transcript stabilization or inhibition of mRNA degradation. The L26 protein resides at the interface of the large and small ribosomal subunit, apparently at or near the ribosomal A site, where it is likely involved in subunit interactions. L26 can be crosslinked to elongation factor 2 (eEF-2), implicating it as a protein involved in forming the region that binds eEF-2 to the 60S ribosomal subunit preceding translocation of peptidyl-tRNA from the A to the P site during peptide bond formation. The link with eEF-2 is interesting because, as will be discussed in detail in Chapter 16, inhibitory control of eEF-2 is one of the important mechanisms involved in suppressing protein synthesis in hypometabolic states. The possibility exists that anoxia-induced up-regulation of L26 could contribute to eEF-2 control. Hence, once again, we have evidence that the control of transcription, via multiple mechanisms, is a critical component of anoxia survival.

Anoxia exposure of *L. littorea* also stimulated the up-regulation of two metal-binding proteins with roles in antioxidant defense, ferritin and metallothionein. Ferritin transcripts rose about twofold in anoxia (Western blots showed that protein also increased), whereas metallothionein transcripts increases three- to fivefold within 12 h of anoxia. As a resident of the intertidal zone at high latitudes, *L. littorea* is also freeze tolerant (see Chapter 17), and freezing is an anoxic/ischemic stress because ice formation halts the circulation of hemolymph to all tissues. Not surprisingly, then, metallothionein transcript levels were elevated just as strongly by freezing as by anoxia (ferritin response to freezing was not assessed). Ferritin sequesters iron and, by doing so, plays an important role in antioxidant defense because free iron is a major catalyst in the production of ROS via the Fenton reaction (see Chapter 12). Metallothionein is generally considered to contribute to the homeostasis of trace metals in the body and, in situations of heavy-metal challenge, to limit toxicity by sequestering metals such as cadmium, lead, and mercury. Much research has correlated metallothionein levels with heavy-metal pollution in aquatic environments, and changes in the levels of the protein are often used as a bioindicator of metal pollution. However, metallothionein is a low-molecular-weight cysteine-rich protein (27 out of 100 amino acids in the *L. littorea* protein are cysteines), and new research suggests that it has antioxidant properties either as a reducing agent (due to its many –SH groups) or due to binding of metals such as copper that, like iron, can catalyze ROS formation. Indeed, transgenic mice that overexpress metallothionein show much reduced damage during cerebral ischemia–reperfusion, whereas cultured embryonic cells that were deficient in metallothionein showed enhanced susceptibility to oxidative stress. Evidence of an antioxidant role in marine mollusks includes the fact that preexposure of mussels to cadmium to enhance their metallothionein levels also increased the subsequent survival of the animals when challenged by iron-triggered ROS production. In general, as discussed earlier, anoxia-tolerant animals maintain constitutively high levels of antioxidant defenses and/or improve these when stimulated by anoxic/hypoxic conditions; for example, we found that anoxia exposure stimulated a strong increase in the amount of reduced glutathione in *L. littorea* tissues. Measurements of conjugated diene contents in lipids suggested that snails are subject to some oxidative stress during the early minutes of recovery after anoxic excursions. Thus, the enhancement of thiol antioxidant defenses and elevated metal-binding capacities under anoxia may prepare the animals for dealing with enhanced levels of ROS when anoxic tissues are reoxygenated.

Some progress has been made in elucidating the regulatory controls on anoxia-induced gene expression in *L. littorea.* When we incubated hepatopancreas explants under aerated versus N_2-bubbled conditions *in vitro*, we found that L26, *kvn*, and ferritin transcripts were all up-regulated under anoxia *in vitro* just as they were *in vivo*. This validated the model for further use in testing the effects of stimulators of protein kinases on gene expression. The effects of dibutyryl cAMP, dibutyryl cGMP, calcium ionophore A23187, or phorbol 12-myristate 13-acetate on gene expression were tested in aerobic incubation. Transcript levels of all three genes were elevated when tissues were incubated with dibutyryl cGMP, whereas only ferritin message responded to any of the other treatments. This implicates a cGMP-mediated signaling cascade in the gene expression response to anoxia in *L. littorea* and agrees with the evidence presented earlier that cGMP-dependent protein kinase (perhaps responding to a nitric oxide signal) regulates a variety of metabolic responses to anoxia in marine mollusks, ranging from enzyme phosphorylation to gene expression.

SUGGESTED READING

Bickler, P. E., and Donohoe, P. H. (2002). Adaptive responses of vertebrate neurons to hypoxia. *J Exp Biol* **205**:3579–3586.

Review of the mechanisms of hypoxia tolerance by neurons including the mechanisms of low oxygen damage and adaptive control over membrane receptors and ion channels.

Boutilier, R. G., and St.-Pierre, J. (2002). Adaptive plasticity of skeletal muscle energetics in hibernating frogs: Mitochondrial proton leak during metabolic depression. *J Exp Biol* **205**:2287–2296. *Review of the concepts of mitochondrial functions in hypoxia and anoxia including proton leak and reverse functioning of the F_1F_0–ATPase.*

DeGracia, D. J., Kumar, R., Owen, C. R., Krause, G. S., and White, B. C. (2002). Molecular pathways of protein synthesis inhibition during brain reperfusion: Implications for neuronal survival or death. *J Cereb Blood Flow Metab* **22**:127–141. *Good review of the mechanisms of protein synthesis regulation with respect to hypoxia challenge.*

Haddad, J. J. (2002). Oxygen-sensing mechanisms and the regulation of redox-responsive transcription factors in development and pathophysiology. *Respir Res* **3**:26 (online at http://respiratory-research.com/content/3/1/26). *Review of oxygen sensing, responses to low oxygen and to oxidative stress, HIF-1 and NFκB transcription factors and the genes they control, and oxygen/redox involvement in apoptosis.*

Hochachka, P. W., and Lutz, P. L. (2001). Mechanism, origin and evolution of anoxia tolerance in animals. *Comp Biochem Physiol B* **130**:435–459. *Good overview of problems and solutions to hypoxia/anoxia by tolerant species.*

Roach, R. C., Wagner, P. D., and Hackett, P. H. (eds.) (2003). *Hypoxia: Through the Lifecycle*. Kluwer/Plenum Academic, NY. *Proceedings of the 13th International Hypoxia Symposium held in February 2003 with state-of-the-art review articles about many aspects of hypoxia adaptation including low oxygen signaling, hypoxia regulation of blood flow, oxidative stress, shock proteins, hypoxia in mammalian hibernation and in fetal development, and high-altitude medicine.*

Storey, K. B. (1996). Metabolic adaptations supporting anoxia tolerance in reptiles: Recent advances. *Comp Biochem Physiol B* **113**:23–35. *A focus on reversible phosphorylation control of enzymes and the protein kinases and phosphatases involved as well as antioxidant defenses.*

Storey, K. B., and Storey, J. M. (eds.) (2002). *Cell and Molecular Responses to Stress, Vol. 3: Sensing, Signaling and Cell Adaptation*. Elsevier, Amsterdam, pp. 27–46. *Includes reviews on brain ischemia tolerance and the role of preconditioning (Ch. 1), hypoxia-induced gene expression (Ch. 2), metabolic and gene expression responses to anoxia in marine snails (Ch. 3), the role of adenosine in tissue protection in ischemia-reperfusion (Ch. 4), and the application of metabolic control analysis to metabolic depression (Ch. 20).*

Storey, K. B., and Storey, J. M (2004). Metabolic rate depression in animals: Transcriptional and translational controls. *Biol Rev., Camb. Philos. Soc.* **79**:207–233. *Review of the principles of metabolic rate depression showing the conservation of mechanisms across phylogeny and between different stresses.*

16

MAMMALIAN HIBERNATION: BIOCHEMICAL ADAPTATION AND GENE EXPRESSION

KENNETH B. STOREY AND JANET M. STOREY

INTRODUCTION

The core body temperature (T_b) of most mammals is maintained remarkably constant (typically 36 to 38°C, depending on the species), regulated by a thermostat in the hypothalamus of the brain and with heating and cooling achieved, as needed, by a variety of physiological and biochemical mechanisms. All mammals, including humans, are endotherms (generating internal heat by metabolic reactions) and most are also good homeotherms (maintaining a highly constant core T_b), and this supports a variety of mammalian successes, including high-speed locomotion, range extension into cold environments, and advanced brain functions. However, endothermy is costly and the metabolic rate of mammals is typically 4 to 7 times higher than that of comparably sized reptiles. This must be supported by equally higher rates of fuel consumption, supplied by foraging or, if food supply is limiting, by food caches or body adipose reserves.

When winter approaches and environmental temperature falls, the metabolic rate, T_b, and food needs of an ectothermic (cold-blooded) organism decline along with it. However, a mammal in the same situation loses body heat faster at colder temperatures and, hence, needs a higher metabolic rate and a greater fuel consumption to support the increased thermogenesis needed to maintain a constant core T_b. Many mammals, even some very small ones such as shrews and some mice, can meet this challenge and remain active throughout the winter by spending as much time as possible in sheltered environments (e.g., under the snowpack and/or in insulated nests), increasing their body insulation (thicker fur, more body fat), and assembling adequate fuel supplies of forage, food caches, and/or body fat. For others, the combination of cold temperatures and lack of food availability in winter makes survival as a homeotherm impossible. The problem is particularly acute for animals such as insectivorous bats or grazing herbivores (e.g., ground squirrels, marmots) that have little or no access to edible food in the winter.

The solution to this problem is hibernation. For 6 to 9 months of the year, many small mammals abandon one of the defining characteristics of mammalian life (homeothermy) and allow their T_b to fall, tracking environmental temperature. By doing so, hibernators gain tremendous energy savings. For example, it has been calculated for ground squirrels that winter hibernation saves 88% of the energy that would otherwise be needed to maintain a euthermic T_b of ~37°C over the winter.

The present chapter explores metabolic regulation as it applies to mammalian hibernation. The field of hibernation research is a huge one that examines the phenomenon at ecological, physiological, and biochemical levels and also includes a huge body of applied research that seeks to use the lessons learned from hibernation in hypothermic medicine, organ preservation (see Chapter 19), and understanding complex brain functions. These subjects fill books of their own so the treatment here is selective and focuses on recent advances in understanding the principles of metabolic regulation as they apply to hibernation.

HYPOTHERMIA AND HIBERNATION

In Chapter 14 environmental stressors were said to affect metabolism in two ways: (a) by direct perturbation of the structure/function of biological molecules and biochemical

Functional Metabolism: Regulation and Adaptation, edited by Kenneth B. Storey.
ISBN 0-471-41090-X

reactions or (b) by jeopardizing the energy currencies of the cell. Mammalian hibernation is an adaptation directed at maintaining cellular energetics over an extended period when the environment is inhospitable for normal life. However, by using this strategy, the animal is then subjected to the perturbing effects of temperature on its metabolism, a situation that is unusual for most mammals. Humans, for example, undergo severe metabolic injuries (often lethal) if our core T_b drops below about 25°C, but small-mammal hibernators readily allow T_b to fall to as low as 0 to 5°C without suffering injury. Hibernating Arctic ground squirrels (*Spermophilus parryii*) are even known to let T_b fall below 0°C; a recorded minimum T_b of −2.9°C is the lowest T_b ever recorded for a living mammal (these squirrels are not frozen, but supercooled; see Chapter 17). Note that the animal that typically comes to mind first when hibernation is mentioned—the bear—actually shows only a very small drop in T_b (to ~35°C) and has a shallow torpor that is easily disturbed. Even though metabolic rate is lowered, the large bulk of the bear, especially its insulating fat and fur layers, prevents its body from cooling very much. Bears are fascinating in that they do not eat, drink, defecate, urinate, or lose muscle mass over the course of many months of winter hibernation. However, physiological data on bear hibernation are still pretty minimal and biochemical data are almost nonexistent, so our discussion here will center on the small mammals that make much better laboratory models.

The injurious effects of hypothermia on nonhibernating mammals arise from two main effects of cold on metabolic systems that have been optimized over millions of years of mammalian evolution for function within a narrow temperature window. The first is the differential effect of temperature change on the rates of thousands of cellular reactions that can culminate in a mismatch between the net rates of reactions that produce and utilize adenosine 5′-triphosphate (ATP). The result is that energy currencies are depleted and the major manifestation of this energy crisis is membrane depolarization, which sets off a range of catastrophic events that are much the same as those caused by anoxia-induced energy failure (see Chapter 15). This is the main biochemical reason for the often lethal effects of hypothermia in humans (see Chapter 19 for the problems that this causes during hypothermic organ storage in transplant medicine). Interestingly, summer-active individuals of hibernating species are just as susceptible to hypothermia-induced membrane depolarization as are nonhibernating species. Hence, the preparation for winter hibernation must include specific adaptations that correct hypothermia-sensitive metabolic processes.

The second main effect of hypothermia is a decrease in lipid fluidity in both membranes and adipose depots as temperature declines. Mammalian membranes and adipose depots are designed to function at about 37°C, and they solidify at about room temperature (think of the difference between olive oil and lard on the kitchen counter). Membrane lipid fluidity is crucial for allowing the protein movements within membranes and protein conformational changes associated with receptor and transporter functions. Impaired fluidity severely disrupts these functions. Hence, hibernation must include mechanisms that sustain or re-establish membrane-associated metabolic functions over a wide range of core T_b values.

Hibernation occurs in multiple mammalian lineages and is believed to have arisen independently several times. For a typical small mammal, the winter hibernation season consists of multiple bouts of torpor interspersed with brief periods of arousal when euthermic T_b is reestablished. The season begins with a series of "test drops" in which the core T_b of the sleeping animal falls by 10 to 15°C for short periods before returning to normal. It is now believed that these test drops are important in triggering the induction of various metabolic adjustments and gene expression changes that support the longer and deeper torpor bouts that follow. Subsequently, the decrease in T_b becomes greater (falling very close to ambient) and the length of each torpor bout increases so that in midwinter T_b is often maintained between 0 and 5°C for 1 to 3 weeks (although an amazing 76 days was recorded in one species of bat). Intervening arousals between torpor bouts generally last 6 to 24 h, depending on the species. During torpor T_b tracks ambient temperature, but with the exception of the Arctic ground squirrels, hibernators do not let T_b fall below 0°C. If ambient temperature decreases to subzero values, the hibernating animal activates a low rate of nonshivering thermogenesis (NST) by its brown adipose tissue (BAT) to avoid any risk of freezing. Hence, despite the deep torpor and the very low T_b, sensory systems are still tracking T_b and the animal is defending a set point temperature.

During torpor all aspects of the animal's physiology slow dramatically. For example, the heart beat of a ground squirrel can drop from 200 to 300 beats per minute to just 5 to 10 beats per minute. Breathing rate similarly declines and in some species changes from rhythmic to intermittent with extended periods of apnea (breath-hold) of an hour or more that are interspersed with brief bursts of breaths. Overall, metabolic rate in torpor at a T_b of 0 to 5°C can be as low as 1 to 5% of the normal resting rate at 37°C. As spring approaches, torpor bouts become shorter and shallower until the animal ceases to hibernate any more. Seasonal hibernation is a longer, deeper, and more dramatic version of the daily torpor that occurs in many small mammals and birds. By allowing T_b to decline by a few degrees during sleep, animals that employ daily torpor can accrue significant energy savings that are often the difference between "making it through the night" and death from starvation.

Metabolism during hibernation is fueled almost exclusively by lipid oxidation; even brain switches to a mixed dependency on glucose and ketone bodies, the latter being produced from fatty acids (see Chapter 9). The whole-body respiratory quotient (RQ; ratio of CO_2 output to O_2 consumption) during steady-state hibernation is typically ~0.7, which is indicative of lipid catabolism. However, studies with ground squirrels have shown that if stressed by ambient temperatures that fall below 0°C (and that necessitate NST to stabilize T_b), RQ values can rise to 0.85. This suggests that other fuels can also be called into play when needed. Fatty acids are particularly important as the fuel for the intense thermogenesis that is needed to rewarm the animal during each arousal from torpor. Heat production comes from two sources: NST due to high rates of uncoupled respiration by BAT mitochondria and shivering thermogenesis by skeletal muscles (see Text Box 16.1). The early minutes of arousal rely

TEXT BOX 16.1 THERMOGENESIS

Metabolism is inefficient and contains many exergonic reactions that release energy, generally as heat. For example, under physiological conditions of ATP, ADP, P_i and H^+ and at cellular pH, the calculated efficiency of energy conservation from the aerobic catabolism of glucose to produce ATP is about 65% so about 35% is lost as heat. This significant heat release is further enhanced by energy release during the subsequent hydrolysis of ATP by ATP-utilizing metabolic reactions. These inefficiencies can be put to use to support either episodic heat production or sustained homeothermy. Thermogenesis can be accomplished by either (or both) of two mechanisms (1) increasing the rate of ATP turnover by speeding up selected ATP-utilizing reactions in a "futile" manner (i.e., no real net benefit except heat output) so that both ATP production and ATP utilization are increased, or (2) uncoupling ATP synthesis by oxidative phosphorylation from the electron transport system so that energy that would otherwise be trapped as ATP is released as heat. Both options are used by animals.

Thermogenesis from Enhanced Rates of ATP Turnover

The high basal metabolic rate of mammals and birds, four- to sevenfold higher than comparably sized reptiles, derives primarily from the first mechanism. Endotherms "waste" high amounts of ATP, supporting much higher rates of transmembrane ion pumping than occurs in ectotherms. All organisms maintain gradients of Na^+, K^+, H^+, Ca^{2+}, and other ions across their membranes, but the membranes of endotherms are much more "leaky" than those of ectotherms due to the presence of higher numbers of ion channels that facilitate ion movements down their concentration gradients. To counteract the dissipation of ion gradients that such leaky membranes cause, endotherms maintain many more ATP-driven ion pumps in their membranes to move ions against their concentration gradients. Although there are positive benefits of high sensitivity and rapid response to stimuli that are supported by this system, such futile cycling of ions across membranes requires high rates of ATP hydrolysis and high rates of substrate oxidation, generating large amounts of heat to support homeothermy. Indeed, as discussed elsewhere in this chapter, strong inhibition of the activities of ion-motive ATPases (and of corresponding ion channels) is one of the major mechanisms of metabolic rate depression that allows T_b to fall during hibernation.

A similar form of futile ATP turnover is the basis of shivering, an episodic mechanism of heat generation used by both vertebrate and invertebrate animals. ATP hydrolysis by the myosin ATPase of skeletal muscles is unhooked from sarcomere shortening (muscle work) so that high rates of fuel and ATP consumption occur with only minor (as in humans) or sometimes no (e.g., honeybees) detectable physical movements of the muscle. For example, various silkmoths need to shiver to warm up their thoracic flight muscles to over 30°C before they can fly and honeybees hold the core temperature of their colony (surrounding the queen) at 34 to 35°C throughout the winter by varying the rate and number of shivering bees and the density of packing of the cluster. The high rate of ATP production needed to support their shivering is supplied by aerobic sugar oxidation in muscle using the massive stores of honey that the colony laid down from summer foraging. Hence, honeybees use a social solution to winter warmth.

Another futile form of ATP turnover that supports thermogenesis occurs in bumblebees, which must also heat their flight muscles before take-off. Here, a futile hydrolysis of ATP occurs from the unregulated function of two enzymes of carbohydrate metabolism: the ATP-dependent 6-phosphofructo-1-kinase (PFK-1) reaction forms fructose-1,6-bisphosphate ($F1,6P_2$), whereas fructose-1,6-bisphosphatase (FBPase-1) immediately hydrolyses it back to fructose-6-phosphate (F6P). Normally, FBPase-1 activity is very low in muscle and the two enzymes are oppositely regulated by allosteric and reversible phosphorylation controls (see Chapter 14). Indeed, bumblebee FPBase-1 is inhibited during flight to allow high rates of unidirectional glycolytic flux (inhibition comes from the high Ca^{2+} released into the cytosol to stimulate contraction). However, during preflight warmup both enzymes function at high rates, achieving a high net rate of ATP hydrolysis that warms the muscle.

Uncoupling Oxidative Phosphorylation from Electron Transport

High rates of heat release from the oxidation of substrates can also be achieved by failing to utilize the proton-motive force to drive the synthesis of ATP; that is, by uncoupling oxidative phosphorylation from the electron transport system (ETS). As noted in Chapter 8, some plants do this by maintaining a second ETS that is never linked to ATP synthesis. In mammals, nonshivering thermogenesis (NST) by brown adipose tissue (BAT) uses an uncoupling protein (UCP) located in the inner mitochondrial membrane to bypass the F_1F_0–ATP synthase.

The mechanism of action of UCP isoform 1 (UCP1), the BAT-specific isoform, is as follows (see Fig. TB16.1). Proton pumping out of the matrix by complexes I, III, and IV of the ETS generates a proton gradient across the inner mitochondrial membrane. Normally this is used to drive ATP synthesis by the F_1F_0–ATP synthase, but, when UCP1 is active, proton reentry bypasses the F_1F_0–ATP synthase and energy that

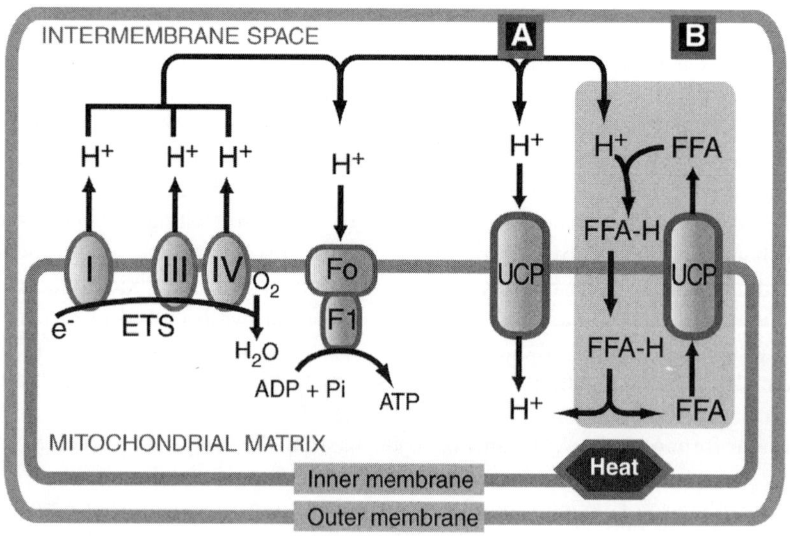

Figure TB16.1 Mode of action of mitochondrial uncoupling protein 1 (UCP1) in heat production by brown adipose tissue. Both the (*a*) original theory and (*b*) current theory of UCP1 action are shown.

would normally be trapped as ATP is released as heat. Two theories of UCP1 action exist. The original theory was that UCP1 was a proton carrier that directly channeled protons back into the matrix. However, recent studies support a different mechanism but with the same net effect. These studies identify UCP1 as one member of a large gene family of mitochondrial inner membrane anion carriers, the physiological substrates of UCP1 being free fatty acid (FFA) anions. Fatty acid anions are transported out of the matrix by UCP1. They are protonated in the acidic intermembrane space, and then neutral FFA-H diffuse across the inner membrane and dissociate again in the more basic pH environment of the matrix. The net effect is that UCP1 functions as a protonophore to return H^+ to the matrix. Although both models can account for the well-known action of FFAs as uncouplers of mitochondrial oxidative phosphorylation, model B also accounts for two other observations: (1) long-chain alkylsulfonates are also transported by UCP1, supporting its role as an anion carrier, and (2) other related mitochondrial transporters (e.g., the dicarboxylate carrier, the ADP/ATP carrier) can also mediate FFA-dependent uncoupling activity.

exclusively on NST by BAT to warm the core thoracic organs, whereas shivering begins once the peripheral skeletal muscles are partly rewarmed. Arousal is a very intense process; for example, in the bat, *Rhinolopus ferrumequinum*, T_b can rise from 7 to 35°C in just 30 min with an increase in oxygen consumption to a level that is 8.7-fold higher than the resting metabolic rate in euthermia. The purpose of the periodic arousals from torpor have frequently been questioned because these brief periods of arousal consume 60 to 80% of the total winter energy budget. This and other key mysteries of hibernation remain unanswered (see Text Box 16.2).

PREPARATORY BIOCHEMICAL ADJUSTMENTS FOR HIBERNATION

Fuel Reserves

The hibernation season can last as long as 9 months for Arctic and alpine mammals, and even with the low metabolic rate in torpor, this still requires a lot of metabolic fuel. Some species cache food in their burrows and eat during their periodic arousals, but many do not. For these latter species, all fuel reserves must be "on board" before hibernation begins. To do this, animals go through a period of intense eating (hyperphagia) in the late summer during which body mass increases by up to 50%, mainly due to the deposition of huge reserves of triglycerides in white adipose tissue. Biochemical adjustments supporting this preparatory phase include elevated activities of lipogenic enzymes that are sustained until autumn fattening is completed. For example, the activity of fatty acid synthase in liver of prairie dogs was 10-fold higher during the prehibernating versus hibernating phase of the year.

Normal hormonal controls on satiety and lipid storage by adipose tissue (see Chapter 9 for a review) are overridden during this period of fattening. In little brown bats (*Myotis lucifugus*) plasma leptin levels decrease during the period of prehibernating fattening, just the opposite of the normal response to increasing adiposity in humans and other nonhibernating species. However, administration of mouse recombinant leptin reduced food intake and body weight gain in Arctic ground squirrels (*S. parryii*) during the fattening period. This suggests that animals remain sensitive to leptin in the prehibernating phase, but the production of leptin has been inhibited or dissociated from the normal signals that indicate rising adiposity probably in response to overriding seasonal controls on body mass set point.

Lipid Fluidity

Direct effects of temperature on the fluidity of triglyceride depots are a key concern for hibernators. To mobilize their fuel reserves when T_b is close to 0°C, lipid depots must be in a fluid state and yet the body fats in most mammals have melting points (MPs) of ~25°C and, hence, are solid near 0°C. The MP of triglycerides depends on the degree of unsaturation of the fatty acids, and the greater the content of mono- and polyunsaturated fatty acids (PUFAs), the lower the MP. Studies have shown that a key element of hibernation success is the presence of sufficient amounts of two PUFAs, linoleic (18 : 2) and α-linolenic (18 : 3) acids, which lower the MP of the depot fats of hibernators to between 5 and −6.5°C. Mammals can synthesize saturated and monounsaturated fatty acids but must get PUFAs from their diet. A natural diet for ground squirrels produces abdominal fat depots with a composition of 16 : 0, 18 : 0, 16 : 1, 18 : 1, 18 : 2, and 18 : 3 fatty acids that

TEXT BOX 16.2 GREAT MYSTERIES IN HIBERNATION

Key mysteries about hibernation remain. One is the nature of the signal that initiates the metabolic suppression to start a torpor bout. During entry into torpor, metabolic rate and T_b drop in parallel, and early studies suggested that it was the drop in T_b that caused the decrease in metabolic rate. However, is has now been proven that the opposite occurs. Careful monitoring of the time course of changes in metabolic rate (measured as oxygen consumption) and T_b have shown that the metabolic rate suppression occurs first. So what triggers cells to reduce their metabolic rate? Part of the puzzle is a lowering of the hypothalamic set point for T_b (the functional equivalent of turning down the thermostat in a home), but it is not yet known how or what the message is that is sent out to all cells/organs of the body to coordinate metabolic rate depression in each.

The existence of a blood-borne hibernation induction trigger (HIT) has been indicated. Some studies have reported that administration of the plasma from hibernating animals induced behavioral or physiological characteristics of hibernation in active animals. However, studies from other laboratories have not been able to confirm this. At present the HIT is believed to be an opiate, but it has never been isolated or purified. The effects of treatment with HIT-containing serum can be mimicked by administration of synthetic opioids, such as D-Ala(2),D-Leu(5)-enkephalin (DADLE), and are opposed by opiate antagonists, but much remains to be done to both identify the natural HIT and determine how it works at the cellular level.

The second great mystery is why hibernators go through periodic arousals during the winter season. These arousals are hugely expensive; it has been estimated that reheating the body and fueling the short periods of interbout euthermia consume, respectively, ~19 and ~52% of the total winter energy budget. Many species neither eat, urinate, or defecate during arousal and actually spend a large portion of their arousal time in REM (rapid eye movement) sleep. Indeed, based on EEG (electroencephalogram) recordings, it is believed that the deep torpor of hibernation is actually a state of sleep deprivation, and this is supported by the measured levels of oleamide, a marker of sleep debt, in ground squirrel brain; oleamide was 2.6-fold higher in brain from hibernating versus euthermic ground squirrels. Hence, periodic arousals may be needed to allow REM sleep in order to reset neuronal homeostasis and also, as winter draws to an end, to allow the animal to sense cues (both endogenous circannual stimuli and exogenous environmental cues) that indicate that it is time to terminate hibernation. Another theory supported by recent experiments is that periodic arousals are needed to reactivate a dormant immune system. Torpid ground squirrels did not activate the normal acute-phase response when injected with bacterial lipopolysaccharide but only developed fever and the immune response during their next arousal period, which they extended to six times longer than normal while they fought the infection.

is approximately 16, 4, 3, 41, 29, and 7% of the total, respectively. Animals fed artificial diets with minimum PUFA contents, resulting in <20% linoleic and 0% linolenic acids in their depot fats, would not enter torpor when placed in the cold. On the other hand, too many PUFAs in artificial diets also inhibited hibernation. When prairie dogs were compared after feeding high-versus-low PUFA diets, those receiving the higher PUFA contents entered torpor earlier, and had lower torpor T_b values and longer bouts of torpor. Hence, a key element in hibernation success in nature is diet selection. However, the natural sensing mechanism involved in evaluating the PUFA content of depot fats remains unknown.

Although the fluidity of depot lipids is increased to allow them to be metabolized by the hibernator at low T_b, it is still not fully understood how the fluidity of phospholipids in cell and organelle membranes is adjusted to maintain membrane functions in hibernation. In ectothermic (cold-blooded) animals, acclimation to temperature change (either up or down) stimulates a rapid restructuring of membrane lipid composition that is termed homeoviscous adaptation (see Text Box 16.3). Interestingly, the quest for homeoviscous adaptation in the membrane lipids of hibernators has produced mixed results when either seasonal or hibernation-induced changes in membrane lipid composition were sought. There may be two reasons for the lack of obvious homeoviscous adaptation in hibernators. One is the metabolic cost. Restructuring all membranes to prevent cold rigidification at the low T_b values in torpor (or alternatively to restore normal membrane fluidity during arousal) would be a huge energetic cost to the animal and a major task to accomplish within a short time frame of entry into or arousal from torpor. Instead of phospholipid restructuring, hibernators might use alternative mechanisms for adjusting membrane fluidity at low T_b. The production of specialized proteins that could affect membrane fluidity during torpor is one untested idea. Another reason for a lack of homeoviscous adaptation in hibernators could be that the rigidification of unadapted membranes that would develop at low T_b values might be

TEXT BOX 16.3 HOMEOVISCOUS ADAPTATION

Lipids in biological membranes normally exist in a "liquid-crystal" state that is intermediate between a highly fluid material and a rather rigid structure. Membrane viscosity is sustained within a narrow window at any T_b by establishing an optimal mix of saturated, monounsaturated, and polyunsaturated fatty acids in the membrane phospholipids. For example, in the phosphatidylcholine fraction of brain membranes the ratio of saturated to unsaturated fatty acids was 0.59 in Arctic fish (sculpin) at a T_b of 0°C, 0.82 in goldfish at 25°C, and 1.22 in rats at 37°C. A temperature increase raises the fluidity of membrane lipids, whereas a temperature decrease causes a more ordered packing of lipids that increases rigidity. Ectotherms respond rapidly to temperature change to modify the composition of their membranes in order to maintain the optimal membrane viscosity, a process called homeoviscous adaptation. During cold acclimation, this involves a rapid increase in the proportion of unsaturated fatty acids that causes a "disordering" of the membrane hydrocarbon interior to offset the cold-induced ordering. This is accomplished by the rapid activation of desaturase enzyme activity. In particular, much attention has been focused on the Δ^9-acyl-CoA desaturase that, in animals, incorporates the first double bond into saturated fatty acids at the C9–C10 position. A double bond at such as central position in a fatty acid chain (C16 and C18 are the most common chain lengths in animals) creates the maximum disordering effect on membrane properties. Figure TB16.2 shows the responses when carp were cooled from 30 to 10°C. The composition of the phosphatidylinositol phospholipid class changed rapidly; monoene content was significantly elevated within 1 day, whereas saturates were reduced. Analysis of chilling effects on the Δ^9-desaturase revealed two mechanisms at work: (a) an early activation of latent enzyme activity (within 1 day) that may result from some sort of posttranslational modification of the enzyme, and (b) up-regulation of gene expression resulting in elevated mRNA transcripts within 2 days and increased Δ^9-desaturase protein levels within 3 to 4 days. The central role of desaturases in homeoviscous adaptation is demonstrated by the fact that the enzymes are induced by low temperature in a wide variety of organisms. For example, when the cyanobacterium *Synechocystis* is shifted from a growth temperature of 34 to 22°C, transcripts of three fatty acid desaturase genes, *desA* (Δ^{12}), *desB* (ω^3), and *desD* (Δ^6), are induced within 30 min and rise rapidly by about 10-fold.

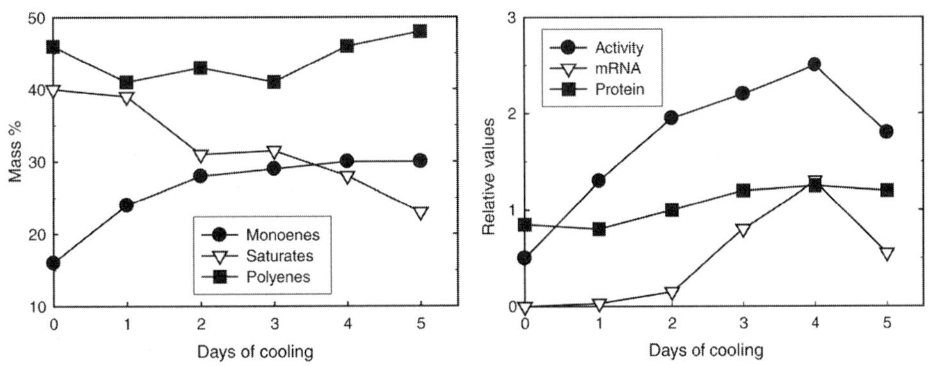

Figure TB16.2 Homeoviscous adaptation. Left panel shows the time course of changes in the composition of the phosphatidylinositol fraction in carp liver microsomes when warm-acclimated (30°C) fish were transferred to 10°C. Right panel shows responses of Δ^9 desaturase. [Redrawn from R. J. Trueman et al. (2000). *J Exp Biol* **203**:641–650.]

exploited as one of the controlling elements in metabolic rate depression. The activities of multiple membrane transporters, channels, and pores are dependent on the fluidity of the membranes around them to permit the conformational changes that are key to protein function. These activities could be reversibly suppressed with ease as T_b cools by simply "trapping" the proteins in increasingly rigid membranes (see section on metabolic rate depression below). Indeed, recent studies with nonhibernating species support this idea. Na^+K^+–ATPase activity correlated positively with the average membrane packing in both rat and toad microsomes. Similarly, Ca^{2+} channel function responded oppositely to detergents that reduce stiffness versus cholesterol that increases the stiffness of the surrounding membrane.

PUFAs, Oxidative Damage, and Antioxidant Defenses

To return to the discussion of depot lipids, there is a serious negative side to the presence of high PUFA contents in lipids and that is that PUFAs are highly susceptibility to autooxidation to produce lipid peroxides, as is explained in detail in Chapter 12. Hence, because of the high PUFA contents of their lipids, hibernators need well-developed antioxidant defenses to prevent oxidative damage (commonly known as rancidification) to their lipid depots over the long winter months. Indeed, ground squirrels fed on a high-PUFA diet showed a 50% increase in superoxide dismutase and a two-fold increase in catalase activities in BAT, compared with squirrels on a low-PUFA diet.

Furthermore, the preparation for natural hibernation in ground squirrels includes an elevation of antioxidant defenses in BAT, including enzymatic (superoxide dismutase, glutathione peroxidase), protein (metallothionein), and metabolite (ascorbate) antioxidants (see Chapter 12 for information on their functions) in order to deal with a surge of reactive oxygen species generation during arousal from torpor when rates of oxygen consumption rise very rapidly. Ascorbate levels are two- to fourfold higher than normal in plasma and cerebrospinal fluid of hibernating ground squirrels but drop quickly during arousal. Furthermore, the up-regulation of uncoupling protein (UCP) isoforms 2 and 3 during hibernation may serve an antioxidant function. Unlike UCP1, these isoforms are not involved in NST (see Text Box 16.1), but recall from Chapter 8 that new studies of UCP2 in mammalian mitochondria have shown that the protein functions to reduce damage by reactive oxygen species (ROS) such as the superoxide that is generated by the electron transport chain. UCP2 shifted the release of ROS from the mitochondrial matrix to the extramitochondrial space and inhibited ROS-triggered apoptosis.

Evidence of oxidative stress associated with hibernation is also found in other organs. During torpor the intestine of ground squirrels shows several signs of oxidative stress, including elevated levels of conjugated dienes (products of lipid peroxidation), a decrease in the ratio of reduced-to-oxidized glutathione, activation of redox-sensitive transcription factors such as the nuclear factor-κB (NF-κB), and expression of the stress protein HSP70.

NONSHIVERING THERMOGENESIS

Metabolic heat production by NST in BAT is one of the hallmarks of hibernation. NST is used to reheat the body during arousal and can often return T_b from near 0°C to 37 to 38°C within minutes. Low-level NST also maintains a minimum T_b just above freezing when ambient temperature in the burrow or den of the hibernator drops below 0°C. BAT is found in large masses in the interscapular region and the perirenal area and surrounds the aorta and heart. BAT differs from white adipose tissue in the presence of many small lipid droplets and a very high number of mitochondria. It has a primary role as an energy-dissipating organ rather than the energy-storing function of white fat. BAT is richly supplied with capillaries that deliver substrates (oxygen, fatty acids) and carry away the product, heat. Brown adipose is not restricted to hibernators but is found in most neonatal mammals (including humans) and persists throughout life in rodents and all hibernating species (except bears). In neonates, NST is critical for maintaining body heat until insulating layers of subcutaneous fat and fur/hair can be grown.

Like white adipose, BAT proliferates during the prehibernation period, for example, increasing in mass by two- to threefold in ground squirrels during late summer and early autumn. Part of the proliferation response to cold exposure in rodents is an increase in vascular endothelial growth factor (VEGF), which stimulates capillary growth (angiogenesis). Recall that VEGF production in hypoxia was under the control of hypoxia-inducible factor (HIF-1) (see Chapter 15), whereas in proliferating BAT, VEGF responds to noradrenaline stimulation, which also triggers NST. Proliferation and differentiation of brown adipocytes is stimulated by a variety of extracellular signals, including noradrenaline, insulin, and insulin-dependent growth factor (IGF-I), each triggering both positive and negative signal transduction pathways. Thyroid hormones are also positive regulators of thermogenic differentiation. The unique protein in BAT that defines its thermogenic capacity is the 33-kD BAT-specific UCP1 that acts as a protonophore to dissipate the proton gradient across the inner mitochondrial membrane and uncouple ATP synthesis from electron transport (see Text Box 16.1 for an explanation of its action). Not surprisingly, then, huge numbers of studies have focused on the induction of UCP1 synthesis under both proliferative and acute signals with multiple signal transduction pathways identified that impact on its synthesis. Major positive signal transduction pathways include protein kinase A (PKA) mediating noradrenaline signals (generally for acute situations) and protein kinase B (Akt) mediating insulin/IGF-I routes (generally for longer term proliferation).

Although most of the work on the molecular mechanisms of BAT thermogenesis has used cold acclimation in nonhibernating species (e.g., mice) as the model system, the mechanisms involved seem identical in hibernators. Acute activation of thermogenesis in brown adipocytes is triggered by noradrenaline released from sympathetic nerves that, in turn, are regulated from the hypothalamus. Noradrenaline activates β_3-adrenergic receptors on the BAT plasma membrane, causing an elevation of intracellular cyclic 3′,5′-adenosine monophosphate (cAMP) and stimulating the dissociation of the active catalytic subunits of PKA. PKA triggers various downstream events, two of the most important being activation of hormone-sensitive lipase (HSL) (see Fig. 9.14) and the up-regulation of UCP1 gene expression via the CREB (cAMP response element binding protein) transcription factor. HSL is activated by PKA-mediated phosphorylation and releases fatty acids from intracellular lipid droplets. These are loaded onto the adipose isoform of fatty acid binding protein (A-FABP) that can form a complex with HSL and are then transported through the cytoplasm to the mitochondria where fatty acids are oxidized by uncoupled respiration to produce heat. Studies in our laboratory showed that there are many additional protein targets of PKA-mediated phosphorylation in BAT. *In vitro* incubation of BAT extracts with cAMP + ^{32}P-ATP showed strong labeling of multiple phosphoprotein peaks that were not labeled when a PKA inhibitor was included in the incubation. In addition, both the pattern and intensity of labeled protein peaks differed between extracts from euthermic-versus-hibernating animals and between incubations done at 37°C versus 5°C, which suggests both hibernation-specific and temperature-specific differences in PKA targets. In particular, strong phosphorylation of proteins in the 40 to 50-kD range occurred in hibernator BAT extracts that were incubated at 5°C.

Brown adipose tissue cells maintain a limited amount of lipid and when this is depleted; lipid fuel can also be derived from circulating triglycerides in chylomicrons or from fatty acids delivered from white adipose tissue. Use of circulating triglycerides depends on the action of lipoprotein lipase (LPL), which is also up-regulated by noradrenaline stimulation in brown adipocytes and exported to the capillary lumen where it attaches to heparan sulfate stalks and hydrolyzes passing triglycerides (see Chapter 9 for more details). Since triglycerides in this form come primarily from dietary intake, the involvement of LPL in NST during arousal from hibernation may be minimal, but the mechanism is important in active rodents in the cold. Fatty acids released by HSL action in white adipose are transported to BAT on plasma albumin and when taken up across the BAT plasma membrane are carried by the heart isoform of FABP (H-FABP) to the mitochondria. Some hibernator organs (BAT, heart) are perhaps unique among mammals in expressing both A- and H-FABP isoforms, reflecting the use of both internal and external fatty acids to fuel NST in BAT and the need for uninterrupted fuel supply to support the rapid increase in heart power output needed to circulate heated blood during arousal. At the mitochondria fatty acids are converted to fatty acyl-CoA moieties, then transported across the inner mitochondrial membrane as carnitine derivatives and catabolized by the normal pathways of β oxidation (to produce acetyl-CoA) and the tricarboxylic acid cycle (see Chapter 9). Electron transport through respiratory complexes I, III, and IV is coupled to the formation of an electrochemical proton gradient across the inner membrane. In conventional cells, this gradient is used to drive ATP synthesis by the F_1F_0–ATPase, but in BAT the presence of UCP1 uncouples ATP synthesis from electron transport (see Text Box 16.1) and energy is dissipated instead as heat.

Highly abundant in BAT, UCP1 can reach 5% of total mitochondrial protein in cold-adapted rats. Interestingly, UCP1 content was increased by 41% and total NST capacity was higher when mice were fed a diet high in PUFAs, suggesting that the elevated PUFA levels (required before animals can enter hibernation) may have a signaling role in optimizing BAT thermogenic capacity. By contrast, the amount of F_1F_0–ATP synthase in BAT mitochondria is typically very low, the ratio of the synthase to the other respiratory chain components being about 10-fold lower than in other tissues. This clearly enhances the ability of UCP1 to "compete" with the F_1F_0–ATP synthase for proton transport. The limiting factor in F_1F_0–ATP synthase expression in BAT is the level of the *c*-F_0 subunit, one of the 10 subunits of the F_0 proton translocator.

Metabolic regulation of UCP1 comes from several sources. Purine nucleotides [ATP, adenosine 5′-diphosphate (ADP), guanosine 5′-diphosphate (GDP), and guanosine 5′-triphosphate (GTP)] are negative modulators of UCP1 that can rapidly change its activity; GDP is particularly important in this regard. UCP1 shows temperature-independent properties with regard to both GDP maximal binding capacity and the dissociation constant (K_d) for GDP, but the K_d is strongly pH-dependent, decreasing with decreasing pH. This would enhance inhibition of UCP1 by GDP under the state of respiratory acidosis that prevails during torpor, but as an initial event in arousal, animals hyperventilate and unload excess CO_2. Cytosolic pH rises and a higher K_d for GDP at higher pH values would reduce the inhibitory effect of GDP on UCP1 and contribute to the activation of NST during arousal. UCP1 is also sensitive to the level of oxidized ubiquinone, and elevated free fatty acids stimulate UCP1 by providing the vehicle for proton translocation (see Text Box 16.1). Once UCP1 is activated, lipid oxidation is no longer inhibited by high levels of ATP and reduced nicotinamide adenine dinucleotide (NADH) that normally integrate the rate of oxidation with organ demand for ATP.

METABOLIC REGULATION IN HIBERNATION

Mammalian hibernation is a complex phenomenon. It includes strong metabolic rate depression with differential suppression of metabolic functions to virtually halt all nonessential functions while continuing to support essential functions with adequate levels of energy. Because hibernating animals can have body temperatures far below euthermic values, temperature effects on metabolism are also a key consideration; both compensation for and exploitation of temperature effects on metabolic reactions can contribute to the metabolic reorganization during hibernation. Hibernation is also a state of long-term starvation, and multiple metabolic adjustments are made to reorganize fuel metabolism for a primary dependence on the oxidation of stored lipids as the primary fuel source. Hibernators also have greatly reduced rates of breathing and heart beat; respiratory acidosis develops and blood flow is reduced to values that would cause severe ischemia at euthermic T_b. Low rates of blood flow increase the risk of spontaneous clot formation, so adjustments must also be made to anticoagulant systems. Long weeks of torpor also mean long weeks of muscle inactivity, so strategies must be in place to minimize disuse atrophy. All of these issues must be dealt with, often by exploiting, modifying, or exaggerating well-known mechanisms of metabolic regulation but also with selected new mechanisms. Regulation can take multiple forms, including exploitation of temperature effects on enzyme/protein functional properties, changes to the structure and/or properties of hibernator proteins/enzymes, posttranslational modification of key proteins, and specific changes in gene and protein expression. The following sections will highlight some examples of these modes of metabolic control, but the reader will notice that there is also considerable overlap between these mechanisms. The mechanisms of metabolic regulation in hibernation are also of great interest to medical science (see Text Box 16.4).

TEXT BOX 16.4 HIBERNATION AND MEDICAL SCIENCE

Medical researchers see multiple applications of hibernation research to humans. The molecular mechanisms that provide long-term stability to hibernator organs at temperatures close to 0°C are of great interest to researchers that want to improve the long-term viability of donor organs while they are transferred on ice to new recipients. Identification of key elements of low-temperature tolerance in hibernators could help to extend hypothermic preservation times from the current limits of about 4 h (heart, lung) to 48 h (kidney) (see Chapter 19). Hibernators are also viewed as possible models for ischemia tolerance because they experience no metabolic damage from very low rates of blood flow in torpor, rates that would normally cause extensive ischemic damage to tissues such as heart and brain. However, the rationale for this ischemia model is faulty. Although experiencing low rates of blood flow, hibernators are not oxygen-limited because their T_b and metabolic rate are also very low; indeed, the fact that lipid oxidation fuels hibernation and lactate does not accumulate demonstrates that metabolism is not oxygen-limited during torpor. But hibernators do appear to be a good model for another ischemia-related problem, that of how to prevent blood clots from forming at extremely low perfusion rates. Hibernators also provide a model for studies of muscle disuse atrophy. Hibernating animals are inactive for many months but lack much of the muscle degeneration that occurs in medical situations of long-term immobility (limb casting, bed confinement) or other situations of disuse such as under the microgravity conditions of long-term space flight. Black bears, for example, lose <20% of their strength over 130 days of hibernation, whereas humans in a similar situation would lose more than 90%. Bears also show no loss of skeletal muscle cell number or size. The mechanism involved is still unknown but may involve periodic shivering or isometric exercise by the torpid animal, limitation of the normally high rates of muscle protein turnover (see Chapter 9), or active muscle protein resynthesis by drawing on labile protein reserves of other organs or recycling urea into new protein synthesis. The space agencies of several countries are keenly interested in hibernation research not only as it relates to muscle atrophy but also to uncover the secrets of metabolic arrest that could allow astronauts to be placed into torpor for long space flights to Mars and beyond.

TEMPERATURE EFFECTS ON HIBERNATOR METABOLISM

Between euthermia and torpor, the core T_b of a hibernator can change by over 30°C. Temperature change affects the rate of virtually all biological reactions and physiological functions, and in most cases reaction rates double with every 10°C increase in temperature (see Chapter 2). This is normally stated as a temperature quotient ($k_{(t+10°C)}/k_{t°C}$), where k is the rate at two temperatures differing by 10°C, abbreviated as Q_{10}. Most metabolic reactions show Q_{10} values of ~2, which over a 30°C change in temperature would mean an 8-fold change in reaction rate. However, Q_{10} values as high as 3 (producing a 27-fold increase in rate over a 30°C range) are quite common, and higher values occur in some cases; for example, the decrease in oxygen consumption of small mammals entering torpor can often have a Q_{10} of about 4 (a 64-fold change in rate over a 30°C range). Indeed, such high Q_{10} values during entrance into torpor have been taken as evidence of the use of active mechanisms of metabolic rate depression that cause the drop in T_b (rather than vice versa). Temperature change also affects the strengths of weak bonds and, therefore, can alter multiple protein/enzyme properties, including conformation, affinity for ligands, substrate or allosteric effectors, subunit dissociation, and binding interactions with other proteins (see Chapter 14). Depending upon the numbers and types of bonds involved, protein/enzyme properties may range from temperature-insensitive to highly affected by temperature change. Hence, differential effects of temperature on reaction rates and protein/enzyme properties could cause havoc with both individual function and the integration of enzymatic pathways and other metabolic processes. This could necessitate the implementation of multiple mechanisms for temperature compensation in order to reestablish homeostasis at the low T_b of the hibernating state. However, the potential havoc could be mitigated by various factors, for example: (a) unusual temperature effects on equilibrium enzymes with low-flux-control coefficients may have little or no effect on the overall activity of the pathway in which they participate, (2) targeted control over specific regulatory enzymes (with high-flux-control coefficients) could effectively override temperature effects and readjust the relative activities of metabolic pathways for function in the hibernating state, (3) temperature effects on membrane properties could alter the temperature sensitivity of membrane-associated enzymes, and (4) temperature effects could alter enzyme-binding interactions with structural proteins. Equally, however, natural effects of temperature change on selected enzymes and functional proteins could also be exploited to assist in the creation of the torpid state by allowing normal or even enhanced temperature sensitivities of various reactions to "effortlessly" shut off metabolic processes that are not needed in torpor (and equally effortlessly reactivate them during arousal). Several examples of temperature effects on enzyme/protein function in hibernating species illustrate the different principles involved.

Maintenance of Low-Temperature Function

Multiple metabolic processes must continue in an integrated fashion during hibernation; for example, aerobic lipid oxidation fuels metabolism in all phases of hibernation (deep torpor, arousal, interbout euthermia), and BAT thermogenesis and the signal transduction pathways that activate it must be functional at all T_b values. Because T_b can range from 0 to 37°C within minutes to hours, a temperature-insensitive design for many proteins would seem to be logical. Indeed, several key enzymes/proteins are designed in this way. As noted above, GDP inhibition characteristics of UCP1 from BAT are temperature-independent. We found that the glycolytic enzyme aldolase from ground squirrel skeletal muscle also showed reduced temperature sensitivity as compared with a nonhibernator. When assay temperature was dropped from 37 to 5°C, the K_m of ground squirrel aldolase for its substrate, fructose-1,6-bisphosphate ($F1,6P_2$), increased by only 50%, whereas the K_m of rabbit aldolase increased threefold.

Temperature-Insensitive Function of Fatty Acid Binding Proteins FABPs also show temperature-independent binding parameters. These intracellular carriers transport fatty acids through the cytoplasm, linking sites of fatty acid import/export (plasma membrane), internal storage (lipid droplets), and oxidation (mitochondria). They are crucial elements in hibernation success because fatty acids are the primary fuel utilized throughout the winter by all organs (even brain uses ketones that are synthesized from fatty acids). Perhaps not surprisingly, then, the genes for FABPs are among those that are strongly up-regulated during hibernation (see section on gene expression below), and key kinetic and structural changes to hibernator FABPs appear to provide the temperature insensitivity of function that is needed of a protein that must function "seamlessly" between low and high T_b values. Indeed, we found that ground squirrel L-FABP shows temperature insensitive dissociation constants for both natural (oleate) and artificial (cis-parinarate) substrates at 5, 25, and 37°C (e.g., K_d for oleate was ~1.5 μM in all cases). By contrast, rat L-FABP showed its lowest K_d (i.e., highest affinity) at 37°C, but K_d values rose about twofold at the lower temperatures (i.e., ligand affinity was reduced by about half). The ability of L-FABP to bind fatty acids of different chain length and degree of unsaturation (16 : 0, 16 : 1, 18 : 1, 18 : 2, 18 : 3) was also analyzed by evaluating their ability to quench the fluorescence of bound *cis*-parinarate. In every case, hibernator L-FABP showed equal or better binding of fatty acids at 5°C compared with 37°C, whereas rat L-FABP showed much poorer fatty acid binding at low temperature (Fig. 16.1). Hence, the properties of ground squirrel L-FABP adapt it for function

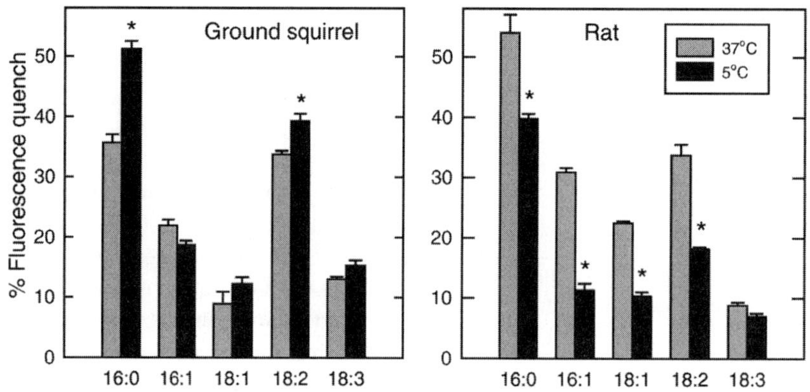

Figure 16.1 Binding of fatty acids to (*a*) ground squirrel and (*b*) rat fatty acid binding protein assessed by the ability of FABP to displace a fluorescent probe *cis*-parinarate at 37°C (shaded bars) or 5°C (black bars); the greater the fluorescence quench, the greater the fatty acid binding. Probe and fatty acids were both 1 μM. Palmitate 16 : 0, palmitoleate 16 : 1, oleate 18 : 1, linoleate 18 : 2, and linolenate 18 : 3. (*) Significantly different from the value at 37°C, $P < 0.05$. Data are means ± SEM, $n = 3$–6. [From J. M. Stewart, T. E. English, and K. B. Storey (1998). *Biochem Cell Biol* **76**:593–599.]

over the entire T_b change experienced by the hibernating animal.

The mechanism of this functional difference between ground squirrel and rat L-FABPs has not been directly determined, but parallel studies of the heart isoform of FABP (H-FABP) indicate that as little as three amino acid substitutions make all the difference. This 133-amino-acid protein is highly conserved with at least 90% sequence identity between hibernating and nonhibernating mammalian species. Of the few substitutions in hibernator H-FABP, three appear to be key. In bat H-FABP these are threonine 83, lysine 73, and leucine 71. Ground squirrel H-FABP shares the threonine 83 substitution but has two other polar amino acid substitutions (lysine 68, asparagine 109) that replace the nonpolar or hydrophobic amino acids present in H-FABP from nonhibernating mammals. The Lys 68 and Asn 109 substitutions in ground squirrel H-FABP occur at turns connecting β sheets in the protein in a "gap" area of FABP that is believed to confer flexibility to the protein. Substitutions in this region could compensate for the effects of low temperature on both protein and ligand rigidity and provide for consistent protein function over the full range of T_b values under which FABP must function.

Improved Low-Temperature Function

Selected enzymes from hibernating species also show temperature effects on substrate affinity that could help them to improve low-temperature function. One example is protein kinase A (PKA), which plays a critical role in the activation of NST. Table 16.1 shows the effects of temperature on the kinetic properties of PKA from ground squirrel BAT versus the rabbit enzyme. The K_m values for both ATP and Kemptide (a peptide substrate) of the hibernator enzymes are sharply reduced at 5°C, compared with 37°C, showing significantly improved substrate affinity at low temperature. By contrast, the substrate affinity of rabbit PKA is unaltered by the temperature change.

A similar situation of improved substrate affinity at low temperature occurs for skeletal muscle creatine kinase (CK) and adenylate kinase (AK) [reactions (16.1) and (16.2)] from hibernators (see more information on these enzymes in Chapter 14). Both are centrally involved in muscle energy metabolism and would have particularly important roles in the supply of ATP for shivering thermogenesis during arousal from torpor:

$$\text{Creatine-P} + \text{ADP} \rightarrow \text{creatine} + \text{ATP} \qquad (16.1)$$

$$2\ \text{ADP} \rightarrow \text{ATP} + \text{AMP} \qquad (16.2)$$

As Table 16.1 shows, the K_m values for all substrates of these enzymes (both forward and reverse reactions) decrease with decreasing assay temperature, an effect that would improve enzyme substrate affinity under the low T_b values of torpor. What is different from the situation with PKA, however, is that CK and AK rabbit muscle also show the same effect, so it seems that the temperature effect on the properties of these enzymes is an inherent property of the mammalian enzymes, perhaps related to the fact that skeletal muscles naturally experience a much wider range of temperatures than do the core organs. The usefulness of enzyme affinities for adenylates that increase

TABLE 16.1 Effect of Assay Temperature on Enzyme Substrate Affinity; Comparison with Hibernation-Associated Changes in Tissue Adenylate Levels[a]

	Hibernator		Rabbit	
	37°C	5°C	37°C	5°C
Protein Kinase A				
K_m ATP.Mg (μM)	49	23	27	29
K_m Kemptide (μM)[b]	50	10	34	35
Adenylate Kinase				
K_m ATP (mM)	0.11	0.06	0.07	0.03
K_m ADP (mM)	0.26	0.11	0.26	0.15
K_m AMP (mM)	0.22	0.10	0.12	0.08
Creatine Kinase				
K_m creatine-P (mM)	2.12	1.76	2.98	2.11
K_m ADP (mM)	0.20	0.01	0.25	0.01
K_m ATP (mM)	0.48	0.20	0.51	0.20
K_m creatine (mM)	3.79	2.29	11.0	2.79
	Prairie Dog		**Jumping Mouse**	
	37°C	5°C	37°C	5°C
Adenylate Levels				
Creatine-P (mM)	—	—	8.1	4.8
ATP (mM)	5.98	3.51	7.2	3.1
ADP (mM)	0.89	1.01	0.8	0.5
AMP (mM)	0.10	0.14	0.05	0.15
Energy charge	0.92	0.86	0.94	0.90

[a]Hibernator enzymes are ground squirrel (*Spermophilus richardsonii*) brown adipose PKA, prairie dog (*Cynomus leucurus*) skeletal muscle AK, and bat (*Myotis lucifugus*) CK. Adenylate levels are for skeletal muscle of prairie dogs and jumping mouse (*Zapus hudsonius*). All K_m values at 5°C are significantly different than the corresponding values at 37°C.

[b]*Note*: Kemptide is a peptide substrate that is widely used for PKA assay; it mimics the phosphorylation site in pyruvate kinase.

Sources: Data are compiled from K. B. Storey (1997). *Comp Biochem Physiol A* **118**:1115–1124; T. E. English and K. B. Storey (2000). *Arch Biochem Biophys* **376**:91–100; J. A. MacDonald and K. B. Storey (1999). *J Comp Physiol* **168**:513–525.

as temperature decreases is apparent when the changes in muscle concentrations of the substrates of these enzymes during hibernation are considered.

As T_b decreases, the creatine phosphate and total adenylate levels in hibernator organs also fall; Table 16.1 shows examples for the skeletal muscle of prairie dogs and jumping mice, but the same effect has been seen in other species and organs. However, despite an approximate 50% decrease in the total adenylate pool size, the energy charge of organs is undisturbed. Energy charge is defined as $[ATP + \frac{1}{2}ADP]/[ATP + ADP + AMP]$ and is effectively a measure of the ATP power available to metabolism (the term $\frac{1}{2}$[ADP] reflects the fact that the adenylate kinase reaction can produce 1 ATP from 2 ADP) (also see Chapter 1). Energy charge in most normal cells is around 0.9 but declines sharply under conditions that cause a rapid utilization of ATP (e.g., muscle exercise) or impair ATP synthesis. A high-energy charge promotes anabolic reactions whereas a drop in energy charge stimulates catabolic reactions to increase ATP synthesis. The decrease in the total adenylate pools of muscle and other organs during hibernation is not accompanied by a change in energy charge or a buildup of end products of AMP breakdown (IMP) such as would act to stimulate increased rates of ATP synthesis. Instead, depletion of the total adenylate pool (with energy charge remaining intact) appears to be one facet of metabolic rate depression. Parallel decreases in the ATP pool size and in the K_m values for ATP of ATP-utilizing enzymes (e.g., PKA, CK, AK) between euthermic and hibernating states help to maintain a near-constant ratio of [ATP] to K_m that may be key to sustaining enzyme function at low temperature. Note, however, that a very strong increase in the affinity of CK for ADP at low temperature (K_m ADP changes by 20-fold) could be an important factor in promoting CK function when rapid ATP generation from creatine phosphate reserves is needed during shivering thermogenesis.

Change or Suppression of Low-Temperature Function

Just as the interactions of temperature with enzyme/protein properties can facilitate the maintenance of consistent enzyme function across a wide temperature range, temperature effects can also be utilized to change or suppress enzyme function at low temperature. A striking illustration of the use of low temperature to change metabolic function is the disaggregation of polysomes in hibernator tissues when T_b falls below 18°C (discussed in greater detail later).

Protein Kinase C A good example of low-temperature-induced enzyme suppression in hibernators is the behavior of the Ca^{2+}-activated and phospholipid-dependent protein kinase, known as protein kinase C (PKC). PKC is a central enzyme of signal transduction in cells (see Chapter 4). The gamma isoform dominates in brain of the bat, *M. lucifugus*, and total activity decreases by about 40% when bats enter hibernation. Arrhenius plots of enzyme activity (log of velocity) versus 1/temperature (in Kelvin) show that activity of the bat brain enzyme declines slowly as temperature initially decreases below 37°C (indeed, Q_{10} is only 1.25 for the 27 to 37°C interval) (Fig. 16.2). Below 10°C, however, activity drops very quickly, with Q_{10} for the 0 to 10°C interval being 3.6 for PKCγ isolated from brain of euthermic bats and 8.1 for the enzyme from torpid bats. By contrast, rat brain PKCγ showed no "break" in the Arrhenius plot over the whole

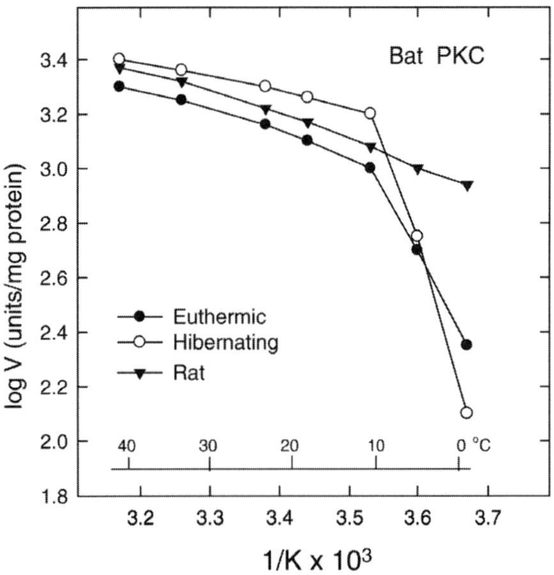

Figure 16.2 Arrhenius plots showing the effect of temperature (in Kelvin) on the activity of PKCγ isoform isolated from brain of euthermic and hibernating bats, *Myotis lucifugus*, and rat brain. Activities were assayed in 5°C increments between 0 and 42°C; note the sharp "break" in the lines for euthermic and hibernator PKC at ~10°C. Inset shows temperature in degrees Celsius. [Data are compiled from H. Mehrani and K. B. Storey (1997). *Neurochem Int* **31**:139–150.]

range. Hence, temperature effects on the activity of bat brain PKCγ could be exploited to suppress signal transduction by this pathway at the low T_b values of the torpid state. Temperature also affected the sensitivity of bat brain PKCγ to phosphatidylserine (PS), a key activator of PKC; the fold activation of PKCγ activity by PS was only 3.5-fold at 4°C compared with 14 to 18-fold at 33°C, a factor that could again aid suppression of PKCγ activity during hibernation. However, bat brain PKCγ showed a feature not found for the enzyme from other mammalian sources. Bat PKCγ was activated by multiple phospholipid types, not just PS. In particular, phosphatidylinositol provided 80 to 90% of the activation seen with PS, whereas phosphatidylethanolamine and phosphatidylcholine were 35 to 45% as effective as PS. This suggests a major modification to the structure of bat PKCγ that imparts to it the possibility of enzyme regulation by different phospholipid types, possibly changing at high versus low T_b values.

Glutamate Dehydrogenase Liver glutamate dehydrogenase (GDH) from the ground squirrel (*Spermophilus richardsonii*) presents another interesting case where the properties of the enzyme in a hibernating species are substantially different from those of the normal mammalian enzyme. Furthermore, a comparison of GDH purified from euthermic (E-GDH) versus hibernating (H-GDH) animals provides strong evidence for a stable modification of the enzyme during hibernation. GDH is a central enzyme of amino acid metabolism; a reversible enzyme found in the mitochondria, it has both an anabolic role in amino acid synthesis and a catabolic role in amino acid degradation and the provision of NH_4^+ for the urea cycle [reaction (16.3)]:

$$\text{Glutamate} + NAD^+ \rightarrow \alpha\text{-ketoglutarate} + NH_4^+ + NADH \qquad (16.3)$$

Purified E- and H-GDH from ground squirrel liver show distinct physical and kinetic differences, particularly when compared at assay temperatures that mimic the euthermic (37°C) and hibernating (5°C) states (Table 16.2). The two forms showed small differences in apparent molecular mass on high-pressure liquid chromatography (HPLC), which suggests that hibernation may induce a posttranslational modification of the enzyme. The mechanism of this modification is not yet known, but mammalian GDH has never been shown to be subject to reversible phosphorylation control. The forward (glutamate-utilizing) reaction may be the functional direction of GDH in liver during hibernation, helping to support gluconeogenesis from amino acids during torpor, and kinetic evidence tends to support this. Both E- and H-GDH showed strong reductions in K_m for glutamate (i.e., increased affinity) at 5°C with H-GDH having the lowest K_m glutamate overall. GDH is sensitive to multiple effectors, the two most important ones being activation by ADP and inhibition by GTP, that place amino acid synthesis versus degradation under control by the energy status of the cell. In the glutamate-utilizing direction, E- and H-GDH both showed strong decreases in K_a ADP (6- and 50-fold, respectively) that greatly increase enzyme sensitivity to ADP activation at low temperature whereas inhibition by GTP was reduced at 5°C in both cases (I_{50} values rose by 2 to 3-fold). Changes in both ADP and GTP effector properties at low temperature would promote the glutamate-utilizing reaction of GDH at low T_b and aid a gluconeogenic function for the liver enzyme during torpor.

METABOLIC RATE DEPRESSION

The decrease in T_b during torpor contributes substantially to the overall reduction in metabolic rate. Indeed, because the Q_{10} value for resting metabolic rate (oxygen consumption) between euthermic and hibernating states is often close to 2, early studies proposed that it was the drop in

TABLE 16.2 Kinetic Properties of Glutamate Dehydrogenase Purified from Liver of Euthermic versus Hibernating Ground Squirrels, *S. richardsonii*, Measured at 37 and 5°C[a]

	Euthermic GDH		Hibernating GDH	
	37°C	5°C	37°C	5°C
Apparent K_m (mM)				
Glutamate	2.03 ± 0.17	0.5 ± 0.05^b	5.2 ± 0.47^c	0.25 ± 0.02^b
NH_4^+	7.0 ± 0.69	24.7 ± 0.01^b	15.8 ± 1.44^c	12.1 ± 0.95^c
α-Ketoglutarate	0.10 ± 0.01	3.66 ± 0.34^b	0.98 ± 0.08^c	$0.43 \pm 0.02^{b,c}$
Effectors of Forward (Glutamate-Utilizing) Reaction				
K_a ADP (mM)	0.181 ± 0.004	0.032 ± 0.002^b	0.58 ± 0.052^c	0.011 ± 0.001^b
I_{50} GTP (mM)	0.019 ± 0.001	0.061 ± 0.0051^b	0.076 ± 0.007^c	0.19 ± 0.013^c

[a] K_m is the Michaelis constant (substrate concentration providing half-maximal enzyme velocity); K_a is the activator constant (activator concentration providing half-maximal activation); I_{50} is the inhibitor concentration that reduces enzyme velocity by 50% under defined substrate concentrations. Data are means ± SEM, $n = 4$. For K_m glutamate, concentration of the co-substrate NAD^+ was 1.5 mM and for K_m values in the reverse direction, NADH was 0.15 mM with either 100 mM NH_4HCO_3 or 7.5 mM α-ketoglutarate.
[b] Significantly different from the value for the same enzyme at 37°C, $P < 0.05$.
[c] Significantly different from euthermic GDH at the same temperature, $P < 0.05$.
Source: Data are compiled from B. J. Thatcher and K. B. Storey (2001). *Biochem Cell Biol* **79**:11–19.

T_b that caused the metabolic depression of torpor. However, it is now known that the opposite is true—a regulated, coordinated, and differential suppression of the rates of multiple metabolic processes results in the drop in T_b. This was confirmed by careful measurements of the decrease in metabolic rate and T_b during entry into a hibernation bout; the former always preceded the latter. Furthermore, Q_{10} for metabolic rate during this entry period is often 3 to 4, indicative of active suppression of metabolic reactions over this period. The overall signal that triggers and coordinates the suppression is still not known (see Text Box 16.2), but one new idea suggests a hypoxia–hypothermia connection (see Text Box 16.5).

Recent studies support the conclusion that metabolic rate depression (MRD) is an active process because many instances of differential regulation of the rates of diverse metabolic processes have now been documented both within and between organs (examples will be discussed throughout this chapter). Protein synthesis provides an interesting case in point. Studies with ground squirrels showed that the rate of ^{14}C-leucine incorporation into protein in brain of hibernating squirrels *in vivo* was only 0.04% of the value in active squirrels, indicating severe suppression of this metabolic function during hibernation (remember, if $Q_{10} = 2$, then a 30 or 40°C change in temperature should result in a rate decrease to 12 or 6% of the euthermic value). Part of this rate suppression was due to an extrinsic factor (the decrease in T_b) and part was intrinsic. Indeed, when measured *in vitro* at a constant 37°C intrinsic suppression was obvious, the rate of protein synthesis in brain extracts from hibernating animals was just 34% of the euthermic value. When comparable assessments using kidney extracts from hibernating squirrels were made, we found an *in vitro* rate at 37°C that was just 15% of the euthermic value, whereas *in vitro* protein synthesis by extracts from BAT were unchanged between torpor and euthermia. These data reinforce the idea that MRD is organ-specific and tailored to achieve the required level of function of each organ in the hibernating state. With its critical role in providing the thermogenesis needed for arousal, BAT clearly maintains its protein biosynthesis capacity even in deep torpor.

Coordinated MRD is one of the hallmarks of hibernation. Over the course of just a few hours, metabolic rate of the hibernator can sink to well below 5% of the corresponding resting rate of the euthermic animal. Although the central trigger that initiates and coordinates MRD is not yet known (see Text Box 16.2), many regulatory controls at the cellular level are now clear. Key among these is reversible protein phosphorylation, a mechanism of metabolic arrest that is widely conserved across phylogenetic lines and is broadly applicable to the control of almost every metabolic function (see Chapter 14). Metabolic functions that are known, to date, to be under reversible phosphorylation control during hibernation include those discussed below.

Carbohydrate Oxidation

Fuel metabolism during hibernation is reorganized so that most organs depend on aerobic lipid oxidation for their energy needs. Carbohydrate catabolism in most organs is spared to save glucose as a fuel for the brain, and gluconeogenesis in liver directs glycerol (from triglyceride hydroly-

TEXT BOX 16.5 THE HYPOXIA–HYPOTHERMIA CONNECTION

The organs of hibernating mammals are hypoperfused and, assessed by the standards of an active mammal at 37°C, would be considered to be severely ischemic; for example, cerebral blood flow is only ~10% of the euthermic value. This has led some researchers to argue that hibernators would make good models for studying ischemia, although this argument does not stand up to scrutiny (Text Box 16.4). Although apnoic breathing patterns may mean that blood oxygen content varies over a considerable range during torpor, their organs are never oxygen-limited; this is confirmed by their reliance on lipid oxidation for ATP production even when extremely high rates of heat production by BAT are required during arousal. However, having said this, there are now at least two lines of evidence that indicate that hypoxia has a role to play in hibernation. These are:

1. An ancient hypoxia–hypothermia interaction may contribute to the mechanism of metabolic rate depression in hibernation. When ectothermic animals are challenged with hypoxia, they will move to a cooler environment to reduce their metabolic rate. Endotherms can achieve the same result in a different way. Hypoxia exposure causes a drop in T_b in many mammalian species. That is, if oxygen is limiting for ATP supply, then a reduction in ATP demand by lowering T_b could keep the system in balance. Hibernating species show a more pronounced drop in T_b in response to hypoxia than do non-hibernating mammals. Furthermore, studies with ground squirrels have shown that metabolic rate was not simply suppressed under hypoxia but was regulated to assist the initial fall in T_b and then acted to stabilize T_b at a new lower level. Indeed, a new set point was established for T_b as long as hypoxia persisted. However, oxygen was not limiting in this situation since a drop in ambient temperature caused the animals to elevate their metabolic rate to maintain the new T_b (this also occurs in hibernation if ambient temperature falls below 0°C). Hence, it is possible that hypoxia signals (perhaps generated from breath-hold episodes) may contribute to initiating and managing the drop in metabolic rate and T_b that occurs during entry into torpor.

2. Hibernating animals show up-regulation of hypoxia-related genes. The hibernation-responsive genes that our laboratory has identified from cDNA array screening included several that code for hypoxia-related proteins in heart and skeletal muscle: HIF-1α, HIF-1β (or ARNT), ORP150 (oxygen-regulated protein) and proline hydroxylase. However, the role that these play is not yet known. HIF is a well-known inducer of glycolytic enzymes, but these do not change in hibernator organs, consistent with a lipid-based metabolism. Thus, it remains to be determined what the HIF signal is doing.

sis) and amino acids (some muscle wasting occurs) into glucose for export. Indeed, the measured rates of gluconeogenesis from lactate and glycerol in isolated hepatocytes were 60 to 100% higher (measured in 37°C incubations) in cells isolated from torpid versus summer-active ground squirrels, documenting one of only a handful of metabolic functions that are enhanced during hibernation. Ketogenesis in liver also augments the fuel supply for brain metabolism. The situation is virtually identical with the human metabolic responses to starvation that were described in Chapter 9.

The primary mechanism responsible for the suppression of carbohydrate catabolism is reversible protein phosphorylation. A key focus of control is the pyruvate dehydrogenase complex (PDC) that commits pyruvate, derived from glycolysis, into the oxidative reactions of the mitochondria. PDC is regulated in an on–off manner by reversible phosphorylation. The active enzyme is the dephosphorylated form, and enzyme activity is typically expressed as the percentage of total pyruvate dehydrogenase (PDH) that is present in the active *a* form (Fig. 16.3). All hibernators examined to date exhibit strong suppression of PDH activity in multiple organs during hibernation, the %PDH*a* typically falling from 60 to 80% active in euthermia to <5% in most tissues of hibernating individuals (Fig. 16.3). Similar strong control of the %PDH*a* also characterizes the shallower daily torpor of hamsters (*Phodopus sungorus*). We found that PDH*a* content in hamster heart and liver was generally only 20 to 40% in torpid animals with low metabolic rates (<40 mL O_2/h) but rose to 60 to 80% in aroused animals with metabolic rates of 75 to 100 mL O_2/h. Hence, a block on pyruvate

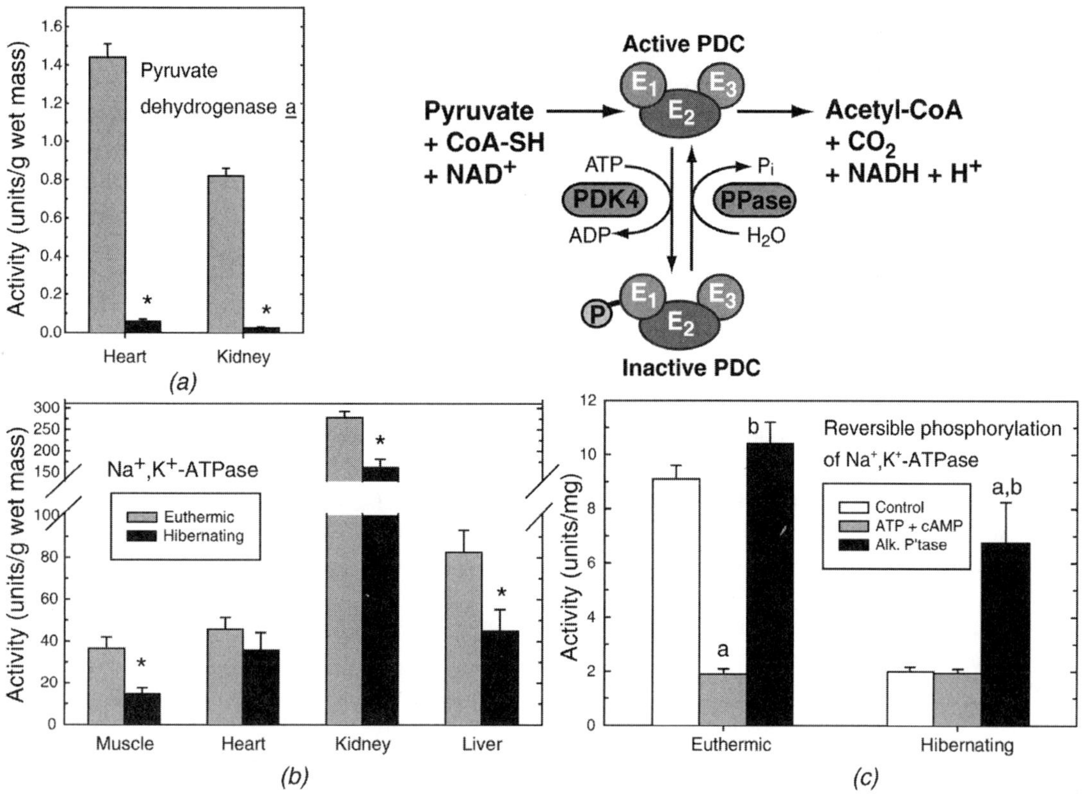

Figure 16.3 Pyruvate dehydrogenase and sodium–potassium ATPase during hibernation in ground squirrel organs. The pyruvate dehydrogenase complex consists of three enzymes: E_1 is pyruvate dehydrogenase, E_2 is dihydrolipoyl transacetylase, and E_3 is dihydrolipoyl dehydrogenase. (*a*) The amount of active pyruvate dehydrogenase a is strongly suppressed during hibernation due to protein phosphorylation by pyruvate dehydrogenase kinase 4 (PDK4) that is strongly up-regulated in torpor. (*b*) Activity of the ion-motive pump, Na^+,K^+–ATPase, is reduced to 40 to 60% of the euthermic value in ground squirrel organs during hibernation. (*c*) The mechanism of Na^+,K^+–ATPase suppression is also protein phosphorylation as shown by the decrease in enzyme activity in extracts of euthermic skeletal muscle when incubated with Mg/ATP + cAMP to stimulate endogenous protein kinase A. Subsequent treatment with alkaline phosphatase reverses the effect and also raises the activity of the enzyme in muscle extracts from hibernating animals. Significantly different from (*) the euthermic value, (a) the untreated control sample, (b) the protein kinase treated sample, $P < 0.05$. [Data are derived from S. P. J. Brooks and K. B. Storey (1992). *J Comp Physiol B* **162**:23–28; J. A. MacDonald and K. B. Storey (1999). *Biochem Biophys Res Comm* **254**:424–429.]

entry into mitochondrial metabolism is one of the central loci of metabolic inhibition in hibernation.

Phosphorylation of PDH is catalyzed by pyruvate dehydrogenase kinase (PDK), and gene screening techniques have repeatedly identified PDK isozyme 4 as one of the genes that is strongly up-regulated during hibernation in multiple organs and different species. For example, PDK4 messenger ribonucleic acid (mRNA) levels were about 10-fold higher in heart of hibernating animals sampled in November to January than in animals sampled in September to November before hibernating; comparable changes in PDK4 protein levels were about sixfold. PDK4 (and PDK2) activities increase in several mammalian organs during starvation, which could indicate that the hibernation response is derived from a preexisting mammalian response that suppresses carbohydrate metabolism under conditions where lipid oxidation is optimized. A new study has shown that rat heart PDK also phosphorylates one of the subunits of NADH ubiquinone-oxidoreductase, also known as NADH dehydrogenase (ND) or complex I of the electron transport system. The result of phosphorylation was reduced ND activity and increased NADH-dependent production of superoxide. PDK phosphorylation of ND would provide a way of coordinating complex I activity with the input of carbohydrate fuel into the tricarboxylic acid cycle and could obviously contribute to metabolic rate suppression during entry into hibernation by targeting the activity of the initial complex of the respiratory chain. Phosphorylation of the enzyme by PKA had exactly the opposite effect, and because PKA mediates signals (e.g., epinephrine) that can increase carbohydrate oxidation by heart, a coordinated activation of both the PDC and the respiratory chain could be achieved. Furthermore, as discussed in Chapters 8 and 12, production of superoxide by the respiratory chain is no longer viewed as a negative by-product of oxygen-based metabolism but is emerging as an important signaling metabolite. A role for superoxide signaling in transitions to and from the hypometabolic state of hibernation needs to be investigated.

Additional phosphorylation controls on liver carbohydrate metabolism closely regulate the disposition of the body's main carbohydrate (glycogen) reserve and the glucose derived from gluconeogenesis (for a review of liver metabolism during starvation see Chapter 9). Inhibitory control of glycolysis targets three loci: glycogen phosphorylase (GP), 6-phosphofructo-1-kinase (PFK-1), and pyruvate kinase (PK). Coordinated suppression of all three is well-illustrated by studies on liver of the jumping mouse, *Zapus hudsonius* (Table 16.3). GP activity is strongly suppressed by two mechanisms, a decrease in total enzyme activity (probably due to a reduced amount of GP protein) and a decrease in the percentage of enzyme present in the phosphorylated active *a* form. Overall, then, the amount of active GP*a* in liver of hibernating mice (0.42 U/g) was only 14% of the value in euthermic animals (3 U/g).

TABLE 16.3 Kinetic Parameters of Liver Glycolytic Enzymes in Euthermic and Hibernating Meadow Jumping Mice, *Zapus hudsonius*

	Euthermic	Hibernating
Glycogen Phosphorylase		
Total activity ($a + b$), U/g	4.5	3.0
Phosphorylase a, U/g	3.0	0.42
Percent active	66	15
6-Phosphofructo-1-Kinase		
K_m fructose-6-P, mM	3.73	3.95
K_a AMP, mM	0.48	0.05
K_a fructose-2,6-P_2, μM	0.14	0.35
I_{50} Mg.ATP, mM	11.4	2.84
I_{50} Mg.citrate, mM	11.7	3.13
Pyruvate Kinase		
$S_{0.5}$ phosphoenolpyruvate, mM	0.47	0.44
K_a fructose-1,6-P_2, μM	0.27	1.18
I_{50} L-alanine, mM	14.0	2.21

Source: Data are compiled from K. B. Storey (1987). *J Biol Chem* **262**:1670–1673.

Entry of hexose phosphates into the "lower" triose phosphate portion of glycolysis is controlled by PFK-1. The enzyme is closely controlled by both energetic signals (ATP inhibition, AMP activation) and fuel supply signals (activation by high fructose-2,6-P_2 signals plentiful carbohydrate that can be used for anabolism; inhibition by high citrate signals plentiful lipid oxidation and, hence, that carbohydrate can be spared). Stable changes to the kinetic constants for these activators and inhibitors occur during hibernation (Table 16.3) and can be interpreted as follows: (a) much greater sensitivity to AMP (K_a decreases by 90%) puts the enzyme under strong control by the energy status of the cell, (b) increased sensitivity to ATP inhibition (I_{50} decreases by 75%) has the same function but also recall that cellular ATP content typically falls by about 50% in torpor (Table 16.2) so part of the change in I_{50} readjusts PFK-1 sensitivity to the prevailing ATP levels in hibernation, (c) reduced sensitivity to fructose-2,6-P_2 (K_a increases by 2.5-fold) makes the enzyme less sensitive to anabolic signals, and (d) increased sensitivity to inhibition by citrate, a signal of high substrate availability in the tricarboxylic acid cycle, enhances the inhibition of glycolysis under conditions of high lipid oxidation.

Pyruvate kinase is the third control point in liver glycolysis. Strong inhibitory control of PK is needed under conditions when gluconeogenesis is needed, a situation that occurs in hibernation. Again, stable changes to PK kinetic parameters were found in liver of hibernating mice: (a) the K_a for fructose-1,6-P_2 rises 4.4-fold, making the enzyme much less sensitive to feed-forward activation by the product of PFK-1, and (b) the I_{50} for L-alanine drops to just 16% of the euthermic value, greatly increasing sensitivity to inhibition by this key gluconeogenic substrate.

In all cases, these kinetic changes to *Z. hudsonius* liver enzymes were stable during enzyme extraction, and studies in this species and others have identified the mechanism of hibernation-induced changes as protein phosphorylation. GP is dephosphorylated and largely returned to its inactive form, whereas PFK-1, PK, and PDH are converted to less active forms by phosphorylation. Not all species show such dramatic and coordinated changes to the properties of their glycolytic enzymes during hibernation (although all show PDH suppression). Such strong regulatory controls are particularly prominent is small species, such as mice and bats, with less consistent changes in larger species such as ground squirrels.

Ion-Motive ATPases

As discussed in Chapter 15, membrane ion pumps consume a huge portion of the total ATP turnover of any cell (5 to 40% depending on the tissue), but their job is crucial for the maintenance of membrane potential difference, which in turn is the basis of multiple activities, including nerve conductance, muscle contractility, the transmembrane movement of many organic molecules, and the sensitivity of multiple receptors. Among these, Na^+K^+–ATPase has a prominent role, especially in endotherms, because high rates of ATP turnover during Na^+/K^+ pumping are a major contributor to thermogenesis (see Text Box 16.1). Hence, Na^+K^+–ATPase activities in mammalian organs are two- to sixfold higher than in the same organs of comparably sized reptiles, and the opposing facilitated flux of Na^+ and K^+ ions through ion channels is similarly greater in mammals. Obviously, then, Na^+K^+–ATPase and other energy-driven pumps as well as the facilitative channels that oppose them represent critical targets for metabolic arrest during hibernation.

Analysis of Na^+K^+–ATPase activities in organs of hibernating ground squirrels confirmed this. We found that activities in extracts from hibernator organs were just 40 to 60% of the comparable values in organs from euthermic animals (all measured at 25°C) (Fig. 16.3). Activity of the sarcoplasmic reticulum (SR) Ca-ATPase that is responsible for Ca^{2+} reuptake into the SR after muscle contraction is similarly suppressed during hibernation. Indeed, in ground squirrel skeletal muscle, the activities of Na^+K^+–ATPase and Ca–ATPase measured under optimal substrate conditions in extracts from hibernating animals were 40 and 41%, respectively, of the corresponding values in euthermic extracts, suggesting coordinate control over these two important ion pumps. Under *in vivo* conditions, suppression may be even stronger because muscle ATP concentration is 30% lower in torpor than in euthermia (reducing ATP substrate availability) at the same time as enzyme affinity for substrates is reduced (e.g., for Ca–ATPase, K_m Ca^{2+} is 75% higher and K_m ATP is threefold higher in muscle extracts from hibernating versus euthermic animals). *In vitro* studies that incubated skeletal muscle extracts under conditions that stimulated protein kinase activity showed us that hibernation-induced suppression of Na^+K^+–ATPase was due to protein phosphorylation whereas Na^+K^+–ATPase activity was restored after alkaline phosphatase treatment (Fig. 16.3). Other proteins involved in SR calcium signaling are also suppressed during hibernation; SR calcium-release channels (ryanodine receptors) decreased by 50%, and most SR calcium binding proteins (e.g., sarcalumenin, calsequestrin) are three- to fourfold lower in hibernating, compared with summer-active, animals. Overall, then, targeted covalent modification of membrane ion pumps and associated proteins strongly suppresses ATP use by these proteins during torpor, but the primary mechanism used (reversible phosphorylation) allows for an immediate reversal and reactivation of ion pumps via stimulation of protein phosphatases during arousal.

Similar controls may regulate other membrane-associated processes during hibernation. For example, rates of K^+ efflux are threefold lower in mitochondria isolated from torpid ground squirrels, compared with euthermic animals, and rates of state 3 respiration are 30 to 66% lower (measured at a constant temperature). By contrast, there are only minor changes in the contents of mitochondrial enzymes or cytochromes between euthermic and hibernating states, and this inhibition is rapidly reversed upon arousal. Again, this implicates reversible controls such as protein phosphorylation in the suppression of mitochondrial activity during hibernation.

Protein Synthesis and Ribosomal Translation Machinery

Protein synthesis is another of the energy-expensive processes in cells and, as noted above, is strongly suppressed during hibernation. As discussed in the previous chapter, two factors could restrict protein synthesis in suppressed metabolic states: (1) mRNA substrate availability and (2) inhibition of the ribosomal translation machinery. Substrate availability does not appear to be a significant factor as there little evidence of reduced mRNA availability in hibernation (similar to the case in other hypometabolic systems; see Chapter 15). No major change in global mRNA transcript levels during hibernation were detected when the question was approached in several ways: total mRNA levels were unchanged, as were the levels of mRNA transcripts of selected constitutively active genes, and evidence from complementary deoxyribonucleic acid (cDNA) array screening showed only a very low percentage of genes whose transcript levels were up- or down-regulated during hibernation. Furthermore, mRNA appears to remain intact during hypometabolism with no reductions in transcript sizes or general shortening of poly(A) tail lengths as typically occurs when transcripts are degraded. Hence, global mRNA substrate availability does not appear to be a factor in protein synthesis inhibition in hibernation.

Instead, organ-specific suppression of the rates of protein synthesis during hibernation is regulated by control over ribosomal translation by two main mechanisms: (1) reversible phosphorylation of proteins in the translational machinery and (2) the state of ribosome assembly.

Reversible Phosphorylation of Ribosomal Proteins Reversible protein phosphorylation regulates a variety of ribosomal proteins. The mechanisms and target proteins are the same as those involved in the suppression of protein synthesis under multiple other states in mammals (notably starvation) (see Chapter 7 for a general discussion) and as also occur in anoxia-induced metabolic rate depression (Chapter 15). As in these other cases, the eukaryotic initiation factor 2 (eIF2), which introduces initiator methionyl–tRNA (transfer RNA) into the 40S ribosomal subunit, is a main target of inhibitory control during hibernation (Chapters 7 and 15). For example, in ground squirrel brain the percentage of total eIF2α that was phosphorylated rose from >2% in euthermic to 13% in hibernation. Hibernator kidney and skeletal muscle also showed severalfold higher levels of phospho-eIF2α than in euthermia.

Protein translation is also controlled at the level of polypeptide elongation. Studies with ground squirrels have shown that mean transit times for polypeptide elongation by ribosomes were twofold longer in extracts from hibernating animals than in control extracts incubated at the same temperature. The mechanism of this effect was traced to the eukaryotic elongation factor 2 (eEF2) (Fig. 16.4). Studies with other systems have established that eEF2 is one of the most prominently phosphorylated proteins in mammalian cells and is the major player in global protein synthesis control at the elongation stage. During hibernation the amount of phospho-eEF2 (the inactive form) increased substantially in brain and liver. Likewise, phospho-eEF2 content in skeletal muscle of hibernating bats was 15-fold higher than in euthermic animals. The mechanism of eEF2 control in ground squirrel brain included an ~50% higher activity of eEF2 kinase during hibernation and a 20 to 30% decrease in protein phosphatase 2A (PP2A) activity (which opposes eEF2 kinase) as a result of a 50 to 60% increase in the levels of the specific inhibitor of PP2A, I_2^{PP2A}.

The eEF2 kinase is Ca^{2+}-dependent, and this initially suggested a way that changes in intracellular Ca^{2+} concentrations could regulate protein synthesis. However, new work shows that this is just a minor part of the whole story for eEF2 kinase is subject to phosphorylation and activation by several of the major cellular protein kinases, including PKA, mitogen-activated protein kinases (MAPKs), $p90^{RSK1}$, and p70 S6 kinase. This puts global protein elongation under regulation by a host of other inputs. New research has also shown that eEF2 kinase is very sensitive to changes in pH. At normal physiological pH in mammals (pH 7.2 to 7.4) the phosphorylation state of the enzyme in liver homogenates was very low, but when the pH of extracts was lowered to 6.6 to 6.8, the amount of phosphoenzyme increased severalfold. This provides a mechanism for a global suppression of protein synthesis in response to cellular acidosis. Strikingly, every known system of natural hypometabolism includes cellular acidification as one of its traits—all facultative anaerobes develop acidosis as a result of a reliance on fermentative metabolism, hibernators and estivators develop respiratory acidosis due to apnoic (breath-hold) breathing patterns, and many other dormant systems are also acidotic as compared with their active states. Hence, exploitation of pH effects on the activities of key enzymes can provide a way of making "effortless" transitions in the activity states of multiple cell functions to facilitate overall metabolic rate depression.

Polysome Disaggregation Translational control in hibernation is also regulated by polysome aggregation state. Studies with brain, liver, and kidney of hibernating ground squirrels have all found a decrease in polysome content and an increase in monosomes in torpor along with a shift of the mRNA for constitutively active genes [e.g., actin, glyceraldehyde-3-phosphate dehydrogenase, cytochrome *c* oxidase subunit 4 (*Cox4*)] into the monosome fraction. This is the same principle that was described for anoxia-tolerant species in Chapter 15, and this confirms that polysome disaggregation is a general principle of metabolic rate depression across broad phylogenetic lines. Figure 16.5 illustrates the principle for ground squirrel tissues. *Cox4* transcripts are retained during hibernation but shifted into the translationally silent monosome fraction so that translation of this constitutively active gene is halted during torpor and its protein levels remain constant.

The trigger for polysome disaggregation during hibernation is not known with certainty, but new evidence indicates that temperature is a factor. When ground squirrels were sampled at multiple T_b values during both entry into (cooling) and arousal from (warming) torpor, the distribution of ribosomal RNA (rRNA) (monitoring ribosomes) and actin mRNA (monitoring transcripts) in liver shifted distinctly when core T_b reached 18°C. Actin mRNA was primarily in the polysome fraction of euthermic squirrels, but below 18°C a large portion of the transcripts, as well as rRNA, suddenly shifted to the monosome fraction, where they remained throughout torpor. Conversely, during arousal, polysome reassembly was first evident when T_b rose to 18°C. Whether this temperature effect derives from a passive influence of temperature on polysome assembly or is due to temperature-stimulated regulation of one or more ribosomal proteins is not yet known.

The principle illustrated above is that an overall suppression of protein synthesis during hibernation is achieved by the dissociation of active polysomes and the storage of mRNA transcripts in the translationally silent monosome fraction. During arousal the reverse transition allows protein synthesis to be rapidly reinitiated without a need for *de novo* gene transcription. Note that by this mechanism an effective "life extension" of mRNA transcripts is achieved. However, this principle does not account for genes that are up-regulated during hibernation, and a growing number of these have been identified (see below). Generally, an increase in mRNA transcript levels goes hand in hand with increased protein synthesis, so

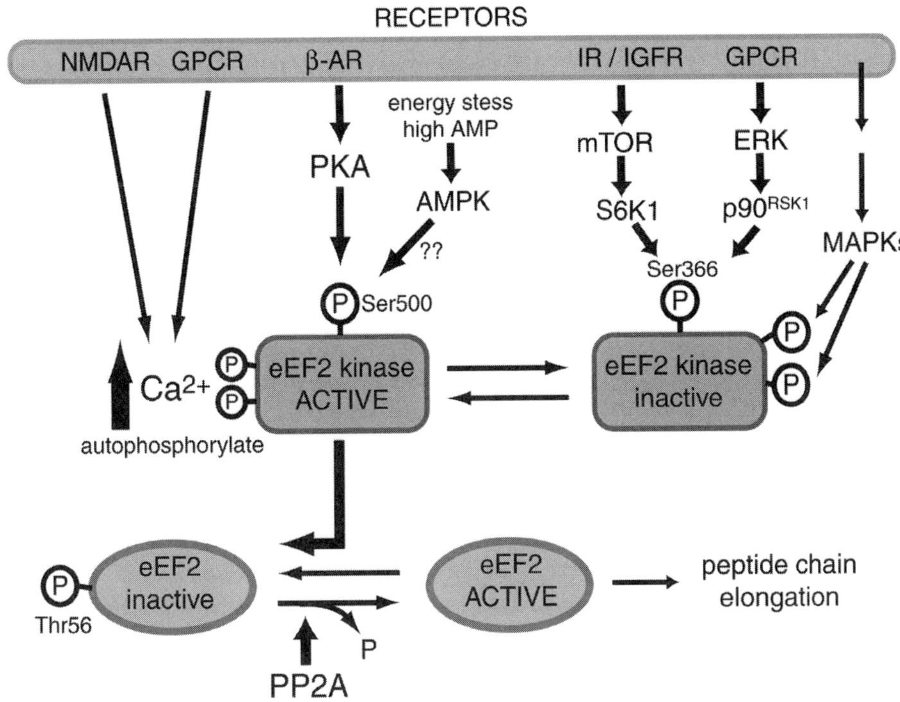

Figure 16.4 Regulation of peptide elongation via controls on the eukaryotic elongation factor 2 (eEF2). eEF2 activity is controlled by reversible phosphorylation of Thr56 with inhibition mediated by eEF2 kinase and activation via dephosphorylation by protein phosphatase 2A (PP2A). eEF2 kinase is the only enzyme that directly phosphorylates eEF2, but the impact of many cellular signals is focused on eEF2 via reversible phosphorylation control over eEF2 kinase at multiple sites. Activation of eEF2 kinase (leading to phosphorylation and inhibition of eEF2) comes from several sources. Initial studies showed that high Ca^{2+} activated the enzyme in a Ca^{2+}/calmodulin-dependent manner, stimulating autophosphorylation of eEF2 kinase at multiple sites. High Ca^{2+} mediates signals from *N*-methyl-D-aspartate receptors (NMDARs) and various G-protein-coupled receptors (GPCRs). Signals such as epinephrine that act via *β*-adrenergic receptors (*β*-ARs) trigger protein kinase A (PKA) to phosphorylate eEF2 kinase on Ser500 and activate the enzyme. Stresses such as exercise or ischemia that deplete cellular ATP levels and elevate AMP also lead to increased phosphorylation of eEF2 by activating the AMP-dependent kinase (AMPK); however, whether AMPK acts via effects on eEF2 kinase or regulates eEF2 dephosphorylation is not yet known. Inhibition of eEF2 kinase (leading to eEF2 activation and the promotion of protein synthesis) comes from phosphorylation of this kinase on different serine residues. Insulin and insulin-like growth factor 1 (IGF1) stimulate their respective receptors (IR, IGR1R) and activate the p70 S6 kinase (S6K1) in an mTOR (mammalian target of rapamycin)-dependent manner to phosphorylate Ser 366. MAPK pathways also lead to inhibitory phosphorylation of eEF2 kinase including ERK-mediated signaling that acts via the $p90^{RSK}$ and SAPK/JNK signaling that phosphorylates other sites. [For a review of this topic see G. J. Browne and C. G. Proud (2002). *Eur J Biochem* **269**:5360–5368.]

how are transcripts that are induced by hibernation handled in a system where ATP availability for protein synthesis is low, polysomes are largely disaggregated, and ribosomal initiation and elongation factors are inhibited? Two cases highlight interesting variations on the general principle.

Differential Distribution of Individual mRNA Species Gene transcripts (*fabp*) for FABP increase severalfold in most organs during hibernation. Figure 16.6 shows the responses by *h-fabp* transcripts in BAT during hibernation; transcript levels are strongly increased and, despite the overall suppression of protein synthesis, H-FABP protein also increases during hibernation. This supports the need for high rates of fatty acid delivery to BAT mitochondria during thermogenesis. How is this accomplished? The answer is differential distribution of *fabp* transcripts in

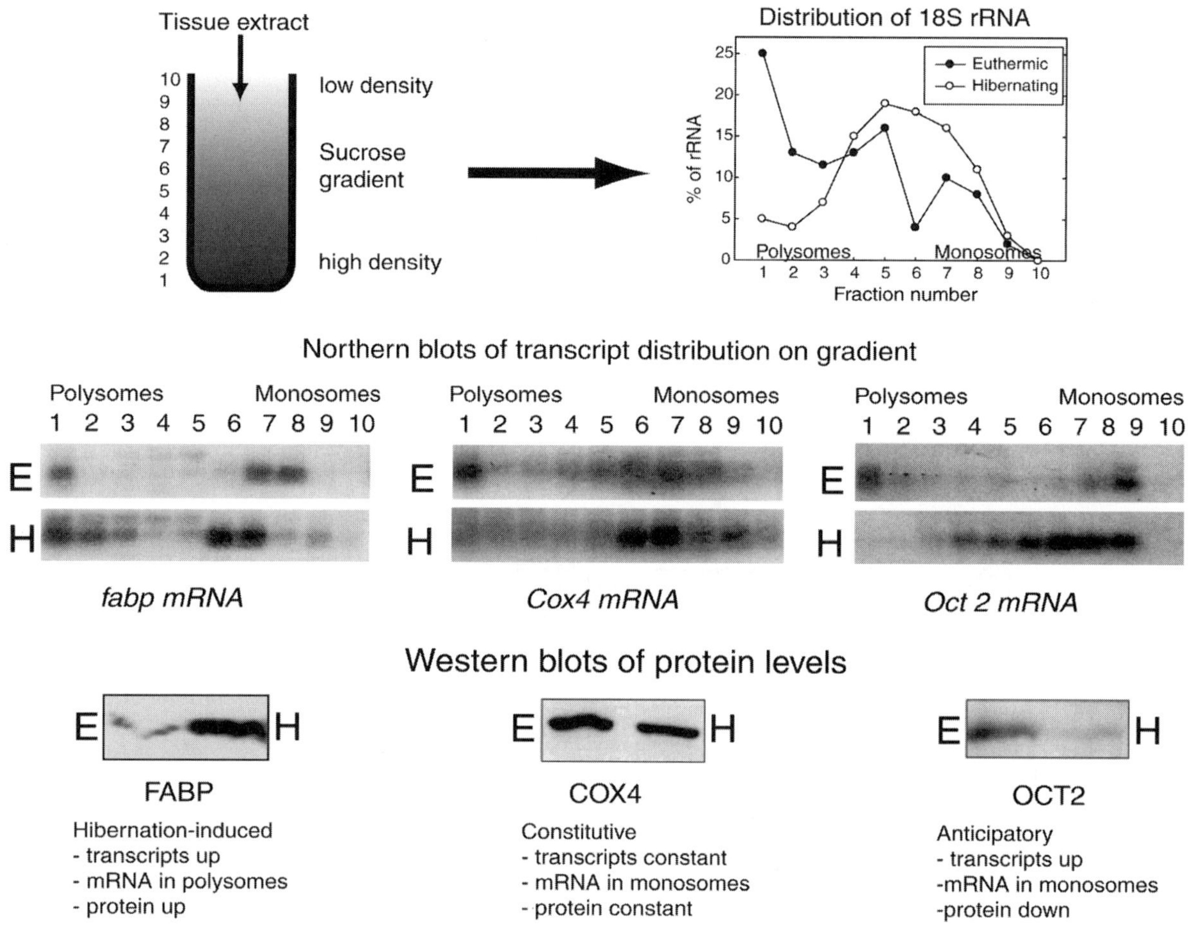

Figure 16.5 Changes in the distribution of mRNA transcripts between translationally active polysomes and translationally silent monosomes during hibernation. Tissue extracts were fractionated on a sucrose gradient and assay of 18S ribosomal RNA content in each fraction shows the relative content of polysomes and monosomes in tissue from euthermic versus hibernating ground squirrels. Northern blots of RNA extracts from each fraction show the position on the gradient and relative abundance of mRNA for three genes: fatty acid binding protein (*fabp*) and cytochrome *c* oxidase subunit 4 (*Cox4*) from brown adipose tissue and the organic cation transporter type 2 (*Oct2*) from kidney. Western blots show the corresponding changes in protein content. The analysis shows (1) enrichment of *fabp* in the polysome fractions and an overall increase in mRNA abundance during hibernation that correlates with the strong increase in FABP protein in hibernator BAT, (2) a strong shift in the mRNA of *Cox4*, a constitutive gene, from the polysome to the monosome fraction during hibernation with unchanged protein levels, and (3) an increase in the abundance of *Oct2* mRNA during hibernation but transcripts are sequestered into the monosome fraction (probably for storage until arousal occurs) while protein content falls during hibernation. [Data compiled from D. S. Hittel and K. B. Storey (2002). *Arch Biochem Biophys* **410**:244–254.]

the monosome versus polysome pools. Whereas transcripts of the constitutively expressed gene *Cox4* are largely sequestered into the monosome fraction during hibernation, we found that *fabp* transcripts were associated with the remaining polysomes. This differential enrichment of *h-fabp* transcripts in the polysome fraction correlated with a threefold increase in H-FABP protein in BAT of hibernating animals. By contrast, COX4 protein levels were unchanged. This illustrates another principle of metabolic control—the rate of translation of individual mRNA species can be altered by differential distribution of transcripts between translationally active and inactive ribosomes. The polysomes in hibernator BAT contain disproportionately higher numbers of those mRNAs (such

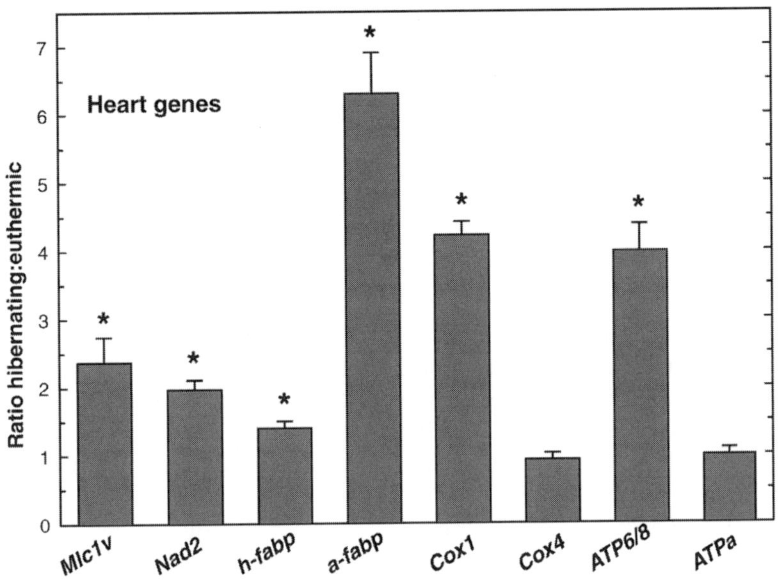

Figure 16.6 Gene up-regulation in ground squirrel heart during hibernation. Relative transcript levels (determined from Northern blots) in hibernating versus euthermic heart are shown for eight genes: *Mlc1*$_v$, myosin light-chain 1 ventricular isoform; *Nad2*, subunit 2 of NADH–ubiquinone oxidoreductase; *FABP-H* and *FABP-A*, heart and adipose isoforms of fatty acid binding protein, respectively; *Cox1* and *Cox4*, subunits 1 and 4 of cytochrome *c* oxidase; *ATP6/8*, ATPase 6/8 bicistronic mRNA; and *ATP*α, α subunit of the mitochondrial ATP synthase. *Nad2*, *Cox1*, and *ATP6/8* are mitochondria-encoded subunits and *Cox4* and *ATP*α are nuclear-encoded subunits of mitochondrial proteins. Data are means $\pm$ SEM, $N = 3$. [Data compiled from D. S. Hittel and K. B. Storey (2002). *J Exp Biol* **205**:1625–1631; A. Falman, J. M. Storey, and K. B. Storey (2000). *Cryobiology* **40**:332–342.]

as *fabp*) that are crucial to the hibernation phenotype, whereas mRNA species that are not needed during hibernation are relegated into the translationally silent monosome fractions.

Anticipatory Up-Regulation Another variation on translational control was identified in studies of the gene for the organic cation transporter type 2 (*Oct2*) in hibernator kidney. *Oct2* transcript levels rose by two- to threefold during hibernation, but OCT2 protein levels actually decreased during hibernation (Fig. 16.5). Analysis of *Oct2* distribution on polysome profiles showed that, although *Oct2* transcripts were much higher in hibernation, they were largely sequestered into the translationally silent monosome fraction. Why would this be? The probable reason for this is so that OCT2 protein can be produced very rapidly from existing *Oct2* transcripts as soon as torpor is broken. Kidney function is virtually shut down during hibernation, and it is possible that this includes an actual degradation of OCT2 protein (rather than a reversible inactivation) since immunoblotting showed much lower levels of OCT2 in hibernator kidney compared with euthermia. In such a situation, resumption of transporter action to support renewed kidney function during interbout arousals would require the rapid synthesis of OCT2. Hence, this suggests another possible principle of translational control—anticipatory up-regulation of selected transcripts during hibernation can support a rapid activation of protein synthesis during arousal that does not depend on enhanced gene expression. This mechanism has a similar functional outcome as the mechanism of ferritin transcript regulation described in the previous chapter. Ferritin transcripts are sequestered until an increase in intracellular iron stimulates their translation, whereas *oct2* transcripts are held in readiness until translation is reactivated during arousal from torpor, possibly as the consequence of polysome reaggregation when T_b rises above 18°C.

Stress Granules The mean half-life of mRNA transcripts in a vertebrate cell is about 3 h, but some transcripts may be stable for only a few minutes and others for a very long time. However, if untranslated mRNAs are to be maintained for days or weeks in hibernator cells, how are they protected from degradation so that translationally competent mRNA transcripts still remain when animals arouse? The poly(A) tail on vertebrate mRNAs is an important protector of the transcript, and a major degradative pathway in eukaryotic cells involves first shortening the poly(A) tail, then decapping the 5′ end and degrading the mRNA in a $5' \rightarrow 3'$ direction. As mentioned previously, analysis of mRNA poly(A) tail lengths has shown no reductions in transcript sizes or general shortening of poly(A) tail lengths during long periods of dormancy, so certainly transcripts are being protected from degradation.

Studies with mammalian systems have linked the suppression of protein synthesis in stress situations with the phosphorylation of eIF2α; recall that increased amounts of phospho-eIF2α is a prominent feature of hibernation. This key initiation factor is phosphorylated by at least four different kinases that mediate different stress signals, and phosphorylation increases binding in an inactive complex with eIF2B, which stops protein synthesis by limiting the availability of the eIF2–GTP–Met–tRNA ternary complex (see Chapter 7). When translation is initiated in the absence of this complex, an eIF2/eIF5-deficient "stalled" 48S preinitiation complex is assembled. This preinitiation complex is joined by RNA-binding proteins, TIA-1 and TIAR, and routed to cytoplasmic foci known as stress granules. In stressed cells, mRNA is in a dynamic equilibrium between polysomes and stress granules. Storage of mRNAs in stress granules during hypometabolism would be an effective protective mechanism that could (a) preserve the valuable pool of untranslated mRNAs until normal conditions were reestablished and (b) provide for a very rapid reinitiation of the translation of key transcripts to provide protein products that are needed immediately during arousal from the hypometabolic state.

Prominent protein constituents of stress granules are poly(A)-binding protein 1 (PABP-1) and T-cell intracellular antigen-1 (TIA-1), a self-aggregating RNA-binding protein. In studies with ground squirrel kidney, Western blotting was used to assess the presence of these proteins in the ribosomal fractions separated on sucrose gradients (as per Fig. 16.5). TIA-1 was restricted to the monosome fractions in trials with kidney extracts from both euthermic and hibernating ground squirrels, but a significant redistribution of PABP-1 occurred in hibernation. In extracts from euthermic animals, PABP-1 was found only in monosome fractions (fractions 8 to 10), whereas in extracts from hibernating individuals PABP-1 was also detected in lower fractions (4, 5, and 7), fractions that also contained substantial amounts of *Cox4* and *Oct2* mRNA (Fig. 16.5). Although the evidence to date is only preliminary, it suggests that a significant amount of the mRNA in the cells of hibernating animals is sequestered into untranslatable stress granules where it is protected from degradation. Furthermore, the presence of ultrastructural changes in the nuclei of hibernating animals and indirect evidence for the binding of PABP to hibernator mRNAs strengthen the argument for the presence of stress granules during torpor.

GENE AND PROTEIN EXPRESSION DURING HIBERNATION

Multiple protein adaptations support hibernation, although, to date, our knowledge of these is still fragmentary. Historically, hibernation research has focused on only a few selected topics, NST and UCP1 being "front and center," with episodic exploration of other events. However, major advances in gene screening technologies over the past few years (cDNA library screening, cDNA arrays, differential display polymerase chain reaction (PCR), reverse transcription (RT)-PCR) are now supporting a more comprehensive analysis of the changes in gene and protein expression that support hibernation. Several studies have examined gene expression differences between euthermic and hibernating states (or hibernating and aroused states) to seek acute changes that support entry into or arousal from hibernation. One consistent result of these studies is that entry to hibernation occurs with very few acute changes in gene expression. This is perhaps not unexpected because (a) entry into a state of cold torpid is not a time to undertake major projects in expensive protein synthesis, (b) cells and organs cannot undergo major modifications because they must remain fully competent to rapidly resume normal body functions during interbout arousals, and (c) many of the required protein changes are put in place well before hibernation begins as part of a strong circannual rhythm that is further cued by photoperiod and ambient temperature changes. Furthermore, it has also been suggested that the test drops in T_b that occur during sleep in the early autumn are used to trigger various metabolic adaptations that support longer full-torpor bouts in the weeks to come.

Hibernation-responsive genes that are up-regulated in ground squirrels include UCP and FABP isoforms in multiple tissues, (α_2-macroglobulin in liver, moesin in intestine, pancreatic lipase in heart, PDK4 in heart, skeletal muscle, and white adipose, the ventricular isoform of myosin light-chain 1 (MLC1_v) in heart and skeletal muscle, OCT2 in kidney, the melatonin receptor in brain, heart, and BAT, and four genes on the mitochondrial genome: NADH ubiquinone–oxidoreductase subunit 2 (ND2), cytochrome *c* oxidase subunit 1 (COX1), and subunits 6 and 8 of the F_1F_0–ATP synthase. Although the genes identified to date represent a wide assortment of cellular proteins, some principles of adaptation are beginning to emerge and will provide interesting directions for new studies over the next few years.

A substantial number of hibernation-responsive genes have been identified in heart (Fig. 16.6). For the mitochondrially encoded genes (*Nad2*, *Cox1*, and *Atp6/8*) this compares with no changes in the transcript levels of nuclear-encoded subunits of these multimeric proteins. The substantial changes in gene responses in heart may be related to the fact that heart must continue to work during torpor and adjustments in gene expression are undoubtedly needed to optimize cardiac function with respect to the changes in T_b, work load, and fuel availability that occur in torpor. Indeed, gene expression studies support substantial myosin restructuring in hibernators. Myosin is made up of two heavy chains (MHC) and four light chains (MLC) that are classified as alkali (MLC1, MLC3) or regulatory (MLC2) light chains; the latter are subject to reversible phosphorylation. Not only is $Mlc1_v$ expression up-regulated in ground squirrel heart and skeletal muscle (the ventricular isoform also occurs in skeletal muscle), but studies with hamster heart have shown that the phosphorylation state of MLC2 decreases from 45% in summer to 23% in torpor. In addition, the proportions of heart MHC isoforms changed from a high content of the β isoform (79% of total) in summer hamsters and winter-active animals at 22°C to near-equal amounts of the α (53%) and β (47%) isoforms in hibernators. Changes in the expression of myosin genes and in the mix of myosin isoforms represented in a muscle are a well-known response for optimizing the myosin motor for function under different conditions. Numerous stimuli (e.g., stretch, electrical stimulation, work load) trigger the response in mammals (see Chapter 11), and myosin restructuring in response to temperature change is well-known in ectotherms. It is not surprising, then, that myosin restructuring occurs in hibernator heart. During hibernation, the heart continues to beat but at a much lower rate and at a much lower T_b. However, peripheral resistance increases substantially, and to compensate for this, the force of contraction actually increases. Increased cardiac contractility in hibernation was originally postulated to result from changes in sarcoplasmic reticulum Ca^{2+} storage and release, but an altered mix of myosin isoforms is another central factor in adjusting heart function for the new work load and thermal conditions of the torpid state.

Another metabolic function that is addressed by adaptive changes in gene expression during hibernation is the clotting capacity of blood. Thrombosis in the microvasculature is a serious risk under low-blood-flow conditions and is a major complication of ischemic conditions in humans (heart attack, stroke). Evidence that hibernators make adjustments to lower the risk of spontaneous blood clot formation during torpor came when studies of liver revealed the up-regulation and export of α_2-macroglobulin during hibernation. α_2-Macroglobulin is a protease inhibitor that binds and inhibits several of the proteases that catalyze steps in the clotting cascade. Other changes during torpor are now known to include reduced levels of several clotting factors as well as reduced platelet numbers (platelets are sequestered into the spleen until arousal), all of which would contribute to minimizing spontaneous clot formation during torpor.

Expression of UCP1 is well known to be induced by cold exposure in both hibernators and nonhibernating rodents. UCP1 levels in BAT are high before animals enter hibernation and remain constant as long as animals are hibernating at ambient temperatures above 0°C. However, when ambient temperature dips below 0°C, the torpid animal increases its metabolic rate and activates NST in order to maintain a constant core body temperature; for example, Arctic ground squirrels could defend a core T_b of 0°C even when ambient temperature fell to −16°C although, to do so, metabolic rate increased about 10-fold. Under this high thermogenic demands, UCP1 transcript levels in BAT rapidly increased. This clearly shows that transcription of at least some genes can go forward at low T_b values and correlates with the protein synthesis capacity of BAT, which appeared to be unaltered in hibernating animals, in contrast to the strong suppression of this function in brain and kidney. The same study also showed that UCP2 and UCP3 transcript levels rose in ground squirrel white adipose and skeletal muscle, respectively, and this suggested that tissues other than BAT could potentially contribute to NST during hibernation.

Supporting the change to a lipid-based economy during hibernation, the capacity for intracellular transport of fatty acids is elevated in most (probably all) tissues during hibernation by increasing the levels of FABPs. Our studies have found that expression of the heart and adipose isoforms of FABP increased in multiple tissues of both ground squirrels and bats during hibernation and up-regulation of other isoforms (at least seven isoforms are known) likely follows suit. Transcripts of A-FABP rose 2-fold in BAT and over 6-fold in heart (Fig. 16.6) of hibernating ground squirrels and were 4-fold higher in BAT of hibernating versus euthermic bats (Fig. 16.7). A-FABP protein levels similarly increased (Fig. 16.7). Transcripts of the heart isoform (that is also expressed in skeletal muscle) were 1.5 to 3-fold higher in BAT, heart, and muscle of hibernating versus euthermic ground squirrels (Fig. 16.6) and nearly 2-fold higher in BAT and skeletal muscle of bats. H-FABP protein also increased strongly, especially in skeletal muscle and BAT (Figs. 16.5 and 16.7).

New studies using cDNA array screening are offering up many more candidate genes that are potentially important to hibernation success. Arrays made with human or rat cDNAs have been used very successfully to analyze gene expression changes in hibernating ground squirrels and bats. Although interspecific sequence differences means that not all hibernator cDNAs will cross-hybridize with the arrays, our studies using human 19K microarrays achieved 85 to 90% hybridization after optimization of hybridization and washing conditions, which means that the responses to hibernation by over 16,000 genes were

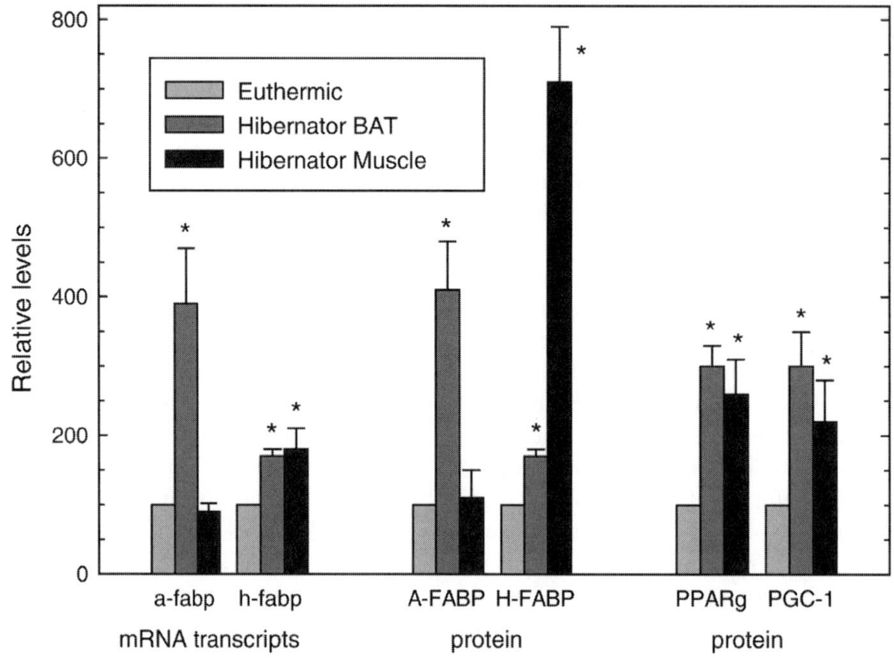

Figure 16.7 Effect of hibernation (36-h reentry into torpor) on transcript and protein levels of the adipose (A) and heart (H) isoforms of fatty acid binding protein and protein levels of the transcription factor, PPARγ and its coactivator, PGC-1, in brown adipose tissue and skeletal muscle of bats, *M. lucifugus*. Transcript levels were measured by Northern blotting; proteins by Western blots. [Data are means $\pm$ SEM, $n = 3$; (*) values are significantly higher in hibernator, compared with euthermic, tissues, $P < 0.05$. [Data on PPARγ and PGC-1 excerpted from S. F. Eddy and K. B. Storey (2003). *Biochem Cell Biol* **81**:269–274.]

assessed. Array screening has two key assets: (1) the opportunity to identify genes (and thereby implicate pathways or functions) that are key to hibernation but have never before been considered as participating in the phenomenon and (2) the opportunity to evaluate the overall responses to hibernation by functionally related groups of enzymes or proteins. For example, in our screening of skeletal muscle extracts from thirteen-lined ground squirrels, *S. tridecemlineatus*, we found that several genes encoding components of the small and large ribosomal subunits were coordinately down-regulated during hibernation, including L19, L21, L36a, S17, S12, and S29. Other insights from array screening are providing avenues for continuing research. For example, analysis of liver and kidney samples from both ground squirrels and bats consistently indicate up-regulation of genes involved in antioxidant defense during hibernation. Glutathione-S-transferase, glutathione peroxidase, and superoxide dismutase mRNA levels were all elevated in hibernator kidney, and these same enzymes as well as peroxiredoxin and metallothionein were up-regulated in liver. These data support our previous discussion of the importance of antioxidant defenses for hibernation success. Array screening of euthermic versus hibernator kidney also showed pronounced increases in transcript levels of selected transport proteins in the hibernator: a twofold increase in aquaporin 3, a fivefold increase in sodium–proton exchanger isoform 2, and a sevenfold rise in the organic cation exchanger isoform 2 (*Oct2*) (discussed earlier).

Array screening revealed very few proteins that were actually down-regulated during hibernation. This correlates well with the results of studies on translational control that indicates that the majority of mRNA transcripts are not degraded during hibernation but are sequestered into translationally inactive monosomes and stress granules. One particularly interesting instance of hibernation-responsive down-regulation is the suppression of insulin-like growth factor (IGF) and its plasma-binding protein (IGFBP-3). The growth-regulatory IGF axis controls somatic growth in skeletomuscular and other tissues, and its suppression during hibernation indicates that energy-expensive growth is curtailed in the hypometabolic state.

Signal Transduction and Transcriptional Control in Hibernation

Relatively little is known about the signaling mechanisms and transcriptional controls that are involved in mediating metabolic and gene expression changes between euthermic and hibernating states. The effects of norepinephrine in the acute activation of BAT thermogenesis during arousal from hibernation have been discussed previously, but other signaling pathways operating to mediate metabolic rate depression and other aspects of cellular responses during hibernation are less well known. Multiple signaling pathways are undoubtedly involved and include well-known mammalian signaling networks that function in starvation to reduce energy expenditure on biosynthesis and shift metabolism to a high dependency on lipid catabolism for ATP generation. Metabolic responses during human starvation are outlined in Chapter 9, and key hormonal signals include a strong decrease in insulin and an increase in glucagon. During hibernation, liver metabolism switches from being an insulin-mediated consumer of glucose (for biosynthesis of glycogen, fatty acids, and many other molecules) to a glucagon-mediated supplier of glucose to other organs (via glycogenolysis or gluconeogenesis) as well as a producer of ketone bodies. Glucagon effects on vertebrate liver include phosphorylation-mediated inhibition of glycolytic enzymes (PFK-1, PK) as well as PDC, and it is likely that glycolytic suppression during hibernation is also mediated in this manner via cAMP as the second messenger and PKA as the kinase. Skeletal muscle $Na^{+}K^{+}$–ATPase activity was also strongly suppressed by PKA treatment, but the enzyme was also inhibited by treatments that stimulated protein kinases C and G activities. Clearly, the activity of this major energy-consuming reaction in cells is responsive to multiple inputs, and it remains to be determined which one mediates the coordinated suppression of this enzyme and other ion-motive ATPases during entry into the hypometabolic state.

Consistent with the starved state, serum insulin levels are very low during hibernation; serum insulin in midwinter hibernating ground squirrels was less than 25% of the values in the prehibernating period. However, insulin transcript levels in the pancreas were just the opposite (highest in hibernation, lowest in prehibernation). Low circulating insulin levels would mean inhibition of intracellular insulin-signaling pathways and that is just what is seen. A variety of insulin effects, especially on carbohydrate metabolism, are mediated by Akt (also called protein kinase B). For example, activation of Akt is linked with increased glucose uptake into muscle cells probably via stimulation of the glucose transporter (GLUT4) and with adipogenesis and lipogenesis in white adipose. Akt also promotes glycogen synthesis in liver by phosphorylating and inhibiting glycogen synthase kinase 3, which prevents the enzyme from inactivating glycogen synthase. During hibernation we found that phospho-Akt (the active form) content in *M. lucifugus* organs was reduced to 30 to 40% of the euthermic value in liver, kidney, brain, and white adipose and was unchanged in heart and skeletal muscle. Furthermore, the total amount of Akt in white adipose also fell to 30% of its euthermic value. This indicates strong suppression of lipogenic pathways during hibernation, as is consistent with the opposite requirement for mobilizing lipid reserves to fuel metabolism during hibernation.

Peroxisome proliferator-activated receptors (PPARs) are ligand-activated transcription factors that play a major role in the regulation of cellular lipid metabolism (see Chapter 9). PPARγ has a primary role in the differentiation and metabolism of adipocytes, both white and brown. All genes involved in lipid catabolism are thought to contain a PPAR response element (PPRE) to which PPAR isoforms bind as a heterodimer complex with activated retinoic acid receptors; this is potentiated by a number of activating factors, including PPAR-gamma coactivator-1 (PGC-1) (Fig. 16.8). Notably, PGC-1 transcription is inhibited by insulin, and this is the mechanism by which the expression of lipolytic enzymes is suppressed when insulin is high. PGC-1 also stimulates the expression of gluconeogenic enzymes, which provides the link between increased expression of both lipolytic and gluconeogenic enzymes during fasting. During fasting, levels of insulin drop and glucagon and glucocorticoid levels increase. Glucagon stimulates an increase in cAMP levels and PKA activity. One of the targets that is phosphorylated by PKA is the CREB protein, a transcription factor. Levels of phosphorylated CREB (the active form) were elevated during hibernation by 2-fold in skeletal muscle of both bats and ground squirrels and by 3.4-fold in heart of squirrels (other organs were not tested). Phospho-CREB binds to the cAMP response element (CRE) that is upstream of the PGC-1 gene (among others) and induces its expression.

Figure 16.7 shows the strong elevation of PGC-1 content in BAT and skeletal muscles of hibernating bats (*M. lucifugus*) that correlates with comparable increases in PPARγ in these organs. Furthermore, coordinated increases in PGC-1 and PPARγ of 1.5 to 2.5-fold were also documented in heart, kidney, liver, and white adipose tissue of bats during hibernation, whereas both were suppressed to about 50% of aroused values in bat brain.

To date, about 50 genes are known to be up-regulated under PPARγ control in mammalian white and brown adipocytes, including large numbers of enzymes involved in both lipogenesis and fatty acid transport and oxidation and genes involved in glucose use for lipogenesis. Of particular relevance to hibernation is PPARγ-mediated up-regulation of the expression of A- and H-FABP, lipoprotein lipase, UCP1, PDK4, carnitine palmitoyl transferase, FATP-1 (the cell surface fatty acid transporter), and several of the enzymes of β oxidation, among others. The gene for leptin is down-regulated by PPARγ. Figure 16.7 shows that the strong up-regulation of PPARγ and

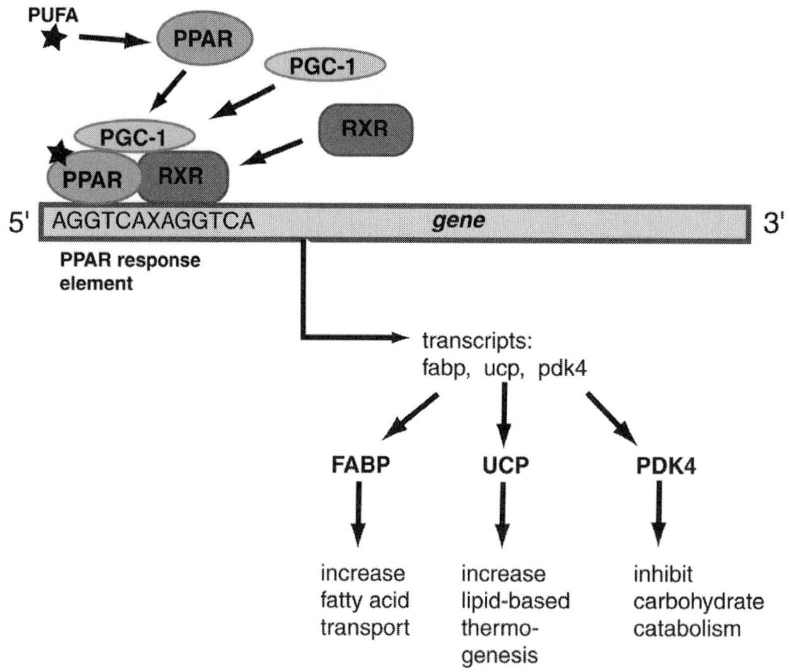

Figure 16.8 When activated by the binding of a ligand such as a polyunsaturated fatty acid (PUFA), peroxisome-proliferator-activated receptors (PPAR) form heterodimers with retinoid X receptors (RXR) and bind to the PPAR response element (AGGTCAXAGGTCA; where X is a variable base) of selected genes. Binding is enhanced/stabilized by the PPAR coactivator, PGC-1. Binding activates transcription of selected genes. In hibernators, these probably include genes for fatty acid binding protein (FABP), mitochondrial uncoupling protein (UCP), and pyruvate dehydrogenase kinase 4 (PDK4).

PGC-1 levels in BAT and muscle of hibernating bats (*M. lucifugus*) correlates with the increase in H- and/or A-FABP mRNA transcripts and protein levels in those organs (skeletal muscle only has trace amounts of A-FABP that do not change in torpor). Thus, we have strong evidence that FABP and undoubtedly other proteins of lipid oxidation (e.g., UCP1, PDK4) are up-regulated in a coordinated fashion under the control of the PPAR transcription factor during hibernation (Fig. 16.8). Furthermore, it is interesting to note that the γ isoform of PPAR is typically described as being abundant in adipose tissue and low in other tissues of nonhibernating mammals, although treatment with PPARγ agonists does up-regulate genes not only in brown and white adipose but also in other insulin-sensitive organs (liver, skeletal muscle). However, in bats the transcription factor was found in all seven tissues tested, the four already mentioned as well as kidney, heart, and brain. This may attest to an enhanced importance of PPARγ and PGC-1 in the regulation of fatty acid catabolism in hibernating species and may represent an adaptive modification of a signal transduction pathway to play a specific role in hibernation.

Mitogen-activated protein kinases also participate in hibernation. MAPKs have widespread roles in regulating gene expression, the p38 MAPKs being particularly involved in cellular responses to stresses. Organ-specific responses by all three MAPK modules [extracellular signal regulated kinases (ERKs), stress-activated kinases (SAPKs), p38 MAPKs] (see Chapter 5 for a review of these modules) were seen in hibernating ground squirrels. ERK1 and ERK2 activities were strongly increased in brain during hibernation, whereas SAPK activity was unchanged in brain but rose in all other organs tested. Western blotting with antibodies specific for the phosphorylated, active p38 showed that levels were twofold higher in heart and over sixfold higher in skeletal muscle of hibernating versus euthermic ground squirrels. Studies focused on brain during arousal from hibernation documented approximately sixfold increases in SAPK and ERK (but no change in p38) during the time that core T_b rose from 7 to 35°C as well as three- and fivefold increases in Akt and PKC (recall both were suppressed in hibernator brain) and elevated levels of glucose-regulated proteins (GRP) and tumor necrosis factor alpha. Hence, multiple signaling pathways are involved in "reactivating" brain function during warming, and the induction of GRP proteins suggests a need to provide protection to multiple cellular proteins under the conditions of rapidly rising T_b and oxygen consumption (undoubtedly generating high levels of reactive oxygen species) over the arousal period.

CONCLUSIONS

In the present chapter we have concentrated mainly on the molecular and biochemical aspects of hibernation, primarily related to fuel metabolism and the control of metabolic rate. The interested reader can explore many other aspects of hibernation; a huge literature is available on the ecological aspects of hibernation, the physiology of tissue and organ functions during torpor including extensive information on neurophysiology, neuroendocrine controls, the relationships between torpor and sleep, circannual rhythms in hibernators, and the regulation of development and proliferation of brown adipose tissue. Our understanding of the biochemical mechanisms that underlie metabolic rate depression in general and hibernation in particular is undergoing very rapid change at the moment due to amazing recent advances in biochemistry and gene expression. For example, gene screening studies (particularly with cDNA arrays) have identified more gene/protein targets of potential interest to hibernation researchers in the last couple of years than have probably been known in total over the past 20 to 30 years of hibernation research. The development of peptide antibodies and of phospho-specific antibodies is vastly improving our capacity to analyze changes in protein content and to trace signal transduction pathways from cell surface to gene or protein target. Overall, this is a hugely exciting time to be involved in hibernation research, and scientists are poised to elucidate many of the remaining mysteries of the phenomenon.

SUGGESTED READING

Boyer, B. B., and Barnes, B. M. (1999). Molecular and metabolic aspects of mammalian hibernation. *BioScience* **49**:713–724. *Easy to read outline of physiological and biochemical changes that support hibernation.*

Heldmaier, G., and Klingenspor, M. (2000). *Life in the Cold.* Springer, Berlin. *Proceedings of the 11th International Hibernation Symposium highlighting multiple advances in the ecology, physiology, and biochemistry of torpor and hibernation.*

Storey, K. B. (ed.) (2001). *Molecular Mechanisms of Metabolic Arrest.* BIOS Scientific, Oxford, UK. *Multiauthor book with 12 chapters assessing metabolic arrest in many forms including hibernation, estivation, anoxia tolerance, diapause, desiccation tolerance with discussion of multiple biochemical mechanisms supporting metabolic arrest.*

Storey, K. B. (2003). Mammalian hibernation: transcriptional and translational controls. In Roach, R. C., Wagner, P. D., and Hackett, P. H. (eds.). *Hypoxia: Through the Lifecycle.* Kluwer/Plenum Academic, New York. *Review of the controls on protein synthesis as they apply to hibernation and exploration of the hypoxia-hibernation connection.*

Storey, K. B., and Storey, J. M. (eds.) (2003). *Cell and Molecular Responses to Stress. Vol. 3: Sensing, Signaling and Cell Adaptation.* Elsevier, Amsterdam. *Includes reviews on uncoupling protein functions from thermogenesis to regulation of reactive oxygen species (Ch. 18), regulation of brown adipocyte proliferation (Ch. 19), hibernators as models for hypoxia-ischemia tolerance in mammalian brain (Ch. 1), and use of cDNA array screening for gene discovery in hibernators (Ch. 22).*

Storey, K. B., and Storey, J. M. (2004). Metabolic rate depression in animals: Transcriptional and translational controls. *Biol Rev Cambridge Philosophical Society* **79**:207–233. *Review of the principles of metabolic rate depression showing the conservation of mechanisms across phylogeny and between different stresses.*

Wang, L. C. H. (1989). Ecological, physiological and biochemical aspects of torpor in mammals and birds. In Wang, L. C. H. (ed.), *Advances in Comparative and Environmental Physiology. Vol. 4: Animal Adaptation to Cold*, Springer, Heidelberg, pp. 361–401. *Still an excellent resource of background information on many aspects of hibernation in a volume that includes many other articles about hibernation and mammalian responses to cold.*

17

COLD HARDINESS AND FREEZE TOLERANCE

JANET M. STOREY AND KENNETH B. STOREY

INTRODUCTION

The Earth is a cold place. About 90% of the water in the oceans is colder than 5°C and surface waters in temperate and polar regions drop to −1.9°C, the freezing point of seawater, for many months of the year. Land is even colder. Winter temperatures of −30°C in temperate zones and −70°C in polar regions are not uncommon. Yet life endures in all cold places. Indeed, many organisms can survive only in the cold; for example, psychrophilic bacteria typically cannot grow above 15°C but grow well at subzero temperatures.

For most multicellular organisms, freezing means death, and, therefore, when winter arrives, strategies of adaptation are needed to survive at subzero temperatures. The fundamental options for protection against the cold are: (a) migration to a warmer climate or microclimate, (b) insulation, (c) thermogenesis, and (d) development of cold hardiness. Migration can be long distance such as birds and butterflies that fly south or short-distances such as frogs that move from their summer haunts in wetlands to spend the winter at the bottom of lakes or garter snakes that travel several kilometers to mass together by the thousands in deep underground dens. Short-distance migration is often, in effect, a way of achieving insulation; animals retreat underground to sites well below the frost line or into deep water that will not freeze to the bottom. Many other organisms also burrow underground (e.g., toads, ground squirrels), seek sheltered sites (e.g., under leaf litter) or dens (e.g., bears, bats), or winter under water (e.g., dragonflies and other insects overwinter in aquatic life stages). Recall also from Chapter 15 that underwater hibernation by turtles, although insulating them from the cold, has required that they optimize their ability to survive without oxygen. Individual insulation consisting of thicker fur or down feathers and layers of fat are key for endotherm survival, as is thermogenesis. Indeed, although hibernating mammals may be exposed to subzero ambient temperatures, they never really have to deal with the possibility of their body fluids freezing because they can always activate nonshivering thermogenesis to maintain a core body temperature, T_b, between about 0 and 5°C. For the vast majority of ectotherms, however, such options are impossible (for an exception see thermogenesis by honeybees in Text Box 16.1). When ambient temperature falls below 0°C, most ectotherms have no physiological or biochemical mechanisms to prevent T_b from falling, and, when T_b falls below 0°C, the risk of body fluids freezing becomes high. For most organisms, freezing is lethal and, hence, their only choice for survival is to develop cold hardiness. Cold hardiness is frequently expressed only in the winter (organisms lose the ability in summer) and is often restricted to one simple life stage (e.g., egg, seed, spore, cyst), which is easier to protect than is a multicellular adult or vegetative form that dies off. Cold hardiness covers a range of biochemical strategies that sustain life at subzero body temperatures. These basically involve multiple options for dealing with the reality that water turns to ice at temperatures below 0°C.

The present chapter explores metabolic regulation as it applies to ectotherm survival at cold temperatures, with a primary focus on survival below 0°C. The subjects of cryobiology and the related medical field of cryopreservation are huge ones that fill books of their own. The focus here will be on the mechanisms of metabolic regulation and biochemical adaptation that contribute to cold hardiness and subzero survival with a major emphasis on the biochemistry of vertebrate freeze tolerance. Advances in understanding these phenomena have provided novel insights about the

Functional Metabolism: Regulation and Adaptation, edited by Kenneth B. Storey.
ISBN 0-471-41090-X

plasticity and adaptability of the mechanisms of metabolic regulation that have been discussed in earlier chapters.

The central theme for preserving life below 0°C is protection. Organisms must arrange protective measures that either prevent their body fluids from freezing or, if managed freezing is allowed to occur, that protect their cells and tissues from the associated physical and metabolic insults. Hence, much of the study of metabolic regulation in the field of cryobiology has centered on the regulation of cryoprotectant production and the synthesis of specific proteins that are involved in ice management or macromolecular stabilization. These will also be our primary focus.

INJURY AND SURVIVAL AT SUBZERO TEMPERATURES

Freezing is lethal for most organisms for several reasons. Ice formation inside cells can destroy both subcellular architecture and metabolic microcompartmentation, and, with only one or two documented exceptions, intracellular ice formation is uniformly lethal for all species and cell types. Both in nature and in the laboratory, freezing typically begins in extracellular spaces and ice propagates through extraorgan spaces (e.g., the abdominal cavity, between skin and muscle layers), through the lumen of the vasculature, and throughout the extracellular fluid spaces surrounding cells, tissues, and organs. Ice does not usually penetrate into cells unless membranes are broken due to physical damage from ice or freezing-related stresses. However, ice growth in extracellular spaces still causes multiple injuries (Fig. 17.1). These include:

1. Ice expansion in the vasculature ruptures capillaries so that upon thawing vascular integrity is lost and there is extensive internal bleeding.
2. The withdrawal of pure water into crystals of extracellular ice elevates the osmotic concentration of remaining extracellular fluid and sets up a powerful osmotic gradient that sucks water out of cells. For example, if the freezing temperature is −10°C, the solute concentration in the unfrozen extracellular liquid of nonacclimated organisms will rise to about 5 osmolar (osM), and more than 90% of the osmotically active water will move out of cells. This extreme cell volume reduction can severely damage membranes and elevates cytoplasmic ionic strength to dangerous levels. Furthermore, during thawing, volume increase can be so rapid as to burst cells.
3. Freezing stops blood circulation so that tissues are deprived of oxygen and blood-borne nutrients as well as of the means to dispose of accumulating waste products. This condition is called ischemia and also characterizes multiple pathological conditions (e.g., heart attack, stroke) and limits the survival time of organs that are removed for transplant (see Chapter 18).
4. Freezing halts multiple vital processes including breathing, heart beat, nerve transmission, and muscle movement that may be impossible to reactivate after thawing.

Because of the dangers of freezing, most ectotherms faced with seasonal exposure to subzero environmental temperatures take steps to avoid ice formation in their bodies, but selected species actually embrace freezing. Ectothermic species use four basic options for survival at subzero temperatures:

1. *Anhydrobiosis* Extensive dehydration that removes all free water from an organism so that there is no water left to freeze
2. *Vitrification* Solidification of water into an amorphous glass state in order to circumvent the physical and osmotic injuries caused by water freezing into ice crystals
3. *Freeze Avoidance* Deep supercooling of body fluids achieved with the use of multiple kinds of antifreezes
4. *Freeze Tolerance* The controlled freezing of body water in extracellular fluid spaces while preserving the liquid state of the cytoplasm

Anhydrobiosis and vitrification strategies are primarily used by microfauna or by "simple" life stages (e.g., eggs, embryos, seeds, cysts), and some unique aspects of these strategies are dealt with in Text Box 17.1. Note, however, that an integral part of both of these strategies, as well as of freeze tolerance and freeze avoidance, is the synthesis and accumulation of high quantities of carbohydrate protectants. Our discussion below of the pathways and regulatory mechanisms that control the synthesis of these protectants applies equally to all four mechanisms of cold hardiness.

At first glance, the freeze avoidance and freeze tolerance options appear to be opposite strategies. The freeze avoidance option prevents freezing by deep supercooling of body fluids, whereas freeze-tolerant organisms allow ice to form in extracellular fluid spaces. However, the two options share at least two key biochemical adaptations: (1) the accumulation of low-molecular-weight carbohydrate protectants and (2) the production of antifreeze proteins. The reason for this is that freeze-tolerant organisms are actually something of a paradox. Ice is allowed to form in extracellular and extraorgan fluid spaces, but intracellular water is prevented from freezing. Hence, freeze-tolerant organisms defend the liquid state of the cytoplasm with the same mechanisms that freeze-avoiding animals use to

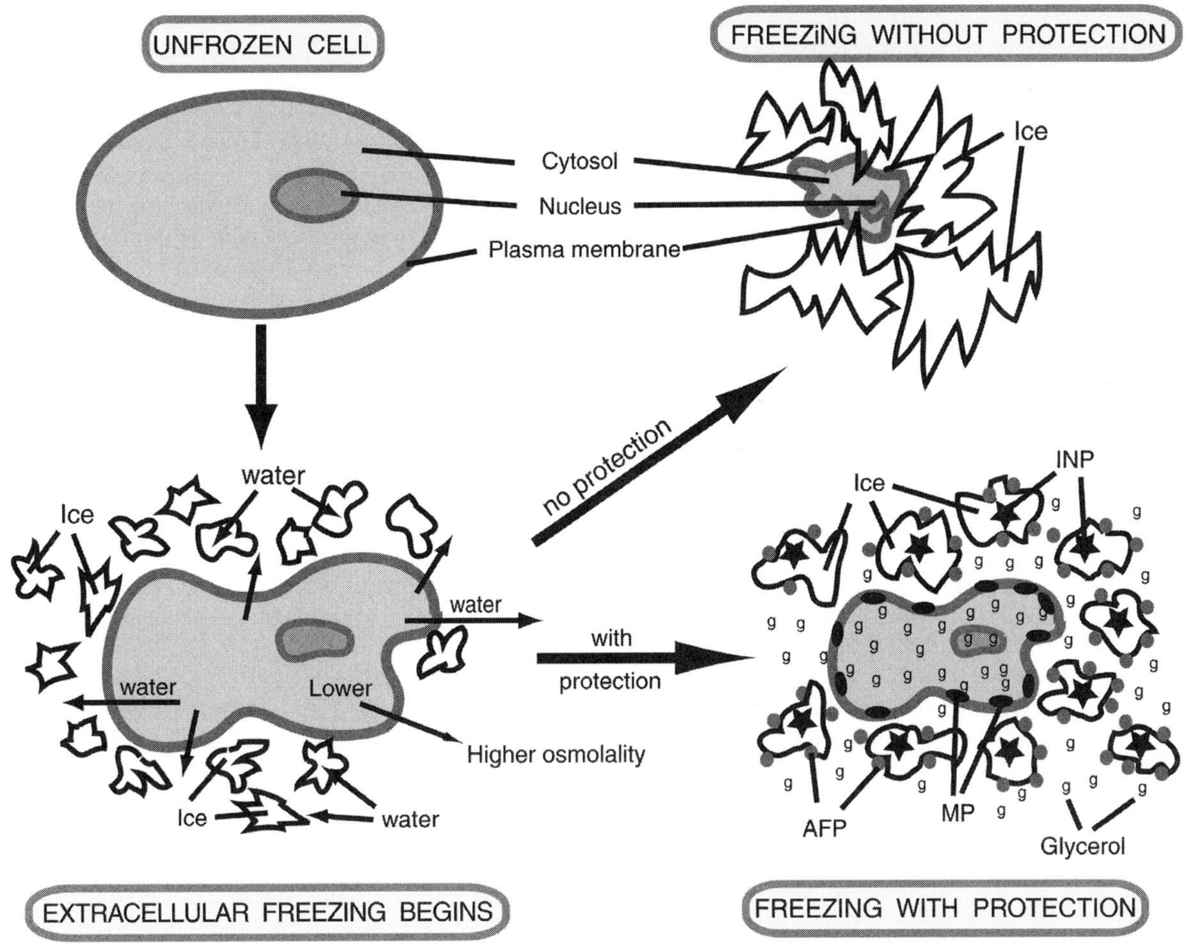

Figure 17.1 Cell responses to extracellular freezing. For an unprotected cell, extracellular ice nucleation results in a rapid growth of ice in large crystals. Solutes are excluded from the crystals and the osmolality of the remaining extracellular fluid rises quickly. Cells respond by loosing water. Cell shrinkage beyond a critical minimum cell volume causes permanent damage to cell membranes so that upon thawing the integrity of the plasma membrane is lost. Cells may also be squeezed and sheared when trapped within growing crystals. Freeze-tolerant organisms use various protective strategies. Freezing is seeded by ice nucleating proteins (INP) at a temperature just under the equilibrium FP of body fluids so that ice growth is slow and controlled. Antifreeze proteins (AFP) regulate the shape of crystal growth and inhibit recrystallization so that crystal size stays small. Low-molecular-weight carbohydrate cryoprotectants such as glycerol (g) or glucose are distributed to all cells and act in a colligative manner to minimize cell volume reduction, whereas others, such as trehalose or proline, act as membrane protectants (MP) to stabilize membrane bilayer structure.

defend the liquid state of the entire organism. Before discussing the biochemical adaptations involved, a further brief explanation of the strategies of freeze avoidance and freeze tolerance are needed.

Freeze Avoidance

Many organisms have mastered the ability to maintain a liquid state at temperatures well below the equilibrium freezing point (FP) of their body fluids. Among animals, the phenomenon has been best studied in two groups: cold-water marine fish and terrestrial insects. Marine fish use this strategy to keep from freezing in surface and inshore waters that chill to -1.9°C in the winter (or year-round in polar regions). This value is well below the -0.5°C FP of the body fluids of teleost fish, and so the potential for inoculative freezing due to contact with environmental ice is a constant hazard. Many insects and

TEXT BOX 17.1 ANHYDROBIOSIS AND VITRIFICATION

Anhydrobiosis

Anhydrobiosis means literally "life without water," and, although this seems a contradiction in terms, it is in fact a critical survival strategy for thousands of species of procaryotes, invertebrate microfauna (e.g., brine shrimp, rotifers, nematodes, tardigrades), some plant seeds, and even the vegetative tissues of a selected group of "resurrection plants." Also called cryptobiosis, the strategy is not just for cold hardiness but allows organisms to endure multiple environmental extremes including very low (−196°C) or very high (100°C) temperatures, desiccation, and oxygen lack. Anhydrobiotic organisms can frequently survive for years, decades, or even centuries. Among animals, the phenomenon has been best-studied in the brine shrimp, *Artemia*. Encysted embryos have as little as 0.1 g of water per gram dry mass (an active animal that is ~80% water would have ~4 g/g dry mass), an amount of water that is insufficient to hydrate intracellular proteins and is therefore incompatible with enzymatic activity or cellular metabolism. Indeed, studies with *Artemia* have found only one indicator of continuing life in dry or anoxic embryos. This is a very slow catabolism of the "energy storage" pool of guanosine polyphosphate (P^1,P^4-diguanosine-5′-teraphosphate) that is built up as animals enter dormancy and appears to support a very slow production of GTP or ATP.

The shell of vicinal water can be stripped off macromolecules during dehydration and, hence, the key to survival in this situation is the choice of effective protectants. Polyhydroxy sugars play a prominent role, not in retaining cellular water as we discussed for freeze-tolerant organisms, but in actually replacing the primary water of hydration surrounding macromolecules (this is called the *water replacement hypothesis*). *Artemia* and many other anhydrobiotes accumulate the disaccharide trehalose; ~15% of the dry mass of encysted Artemia is trehalose. Trehalose is an excellent stabilizer of both biological membranes and proteins (Fig. 17.3) and also has the ability to form an amorphous glass at low water contents (as does the sucrose accumulated by desiccation-tolerant plants). An important feature of such glasses is the huge reduction in diffusion coefficients of metabolites and macromolecules within the glass, which can effectively bring metabolism to a stand-still.

Protection is also provided by the synthesis of stress proteins that prevent protein unfolding or aggregation. In *Artemia*, a protein called p26 plays this chaperone role and makes up 10 to 15% of the nonyolk protein in the dry embryos. The protein is a member of the small heat shock/α-crystallin family. It polymerizes into oligomers of ~500 kD and displays a GTPase activity that may aid the extensive translocation of p26 to the nucleus during stress. A second stress protein found in encysted *Artemia* is artemin. It shows derivation from the iron storage protein, ferritin, but artimen multimers lack the central space used for metal storage. Artemin is enriched in cysteines, and its proposed function is in the protection of encysted embryos from oxidative damage.

Vitrification

Vitrification is a process by which water is solidified, not into a crystal but into an amorphous glass. As mentioned above, trehalose or sucrose glasses are an integral part of anhydrobiosis so the "line" between anhydrobiosis and vitrification as a winter cold-hardiness strategy is sometimes vague. The difference is that various organisms can solidify water into sugar glasses without having to undergo the massive dehydration of the anhydrobiotes. Sugar glasses form in many plant seeds and are key to subzero survival in the twigs of various subarctic woody plants such as poplar and birch. In poplar, for example, sugar glasses form below −20°C, and twigs that are cold-hardened at −20°C to optimize sugar production can subsequently endure exposure to liquid nitrogen. The advantage of vitrification as a strategy for cold hardiness is that a sugar glass incorporates all of the dissolved solutes present in the extra- and/or intracellular water of an organism and, hence, vitrified cells are not under osmotic, ionic strength, or volume stresses as frozen ones are.

Vitrification has been widely explored as a strategy for the applied cryopreservation of cells, tissues, and organs, and, although successful in some cases, researchers have often found that the requirements for achieving a glass transition can be daunting. Necessary conditions can include the need to add extremely high concentrations of solutes (about 40% solutions), to achieve rapid cooling to the glass transition temperature (T_g), that is often well below −30°C, and to use warming rates in the order of 30 to 50°C/min to prevent devitrification, the instantaneous crystallization of ice that can occur during warming at any temperature between the T_g and the melting point (MP) of the solution. How, then, do organisms manage to achieve vitrification in nature?

Natural vitrification occurs under much less rigorous conditions due to at least two factors that cannot always be used in cryomedical situations: (1) vitrifying systems undergo at least partial dehydration, and (2) cells produce and accumulate high concentrations of sugars that have particularly high glass transition temperatures (e.g., trehalose in animals and sucrose, raffinose or stachyose in plants). High glass transition temperatures are very important so that organisms can make the transition to the vitrified state before there is any risk of spontaneous freezing. Indeed, work with both resurrection plants and *Artemia* shows evidence that vitrification occurs during drying at temperatures well above 0°C.

Vitrification is now used quite widely for the ultralow storage of medically important cells and some tissues, but successful preservation of vitrified organs remains elusive. There are several reasons for this including the low tolerance of mammalian organs for substantial dehydration, the requirement for extremely high amounts of cryoprotective agents (such as dimethylsulfoxide, ethylene glycol, glycerol), which may have cytotoxic effects, the impermeability of mammalian cells to the best protective agents with high T_g values (e.g., trehalose), and the need for very fast and even cooling and warming throughout the entire organ mass to both induce vitrification during cooling and avoid devitrification during warming.

other invertebrates are even more adept at freeze avoidance and can remain unfrozen down to −40°C or even lower (see Text Box 17.2). Deep supercooling also occurs within the primordium tissue of the buds and xylem of many woody plants.

The freeze avoidance strategy exploits two of the physical properties of water solutions. The first is a colligative property: the greater the concentration of dissolved solutes, the lower the FP of a solution. The second is the phenomenon of supercooling: the ability of water solutions to chill to temperatures below their equilibrium FP without freezing. All solutions can supercool to some extent, but prolonged supercooling by more than a few degrees below the FP does not normally occur in nature because crystallization is triggered either by contact with environmental ice or by the presence of heterogeneous nucleators such as proteins, food particles in gut, or bacteria or surfaces that can orient water molecules into the crystal lattice and trigger freezing. Freeze-avoiding organisms exploit both of these properties using an accumulation of high concentrations of solutes to achieve strong colligative suppression of the temperature at which body fluids freeze as well as mechanisms that stabilize the supercooled state. The implementation of both of these principles present us with interesting case studies in metabolic regulation and biochemical adaptation.

Freeze Tolerance

The negative side to freeze avoidance is that the supercooled state is not stable, and, if a supercooled organism is seeded by random contact with ice or another nucleator, it will flash freeze and die. Therefore, some organisms have made the "choice" to let themselves freeze but dictate the conditions of the freeze. Freezing is initiated at a high subzero temperatures (usually well above −10°C) so that the rate of ice formation is slow, crystal size is regulated, and ice growth is limited to extracellular and extraorgan spaces only. The best-studied groups of freeze-tolerant animals are certain insects (Text Box 17.2), some intertidal marine mollusks, and several kinds of frogs that hibernate

TEXT BOX 17.2 ONE HABITAT—TWO INSECTS—TWO STRATEGIES

Many kinds of insects make their homes inside galls—hollow nodules on plant stems, leaves, or roots that provide the insects with shelter and food while doing little or no harm to the plant. Life in a gall is restrictive, and insects that live in stem or twig galls may be imprisoned in them throughout the winter and exposed to deep subzero temperatures. The woody stems of goldenrod are home to two gall insects with contrasting strategies of winter survival. Each of these species has become a major model animal for studies of insect cold hardiness.

Caterpillars of the gall moth *Epiblema scudderiana* live inside elliptical galls. They have perfected the freeze avoidance strategy that allows them to stay liquid down to nearly −40°C (see Fig. TB17.1). Triggered when autumn temperatures fall to ~5°C, the insects convert their huge stores of glycogen to glycerol, which reaches levels of over 2 M or ~19% of larval body mass by early winter. High glycerol plus the production of antifreeze proteins results in a suppression of the insect's supercooling point (SCP) from −14°C in September to −38°C by December, a value well below the lowest expected environmental temperature. Freeze avoidance is also aided by a decrease in total body water content and by the presence of a waterproof silk cocoon that lines the interior of the stem cavity to prevent inoculative freezing by contact with environmental ice.

Larvae of the gall fly *Eurosta solidaginis* live inside ball galls, frequently sharing the same stem with a gall of *Epiblema*, but they use the opposite strategy of winter survival—freeze tolerance. Gall fly larvae actually raise their SCP to about −8°C during the winter and, lacking a cocoon, they are susceptible to freezing at even higher temperatures due to contact with frozen detritus or ice in the gall interior. Freeze-tolerant animals encourage ice formation at high subzero temperatures because this allows a slow rate of ice formation that gives cells the greatest amount of time to adjust to the stresses imposed by ice growth through their tissues. Gall fly larvae use a dual system of colligative cryoprotectants: glycogen is converted to ~300 mM glycerol and ~150 mM sorbitol each protectant accumulating on different time frames. Glycerol synthesis is triggered when ambient temperatures cool below 15°C and sorbitol by 5°C exposure. The two protectants are also catabolized separately with sorbitol being reconverted to glycogen, whereas glycerol is probably oxidized as an aerobic fuel during pupal/adult development in the spring. Larvae also accumulate 50 to 60 mM proline, which contributes to membrane stabilization during freezing.

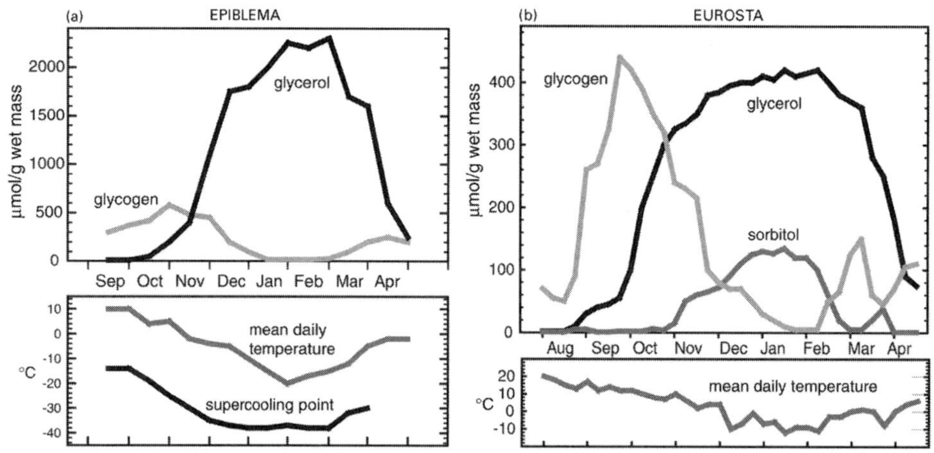

Figure TB17.1 Seasonal synthesis of cryoprotectants by goldenrod gall insects, (a) the freeze-avoiding *Epiblema scudderiana* and (b) the freeze-tolerant *Eurosta solidaginis*. In *Eurosta* note the mid-February loss of sorbitol during a warm spell (with carbon returned to glycogen) but renewed production during colder weather in March. [Data are compiled from J. Rickards, M. J. Kelleher, and K. B. Storey (1987). *J Insect Physiol* **33**:443–450 for *Epiblema* and from J. M. Storey and K. B. Storey (1986). *J Insect Physiol* **32**:549–556 for *Eurosta*.]

on land (Text Box 17.3). Freeze tolerance is also widely studied in agricultural science since the mechanisms of inducing or enhancing frost hardiness have great benefit for the engineering of plant cultivars that survive in cold climates. Freeze-tolerant organisms typically regulate the conversion of up to 65% of their total body water into extracellular ice while using high concentrations of colligative protectants within their cells to prevent the cytoplasm from freezing. Multiple biochemical adaptations address issues associated with freezing. These include:

1. *Ice Management* Ice nucleating agents (often specific proteins) are used to trigger and regulate ice crystal formation in extracellular compartments. Antifreeze proteins help to manage crystal size and prevent recrystallization. Enhanced synthesis of blood clotting proteins in vertebrates also helps to repair any internal bleeding injuries caused by ice when animals thaw.

2. *Cell Volume Reduction* The growth of ice (a crystal of pure water that excludes solutes) in extracellular spaces creates a powerful osmotic force that sucks water out of cells and causes a major decrease in cell volume and a large increase in intracellular ionic strength (Fig. 17.1). Volume changes and the structural pressure that they place on cellular membranes must be managed. The action of membrane protectants stabilizes bilayer structure in response to the compression stress imposed by cell volume loss, whereas low-molecular-weight carbohydrate

TEXT BOX 17.3 FROZEN AMPHIBIANS AND REPTILES

There is nothing quite so amazing as removing an ice-coated, stiff and frozen solid frog from a freezer, setting it on the lab bench, and then watching as ice melts away. The frog relaxes into a seemingly lifeless blob, and then after about an hour a flutter of life appears as a heart beat becomes discernible through the chest wall. Minutes later, the frog takes a first gulp of air, then an eyelid blinks, then legs shuffle and push the frog up into a sitting position, and then it leaps away!

Freezing and thawing several times a winter is part of normal life for several species of woodland frogs that hibernate on the forest floor. They are insulated by layers of leaf detritus and snow from the harshest air temperatures above the snowpack, but under the snow, midwinter temperatures may still fall to $-5°C$ or lower for days or weeks. Frog skin is no barrier to water or ice, and so, when ice forms on the skin surface, body water is quickly seeded and ice propagates through the frog's body over several hours. Ice nucleation triggers an immediate outpouring of glucose from the liver, and sugar is packed into all cells to provide intracellular cryoprotection at the same time as 50 to 70% of the frog's total body water turns to ice in extracellular spaces. When fully frozen, ice fills all extracellular compartments including the abdominal cavity, the ventricles of the brain, the bladder, and runs in sheets between the skin and skeletal muscle. Breathing and heart beat stop, muscles cannot move, no electrical activity can be detected in the brain, and the frog waits, in an icy suspended state, until temperature rises and the melt begins.

The best-studied of the freeze-tolerant frogs is the wood frog, *Rana sylvatica*, the subject of most of the discussion in this chapter, but others include the spring peeper (*Pseudacris crucifer*), the chorus frog (*P. triseriata*), and the tree frogs (*Hyla versicolor*, *H. chrysoscelis*). Several reptiles also show sufficient freezing survival to indicate that freeze tolerance contributes to their winter survival. These include the European common lizard (*Lacerta vivipara*), the box turtle (*Terrapene carolina*), and hatchlings of the painted turtle (*Chrysemys picta*). Various other reptiles and amphibians have a capacity to endure brief freezing exposures but succumb during long-term exposures where ice content rises above 50% and vital processes such as heart beat are interrupted.

protectants provide colligative action to increase cellular osmolality and keep cell volume from falling below a critical minimum.

3. *Ischemia* The freezing of blood plasma halts circulation so that while frozen organs must survive without supplies of oxygen and blood-borne fuels. Survival requires good anoxia tolerance as well as antioxidant defenses that prevent damage due to the formation of reactive oxygen species when oxygen is reintroduced upon thawing. Mechanisms of metabolic rate depression may also be employed to minimize cellular energy requirements and, hence, maximize survival time in the frozen state.

4. *Vital Signs* Freezing halts all vital signs including heart beat, breathing, muscle movement, and nerve transmission. All are sequentially reactivated during thawing, but the molecular mechanisms underlying these processes are still largely unknown.

BIOCHEMISTRY OF CARBOHYDRATE PROTECTANTS

Common to most cold-hardy organisms, both animal and plant, is the accumulation of high concentrations of low-molecular-weight sugars and polyhydric alcohols as protectants. For anhydrobiotic organisms these stabilize macromolecular structure and substitute for the hydration shells that normally contribute significantly to stabilizing protein conformation. For vitrification, high concentrations of protectants allow the formation of sugar glasses in which water is solidified but not crystallized. For freeze-avoiding organisms, high levels of carbohydrate protectants provide the colligative suppression of FP and supercooling point (SCP) of biological fluids that is integral to maintaining body water in a liquid state at subzero temperatures. This use of biological protectants is the functional equivalent of adding ethylene glycol to the water in the radiator of a car (indeed, some insects actually accumulate ethylene glycol). Concentrations of 2M or more are often achieved in the body fluids of freeze-avoiding insects (see Fig. TB17.1). For freeze-tolerant organisms, carbohydrate protectants preserve the liquid state of the cytoplasm and defend a minimum cell volume by limiting water loss into extracellular ice. Although total cryoprotectant levels, measured per gram wet mass (gwm), are lower in freeze-tolerant animals (typically 300 to 500 μmol/gwm) (see Text Box 17.2 for freeze-tolerant insects), once 65% of body water is frozen as ice, protectant concentrations in remaining intracellular water will rise to 1M or more.

Glycerol is by far the most common protectant used by insects and other terrestrial arthropods, but other carbohydrates include sorbitol, mannitol, ribitol, xylitol, erythritol, ethylene glycol, glucose, trehalose, and sucrose

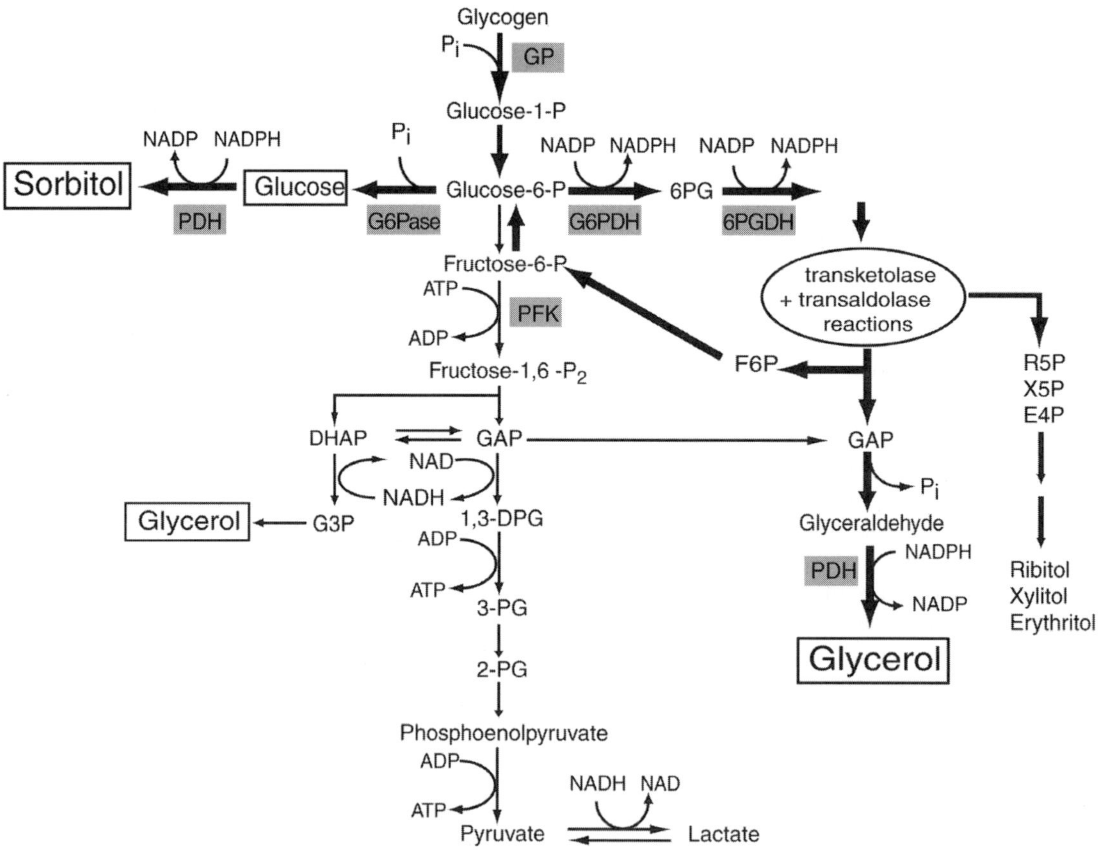

Figure 17.2 Biosynthetic pathways for glycerol, sorbitol, and glucose cryoprotectants. Glycogen is the carbon source and enzymes of glycolysis and the pentose phosphate cycle (PPP) are used. Glycerol in insects is primarily made from the glyceraldehyde-3-phosphate (GAP) output of the PPP using NADPH reducing power that is also generated by the PPP. An alternative output of glycerol from glycerol-3-phosphate (G3P) is shown and is the route used in vertebrate liver. Outputs of four and five carbon sugar phosphates from the PPP can also be converted to their corresponding polyols after phosphate removal and then conversion of sugar to polyol by the enzyme polyol dehydrogenase (PDH). Other enzymes in shaded boxes are: GP, glycogen phosphorylase; G6Pase, glucose-6-phosphatase; G6PDH, glucose-6-phosphate dehydrogenase; 6PGDH, 6-phosphgluconate dehydrogenase; PFK, 6-phosphofructo-6-kinase.

(Fig. 17.2). A few marine fish of the smelt family also produce large amounts of glycerol in the winter to raise their body fluid osmolality to match that of seawater. This is a expensive proposition for these fish because glycerol passes easily across biological membranes, and about 10% of the glycerol pool is lost each day to the surrounding seawater, necessitating continuous repletion. Glucose is used as the protectant by three kinds of freeze-tolerant frogs, but one species (*Hyla versicolor*) accumulates glycerol. Freeze-tolerant marine mollusks and barnacles do not make carbohydrate protectants, but they do maintain constant high levels of free amino acids in their cells to balance their internal osmolality with that of seawater. During freezing, these amino acids provide the same colligative resistance to cell water loss as is provided by carbohydrates for most other species. Sucrose is the most common carbohydrate protectant found in cold-hardy plants but raffinose, sorbitol, and others have been reported.

Most of the protectants mentioned above function as colligative protectants, their actions being primarily a function of their high concentration in body fluids, but some have main actions as membrane protectants. These include the disaccharide, trehalose, and the amino acid, proline. Multiple forms of damage to membranes can occur during freeze/thaw; for example, in plants these include expansion-induced lysis, lamellar to hexagonal-II phase transitions, and freeze fracture jump lesions. Protective metabolites act to stabilize the bilayer structure of membranes. For example, trehalose appears to inhibit the phase transition to the gel phase that is induced by

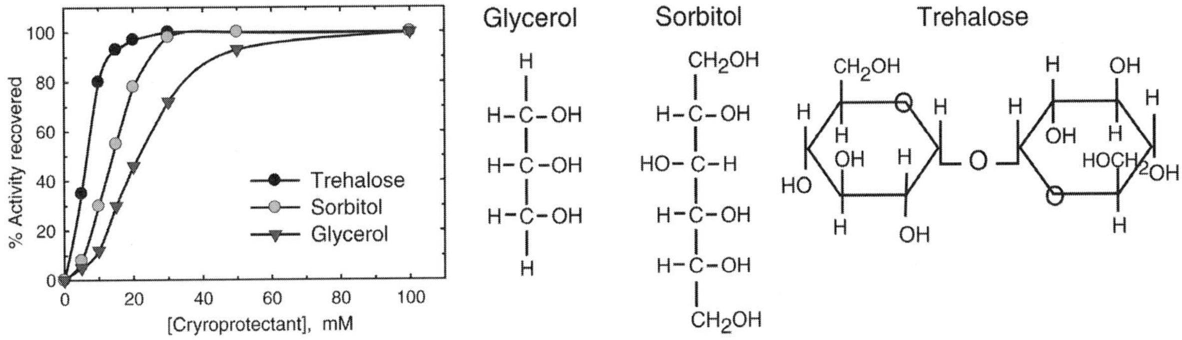

Figure 17.3 Chemical structures of the three most common carbohydrate cryoprotectants and their effects in protecting an enzyme from freeze denaturation. Graph shows the effect of increasing cryoprotectant concentration on the activity of *Eurosta solidaginis* glucose-6-phosphate dehydrogenase recovered after 1 h of freezing at −77°C. Half-maximal protection occurs with 7 mM trehalose, 14 mM sorbitol, or 20 mM glycerol. [Data are from K. B. Storey, D. Keefe, L. Kourtz, and J. M. Storey (1991). *Insect Biochem* **21**:157–164. Used with permission.]

low water stress (freezing or dehydration) (also see Text Box 17.1) apparently by hydrogen bonding with polar head groups on phospholipids to spread the phospholipid monolayers. The importance of these membrane protectants has been emphasized in studies with the freeze-tolerant plant *Arabidopsis thaliana*. Multiple genes associated with proline metabolism are among the suite of cold-responsive genes in this species and transgenic studies traced the constitutive freeze tolerance in a mutant, *eskimo1* (*esk1*), to the accumulation of proline even at warm temperatures; levels of proline were 30-fold higher in *esk1* mutants than in wild-type plants. In recent years a number of small hydrophilic proteins have also been found in plants that are cold/freeze-responsive and appear to act as membrane stabilizers (more about these in the section on gene expression later). Analysis of the damage caused to membranes by freezing (and/or dehydration) and the molecular mechanisms of membrane stabilization by protectants (both trehalose/proline and small proteins) is a huge field of its own that we cannot do justice to in this chapter, but the interested reader is directed to reviews cited in the Suggested Reading list at the end of this chapter.

Why Choose Carbohydrates for Colligative Cryoprotection?

Polyhydric alcohols such as glycerol are a common choice for a protectant role for several reasons: (a) They are highly soluble and therefore can be accumulated in very high concentrations; for example, glycerol levels reach 2 M or more in cold-hardy insects to provide strong colligative suppression of FP and SCP (see Text Box 17.2); (b) they are easily synthesized from glycogen (or from starch in plants) with near-perfect stoichiometry and with little or no adenosine 5′-triphosphate (ATP) input needed; (c) they are synthesized as offshoots from central pathways that are typically present in all cells and organisms; (d) they are easy to remove when no longer needed either by reconversion to glycogen (or starch) or by oxidation as an aerobic fuel; (e) they are good stabilizers of protein conformation in the face of multiple stresses including dehydration, dilution, heat/cold, and freezing; and (f) they are "compatible solutes," which means that their presence in high concentrations has little or no effect on enzyme kinetic parameters. The use of glucose as a cryoprotectant by frogs also fits most of these criteria, particularly its ease of synthesis in three ATP-independent steps from glycogen, although the potential injurious effects of extremely high glucose levels (as are seen in the disease diabetics mellitus) must be addressed.

The stabilizing effects of sugars and polyols in the preservation of enzyme activity are illustrated in Figure 17.3. The figure shows that glycerol, sorbitol, and trehalose protect the enzyme glucose-6-phosphate dehydrogenase (G6PDH) from freeze denaturation. As little as 40 to 50 mM protectant was needed to provide full recovery of enzyme activity after freezing, whereas in the absence of protectant, no enzyme activity could be recovered. Similar results have been documented for many other enzymes and also for isolated cell systems. The actions of polyols as compatible solutes that have little effect on enzyme properties were discussed in Chapter 14. Table 14.6 showed that glycerol and sorbitol themselves have little effect on enzyme properties and also stabilize enzyme kinetic properties against the disruptive influences of low temperature and high salt. For freeze-tolerant animals, this latter effect of minimizing the negative consequences of high ions (e.g., K^+, Na^+, Ca^+) on metabolic reactions is probably very important. By introducing high concentrations of polyhydric alcohols as solutes to minimize the amount of ice that can form at any given subzero temperature, the increase in cytoplasmic ionic strength during freezing is also minimized.

Biosynthesis of Polyhydric Alcohols

Glycerol can be synthesized in two steps from one of the triose phosphate intermediates of glycolysis, dihydroxyacetone phosphate (DHAP). DHAP is converted to glycerol-3-phosphate (G3P) by the enzyme glycerol-3-P dehydrogenase (G3PDH) (Fig. 17.2); this enzyme is found in all cells because of its role in providing G3P as the "backbone" to which fatty acids are esterified in the formation of mono-, di-, and triglycerides. Phosphatase action on G3P produces glycerol. This route appears to be used by vertebrates (e.g., marine smelt, gray tree frogs) that accumulate glycerol as a protectant. In support of this, these species show much higher activities of G3PDH in liver than do comparable species in the same environment that do not synthesize glycerol. For example, we found that G3PDH activity in liver of cold-acclimated gray tree frogs (*H. versicolor*) was four- to fivefold higher than in wood frogs (*Rana sylvatica*), reflecting the accumulation of glycerol versus glucose, respectively, by these freeze-tolerant frogs. G3PDH activity in smelt liver also peaks during the winter months when glycerol biosynthesis is highest. The carbon source for glycerol synthesis is liver glycogen that is accumulated during summer/autumn feeding or, in the case of smelt, dietary carbohydrate and amino acids are also used, the latter being processed by gluconeogenic reactions in liver.

The synthesis of glycerol (or other polyols) in insects seems to be accomplished by a more specialized coupling of two metabolic pathways: glycolysis and the pentose phosphate pathway (PPP). In the insect fat body (its liver-like organ), glycogen breakdown proceeds to the level of glucose-6-phosphate (G6P), and then carbon flow is diverted into the PPP from which both sugar phosphate precursors (hexose, pentose, tetrose, or triose phosphates) and reduced nicotinamide adenine dinucleotide phosphate (NADPH) emerge (Fig. 17.2). After removal of the phosphate group, sugars are converted to sugar alcohols by the enzyme polyol dehydrogenase, whose activity increases strongly in the autumn during the time of active polyol synthesis:

$$\text{Sugar} + \text{NADPH} + \text{H}^+ \rightarrow \text{polyol} + \text{NAD}^+ + \text{H}_2\text{O}$$
$$\text{Glyceraldehyde} + \text{NADPH} + \text{H}^+ \rightarrow \text{glycerol} + \text{NAD}^+ + \text{H}_2\text{O}$$
$$\text{Glucose} + \text{NADPH} + \text{H}^+ \rightarrow \text{sorbitol} + \text{NAD}^+ + \text{H}_2\text{O}$$

Glycerol and sorbitol are the two most common polyol products in insects, but, as Figure 17.2 shows, other sugar phosphates can be drained from the PPP to synthesize alternative polyols. Because 2 mol of NADPH are produced for every G6P moiety that enters the PPP, the synthesis of sorbitol is balanced when about half of the G6P first cycles through the PPP, whereas the other half is converted directly to glucose and then to sorbitol. For the synthesis of glycerol, however, almost all the carbon must cycle through the PPP to achieve a near equal output of 2 glyceraldehyde-3-P (GAP) and 2 NADPH per G6P that enters the PPP. We say "near equal" because each cycle of the PPP leads to the loss of 1 CO_2 at the 6-phosphogluconate dehydrogenase reaction so that the input of 6 G6P results in the net output of 10 GAP and 12 NADPH. Hence, to use up all the NADPH and maintain redox balance, one-sixth of the G6P must be converted to GAP (or G3P) via the nonoxidative reactions of glycolysis that require ATP input at the 6-phosphofructo-1-kinase (PFK-1) reaction (Fig. 17.2). Furthermore, the ATP requirement of this route has to be satisfied by utilizing a small portion of the G6P to generate ATP via oxidative phosphorylation. When all demands are accounted for, the theoretical efficiency of carbon conversion from glycogen to glycerol can be calculated as 84 versus 95% if sorbitol is the product.

Documenting the Role of the PPP

The obligatory coupling of the PPP and glycolysis during polyol synthesis in insects is another excellent example of the reorganization or adaptation of central metabolic pathways to achieve a very specific goal (recall the multiple options for modifying glycolytic outputs during anaerobiosis discussed in Chapter 15). This coupling was confirmed in a novel way in studies that we undertook with freeze-tolerant gall fly larvae (*Eurosta solidaginis*), and these illustrate how the use of a simple stress situation (e.g., anoxia) can be employed to gain information about the integration of metabolic pathways. As discussed in Text Box 17.2, gall fly larvae produce both glycerol and sorbitol as their cryoprotectants. Glycerol synthesis is triggered during early autumn cooling when ambient temperatures drop below $\sim 15^\circ C$, whereas sorbitol synthesis is triggered by colder temperatures below $5^\circ C$ (see Fig. TB17.1). Since glycerol synthesis has an ATP requirement because a portion of the carbon flow must be routed via PFK-1 to reach GAP (or G3P) (Fig. 17.2), we wondered if glycerol synthesis could go forward under anoxic conditions.

It is obvious from looking at Figure 17.2 that if glycerol was produced using *only* glycolysis and the G3PDH reaction (with no PPP involvement), then when insects were placed under anoxia (nitrogen gas atmosphere) glycerol synthesis could only continue if G3PDH used the NADH that was produced by the glyceraldehyde-3-P dehydrogenase (GAPDH) reaction of glycolysis. This would necessitate a 1 : 1 split of triose phosphates into the G3PDH versus GAPDH reactions and, furthermore, the 50% of carbon that continued through glycolysis to reach pyruvate could not be converted to lactate because the NADH needed by the lactate dehydrogenase reaction would have already been used by G3PDH. Pyruvate would have to be converted to a product that does not require reducing equivalents, such as transamination to form alanine. Hence, without PPP participation, glycerol synthesis under anoxia would result in a 1 : 1 accumulation of glycerol and alanine; notably ATP production from glycolysis ending in alanine would provide the necessary ATP for PFK-1. However, when we actually evaluated the metabolic responses of *Eurosta* to anoxic conditions at 13°C, a very different result was seen (Table 17.1). Under anoxic conditions, glycerol accumulation by the larvae at 13°C was still substantial (about half as much as under aerobic conditions), both lactate and alanine accumulated as glycolytic end products, glycerol accumulation was about four times greater than alanine, and a considerable amount of sorbitol was also made. What does this mean? First of all it means that the PPP must be at work because the total reducing equivalents needed to produce the observed glycerol + sorbitol output was 85 μmol NAD(P)H per gram wet mass, a value over 5 times greater than the amount of NADH that would have been available from anaerobic glycolysis alone, based on alanine accumulation. At 3°C the imbalance was even greater, an 8-fold higher output of sorbitol, compared with alanine.

However, when viewed in another way, what is interesting about this partnership between glycolysis and the PPP in cold-hardy insects is that having created a system where G6P must cycle through the PPP to optimize polyol synthesis, insects appear to be "stuck" with this system for all their metabolic needs during the period of autumn cold hardening. Thus, under anoxia, organisms typically suppress biosynthesis and focus their carbohydrate use into glycolytic ATP production (Chapter 15). There is no energetic benefit to be gained from sorbitol or glycerol synthesis in anoxia. So, why does sorbitol biosynthesis go forward at high rates in anoxia? Note that sorbitol is not normally produced at 13°C in *Eurosta* and that under anoxia at 3°C sorbitol was accumulated at 60% higher levels than under aerobic conditions. The answer is that if the route of carbohydrate degradation in autumn larvae is obligately linked to channeling a very high percentage of carbon through the pentose phosphate cycle, then an anoxia-induced activation of carbohydrate catabolism for the purpose of anaerobic ATP synthesis would elicit an enormous overproduction of NADPH. This is because hexose phosphates have to be channeled through the NADPH-generating reactions of the PPP in order to deliver GAP to the ATP-generating reactions of the lower half of glycolysis. For example, look at the situation during anoxia at 3°C. If carbon flowed directly through glycolysis, then an accumulation of 10 mol alanine as the glycolytic end product would have come from 5 mol of G6P (derived from glycogen) and would have produced 10 mol of NADH at the GAPDH reaction and an output of 15 mol of ATP. However, if carbon must go through the PPP instead, we calculated that 10 mol of alanine would have come from the mobilization of 6 mol of G6P (one-sixth of carbon is lost as CO_2) and would be accompanied by the production of 12 mol of NADPH in the PPP plus 10 mol of NADH at the GAPDH reaction and produce 20 mol of ATP (since the PFK-1 reaction is circumvented). Although the ATP yield of this routing is seemingly marginally better than for glycolysis alone, the output of reducing equivalents is more than doubled, and redox balance can only be maintained by the production of large amounts of sorbitol. Hence, this example shows how the "choice" of metabolic pathways needs to consider various pros and cons. Highly efficient polyol synthesis can be accomplished using the PPP to output both NADPH and sugar phosphates. Under anoxic conditions, however, carbon routing through the PPP is highly inefficient because a high percentage of the carbohydrate substrate ends up in a "dead end" product, sorbitol, that cannot be mobilized to contribute to the anaerobic generation of ATP.

TABLE 17.1 Effect of Anoxia Exposure on Balance of Carbon Compounds, Hydroxyl Equivalents, and ATP Produced by Freeze-Tolerant Insect *Eurosta solidaginis* at 13 vs. 3°C [a]

	13°C Exposure		3°C Exposure	
	Aerobic	Anoxic	Aerobic	Anoxic
Net synthesis of carbon compounds				
C6				
Sorbitol	0.2	21.3	49.7	79.3
Glucose	0.2	4.0	2.2	27.0
C3				
Glycerol	112	64	15	0
Glycerol-3-phosphate	14.4	23.7	6.8	0
Lactate	1.4	15.0	0.5	0
Alanine	0.5	15.8	1.9	10.4
Total products in C6 equivalents	64.6	84.5	64.0	125

[a]All data are μmol/g wet weight.
Source: Compiled from K. B. Storey and J. M. Storey (1992). *Adv Low Temp Biol* **1**:101–140.

Enzymatic Regulation in Polyol Biosynthesis

The production of polyols in insects also involves some interesting variations in metabolic regulation. With polyols representing as much as 20 to 25% of the insect's body mass in midwinter, it is obvious that major preparations are needed. These include the accumulation of huge amounts of glycogen in the fat body during summer/autumn feeding and the synthesis of increased amounts of key enzymes. For example, in caterpillars of the glycerol-producing *Epiblema scudderiana* (see Text Box 17.2), we found that the amount of active glycogen phosphorylase (GP*a*) rose from about 0.4 units/g wet mass in early autumn (total GP $a + b = 4$ U/g with 10% *a*) to 7.2 U/g during peak glycerol synthesis in November (total GP = 12 U/g with 60% *a*). At the same time, polyol dehydrogenase activity rose over sevenfold and G3PDH increased threefold.

Once preparatory measures are made, the actual initiation of polyol synthesis relies on low-temperature triggers that are positioned such that polyol accumulation can be complete before the insect experiences the subzero temperatures that require cryoprotection. For species that produce only glycerol, synthesis is initiated by exposure to ~5°C (in nature three to four consecutive nights with lows in this range are needed), and glycerol accumulates at the highest rate at temperatures between 0 and −5°C (insects are not frozen in this range because SCPs of cold-hardy species are typically −8°C or lower). Species that accumulate two protectants, such as glycerol and sorbitol, do so with two independent synthesis events; for example, in *Eurosta* the triggers are 15°C for glycerol and 5°C for sorbitol (Text Box 17.2 and Table 17.1). The primary metabolic focus of these temperature triggers is GP. The percentage of GP in the active *a* form rises from low to high values within 1 to 2 h when insects are transferred from warm to cold temperatures. The basis of this effect has been traced to differential temperature effects on the activities of the two enzymes that regulate GP, phosphorylase kinase and phosphorylase phosphatase. Over the range between 30 and 0°C the kinase undergoes a normal reduction in activity (Q_{10} ~2) and the K_m value for phosphorylase *b* is conserved. The phosphatase, however, is rapidly inactivated at temperatures below ~8°C with a functional Q_{10} of about 23 at low temperatures. Thus, when silkmoth fat body GP was incubated *in vitro* with partially purified preparations of its phosphorylase kinase (+ATP) and phosphorylase phosphatase, the phosphatase was unable to prevent GP*a* content from soaring to maximum values when incubations were at 0°C, whereas at 25°C competition between the two enzymes maintained GP*a* at about 25% of its maximum value. Although other factors may also be involved *in vivo*, this elegant experiment showed that simple temperature effects on a single enzyme can have a huge effect on a major metabolic function and provides a very simple way of modulating the rate of polyol synthesis in direct response to environmental temperature change.

Cold activation of GP*a* is further potentiated by the effects of low temperature on the kinetic properties of the enzyme and by oppositely directed controls of glycogen synthase (GS), which ensure that no recycling of polyols to glycogen places in the cold. As Table 17.2 shows the affinity of *Epiblema* GP*a* for both glycogen and inorganic phosphate (P_i) increased strongly at low temperature (K_m values decreased to just 20% of the values at 22°C), as did sensitivity to adenosine 5′-monophosphate (AMP) as an activator (K_a dropped just 30% of the value at 22°C). Notably, the addition of 500 mM glycerol, at about 25% of the maximum level accumulated in *Epiblema*, further potentiated GP*a* affinity for its substrates. However, low temperature had just the opposite effect on GS. Like GP, GS is interconvertible between an active, phosphorylated, G6P-independent (I) form and an inactive, dephosphorylated form (D) that is dependent on G6P to show activity *in vitro*. Only the inactive D form could be found in autumn-collected *Epiblema*, which fits with the idea that the larvae are poised for unidirectional glycogenolysis during the season of cold-hardening. Low-temperature effects on GS kinetic properties further show an enzyme that is much less active in the cold. At 5°C our studies revealed that the enzyme showed much lower affinities for both substrates; K_m values for glycogen and uridine diphosphoglucose were 4.4- and 1.8-fold higher at 4°C compared with 22°C. The enzyme was also much less sensitive to activation by G6P at 4°C.

Regulation of GP activity controls the input of carbon into polyol synthesis but cannot determine which polyol product is made. How is that decision made? This question is particularly interesting for a species such as *Eurosta* that produces glycerol as the result of glycogenolysis at cool temperatures (below ~15°C) but accumulates sorbitol when a second round of glycogenolysis is activated by

TABLE 17.2 Effect of Assay Temperature on Kinetic Properties of Glycogen Phosphorylase and Glycogen Synthase from *Epiblema scudderiana*[a]

	22°C	5°C
Phosphorylase a		
K_m glycogen, μg/ml		
No additions	120 ± 4	24 ± 1[b]
+ 0.5 M glycerol	73 ± 2[b]	15 ± 1[b]
K_m inorganic phosphate, mM	6.5 ± 0.1	3.7 ± 0.1[b]
K_a AMP, nM	176 ± 4	53 ± 1[b]
Synthase D		
K_m glycogen, μg/mL	1170 ± 350	5160 ± 160[b]
K_m UDPG, μM	48 ± 4	88 ± 3[b]
K_a G6P, mM	2.1 ± 0.2	6.2 ± 0.2[b]

[a]Note that addition of glycerol further increases the affinity of phosphorylase for glycogen. UDPG is uridine diphosphoglucose.

[b]Significantly different from the corresponding value at 22°C with no glycerol added.

Source: Compiled from C. P. Holden and K. B. Storey (1993). *J Comp Physiol B* **163**:499–507, and A. M. Muise and K. B. Storey (1999). *Cryo-Lett* **20**:223–228.

5°C exposure. Sorbitol synthesis is a less efficient use of the carbon pool for the production of colligatively active molecules (1 G6P makes 1 sorbitol versus 2 glycerol) so we reasoned that there must be a metabolic block on the production of triose phosphates by the larvae at low ambient temperatures. To determine where this block might be, we used a technique called crossover analysis (the theory is explained in Chapter 1) that looks at the changes in levels of metabolites in a pathway as the result of a metabolic stress and evaluates the relative change, compared with control values, of the substrate versus product concentrations of the each enzyme. A crossover occurs when a stress has a differential effect on substrate versus product concentrations of an enzyme; that is, the ratio of experimental/control concentrations for substrate rises to >1.0, whereas the ratio for product concentrations decreases to <1.0, or vice versa. Relative changes in substrate versus product levels "cross over" the line where experimental/control concentrations = 1. This analysis was applied to assessing the changes in glycolytic intermediates in *Eurosta* when the larvae were acutely transferred from 13 to 3°C, a temperature shift that rapidly initiates sorbitol synthesis (Table 17.1). Within 1 h of the shift to the cold temperature, an activation of glycogenolysis was obvious from the significant rise in G6P levels, and by 2 h G6P had tripled, glucose had risen by 15-fold, and sorbitol accumulation had started (Table 17.3; see also Fig. 17.2 for the reactions involved). When G6P rose, so did fructose-6-phosphate (F6P), the substrate of PFK-1; this is because phosphoglucoisomerase is an equilibrium enzyme. The ratio experimental/control [F6P] rose to 38 within the first 2 h of cold exposure, but, despite this massive proportional increase in the substrate of PFK-1, the experimental/control concentrations of the product of PFK-1, fructose-1,6-bisphosphate ($F1,6P_2$), behaved oppositely and fell to 0.33 within 2 h and to 0.50 by 4 h of cold exposure. This indicates a very strong inhibition of PFK-1 because, despite the huge increase in F6P substrate concentrations, $F1,6P_2$ product was depleted because it was consumed faster by the rest of glycolysis than it was produced by PFK-1. By contrast, when the same analysis was applied to pyruvate kinase, a regulatory enzyme in the lower part of glycolysis, no such changes in its substrate and product concentrations were seen; levels of both the substrate (phosphoenolpyruvate) and product (pyruvate) remained constant (Table 17.3). The crossover theorem can be applied to many different pathways and effectively pinpoints rate-limiting steps in pathway response to different stresses. In this case, it indicated strong negative control of glycolytic flux at the PFK-1 reaction as a result of the decrease in temperature. Such a block on glycolysis would promote the diversion of carbon into the reactions of sorbitol synthesis. The mechanism of PFK-1 inhibition by cold was subsequently explored, and we found that it was due to multiple factors: (1) a large negative effect of low temperature on enzyme activity ($Q_{10} = 3.6$), (2) a decrease in enzyme affinity for F6P (K_m increased) and a decrease in enzyme sensitivity to activators, AMP, and fructose-2,6-bisphosphate ($F2,6P_2$) when assayed at low temperature, and (3) elevated levels of PFK-1 inhibitors (G3P, sorbitol) and decreased levels of PFK-1 activators ($F2,6P_2$) *in vivo* at low temperature.

TABLE 17.3 Changes in Concentrations of Intermediates of Sorbitol Synthesis When Production is Stimulated by Acute Temperature Decrease from 13 to 3°C[a]

	Control (time zero)	2 hours cold	Ratio Cold : Control
Glycogen	304	299	
Glucose	0.1	1.5	15
Sorbitol	2	4	2
G6P	0.15	0.45	3
F6P	0.002	0.076	38
$F1,6P_2$	0.105	0.07	0.67
PEP	0.07	0.08	1.1
Pyruvate	0.055	0.055	1.0

[a]All data are μmol/g wet weight.
Source: Data are excerpted from J. M. Storey and K. B. Storey (1983). *J Comp Physiol* **149**:495–502.

Biosynthesis of Glucose as the Cryoprotectant in Frogs

The glucose used for cryoprotection by woodland frogs is produced in liver using the same three-step pathway present in the liver of all vertebrates (Fig. 17.2):

$$\text{Glycogen} \rightarrow \text{glucose-1-phosphate} \rightarrow \text{glucose-6-phosphate} \rightarrow \text{glucose}$$

The enzymes involved are GP, phosphoglucomutase, and glucose-6-phosphatase. Preparatory measures for winter cryoprotectant synthesis include the accumulation of large amounts of glycogen in liver during summer/autumn feeding, so that about 20% of liver mass is glycogen when hibernation begins, and the presence of high activities of GP (e.g., activity in wood frog liver is 12-fold higher than in liver of freeze-intolerant leopard frogs). Unlike the situation in insects, however, frogs do not show anticipatory accumulation of their cryoprotectant during an autumn cold-hardening period but only trigger glucose synthesis when the frog's body actually begins to freeze. The reason for this may be to minimize the time that tissues are exposed to extreme hyperglycemia because, unlike glycerol or other polyols, glucose is not a chemically or metabolically inert carbohydrate.

Indeed, diabetes research has shown that many of the pathophysiological features of the disease are directly related to chronic hyperglycemia. For example, the nonenzymatic attachment of glucose to long-lived proteins underlies diabetic microvascular disease (due to glycation of collagen and other basement membrane proteins) and diabetic cataract (glycation of lens crystallins). Glucose and glycated proteins are also effective pro-oxidants. Free glucose is prone to auto-oxidation in the presence of transition metals (iron, copper) to form protein-reactive dicarbonyl compounds and hydrogen peroxide (see Chapter 12). The potential for hyperglycemic damage to frog tissues can be appreciated from the glucose concentrations involved. Glucose is normally regulated at 4 to 5 mM in vertebrate blood but may rise to 7 to 8 mM after a meal with insulin secretion typically acting to prevent higher values (see Chapter 9). Multiple metabolic injuries arise in uncontrolled diabetes when glucose ranges between 10 to 50 mM. However, freeze-tolerant frogs elevate glucose to 150 to 300 mM, and, when as much as two-thirds of total body water freezes out as extracellular ice, the resulting concentration of glucose in the cytoplasm can be nearly 1 M!

The production of glucose for cryoprotection is triggered within seconds when the frog's body begins to freeze. The initial freezing event for frogs is typically ice nucleation on the skin surface, either seeded by contact with environmental ice crystals (at any temperature below of -0.5°C FP of frog body fluids) or as a result of the ice nucleating activity of skin bacteria that stimulate freezing at -2.5 to -3°C (see section below on ice nucleators). Within 2 to 4 min after nucleation, liver glucose levels have doubled, and significant export into the blood can be detected within 4 to 5 min. Glucose output continues at high rates so that levels in blood and other organs rise from <5 to over 100 mM within just a few hours (Fig. 17.4), quite a feat when you remember that the frog's body temperature is below 0°C!

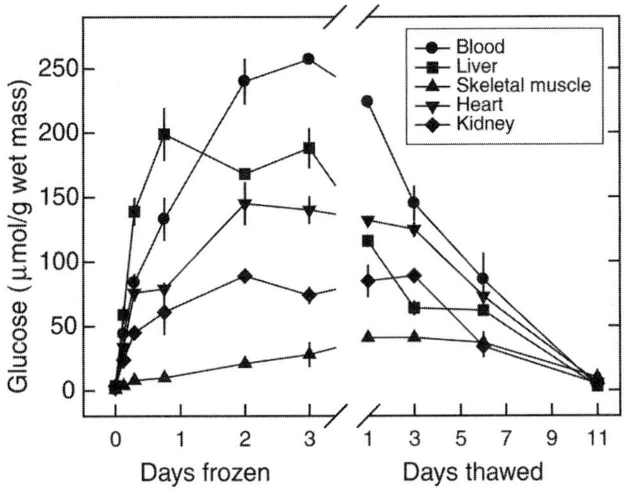

Figure 17.4 Freezing triggers high rates of glycogenolysis in frog liver, and glucose builds up quickly and is exported via the blood and taken up by all other organs. Shown is the time course of changes in glucose levels during freezing at -2.5°C and thawing at 5°C; data are means + SEM, $n = 3$.

Signal Transduction and Enzymatic Regulation

The regulation of freeze-induced glucose production has been extensively studied in wood frogs, *R. sylvatica*. Much is known but some intriguing questions remain. One is the nature of the signal that links ice formation at a peripheral site on the skin with the immediate activation of glycogenolysis in the liver. The signal may be nervous or hormonal, but its action is a stimulation of β_2-adrenergic receptors on the plasma membrane of liver cells. β-Adrenergic reception was identified when we showed that injections of β-adrenergic blockers (propranolol) impaired the hyperglycemic response to freezing, whereas α-adrenergic blockers (phenolamine) did not. β_2-Receptor activation triggers the production of cyclic 3′,5′-adenosine monophosphate (cAMP) in liver, whose levels rise twofold within 2 min postnucleation, and this, in turn, activates cAMP-dependent protein kinase (PKA). Binding of cAMP to regulatory subunits on PKA triggers the dissociation of the inactive PKA tetramer to release the two catalytic subunits and the percentage of PKA present as the active subunit in frog liver rises from ~7% in controls to over 60% within 5 min after freezing begins. PKA phosphorylates glycogen phosphorylase kinase, which in turn phosphorylates the inactive *b* form of GP to convert it to the active *a* form. The percentage of GP in the *a* form rises quickly to nearly 100% and the production of glucose cryoprotectant is underway. This cascade of events is well-known as the standard way of activating glycogenolysis in vertebrate liver in response to glucagon, epinephrine, or other hyperglycemic stimuli (see Chapters 9 and 10).

However, as we learned in Chapters 9 and 10, blood glucose levels are normally under tight homeostatic controls in vertebrates due mainly to the actions of insulin versus glucagon and their oppositely directed effects on the activities of GP versus GS. Indeed, this system of homeostatic regulation of glucose is fully functional in unfrozen frogs, but, obviously, one or more novel regulatory mechanisms must come into play to allow the massive freeze-induced hyperglycemia. The full picture is not yet known but significant elements are.

Once again, similar to the situation in cold-hardy insects, one key factor in GP control is the behavior of protein phosphatase-1 (PP-1) that acts to dephosphorylate active GP*a* and return it to its inactive *b*. In normal vertebrate liver, PP-1 activity rises as glucose levels increase and the enzyme dephosphorylates GP*a* to halt further glycogenolysis. PP-1 activity also rises in wood frog liver during cryoprotectant production (a 70% increase over the first 20 min) but GP*a* activity remains high and unaffected. Why is this? The answer seems to be that during freezing PP-1 is kept out of physical contact with liver glycogen granules. In normal vertebrate liver, a high proportion of PP-1 catalytic subunits occur as dimers, bound 1 : 1 with a G protein that confers glycogen-binding ability (see Fig. 4.14). The α and δ isoforms of PP-1 are the main ones that associate with glycogen. Subcellular fractionation and immunoblotting were used to follow PP-1 isoform distribution in wood frog liver. The results showed that in control frogs PP-1 α and δ were overwhelmingly associated with glycogen particles with little enzyme free in the cytosol (Fig. 17.5). However, when frogs were frozen for 12 h at −2.5°C, the opposite response was seen. PP-1 was translocated to the cytosolic fraction with virtually no enzyme left bound to glycogen particles. Hence, PP-1 cannot halt glycogenolysis in liver during freezing and glucose output continues unabated. The mechanism of PP-1 translocation may be linked with PKA-mediated phosphorylation of the G subunit, which in mammals increases the rate of PP-1 inactivation by inhibitor-1 and promotes PP-1 release from glycogen. But, whereas PKA activation is transient under normal circumstances in vertebrate liver, the percentage of PKA present as the active catalytic subunit was

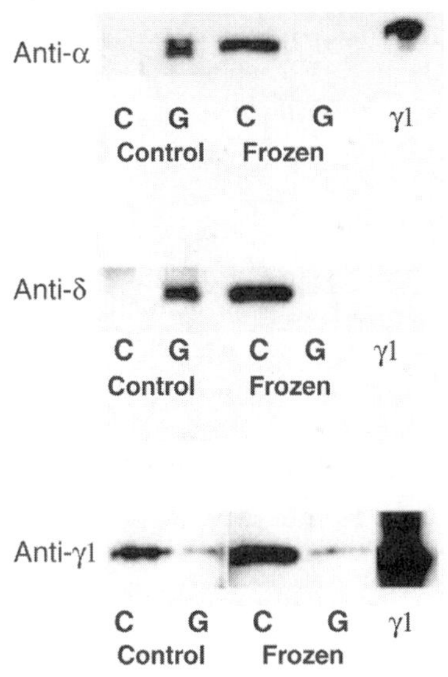

Figure 17.5 The distribution of protein phosphatase-1 (PP-1) isozymes between the cytosolic (C) and glycogen particle (G) fractions of wood frog liver was investigated by using Western blotting with antibodies to rat liver α, δ, and $\gamma 1$ PP-1. The blots compare isozyme distribution between control and 12-h frozen wood frogs, clearly showing a shift by the α, δ isozymes from the glycogen fraction in control frogs to the cytosolic fraction when frogs were frozen. Recombinant PP-1-$\gamma 1$ is present in the fifth lane as a positive control. [From J. A. MacDonald and K. B. Storey (1999). *Cryo-Lett* **20**:297–306. Used with permission.]

sustained at high levels in frog liver for at least the first 8 h during freezing.

A rise in blood glucose levels beyond ~7 mM in vertebrate blood is typically aggressively reversed by the secretion of insulin to promote uptake and storage of the sugar by insulin-sensitive organs. How then does blood glucose continue to rise to over 200 mM in frogs during freezing? Freezing must interfere with the normal homeostatic control of glucose perhaps by interfering with insulin action. Several possibilities exist but the answer is still not known. Insulin secretion by the pancreas could be blocked, although, in fact, we found that plasma insulin levels do double during freezing. Alternatively, insulin interaction with its receptors on hepatocyte membranes could be blocked. However, one of the main intracellular responses that is activated by insulin binding to its receptor is the activation of Akt (protein kinase B) (see Chapters 5 and 9), and Akt activity rises by threefold in the liver of freezing frogs. This suggests that the initial steps of insulin signal transduction may continue during freezing. However, subsequent actions of Akt are clearly disrupted during cryoprotectant synthesis. A main action of Akt signaling in liver cells is to activate GS and thereby stimulate glycogen synthesis when glucose is high. This clearly does not occur during freezing; we have shown that GS activity remains very low throughout freezing (exactly the opposite of GP) and only rises when frogs thaw. Akt phosphorylates glycogen synthase kinase 3 (GSK3) to inactivate it and prevent it from phosphorylating and inactivating glycogen synthase. Hence, the disruption of insulin action may occur in the link between Akt, GSK3, and GS.

A structural modification of wood frog insulin could also reduce its effectiveness in regulating glucose levels in freezing frogs and, indeed, sequence analysis of the wood frog hormone showed some anomolous features that could limit its effectiveness (Fig. 17.6). One unusual feature was the presence of a two-amino-acid extension (lysine–proline) on the N terminus of the A chain. Although shared by other ranid frogs, this extension does not occur in any other vertebrates and its role remains unknown. The rest of the A chain, as well as the B chain, are highly conserved in other frogs as compared with the mammalian hormone, but wood frog insulin showed two anomolies. The serine residue at position A23 in wood frog insulin (A21 of human) is an asparagine in all other species, and the aspartic acid at B13 in wood frog insulin is glutamic acid in nearly all tetrapods. Both of these residues have been shown to play important roles in insulin action. A21 is key to the maintenance of the biologically active conformation due to its bonding to B22/23, whereas B13 aids binding to the insulin receptor. Both of the amino acids substitutions in wood frog insulin, as minor as they may seem, could significantly impair its function. Indeed, the only other known instance of a Glu to Asp substitution at B13 results in a

	Insulin A-chain
Rana sylvatica	KP GIVEQ CCHNM CSLYD LENYC S...
Rana catesbeiana	-- ----- ----T ----- ----- N...
Rana ridibunda	-- ----- ----T ----- ----- N...
Xenopus laevis	----- ---ST --F-- --S-- N...
Homo sapiens	----- --TSI ----Q ----- N...

	Insulin B-Chain
Rana sylvatica	FPNQH LCGSH LVDAL YMVCG DRGFF YSPRS...
Rana catesbeiana	----Y ----- --E-- ----- ----- -----...
Rana ridibunda	----Y ----- --E-- ----- E---- -----...
Xenopus laevis	LV--- ----- --E-- -L--- ----- -Y-KV...
Homo sapiens	-V--- ----- --E-- -L--- E---- -T-KT...

Figure 17.6 N-terminal sequences of insulin A and B chains in wood frog, bullfrog, green frog, African clawed frog and humans. [From J. M. Conlon, K. Yano, N. Chartrel, H. Vaudry, and K. B. Storey (1998). *J Mol Endocrinol* **21**:153–159.]

low potency insulin. Interestingly, although wood frog insulin displays novel features, its glucagon is identical with the hormone from the bullfrog and has only one amino acid substitution as compared with human glucagon.

Control over GP regulates the production of glucose-1-P (which is interconverted with G6P) but, as mentioned previously, cannot determine the ultimate product of glycogenolysis. Multiple metabolic fates are open to G6P including (a) hydrolysis to glucose by glucose-6-phosphatase (G6Pase), (b) entry into the PPP via the G6PDH reaction, and (c) catabolism via glycolysis (Fig. 17.2). In the liver of cold-acclimated wood frogs, competition between G6Pase and G6PDH seems to be ruled out by G6Pase activities that are >200-fold higher than G6PDH. In addition, freezing stimulates a further 70% increase in G6Pase activity while reducing G6PDH activity to 20% of its former value. G6P use by glycolysis is also effectively shut down during freezing. Once again we demonstrated that inhibition is centered on PFK-1. The maximal activity of PFK-1 in wood frog liver is similar to that of G6Pase, and PFK-1 activity does not change during freezing, so a different approach to PFK-1 control is needed. When levels of glycolytic metabolites were measured over the initial minutes of freezing, all hexose phosphates (G1P, G6P, F6P) were strongly elevated within 5 min postnucleation, whereas $F1,6P_2$ levels were unaffected; when crossover analysis was applied, strong negative control of glycolytic flux was again indicated at the PFK locus. The mechanism of PFK-1 inhibition was traced to significant changes in the kinetic and regulatory properties of the frog liver enzyme occurring over the first 70 min postnucleation, changes that are characteristic of protein phosphorylation of the vertebrate liver enzyme. Furthermore, levels of the PFK-1 activator, $F2,6P_2$, dropped sharply during freezing as a result of freeze-induced changes in the activity and properties of 6-phosphofructose-2-kinase (PFK-2), which produces $F2,6P_2$ (see Chapter 14 for a discussion of PFK-1 and PFK-2 relationships). Kinetic changes to PFK-2 (e.g., at 10-fold increase in K_m for F6P) are also consistent with freeze-stimulated phosphorylation of this enzyme to produce a less active form. Indeed, the activation of GP and the inhibition of PFK-2 are linked because they are both regulated by PKA; the same activation of PKA that stimulates glycogen phosphorylase kinase inhibits PFK-2, causing the fall in $F2,6P_2$ levels that inhibits PFK-1 during freezing.

Glucose Distribution

Once synthesized, glucose needs to be quickly exported to other organs of the frog to provide them with cryoprotection. This must be accomplished within just a few hours because, as ice grows inward from the peripheral regions of the frog's body and into the core, blood flow is progressively cut off and, with it, the ability to distribute glucose. Hence, an organ-specific pattern of glucose distribution results with lowest levels in skeletal muscle and skin, intermediate levels in organs such as brain and kidney, and highest amounts in liver and heart (Fig. 17.4). Significantly, the very high levels of glucose in liver and heart prove to be a bonus to the frog during thawing because these organs have the highest concentrations of colligative solutes and, therefore, the lowest melting points. Hence, the heart and liver melt first, and this allows for the resumption of heart beat as the first vital sign detectable in the thawed animal and means that circulation of oxygenated and nutrient-laden blood can resume to each other organ as soon as it thaws.

The uptake of glucose into vertebrate cells depends on the carrier-mediated transport of glucose across the plasma membrane by glucose transporters (GLUT), and this transport system is also the focus of adaptive change in freeze-tolerant frogs. Our studies suggest that the mechanism is to be purely quantitative with no evidence of altered kinetic properties, at least for the liver isoform of the transporter (GLUT2). Thus, we found that the number of glucose transporters in liver plasma membrane of freeze-tolerant wood frogs is 5 times greater than in freeze-intolerant leopard frogs (*Rana pipiens*) and, between summer and autumn, the number of GLUT2 transporters in wood frog liver plasma membranes increases by 8.5-fold. Glucose transporter numbers in the receiving organs are also high in freeze-tolerant frogs; for example, the maximal rate of glucose transport by skeletal muscle of wood frogs was 8-fold higher than in leopard frogs. New data from complementary deoxyribonucleic acid (cDNA) array screening of wood frog heart has also identified the GLUT4 transporter, the insulin-responsive form, as one of several prominently up-regulated genes in 24-h frozen frogs; transcript levels are 2- to 3-fold higher in heart of frozen versus control frogs. This indicates that an acute (as well as seasonal) up-regulation of glucose transport capacity helps frog organs to rapidly accumulate cryoprotectant.

Once glucose is transported into the cells of freezing frogs, the next question is: Why is it not catabolized? The normal response of most organs after glucose uptake is immediate phosphorylation via the ATP-dependent hexokinase reaction followed by the use of G6P by one of several pathways. Indeed, glucose would be a valuable substrate for anaerobic glycolysis under the ischemic state imposed by freezing. Although the mechanism is not yet known, it is clear that glucose catabolism by wood frog organs is specifically inhibited at low temperatures. This was shown in our studies with wood frog erythrocytes. Red blood cells incubated with [U-^{14}C]-glucose took up glucose just as effectively at 4°C as they did at 12, 17, or 23°C. However, when ^{13}C nuclear magnetic resonance (NMR) was used to track the change in intracellular glucose levels over time, red blood cells at 17 or 12°C showed constant temperature-dependent rates of glucose depletion, whereas cells incubated at 4°C showed no glucose catabolism. The mechanism of this inhibition may be control over the enzyme, hexokinase. Although total hexokinase activity is not altered in most organs during freezing (although a 40% decrease occurred in kidney), the physical association of hexokinase with subcellular binding sites is key to its function (see Chapter 14), and these physical interactions may be disrupted in the frozen state to prevent glucose catabolism.

Development of the Cryoprotectant Response in Freeze-Tolerant Frogs

How did the use of glucose as a cryoprotectant get started in frogs? As we have seen in other instances in these chapters on biochemical adaptation, novel adaptive strategies typically develop from some underlying metabolic capacity. In this case, the origin of glucose use as a cryoprotectant probably grew out of an underlying hyperglycemic response to dehydration that is shared by both freeze-tolerant and freeze-intolerant frogs. Amphibians as a group have highly water-permeable skins. This tends to restrict them to living in moist habitats and has required them to develop the greatest tolerance of all vertebrates for variations in body water content and body fluid ionic strength. For example, mammals manifest nervous system dysfunction when plasma sodium rises by 30 to 60 mM, whereas amphibians can tolerate sodium at 90 to 200 mM above normal. Many amphibians can readily endure the loss of 50 to 60% of their total body water, and desert, arboreal, and freeze-tolerant frogs and toads are all at the top of the list. From the point of view of a cell, intracellular dehydration due to evaporative water loss is really no different from intracellular dehydration due to water loss into extracellular ice masses.

To experimentally test the idea that freeze tolerance and desiccation tolerance were related, we analyzed the metabolic responses of wood frogs to dehydration. Frogs were placed in dry containers where they lost body water at a rate of ~1% of total per hour to a final total of 50% lost (both the rate and the final maximum are fully survivable by wood frogs). Dehydration stimulated hyperglycemia in wood frogs just as effectively as freezing did, and glucose levels rose to values just as high as in frozen frogs (e.g., a 300-fold increase in liver) and with the same β_2-adrenergic signal transduction pathway activated in liver. Amazingly, however, the hyperglycemic response to dehydration also occurred in a freeze-intolerant frog, the leopard frog *R. pipiens*, although the response was more muted (a 24-fold increase in liver glucose). These results strongly indicated that the cryoprotectant biosynthesis response to freezing grew out of a preexisting hyperglycemic response by frogs to water stress. Notably, however, wood frogs given anoxia exposure at 5°C (to mimic the oxygen deprivation that is another component of freezing) did not show elevated glucose. As will be discussed in the later section on gene expression, many of the genes that are up-regulated in response to freezing also respond to either dehydration or anoxia stresses, and this information has provided valuable clues, particularly for novel genes of unknown function, as to the possible functional roles of gene products and to the signal transduction pathways that may be involved in regulating expression.

ICE MANAGEMENT BY ANTIFREEZE PROTEINS AND ICE NUCLEATING PROTEINS

Antifreeze proteins (AFPs) and ice nucleating proteins (INPs) are superb examples of novel proteins that have been developed in selected organisms for very specific purposes. AFPs are widely distributed in coldwater marine fishes, terrestrial arthropods, and many plants where they act to inhibit the growth of microscopic ice crystals and thereby prevent the freezing of body fluids among organisms using the freeze-avoiding strategy of cold hardiness. INPs have the opposite action—they act to order water molecules into the ice lattice structure and thereby trigger crystallization in freeze-tolerant species. Interestingly, some freeze-tolerant insects have both INPs and AFPs, and the postulated role of an AFP in a freeze-tolerant organism is to inhibit recrystallization, the process whereby small ice crystals regroup over time into larger and larger crystals that could be physically damaging to tissues (Fig. 17.1). Hence, AFPs are employed to provide long-term stability of crystal size and shape in organisms that could be frozen for weeks or months. Many reviews cover the protein chemistry of AFPs and INPs and their mechanisms of physical interaction with ice (see Suggested Reading at the end of this chapter), and in Chapter 14 fish AFPs were presented as an excellent example of gene/protein evolution. Our remaining discussion here will focus mainly on physiological function and the metabolic regulation of AFP and INP synthesis.

Antifreeze Proteins

Antifreeze proteins come in many types. Five different groups have been described in coldwater fish and many other kinds occur in insects and plants. Although each type of AFP is structurally distinct, the mechanism by which they inhibit the growth of ice is much the same in all cases. AFPs adsorb onto the surface of seed crystals via hydrogen bonding to the surface water molecules and thereby prevent crystals from growing beyond a microscopic size. Bonds form with the polar side chains of amino acids on most AFP types or with the hydrophilic groups on the sugar side chains of antifreeze glycoproteins. This forces crystal growth into highly curved (high surface free energy) fronts rather than the preferred low curvature fronts (low surface free energy) and growth stops (although growth will resume with the impetus of a lower temperature). The presence of AFPs causes a reduction in the apparent FP of body fluids without changing the melting point; this thermal hysteresis is exploited for AFP detection and assay. The thermal hysteresis provided by AFPs to the blood of marine fish is only about 0.7 to 1.5°C, but this is enough to lower plasma FP (about −0.5°C for most teleost fish but higher in glycerol-accumulating smelts) below the FP of seawater (−1.9°C). Insect and plant AFPs are much more powerful because they must act in much colder terrestrial environments. Thermal hysteresis values for the AFP-containing hemolymph (the blood equivalent) of terrestrial arthropods in midwinter are generally 3 to 6°C, and together with high levels of carbohydrate protectants and a nucleator-free environment this allows many insects to chill to −20 to −40°C without freezing.

Whereas polar fish retain AFPs year-round, temperate-zone fish show a seasonal cycle of blood AFP levels that peaks in midwinter. Multiple influences on AFP production have been demonstrated using species that inhabit the waters off the Northeast coast of North America. These include (a) geographic variation—AFP gene dosage and AFP levels in plasma are higher in more northerly populations of individual species (e.g., the technique of Southern blotting showed that ocean pout from Newfoundland have 150 AFP gene copies, whereas New Brunswick pout have only 30 to 40 copies; correspondingly, plasma AFP levels that are 5- to 10-fold higher in the more northerly populations), and (b) environmental niche—inshore species that could frequently encounter surface ice show an anticipatory increase that elevates blood AFPs before water temperature drops to subzero values, whereas deep-water fish synthesize AFPs only if exposed to subzero temperatures. Deep-water species also show lower gene dosage and lower maximal AFP levels than do closely related inshore species.

Recent studies with the AFPs of winter flounder have contributed information about the metabolic regulation of AFPs. Liver is the production site for the plasma AFPs. The type I AFP of the flounder is encoded as a preproprotein; the presequence is removed cotranslationally and the pro-AFP is secreted into the blood where the prosequence is removed within 24h to yield a 37-amino-acid mature AFP. AFP messenger ribonucleic acid (mRNA) appears in the liver in October, about one month prior to AFP appearance in the blood, which occurs when water temperature drops below 8°C. Low temperature appears to be needed to allow AFP mRNA to accumulate (the mRNA is destabilized at warmer temperatures) but does not regulate AFP gene expression. Photoperiod is the main trigger of gene expression; long day lengths (>14h) strongly delay AFP production, whereas short days (4 to 8h) are without effect. Insect AFPs also respond to photoperiod as the primary trigger for AFP synthesis (critical daylength is between 10 and 11h), but other cues include temperature, thermoperiod (lengths of warm versus cold phases each day), and low relative humidity. In insects, information from environmental cues is integrated in the central nervous system and results in the release of juvenile hormone that stimulates the production of AFPs in the insect fat body.

In fish, photoperiod effects also act via the central nervous system to control the release of growth hormone (GH) from the pituitary. High GH levels repress AFP gene transcription, apparently via the action of insulin-like growth factor (IGF-1) (see Chapter 10) as the intermediary. The annual life cycle of winter flounder includes the cessation of feeding and growth from about October through

March, which leads to reduced GH and IGF-1 levels over the winter months. This releases the repression of AFP genes and allows the synthesis and accumulation of AFPs. At the molecular level, IGF-1 can inhibit liver AFP gene enhancer activity. Analysis of the liver AFP gene promoter showed DNA-binding motifs for the C/EBPα transcription factor and a novel AP-1-binding complex that has been termed the antifreeze enhancer-binding protein (AEP). In other systems, IGF-1 causes a dephosphorylation and deactivation of the cCAAT enhancer binding protein (c/EBPα), and it is proposed that this effect and/or an effect on AEP expression causes transcriptional inhibition of the liver AFP genes. With the loss of long day length in the fall, GH production is inhibited, IGF-1 levels fall, and C/EBPα and AEP are activated to enable AFP gene expression.

Whereas liver produces the secreted form of fish AFP, other studies have recently shown that AFP production is not limited to the liver in winter flounder. When a liver cDNA probe was used to screen a skin cDNA library, other AFP clones were found. Distinct but homologous alanine-rich AFPs are produced in skin, their main differences from liver AFP being a lack of the signal and prosequences that allow export of liver AFPs. Hence, skin AFPs remain *in situ* to protect the integument from intracellular freezing. Northern blot analysis showed that the skin type AFP was also strongly expressed in other exterior tissues such as scales, fin, and gills. Hence, two multigene AFP families occur in winter flounder, both with the same functional amino acid repeats. One produces the secretory liver type of AFP and the other nonsecretory AFP that provides antifreeze protection to surface tissues of the fish.

Ice Nucleating Agents

Ice nucleating agents are found in most freeze-tolerant organisms where they are employed to trigger ice formation in the extracellular body fluids. By their action, nucleation is triggered at mild subzero temperatures (usually well above $-10°C$), extensive supercooling is avoided, and slow ice growth through extracellular spaces is managed so as to minimize injury to the organism. Ice nucleating agents are sometimes novel INPs that apparently evolved specifically for this purpose, but other agents can have ice nucleating activity including heterogeneous material in the gut or in the external environment around the organism, bacteria, and in the case of insects, various mineral crystals (e.g., calcium phosphate) that are stored in fat body or Malpighian tubules. The presence of such material is the reason that freeze-avoiding species typically clear their gut and/or spin a protective cocoon to minimize the risk of nonspecific nucleation over the winter months, but freeze-tolerant species can exploit these actions.

Wood frogs, *R. sylvatica*, appear to have three modes of ice nucleation. When cooled on a wet substrate (typical of their forest floor hibernaculum), freezing can begin at any temperature below the $-0.5°C$ FP of frog blood due to seeding across the skin by contact with environmental ice crystals. If cooled on a dry substrate, however, wood frogs chill to a characteristic SCP of -2.5 to $-3°C$ before freezing is stimulated by the nucleating activity of *Pseudomonas* and *Enterobacter* bacteria on the surface of the frog's skin or in the gut. The third option is the presence of a plasma INP. We have found that the INP appears seasonally, triggers freezing *in vitro* at -6 to $-8°C$ (as assessed by differential scanning calorimetry), and is susceptible to denaturation by heat and various chemical treatments that confirm that it is a protein. However, it is difficult to envision the role of the plasma INP if it acts only at these low temperatures *in vivo* for such temperatures are near the survival limits of wood frogs, but it is possible that within the body the protein is effective at higher temperatures. Furthermore, this plasma INP may not actually trigger ice formation but, instead, may aid the propagation of ice through the frog's body, perhaps by modulating or regulating crystal growth through the vasculature.

Many bacteria show powerful ice nucleating activity that is a major cause of frost damage to crop plants. From the bacterial point of view, their INP activity is beneficial and has functions that could be nutritive (frost injury to plant cells causes them to leak nutrients that can be used by the bacteria) or protective (INPs can provide resistance to desiccation for the bacteria by sequestering a reserve of water close to the cell surface as well as dispersing ice crystals away from the cell to limit the chance of intracellular inoculation). The structure and function of bacterial INPs has been extensively studied for two reasons: (1) to find a way of impeding their action to reduce frost damage to crops, and (2) to optimize their function as nucleators in commercial snow-making operations. The INP monomers of *P. syringae* have a molecular mass of 120 to 180 kD, but the relationship between nucleation temperature and molecular mass predicts that a nucleator must be ~20,000 kD to trigger crystallization at $-2°C$. Not surprisingly, then, huge aggregates of INP monomers are found on the bacterial cell membrane, and a phosphatidylinositol (PI) component of the protein appears to anchor them in groups.

Insect INPs Hemolymph INPs have been found in a number of freeze-tolerant insects where they act to raise the SCP during the winter months so that the insects freeze at relatively high subzero temperatures. Only the INP of the cranefly, *Tipula trivittata*, has been extensively characterized. It is an 800-kD globular lipoprotein (45% protein, 4% carbohydrate, 51% lipid) with two protein subunits (265 and 80 kD) that is present year-round and also functions as a normal lipophorin in the transport of lipids through the hemolymph. A distinguishing feature of the

INP compared with other lipophorins is the presence of PI (not previously found in insect lipophorins), which accounts for 11% of the bound phospholipids. Surface phospholipids on insect lipophorins can interact with water, and the PI content of the protein appears to be key to the ice nucleating action. Thus, treatments that affect the PI head groups inactivate INP function, and the data suggest that the hydroxyl groups on the inositol rings form the active water-organizing site. As with bacterial INPs, function of the *T. trivittata* INP also requires protein aggregation, in this case into long chains with two to three chains closely associated.

ENERGETICS, ISCHEMIA RESISTANCE, AND METABOLIC RATE DEPRESSION

Freezing is an ischemic event. When blood plasma freezes, each cell in each organ is isolated and cut off from the delivery of oxygen and nutrients, the removal of wastes, and interorgan communication via hormones and other signals for the duration of the freeze. Although metabolic rate is very low in a frozen organism, freezing may extend for many weeks, and hence, each cell of a freeze-tolerant organism must have fermentative fuel reserves and maintain a viable basal metabolism throughout the freeze. Indeed, the injurious effects of ischemia on human organ explants are one of the biggest challenges faced by researchers working in the field of organ transplantation (Chapter 19) and the main driving force behind continuing research that aims to extend cryopreservation technology for ultralow temperature storage (−196°C) to more and more types of tissues and organs.

Freeze-tolerant animals show well-developed anoxia tolerance; in laboratory trials we have shown that wood frogs readily endure at least 2 days under a nitrogen gas atmosphere at 5°C, whereas *Eurosta* survived an 18-day exposure. Known freezing survival times extend to at least 4 weeks for frogs at −3°C and 16 weeks for *Eurosta* at −15°C. Over the course of a freezing episode, both frogs and insects show typical animal responses to oxygen limitation: a slow progressive depletion of organ glycogen, phosphagen, and ATP over time and an accumulation of lactate and alanine as glycolytic end products. No unusual anaerobic end products are made although, as noted earlier, the peculiarities of carbohydrate pathways in *Eurosta* during the autumn cold-hardening period mean that large amounts of sorbitol accumulate under anoxic conditions. All parameters normalize again within 3 to 11 days postthaw. Although the glucose that frogs accumulate as a cryoprotectant is a fermentable fuel, frog organs appear to deplete only glycogen in support of anaerobic metabolism while frozen. As mentioned earlier, this may be due to strong inhibition of hexokinase in the cells of frozen frogs that prevents cryoprotectant pools from being catabolized.

When we analyzed the effects of freezing on many enzymes of glycolysis, gluconeogenesis, the tricarboxylic acid cycle, amino acid metabolism, fatty acid metabolism, and adenylate metabolism in wood frog organs we found that, in general, there are very few changes in enzyme activities in response to freezing (apart from those involved in cryoprotectant production in liver); equivalent studies assessing responses to anoxia exposure concurred. Enzyme changes that did occur during freezing generally involved a decrease in activities (in 75% of cases), often of biosynthetic enzymes, that was reversed after thawing. Hence, freeze-tolerant animals, like anoxia-tolerant animals, are well-prepared with constitutive activities of enzymes for dealing with aerobic-anoxic transitions during freezing. Interestingly, our recent application of cDNA array screening to identify freeze-responsive genes in wood frog organs identified the hypoxia-inducible factor 1 (HIF-1) as a freeze up-regulated gene but did not find up-regulation of any of the glycolytic enzymes that are usually under HIF-1 control. HIF-1 is also well-known to stimulate the up-regulation of the GLUT1 transporter (the ubiquitous form) so it may be that during freezing in wood frog organs HIF-1 triggers an acute up-regulation of glucose transporter levels to aid cryoprotectant uptake without triggering accompanying changes in glycolytic enzymes. The regulatory mechanisms involved in selectively activating just a subset of genes that are normally under HIF-1 control are not yet known, but we are sure that they will provide a fascinating study in adaptation that is likely intertwined with other modified regulatory mechanisms to accomplish and sustain the extreme hyperglycemia of the frozen state.

Freezing survival seems to be associated with metabolic rate depression (MRD), although there have been few studies that specifically address the issue. Many freeze-tolerant organisms are in preexisting hypometabolic states (dormancy, diapause, anhydrobiosis) when they are exposed to freezing, and this would clearly facilitate long-term freezing survival. For example, many insects undergo a winter diapause with metabolic rates typically <10% of normal values at the same temperature. The strong suppression of aerobic metabolism during diapause is associated with decreased activities of mitochondrial enzymes. For example, Figure 17.7*c* shows that activities of two enzymes of the tricarboxylic acid cycle, citrate synthase and NAD-dependent isocitrate dehydrogenase, decrease by about 50% in *Eurosta* over the midwinter months and then rebound in the spring; the activity of glutamate dehydrogenase, another enzyme located in the mitochondrial matrix, follows suit. The same thing happens in *Epiblema*. In some species, an actual degradation of mitochondria is seen so that only low numbers of the organelles remain throughout the dormant period.

The probability that freezing stimulates MRD in wood frogs is supported by data showing a freeze-induced

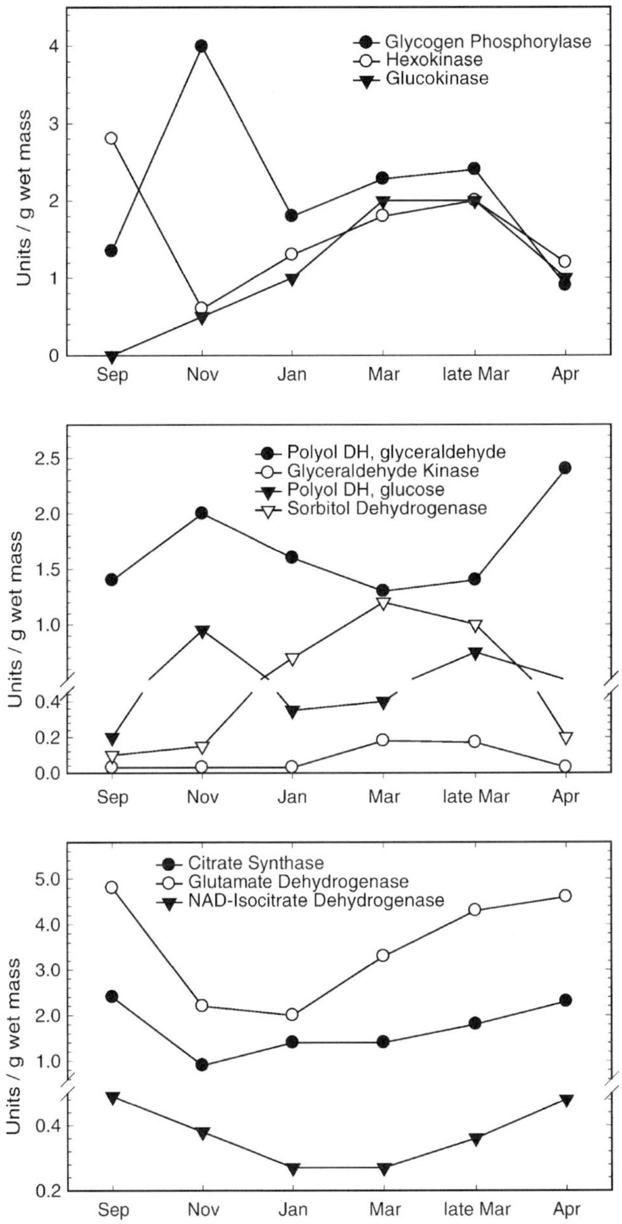

Figure 17.7 Changes in the activities of selected enzymes in goldenrod gall fly larvae, *Eurosta solidaginis* over the winter months. For polyol dehydrogenase, activity was measured with each of two substrates: with glyceraldehyde the product is glycerol, whereas with glucose the product is sorbitol. Animals were sampled within the first week of the month except for a late March sampling point, and the final sample was on April 21, just before larvae pupated. [Data compiled from D. R. Joanisse and K. B. Storey (1994). *J Comp Physiol B* **164**:247–255 and D. R. Joanisse and K. B. Storey (1994). *Insect Biochem Mol Biol* **24**:145–150.]

increase in the content of phosphorylated eIF2α in organs of frozen frogs. Recall that a high phospho-eIF2α is indicative of an inhibition of protein synthesis and commonly accompanies MRD in other systems. We have also found that freeze exposure suppresses the activities of ion-motive ATPases in wood frog skeletal muscle, again indicating MRD. Na^+K^+–ATPase and sarcoplasmic reticulum (SR) Ca^{2+}–ATPase activities were reduced by 30 to 50% during freezing as was calcium binding by SR fractions; furthermore, oxalate-stimulated calcium uptake was reduced by 92% in SR fractions from frozen frogs. All rebounded partially or fully after 24 h thawing. This strongly indicates that active steps are taken to conserve organ energy expenditure on ATP-dependent ion pumping in the frozen state.

Once again, our new data from cDNA array screening provide additional indications that freeze-induced MRD occurs, probably as a mechanism of ischemia resistance. Screening identified adenosine A1 and A2A receptors and the enzyme 5′-nucleotidase (that synthesizes adenosine from AMP) as putatively up-regulated during freezing in frog heart. Recall from Chapter 15 that adenosine accumulates quickly in response to anoxia in organs (particularly brain) of anoxia-tolerant turtles and acts via adenosine A1 receptors to trigger ion channel arrest. Overexpression of adenosine A1 receptors is also known to increase myocardial tolerance of ischemia in transgenic mice. Hence, the putative up-regulation of adenosine receptors and 5′-nucleotidase in frog heart provides evidence that as wood frogs freeze they activate a set of ischemia-protective signals and receptors and implement hypoxia/anoxia-induced suppression of ATP-expensive metabolic processes. Interestingly, in the study with transgenic mice, overexpression of adenosine A1 receptors was also associated with enhanced GLUT4 expression, suggesting yet another mechanism that frog organs adapt and use to acutely enhance cryoprotectant uptake during freezing.

ANTIOXIDANT DEFENSES

All organisms maintain antioxidant defenses, including enzymes, metabolites (e.g., glutathione, ascorbate) and proteins (thioredoxin), that deal with the continual assault to cellular metabolism by reactive oxygen species such as superoxide, hydrogen peroxide, hydroxyl radicals, and peroxynitrite (see Chapter 12). The latter two, in particular, are highly reactive species that can cause serious damage to cellular lipids, proteins, and DNA. Studies in mammalian systems have linked a substantial portion of ischemic damage to organs, not to the period of oxygen deprivation itself, but to the reperfusion phase when oxygen is reintroduced and a burst of oxyradical production can temporarily overwhelm the antioxidant defenses of an organ. Ischemia-linked oxidative damage is also a serious issue in hypothermic organ preservation and is lessened by the addition of antioxidants to the perfusion medium during organ retrieval and storage (see Chapter 19). Freeze–thaw is an ischemia-reperfusion event that has the potential to result in oxidative damage to organs when heart beat and breathing are restored during thawing.

Furthermore, in the case of freeze-tolerant frogs that use glucose as a cryoprotectant, the potential for glucose-mediated oxidative damage is high since glucose-related oxidative damage is known to play a role in tissue damage in diabetes. Persistent high glucose causes damage to tissues for two main reasons: (1) the nonenzymatic glycation of proteins impairs their function and leads to the accumulation of damage products called advanced glycation end products (AGEs), and (2) the pro-oxidant actions of glucose, glycated proteins, and AGEs stimulate the production of reactive oxygen species that cause further damage to macromolecules. Indeed, we suspect that the damage that could accrue as a result of sustained high glucose is a major reason why cryoprotectant biosynthesis by frogs is triggered only when the animals actually begin to freeze.

Two strategies contribute to the defense against oxidative stress associated with freeze–thaw in freeze-tolerant animals. The first and most important is elevated constitutive defenses. For example, the activities of six antioxidant enzymes (superoxide dismutase, catalase, glutathione *S*-transferase, glutathione reductase, and total and Se-dependent glutathione peroxidases) were generally at least 2-fold higher in organs of wood frogs compared with the same organs in freeze-intolerant leopard frogs that hibernate underwater. Furthermore, the levels of reduced glutathione (GSH) were up to 10-fold higher in wood frog organs. Indeed, the antioxidant defense capacity of wood frog organs is very close to that of anoxia-tolerant turtles, and both are high for ectothermic vertebrates and closer to the antioxidant capacities of mammalian organs. This indicates that good antioxidant defenses are a necessary aid for freezing survival. Not surprisingly, then, wood frog organs showed no signs of oxidative damage to lipids over a course of freeze–thaw. The freeze-tolerant insect, *Eurosta*, similarly showed no accumulation of oxidative damage products as a result of freeze–thaw, although a 5-fold increase in the amount of oxidized glutathione (GSSG) suggested that some oxidative stress occurred during freezing that consumed GSH.

The second principle of antioxidant defense in freeze-tolerant animals is selective changes to the activities of some enzymes in response to freezing. *Eurosta* showed no change in the activities of antioxidant enzymes in response to either freezing or anoxia exposure but wood frogs did. Both total glutathione peroxidase and selenium-dependent glutathione peroxidase activities increased in frog organs during freezing and returned to near control values after 24 h thawing. This suggests that the need for enhanced defenses against peroxidative damage is greatest during freezing or immediately after thawing, a time when glucose is also very high. These changes may minimize the potential for glucose-mediated oxidative damage to macromolecules, which is a significant problem associated with sustained high glucose levels in diabetes.

COLD HARDINESS AND GENE EXPRESSION

Polar species are typically cold-hardy all year round, but organisms living in temperate zones of Earth acquire their cold hardiness on a seasonal basis. Acquisition of cold hardiness is often linked with photoperiod cues and is accelerated by decreasing environmental temperatures (similar to the situation for AFP induction, discussed earlier). Insects that have only one generation a year (univoltine) often show an obligatory acquisition of cold hardiness with the transition to the overwintering life stage (e.g., in *Eurosta* this is the third instar larva) as well as an obligatory period of diapause (arrested development) before they can resume growth and development. Thus, the goldenrod gall formers, *Eurosta* and *Epiblema*, need a minimum of about 3 months of cold exposure before they can break diapause. They will not resume development if removed from cold storage earlier, and after diapause ends (usually in early February in Canada), they remain in a quiescent state until warming temperatures in the spring facilitate pupation. Insects with two or more generations over the year (bi- or polyvoltine) are typically cold-hardy only in the last generation that overwinters. This generation is usually preprogrammed by photoperiod triggers (day shortening) received by the female in the previous generation that cause the production of diapause hormone that acts on the developing eggs of the next generation to set them on the course for diapause and cold hardiness. For example, diapause-bound eggs laid by silkmoths *Bombyx mori* already contain higher glycogen levels than nondiapause eggs as well as the enzyme complement to immediately begin glycerol and sorbitol synthesis. After these eggs are laid, development arrests at an early embryonic stage, and in eggs stored at 5°C diapause persists for several months before an induction of sorbitol dehydrogenase (to catabolize sorbitol) and other enzymes signals its end.

The seasonal acquisition of cold hardiness involves changes in the expression of multiple genes. Gene expression responses can occur on two time frames: (1) seasonal and/or cold acclimation responses that accrue over days or weeks prior to exposure to potentially damaging subzero temperatures, and (2) rapid responses triggered within minutes to hours when temperature drops (cold shock) or freezing begins. For example, the appearance of AFPs, INPs, and increased activities of enzymes involved in cryoprotectant synthesis all result from seasonal changes in gene expression. Other factors can also come into play such as changes in food availability; goldenrod gall forming insects also seem to take cues from the senescence of the plant galls in which they live.

Figure 17.7 shows the seasonal changes in activities of selected enzymes of polyol metabolism by freeze-tolerant gall fly larvae; these changes can be compared with the acquisition of cryoprotectants by the larvae as shown in Figure TB17.1. Figure 17.7*a* shows that hexokinase, which is involved in the uptake, phosphorylation, and storage of dietary sugars as glycogen is highest in early autumn, coincident with the time of major glycogen storage. Glycogen phosphorylase *a* activity peaks later in the fall, corresponding with high rates of glycerol and sorbitol synthesis (see Fig. TB17.1) and with peaks of polyol dehydrogenase activity using either glyceraldehyde (making glycerol) or glucose (making sorbitol) as the substrate (Fig. 17.7*b*). However, enzymes that catabolize polyols do not appear until later in the winter. Sorbitol dehydrogenase (SDH), which converts sorbitol to fructose, is induced in January and provides the larvae with flexibility in their sorbitol pools so that these can be catabolized whenever with weather warms up. Notably, SDH induction is linked with the end of diapause in silkmoth eggs, and this may also be so in *Eurosta*, the larvae entering a quiescent state after diapause ends, which endures until warming temperatures trigger massive losses of polyols and the beginning of renewed development. Rising glucokinase and hexokinase activities over the late winter months would phosphorylate the fructose produced by SDH so that it can reenter the hexose phosphate pool for reconversion to glycogen. Glyceraldehyde kinase, which catabolizes the glycerol pool, does not appear till March and this correlates with the beginning of glycerol clearance at this time (see Fig. TB17.1).

Thus, for many insects the acquisition of cold hardiness and/or freeze tolerance is an obligate seasonal event, often intertwined with development and virtually always completed before subzero temperatures are experienced in nature. For this reason, insects are generally poor models for studying acute effects of cold and/or freezing on gene expression. Indeed, when we tried such experiments with *Eurosta* and *Epiblema*, we found very few gene expression changes that are directly triggered by cold or freezing exposures. For example, a search for freeze-induced gene expression in *Eurosta* using differential display polymerase chain reaction (PCR) found only eight transcripts that were putatively up-regulated by freeze exposure, only one of which could be tentatively identified by its high sequence identity with the *Drosophila* ribosomal protein, rpA1. It appears that the program of protein changes needed for cold hardiness in this univoltine species is so "hard-wired" into the developmental and seasonal patterns of the species that few, if any, gene expression and protein synthesis changes are needed when tissue freezing actually begins. With the small body size of insects (gall fly larvae are 50 to 60 mg), this result is not surprising for these insects freeze very fast, probably without sufficient time for significant gene induction and protein synthesis responses.

However, other models offer good opportunities to study cold- and freeze-induced gene expression, and we will consider recent advances with two of these: the model plant *Arabidopsis thaliana* and the wood frog *R. sylvatica*. Recent work with both of these models has employed state-of-the-art gene screening techniques (e.g., cDNA library screening, differential display PCR, cDNA array screening, reverse transcription (RT)-PCR) to provide a wide view of the many different types of cellular proteins and enzymes that respond to cold or freezing stress. Our focus below, then, is on cold-induced gene expression in *Arabidopsis* and freeze-induced gene expression in wood frogs.

Cold-Induced Gene Expression in *Arabidopsis*

Arabidopsis is the "*Drosophila*" of the plant world, a very widely used model whose metabolism and genetics have been extensively characterized and that can be easily manipulated with deletion or overexpression mutants to characterize the phenotypic effects of genes. Until recently, about 40 cold-induced genes had been identified in *Arabidopsis*, representing a wide variety of cell functions (with additional novel unidentified genes indicated from cDNA array screening). The plant is freeze-tolerant so these include a mixture of genes that may have primary functions in dealing with either cold or freezing stress. Many of these genes are responsive to abscisic acid (ABA), and it appears that this plant hormone is the major mediator of the cold or freezing signal in all plants. In *Arabidopsis*, treatment with ABA at normal growth temperatures improved survival in subsequent tests of frost hardiness, whereas ABA-deficient mutants were impaired in developing freezing tolerance during cold acclimation. Among the first genes shown to be cold-inducible were several that encoded enzymes/proteins involved in membrane protection including (a) desaturases that increase the extent of lipid unsaturation to adjust membrane fluidity for low-temperature function, (b) enzymes of proline biosynthesis that produce the low-molecular-weight membrane-protectant, proline, and (c) dehydrins that appear to function as membrane protective small proteins. Alcohol dehydrogenase (ADH) was also up-regulated in the cold to increase the capacity for glycolytic energy production during freezing since plants produce ethanol, not lactate, as their glycolytic end product. Other cold-responsive genes included those encoding sugar transporters, water channel proteins, Ca^{2+} and iron-binding proteins, a wound-inducible protein, and several antioxidant enzymes. Several kinases are also cold-induced, including members of the mitogen-associated protein kinase (MAPK) signaling pathway and the ribosomal S6 kinase, implicating their roles in regulating the functions of other proteins in the cold. Most of these genes are constitutive at normal growth temperatures and undergo 2- to 5-fold induction during cold acclimation. However, one class of cold-regulated (*COR*) genes is strongly induced by cold, typically by 50- to 100-fold, and because of this strong induction response they have been the target of many regulatory studies.

COR genes (and a few others) are regulated by the CBF (CRT binding factor) transcription factors (three isoforms are known) that are strongly up-regulated by cold exposure in *Arabidopsis*. This transcription factor binds to the cold-repeat element (CRT) in the 5′ promoter of CBF-responsive genes. Within 15 min of transfer to low temperature, CBF transcripts begin to accumulate, followed after 1 to 2 h by accumulation of transcripts of the CRT-regulated genes. Deletion analysis of the promotors of some of the *COR* genes determined the sequence of the CRT to which CBF binds. The surprising result of this was that the CRT was identical with a drought-responsive element (DRE) that is present in the genes of many other plant species and is regulated by the DREB (DRE binding factor) transcription factors. Thus, two different stresses (freezing, drought), acting via two different transcription factors (CBF, DREB), can activate a common set of genes. As discussed previously in Chapter 14 (see Fig. 14.12), this system illustrates an important principle of biochemical adaptation. Adaptive mechanisms that work well in one situation are often "borrowed" and put to use in organismal responses to a related stress. In this case, with the simple creation of a new transcription factor (CBF), a whole group of preexisting dehydration-responsive genes/proteins could be put to work to aid freeze tolerance and address problems that are common to both stresses (e.g., cell volume reduction, cell water loss, membrane stress). Interestingly, as we will see in the next section, quite a few of the freeze-responsive genes in wood frogs are also induced by dehydration in both freeze-tolerant and freeze-intolerant amphibians, indicating that their origin was also as dehydration-responsive genes.

A recent study has hugely expanded the database on cold-responsive genes in *Arabidopsis* and provides a superb model for the way in which cDNA array screening can be used in studying biochemical adaptation (see Suggested Reading list). Plants grown at 22°C were acutely transferred to 4°C and sampled after 0.5, 1, 4, 8, and 24 h and 7 days. Gene expression at each time point was assessed on gene chips containing 8296 different cDNAs with a 3-fold change in expression considered significant. Gene expression was also assessed in warm-grown plants that constitutively expressed CBF isoform 1, 2, or 3 to identify those genes that were controlled by CBF regulons. Screening found 306 genes that were cold-responsive, 218 that were up-regulated, and 88 that were down-regulated by cold exposure. Only 12% of these genes could be clearly assigned to the CBF regulons, which included subregulons controlled by the RAP2.1 and RAP2.6 transcription factors. Genes for at least 15 other putative or known transcription factors were CBF-independent and were expressed in different time patterns, which indicates the existence of several other regulons that control groups of cold-responsive proteins. Multiple patterns of gene expression were identified. Some genes were activated early, whereas others responded later. Early responders included CBF transcripts that rose within the first hour of cold exposure and were followed within 4 h by increased expression of several *COR* genes. About 50% of genes were only transiently induced (up-regulated at only 1 or 2 time points), whereas others were expressed for longer times, including 64 genes that were expressed over the full 7-day time frame. "Waves" of gene expression occurred with new groups of genes up-regulated at each time point. Cold-responsive genes belonged to many different groups with identified genes encoding proteins involved in transcription regulation, intracellular signaling, sugar metabolism, antioxidant defenses, COR and late embryogenesis proteins that are believed to provide cryoprotection, and many others. Overall, then, the results from this transcriptome screening indicate the highly complex nature of the metabolic responses that make up cold acclimation (and our fractional knowledge about these to date) and also emphasize the existence of regulated patterns of coordinated responses that provide both acute responses that are involved in immediate relief from cold shock and long-term responses that readjust structural and functional components of cells to reestablish homeostasis at low temperature.

The dehydrin and COR proteins in plants deserve some further comment. Dehydrins are induced by environmental stresses that result in cellular dehydration (e.g., drought, high salt, freezing, seed desiccation). At least five different dehydrins are up-regulated by cold exposure in *Arabidopsis*, one of which is also a COR protein. Dehydrins are glycine-rich, hydrophilic, heat-stable proteins that include a characteristic lysine-rich sequence near the C terminal. The polypeptide is believed to form an amphipathic helix that could interact with membrane lipids or with exposed hydrophobic patches of partially denatured proteins in a manner similar to chaperones. They are believed to serve a protective function during stress. Studies with the peach dehydrin PCA60 showed that it protected the activity of lactate dehydrogenase during freeze–thaw cycles *in vitro*, and it also exhibited antifreeze activity as evidenced by thermal hysteresis and effects on ice morphology. Similarly, the COR proteins are not enzymes, but many are very hydrophilic and remain stable upon boiling, which suggests that they act as cryoprotective proteins similar to dehydrins. Overexpression of one *COR* gene, *COR15a*, improved freezing survival by chloroplasts and protoplasts in warm-acclimated transgenic plants. Localized in the chloroplast stroma, it is proposed that the 15-kD COR15am mature protein functioned to stabilize membranes by decreasing the propensity of membranes to form deleterious hexagonal II phase lipids upon freeze-induced dehydration. To do this the proteins may alter the intrinsic curvature of the inner membrane of the chloroplast envelope. While these protective proteins in plants seem to play roles in membrane or protein stabilization, a group of cold-shock proteins in bacteria appear to have primary roles in preserving nucleic acid stability (see Text Box 17.4).

Agricultural scientists are very interested in the possibility of using transgenic technology to enhance freeze tolerance or to induce constitutive freeze tolerance in various crop species, thereby reducing the economic cost of frost damage. In one study, for example, *Arabidopsis* were transfected with insect AFP genes (with or without the signal peptide for AFP export). Plants expressed the protein and plants with AFPs in the apoplast fluid froze at significantly lower temperatures than did wild-type plants, especially in the absence of extrinsic nucleation events. However, transgenic lines showed no improvement in freeze tolerance. In other studies, transgenic *Arabidopsis* lines that constitutively express the CBF transcription factor showed the continuous presence of various proteins that are normally cold-inducible. One line showed a level of freeze tolerance comparable to that of the cold-acclimated wild type. However, in another line the transgenic plants were all dwarfs. Hence, there can be a growth penalty for the constitutive maintenance of the many metabolic adaptations needed for freeze tolerance, and this can affect both yield and productivity, characteristics that are generally positively selected in nature and in agriculture. This is probably the reason that frost hardiness is only seasonally expressed in most plants.

TEXT BOX 17.4 BACTERIAL COLD SHOCK PROTEINS

Both heat and cold induce shock responses by organisms. The heat-shock response has been extensively studied, and its prominent feature is the synthesis of protein chaperones that aid protein refolding and act to protect/rescue proteins whose active conformation is disrupted/denatured by heat exposure. A cold-shock response has been characterized in mesophilic bacterial species, such as *Escherichia coli*. A decrease from warm to cold temperature immediately halts growth, primarily because of cold effects on ribosomes that cause a break-up of polysomes into monosomes. However, after a lag period during which cold-shock proteins (Csps) are synthesized, protein translation and bacterial growth resumes. Csps are typically small ($\sim$7 kDa) proteins that have roles in gene expression, mRNA folding, transcriptional initiation and regulation, and/or freeze protection. Different bacterial species may show 12 to 37 types of Csps. Some are known cellular proteins whose levels are up-regulated by cold, whereas others are apparently unique to cold shock. CspA, for example, appears to function as an RNA chaperone to guide efficient translation by either (a) preventing formation of secondary structure in mRNA that would make it inaccessible to the translation apparatus or (b) protecting the message from enzymatic degradation. The responses by psychrotolerant and psychrophilic bacteria to cold shock is different. The number of proteins that respond to cold is greater, but their level of expression is weaker than the mesophilic response. This is not unexpected as these species are better prepared constitutively for dealing with low temperature. A subgroup of these proteins, termed cold-acclimation proteins (Caps), is continuously produced during low-temperature growth and may have roles key to life in the cold.

Effects of Csps on freezing survival have been assessed. A single *cspB* deleted mutant of *Bacillus subtilis* showed lower freeze survival than wild-type cells. A freeze-sensitive phenotype of *Lactococcus lactis* was observed upon deletion of the *cspA*, *cspB*, and *cspE* genes, whereas overproduction of the protein products of these genes increased freezing survival. Since some Csps are recognized as nucleic acid binding proteins, they might protect RNA and DNA during the freezing process by binding to these nucleic acids and, hence, increase the survival of the bacterial cells.

Freeze-Induced Gene Expression in Wood Frogs

The frozen state is an ischemic state; as the ice front moves through an organism, freezing extracellular fluid and plasma, the delivery of oxygen and substrates to tissues is cut off. Hence, the transition to the frozen, ischemic state would not be the time to undertake major ATP-expensive biosynthetic projects. Indeed, studies with wood frog heart have shown that a general suppression of transcription and translation takes place and comparable events probably occur in all other organs. To assess the transcriptional capacity of wood frog heart, we used the nuclear run-off assay (see Fig. 14.11 for an example of the technique). The rate of ^{32}P-uridine triphosphate (UTP) incorporation into preinitiated RNA transcripts in fresh nuclei isolated from the hearts of 5°C-acclimated (control) frogs was compared with the rate in frogs frozen for 24 h at -2.5°C. After Trizol isolation of total RNA, the amount of ^{32}P-UTP incorporated by transcripts in nuclei from frozen frogs was only 47% of the amount in control hearts, indicating a substantial suppression of net transcriptional activity during freezing.

Similarly, translational activity in frozen hearts was strongly suppressed as evidenced by the phosphorylation status of the eukaryotic initiation factor 2α alpha (eIF2α); phospho-eIF2α was threefold higher in hearts of frozen frogs compared with unfrozen controls. Thus, against a background of a general suppression of transcription and translation during freezing, the up-regulation of selected genes suggests that they have important roles to play in freezing survival. Some of these are discussed below.

Over the last 8 years, studies in our lab at Carleton University have focused on the discovery and analysis of freeze-responsive genes in the organs of wood frogs. As with many other fields of biology, the recent "explosion" of technique choices for gene discovery have allowed a whole new approach to be taken in identifying the genes/proteins that support freeze tolerance. Where previously we needed to identify a physiological or biochemical trait of freeze-tolerant animals (e.g., cryoprotectant production) and then work backward to elucidate the protein mechanisms underlying it, we can now work in the opposite direction beginning with up-regulated genes and discovering the metabolic and physiological consequences of the expression of these genes. One might think that there would be a lot of overlap in the results of these two approaches but, to date, there has been very little. Indeed, one key result of gene screening technology has been the identification of dozens of genes/proteins, both previously known and novel, that have never before been linked with freeze tolerance (several of these are discussed below). Each of these genes has implicated a previously unsuspected area of cellular metabolism as contributing to freeze tolerance. Using techniques of cDNA library construction and screening and differential display PCR, a variety of freeze-responsive genes have been identified. Note that these studies were done with cDNA isolated from 5°C acclimated frogs as controls and frogs frozen at −2.5°C (usually for 24h) as the experimental condition and that, therefore, they represent freeze-induced genes, not cold-induced genes as for the *Arabidopsis* studies described above. Freeze-responsive genes that have been confirmed as up-regulated in wood frogs include fibrinogen, mitochondrial membrane proteins [adenosine 5′-diphosphate (ADP)/ATP translocase, the inorganic phosphate carrier, F_1−ATPase subunits 6 and 8], myosin light-chain subunit 2, atrial natriuretic peptide, the glycolytic enzyme phosphoglycerate kinase 1, and several proteins involved with protein synthesis (ribosomal phosphoprotein P0, ribosomal protein L7, and the eukaryotic elongation factors 1 and 2). In addition, three novel proteins (FR10, FR47, Li16) were found.

Very new studies have employed cDNA array screening with 19,000 human cDNA gene chips to further explore freeze-responsive gene expression in wood frogs. Although cross-species reactivity was only 60 to 80% between frog cDNA and human chips, the responses of thousands of genes could still be evaluated. Gene expression changes in nine different organs have been assessed. Results for wood frog heart provide an instructive example. Over 200 genes showed a greater than twofold increase in expression in heart of frozen frogs, compared with 5°C controls. Many of these fell into six general categories: (1) signal transduction: serine/threonine protein kinase, protein phosphatases, and their related regulatory proteins; (2) shock proteins with known or probable chaparone functions such as heat-shock protein 27 and glucose-regulated proteins; (3) membrane transporters including several solute carriers such as the inorganic phosphate carrier and the GLUT4 glucose carrier; (4) genes encoded on the mitochondrial genome; (5) antioxidant defenses including cytochrome P450, aldo-ketose reductases, superoxide dismutase; and (6) genes related to hypoxia/ischemia defense including HIF-1, adenosine receptors, and 5′-nucleotidase. Genes belonging to the last two categories suggest coordinated control during freezing in heart by genes containing either the hypoxia response element or the antioxidant response element. Genes containing these response elements are well-known to be involved in mammalian organ responses to ischemia-reperfusion, which suggests that a substantial part of the heart response to freezing is the implementation of defensive or protective strategies for dealing with long-term ischemia. This makes sense when we remember that heart is one of the last organs to freeze and the first to thaw and that despite ischemic insults to its metabolism, the heart must continue to do physical work as the frog is freezing in order to distribute cryoprotectant and must resume function quickly after thawing in order to reestablish perfusion of other organs as soon as possible. Selected freeze-responsive genes are discussed more fully below.

Fibrinogen Fibrinogen plays an important role in wound repair. This acute-phase plasma protein is synthesized by liver and secreted into the plasma with increased production stimulated by stresses including infection, inflammation, and tissue injury. The protein has two halves, each made of three subunits (Aα, Bβ, and γ) and as the final step in the coagulation cascade, thrombin cleaves near the N termini of the Aα and Bβ chains to release the A and B fibrinopeptides and expose sites for polymerization into the fibrin mesh of a growing blood clot. Screening of a cDNA library made from liver of frozen wood frogs retrieved freeze up-regulated clones with inserts encoding the α and γ subunits of fibrinogen. Northern blots showed that both transcripts were coordinately expressed over a time course of 24h freezing at −2.5°C followed by 24h thawing at 5°C; transcript levels rose by more than threefold after 8h freezing, remained high with longer freezing, and then returned to near control values within 24h after thawing. In a subsequent study, the β- and γ-fibrinogen subunits were retrieved from screening a liver cDNA library made

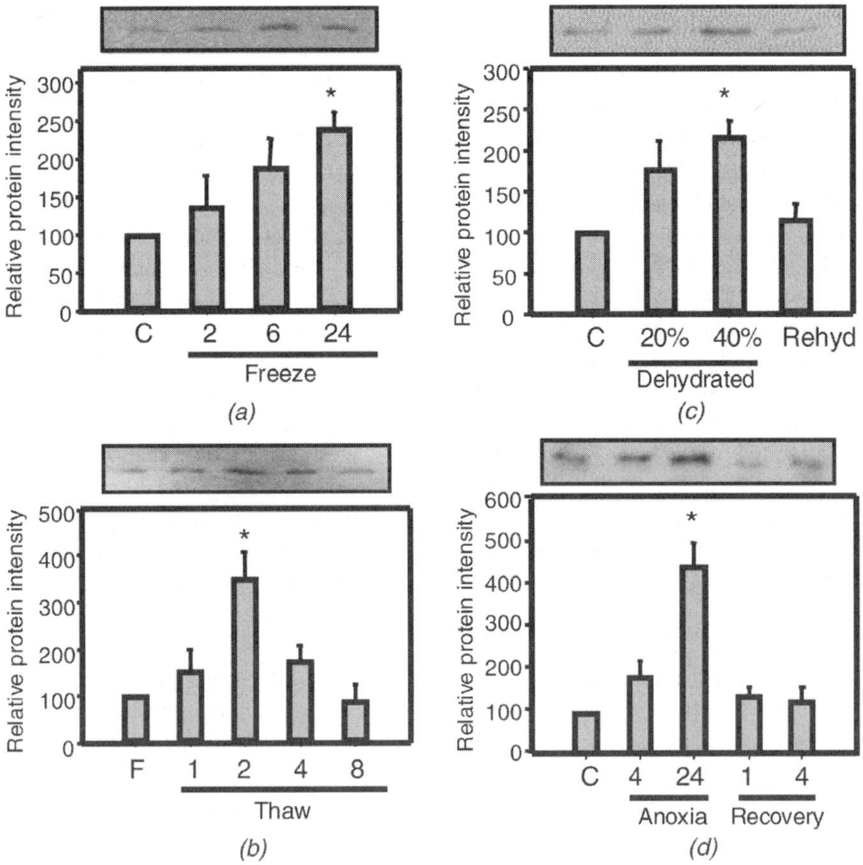

Figure 17.8 Responses of Li16 protein levels in liver of wood frogs, *Rana sylvatica*, under different conditions: (*a*) freezing at $-2.5°C$ for 2, 6, or 24 h, (*b*) thawing at 5°C after 24 h frozen for 1, 2, 4, or 8 h; note that on this graph changes during thawing are expressed relative to protein levels in 24-h frozen (F) frogs, (*c*) dehydration at 5°C to 20 or 40% of total body water lost followed by 24 h rehydration, (*d*) anoxia exposure under a nitrogen gas atmosphere at 5°C for 4 or 24 h followed by aerobic recovery for 1 or 4 h. Example Western blots are shown along with histograms presenting mean values for relative band intensities for $n = 3$ independent trials. C, control at 5°C, *, significantly different from zero time value, $P < 0.01$. See Figure 14.10 for more information about Li16 expression in frozen frogs. [Data from J. D. McNally, S. Wu, C. M. Sturgeon, and K. B. Storey (2002). *FASEB J* **16**:902–904.]

from glucose-loaded wood frogs; glucose loading was used to assess genes that were responsive to the extreme hyperglycemia of the frozen state. Hence, coordinate expression of the three subunits undoubtedly occurs during freezing, as is the well-known response by fibrinogen subunits in other vertebrates. The probable function for fibrinogen up-regulation during freezing is to improve the frog's damage repair capabilities so that any internal bleeding injuries caused by ice crystal damage to the microvasculature can be rapidly addressed when organs thaw. Since the normal mechanisms of interorgan signaling (hormone transport by blood, nervous stimulation) are cut off during freezing and impaired for a considerable time postthaw, an anticipatory up-regulation of damage repair mechanisms makes sense so that newly synthesized clotting proteins can be available immediately upon thawing.

To explore the functions of freeze-responsive genes, we routinely examine the parallel response of each gene to two of the component stresses of freezing: anoxia and dehydration. From these studies, we can gain a good idea of whether freeze-responsive genes are responding to the ischemia or the cellular dehydration components of freezing. Recall from the section on cryoprotectant biosynthesis that glucose production by wood frog liver was stimulated just as strongly by dehydration as by freezing but did not respond to anoxia. Similarly, all freeze-responsive genes that have been extensively characterized to date in wood frogs also respond to either anoxia or dehydration (but not both). In the case of fibrinogen, transcript levels were elevated under dehydration stress but not in anoxia. This suggests that it is the structural strain on cells that necessitated the need for defenses against bleeding injuries. The endothelial cells lining blood vessels are particularly vulnerable to freezing injury for they face not only volume changes (shrinking during freezing, rapid swelling during thawing) but also the possibility that ice expansion within the lumen of capillaries will cause them physical damage or breach their walls. Indeed, studies of mammalian liver preservation have identified the endothelial cells of the vasculature as primary targets of freezing damage and much more sensitive than the hepatocytes.

Mitochondrial Transporters The mitochondrial inner membrane is highly selective as to which metabolites can enter the organelle; for example, glucose cannot, which leads to the question of how mitochondrial volume is regulated when frog cells dehydrate during freezing! The mitochondrial ADP/ATP transporter (AAT) was one of the first freeze-responsive genes identified in wood frog liver, and recent studies have identified two others, the inorganic phosphate carrier (PiC) and the mitochondrially encoded subunits 6 and 8 of the F_1F_0–ATP synthase. However, transcript levels of two other transporters (the dicarboxylate carrier and the oxoglutarate transporter) did not change during freezing, which indicates that only selected functions need modification to deal with the consequences of freezing. The three freeze-responsive transporters showed somewhat different patterns during freezing. Transcripts of the bicistronic mRNA encoding F_1F_0–ATP synthase subunits 6 and 8 (also known as the β and γ subunits of the F_1–ATPase) were elevated by 7-fold in liver of 24-h frozen frogs and remained high after 24 h thawing at 5°C. *Aat* transcripts reached a peak of 4.5-fold higher than controls after 8 h freezing but declined with longer freezing and during thawing. The levels of *pic* transcripts increased from nearly undetectable amounts in control and 2-h frozen frogs to reach >70-fold higher than control values after 24 h frozen. AAT and PiC protein levels were also quantified; these lagged behind the changes in mRNA transcripts but were ~2-fold and 3.3-fold higher than controls after 24 h frozen. PiC protein returned to near control values by 8 h thawed, whereas AAT was reduced to about 1.5-fold higher than controls after 24 h thawed. The fact that peak protein levels for both genes were reached during freezing suggests that they have a role to play in the survival of liver mitochondria in the frozen state. What that role is remains to be determined, but it may be related to stabilizing mitochondrial energetics under the ischemic conditions during freezing. Indeed, we found that both AAT and ATPase 6/8 also responded strongly to anoxia exposure (but not to dehydration), which further supports the idea that their expression may respond to signals coming from low oxygen or low energetics. However, PiC responded to dehydration stress, which suggests that changes in inorganic phosphate carrying capacity may be needed in order to adjust ionic, osmotic, or volume regulatory parameters between the mitochondrial and cytoplasmic compartments during freeze/thaw-induced changes in cell volume. Indeed, liver lost ~58% of its water during freezing at −2.5°C (when 65% of total body water was converted into ice), and this would cause major changes in ionic strength and osmolality of liver intracellular fluids, necessitating regulatory adjustments of these parameters by the various subcellular organelles.

Novel Proteins To date, screening of wood frog liver has revealed three unidentified freeze-responsive genes that have been named *fr10*, *li16*, and *fr47*. They code for proteins with molecular weights of 10, 14, and 47 kD, respectively. The three proteins show no sequence similarity to any known gene (or reported expressed sequence tag) when subjected to searches of gene banks nor do they appear to be related to each other. Apart from being up-regulated during freezing in liver, the only other shared characteristic that we know of to date is the presence in all three proteins of a hydrophobic region of 21 amino acids in length. In FR10 and Li16 this is an N-terminal region, whereas in FR47 the hydrophobic region is near the C terminus. In both FR10 and Li16 this region contains four leucine and four valine residues. Such hydrophobic regions are not uncommon in proteins and may represent a transmembrane segment.

To characterize the function of an unknown protein, pieces of information can be gathered from a variety of sources (these are summarized in Table 17.4). Also refer to Figure 14.10 for a diagram showing the various techniques that can be used for analysis of a novel gene and its protein product. Some of the approaches that can be used are as follows. First is the characterization of the response by these proteins to the primary stress itself, freezing. Gene transcript levels of all three increased by 3- to 5-fold in liver after 24 h of freezing at −2.5°C, and for FR10 these had returned to control values after 24 h thawed. Antibodies were raised against the C-terminal peptide of Li16 and FR47, and these were then used in Western blotting to assess protein responses to freeze/thaw. Western blots confirmed that both Li16 and FR47 proteins were also elevated in liver of frozen frogs (by 2.4- and 1.6-fold, respectively, after 24 h) (see Fig. 17.8 for Li16). Levels of both proteins continued to increase over the first 2 h when frogs were returned to 5°C to thaw reaching final maxima of 8.4- and 3.5-fold higher than unfrozen controls. However, by 8 h thawed levels had returned to about the same levels as in 24-h frozen frogs (Fig. 17.8). The peak protein level after 2 h of thawing corresponds with a time when only about half of the ice has melted in the frog's body.

TABLE 17.4 Comparison of Properties of Three Novel Proteins That Are Freeze-Responsive in Wood Frog Liver

	FR10	*Li16*	*FR47*
Molecular weight, kD	10	12.8	45.7
Number of amino acids	90	115	390
Sequence characteristics	N-terminal hydrophobic region, aa 1–21, possible nuclear exporting signal	N-terminal hydrophobic region, aa 1–21	Hydrophobic region near C terminal, aa 350–370
Transcript response to stress (relative to unfrozen control = 100)[a]			
24-h freeze (2-h thaw)	320 (110)	370	510
20% dehydration (full rehydration)	↑↑↑ (↑↑)	80	220
24-h anoxia (1-h recovery)	↓↓↓↓ (n.a.)	760	620
Protein response to stress (relative to unfrozen control = 100)[a]			
24-h freeze (2-h thaw)	n.a.	240 (840)	160 (350)
20% dehydration (full rehydration)	n.a.	175 (110)	65 (125)
24-h anoxia (1-h recovery)	n.a.	440 (130)	60 (90)
Transcript response to *in vitro* tissue incubation with second messengers[b]			
Dibutyryl-cAMP	n.a.	n.s.	n.s.
Dibutyryl-cGMP	n.a.	↑↑	n.s.
PMA	n.a	n.s.	↑↑
Organ distribution	All organs tested	Liver, gut, heart	Liver
Genbank accession number	U44831	AF175980	AY100690

[a]Values for liver mRNA transcript and protein levels are relative to levels in controls (set to 100); data show effects of the stress with numbers in brackets for the relative level in recovery (2-h thaw, full rehydration sampled after 24 h, or 1 h back under aerobic conditions); the numbers are mean values from quantification of band densities on 3–5 blots whereas for *FR10* responses to freezing and anoxia, the arrows show results of $n = 1$ trials.

[b]For *in vitro* incubations, liver slices from control frogs were incubated with varying concentrations of second messengers (20–2000 μM db-cAMP or db-cGMP or 2–200 μM PMA) or for varying times (1–10 h) with a set concentration (200 μM) of second messenger. PMA is phorbol 12-myristate 13-acetate. n.a. = not available. n.s. = no significant change.

Source: Data are compiled from Q. Cai and K. B. Storey (1997). *Gene* **198**:305–312; J. D. McNally, S. Wu, C. M. Sturgeon, and K. B. Storey (2002). *FASEB J* **16**:902–904; J. D. McNally, C. M. Sturgeon, and K. B. Storey (2003). *Biochim Biophys Acta* **1625**:183–191.

Organs are still shrunken, and heart beat has not yet resumed so the liver is still under ischemia and volume stresses, but their suppression by 8 h thawed (when heart beat and breathing have resumed and liver appears to be visibly restored to normal size) indicates that they are no longer needed once recovery is well-advanced. Overall, then, this argues for important roles for the three novel proteins in dealing with the stresses associated with freezing.

Next we looked at the organ distribution of the transcripts to give us further hints as to protein role. In this regard the three proteins are very different. Transcripts of *fr47* were found only in liver, *li16* was prominent and strongly up-regulated in liver, heart, and gut whereas *fr10* appeared in all eight organs that we tested. Hence, the functions of FR47 can be interpreted as liver-specific, whereas Li16 is restricted to selected organs (which may share a common characteristic), and FR10 appears to have a universal function. Furthermore, FR47 protein was found in liver of two other freeze-tolerant frogs, *Pseudacris crucifer* and *Hyla versicolor* but not in freeze-intolerant *Rana pipiens* or *Scaphiopus couchii*, which again indicates a definite freeze-specific role.

Our next approach was to find out how these proteins responded to two of the component stresses of freezing—ischemia and cell dehydration. We found that transcripts of *fr10* were strongly up-regulated by dehydration, whereas both *li16* and *fr47* transcripts responded strongly to anoxia exposure. The responsiveness of *fr10* to dehydration and its presence in all organs leads us to think that this protein may have a protective role in dealing with the consequences of freezing on cell membranes, perhaps similar to the functions of the dehydrin or COR proteins in plants. Studies will next be designed to see if the overexpression of FR10 in cell lines from freeze-intolerant organisms can improve their freezing survival. Li16 protein levels rose strongly under anoxia (24 h under N_2 gas at 5°C) to very high levels (Table 17.4), suggesting a function for Li16 in the response to freeze-associated ischemia. Note that a significant response by Li16 to 40% dehydration (Fig. 17.8) is probably related to the hypoxia that occurs at high dehydration levels when blood viscosity becomes high and circulation is impaired. However, although *fr47* transcripts responded to anoxia, the protein did not. This difference between transcript and protein expression suggests that additional controls are involved in FR47 protein production; an anoxia/ischemia signal may have been "co-opted" to trigger expression of the *fr47* gene during freezing, but the protein is actually produced only when the frog is freezing.

Finally, if we can identify the signal transduction pathway that regulates the expression of a gene, we can get a further idea of the possible metabolic role of the protein or the possible initiating signal, such as a hormone, that may have triggered the signal transduction cascade involved. Recall that freeze-induced cryoprotectant synthesis in frogs was traced to the activation of β-adrenergic receptors and cAMP-dependent protein kinase. To gain similar information about *li16* and *fr47* expression, we incubated liver slices *in vitro* with different second-messengers (Table 17.4). Interestingly, neither gene responded to incubations with the cAMP analog, dibutyryl cAMP, but *li16* transcripts were up-regulated by exposure to dibutyryl cGMP (guanosine 3′5′ cyclic monophosphate). On the other hand, *fr47* transcript levels responded only to phorbol 12-myristate 13-acetate, a stimulator of the Ca^{2+} and phospholipid-dependent protein kinase C. This data provides additional evidence that the two proteins have quite different functions *in vivo* since they respond to different signals.

Furthermore, the results also support the idea that multiple signaling pathways are involved in implementing protective responses during freezing. We now know that protein kinases A, G, and C all participate in mediating different known metabolic responses to freezing and, in addition, we have measured freeze- or thaw-specific changes in the activities of two mitogen-activated proteins kinases: the stress-activated protein kinase (SAPK) and p38 (see Chapter 4 for information on these kinases) in frog organs. This provides strong evidence that natural freezing survival is not just a matter of packing cells with cryoprotectants and hoping for the best (as is still the practice in the pseudoscience field of cryonics) but involves coordinated responses by many aspects of cellular metabolism, triggered and regulated by multiple signal transduction pathways.

To return to the signal transduction pathways implicated in *fr47* and *li16* control, we had a previous indication that the PKC pathway is active in liver during freezing. Analysis of changes in second-messenger levels for both PKA and PKC in liver over the course of a freeze–thaw exposure showed that inositol 1,4,5-trisphosphate (IP_3), an intracellular second messenger of PKC, rose slowly over time during freezing, reaching about 3- and 11-fold higher than control values after 4 or 24h of freezing. Under anoxia stress, however, IP_3 levels rose by 8-fold within 30min in frog liver, but IP_3 was much less responsive to dehydration (about 3-fold higher in 20% dehydrated animals). The strong response of IP_3 to anoxia coupled with the graded accumulation during freezing (also seen for *fr47*) suggests that IP_3 accumulation in liver of freezing frogs mirrors the progressive development of anoxia/ischemia as the animal freezes and that one of the actions of IP_3 and PKC is the up-regulation of the *fr47* gene.

We do not have the same correlation with second-messenger levels for *li16*, but the putative link between *li16* and cyclic guanosine 5′-monophosphate (cGMP) fits very well with some of the new data that we have retrieved from cDNA array screening. Array screening indicated putative up-regulation of the adenosine receptor signaling pathway in heart and other organs of frogs during freezing: both A1 and A2 adenosine receptors were up-regulated as well as 5′-nucleotidase, the enzyme that synthesizes adenosine from AMP. Adenosine receptor signaling is mediated intracellularly by cGMP. Recall from Chapter 15 that adenosine accumulates quickly in the brain of anoxia-tolerant turtles when animals are exposed to anoxia and acts via adenosine A1 receptors to suppress the activities of ATP-dependent ion channels as part of the overall metabolic rate depression that ensures anoxia survival. Overexpression of adenosine A1 receptors is also known to increase myocardial tolerance of ischemia in transgenic mice. Hence, taken together, these data give a clear indication that as wood frogs freeze they activate a set of ischemia-protective signals and receptors and suggest that the Li16 protein has a role in this capacity. Overall, then, we have assembled a considerable amount of information about each of the three novel genes and, although we do not yet know the function of their protein products, we are continuing to move forward on several fronts to elucidate their responses and actions.

CONCLUSIONS

The fields of cold hardiness, cryobiology, and freeze tolerance are huge ones and we have only scratched the surface in this chapter with our focus on some of the metabolic and gene expression responses by freeze-tolerant organisms. Principles of biochemical adaptation that we have stated before have emerged again including (a) bending preexisting metabolic capacities to new functions and to new triggers, (b) producing minimum numbers of truly new protein adaptations, (c) employing low-molecular-weight carbohydrate protectants and selected small proteins to stabilize macromolecular structures under stress, (d) modulating regulatory mechanisms and signal transduction pathways to achieve new stress-specific outcomes (such as producing an uninhibited extreme hyperglycemia). Much, much more remains to be learned about natural freezing survival and with the help of amazing new genomics and proteomics technologies, we hope to continue the search for many years to come. Remember, as Albert Einstein once said "Science is the game we play with God to find out what his rules are" and it is, perhaps, the most amazing game around.

SUGGESTED READING

Benson, E., Fuller, B., and Lane, N. (2004). *Life in the Frozen State*. CRC Press, Boca Raton. (Includes as one chapter, K. B. Storey and J. M. Storey, Physiology, biochemistry and molecular biology of vertebrate freeze tolerance: The wood frog). *Up-to-date analysis of all aspects of freezing survival: biophysics of ice and water, mechanisms of ice damage, natural freeze tolerance in animals and plants, medical cryopreservation, cryobanking of biological materials, vitrification, and cell preservation by drying.*

Duman, J. G. (2001). Antifreeze and ice nucleator proteins in terrestrial arthropods. *Ann Rev Physiol* **63**:327–357. *Function, structure, and regulation of these specialized proteins in insects and other terrestrial invertebrates. See also suggested reading in Chapter 14 for a review on fish antifreeze proteins.*

Fowler, S., and Thomashow, M. F. (2002). *Arabidopsis* transcriptome profiling indicates that multiple regulatory pathways are activated during cold acclimation in addition to the CBF cold response pathway. *Plant Cell* **14**:1675–1690. *Superb example of state-of-the-art usage of cDNA array screening technology for identifying cold-responsive genes that shows how "waves of changes" in gene expression provide both early versus late and transient versus sustained responses during cold acclimation.*

Margesin, R., and Schinner, F. (1999). *Cold-adapted Organisms: Ecology, Physiology, Enzymology and Molecular Biology*. Springer, Heidelberg. *Multiauthor book dealing with many different topics in natural cold hardiness of animals, plants, and microorganisms.*

Storey, K. B. (1997). Organic solutes in freezing tolerance. *Comp Biochem Physiol A* **117**:319–326. *Review of the functions and regulation of low-molecular-weight cryoprotectants used by cold-hardy animals.*

Storey, K. B., and Storey, J. M. (1996). Natural freezing survival in animals. *Ann Rev Ecol System* **27**:365–386. *General review emphasizing the types of freeze-tolerant animals and the principles of survival.*

Storey, K. B., and Storey, J. M. (2001). Hibernation: poikilotherms. In *Encyclopedia of Life Sciences*. MacMillan Reference, London. Online at www.els.net. *Easy-to-read review of how animals survive in the cold.*

Storey, K. B., and Storey, J. M. (eds.) (2001). *Cell and Molecular Responses to Stress. Vol. 2: Protein Adaptations and Signal Transduction*. Elsevier, Amsterdam. *Multiauthor book with review on gene expression in animal freezing survival, transgenic expression of antifreeze proteins, cold-adapted enzymes, cold-shock proteins, and antioxidant defenses, among others.*

Thomashow, D. F. (1999). Plant cold acclimation, freezing tolerance and regulatory mechanisms. *Ann Rev Plant Physiol Plant Mol Biol* **50**:571–599. *A good recent review of the principles of cold hardiness and freeze tolerance in plants touching on membrane protection, gene expression, and signal transduction.*

18

ASPECTS OF BLOOD CELL BIOCHEMISTRY: ERYTHROCYTES, PLATELETS, AND STEM CELLS

JEAN E. GRUNDY

INTRODUCTION

Blood is a fluid tissue comprising 7 to 8% of the total body mass and containing various chemicals in solution and cells in suspension. It provides oxygen and nourishment to all organs, clears away waste products, and controls infection. The extracellular component of blood, representing ~55% of blood by weight, performs two major activities that are closely linked: (a) the response to infection, mediated by the complement system, and (b) hemostasis, or the balance between blood clot formation (coagulation) and dissolution (fibrinolysis). The cell component of blood is derived from eight different lineages, differentiated to provide oxygen delivery, clot formation, and a broad spectrum of cells that act at different levels to defend the host from infection (Fig. 18.1).

Since blood is easily collected with little harm to the host organism and there is no need to disrupt tissue structure in order to isolate cells, blood cells have been widely used for studies of differentiation and metabolism. As well, cells with very different functions are similar in terms of their intracellular organization, metabolism, and chemistry, which allows information from intracellular mechanisms in blood cells to inform our understanding of other, less accessible cells. Most of the metabolic pathways found in blood cells are universal so that a metabolic step can be discussed without having to identify the cell it came from.

The pluripotent stem cell, from which other blood cells are derived, differentiates in response to specific signals into one of the cell lineages, undergoing replication and producing many mature differentiated cells. In the process of differentiation, many of the mature cell systems in blood lose certain subcellular organelles, so their metabolic systems are simplified compared with the cells of most organs. Ranked from most to least complex are the leukocytes, platelets, reticulocytes, and erythrocytes. Erythrocytes, for example, are simplified yet highly specialized cells. They have no nucleus or mitochondria and, therefore, lack numerous complex regulatory interactions seen in most other cells. They are vulnerable to damage because they cannot synthesize new proteins and are completely dependent on anaerobic glycolysis for energy generation. Platelets have mitochondria and, therefore, can synthesize mitochondrial protein, metabolize fat and synthesize glucose, but lack a nucleus and so cannot synthesize most proteins. However, their most interesting characteristic is the stunning change they undergo as they become activated and participate in clot formation. These terminal cells, with simplified metabolic systems, can provide information about metabolic systems in the absence of other competing pathways.

The process by which blood cells mature and ultimately reach the circulation is called hematopoiesis. Mature blood cells are derived from hematopoietic organs, namely bone marrow, spleen, liver, and lymph nodes. Cell composition of the hematopoietic organs and the peripheral blood is extremely diverse, with a multilayered organization of the hematopoietic system from stem cells to committed progenitor cells and precursor cells, to final release of mature cells into the blood, lymph nodes, or tissue. This chapter discusses metabolic aspects of three cell types: erythrocytes, platelets, and stem cells. Erythrocytes and platelets are mature differentiated cells responsible for oxygen delivery to tissues and for blood clotting, respectively. Hematopoietic stem cells differentiate into their progeny,

Functional Metabolism: Regulation and Adaptation, edited by Kenneth B. Storey.
ISBN 0-471-41090-X

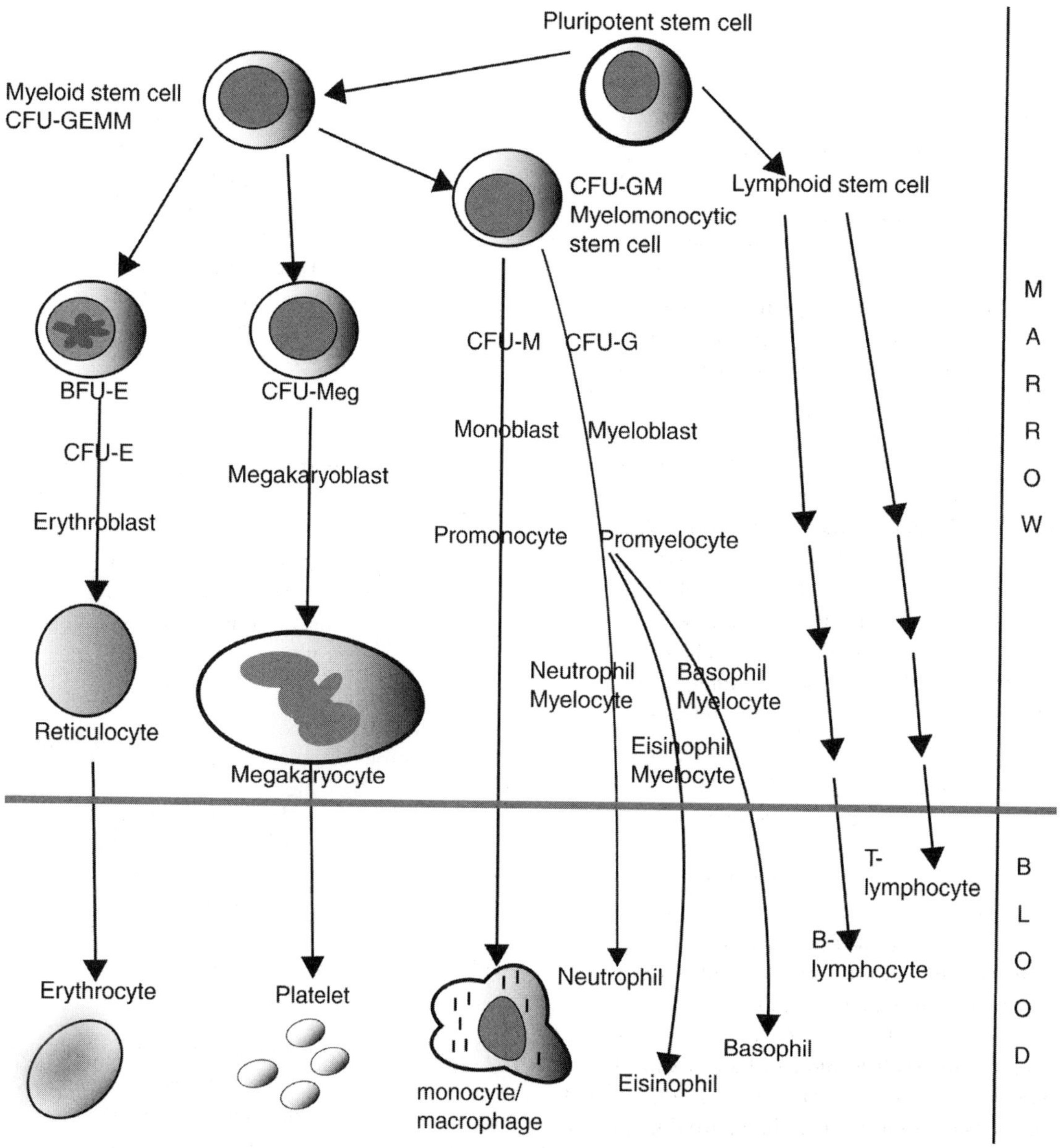

Figure 18.1 Hematopoietic cell lineages in humans. Hematopoietic stem cells differentiate into cells of eight different lineages in response to different cytokines. The different compartments of differentiation are the stem cell compartment, the committed progenitor cell compartment, and the precursor cell compartment. Mature cells that do not contain a nucleus (platelets and erythrocytes) have lost their nuclear material at the last step.

mature differentiated blood cells, and thus can be termed the "mother" of all blood cells.

ERYTHROCYTES

Physiology

Erythrocytes, or red blood cells (RBCs), are produced at a rate of 250 billion per day in humans and have a lifetime in circulation of approximately 120 days. During this time, RBCs traverse the circulation over 100,000 times, frequently deforming their biconcave disk shape to squeeze through microcapillaries smaller than their 7.5 μm diameter. RBC membranes are pliable yet resilient to resist shear-induced fragmentation, distensible to allow swelling or shrinkage in response to osmotic challenge, and resistant to oxidation and proteolysis. The red color of the cells is derived from the enormous load of the intracellular oxygen-binding protein, hemoglobin (Hb), which is present at levels close to its limit of solubility. This high protein concentration creates a powerful osmotic pressure gradient across the cell membrane so that RBCs swell in isotonic solutions and shrivel in hypertonic solutions. During circulation through the kidney, RBCs pass through zones of extracellular fluid concentrations that range from isotonic to 6× isotonic and back in milliseconds. Iron and hemichromes released from damaged Hb trigger oxidative stress, intensified by marked pH and pO_2 changes that are endured in the kidney, spleen, and muscle. The erythrocyte membrane is durable and resilient, adapted to meet these challenges.

Anion exchanger 1 (AE1), also known as protein 3 or band 3, is a highly abundant protein in erythrocytes and the major anion (HCO_3^-/Cl^-) exchange protein. It is composed of a 43-kDa amino-terminal cytosolic domain that binds the cytoskeleton, hemoglobin, and glycolytic enzymes and a 52-kDa carboxyl-terminal domain that mediates anion transport. The link between the domains is flexible, which permits rotation of AE1 despite its attachment to the cytoskeleton. AE1 suppresses metabolic pathways by sequestering key pathway enzymes, such as glyceraldehyde-3-phosphate dehydrogenase (GAPDH), phosphoglycerate kinase (PGK), and aldolase, as well as carbonic anhydrase II, which bind to its highly acidic cytoplasmic region. The same region also binds hemoglobin and hemichromes (generated upon denaturation or oxidation of hemoglobin). Hemichrome binding appears to stimulate the formation of clusters of AE1 (discussed further below) that are recognized by an AE1 isoantibody, leading to removal of the cell from the circulation by the reticuloendothelial system.

Metabolism

Human RBCs are devoid of a nucleus, mitochondria, polyribosomes, and nucleic acids. They cannot synthesize new proteins or repair damage to existing ones. RBCs rely solely on sugar fermentation for maintaining energy and redox balance, supplying adenosine 5′-triphosphate (ATP), 2,3-diphosphoglycerate (2,3-DPG), reduced nicotinamide adenine dinucleotide (NADH), and NADH phosphate (NADPH) and reduced glutathione (GSH) as needed. Among other activities, energy currencies are used in the RBC to support osmotic stability through ATP-dependent membrane pumps, maintain hemoglobin iron in its ferrous (Fe^{2+}) form via the NADH reducing system, and protect membrane proteins and enzymes from oxidative damage via NADPH/glutathione-dependent mechanisms. Glucose is the main sugar substrate for metabolism, but RBCs can also use galactose, fructose, and mannose; however, pentoses and disaccharides are not metabolized. Glucose uptake into erythrocytes is mediated by the glucose transporter isoform (GLUT1) of which there are about 300,000 copies per cell. The high glucose transport capacity of the human erythrocyte is consistent with its dependence on glucose for energy. However, mature human RBCs actually consume glucose at a slow rate compared with the cells of other tissues; for example, glucose consumption by white blood cells is about 500 times higher than that of RBCs.

Fructose entry into RBCs is mediated through two other glucose transporter isoforms: GLUT5, which does not transport glucose but has a high affinity for fructose, and GLUT2, which has a weak affinity for both glucose and fructose. Both glucose and fructose enter the glycolytic pathway after phosphorylation by the enzyme hexokinase to produce glucose-6-phosphate (G6P) or fructose-6-phosphate (F6P), respectively. The primary substrate of hexokinase is glucose but it has a low affinity for fructose. While the liver is considered to be the primary consumer of fructose, RBCs may be a significant but underrecognized contributor to nonhepatic fructose clearance.

Since RBCs have minimal glycogen stores, they are completely dependent upon plasma sugars for fuel. Although the normal enzymes for glycogen production and breakdown are present, and glycogen turnover does occur, the activity of glycogen synthetase is low. In some disorders, the activity of glycogen breakdown enzymes is low, resulting in a net accumulation of glycogen in the erythrocyte (glycogen storage disease). In forms of the disease where phosphofructokinase or phosphohexose isomerase are affected, preventing glycogen breakdown, excessive glycogen accumulation can occur to the point of red cell rupture (hemolytic anemia). Disorders of other enzymes of glycogen breakdown do not greatly affect erythrocytes, other than increasing glucose consumption.

Over 95% of glucose consumed by erythrocytes is processed through the Embden–Myerhof pathway (EMP) of anaerobic glycolysis. The remaining 5% passes through the pentose phosphate pathway (PPP) to generate the NADPH reducing power needed by the cell. The enzymes of glycolysis are associated with the plasma membrane in RBCs and regulate the three most important intermediates in erythrocyte metabolism: 2,3-DPG, ATP, and oxidized nicotinamide adenine dinucleotide (NAD^+). Glycolysis in erythrocytes is the same as in all other cells but carbon can be diverted at one key point. This is the Rapoport–Leubering shunt, which uses 1,3-diphosphoglycerate to synthesize 2,3-DPG via the enzyme diphosphoglycerate mutase (Fig. 18.2).

2,3-Diphosphoglycerate is an important regulator of oxygen binding to Hb. Hb is a tetramer of two α-β heterodimers, each subunit containing a heme group capable of binding oxygen. Oxygen binding is cooperative, which means that as one subunit binds oxygen, the conformation of adjacent subunits changes in a way that makes them more likely to bind oxygen. This results in a sigmoidal

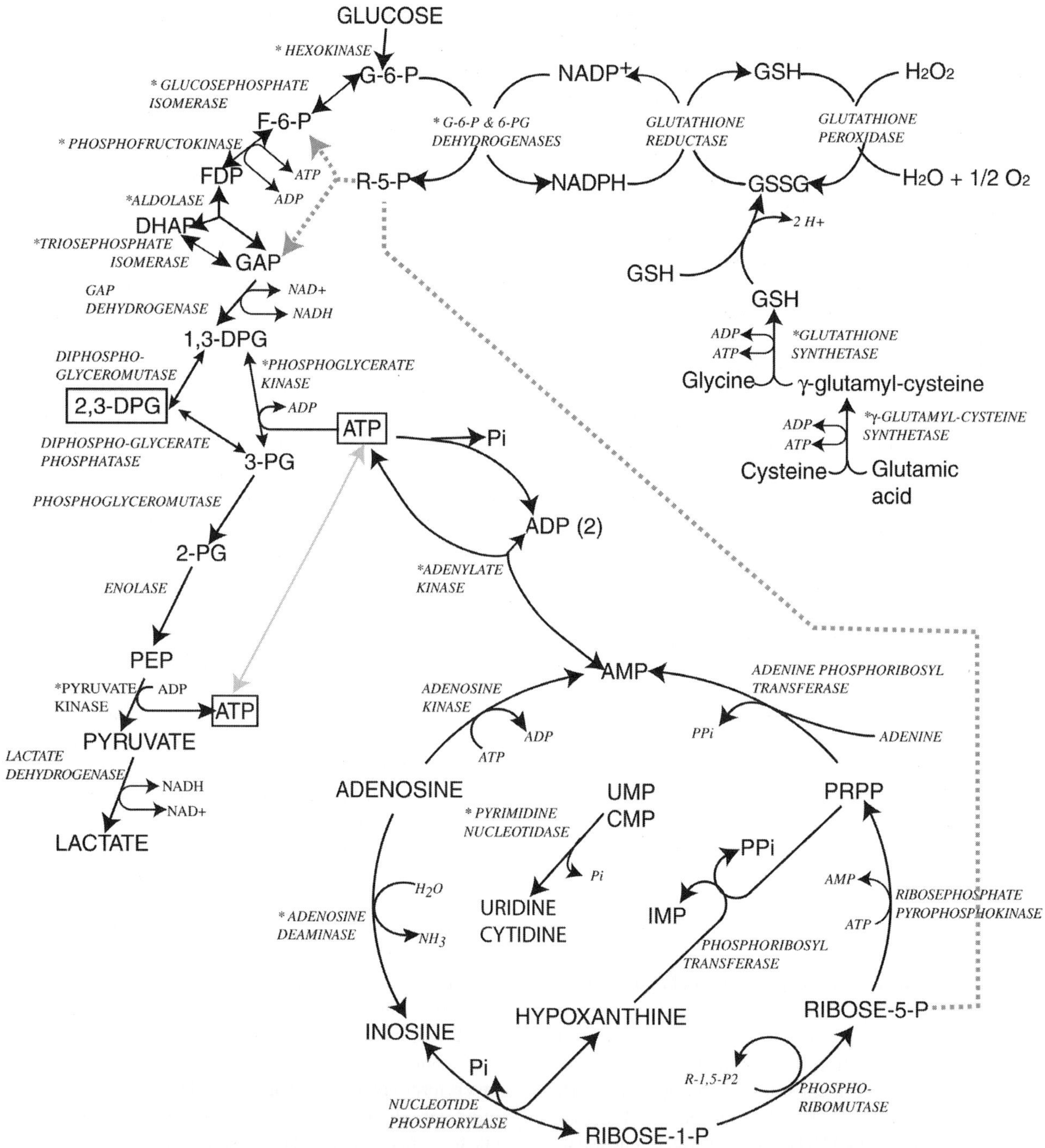

Figure 18.2 Glycolytic and related processes in the human red cell. Stars indicate enzymes found to be defective in association with hereditary hemolytic anemias. [From Scriver et al. (eds.) (1989). *The Metabolic Basis of Inherited Diseases*, McGraw-Hill, New York, p. 1608. With kind permission from the publishers McGraw-Hill, Inc.]

binding curve. Hb exists in two conformations: the taut (T) conformation, which is deoxygenated and has a relatively low oxygen affinity, and the relaxed (R) conformation, which has an oxygen affinity 500 times higher. 2,3-DPG binds between the β subunits of T-Hb to stabilize the T conformation, reduce oxygen affinity, and promote oxygen unloading. The binding of oxygen and 2,3-DPG to Hb is mutually exclusive. As Hb picks up oxygen in the lungs, its conformation changes from T to R, the intersubunit space increases, 2,3-DPG is displaced, and oxygen is bound. On reaching the peripheral tissues where there is a low partial pressure of O_2, 2,3-DPG reinserts itself, O_2 is released from the Hb [reaction (18.1)], and Hb returns to the T conformation.

$$\text{Hb-DPG} + 4\text{O}_2 \longleftrightarrow \text{Hb-4O}_2 + \text{DPG} \qquad (18.1)$$

Amino acid substitutions in the globin chains that govern heme-oxygen interactions may impact negatively on 2,3-DPG binding to Hb. The result is increased Hb affinity for oxygen and a lower release of O_2 in the peripheral tissues, creating a situation of low O_2 (hypoxia). Such tissue hypoxia stimulates erythropoietin synthesis, mediated via the hypoxia-inducible transcription factor (HIF) (discussed in Chapters 6 and 15), and this stimulates increased red cell production. The same effect is also associated with a rare enzymatic disorder affecting 2,3-DPG mutase, in which insufficient 2,3-DPG is produced.

Modulation of 2,3-DPG levels is an important regulatory mechanism for improving O_2 release from hemoglobin under various stress conditions. Elevated 2,3-DPG is found in the RBCs of people that are experiencing hypoxia whether it be as a result of a lung ailment or due to high altitude. Increased levels of 2,3-DPG allow hemoglobin to release O_2 at lower partial pressures. Oppositely, fetal hemoglobin is "engineered" to have an increased affinity for O_2. γ Subunits replace β subunits in the fetal Hb tetramer ($\alpha_2\gamma_2$ instead of $\alpha_2\beta_2$), the γ subunit binding 2,3-DPG less strongly than the β subunit. This ensures that the fetus, which has to obtain O_2 from the maternal blood supply at lower partial pressures than outside the womb, is able to take O_2 from maternal Hb.

The primary energy currency of the erythrocyte is ATP and is associated with three reactions: (1) It can substitute for 2,3-DPG in affecting the position of the Hb–oxygen dissociation curve, (2) the preservation of red cell shape and flexibility is closely dependent on the concentration of ATP in the cell, and (3) the active transport of cations across the cell membrane, maintaining specific intracellular cation concentrations, requires energy provided by ATP.

The ATP and redox equivalents must be continually regenerated by erythrocyte metabolism. If an enzyme deficiency or metabolic stress disrupts the ATP or NADPH generation system, distinct membrane changes result, particularly to AE1, and these cause recognition of the RBCs as "damaged," promoting their removal by the spleen and the macrophage system.

Glycolysis converts glucose into two molecules of pyruvate, generating two molecules each of ATP and NADH in the process. The NAD^+ consumed by GAPDH must be regenerated for glycolysis to continue. In erythrocytes, which lack mitochondria, this is accomplished by the reduction of pyruvate to lactate, making lactate the primary end product of RBC energy metabolism. Hence, RBCs have continuous lactate output, and alterations in RBC metabolism may be observed as changes in blood lactate levels. Lactate is efficiently returned to the liver for gluconeogenesis.

Redox equivalents maintain the reduced state of hemoglobin (Fe^{2+}) and reduced sulfhydryl groups of enzymes, glutathione, and membrane components. The most important metabolite to redox metabolism in erythrocytes is the tripeptide glutathione (γ-glutamyl cysteinyl glycine, abbreviated GSH), which is one of the most abundant short peptides in eukaryotic cells and the most prevalent intracellular thiol. The oxidized form of glutathione, GSSG, is composed of two molecules of glutathione linked by a disulfide bond. GSH can be directly oxidized to GSSG by oxygen-centered free radicals or can act as the mandatory substrate for the enzymes glutathione peroxidase and glutathione transferase. GSSG is recycled back to two molecules of GSH by the enzyme glutathione reductase, which uses the NADPH product of the PPP as its source of reducing equivalents [reaction (18.2)] (see Fig. 18.2). Antioxidant defense, especially provided by renewable GSH, is of great importance to RBCs because of the presence of high levels of both iron and oxygen in the cells. Oxidation of Fe^{2+} to Fe^{3+} creates a catalyst for generation of harmful reactive oxygen species:

$$\text{GSSG} + 2\ \text{NADPH} + 2\text{H}^+ \longrightarrow 2\ \text{GSH} + 2\ \text{NADP}^+ \qquad (18.2)$$

Regulation of Glycolysis

The maximal activities of glycolytic enzymes of the EMP range in RBCs over three orders of magnitude, with glucose phosphate isomerase, triose phosphate isomerase, phosphoglycerate kinase, and lactate dehydrogenase having the highest activities, whereas hexokinase and aldolase have the lowest. Glycolytic flux typically operates at a rate well below the maximum possible rate. The most important enzymes regulating the rate of glycolysis in RBCs are hexokinase, phosphofructokinase (PFK), and pyruvate kinase (PK).

The control of glycolysis by these three key enzymes in RBCs may be considered by examining two segments of the pathway: the section containing hexokinase and PFK and the section from GAPDH onward. For the hexokinase–phosphofructokinase section, the relevant reactions are as follows:

$$\text{Glucose} + \text{ATP} \longrightarrow \text{G6P} + \text{ADP} \quad \text{(hexokinase)} \qquad (18.3)$$

$$\text{G6P} \longleftrightarrow \text{F6P} \quad \text{(phosphoglucoisomerase)} \qquad (18.4)$$

$$\text{F6P} + \text{ATP} \longrightarrow \text{F1,6P} + \text{ADP} \quad \text{(PFK)} \qquad (18.5)$$

where F1,6P is fructose-1,6-phosphate.

An increase in the concentration of ATP will increase the level of Mg–ATP as the substrate for hexokinase and PFK. Of the two enzymes, PFK has higher affinity for Mg–ATP, but ATP inhibits PFK activity at higher levels (substrate inhibition). Hexokinase flux increases with increasing ATP, but the product, G6P, inhibits hexokinase activity. The net result is a slowdown of glycolysis as a result of an increase in cellular ATP. Oppositely, a decrease in ATP content releases PFK inhibition, increases the consumption of G6P so that hexokinase is activated, and speeds up the general rate of glycolysis.

The second segment of the pathway, from GAPDH onward, produces ATP from ADP, recycles NADH to NAD^+, and produces 2,3-DPG. The role of PK in limiting this pathway is to regulate this section of the pathway and to distribute the flow of substrate into the 2,3-DPG bypass. PK is inhibited by ATP in the same manner as PFK, forcing higher production of 2,3-DPG when PK is inhibited. PK is also inhibited by 2,3-DPG, although the reason for this inhibition is not clear.

Other factors contribute to glycolytic flux, so it is important not to assume that simple ATP concentrations control glycolysis in red cells. For instance, adenosine 5′-monophosphate (AMP), which plays a key modulatory role in muscle cell glycolytic regulation, may play an as yet unrecognized role in blood cells.

Erythrocyte Enzymopathies

Diseases caused by deficiencies of selected enzymes in RBCs are widespread and of major consequence to human populations. Enzymopathies can be divided into three groups. The first are those that involve the main glycolytic pathway and, hence, impact on cellular energy production. Defects in glycolysis generally cause a nonspherocytic hemolytic anemia. The most commonly found and best-studied of these defects is pyruvate kinase deficiency. The second group involves defects in the PPP, which produces the reducing equivalents needed for antioxidant defense. The most common erythrocyte enzymopathy belongs to this group: glucose-6-phosphate dehydrogenase (G6PDH) deficiency exists in ~400 million people worldwide. However, only in severe cases is the deficiency linked to hemolytic anemia. The third broad category of RBC enzymopathy involves defects of the various linked reactions, the most important of which are deficiences in the enzymes that synthesize or catabolize 2,3-DPG or in methemoglobin reductases that catalyze the reduction of Fe^{3+} (ferric heme iron, capable of catalyzing production of reactive oxygen species) to Fe^{2+} (ferrous heme iron, which carries oxygen in heme).

Glucose-6-Phosphate Dehydrogenase Deficiency Glucose-6-phosphate dehydrogenase (G6PDH) catalyzes the first reaction of the PPP [reaction (18.6)] and is one of its two NADPH-producing reactions:

$$G6P + NADP^+ \longrightarrow \text{6-phosphogluconate} + NADPH + H^+ \tag{18.6}$$

The PPP is regarded as a shunt since the pentose sugars it produces can be recycled by transketolase and transaldolase to fructose-6-phosphate and triose phosphate. The pentose sugars can also be produced from fructose-6-phosphate by the reverse reaction, bypassing the first part of the PPP, and making it redundant for pentose production. The indispensable product of the PPP is NADPH, which is not produced elsewhere in carbohydrate metabolism. NADPH is a coenzyme in many reductive biosynthetic pathways, such as production of fatty acids, cholesterol, purines, and pyrimidines. In RBCs, most of these pathways have been lost during maturation, and the primary role of the pathway is in NADPH generation for antioxidant defense.

NADPH is the substrate for GSH regeneration by glutathione reductase (GR). GR maintains GSH concentrations for protection against damage by oxygen radicals, including as a reactive oxygen detoxifying substrate for peroxidase enzymes (as described above). NADPH itself is important for the activity of catalase, which directly reduces hydrogen peroxide.

Glucose-6-phosphate dehydrogenase has a high affinity for $NADP^+$ (K_m is approximately 3 to 13 μM) and is inhibited by NADPH (K_i is approximately 16 μM). In normal red cells in the steady state, $NADP^+$ concentration is very low, and NADPH concentration is about 35 μM. Thus G6PDH in red cells is normally almost completely inhibited, estimated to function at 1/60 of its maximal capacity. This provides a large reductive potential for rapid NADPH generation, without any need to synthesize new protein. When an oxidative insult occurs, the rapid result is GSH depletion as it is used directly and by glutathione-utilizing enzymes to neutralize hydrogen peroxide and organic peroxides. As GSH is replenished, NADPH is used, and $NADP^+$ is formed. G6PDH inhibition is decreased and glucose flux is diverted through the PPP to produce more NADPH.

The most common erythrocyte enzymopathy is G6PDH deficiency. In G6PDH-deficient cells, the enzyme is present in an inactive or much lower activity form, and under normal conditions the enzyme activity is at or close to its maximal capacity. The large reductive potential does not exist, and, since G6PDH cannot work any faster when challenged by oxidative stress (e.g., oxidant chemicals, bacterial/viral infections), G6PDH-deficient RBCs cannot sufficiently raise the rate of their PPP to prevent oxidative damage. Oxygen-centered free radicals are formed and catalyze multiple damaging effects (see Chapter 12).

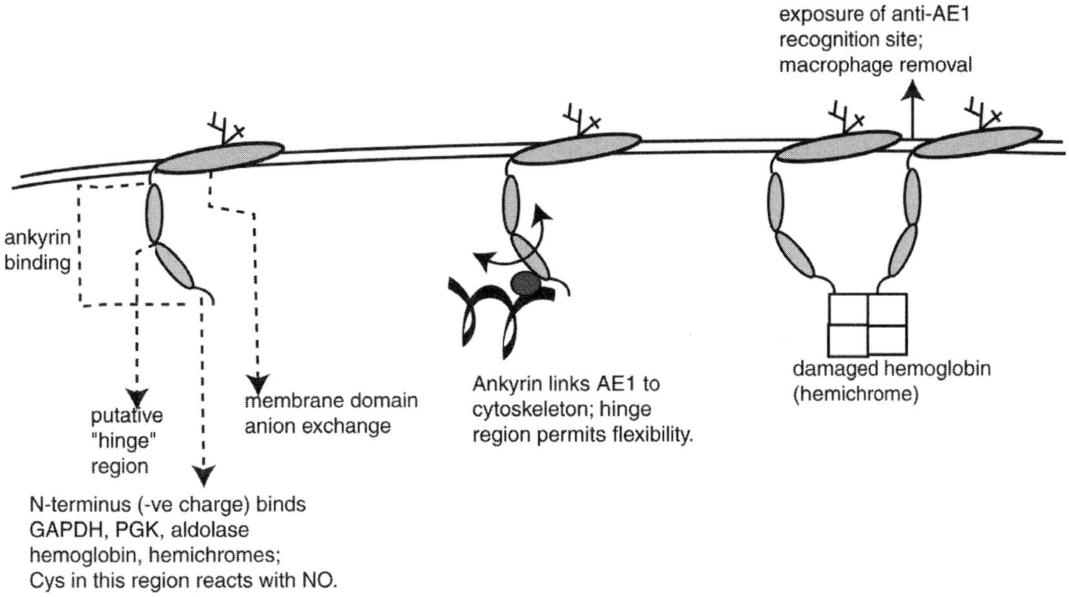

Figure 18.3 AE1 (Anion exchange protein, band 3). The N-terminal end of AE1 binds proteins and the C-terminal end mediates anion exchange. Point mutations in the so-called hinge region of the N-terminal domains cause rigidity at this region and loss of deformability of the erythrocyte. Hemichrome binding and subsequent clustering of AE1 results in extracellular exposure of a recognition site for naturally circulating anti-AE1 antibodies, followed by macrophage recognition and removal.

In particular, oxygen free radicals increase the formation of hemichromes (a mixture of heme and denatured globin). The iron in hemichromes catalyzes the formation of other free radicals that accelerate oxidative damage, especially of the red cell membrane. The cytoplasmic domain of the membrane protein AE1 binds hemichromes, and two sensitive cysteine residues are oxidized to form a disulfide (Fig. 18.3). This results in clustering and formation of high-molecular-weight AE1 aggregates that form an epitope for naturally circulating anti-AE1 antibodies, permitting high-affinity antibody binding. AE1 is present in many tissues of the body: Presentation of the AE1 antigen in response to damage to the protein is common to most cells, and it has been termed *senescent cell antigen* (Fig. 18.3). The end result is removal of the circulating red cell by the spleen, recognition of the damaged erythrocyte by macrophages, and its destruction. During an oxidative crisis, anemia due to the destruction of RBCs occurs in G6PDH-deficient individuals. Another common symptom is jaundice, as the liver struggles to break down massive amounts of hemoglobin liberated by macrophage-induced erythrocyte removal.

Nearly 70 variants of G6PDH mutation exist, most being point mutations that result in an amino acid substitution. Many of these variants cause a small level of G6PDH deficiency but are asymptomatic; in one case it is actually adaptive in dealing with malaria (see Text Box 18.1). The most commonly known G6PDH deficiency is the variant called Mediterranean. Sufferers react to the oxidant, divicine, found in fava beans with sudden, painful crises after a meal of beans or a walk through a field of beans in

TEXT BOX 18.1 G6PDH DEFICIENCY AND MALARIA RESISTANCE

The geographic distribution of G6PDH deficiency corresponds to an area of the world with high malaria risk. Hemizygous males and heterozygous females show increased resistance to malaria caused by the parasite *Plasmodium falciparum*. Although other factors make contributions, the advantage of G6PDH deficiency is as follows. Parasite invasion into RBCs induces an oxidative stress that the infected cell cannot overcome because it is unable to increase its rate of NADPH production by the PPP due to its limiting G6PDH activity. Hence, GSH levels drop rapidly, oxidative damage accumulates, AE1 clustering and antigen presentation occurs, and the infected cells are destroyed by macrophages. Hence, parasitized cells are removed from circulation, preventing or minimizing the development of the disease.

flower. Crises can last for days. Research has also shown that primaquine, and other oxidative drugs, induce the same symptoms in affected individuals within a few hours.

Other Enzyme Deficiencies Pyruvate kinase deficiency has been implicated in chronic hemolytic anemia, a condition in which ATP levels are decreased and the concentration of 2,3-DPG is increased twofold. Low pyruvate kinase activity limits the rate of ATP production by this reaction of glycolysis and causes a "backup" of carbon in previous steps and an "overflow" into 2,3-DPG production. 2,3-DPG inhibits pyruvate kinase and also inhibits hexokinase, which may exacerbate the decrease in glycolytic ATP production caused by the pyruvate kinase deficiency. Other rarer enzymopathies are hexokinase, phosphoglucoisomerase, phosphofructokinase, aldolase, triosephosphate isomerase, PGK, or enolase deficiencies. The pattern of inheritance for most of these disorders is autosomal recessive and metabolic depletion, and failure of energy generation is the primary pathogenic mechanism of hemolysis in this group of enzymopathies. Treatment is usually splenectomy, which is not curative, but hemoglobin levels improve or transfusion requirements decrease. This may be because the spleen normally sequesters blood cells, and once it is removed, all blood cells must circulate.

IRON METABOLISM

Iron is a key element in the metabolism of all organisms, particularly as a part of heme. Heme is an integral part of many proteins including cytochromes and cytochrome oxidase of the electron transport chain, enzymes of the Krebs cycle, the oxygen-binding proteins myoglobin and hemoglobin, and various peroxidases that protect cells from oxidative injury by reducing peroxides to water. Nearly half of the enzymes and cofactors of the Krebs cycle either contain iron or require its presence.

Approximately 2.0g of body iron in men and 1.5g in women is contained in hemoglobin, which is 0.34% iron by weight. One milliliter of packed erythrocytes contains approximately 1mg of iron. From the point of its entry into the plasma, iron is kept in a tightly controlled system, protein-bound and reduced. It is transported in plasma by transferrin to the erythroblasts of the marrow. There transferrin binds to the transferrin receptor (TfR) on the erythroblast membrane. The membrane of a reticulocyte (the final precursor to an erythrocyte) can bind up to 50,000 diferric transferrin molecules per minute. Transferrin bound to its receptor forms clusters in pits on the cell membrane, and these are internalized by endocytosis. Iron release is achieved by acidification when the internalized vesicle fuses with an endosome and the apotransferrin-transferrin receptor complex then returns to the cell membrane, where at neutral pH, apotransferrin is released back into the plasma. Within the vesicle, another protein called the divalent metal transporter 1 (DMT1) releases Fe^{3+} into the cytosol, where it is taken up by mitochondria for heme synthesis. Within mitochondria, iron is inserted into protoporphyrin by heme synthetase. When heme synthesis is impaired, as in lead poisoning or in the sideroblastic anemias, the mitochondria accumulate excessive amounts of amorphous iron aggregates.

Iron is rigorously controlled in all tissues because the free metal ion participates in the Haber–Weiss and Fenton reactions that generate highly reactive hydroxyl ions (see Chapter 12). Oxidative damage arising due to high levels of free iron has been linked to inflammatory diseases such as arthritis and to the development of atherosclerosis and cancer. To prevent excessive iron release, senescent, damaged, or malformed red blood cells are cleared by macrophages. The safe catabolism of hemoglobin, formation of bile pigment, and iron recycling takes place outside of the circulation. If intravascular hemolysis occurs, free hemoglobin is released and is stoichiometrically complexed to haptoglobin. This complex is recognized and internalized by tissue macrophages via the CD163 receptor.

However, since macrophages themselves contain reactive oxygen species, there is a danger that heme and complexed iron taken up by these cells can potentiate macrophage-derived reactive oxygen metabolites. These reactions are thought to contribute to tissue injury and pathological processes such as atherosclerosis. This emphasizes the need for tight physiological controls over iron since it can catalyze formation of tissue-damaging species even from within the cells that remove it from the circulation.

Within the closed system of iron use and degradation, about 80% of iron is rapidly reincorporated into hemoglobin with some remainder routed into storage pools associated with the proteins ferritin or hemosiderin. There is no mechanism for iron excretion, and it is lost only adventitiously due to bleeding (including menstruation) or as skin cells are sloughed off.

Another Role for Red Blood Cells: Nitric Oxide Delivery

Nitric oxide (NO) is a signaling molecule that within the blood system is particularly important as a platelet activation inhibitor and as a vasodilator. It is secreted by endothelial cells but has to be carried to its target via the blood. Until recently it was believed that passive diffusion was the only way to get NO to its smooth muscle target cells. NO reacts rapidly with heme and is immediately converted to nitrate, suggesting that endothelium-derived NO would be rapidly consumed in the blood and unavailable to smooth muscle. However, recent studies suggest that contrary to the dogma of passive diffusion of NO from endothelial cells to target smooth muscle cells, NO delivery is not simple, random, or mediated solely by diffusion. Studies showed that erythrocytes that were pretreated with NO were able to relax smooth muscle cells, suggesting that NO activity can be salvaged from red blood cells. While most NO is destroyed by heme, a small portion is transferred intramolecularly to the 93-β-cysteine residue of "relaxed" R-Hb, binding to the cysteine thiol to form *S*-nitrosothiol-hemoglobin. When the erythrocyte enters an area of the circulatory system with a low oxygen saturation, oxygen is released from hemoglobin and the quaternary structure of the protein reverts to the deoxygenated form. This promotes binding of NO-bound hemoglobin to the anion exchange protein AE1 and transfer of NO to a thiol on AE1. *S*-nitrosothiol-AE1 can transfer NO activity out of the cell and into the vessel wall, possibly by transport through AE1, diffusion across the membrane, or as the erythrocyte bumps along the wall itself. The precise method for delivery of NO from the erythrocyte to smooth muscle cells in capillaries has not yet been elucidated. Whatever the method of delivery, NO then acts as the endothelial-derived relaxing factor (a term that was originated to describe this activity before it had been identified with NO) to allow widening of the capillaries and increased blood flow and oxygen delivery to tissues.

PLATELETS

Physiology

Resting human platelets are anucleate cells of 2 to 3 μm in diameter, with a cell volume of 10 fL (Fig. 18.4). Their prevalence in blood is 250,000 $\pm$ 100,000 cells/μL of blood, and they are produced at a rate of approximately 40,000/μL of blood per day with a life span of 7 to 10 days. Platelets are essential parts of the reactions of blood coagulation, inflammation, and wound healing. Their primary function is to stop the bleeding that results from injury to blood vessel walls by (a) forming a hemostatic plug at the site of injury and (b) providing a catalytic site for the coagulation process. Platelets can also participate in pathophysiology as they help form the intravascular clots that are responsible for the morbidity and mortality associated with arterial vascular disease. Depletion of these short-lived minuscule cells occurs as a result of their use in hemostatic reactions and *senescence* (aging platelets expose signals that trigger their removal from the circulation by macrophages). Platelets exist in excess of their normally required numbers and, in general, spontaneous bleeding does not occur until platelet numbers are reduced to at least one-tenth of normal.

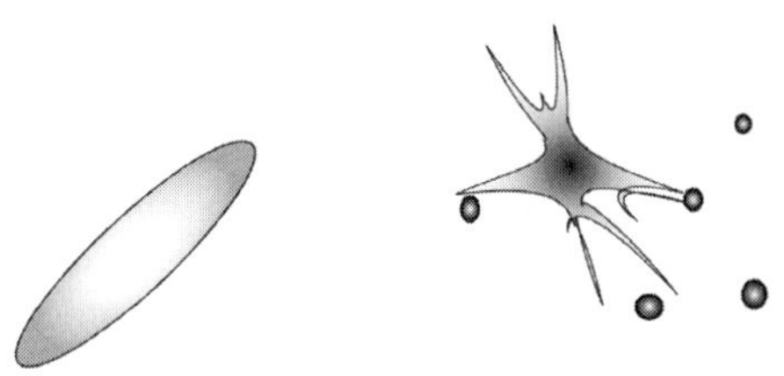

Figure 18.4 Resting and activated platelets. Platelets change shape upon activation, releasing granule contents, extending pseudopods from the plasma membrane and producing anionic-phospholipid-rich microparticles, which themselves contribute to clot formation.

Platelets are discoid, having a smooth surface interrupted periodically by small pits. These pits connect the surface to a networked system of membrane-bound tubes, called the surface-connected canalicular system (SCCS), also called the open canalicular system. This network is a reservoir of plasma membrane, membrane receptors, and proteins to be used when the cell activates and a conduit to the extracellular milieu into which intracellular granules release their contents for secretion after activation. Platelets contain two important types of secretory granules: α granules and dense granules. α Granules contain clotting factors, such as factor V, factor VIII, fibrinogen, high-molecular-weight kininogen, von Willebrand factor, thrombospondin, glycoprotein CD62 (P-selectin), platelet factor IV, β-thromboglobulin, platelet-derived growth factor, and low concentrations of all plasma proteins (e.g., immunoglobulin G and albumin). Dense granules contain Ca^{2+}, Mg^{2+}, adenosine 5′-diphosphate (ADP), and ATP plus several vasoactive amines, notably 5-hydroxytryptamine (serotonin). There is also a membranous compartment that is randomly woven through the cytoplasm, which serves as the major calcium store for the platelet: This is the dense tubular system (DTS). The DTS is similar to the sarcoplasmic reticulum in other cells. The membranes forming the DTS contain an inwardly directed Ca^{2+}/Mg^{2+}–ATP pump that maintains the cytoplasmic concentration of Ca^{2+} in the nanomolar range in the resting platelet. The DTS is also the major site of prostaglandin and thromboxane A_2 (TxA_2) synthesis in the platelet. Platelets also contain lysosymal granules, similar to those of other cells, and inclusions composed of glycogen, which can be rapidly mobilized to form glucose when the platelet is stimulated or used as energy sources in the absence of glucose (e.g., in storage solutions without glucose).

The platelet cytoskeleton is an elaborate framework of a planar spectrin network surrounding the cytoplasm, crosslinked by long actin filaments in a rigid structure that fills the cytoplasmic space. Actin in the cytoplasm is crosslinked by filamin and α-actinin. Most of the platelet filamin is complexed to a platelet surface membrane receptor, GPIb/IX/V complex, which is the receptor for a large plasma protein called von Willebrand Factor (vWF). This effectively links the vWR receptor to actin and also makes the spectrin network more rigid as GP1b/IX/V–filamin units are trapped on the surface in the spectrin web. The spectrin surface confines platelet membrane receptor movement and is a scaffold to which membrane and cytoplasmic signaling proteins attach. Tubulin, a cytoskeletal protein used to transmit force for organelle transport and chromosome separation, is also found in platelets, tightly coiled. It is a matter of discussion at present whether tubulin helps to maintain the discoid shape of the platelet, is simply a remnant of platelet production, or helps to form filipod extensions of the plasma membrane that are extended during cell activation and spreading.

Resting platelets circulate until they encounter an external stimulus, such as a nonendothelial surface, thrombin, or collagen. Contact with the stimulus induces a signal transduction cascade that results in drastic and usually irreversible changes, termed platelet activation. Platelet activation can also be induced by altered conditions, such as cooling, which induces a membrane phase transition at about 15°C. The following changes occur rapidly, and often concomitantly, as platelets activate and participate in clot formation. Anionic phospholipids normally sequestered on the inside of platelet membranes flip to the exterior, providing a scaffold for assembly of coagulation enzyme complexes (see Text Box 18.2). Platelet membrane extrusions (pseudopodia) extend from the plasma membrane and separate into microparticles rich in anionic phospholipid which themselves are strongly implicated in clots (Fig. 18.4). Platelets change shape from discoid to a spiky sphere and stick to the vessel wall (adhesion) and to each other (platelet aggregation), forming tight bunches and binding fibrinogen and fibrin through a membrane signaling protein (GPIIb/IIIa). Alpha and dense granules merge with the SCCS and the plasma membrane, and the contents are released into the surrounding milieu, speeding thrombin formation as platelet-derived calcium and clotting factors are trapped within the spreading clot. Thrombin, the terminal enzyme of coagulation, is produced and cleaves fibrinogen to produce fibrin strands; further crosslinking between strands supports the platelet plug and forms the clot. Finally, trapped platelets within the clot, still bound tightly to the fibrin web by GPIIb/IIIa mediate clot retraction by contractile tension applied across the interior platelets by actomyosin to contract the clot, reducing its volume, making it less sensitive to clot breakdown (fibrinolysis) and efficiently protecting the initial site of injury.

The diverse events of platelet activation and clot formation are very complex and the subject of several excellent in-depth articles and chapters. A general view of some key pathways will be discussed here including inhibition of platelet activation, initiation of platelet activation, and phospholipid signaling pathways.

TEXT BOX 18.2 PROCOAGULANT SURFACE ON PLATELETS

All cells including platelets maintain an asymmetric distribution of surface phospholipids on their plasma membrane, with anionic phospholipids, such as phosphatidylserine and phosphatidylethanolamine on the cytoplasmic face, and neutral phospholipids such as phosphatidylcholine and sphingomyelin on the extracellular face. The transmembrane distribution of phospholipids is especially important for platelets, as they expose anionic phospholipid on the outer leaflet of their plasma membrane when stimulated to do so by external signals. The provision of the procoagulant surface was previously known as *platelet factor 3 activity*. Exposed anionic phospholipid, especially phosphatidylserine (PS), is a requirement for blood coagulation and acts as a scaffold upon which complexes of blood coagulation proteins assemble in a calcium-dependent manner. The ability of platelets to support the tenase and prothrombinase reactions of coagulation is correlated with the extent to which lipid asymmetry becomes perturbed and is related to the type of agonist. Anionic phospholipid distribution is controlled by three different enzymes. The first is an ATP-dependent aminophospholipid translocase, termed a *flipase*, that rapidly transports anionic phospholipids to the interior leaflet. Flipase activity is mediated by a 130-kDa integral membrane protein that is a member of the Mg^{2+}-dependent phosphoglycoprotein ATPases. The second, an ATP-dependent *floppase*, is slower and shuttles neutral phospholipids from the inner to the outer leaflet. Both of these enzymes operate normally to maintain membrane lipid asymmetry. During activation, an increase in intracellular Ca^{2+} blocks flipase and floppase activities and activates a Ca^{2+}-dependent *scramblase*. Scramblase induces rapid mixing of phospholipids in the plasma membrane bilayer, resulting in the dissipation of plasma membrane asymmetry. The appearance and clustering of anionic phospholipid on the exterior leaflet of the plasma membrane is correlated with the appearance of small, PS-rich *microparticles*, vesicles much smaller than platelets that are expelled from the plasma membrane. Although the anionic phospholipid-rich new surface of the platelet is well-known to support thrombin production, microparticles have not been extensively studied and are surmised to also provide a procoagulant surface.

The significance of the platelet procoagulant response is highlighted by a rare bleeding disorder, Scott syndrome, that is characterized by decreased Ca^{2+}-induced lipid scrambling and decreased production of microparticles, while all other platelet functions remain normal. Despite the availability of other cells (e.g., leukocytes trapped in the clot) expressing procoagulant phospholipid, Scott syndrome is characterized by a bleeding diathesis.

Inhibition of Platelet Activation To better understand platelet activation and participation in clot formation, it is important to understand that a general *clot prevention* (antithrombotic) state normally exists in the blood. Blood has to remain liquid, clot formation is suppressed, and platelet activation is inhibited until conditions arise that overwhelm the protective mechanisms in place. To prevent adventitious clot formation, circulating platelets are maintained in a nonreactive state by intrinsic and extrinsic factors.

Intrinsic factors contributing to maintenance of the nonreactive state are:

1. Platelets have a net negative charge, which prevents their interaction with red cells, leukocytes, endothelial cells, or other platelets.
2. The key platelet plasma membrane receptor GPIIb/IIIa exists in an inactive state and must undergo a conformational change before it can bind to plasma fibrinogen, and, if only minimal binding occurs, the GPIIb/IIIa complex may be cleared from the surface of the platelet, attenuating the signal.
3. An increase in intracellular calcium is the trigger for activation of many processes involved in platelet activation (this is described in more detail later). Cytosolic Ca^{2+} is normally maintained at a low level by sequestration in the DTS via ATP-driven pumps, and platelet activation is inhibited by Ca^{2+} antagonists [e.g., chelators such as citrate or ethylenediaminetetraacetic acid (EDTA) are routinely used to prevent platelet activation in collected blood].
4. ADP, a potent platelet agonist secreted by platelets from dense granules, and by other cells, is rapidly hydrolyzed to AMP by platelet surface ADPases.
5. TxA_2, a potent platelet activator produced and secreted by platelets to amplify the platelet activation response, is labile and rapidly converts to the inactive TxB_2.
6. Phosphatases, enzymes that dephosphorylate proteins, limit the signaling by intracellular protein kinases that drive platelet activation. However, not all kinases are stimulatory for activation; for example, protein kinases A and G (PKA and PKG) are implicated in inhibition of platelet activation (see below).
7. Platelet receptors coupled to a superfamily of signal transducing proteins, called G proteins, are rapidly desensitized by serine/threonine phosphorylation.

Extrinsic factors that inhibit platelet activation include:

1. Rapid inactivation of activated clotting factors (which can activate platelets) by normally circulating anticoagulant plasma proteins.
2. Physical barriers to activation include the protection of platelets from the connective tissue matrix within the vascular wall by a continuous barrier of endothelial cells.
3. Compounds secreted by endothelial cells to inhibit platelet activation, among them an ecto-ADPase, NO, and prostaglandin I_2 (PGI_2). These act inside the platelet to stimulate the inhibitory PKA and PKG pathways. NO diffuses across the platelet membrane, binding to guanylyl cyclase and enhancing cyclic guanosine 5′-monophosphate (cGMP) production. PGI_2 binds to a G-protein coupled receptor and activates adenylyl cyclase, resulting in cyclic AMP (cAMP) production. The inhibitory effects of cAMP and cGMP are principally mediated through cAMP- and cGMP-dependent protein kinases. One example of this is that high cAMP levels, through activation of the cAMP-dependent protein kinase (PKA), stimulates Ca^{2+} uptake by the DTS, thereby diminishing any stimulatory Ca^{2+} signal. NO and PGI_2 may act synergistically.

Cyclic-AMP levels are controlled by G-protein linked mechanisms since the activity of adenylate cyclase, which forms cAMP from ATP, is regulated by G proteins (see Chapter 4). G proteins are molecular switches containing a guanosine 5′-triphosphate (GTP)-binding α subunit together with a β-γ heterodimer, which are "on" when the α subunit is bound to GTP and dissociated from the β-γ subunits, and "off" when it is bound to guanosine 5′-diphosphate (GDP) and reassociates with the β-γ subunits. Activated receptors serve as guanine nucleotide exchange factors, promoting the replacement of GDP with GTP and thereby activating the G protein. An intrinsic GTPase activity of the α subunit ensures these receptors only transmit their signal for a short time.

Adenylate cyclase has low activity in the absence of the stimulatory guanine nucleotide binding protein termed G_s. The α subunit of this G protein binds to and activates adenylate cyclase when the β and γ subunits are not bound. Inhibition may occur when the α subunit of G_i, an inhibitory G protein that shares the β and γ subunits with G_s, dissociates and the subunits rebind to G_s, or when G_s-bound GTP is converted to GDP, and the α subunit of G_s dissociates, ending the activation.

The other important enzyme involved in regulating cAMP levels is cyclic nucleotide phosphodiesterase. Many inhibitors of platelet function, such as the adenylate cyclase stimulators PGE_1, PGI_2 and PGD_2 act by raising cAMP levels. Since turnover of cAMP in platelets has a half-time of less than 1 s, perturbations in adenylate cyclase or phosphodiesterase may rapidly affect platelet activation status. Several platelet agonists (e.g., ADP and epinephrine) inhibit adenylate cyclase in cell-free preparations, all through receptors coupled to G_i.

Initiation of Platelet Activation—Binding to Receptors After exposure to a stimulus such as collagen exposed by vascular wall injury, platelets adhere to the stimulating surface, flatten, and spread on the surface. In the process their shape changes from a disk to a "sticky" spiny sphere with long filopodia, derived from a rearrangement of the membrane of the SCCS. Shape changes are also induced by agonists binding to one or several of the thousands of receptors studding the surface of platelets. The receptors detecting collagen or the agonist transmit signals from the outside to the inside of the platelet, directing signaling pathways that induce cytoskeletal reassembly (shape change), release of intracellular Ca^{2+}, secretion of granule contents, activation of other receptors, and mediation of platelet adhesion and aggregation.

Platelet receptors include the von Willebrand factor receptor, "serpentine" receptors linked to G proteins, integrins composed of an α and a β subunit, and immunoreceptor tyrosine-based activation motif (ITAM)-containing receptors. The best understood of these are the serpentine receptors, such protease-activated receptor (PAR), receptors that respond to thrombin. Agonist binding to receptors induces signal transduction.

Agonists that induce platelet activation include conformationally altered von Willebrand factor, thrombin, collagen, ADP, thromboxane A2 (TxA_2), epinephrine, vasopressin, and platelet-activating factor. Of these, thrombin, epinephrine, platelet-activating factor, vasopressin, ADP, and TxA_2 bind to G-protein-coupled receptors. These receptors (so-called serpentine receptors) have seven transmembrane domains and an extracellular N terminus (Fig. 18.5). The receptor signal is passed to a G protein inside the cell (of the G_q subfamily—there are several subfamilies of G proteins in human platelets), which transduces the signal. G-protein-coupled receptors typically have a

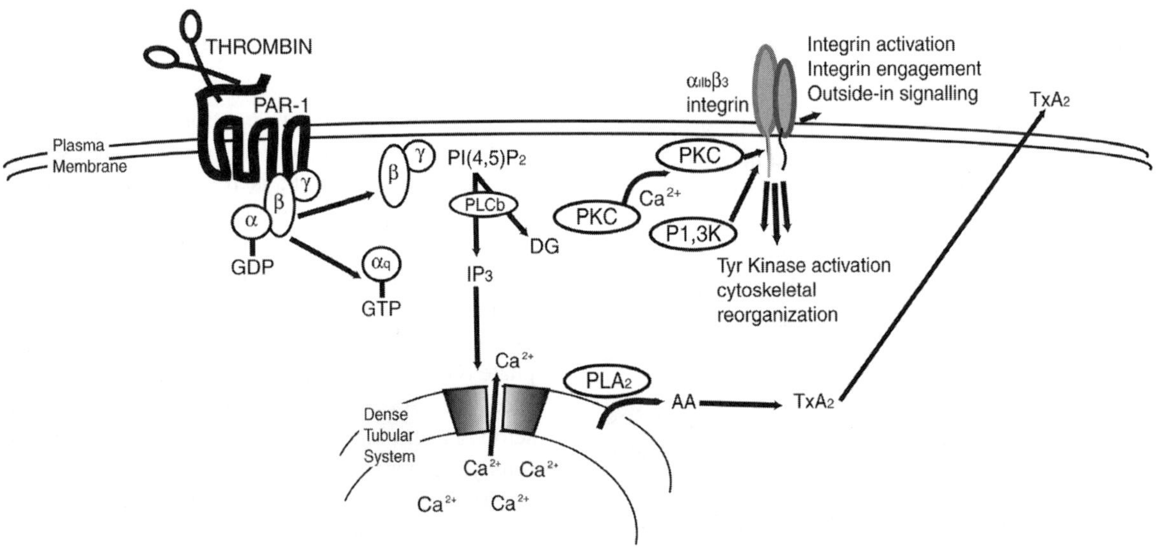

Figure 18.5 G-protein-mediated platelet activation by thrombin. A simplified diagram of platelet activation in response to thrombin (PAR-1) receptor stimulation. Thrombin cleavage of protease-activated receptor-1 (PAR-1) causes G-protein phosphorylation and dissociation, and downstream activation of phospholipase C (PLC), which produces inositol triphosphate (IP_3) and diacylglycerol (DG). IP_3 mediates Ca^{2+} release from intracellular stores in the dense tubular system (DTS). Protein kinase C (PKC) is mobilized to the plasma membrane by DG and released Ca^{2+}. PKC action results in signaling through α_{IIb}-β_{III} integrin, resulting in cytoskeletal reorganization and tyrosine kinase activation. Phospholipase A2 (PLA2) activation results in production of thromboxane A2 (TxA2). TxA2 leaves the platelet and through its own, similar, receptor-mediated action, activates other nearby platelets. [Modified from C. S. Abrams and L. F. Brass (1999). The molecular basis for platelet activation. In E. Benz, H. Cohen, B. Furie, R. Hoffman, and S. Shattil (eds.). *Hematology: Basic Principles and Practice*, 3rd ed. Churchill Livingstone, New York, p. 1757.

limited duration of activity after which they are turned off either by serine/threonine phosphorylation or cleared from the cell surface by endocytosis. In several cases there are multiple receptors for one activator; for instance, there are two different G-protein-coupled protease-activated receptors (PAR-1 and PAR-4) on human platelets that respond to thrombin and at least one glycoprotein (GPIb of the glycoprotein complex GPIb/IX/V) that binds thrombin, although this has not been shown to lead to intracellular signaling.

Thrombin is the terminal enzyme of coagulation, a serine protease that cleaves fibrinogen to form insoluble fibrin strands, which make the clot. However, thrombin has several other roles, including platelet activation, and is considered to be the most potent platelet activator *in vivo*. PAR-1 is considered to be the most important thrombin receptor on platelets, based on its sensitivity and expression level. Thrombin as low as 0.1 nM activates platelets, cleaving PARs and inhibiting synthesis of cyclic-AMP, by inhibiting adenylyl cyclase. Cleaved PAR receptors are internalized, attenuating further signal transduction, but the strength of the signal even for a short time is enough to activate phospholipase C, which liberates diacylglycerol (DAG) and inositol trisphosphate (IP_3) from phosphatidylinositol-4,5-bisphosphate (Fig. 18.5). Activation of phospholipase C is a common result for signaling by G-protein-coupled receptors. Further signal transduction by this pathway is discussed in the section on phospholipid signaling pathways below.

If thrombin is the strongest platelet activator, collagen is the "classic" activator. Collagen in the subendothelium is exposed after damage to the vascular endothelium and rapidly stimulates platelet adhesion. At sites of low shear stress, or *in vitro*, adhesion to collagen can occur without the help of accessory molecules; however, at sites of high blood turbulence *in vivo*, platelet adhesion to collagen additionally requires participation of a plasma protein called von Willebrand factor (vWF) that strengthen the bond between collagen and platelet in order to resist the shear forces caused by flowing blood. Circulating vWF binds to collagen or immobilized fibrinogen in the subendothelial matrix and in response to high shear stress linearizes. Platelet GPIb/IX/V receptor binds transiently to the remodeled von Willebrand factor, slowing the platelet down so it appears to "roll" along the site of vascular damage. Platelet GP1b/IX/V sends intracellular signals that trigger activation of membrane GPIIb/IIIa, a receptor for vWF or fibrin/fibrinogen. vWF bound to the subcellular matrix binds more strongly to the activated GPIIb/IIIa, anchoring the platelet to the site of damage. In this way platelets form a discontinuous carpet over the subendothelium. Nonintegrin receptors such as glycoprotein VI (GpVI) are then engaged. GVI signaling occurs by physical association with a critical ITAM-containing Fc(γ) receptor [Fc(γ)R] chain, used to dock proximal transducers of receptor signaling pathways. Dimerization or aggregation of the occupied or crosslinked receptors induces the proximal transducers to generate an intracellular signal, such as activation of phospholipase C, initiating phospholipid signaling (see below). vWF/collagen receptor signaling results in downstream activation of multiple transducers such as Src family kinases, syk tyrosine kinase, PI-3 kinase, Btk/Tec kinase family, and phospholipase C. Irreversible platelet activation ensues, linking these surface receptors to the release of intracellular calcium, cytoskeletal rearrangement, and granule secretion. Granule secretion releases ADP and fibrinogen, required for platelet aggregation, as well as Ca^{2+} and other compounds that propagate the activation reaction. As activation proceeds, thrombin is produced on platelets as a result of exposure of procoagulant phospholipid on platelet membranes, and thrombin signaling occurs as described in the previous paragraph (see Text Box 18.3).

Integrin receptors of the β_3 (CD61) subclass mediate cell adhesion and signaling. Platelet integrin receptor GPIIb/IIIa (also termed $\alpha_{IIb}\beta_3$) responds to signals from within the platelet (inside-out), such as phosphorylation by tyrosine kinases, by altering conformation to an active form, and binding fibrinogen and/or vWF. Fibrinogen and fibrin have several binding sites for this receptor, and they act as a "bridge" between two activated platelets, resulting in platelet aggregation. Outside-in signaling then sends signals through other pathways triggering other responses, including cytoskeletal reorganization, shape change, and granule movement to the center of the platelet, followed by granule secretion. Secretion involves a rapid reorganization of the actin cytoskeleton as actin filaments are uncapped, severed, and rebuilt in response to signals that involve changes in cytosolic Ca^{2+} concentration and the sequential activation of low-molecular-weight GTPases of the RhoA family. Activation of these small GTPases is central in the assembly of actin and cell architecture. These include rac, a GTPase implicated in cell spreading onto the extracellular matrix and fibrinogen, and rho, a GTPase implicated in mechanotransduction, which mediates formation of stress fibers and focal contacts. Increased Ca^{2+} (to micromolar levels) activates gelsolin to fragment actin in the cytoskeleton and myosin light-chain kinase.

Platelet aggregation requires ADP and fibrinogen, either secreted by the platelet or exogenously derived. Whereas vWF appears to link platelets to the subendothelium, fibrinogen appears to form part of the mechanism linking platelets to each other. The growing plug of sticky platelets is eventually stabilized by crosslinked fibrin in the clot and platelet–platelet contact is further stabilized by binding of fibrin or fibrinogen to the integrin GPIIb/IIIa on adjacent platelets. Glanzmann thrombasthenia is characterized by a severe reduction in, or absence of, platelet aggregation in response to all platelet agonists thought to operate *in vivo*

TEXT BOX 18.3 SIMILARITY BETWEEN PLATELET ACTIVATION AND APOPTOSIS

Several of the features of platelet activation, particularly the formation of microparticles and exposure of anionic phospholipids such as phosphatidylserine, are hallmarks of cells undergoing programmed cell death, or apoptosis. Apoptosis is a physiological mechanism for removing unneeded or damaged cells and plays a key role in embryogenesis, morphogenesis, normal cell turnover, and other processes. Even though nuclear collapse and DNA fragmentation are important hallmarks of apoptosis, the nucleus itself is not required. Proper exposure of a surface signal that ensures rapid and complete recognition and removal of the dying cell is important to prevent the release of the inflammatory cell contents upon lysis. One of the recognition signals is increased surface anionic phospholipid, which is recognized by macrophages in a reaction promoted by β-2-glycoprotein I. During apoptosis, signaling via *death domain* receptors results in proteolytic activation of a family of proteases termed caspases, which are centrally involved in apoptotic signaling and execution. The presence of messenger ribonucleic acid (mRNA) coding for death receptors and caspases has been shown in platelets. The contribution of caspase activity, especially of the *effector caspase*, caspase 3, to a specific subset of platelet activation responses (platelet phosphatidylserine exposure, microparticle release, and cleavage of a cytoskeletal-membrane linker protein) has accordingly been demonstrated. Caspase inhibitors inhibit these responses. Shape change, α-granule secretion, and aggregation were shown to be independent and unaffected by caspase inhibitors. Interestingly, other studies have shown that platelet senescence (aging resulting in removal from circulation) is caspase-independent and possibly related to exogenous plasma-derived factors. However macrophage removal of platelets is thought to be at least partially mediated by phosphatidylserine exposure, which the previous reports ascribed to caspase activity. The role of caspases, the "effectors of apoptosis" in platelet processes is still under investigation.

(ADP, epinephrine, thrombin, collagen, thromboxane A_2) due to qualitative or quantitative abnormalities of platelet glycoprotein IIb (α_{IIb}, CD41) and/or GPIIIa (β_3, CD61). As this integrin is the membrane link for mechanical force transduction, Glanzman thrombasthenia is also marked by a lack of clot retraction.

Phospholipid Signaling Pathways Membrane phospholipids are substrates for two related signaling pathways mediating platelet activation. These are the phosphoinositide pathway and the arachidonate (eicosanoid) pathway.

Phosphoinositide Pathway Many of the initial signaling responses discussed above include activation of phospholipase C. In fact, all platelet responses after agonist binding center on the metabolism of phosphoinositides, which results in eventual calcium release. Of the primary platelet membrane phospholipid, 5% is phosphatidylinositol. Phosphorylation of phosphatidylinositol (PI) yields phopsphatidylinositol-4-phosphate and phosphatidylinositol-4,5-bisphosphate (PIP_2). Activation of specific G-protein-coupled receptors results in the activation of plasma membrane phospholipase C (PLC), which hydrolyzes PIP_2 to two second messengers: diacylglycerol (DAG) and inositol trisphosphate (IP_3). IP_3 diffuses from the plasma membrane to the DTS. There it binds to IP_3 receptors, rendering the membranes of the DTS permeable to calcium and releasing Ca^{2+} into the cytoplasm. Ca^{2+} is also elevated by the opening of plasma membrane channels but, interestingly, Ca^{2+} sequestered in dense granules is expelled to the extracellular milieu and is not thought to contribute to elevated intracellular Ca^{2+}. Increased levels of intracellular Ca^{2+} triggers secondary activation of phospholipase A2, as well as inducing membrane translocation of protein kinase C (PKC), where it is activated by membrane-associated DAG. PKC phosphorylates target proteins that participate in granule secretion and the up-regulation of the fibrinogen receptor, GPIIb/IIIa. The Ca^{2+} signal also serves to activate myosin light-chain kinase (MLCK). MLCK phosphorylates the myosin light chain, mediating the actomyosin-based force generation that drives granule centralization for secretion and clot retraction. Other reactions stimulated by Ca^{2+} include activation of mitogen-activated protein kinases (MAPK) and of proteolysis by a Ca^{2+}-dependent protease, calpain, which is found in the platelet cytosol. Glycogenolysis is also activated within seconds of platelet activation.

Arachidonate Pathway The oxygenated derivatives of arachidonic acid (AA) are the eicosanoids, a large family of signaling molecules (Fig. 18.6). When a specific stimulus acts on platelets, cytosolic phospholipase A2 is activated, and liberates arachidonic acid from platelet membrane phospholipids such as phosphatidylethanolamine and phosphatidylcholine. Arachidonate is converted by platelet cyclooxygenase 1 to the labile cyclic endoperoxides prostaglandins G_2 and H_2 (PGG_2 and PGH_2) [reaction (18.7)]. PGH_2 is then converted to thromboxane A_2 (TxA_2), a potent but short-lived platelet agonist, by thromboxane synthetase, or the cyclic endoperoxides spontaneously degrade. TxA_2 exits the platelet and binds to a G-protein-

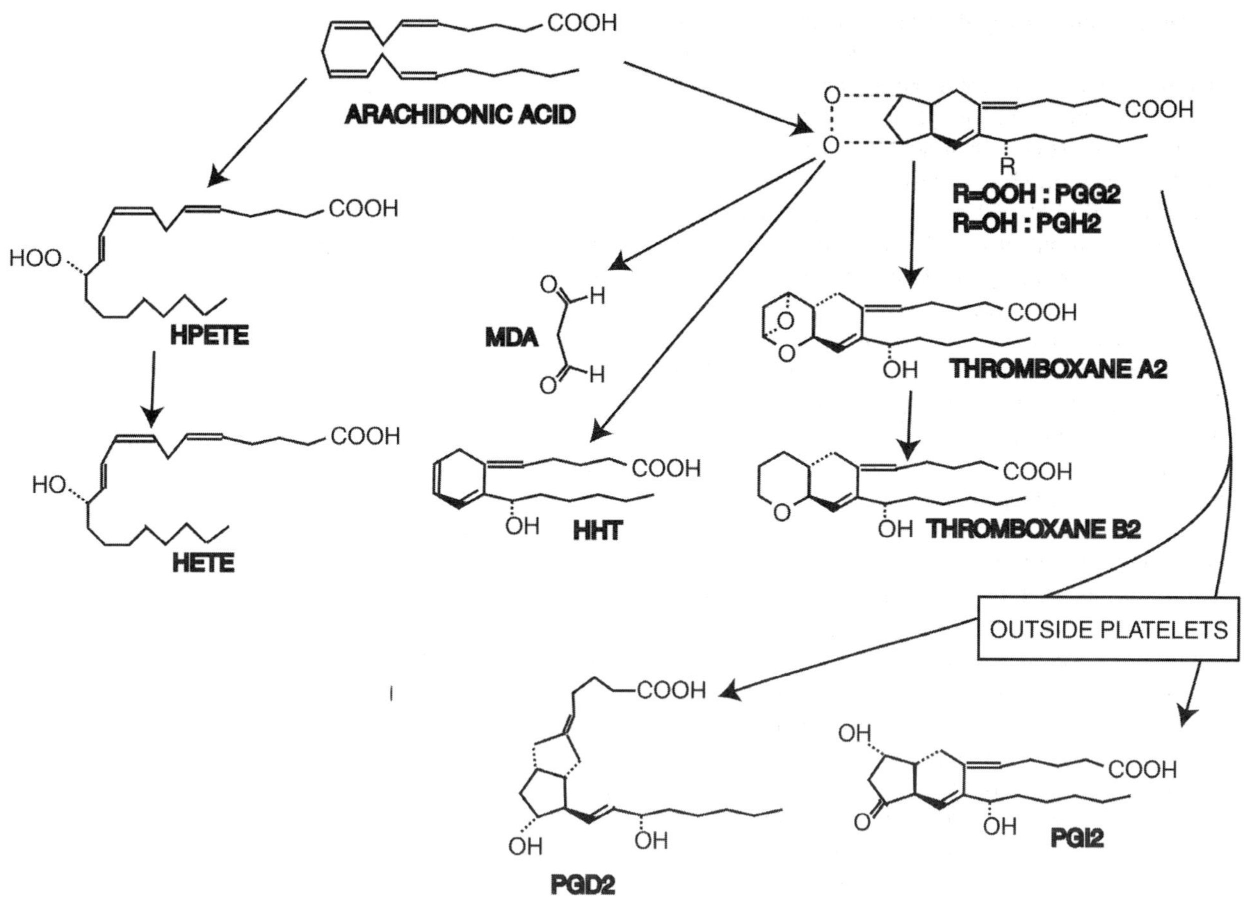

Figure 18.6 Arachidonic acid metabolism in platelets. Arachidonic acid is the precursor to the potent platelet activator thromboxane A_2 (TxA_2). Conversion of PGH_2 to PGI_2 does not occur within platelets. TxA_2 and PGH_2 stimulate platelet aggregation and PGI_2 and PGD_2 inhibit it. Endothelial cells produce PGI_2, which inhibits platelet activation. Aspirin and fish fatty acids inhibit the first step of the pathway, at cyclooxygenase. [From P. Majerus (2001). Platelets. In Stamatoyannopoulos, Majerus, Permutter, and Varmus. *The Molecular Basis of Blood Disease*, 3rd ed. W. B. Saunders, p. 771. Used with permission.]

coupled receptor on the same or on nearby platelets. TxA_2 binding activates phospholipase C, activating the phosphoinositide pathway and IP_3/DAG production. In endothelial cells, the same pathway has a different result. PGH_2 is converted by prostacylin synthase to PGI_2, and specific synthetases also produce prostaglandins D_2 and E_2, which bind to platelet receptors and stimulate adenylyl cyclase. Hence, they are inhibitors of platelet aggregation.

$$\text{Arachidonate} \xrightarrow{\text{Cyclooxygenase}} \text{Prostaglandin G2 (PGG2)} \tag{18.7}$$

Two compounds shown to decrease blood clotting, eicosapentanoic acid and aspirin, work by acting on cyclooxygenase to inhibit the first stage of TxA_2 production and impair platelet aggregation. Eicosapentanoic acid, a fatty acid found in cold-water fish, is a competitive inhibitor of PGG_2 and PGH_2 formation, competing with arachidonic acid for cyclooxygenase. Decreased production of PGG_2 and PGH_2 affects subsequent TxA_2 formation. People with a diet rich in this fatty acid may have a mild bleeding disorder but may be protected against atherosclerosis. Aspirin has the same net effect but acts by acetylating a serine residue on cyclooxygenase that irreversibly inhibits the enzyme. The effect is permanent since platelets cannot produce new protein. Much lower levels of aspirin are required for platelet cyclooxygenase inhibition than for other pharmacologic effects of the drug and, hence, repeated low doses of aspirin (<80 mg/day) can yield complete platelet inhibition without affecting other functions. This is the basis for the prescription of low doses of aspirin to patients at risk of heart attack. Other platelet cyclooxygenase inhibitors, indomethacin and sulfinpyrazone, are competitive inhibitors of the enzyme.

Energy Metabolism

Unlike erythrocytes, platelets are not limited to glycolysis as their ATP-generating pathway but are capable of carrying out most reactions of carbohydrate and lipid metabolism. Platelets have a total ATP turnover rate of 3 to 8 times that of resting mammalian muscle. Glycolysis accounts for the majority of the ATP produced, since plasma glucose is readily available as a fuel source. There is additional evidence that in a hypoxic situation, such as platelet storage, the lactate produced by glycolysis can also be used for mitochondrial ATP production. Platelets also have glycogen stores that can be rapidly mobilized. A rationale for the need for high-energy turnover in platelets is that they must maintain continuous high activities of ATP-dependent intra- and extracellular pumps (such as the Ca^{2+}/Mg^{2+}–ATPase pump of the DTS) and must be able to undergo a rapid shape change with redistribution and secretion of intracellular contents during activation. In support of this, the metabolic inhibitors antimycin A and 2-deoxyglucose have been shown to decrease agonist-induced α-granule secretion.

Platelet Disorders

Platelets and Atherosclerotic Plaques Atherosclerotic plaques are composed largely of lipids and cellular material from fibrous tissue and smooth muscle fibers. The contribution of platelets and thrombotic material to atherosclerosis has been suggested by the fact that fibrin and material from platelets can be found in sandwiched layers within the plaque. It has been suggested that initial formation of the plaque is due to formation of a clot on an injured vessel, followed by regrowth of endothelial cells to cover the thrombus. The entry of platelets into atherosclerotic plaque could be during the formation of this initial thrombus, or by hemorrhage within the plaque if turbulent blood flow pulls it away from the vascular wall. Platelets likely stimulate plaque growth and proliferation through the release of, among other effectors, platelet-derived growth factor (PDGF) and transforming growth factor-β. PDGF stimulates growth and proliferation of smooth muscle cells and fibroblasts, which are the major sources of fibrous tissue within the plaque. Endothelial injury is followed by platelet fibrin deposition and monocyte infiltration. Monocytes differentiate into macrophages, which then absorb lipids in the plaque to become foam cells. Death of the foam cell liberates cholesterol esters and oxidation products—which reinjure the endothelial surface and start the process again.

von Willebrand factor (vWF) is a large multimeric protein with two main functions: (a) it carries factor VIII, the precursor molecule of a cofactor for a coagulation complex, and (b) it promotes platelet adhesion to damaged vessel walls, particularly under high shear stress, for instance, in areas of high blood turbulence. These regions are where atherosclerotic plaques grow and develop in response to a small initial injury. As described above, vWF binds to subendothelium, undergoes a conformational change at high shear rates, leading to platelet binding and activation. Activated GP IIb/IIIa binds to vWF under conditions of high shear stress, leading to platelet aggregation. At low shear stress, aggregation is mediated through binding of activated GP IIb/IIIa to fibrinogen. Elevation of vWF or fibrinogen is thus associated with an increased tendency to thrombosis.

Quebec Platelet Disorder—An Inherited Platelet Disorder The clotting factors contained in platelets are also in plasma, which could be seen as a redundancy. Why would an individual need cells containing clotting factors when they are already present in plasma? The answer is that platelets provide targeted clotting factor delivery that cannot be achieved using plasma alone. Individuals with platelets that are deficient in clotting factors show bleeding and impaired clotting, despite normal amounts of all clotting factors in plasma. Quebec platelet disorder (QPD) is one such deficiency that has manifested itself in two families. It is characterized by a bleeding diathesis associated with platelets that do not contain intact clotting factors. The description of the defect underlying QPD and the explanation of why it causes impaired clotting have only been described very recently. Breakdown of fibrinogen and factor V contained in platelet α granules has been ascribed to paradoxically high levels of urokinase plasminogen activator (u-PA) in these granules; this acts to break down the factors in the platelets before they can be released. This implies that, at the site of a clot, plasma clotting factors are not enough for efficient clot formation, and the release of platelet α-granule contents are a requirement for full clot formation in humans. An animal model approximates one feature of QPD: Cows have only 1% of the essential coagulation cofactor factor V (FV) sequestered in their platelets. To efficiently form clots, their plasma must deliver high levels of FV to a forming clot, and indeed, plasma levels of bovine FV are five times that found in human plasma. Many other platelet disorders, resulting from both qualitative and quantitative defects, have been described, such as Glanzmann's thrombasthenia (described previously in the text).

STEM CELLS

During embryogenesis, a single fertilized oocyte becomes a multicellular organism with cells committed to different tissues and to the performance of different functions within the body. As the embryo grows, cells with a predetermined fate proliferate to enable tissues and organs to grow. Even after an animal is fully grown, however, many tissues and organs in the adult continue a precise system of death and rebirth wherein mature cells die, either naturally or due to injury, and are replenished. Accordingly a lifelong supply of pliable stem cells is necessary. Adult stem cells are often relatively slow-cycling cells, with a lifetime of about 3 months, that are able to respond to stimuli and produce daughter cells to self-renew or to enter committed differentiation. The pattern of stem cell proliferation and terminal differentiation dates back to the most primitive animals (e.g., sponges and hydrozoans), but higher organisms still show remarkable plasticity in many cases (e.g., the limb or tail of a lizard can regenerate after being severed). Although mammals do not show such remarkable characteristics, some organs and tissues can regenerate when injured, a primary example being the liver (see Chapter 19). The epidermis, hair, small intestine, and blood system are all examples of adult cells with a naturally dynamic existence. Even in the absence of injury, these structures constantly produce new cells that transiently divide, terminally differentiate, live for a set period, and then die.

Self-renewal is essential in the hematopoietic system as humans lose and replace over a billion red blood cells per day. Hematopoietic stem cells are probably the best-characterized stem cell population and have contributed to many of the concepts and generalizations about other stem cells. The best description of the hematopoietic system is as a continuum of functional compartments: the stem cell compartment, the progenitor compartment, and the precursor compartment.

The primitive, rare cells in the hematopoietic stem cell compartment provide more differentiated cells through a process termed *commitment* downstream into the next compartment, allowing sustained, lifelong production of all mature blood cell types. The population of stem cells maintains itself by asymmetric division into a committed colony forming unit (CFU) and another stem cell. The primary location of stem cells in adults is in bone marrow, but some cells are also released into circulation and allow cell differentiation to occur in places other than marrow. Through a stepwise series of extrinsic and intrinsic cues, hematopoietic stem cells differentiate to give rise to progenitor cells with progressively more limited options, such as lymphoid cells resulting in T and B lymphocytes or myeloid progenitor cells resulting in basophils, eosinophils, neutrophils, macrophages, platelets, and erythrocytes (Fig. 18.1). Experiments in the early 1960s showed that a population of clonogenic bone marrow cells could generate myeloerythroid colonies in the spleens of lethally irradiated hosts. In some cases, these cells gave rise to others that could be transferred to secondary hosts and there reconstitute all blood cell lineages, an important initial result that showed that stem cells could be transplanted.

Self-renewal of stem cells is correlated with a high proliferative potential. The stem cell compartment must be multipotential, given the existence of at least eight distinct lineages of mature blood cells. The functional abilities of the stem cell population must be rigorously controlled to ensure the correct balance of self-renewal versus commitment to differentiation. Control mechanisms for self-renewal or production of mature cells must also be flexible to allow response to situations of hematological stress. Control at an even higher level (a stem cell that can differentiate into all cell types, such as is found in the developing embryo) is suggested but not evident in most adult animals. Animal model studies (using irradiated animals or those genetically altered to have few stem cells) have shown a link or possibly a common ancestor cell for endothelial cells and hematopoietic cells; this is definitive in avian models, but less clear in mammals.

Human fetal stem cells that can differentiate to many fates (totipotent) are very similar to those of lower vertebrates; these coexist with stem cells such as hematopoietic stem cells, which are pluripotent—able to form many cell types but within one tissue type (blood) only. During mammalian development, the first blood cells, called embryonic nucleated erythrocytes, appear in blood pools in the extra-embryonic yolk sac and express certain transcription factors that specify these cells to their hematopoietic fate. In the developing embryo itself, precursor cells migrate to an environment permissive to hematopoiesis that is located in the aorta, genital ridge, and mesonephros (AGM) region. Vascular endothelial growth factor (VEGF) and its tyrosine kinase receptor (Flk1) as well as the vascular endothelial growth factor receptor 2 (VEGFR2) signal transduction pathway appear essential in regulating migration of embryonic blood cells and progenitors to the AGM. As development proceeds, fetal liver becomes the focal point of hematopoiesis. During this phase, some hematopoietic stem cells can differentiate to lymphoid and myeloid progenitor cells. Over time, hematopoiesis is transferred to the spleen and by the time of birth is completely transferred to the bone marrow, where blood formation is maintained throughout the life of the adult. A key early stem cell receptor, c-kit (described in greater detail later) is present in the yolk sac 8 to 9 days postconception and remains an essential receptor for early hematopoietic stem cell differentiation in the adult human.

The ability to regenerate and continuously generate large numbers of cells throughout life is shown also by the intestinal epithelium, the skin epithelium, and the male germline. Other tissue systems such as the liver, the muscles, the vasculature, and the nervous system can regenerate mature cells in response to injury or stress. Although they are less well developed, possible stem cell populations have been proposed for muscle cells, central and peripheral nervous system cells, and mesenchymal stem cells. Hematopoietic stem cells differ from epithelial and embryonic stem cells in that they do not adhere tightly to one another, which may account for their difficulty in settling down to a particular place. Even in the adult, populations of stem cells and their progeny with more restricted potential are detectable in peripheral blood, spleen, and liver, as well as in bone marrow, periodically leaving and returning to one bone marrow niche where they are "nursed" by bone stromal cells and a complex array of other cells in the marrow environment that produce a large array of extrinsic molecules. This environment imparts to stem cells many of the properties that they display and cannot so far be replicated by a growth medium. Despite efforts to characterize these molecules, stem cells still survive better when cultured in bone marrow stroma than in defined medium supplemented with characterized bone marrow components.

Differentiation in Hematopoiesis

Growth factors, cytokines, and chemokines regulate hematopoiesis by expanding progenitors and regulating their cell cycle, killing off cells by mediating apoptosis and regulating the function of differentiated cells. The study of these regulatory molecules has led to the identification of literally hundreds of factors and the receptors that mediate the effects. Understanding the entire effect, that is, the full cellular consequences of receptor–ligand interaction, and the entire signal transduction pathway, has occurred more slowly. There are several classes of hematopoietic growth factors (HGFs):

1. Multilineage HGFs, such as interleukin (IL)-3, stimulate the survival, proliferation, and differentiation of a broad range of progenitors.
2. Lineage-specific HGFs, such as erythropoietin (EPO) and thrombopoietin (TPO), stimulate the proliferation and differentiation of more mature progenitor cells, in this case erythroid and megakaryocyte progenitor cells, respectively.
3. Two other early-acting factors, stem cell factor (SCF) and Flt3 ligand (Flt3L), are important for stem cell survival and act in synergy with other factors.
4. Indirect stem cell stimulation, possibly requiring other cytokines, is provided by IL-1, IL-6, and IL-11.
5. Other ligands affect stem cell self-renewal: the ligands delta, jagged 1, and jagged 2 act through the notch receptors, sonic hedgehog acts through patched and smoothened receptors, and Wnt acts through frizzled and lipoprotein receptor-related families.

Receptors for groups 1 and 2 above include the large cytokine receptor superfamily, which show similarities in their extracellular domain but are different intracellularly. This contrasts with the SCF receptor and Flt3, which are class III receptor tyrosine kinases.

Cytokine Receptor Signaling Cytokine receptors do not contain intrinsic tyrosine kinase activity; however, after stimulation of cytokine receptors such as the IL-3 receptor, tyrosine phosphorylation is evident. This is thought to be accomplished by one of the four Janus family kinases (JAK) (Fig. 18.7). Upon stimulation of the IL-3 or the EPO receptor, JAK-2 is autophosphorylated, recruited to the site of the receptor, and phosphorylates the receptor. This action recruits one of a family of STAT (signal transduction and activator of transcription) proteins, which are themselves phosphorylated, dimerize, and enter the nucleus to bind deoxyribonucleic acid (DNA) to activate or repress transcription. Activated JAK/STAT pathways are opposed by phosphatases, such as shp1, which binds to the EPO receptor and regulates erythropoiesis. STAT proteins are also down-regulated by phosphatase and ubiquitin pathways, and further regulation takes place at the nuclear level. It is important to note that this is only one of many signaling events emanating from the site of the activated cytokine receptor; the signal is also transmitted through Ras and PI3 kinase to a series of downstream kinases, also resulting in nuclear activation.

Tyrosine Kinase Signaling—SCF Stem cell factor (SCF), also termed c-kit ligand, mast cell growth factor, and Steel factor, is a hematopoietic growth factor acting on early-stage hematopoietic cell lineages. It cannot act alone but acts synergistically with other cytokines to stimulate growth of hematopoietic progenitors specific for different lineages *in vitro* and to stimulate blood cell production *in vivo*. In addition to its *in vitro* applications (stem cell culture), therapeutic evaluation of the molecule to aid mobilization of peripheral blood progenitor cells for autologous transplantation patients is under evaluation.

Stem cells are arrested in the growth cycle at the G_0 stage. The G_0 stem cell is moved into its active proliferative state by SCF. Stem cell factor exists in a transmembrane and a soluble (truncated) form; it likely plays a role in stromal cell–stem cell interactions. SCF binds to a receptor (c-kit) on hematopoietic progenitor cells, which is the product of a protooncogene related to a feline sarcoma virus oncogene. As mentioned above, c-kit is a tyrosine kinase receptor.

One of the largest families of cellular ligand receptors, the protein tyrosine kinase receptors have been identified in early hematopoietic stem cells and are implicated in the regulation of hematopoiesis, as well as cell survival, proliferation, transformation, and differentiation. In this case,

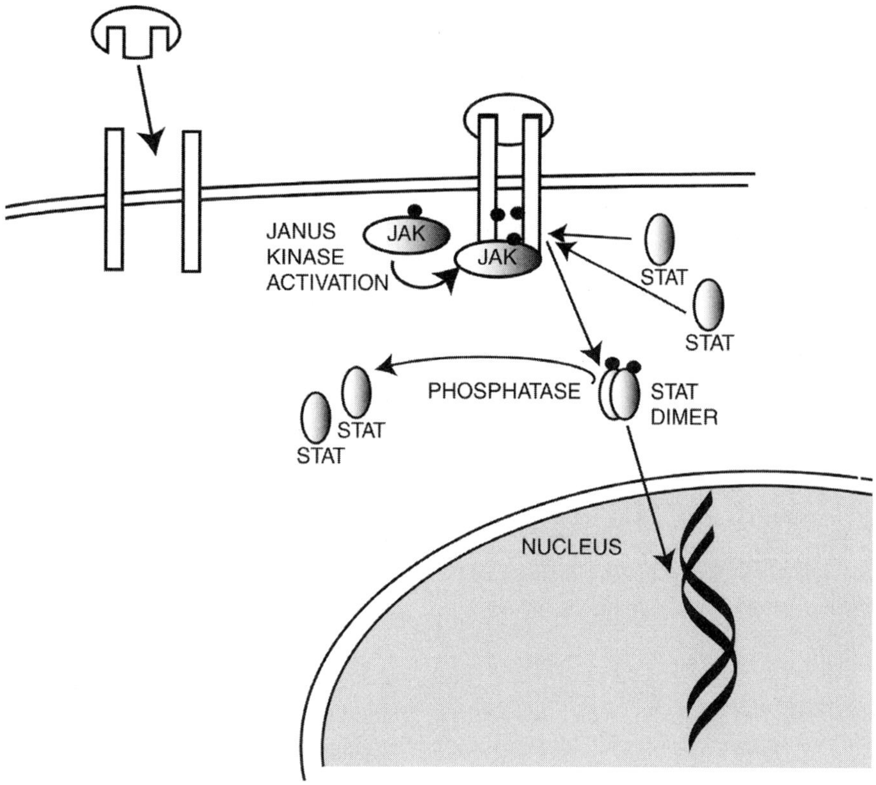

Figure 18.7 JAK-STAT signaling pathway. Members of the cytokine receptor superfamily lack intrinsic tyrosine kinase activity. Upon ligand binding to their specific receptors, the receptors dimerize and signal cytoplasmic kinases such as Janus kinase (JAK). JAK stimulation results in autophosphorylation, binding to and phosphorylation of the receptor and recruitment of signal transducer and activator of transcription (STAT) proteins, which are then phosphorylated. STAT dimerizes and moves to the nucleus, binding and activating transcription.

c-kit and SCF are implicated in expansion of early hematopoietic progenitor cells. Tyrosine kinase receptors have a single transmembrane spanning domain and a cytoplasmic domain that has a protein tyrosine kinase catalytic domain (Fig. 18.8; see Chapter 5 for more information). Further classification of subfamilies is based on the structure of the extracellular domains.

Four general steps are involved in c-kit tyrosine kinase receptor signaling: ligand-induced dimerization of the receptor, transphosphorylation of the activation loop, phosphorylation of additional sites and recruitment of proteins to the receptor complex, and phosphorylation of substrates. The key function of the ligand is to trigger oligomerization of the receptor. This can occur by several methods; for example, the ligand itself can exist as a dimer or the ligand may have two binding sites. Following oligomerization, a transphosphorylation of a tyrosine residue in the kinase domain results in receptor activation. The phosphorylated c-kit receptor provides binding sites for SH2-domain-containing proteins, such as the p21Ras–MAPK pathway, the phosphatidylinositol-3-kinase (PI3K) pathway, Src family kinases, phospholipase C-γ, and phosphatases SHP1 and SHP2, among others. This is followed by transduction of the signal to a variety of sites within the cell.

Recall, however, that SCF acts in concert with other cytokines. To promote cell differentiation, synergy and amplification of the SCF-R activation pathway are provided by the early acting cytokines that are lineage nonspecific: IL-3, IL-6, and Il-9, and by G-CSF (granulocyte colony-stimulating factor) and GM-CSF (granulocyte-macrophage colony-stimulating factor). In response to these stimuli, the cell product derived is the multipotent CFU–GEMM (colony forming unit–granulocyte, erythrocyte, macrophage, megakaryocyte), which is the progenitor of the myeloid, monocyte, megakaryocyte, and erythroid cells (Fig. 18.1).

Further differentiation is under the control of the growth and differentiation factors, which often overlap between cell types. EPO, G-CSF, and/or GM-CSF are necessary not only for differentiation but also for cell survival. Withdrawal of these cytokines results in apoptosis. For erythrocyte production, IL-11 and IGF-1 stimulate CFU–GEMM to form

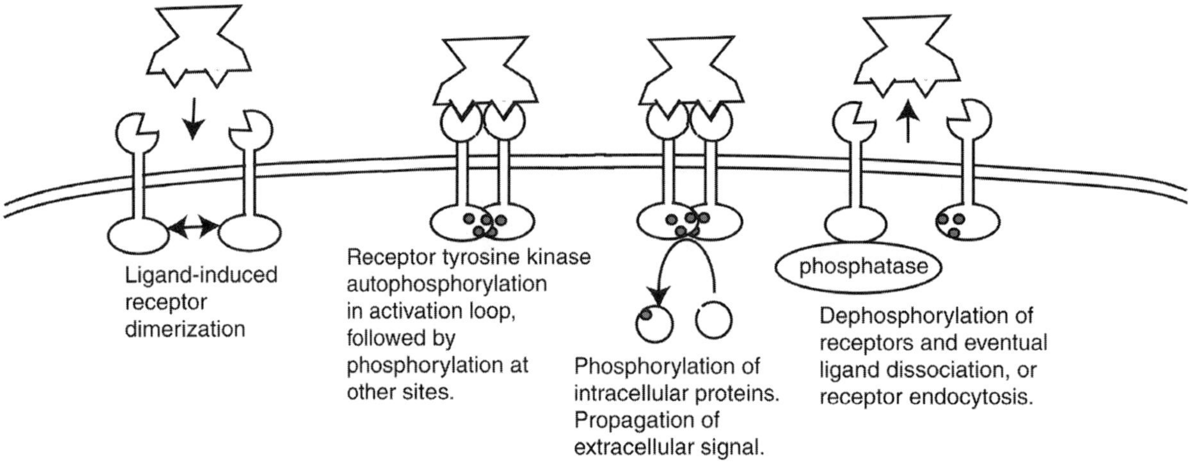

Figure 18.8 Tyrosine kinase receptor activation. Binding of specific ligands such as stem cell factor induces receptor dimerization and autophosphorylation in the activation loop forming a functional tyrosine kinase. The membrane-bound kinase can then phosphorylate cytosolic proteins, including many kinases, amplifying the signal.

the burst forming unit–erythroid (BFU-E), which are then further differentiated by GM-CSF, IL-9, insulin-like growth factor 1, and EPO to form the colony-forming units for erythrocyte precursors (CFU-E). Both BFU-E and CFU-E are considered committed progenitor cells. The further production of erythrocytes from CFU-E is under EPO control. Formation of the erythroblast, enucleation, and release of the erythrocyte into the bloodstream all follow.

Interleukin-11 stimulates CFU–GEMM to produce the colony-forming unit–megakaryocyte (CFU–Meg), which ultimately results in platelet production. Further differentiation is under the control of TPO, a relatively recently discovered cytokine produced in liver and kidney. TPO action results in maturation of the megakaryocyte, uncoupling of nuclear replication, and cell division, which produces a large polyploid cell, and production of platelets from extrusions of the megakaryocyte membrane. Circulating platelets and megakaryocytes contain TPO receptors, which internalize and degrade TPO; it is surmized that this controls platelet population since when platelet levels drop, TPO levels rise, and platelet production is stimulated.

Negative Regulators of Hematopoiesis Prostaglandins, interferon, and transforming growth factor β all inhibit stem cell proliferation. Integrin-mediated direct cell contact between hematopoietic stem cells and stromal cells has been shown to inhibit differentiation of stem cells in some culture systems.

Stem Cell Mobilization Integrins are cell surface glycoproteins that interact with complementary adhesion molecules (recall that platelets also use integrin signaling for aggregation). They are heterodimers of noncovalently associated α and β subunits, with a large extracellular binding domain, a transmembrane segment and a short cytoplasmic domain. Integrins transmit signals bidirectionally from inside to outside the cell and then back again. The β_1 integrin family may be responsible for localization of hematopoietic cells within bone marrow niches. These bind to fibronectin, collagen, thrombospondin, and vascular cell adhesion molecules. Members of the β_2 integrin family interact with intracellular adhesion molecules and fibrinogen to regulate the exit of mature cells from the bone marrow.

Stem Cell Transplantation to Regenerate the Hematolymphoid System

The regeneration of the hematolymphoid system following an otherwise lethal dose of whole-body irradiation or chemotherapy has become the basis for the use of bone marrow transplantation. Hematopoietic stem cells can be obtained from bone marrow, peripheral and umbilical cord blood, and from autologous, syngeneic, and allogeneic donors. Specific separation methods for the isolation of human hematopoietic stem cells by phenotype allows these cells to be harvested for transplant (Fig. 18.9). Hematopoietic cells constitute 0.05 to 0.1% of human marrow and circulating hematopoietic cells and are capable of repopulating hematopoietic cells of multiple lineages. Allogeneic bone marrow transplantation leads to a graft versus host disease caused by contaminating T lymphocytes. However, transplantation of purified hematopoietic stem cells leads

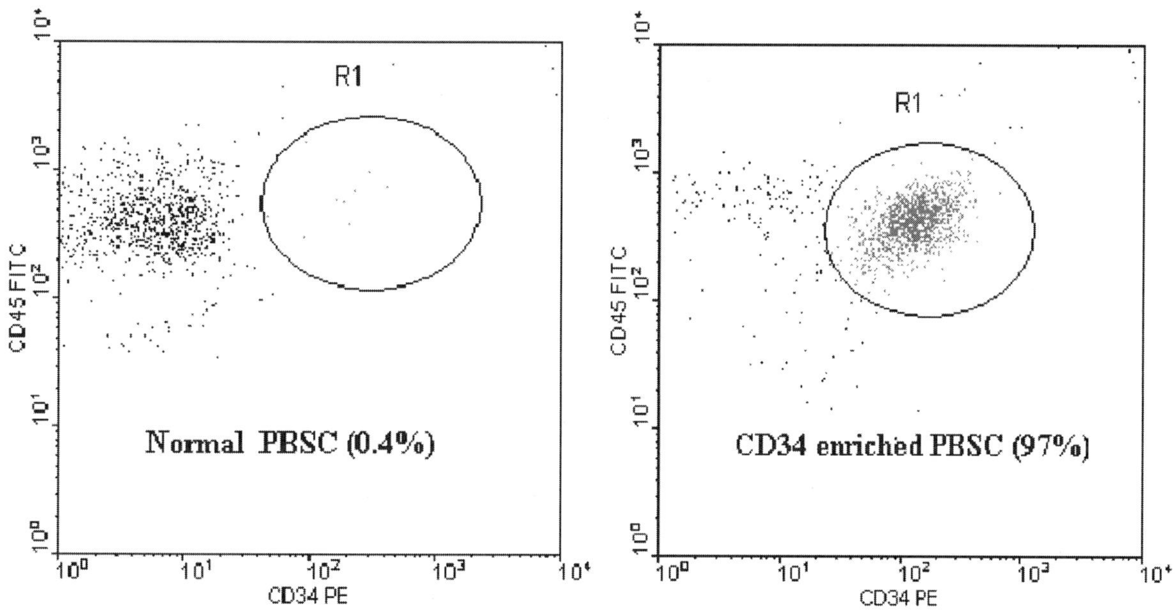

Figure 18.9 Peripheral blood stem cells (PBSC) detected by flow cytometry. (*a*) PBSC in a sample of blood prior to purification. (*b*) PBSC after purification on the basis of CD34, a surface antigen of stem cells. (Image courtesy of Dr. Doug Palmer and Mr. Paul Birch, Canadian Blood Services.)

to donor blood engraftment without the appearance of this disease, as well as rapid engraftment of myeloid cells and platelets. To avoid marrow harvest and anaesthesia, as well as expanding the pool of eligible patients to those for whom marrow harvest could be contraindicated, medical centers are now more frequently using hematopoietic stem cells obtained from peripheral blood by leukapheresis after mobilization with chemotherapy and/or hematopoietic growth factors. Since the 1980s, transplantation of hematopoietic stem cells from peripheral blood has almost supplanted conventional bone marrow transplantation because it offers more rapid hematopoietic recovery, less morbidity, decreased medical support, and lower total costs.

In addition to the well-publicized use of hematopoietic stem cells to replace bone marrow destroyed during cancer treatment, stem cells are widely used to treat nonmalignant diseases such as in cases of genetically determined primary immunodeficiency and genetic metabolic and hematologic disorders. Stem cells have been successfully used to treat Gaucher's disease (an abnormality of the enzyme glucocerebrosidase), Hurler's syndrome (a deficiency of α-iduronidase), and osteopetrosis. In all these cases, the disease continues to progress for about 6 months until donor tissue macrophages or osteoclasts are produced in sufficient numbers to help the patient. There are additional implications for autoimmune diseases, such as diabetes. Transplantation of donor cells from normal mice into mice bearing mutations that predispose them to an autoimmune, T-cell-mediated type I diabetes not only induces tolerance to donor tissues but also blocks the progression of the autoimmune disease. Thus, allogeneic stem cells can induce tolerance of other tissue-specific stem cells from the same donor.

Totipotent Stem Cells

Following fertilization of the egg and formation of the blastocyst, *totipotent* stem cells are generated within the inner cell mass. These embryonic stem cells are thought to segregate into three major lineages within the three primary germ layers (endoderm, mesoderm, and ectoderm). It has been assumed that the capacity of stem cells to generate all cell types becomes narrowed as they differentiate into tissue- and organ-specific committed stem cells. These compartmentalized stem cells then become responsible for the subsequent growth of organs and tissues during development and for tissue homeostasis and repair in adulthood. However, there is now evidence that suggests that lineage differentiation of stem cells may be considerably broader. For example, hematopoietic stem cells have been shown to contribute to muscle development, brain stem cells and cultured central nervous system stem cells can give rise to blood cells in irradiated mice, transplanted bone marrow can participate in the repair of myocardium, and adult neural stem cells have been transformed into cells that resemble embryonal stem cells. This raises the question of whether any so-called organ-specific stem cell has the capability to "dedifferentiate" and become totipotent. The paradigm of stem cells, even stem cells derived from adult tissue, as committed to a single organ or tissue type, has changed in light of new evidence that stem cells from a particular lineage, when injected into different tissue, can produce cells characteristic of their new environment. Both intra- and intergerm layer transdifferentiation has been demonstrated (e.g., mesoderm-derived bone marrow stem cells normally undergo myogenic differentiation but can also be induced to produce endoderm-derived hepatocytes and neuroectoderm-derived glia). This challenges the dogma of cells or cell nuclei as being committed to a particular fate.

By removal of a stem cell from its somatic tissue and transplantation to a different microenvironment, researchers have been able to show differentiation of the transplanted cell in response to its new environment. These and other findings, although preliminary, show the dominant contribution of the niche to the developmental stage-specific gene expression program of the stem cell. But because the potential of stem cells has usually been revealed by forcing the cells to mature in altered environments or in regions and developmental stages other than those of their origin, the implications of these findings for understanding and application to normal development remain areas of intense speculation. Skepticism remains since the functional properties of the newly derived cells have not been clearly shown, and the frequency of these "transdifferentiating" cells appears to be very low. Recently so-called transdifferentiated stem cells were shown not to produce new cells but instead to fuse with surrounding cells, forming diploid or tetraploid hybrid cells. However, the therapeutic potential of a ready-made "cell bank" that can produce any cell or tissue with the right stimulus is enormous.

Problems with transdifferentiating stem cells also derive from the possibility that an even more primitive cell may instead be responsible for the observed effect. Adult neural stem cells, when injected into the blastocyst, have been shown to contribute to embryonic tissues of endodermal, ectodermal, and mesodermal origin in the same manner as embryonic stem cells. Another explanation for these results is that the donor stem cell populations used for regeneration are themselves contaminated with extremely rare totipotent stem cells. In this scenario, neither dedifferentiation nor transdifferentiation occurs, but rather totipotent stem cells (which are known to migrate) in unexpected locations are responsible for successful donor–host engraftment and new growth.

What are the implications for human treatment in clinical settings? At present the concept of adult stem cells as equivalent to embryonic stem cells is not solidly founded. However, the new demonstrations of heretofore unimagined plasticity will lead to further investigation of adult stem cells as an alternative to embryonic or fetal stem cells, thereby avoiding the ethical considerations normally not applied to adult tissue use (see Text Box 18.4).

TEXT BOX 18.4 BLOOD CELL SUBSTITUTES

Blood transfusion is fraught with difficulties. Multiple incompatibilities can occur, the most common being due to differences in cell surface protein glycosylation that results in type A, B, or O (no glycosylation) blood types as well as the Rh factor incompatibility. Transfusion can also cause transmission of blood-borne diseases, including the human immunodeficiency virus (HIV), and hepatitis C, as well as a theoretical risk of transmission of unusual infections such as Creutzfeld–Jakob disease. Identification of each new risk factor also limits the pool of eligible donors and, hence, the available blood supply. In 2001, in the United States, the risk of infection from an infected unit of blood was 1 : 30,000 to 1 : 150,000 for hepatitis C and 1 : 200,000 to 1 : 2,000,000 for HIV. Clerical errors lead to mismatched blood administrations with a rate of fatal ABO blood transfusions of 1 : 500,000 to 1 : 2,000,000, and more than 2% of U.S. military identification tags for blood still have ABO errors. Finally, patients may refuse blood transfusions (e.g., Jehovah's witnesses) on religious grounds but may accept a synthetic blood substitute. For all these reasons, a reliable substitute for blood would be the ideal solution.

Oxygen Delivery—Erythrocyte Substitutes

The search for a blood substitute has focused almost exclusively on the ability of the blood to carry oxygen since this is the function of blood that is the most rapidly fatal when lost. Two major classes of compounds are currently under investigation as oxygen-carrying substitutes.

Perfluorochemicals (PFCs) are created by complete or nearly complete fluorine substitution for hydrogen in hydrocarbons, yielding a virtually inert compound. PFCs dissolve larger quantities of oxygen and carbon dioxide than water-based liquids such as blood plasma. The ability to hyperoxygenate PFCs outside the body allows a directed flow of oxygenated fluid to tissues in danger of anoxia such as the distal tissue during balloon angioplasty. The first challenge in the use of PFCs is that they are not miscible with water and require suspension as microdroplets with the use of emulsifying agents. The second challenge is that oxygen solubility in PFC has a linear relationship with the partial pressure of oxygen in the atmosphere, unlike the sigmoidal dissociation curve for hemoglobin that favors full loading at normal atmospheric oxygen levels. Hence they are not the ideal oxygen carrier substitute for Hb.

Other strategies have focused on using free hemoglobin (as opposed to RBC-enclosed Hb) as an oxygen carrier substitute. For this, methods are needed to increase the "size" of the Hb molecule to prevent it being cleared by the kidney. These have included cross-linking the subunits within tetramers, adding large molecules to hemoglobin (e.g., covalent attachment of polyethylene glycol to hemoglobin), polymerization of hemoglobin, encapsulation of hemoglobin in artificial red cells or liposomes, and using microsphere technology to form stable million-molecule microbubbles. However, the sources of the hemoglobin used for this work has been human outdated banked blood, bovine blood, transgenic swine, or transgenic *Escherichia coli*. Each of these sources has its own potential for contamination with a variety of infectious agents as well as antigenic problems when nonhuman blood is used. Research and development of effective oxygen carrier substitutes remains an active field.

Clot Delivery—Platelet Substitutes

Another crucial role of blood is clot formation, in which platelets play a central role. Individuals with thrombocytopenia (platelet deficiency) experience prolonged bleeding times that require transfusions with donor platelets. However, maintenance of adequate supplies of donor platelets is much more difficult than for RBCs. Unlike RBCs that can be kept for months in cold storage, platelets can only be stored at room temperature because cold exposure triggers platelet activation. This increases the risk of microbial contamination and viability drops over time so that donor platelets can only be stored for a maximum of 5 days. Hence, platelet transfusions are limited in several ways, including limited supply, loss of functional properties during storage, donor availability, transmission of infection, and cost. Platelet preservation or the use of platelet substitutes are consequently of great interest to blood providers.

Several types of artificial platelet substitutes have been developed including thromboerythrocytes, plateletsomes, infusible platelet membranes, sythocytes, and fibrinogen-coated microcapsules. The goal for synthetic platelets is to contribute to clotting without triggering adventitious clot formation in the absence of clot-stimulating conditions. Plateletsomes, for example, were developed from a liposome preparation with platelet proteins incorporated into the lipid bilayer. The goal of this artificial platelet is to enhance primary hemostasis and ameliorate bleeding despite the absence of metabolic processes in the plateletsome. In tests on animals deficient in platelets, the importance of the presence of negatively charges phospholipids was shown, as well as a dependence on platelet glycoproteins. Plateletsomes were shown to be effective in reducing bleeding, showing that despite formidable obstacles such as the absence of active metabolic processes, a platelet substitute is feasible.

Fibrinogen-coated microcapsules, by contrast, are formed from albumin formed into microcapsules by freeze-drying, then coated with fibrinogen. The size of the microcapsules is about 3.5 to 4.5 μm. Both plateletsomes and fibrinogen-coated microcapsules have been shown to stop active bleeding. However, in the absence of bleeding, they do not cause clotting, suggesting that they can be used as emergency measures to stop bleeding but not to regulate or control thrombocytopenia.

One method of platelet preservation has been the development of lyophilized platelets. Rehydrated, lyophilized platelets retain a similar ultrastructure and many of the surface membrane functions of fresh platelets. Their actions are similar to those of normal platelets: They spread on foreign surfaces, they adhere to denuded subendothelium and bind fibrinogen, and they support *in vitro* clot lysis. However, intracellular processes are only partially operational in reconstituted platelets, and these platelets also show depleted nucleotides, with lower levels of AMP, ADP, and ATP. Signaling and the response to activation in rehydrated platelets is also diminished. However, in comparison to artificial platelet substitutes, which mainly provide a surface of anionic phospholipid, the lower response level of lyophilized platelets is still sufficient to provide positive feedback amplification in coagulation processes, analogous but not equivalent to native platelets.

SUGGESTED READING

The following three textbooks are excellent general references on stem cells, erythrocytes, and platelets.

Beutler, E., Lichtman, M. A., Coller, B. S., Kipps, T. J., and Seligsohn, U. (eds.) (2001). *Williams Hematology*, 6th ed., McGraw-Hill Medical, New York. *This is a hematology textbook aimed at medical students, presenting the basics and then exploring aspects of blood diseases. Useful section on hematopoiesis.*

Israels, L. G., and Israels, E. D. (eds.) (2002). *Mechanisms in Hematology*, 3rd ed., Core Health Services, Concord, Ontario. *An unusual and acclaimed textbook that presents the basics of hematology and signal transduction, using pathways described and diagrammed very simply and clearly. An easy-to-understand and useful book, especially the sections on the regulation of hematopoiesis (Section 5) and platelet structure and function (Section 30).*

Stamatoyannopoulos, G., Majerus, P. W., Permutter, R. M., and Varmus, H. (eds.) (2001). *The Molecular Basis of Blood*

Diseases, 3rd ed., W. B. Saunders, Philadelphia. *Presents the basics of blood cell biochemistry and molecular biology, then relates them to specific disorders. Good chapters on signal transduction, iron metabolism, red cell membranes, and platelets.*

SPECIFIC FOR ERYTHROCYTES

Grimes, A. J. (1980). *Human Red Cell Metabolism.* Blackwell Scientific, Oxford, UK. *Comprehensive book about many aspects of red cell metabolism, particularly useful for its discussion of anaerobic glycolysis in Chapter 3.*

SPECIFIC FOR PLATELETS

Abrams, C. S., and Brass, L. F. (2001). Platelet signal transduction. In Coleman, R. W., Hirsh, J., Merder, V. J., Clowsd, A. W., and George, J. N. (eds.), *Hemostasis and Thrombosis: Basic Principles and Clinical Practice*, 4th ed. Lippincott Williams and Wilkins, London. *Reviews the current understanding of the initiation of platelet activation, propagation of the signal to amplify the response and signal other platelets, and signals that prevent platelet activation.*

SPECIFIC FOR STEM CELLS

Fuchs, E., and Segre, J. A. (2000). Stem cells: A new lease on life. *Cell* **100**:143–155. *An overview of the stem cell describing adult and embryonic stem cells, epithelial, neuronal, intestinal, and hematopoietic stem cells, as well as how stem cells exist in particular "niches" within the body, and emerging stem cell therapeutic potential.*

19

ORGAN PRESERVATION FOR TRANSPLANTATION

THOMAS A. CHURCHILL

INTRODUCTION

Transplantation: Not a Novel Concept

The concept of organ transplantation from one individual to another or from one species to another is not a new one. Sanskrit texts from India in the second and third century BC describe the autotransplantation (donor and recipient are the same individual) of pedicled (with vessels intact) grafts from the forehead, neck, and cheek to restore mutilations of the nose, ear, and lip. A Greek legend from AD 285 to 305 holds that Cosmos and Damian, the patron saints of medicine, transplanted the leg of a recently dead Moor to a patient whose leg had been amputated. Modern transplantation began with the experiments of John Hunter in the eighteenth century who transplanted a human tooth to a cock's comb and later a hen's spur to a cock's comb.

In the years following the initial experiments of Hunter, many obstacles remained to successful organ transplantation, most of which have been overcome within the last century by medical and surgical science. In the late 1800s, Lister described techniques of proper asepsis that allowed for a decrease in the infection rate. At the turn of the century, Carrel and Guthrie pioneered the technique of proper vascular anastomosis (the suturing of two vessels, bringing the vessels together) using fine sutures that penetrated all layers of the vessel wall. The problem of revascularizing explanted organs through effective vascular anastomosis without thrombosis (blood clotting), stenosis (partial blockage), and hemorrhage was therefore overcome, making solid visceral transplants possible.

While the rejection of allograft transplants remains a problem in modern transplantation, it is no longer an absolute barrier to success. Over the past 40 to 50 years, our knowledge of immunology and its role in tissue rejection has grown tremendously. This has led to the development of pharmaceutical agents capable of suppressing the immune response to allografted tissue and to the application of strategies such as tissue typing that lessen the incidence of rejection. While rejection has not been eliminated, the incidence has been reduced to acceptable levels, and when it does occur, it can be dealt with successfully in most cases.

Modern Advances in Organ Transplantation

The ultimate goal of organ transplantation is life extension for patients whose life expectancy would otherwise be measured in weeks or months, as well as an improvement in the quality of life. This applies to many people who suffer from end-stage disease of liver, kidneys, small bowel, heart, or other organs. While the concepts outlined in the following discussion deal mainly with specific organs, many of the concepts and principles are applicable to other solid viscera such as liver, kidneys, pancreas, small bowel, and lungs. Important advances in the last 30 years have contributed to the enormous success of transplantation of many organs. The major factors are:

1. Improved surgical techniques
2. Improved immunosuppression
3. Increased willingness of families to donate the organs of deceased loved ones
4. Development of methods to successfully preserve organs

Functional Metabolism: Regulation and Adaptation, edited by Kenneth B. Storey.
ISBN 0-471-41090-X

PRESERVATION

Problems Associated with Low-Temperature Organ Storage

When an abdominal or thoracic organ is harvested, its main arterial vessels are used to flush the organ with a preservation solution. A number of established preservation solutions are available commercially, and the one selected is dependent on organ type and on surgeon/hospital preference. The flush of cold solution (usually 2 to 4°C) performs two general functions. First, the blood is cleared from the organ to prevent blood clots and to ensure proper reperfusion of the organ once blood flow is reestablished in the recipient patient following transplantation. Second, the organ is cooled to reduce organ metabolic rate and slow down the rate of various degenerative processes. Once the organ has been flushed and excised from the donor, it is then placed in additional cold preservation solution and placed on ice inside a cooler for easy transport to the site of transplantation. As soon as blood flow is cut off from the organ during this harvesting procedure, the "clock starts ticking." There is only a finite period that the organ can remain outside the body in this state of cold storage; this period is organ-dependent but generally ranges from 4 h (heart, lung) to 48 h (kidney). Obviously, the longer the period, the greater the time available for surgical team preparation, tissue matching, and transportation.

Freezing whole organs is not presently an option because large, piercing ice crystals cause physical injuries to frozen organs, and unless storage temperatures are very low (e.g., at −196°C in liquid nitrogen) degenerative metabolism continues, albeit at a reduced rate. Currently, cryopreservation (preservation with freezing) is widely used for many kinds of single cells (i.e., blood, sperm, hepatocytes) or small/thin pieces of tissue (cornea, skin, pancreatic islets), which can be rapidly frozen in liquid nitrogen under conditions that promote intracellular water solidification into a "glass" (not ice crystals), which does not damage the tissue as ice crystals would. The often-quoted term *suspended animation* with respect to organ preservation refers to the use of pharmacologic agents (and temperature) to reversibly halt all metabolic activity. This is a great idea in theory but extremely difficult to put into practice. Currently, the only solution for solid organs is storage at near-zero, nonfreezing temperatures with preservation solutions that hopefully mitigate at least some of the primary problems associated with organ removal from the body.

Humans and most other mammals have evolved over millions of years to optimize their metabolism for function within a very narrow temperature window around 37°C. Having optimized metabolism for homeothermy, they are now incapable of surviving deep hypothermia because the differential effects of cold on factors including biochemical reaction rates, protein conformations, and lipid fluidity severely impair key functional aspects of intermediary metabolism, ion regulation, membrane transport, and excitation-contraction processes, to name but a few. Generally, most enzymes exhibit a 2.0 ± 0.2-fold decrease in rate for every 10°C drop in temperature; this is a direct result of a negative shift in the Arrhenius energy distribution. If you recall from basic physical chemistry, the number of molecules equal to or greater than the activation energy for the process of interest drops by ~50% for every 10°C decrease in temperature (Fig. 19.1). An added complication for organs removed from the body is a severe limitation of oxygen availability, which rapidly compromises the capacity of adenosine 5′-triphosphate (ATP)-generating processes even though most ATP-utilizing processes are still in operation. Once oxygen becomes limiting, cells switch to a greater reliance on anaerobic pathways, and glycolysis becomes the primary route of ATP production in excised organs. Unfortunately, glycolysis can only continue for a few hours during organ storage before problems arise. Substrate (endogenous glycogen, exogenously supplied glucose) can run out but, more importantly, end product accumulation becomes prohibitive. In particular, the accumulation of protons (acid) is the principal impediment to continued glycolytic energy production; acidification strongly inhibits the main regulatory enzyme of glycolysis, 6-phosphofructo-1-kinase (PFK). Consequently, rapidly precipitated imbalance arises between energy-generating and energy-consuming processes during cold organ storage.

In terms of metabolism, one of the primary defining characteristics of a living cell is the maintenance of ion gradients across the cell membrane. Specifically, potassium (K^+), sodium (Na^+), and calcium (Ca^{2+}) concentration

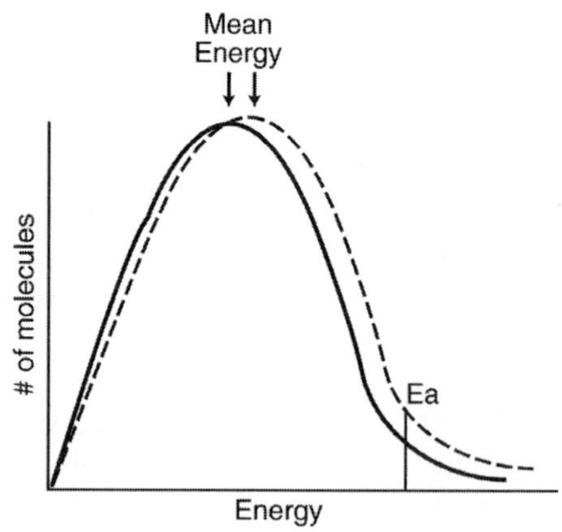

Figure 19.1 Arrhenius profile of a population of molecules with varying energy levels. E_a, activation energy of a given process. The mean energy of a population may decrease by a small amount as shown; however, the number of molecules with an energy > E_a drops by a much greater percentage.

gradients are of prime importance to the maintenance of cellular homeostasis. In addition to a multitude of sensitive cellular processes, these ions also regulate cell volume. Shifts in ion concentrations during organ storage influence water uptake (edema) and uncontrolled or excessive ion movements will result in damage due to edema. Consequently, the regulation of ion gradients is vital to maintaining cell/organ viability during organ storage at a number of biochemical levels (Fig. 19.2). Due to the energy dependency of ion pumps, changes in ATP availability will directly affect the intracellular and extracellular concentrations of these ions. One of the consequences of diminished ATP production via anaerobic pathways (glycolysis) in stored organs is a reduction in sodium–potassium ATPase activity. As a result, intracellular Na^+ levels rise. This is also exacerbated by proton accumulation as a by-product of glycolysis because increased activity of the Na/H exchanger also results in increased intracellular Na^+. There are two deleterious consequences of these events: (1) edema and (2) autolysis. First, since Na/K–ATPase activity is limited by dwindling ATP availability (this enzyme is also inhibited by low temperature), Na^+ extrusion from the cell is largely halted. Since the intracellular compartment is negatively charged with respect to the extracellular space (due to ion gradients as well as intracellular proteins that carry negative charges), the reduction

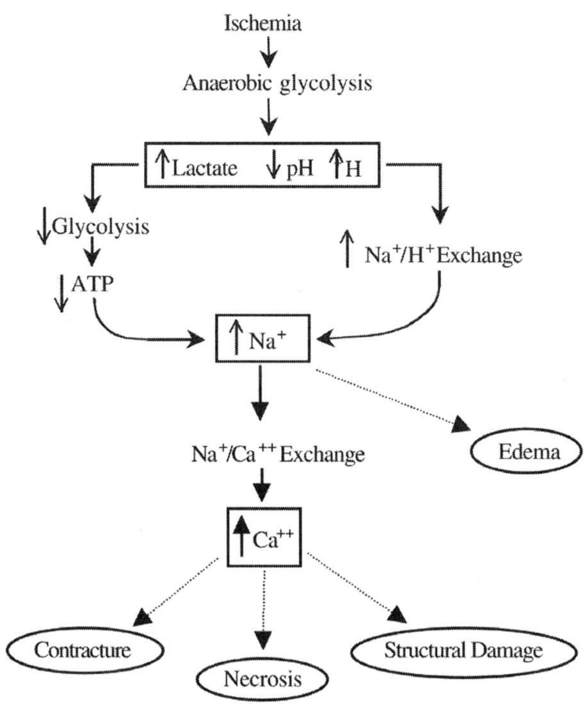

Figure 19.2 Events of ionic dyshomeostasis. Outlines the metabolic events that occur as a result of ischemia eventually leading to cell death.

in negative charge due to increased Na^+ retention leads to an influx of chloride (Cl^-) ion and a net uptake of water. This, in turn, can lead to cell swelling, lysis, and death. In an environment of low temperature and low ATP, cellular edema is unavoidable unless an impermeant molecule (e.g., dextran, starch, albumin, or selected impermeant sugars) is introduced into the extracellular milieu. For this reason, one or more impermeant constituents are included in commercial preservation solutions to limit tissue edema.

Although edema can be easily combated by modifying the composition of the preservation solution, a second effect of reduced Na^+ efflux cannot be so easily remedied. Uncontrolled Na^+/Ca^{2+} exchange across the cell membrane or mitochondrial membrane leads to increased intracellular calcium. Although the mean calcium concentration in cells is ~2 mM, cytosolic free calcium levels are only 2 μM! Calcium is sequestered in subcellular stores (chiefly in the endoplasmic reticulum, sarcoplasmic reticulum, and mitochondria), and regulation of calcium release is a fundamental concern in preserving the metabolic integrity of all cells. Controlled Ca^{2+} release from intracellular stores has a role in many intracellular processes (e.g., signal transduction, muscle contraction), but unregulated increases in cytosolic Ca^{2+} can have devastating consequences. High cytosolic Ca^{2+} activates calcium-dependent phospholipases, leading to the destruction of cellular membranes including those surrounding lysosomes that then release their contents to initiate uncontrolled autolysis (lysosomes contain multiple degradative enzymes including lysozyme, acid phosphatases, ribonucleases, and others). These degenerative processes are common to all systems of static (without perfusion) organ storage, although their progress is faster in some organs than others. For clinical organ storage the primary way to retard metabolic degeneration is to reduce the storage temperature and slow down all of these processes as a result of temperature effects on reaction rates. Other metabolic factors are also addressed in the design of various organ preservation solutions. The development of one such solution in the mid-1980s at the University of Wisconsin has been the most successful to date.

UNIVERSITY OF WISCONSIN COLD-STORAGE SOLUTION

The University of Wisconsin (UW) solution (Table 19.1) was developed by Folkert Belzer and James Southard and was originally formulated for preservation of the pancreas. However, it was quickly adopted by many centers as an excellent solution for the cold preservation of liver and kidney as well. The UW solution has several components that were added on theoretical grounds, the most significant being the addition of large impermeant molecules in the form of lactobionate, raffinose, and hydroxyethyl-starch;

TABLE 19.1 Composition of Two Major Preservation Solutions: University of Wisconsin Solution and Bretschneider's Solution

Component	UW Solution (mM)	Bretschneider's Solution (mM)
Lactiobionate	100	—
Raffinose	30	—
K^+	125	9
Na^+	15	15
Mg^+	5	4
SO_4	5	—
$(H_2PO_4)^-$	25	—
Allopurinol	1	—
Gutathione	3	—
Adenosine	5	—
Dexamethasone	8 mg/L	—
Hydroxyethyl starch (HES)	5%	—
Insulin	100 U/L	—
Cl^-	—	50
Histidine	—	198
α-Ketoglutarate	—	1
Tryptophan	—	2
Mannitol	—	30
Osmolality	320	290
pH	7.45	7.1

as discussed above, these limit edema during storage. The tripeptide antioxidant, glutathione (GSH: γ-GluCysGly) was added too minimize damage from oxygen radicals (see Chapter 12), and adenosine was added to enhance the ability of the cells to regenerate ATP upon reperfusion. Although the original UW formulation was initially constructed based on sound theory (most of which was borne out in experimentation), there are new aspects of organ preservation/reperfusion that are currently under investigation. Interestingly, some of these new factors were coincidentally being addressed with the original UW formulation, although they were not appreciated at the time. For example, lactobionate is an effective chelator of divalent cations such as Ca^{2+}; at the lactobionate concentrations present in the UW solution 85% of Ca^{2+} is chelated and 15% is free. Not only is control of free calcium generally desirable to minimize the activation of various calcium-sensitive cascades (in addition to preventing autolysis), the consequent inhibition of metallo-matrix proteinases (MMPs) as a result of Ca^{2+} chelation is also a beneficial outcome of the incorporation of lactobionate. MMPs are enzymes responsible for the degradation of the connective tissue that holds organs together. The UW solution remains the solution of choice for preserving liver, pancreas, and kidney and exhibits a significant extension of the safe storage times for these organs. However, although the UW solution has also been used for preservation of heart, small bowel, and lung, its success in extending storage times or improving organ viability after storage has not been anywhere near the success that has been achieved with organs such as liver and pancreas. Additional organ-specific factors must be required for preservation success, and this is the basis for much of the current research in this field.

Pathophysiology of Ischemia

When a tissue or organ is removed from the body, it is immediately cut off from the blood circulation (a condition called ischemia) and, hence, is deprived of its supply of oxygen and blood-borne fuels. Ischemia caused by organ excision initiates a series of biochemical events that lead to cellular and membrane dysfunction, intracellular edema, and cell death. The metabolic injuries initiated by organ excision are virtually equivalent to the ischemic damage seen in heart attack or stroke, and the reader is referred to the additional discussion of ischemic/anoxic damage in Chapter 15. Many, but not all, of the metabolic injuries associated with ischemia arise due to the interruption of oxygen supply, leading first to hypoxia (low oxygen) and then to anoxia (no oxygen). As discussed in Chapter 9 the majority of human organs rely on an uninterrupted supply of oxygen in order to catabolize fuels via aerobic pathways in the mitochondria and produce ATP via oxidative phosphorylation. Because ischemia compromises aerobic metabolism, it rapidly disrupts the balance between ATP production and ATP utilization in cells and is rapidly damaging especially to organs with high metabolic rates or low endogenous reserves of fermentable fuels or both (e.g., brain). All organs have some capacity for ATP generation via fermentative glycolysis when oxygen is limiting, but this is a less efficient mode of ATP production and can only be used for the catabolism of carbohydrate fuels. Under ischemic conditions, hypoxia develops rapidly, and, although normal cellular functions initially continue (but often for no more than a few minutes at physiological temperatures), cellular ATP reserves typically become rapidly depleted as demand for ATP by cellular reactions outstrips its supply.

Energy Metabolism: What Can Nature Teach Us?

Mammals, including humans, are endothermic homeotherms, which means that core body temperature is typically held within a very narrow range (about 37 to 38°C) fueled by internal metabolic heat production. Most mammals can tolerate very little deviation in core body temperature either up or down. Although hibernating species can survive for extended periods of time with body temperatures near 0°C (see Chapter 16), hypothermia is highly damaging for most mammals, including humans. Hence, organ explants, although benefiting from a reduced rate of metabolic decay when stored on ice, also undergo metabolic injuries as a result of the low temperature. Reasons for this include the differential effects of temperature change on the rates of multiple biochemical reactions that can result in severe imbalances and impairments of key metabolic pathways, ion regulation, membrane fluidity, and excitation-contraction processes. Again, as with hypoxic/ischemic injury, a rapidly precipitated imbalance between energy-generating and energy-consuming processes arises during periods of induced hypothermic ischemia (i.e., organ storage). Nevertheless, it is necessary to expose the target organ to hypothermic storage in order to reduce the deterioration of cellular homeostasis caused by hypoxia. How can this "catch 22" situation be resolved?

Advances in organ preservation have been made by identifying and correcting various metabolic deficiencies by inhibiting, supplementing, and stabilizing various cellular and metabolic parameters at low temperature. Most advances have been derived chiefly from empirical experimentation with an understanding of the biochemical mechanisms involved sometimes lagging behind. In subsequent sections, we will analyze some of the strategies for extending organ viability in hypothermia, examining the underlying biochemistry that causes the injury and the metabolic effects of the applied treatments.

Key elements of hypoxic/hypothermic survival can be derived from studies of the physiological and biochemical adaptations of hibernating mammals and of anoxia-tolerant lower vertebrates and invertebrates. In Chapters 15 and 16 we learned that the overriding critical adaptation for survival of these stresses was metabolic rate depression. By coordinating a strong suppression of the rates of ATP-utilizing reactions, metabolic rate could be reduced to a level that was sustainable by the ATP output from anaerobic glycolysis alone (during anoxia) or that could be sustained over many months by the slow oxidation of internal stored fuel reserves (hibernation). For hibernating mammals, additional energy savings are accrued by abandoning homeothermy and allowing body temperature to fall to ambient, thereby using the Q_{10} effect of temperature on reaction rates to slow the rate of ATP demand. As a result of metabolic rate depression and a readjustment of the hypothalamic set point for body temperature, hibernating mammals can sustain a *reversible hypothermic state for several months.* During hibernation, breathing and heart beat also slow to just a fraction of normal so that organ perfusion rates fall to less than 10% of euthermic values, a rate that would cause ischemic injury in euthermic animals. It is apparent, then, that hibernating mammals as well as many types of lower animals have the capacity to adjust their organ metabolism to endure ischemia, hypoxia, and hypothermia, and this has two important implications for human organ preservation technology: (a) mammalian organs can, with the proper preparation, endure these stresses, and (b) identification of the metabolic regulatory mechanisms that aid stress survival in natural systems will allow the development of applied strategies that can be used to impose metabolic rate depression upon organ explants to improve their long-term preservation. Based on such animal studies, it is already known that two strategies are particularly important to the maintenance of cellular energy homeostasis during cold storage: (a) maintenance of an adequate rate of ATP production, and (b) reduction of the rates of energy-consuming processes (Fig. 19.3). Applied strategies for organ preservation have incorporated both of these principles, and the following sections will discuss the progress to date in improving the preservation of organ explants.

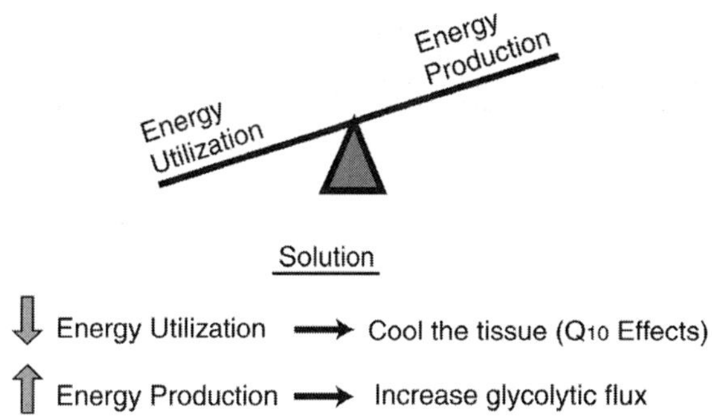

Figure 19.3 A metabolic problem: the energetics of cold storage. This figure depicts the imbalance in tissue energetics that occurs as a result of cold hypoxic organ storage and the potential actions that can remedy the effects of this insult.

Transition to Anaerobic Metabolism

The period between harvesting a donor liver and transplantation of the organ into the recipient involves cold storage of the isolated organ. During this period, the supply of blood, and consequently of oxygen, is cut off from the organ. With present technology and practice, this period of cold ischemia is unavoidable and, unfortunately, results in a gradual deterioration of organ function, eventually progressing to irreversible damage. As soon as the oxygen supply is halted, the organ switches to a greater reliance on anaerobic energy-producing pathways, glycolysis becoming of primary importance (Fig. 19.4). Energy levels within the tissue are soon compromised during ischemic storage since many energy-utilizing pathways/steps are still in operation, even though cold storage reduces their rate. Consequently, there is a rapid decline in organ ATP levels as well as a corresponding rise in lactate (Fig. 19.5).

In addition, intracellular acidification progresses quickly during cold storage. For example, pH values during liver storage can fall from 7.6 to 6.8 within 10h of cold storage.

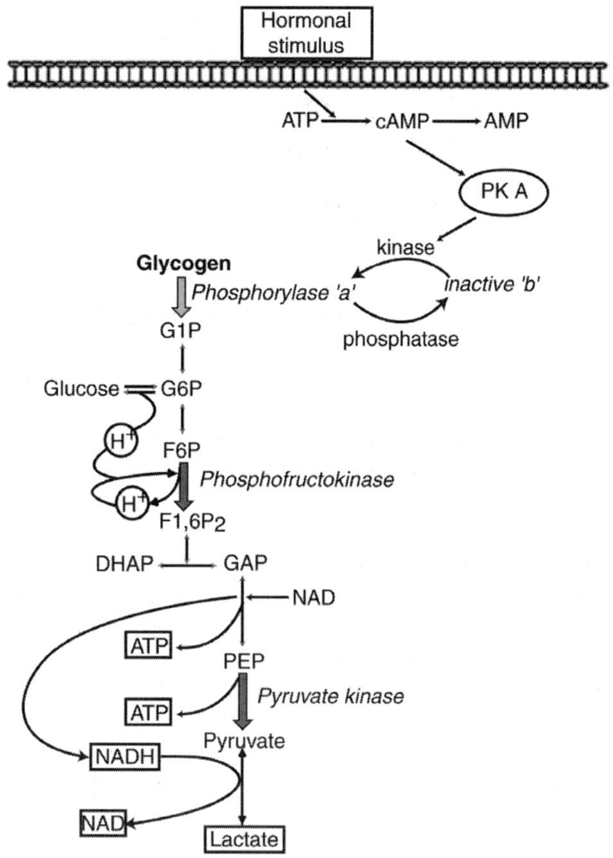

Figure 19.4 The glycolytic pathway. Various aspects of metabolic regulation depicted in this figure are discussed with respect to their involvement in limiting energy production during organ storage.

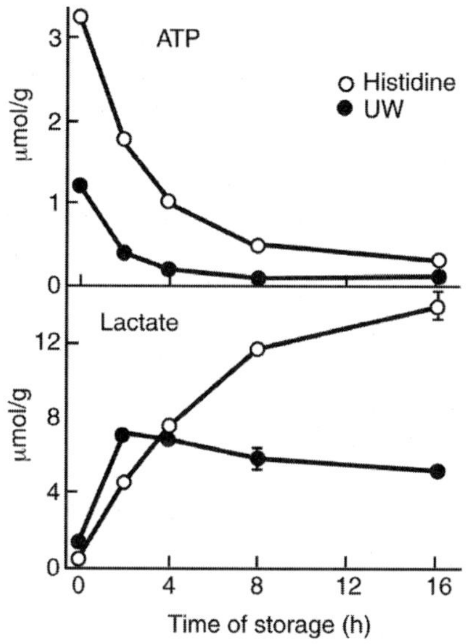

Figure 19.5 ATP and lactate levels in rat liver during cold storage. ATP levels in livers stored in UW solution drop off rapidly over the first 2 to 4h; an initial attempt to replenish dwindling ATP stores via anaerobic glycolysis is made, but lactate levels only increase over the first 2h. With a histidine-based solution (90mM histidine), the superior buffering afforded by the histidine results in superior ATP levels throughout the 16-h storage. The mechanism involved is alleviation of a pH-mediated inhibition of glycolysis; lactate levels rise continuously throughout most of the time course shown.

Acidification arises because of an imbalance in proton metabolism between ATP-producing and ATP-consuming reactions. ATP hydrolysis by cellular ATPases produces protons [the products of an ATPase are adenosine 5′-diphosphate (ADP), inorganic phosphate (P_i), and H^+] that are normally balanced by proton consumption during oxidative phosphorylation. However, this balance breaks down under anoxia or ischemia because the combination of ATP production by anaerobic glycolysis + cellular ATP consumption by ATPases is actually a net producer of H^+. As an aside, it is true that the net reaction mediated by glycolysis under anerobic conditions is: Glucose $\rightarrow$ 2 lactate + $2H^+$. However, it must be emphasized that the terminal production of lactate is not the direct source of protons; the pyruvate to lactate conversion actually consumes H^+. It is the hydrolysis of ATP used to "prime" the glucose molecule that produces $2H^+$. A second source of protons is the generation of reduced nicotinamide adenine dinucleotide (NADH) from oxidized NAD (NAD^+), which is accompanied by a proton (H^+) (charge must be conserved). During ischemic organ storage, the NAD^+ + NADH pool converts almost entirely (99%) to NADH and consequently contributes to the proton pool. The drop in cellular pH during organ storage is large enough to have substantial effects on a number of pH-sensitive enzymatic reactions. In particular, low pH inhibits PFK, the most pH-sensitive enzyme of glycolysis, and this has important implications because, of course, cellular energy production during cold storage of most organs relies almost entirely on glycolysis. Although this is true for most cases, novel strategies for the preservation of small bowel (outlined later in this chapter) rely instead on the limited amount of residual oxygen present in the tissue/solution to facilitate energy production.

The regulation of flux through the glycolytic pathway typically occurs at several irreversible enzyme loci (which show high flux control coefficients; see Chapter 1): glycogen phosphorylase, phosphofructokinase (PFK), and pyruvate kinase (PK). Note that hexokinase is not of concern in the ischemic model since blood glucose is not available as a fuel. Other enzymes of glycolysis catalyze reversible reactions. Regulatory control of the glycolytic pathway is extremely sensitive to cellular energy levels, as well as to pH, ion concentrations, substrate/product levels, and numerous enzyme-specific effector molecules. These regulatory enzymes are also typically targets of reversible phosphorylation that alters enzyme conformation and changes kinetic parameters (K_m, $V_{\max}$, K_a, I_{50}). In the case of liver PFK, for example, the high-phosphate form of the enzyme is a less active form and the low-phosphate form is the more active. The low-phosphate form of PFK is much less sensitive to inhibition by ATP and more sensitive to activation by adenosine 5′-monophosphate (AMP). The low-phosphate form of PFK would be expected to predominate during ischemia since low ATP levels limit kinase-mediated enzyme phosphorylation. Additionally, the inhibitory effects brought about by low pH severely impair PFK reaction rates. However, the low-phosphate form of liver PFK is inhibited to a lesser degree than the high-phosphate form. Thus, the inhibitory effect of pH may not be predominant in the regulation of PFK under these unique conditions, and the net activating effect of increased sensitivity to activators combined with decreased sensitivity to inhibitors could become more important with respect to final activity state.

So what actually happens at PFK in "real life"? Exposure of a mammalian organ to the insult of extreme hypothermic hypoxia is not a "typical" event. Hence, the metabolic responses cannot be regarded as a concerted effort to combat the stress as is often the case with well-adapted animals in nature. Interestingly, the net effect of all influencing facets on PFK regulation in mammalian liver is an activation of the enzyme after 8 to 10 h of cold storage; the substrate and product of this reaction suddenly drop and rise (analogous to a dam in a river opening, allowing the flow of water through to the next block in the river). Unfortunately, glycolysis is inefficient with respect to maximizing ATP production per molecule of glucose during anaerobic conditions (only 5 to 8% compared to oxidative phosphorylation) and even to "break even," the glycolytic rate must be increased considerably to maintain the delicate balance in cellular energetics. Hence, despite attempts to promote flux through glycolysis, there is clearly an inhibition of the pathway early during cold storage and a consequential failure to maintain energy pools.

A primary limitation of continuous flux through the glycolytic pathway is maintenance of *redox* (reduction/oxidation) balance; in the cell, this is defined as the ratio of $NADH/NAD^+$. NADH is involved in numerous redox reactions in which a substrate receives the proton from NADH and is reduced (the converse is true for NAD^+); consequently NADH [and NADH phosphate (NADPH)] is referred to in terms of *reducing equivalents*. Glycolysis can only continue so long as the NAD^+ that is reduced to NADH by the glyceraldehyde-3-phosphate dehydrogenase (GAPDH) reaction is regenerated. Under aerobic conditions, this is accomplished by shuttling reducing equivalents into the mitochondria; but, under hypoxic/anoxic conditions, NADH is consumed via the conversion of pyruvate to lactate by the lactate dehydrogenase reaction. Hence, lactate typically accumulates under hypoxic/ischemic conditions. During flush/storage of liver, the $NAD^+/NADH$ pool is rapidly converted to >50% NADH during the flushing process and ~100% NADH by 25 min postflush. The implication with respect to NAD^+-requiring reactions is that there may be a block in glycolysis at the GAPDH reaction as a result of the imbalance in $NADH/NAD^+$ ratios. Therefore, even though PFK (the usual rate-limiting enzyme) may be functional, glycolysis may be severely impaired at the level of GAPDH by an imbalance in redox potential.

In addition to problems of low oxygen and the limited ATP production capacity available from glycolysis in the ischemic excised organ, another of the major problems of static storage of organs is the accumulation of anaerobic end products such as lactate and H^+. End products eventually accumulate to inhibitory concentrations and result in feedback inhibition of the glycolytic pathway that suppresses further ATP production. The practicalities of clinical organ harvesting dictate that a static method of organ storage is much more feasible than its alternative, which is much better at maintaining viability, a continuous perfusion method. Furthermore, observation of the techniques involved in clinical organ harvesting is that once the organ is flushed and transferred to the storage container, preservation solution is permitted to escape due to vascular collapse. At a metabolic level, even though lactate and H^+ are continuously removed from the intracellular component via H^+/lactate pumps, the critical inhibitory concentration will be reached more quickly when the vascular system is collapsed than if the vascular space was maintained in a filled state.

Regeneration of ATP The decline in organ ATP levels during storage does not simply mean a corresponding increase in ADP and AMP. Further degradation of AMP proceeds to yield elevated IMP, inosine, hypoxanthine, and xanthine (Fig. 19.6). Consequently, the total adenylate (TA) pool (ATP + ADP + AMP) declines. Not only do these events have direct implications for maintaining energy-utilizing processes throughout static storage, but a reduced TA pool also limits the ability to replenish ATP once oxygen is reintroduced to the organ. The term *reperfusion* refers to the period immediately following transplantation of the organ once blood flow is reestablished. The regeneration of TA stores is critical to organ viability following reperfusion in the recipient patient. The length of ischemia during storage directly influences the extent of ATP depletion, and the potential for full adenylate recovery depends on whether ATP levels and/or the TA pool have fallen below a "critical" minimum value during the storage period.

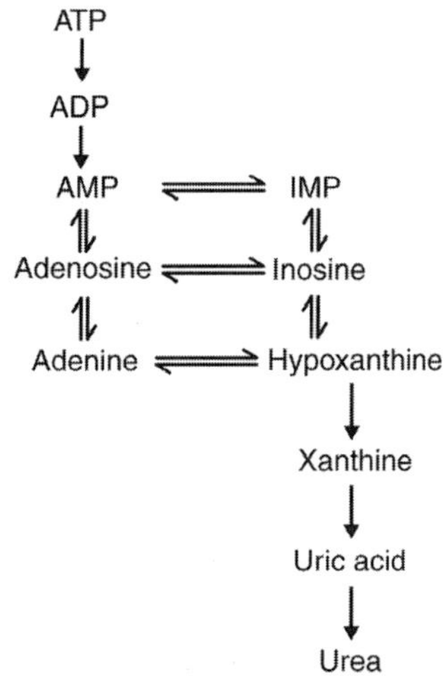

Figure 19.6 The purine (adenine) pathway.

Under normal circumstances in organs, ATP hydrolysis by enzymatic reactions produces ADP, which is rapidly rephosphorylated, generally by oxidative phosphorylation but also from creatine phosphate pools (in muscle). However, when ATP-producing reactions are limiting, accumulated ADP is processed by other reactions. The adenylate kinase reaction can temporarily boost ATP production ($2\ \text{ADP} \rightarrow \text{ATP} + \text{AMP}$) but this rapidly elevates AMP. When AMP levels become too high, AMP is withdrawn from the adenylate pool by its catabolism to either IMP or adenosine (depending on the tissue or circumstances). Removal of AMP from the TA pool allows cellular energy charge, $\text{EC} = (\text{ATP} + 0.5\ \text{ADP})/(\text{ATP} + \text{ADP} + \text{AMP})$, to be maintained at a high value for as long as possible. Subsequently, IMP and adenosine can be further catabolized to form products including inosine, hypoxanthine, and xanthine.

Restoration of tissue adenylate status during the reperfusion period is an energy-expensive process that can involve two routes, the so-called *salvage* pathway and the *de novo* pathway. The *salvage* pathway is a much simpler and less energy expensive process than the *de novo* pathway and involves a single reaction in which a free purine (in this case adenine) reacts with 5-phosphoribosyl-1-pyrophosphate (PRPP) to yield the corresponding adenine nucleotide, AMP. The *de novo* pathway, as the name suggests, forms the purine ring system (on ribose-5-phosphate) in a lengthy step-by-step series of reactions. Cold ischemic storage may reduce the capacity to regenerate adenine nucleotides and, thus, further reduces the resuscitation possible upon reperfusion. However, one of the strategies being tested as a means of aiding organ recovery during reperfusion is the addition to the perfusate of substrates that enhance both the *salvage* pathway and *de novo* pathway of adenylate synthesis (e.g., fructose, ribose, adenine, adenosine, glutamine, glycine, aspartate). These may improve the overall yield of adenine nucleotides and the recovery of energy metabolism during reperfusion.

In summary, it is becoming increasingly apparent that the compromised efficiency of energy metabolism during cold, ischemic storage of organs is multifactorial and includes: (a) inhibitory effects of declining pH on glycolysis, (b) imbalanced redox state, (c) covalent and allosteric modulation of key regulatory enzymes, (d) reduced transport of inhibitory end products into the vascular space, and (e) capacity to regenerate high-energy phosphates upon reperfusion.

Obvious Solution: Continuous Perfusion It has long been recognized that the injuries to excised organs that are caused by cold ischemic storage could be easily minimized or fully alleviated by storing organs under conditions of continuous hypothermic perfusion. Particularly for kidney and liver, studies have shown that perfusion under optimal conditions of oxygenation, perfusate flow, and substrate supply can yield excellent long-term preservation. For example, in studies with liver, good recovery of hepatic function and survival after reimplantation can easily be achieved from organs that have undergone hypothermic perfusion for up to 72 h compared with static storage that provides only 12 to 24 h of viability. One of the primary events of static organ storage is the progressive impairment of mitochondrial function by an increase in uncoupled respiration; this can be avoided when the mitochondria are maintained in an active, oxidative state by continuous perfusion, even at hypothermic temperatures. It has been suggested that it is specifically the hypoxia stress due to the static nature of organ storage that is responsible for the opening of the mitochondrial transition pore, which causes uncoupling of respiration by collapse of the membrane potential and permeabilization of the inner mitochondrial membrane. However, at present, the technique of continuous perfusion is logistically difficult to transfer to the clinical arena, and, although there is much sound evidence that clearly shows the benefit of a continuous perfusion system, simplicity is often the factor that dictates clinical practice. Consequently, scientists working to develop improved strategies for organ preservation must take heed of the confines of the clinical arena, and, currently, for the preservation of solid organs this means static storage is the only realistic option.

Role of Buffers One preservation solution used mainly in Europe was developed by Bretschneider in Germany in the early 1970s for the purpose of cardioplegia ("stopping the heart") during open-heart surgery (Table 19.1). Since 1985, Bretschneider's solution has also been used during heart transplantation. Bretschneider recognized the problem of the development of intracellular acidosis as a result of ischemia. To address this problem he designed a flush solution that had a high buffering capacity provided by the addition of 198 mM histidine. Histidine is a good physiological choice for a buffer because the imidazolium groups on histidine residues (on proteins, in small peptides, and free histidine) are actually the major intracellular buffering agents in animal cells. The data in Figure 19.7 illustrate the relative buffering capacities of two organ preservation solutions, the UW and the St. Thomas Hospital solution (STHS), along with histidine and other buffers [bicine, bis(2-hydroxyethyl)-2-aminoethanesulfonic acid (BES)], the buffers being present at 90 mM each.

As the figure shows, STHS has a very poor buffering capacity, and a very small acid load results in a steep drop in pH. UW fairs slightly better due to the presence of 25 mM phosphate as its primary buffer, but its buffer capacity is still much lower than the three buffers. Histidine has a strong buffering capacity within the physiological pH range, as depicted for 90 mM histidine on the graph, and Bretschneider's solution with 198 mM histidine will have a 2.2-fold greater buffering capacity (producing an x intercept of $\sim$120 mmol/L) The properties of an effective buffer for use in biological systems are further discussed below.

In recent years, Bretschneider's histidine-based HTK cardioplegia solution has been tested experimentally and used clinically as an alternative to the UW solution for both hepatic and cardiac preservation. No apparent differences in the ultimate survival rates of the organs after storage and implantation have been seen, indicating that both are effective preservatives. Interestingly, a variety of solutions have been promoted as being appropriate for cardiac preservation, approaching the problem of graft preservation from several different perspectives. For example, the UW solution addresses the problem of cell swelling with less emphasis on tissue acidosis, whereas the opposite is true for Bretschneider's solution.

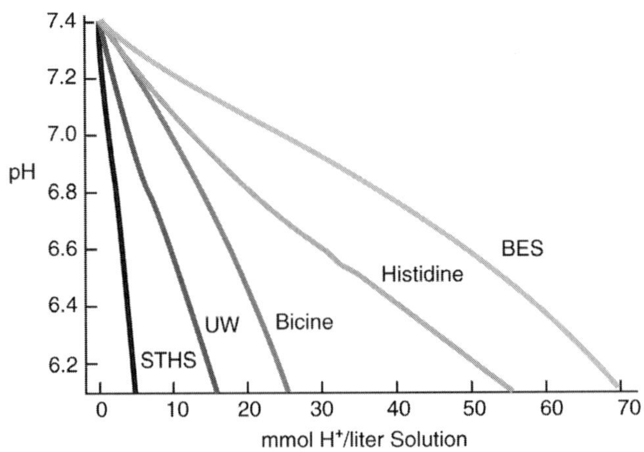

Figure 19.7 The relative bufferng capacities of various solutions: STHS (St. Thomas Hospital solution), UW, bicine, histidine, BES (modified UW solution supplemented with 90-mM bicine, histidine, or BES).

Given that Bretschneider's and UW solutions address different aspects of preservation, it is obvious that a combination of the two could be even better. Indeed, this has proven to be the case. A considerable improvement in CIT (cold ischemic time) is possible by combining the osmotic impermeants of the UW solution (lactobionate and raffinose) with the increased buffering capacity (provided by histidine) of Bretschneider's solution. Using rat livers, 85% survival after reimplantation was reported after 24 h cold storage when the organ was initially flushed with a combined histidine–lactobionate (HL)-based solution as compared with 29% with the UW solution. One of the effects of this novel formulation was a lesser decline in ATP levels during storage as compared to livers stored in UW. In addition to superior energetics, increased lactate production (i.e., better maintenance of glycolysis), decrease pH disturbance, and maintenance of the activities of regulatory enzymes were subsequently established as part of the underlying mechanism of the novel HL formulation. Clearly, the added buffering capacity of the HL solution explains, in part, the improvement in tissue energetics. However, despite positive effects of histidine-based or histidine-lactobionate-based preservation solutions with organs of small animals, studies with large-animal systems (dog, pig, human) have not found an improvement in protection. This may indicate a problem of suboptimal type of buffer. Although histidine is an excellent choice for clinical solutions (i.e., low toxicity, moderate buffering capacity, high solubility), there are many compounds with markedly superior buffering capacity that may be better suited to larger organs (e.g., BES). Overall, there are several important selection criteria that must be considered for an effective buffering agent:

1. The pK value for the buffer must be near to the desired pH of the storage solution to provide maximal buffering capacity; the most effective buffering occurs at pK $\pm$ 0.5 to 0.8 pH units.
2. A low molecular weight is needed to aid in transmembrane diffusion. One frequent, but not required, characteristic of a buffering agent is that it be a compound capable of zwitterion activity (both a positive and negative charge on the same molecule). Thus, at no time will the buffering molecule be uncharged (it will have a net charge of zero as a result of the generation of a zwitterion when pK = pH).
3. A structure that is similar to that of an amino acid (or other natural biological molecule) is advantageous

so that it might be possible to facilitate passage of the buffer into cells using existing membrane transporters.

4. High solubility, low chemical reactivity and low toxicity minimize damage to cells by buffers that are not natural cellular constituents.
5. A similar effect of temperature change on buffer pH (ΔpH/°C) as that displayed by water or histidine (-0.018 pH/°C) is advantageous so that when organs are cooled, the pH change with temperature of the buffering system mimics the natural behavior of body fluids when cooled.

According to the "α stat" theory of pH control, as temperature changes and causes a shift in intracellular pH, the histidine moieties (the primary amino acid responsible for enzyme catalysis) of intracellular proteins are able to maintain their ionization status at an ideal value required for effective reversible enzyme catalysis (for an equilibrium type enzyme). The α stat refers to the state of α-imidazole groups of histidine residues found in the catalytic region of an enzyme. The mathematical term for the state of the α-imidazole groups is simply the ratio of unprotonated imidazole groups to total imidazole groups, or [imid]/([imid·H^+] + [imid]). The ideal status of catalytic histidines where reversible protonation/deprotonation events are required is for this ratio to be equal to 0.50. Hence, 50% of the imidazole groups (of catalytic histidine residues) are protonated and half are unprotonated. The result is that an enzyme is equally capable of catalyzing a forward or reverse reaction. α-Stat regulation is believed to play a fundamental role in enzymatic catalysis, the three-dimensional structure of proteins, and buffering transient changes in pH.

Clinically, only a histidine-containing preservation solution (Bretschneider's solution) has been explored as an agent for improving the problem of pH buffering during organ storage, but *in vitro* studies suggest that other buffers such as bicine, tricine, and BES are superior with respect to maintaining the activity state of key regulatory enzymes and sustaining ATP generation by glycolysis. In liver, buffering capacity correlates well with the length of time that glycogen phosphorylase is maintained in its active *a* form. Although Figure 19.7 suggests that bicine is not a great buffer, one must consider the starting pH (7.4) of the titration curve. If the initial pH is 8.0, then the buffering capacity for bicine is much superior to that of a comparable histidine-containing preservation solution. Interestingly, the ideal pH of preservation solution has been studied to some extent, with conflicting results. Some studies suggest that higher initial solution pH is beneficial, while others have found improved preservation of certain organs at a lower pH.

Buffering agents such as bicine, tricine, and BES are amine-type buffers that were first described by N. E. Good. These "Good" buffers are a series of compounds that exhibit excellent buffering capabilities across a wide pH range. Although the Good buffers are widely recognized and used in laboratory settings, one potential problem that is characteristic to many low-molecular-weight amine buffers is that these compounds have a tendency to uncouple mitochondrial oxidative phosphorylation owing in part to their ability to traverse cell membranes. Imidazole, despite having excellent buffering abilities, is such a compound and is know to be a very strong uncoupler of mitochondrial oxidative phosphorylation. When imidazole is added to UW solution, there is a rapid depletion of organ ATP content, even during the time that it takes to flush 5 to 10 mL of solution through a 15 to 20-g rat liver. Within 60 min, ATP drops by >97%! Hence, although favorable characteristics are outlined above, there are no "hard and true" rules when selecting a suitable buffer, and caution must always be used when incorporating any new buffer into a potentially novel preservation solution.

Role of Protein Kinases *In vivo*, whole organs are subject to hormonal regulation with hormones binding to specific membrane receptors and subsequently eliciting second-messenger responses. These trigger cascades of protein kinase and/or protein phosphatase enzymes so that a single hormone-receptor binding event is amplified 10^2- to 10^3-fold in terms of final kinase/phosphatase activity on the target protein (Fig. 19.8) (see also Chapter 4). This

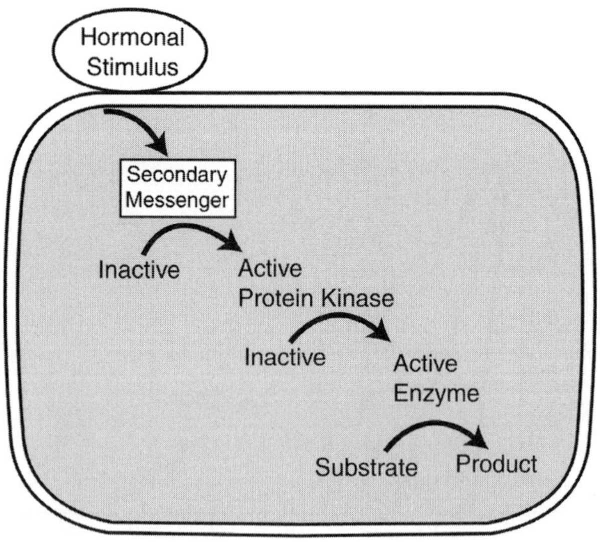

Figure 19.8 Amplification cascade involving enzyme phosphorylation. The consequence of this cascade is that from a single hormone binding event, a 10-fold increase in activity can occur at each step, resulting in an overall 10^2 to 10^3 increase in substrate to product conversion.

type of enzyme regulation is constantly in operation with many regulatory proteins and is not confined solely to the regulation of energy production; many, if not all, energy-utilizing processes are also controlled in a similar manner. The role of many protein kinase and protein phosphatase systems in metabolic control has been well-documented *in vivo* for mammalian systems (see Chapters 4, 5, and 10). Such an all-encompassing mechanism of cellular control that can potentially regulate both energy-producing and energy-consuming processes is viewed as an important target for improving preservation techniques. For example, a single extracellular signaling molecule could be employed to initiate reversible phosphorylation/dephosphorylation regulation over multiple enzymes to activate anaerobic metabolism in excised organs or to reduce ATP use by selected energy expensive processes (i.e., ion channel regulation, protein synthesis).

During cold hypoxic storage, organs need a continuous supply of ATP produced by the catabolism of endogenous glycogen reserves via anaerobic glycolysis. Despite this, the observed occurrence in excised organs is a progressive and continuous decline in the activity of the main enzyme responsible for glycogen breakdown, glycogen phosphorylase (GP). As discussed in previous chapters, GP is interconverted between an active *a* form (phosphorylated enzyme) and an inactive *b* form (dephosphorylated); activation is mediated via cyclic-AMP-dependent protein kinase (PKA), which activates GP kinase, which in turn phosphorylates GP. During clinical isolation of the donor organ, all hormonal controls on the phosphorylation states of cellular proteins (including GP) are lost. As a result, the amount of active GP*a* falls, which results in a decrease in glycogen degradation, at the same time as the isolated organ needs an increase in anaerobic glycogenolysis to support the ATP needs of the organ during prolonged storage.

How can this problem be corrected in the applied situation of an organ explant? Researchers turned to the use of a cell-permeable analog of cyclic-AMP (cAMP). By providing dibutyryl-cAMP in the flush solution, a high level of PKA activity, and hence of GP*a* activity could be maintained throughout the hypothermic storage period. With sustained high GP*a*, lactate production increased and, consequently, tissue ATP was better maintained. Figure 19.9 shows results from a study on liver that utilized dibutyryl-cAMP combined with the phosphodiesterase inhibitor, dipyridamole, that blocks the enzymatic degradation of the cAMP analog. In combination these two treatments produced an elevation of GPa activity compared to untreated UW flushed/stored liver. As a result, net lactate accumulation increased by 2- to 3-fold over 10 to 24h, and ATP was significantly higher than in organs stored in UW- or histidine-based solutions.

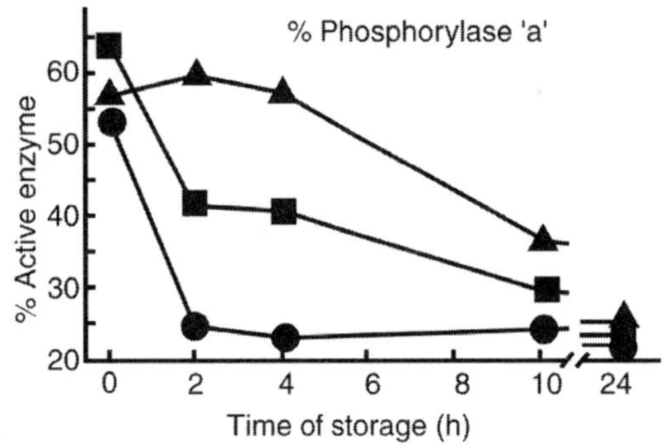

Figure 19.9 Cyclic-AMP activation of hepatic glycogen phosphorylase. The percentage of glycogen phosphorylase *a* (active form) in livers flushed with UW solution (circles) dropped by 50% within 2-h storage. In a histidine supplemented modified UW solution (squares), there still was a decrease, but the magnitude was less than with UW alone. However, with the addition of a cell-permeable analog of cAMP, dibutyryl-cAMP, plus a phosphodiesterase inhibitor (dipyridamole), the active *a* form of the enzyme was preserved for the first 4h of storage (triangles) before beginning to decline.

Importance of End Product Clearance One of the problems routinely encountered with current clinical harvesting techniques is the limited vascular volume left after initial flushing is complete. Preservation solution escapes due to vascular collapse caused simply by gravity and the weight of the organ. As a result there is very little extracellular fluid volume remaining within the excised organ into which end products can be excreted. To determine whether improved end product clearance would allow for better maintenance of energy levels during static organ storage, we used a unique experimental setup that involves increasing the volume of preservation solution in direct contact with the organ. The organ is allowed to take up preservation solution under a physiological pressure head and undergoes a "ballooning" effect. All inlets are then tied off to maintain the ballooned state of the organ throughout storage. The pressure must be carefully controlled since too great a pressure will cause physical damage, especially to the microvasculature, whereas too little pressure will not counteract the weight of the organ and it will maintain collapsed. We have used this technique of organ ballooning with both liver and small bowel to increase the volume of preservation solution contained within the organ and consequently permit greater tissue interaction with the preservation solution. The resulting energy profiles in liver and bowel are markedly superior in ballooned organs as compared with the normal state of vascular collapse. Ballooned organs also show superior recovery of high-energy adenylates upon reoxygenation. Furthermore, with the greater perfusate volume to tissue ratio allowed by the ballooning technique, greater benefits from the supplemental agents added to the preservation solution (e.g., buffers, substrates, agonists/antagonists of protein kinases or phosphatases) are also possible. When used in concert with oxygenated solution, greater amounts of oxygen can also be provided to the organ during static storage. The ballooning technique has been remarkably successful for preserving the metabolic integrity of organs. In particular, by providing a much greater extracellular volume into which products can be excreted, the rise in intracellular lactate and protons is retarded and longer storage times are allowed before intracellular acidification and end product accumulation become major problems. One of the attractive features of this technique is its simplicity. It requires no additional equipment during organ harvesting, only several well-placed ties. This simple, yet elegant technique combines some of the benefits of continuous perfusion with the simplicity and practicality of static storage.

Clearly, the problems associated with organ preservation cannot be attributed to a single mechanism. As a result of the obvious complexities involved in designing an "ideal" organ preservation solution, any novel solution should be carefully formulated so that all of the relevant factors are addressed.

REPERFUSION

Ischemia–Reperfusion Injury

As discussed above, the ischemia imposed by static storage of organs causes a variety of metabolic injuries. Paradoxically, so does the reperfusion of ischemic tissues. Although reperfusion quickly supplies oxygen and fuels to tissues so that normal metabolism should resume, a number of negative events also occur. The phenomenon is called ischemia–reperfusion injury (IRI) and is well-known both in organ preservation and in situations such as recovery after blood vessel occlusion in heart attack or stroke. IRI consists of dedicated sequence of events at multiple cellular levels, some of which have been discussed earlier (see Chapter 15). With respect to the recovery of organs after implantation, three categories of IRI events are important (Fig. 19.10):

1. Production of reactive oxygen species (ROS)
2. Cell-specific events targeted at the endothelium
3. Reduced bioavailability of nitric oxide (NO)

Reactive Oxygen Species

Reacitve oxygen species (ROS) including superoxide radical, hydroxide radical, and hydrogen peroxide cause a variety of cytotoxic effects (see Chapters 12 and 13). ROS are the continuous products or by-products of a variety of cellular processes and are routinely managed by the antioxidant defenses of all organs. However, numerous situations arise where an overproduction of ROS can overwhelm existing antioxidant defenses and lead to multiple forms of ROS damage by macromolecules including deoxyribonucleic acid (DNA), lipids, and proteins (see Chapter 12 for an in-depth discussion of ROS-mediated damage). It is well-known that the reintroduction of oxygen after an ischemic event results in a brief burst of ROS production and that the damage incurred by these ROS is a fundamental mechanism of IRI in all organs. This has been confirmed for IRI-associated hypothermic preservation in small bowel, kidney, lung, heart, and liver. Hence, the development of applied techniques that prevent or minimize damage by ROS during organ reperfusion is a major goal of transplantation technology.

The extent and type of ROS-mediated stress has a strong organ-specific component to it, and, hence, preservation strategies may need to include treatments that deal with organ-specific concerns. For example, the intestine is extremely susceptible to ROS-mediated IRI due to an abundance of the enzyme xanthine oxidase (XO) in mucosal epithelial cells. XO and xanthine dehydrogenase (XDH) are identical proteins, but in multiple pathologic states, including transient ischemia, XDH is converted to XO. During ischemia, the breakdown of ATP causes an accumulation of the degradation products, hypoxanthine and xanthine, which are

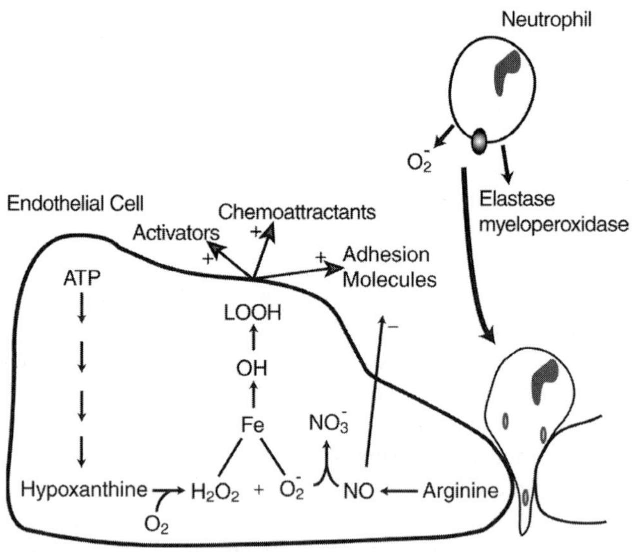

Figure 19.10 Events involved in ischemia–reperfusion injury. ATP degrades during storage, eventually leading to the generation of reactive oxygen species. These ROS can lead to the scavenging of nitric oxide and the "attraction" of neutrophils to the site of injury. The infiltration of recruited neutrophils results in further damage due to degradative enzyme release.

substrates of XO. Elevated substrate availability, especially when oxygen is reintroduced during reperfusion can lead to a rapid production of superoxide radicals via the XO reaction. Interestingly, activities of XO are also proportional to the duration of ischemia. Hence, treatments that specifically target XO could be useful for improving the preservation of intestine.

Involvement of Neutrophils

The vascular endothelium is an active site in the mediation of IRI, both as a source and target of destructive mediators. Vascular endothelial cells release hormones and endothelial factors that regulate vascular tone, platelet aggregation, and neutrophil infiltration. Neutrophils elicit their damaging effects predominantly by local infiltration of ischemic tissue upon reperfusion followed by the release of multiple degradative enzymes (myeloperoxidase, elastase, collagenase) that then potentiate cell damage. A controlled destruction of cells at the site of an injury or microbial invasion is a necessary step in the repair of injured tissue. Consequently, neutrophils produce and release large amounts of superoxide in order to facilitate this directed disintegration of tissue. There can be as many as 10 million neutrophils per gram of tissue (intestine has the highest numbers), and, thus, the contribution of neutrophils to the cytotoxic production of superoxide is responsible for much greater ROS production than the xanthine oxidase system.

Reactive oxygen species clearly play a role in the recruitment of neutrophils to postischemic tissue, and it is hypothesized that XO-derived ROS produced by endothelial cells initiate the production and release of proinflammatory agents that cause the subsequent attraction and activation of neutrophils. Neutrophil chemoattraction, migration, and infiltration at sites of inflammation is a complex process. In brief, neutrophil chemoattractants, including leukotriene B_4 (LTB_4), platelet-activating factor (PAF), and interleukin-8 (IL-8), are important in directing the migration of neutrophils and activating local adhesion molecules on endothelial cell surfaces. Chemotaxis occurs when a soluble chemoattractant molecule diffuses away from the site of inflammation and induces neutrophil movement in the direction of increasing concentration of the chemoattractant. Neutrophils express integrins, which are ligands of endothelial adhesion molecules, and these interact with selectins and intercellular adhesion molecules on the surface of endothelial cells at the site of inflammation. Selectins (E-selectin and P-selectin) are adhesion molecules responsible for the initial attachment of leukocytes to endothelial cells and mediate the phenomenon of leukocyte rolling along the vessel wall. Firm attachment of the neutrophil to the endothelial wall occurs by the interaction of a specific neutrophil integrin (CD11/CD18) with an intercellular adhesion molecule (ICAM-1) on the endothelial surface. Transendothelial migration from the blood vessel lumen to the extracellular space is also mediated by chemoattractants, including IL-8 released from endothelial cells at inflammatory sites. The entire process of neutrophil migration and activation is the topic of considerable research; this is not entirely unexpected when one considers the impact that blocking the inflammatory process would have on modern medicine.

Activated neutrophils mediate tissue damage by three main mechanisms during IRI:

1. Release of proteolytic enzymes that directly degrade the endothelial basement membrane and extracellular matrix, including elastase, collagenases, and proteases
2. Generation and release of ROS, via NADPH peroxide (myeloperoxidase) reduction of O_2
3. Impeding blood flow in mucosal capillaries and postcapillary venules, causing the no reflow phenomenon

Nitric Oxide

Prior to 1987, nitric oxide (NO) was known as endothelium-derived relaxing factor. NO is formed endogenously in endothelial cells via the reduction of L-arginine in the presence of molecular oxygen via the nitric oxide synthase (NOS) reaction (L-arginine $\rightarrow$ L-citrulline + NO). Physiological functions of NO include the regulation of endothelial integrity, permeability and proliferation, vascular tone, antithrombosis, and, in the case of myocardium, regulation of contractility (Ca^{2+} entry into myocytes). Although the only definitively accepted receptor for NO is soluble guanylate cyclase (sGC) and most NO-mediated effects are mediated via this cascade pathway, there are other non-sGC-related effects of NO that are less well-clarified. NO is an essential component in the maintenance of organ homeostasis. Many lines of evidence indicate that NO contributes to the cytotoxicity required for the elimination of tumors and invading microorganisms. More recently, many investigators have taken an interest in the ability of NO to suppress lymphocyte proliferation, antigen-specific T-cell proliferation, antibody production, graft-versus-host reaction, mucosal damage, and hemorrhage. IFN (interferon gamma) is a potent inducer of the enzyme responsible for endogenous NO production, nitric oxide synthase (NOS) in macrophages, and induction is enhanced by tumor necrosis factor α (TNFα) and other agents. Besides the role of NO in immune defense, NO is involved with inflammation and autoimmune-mediated tissue destruction. Inhibition of hypoxia-induced vasoconstriction and inhibition of platelet adhesion and aggregation are both key processes of particular relevance to ischemia/reperfusion injury following organ transplantation. Studies

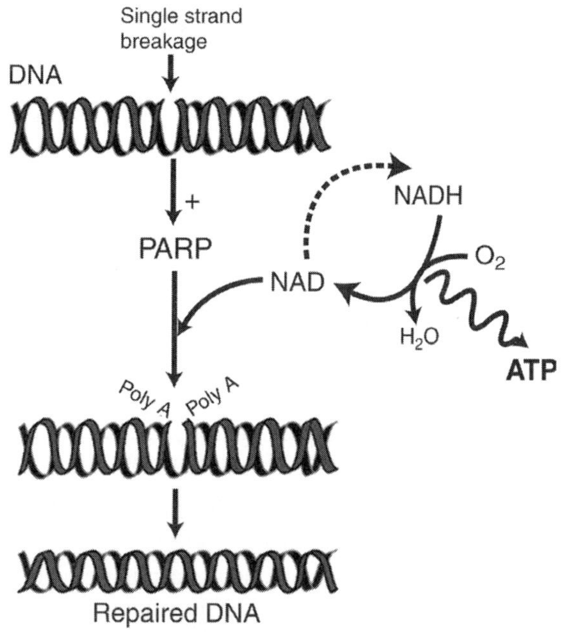

Figure 19.11 PARP-mediated consumption of cellular NAD. A break in one of the strands of DNA due to reactive oxygen species damage activates the repair enzyme, poly(ADP-ribose)polymerase (PARP). The mechanism of repair involves polyADP-ribosylation of both ends of the broken DNA strand. For each ADP-ribose of the polyADP-ribose tail, a molecule of NAD is used. This consumption of cellular NAD, interferes with oxidative phosphorylation responsible for the generation of ATP in oxygenated tissue.

have suggested that exogenous supplementation with L-arginine or the administration of exogenous NO (via a NO donor) can be cytoprotective in relieving IRI via four general mechanisms:

1. Blocking neutrophil infiltration
2. Scavenging oxygen-derived free radicals
3. Direct vasodilation of the coronary circulation
4. Decreasing oxygen demands through vasodilation-induced hypotension

Unfortunately, despite the benefits of NO under physiologic conditions, in situations of ischemia–reperfusion NO can also facilitate damage. Upon reperfusion, a burst of ROS production occurs, and in this situation NO can combine with superoxide and form a very reactive species, peroxynitrite (see Chapter 1). Peroxynitrite can attack and damage multiple cellular macromolecules as well as strands of DNA. If ROS are produced in an uncontrolled manner, then NO can be depleted during the initial phase of reperfusion. Consequently, in the latter phase of reperfusion there is a limited capacity for NO production and NO-mediated cell functions that are involved in restoring cell homeostasis are impaired. Hence, in the realm of organ preservation a combined approach of antioxidant therapy and NO donor may provide protection from ROS in conjunction with the benefits of NO.

DNA Repair Mechanisms

Peroxynitrite-mediated DNA single-strand breakage activates nuclear poly(ADP-ribose)polymerase (PARP) (EC 2.4.2.30). PARP binds to damaged DNA and consumes NAD^+ to synthesize poly(ADP-ribose) chains on the broken DNA strand in a futile cycle that depletes cellular NAD^+ (Fig. 19.11). This depletion leads to a suppression of mitochondrial respiration and results in the exhaustion of both cellular NAD^+ and ATP. In a cell with extensive DNA damage and a consequent massive activation of PARP, depletion of NAD^+ can have death-inducing consequences. In a mildly damaged cell, NAD^+ depletion does not appear to be significant enough to cause death, but it may affect cellular energy dependent processes. A limitation of ATP production via this mechanism could be of significance in tissues during reperfusion following transplantation because during this period the organ requires high ATP availability to support the many metabolic processes involved in regeneration.

Microcirculation and the No-Reflow Phenomenon

The phenomenon termed *no reflow* is a serious problem during reperfusion of organs and contributes significantly to much of the damage following ischemia and reperfusion. The basis of this phenomenon is blocked capillaries. As many as 70% of capillaries may be blocked in an organ subjected to cold ischemic storage followed by transplantation so that the microcirculation of the organ is severely compromised upon reperfusion. The severity of the no-reflow phenomenon increases with increasing length of cold ischemia. The primary mechanism implicated in diminished blood flow is the accumulation of neutrophils in the microcirculation causing a mechanical blockage of capillaries and promoting platelet aggregation that further plugs the microcirculation (see Chapter 18 for more about platelet aggregation). Another mechanism involved is neutrophil-mediated injury to the endothelial lining and basement membrane by the release of proteases, collagenases, and elastases, which causes an increase in capillary permeability and leakage from the plasma to the interstitium. This results in increased viscosity with a consequent decrease in blood velocity and ultimately complete cessation of blood cell flow through the microcirculation.

TABLE 19.2 Current Status of Organ Transplantation: Estimates Based on 2000–2001 Data from the United States[a]

Organ	Number of Transplants	Number Waiting for Transplant	New Patients	Mortality Rate (%)	Survival: Adult Graft (3 yr) (%)	Survival: Adult Patients (3 yr) (%)
Heart	2,185	4,091	3,429	15	78	78
Liver	4,621	15,695	11,236	11	72	81
Kidney	8,069	45,843	22,482	6	82	96
Kidney–pancreas	896	2,382	1,913	9	84 (K), 77 (P)	90
Heart–lung	39	209	135	16	46	47
Lung	974	3,622	2,029	14	58	59
Intestine	103	135	201	16	58	61

[a]The organ shortage is highlighted by the large percentage of patients waiting for a transplant and the number of new cases that also put a drain on the limited number of organs available.
Taken from UNOS (United Network for Organ Sharing) database.

Interventions to Improve Ischemia–Reperfusion Injury

Supplementation or administration of chemical or enzymatic antioxidants has been the focus of much research over the past several decades, not just with respect to IRI in organ preservation but also in other situations of ischemic stress. Several studies have shown successful attentuation of ROS-mediated events (e.g., increased microvascular permeability, inflammatory cell response) with the use of a specific inhibitor of XO, allopurinol, added to the perfusion medium (inhibition constant, K_i, is $\sim 50\,\mu M$). Other additives shown to reduce ROS-mediated damage include the iron chelating agent deferoxamine (or desferrioxamine) (recall that iron triggers hydroxyl radical production via the Haber–Weiss reactions; see Chapter 12), the iron-binding protein apo-transferrin, hydroxyl radical scavengers such as dimethyl sulfoxide (DMSO), as well as the direct addition of SOD and/or catalase enzymes that convert superoxide radicals to water.

Additives can also be used with the aim of supplementing intracellular antioxidant defenses. Treatment with chemical antioxidants, such as α-tocopherol or n-acetylcysteine, has been beneficial in attenuating ischemia and reperfusion-related histologic injury, neutrophil infiltration, and cellular peroxidation (lipid, protein, nucleic acids). The tripeptide, glutathione, is a primary intracellular reducing agent that is both a substrate of antioxidant enzymes (e.g., glutathione peroxidase) and an antioxidant itself (see Chapter 12), and studies have shown that augmenting the levels of GSH in cells can lessen IRI. Glutathione itself cannot cross the cell membrane, but studies have shown that glutamine/glutamate supplementation (glutamate being one of the constituent amino acids of GSH) can elevate intracellular GSH levels during ischemia–reperfusion and reduce ROS-mediated injury.

There is little doubt that ROS mediate IRI by a number of mechanisms, including direct cytotoxicity. However, the most important effect of early ROS generation is their role in the activation and infiltration of neutrophils, which are the primary mediators of tissue damage and dysfunction in IRI. Interestingly, strategies of IRI relief also exist that involve administration of monoclonal antibodies directed against neutrophil adhesion glycoproteins as well as antineutrophil serum. Both strategies have resulted in a significant reduction in inflammation and subsequent tissue injury (although details are beyond the scope of this chapter).

ORGAN-SPECIFIC CONSIDERATIONS

Table 19.2 outlines the status of solid-organ transplantation, describing the number of transplants performed per year, number of patients waiting for a suitable organ, and the success rates for each type of organ transplant. The "safe" window of storage for a donor organ is highly tissue-specific (Table 19.3). For some organs, such as the heart, which is highly metabolically active, it is reasonable to expect a short storage time. However, why kidney can be stored for such a long period (up to 72 h) with only minor clinical consequences and the liver cannot remains the focus of much investigation.

TABLE 19.3 Mean Clinical Storage Times of Human Organs Removed for Transplant

	Hours
Heart	6
Liver	15
Kidney	41
Kidney–pancreas	22
Heart–lung	6
Lung	6
Intestine	10

Liver

Liver transplantation is now of proven value in the treatment of various end-stage hepatic diseases. Improvements in patient selection, anesthesia, operative technique, intensive care, and newer schedules of immunosuppression have led to reported one-year survival rates of 80 to 85% for the 10,000 patients yearly that receive liver transplants around the world. Development of organ preservation techniques have extended the safe cold ischemic time for human hepatic grafts to beyond 12h. However, concern about the limited number of organ donors and the quality of early graft function in stored livers remains. The two most critical factors predicting outcome of the transplant are: (a) degree of patient illness and (b) organ function following preservation. With the cost of retransplantation, including hospital time and care, at $250,000 per patient (5 times that of the initial transplant), the design of effective preservation strategies is an important facet of the transplantation process.

Use of the UW solution that eliminates the problem of cellular edema during storage has increased the viability of liver grafts to at least twice the storage time that was seen in original trials that used a citrate-based preservation solution. However, even though the UW solution is widely accepted as the "gold standard" for liver preservation, the maximum storage time remains relatively brief (10 to 24h). Fortunately, the liver has an amazing capacity for regeneration and as much as 85% loss of the parenchymal tissue can be tolerated. Regenerative growth can then restore initial liver mass and function.

Living Related Liver Transplantation Since the first successful living donor liver transplantation (LDLT) in 1989, this surgical innovation has had a substantial, positive impact on pediatric liver transplantation. As surgical skills improve and propagate throughout the surgical community for this demanding procedure, LDLT will overcome the

TEXT BOX 19.1 ARTIFICIAL LIVER SUPPORT (ALS)

At present, the treatment of a patient in fulminant hepatic failure (FHF) (e.g., often as a result of drug overdose) is based upon scrupulous intensive care and often when patient condition deteriorates, emergency liver transplantation must be considered. Treatments that stabilize the condition of a patient in FHF would be of great benefit, particularly those that could provide an artificial support that reduce the metabolic burden on the patient's own liver and allow it time to regenerate itself. Development of artificial liver support (ALS) systems capable of supplementing hepatic function in patients with FHF has been ongoing for more than 40 years. There are obvious benefits to a bridging period during which a patient can be supported via an extracorporeal liver support system that detoxifies the patient's plasma by using porcine hepatocytes and allows sufficient time for either the regeneration of the patient's own damaged liver or to obtain a suitable donor liver. An ALS device consists of a physical column in which porcine hepatocytes are attached to an insoluble matrix. The patient's blood is separated into red blood cells (RBCs) and plasma. The plasma is then passed through the column and the column removes the toxins from the plasma. Once the plasma is detoxified, the cells and plasma are recombined and returned to the patient (Fig. TB19.1).

Liver is a multifunctional organ with synthetic, metabolic, and detoxification roles. There is a sound basis in the literature for the idea that if the detoxification role can be assumed by an ALS (utilizing porcine hepatocytes) for a minimum of 24 to 48h (with requirements for synthetic and metabolic functions addressed by the attending physician), then a patient's own liver can regenerate sufficiently. The liver is the only human organ capable of significant regeneration and does so in response to partial hepatectomy (amazingly up to 85% can be resected) or toxin insult. Indeed, this is the basis for living related liver donors, where a parent can provide a lobe of their own liver for their child; both livers then respond to environmental parameters and grow according to size of the host. At a cellular level, within minutes of insult, hepatocyte growth factor (HGF) receptors on hepatocyte membranes are up-regulated, and within hours messenger RNA levels in the liver expressing HGF are elevated. Within 1 to 2 days the liver has started its regenerative process, and within 6 to 12 months the injured liver has returned to its original size and state (Fig. TB19.2).

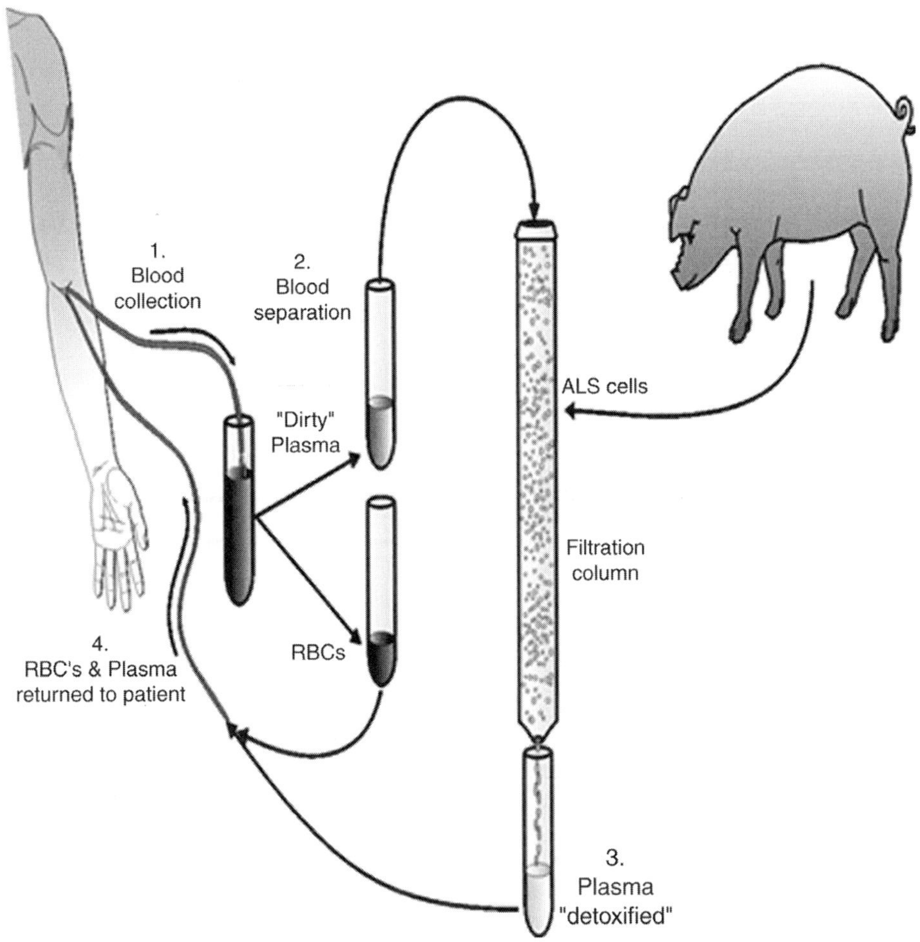

Figure TB19.1 Diagram of an artificial liver support system. (1) Blood is collected from the patient. (2) Blood separated into red blood cells and the plasma requiring treatment. (3) The plasma is filtered through a column of immobilized hepatocytes (commonly isolated from pig) and is eluted in a detoxified form. (4) The red blood cells and "clean" plasma are reconstituted and returned to the patient.

Experience with ALS has been limited to date due to the necessary combination of biological and physical or architectural aspects of extracorporeal devices. Nevertheless, in the clinical realm, studies have shown significant reductions in intracranial pressures (ICP) as a direct result of reductions in ammonia levels; the accumulation of ammonia in the serum can cause an increase in intracerebral edema leading to irreversible brain injury. Decreases in other metabolites (lactate and bilirubin) as well as enzyme markers of cell damage released from the patient's liver, reflect the potential effectiveness of an ALS on several key biochemical and physiological parameters. This reduction in metabolite, ICP, and tissue injury is transient and can provide a brief "bridging" period for patients in FHF prior to liver transplantation (Table TB19.1). In one set of patients subjected to 6h on ALS, levels of HGF rose dramatically from 7.9 to 30.5ng/ml serum HGF, indicating that liver regeneration was underway and emphasizing the important role that an ALS can play in the bridging period while a damaged liver (e.g., as a result of acetominophen overdose) begins to regenerate. Thus, current attempts at ALS in clinical trials have produced positive outcomes with respect to several critical functions of the liver (ammonia clearance, bilirubin conjugation) as well as an increase in the potent hepatotrophic factor, HGF. Unfortunately, the survival of patients supported by ALS after subsequent liver transplantation has not been significantly greater than patients with no

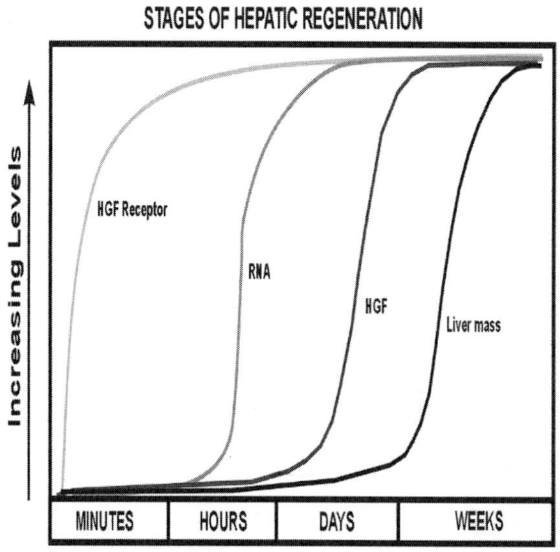

Figure TB19.2 Stages of hepatic regeneration. Although growth in liver mass following hepatic resection can take several weeks, the molecular mechanisms involved in hepatic regeneration are triggered within minutes to hours of the insult.

extracorporeal support. However, both hepatocyte mass and time of ALS treatment are critical factors when determining effectiveness of the treatment in both experimental and clinical settings, and ALS use to date has been inconsistent.

TABLE TB19.1 Benefit of Artificial Liver Support (ALS) in Clinic: Biochemical and Physiological Parameters of Patients before and after ALS Treatment[a]

	Pre-ALS	Post-ALS
AST (U/L)	1,325	880
ALT (U/L)	1,240	770
Ammonia (μM)	165	120
Bilirubin (μM)	440	310
ICP (mmHg)	20	9

[a]Serum enzymes and metabolites were measured in patients prior to ALS treatment (pre-ALS) and after ALS (post-ALS). Aspartate transaminase (AST) and alanine transaminase (ALT) levels decrease following ALS treatment; this reflects a reduction in cell/tissue injury. Ammonia and bilirubin also decrease after ALS treatment, reflecting the clearance of potentially toxic end products of metabolism. Bilirubin is a breakdown product of heme degradation and is extremely lipid soluble (so it can easily pass into cells) and also toxic (it can uncouple oxidative phosphorylation), particularly to nervous tissue. An important hepatic function is the conjugation of bilirubin with glucuronic acid, so that excretion is possible; hence, a reduction of bilirubin is beneficial to liver viability. Intracranial pressure (ICP) can increase as a consequence of end product accumulation and eventually lead to irreversible brain damage; a reduction in ICP represents an important benefit of this novel technology.

donor organ shortage in the pediatric population. Despite the main ethical issue of subjecting healthy individuals (donors) to the risks of liver resection and general anesthesia, any parent can easily relate to the sacrifices that one would make for his or her own child. The advantages of LRLT are:

1. Optimal selection of donors (usually related) with excellent liver function
2. Little graft injury because simultaneous donor/recipient operations minimize storage time
3. An elective operation can be performed before the recipient's health is compromised. In other situations it is usually the sickest patients that are at the top of the transplant list, and their poor health often leads to complications during recovery.

Pancreas

The pancreas is responsible for the production of hormones (e.g., glucagon, insulin, somatostatin) that regulate fuel metabolism as well as enzymes (e.g., chymotrypsin, carboxypeptidase, trypsin), which are secreted into the intestines to aid in the digestion of food. The cells responsible for the production of enzymes are the exocrine ("secreting out") component of the pancreas and the cells that produce hormones are the endocrine tissue. Distributed throughout the pancreas are clusters of cells called the islets of Langerhans, which comprise ~1 to 2% of the total pancreatic mass. (See Text Box 19.2). Islets contain several types of endocrine cells, the two most important for transplant medicine being the α cells that synthesize glucagon, a hormone that elevates blood glucose, and the β cells that produce insulin, a hormone that lowers blood glucose (also see Chapter 10). In type 1 diabetes, insulin shortage is caused by an autoimmune process in which the β cells are destroyed by the body's own immune system. Although the β cells are destroyed, immature precursor cells appear to survive the attack; this is particularly relevant when considering potential remedies. In type 2 diabetes the β cells do not function at 100% efficiency, and furthermore there is peripheral resistance by organs such as skeletal muscle to the insulin that is produced. Consequently, people with type 2 diabetes do not produce enough insulin to overcome the combined stress of insulin resistance and low insulin production in response to a glucose load.

Currently, diabetes is among the top five causes of death in North America. Historically, patients with diabetes have needed to pursue a return to normoglycemia since high fluctuations in blood glucose can culminate in renal failure, blindness, heart disease, neuropathy, and artherosclerosis. Although intensive long-term insulin therapy is associated with reduced risk of developing related complications,

> **TEXT BOX 19.2 ISLET TRANSPLANTATION**
>
> Since most of the pancreatic tissue of a diabetic patient is completely healthy, it makes sense to transplant only the "diseased" tissue, the β-cell-containing islets themselves. There are numerous problems that must be considered when carrying out the transplantation of islets, and these include low recovery of islets from the donor pancreas, poor engraftment, poor function, and tissue rejection by the recipient's immune system. Now that medicine has provided newer (and less toxic) immunosuppressive agents, islet cell transplantation is becoming a real alternative to whole-organ pancreas transplantation. Approximately 1 million islets have to be transplanted to make enough insulin so that insulin injections will no longer be necessary. Fortunately, islets can be cryopreserved, thereby allowing the use of multiple donors for a single recipient. Typically, multiple (usually 2 or 3) donor pancreases are needed for every one recipient in order to allow for a less than 100% efficiency in isolation and purification. Islet yields generally range from 350,000 to 650,000 per donor depending on donor age, pancreatic storage time, and other variables.
>
> Compared to vascularized pancreatic grafts, transplantation of isolated islets offers a number of advantages. For example, donor islet tissue can be tested and/or pretreated before implantation, thus allowing the possibility of using grafts with defined (and reliable) metabolic and immunologic characteristics. The transplantation process is simplified because islets are implanted into the vasculature of the liver where they are "sieved out," becoming lodged in the liver where they release insulin in response to a glucose challenge.
>
> Researchers are trying to develop methods of transplanting islet cells to reduce or eliminate the need for immunosuppression and the risk of rejection. Immunoisolation methods separate the islets from the immune system of the patient by coating cells or encapsulating them into microscopic containers (often synthesized from alginate polymers). Immunoalteration of the islets themselves is also being attempted so that the surface receptors/molecules are not recognized by the patient's immune system.

harmful side effects due to recurrent hypoglycemia can also be a problem.

Pancreatic Transplantation Pancreatic transplantation was first used for the treatment of type 1 diabetes in humans in 1966. In that early era, the rates of graft and patient survival were low, so very few procedures were performed until 1978. Important steps toward improving surgical results included the introduction of improved immunosuppressive regimens, especially the use of cyclosporin and anti-T-cell agents, new surgical techniques, and the selection of healthier recipients. In the past decade, the number of procedures performed has steadily increased each year. The objectives in this instance are to render the patient free of exogenous insulin therapy, to arrest the progress of ongoing secondary complications, to protect the cotransplanted kidney from hyperglycemia, and to improve quality of life.

As of 2000, nearly 14,000 pancreas transplants had been done, and the success rate is comparable to those achieved with other solid-organ transplants. Since transplantation of any organ requires long-term immunosuppression, the main accepted indication for pancreas transplant is type 1 diabetes with kidney transplant for renal failure.

Heart

In North America, heart disease assumes a particularly prominent position because of its widespread prevalence, and transplantation for end-stage heart failure offers a form of surgical intervention that, in many cases, is the only hope of significantly extending life with excellent prospects for rehabilitation. The first reported cardiac transplant, was performed in 1905 by Carrel and Guthrie, who heterotopically transplanted a canine cardiac allograft into the neck of a recipient dog. With the development of cardiopulmonary bypass in 1953, the groundwork necessary for cardiac transplantation was laid, and in the 1960s, Shumway and Lower at Stanford University clearly demonstrated that a denervated (without nerves) dog heart could support a canine recipient's circulation. Based upon the experiments of the Stanford group, Barnard performed the first human orthotopic heart transplant in South Africa in December 1967. Despite the fact that his patient died of sepsis less than 3 weeks later, this landmark event served as a catalyst for many other surgeons who collectively performed over 100 transplants in more than 60 centers in the next year. Early results were generally poor due to problems with graft rejection and immunosuppression. However, in the early 1980s cyclosporin was introduced and was subsequently shown to improve survival with decreased morbidity in patients undergoing cardiac transplantation. By the end of 1984, 350 cardiac transplants had been performed, and currently the one-year survival of recipients exceeds 80%. Cardiac transplantation has become a very real option in the treatment of end-stage cardiac disease.

At present, the supply of hearts still falls far short of demand. With this in mind, the immense importance of optimal donor organ preservation becomes very apparent. At present, cardiac preservation techniques are not optimal. While research with liver and pancreas have seen a doubling of safe storage time over the past 1 to 2 decades to times as long as 24 to 30h, the safe storage time limit for hearts remains only 4 to 6h. Improvement of cardiac graft preservation will benefit cardiac transplant programs in several ways including:

1. Increasing the time available for graft transport between donor and recipient hospitals; this would allow organ procurement from more distant centers, and thereby increase the size of the potential donor pool.
2. Increasing storage times also allows for improved tissue matching, which would lead to fewer problems with rejection, and also allows more time for the better preparation of the recipient and the surgical team.
3. Better organ preservation would have favorable impact on postoperative graft function, which should make the posttransplant period less eventful and more cost effective.

Optimizing preservation techniques for the heart is critical, perhaps more so than for any other organ. The reason for this is that the allowable recovery period for the heart after transplantation is extremely short. The heart must be fully capable of supporting the patient's circulation immediately following transplantation with very little tolerance for compromised function. By contrast, patients with transplanted kidneys can be supported with dialysis until recovery is optimized, and patients with liver transplants take several hours to resynthesize fuel reserves, correct tissue edema, and reestablish normal electrolyte concentrations. However, the heart must resume optimal function virtually immediately, including all aspects of its function—energy metabolism, contractile function, regulation of the fluxes of calcium and other ions. The most effective way of accomplishing this is to ensure that the best possible preservation techniques are fully utilized. (See Text Box 19.3.)

Lung

During the past two decades lung transplantation has evolved from an experimental stage to become the mainstay of therapy for end-stage lung disease. Despite the success of lung transplantation (to date, >10,000 have been performed worldwide), rejection, infection, and high IRI contribute to poor immediate graft function. In 10 to 20% of patients the lung is so severely damaged that the patient requires extended support with positive-pressure ventilation, pharmacologic therapy, and occasionally extracorporeal membrane oxygenation to support gas exchange. In addition to a high mortality rate in the early postoperative period, there is increasing evidence that severe reperfusion injury

TEXT BOX 19.3 MYOCARDIAL STUNNING

The term *myocardial stunning* refers to the mechanical dysfunction that persists after an ischemic insult despite the absence of irreversible damage. An essential point in this definition is that the dysfunction is completely reversible and, therefore, the term does not apply to myocardium that is infarcted and nonviable. This concept is important in all settings where the heart is subjected to reversible nonlethal ischemia, such as coronary bypass and transplantation, because it predicts a certain degree of functional impairment of the postischemic myocardium, even when the cells are completely viable. This implies that enhancing viability through appropriate preservation solutions and techniques may not necessarily lead to better function, at least in the short term, unless this issue of stunning is addressed.

While ATP levels must be maintained for cellular viability, simply preserving ATP alone is not sufficient to prevent myocardial stunning. This is consistent with the statement earlier that maintaining viability is important for function, but viability in and of itself does not necessarily ensure a good graft function, at least in the short term. Pharmacological agents can be utilized to minimize stunning based upon these recognized mechanisms. Treatment with SOD/catalase or an inhibitor of Ca^{2+} release from the sarcoplasmic reticulum (SR) help alleviate episodes of stunning. Hence, the two major factors involved in myocardial stunning are: (a) generation of oxygen free radicals and (b) excitation-contraction uncoupling due to Ca^{2+} release from SR.

is also associated with an increased risk of acute rejection and graft dysfunction in the long term.

Preservation strategies aim to limit injuries occurring before lung procurement, during ischemia, and after reperfusion. Mechanisms leading to the development of graft failure are the result of a cumulative effect of these three phases: (a) injurious events related to brain death, (b) ischemia during preservation followed by reperfusion, and (c) mechanical ventilation of damaged lungs.

Small Bowel

At the turn of the twentieth century, Alexis Carrel performed heterotopic (Greek translation, *hetero* = other and *topic* = place; hence at another place or not in the normal place) intestinal transplantation into the necks of dogs. The first clinical attempts at small-bowel (SB) transplantation between 1964 and 1970 were fraught with problems of sepsis, rejection, and technical complications. After the introduction in 1968 of intravenously administered nutrition, called total parenteral nutrition (TPN), interest in clinical intestinal transplantation ebbed. As long as nutritional support for the patient could be administered intravenously, there was no need to undergo a major surgical procedure with a significant likelihood of failure. However, with the eventual realization that TPN can be associated with serious life-threatening complications (loss of venous access, sepsis) and with the development of safer and more potent immunosuppression (cyclosporin A) in the 1980s, small-bowel transplantation has returned to the clinical arena. (See Text Box 19.4.)

TEXT BOX 19.4 NEW FRONTIERS IN BOWEL PRESERVATION

Clinical organ procurement involves storing the organ in a static manner; unfortunately, once the organ is flushed, the preservation solution is permitted to escape due to vascular collapse under the weight of the tissue itself. This limits the amount of preservation solution in intimate contact with parenchyma of the bowel and hence minimizes solution–cell interactions. However, due to the physical nature of the bowel, multiple factors can be addressed in a simple manner by not only flushing the vasculature but by also flushing the lumen and gently inflating the bowel with preservation solution.

Although oxygen tensions drop quickly during storage, there is enough oxygen remaining (either in the tissue or delivered in the aqueous preservation solution) to facilitate fuel catabolism for the production of much needed ATP. The small bowel primarily draws its energy from the catabolism of three amino acids; 98–99% of the dietary glutamate and aspartate and approximately two-thirds of glutamine are metabolized within the mucosal cells. By supplying these amino acids in a tailored preservation solution (in addition to the other amino acids, high-molecular-weight impermeants and buffer) to the bowel via luminal flush, the intestine is capable of maintaining oxidative phosphorylation for a short time after organ harvest. The ATP synthesized is vital to maintaining the mucosal barrier and absorptive function throughout cold storage (Fig. TB19.3). The reason for the improvement in tissue viability via this novel luminal flushing technique is due to a combination of increased substrate delivery to the mucosal cells as well as the efficient clearance of end products.

The benefits of luminal flushing/ballooning include:

1. Removal of toxic enteric contents (proteases, digestive enzymes)

2. Supplying nutrients directly to the epithelium
3. Removal of toxic end products (lactate and ammonia)
4. Direct contact with buffering agents (reducing pH shifts)
5. Increased oxygen supply

The simple, yet elegant technique of luminal flushing with a novel amino-acid-based preservation solution (AA) may provide the protection needed to ensure an intact and fully functional organ upon transplantation in the clinic.

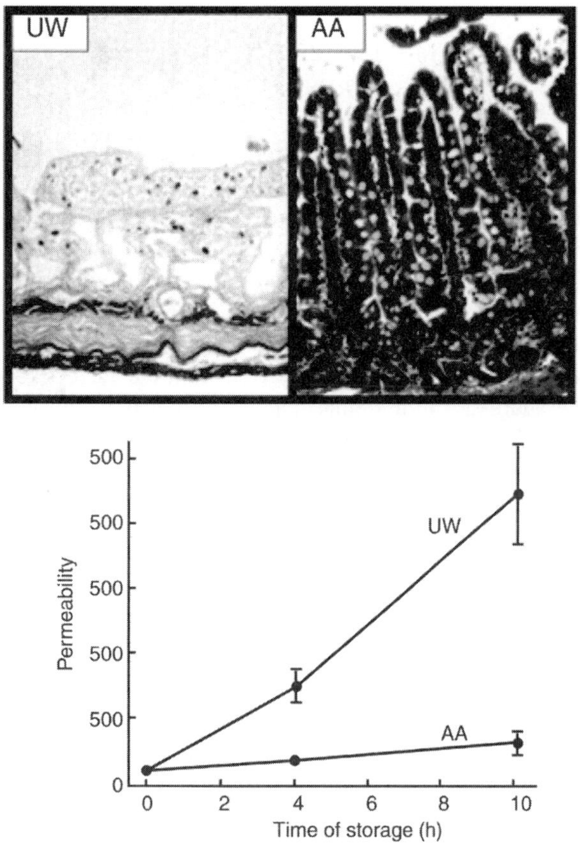

Figure TB19.3 Mucosal protection by luminal administration of a nutrient-rich preservation solution. After 10h cold storage of rat small bowel, tissues flushed vascularly with UW solution were severely damaged with complete denudation of the villi and crypt infarction (upper left picture). With the addition of a luminal flush with an amino-acid-rich preservation solution, the villi remained largely intact with only minor lifting of the epithelium (upper right picture). The "leakiness" of the mucosal barrier was also assessed. After 10h storage, UW-treated tissues were extremely permeable compared to the AA-treated tissues; permeability values were >400 units in UW-treated tissues, compared to ~60 units in AA-treated specimens.

As of 1999, there had been a total of 474 small intestinal transplants carried out worldwide. Small-bowel transplantation (SBT) is currently limited to patients with irreversible intestinal failure who have developed life-threatening complications or have otherwise failed with intravenous nutritional support via TPN. Unfortunately, this limits the recipients to only the sickest of patients with intestinal failure, those for whom transplantation offers the only hope for survival, and indeed many patients (usually children with malformed/malfunctioning bowel) do not survive while waiting for a suitable donor. Consequently, the success rate of SBT is currently nowhere near the success rates of the other commonly transplanted organs. Overall survival rates up to 1999 have remained low (<40% five-year survival). However, encouraging new data indicate that post-transplant survival is steadily improving; as of 2003, rates of 71% and 88% (3 year survival) for graft and patient survival have been reported. Although the major obstacles to the development of reliable and safe SBT have been quite high [bacterial infection (55%) and allograft rejection (12%)] up to 1999, the incidence of these complications are dropping as surgeon experience, medical management (immunosuppression), among other clinical factors improve. Major obstacles to the development of reliable and safe SBT are bacterial infection (55%) and the control of allograft rejection (12%).

The most important advance in solid-organ preservation in recent years has been the development of the UW solution, used to vascularly flush and store the donor organ during the retrieval operation. Despite the benefit of UW for preservation of other intra-abdominal organs (primarily liver, pancreas), the maximum storage time for SB remains relatively brief (6 to 8h); furthermore graft quality is often compromised. No single preservation solution has proven effective for SB, and equivalent results can be achieved with simple crystalloid solutions (normal saline) or with complex solutions such as UW solution. Thus, it is by default that the current clinical standard for SB consists of a vascular flush with UW solution as part of multivisceral organ procurement.

Ischemia and Preservation Injury The small bowel has two primary functions that must be maintained in order to perform its function once reimplanted into the recipient. First, absorptive function is the main function of the bowel that facilitates the small bowel to absorb nutrients from ingested food and transports these essential building blocks and cofactors of metabolism. Second, barrier function to prevent bacterial invasion is critical. The inside of the bowel (the lumen) contains many species of bacteria that if allowed into the bloodstream (due to degradation of the tissue) result in a generalized infection (sepsis). The preservation of both of these functions is

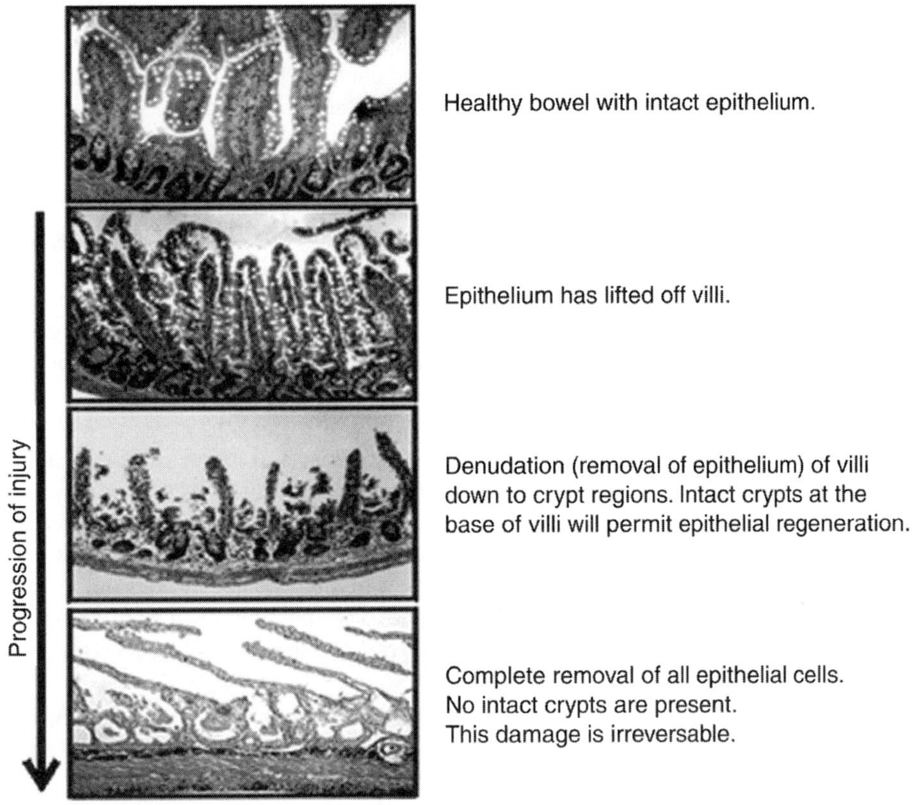

Figure 19.12 Ischemic injury in the small intestine. Healthy intestine possesses thick fingerlike projections (villi) that are covered in an outer layer of epithelial cells. This epithelial barrier is critical in keeping bacteria captive in the lumen of the bowel. After a brief period of ischemia, the epithelial layer begins lifting off the underlying villus tissue. As ischemia progresses, the epithelium lifts more and is completely removed (denudation). As long as the cryptal regions of the mucosa remain intact, regeneration of the epithelial layer is possible. However, once the epithelial cells in the crypts are disrupted, the damage incurred is irreversible. At all stages of epithelial disruption, there is an increased probability that bacteria can migrate into the mucosa and enter the bloodstream, potentially leading to sepsis in the recipient patient.

fundamental to a successful transplant outcome for the recipient.

Unfortunately the small bowel is extremely susceptible to ischemic injury; even brief periods of reduced oxygen initiate tissue injury. The luminal surface of the bowel is covered in fingerlike projections (singular, villus; plural, villi), which effectively increase the absorptive surface area of the intestine. These villi are covered by a layer of epithelial cells, which act as the primary barrier for the bowel. Within 30 min of partial occlusion of the main oxygen-supplying artery to the small bowel (or 4 to 8 h of cold storage), this epithelial layer begins lifting off the underlying villus tissue. As long as the epithelium along the upper sides and tips of the villi are only affected, then the injury can be reversed by restoring blood supply to the tissue. However, if the epithelial layer extending down into the troughs of the villi (the correct term is *crypts*) is affected, then the damage becomes irreversible. Although the bowel is subject to bacterial invasion at all stages of injury, a translocation event of a single bacterium across the epithelium into the villus vasculature becomes more probable as the degree of injury increases (Fig. 19.12). The cryptal regions are fundamental to maintaining or regenerating the epithelial layer over the villus since this region is highly proliferative and is the source of the epithelial cells.

ORGAN SHORTAGE: PROBLEMS AND SOLUTIONS

Organ Shortage Crisis

Building on the important contributions of individual scientists such as Carrel (vascular anastomosis), Medawar and Burnett (immunological tolerance), and Calne (cyclosporin), allotransplantation (transplantation within a species) has had considerable success in recent years. To date, there have been over a million organ and tissue transplants performed worldwide. In 1999 alone there were over 40,000 cadaveric solid-organ transplants performed, mostly in the United States. Currently, there is a severe shortage of donor organs that is only expected to get worse (Table 19.2).

Consent for donor organ harvesting remains an important and controversial issue. Generally, there are two categories of consent: informed/expressed consent and presumed consent. *Informed or expressed consent* means that people actively understand that they are volunteering for organ donation and do so by signing a donor card that they carry with them in case of an unfortunate accident. Countries that function strictly on informed consent include the United States, Canada, Latin America, the United Kingdom, Ireland, Denmark, the Netherlands, and Germany. The term *presumed consent* means that organs are harvested in every possible case unless one specifically refuses. Countries that have presumed consent include Finland, Portugal, Austria, Sweden, Czech Republic, Slovak Republic, Hungary, and Poland. Some countries theoretically have presumed consent but, in practice, use informed consent; these include Spain, Italy, Greece, Belgium, and France. Overall, although it might be predicted that presumed consent would increase the number of available donor organs, the current statistics from these different countries actually show no significant differences in the percentage of actual organ donations retrieved from the pool of potential organ donors. However, one has to wonder if there are not other complicating factors such as financial and expertise resources in some of the countries listed under presumed consent that lower the actual capacity of these countries to make use of available organs. In some countries such as Japan, organ harvesting from brain-dead individuals has been culturally unacceptable until just recently. However, living organ donors are common, especially for kidney transplants, a procedure that is also common in most of the Western world. Recently, living donor liver transplants have also become widely used, a procedure that makes use of the high regenerative capacity of liver, as discussed earlier. However, the use of living people as organ donors is always complicated by the potential risks to the donor.

Exciting new work in genetics and cloning may also help with the organ shortage crisis. Researchers are now able to clone animals (sheep, pigs) from an adult cell and are also able to grow tissues such as skin and pancreatic islets *in vitro*. It may be possible to combine all these technologies to clone human organs from the patient's own organ cells or animal organs that are genetically modified not to carry immune activating antigens. Obviously, there is the logistical problem of the time required to grow an entire new organ, which would limit the use of the technique for treating acute organ failure. However, organ cloning could be beneficial for treating various types of chronic organ failure (e.g., cystic fibrosis, chronic heart failure) in which the cloned organs containing the defective gene(s) could be repaired during development. Newer research in the area of treating heart failure deals with stem cell research. Stem cells are undifferentiated cells that have the potential to develop into any type of cell. Researchers are looking at the potential of using stem cells to help regenerate an entire functional organ. Although this is in its early stages, this form of therapy may have a significant impact especially for people who are high surgical risks. (See Text Box 19.5.)

Controversial sources of organs and tissues include using organs and tissues from executed prisoners, aborted or anencephalic (without brain) fetuses, and paying people to donate sparable organs (i.e., a kidney). These methods have significant ethical and practical concerns. All of the above are possible sources of organs and possible alternatives to organ transplantation. However, no one strategy is without problems. Many methods are in place or are being explored to increase the organ donor pool. These can be categorized as regularly practiced methods (public awareness campaigns, organ donor cards, presumed consent), new clinical methods (accepting older donor organs), experimental methods (cloning, xenotransplantation), and controversial methods ("alternative" human sources).

XENOTRANSPLANTATION: A BRIEF HISTORY

Xenotransplantation is the transplantation of organs and tissues between different species. The idea of xenotransplantation dates back centuries as can be seen in stories of mythologcial beings that were more powerful when they had the advantage of being part animal as well as part human. The first surgical attempt at xenotransplantation was reported in 1804 when Boronio described the failure of a cow and horse skin graft exchange. In the early 1920s, Russian surgeon Serge Voronoff transplanted 51 older human males with baboon testicles in an attempt to "rejuvenate their physical, mental, and sexual condition." Voronoff writes about how successful his procedure was, but, with today's knowledge, we realize that these grafts would not have survived very long and that any "benefits" were likely due to a placebo effect.

Of more practical relevance are attempts to transplant chimpanzee or baboon kidneys into human recipients. Chimpanzee kidneys were typically rejected within 2 months (but one lasted for $\sim$9 months) and baboon kidneys were rejected even faster. The baboon is more phylogenetically disparate (further away) from humans than are

TEXT BOX 19.5 CHROMOSOMAL MODIFICATION

This technology does not refer to genetic manipulation for the purpose of transfer of genetic information, but rather to the modification of existing chromosomes (and the contained information) so that the tissue will remain viable for the entire lifetime of the organ recipient. As organ transplantation is becoming more common, it is becoming increasingly apparent that an organ from a 55-year-old person may not possess the same capacity for long-term viability as that of a 20 year old. Hence, donor organs have a finite life span whether they are transplanted into a young or an old person. If the organ life span can be lengthened, then the drain on the donor organ pool can be avoided in 20 to 30 years time when, for example, an organ donated by a 55 year old to an 18 year old begins to fail due to this phenomenon.

Telomeres are specialized DNA–protein complexes located at the ends of linear eukaryotic chromosomes. These nucleoprotein complexes have important functions, primarily in the protection, replication, and stabilization of the chromosome ends. They are essential for the proper maintenance of chromosomes and may play a role in aging and cancer. In most organisms studied, telomeres contain lengthy DNA sequences of tandem repeats composed of a G-rich strand and a C-rich strand (called terminal repeats). These terminal repeats are highly conserved and, in fact, all vertebrates appear to have the same simple G-rich sequence repeat in telomeres: (TTAGGG)*n*. Telomerase is a novel ribonucleoprotein, or reverse transcriptase (RT), that adds nucleotides to the ends of chromosomes during DNA replication. Unlike other known reverse transcriptases, all of which are associated with viruses, the telomeric RT is the only one known to date that is associated with a normal vertebrate genome. In the absence of such an enzyme, chromosomes get progressively shorter with each cell division and consequently, the life span of the organ shortens. The age span of cultured cells is normally limited to about 50 cell divisions. This is known as the Hayflick limit, named for the scientist who first observed that the life span of cultured cells was finite. However, this limit can be more than doubled by transfecting the cultured cells with telomerase genes. This finding proved what was a highly controversial model linking telomeres to cellular aging. This area of research is attempting, through genetic modification, to increase organ life span, so that transplanted organs will not become senescent in their recipients.

chimpanzees and as a rule "the more disparate the species, the more humoral the rejection response." Between 1906 to 1992, there have been over 40 attempted xenotransplants into humans and all have failed; however, all have provided useful information about the xenorejection process.

Solid-Organ Xenografts

Xenografts (organs from different species) can be categorized into two types: concordant and discordant. Discordant grafts are between phylogenically disparate combinations (e.g., pig to human). Higher primates and humans have naturally occurring antibodies (reactive to galactose-α-1,3-galactose and related epitopes) against the proteins of more distant mammalian species. Because of these antibodies, a discordant xenograft undergoes a very rapid (minutes to hours) form of rejection that is termed hyperacute rejection (HAR). Concordant xenografts are grafts between phylogenetically close species such as mouse to rat or chimpanzee to human. These combinations do not undergo HAR but display a less severe form of rejection termed acute vascular rejection or delayed xenograft rejection (as do discordant xenografts if HAR is overcome). This form of rejection develops after a few days of transplantation and the induction of new antibody synthesis is important in its development.

Hyperacute Rejection

Naturally occurring antibodies (or xenoantibodies) are believed to originate from exposure of the host to certain α-galactose (α-gal) expressing pathogens. These xenoantibodies aid in defending the host and warding off infection. When a pig organ is transplanted into a human recipient, a vigorous rejection due to the xenoantibodies present will occur within minutes of implantation. However, if the circulating anti-α-gal antibodies are competitively bound with infused α-gal oligosaccharide, or the α-gal antigens are enzymatically cleaved from the organ's endothelium, then the organ will survive for a prolonged period in its new host. The following summarizes some of the important points known about the α-gal antigen and the anti-α-gal antibody:

1. Anti-α-gal antibody is present in humans, apes, and old world monkeys. Other animals/organisms (e.g., pigs, bacteria) express the α-gal antigen.
2. The α-gal antigen is heavily expressed ($\sim 10^6$ to 10^7 epitopes per cell).
3. The α-gal antigen is expressed by some diseased human cells/tissues and senescent red blood cells.

4. The α-gal antigen is very similar in structure to the blood group B antigen.
5. The physiological purpose of anti-α-gal antibody is for protection against pathogens, cancerous cells, and removal of senescent red blood cells.
6. Animals that do not express α-gal antigen, do express other xenoantigens (e.g., β-gal); humans also have a β-gal antibody. Although attempts to remove or block the α-gal antibody : antigen response have been successful, several other problems can arise, including treatment toxicity, antibody rebound, reactions by antibodies other than gal-α-gal, and cell-mediated immune responses that may limit the practical usage of this strategy. Of particular interest to this field, a genetically modified pig strain has been developed recently in which the porcine cells lack the α-gal antigen. This has given new hope to the potential for xenotranplantation from this alternative organ source.

TABLE 19.4 Index of Dissimilarity of Some Mammalian Species (Based on Immune Reactivity): as the Values Deviate from 1.0, the Degree of Dissimilarity Increases

Species	Index
Primates	
Humans/apes	
Human	1.0
Gorilla	1.09
Chimpanzee	1.14
Orangutan	1.22
Old-world monkeys	2.44
New-world monkeys	3.85
Prosimians	13.3
Nonprimate	
Pig	>35

Choice of Organ Donor

Based on immune reactivity, a dissimilarity index has been developed for multiple species (Table 19.4). Obviously, the optimal animals to use for xenotransplantation (based on the least dissimilarity) are the great apes, particularly chimpanzees and gorillas. Typically, organs from nonhuman primates do not undergo HAR but are subject only to a less severe form of antibody-mediated rejection called acute vascular rejection. Of course, the more similar the species is to humans, the higher the risk of xenosis (diseases that animals can pass on to other species) and the greater the moral issue that we face in trapping/breeding and then killing these animals for our own use. A further practical concern is the limited numbers of many of the nonhuman primates, their slow rate of breeding and maturation, and the lack of organ size matching between most of these species and humans. Hence, despite the problems of discordant xenograft rejection, pigs remain the experimental animal of choice at the moment due to the low cost associated with breeding, our few moral concerns with the use of a species that is also a major food item, and many similarities in organ size and functionalities between pigs and humans.

Xenosis: Cross-Species Transmission of Pathogens

Humankind has always been in contact with animals capable of transmitting a variety of pathogens and this is not likely to change. Some prime examples of animal-to-human pathogen transmission (xenosis) resulting in disease are: swine/avian influenza, human immunodeficiency virus (HIV), bovine spongiform encephalitis ("mad cow" disease), West Nile encephalitis, and malaria. This leads one to wonder how many other viruses/prions have been transmitted across species (particularly to humans) without any pathologic consequence or disease. Of particular relevance, porcine endogenous retrovirus (PERV) appears to go dormant in mice and becomes latent with no evidence of spreading or amplification. Human recipients of xenotrans-

TABLE 19.5 Pros and Cons of Xenotransplantation

Advantages	Potential Disadvantages
Unlimited number of donor organs	Difficulty in overcoming HAR
Minimal storage injury potentially reducing rejection	Biochemical and physiological dissimilarities (hormone sensitivities, synthesis of bioactive proteins, differences in basal body temperature, differences in electrolyte concentrations)
Potential for genetic modification to eliminate HLA antigens and reduce rejection	Moral issues of sacrificing evolutionarily advanced animals
Potential to genetically create organs that are disease-resistant	Xenosis—interspecies transfer of pathogens (PERV, prions)
Few moral issues	Expense—industry will attempt to recover past research costs (should society or the individual pay?)
	Emotional attitudes—recipients may feel not completely human

plants from pigs will most likely become infected with PERV, but the absence of replication will prevent viral adaptation and spread; as a consequence, the infection will not lead to any pathology or disease. However, the fear remains that there may be many other viruses in pigs or other species that are inconsequential in their native hosts but could be pathological in humans. Table 19.5 summarizes the pros and cons of xenotransplantation.

SUMMARY

Despite the wealth of knowledge encompassing the entire field of organ preservation and transplantation, current clinical practice stills relies heavily on a relatively simple cold preservation solution to: (1) clear the blood from the vascular bed in order to facilitate efficient reperfusion; (2) cool the organ to near zero temperatures in order to slow down tissue degeneration; and (3) reduce/eliminate cellular/tissue edema. Pharmacologic interventions targeted at reducing adenylate depletion, ion gradient conservation, combating reactive oxygen species, maintaining vascular tone, and reducing immune cell activation have all been attempted, and research into these areas is ongoing. The theory regarding the importance of these multiple processes is not in question. However, for there to be any change in clinical practice, there must be substantial improvement in both experimental and preclinical outcomes. Because the problems associated with organ preservation are numerous and complex, it is indeed difficult to demonstrate improvements in viability outcomes in an experimental setting when the implementation of sound scientific process dictates that only a single parameter can be adjusted at a time in order to determine true cause-and-effect relationships. In the future, major advances in organ preservation technology will likely be derived from:

1. Aggressive development of multipronged approaches to preservation solution design in order to target multiple aspects of low-temperature organ storage
2. Modification of existing static methods of storage to continuous/pulsatile perfusion so that the problems of limited oxygen are circumvented
3. Major advances in genetic engineering either in the field of stem cell research or, perhaps more probable in the foreseeable future, the development of genetically altered pigs devoid of the xenoantigens

SUGGESTED READING

Bagley, J., and Iacomini, J. (2003). Gene therapy progress and prospects: Gene therapy in organ transplantation. *Gene Therapy* **10**:605–611. *A new frontier in organ transplantation.*

Hon, W. M., Lee, K. H., and Khoo, H. E. (2002). Nitric oxide in liver diseases: friend, foe, or just passerby? *Ann NY Acad Sci* **962**:275–295. *Examines the complex and sometimes apparently paradoxical role of NO.*

Jaeschke, H. (2003). Molecular mechanisms of hepatic ischemia-reperfusion injury and preconditioning. *Am J Physiol* **284**:G15–G26. *This review examines cellular mechanisms of injury, formation of pro- and anti-inflammatory mediators, expression of adhesion molecules, and the role of oxidant stress during the inflammatory response. The role of NO is also examined in liver ischemic-reperfusion injury in the surgical realm.*

Jahania, M. S., Sanchez, J. A., Narayan, P., Lasley, R. D., and Mentzer, R. M. (1999). Heart preservation for transplantation: Principles and strategies. *Ann Thoracic Surg* **68**:1983–1987. *This review focuses on the physiologic principles underlying the use of hypothermia and the chemical components of preservation fluids for heart grafts.*

Jassem, W., Fuggle, S. V., Rela, M., Koo, D. D., and Heaton, N. D. (2002). The role of mitochondria in ischemia/reperfusion injury. *Transplantation* **73**:493–499. *Mitochondrial ischemia-reperfusion injury is examined focusing on a complex chain of events involving depletion of energy substrates, alteration of ionic homeostasis, production of reactive oxygen species, and cell death by apoptosis and necrosis is examined.*

Laskowski, I., Pratschke, J., Wilhelm, M. J., Gasser, M., and Tilney, N. L. (2000). Molecular and cellular events associated with ischemia/reperfusion injury. *Ann Transplant* **5**:29–35. *Organ preservation and transplantation initiates a cascade of molecular and cellular events triggering the release of proinflammatory mediators and immune cell infiltrates. This inflammatory event influences both acute functional and structural changes in the organ, which contributes to reduced graft survival.*

Rangan, U., and Bulkley, G. B. (1993). Prospects for treatment of free radical-mediated tissue injury. *Br Med Bull* **49**:700–718. *Examines the role of antioxidant therapy in experimental and clinical situations of ischemia-reperfusion injury.*

Southard, J. H., and Belzer, F. O. (1988). Principles of solid-organ preservation by cold storage. *Transplantation* **45**:673–676. *Short yet concise review article focusing on importance of UW solution.*

Southard, J. H., and Belzer, F. O. (1989). Organ Preservation (Chapter 10). In Wayne Flye, M. (ed.). *Principles of Organ Transplantation.* W. B. Saunders, Philadelphia, pp. 194–215. *This and the previous review are by two of the "fathers" of organ preservation. The team of Southard and Belzer was a great example of "when science meets medicine."*

Southard, J. H., and Belzer, F. O. (1995). Organ preservation. *Ann Rev Med* **46**:235–247. *Review.*

20

METABOLIC EVOLUTION AND THE ORIGIN OF LIFE

TOLGA BILGEN

As a topic, the origin of life might be appropriate for a book's opening chapter—after all, every book ought to begin by considering "how it all started." However, having waded through the complexities of contemporary cellular metabolism and metabolic regulation over the last 19 chapters, the reader might now better appreciate some of the ideas surrounding how such metabolism may have emerged in the first place.

WHAT LIFE WANTS OUT OF LIFE

Life is order's way of maintaining order. The organization of any living organism, its ordered structures, must persist in the face of short- and long-term environmental flux. Otherwise, functional efficiency may decline and the organism may die. Metabolism generally operates in the short term, adjusting the activities of a single individual throughout its life. Replication, meanwhile, can be thought of as a response to long-term environmental changes that persist beyond the individual's life. In this case, order can only be maintained if an individual's progeny bear advantageous modifications.

Despite their limitations, every definition of life requires a given system to carry out metabolism and replication. But feedback is also crucial for the maintenance of order, within both metabolic and replicative timescales. Polymer-driven feedback loops may be considered the first "animate" matter on Earth. Operating at all organizational levels, from enzyme pathways and organelles up to ecological communities, feedback loops may be considered a primary selective unit.

Feedback also illustrates mutual dependence; no single polymer, no single catalytic cycle, can provide all the necessary adjustments required for the persistence of order. Consider a system composed of several distinct, co-dependent subsystems, each contributing some crucial product to another. Each subsystem has its own co-dependent components, each of which provides crucial products to other components.

Another theme is that innovation tends to arise from the liberal recombination of a conserved assortment of component parts. A novel subsystem, for instance, is more likely to be assembled out of preexisting components, rather than from new components arising *de novo*. Similarly, preexisting subsystems are often co-opted in the formation of novel systems, with novel functions as the final outcome.

This conservatism, or continuity, also highlights the relative likelihood (and advantage) of evolving biopolymers to catalyze slow, abiotic reactions.

A consideration of the central progression from abiotic monomers and polymers to compartmentalized lifelike systems will provide context for the particulars of metabolic origins.

INITIAL CONDITIONS AND THE MONOMER WORLD

Polymer systems require a steady supply of monomers for replacement and growth. Present-day organisms can gain their monomers from three sources: *de novo* synthesis of monomers, the recycling of their own polymers into monomers, and obtaining monomers from other organisms. However, what was the abiotic source of organic monomers on the primordial Earth?

Functional Metabolism: Regulation and Adaptation, edited by Kenneth B. Storey.
ISBN 0-471-41090-X

Appetite for Reduction

Ultraviolet or electrical energy input to a reducing gas mixture of NH_3, CH_4, H_2O, and H_2 readily produces amino acids and other organic compounds (see Text Box 20.1). A primordial atmosphere of similar composition may have allowed for the precipitation of a variety of organic monomers. Accumulating in oceans or other bodies of water, these would have formed a rich "prebiotic soup," ripe for the initiation of lifelike polymer systems.

However, current theories about the composition of the primordial atmosphere suggest that no such soup would have been served. Volcanic outgassing is likely to have supplied the bulk of atmospheric gases, making for a nonreducing mixture of CO_2, N_2, and H_2O. Any ammonia would have undergone photodegradation, and methane

TEXTBOX 20.1 THEORIES OF PREBIOTIC MONOMER SYNTHESIS

PREBIOTIC SOUP

This theory assumes that the primordial atmosphere was a mixture of reducing gases. With an input of ultraviolet or electrical energy, carbon can be reduced into various classes of organic compounds. In 1953, Stanley Miller and Harold Urey's experiments produced amino acids and other biologically relevant compounds from a starting mixture of NH_3, CH_4, H_2O, and H_2. Some of the organic compounds produced included urea, the amino acids glycine, alanine, aspartate, and glutamate, and the organic acids lactic acid, acetic acid, and formic acid.

Subsequent experiments, employing a variety of gas mixtures and conditions, led to the following conclusions:

- Hydrogen cyanide (HCN) and aldehyde were crucial intermediates in prebiotic amino acid synthesis.
- Even mildly reducing conditions (i.e., as long as some H_2 is available) will produce organic compounds in varying yields.
- Although the presence of ammonia boosts both product yield and diversity, abiotic synthesis of organic compounds will still occur in its absence using nitrogen gas as a substitute.
- Methane is a crucial carbon source; without it, product variety declines, and glycine is the major compound formed.
- Recent experiments using a $CO{-}CO_2{-}N_2{-}H_2O$ gas mixture suggest that such an atmosphere, bombarded by cosmic radiation, can support the synthesis of organic compounds comparable to the yields estimated for a strongly reducing atmosphere.

IRON–SULFUR WORLD

In the late 1980s, Gunter Wachtershauser proposed that, in a nonreducing atmosphere comprised mostly of CO_2, CO, and N_2, the oxidative formation of pyrite (FeS_2) would provide a steady source of energy and reducing power for the synthesis of organic compounds:

$$FeS + H_2S \rightarrow FeS_2 + H_2 + 2e^- \quad (\Delta G = -38.6 \text{ kJ/mol}) \qquad \text{(TB20.1)}$$

Under anaerobic conditions, protons have been shown to act as electron acceptors in the above reaction, although carbon can also successfully compete for electrons. In this way, pyrite formation can fix inorganic carbon into reduced organic compounds. Experiments using iron–sulfur slurries and inorganic carbon have demonstrated the synthesis of methane and saturated hydrocarbons, in various yields, and molecular nitrogen has been reduced to ammonia. Experimental reactions using Fe, S, and CO-rich slurries have also produced acetic acid ($CH_3{-}COOH$):

$$CO + FeS + 3H_2S \rightarrow CH_3{-}SH + FeS_2 + H_2O + S^{O} \quad (TB20.2)$$

$$CH_3{-}SH + CO \rightarrow CH_3{-}CO{-}SH \quad (TB20.3)$$

$$CH_3{-}CO{-}SH + H_2O \rightarrow CH_3{-}COOH + H_2S \quad (TB20.4)$$

In reactions (TB20.2) and (TB20.3), CO (or CO_2) is first reduced to methyl mercaptan ($CH_3{-}SH$), which then reacts with excess CO to give thioacetic acid ($CH_3{-}CO{-}SH$). The thioacetic acid is then hydrolyzed to acetic acid [reaction (TB20.4)]. The generated pyrite appears as a precipitate. Further work has also demonstrated the synthesis of pyruvate ($CH_3{-}CO{-}COOH$) from the fixation of carbon dioxide into thioacetic acid:

$$CH_3{-}CO{-}SH + CO_2 + FeS \rightarrow CH_3{-}CO{-}COOH + FeS_2 \quad (TB20.5)$$

The formation of pyrite [reaction (TB20.1)] liberates more energy at higher temperatures and pressures, and at lower pH, conditions presumed to be characteristic of hydrothermal vent systems. Also, CO would have been an ample carbon source, as it is a common component of hydrothermal waters. Since hydrothermal vents may have existed since prebiotic times, it is suggested that they were the actual cradles of life on Earth.

Further speculations arise from pyrite's removal from the reaction as a precipitate; the mineral's positively charged surface can act to localize any negatively charged organic compounds through electrostatic interactions. It is predicted that the adsorbed organic compounds would react with one another, generating an expanding network of autocatalytic pathways. This so-called surface metabolism would then generate a wide variety of products, including sugars, amino acids, and lipids, in a carbon fixation cycle similar to a reductive TCA.

This theory also offers mechanistic continuity with extant metabolism in that redox functions are frequently carried out by proteins bearing iron–sulfur clusters.

The predictions of and the experimental evidence supporting the iron–sulfur world scenario are manifold, though some problems persist. First, the experimental reaction conditions frequently favor the synthesis of one organic compound while prohibiting the production of others; this would appear to preclude the establishment of autocatalytic cycles and a reductive protometabolism. Second, the concentrations of H_2S used in the experiments may have been unrealistically high, compared to H_2S availability in nature. Lastly, although amino acids are a predicted product of this mineral-catalyzed system, they have yet to be synthesized experimentally.

EXTRATERRESTRIAL SOURCES

Precluding or enhancing Earth-bound syntheses, preformed organic compounds from outer space could have been deposited by meteor or comet impact events. A wide variety of organic compounds are available on carbonaceous meteorites, including alcohols, amino acids, purines, pyrimidines, various hydrocarbons, and amphiphiles. The Murchison meteorite is a well-studied example and among the organic compounds found in it were several amino acids (alanine, aspartate, glutamate, glycine, proline, valine) as well as a peptide (diglycine) and assorted aliphatic compounds including monocarboxylic acids. More recently, over 100 aliphatic hydrocarbon compounds were identified on the Tagish Lake (British Columbia) meteorite. In addition to meteorites, comets and interplanetary dust particles (also called micrometeorites) have also been suggested as delivery systems for extraterrestrial organic compounds.

and hydrogen gas would have been present only in trace amounts. Under these conditions, the synthesis of organic compounds in the atmosphere would have been minimal. Recent work with CO-rich gas mixtures, however, suggests that even nonreducing conditions could have supported the production of organic compounds (see Text Box 20.1).

Fool's Gold

An alternative to the atmospheric synthesis of monomers was proposed in the late 1980s. This theory uses the reducing power generated during the formation of pyrite (FeS_2), or other iron–sulfur minerals (marcasite, pyrrhotite), to synthesize organic compounds. The exergonic formation of pyrite from FeS and H_2S occurs under conditions typical of hydrothermal vents. Indeed, under simulated hydrothermal conditions, slurries of iron and sulfur have been found to reduce inorganic carbon into pyruvate and hydrocarbons, among other compounds (see Text Box 20.1).

Furthermore, pyrite had been predicted to supply more than an energy source for the synthesis of organic compounds. The polymerization of monomers has also been attributed to iron–sulfur chemistry, and the mineral's charged surface has been proposed as a substrate to keep organic compounds and polymers localized, allowing for the emergence of lifelike polymer systems carrying out a *surface metabolism* (see discussion under Polymer Worlds).

Special Delivery

Outer space offers another source of organic compounds. Carbonaceous meteorites have carried amino acids, lipids, and other biologically relevant products to Earth (see Text Box 20.1). Given the frequency of impact events early in Earth's history, extraterrestrial compounds may have been delivered in sufficient quantity to have initiated biogenesis. Meteor-borne materials also reflect the robustness of abiotic chemistry. Thus, if such a variety of reduced compounds could be formed in space, where starting material and energy sources are rare or widely dispersed, then abiotic synthesis on a primordial Earth, with its abundance of inorganic precursors and energy, would appear quite plausible.

Nucleotides from Where?

With its informational and catalytic capabilities, ribonucleic acid (RNA) is favored as the original lifelike polymer (see Polymer Worlds below). Montmorillonite clay has been shown to catalyze nucleotide formation under drying reactions, but abiotic synthesis of nucleotide building blocks is problematic for several reasons. One of these is that even though a wide variety of sugars can be generated by abiotic processes, only meager yields of ribose (the requisite sugar for RNA synthesis) have been achieved (see Text Box 20.2). Furthermore, sugars have short half-lives in aqueous solution, and this may preclude the accumulation of enough pure ribose to allow nucleotide synthesis.

The abiotic production of purine and pyrimidine bases poses additional challenges. One route to purine synthesis imagines a traveling reaction, encountering varied environmental conditions (i.e., hydric and thermal gradients) as it trickles down a mountain. Abiotic pyrimidine synthesis, traditionally regarded as impossible, may have started with the oligomerization of hydrogen cyanide (HCN); the resulting compound, orotic acid, is a uracil precursor. Pyrimidines pose a further problem. While a solution of purine bases and carbohydrates will, upon evaporation, form nucleotides, pyrimidine bases will not. Also, compared to adenine or uracil, the pyrimidine cytosine is known to have a short half-life. This raises the intriguing notion that the first oligonucleotides were composed of purines alone (see Text Box 20.3).

All You Can Eat

Even weak reducing conditions initiate amino acid synthesis. Hydrothermal vents have likely existed from prebiotic times, and meteor impact events occurred at greater frequencies on the primordial Earth. Therefore, it is possible that all three of these sources, and possibly other as yet uncharacterized processes, contributed to the accumulation of organic compounds to the prebiosphere. By combining the organic monomers available from all three sources, there may not then have been a major stumbling block in the path to the first living systems.

POLYMER WORLDS

While the abiotic mechanism to link monomers together is a problem in itself, the nature of the first lifelike polymers—what they were and what they actually did—is a fundamental question in the origin of life. Did life spring from a peptide- or RNA-dominated assortment of polymers? These are the so-called peptide world and RNA world theories. Furthermore, did these polymer worlds specialize in metabolic catalysis (the harnessing of chemical energy or monomer production) or replication (the production of more polymers)?

Dry up, the Water's Fine

Numerous experiments have demonstrated that peptides and oligonucleotides can form without the aid of enzymes (see Text Box 20.3). With dehydration (or condensation) being the key to biopolymer synthesis, various scenarios have been proposed.

TEXT BOX 20.2 PREBIOTIC RIBOSE AND OTHER CARBOHYDRATES

The inherent instability of sugars in aqueous solution has been cited as a major problem for explaining the origin of life. How were carbohydrates established as a metabolic nexus if they were scarce or unavailable? How could an RNA world be initiated in the absence of sufficient ribose? The formose reaction provides a plausible answer. Starting with formaldehyde, the formose reaction has been shown to produce triose, tetrose, pentose, and hexose sugars:

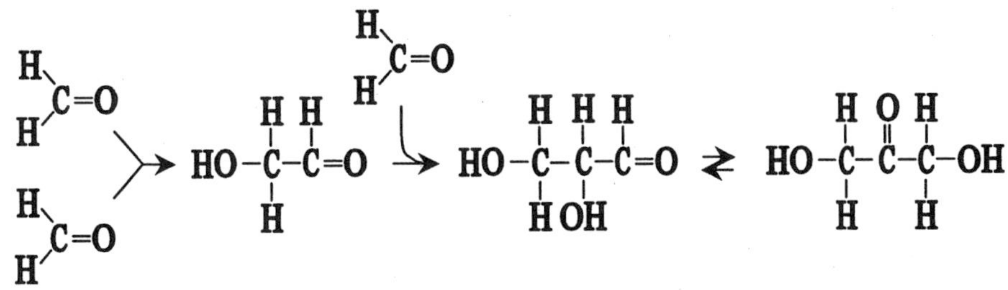

Formaldehyde **Glycoaldehyde** **Glyceraldehyde** **Dihidroxyacetone**

From the above example, two glycoaldehydes can combine to produce a tetrose. The trioses, glyceraldehyde and dihidroxyacetone, can combine with glycoaldehyde to form pentoses, or with other trioses to generate hexoses.

Synthesis could have occurred as follows. In the upper atmosphere, UV rays can break down H_2O and CO_2. The resulting free radicals could combine to form formaldehyde, CH_2O. The solubility of formaldehyde in water would allow it to rain down to Earth, where it would accumulate and become concentrated through evaporation. When formaldehyde concentration reaches 1 to 2%, the formose reaction can proceed autocatalytically at 55°C in the presence of calcium hydroxide, $Ca(OH)_2$. Within minutes, a broad array of sugars (straight-chained and branching) and other compounds can be produced. In a nonoxidizing atmosphere, sugars could have been available for use by the first lifelike systems. One problem with this proposal concerns ribose production; yields of this sugar are usually $<1\%$ of total sugar output. Furthermore, ribose half-life is very low at the alkaline pH ($\sim$pH 11) required for the $Ca(OH)_2$ catalyst to work.

However, with magnesium hydroxide, $Mg(OH)_2$, as a formose catalyst, conditions change. Although the percentage of aldopentoses formed from the reaction is lower than with calcium hydroxide as the catalyst, the reaction takes place at a lower pH (pH 9.4) that improves aldopentose stability. Furthermore, with $Mg(OH)_2$ catalysts, the conversion of preformed glucose to aldopentoses has been demonstrated. Plumbous salts have also been found to act as catalysts in sugar formation from formaldehyde. In combination with magnesium hydroxide, lead salts have also been shown to promote the conversion of other sugars into aldopentoses, favoring ribose production.

Although carbohydrate synthesis might not have been as limiting a factor as previously thought for the construction of abiotic RNA, there are still some interesting difficulties associated with the use of ribose. For example, among all of the prebiotically formed aldopentoses, how did ribose come to be singled out for use in nucleotide production? In addition, since ribose will spontaneously convert to other aldopentoses, how was ribose obtained in pure enough form to support extensive RNA synthesis?

TEXT BOX 20.3 PREBIOTIC POLYMERS

Numerous studies have demonstrated abiotic synthesis of peptides and oligonucleotides, under various conditions, using different catalysts.

PEPTIDES

Several ways of producing amino acid polymers have been demonstrated under both wet and dry conditions. These include:

- In solution, activated amino acids show limited polymerization, although glycine has been shown to facilitate the formation of di- and tripeptides from free alanine in dilute solution. (See Peptide World for more on the catalytic capacity of free amino acids.) In the presence of chemical condensing agents, such as polyphosphates or cyanamide, the formation of peptides up to 50 residues in length has been achieved in solution. Recent attention has also been paid to inorganic salts as condensing agents; NaCl and $CuCl_2$, in concentrations typical of evaporative cycles, have been shown to catalyze the formation of oligopeptides from free amino acids.
- Under dry-state conditions, trimetaphosphate, ATP, or Mg^{2+} can catalyze the polymerization of glycine at 65°C. However, overheating of amino acid mixtures results in the formation of carbonized tar.
- Wet–dry cycling has proven most effective in polymerizing amino acids, as shown by Fox and Harada in 1958. Amino acid solutions were heated to dryness, forming polymers that were dubbed proteinoids. When rehydrated, these proteinoids displayed self-assembly and a variety of weak catalytic properties. The proportions of amino acids incorporated into proteinoids was found to differ from that of the starting mixture; for example, aromatic amino acids were overrepresented in the proteinoids. This suggests that nonrandom chemical forces contributed to the selection of repeatable sequence motifs. However, problems occurred with the amino acid linkages that were found in proteinoids because not all were true peptide bonds. Furthermore, the high concentrations of pure amino acids used in Fox's experiments are very unlikely to reflect prebiotic monomer abundance.
- Iron–sulfur mineral formation has been shown to polymerize amino acids. Under hot (100°C), anaerobic, aqueous conditions, slurries that include FeS, NiS, and H_2S have been shown to catalyze di- and tripeptide formation from phenylalanine, tyrosine, and glycine. However, peptide half-life is short under these reaction conditions.
- Montmorillonite clay has been shown to order amino acids, catalyzing peptide synthesis at temperatures below 100°C. Clay surfaces may also have adsorbed preexisting short peptide chains (formed previously in solution), subsequently catalyzing their linkage into longer polymers.

OLIGONUCLEOTIDES

There are also multiple ways of producing oligonucleotides under abiotic conditions. These include:

- Evaporation of a nucleotide solution will result in oligomer formation.
- Polymerization of free nucleotides in solution is influenced by condensing agents such as trimetaphosphate or cyanimide.
- Activated monomers (nucleotides) polymerize in solution, although high rates of hydrolysis limit oligomer length to 20 to 40 nucleotides.
- Nucleotide polymerization has been demonstrated at subzero temperatures, producing oligomers of up to 11 bases. However, the presence of inorganic ions greatly inhibit oligomerization.
- Montmorillonite clay has been shown to order nucleotides in solution. Purines and pyrimidines, adsorbing to the positively charged clay surface, form ordered monolayers. These can subsequently polymerize producing oligonucleotides of up to 55 bases.
- A single strand of RNA can act as a template, promoting base pairing by free nucleotides in solution. Magnesium and zinc have been shown to be effective catalysts in this base pairing. Notably, both ions are required by modern nucleic acid polymerases.

The aqueous medium in which life arose made dehydration reactions thermodynamically unfavorable. Wet–dry cycles, however, would have allowed the periodic evaporation of monomer solutions. Monomer concentrations would increase with decreasing solution volume, and anhydrous heating would encourage polymer formation. Shallow tide pools, lagoons, or waves splashing over exposed mineral surfaces could have been ideal polymer reactors. Furthermore, the Moon was likely much closer to Earth

4 billion years ago. This would have created stronger gravitational pull on the oceans that would have exposed much larger surface areas to dry out during low tides. With the rising tide, the newly formed polymers might have washed out to sea, might have hydrolyzed back into their component monomers, or remained localized, contributing to a growing assortment. Adsorption to charged mineral surfaces may have contributed to this localization (see Compartment of One's Own below).

In the absence of wet–dry cycles, energy for polymer synthesis may have been supplied by chemical condensing agents, such as polyphosphates, or HCN-derived compounds (e.g., cyanamide). Mineral surfaces, such as pyrite and clays, may also have been instrumental in polymer formation. Their positively charged surfaces adsorb free monomers, bringing them into close proximity and promoting their linkage (see Text Box 20.3).

Peptide World

The relative ease of abiotic amino acid synthesis, and the variety of conditions under which amino acids polymerize, have made peptides an attractive starting point for the development of lifelike polymer systems. A peptide world scenario predicts that the essential functions associated with life, including metabolism, growth, and replication were performed by peptides or proteins.

Sidney Fox found that wet–dry cycling of amino acid mixtures caused the formation of polymers up to 50 residues long (see Text Box 20.3). Upon rehydration, hydrophobic interactions caused these *proteinoids* to form microspheres averaging 1 to 2 μm in diameter. Importantly, these proteinoids possessed weak catalytic activities, including the ability to catalyze peptide formation from free amino acids (see Table 20.1). This presents the possibility of a peptide-based protometabolism.

Amino-acid-based polymers are natural chemical catalysts, but proteins are not well-known for their template-based replication. Even if a chance set of peptides could self-organize into a functional lifelike system, its persistence would be problematic. Could abiotic, apparently random, peptide generation provide polymer replacements over time?

Abiotic peptide sequences might not have been entirely random, for two reasons. First, the limited variety of available amino acids could have restricted peptide sequence diversity. Second, proteinoid sequences have shown a nonrandom quality; certain residues and sequences are favored over others, regardless of the proportions of different amino acids in the starting mixture. This might reflect differential bond stabilities between various amino acids or other as yet unknown ordering principles. The charged surfaces of the reaction vessels used in the experiments may have brought some repeatable order to adsorbed amino acids. Thus, nontemplate-driven, abiotic synthesis could have produced multiple copies of useful peptides.

Another argument favoring a peptide world scenario is the demonstrated ability of even short peptide sequences to have catalytic activity. Indeed, free amino acids in solution have been shown to cluster around and hydrolyze substrate molecules, reproducing the functions of some of the simpler enzyme activities such as carbonic anhydrase, catalase, and β-galactosidase. Substrates, themselves, might also have acted as templates, further reducing the randomness of peptides:

$$CO_2 + H_2O \longleftrightarrow H^+ + HCO_3^- \quad \text{(carbonic anhydrase)}$$
$$2H_2O_2 \longrightarrow 2H_2O + O_2 \quad \text{(catalase)}$$
$$\text{Lactose} \longrightarrow \text{galactose} + \text{glucose} \quad (\beta\text{-galactosidase})$$

Therefore, peptide world theories are supported by several engaging lines of evidence, including the recent *in vitro* development of self-replicating peptides (see Text Box 20.4). The discovery of self-splicing RNA (ribozymes) in the early 1980s, however, introduced another candidate for the first lifelike polymers.

TABLE 20.1 Observed Catalytic Activities of Proteinoids and Ribozymes

Proteinoids	Ribozymes
Deamination	Acyl transfer
Decarboxylation	Aminoacylation
Esterolysis	DNA and RNA cleavage
Oxidation	DNA and RNA ligation
Photochemical decarboxylation	Peptidyl transfer
Amino acid polymerization	Ribonucleotide synthesis
	RNA-templated RNA synthesis
	Transesterification

Oligonucleotide World

The RNA world theory predicts a population of catalytic oligonucleotides that could establish basic metabolic networks while simultaneously providing the information storage necessary for replication. This scenario has won tacit acceptance among origin of life theorists, despite problems with abiotic ribose and nucleotide generation, as discussed previously. These latter have prompted speculation about alternative oligonucleotide-like polymers making up a so-called pre-RNA world. Hexose sugars, for instance, or threose (a four-carbon sugar), or even glycine may have provided the original backbone for some lost pre-RNA molecule.

TEXT BOX 20.4 PEPTIDE SELF-REPLICATION

Experiments in abiotic synthesis never fail to yield amino acids. The polymerization of amino acids occurs under such a wide range of conditions that the primordial Earth was certain to support abiotic peptide synthesis. The diversity and versatility of peptides and proteins, in terms of their self-organization, structural motifs, stability, and their catalytic activities, would seem to make them obvious candidates for the first life-like polymers. This putative peptide world only lacked information storage and replicative capability. Or did it?

AUTOCATALYTIC PEPTIDES, *IN VITRO*

The basic model system consists of a template peptide that produces copies of itself by aligning and ligating two peptide fragments identical to portions of its own sequence:

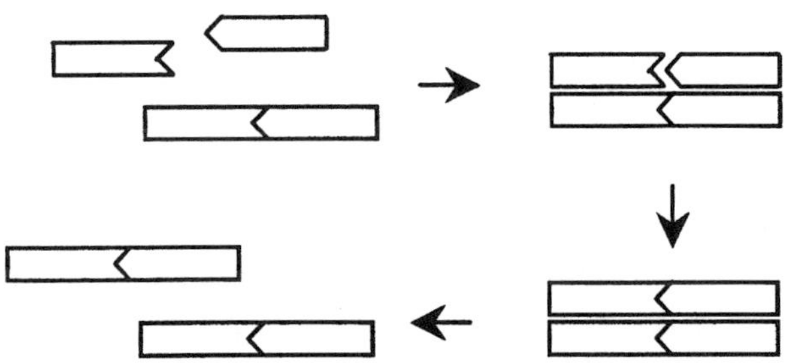

This model was realized in 1996 by the Ghadiri group, which based its autocatalytic peptide on the coiled-coil helical region of a known transcription factor. The template forms a ternary complex with fragment peptides through interhelical hydrophobic interactions, facilitating their ligation. A number of experiments with this system and variants have revealed the following:

- Self-replication of the template increases exponentially, although product inhibition eventually occurs.
- The template peptide is chiroselective, preferentially utilizing homochiral peptide fragments in a racemic mixture. Fragment peptides possessing even a single stereochemical mutation (i.e., one amino acid of opposite chirality) were not utilized by the template for self-replication. Stereochemical mutations may sterically disrupt coiled-coil structures, which inhibits proper association between template and fragment peptides.
- Template peptides are sensitive to sequence mutations in the peptide fragments. The alignment of the fragment peptides by the template is mediated not by every amino acid residue but only by those involved in the helix–helix interactions. Mutations in these key positions likely lower the stability of the ternary structures, as competition assays demonstrate that template peptides ligate "wild-type" fragments over mutant fragments. The template peptide's fidelity is not entirely rigid, however, and templates possessing a single mutation can still coordinate and ligate wild-type fragment peptides.
- Cross-catalysis has been observed, where one species of template peptide organizes and ligates the fragment peptides of a second template species, and vice versa. Helix heterodimers form via electrostatic interactions, for instance, between lysine residues on one coil and glutamic acid residues on another. This electrostatic complementarity allows for selectivity, that is, peptide 1 will preferentially support the formation of peptide 2, and vice versa. Also, the system's activity can be regulated by environmental conditions. For example, at low pH, the protonation of glutamic acid residues lowers electrostatic interactions with lysine and, hence, heterodimer formation and mutual cross-catalysis is inhibited. At neutral pH, self-replication of each individual template species is favored.
- Cross-catalyzing systems produce not only new copies of the original template peptides but also produce hybrid templates consisting of combinations of each species' fragment peptides. This is also influenced by environmental conditions, including pH and salt concentrations.

The abiotic formation of short peptides (i.e., fragment peptides) in multiple copies is plausible (see Text Box 20.3). The *in vitro* systems described above suggest that combinations of short peptides could give rise not only to homochiral, self-replicating ligases but also to mutually supporting hypercycles that increase peptide diversity by the recombination of available fragments. A peptide world may been self-supporting after all.

The catalytic potential of an abiotic RNA world is reflected by observations of natural and artificial ribozymes. Cellular ribozymes, including spliceosomes and RNAse P, are crucial for messenger, ribosomal, and transfer RNA (mRNA, rRNA, and tRNA) processing. The pivotal role of rRNA in peptide bond formation is also well-known. Experimentally, ribozymes have been "evolved" *in vitro* from pools of random oligomers or from known ribozymes; successive cycles of *in vitro* mutation and selection have generated a variety of activities including cleavage/ligation of RNA, peptide bond formation, and nucleotide synthesis (Table 20.1 and Text Box 20.5).

Not only do ribozymes display catalytic diversity, but recent work also demonstrates their catalytic promiscuity, meaning a single oligonucleotide can adopt more than one tertiary conformation, each with its own distinct capability. For example, one *in vitro* study employed a viral

TEXT BOX 20.5 RIBOZYMES NATURAL AND ARTIFICIAL

Ribozymes are RNAs that possess enzymelike activity, some possessing rate constants approaching those of their protein enzyme counterparts. Since the first discovery of a self-splicing ciliate pre-rRNA, much attention has been given to ribozymes, including those found in nature and those developed in the laboratory.

NATURALLY OCCURRING RIBOZYMES

- Many naturally occurring ribozymes are self-splicing introns that excise themselves out of pre-mRNA, pre-rRNA, and pre-tRNA molecules, via the hydrolysis of phosphodiester linkages. Some of these ribozymes require accessory protein components for full activity. Thus, deproteinized ribozymes show weaker activity, although this can be partially regained under nonphysiological conditions (high salt or temperature).
- Ribonuclease P (RNase P), a ribonucleoprotein, catalyzes site-specific cleavage of certain RNA precursors, namely pre-tRNA. A complex of RNA and protein, the catalytic activity of RNase P resides in its RNA component. This ribozyme possesses true enzymelike activity, as it is a multi-turnover catalyst that remains unchanged after processing its substrate.
- The ribosome is a ribozyme. Contrary to what was once believed, the rRNA is the catalytic portion of the ribosome (carrying out peptidyl transfer), whereas the numerous accessory proteins appear to offer physical support.
- Like protein enzymes, many ribozymes use cofactors such as imidazole, metal ions, and other small allosteric effector molecules. Unlike the case in proteins, however, some ribozymes might employ divalent metal cations to help stabilize tertiary folding, as opposed to exploiting their catalytic properties.
- RNA tertiary conformations can be complex and varied, providing sites that coordinate substrates, metal ions, and catalytic nucleotide residues. Like protein enzymes, ribozymes can "force" unfavorable contacts between substrates and catalytic groups within a catalytic pocket or groove.

IN VITRO SELECTION WITH RIBOZYMES

In vitro selection experiments on ribozymes can be viewed as a model of Darwinian evolution acting in a putative RNA world. In such experiments, ribozymes exhibiting a selected catalytic activity outcompete others in a "survival of the fittest" scenario: A population of RNAs is assayed for a given catalytic activity (the selection pressure). Those few RNAs with the greatest activity are amplified, and random mutations are introduced into the new population. This new pool of RNAs, enriched for the sought-after activity, is then tested in a second round of assays. The strongest ribozymes are once again selected to be amplified, with mutations, into a third round, and so on. This cyclic selection–amplification–mutation typically results in greater and greater catalytic activity. The starting population of oligonucleotides can be (1) a large assortment of random sequences or (2) based on a known ribozyme.

Catalytic activities that have been selected for include:

- Nucleotide synthesis: Starting from a pool of random sequences, a ribozyme has been "evolved" that catalyzes glycosidic bond formation between thiouracil and ribose, making thiouridine.
- Peptide bond formation: From random starting pools, ribozymes have arisen that catalyze the same peptidyl transfer reaction carried out by natural ribosomes. Mg^{2+} is a necessary cofactor, for structural or catalytic reasons.
- RNA-templated primer extension: Starting from a sequence pool based on an RNA-ligase ribozyme (itself developed from random sequences, also see below), a ribozyme was derived that can synthesize complementary copies of any template RNA. Primer extension occurs accurately (the ribozyme adds the correct nucleotide 95.8% of the time) to a length of up to 14 nucleotides. This RNA polymerase ribozyme employs Mg^{2+} as a cofactor.
- Other activities successfully selected for include the cleavage of DNA–RNA hybrids, DNA cleavage, DNA ligation, aminoacylation, acyl transfer, and porphyrin methylation.

SELF-REPLICATING RIBOZYMES

In vitro experiments have also "evolved" oligonucleotides that can replicate themselves. One model consists of a template RNA that coordinates two RNA fragments and ligates them to make another copy of itself. This is similar to the autocatalytic peptide described in Text Box 20.4. The two substrate (or fragment) oligos are aligned by complementary base-pair binding along the template:

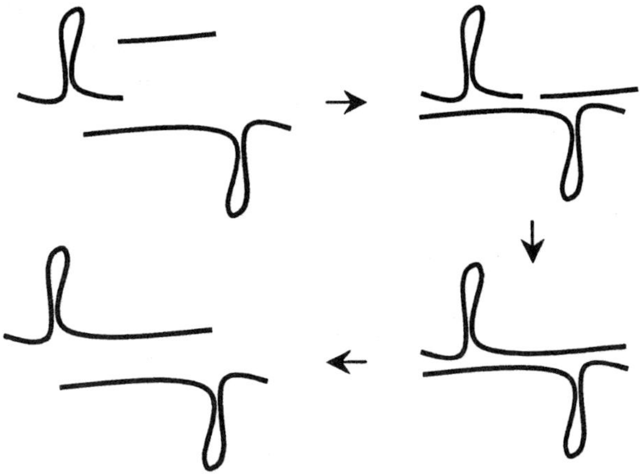

This replicating ligase was developed by the Joyce group in 2002 and was based on an RNA-dependent RNA ligase ribozyme, itself previously "evolved" *in vitro* from a random sequence pool. The ribozyme uses Mg^{2+} as a cofactor.

Further experiments with this system and variants have revealed the following:

- The replicating ligase has been successfully "re-evolved" without cytosine, that is, it can function while being composed only of A, G, and U. Its tertiary structure is maintained by A–U base pairing and by G–U wobble pairing. Subsequent *in vitro* selection with this C-less ribozyme demonstrated further tolerance to missing nucleotides: Even when composed of only two bases, uracil (U) and diaminopurine (D), it retained its functionality.
- In a separate experiment, another C-less, self-replicating ligase was developed *in vitro* from a starting pool of random sequences.

The RNA world theory predicts a lifelike polymer system in which RNA provides catalytic, information storage, and replicative functions. The sheer variety of catalytic activities found in nature and in selection experiments, including nucleotide synthesis and self-replication, bolster this concept.

RNA-cleaving ribozyme, and a ligase developed through *in vitro* selection, each possessing its own specific sequence. An intermediate nucleotide sequence was determined between these two ribozymes, that is, one possessing combined elements of both. When synthesized, this intersection sequence was found to fold into either of two distinct tertiary conformations, one allowing for cleavage activity, the other promoting ligation. This implies that, in a putative RNA world, a single polymer sequence could have been quite versatile and that only slight changes in its sequence would have been enough to push it in one catalytic direction or another.

Crucially, recent *in vitro* selection studies have developed a ribozyme with RNA polymerase activity (see Text Box 20.5). Approximately 200 nucleotides in length, this polymerase employs RNA templates and displays proofreading activity. This potential for replication is a boost for the RNA world scenario in that multiple copies of useful polymers could have been produced through template-driven synthesis. Problems arise, however, over oligonucleotide length. First, known abiotic processes cannot account for the production of a ribozyme of 100 to 200 nucleotides. Second, this ribozyme's own length precludes its self-replication; the polymerase activity is far too slow to produce 100- to 200mers. Overall, both points illustrate the problems arising when synthesis cannot keep pace with degradation.

Multimeric ribozymes have been invoked to counter these difficulties. The first self-replicating ribozyme may have consisted of numerous shorter RNA strands, which were more feasible in an abiotic world and easier to replicate.

Cooperative World

In extant biology, nucleic acids are rarely found without some proteins bound to them, and many proteins require nucleotide-derived coenzymes to function. The ribozyme component of RNase P, for example, displays catalytic activity when deproteinized, but when complexed with its protein accessories, its activity increases by approximately 20-fold. In the case of the ribosome, the catalytic rRNA retains only some activity if partially deproteinized and becomes inactive without its protein complement. This mutual support between proteins and nucleotide polymers may reflect relationships that date back to the original lifelike polymers. Nevertheless, proposed models place the initial onus on either ribozymes or peptides.

Most scenarios generally favor a ribozyme-dominated origin to life in which associated peptides gradually took over catalytic roles. This seems plausible because of the efficiency of enzymes, which can far outperform ribozymes in catalyzing the same reaction. It is unlikely that any natural ribozyme existing today replaced a primordial enzyme.

One model posits primordial ribozymes using available amino acids as structural supports, electrostatic anchors, or as catalytic cofactors. A ribozyme that polymerizes amino acids, even in a nontemplated fashion, might introduce a significant competitive advantage, perhaps by providing itself with a more stable structural support.

The limited variety of amino acids available at that time would make for peptides of similar sequences, likely glycine-rich. Three-dimensional folding of the nascent peptide chains would also terminate their synthesis at comparable lengths, so that they would be of similar size. Variety in the growing peptide population would result from the occasional introduction of different amino acids into the available pool. The properties of these newly emerging peptides would allow them to adopt numerous functions such as ribozyme stabilization (polyglycine, e.g., would likely have served a structural role) or even RNA-templated RNA polymerization. This latter function would result in the amplification of any available RNA, including the amino acid polymerase that gave rise to the peptide population in the first place.

Any RNA conformations hindering oligonucleotide synthesis could be overcome by daily temperature fluctuations; daytime high temperatures would linearize ribozymes sufficiently to allow their use as templates by protein polymerases, whereas nighttime low temperatures would allow ribozyme refolding and catalytic reactivation. Imagining these events to occur near a hydrothermal vent would also allow for temperature cycling, as the waters surrounding the vents offer a steep temperature gradient, and convection currents may allow for polymers to cycle through numerous folding/melting regimes. In this manner, the growing peptide pool would slowly take over ribozyme functions, while the oligonucleotides gradually adopted a more informational role (see More Leftovers—The Nucleotide Coenzymes below).

What of the reverse scenario, where peptides promote the formation of the first complex ribonucleotides? Recent *in vitro* work with a viral RNA-binding protein raises this possibility. When synthesized in isolation, the protein's RNA-binding portion (a helical peptide 17 residues in length) still recognizes and binds an RNA of specific sequence. When this RNA is experimentally cleaved, and its 5′ and 3′ ends are activated with cyanogen bromide, the peptide is able to align the fragments, and promote their ligation. Though the peptide does not catalyze the formation of the phosphoanhydride bond, it does act as a template for the oligonucleotide fragments.

Whichever model best approximates the origin of lifelike polymers, it is clear that mutual support between ribozyme and peptide was plausible. In light of extant biology, perhaps their interdependence was essential.

Fearful Symmetry

Abiotic processes generally result in a racemic mix of monomers and so the establishment of the homochirality so common in biological systems remains a controversial topic.

Enantiomeric selection of L amino acids and D sugars may have occurred by multiple mechanisms. For example, differential photodegradation by ultraviolet (UV) light may have selected one enantiomer over another. Mineral surfaces might have exerted further selection by favoring the adsorption of different enantiomers.

Homochirality may have been encouraged by the efficiency of one chimeric form over another. For instance, free L amino acids in solution display greater catalytic activity (mimicking carbonic anhydrase, catalase, and β-galactosidase activities; see Peptide World, above) than D enantiomers. Stereochemical cooperativity may also play a role, in which a homochiral polymer's catalytic efficiency is enhanced over polymers with mixed chirality. This has been demonstrated recently by the development of a chirosensitive, self-replicating peptide (see Text Box 20.4). This peptide synthesizes homochiral copies of itself from a racemic mixture of peptide fragments. If the pool of component fragments possess chiral mutations, the rate of replication diminishes significantly. This suggests that the earliest self-replicating polymers were themselves key in the development of biological homochirality.

EARNING A LIVING

Whether populated by catalytic peptides, ribozymes, or both, how did the polymer world establish ordered, lifelike systems? One strategy describes an initially simple system that steadily increased in complexity. Another model assumes a more chaotic start where complexity was inherent from the beginning.

Replicators—Simplicity First

Among a population of abiotically produced polymers, chance allows a self-replicator to arise; such a polymer uses itself as a template to copy itself. Assuming the replicator works faster than abiotic processes, copies of the replicator would soon outnumber abiotically produced polymers. But, would 100% fidelity, the production of error-free copies, assure the persistence of this self-replicator? Only in the short term. Abiotic monomer production could not keep up with the demand of an ever-growing number of replicators. The replicator population, all identical, would be unable to respond to monomer depletion or to any other environmental change. Hydrolyzed replicators might supply some monomers, but this would effectively halt any further population growth; a grim outcome for the perfect self-replicator.

In another scenario, the abiotic process produces a less-than-perfect self-replicator, one that introduces occasional errors into its copies. This invites what Freeman Dyson called *error catastrophe*, where inaccuracies in progeny lead to functional degeneration and the eventual collapse of the population. But imperfect replication would also open the door to *error opportunity* where mutants of the original replicator might possess novel and useful catalytic activities. Indeed, *in vitro* selection experiments with ribozymes and peptides appear to support this idea; a polymer's catalytic activity typically increases hyperbolically with each successive mutation-selection cycle. Greater polymer variety, resulting in a wider range of catalytic activities, would allow for versatility in the face of environmental challenge. Environmental depletion of a crucial monomer, for instance, might not signal disaster if that monomer can be synthesized by a new catalyst within the functionally diversified population. Such diversity would be more likely to support the long-term persistence of the population.

Hypercycles—Complexity First

An alternative polymer world model excludes self-replicators. Instead, abiotic synthesis increases the variety of random polymers until a complexity threshold is surpassed. At this point, the population possesses so many different catalytic activities that a minimal lifelike metabolism emerges, virtually intact. Catalysts promote their own persistence by contributing toward the maintenance of the whole; each species of polymer is synthesized by another, making a closed autocatalytic loop. The entire system is collectively self-sustaining and self-reproducing.

Such a hypercycle depends from the outset on high polymer diversity, suggesting an "order out of chaos" effect. Any polymers that accelerate the increase in variety—for instance, a ligase or a template-directed polymerase—would be particularly useful in hypercycle initiation. This model takes into account the fact that no single polymer species is multifunctional enough to catalyze all the functions required to maintain order in a living system.

Hypercycles also depend on mutual support, suggestive of an ecological relationship between polymer species. For instance, catalyst 1 produces the substrate for catalyst 2, which produces the substrate for catalyst 3, and so on, until catalyst *n* closes the loop by synthesizing the substrate for catalyst 1. Furthermore, one catalytic loop might synthesize a by-product that feeds into a second catalytic loop, which in turn synthesizes a nutrient for the first loop. Chemical symbiosis between hypercycles would support further diversity.

But this raises the possibility of a hypercycle analog to error catastrophe. One catalyst might not return any useful products to the cycle, even though it consumes the substrates a cycle provides. Similarly, a polymer might not feed back into the loop that contributes to its own synthesis. In both situations, a collaborative population of catalysts would likely collapse.

Lesson of the Selfish Polymer

While either initiation model could establish a set of cooperative catalysts, both are vulnerable to parasitic elements. Selfish replicators (see Replicators—Simplicity First, above) would choke polymer diversity. Selfish catalysts (see Hypercycles—Complexity First, above) would disrupt cooperative networks by consuming a hypercycle's resources without contributing any useful products back into it. How could early lifelike polymer systems overcome such parasitic competitors?

One possibility is that the only polymer systems to persist were those that avoided generating selfish polymers in the first place. By crippling or destroying the populations they parasitized, selfish polymers weeded themselves out. This might account for the absence of selfish replicators or catalysts in modern organisms, but it assumes a number of separate polymer populations at the start.

Another possibility is that only those polymer systems possessing regulatory mechanisms could persist. By modulating the activity of the selfish polymer or of other catalysts in the system, one can compensate for imbalanced or disruptive production. The ability to control anabolic and catabolic activities would also have been crucial in order to respond to environmental fluctuations.

The selective pressures of selfish polymers and environmental flux would have favored those polymer systems that could regulate their metabolic and replicative activities. But how could Darwinian selection operate on loose sets of catalytic polymers? Why should an assortment of hypercycles compete or establish symbiotic relationships?

A COMPARTMENT OF ONE'S OWN

Compartmentation—the localization of catalysts, their substrates and products—is key to the transition from lifelike to living systems. This is evidenced by the cellular nature of all known life, including the viruses, which only become "alive" within a cellular milieu.

Problems with Solutions

A multicomponent, lifelike system faces the danger of dilution, in which catalysts or their products are lost to the surrounding medium. A self-sustaining system that develops a new synthetic pathway, for example, might not benefit from its innovation if the products drift away to feed neighboring systems. Selection would have been impaired. A free-floating polymer world may have solved these problems in a number of ways.

Polymer Scaffolding

One way to keep a set of catalysts localized is for them to associate physically with one another or with some stabilizing polymer structure. A fledgling hypercycle in the polymer world would have gained a significant advantage over other hypercycles if it acquired the means to produce long polymer scaffolds. This system would then possess increased catalytic efficiency, as catalysts bound to the scaffold structure would be more likely to interact chemically. The eukaryotic cytoskeleton is an extant example of a polymer scaffold system. It organizes cellular components in space and allows the existence of larger and more complex cooperative systems.

In a peptide world, the presence of proteinoid microspheres would have added structure. The diverse catalytic activities of individual proteinoids could have initiated a hypercycle, while the hydrophobic interiors of microspheres would have provided a sheltered environment in which (1) new activities or products could develop and (2) more sensitive molecules could be protected. By contrast, in an RNA world, noncovalent or short-lived covalent interactions could have localized RNAs and various small organic molecules into *organizing centers*, analogous to the modern ribosome. Ribosomal proteins, for instance, are presumed to stabilize or orient what may be a *floppy* ribozyme into a specific tertiary structure, promoting its full catalytic activity.

Sink, Don't Swim—Mineral Surfaces

Clay and pyrite, with their positive surface charges, could have produced and/or localized a cooperative set of polymers through ionic interactions. Compartmentation on mineral surfaces has also been proposed as a selective mechanism, explaining the anionic nature of biopolymers. Hence, electrostatically neutral or cationic molecules would have been lost to the surrounding environment. Other scenarios posit microscopic pores in rock surfaces as providing the first reaction vessels in which lifelike polymer systems could develop.

Lipid World and Protocells

As the cell is the functional unit of all known life, it is obvious that amphiphilic lipids ultimately offered the polymer world its most versatile compartment. What is not immediately obvious, however, is the abiotic source of amphiphiles (see Text Box 20.6).

> **TEXT BOX 20.6 PREBIOTIC LIPIDS**
>
> The first membranes could have been assembled from a diverse assortment of abiotically generated amphiphiles or lipids. These include glycerol, glycolipids, chlorosulfolipids, oxidized cholesterol, sphingolipids, and sterol. The abiotic source of these compounds is still in question, though recent experiments lend support to both terrestrial and extraterrestrial synthesis.
>
> Phospholipids characteristic of prokaryotic and eukaryotic membranes have been synthesized abiotically using 12-carbon fatty acids, glycerol, and phosphates. Dehydration produced phosphatidic acids. The abiotic production of the necessary precursor single-chain fatty acids is problematic, however.
>
> Prebiotic synthesis could have begun via the reductive power of pyrite formation (see Text Box 20.1). This process has been shown to synthesize saturated hydrocarbons, which are precursors of membranogenic molecules.
>
> Other experiments in lipid synthesis used formic and oxalic acid as a source of carbon. Aqueous solutions of these acids, heated to 150 to 175°C, yield lipid compounds with chain lengths ranging from C2 to C35. These experiments did not involve mineral catalysts, but nevertheless implicate a hydrothermal origin of lipids.
>
> Extraterrestrial lipid input has also been supported since amphiphiles have been recovered from the Murchison meteorite (see Text Box 20.1). These were found to form bilayered vesicles in water that could capture and retain polar dyes. Interplanetary clouds and comets are also plausible sources of amphiphilic compounds; under simulated interstellar conditions, mixtures of water, methanol, carbon monoxide, and ammonia (material consistent with the contents of interstellar ice) form complex mixtures of lipid compounds when bombarded with UV radiation. Among the synthesized products are amphiphiles, which form bilayered vesicles in water.

Whatever the rate of their abiotic synthesis, the remarkable stability of free fatty acids—at extreme temperatures, even over millions of years—would have assured their eventual accumulation. With a supply of amphiphilic compounds, polymer systems could have been encapsulated within vesicles. Several scenarios have been suggested to explain the formation of protocells:

1. Wet–dry cycling has been demonstrated to cause the capture of biopolymers within vesicles. When dried,

vesicles collapse into layers, "sandwiching" other solutes between them. When vesicles reform upon rehydration, solutes, including polymers, become encapsulated. Freeze–thaw cycles are predicted to give similar results.

2. Hydrophobic or electrostatic interactions could have attracted and attached drifting amphiphilic molecules to polymers. In this way, a polymer system may have assembled a layer of lipids around itself. Lipid association may have stabilized new polymer conformations, promoting novel or more efficient catalytic function. A variation of this idea posits that polymers only acquired useful catalytic function within vesicle membranes. This *cells first* scenario suggests that cooperative polymer systems were only initiated after their localization within membranes.
3. A polymer system could have evolved the ability to synthesize lipids. These may have initially represented a form of nutrient storage, with compartmentation arising secondarily.
4. Pyrite-catalyzed lipid synthesis could have resulted in a steadily growing membrane-like layer bound to the mineral surface. The hydrophobic microenvironment beneath this lipid layer could have promoted monomer and polymer synthesis on the mineral surface. Vesicles blebbing off the ever-growing lipid envelope could then capture and carry away assorted polymer catalysts, giving rise to lifelike protocells.

The likelihood of an itinerant vesicle capturing all the polymer components of a hypercycle is questionable. Capturing a self-replicating polymer might be more feasible. Regardless of the initial polymers caught, gaining subsequent access to monomers might have caused further difficulties. Diffusion of polar solutes through C18 hydrocarbon bilayers is highly restricted and leads to the question: Did vesicles offer polymer systems compartmentation in exchange for starvation?

Lipids of C12–C14 lengths have been shown to encapsulate an RNA polymerase through wet–dry cycling. The bilayered vesicles prevented escape of the enzyme, but allowed nucleosides to diffuse in from the surrounding medium. This simple system was subsequently shown to synthesize RNA oligomers, which remained localized inside the vesicle (see Text Box 20.7).

One problem with abiotic vesicle formation, however, is the presence of divalent cations in marine salts. In high concentrations, magnesium, calcium, and ferrous iron precipitate amphiphilic compounds. Other salts present in seawater have also been demonstrated to interfere with vesicle stability. It is possible, then, that cellular life first evolved in fresh water.

TEXT BOX 20.7 PERMEABILITY WITHOUT MEMBRANE CHANNELS

Would the earliest protocell membranes have allowed the transmembrane passage of amino acids, nucleotides, and other polar nutrients? Membrane permeability is affected by a number of parameters.

LIPID CHAIN LENGTH

David Deamer and his group (1994) conducted experiments using lipids of various lengths. They found that while C6–C8 hydrocarbon chains would allow the diffusion of polar solutes, the vesicles they formed were unstable. Hence, polymer catalysts could have escaped encapsulation. C16–C18 chains offer structural integrity, but prohibitive permeability. Hence, vesicle-bound polymer systems would have starved. However, hydrocarbon chains of C12–C14 in length have been shown to offer a compromise between vesicle stability and nutrient permeability. Such vesicles retain enzymes while allowing the influx of polar molecules, like nucleosides.

COMPOSITE MEMBRANES

An assortment of abiotically generated lipids (see Text Box 20.4) may have contributed collectively to the first membranes. Combinations of these offer various membrane properties, effecting stability and permeability.

CO-SURFACTANTS AND MINERALS

Compounds such as cholate (abiotically formed or synthesized by polymer catalysts) and minerals such as calcium could have contributed to membrane permeability.

DIVIDE AND REGULATE

The establishment of vesicle-bound cooperative polymer systems opened the way for Darwinian selection; it also created a need to control anabolic, catabolic, and replicative activities within these protocells.

The protocell membrane would augment itself by acquiring lipids either from the environment or from polymer-mediated synthesis. Increased vesicle size would cause structural instability, relieved only by vesicle division. Persistence of the polymer system would require that polymer components be distributed within the daughter protocells.

Among other problems, the polymer system's rate of replication would have to keep pace with the rate of vesicle division, or else new protocells might lack crucial elements of the cooperative hypercycle. Those systems with greater control over metabolic and replicative activity would be more likely to achieve long-term continuation, hence a selection regime can be established.

A Toy Protocell—Need for Metabolic Regulation

Consider a population of protocells, all of which are dependent on nutrient A. Nutrient A could be a precursor of a monomer, lipid, or energy carrier. With sufficient A available in the medium, all protocells will grow and replicate.

Nutrient A is derived from another substrate, N, also available in the environment. However, the abiotic $N \rightarrow A$ conversion is slow, and ever-increasing protocell numbers will soon deplete the supply of A. Should an enzyme, E1, arise that catalyzes the $N \rightarrow A$ conversion, the protocell possessing it will gain advantage. With a steady supply of A at its disposal, this protocell will outcompete the rest, and its descendants will eventually comprise the majority of the population.

Subsequently, as the environmental availability of N fluctuates, the advantage of E_1 may be lost. Within a new population, however, is a mutant protocell that possesses a second enzyme, E_2, that reversibly catalyses $N \leftrightarrow B$. B is the storage form of N and is restricted to the protocell's interior. Whenever N is plentiful, the protocell will be able to accumulate B while still supplying itself with nutrient A. During periods of N scarcity, B can be reconverted to N, again maintaining a steady supply of A. This mutant has a greater resistance to environmental flux and, hence, a steadier supply of A than its competitors. Due to this advantage, the majority of the protocell population will soon possess both E_1 and E_2.

During times of high N abundance, however, overproduction of B would be disadvantageous, as this might hinder the production of nutrient A. Simultaneously, the overaccumulation of A could be dangerous or wasteful: Nutrient A might be toxic at high concentrations, or it may degrade before it can be converted into a useful product.

This problem could be solved in various ways:

1. Protocells can be selected in which E_2 has a lower K_m or higher $V_{\max}$ than E_1. During times of plentiful N, the slow-acting E_1 will provide a steady supply of nutrient A, while fast-acting E_2 will convert any excess N to its storage form B.
2. Protocells can be selected where nutrient A provides negative feedback to E_1; E_1 inhibition would prevent the overproduction of A.
3. Nutrient A could provide positive feedback to E_2, accelerating B synthesis to prevent A overproduction.
4. E_2 activity could be regulated by N or B concentrations in the protocell. For instance, the $B \rightarrow N$ reaction would predominate if $[N] < [B]$. This would promote the $B \rightarrow N \rightarrow A$ pathway during times of N scarcity.

This toy model illustrates not only *the need for metabolic regulation* but also the advantage conferred to an organism that can accelerate abiotic reactions (see Table 20.2 for a summary and Chapters 1 and 2 for comparable concepts in enzyme regulation in extant biochemistry). The model also assumes a heterotrophic lifestyle, which is thought to have been key in the evolution of glycolysis and the tricarboxylic acid (TCA) cycle (also called the Krebs cycle). However, modern metabolism is not merely a suite of enzyme-catalyzed abiotic reactions since biosynthetic processes do not always parallel abiotic chemistry. Nevertheless, it would be easier to evolve an enzyme catalyst for a sluggish, nonenzymatic abiotic reaction than for an abiotic reaction that does not occur at all. Ultraviolet photolysis of some TCA cycle intermediates, for example, has demonstrated that a number of this pathway's reactions can occur spontaneously. Primitive pathways can be imagined then, where some but not all steps were enzyme-mediated.

CENTRAL CARBON PATHWAYS

The central carbon pathways are key to the generation of biological energy and of material precursors necessary for cellular maintenance and growth. These critical roles suggest that the earliest organisms must have possessed

TABLE 20.2 How the Protocell Got Its Regulation

Problem	Solution
1. Slow abiotic conversion of $N \rightarrow A$	Catalyst E_1 to convert $N \rightarrow A$
2. Periodic scarcity of N	Catalyst E_2 to convert $N \leftrightarrow B$, a polymer storage form of N
3. Balancing $N \leftrightarrow B$ and $N \rightarrow A$ conversions	Modulate E_1 and E_2 activities through allosteric binding or other mechanisms

these pathways, in some form, prior to any significant phylogenetic divergence. As evidence of their antiquity, variations of the central carbon pathways occur in each of the three existing domains of life, the Archaea, the Bacteria, and the Eukarya.

Regarding the origin of two of the most ubiquitous carbohydrate pathways, glycolysis and the TCA cycle, two questions are worth asking:

- What good is half a pathway? To understand how an extant system came to be, it is often useful to consider how an incomplete version of the system, a portion, could have offered an advantage to some ancestral organism. Each portion of a system can then be examined to determine (a) how the individual parts could have been beneficial, and (b) what drove their assembly into a novel, more complex whole. This question assumes the gradual, stepwise assembly of metabolic pathways, or the linkage of shorter pathways together into larger systems. This also takes into account the possibility that the function of ancient, partial pathways (e.g., portions of glycolysis or half of a TCA cycle) may have differed from their modern counterparts.
- What can be learned from extant variations of these pathways? All known forms of cellular life carry out carbohydrate metabolism. By comparing alternate forms of glycolysis and the TCA cycle, the core elements of these pathways, portions that may have been present in the earliest organisms, can be discerned.

TEXT BOX 20.8 EVOLUTION OF GLYCOLYTIC ENZYMES

Striking structural similarities exist among the 10 glycolytic enzymes. For example, the TIM-barrel domain, a core of β strands flanked by α helices, is possessed by all glycolytic enzymes. Several enzymes also share similar nucleotide- and cofactor-binding structures. However, despite these topographical resemblances, it is unlikely that the enzymes diverged from a common ancestral protein. Low sequence homologies indicate convergent evolution instead, where several protein ancestors independently evolved the same stable and useful tertiary structures. Furthermore, convergence is suggested by the high protein : structural motif ratio among modern proteins—there are more extant proteins than there are common motifs.

Common ancestry is more likely between enzymes of similar catalytic activity. For example, kinases share greater homology with other kinases than with isomerases. One interesting example of this is a bifunctional enzyme, with both hexokinase and PFK activity, recently found in a hyperthermophilic archaeon. Competitive inhibition studies suggest that the enzyme uses the same binding site for both glucose and fructose-6-phosphate. If this bifunctional enzyme is ancestral to the hexokinases and PFKs of other Archaea (and this is supported by sequence similarities,) a gene duplication/divergence event would account for its evolution into two monofunctional enzymes. Although this is an example of two glycolytic enzymes sharing a common ancestor, recall that the hexose-intermediate enzymes of the pathway's upper half are likely to have arisen independently in the Archaea.

Glycolysis as Living Fossil

Glycolysis is present, at least in part, in all living organisms. Hence, it is thought to have been established early in the history of life. The pathway's great age is also suggested by several of its characteristics:

- Glycolysis is structurally simple. Enzymes operate independently within an aqueous medium, and adenosine 5′-triphosphate (ATP) production does not require a membrane.
- Glycolytic enzyme sequences are highly conserved across all kingdoms of life (see Text Box 20.8).
- Glycolysis functions independently of oxygen and would have been well-suited to an ancient Earth, which is predicted to have had an oxygen-poor atmosphere.
- The two substrate-level phosphorylation reactions of glycolysis would have provided enough energy to meet the needs of a primitive cell.
- With pyruvate as the terminal electron acceptor, redox balance of the pathway can be maintained.

Finding Purpose in Sugar-Free World

It is likely that glycolysis is quite ancient, but this in itself presents a complication, as hexose sugars were not likely to have been abundant. The notoriously short half-lives of sugars would have prohibited their accumulation in a prebiotic world (see Text Box 20.2). Why then, would a hexose degradation pathway have evolved so early on, if there was little or no sugar available? What advantage did glycolysis offer ancestral life?

Apart from energy production, glycolysis also supplies three-carbon compounds that can be channeled into a variety of synthetic pathways, including fatty acid or amino acid production. Indeed, the role of glycolysis in providing biosynthetic outputs remains constant in all extant organisms, yet its role in energy production varies. Some microorganisms, for instance, generate ATP independently of central carbon metabolism, even operating glycolysis without net ATP production. Hence, biosynthesis may have been a stronger factor in shaping the evolution of the glycolytic pathway, offering a greater advantage to organisms than energy production (see Evolution in Reverse and TCA's Lucky Horseshoe below).

But some substrate for glycolysis to work upon would still have been required. While abiotic reactions probably could not have supplied sufficient hexose sugars, they may have produced amino acids in relative abundance (see Text Box 20.1). Given this, the original function of glycolysis may have been to derive energy and biosynthetic precursors from abiotically formed glycine; converted to 3-phosphoglycerate, it could be catabolized to yield pyruvate, acetyl-CoA and ATP. Alanine, also an abundant product of abiotic synthesis experiments, could have been converted to pyruvate. Other feasible prebiotic compounds, including glycerol and glyceraldehyde, may have been substrates. The abiotic phosphorylation of glycoaldehyde, experimentally demonstrated, suggests that phosphorylated glycerol or glyceraldehyde may have been available to early life; these phosphorylated intermediates could subsequently have fed a primordial glycolytic pathway. That glycolysis may have originated as a means to metabolize amino acids accounts for a portion of the pathway, the lower half that deals with triose intermediates. Might this portion represent an original metabolic core?

Glycolysis—The Trunk

All cellular life can convert glucose to pyruvate, though not through identical glycolytic pathways. The Embden–Meyerhof pathway, perhaps the one most familiar to students of biochemistry, is present in both bacteria and eukaryotic organisms, but only incomplete versions have been detected in archaeans (more on this later). Compare this to the Entner–Douderoff pathway, common to Bacteria and Archaea; both pathways share a common "trunk" of reactions interconverting triose intermediates, that is, from glyceraldehyde-3-phosphate to pyruvate (see Fig. 20.1). Though not shown, the pentose phosphate pathway also shares this same trunk. Might this portion of glycolysis have been in place among the earliest organisms?

Enzyme sequence similarities suggest so. The enzymes mediating the trunk reactions, in particular triosephosphate isomerase (TPI), glyceraldehyde-3-P dehydrogenase (GAPDH), phosphoglycerate kinase (PGK), enolase, and pyruvate kinase (PK), are widely distributed and are extremely well preserved. Sequence analysis supports the idea that each of these enzymes is of monophyletic origin—they were all present within a common ancestor prior to any significant phylogenetic divergence.

But perhaps the "root" of the trunk came first. The nonphosphorylated Entner–Douderoff pathway (see Fig. 20.1) is prominent among hyperthermophilic Archaea and is also present in some Bacteria. Regarded as an example of *paleometabolism*, this pathway does not yield net ATP from glucose and shares only two steps with the other two variants of glycolysis, those mediated by enolase and PK. Indeed, enolase and PK are considered to be the oldest glycolytic enzymes. Present in all three kingdoms, they show the greatest conservation.

To place these three alternate glycolytic pathways together into an evolutionary scheme, life's growing biosynthetic and energetic requirements might be taken into account. The first pathway to arise would have been the nonphosphorylated Entner–Douderoff, yielding biosynthetic intermediates without net energy yield. The Entner–Douderoff pathway would come next, given its complement of trunk enzymes, its production of useful intermediates and a net yield of 1 ATP from glucose. Finally, the Embden–Meyerhof pathway, with its net yield of 2 ATP, would be last on the scene.

Yet this scenario raises another question: Why should PK and enolase represent the oldest, most conserved steps of glycolysis? What functional advantage might they have offered, and what about the hexose-converting steps of the glycolytic pathway?

Glycolysis—The Upper Branches

As already noted, the differences among the extant glycolytic pathways occur predominantly among their upper portions (see Fig. 20.1), suggesting that the hexose conversion steps evolved after some degree of divergence from a common ancestor. This is supported by phylogenetic comparisons of enzymes catalyzing these hexose conversion steps (see Text Box 20.8) and by certain characteristics of the hexose degradative pathways among the Archaea:

- Archaeans degrade glucose predominantly via the Entner–Doudoroff pathway. This pathway differs significantly from the Embden–Meyerhof pathway in that it does not require enzymatic steps catalyzed by phosphoglucoisomerase (PGI), phosphofructokinase (PFK), aldolase, or triose phosphate isomerase (TPI).
- The Embden–Meyerhof pathway is found among the Archaea but always in a variant or incomplete form. Some variants employ adenosine 5′-diphosphate (ADP) as the phosphate donor, while others use pyrophosphate (PPi) (see Why ATP below). The enzyme glyceraldehyde-3-phosphate ferrodoxin oxidoreduc-

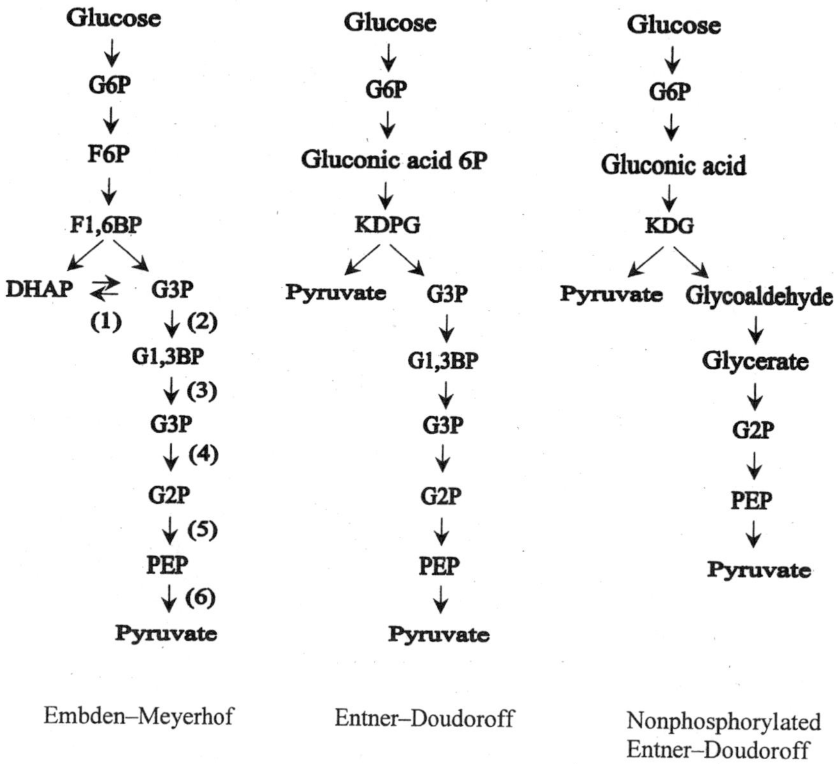

Figure 20.1 Three modern variations of the glycolytic pathway. Left to right, they are: the Embden–Meyerhof pathway, the Entner–Doudoroff pathway, and the nonphosphorylated Entner–Doudoroff pathway. Glucose is processed to pyruvate in each, and note the conservation of the five triose conversion steps in the Embden–Meyerhof and the Entner–Doudoroff pathways (from G3P to pyruvate). The last two steps in glycolysis, converting G2P to PEP and then to pyruvate, are common to all three variants. Abbreviated intermediates: G6P, glucose-6-phosphate; F6P, fructose-6-phosphate; F1,6BP, fructose-1,6-bisphosphate; DHAP, dihydroxyacetone phosphate; G3P, glyceraldehyde-3-phosphate; G1,3BP, glycerate-1,3-bisphosphate; G3P, 3-phosphoglycerate; G2P, 2-phosphoglycerate; PEP, phosphoenolpyruvate; and KDPG, 2-keto-3-deoxy-6-P-gluconic acid; and KDG, 2-keto-3-deoxygluconic acid. Numbered enzymes: (1) triosephosphate isomerase (TPI), (2) glyceraldehyde-3-P dehydrogenase (GAPDH), (3) phosphoglycerate kinase (PGK), (4) phosphoglyceromutase, (5) enolase, and (6) pyruvate kinase (PK).

tase catalyzes the conversion of glyceraldehyde-3-phosphate to 3-phosphoglycerate, taking the place of two enzymes of the canonical Embden–Meyerhof pathway, GAPDH and 3-phosphoglycerate kinase.

- While the bacterial and eukaryotic Embden–Meyerhof pathway enzymes show significant conservation, they are phylogenetically far removed from their archaean counterparts. Amino acid sequences of archaean hexokinase, PFK, aldolase, and TPI all differ markedly from those found in the other kingdoms.

As the Archaea are frequently considered to be the least-evolved among extant organisms, their variations of the glycolytic pathway may reflect those of the Earth's most ancestral life. Indeed, one sulfur-reducing archaean is able to metabolize glucose despite its apparent lack of genes encoding any recognizable glycolytic enzymes. Alternatively, their hexose-degradative pathways reflect those portions that were "reinvented" after a split from an ancestral lineage.

Evolution in Reverse

Some questions about the origin of glycolysis may be answered by another question: What if glycolysis is actually gluconeogenesis running in reverse? As mentioned earlier, functional adaptability can contribute to the longevity of ancient systems and life has a knack for finding novel applications for preexisting tools. Most commonly regarded as a catabolic pathway, glycolysis may have initially served life as an engine of hexose synthesis. Some evidence supports this notion:

- Gluconeogenesis is more widespread than glycolysis, being present in all three kingdoms. Archaeans, for instance, typically employ the Entner–Doudoroff pathway for hexose catabolism, but synthesize hexoses using a modified Embden–Meyerhof pathway running in the anabolic direction.
- Seven of the 10 steps of glycolysis (Embden–Meyerhof pathway) are reversible. This means most of the pathway enzymes can operate in both anabolic and catabolic directions. The three unidirectional steps involve ATP–ADP conversions. Hence, only three loci (hexokinase, PFK, PK) need be bypassed during gluconeogenesis. Furthermore, some archaean PFKs readily catalyze the reversible phosphorylation of fructose-6-phosphate (F6P) interconverting it with frucose-1,6-bisphosphate ($F1,6P_2$). Alternatively, only these three enzymes had to be developed in order to switch from an ancient gluconeogenic function of the pathway to a glycolytic function.
- From the standpoint of survival value, the synthesis of glucose (or other hexose sugars) would have provided stored energy and biosynthetic precursors in a readily accessible form. Also, it would have been useful in the construction of primitive cell walls.

Considering first the shared trunk of glycolysis found in the Embden–Meyerhof and Entner–Doudoroff pathways (see Fig. 20.1), four of its five steps (except PK) are reversible. An ancestral organism employing this reversible pathway could benefit from the production of either glyceraldehyde-3-phosphate (for subsequent amino acid and glycerol synthesis) or pyruvate (for amino acid or fatty acid synthesis from acetyl-CoA), depending on the carbon input available.

The "upwards" evolution of the trunk pathway (i.e., starting from pyruvate) might explain the ubiquity and antiquity of enolase and PK but would depend on an abundance of alanine as a starting material. This raises the possibility that the earliest organisms were autotrophic, fixing CO or CO_2 into acetate, then synthesizing pyruvate. Indeed, the iron–sulfur world theory (see Text Box 20.1) predicts that the first metabolic pathway was a reductive TCA-like cycle, fixing inorganic carbon into organic compounds. Glycolysis is thought to have extended out from this central reductive pathway, in the anabolic direction.

Since glycine and alanine are common products of abiotic synthesis, perhaps the trunk pathway extended from both ends to meet somewhere in the middle. The conversion of glycine to glyceraldehyde-3-phosphate would have started the pathway off in the catabolic direction, whereas the anabolic direction would have started from the conversion of alanine to pyruvate. These separate pathways may have joined at the phosphoenolpyruvate (PEP) step, an end product of potential import to early organisms. PEP can be converted to oxaloacetate, a precursor of the aspartate family of amino acids, so a primordial organism could have benefited from possessing either portion of the trunk pathway.

Considering the upper portion of glycolysis, the hexose-intermediate steps, most steps in this pathway are reversible, and the anabolic end products, glucose-6-phosphate and glucose, can offer significant advantage in the biosynthesis of ribose and cell walls, respectively. It seems plausible then that these steps could have arisen as a part of an anabolic pathway, extending upward from glyceraldehyde-3-phosphate.

Taking Control—Enzyme Modulation

The toy protocell model emphasized the need to adjust metabolic flux to meet an organism's immediate requirements. Glycolysis offers clues to the evolution of a few control mechanisms. The glycolytic enzymes are shared by different phylogenetic lineages, but their regulation within each lineage differs markedly, suggesting that control mechanisms arose after the establishment of the pathway.

Gene duplication–fusion events, producing doublet proteins, appear to be a common motif in the evolution of enzyme control. Redundant, duplicated structures are free to accumulate changes that may lead to novel functions. Hexokinase provides an example. In yeast, hexokinase subunits are not inhibited by physiological concentrations of their product, glucose-6-phosphate (G6P). However, allosteric inhibition is observed in the mammalian enzyme. At approximately twice the size of the yeast homolog, the mammalian subunits likely arose through duplication–fusion. The N-terminal half of the mammalian enzyme subunit evolved into an allosteric effector site, capable of binding G6P. Similar regulatory divergence is observed in PFK. Bacterial and yeast PFKs show no allosteric inhibition by the enzyme product, $F1,6P_2$, whereas the mammalian enzyme does. This arises from a duplicated active site that evolved to bind $F1,6P_2$.

Taking Control—Translation

Apart from modulating enzyme activities, modern organisms also affect their metabolic activity by altering enzyme quantity. This requires transcriptional and translational control, often mediated by transcription factors and effector proteins responding to signals or substrate concentrations. Recent evidence suggests the existence of a translational feedback mechanism that operates independently of proteins.

In bacteria, high thiamine concentrations in the medium inhibit the expression of those enzymes involved in thiamine synthesis. The mRNA transcripts encoding these enzymes are thought to sense cellular thiamine concentrations directly. By binding thiamine, the mRNA undergoes conformational changes that prevent its translation by the ribosome. In addition, RNAs have been designed *in vitro* to recognize and bind various types of small molecules with high affinity. Hence, it is possible that the cellular mRNA transcripts of at least some proteins may also possess such capabilities, enabling them to regulate their own expression by binding to the products of the proteins they encode.

This feedback, acting independently of proteins, raises the possibility of a convenient regulatory mechanism in a time before translation. Hence, the ribozyme-catalyzed pathways of a primitive polymer world may have been controlled by product inhibition, similar to that observed in modern enzymes.

TCA's Lucky Horseshoe

Pyruvate fermentation to lactate or ethanol allows for the regeneration of oxidized nicotinamide adenine dinucleotide (NAD^+) which is consumed by the GAPDH step of glycolysis; and, thereby, early anaerobic life could maintain redox balance during the catabolism of glucose or other sugars. However, although this is an effective redox solution when the purpose of glucose catabolism is ATP synthesis, it ruins the biosynthetic potential of pyruvate and all triose phosphate intermediates produced between the GAPDH and pyruvate kinase reactions. Any of the intermediates siphoned off into biosynthesis would upset redox balance in glycolysis. Selective pressure to satisfy both redox and anabolic needs likely lead to the development of alternative pathways, stemming from pyruvate.

Assuming that a primitive organism has glycolysis in place, producing two molecules of pyruvate from a hexose, the conversion of pyruvate to oxaloacetate (catalyzed by pyruvate carboxylase in extant Archaea, Bacteria, and Eukarya) would have conferred an immediate advantage, for not only is oxaloacetate an amino acid precursor, but it can also act as an electron acceptor. Oxaloacetate reduction produces malate, and this frees some pyruvate from redox duty. The elaboration of the reductive pathway from pyruvate → succinate not only provides redox balance for glycolytic oxidation but also offers increased biosynthetic potential (see Fig. 20.2).

The increasingly available pyruvate, in turn, provides a biosynthetic opportunity. Pyruvate is a precursor for amino acid synthesis, for instance, and acetyl-CoA formation from pyruvate provides the starting material for fatty acid biosynthesis (pantetheine, the active component

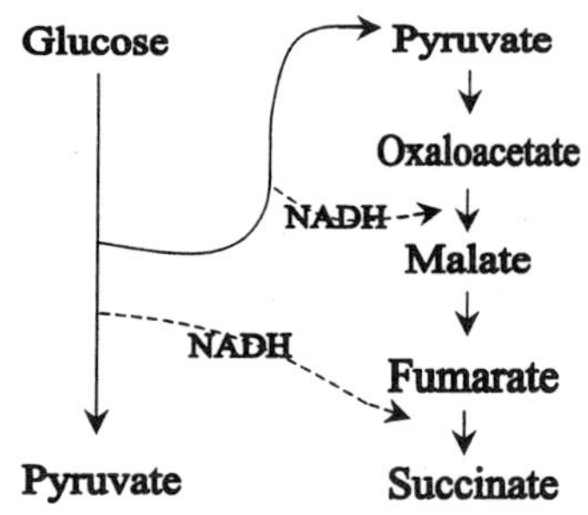

Figure 20.2 A reductive pathway stemming from pyruvate can offer redox balance for glycolysis, while increasing the biosynthetic potential of pyruvate.

of CoA, has been synthesized in simulated early Earth conditions, so it is plausible that CoA was a component of the prebiosphere). Acetyl-CoA condensation with oxaloacetate forms citrate, which is further converted to α-ketoglutarate, a precursor to glutamate and other amino acids.

The advantages of increased biosynthetic precursors derived from pyruvate selected for the evolution of two anabolic pathways, each stemming from pyruvate and each representing half of the modern TCA cycle (also called the Krebs cycle). Together they can achieve redox balance in an anaerobic world. Oxidations in the pyruvate → α-ketoglutarate arm satisfied the reductive steps of the pyruvate → succinate arm. This "horseshoe" design (see Fig. 20.3) may have also prompted the selection for an alternative means to balance glycolytic oxidations; electrons from glycolysis could have been made available for reductive biosynthesis and for the generation of ATP (see pH Poisoning below).

In this manner, when free oxygen increased to levels where it could be exploited as a terminal electron acceptor, a complete TCA cycle was almost ready. All that remained to be done was to link the reductive and oxidative pathways into a cyclic one. These two pathways could have met in at least two possible ways, depending on how far each pathway extended.

- In one scenario, the oxidative arm may have extended from pyruvate, via acetyl-CoA and citrate synthesis, through to succinyl-CoA. This possibility is supported by the fact that modern pyruvate dehydrogenase possesses some α-ketoglutarate dehydrogenase activity. Perhaps an ancestral enzyme carried out dual functions, converting pyruvate to acetyl-CoA or α-ketoglutarate to succinyl-CoA. In this case, the conversion of succinyl-CoA to succinate would unite the oxidative and the reductive pathways, meaning that succinyl-CoA synthase was the last enzyme activity to be added to the system. A reversal of the fumarate reductase and malate dehydrogenase reactions to instead catalyze the oxidation of succinate and malate would then allow the reductive arm of the cycle to operate in an oxidative manner (Fig. 20.3).

- In an alternative scenario, the oxidative pathway may have extended only from pyruvate → α-ketoglutarate. In this case, α-ketoglutarate production would allow for amino acid biosynthesis, and the reductive pathway, from pyruvate → succinyl-CoA, would be used for the synthesis of porphyrins (possibly for membrane-bound electron transfer; see pH Poisoning below). This would mean that α-ketoglutarate dehydrogenase would have been the last enzyme required to link the two pathways together into a cycle. Interestingly, modern *Escherichia coli*, a facultative anaerobe bacterium, employs either the horseshoe or the cyclic arrangement of these pathways, depending on oxygen availability. Under aerobic conditions, the TCA cycle runs oxidatively, whereas under anaerobic conditions, α-ketoglutarate dehydrogenase synthesis is repressed. This breaks the cycle into the two parallel linear pathways. Fumarate reductase is also upregulated, not only allowing the reductive arm to proceed from pyruvate but also contributing to redox balance (see pH Poisoning below).

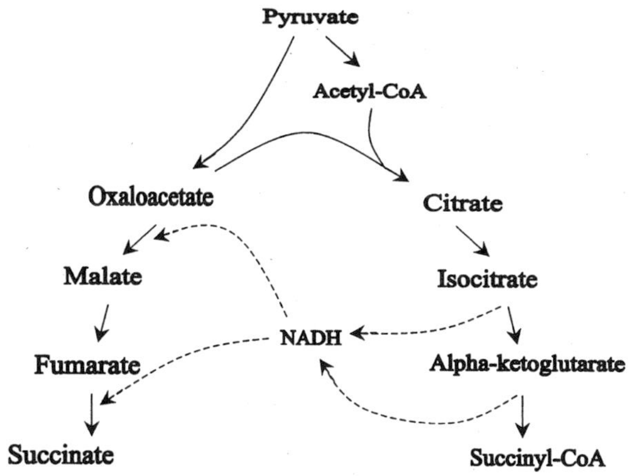

Figure 20.3 An incomplete TCA cycle, consisting of two pathways stemming from pyruvate. The oxidative and reductive arms both provide biosynthetic intermediates and redox balance.

pH Poisoning, Electron Transfer Proteins, and Respiration

What would favor the closing of the horseshoe into a cycle? How did early anaerobes realize the potential of the TCA cycle's synthetic and energy-producing capacities? A possible explanation is as follows. Utilizing glycolysis alone, the first anaerobic life forms faced internal acidosis as a side effect of a fermentative metabolism. As a side note and contrary to popular thought, H^+ production during fermentative metabolism is not actually due to the formation of lactic acid but rather the protons derive from ATP hydrolysis (the products of an ATPase are ADP, P_i, and H^+). The net system (glycolysis + ATPase) is a net producer of H^+, whereas when oxidative phosphorylation and cellular ATPases are coupled, there is no net acidosis. Hence, in ancient cell systems that relied on glycolysis for ATP generation, the need arose for a mechanism of pumping protons out of the cell. Proton efflux would relieve the pH stress and favor the evolution of a membrane channel to export H^+. Initially, this channel could have been non-specific, allowing various cations to pass through. The *E. coli* H^+ channel, for instance, also allows the passage of Na^+. (Hence, a bacterium with a well-developed proton pumping system would make an excellent candidate for a mitochondrion. See Energy for Evolution below.)

Accumulation of protons in the surrounding medium would eventually create an electrochemical gradient, impeding further proton export. Any cells able to couple ATP hydrolysis to channel function could then actively pump protons out against the gradient. This modular evolution is suggested by observations of the modern H^+–ATPase, where the H^+ channel has been shown to function independently of the ATPase activity.

Meanwhile, biosynthetic pressures described earlier would have been selecting for cells that can spare oxidized substrates for anabolism by minimizing fermentation. One electron sink is the reductive pathway starting with pyruvate and ending in succinate. Porphyrins, derived from succinyl-CoA, contribute to electron transfer systems in extant biology and may have done so early in life's history. Membrane proteins could also act as electron acceptors; in doing so, these proteins extruded more protons from the cell.

Arising in an anaerobic world, these protoelectron transport proteins also had to reduce other available oxidants:

1. *Inorganic oxidants such as elemental sulfur or NO_3^-.* Many extant Bacteria and Archaea take advantage of protons as an abundant electron sink; a membrane-bound hydrogenase reduces H^+ to evolve H_2 gas and also pumps protons out of the cell. Similar proteins, called sufhydrogenases, reduce elemental sulfur as well. Geological evidence, in the form of sedimentary sulfides, suggest the presence of sulfate-reducing microbes up to 3.4 billion years ago, indicating the great antiquity of respiration.
2. *Organic oxidants such as fumarate.* Some prokaryotes express fumarate reductase during anaerobic metabolism. This membrane-bound enzyme not only produces succinate but also extrudes protons from the cell. (Recall that this is part of the reductive arm of the interrupted TCA cycle.)

Possessing even a simple electron transfer system, the organism (a) achieved redox balance, (b) produced useful biosynthetic precursors, and (c) expelled more protons into the environment. Additional electron transfer proteins would not only augment the proton gradient but would relieve the proton channel of its efflux duty. Influx through the channel could then occur, and the H^+–ATPase could now operate as an H^+–ATP synthase. This anaerobic electrophosphorylation persists in modern prokaryotic anaerobes, supplementing the ATP generated by substrate-level phosphorylation. Interestingly, the modern-day eukaryotic mitochondrial F_1F_0–ATP synthase is also bifunctional; under normal aerobic conditions it is an ATP synthase, but, during hypoxia, when proton pumping by the electron transfer chain ceases, the enzyme reverses its function and becomes an ATP-dependent proton pump. In this way, during oxygen stress, the enzyme helps maintain mitochondrial membrane potential.

Once this system was in place, relatively minor changes were required for the organism to exploit increased atmospheric oxygen as a terminal electron acceptor. Various parts of the whole were already of considerable use and were adopted for new purposes with the addition of minimal new materials. Phylogenetic analysis suggests that the various respiratory systems, including aerobic, nitrate, and sulfur, share both great antiquity and common ancestry (see Lucky LUCA below).

WHY ATP?

The above schemes stress the importance of generating biosynthetic precursors, but the development of ATP was also crucial in evolving metabolic pathways. For instance, the hydrolysis of high-energy phosphoanhydride bonds in ATP is directly coupled to many biosynthetic reactions. Consider also that most glycolytic intermediates are phosphorylated. This may have originally evolved as a mechanism to reduce their membrane permeability and retain them within protocells or, in a polymer world, to maintain them localized by adsorption onto a positively charged mineral surface. ATP must have been established as a universal energy-conversion molecule very early in life's history, but how?

Phosphorus Made Them Do It

The central role of phosphorus in energy transfer might have to do with its unique chemical characteristics. Phosphorus forms weaker bonds than other available elements, and it can also form double bonds and bonds with more than four covalent linkages.

Today, apatites and phosphates make most phosphorus inaccessible to the biosphere, but its distribution may have been different on prebiotic Earth. Volcanic outgassing could have transported great quantities of elemental phosphorus to the planet's surface, and high concentrations may have been present in the first water to condense.

PPi—The Poor Man's ATP

Inorganic diphosphate, called pyrophosphate (PPi), may have been the predecessor of ATP for early membrane-bound life or even for polymer world systems. Wet–dry cycling of phosphate in solution yields PPi. Volcanic magma has been shown to produce PPi as well as polyphosphate (poly-P), a linear polymer of PPi. These prebiotic processes as well as the stability of PPi suggest that it may have saturated primordial Earth.

The phosphoanhydride bonds of PPi and poly-P store one-half to two-thirds of the energy compared to that stored in ATP. Polyphosphates have been shown to be effective condensing agents, facilitating both peptide and oligonucleotide formation in solution (see Text Box 20.3). Assuming the abundance of PPi and the low energy demands of early life, PPi was probably established as the first energy donor.

Many modern organisms still use PPi and poly-P as energy donors and acceptors, suggesting possible holdovers from a world before ATP. These include:

- Phototrophic bacteria possess membrane-bound PPi-synthases, which photophosphorylate orthophosphate to organic PPi.
- A bacterial PPi-dependent proton pump has been found; this membrane-bound H^+–PPase may share a common ancestor with the H^+–ATPase.
- A PPi-dependent PFK (PPi–PFK) occurs in some Archaea, protists, and plants. It catalyzes the reversible phosphorylation of fructose-6-P and likely shares a common ancestry with ATP–PFK. Indeed, using site-directed mutagenesis, a protozoan PPi–PFK has been converted into a functional ATP–PFK with a single amino acid substitution. This suggests that only minor changes were required for PPi-dependent organisms to switch to an ATP-dominated energy economy.
- A poly-P kinase is known, which uses ATP to catalyze the phosphorylation (elongation) of polyphosphate chains.
- Bovine and rat mitochondria have been shown to possess a membrane-bound pyrophosphatase, which acts as a coupling factor in electron transfer to drive PPi synthesis.

The ATP Takeover

Energy storage in the triphosphate tail of ATP and in poly-P is comparable, and studies show that the adenosine moiety of ATP contributes little to the reactivity of the triphosphate group. Given the similarities in energy availability, and the efficiency of PPi and poly-P as energy carriers, why should ATP have replaced them?

Ultraviolet exposure and dry heating has been shown to produce adenosine 5′-monophosphate (AMP), ADP, and ATP from adenosine and inorganic phosphate. Assuming the availability of purines and ribose (see Monomer World and Text Box 20.2), early life may have developed an alternative source of phosphoanhydride that could compete with PPi.

Early life that used ATP may have had a selective advantage, perhaps because of ATP's functional versatility compared to PPi.

- ATP phosphate bonds store more energy than the bonds in PPi or poly-P.
- ATP can contribute directly to oligonucleotide synthesis.
- While ATP can phosphorylate and pyrophosphorylate, its ability to adenylate also made it a prerequisite in various synthetic pathways.
- ATP offers a more defined structure than PPi or poly-P, making it more recognizable by enzymes. The adenosine structure can be oriented and bound during substrate binding by kinases in order to optimally position the phosphate tail for catalytic reaction.

Thus, despite being simpler and more stable molecules, neither PPi nor poly-P could compete with the greater versatility of the new phosphoanhydride carrier, ATP.

Another consideration is cell membrane viscosity. The experimental manipulation of rat mitochondria demonstrates that increasing membrane viscosity inhibits ATP production in favor of PPi synthesis. Early membranes may have been very rigid, favoring the initial production and use of PPi. The switch to ATP production may have followed the evolution of more fluid membranes, probably by a change in lipid composition.

Again, the adoption of ATP demonstrates how minor changes in existing processes can spur a new way of life. A phosphoanhydride energy economy was certainly in place by the time of the first protocells, and they were almost ready for an ATP-driven world.

MORE LEFTOVERS—THE NUCLEOTIDE COENZYMES

Metabolic evolution favored increased catalytic efficiency and efficient control over redox reactions, the latter probably necessitated to prevent damage to vital components of protometabolic systems by unwanted or dangerous reductions during sugar oxidation. Metabolic intermediates may have supplied the first electron sinks (see TCA's Lucky Horseshoe above), but dedicated electron shuttles would have provided a greater advantage, not only for protection but also for the reductive synthesis of new compounds.

The initial electron sinks or shuttles may have been inorganic compounds, which were later incorporated into emerging organic molecules. The redox functionality of transition metals, for instance (see Text Box 20.1 for an outline of the iron–sulfur world theory of biogenesis and Chapter 12), may have offered such an advantage that metal clusters were incorporated into primitive peptides to enhance their catalytic activities. The success of this acquisition is reflected by the ubiquity of metalloenzymes in modern biology, where transition metal clusters are employed for redox purposes. The membrane-bound hydrogenase (MBH), for example, is present in many microorganisms, including an archaean hyperthermophile. The MBH coordinates both FeS and NiS clusters to eliminate electrons by reducing protons to H_2 gas (see pH Poisoning above). Ferredoxin provides another example, being a protein that coordinates FeS in its role as a redox carrier. In plants, ferredoxin helps generate reduced adenine dinucleotide phosphate (NADPH) during photosynthesis. In some archaean microbes, ferredoxin replaces NAD as an electron shuttle in oxidative metabolism, passing electrons to MBH.

But given that RNA world scenarios tend to dominate current thinking, and that nucleotide-based catalysts are unlikely to have replaced existing amino-acid-based catalysts, it would seem likely that modern nucleotide coenzymes arrived on the scene very early on in life's history.

First Coenzymes

The nucleotide coenzymes could have been the metabolic by-products of early lifelike systems. Abiotic processes could have supplied biosynthetic precursors for early life to work with. Quinoline, for example, is a NAD. Nonenzymatic quinoline production may have supplied an early protocell with substrate for NAD synthesis. The utility of coenzymes in enhancing catalysis and electron shuttling may have allowed them to become quickly established.

First Ones On, Last Ones Off

The structure of some nucleotide coenzymes, including coenzyme A and NAD, suggest that they may have been situated at the 5′ end of RNA molecules. Perhaps they began as part of polymer world's oligonucleotide complement. In this light, two scenarios depict an evolutionary progression from ribozyme to ribonucleoprotein (RNP) to enzyme.

Early catalytic ribozymes, with active sites resembling coenzymes, could have bound amino acids or short peptides as cofactors. The structural or catalytic advantage gained in this association could have selected for the steady elaboration of the protein component, until it took on a catalytic role. The ribozyme, rendered increasingly redundant, would have atrophied until only its original active site remained as a coenzyme.

An alternative scheme begins with an assembly of prebiotically formed coenzymes and amino acids. Their combined catalytic activities may have initiated the synthesis of peptides and oligonucleotides. Although initially ribozyme-driven, the use of cofactor peptides would have led to a RNP-to-enzyme progression, described above.

SAVING UP FOR A RAINY DAY

Storage of energy and nutrients allows an organism increased independence from the environment. A period of nutrient scarcity or of environmental stress can be endured more readily with onboard supplies.

Versatility Wins Again

Polyphosphate (poly-P) may have been an early energy storage molecule. Its availability, utility, and simplicity make it a plausible candidate. Poly-P might still be used by some photosynthetic bacteria to synthesize ATP during overnight dark periods. But overall, poly-P has been replaced by carbohydrates for energy storage, probably for the same reason PPi was succeeded by ATP—versatility. Carbohydrates can be used to synthesize ATP as well as many other cellular components, whereas poly-P can only transduce energy. One near-universal storage molecule is glycogen, a glucose polymer.

Glycogen's Lesson

The structure of glycogen maximizes the storage and rapid mobilization of glucose. Two enzymes are largely responsible for that structure: Glycogen synthase increases the length of glucose chains, whereas branches in the chains are introduced by the branching enzyme. The latter enzyme's activity is 200-fold higher than that of glycogen synthase. This makes for highly branched glucose polymers of uniform length, arranged in spherical tiers.

Branching increases the solubility of glycogen and, hence, its storage capacity in a limited space. A greater number of terminal glucose residues also allows for rapid glycogen synthesis and degradation while still leaving enough room between branches for enzymes to operate. Common to bacteria, yeast, and animals, glycogen structure must have been optimized early in metabolic evolution (plants developed a similar polymer, starch, instead).

Glycogen also demonstrates the optimization of a complex structure *automatically* at the enzyme level. No instruction or direction or templating from any higher level is required. This is further evidence that glycogen became the storage molecule of choice early in life's history.

LUCKY LUCA

Life's three extant domains—the Archaea, the Eubacteria, and the Eukaryota—display commonalities and disparities that raise the question of what the last universal common ancestor (LUCA) was like.

Vesicular Jackpot

One problem, already raised, starts with the assumed encapsulation of polymer world catalysts by itinerant lipid vesicles. The probability of one lucky vesicle capturing enough catalysts to become a lifelike protocell would seem small. Could a single vesicle have given rise to all three lineages?

Further inconsistencies stem from phylogenetic comparisons among the three domains. Some bacterial genes resemble eukaryotic or archeaen homologs more than they do other bacterial versions. Similarly, many metabolic genes shared among the Archaea and Eubacteria are absent from the Eukaryota. These problematic gene similarities and distributions further cloud the picture of the progenitor organism.

Cell and Supercell

The divergent natures of the three domains, and the metabolic and genetic characteristics unique to each, might suggest a totipotent LUCA. Rather than being a primitive, inefficient protoorganism, this sophisticated LUCA would have been able to cope with a variety of environmental challenges.

Phylogenetic comparisons suggest that the LUCA possessed at least four respiratory chains, making it capable of oxygen-, nitrate-, sulfate- and sulfur-based respiration. Some Archaea and Eubacteria still possess more than one respiratory system, up-regulating the most useful one depending on environmental circumstances. The LUCA is also proposed to have possessed a complete TCA cycle and carbohydrate metabolism, antioxidant proteins, and even flagellar motility.

Every übercell, however, has at least one tragic flaw and this postulated LUCA suffers from timescales. Earth is estimated to be approximately 4.5 billion years old. Microfossil evidence suggests the presence of bacterial life as early as 3.8 billion years ago. Could something as sophisticated as the super-LUCA have evolved from lifeless matter in so short a time? Also, can populations of this ancestral organism have diverged into three unique domains so quickly afterwards?

Promiscuous Communes

Rather than a single protocell ancestor, the LUCA may have been a loosely knit association of protocells. A combination of (a) error-prone replication and (b) frequent horizontal transfer would have allowed independent mutations and innovations to spread. This collective of protocells, cross-feeding each other genetically and metabolically, could have possessed all the characteristics ascribed to the super-LUCA. This scenario could account for inconsistent gene distribution among domains, in which case phylogenetic comparisons may offer the history of individual genes, rather than of organisms.

In this scenario, the horizontal transfer of more modular subsystems would have been favored. Universally useful elements, as well as those that can operate independently of other subsystems, would have successfully passed throughout the protocell community. This would include certain metabolic genes and translational mechanisms. In this way, early steps in cellular evolution would have been driven in large part by the recombinative power of horizontal transfers.

Over time, the exchange of simpler, independent subsystems would allow for the evolution of subsystems of greater complexity and interconnectedness. Such interdependent subsystems would have become increasingly refractory to horizontal transfers, initiating divergence among the protocells. For instance, portions of an incompletely transferred subsystem might not confer any advantage to the receiving protocell. Also, if one transferred subsystem depends on a second for substrate, only those protocells possessing both would gain advantage.

The development of self-recognition mechanisms would also have intensified the divergence, until the affiliation of protocells formed separate lineages. The isolation of each protocell line would allow independent innovations to occur, until they each gave rise to one of the currently recognized domains of life.

In this model, an initially chaotic, fluid exchange system among simpler organisms led to the establishment of more complex, exchange-resistant organisms. Such a collective evolutionary mechanism requires a common language, suggesting an early origin of the genetic code.

ENERGY FOR EVOLUTION—MITOCHONDRIA

Eukaryotes have displayed a startling degree of innovative change, including multicellularity, since acquiring mitochondria 1.9 billion years ago. Contrast this to the Eubacteria and Archaea, which remain largely unchanged after 3.8 billion years. How did the advent of mitochondrial respiration allow for such rapid eukaryotic evolution?

The symbiotic association of an ancestral eukaryote with the eubacterial protomitochondrion (perhaps the most extreme example of horizontal transfer) is a familiar model (see theories of the origin of mitochondria in Chapter 8), but questions remain. These include:

- How did the host cell engulf the eubacterium? The host organism was not likely to possess the cytoskeletal proteins required for endocytosis. Also, how did the eubacterium avoid digestion to become a symbiont?
- An alternate proposal imagines a predatory eubacterium that, after forcing its way into a prey cell, digests it from the inside. The question then becomes, how did the prey avoid digestion to become a symbiont of the predator?
- While the advantage for the host cell is clear, what did the symbiont gain from the association?

Regardless of how the symbiosis came to be, the advantages of increased energy production available through mitochondrial ATP generation can be inferred from both bacterial and yeast examples. In comparison to eukaryotic genomes, bacterial genomes are remarkably small and streamlined. They are among the smallest known, are haploid, and generally consist of intronless, single-copy genes. Smaller genomes offer the advantage of faster replication, hence, the short generation times of bacteria under optimal conditions. But a minimalist genome might also have been the consequence of limited energy supply. Deoxyribonucleic acid (DNA) synthesis is ATP-dependent, and bacterial metabolism may be unable to support the maintenance and replication of larger genomes. This would limit the possibilities offered by gene duplication and divergence events. In this way, bacterial evolution may have been held back by limited genome size, which in turn was limited by energy availability.

In a similar vein, yeast maintain diploid genomes in nutrient-rich media but reduce to a haploid genome when resources become scarce. Again, energy production affects how much DNA a cell can afford to maintain. The addition of an organelle (the mitochondrion) dedicated to energy production may have provided the impetus for the expansion of the ancestral eukaryotic genome. Large-scale duplications and multiple chromosomes provide the redundancy for mutational tampering and potential innovation. A greater energy budget may also have prompted the evolution of specialized nuclear structures to house the expanding genome.

An increase in toxic oxygen levels in the atmosphere may have been another spur for symbiosis. The ancestral eukaryote could have gained additional advantage by hosting an oxygen-consuming symbiont. The latter would have maintained oxygen levels inside the host to within tolerable limits.

TOO MANY ANSWERS?

The question of life's origins has inspired so many competing ideas and explanations, it is tempting to think that the problem is just too complex and remote to ever solve for certain. If we will never know for sure how life arose, then where does that leave "the truth"? Perhaps the remarkable variety of origin scenarios might instead indicate life's inevitability. Monomer and polymer syntheses, for instance, are encouraged by a variety of plausible starting conditions. The catalytic potential invariably displayed by biological polymers, coupled with the resilience of membrane-forming lipids, would suggest that the emergence of life on Earth was beyond doubt, later if not sooner. Perhaps the truth is that, given permissive conditions, life cannot be denied.

RECOMMENDED READING

Dandekar, T., Schuster, S., Snel, B., Huynen, M., and Bork, P. (1999). Pathway alignment: Application to the comparative analysis of glycolytic enzymes. *Biochem J* **343**:115–124. *A phylogenetic and functional comparison of the major variants of the glycolytic pathway.*

Huynen, M. A., Dandekar, T., and Bork, P. (1999). Variation and evolution of the citric-acid cycle: A genomic perspective. *Trends Microbiol* **7**:281–291. *Taking advantage of the growing number of completely sequenced prokaryotic genomes, this review offers a phylogenetic perspective on TCA cycle evolution.*

Issac, R., Ham, Y. W., and Chmielewski, J. (2001). The design of self-replicating helical peptides. *Curr Opin Struct Biol* **11**:458–463. *Comprehensive review of* in vitro *self-replicating peptides, with ramifications for a peptide world.*

Joyce, G. F. (2002). The antiquity of RNA-based evolution. *Nature* **418**:214–221. *Extensive review of the prebiotic origins of putative pre-RNA molecules, RNA, and the rise of an RNA world. This issue of* Nature *also contains other useful reviews about ribozymes, both natural and artificial.*

Lazcano, A., and Miller, S. L. (1999). On the origin of metabolic pathways. *J Mol Evol* **49**:424–431. *This review considers all major origin of life issues, although its focus is on the enzymatic takeover of prebiotic chemical reactions.*

Martin, W., and Russell, M. J. (2003). On the origins of cells: A hypothesis for the evolutionary transitions from abiotic geochemistry to chemoautotrophic prokaryotes, and from prokaryotes to nucleated cells. *Phil Trans Roy Soc London, Ser B Biol Sci* **358**:59–83. *A whopping, integrative review that begins with the iron–sulfur world and ends with the rise of all three branches of cellular life.*

Melendez-Hevia, E., Waddell, T. G., and Cascante, M. (1996). The puzzle of the Krebs citric acid cycle: Assembling the pieces of chemically feasible reactions, and opportunism in the design of metabolic pathways during evolution. *J Mol Evol* **43**:293–303. *A theoretical analysis, in which the modern TCA, portions of the TCA, and hypothetical, TCA-like pathways are compared in light of energy and chemical efficiency.*

Monnard, P. A., and Deamer, D. W. (2002). Membrane self-assembly processes: Steps toward the first cellular life. *Anat Rec* **268**:196–207. *A membranes-first perspective of life's origin, with surveys of prebiotic lipid synthesis and David Deamer's pivotal studies with lipid vesicles and catalytic polymers.*

Romano, A. H., and Conway, T. (1996). Evolution of carbohydrate metabolic pathways. *Res Microbiol* **147**:448–455. *This review serves as a fine introduction to the various modern glycolytic pathways, the common trunk pathway, and considers their origins.*

Zubay, G. (2000). *Origins of Life on the Earth and in the Cosmos*, 2nd ed. Harcourt Academic. *A college-level textbook that thoroughly covers all aspects of the origins of life problem, with an emphasis on extant cell biology and experimental evidence.*

Two journals that are particularly useful in this field include *Astrobiology* and *Origins of Life and Evolution of the Biosphere*.

INDEX

Functional Metabolism: Regulation and Adaptation, edited by Kenneth B. Storey.
ISBN 0-471-41090-X